TIME-SAVER STANDARDS
FOR BUILDING MATERIALS & SYSTEMS
DESIGN CRITERIA AND SELECTION DATA

TIME-SAVER STANDARDS
FOR BUILDING MATERIALS & SYSTEMS
DESIGN CRITERIA AND SELECTION DATA

Donald Watson, EDITOR

McGraw-Hill

New York ▪ San Francisco ▪ Washington, D.C. ▪ Auckland ▪ Bogotá ▪ Caracas ▪ Lisbon
London ▪ Madrid ▪ Mexico City ▪ Milan ▪ Montreal ▪ New Delhi ▪ San Juan
Singapore ▪ Sydney ▪ Tokyo ▪ Toronto

McGraw-Hill

A Division of The McGraw·Hill Companies

1 2 3 4 5 6 7 8 9 0 KGP/KGP 0 6 5 4 3 2 1 0

ISBN 0-07-135692-4

SPONSORING EDITOR

Wendy Lochner

BOOK DESIGN & PRODUCTION

Susann L. Gierman / Suzani Design
Todd S. Gordner
Sandra K. Littell

This book was set in Syntax, Helvetica, and Zapf by ITC.
Printed and bound by Quebecor/Kingsport.

Disclaimer:

The publisher, editors and contributors have made every reasonable effort to make this reference work accurate and authoritative, but make no warranty and assume no liability for the accuracy or completeness of the text, tables or illustrations or its fitness for any particular purpose. The appearance of technical data or editorial material in this publication does not constitute endorsement, warranty or guarantee by the publisher, editors or contributors of any product, design, service or process. Neither the editors or McGraw-Hill shall have any liability to any party for any damages resulting from the use, application or adaptation of information contained in Time-Saver Standards, whether such damages be director or indirect or in the nature of lost profits or consequential damages. Time-Saver Standards is published with the understanding that McGraw-Hill is not engaged in providing architectural, engineering design or other professional services.

PREFACE

This inaugural volume of *Time-Saver Standards for Building Materials & Systems* is a revised edition of technical data sheets developed over the last twenty years and previously published as SELECTION DATA, a volume of SWEET'S Catalog Files. SELECTION DATA was designed to respond to the expressed need of construction professionals for a single source of unbiased and generic product information. The present volume continues that objective and has been reformatted, updated and augmented with new sections and pages. Its purpose is to assist the design and building professional in design, specification and product selection.

Time-Saver Standards for Building Materials & Systems is a guide and data source for building materials and systems design, evaluation and selection. The Table of Contents provides a comprehensive listing of construction materials, components and systems of a building. Text and illustrations highlight design criteria and selection data of the entire assembly of a building construction project.

The format and sequence of the Table of Contents conforms to UNIFORMAT II, the widely adopted classification for preliminary cost estimating that places building elements in the order of construction and location in a completed building. This follows the order of building design and construction, in the same manner and general sequence that architects and building designers think about design and construction as a process.

Time-Saver Standards for Building Materials & Systems consists of brief summaries, bulleted text and illustrations that lead the reader through the design, specification and product selection process necessary for informed judgment in design and selection of all building elements, components and systems. Comparisons and evaluations can be made by referring to charts and tables that summarize various properties, uses and choices of construction materials, product and system types. Selection checklists provide information on materials capability, durability, and other design and performance criteria. These data are typical of the key information needed for design development, detailing and construction documents phases of architectural services.

A companion to this present volume is *Time-Saver Standards for Architectural Design Data (7th edition 1997)* which compiles detailed design data and fundamentals in the form of archival reference articles. It provides a second level of detail, design data and reference for readers of the book. It is referred to throughout this book as "Reference 1" and is an appropriate follow-up for the more detailed, quantitative data, required for completion of design.

Originally developed as a volume in SWEET'S Catalog File, *Time-Saver Standards for Building Materials & Systems* provides a useful introduction and review to building product manufacturers' catalogs. It allows designers to comprehensively evaluate manufacturer's product literature and product specific data in SWEET'S Catalog File. The Table of Contents is cross-referenced to MasterFormat and the filing numbers used in SWEET'S Catalog File.

The format of *Time-Saver Standards for Building Materials & Systems* also allows for easy and comprehensive reference, as well as an overview of the elements of building design and construction, appropriate for classroom, training and educational venues.

Donald Watson,
FAIA, NCARB, Editor

CONTRIBUTORS

The contributions of the following individuals is gratefully acknowledged in the historic development of SWEET'S Selection Data volume:

Douglas Behrens, PE
Algirdas Brazinskas, AIA
Carl T. Grimm, CSI, CCS
Anna M. Halpin, FAIA, CSI
Michael R. Kretschmann, PE
JoAnne Lindsley, IALD, MIES

In addition, the following individuals provided contributions for *Time-Saver Standards for Architectural Design Data (7th ed.)* also included in the present volume:

Donald Baerman, AIA
Everett M. Barber, Jr.
Elmer E. Botsai, FAIA
John C. Carmody
Robert P. Charette, PE
Walter Cooper, PE
Arturo De La Vega
Robert DeGrazio
M. David Egan, PE
Philip Fairey
Martin Gehner, PE
William Hall
Rita M. Harrold, IESNA
Joseph Lstiburek, P. Eng.
Nadav Malin
Fred Malven, PhD.
Murray Milne
James. C. Myers
Richard Rittelmann, FAIA
Stephen S. Ruggiero
Stephen Selkowitz
Peter R. Smith, FRAIA
Benjamin Stein
Timothy T. Taylor, AIA, ASTM
John Templer
Joel Ann Todd
Alex Wilson

THE EDITOR

DONALD WATSON, FAIA, NCARB, is an architect and author and is editor-in-chief of *Time-Saver Standards for Architectural Design Data (7th Edition)*. He is Professor of Architecture and former Dean, School of Architecture, Rensselaer Polytechnic Institute, Troy, NY.

HOW TO USE THIS BOOK

The purpose of *Time-Saver Standards for Building Materials and Systems* is to provide generic **selection guidelines and criteria for design and specification** of architecture and construction materials, systems and products.

The **TABLE OF CONTENTS is a summary of topics,** organized in a logical sequence that follows the process of construction. The first chapter provides an introductory overview of the physical principles of design and building construction.

Chapter topics and subtopics follow the process of construction from "the ground up," that is, foundations, structure, building enclosure, interior and services. This sequence, along with topic numbering, conforms to the Uniformat II classification system, the industry-adopted reference for preliminary design and cost estimating formats.

The **INDEX** provides an easily referenced tool **to search for a particular topic,** and includes the page numbers to find **key definitions.**

Within this outline, each page is organized around a specific topic, highlighted to be easily read and comprehended.

The topics are presented in brief and usable form, as checklists, tables and summaries. Where appropriate, references are indicated for further and more detailed information. Indicated in the text as **Reference 1,** *Time-Saver Standards for Architectural Design Data, 7th Edition, 1997* contains **extended and detailed reference articles** on the topics in this book and is thus recommended as a companion to this volume.

The contents and topics in this book are indicated in the accompanying Table in **both UniFormat and MasterFormat classification formats. This enables cross-referencing to SWEETS Catalog Files,** as well as industry literature numbering systems which follow MasterFormat.

UniFormat		General Conditions (1)	Site Construction (2)	Concrete (3)	Masonry (4)	Metal (5)	Wood and Plastics (6)	Thermal & Moisture Protection (7)	Doors & Windows (8)	Finishes (9)	Specialties (10)	Equipment (11)	Furnishings (12)	Special Construction (13)	Conveying Systems (14)	Mechanical (15)	Electrical (16)
A	**FOUNDATIONS**			√	√			√									
B	**BUILDING SHELL**			√	√	√	√	√	√	√	√						
B1	**STRUCTURE & MATERIALS**																
B1.1	STRUCTURE			√	√	√	√										
B1.2	CEMENT/CONCRETE			√													
B1.3	STONE/MASONRY				√												
B1.4	METALS					√											
B1.5	WOOD/WOOD PRODUCTS						√										
B2	**EXTERIOR WALL ASSEMBLIES**																
B2.1	INSULATION							√									
B2.2	WATERPROOFING/DAMPROOFING							√									
B2.3	EXTERIOR WALLS			√	√	√	√	√		√							
B2.4	BEARING WALLS			√	√	√	√										
B2.5	CURTAIN WALLS					√		√		√							
B2.6	WALL FACINGS					√				√							
B2.7	GLASS/PLASTICS								√								
B2.8	WINDOWS						√		√								
B2.9	ENTRANCES/DOORS					√	√		√		√						
B2.10	COATINGS			√		√	√			√							
B2.11	SEALANTS			√		√	√			√							
B3	**ROOFING ASSEMBLIES**																
B3.1	ROOFING							√									
B3.2	SKYLIGHTS							√				√					
C	**INTERIORS**			√		√	√			√				√	√		
C1	**INTERIOR PARTITION SYSTEMS**					√	√			√							
C2	**STAIRCASES**			√		√	√							√	√		
C3	**INTERIOR SYSTEMS & FINISHES**																
C3.1	CEILINGS									√							
C3.2	FLOORING									√							
D	**SERVICES**			√	√	√	√	√	√	√		√			√	√	√
D1	**CONVEYING SYSTEMS**														√		
D2	**PLUMBING**															√	
D3	**HVAC**							√						√		√	√
D4	**FIRE PROTECTION**			√	√	√	√	√	√	√						√	√
D5	**ELECTRICAL**							√				√					√

TABLE A: TABLE OF CONTENTS (listed vertically) conforms to Uniformat II classification. Each section topic is cross-referenced in this table to its corresponding Masterformat Division, listed horizontally.

TABLE OF CONTENTS

PART 1

DEFINITIONS & PHYSICAL PRINCIPLES

A • FOUNDATIONS

A1 • SUBSTRUCTURE

B • BUILDING SHELL

B1 • STRUCTURE & MATERIALS

B2 • EXTERIOR WALL ASSEMBLIES

B3 • ROOFING ASSEMBLIES

C • INTERIORS

C3 • INTERIOR SYSTEMS & FINISHES

D • SERVICES

DEFINITIONS & PHYSICAL PRINCIPLES

BUILDING

BUILDING=

= MEANS TO ALLOW UNIMPEDED HUMAN ACTIVITY IRRESPECTIVE OF THE CONDITIONS OF THE NATURAL ENVIRONMENT

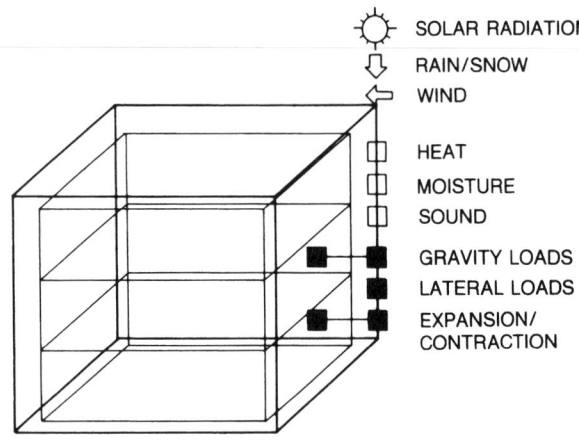

SOLAR RADIATION
RAIN/SNOW
WIND

HEAT
MOISTURE
SOUND

GRAVITY LOADS
LATERAL LOADS
EXPANSION/
CONTRACTION

ENCLOSURE may be defined as being:
■ **Means of containment and articulation of habitable space, selected and assembled to resist and/or control effects of:**
• Factors of environment external to such means of containment and/or factors of the environment thus contained.
• Forces acting upon and forces developing within such means of containment.
■ **While maintaining its intended shape and integrity under all possible combinations of such factors and forces.**

ENCLOSURE

SERVICES

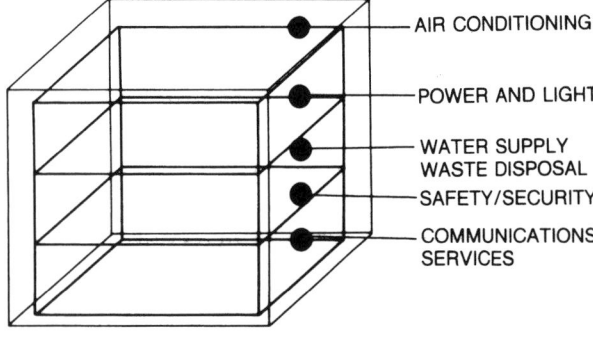

AIR CONDITIONING

POWER AND LIGHT

WATER SUPPLY
WASTE DISPOSAL
SAFETY/SECURITY

COMMUNICATIONS
SERVICES

SERVICES may be defined as being:
■ **Means of maintaining the environment internal to the means of containment at:**
• Comfort and convenience levels acceptable to the occupants.
• Safety levels required by the occupants.

DEFINITIONS & PHYSICAL PRINCIPLES

ENCLOSURE

ENCLOSURE=

MEANS OF SUPPORT

to maintain any enclosure in the shape intended while resisting all external and internal forces acting upon such enclosure, and transmitting all such forces to the ground.

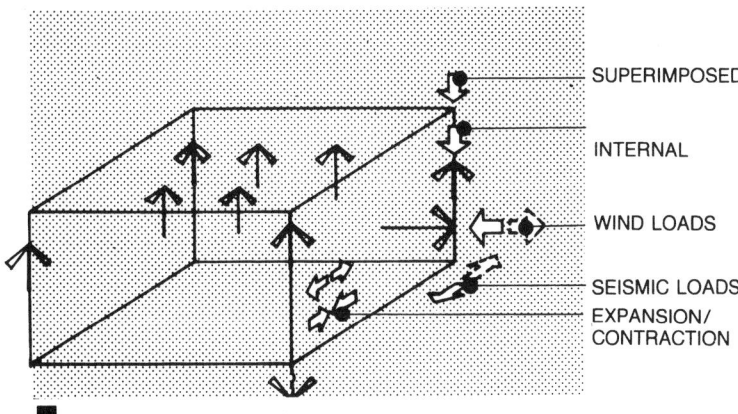

SUPERIMPOSED
INTERNAL
WIND LOADS
SEISMIC LOADS
EXPANSION/
CONTRACTION

ENVELOPE

to separate environment internal to the enclosure from that external to it while acting as barrier and/or selective filter to all external or internal environmental factors acting upon such envelope.

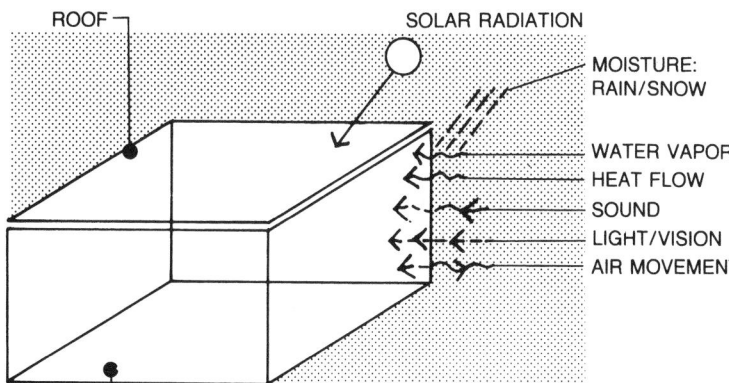

ROOF
SOLAR RADIATION
MOISTURE:
RAIN/SNOW
WATER VAPOR
HEAT FLOW
SOUND
LIGHT/VISION
AIR MOVEMENT
WALLS

ELEMENTS OF INTERIOR

to articulate, define, separate, or connect space or spaces within the enclosure while resisting any environmental factors and/or forces acting upon them, and transmitting all such forces to means of support.

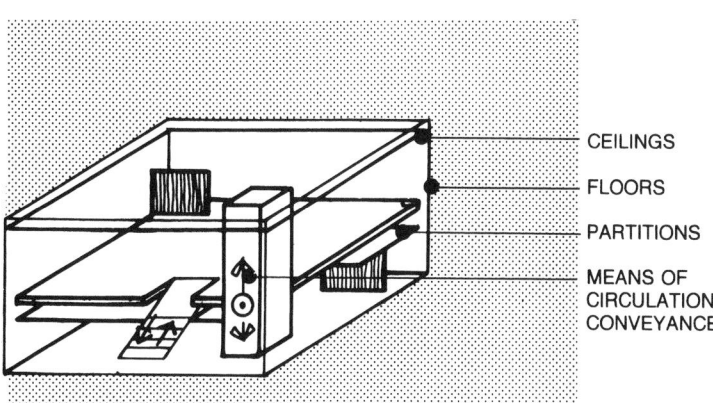

CEILINGS
FLOORS
PARTITIONS
MEANS OF
CIRCULATION/
CONVEYANCE

DEFINITIONS & PHYSICAL PRINCIPLES

FORCES

GRAVITY LOADS

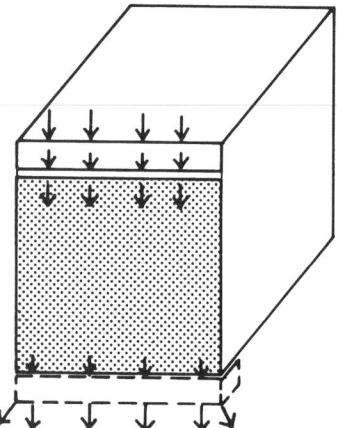
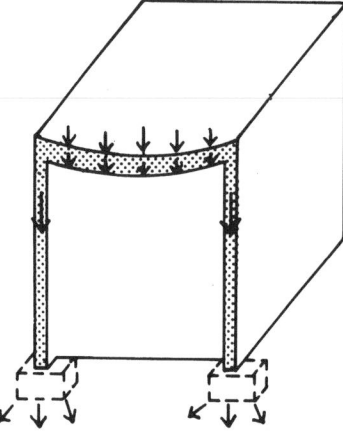

LOAD TRANSFERRED
CONTINUOUSLY

POINT TRANSFER
OF LOADS-
SIMPLY SUPPORTED

The envelope of any enclosure, whether such enclosure is exposed to the elements or contained within the controlled environment of a large enclosure, will always be subject to gravity load of its own weight, permanent and fixed as to magnitude and location; generally referred to as dead load. An envelope may also have to sustain randomly superimposed gravity and wind loads of varying magnitude and/or duration; generally referred to as live loads.

All gravity loads are transferred through the envelope or its components to the ground in rigid envelopes, or through rigid components of rigid and flexible envelopes; resisted through internal air pressure in air supported flexible ones, with the ground then acting as counterweight to uplift.

- Flexible envelopes, and flexible components of envelopes will always act in tension, irrespective of their shape and distribution of loads.
- Stresses in rigid envelopes will be essentially determined by: distribution of loads; paths of transfer of loads; shape and/or position relative to the ground of the rigid envelope or of its rigid components.

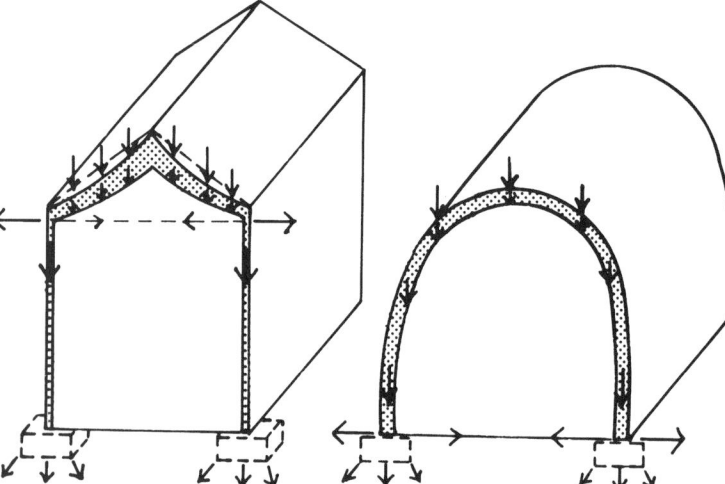

POINT TRANSFER
OF LOADS
OUTWARD THRUST
AT SUPPORTS

CONTINUOUS TRANSFER
OF LOADS -
OUTWARD THRUST
AT FOUNDATION

Gravity loads will cause deformations to develop in the envelope:
- deformations in rigid envelopes due to dead loads only will be permanenet but not necessarily unchanging: certain materials tend to continue to deform over time under sustained load even when there is no change in the magnitude of such load.
- deformations due to dead load will also be randomly amplified through superimposed live loads, with the deformations generally reverting to their previous position after removal of such live loads, as long as stresses resulting from combined loading do not exceed the elastic limit of the material.
- any horizontal or inclined component of any envelope except when held up by internal pressure will be subject to deformation, however slight, due to its own weight alone.

DEFINITIONS & PHYSICAL PRINCIPLES

WIND/SEISMIC LOADS

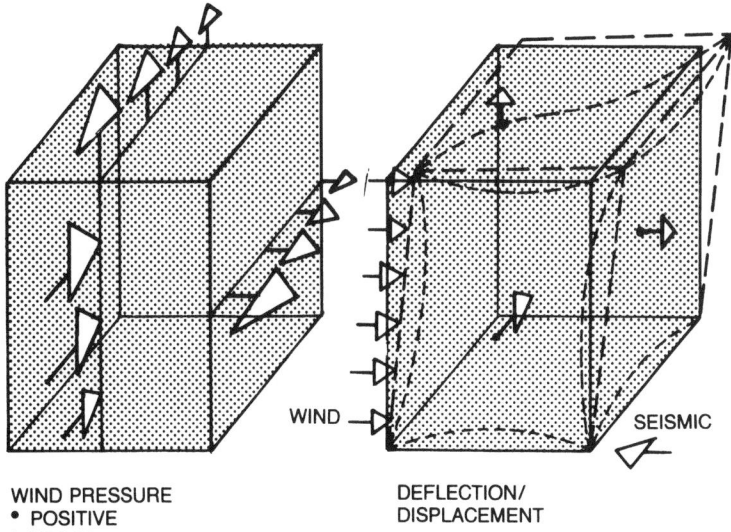

WIND PRESSURE
• POSITIVE
• NEGATIVE

WIND

DEFLECTION/
DISPLACEMENT

SEISMIC

Wind forces - the flow of air against, around, and over an envelope of an enclosure - will affect tthe stability of the envelope:
• vertical components facing into the wind will be under positive pressure which will cause deformation in such components.
• vertical components parallel to or facing away from the direction of wind will be subject to negative air pressure, as will all horizontal or nearly horizontal surfaces.

Lateral deflection or deformation of the vertical frame of an envelope is resisted by horizontal components of the enclosure, roof and floor assemblies, acting as diaphragms.

The dead load of flat or nearly flat roof assemblies will counteract the negative wind pressure proportionately to the weight of each component: light horizontal envelopes may be deflected upwards; light membranes attached to heavy assemblies may be lifted off.

EXPANSION/CONTRACTION

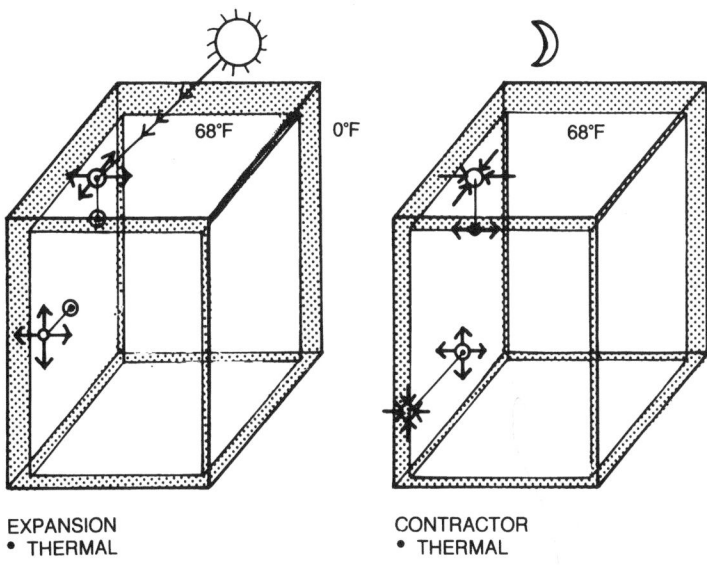

68°F   0°F

68°F

EXPANSION
• THERMAL

CONTRACTOR
• THERMAL

Movement will occur in all components of the envelope due to variations in their internal temperature:
• Components of envelopes exposed to solar radiation even when air temperatures are low, will gain heat and expand proportionately to their individual coefficients of expansion. Components adjacent to them but not thus exposed may remain at constant temperature and not expand at all: when such components are continuously attached to each other, they may fail due to differential movement between them.

Components of an envelope may also swell and shrink due to changes in their internal moisture content, and such movement may also be differential.

All components of an enclosure are in almost constant movement, interacting between each other based solely on their physical state and properties, and not necessarily on the intent of the designer.

DEFINITIONS & PHYSICAL PRINCIPLES

ENVIRONMENTAL FACTORS

DEFINITIONS & PHYSICAL PRINCIPLES

HEAT

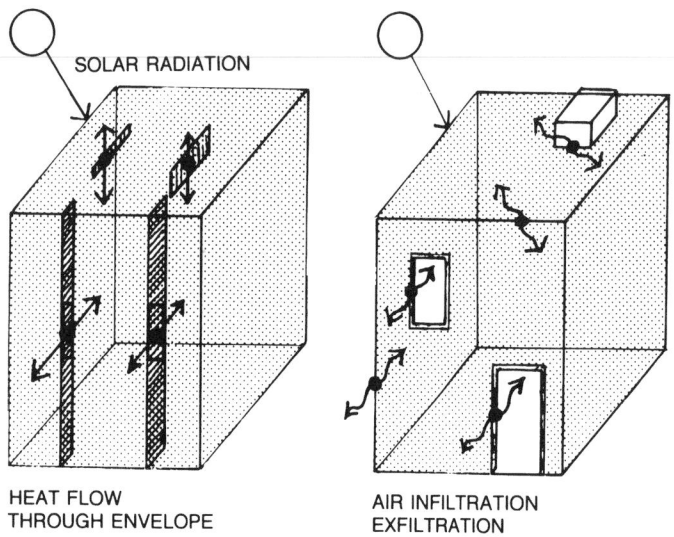

SOLAR RADIATION

HEAT FLOW
THROUGH ENVELOPE

AIR INFILTRATION
EXFILTRATION

Heat will flow through the envelope whenever a temperature differential exists between its outer surfaces, and such flow of heat must be controlled whenever the interior environment of an enclosure has to be maintained within limits of comfort for the occupants:
*   Flow of heat cannot be stopped, it can only be impeded by components of the envelope when selected to reduce the total amount of heat being transferred through them during a given unit of time.

Heat will also be gained by, or lost from, a contained space of an enclosure by air leakage through the envelope whenever temperature and/or pressure differentials between interior and exterior environments exist, and such gains or losses may significantly affect the comfort of the occupants:
*   air leakage can be effectively minimized by making the envelope airtight; completely preventing air leakage can seldom, if ever, be achieved.

MOISTURE

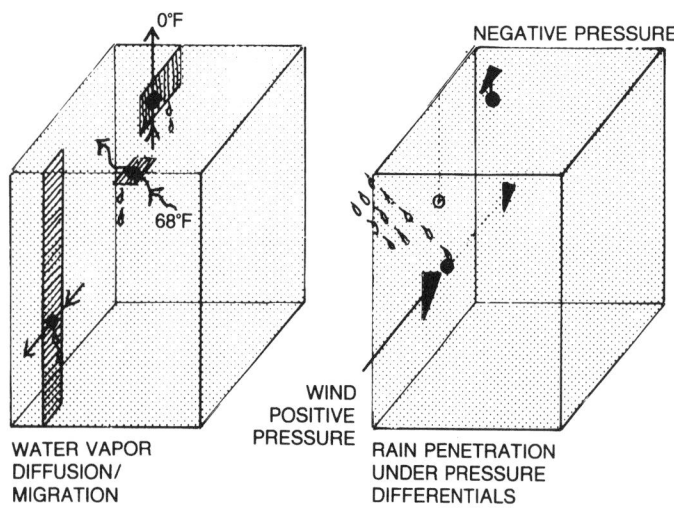

0°F

68°F

NEGATIVE PRESSURE

WIND
POSITIVE
PRESSURE

WATER VAPOR
DIFFUSION/
MIGRATION

RAIN PENETRATION
UNDER PRESSURE
DIFFERENTIALS

Air leaking through the envelope will also carry with it water vapor suspended in it. Water vapor condenses out of the air/vapor mixture when the mixture drops below a specific temperature - the dew point - at which the air is fully saturated, i.e., at 100 percent relative humidity. Water vapor will also migrate from an area of higher vapor pressure to an area with lower vapor pressure, and will condense upon reaching the dew point:
*   condensation may occur within the envelope, which may lead to premature failure of the envelope.
    Rain water may be drawn into or through an envelope by differences in air pressure across the skin:
*   wind pressures against the exterior surfaces will be greater than interior air pressures, and such difference then becomes the driving force for water and air penetration into the interior.

SOUND

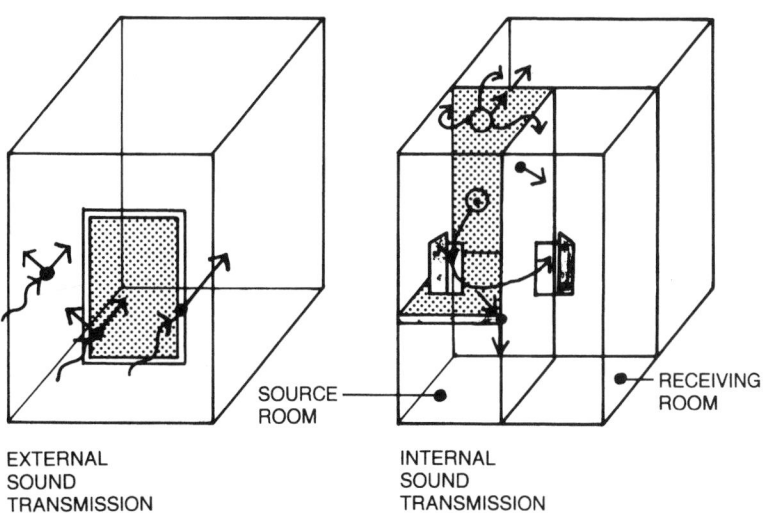

EXTERNAL
SOUND
TRANSMISSION

SOURCE
ROOM

INTERNAL
SOUND
TRANSMISSION

RECEIVING
ROOM

Transmission of external air borne sound through an envelope may have to be controlled for the comfort of the occupants:
* transmission of sound through a barrier is inversely proportional to the mass of the barrier; light envelopes will be less effective than heavy ones.
* any opening in the envelope will effectively destroy the integrity of the envelope as a barrier to sound transmission.

Interior components of an enclosure, such as floor assemblies, partitions, ceilings may also be required to control sound:
* air borne sound within a space by absorbing it to reduce its intensity; reflecting and/or scattering it to improve its diffusion.
* air borne sound transmission from one space to adjacent ones.
* structure-borne sound by damping it, or by isolating its source.

FIRE

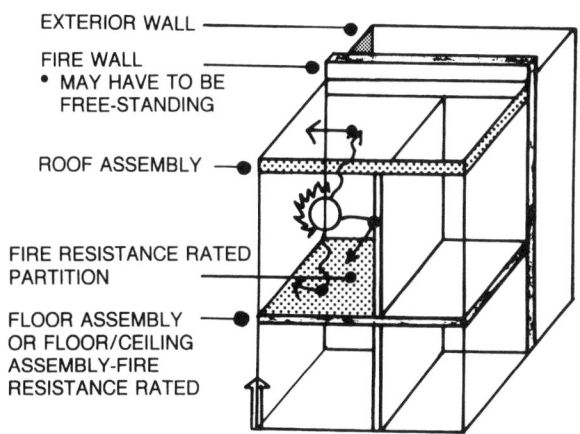

EXTERIOR WALL

FIRE WALL
* MAY HAVE TO BE FREE-STANDING

ROOF ASSEMBLY

FIRE RESISTANCE RATED PARTITION

FLOOR ASSEMBLY
OR FLOOR/CEILING
ASSEMBLY-FIRE
RESISTANCE RATED

Envelope of an enclosure and/or its components may be required to resist the effects of fire for a specific minimum of time without a significant reduction of structural strength and/or stability to ensure the safety of the occupants:
* generally the interior structural assemblies of an enclosure are required by building codes to be fire-resistant rated for a specific time interval.
* walls to prevent the spread of interior fire, commonly referred to as fire walls, may be required to compartmentalize large spaces, or to separate different activities within a single enclosure.
* exterior walls may have to be fire-resistant rated when separation from adjacent enclosures is less than a specific minimum.
Fire-resistant ratings are established based on tests of full scale assemblies, constructed of specific materials in a particular manner, and such construction may generally not be modified.

DEFINITIONS & PHYSICAL PRINCIPLES

MEANS OF SUPPORT

DEFINITIONS & PHYSICAL PRINCIPLES

COLUMNS AND GIRDERS

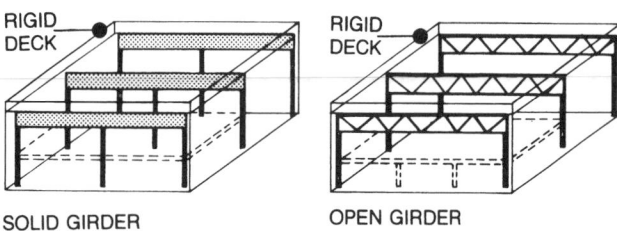

SOLID GIRDER

OPEN GIRDER

COLUMNS AND GIRDERS/BEARING WALLS

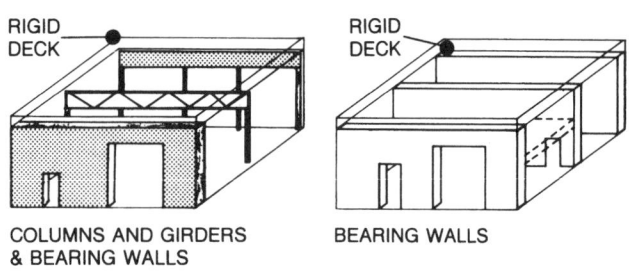

COLUMNS AND GIRDERS
& BEARING WALLS

BEARING WALLS

CURVED GIRDER

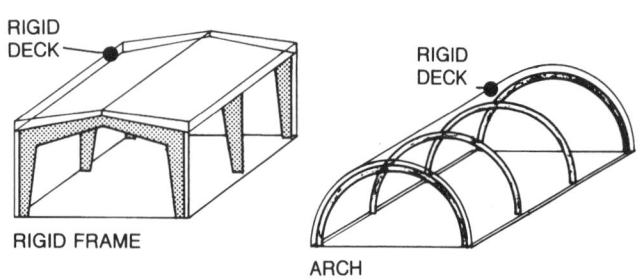

RIGID FRAME

ARCH

Means of support for the envelope and/or interior elements of an enclosure may consist of:
■ Horizontal elements to safely resist:
• gravity loads of a roof deck, or also of floor deck or decks, including all loads superimposed or likely to be superimposed upon such decks.
• gravity loads of walls when supported on such horizontal elements.
• lateral loads either acting directly on such elements or transmitted to them through walls which they brace.

■ Vertical elements to transmit gravity and lateral loads imposed upon them by horizontal elements and/or walls to the foundation/ground.

Horizontal elements may be classified as:
■ Primary: when the deck assembly or the decking component of such assembly bears directly on them.
■ Secondary: when supporting the decking component of a deck assembly between widely spaced primary supports.

PRIMARY horizontal elements will be referred to as GIRDERS in all instances;
VERTICAL elements which support them as COLUMNS. SECONDARY vertical elements as FRAMING. Decking of a roof or floor assembly, whether alone or combined with framing to support it, will be referred to as DECK.

■ GIRDERS may be:
• Solid web also often referred to as beams of various materials, such as structural steel; solid or laminated wood; reinforced concrete.
• Open web commonly referred to as trusses of various materials, such as structural steel; wood.
• Pitched or curved such as trusses with pitched or curved top chords supported on columns.
• Curved in different configurations, with the girder and columns being combined into one element; generally referred to as arches.

Primary horizontal supports may also be walls which combine girders and columns into one element called bearing walls, or portions of walls may function as columns commonly referred to as pilasters to support girders.

CURVED DECK

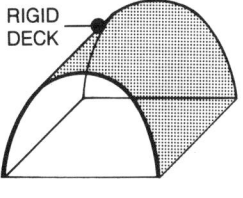

RIGID DECK

BARREL ARCH   DOME

Means of support for the envelope of an enclosure, but not for interior elements, may consist of:
Curved surface monolithic deck only commonly referred to as structural membrane or shell which combines girders and deck into one:
- Usually capable of transmitting loads in more than two directions to foundation/ ground.
- Highly efficient structurally when it is so shaped, proportioned, and supported that it transmits gravity and lateral loads acting upon it essentially in compression; without bending or twisiting.
- Curvature is principally influenced by requirements of load transfer; shapes may be barrel arches, domes, cones, hyperbolic paraboloids.
- Reinforced concrete is most commonly used in building construction.

POINT SUPPORTED

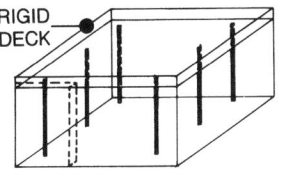

RIGID DECK   FLEXIBLE MEMBRANE

FLAT PLATE   TENSILE: NONPRESTRESSED

Means of support for the envelope of an enclosure may consist of individual columns and:
■ Flat monolithic deck only commonly referred to as two way slab which combines girders and deck into one:
- Of concrete reinforced for flexure in more than one direction, of uniform thickness throughout, or with drop panels at columns only, supported by any combination of columns.
- Columns may be arranged so that their center lines divide the deck into square, rectangular, triangular, or irregular panels; bearing walls may also be used.
■ Flexible membrane suspended or stretched between columns commonly referred to as posts and subject to tensile stresses only:
- flexible membranes are completely incapable of resisting bending stresses.
- nonprestressed when suspended; prestressed when stretched.

AIR PRESSURE

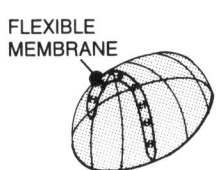

FLEXIBLE MEMBRANE   FLEXIBLE MEMBRANE

AIR SUPPORTED   AIR INFLATED

Means of support for an envelope consisting of a flexible membrane, which generally functions as the complete enclosure, may be air pressure only:
■ Air supported: when the interior is sufficiently pressurized to counteract the effects of gravity load of the membrane itself as well as all superimposed gravity and lateral loads.
- Interior is always under positive pressure and provisions to maintain such pressure are required at all penetrations through the membrane.
■ Air inflated: when completely supported by pressurized air entrapped within the membrane.
- Interior of air inflated enclosure is at atmospheric pressure.
■ Both types have to be anchored to a foundation or directly to the ground against displacement by wind forces, and/or to resist uplift of pressurization.

DEFINITIONS & PHYSICAL PRINCIPLES

DEFINITIONS & PHYSICAL PRINCIPLES

ENVELOPE

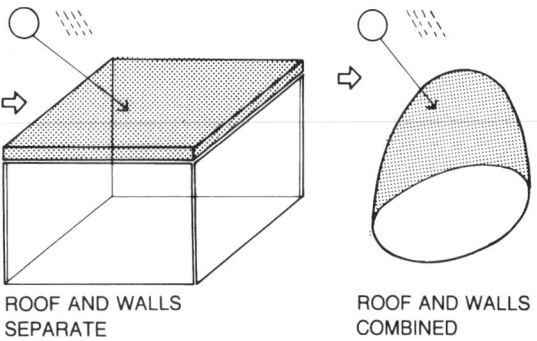

ROOF AND WALLS
SEPARATE

ROOF AND WALLS
COMBINED

ENVELOPE of an enclosure is a continuous air and watertight barrier, maintained in a given form by means of support to separate the contained environment from that external to it. The barrier or envelope consists of a roof assembly covering the contained space and wall assemblies surrounding it. Roofs and walls may be separate distinct elements, or essentially one, without any clear differentiation between them.

Envelope may be:
■ Flexible fabric membrane functioning as a complete enclosure.
■ Flexible roof assembly and rigid wall assemblies.
■ Rigid roof and wall assemblies of components selected to resist and/or control some or all environmental factors.

ROOFS

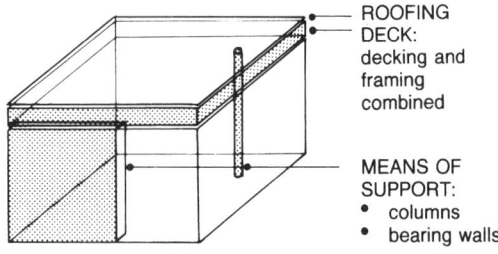

ROOFING
DECK:
decking and
framing
combined

MEANS OF
SUPPORT:
• columns
• bearing walls

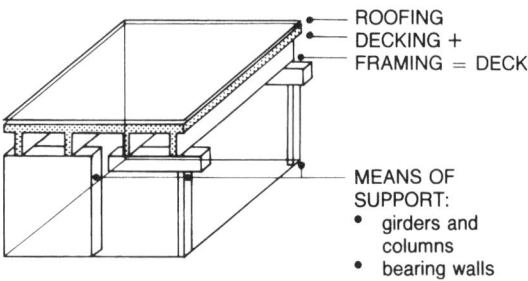

ROOFING
DECKING +
FRAMING = DECK

MEANS OF
SUPPORT:
• girders and
columns
• bearing walls

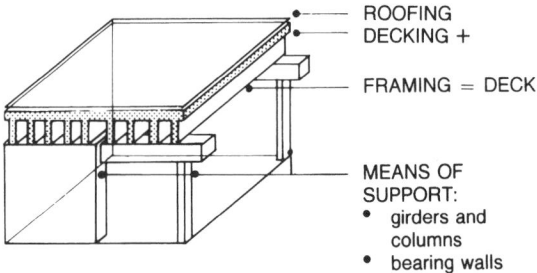

ROOFING
DECKING +

FRAMING = DECK

MEANS OF
SUPPORT:
• girders and
columns
• bearing walls

ROOF ASSEMBLIES commonly include:
■ Roofing: to resist the effects of environmental factors.
■ Deck: a substrate for the roofing which not only carries the roofing but also resists the effects of all forces acting on the assembly.
■ Means of support for the deck: such as girders, bearing walls, columns.

Roof decks may be:
■ Decking or substrate, only when such decking is capable of spanning between widely spaced primary supports without the need for any secondary framing:
• Long-span decking may be considered as combining decking and framing in one when its span exceeds an arbitrary maximum of eight feet.
■ Decking and widely spaced framing, eight feet or less on centers, with the framing spanning between widely spaced primary supports:
■ Decking and closely spaced framing, two feet or less on centers, with the framing spanning between primary supports.
■ Decking such as rigid panels or flexible membrane supported by a cable network.
• Flexible membranes function as decking and roofing combined.
■ Flexible membranes functioning as decking and roofing combined either suspended from or stretched over rigid vertical supports:
• Flexible membranes may also be air supported or air inflated. Rigid roof assemblies may be flat, pitched, curved; or any combination thereof.

WALLS

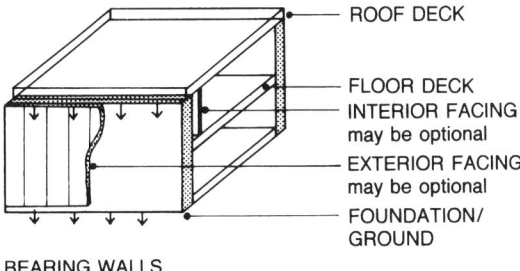

BEARING WALLS

- ROOF DECK
- FLOOR DECK
- INTERIOR FACING may be optional
- EXTERIOR FACING may be optional
- FOUNDATION/ GROUND

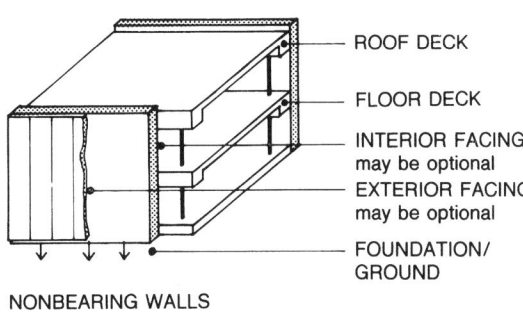

NONBEARING WALLS

- ROOF DECK
- FLOOR DECK
- INTERIOR FACING may be optional
- EXTERIOR FACING may be optional
- FOUNDATION/ GROUND

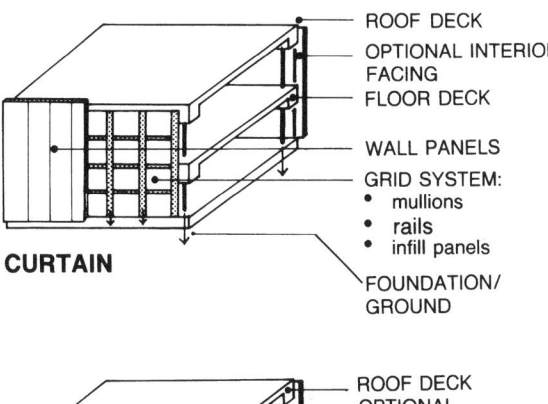

CURTAIN

- ROOF DECK
- OPTIONAL INTERIOR FACING
- FLOOR DECK
- WALL PANELS
- GRID SYSTEM:
  - mullions
  - rails
  - infill panels
- FOUNDATION/ GROUND

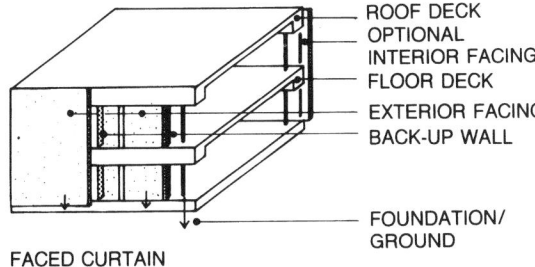

FACED CURTAIN

- ROOF DECK
- OPTIONAL INTERIOR FACING
- FLOOR DECK
- EXTERIOR FACING
- BACK-UP WALL
- FOUNDATION/ GROUND

WALL ASSEMBLIES commonly include:

- Structural core: to resist gravity loads of the assembly itself, those that might be superimposed upon it, and lateral loads:
  - Structural core may be a separate component such as framing or the core may function as the complete wall assembly.
- Exterior facing: to resist the effects of environmental factors:
  - Exterior facing may be a separate component attached to and supported by the structural core, or it may be an integral part of such core.
- Interior facing: either as a required component to complete the wall assembly such as over framing or as an optional component added to satisfy functional and/ or visual requirements.
- Means of support against lateral forces, either wind or seismic: columns or pilasters when span of wall is horizontal, floor and roof assemblies when loads are transferred vertically.

WALL ASSEMBLIES may be classified as:

- Bearing: walls carrying superimposed gravity loads in addition to their own weight.
- Nonbearing: walls not carrying superimposed gravity loads in addition to their own weight, whether capable of carrying such loads or not, and supported directly on foundations/ground.
- Curtain: nonbearing walls secured to and supported by the structural frame of an enclosure:
  - Grid type: of vertical and horizontal framing members supported by floor or roof assemblies and holding between them and supporting various in-fill panels.
  - Wall panel type: of prefabricated panels spanning between floor and roof or between floors and functioning as the complete wall assembly.
- Faced: walls functioning as continuous backup and support for various types of facings:
  - Backup walls may be bearing, nonbearing, or curtain.
  - Facings may be: off-site fabricated panels or units such as metal-faced composite panels, ceramic tile units; or on-site made, such as stucco.
- Composite: when consisting of two or more wythes of masonry where at least one wythe is dissimilar to other wythes.
- Cavity: of two wythes of masonry built to provide an air space within the wall.
- Shear: when wall is to resist horizontal forces in the plane of the wall.

DEFINITIONS & PHYSICAL PRINCIPLES

PART 1

ELEMENTS OF INTERIOR SPACE

FLOORS

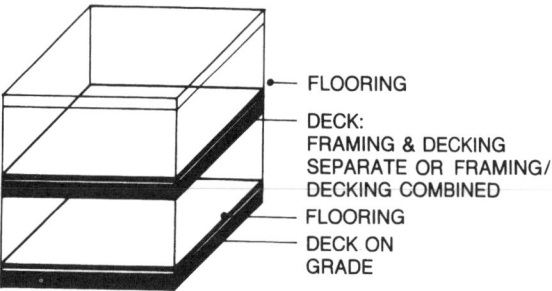

FLOORING

DECK:
FRAMING & DECKING
SEPARATE OR FRAMING/
DECKING COMBINED

FLOORING

DECK ON
GRADE

Floors are flat, commonly horizontal surfaces within the envelope of an enclosure.

Floor assemblies commonly include:
■ Flooring: to resist the effects of traffic over the surface of the floor deck.
■ Deck: to support all loads imposed on the floor assembly.
■ Means of support for the deck.

Floor decks may be:
■ Decking only when capable of spanning between widely spaced primary supports without the need for any secondary framing, or when supported by the ground.
■ Decking and secondary framing spanning between widely spaced primary supports.

CEILINGS

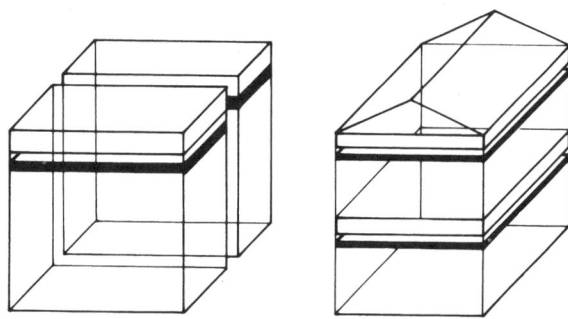

Ceiling membranes are nonstructural components of an enclosure, depending on their support on floor or roof assemblies. ceilings may be:
■ Visual screens and/or functional separation between an inhabited space and the underside of a floor or roof assembly above: minimizing dust accumulation, concealing service lines, minimizing sound reflection when of sound absorbing materials, containing air within the space above them when such space is used to return or supply conditioned air in lieu of ducts.
■ Integral components of floor or roof assemblies when such assemblies are required to be fire-resistant rated: protecting the structural framing and/or decking from effects of fire when the structural components are not of themselves fire-resistive.

PARTITIONS

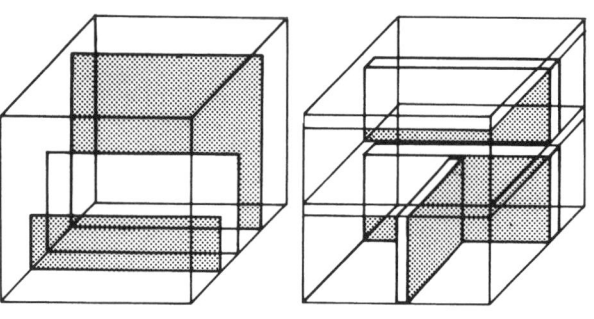

The space within an envelope may be defined or divided vertically by partitions to:
■ Control movement through enclosed space.
■ Provide visual and/or speech privacy to the occupants.
■ Enclose different environments within a single envelope.
■ Separate or isolate different activities.
■ Prevent the spread of fire within the enclosed space.

Partitions may be:
■ Of different heights: below eye level, to above eye level, to ceiling, or to underside of floor or roof assembly above:
■ Fixed, relocatable, or operable; supported on, or suspended from floor or roof assemblies:
• When supported, they are capable of carrying their own weight, but generally not superimposed loads.

MEANS OF CIRCULATION

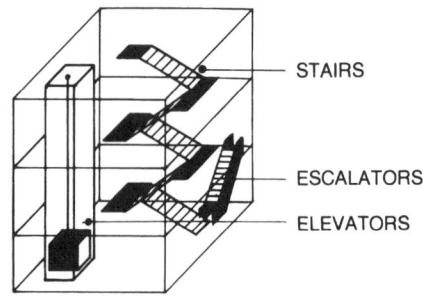

STAIRS

ESCALATORS

ELEVATORS

• CIRCULATION
• CONVEYANCE

Means of conveyance/circulation between two or more floors or levels may include:
■ Ladders for limited access
■ Stairs for foot traffic
■ Ramps for foot traffic, handicapped person's access, vehicular traffic
■ Escalators for continuous movement of large number of persons
■ Elevators for intermittent rapid movement of persons, goods or vehicles
■ Dumbwaiters for continuous or intermittent movement of goods or records
■ Various systems of conveyors for continuous transport of goods
■ Moving sidewalks

SERVICES

AIR CONDITIONING

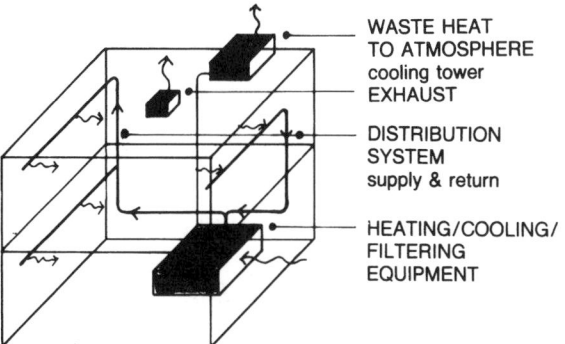

WASTE HEAT
TO ATMOSPHERE
cooling tower
EXHAUST

DISTRIBUTION
SYSTEM
supply & return

HEATING/COOLING/
FILTERING
EQUIPMENT

Air Conditioning consists of simultaneous control of temperature, humidity and the quality of air within an enclosure, at comfort levels acceptable to the occupants.

Air conditioning systems generally include:
- Boiler to supply sufficient heat to replace that transmitted and lost to the exterior through the envelope.
- Equipment to cool and dehumidify the air: chiller, condenser, fans, pumps.
- Filters to clean the air of pollutants.
- Humidifier to maintain the air at desired level of relative humdity.
- Distribution system: supply and return.
- Fresh air supply to air conditioning equipment to replace air exhausted or lost through exfiltration.
- Exhaust systems to rid the interior of polluted air.

WATER/WASTEWATER SYSTEMS

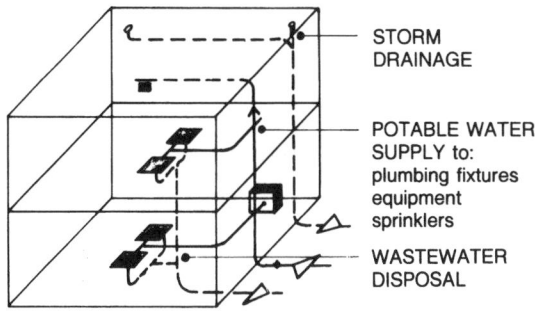

STORM
DRAINAGE

POTABLE WATER
SUPPLY to:
plumbing fixtures
equipment
sprinklers

WASTEWATER
DISPOSAL

All buildings housing human activity must be provided with portable water in quantities sufficient to meet the needs of the occupants and of all related activities:
- Cold and hot water for sanitation, food preparation, cleaning.
- Cold water for sprinkler systems, service equipment.

Water supply systems distribute water to fixtures or devices which serve as the terminals of such system:
- Distribution systems are designed based on the probable maximum water demand, taking into account anticipated rate and duration of flow through all fixtures and frequency of use.

Waste water systems are largely based on the anticipated quantities of water flow through all fixtures.

Storm drainage systems are sized based on: area to be drained; amount of rainfall per unit of time.

POWER/LIGHTING/COMMUNICATIONS

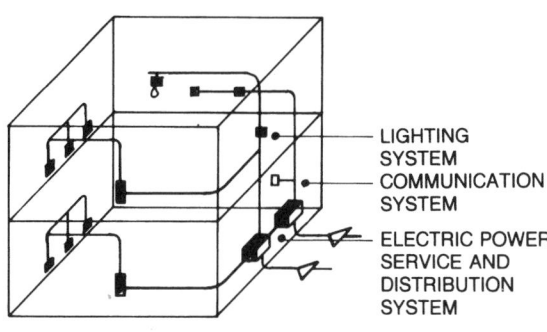

LIGHTING
SYSTEM
COMMUNICATION
SYSTEM

ELECTRIC POWER
SERVICE AND
DISTRIBUTION
SYSTEM

Electric service: conductors and equipment for delivering energy from external or internal electricity supply system to the wiring system of an enclosure being served.
- Service equipment usually consists of a circuit breaker and accessories located near the point of entry.
- Wiring system: distributes power from service equipment to the point of use.

Electric service is commonly provided for several different kinds of loads: lighting, various plug-in equipment and appliances, air conditioning, elevators, process equipment, communications or security systems.
- These loads may vary in voltage and times of service, and for the determination of the probable maximum load a demand factor - the ratio of probable maximum demand for power to the total connected load - is generally used.

SAFETY/SECURITY SYSTEMS

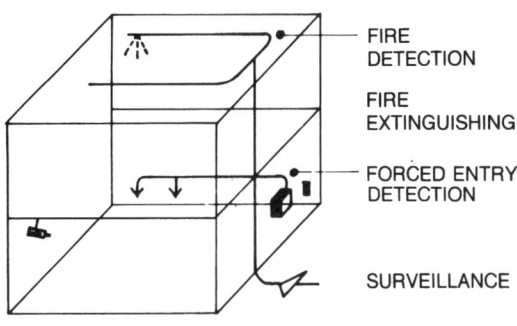

FIRE
DETECTION

FIRE
EXTINGUISHING

FORCED ENTRY
DETECTION

SURVEILLANCE

Safety systems generally include fire suppression and/or detection systems such as:
- Automatic sprinklers: either wet systems when spaces served are maintained at temperatures to prevent freeze-up, or dry systems where exposed to freezing temperatures.
- Chemical systems: used when materials contained may not be readily extinguished with water alone. Some of the more commonly used are:
  - Foamed chemicals: produced by mixing chemical and water.
  - Gas: such as carbon dioxide or Halon discharged from pressurized storage cylinders.
  - Dry chemicals: discharged from central supply through piping or from portable equipment.
- Fire detectors: fixed-temperature; rate-of-rise; photoelectric; combustion-product sensing; flame sensing devices.

DEFINITIONS & PHYSICAL PRINCIPLES

PRESSURE

DEFINITIONS & PHYSICAL PRINCIPLES

WIND PRESSURES

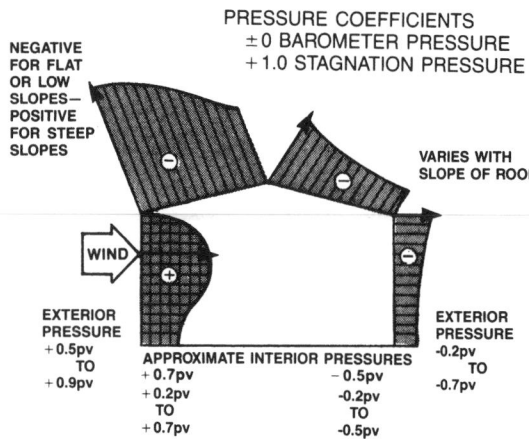

PRESSURE COEFFICIENTS
±0 BAROMETER PRESSURE
+1.0 STAGNATION PRESSURE

NEGATIVE FOR FLAT OR LOW SLOPES— POSITIVE FOR STEEP SLOPES

VARIES WITH SLOPE OF ROOF

WIND

EXTERIOR PRESSURE
+0.5pv TO +0.9pv

EXTERIOR PRESSURE
-0.2pv TO -0.7pv

APPROXIMATE INTERIOR PRESSURES
+0.7pv   -0.5pv
+0.2pv   -0.2pv
TO   TO
+0.7pv   -0.5pv

WIND PRESSURES ON BUILDINGS OF SIMPLE RECTANGULAR SHAPE

Wind pressure on the windward wall of an enclosure builds up because the velocity of the air stream is reduced upon reaching any obstruction in its path. If air flow could be stopped completely at such wall, then all of the kinetic energy of the wind would be translated into pressure on the wall, resulting in maximum pressure at a given velocity, also referred to as stagnation pressure:
- stagnation pressure on unit area equals the square of wind speed times a constant related to the mass density of air.
■ Pressure distribution patterns do not vary with the wind velocity, and the relationship of pressures at one point relative to another is constant for envelopes of different sizes but of similar proportions.
■ Pressure distribution on walls will vary with wind directions: pressure will approach a coefficient of 1.0 in the central portion when acting perpendicular to a wall surface, a coefficient of 0.8 when at 45 degrees to the plane of the wall.
■ Pressure distribution on roofs will vary with shape of roof and wind direction: • flat roofs will be subject to negative pressures; • windward surfaces of sloping roofs may be under positive pressure, with the critical angle being: 5 percent for small low enclosures, up to 25 to 30 percent for higher, larger enclosures.

NEGATIVE PRESSURE ON ROOFS

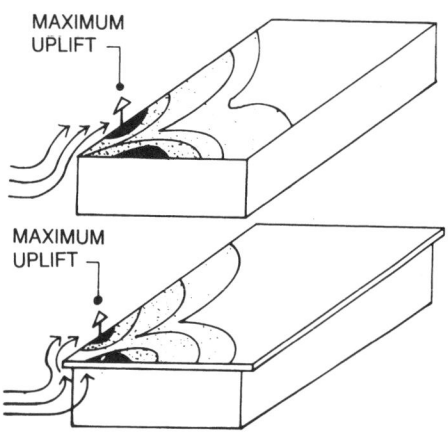

MAXIMUM UPLIFT

MAXIMUM UPLIFT

Flat or nearly flat roofs will always be under negative pressure whenever there is air flow over them:
- when direction of wind is perpendicular to a wall, maximum negative pressure on the roof, with a coefficient of about -0.8, will occur over the center portion of the windward wall, diminishing in the direction of air flow over the roof to a coefficient of -0.2.
- when direction of wind is at 45 degrees to the plane of a wall, maximum negative pressure will occur at the windward corner, reaching a coefficient of -3.0 or more.
- when the roof overhangs walls, the flow of air is stopped at such overhang, resulting in positive pressure with a value close to stagnation pressure under the overhang. Air is still forced to rise over the leading edge of the overhang, resulting in very high negative pressure: the total uplift on the overhang is then the sum of the pressure below and the suction above.
- when parapets are provided, the negative pressure at a leading corner will increase slightly when the parapet is low, and will be significantly reduced when parapets are high. The effective height of a parapet for reducing pressure is a function of building height.

PRESSURES ON WALLS

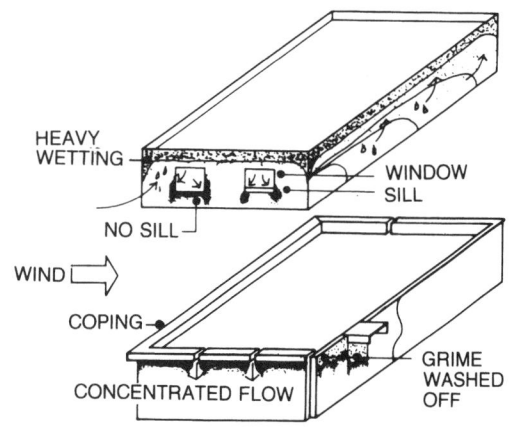

HEAVY WETTING

WINDOW SILL

NO SILL

WIND

COPING

CONCENTRATED FLOW

GRIME WASHED OFF

Positive air pressures against walls affect air and water penetration through them. Wind also influences the deposition and migration of rain on walls of an enclosure:
- upper surfaces and edges of windward walls will be wetted more than center and lower portions.
- parapets at roof level are likely to be the most wetted parts.
Flow of rain water over a wall surface will generally not be even:
- concentration at irregularities on the surface will occur.
- amounts will vary due to differences in air pressures over a single wall plane.
Rain water flowing over a wall surface will tend to wash off and will often redeposit grime which had accumulated:
- rate of flow over glass will be faster than over porous materials: when they are adjacent to each other, a concentration of water at the bottom of a glass panel will occur.
- horizontal projection will shelter surfaces right below them, thus minimizing wash-off of grime on such surfaces.
In both instances, differential streaking is likely to result.

MOVEMENT

MOVEMENT TYPE	DESCRIPTION

LINEAR

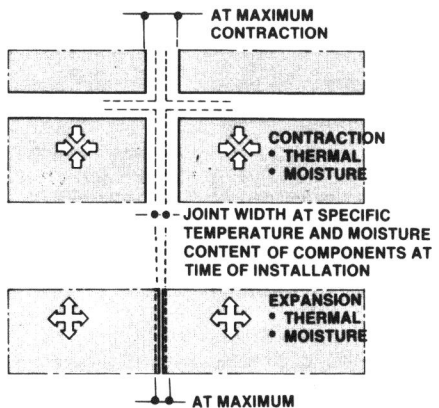

Linear movement - dimensional change - in components is due to:

■ **Variations in internal temperature** of the components:
* all components, regardless of the material used, are subject to thermal movement.
* exterior temperatures change all the time; thus thermal movement in components is a continuous process.

■ **Variations in internal moisture content:** porous materials may absorb and release moisture; this, in turn, may cause expansion/contraction, or shrinking/swelling in the components.

■ **Dimensional changes** in porous materials due to thermal and moisture movement may be cumulative or off-setting: a rise in temperature may be simultaneous with either an increase or a decrease in the relative humidity of surrounding air.

■ **The design width of a joint should be based on:**
* amount of **movement** to be expected: thermal only for non-porous materials; thermal and moisture for porous materials;
* **temperature,** or temperature and moisture content, of the components at the time of installation in relation to in-service conditions to be expected.

DIFFERENTIAL

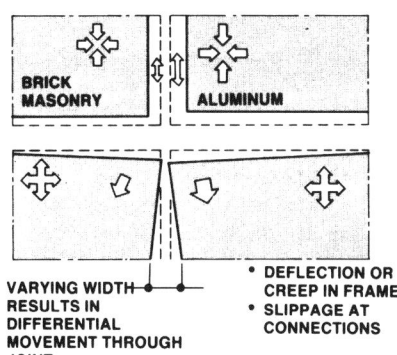

Differential movement - differential dimensional changes - between components may be due to:

■ **Different coefficients of expansion** of the components:
* aluminum has a coefficient of expansion of 13×10^{-6} inch per inch per degree F, and a 10 foot square panel will contract 0.155 inch for a temperature drop of 100°F.
* an adjacent masonry panel of the same size will only contract 0.037 inch under the same conditions.
* coefficient of expansion for wood perpendicular to grain varies from 19 to 32×10^{-6} in/in/°F, while the coefficient of expansion parallel to grain is only 2.1 to 3.6×10^{-6}.

■ **Different rates of shrinkage/swelling** due to changes in internal moisture content between components, or within the component.
* dimensional changes in wood during loss or gain of moisture vary depending on whether expansion is tangential in the direction of growth rings, or radial across growth rings.
* a piece of green Douglas fir will shrink about 1.5% tangentially and about 2.5% radially when dried to 20% moisture content, the volumetric change being about 3.7%

■ Differential movement may also result from **movements in supporting frame** acting simultaneously with thermal and/or moisture movements.

TRANSVERSE

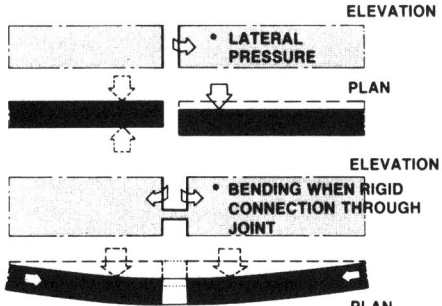

Transverse movement or movements perpendicular to the plane of components may result from:

■ **Differences in the magnitude of lateral loads** acting on adjacent components; or from bending stresses.
* differences in pressures on vertical components, such as wall panels;
* moving loads over horizontal components, when the edge of one is free to deflect.
* deflection due to thermal or moisture movement in a component with two edges restrained which is adjacent to an unrestrained component.
* deflection due to horizontal loads in a free edge of a component next to an attached edge of another component.
* differences in deflection between a relatively stiff component next to a flexible one.
* deflection in two components with rigid, spaced connections between them, when they are restrained at their edges or when under lateral load.
* deformation in components due to differentials in temperature between opposite faces of one component, or differences in moisture gain/loss, such as warping in wood.

DEFINITIONS & PHYSICAL PRINCIPLES

FORCES/FACTORS

MOVEMENT DUE TO CHANGES IN TEMPERATURE

■ **Deformations induced by changes in temperature** may result in the cracking of inelastic materials, such as concrete or stone, even under moderate changes in temperature:

- concrete will expand or contract 0.05 percent for a change in temperature of 80°F;
- if free to contract, it should not fail in compression since deformation required for it to fail in compression is 0.10 percent;
- if tensile stresses develop due to restraint, cracking may result, since a deformation of only 0.01 percent in tension will cause failure.

■ **Corresponding values** for some other typical inelastic facing materials subject to 80°F temperature changes are:

- brick: 0.024 percent deformation: deformation required to cause failure in compression: 0.20 percent; in tension: 0.016 percent.
- marble and dense limestone: 0.024 percent deformation; deformation required to cause failure in compression: 0.25 percent; in tension: 0.006 percent.

■ **Facings, when of thin sections, will absorb and lose heat quickly** with resultant wide dimensional fluctuations, as illustrated in the table below for the more commonly used facing materials:

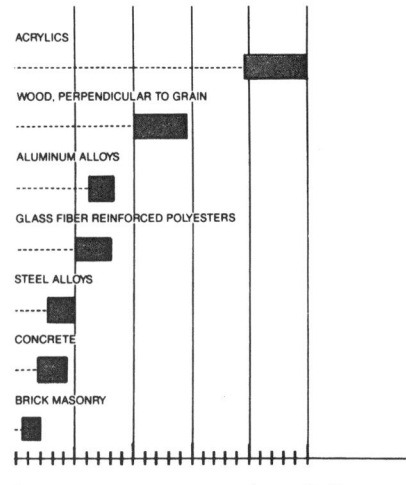

ACRYLICS

WOOD, PERPENDICULAR TO GRAIN

ALUMINUM ALLOYS

GLASS FIBER REINFORCED POLYESTERS

STEEL ALLOYS

CONCRETE

BRICK MASONRY

0   10   20   30   40   50 60×10^{-6} per °F

■ **Thus a 10 foot long section of acrylic plastic** with a coefficient of expansion of 40×10^{-6} will elongate by 0.72 inches for a 150°F rise in temperature, while a section of the same length of aluminum will expand 0.232 inches and stainless steel only 0.104 inches under the same conditions.

■ **Materials should be assembled with a minimum of restraint** to allow for volumetric changes:

- failures occur when clearances are insufficient, fasteners do not allow for movement, or deformations are greater than sealants or gaskets can accommodate.
- when materials are restrained by connection to other materials, deformations from changes in temperature may induce stresses in them, and if these exceed the strength of such materials they will crack or deform permanently.

■ **When facings are restrained either at their edges or by fasteners,** local deformations such as buckling may result.

■ **Temperature differential across a heavy section results in differential thermal expansion** which causes warpage or deflection:

■ **Precast concrete units of a roof deck** were observed to rise and fall at the center of the span on a daily cycle corresponding to the hours of sunlight:
- the units were 20 feet long and the movement due to thermal expansion/contraction amounted to about ¼ inch.
- in this particular instance, a plaster ceiling was directly applied to the underside of the precast units; cracks in the ceiling could not be patched since they opened and closed almost daily.

■ **Thermal movement may be differential** even in facings of the same material when small or thin parts of low thermal capacity are interconnected with larger sections which will tend to heat up much more slowly.

■ **Light colored and/or reflective surfaces** will minimize radiant heat gain and thus also reduce thermal movements.

MOVEMENT DUE TO CHANGES IN MOISTURE CONTENT

Materials capable of absorbing water generally expand as they do so and contract again on drying, but in some instances moisture deformation is not reversible:

- concrete, mortars, and plasters generally contract more during initial drying than during any following reversible deformation.

■ **Deformations induced by wetting from dry to saturated state,** or drying from saturated state for some inelastic materials are:

- concrete: 0.03 percent.
- brick: 0.007 percent.
- marble: 0.001 percent.
- deformations required to cause failure are the same as for temperature-induced deformations.

■ For information on moisture induced deformations in wood, refer to WOOD/WOOD PRODUCTS.

Differential shrinkage may result in differential stresses and deflection:

- assuming 0.03 percent shrinkage of concrete, a 4 inch thick simply supported slab with reinforcing at one face only and spanning 10 feet would deflect about ⅛ inch or more at the nonreinforced face.

VAPOR MIGRATION

Moisture content of warm air within a heated building is almost always higher than that of cold outdoor air. The resulting difference in water vapor pressures tends to move the vapor from the high pressure interior zone to the low pressure outdoor zone, even against a reverse air pressure.

■ **Water vapor migrates:**

- by diffusion through the components of the wall or roof assemblies.
- by air leakage through cracks in wall/roof assemblies, when outdoor air pressures are lower than interior pressures.

■ **The rate of water vapor migration by diffusion** depends upon:

- difference in vapor pressure.
- permeability of components of the building envelope.
- length of the flow path.

METHODS OF ASSEMBLY

SITE ASSEMBLED

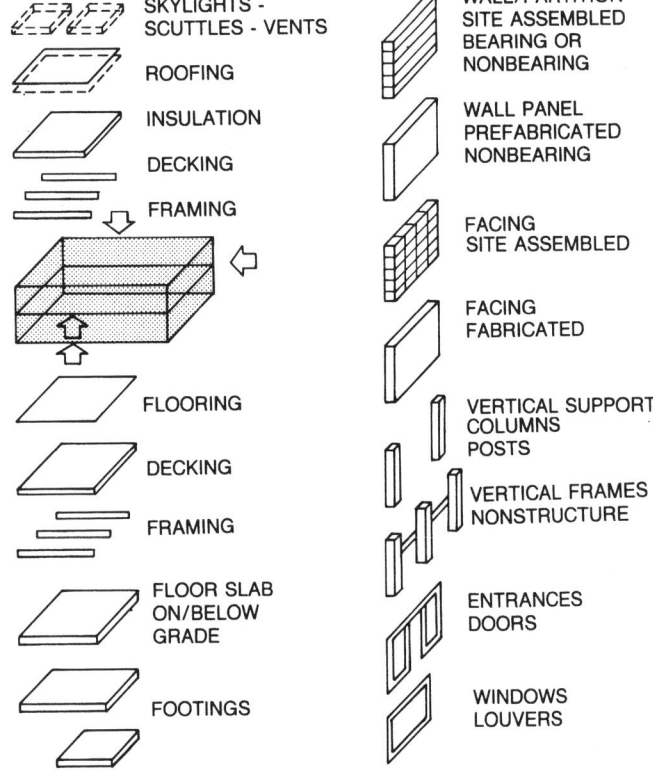

SKYLIGHTS - SCUTTLES - VENTS

ROOFING

INSULATION

DECKING

FRAMING

FLOORING

DECKING

FRAMING

FLOOR SLAB ON/BELOW GRADE

FOOTINGS

WALL/PARTITION SITE ASSEMBLED BEARING OR NONBEARING

WALL PANEL PREFABRICATED NONBEARING

FACING SITE ASSEMBLED

FACING FABRICATED

VERTICAL SUPPORTS: COLUMNS POSTS

VERTICAL FRAMES NONSTRUCTURE

ENTRANCES DOORS

WINDOWS LOUVERS

DEFINITIONS & PHYSICAL PRINCIPLES

The process of contructing any enclosure consists of assembling and joining together diverse individual parts into an integrated whole.

The commonly used method, which allows for the widest choice of materials and components as well as for a variety of sizes and shapes for a given enclosure, is the site assembly of a number of specifically selected constituent parts such as:

■ **Basic ingredients of heterogeneous materials, fully processed into the finished product at the site; such as:**
* cement, aggregate, and water into concrete
* gypsum and water into plaster.

■ Materials processed/fabricated offsite into various forms and sizes and joined on-site into their final configuration; such as:
* masonry units into walls.
* felts and bitumen uinto roofing or waterproofing membranes.

■ Materials fabricated into units and assembled into their final form on-site; such as:
* resilient tile or sheets into flooring.
* concrete or wood panels into decking.
* structural steel girders, beams, columns into framing systems.
* wall panels, wall facing panels.

■ **Components of one or several different materials fabricated and preassembled off-site into self-contained units of diverse complexity and installed in their final location on-site**
Components may include:
* windows, doors and frames or prehung door units.
* skylight units.
* movable partitions.
* prefabricated modules, such as shower compartments.
* mechanical equipment: air conditioning units, boilers, heaters, fans, pumps.
* electrical equipment : switchgear, lighting fixtures, motors.
* conveying equipment: elevators, escalators.

SITE ASSEMBLED - PREENGINEERED

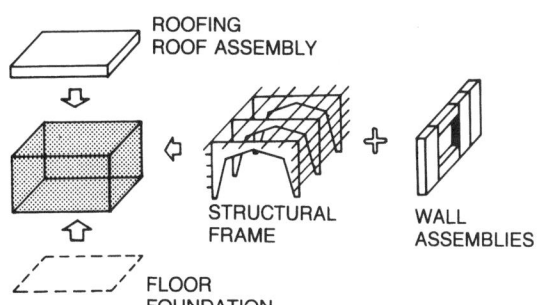

ROOFING
ROOF ASSEMBLY

STRUCTURAL
FRAME

WALL
ASSEMBLIES

FLOOR
FOUNDATION

Pre-engineered enclosures are envelopes and means of support only, assembled on-site, of standardized materials and components:
■ Forms of enclosure are generally limited to several types.
■ Means of support are available in standard types, such as solid girders, open girders, arches, of several standard spans; spaced at several fixed intervals.
• means of support are commonly designed to function under specific loading conditions only.
■ Envelopes consist of standardized materials supplied as part of the system.
• optional components such as doors, windows, louvers are generally available.

PREFABRICATED - SITE ASSEMBLED

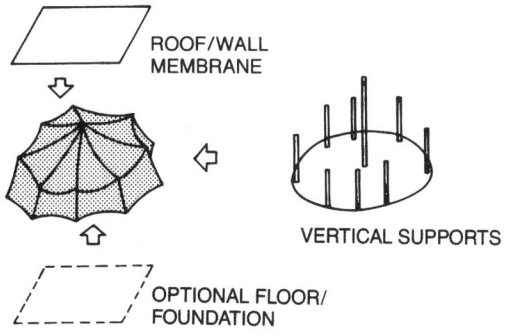

ROOF/WALL
MEMBRANE

VERTICAL SUPPORTS

OPTIONAL FLOOR/
FOUNDATION

Prefabricated enclosures consist of flexible fabric membranes and means of support: the membranes are shipped to the site completely fabricated, ready to be suspended from or stretched over site erected supports:
■ Membranes are custom made for each specifc installations and are only limited as to shape, size, or span between supports by properties of the membrane itself and/or manufacturing processes.

PREFABRICATED - SITEERECTED

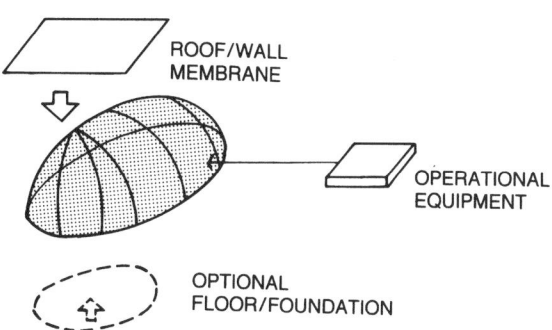

ROOF/WALL
MEMBRANE

OPERATIONAL
EQUIPMENT

OPTIONAL
FLOOR/FOUNDATION

Site-erected prefabricated enclosures are shipped to the site as complete units, ready to be erected:
■ Site work is commonly limited to anchoring the flexible fabric membrane to the ground or to a foundation, and inflating it.
■ Enclosure are custom made for each specific installation, and sizes are only limited by manufacturing processes.
■ Components - such as blowers to maintain pressure and/or to air condition the interior, entrance assemblies - are generally available from the manufacturer of the enclosure.

DEFINITIONS & PHYSICAL PRINCIPLES

SITE-ASSEMBLED ENCLOSURE

DEFINITIONS & PHYSICAL PRINCIPLES

FORM	DESCRIPTION
RECTANGULAR/CURVED: flat roof	**Rigid flat or nearly flat roof decks covering just about any arrangement of walls:** • area within exterior walls may be of any size and-shape: square, rectangular, oval, circular, serpentine, or any combination thereof. • parts of the same enclosure may be of varying heights, sizes, and shapes. • may be combined with sections having pitched or curved roofs. • may be constructed to be expanded horizontally and/or vertically at a later date.
RECTANGULAR/CIRCULAR: pitched roof	**Pitched roofs either double or single pitch covering various arrangements of walls:** • area within exterior walls is generally limited in size for double-pitched roofs when space within the roof construction is not being utilized, primarily due to economic considerations. • shape of the total enclosure is influenced by the shape of the roof, but a diversity of arrangements is possible. • may be used as the top section of a multi-level enclosure. • may be expanded horizontally but generally not vertically.
RECTANGULAR: curved roof	**Rigid roofs curved in one direction either linear or intersecting covering various square or rectangular arrangements of walls;** • area within exterior walls generally limited to maximum span of roof assembly. • shape of enclosure largely determined by shape of roof, limited number of possible arrangements. • may be used as the top section of a multi-level enclosure. • may expand horizontally but not vertically.
RECTANGULAR: curved roof/walls	**Rigid roofs curved in one direction and functioning as the envelope, defined by walls in the other direction, commonly one story in height:** • area limited by span of roof/walls across the enclosure, length not limited. • shape limited to square or rectangular; sections may be offset laterally. • roof and walls may also be separate elements in the long dimension, such as in rigid frames.
CIRCULAR: curved roof/walls	**Rigid combined roof/walls, or a combination of roof/walls and walls, over circular or modified circular arrangement of walls:** • area within exterior walls limited by maximum span of means of support. • shape limited to hemispheres with different span-to-rise ratios; hemispheres may be modified by slicing off lower portion on three or more sides vertically. • may be raised above ground and supported by columns or bearing walls. • may be incorporated into a flat roof assembly. • cannot be expanded once constructed.

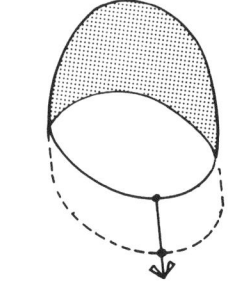
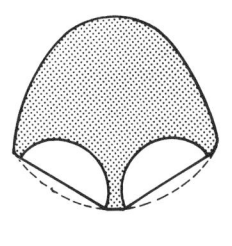

Means of support may be:
- Columns only for monolithic roof/floor decks.
- Columns and girders of either solid or open cross section for all types of decks:
  - Girders for roofs may be slightly pitched to provide for positive drainage.
- Bearing walls when enclosures are of limited height.
- Combination of bearing walls, columns, and girders.

Envelope:
- Roofs: various types of roof decks with built-up or single-ply roofing.
- Walls: bearing, nonbearing, or curtain.
- curtain walls may be grid or wall panel types.

Elements of interior space commonly include:
- Floors: which may or may not be of the same construction as roofs.
- Partitions.
- Ceilings.
- Means of circulation and conveyance.

Means of support may be:
- Columns and pitched girders of solid cross section for relatively short spans.
- Columns and girders of open cross section or trusses with pitched top chords for long spans and/or heavy loads.
- Bearing walls supporting light secondary framing, such as rafters or light trusses.

Envelope:
- Roofs: various types of decks with roofing of shingles, flat metal, metal roof panels functioning as decking and roofing combined, monolithic decks such as folded plate.
- Walls: bearing walls for enclosures of limited size; nonbearing, or curtain.

Elements of interior space commonly include:
- Floors: which may be of similar or different types of construction as roofs:
  - generally column and solid girder types when roofs are supported by solid or open girders.
  - commonly of light framing members such as joists or light trusses when roofs are supported by bearing walls.
- Partitions.
- Ceilings are optional, not used with certain types.
- Means of circulation and conveyance when multi-level.

Means of support may be:
- Curved monolithic decks: barrel vaults, either single, in series, or intersecting.
- require securing against outward thrust at supports.
- Columns and curved girders of solid cross-section.
- generally require securing against outward thrust at supports.
- Columns and open cross section girders with straight bottom cord and curved top cord, commonly referred to as bow string trusses.

Enclosure:
- Roofs: various decks with built-up, single-ply, or liquid-applied membrane roofing.
- Walls: nonbearing or curtain.
- curtain walls more commonly of the grid type.

Elements of interior space commonly include:
- Floors: of different types of construction than roofs.
  - commonly of columns and solid or open girders, may have additioal interior columns to reduce spans.
- Partitions.
- Ceilings may be used with trusses; commonly not used for barrel vaults.
- Means of circulation and conveyance when multi-level.

Means of support:
- Curved monolithic decks: thin concrete shells.
- Curved girders: either solid or open of constant or nearly constant cross section.

All require secure anchorage at foundation to counteract outward thrust.

Envelope:
- Roofs: monolithic or composite decks with adhered single-ply or liquid-applied elastomeric membranes, flat seamed metal, curved formed metal panels.
- Walls: nonbearing or curtain end walls:
- may require additional framing, vertical or vertical and horizontal, to resist lateral loads.

Elements of interior space:
- Floors: are generally slabs on grade; when envelope is raised above grade, floors commonly are independent structures within the enclosure.
- Partitions of various heights may be incorporated into the enclosure.

Means of support:
- Curved monolithic decks: thin concrete shells.
- Curved girders: either solid or open of constant or nearly constant cross section, commonly for fully circular forms only.

All required provisions at foundation to counteract outward thrust such as butt-ressing, or tension rings for fully circular forms.

Envelope:
- Monolithic or composite decks, generally with roofing of: liquid-applied elastomeric membranes, flat-seamed metal, plastic panels, shingles.
- Walls when used are commonly curtain, grid type.

Elements of interior space:
- Floors: are generally slabs on grade; when envelope is raised above grade, floors commonly are independent structures within the enclosure.
- Partitions of various heights but commonly not full height may be incorporated into the enclosure.

DEFINITIONS & PHYSICAL PRINCIPLES

TOLERANCES/CLEARANCES

ACCURACY IN CONSTRUCTION

The process of constructing or assembling an enclosure is one of combining diverse separately manufactured parts to produce a finished whole:

■ **The parts to be combined generally include:**

- materials fabricated on-site: such as concrete placed and finished in its final location.
- materials fabricated off-site and then finished on-site: such as wood studs cut to required size during installation.
- materials fabricated off-site and assembled on-site: such as masonry units, roofing, facing panels.
- materials fabricated and assembled into components off-site, and secured in their final location on-site: such as wall panels, windows, doors, equipment.

■ **The parts to be combined commonly are:**

- manufactured of various materials by different manufacturers using diverse processes.
- assembled into components using different means and methods.

■ **The parts manufactured off-site under the relatively controlled conditions of a factory should be, and generally are, made to be reasonably uniform:**

- even so, there is usually a difference between the actual size and that specified; and such difference may often be much larger than anticipated.

The conformance of a finished enclosure to its theoretical design values will principally depend on the extent of:

■ Permissible deviations from nominal dimensions generally referred to as dimensional tolerances of components and materials during manufacture.

■ Permissible deviations from theoretical values during site assembly of materials and components.

■ Allowance or clearances provided during design and detailing to accommodate inaccuracies which may be expected to occur during both the manufacture and erection.

■ Absolute precision is seldom re-quired and is almost never attainable in site-assembled enclosures.

TOLERANCES

Tolerances for manufactured products are commonly included in standards or specifications for such products or their assemblies which are issued by various agencies, institutes, or industry associations.

Standards or specifications may establish acceptable limits for:

■ **Deviations from:**

- standard size, shape, thickness.
- straightness or squareness.

■ **Variations in properties:**

- strength, hardness, composition.
- color or surface finish.

Tolerances vary for different components and materials, and such differences may become a factor when diverse products are joined or combined into a single assembly at the site:

■ Deviations from the standard may be cumulative or offsetting:

- it is not very likely that all deviations in different products obtained from various sources will be cumulative, but it should be realized that such may be the case, and proper fit may not be achieved despite the fact that tolerances for all constituent parts were within acceptable limits.

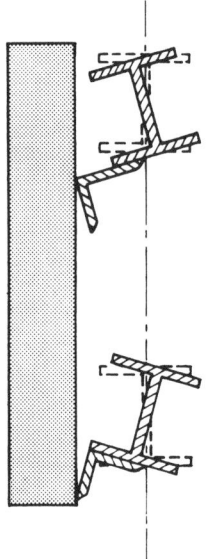

- beams and angles shown above with deviations from nominal shape exaggerated may be within standard rolling mill tolerances, but the end result will not be acceptable if straight vertical elements are to be properly attached to such an assembly: only one point on each angle can be aligned during erection.

TOLERANCES FOR STRUCTURAL FRAME

The means of support of an enclosure - the structural frame - is the part which is assembled first, and the level of accuracy achieved, or achievable, will influence the level of accuracy of the complete enclosure.

The accuracy of the structural frame is affected by:

■ **Proper layout:** the establishment of an accurate and maintainable system of reference points for controlling the alignment and levels of the principal elements of the structural frame as well as that of the entire enclosure.

■ **Maintaining close dimensional tolerances of the structural frame during erection,** whether such frame is fabricated on or off site.

The standards of accuracy generally referred to as erection tolerances have been established for both structural steel and reinforced concrete frames:

■ **Tolerances for structural steel frames** should conform to those given by the American Institute of Steel Construction under their Code of Standard Practice for Steel Buildings and Bridges.

- Girders and secondary framing of a steel framed roof or floor assembly may be erected level within specified tolerances but will permanently deflect between vertical supports when dead loads such as concrete floor fill are superimposed upon them: when floors must be level within narrow limits, provisions must be made to effectively counteract effects of deflection in framing members.

■ **Tolerances for reinforced concrete structures** should conform to requirements of Tolerances for Concrete Construction and Materials issured by the American Concrete Institute:

* Tolerances for concrete floors of ⅛ inch to 10 feet, presently given in above standard, require highly specialized placing and finishing techniques, and the indiscriminate use of this tolerance when not needed may result in unnecessary contractual and legal disputes: The American Concrete Institute cautions against the use of this tolerance except where definitely required, such as in the drive aisles of high lift fork truck warehouses.

■ It should be realized that the above standards namely for structural steel and reinforced concrete apply only to respective types of structural frames and not to composite ones, such as concrete floor slabs supported by steel columns.

* tolerances for composite construction should be established by the designer, and such tolerances should be realistic:
* the standard of accuracy thought to be attainable in construction is often much higher than that actually attained in practice.
* close tolerances will increase costs.
* ultimately may not be fully achievable in whole or in parts.
* designs and details that demand higher accuracy than can normally be achieved will inevitably be more troublesome to construct: present general level of accuracy attainable should be viewed as a fact of life, and design and detailing should include provisions to accommodate inaccuracies and/or errors which can reasonably be expected to occur.

CLEARANCES

Site assembly of enclosures can seldom if ever be accomplished under ideal conditions, and reasonable allowances should be made for environmental and human factors affecting such process:

■ **Selection and detailing of materials and components to be incorporated into an enclosure** should consider the anticipated erection sequence and the cumulative effect of tolerances in preceeding parts upon those to be erected later:
• detailing based on tolerances which

are too close to be consistently achievable or with no allowance for such tolerances will inevitably result in problems during construction.

■ **Off-site fabrication of components,** therefore, should not be based on design dimensions only unless proper clearances have been incorporated into the design:

* clearances are spaces or distances purposely provided between adjacent components of an assembly to facilitate their proper alignment during erection.

The extent of clearances required will be influenced by degree of accuracy reasonably attainable during fabrication of components and their erection with the latter often being more critical.

■ **Reference points and dimensions change during erection as materials expand and contract with variations in temperature and/or their moisture content:**

* dimensions of a wall taken in early morning on a cold day will be different from those taken at noon when the wall has been exposed for several hours to solar radiation.
* levels established for a multistory reinforced concrete structure shortly after removal of formwork are likely to be different from those taken a month or so later, due to plastic flow and initial drying shrinkage occuring in concrete during that period of time.
* accuracy in layout will depend on the skill of persons performing it, and on the types of instruments used.
* working conditions will also affect the accuracy of layout; i.e., measuring tapes exposed to sunlight are subject to thermal expansion/contraction just as all other materials.

Considerations of proper clearances should also include the effects of environmental factors upon the complete enclosure:

■ **Thermal expansion and contraction** for all materials.

■ **Shrinking and swelling** due to changes in moisture content of materials thus affected.

■ **Deformations** due to temporary or permanent gravity loads.

■ **Movement and resulting deformations** due to lateral loads.

Clearances may be provided by:

■ **Connections designed and detailed** to be adjustable in one or several directions during erection and the incorporation of sufficient space for such adjustment in their design.

■ **Special components** such as corner trim or joint covers detailed to allow for field cutting of adjacent components should such cutting become necessary.

■ **Joint design** which allows for a certain amount of inconsistancy in width without becoming visually objectionable.

OTHER CONSIDERATIONS

The relative location of materials and components in an enclosure may also have an effect on their selection:

■ **Finishes for components normally seen at oblique angles** should be selected to hide rather than highlight minor variations in flatness which may exist in such components:

* sunlight striking a flat, smooth, and glossy surface at an oblique angle will highlight even slight imperfections which would not normally be noticeable at different angles of illumination.

■ **Components fabricated off-site** to close tolerances when placed directly adjacent to on-site fabricated ones may emphasize any inaccuracy in the latter.

■ **Identical coating applied to adjacent surfaces** with different properties such as porosity, texture will not appear to be the same.

DEFINITIONS & PHYSICAL PRINCIPLES

RESISTANCE TO FIRE

DEFINITIONS & PHYSICAL PRINCIPLES

Building codes regulate the construction and use of buildings and structures by establishing minimum requirements to insure public safety, health, and welfare:

■ **Type of construction is controlled by defining requirements for:**

• performance characteristics of materials permitted to be used in specific components: such as noncombustibility, resistance to decay, allowable stresses.
• fire-resistance requirements for major components of an enclosure such as for framing, columns, walls.

When resistance to fire is a requirement, fire resistance ratings of specific assemblies should be ascertained prior to their selection:

■ **Assemblies:** - such as roof/floor decks, girders and columns, walls, partitions - are assigned a rating based on tests carried out on identical assemblies under laboratory conditions, the most common test being ASTM E119 - Standard Methods of Fire Tests of Building Construction and Materials:

• The test consists of three criteria and ends when any one of the three is reached:
  1. the test assembly fails to sustain applied loads and collapses.
  2. cotton waste placed on unexposed side of assembly is ignited through cracks or fissures.
  3. temperature of unexposed surface rises an average of 250°F above its initial temperature and/or fails under a hose stream test.

■ **While these tests provide a measure of relative fire resistivity, they must not be seen as indicative of the actual behavior of an assembly in the presence of fire:**

• all three criteria are assigned equal value where structural integrity is the most critical of the three.
• standard tests are conducted mostly under negative pressures while positive pressures may prevail in the immediate vicinity of the fire.
• amount of fuel used in test before failure occurs is not a criterion in the test whereas in real life situations fuel contributed by assemblies intensifies the fire.

• rational analytical procedures can replace ratings where these are not readily available, especially in evaluating concrete components.

■ In order to be approved by building officials the rated assembly and all of its components must be installed exactly as tested. At the discretion of building officials some exceptions may be approved, such as:

• greater height of thickness of assembly.
• larger structural members with thicker flanges, webs or diameter of chords.
• reducing compressive strength of concrete by not more than 500 psi.
• increasing compressive strength of concrete.
• substituting continuous span design for a standard simple span design as tested.
• substituting fire-retardant treated wood for untreated in partition assemblies.
• addition of roof covering of saturated felt and hot-mopped asphalt to any floor design to be used as a roof.
• reduction of thickness of roof insulation.

■ Fire-rated assemblies commonly may not be altered by:

• substituting lightweight concrete for normal weight concrete.
• substituting normal weight for lightweight concrete.
• increasing spacing of ceiling suspension system supports, although they may be reduced.
• addition of insulation to a floor assembly, to be used for a roof, as it may cause higher temperatures within the assembly.
• increasing the thickness of insulation.

■ **Many structural elements of roof and floor assemblies have been fire-tested under restrained conditions wherein the surrounding or supporting structure is capable of resisting substantial thermal expansion through out the range of anticipated elevated temperature.**

• Construction not complying with this definition is assumed to have load bearing elements free to rotate and is considered to be unrestrained.

• Generally, restrained assemblies have a higher fire-resistance rating than those that are unrestrained, although both may be essentially the same assembly of components.
• Refer to UL "Fire Resistance Directory" and to ASTM E119 for more information.

■ **Providing automatic sprinkler systems will:**

• increase safety of occupants.
• reduce insurance premiums.
• allow an increase of lengths of travel from a habitable space to a means of egress.
• generally will allow an increase in area permitted for a given construction type.

■ **Penetrations through fire-rated floor/roof, floor/ceiling assemblies must be protected and/or limited in area and type.**

■ **Consideration should be given to durability of fireproofing materials to resist:**

• construction abuse.
• moisture from leaks and condensation.
• water from sprinkler systems.
• vibration, which could cause dusting and flaking.
• erosion due to air movement.

Building codes commonly require Class A or B roof coverings within the fire limits of cities:

■ **Class A coverings:**

• Effective against severe fire exposures.
• Are not readily flammable.
• Do not carry or communicate fire.
• Offer high degree of fire protection to roof deck.
• Do not slip from position.
• Possesses no flying fire brand hazard.
• Do not require frequent repairs to retain fire retardant properties.

■ **Class B Coverings:**

• Effective against moderate fire exposures.
• Offer moderate degree of fire protection.
• May require infrequent repairs.

ACOUSTICS

SOUND AND VIBRATIONS

Sound is a vibration in an elastic medium such as air, water, most building materials, and the earth. Noise can be defined as unwanted sound; that is, annoying sound made by others or very loud sound which may cause hearing loss.

Sound energy progresses rapidly, producing extremely small changes in atmospheric pressure, and can travel great distances. However, each vibrating particle moves only an infinitesimal amount to either side of its normal position. A full circuit by a displaced particle is called a cycle. The time required for one complete cycle is called the period and the number of complete cycles per second is the frequency of vibration. Consequently, the reciprocal of frequency is the period.

A pure tone is vibration produced at a single frequency, such as the variation in pressure caused by striking a tuning fork, which produces an almost pure tone by vibrating adjacent air molecules.

■ **Frequency of sound is the rate of repetition of a periodic event.** The frequency of a sound wave is determined by the number of times per second a given molecule of air vibrates about its neutral position. The greater the number of complete vibrations or cycles, the higher the frequency. The unit of frequency is the Hertz (Hz). Pitch is the subjective response of human hearing to frequency.

■ **Wavelength is the distance a sound wave travels during one cycle of vibration.** It also is the distance between adjacent regions where identical conditions of particle displacement occur.

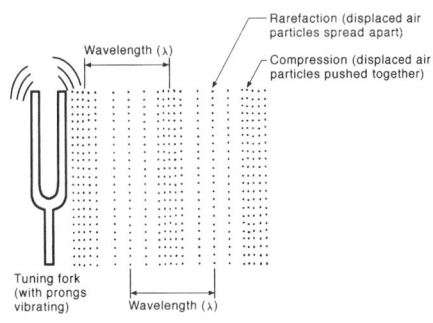

Fig. 1. Wavelength in air from a vibrating tuning fork. (Egan 1988)

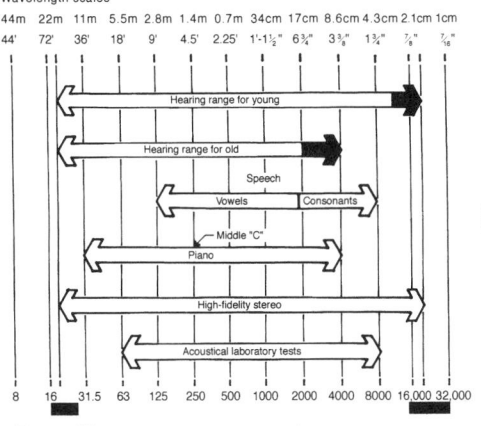

Fig. 2. Wavelength scales. Vibrations below 20 Hz. are not audible by humans, but can be felt. (Egan 1988)

■ **Velocity of sound.** Sound travels at a velocity that depends primarily on the elasticity and density of the medium. In air, at normal temperature and atmospheric pressure, the velocity of sound is approximately 1,130 feet per second (ft/s), or almost 800 mi./h.

SENSITIVITY OF HEARING

The adjacent graph shows the tremendous range of sound levels in decibels (dB) and frequency in Hertz over which healthy young persons can hear. Also shown is the frequency range for "conversational" speech, which occurs in the region where the ear is most sensitive.

Inverse-square law

Sound waves from a point source outdoors with no obstructions (called "free-field conditions") are virtually spherical and expand outward from the source.

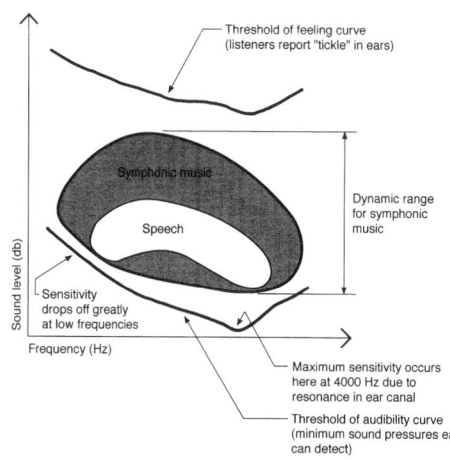

Fig. 3. Human audible sound level and frequency (Egan 1988)

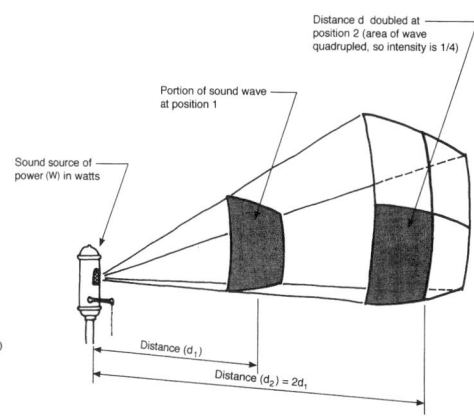

Fig. 4. Inverse-square law (Egan 1988)

Power is a basic quantity of energy flow. Although both acoustical and electric energies are measured in Watts, they are different forms of energy and cause different responses. For instance, 10 Watts (abbreviated W) of electric energy at an incandescent lamp produces a very dim light, whereas 10 W of acoustical energy at a loudspeaker can produce an extremely loud sound.

REFERENCES:

D. Egan *"Acoustics"* in Reference 1; also: Egan, M. David. 1988. Architectural Acoustics. New York: McGraw-Hill.

DEFINITIONS & PHYSICAL PRINCIPLES

SOUND: units of measure

SOUND PRESSURE LEVELS

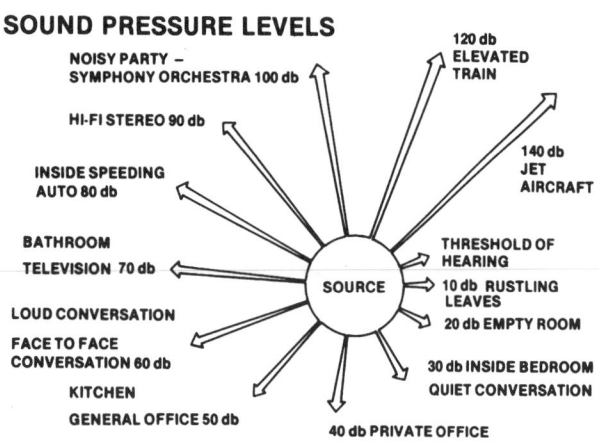

Sound is a vibration at various frequencies in any elastic medium:
- it is generated by a source;
- it requires a path for transmission: gaseous, liquid, or solid.
- to reach the receiver: usually the ear of a living being.

■ **Sound pressure level** (or loudness) is given in **decibels (db):** a ratio of the intensity of sound measured to a reference intensity roughly equivalent to the threshold of hearing.
- a change of **10 db** in pressure will result in perceiving the sound as twice as loud, or half as loud.
- a change of **3 db** in pressure will be just perceptible.

FREQUENCY

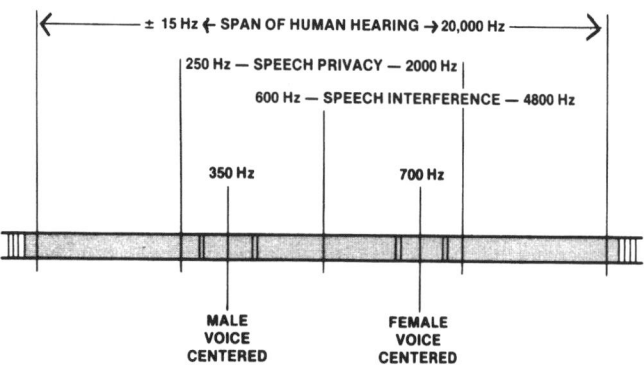

■ **Resistance of materials to sound transmission** and sound transmission loss varies with different frequencies. To establish the STC rating sound transmission loss is tested at all frequencies from 125 to 4000 Hz.

■ **Sound absorption of materials** will also vary with different frequencies; the most commonly used frequencies in testing materials are 125, 250, 500, 1000, 2000, and 4000 Hz.

■ **Frequency band:** a division of audible frequency band into more convenient sections or octave bands centered at the following frequencies: 31.5 at 1000 Hz; 63 at 2000 Hz; 125 at 4000 Hz; and 250 at 8000 Hz.

NOISE REDUCTION COEFFICIENT

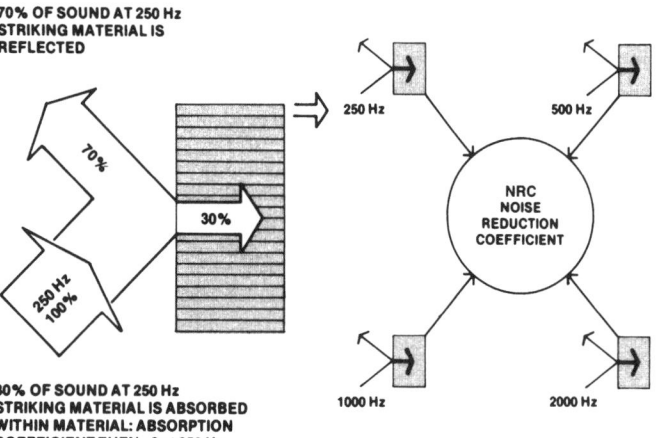

■ **Noise Reduction Coefficient** is an arithmetic average–to the nearest .05- of four sound absorption coefficients; or the ratio of sound absorbing effectiveness of a material at four specific frequencies to the effectiveness of a perfectly sound absorbing material at the same frequencies.

■ **Most materials effective in absorbing sound** are porous and lightweight and do not effectively resist transmission of sound through them. Thus a suspended acoustic tile ceiling will allow sound from one space to pass through it into the plenum and down again into an adjacent space.

■ **Sound attenuation factors** have been established by manufacturers to indicate the effectiveness of commonly used acoustic materials in resisting sound transmission to adjacent spaces via the plenum space. Refer to "Performance Data of Architectural Acoustical Materials" published by Acoustical Materials Association.

■ **Sound reverberation** is the continued multiple reflection of sound after it has been stopped at the source:

■ **The amount of reverberation** in a space is measured by reverberation time, or the time required to reduce the energy of reflected sound to one millionth of the level it had when the source was stopped.

■ **The STC of the ceiling** should equal the STC of partitions between adjacent spaces.

DEFINITIONS & PHYSICAL PRINCIPLES

SOUND TRANSMISSION CLASS

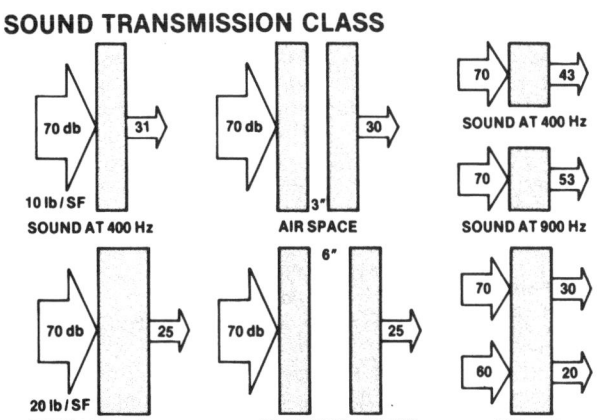

- **A single number rating system** which compares the Transmission Loss test curve at all frequencies of a given construction with a "Standard Contour" which reflects known subjective responses to the Transmission Loss performance.

- Theoretically, **the doubling, or halving of the weight of construction** will change the Transmission Loss by 6 db at any given frequency; the Transmission Loss of the assembly, however, will vary at different frequencies.

- Air spaces, resilient attachment of surfacings, sound absorbing materials (such as insulation within the air space) will tend to **increase Transmission Loss of the assembly** at all frequencies. Minimum effective air space is 2 inches.

IMPACT NOISE

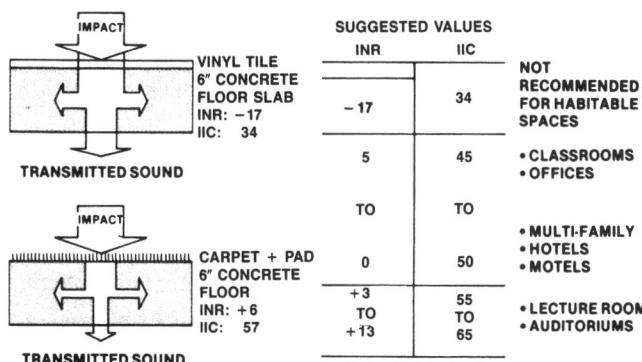

	SUGGESTED VALUES		
	INR	IIC	
VINYL TILE 6" CONCRETE FLOOR SLAB INR: –17 IIC: 34	–17	34	NOT RECOMMENDED FOR HABITABLE SPACES
	5	45	• CLASSROOMS • OFFICES
	TO	TO	
CARPET + PAD 6" CONCRETE FLOOR INR: +6 IIC: 57	0	50	• MULTI-FAMILY • HOTELS • MOTELS
	+3 TO +13	55 TO 65	• LECTURE ROOMS • AUDITORIUMS

- **Impact noise** caused by impact of walking, dropped objects, scraping furniture.

- **Impact noise insulation** - or isolation is expressed as either impact noise rating INR or impact insulation class IIC. It is a measure of the effectiveness of a given floor or floor and ceiling assembly to prevent impact sound transmission to spaces below such assemblies.

- **Both INR and IIC** are single number rating systems. IIC values are the ones currently used, and can be converted to previously used INR values by subtracting 51 points, e.g. IIC 45 equals about INR-6.

STRUCTURE BORNE SOUND

- **Sound will travel** quite efficiently -i.e., without much loss of energy- through structural components. If the level of sound energy is strong enough, sound will be re-radiated into surrounding air.

- **Air borne sound at high levels** in space will induce structure borne sound waves which may re-radiate into adjacent spaces.

- **Common structure borne sound** sources are generally those which act directly upon the structure, such as footfalls, vibrations transmitted from mechanical equipment directly connected to the structural elements.

- **Control of structure borne sound** is primarily a question of effectively isolating the sound sources from the structure, and of providing cut-offs or breaks that confine structure-borne sound travelling from one space to another.

MASKING SOUND

- **Whether a transmitted sound will be heard in a receiving space** will depend upon the level of sound present within that space. If the sound within the space is at 20 db level, a transmitted sound at the same pressure will not be heard, but a transmitted sound at 30 db will be quite noticeable. Thus a sound constantly present within a space may mask sound transmitted into it: the higher the level of the masking sound, the lower the requirements for sound isolation of the space.

- **Generally, any space will have some background noise,** generated by HVAC systems, activities in other parts of the structure, traffic or wind outside. Some of these background noises are intermittent and may not be relied upon to provide a constant masking sound. In certain areas, such as large open offices, maintaining a constant level of masking sound may be important enough to warrant an independent masking sound system.

NOISE CRITERIA CURVES

- **Noise criteria curves** give permissible sound pressure levels or background noise within a space; they are given at eight octave bands based on subjective human response to sound pressure and frequency.

- **The range recommended** is from NC-20 for a broadcast studio to NC-40 to NC-65 for factories.

- **The generally used classification** of background noise levels is:
  • under NC-25 low level: quiet location, no HVAC noise.
  • NC-25 to under NC-35 average level: some traffic noise, some HVAC noise.
  • over NC-35 high level: street and traffic noise; noisy HVAC systems.

DEFINITIONS & PHYSICAL PRINCIPLES

HEAT: modes of transfer

RADIATION, reflectivity

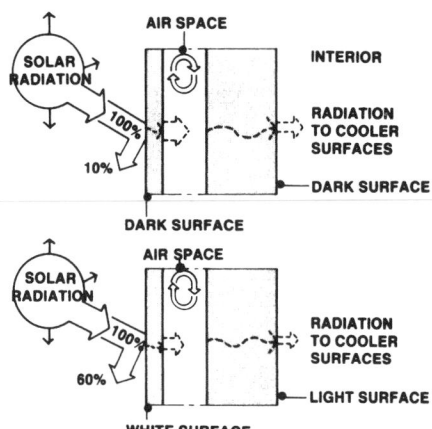

Radiation is the phenomenon of heat transfer by radiant energy through space (without the need of a medium of transfer) from a body at a higher temperature to all bodies at lower temperatures which are in its line of sight.

■ Heat transfer by radiation increases significantly as the temperature of the emitting surface rises.

Solar energy striking a surface will be partially reflected and partially absorbed, with the fractions primarily dependent on the color of the surface:
* a dark surface may absorb about 90 percent, while a white one will absorb from 20 to 40 percent, reflecting the balance.
* the difference between maximum temperatures of white and dark surfaces exposed to solar radiation may be as much as 50 to 60°F as a result of their surface reflectance.

■ Even though reflectance of two light surfaces may be similar, their emissivity may differ resulting in significantly higher temperatures in those with lower emissivity under the same exposure.

RADIATION, emissivity

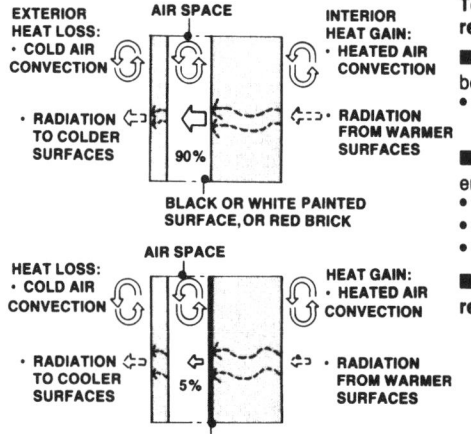

Temperature and emissivity of the surface will determine how heat gained is reradiated.

■ Painted surfaces have a higher coefficient of emissivity: most of the heat absorbed will be reradiated faster.
* the color of a painted surface has little effect on emissivity: black and white lacquers at 100° to 200°F both have an emissivity range of 0.80 to 0.95.

■ Metallic surfaces, especially when polished, have much lower coefficients of emissivity:
* bright aluminum foil is 0.04 to 0.05
* commercial grade emeried, polished copper 0.03, but 0.78 when heavily oxidized
* aluminum coated roofing 0.1 to 0.2.

■ Metallic surfaces therefore will reradiate more slowly than painted surfaces and remain hotter.

CONVECTION, surface conductance

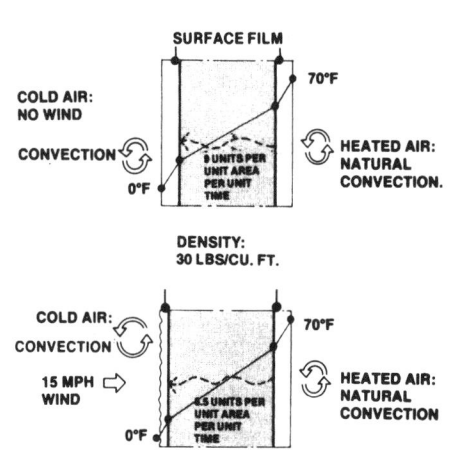

Heat will flow through a solid body when there is a temperature differential between air on opposite sides of it.

■ Heat gain and heat loss by and from the body will be by convection: air in contact with the surfaces will either give up heat to the body, or pick up heat from it:
* natural convection will take place when the motion of air is due entirely to differences in density
* forced convection occurs when the motion is amplified by external forces.

The transfer of heat from or to air is affected by a layer of air adjacent to the surfaces of a body, or the surface film:

■ Surface film is a layer of stagnant air which clings to the surface of any object and offers resistance to the flow of heat.

■ The heat flow through a surface film, the convective surface conductance, in general use is:
* design value for interior surface: still air generally assumed at .65 Btu/hr. sq. ft. °F. Since this varies somewhat depending on surface material, relative position of surface, direction of heat flow and temperature, other design values are sometimes used.
* design value for exterior surface, whether vertical or horizontal, with 15 MPH wind: 6.00 Btu/hr. sq. ft. °F.
* these design values are incorporated into the temperature gradient calculations shown on following pages

CONVECTION, RADIATION, vertical air space

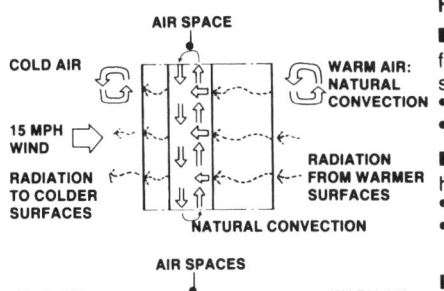

Heat transfer through an air space incorporated into a vertical assembly will be by:

■ **Natural convection within the air space:** temperature differences between the surfaces of components facing the air space will set up convective currents within the air space. **Amount of heat** transferred will:
• **increase** with increase in temperature differences of the two surfaces
• **will not be** significantly affected by the temperature level.

■ **Radiation through the air space from the warm surface to the cold one.** Amount of heat transferred by radiation will vary:
• with the temperature **difference**
• also with temperature **level,** increasing rapidly with increase in surface temperature levels.

■ **At low temperature levels,** convection will be the controlling factor; at very high temperatures, radiation.

■ **When vertical air space** is broken up into a number of horizontal cells:
• heat transfer by convection is reduced by minimizing convective currents
• transfer by radiation will remain unchanged
• the horizontal divisions will allow some heat transfer through them by conduction.

CONVECTION, RADIATION, horizontal air space

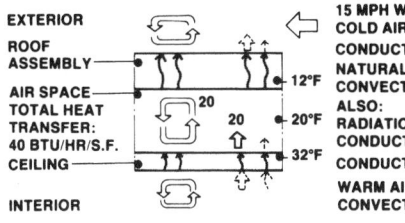

Heat transfer through horizontal air spaces will differ depending on the direction of heat flow:

■ **Upward heat flow** through an air space will be by convection and radiation, similar to that for vertical air space.

■ **When the flow of heat is downward,** the air in contact with the upper warmer component will also be warmer and less dense than air in contact with the lower colder component:
• heat transfer will be by radiation
• convection will be at a minimum
• a small amount will be transferred by conduction.

■ **Transfer by radiation** is the same through vertical and horizontal air spaces:
• assuming that at normal temperatures the emitting surface has a coefficient of 0.9, such as for painted surfaces or red brick, transfer by radiation might be about 50 percent of the total heat transferred.
• if a bright metallic surface is substituted, heat transfer by radiation may be reduced to about 5 percent of the total.

CONDUCTION

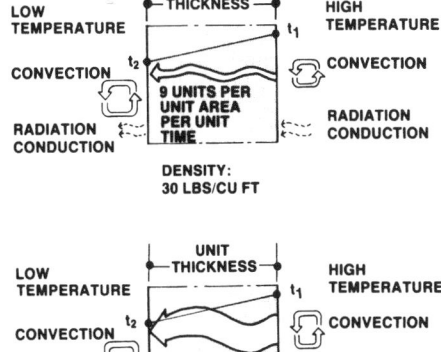

■ **Conduction is the transfer of heat** from one part of the body to another part, or from one body to another which is in physical contact with it, without any appreciable displacement of the particles of the body or bodies.

■ **Heat continues to flow as long** as a temperature difference exists within the body, or between bodies in contact with one another.

The rate of heat flow depends upon the conductivity of the body:

■ **Conductivity of materials varies** with differences in their densities: low density materials have voids in them, which contain air or other gaseous substances and which impede the transfer of heat by increasing the cross sectional area or length of travel:
■ Regular weight concrete with a density of 140 lbs./cu. ft. has a coefficient of conductivity of $k = 9.09$.
■ Cellular concrete with a density of 30 lbs./ cu. ft. has a coefficient of $k = .90$.
• transfer of heat will be 10 times less per unit area per unit time for the less dense material under the same difference in surface temperatures.

DEFINITIONS & PHYSICAL PRINCIPLES

HEAT FLOW: temperature variations,

EQUIVALENT TEMPERATURE

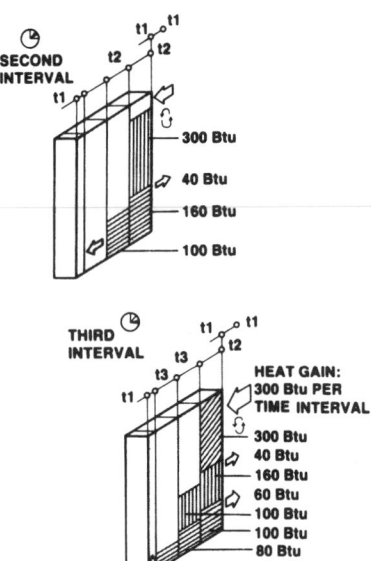

Heat gain and heat loss in assemblies is normally calculated at the time of greatest heat flow which implies that such conditions remain the same at all times.

■ This approach is referred to as **the steady state of heat flow;** it assumes that:
• the rate of heat flow through the assembly will not vary with time
• temperature differentials within the assembly, and outside of it, will remain constant.

■ **For typical heat flow calculations** refer to Temperature Gradients on following pages.

■ **Actual conditions do change** almost constantly, especially when an assembly is exposed to variable solar radiation, resulting in an unsteady state of heat flow.

■ **An assembly may be exposed to instantaneous heat gain** through solar radiation, which will first be absorbed by the surface layer. The temperature of this layer will rise above the temperature of the remainder of the assembly, and also above the temperature of outdoor air:
• heat flow will occur into both regions of lower temperatures
• the amount of heat flowing in either direction will depend on the resistances of the assembly and the surface film coefficient.

The unsteady flow of heat or dynamic response generally is accounted for by using the equivalent temperature difference:

■ **The temperature difference** which reflects the total heat flow through an assembly caused by variable solar radiation and outdoor temperature.

■ **The information on solar radiation** required to establish its amount at a given location can be found in Selection Data SOLAR ENERGY.

■ **Design temperature differentials** for a given location are available from the local U.S. Weather Bureau.

■ **The heat flow through** the assembly is then calculated using the steady state heat flow equation.

VARYING OUTDOOR TEMPERATURES

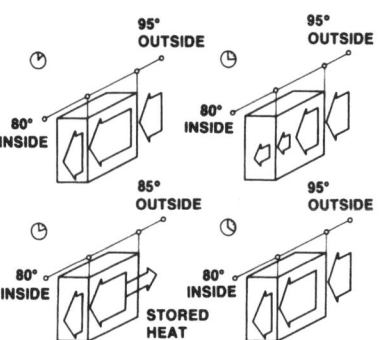

The equivalent temperature difference must take into account the duration of the exposure during various times of the day. Outside temperatures vary with a resultant immediate effect on the flow of heat:

■ **If outdoor temperature suddenly drops** from 95°F to 85°F:
• heat continues to flow from the interior surface of the assembly into an interior space at 80°F
• there is also heat flow from the outer surface of the assembly to now cooler outside air
• therefore the amount of heat stored within the assembly is reduced.

■ **If the outside temperature rises again** to 95°F after several hours and the outer surface of the assembly begins to gain heat, the flow of heat from the inner surface of the assembly into the interior space does not immediately rise to its previous level:
• the inner surface remains slightly above the temperature of the interior air due to negligible heat flow when outdoor temperature was at 85°F.
• heat flow into the interior increases gradually returning to the previous level only after the temperature of the entire assembly has risen to a point where the steady state condition is re-established.

■ **The interval between the change in outdoor temperature and the temperature of the inner surface is known as the time lag;** it is due mostly to the heat required to raise the temperature of the assembly itself.

TIME LAG

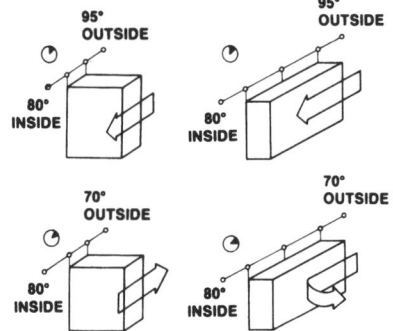

Time lag is the time required to establish steady state condition through an assembly: for heat to travel through an assembly from the warm surface to the colder one:

■ **Thin lightweight assemblies** have little mass and do not require large amounts of heat to raise their temperature:
• the steady state temperature distribution is reached soon after the temperature of their outer surface rises.
• since little heat is stored in such assemblies, the temperature of the inner surface drops quickly after a drop in outside temperatures: the time lag is short.

■ **Dense, thick assemblies** have a large heat storage capacity:
• a considerable amount of time may be required for the heat being absorbed at the outer surface to reach the inner surface.
• should the temperature at the outer surface drop before the heat reaches the inner surface, the flow will reverse and heat will flow back to the outer surface; and from there to the cooler outside air.
• heat will be stored in the assembly, some of it being released when outdoor temperature falls below the temperature of the assembly, then replenished as outdoor temperature rises.

■ **The thermal capacity of an assembly** is determined by the volume × the density of the materials incorporated into the assembly × the specific heat of the material.

DEFINITIONS & PHYSICAL PRINCIPLES

units of measure

CONDUCTIVITY

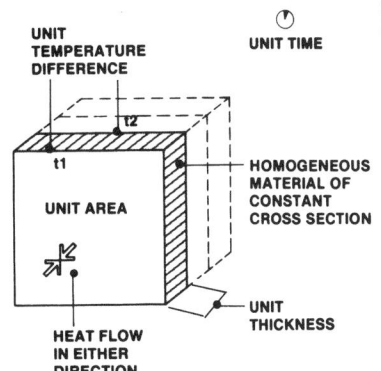

Thermal Conductivity designated k, is a property of homogeneous material:

■ **It is measured by** the quantity of units of heat passing through a unit thickness, per unit area, in unit time, when a unit temperature difference is maintained between the outer surfaces of the material.

■ **Generally used units are:**
• units of heat given by British Thermal Unit, or Btu, which is the amount of heat required to raise the temperature of one pound of water from 63°F to 64°F.
• unit thickness: one inch.
• unit area: one square foot.
• unit time: one hour.
• unit temperature difference: one degree F.

■ Coefficients of conductivity are **not additive.**

Resistivity designated by r or 1/k is the reciprocal of conductivity:

■ **It is measured by** the temperature difference in degrees F between smooth parallel outer surfaces of one inch thick material that are required to cause one Btu to flow through one square foot per hour: r = temperature difference in degrees F per inch per one square foot per hour, divided by Btu

$$k = \frac{Btu}{HR./Sq.Ft./°F/inch} \qquad r = \frac{HR/Sq.Ft./°F/inch}{Btu}$$

■ Coefficients of resistivity **are additive.**

CONDUCTANCE

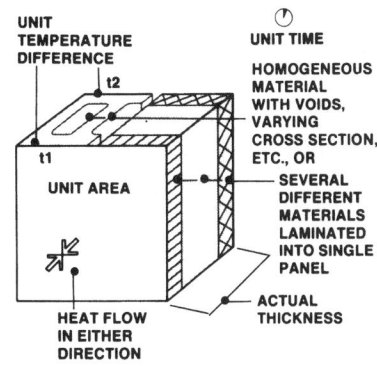

Thermal Conductance designated C, measures the rate of heat flow through the actual thickness of homogeneous, nonhomogeneous, or composite materials:

■ **Composite materials** are those where the cross sectional area is not identical throughout, such as in hollow core concrete block, or where a product consists of several layers of similar or different material, such as plywood or built-up roofing.

■ **Conductance is defined** as the heat flow in Btu per hour through one square foot area of given thickness for one degree F difference in temperature between the outer surfaces.

■ Coefficients of conductances **should not be added.**

Thermal Resistance, designated R or 1/C, is the reciprocal of conductance:

■ **It is a unit for the resistance to heat flow** through a given thickness of a homogeneous, non-homogeneous, or composite material.

■ **It is measured by** the temperature difference in degrees F between the outer surfaces required to cause one Btu to flow through one square foot per hour: R = temperature difference in degrees F divided by Btu per one square foot per hour.

■ Resistances **may be added.**

TRANSMITTANCE

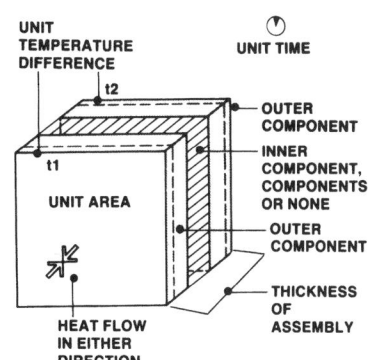

Thermal Transmittance, designated as U-value, is the measure of heat flow through a component of the building, whether vertical or horizontal, when a difference between air temperatures on either side of such component exists:

■ **The effect of air spaces** ¾ inch and wider, incorporated into the assemblies, and that of surface air films is included in the coefficient of thermal transmittance.

■ **Thermal transmittance is measured** by Btu per hour through one square foot, when the temperature difference is one degree F between the air at the two surfaces of the assembly.

■ **The U-value is the reciprocal,** or 1/ΣR, of the sum of all thermal resistances of the components, or the total resistance to heat flow through a complete assembly: Σ R = R of surface film plus R of outer component or components plus R of air space or spaces plus R of inner component or components plus R of surface film.

■ **U-values are not additive:** when modifications of an assembly are investigated, thermal resistances should be used.

DEFINITIONS & PHYSICAL PRINCIPLES

PART 1

HEAT FLOW: temperature gradients

SOLID ASSEMBLY

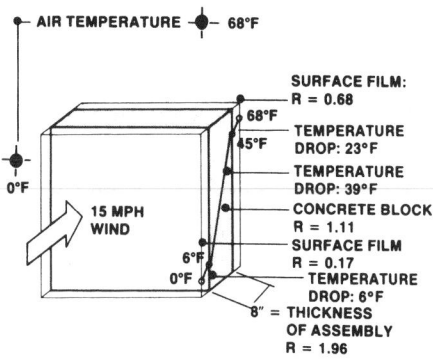

AIR TEMPERATURE — 68°F

SURFACE FILM: R = 0.68
68°F
TEMPERATURE DROP: 23°F
45°F
TEMPERATURE DROP: 39°F
CONCRETE BLOCK R = 1.11
SURFACE FILM R = 0.17
6°F
0°F
TEMPERATURE DROP: 6°F
8" = THICKNESS OF ASSEMBLY R = 1.96

0°F

15 MPH WIND

TRANSMITTANCE: U = 1/R = 1/1.96 = 0.51 BTU/SF/°F/HR

TEMPERATURE GRADIENTS

The temperature distribution, or gradient, of a given point in an assembly under different outside and inside temperature and humidity conditions **is of major significance. It will:**
* indicate the magnitude of heat flow to be expected at such a point
* locate condensation and freezing planes within the assembly.

■ **Temperature gradients under steady state heat flow** are determined assuming that:
* all parallel paths for heat flow have the same conductivity
* all components of the assembly have reached steady temperatures, neither storing nor releasing heat.

■ **Therefore all the heat that enters** through the warm surface of the assembly must flow through each component to be carried away from the cold surface.

■ **The rate of flow is:**
* directly proportional to the driving force, the difference in temperature
* inversely proportional to the thermal resistance of the components.

■ **Thus the temperature drop** through each component is proportional to its thermal resistance.

■ **Thermal resistances** of many building components and assemblies may be found in "Time Saver Standards" and "Architectural Graphic Standards."

ASSEMBLY WITH AIR SPACE

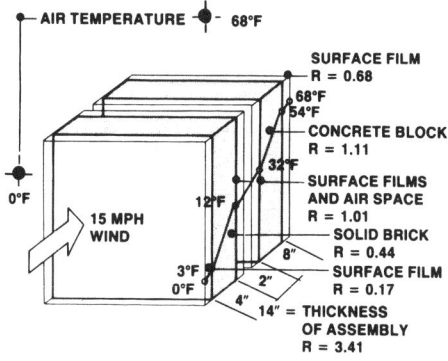

AIR TEMPERATURE — 68°F

SURFACE FILM R = 0.68
68°F
54°F
CONCRETE BLOCK R = 1.11
32°F
SURFACE FILMS AND AIR SPACE R = 1.01
12°F
SOLID BRICK R = 0.44
8"
3°F
SURFACE FILM R = 0.17
0°F
2"
4"
14" = THICKNESS OF ASSEMBLY R = 3.41

0°F

15 MPH WIND

TRANSMITTANCE: U = 1/R = 1/3.41 = 0.29 BTU/SF/°F/HR

CALCULATING TEMPERATURE GRADIENTS

Temperature gradients are calculated based on the total temperature drop through the assembly, distributed among the individual components in proportion to their resistances.

■ **The temperatures at interfaces** of the components can then be determined, and the plane at which condensation may occur at a given relative humidity determined.

Temperature gradients may be established by graphic or arithmetical methods:

■ **The arithmetical method** might be best illustrated with the solid assembly shown as an example:
* the temperature drop is 68°F, and the resistance of the entire assembly, including surface films, is R = 1.96
* the temperature drop through the interior surface film will be: 68 × 0.68 over 1.96 = 23.6°F
* through the concrete block; 68 × 1.11 over 1.96 = 38.5°F
* the exterior film: 68 × 0.17 over 1.96 = 5.89°F.

■ **Graphical Method** consists of drawing components with their thickness proportional to their resistance to heat flow and connecting the points of outside and inside temperature with a straight line.

INSULATED ASSEMBLY

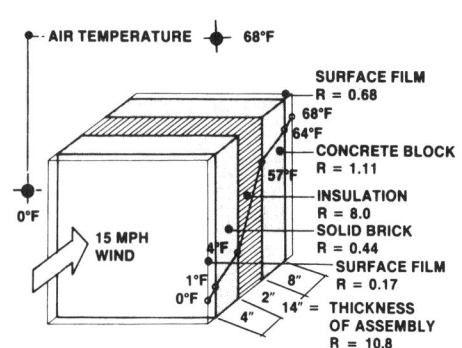

AIR TEMPERATURE — 68°F

SURFACE FILM R = 0.68
68°F
64°F
CONCRETE BLOCK R = 1.11
57°F
INSULATION R = 8.0
SOLID BRICK R = 0.44
SURFACE FILM R = 0.17
1°F
0°F
8"
2"
14" = THICKNESS OF ASSEMBLY R = 10.8
4"

0°F

15 MPH WIND

TRANSMITTANCE: U = 1/R = 1/10.3 = 0.09 BTU/SF/°F/HR

IMPROVING RESISTANCE TO HEAT FLOW IN ASSEMBLIES

The solid assembly illustration above has poor resistance to heat flow, which could be improved by:
* adding rigid insulation to either one of the surfaces of the assembly
* by filling the voids of the concrete block units with loose fill or foamed-in insulation.

■ **It could be modified into an assembly with air space** by adding a new component in front of either the outside or the inside surface, separated by an air space.

■ **Improved resistance** to heat flow would be realized more through the **incorporation of the air space** into the assembly than through the added component:
* face brick four inches thick has a coefficient of resistance R = 0.44, while a two inch air space has an R = 1.01.

■ The resistance to heat flow of the new assembly, even though improved, may still be too low to meet energy conservation requirements **and an insulated assembly might be needed.**

■ **Several options are available** to improve the thermal performance of the assembly
* insulation may be used to fill the air space, as illustrated
* the voids in the concrete masonry units could also be filled with insulation
* the assembly could be widened to provide an air space between the inner component and the added insulation in the air space
* reflective surface could be added to the new air space in the modified assembly.

CORROSION OF METALS

CORROSION OF METALS

Corrosion in general

Even if only one metal is involved, small differences in electric potential cause electrons to flow in the metal. If adjoining metals are different, and especially if they are far apart on the galvanic series, the reaction will be more rapid. In a fresh water environment, the following galvanic series, from "more noble" to "less noble" applies (in a wet environment, metals toward the bottom of the series will corrode in contact with those toward the top):

Monel
Copper
Stainless steel
Lead
Tin on steel
Galvanized iron or steel
Zinc
Aluminum
Magnesium

Biological corrosion is corrosion in which living organisms affect the corrosion. Examples are destruction of protective coatings and changing the pH of the environment.

Iron and steel corrosion

The corrosion of iron and steel is among the most harmful natural environmental forces acting on buildings. Iron can corrode under a number of different conditions in damp environments.

OTHER TYPES OF CORROSION OF CONCERN TO ARCHITECTS

■ **In building locations subject to acid deposition (acid rain),** a residue of acid may remain on roofs after the rain dries and acid aerosols may settle there.

• **When the acid rain dries on chemically inert surfaces,** and when additional acid aerosols settle, the acids become concentrated.

• **Rain will dilute the acid and wash it off,** but mist and dew on the roof may dissolve the acid residue without diluting it much.

• **The resulting acid may attack** copper, lead-coated copper, and some other roofing metals, removing the protective patina.

• **Without a coating of protective patina**, copper's durability is reduced. If the copper is designed not to accept runoff from other surfaces, or if it is protected by zinc sacrificial anodes, it is very durable.

■ **Intergranular corrosion (corrosion of one alloy component) may occur in alloys.** One type is dezincification of brass, commonly called "crystallization."

■ **Lead** may leach out of domestic water lines, solder, and fixtures, and make the domestic water toxic.

■ **Stainless steel in an environment with halide ions suffers "autocatalytic corrosion."** The corrosion occurs in small pits and may cause premature failure. Thus, stainless steel ceiling hangers over chlorinated swimming pools are not a wise design choice.

■ **Aluminum will corrode in the presence of hydroxyl ions.** Hydroxyl ions are found in concrete and masonry mortar. This is a rapid reaction. If aluminum is embedded in fresh concrete, an area of hydrogen bubbles can sometimes be seen at the surface above it. Since concrete is rich in hydroxyl ions, wetting of the concrete-aluminum boundary may continue to cause corrosion in the completed building.

SOME METHODS OF CORROSION PREVENTION

In general, subvert any part of the corrosion reaction. Following are some of the strategies:

■ **Since the reactions listed above are aqueous reactions,** prevent the metals from getting wet. This protection must be highly reliable. If a leak occurs and the steel gets wet, corrosion will proceed.

■ **Select proper materials.** For severe service, consider high-silica

cast iron, certain types of wrought iron (if available), stainless steel, bronze, or other noncorroding metals. Where materials vulnerable to corrosion are used in a damp environment, protect them by one of the methods listed below.

■ **Provide anodic protection.** This is based on a property of iron and some other metals to "passivate" at an intermediate level of oxidizing power. Paint coatings such as zinc chromate and red lead protect by this mechanism.

■ **Provide cathodic protection.** This is similar to the galvanic corrosion noted above, but the protective metal is less noble than the material being protected. The protected base metal becomes the cathode, and the less noble metal—such as zinc or aluminum—becomes the sacrificial anode. This is the basis for protection by the zinc coating on galvanized steel. However, since the protective anodic metal is sacrificed, there is a time limit to this protection. In some cases the sacrificial anode may be replaced, but galvanized steel inside construction cannot easily be replaced.

• **Blocks of sacrificial anodic metal can provide cathodic protection.** Cathodic and anodic protection can also be provided by applying a direct current to the metal and its environment. Protection using these methods should be designed by a corrosion engineer. A commonly seen use of cathodic protection is the zinc blocks coupled to underground steel fuel tanks.

• **Maintaining a high pH (hydroxyl-rich) environment will protect steel unless there are halide ions present.** ("pH" is a measure of the acidity and alkalinity of a solution. "7" is neutral, below 7 is acid, and above 7 is alkaline. The numbers are on a logarithmic scale. Each unit above 7 is 10 times the previous unit, and every unit below 7 is 1/10 the previous unit. Thus, steel framing and reinforcing steel encased in concrete and masonry are generally not subject to corrosion.

• **The embedment of cathodic protection must be great enough to**

DEFINITIONS & PHYSICAL PRINCIPLES

PART 1

CORROSION OF METALS (cont.)

avoid neutralizing the alkaline environment by acid deposition. Special quality control is needed to maintain the proper separation of embedded steel from exterior surfaces. Using "chairs" between the reinforcing and the side and bottom forms is one method of maintaining proper separation. Vigilant inspection with a mirror or flashlight is another.

• **The concrete or masonry must be sound and relatively uncracked.**

• There must be no penetration by chloride and other halide ions. **Chlorides and other halides "depassivate" the surface of the metal.** The chemical mechanism by which they do this is apparently not known with certainty, but appears to concern complex reactions with the passivation layer. Therefore some methods which protect steel in a salt-free environment don't work in an environment rich in chlorides and other halides.

• **Corrosion-inhibiting admixtures for concrete are available.** Also, highly impermeable concrete and concrete with durable coatings will resist the penetration of halide ions.

■ **Apply coatings.** This is the predominant method of corrosion protection, but has its problems. Most paints of most types are somewhat porous, and many paints do not contain corrosion-inhibitive pigments. Corrosion-protective coatings for each specific environment should be selected through consultation with a technical representative of the selected paint manufacturer or, for critical applications, a corrosion engineer. The Steel Structures Painting Manual (1993) contains recommendations for preparing, priming, and painting steel structures under numerous service conditions. Preparation, priming, and painting may be specified by reference to SSPC standard specifications.

• **A few coatings, including coal tar enamel,** are totally nonporous, but it is prudent to assume that paint coatings are somewhat porous.

• **Most organic corrosion-resistant coatings (paints),** such as red lead and zinc chromate, protect by creating an oxidizing polarized layer at the metal surface. See "anodic protection" above. They are not recommended for total or frequent immersion, however.

• **A few organic corrosion-resistant coatings (paints),** such as zinc dust primer, protect by becoming sacrificial anodes and thus making the steel the cathode. The zinc must be tightly packed and within a few angstrom units of the surface of the metal (the zinc-rich primer must be applied almost immediately after abrasive blasting). These paints are called "zinc rich primers and paints."

• **Epoxy coatings** are commonly applied to concrete reinforcing bars in critical environments. An alkaline environment does not harm epoxy coatings, while most alkyd and oleoresinous paints are saponified and loosened in that environment.

• Applying most finish paints directly to the metal does not give good protection; they need a proper primer. The preparation must also be proper.

• There is a **synergistic effect** with the use of coatings and catalytic protection. There is an advantage to using zinc dust polymer coatings over abrasive-blasted steel, in that both cathodic protection and an impermeable layer are employed. Applying a proper protective coating over galvanized steel protects the zinc coating and thus prolongs the life of the coating and its protection.

RECOMMENDED DESIGN PRACTICE

While specialized corrosion protection may require the services of a corrosion engineer, the following good design practices can be implemented by the architect, and can be highly effective.

■ **Protect corrosion-vulnerable components from water.** Design details to be free draining.

■ **Maintain a proper environment near corrodible metals.** For example, don't locate lead-acid batteries near structural steel. Don't locate steel structures in a place formerly or presently used to store salt. Low temperatures retard corrosion. Hot, steamy environments hasten corrosion.

■ **Protecting steel from corrosion in a halide-rich environment is difficult and often unreliable.** Parking garage decks in areas where melting salts are used in winter, especially the floors, which are not exposed to rain washing, are highly vulnerable to corrosion. Cathodic and anodic protection is currently being used for bridge structures, and such protection can be used for parking structures as well. The design of such protection is performed by corrosion engineers, but it is the architect's role to request consultation with the corrosion engineer when appropriate. One beneficial maintenance operation is to wash the lower floors of parking garages with clear water in the spring, but the designer can't be sure that this maintenance will be performed.

■ **Increase wall thickness to allow some corrosion without affecting required strength.** Example: cast iron roof drains corrode, but their thickness, together with the slow corrosion rate of cast iron, allows them to function for the life of the roof.

■ **Avoid open joints.** Welded joints and rolled sections resist internal corrosion, while riveted and bolted connections are more prone to such corrosion. Where it is appropriate to use riveted or bolted connections exposed to a corrosive environment, provide special protection.

■ **Design to facilitate cleaning, maintenance,** and replacement. Design inspection panels at critical joints.

■ **Avoid corrosive conditions at stress concentrations** (or, avoid stress concentrations in corrosive environments). Stress concentrations cause differences in potential and thus galvanic corrosion.

■ **Avoid electrical circuits in metal in corrosive environments.** Electrical circuits cause differences in potential and thus galvanic corrosion.

■ **Avoid heterogeneity of metals,** especially metals far apart in the galvanic series.

■ **Observe and learn from experience in the locale where you practice.** What works in Houston won't necessarily work in Montpelier. When practicing outside of your familiar area, confer with architects and engineers familiar with the environmental factors specific to the locale of the building project.

■ **Insulate galvanic couples.** For example, isolate steel from copper pipe in water supply lines.

■ **Keep the exposed area of the cathode small, and keep the exposed area of the anode large.** A copper nail in a steel sheet is not very harmful, but a steel nail in a copper sheet is harmed. If you can only protect half the system, protect the cathodic part.

■ **Note that "weathering steel" ("CorTen" and "Miari-R") resists corrosion** except in salt atmospheres and where exposed to standing water. Follow the manufacturer's precautions.

■ **In metals imbedded in concrete:**

• **Maintain adequate cover,** following the recommendations of the American Concrete Institute and increasing the cover where practicable. Exercise tight quality control regarding this. Inspect the formwork, using a good flashlight on dark days and a mirror on bright days. Do not permit the concrete to be placed until certain that there is no metal near the outside forms. This applies to tie wires as well as reinforcing. Specify "chairs" between the reinforcing and the outside forms.

• **Maintain a high pH.** Note that acid deposition can neutralize the concrete, especially if the concrete cover is thin. To some extent, coating old concrete with lime wash (calcium hydroxide plus binders and other admixtures) may stop neutralizing by acid deposition.

• **Concrete should be of high quality and nonporous.** Avoid excess water in the mix. Use water-reducing admixture or other means to lower the water content. Consider silica fume precipitate. Remember that nonstructural concrete must be treated with the same care as structural concrete.

• **Avoid chlorides.** Don't add them, and don't allow use of aggregates and water containing chlorides (salt sand; salt used to melt ice in the mixer, etc.). Note that some proprietary products may contain chlorides without saying so. Consider having plastic concrete tested for chlorides. The chloride content allowed under codes may not be safe; consider using a lower limit. A desirable limit, but hard to attain, is 0.1%. by weight of cement. ACI permits 2%, with less for prestressed work.

• **If the concrete will be exposed to chlorides** (near ocean, in parking garage, in swimming pool, etc.), the integral or applied protection to exclude chlorides must be in place before exposure. Such methods do not work after the chlorides have already penetrated the concrete. Methods of sealing the concrete include coatings, penetrating water repellents, and very nonporous concrete. Some success has been reported regarding removal of chlorides by setting up an electrical current. A positive charge is imposed on the top of the concrete, drawing the chloride anions to the top where they can be washed off.

• **Avoid cracking.** Discuss methods with the structural engineer. Remember that nonstructural concrete must be protected from cracking as well as structural concrete. Cracking is especially likely at areas of tension and thermal movement concentration.

■ **Galvanize reinforcing, or use epoxy-coated reinforcing, or use stainless steel, or use nonmetallic reinforcing such as alkali-resistant fiberglass and aramid fiber.** These methods should be used in conjunction with good design; they may not be adequate by themselves.

• **Patches in concrete at reinforcing can create a galvanic cell,** which makes other parts of the reinforcing anodic. For this reason the steel should be coated with an electrically insulating coating before the concrete is patched.

• **Salt splash zones in concrete and masonry are vulnerable to corrosion.** Salt splash zones, which are covered and not washed by rain, are especially vulnerable.

• **Apply phosphate or other pre-treatment of steel surfaces.** This offers corrosion protection, and it aids in paint adhesion.

■ **On every project consider possible corrosion problems, and seek professional advice when in doubt. For example:**
• Parking garages in climates where melting salt is used.
- Swimming pools.
- Structures near highways where melting salt is used.
- Structures near salt water.
- Structures with strong underground electrical currents.
- Structures in contaminated soil.
- Structures with critical metal components which cannot be inspected and repaired in the future.

REFERENCES:

D. Baerman *"Corrosion of Metals"* in Reference 1.

National Association of Corrosion Engineers (NACE), 1440 South Creek Drive, Houston, TX 77084-4906.

DEFINITIONS & PHYSICAL PRINCIPLES

WATER VAPOR MIGRATION

DEFINITIONS & PHYSICAL PRINCIPLES

SOLID ASSEMBLY

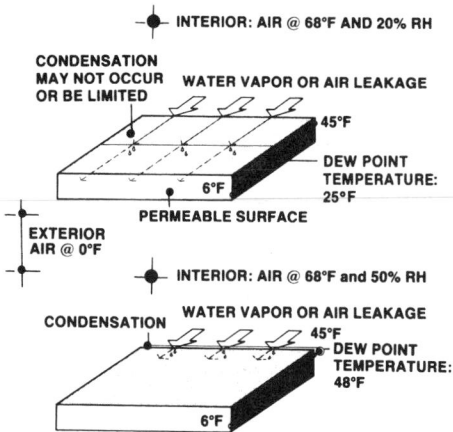

Water vapor is one of the several gaseous constituents of air.

■ **When a difference in concentration** of any one of these gases exists between two points, there will also exist a difference in partial pressure of the gas, causing the gas to flow from the point of higher concentration to the point of lower concentration, through air or through any permeable material.

■ **The maximum amount of water in the water vapor/air mixture** is limited by temperature: this amount will increase or decrease proportionately to the increase or decrease in the temperature of the mixture.

■ **The ratio between** the weight of water vapor actually present in the air and the maximum weight the air can contain is referred to as the **relative humidity,** expressed as a percentage: at 100 percent relative humidity the air is saturated.

■ If the temperature at which air becomes saturated, referred to as the **dew point, is further lowered, condensation occurs:** this process will continue until the ratio of the water vapor to air is stabilized at the new temperature level.

■ **Dew point temperature for a given air temperature** and relative humidity can be determined using a psychrometric chart, a version of which may be found in "Architectural Graphic Standards".

ASSEMBLY WITH AIR SPACE

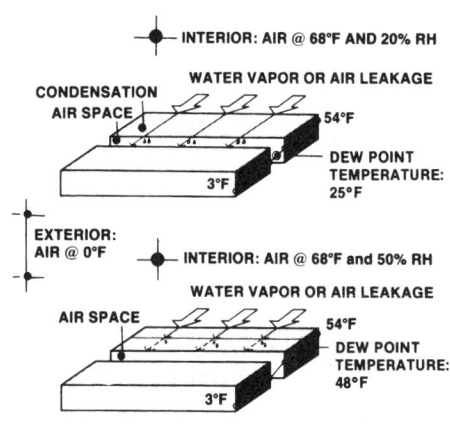

Water vapor migrates from higher pressure zones to those of lower pressure by diffusing through materials in direct proportion to their permeability: porous materials generally have a high degree of permeability.

■ **Water vapor may also penetrate** into an assembly as part of the water vapor/air mixture through air leakage.

■ **Water vapor starts condensing** upon reaching the dew point temperature, except that:
• though the dew point is reached within a component, condensation may only occur at **its interface** with another component, even when that interface is at a lower temperature than the dew point temperature.
• water vapor diffusing through a porous outer component **may not condense at all** or only to a limited extent, provided its passage to the lower vapor pressure zone outside is not impeded; applying a vapor impermeable coating over the exterior surface of a porous substrate blocks the passage of vapor.

■ **Vapor migration by diffusion** can be controlled by the use of vapor retarders; but **air leakage** generally will be the **prime cause of most of the condensation** problems in vertical and horizontal assemblies.

INSULATED ASSEMBLY

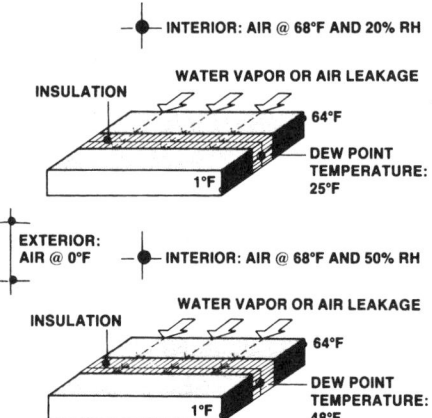

Vapor migration by diffusion only or by diffusion and air leakage is **shown for the same assemblies** for which temperature gradients were shown on the opposite page, with the relative humidity for interior spaces added. The two values for relative humidity used represent a low level at 20 percent, and a comfort level at 50 percent, steady state of heat flow was assumed in all examples.

■ **Vapor migration through a solid assembly** may not present a problem at 20 percent relative humidity, since in a porous substrate no, or only slight, condensation may occur, provided vapor is allowed to escape to the outside; at 50 percent relative humidity, condensation is likely to take place on the cold surface.

■ **With an air space** and 20 percent relative humidity, condensation should occur on the cold surface facing the air space, especially since brick has low permeability to vapor; at 50 percent, condensation should take place on the warm surface facing the cavity.

■ **The dew point temperature for the insulated assembly** is reached within the insulation at both values for relative humidity. If the insulation is permeable to water vapor, condensation should occur at the interface of insulation with the colder component.

TEMPERATURE GRADIENTS

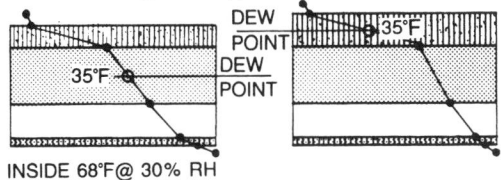

OUTSIDE AIR @ 0°F.15 MPH WIND

DEW POINT

35°F

DEW POINT

35°F

INSIDE 68°F @ 30% RH

ASSEMBLY A   ASSEMBLY B and C

ASSEMBLY A

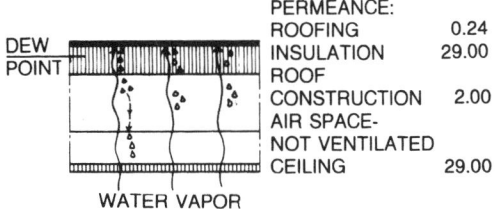

DEW POINT

WATER VAPOR

PERMEANCE:
ROOFING   0.24
INSULATION   29.00
ROOF
CONSTRUCTION   2.00
AIR SPACE-
NOT VENTILATED
CEILING   29.00

ASSEMBLY B

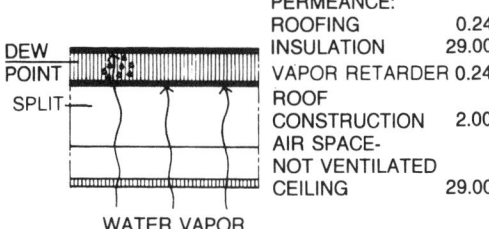

DEW POINT

SPLIT

WATER VAPOR

PERMEANCE:
ROOFING   0.24
INSULATION   29.00
VAPOR RETARDER   0.24
ROOF
CONSTRUCTION   2.00
AIR SPACE-
NOT VENTILATED
CEILING   29.00

ASSEMBLY C

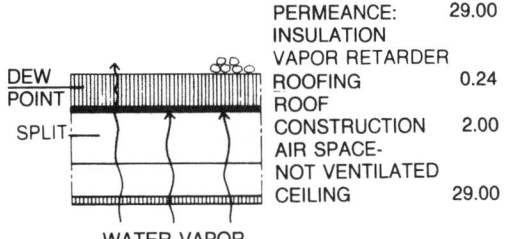

DEW POINT

SPLIT

WATER VAPOR

PERMEANCE:
INSULATION   29.00
VAPOR RETARDER
ROOFING   0.24
ROOF
CONSTRUCTION   2.00
AIR SPACE-
NOT VENTILATED
CEILING   29.00

Water vapor migration and condensation may be illustrated by considering conditions which may develop in several hypothetical roof assemblies under cold, dry outside with warm and humid inside conditions:
* outside and inside conditions and dew point temperature as shown under Temperature Gradients through assemblies.
* permeance of components remains the same for all assemblies shown.

ASSEMBLY A:
The assembly is poorly insulated, and the dew point occurs within the roof construction:
■ Vapor will start condensing within the roof construction, and will condense on the underside of the low permeability roofing:
* roof insulation may become saturated, losing some of its insulating value.
* reevaporating liquid may lift or blister the roofing membrane when exposed to intense solar radiation.
* condensed vapor may run back through the roof construction and drip on the ceiling.
■ Adding sufficient insulation to shift the dew point beyond the cold side of the assembly should prevent condensation within the assembly.
■ Alternately, a vapor retarder could be installed over the top of the roof construction, and the thickness of the insulation increased sufficiently to shift the dew point beyond the cold side of such a retarder.

ASSEMBLY B:
Adding a vapor retarder on top of the roof construction and increasing thickness of insulation to shift the dew point beyond the cold side of the vapor retarder may create other problems:
■ Vapor may penetrate the retarder:
* through voids or punctures resulting from poor installation.
* through splits in the retarder caused by cracks developing in roof construction.
■ Then the following may happen:
* upon reaching the dew point, vapor will start condensing as it continues to migrate to the underside of the roofing membrane.
* water, either in gaseous or liquid state will be then trapped between two impermeable surfaces, unable to evaporate to the outside, nor to run back into the heated space.
* any moisture penetrating into the insulation before or during the installation will also be trapped, and may deteriorate the insulation even if no vapor penetrates the retarder.
■ Providing means of ventilating the underside of the insulation, such as a ventilating base sheet, will help if continuous passage of vapor to the outside can be assured, but effectiveness diminishes as the area to be ventilated increases.
■ In such cases roof vents may be used:
* roof vents will be effective if vapor can freely migrate to them through the insulation.
* unless special self-regulating types are used, they will also admit vapor migration into the insulation when outside vapor pressures exceed those within the insulation.
Additional approach used is to delete the roofing, and combine it with the vapor retarder on top of roof construction, under the insulation, as shown under:

ASSEMBLY C:
■ Vapor migration would still be up to the vapor retarder/roofing, and no condensation should occur within the roof construction:
■ Cracks in the vapor retarder/roofing would still allow passage of vapor, but it would no longer be trapped in the insulation.
■ Insulation would:
* minimize thermal expansion/contraction in the vapor retarder/roofing.
* protect the vapor retarder/roofing from effects of ultraviolet radiation, hail, rain, snow and ice.
* the insulation should be resistant to the exterior environment, including airborne pollutants.
■ It should be kept in mind that repair of the vapor retarder/roofing would be difficult and costly, as not only aggregate ballast used to hold down the insulation, but the insulation as well, would have to be removed.

DEFINITIONS & PHYSICAL PRINCIPLES

AIR/WATER PENETRATION

WIND PRESSURE

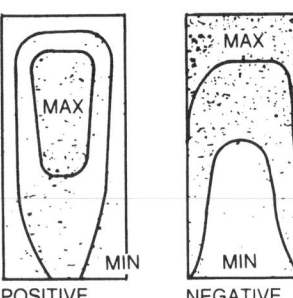

AIR PRESSURE DISTRIBUTION ON
SAME WALL UNDER DIFFERENT
WIND DIRECTIONS

Air infiltration/exfiltration will occur through joints and cracks between exterior wall components due to exterior and interior pressure differentials which result from air flow around and over an enclosure :

Exterior pressures vary even under constant wind velocity depending on terrain, exposure and orientation, also on height and shape of structure.

Interior pressures must adjust to equalize exfiltration and infiltration; generally pressures will vary with the number and location of openings in relation to a given wind direction: if most openings occur on the windward, interior pressures will increase and approach the value of exterior pressures; openings on the leeward side will have an opposite effect.

Infiltration may result in drafts and increase of heating load; **exfiltration may** carry air and water vapor to cold, hidden parts of walls and roofs, where the vapor may condense.

Rain penetrates through joints and cracks between and within the components of windward walls under pressure differentials between the exterior and interior spaces:

Wind driven rain will generally form a film over wall surfaces of low absorptivity; this film increases in thickness towards lower levels of multi-story buildings as it migrates laterally with a downward flow concentrated at vertical irregularities (such as joints).

CHIMNEY EFFECT

EXTERIOR AIR: – 10° F

APPROXIMATE TOTAL
PRESSURE DIFFERENCE
IN INCHES OF WATER
COLUMN FOR BUILDINGS
OF VARIOUS HEIGHTS:
100 FT – 0.3
200 FT – 0.56
400 FT – 1.06
600 FT – 1.6pv

NEUTRAL
PRESSURE
PLANE:
POSITION
VARIES

INTERIOR: + 70°F

PRESSURE DIFFERENCE

INFILTRATION/EXFILTRATION OF AIR
DUE TO TEMPERATURE/PRESSURE DIFFERENCES

Air infiltration/exfiltration will occur, especially in multi-story buildings, as a result of pressure differences, that is differences in the density of air when interior temperatures differ from outside temperatures:

Warm air will rise through openings in floor assemblies, elevator and stair shafts; it will produce:
* **negative** pressures in relation to outside air pressures at lower levels of the structure
* **positive** pressures at upper levels.

The number, size, and distribution of openings through which air can penetrate will determine the rate of infiltration/exfiltration and affect the relative location of the **neutral pressure plane,** at which inside and outside pressures are equal.
* in a tall structure with no internal separations and a single large opening at the bottom, the neutral pressure plan will be at the bottom of the structure: **exfiltration only will occur above the opening.**
* if the opening is relocated to top of the structure, the pressure inside will be lower than the pressure outside at all levels: **infiltration only would occur below the opening.**

Negative pressures inside the structure at lower levels will tend to draw in rain water.

Positive pressures at upper levels will result in exfiltration (or air leakage), this may cause:
* loss of heated air
* water vapor, carried by the heated air, to condense on cold surfaces within the wall and/or roof assemblies.

PRESSURE DIFFERENCES

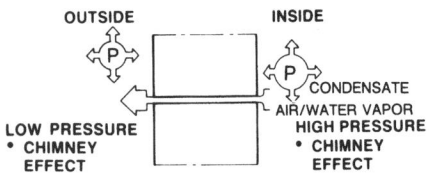

AIR/WATER VAPOR EXFILTRATION TO EQUALIZE
PRESSURE THROUGH JOINTS AND/OR CRACKS

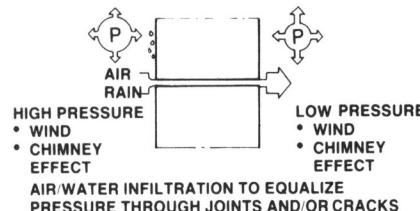

AIR/WATER INFILTRATION TO EQUALIZE
PRESSURE THROUGH JOINTS AND/OR CRACKS

Air will move from low pressure to higher pressure zones and may carry with it suspended water vapor; or water which is either moving through the air as drops of rain or is blocking its free passage to the lower pressure zone:
* so long as a difference in pressure exists on opposite sides of an assembly and there are openings in it, such as joints or cracks, **infiltration or exfiltration of air will continue;**
* water vapor may move as an air/vapor mixture under differences in air pressures; it may even **diffuse from a zone of low air pressure to one with high pressure** if the vapor pressure is lower in the high air pressure zone.

Water will penetrate through an assembly under gravity, capillary action, or under differences in air pressures:
* water penetration will **not occur** when **no pressure differences** exist and no other driving forces, such as gravity, are present;
* air pressure differences **may combine** with forces of gravity and/or capillarity **to increase** the rate of water penetration;
* water may **not penetrate** under gravity or capillarity if **positive air pressure** exists to counteract its flow;
* water suspended in a high velocity air stream may have sufficient **kinetic energy** to penetrate into a zone of **higher air pressure.**

DEFINITIONS & PHYSICAL PRINCIPLES

SOLAR RADIATION

HEAT FLOW, time lag, thin section

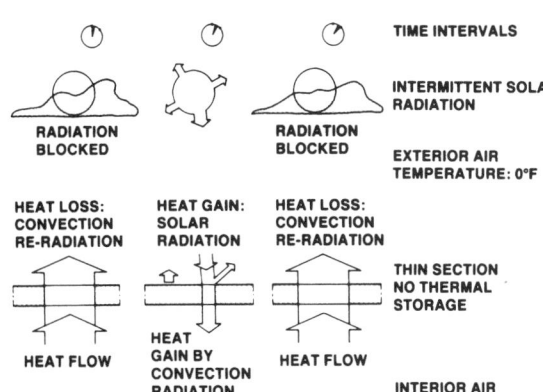

TIME INTERVALS

INTERMITTENT SOLAR RADIATION

EXTERIOR AIR TEMPERATURE: 0°F

RADIATION BLOCKED

RADIATION BLOCKED

HEAT LOSS: CONVECTION RE-RADIATION

HEAT GAIN: SOLAR RADIATION

HEAT LOSS: CONVECTION RE-RADIATION

THIN SECTION NO THERMAL STORAGE

HEAT FLOW

HEAT GAIN BY CONVECTION RADIATION

HEAT FLOW

INTERIOR AIR TEMPERATURE: 68°F

Solar radiation striking the outer surface of a section may raise its temperature above that of the inner surface and start heat flowing through it.

■ **For the flow of heat to reach the inner surface,** the temperature of the section itself must first rise to above the temperature of the surface. The time required to raise the temperature of the section to such a level is known as the time lag.

■ **The heat absorbed** by the section to raise its temperature is then referred to as **stored heat.**

■ **Thin sections** do not require large amounts of heat to raise their temperatures, and **have short time lags.**

REFLECTION, ABSORPTION, TRANSMISSION, transparent materials

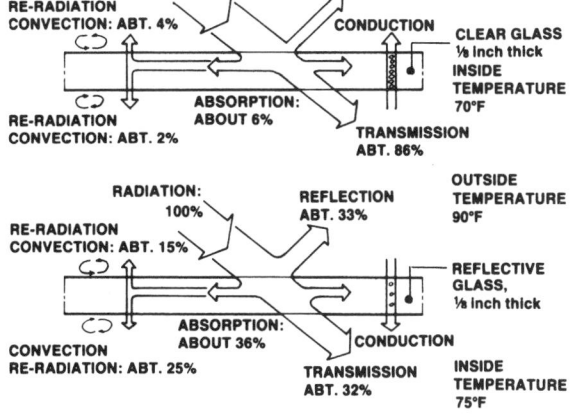

RADIATION 100%

REFLECTION ABT. 8%

OUTSIDE TEMPERATURE 25°F

RE-RADIATION CONVECTION: ABT. 4%

CONDUCTION

CLEAR GLASS ⅛ inch thick

INSIDE TEMPERATURE 70°F

RE-RADIATION CONVECTION: ABT. 2%

ABSORPTION: ABOUT 6%

TRANSMISSION ABT. 86%

RADIATION: 100%

REFLECTION ABT. 33%

OUTSIDE TEMPERATURE 90°F

RE-RADIATION CONVECTION: ABT. 15%

REFLECTIVE GLASS, ⅛ inch thick

ABSORPTION: ABOUT 36%

CONDUCTION

CONVECTION RE-RADIATION: ABT. 25%

TRANSMISSION ABT. 32%

INSIDE TEMPERATURE 75°F

Transparent materials, such as glass or clear plastics, not only reflect and absorb radiation, but also transmit it.

■ **Clear glass will reflect** ultraviolet radiation, but will allow most of the visible and infrared to pass through.

■ **During the heating season,** radiation admitted through glazed openings to the interior may supply a significant part of the total energy demand, provided the reverse flow of heat is effectively controlled.

■ Even though **the percentage** of radiation transmitted through clear glass **does not change between winter and summer,** it has been determined that about **twice the amount is gained in winter:**
• south facing openings are exposed to radiation for longer periods of time in winter;
• the angle of incidence is closer to the perpendicular, increasing the amount per unit area received;
• though length of travel through the atmosphere increases in winter, it is largely offset by the sun being closer to earth during that season.

■ **Reflective glass will reduce heat gain** in all seasons and generally only offers advantages by reducing summer cooling load.

REFLECTION, ABSORPTION, EMITTANCE, solid materials

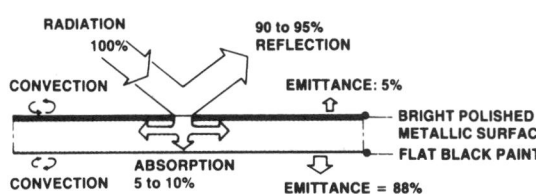

RADIATION 100%

90 to 95% REFLECTION

CONVECTION

EMITTANCE: 5%

BRIGHT POLISHED METALLIC SURFACE

FLAT BLACK PAINT

CONVECTION

ABSORPTION 5 to 10%

EMITTANCE = 88%

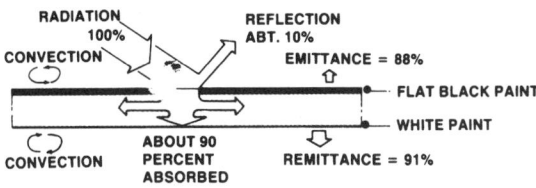

RADIATION 100%

REFLECTION ABT. 10%

CONVECTION

EMITTANCE = 88%

FLAT BLACK PAINT

WHITE PAINT

CONVECTION

ABOUT 90 PERCENT ABSORBED

REMITTANCE = 91%

The total solar radiation striking a plane surface is the sum of direct radiation and diffuse sky radiation, and may also include radiation reflected from surrounding surfaces.

■ **All substances will reflect** a certain percentage of the solar radiation striking them and absorb the balance. The sum of the values for reflectivity and absorptance always equal to 1.
• warm surfaces also emit energy as long wavelength radiation.
• reflectivity, absorptance, and emittance are determined by the surface characteristics of a given material.

■ **When heat gain is to be minimized,** surfaces selected should have high value of reflectivity to reduce the amount of radiation absorbed, and also a high value of emittance to re-radiate the maximum amount of heat to the outside.

■ **When heat gain is to be maximized,** the receiving surface should have a high value of absorptance, with a low value of emittance:

■ **For additional information** on surface characteristics, refer to INSULATION

DEFINITIONS & PHYSICAL PRINCIPLES

SOLAR RADIATION (cont.)

DEFINITIONS & PHYSICAL PRINCIPLES

SOLAR RADIATION

Solar energy reaching the earth in the form of short-wave radiation fuels the entire life process: it heats the atmosphere and the surface of the earth; it supplies the heat for all biological activity; it illuminates the earth.

■ **Solar radiation is a constant yet variable source of energy:** it reaches the upper atmosphere of the earth at a nearly constant rate, but only a portion of the upper atmosphere is exposed to it at any given time as the earth rotates.

At the surface of the earth it is available as:

■ **Direct** solar radiation;

■ **Reflected** radiation: direct radiation reflected by surrounding surfaces;

■ **Diffuse** radiation, or skylight: direct radiation scattered during its passage through the atmosphere.

■ **Stored energy** in two forms:
• direct radiation converted to heat and absorbed in the atmosphere, water, objects on the surface of the earth, and the earth itself.
• direct radiation photosynthesized and stored in: living plants, or biofuel & in fossil fuels.

Solar radiation reaching the surface of the earth consists of:

■ **Visible light.**

■ The invisible **infrared** wavelengths.

■ The third, **the ultraviolet band,** even though important, constitutes less than 10% of solar radiation.

Any structure located upon the surface of the earth will be subject to a certain amount of solar radiation: either directly, by being exposed to it, or indirectly through its effects:

■ **When the structure is directly exposed,** it will absorb varying amounts of the radiation striking its envelope, functioning as a collector of solar energy whether thus intended or not: **solar radiation cannot be excluded from the design process, it can only be ignored.**

The total amount of radiation received by the earth in the course of a full orbit around the sun **is virtually constant, but its distribution over the earth's surface varies considerably:**

■ **It is available** only at specific times and from specific directions, during a 24-hour period;

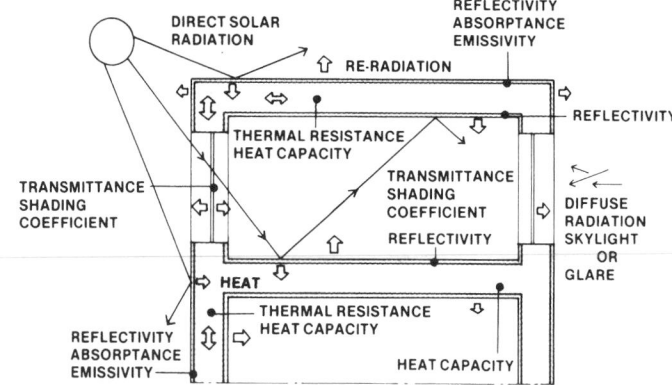

PROPERTIES OF COMPONENTS OR ASSEMBLIES WHICH DIRECTLY AFFECT CONTROL AND/OR UTILIZATION OF SOLAR RADIATION

■ **The amount** of solar radiation at any given time depends on atmospheric conditions, such as cloud cover and level of pollutants.

■ **The intensity** of radiation striking a surface fluctuates with daily and seasonal variations in the position of the sun, and also depends on the position of the surface in relation to the sun:

■ **The orientation and shape** of a structure will determine the amounts of radiation received at the envelope.

■ **The extent and distribution** of opaque and transparent components of the envelope will influence both:
• heat gain/loss
• penetration of visible light to the interior.

CONTROL AND UTILIZATION

Solar energy available to a building needs to be effectively **controlled and utilized for optimum contribution to the interior climate created.**

While the infrared band contributes heat only and may or may not be desirable, the visible light band generally is. **Control and utilization of solar energy should therefore consider:**

■ **Heat** at opaque surfaces.

■ **Heat and light** at transparent or translucent surfaces.

HEAT absorbed at opaque surfaces of the envelope may be:

■ **Allowed free flow** to the interior, with the reverse flow controlled;

■ **Controlled** to reduce its flow to the interior at all times;

■ **Stored** in solid assemblies for gradual release into the interior.

■ **Properties of the exterior** components determine the amount of radiation rejected, absorbed, or admitted.

■ Those properties of components and assemblies which **directly affect control and/or utilization of solar radiation are illustrated above.**

■ **Direct radiation may be:**
• reflected to strike surfaces not directly exposed to it;
• selectively reflected and admitted;
• diffused over a large area;
• concentrated over a small area;
• excluded at certain angles of incidence, admitted at others.

■ **Surfaces receiving** direct radiation may be:
• controlled to maintain them at an angle for optimum gain;
• selectively coated for maximum absorption and minimum re-radiation.

■ **Radiation absorbed** and converted to heat in the absorbing medium may need to be concentrated for efficient utilization.

Light and heat of solar radiation received at or through a glazed opening in the envelope may be:

■ Controlled **before** reaching the opening;

■ Controlled **at the opening** either by the glazing of or by shading devices incorporated into a window assembly;

■ Controlled **after being transmitted** through the opening by operable shading devices, such as blinds, shades.

Control Devices

Control devices are used to manage the flow of solar energy through the openings in the building envelope.

The internal environment to be maintained under varying external conditions will influence the type and extent of control required:

■ **An opening intended to admit direct** solar radiation in winter, in order to help heat the interior, may have to be shaded to exclude radiation during warm weather to prevent overheating.

■ **The same opening** may have to be controlled to prevent heat loss to the outside at night or during cloudy days during the heating cycle;

■ **Low-angle early morning or late afternoon radiation** may have to be blocked to prevent visual discomfort while during the balance of daylight hours such openings should be free to admit usable daylight.

Solar-heat can be invited in the winter and guarded against in the summer through any/all of the following means:

■ **orientation** of building on site

■ appropriate landscape **plantings**, for example
• evergreens where year-round protection is needed
• deciduous trees where summer shade, winter sun are wanted
• ground covers on driveways and parking lots to keep the ground cool.

The control of solar radiation at transparent openings is achieved through the use of various shading devices, such as:

■ **Fixed exterior** devices that block solar radiation at specific angles of incidence:
• overhangs, vertical or horizontal louvers
• screens
• grilles, trellises.

■ **Variable exterior** devices that control solar radiation selectivity at most or all angles of incidence:
• awnings
• adjustable louvers, vertical or horizontal
• shutters.

■ **Variable interior** devices that control visible light selectively at different angles of incidence:
• vertical or horizontal blinds
• plain and insulating shades
• drapes.

■ **Fixed interior devices,** to reduce amount of solar radiation transmitted:
• reflective glass and tinted glass
• reflective films on glass.

Utilization Methods

Methods of utilization more commonly employed use all the forms of solar radiation available: direct, reflected, diffuse and stored. **Direct and/or reflected radiation is used in:**

■ **Passive space heating systems:** function by absorbing energy at the surface of and storing the heat in exterior/interior assemblies for gradual release.
• flow of heat is controlled to minimize losses to the outside.
• auxiliary energy source is generally provided.

■ **Active space heating/cooling systems and domestic water heating systems:** function by absorbing energy at the surface of a collector, flat or concentrating, and transferring the heat collected by means of a transport medium to the point of use:
• provisions for storing heat usually are incorporated into space heating systems.
• auxiliary energy source generally provided.

■ **Active process heat systems** for low temperature applications: function by absorbing energy at the surface of a collector and transferring it to the point of use:
• auxiliary energy source may be required.

■ **Photovoltaic generation:** through photovoltaic cells which convert solar energy directly into electricity.
• in bright sunlight they can produce direct current at 0.6V and approximately 0.030 amps per square centimeter of exposed cell area.
• conversion efficiency is about 16 percent. Direct solar radiation is also used in power generating and industrial installations, such as:

■ **Solar furnaces:** function by concentrating solar radiation over a small area through the use of a large array of sun tracking reflective surfaces:
• temperatures in excess of 6,000°F have been generated.

■ **Power generating plants:** function by:
• concentrating solar radiation via an array of heliostats—suntracking reflective surfaces—on a central receiver.
• there it is absorbed, converted to heat.
• next it is transferred through a liquid transport medium to drive a steam or gas turbine.

• process is still at an early stage of development.

Natural daylight can supplement or even replace artifical lighting of a building.

■ **Clearstory strip windows** increase light penetration into the interior.

■ **Reflective surfaces** such as light shelves and heliostats are used to improve penetration of daylight into the interior for maximum amount and better distribution of usable daylight.

■ **Skylights and roof monitors** are installed in single story structures or top stories for wide area lighting.

Skylights as a means for energy conservation may be adopted in any building space which is covered by a roof. You should compare:

☐ **the energy consumed** by heat loss through them to:

☐ **the energy saved** by not consuming electricity due to natural illumination.

☐ **The increased heat** from the sun is largely offset by the elimination of the heat from the electric lights. Thus, the air conditioning annual demand is not substantially changed, and the electrical light energy saved by the use of solar lighting represents virtually **all a net energy saving**.

☐ It takes over three times as much fossil fuel to create a BTU of heat. Thus, when solar energy replaces electricity (as in lighting), it is **three times as effective** as when it replaces heating.

☐ A design that provides adequate lighting for 50% of the working hours should **pay out in energy savings** in from 5 to 10 years (1978 energy rates).

Radiation absorbed and stored as heat in the atmosphere may be utilized by:

■ **Heat pumps** to concentrate and transfer it to interior spaces.

■ **Wind driven** electric power generators or water pumps.

Radiation absorbed and stored in water may be utilized by:

■ Using small bodies of water **as heat storage/heat sinks.**

DEFINITIONS & PHYSICAL PRINCIPLES

SOLAR RADIATION (cont.)

DEFINITIONS & PHYSICAL PRINCIPLES

SUN AND EARTH ORBIT

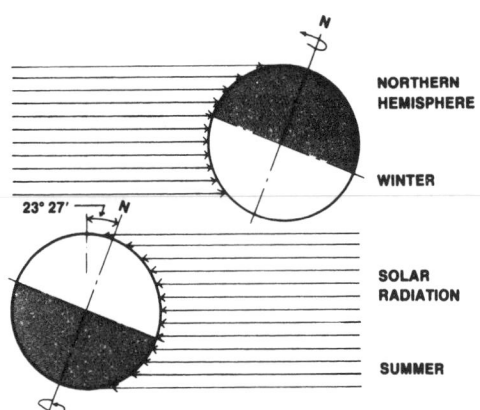

The sun radiates energy at a constant rate of 43,000 kw/m² of its surface. Only a tiny fraction of this energy, about 7×10^7 kw/hr reaches the upper atmosphere of the earth, but that amounts to 30,000 times more than is used in all man-made devices. **The actual amount of radiation striking a particular point on the earth's surface depends on its position in relation to the sun:**

■ **Intensity of radiation varies** depending on whether it strikes a surface at right **angles** or oblique ones, as a result of the curvature of the earth.

■ **Intensity is further modified** because the earth **is tilted** at an angle of about 23½° in relation to the plane of its orbit around the sun.

■ **Solar radiation reaching the earth's surface on a slanting path** due to curvature and tilt of the earth must travel through more of the earth's atmosphere, thus losing more energy through absorption and scattering within the atmosphere.

EARTH ATMOSPHERE, TEMPERATURES

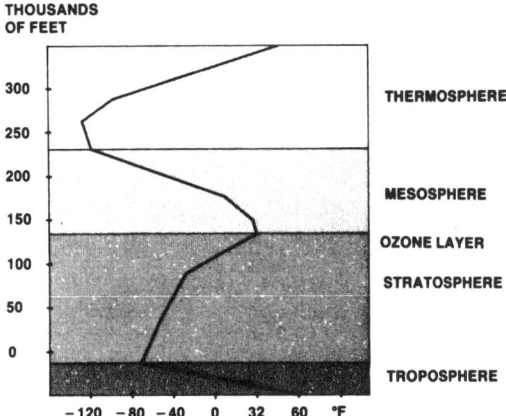

To reach the earth's surface, solar radiation must penetrate through the atmosphere: a mixture of gases of varying densities surrounding the earth in concentric layers:

■ **The atmosphere is composed mainly** of oxygen and nitrogen with minute quantities of carbon dioxide and hydrogen, and with varying amounts of water vapor, ozone, dust and smoke.

■ **Ozone,** a heat absorbing form of oxygen in the stratosphere, **blocks the passage** of most of the harmful ultraviolet band of radiation; this also accounts for the higher temperature of this layer.

■ **The upper atmosphere is largely transparent** to solar radiation and absorbs about one-seventh of the radiation passing through it; it is **practically opaque** to terrestrial long-wave radiation, absorbing 85 to 90 percent of it. Such long-wave radiation is the primary source of heat for the lower atmosphere.

■ **Incoming solar radiation exceeds** outgoing terrestrial radiation about between latitudes 37°N and 37°S. North and south of 37° **the opposite holds true.** This latitudinal imbalance sets up the general circulation of the atmosphere.

WAVELENGTHS OF RADIATION

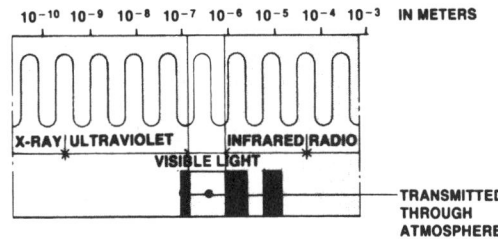

Solar energy is transmitted through a wide band of electromagnetic radiation, with only a narrow segment of it, ranging from about 0.29 to about 3.3 microns, reaching the surface of the earth:

■ **Appr. 10 percent is the invisible band of ultraviolet radiation,** with wavelengths below 0.4 microns. Ultraviolet radiation destroys cells of the human skin, but also aids the body in synthesizing Vitamin D;

■ **Visible light, the band of wavelengths** perceived by the human eye, ranges from about 0.4 microns to about 0.7 microns, and comprises about 40 percent of incoming radiation;

■ **Invisible infrared waves** beyond 0.7 microns in lengths account for the remaining 50 percent.

■ Some of the solar radiation absorbed at the surface of the earth is in turn **reradiated back into space** in the form of long wavelength infrared radiation: the earth emits about 114 units of infrared radiation: the net infrared emission is the difference between this value and the 93 units absorbed by the atmosphere and emitted back to the surface of the earth.

ANGLE OF INCIDENCE, SOLAR CONSTANT

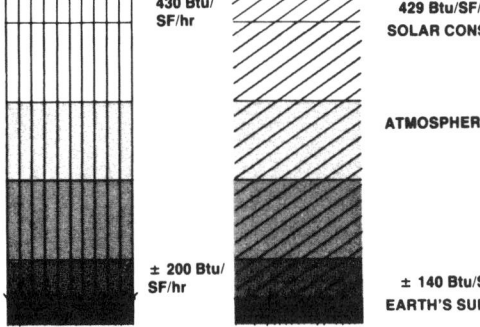

430 Btu/SF/hr

± 200 Btu/SF/hr

429 Btu/SF/hr = SOLAR CONSTANT

ATMOSPHERE

± 140 Btu/SF/hr
EARTH'S SURFACE

Solar radiation reaches the outer edge of the atmosphere at a nearly constant rate which is referred to as the solar constant.

■ **Solar constant may vary** by about 2 percent due to changes in the energy output of the sun, and by about 3.5 percent due to variations in distance between the earth and the sun. The amount of radiation striking a unit area of the upper atmosphere **also depends on the angle of incidence,** which changes constantly as the earth rotates.

■ **The values generally used for the average intensity** of solar radiation perpendicular to the surface of the atmosphere vary from 442 Btu/SF/hr to 420 Btu/SF/hr depending on the source.

■ **The amount of radiation reaching a horizontal unit area** on the surface of the earth at a particular point depends on the **angle** at which it strikes such area.
• it is **maximum** up to 300 to 350 Btu/SF/hr on a clear, cloudless day, when perpendicular to the surface of the unit area.
• it **decreases** by the cosine of the angle by which it deviates from the perpendicular, dropping to zero at sunset and sunrise.
• **curved surfaces** receive the same amount of radiation as flat surfaces of equal projected area.

DIFFUSE RADIATION

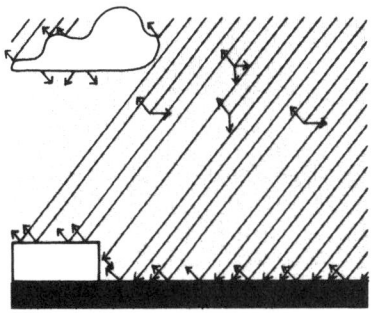

TROPOSPHERE

EARTH'S SURFACE

Diffuse radiation is direct radiation scattered during its passage through the atmosphere, and received from all parts of the sky. It is also often referred to as luminance, or skylight.

■ **Diffuse radiation may** only contribute as little as 10 percent of the total radiation striking a surface on a clear day, but may reach 100 percent in cloudy weather.

■ **Even in cloudy weather,** as long as the sun still casts a shadow, up to 50 percent of possible radiation may be received.

■ **The amount of diffuse** radiation received by a surface does not depend on its orientation, as with direct radiation, but is determined by its position: a flat roof which faces the sky will receive about twice the amount per unit area than a vertical wall.

REFLECTION, ABSORPTION

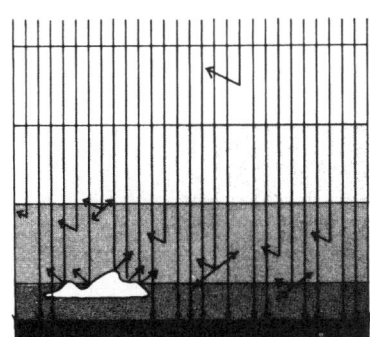

430 Btu/SF/hr = 100%

ATMOSPHERE

LOSS THROUGH:
REFLECTION = ± 35%
ABSORPTION = ± 19%

± 200 Btu/SF/hr = ± 46%
EARTH'S SURFACE

Only about half of the radiation striking the upper atmosphere is finally absorbed and converted to heat at the surface of the earth: the rest is reflected back into space by the atmosphere and by ground level reflective surfaces, such as polar ice caps; or is absorbed in the atmosphere:

■ Cloud droplets, microscopic particles of dust, smoke particles, air molecules and ground surfaces reflect and scatter solar radiation: from 25 to 35 percent of all incoming radiation is lost back into space;
• backscattering by air: about 6 percent.
• reflected by clouds: about 20 percent.
• reflected by ground surface: about 4 percent.

■ **Dust, water vapor,** ozone, and carbon dioxide in the atmosphere also **absorb solar radiation:** from 15 to 20 percent of all incoming radiation is lost through absorption, mostly in the upper layers of the atmosphere.

■ **More solar radiation is lost by absorption at low sun angles** because the length of travel through the atmosphere is increased.

DEFINITIONS & PHYSICAL PRINCIPLES

DAYLIGHTING

SUNLIGHT

Solar radiation is absorbed, reflected, and scattered in the atmosphere so that the visible band of radiation reaches the surface of the earth not only as direct radiation, but also as reflected or diffuse light from the sky.

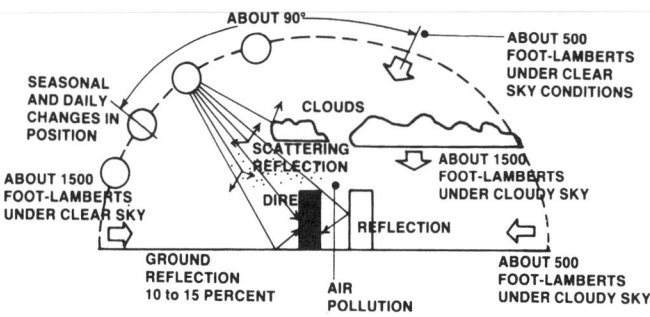

■ **The amount of daylight received** from the sky depends on the position of the sun and on atmospheric conditions.

■ **Generally one or more of three basic conditions** of incident light are considered in design evaluations, depending on climatic and site conditions and on orientation:

* **light from an overcast sky only:** this varies with the density and uniformity of the overcast. A uniformly overcast sky is usually two-and-half to three times as bright overhead as near the horizon.
* **light from a clear sky only:** is brightest around the sun with the darkest area approximately 90 degrees from across the sun's position, and then generally brighter at the horizon than overhead.
* **light from a clear sky plus direct sunlight:** the effect of direct sunlight on surfaces exposed to it is added to clear sky conditions.

INCIDENT LIGHT ON STRUCTURES, northern hemisphere

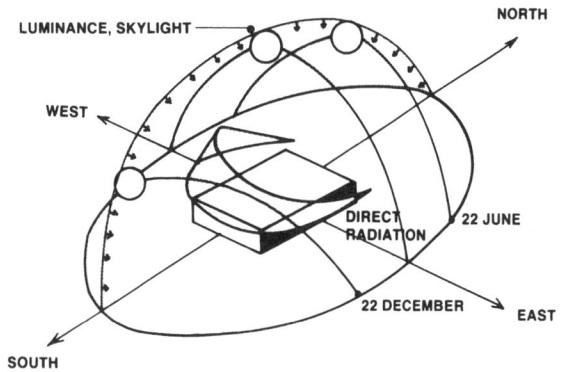

Daylight is a dynamic visual phenomenon varying with location time and seasons: it is an instantaneous event and cannot be stored.

■ **Computational and graphic analysis methods exist,** but none provide the combination of reliability, accuracy and dynamic response which daylight design requires;

■ **Physical modeling of daylight design** for each specific project in actual site conditions is the recommended approach:
* it predicts both the quality and quantity of daylight,
* it can reveal lighting gradients, specific glare problems and the effects of building forms and finishes.

■ **Numerical calculations fall into two main categories:**
* **the lumen method** which allows for a comparison of various window-wall schemes on a center line perpendicular to them. Effects of surface reflectances and of shading devices are included in calculations.
* **the daylight factor method** which predicts the daylight anywhere in a space for a standard sky condition without direct radiation striking the window. The daylight factor method is essentially an analytical process.

DEFINITIONS & PHYSICAL PRINCIPLES

SOURCES OF ILLUMINATION

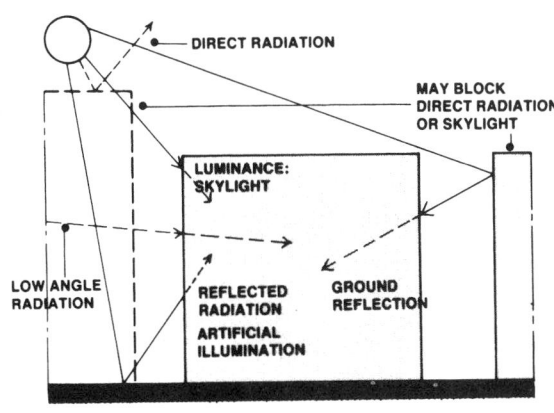

Daylighting depends not only on atmospheric conditions, but also on:

■ **Daily and seasonal** changes in the sun's position;

■ **The orientation** and location of the structure;

■ **Daylighting** generally is only readily available at the outer surfaces of a multi-story structure except where special design features, such as light wells, are used.

■ **Provisions for daylighting,** therefore, are not added to, but integrated into the total design of a structure.

■ **Orientation is an important consideration** in daylight design: a predominantly north/south orientation is generally desirable as it lends itself to seasonal, or fixed, controls.

■ Despite such limitations, **maximizing the use of daylighting is usually desirable,** especially because daylighting, using skylight as the source, will generally deliver more illumination per unit of heat gained in the space than artificial sources:
• artificial illumination not only consumes energy, but also **adds to internal heat load,** the control of which requires additional energy during the cooling period.
• artificial illumination in commercial structures, such as office buildings or stores **uses from 35 to 50 percent** of the total energy demand.

■ Artificial illumination on the other hand **is easily controlled, predictable,** and can be provided at any point within the structure.

SIDELIGHTING

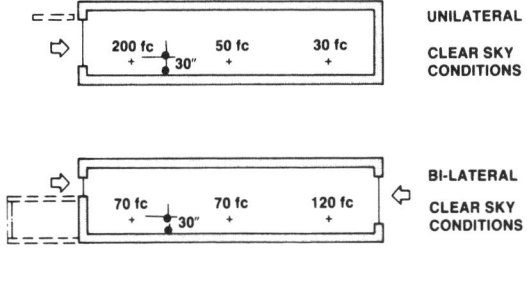

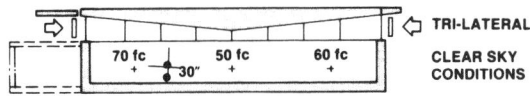

The most familiar approach to daylighting is sidelighting.

■ **Generally only diffuse radiation,** or skylight, is considered: direct solar radiation may lead to visual and thermal discomfort and should be excluded.

■ **Windows are more effective when horizontal:** vertical windows set up strong patterns of contrasting brightness:
• **the higher the location** of a window in wall and the higher the ceiling, the better the distribution of light. A typical practical limit for unilateral daylight penetration into an office is 15 to 20 feet.
• **penetration** of daylight may be improved through the use of clearstory windows. For additional information on clearstory windows, refer to Time Saver Standards.

■ **Reflectances of wall,** floor, and ceiling surfaces will also influence light levels within a space for any type of illumination.

DEFINITIONS & PHYSICAL PRINCIPLES

PART 1

DAYLIGHTING (cont.)

TOPLIGHTING

Toplighting is a method of daylighting large single story structures; more appropriate for general illumination than for task lighting.

■ **Toplighting may be provided by:**
* roof monitors, glazed sawtooth roofs, skylights;
* the entire roof may be translucent, as in fabric covered tensile or air-supported structures.
* for additional information on types and methods of toplighting, refer to Time Saver Standards and Architectural Graphic Standards.

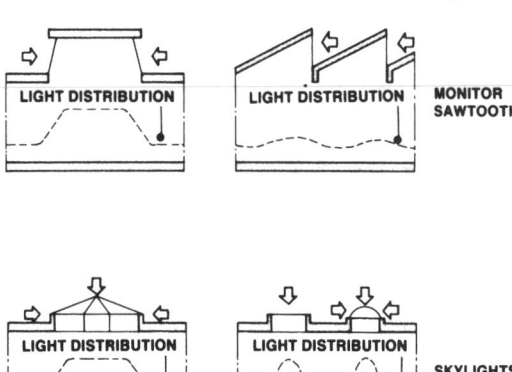

■ **Monitors admit light** from opposite compass directions and:
* when oriented north/south they may require control of direct solar radiation on the south facing side.
* sawtooth roofs should be faced north to exclude direct solar radiation; the exception possibly being buildings located in the northern latitudes where south facing opening will admit more daylight in winter.
* reflective roof surfacing may be used to increase level of daylight admitted.

■ **Skylights,** like south facing monitors, may need provisions to control heat gain through them; translucent roofs are very difficult to control.

REFLECTIVE SURFACES

Sidelighting of large multi-story structures generally is unilateral and confined to the perimeter of the structure; toplighting can only be used for the upper story.

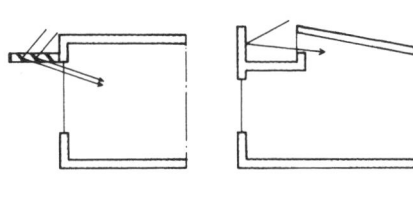

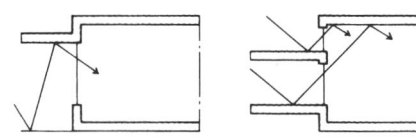

■ **Light wells** may be used to improve penetration of daylight into large structures.

■ **Some of the other approaches** to allow daylight to penetrate deeper into the structure are:
* **reflective surfaces on exterior** louvers and interior blinds: when fixed they may reflect skylight only, and should be automated to follow the daily path of the sun when direct radiation is to be reflected.
* **light shelves** located below full height and/or clearstory windows;
* **fixed overhangs** to re-reflect ground reflection into the space;
* **parapets** facing a north oriented clearstory window or monitors to reflect light from the sun and the southern sky through them to the interior.
* **upward facing** reflective surfaces are subject to dust collecting on them, and will not perform as intended unless maintained.

DEFINITIONS & PHYSICAL PRINCIPLES

DAYLIGHTING: preliminary selection

The ratio of light output to heat gain from a light source is generally expressed as lumens per watt: the higher the value of such a ratio, the less heat is associated with a given level of illumination:

■ **Solar radiation as a source** of illumination produces much less heat than artificial light sources:
- direct radiation has a value of about 125 lumens per watt.
- diffuse clear sky luminance about 150 to 170 lumens per watt.
- fluorescent lamps have 60 to 70 lumens per watt.
- incandescent lamps between 10 and 20 lumens per watt.

■ **Reducing heat gain** from sources of illumination within a space through increased use of daylighting results in energy savings for space which requires cooling to maintain comfort levels.

■ **Daylighting can be defined** as providing usable levels of daylight over maximum areas of interior spaces. This involves control as well as utilization since in order to be usable, daylight may have to be controlled.

Daylighting and Electric Lighting

■ Direct sunlight striking a bright surface **can cause glare with attendant visual discomfort** when the level of illumination of surrounding surfaces is markedly lower.

■ A high level of direct solar illumination, therefore, **requires a concomitant high level of artificial illumination** to reduce sharp contrasts in brightness.

Maximizing the use of daylight and reducing energy consumed by artificial illumination **may require more sophisticated lighting controls** than the on/off switches generally used:

■ **Dimmable controls** should be considered for better response to changing levels of daylight.

■ **Sensors for user-independent control** of dimmable artificial illumination systems in large open office spaces may be required to adjust the level of light automatically and maximize conservation of energy.

■ **Daylighting to provide ambient lighting,** with artificial task lighting individually controlled, should also be considered.

■ **Control of daylighting** by excluding direct radiation, or diffusing it evenly over the entire space:
- will improve visual comfort
- but may also interfere with the use of such radiation for heating the space when needed.

Available Daylight

Daily conditions cannot be predicted with any degree of accuracy, but the average annual amounts of sunshine have been established: **data compiled by the U.S. Weather Bureau provide a statistically reliable guide** to the amount of daylight which may be anticipated at a given location.

External factors which affect daylighting are:

■ **Solar positions,** luminance of the sky and the distribution relative to the specific orientation considered.

■ **Local atmospheric** conditions and their effect on incident daylight: clear or cloudy; clean or polluted.

■ **Seasonal changes** and their effect: e.g. the presence of snow on the ground for extended periods of time.

■ Solar radiation is **a dynamic source of energy:** methods of utilization and control must be responsive to its nature, or they will fail.

Sidelighting

When solar radiation is not needed for heating a space, glazed openings may be proportioned and located:

■ **To minimize** heat gain to a space.

■ **To increase** penetration of daylight into it.

■ Generally, **the higher** a continuous horizontal glazed opening is above the floor, **the deeper the penetration.**

■ Unless the glazed openings are made larger by extending them closer to the floor with resultant increase in heat gain, **visual contact with the outdoors may be largely lost.**

■ Use of **narrow vertical openings** rather than narrow horizontal ones restores visual contact with the outdoors, but also **creates sharp contrasts** in brightness resulting in visual discomfort.

Direct penetration of daylight at usable levels into a side-lighted space is generally limited, and **means for im-** proving it should be considered, including:

■ **Reflective surfaces attached** to or incorporated into the exterior assemblies, (such as light shelves) which reflect and diffuse direct solar radiation over the space.

■ Reflective surfaces **adjacent to the structure,** (e.g. crushed stone ground surfacing) effectively utilized, light reflected from the ground particularly at points well removed from the windows:
- on non-sun exposures, light reflected from the ground may account for more than half of the total light reaching a window.

■ **Sun-tracking mirrors** distribute sunlight through the interior spaces.
- the use of heliostats is still largely in an experimental stage.
- light levels will vary quickly and extensively under partly cloudy conditions and elaborate controls for artificial illumination may be required to compensate for such variations.

■ **Interior reflective** surfaces allow for deeper penetration of daylight received at the perimeter.

■ **Light colored** floors, ceilings, draperies, blinds will also amplify interior light.

Toplighting

Similar trade-offs between heat gain and improved daylighting apply to the use of toplighting, by means ranging from single skylights to elaborate skylight systems.

DEFINITIONS & PHYSICAL PRINCIPLES

SHADING

SOLAR ANGLES, northern hemisphere

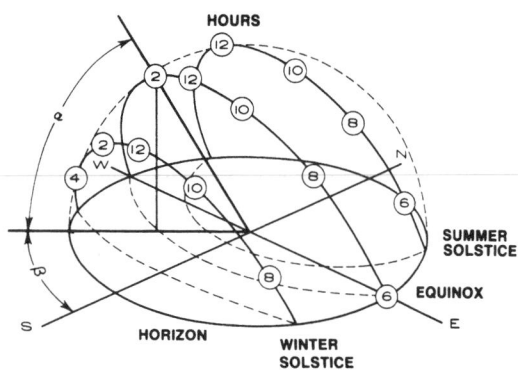

The position of the sun in relation to a specific geographic location, season and time of day determine the exposure to solar radiation of a given surface at that location.

■ **Position of the sun and resultant shading may be determined** by combining site-specific shading data with published solar radiation figures for the given locale;

■ **Methods used to obtain site-specific data are:**
- surveyor's transit to determine the angle of each shadow-casting object; the process is slow and the recording of data time consuming.
- sighting through wide angle "fisheye" lens. The sun's path for selected seasons or times is then printed on a transparent grid. Presently the method is still limited.
- reflective devices consisting of either a plane mirror, or a contoured dome.

ORIENTATION

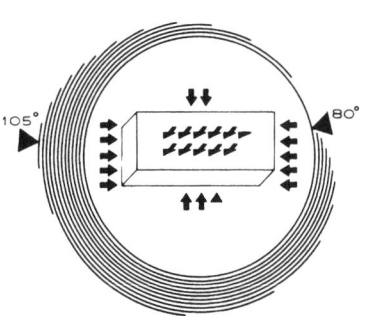

THERMAL IMPACT
ON DIFFERENTLY
EXPOSED SURFACES

JULY
PHOENIX, ARIZONA

Orientation includes both physiological and psychological considerations:

■ **Physiological considerations generally include:**
- **climatic** factors such as: wind intensity and direction; air temperatures;
- **seasonal** variations in thermal impact of solar radiations on a structure, as illustrated: with each arrow representing 250/Btu/SF/day, the north and south walls will be receiving considerably less radiation than east and west walls, with the maximum amount striking the horizontal roof plane.
- **environmental** factors include: noise, dust, smoke, and smell.

■ **Psychological considerations** may include view, privacy, relationship to other structures or to natural features of the environment.

SUN TIME

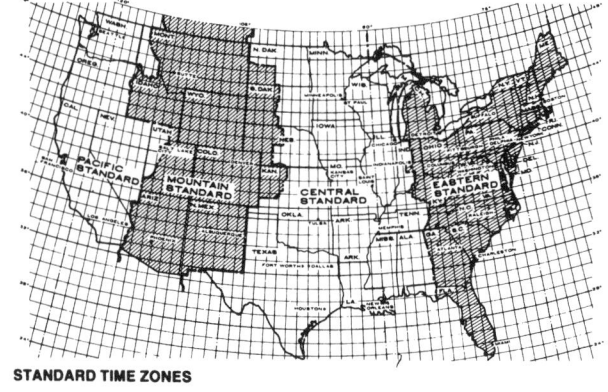

STANDARD TIME ZONES

Determination of solar angles and positions, as well as the calculation of shading, shading devices, and the construction of shadows **requires an accurate determination of sun time** at a given location as opposed to the standard time used in the time zone of such particular location:

DEFINITIONS & PHYSICAL PRINCIPLES

SOLAR RADIATION, EFFECTS AND CONTROL

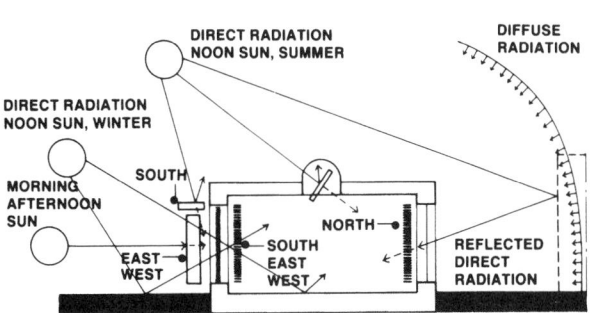

Direct solar radiation needs to be controlled to:

■ **Reduce** heat gain through openings during warm weather;

■ **Maximize** heat gain when space heating is required;

■ **Reduce visual discomfort** of glare or sharp contrasts and improve level & diffusion of illumination.

■ Reduce visual discomfort from **sunlight penetrating at low angles.**

■ **Reflections of direct radiation** off the ground, water, snow, or adjacent vertical surfaces may cause glare.

■ **Diffuse radiation levels** at the horizon vary under different atmospheric conditions and their intensity may reach levels sufficient to result in glare.

■ **The means of control** used must not prevent achievement of
• proper levels of daylight
• visual contact between the inside and outside.

SHADING DEVICES

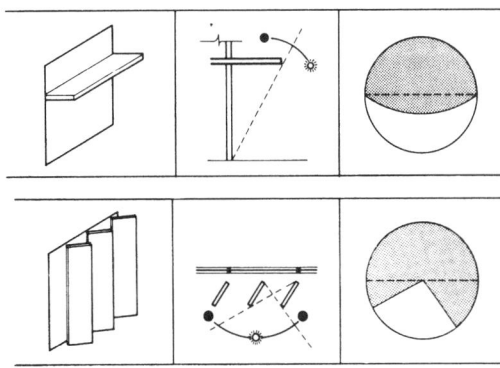

Shading devices may be grouped by location into:

■ **Exterior:** fixed or operable
• exterior shading devices generally offer better control of heat gain, since heat absorbed is dissipated into the surrounding air.

■ **Within window assembly:**
• operable or fixed blinds between two panes of glass; not as effective as exterior devices in preventing heat gain, and difficult to clean or repair.
• glazing itself may provide shading and heat absorption when coated with reflective films or tinted; once treated, it will not respond to changes in conditions.

■ **Interior:** operable blinds, shades, drapes.
• not effective in preventing heat gain
• easy to operate and maintain.

SHADING AND ORIENTATION

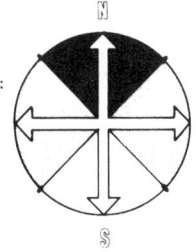

CLEAR GLASS, SINGLE OR DOUBLE
BLINDS, VERTICAL OR HORIZONTAL,
TO CONTROL SKY GLARE

REFLECTING OR
ABSORBING GLASS; OR:
CLEAR CLASS, SINGLE,
DOUBLE, AND:
VERTICAL LOUVERS,
VERTICAL BLINDS
SHADES, DRAPES;
ALSO:
HORIZONTAL BLINDS

REFLECTING OR ABSORB-
ING GLASS; OR:
CLEAR GLASS, SINGLE,
DOUBLE, AND:
VERTICAL LOUVERS,
VERTICAL BLINDS,
SHADES, DRAPES;
ALSO:
HORIZONTAL BLINDS

CLEAR GLASS, SINGLE OR DOUBLE
AND SHADING DEVICE:
OVERHANGS, HORIZONTAL LOUVERS,
HORIZONTAL BLINDS, SCREENS,
SHADES, DRAPES; OR TRELLIS;
OR:
REFLECTING OR INSULATING GLASS

Selection of a shading device depends on its shading mask and orientation:

■ **Fixed shading devices** have the advantage of simplicity, but lack dynamic response to changing solar positions.

■ **Orientation between principal** compass directions may require a combination of several types to provide full shading.

■ **The extent of shading** a shading device or an element of the structure will provide can be determined by the use of solar path or solar angle diagrams and of a shading mask protractor:
• shading masks are used to evaluate the need for shading devices, their effectiveness
• and may also be used to design them.

DEFINITIONS & PHYSICAL PRINCIPLES

SHADING (cont.)

HEAT FLOW, time lag, heavy section

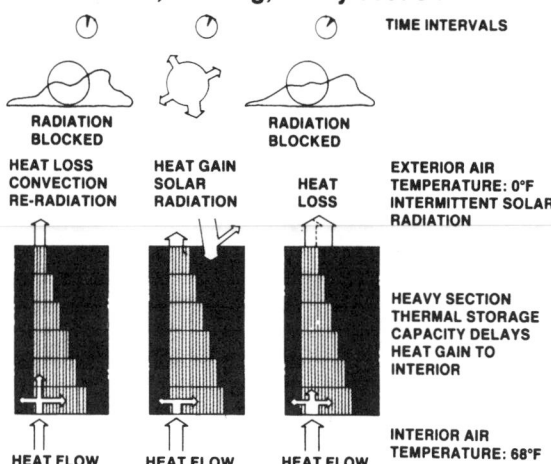

TIME INTERVALS

RADIATION BLOCKED

RADIATION BLOCKED

HEAT LOSS CONVECTION RE-RADIATION

HEAT GAIN SOLAR RADIATION

HEAT LOSS

EXTERIOR AIR TEMPERATURE: 0°F INTERMITTENT SOLAR RADIATION

HEAVY SECTION THERMAL STORAGE CAPACITY DELAYS HEAT GAIN TO INTERIOR

HEAT FLOW   HEAT FLOW   HEAT FLOW

INTERIOR AIR TEMPERATURE: 68°F

Massive sections require large amounts of heat to raise their temperatures: storage capacity is high and time lag is long.

■ **When a massive section,** such as a heavy concrete wall, is **exposed to intermittent** solar radiation, the heat gained at the outer surface will start flowing into the section, gradually raising its temperature.

■ **The outer surface temperature will rise first,** and more heat will be lost to the outside because of higher temperature differential.

■ **Exposure to solar radiation may not be long enough** for the entire section to absorb a sufficient amount of heat to raise its temperature above that of the inner surface; in this case no heat gain to the interior will result.

■ **When heat gain at the outer surface ceases,** heat loss from it will gradually return to a steady state flow.

■ **During the heating season** radiation admitted through glazed openings to the interior may supply a significant part of the total energy demand provided the reverse flow of heat is effectively controlled.

HEAT STORAGE, mass effect

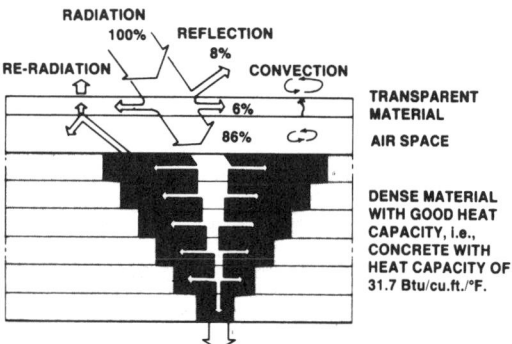

RADIATION 100%   REFLECTION 8%

RE-RADIATION   CONVECTION

6%

86%

TRANSPARENT MATERIAL

AIR SPACE

DENSE MATERIAL WITH GOOD HEAT CAPACITY, i.e., CONCRETE WITH HEAT CAPACITY OF 31.7 Btu/cu.ft./°F.

Massive sections require considerable time to raise their internal temperature and may absorb and hold considerable amounts of heat.

■ A massive section such as a heavy concrete or masonry wall, or a concrete floor, **may be used as heat storage:**
• to moderate heat gain to a space,
• to release heat stored when solar radiation is not available.
• this is one of the principal methods used in passive solar heating systems.

■ **Heat may also be stored** in various other substances, such as in containers filled with water.

■ **In all instances,** heat storage should be protected from heat loss to the outside. Protection usually consists of:
• transparent materials installed to prevent re-radiation when the storage is exposed to solar radiation;
• movable opaque covers are generally used to prevent reverse heat flow during cloudy weather and at night.

■ **Heat transfer** to the space is by natural or forced convection.

HEAT TRANSFER

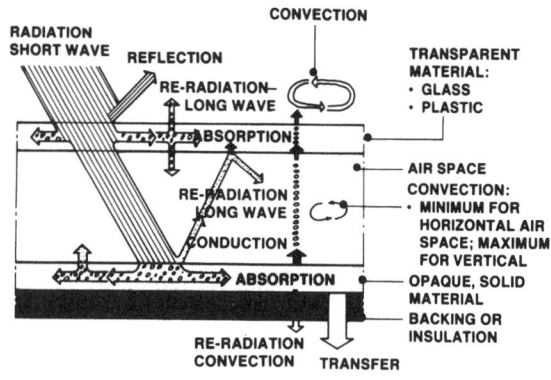

CONVECTION

RADIATION SHORT WAVE   REFLECTION

RE-RADIATION— LONG WAVE

ABSORPTION

RE-RADIATION LONG WAVE

CONDUCTION

ABSORPTION

RE-RADIATION CONVECTION   TRANSFER

TRANSPARENT MATERIAL:
• GLASS
• PLASTIC

AIR SPACE CONVECTION:
• MINIMUM FOR HORIZONTAL AIR SPACE; MAXIMUM FOR VERTICAL

OPAQUE, SOLID MATERIAL

BACKING OR INSULATION

Clear glass transmits most of the short-wave radiation striking it, but is almost totally opaque to long-wave thermal radiation emitted by surfaces with temperatures between 200°F to 400°F:

■ Radiation which passes through glass and is absorbed by an opaque surface is **prevented** by the same glass from **being re-radiated to the outside.** This phenomenon is known as the **Greenhouse Effect.**

■ **To minimize heat losses** from the absorbing surface to the outside, transparent materials are used to take advantage of the greenhouse effect: long wave re-radiation is trapped by the transparent cover, and heat loss through convection is also minimized.

SOLAR DESIGN

SYSTEMS

All systems for direct utilization of solar energy to heat interior spaces:
- **Collect** solar energy available.
- **Transport it** to the point where it is to be released.
- With/without intermediate **storage.**

The two basic methods generally employed are:
- **Passive:**
  - when mechanical equipment is not used, or when its use is very limited in transferring heat from surfaces at which it is absorbed to surfaces at which it is to be released.
  - passive systems generally are intrinsic components of a structure, not separate entities.
- **Active:**
  - when most or all of the transfer of heat is by mechanical means.
  - some components of an active system may be incorporated into the structure, but the system basically is a separate entity.

PASSIVE SYSTEMS

Any structure exposed to solar radiation will absorb it, convert it to heat, and transfer it to the interior when a difference in temperature exists between the outside and inside surfaces of its envelope.

- **A passive system is essentially a structure,** or assemblies within such structure, designed to:
  - absorb maximum amount of the available solar radiation
  - to retain, or store, the heat absorbed to be used when direct solar radiation is not available.

- The principle is based on the **use of materials with high heat storage capacities,** such as concrete, masonry, or rocks in assemblies exposed to maximum solar radiation during the heating season.

- Heat may also be **absorbed by and stored in elements independent of the structure** itself, such as containers filled with water or other substances with high heat storage capacities.

- The heat absorbed by such assemblies must **be prevented from being lost to the outside** through convection and re-radiation so that it can be used to heat the space.

- **A reversal of this principle** is used in localities where high daytime and low nighttime temperatures prevail to keep the interior cool:
  - The heat generated in the interior and absorbed during the day
  - it is then removed at night by flushing the interior with the cool night air.

Some of the methods employed in passive systems are:

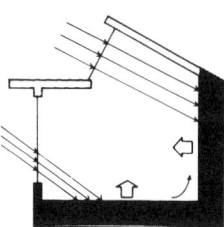

Internal Mass

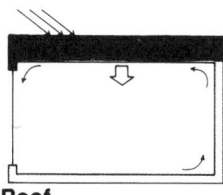

Roof

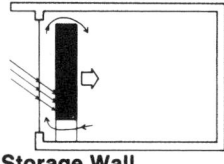

Storage Wall

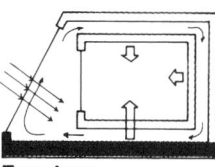

Envelope

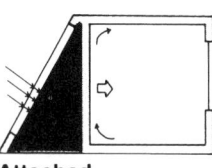

Attached

- **Materials used** for storing heat should have high specific heat coefficients.

- **Transparent covers** used to minimize heat loss from storage assemblies should have high transmittance to short wave radiation, and be opaque to long wave re-radiation.

- **Exterior envelope** should be well insulated to minimize heat losses to the outside. For additional information refer to INSULATION.

ACTIVE SYSTEMS

All active systems consist of three components: collector(s), heat storage and heat distribution; they may also include transport devices, controls and an auxiliary energy system.

The selection of an active system should start with its principal component: the collector.

- **The collector converts** the incident solar radiation to thermal energy by absorption on a suitable surface.

- **The method of transporting heat** absorbed by the collector, and the type of collector selected will largely determine the selection of other components of the system, such as:
  - heat storage
  - piping or ductwork
  - heat exchangers, pumps, or fans
  - controls.

- The first consideration in the selection of a collector and the design of the entire system is the percentage of the total annual heat demand the system is expected to satisfy:

- **The efficiency** of the system will be reduced with an increase in such percentage, since the load factor for the system will be decreased:

- A system designed to **satisfy the total annual heat demand** will have to be sized for the peak heating period. This means that it will be oversized for the months preceeding and following that period.

DEFINITIONS & PHYSICAL PRINCIPLES

BIOCLIMATIC DESIGN

BIOCLIMATIC DESIGN

Bioclimatic design is the design of buildings to create comfort and energy-efficiency based on analysis of the climate and ambient energy represented by sun, wind, temperature and humidity.

In using the term "bioclimatic," architectural design is linked to our biological, physiological and psychological need for health and comfort. Bioclimatic approaches to architecture attempt to create comfort conditions in buildings by utilizing the microclimate and resulting design strategies that include natural ventilation, daylighting, and passive heating and cooling.

CHARACTERIZATION OF CLIMATES

Indicated in Fig. 1, regions exceeding 8,000 annual heating degree days (HDD) are defined as predominately "underheated," that is, case the need for heating predominates, such as through direct solar gain and energy conservation. The large temperate area between 2,000-8,000 HDD has both heating and cooling requirements that must be balanced to assure that design techniques favored for one condition are compatible with all others. Sun-tempering (modest but careful use of south-facing windows) may provide a substantial portion of winter heating, but must also be dimensioned to provide summer shading. Regions with less than 2,000 HDD require little heating in comparison to cooling and are thus defined as "overheated."

The relative effectiveness of passive cooling strategies follows in part the climatic characterization from "arid" to "humid." The suitability of ventilation and evaporative cooling as cooling strategies are related to atmospheric humidity during summer (overheated) months. Those with dew points averaging less than 50F may be considered "arid." Regions having a combined July and August average dew point temperature greater than 65F (18.3°C) may be considered "humid." The entire southeast quadrant of the U.S. has mean daily humidity readings exceeding comfort limits under still air conditions. The main bioclimatic strategy of this region is thus to use shading and ventilation, to minimize if not to replace mechanical dehumidification and air conditioning, which may be required as a function of building type and climate.

The 50F (10°C) dewpoint temperature is an arbitrary way of defining the upper limit of arid conditions, but is convenient since it produces an outdoor daily temperature range of roughly 30F (-1.1°C) dry-bulb. Arid and semiarid conditions favor evaporative and radiative cooling and generally discourage summer daytime ventilation, since the air is both hot and dry. Thermal mass is especially effective in arid regions with extremely high daily maxima with nighttime lows that fall within the comfort range.

BIOCLIMATIC DESIGN PRINCIPLES

■ Bioclimatic design strategies are effective for "envelope-dominated" structures, to provide a large portion if not all of

Fig. 1. U. S. regions based on bioclimatic design conditions.

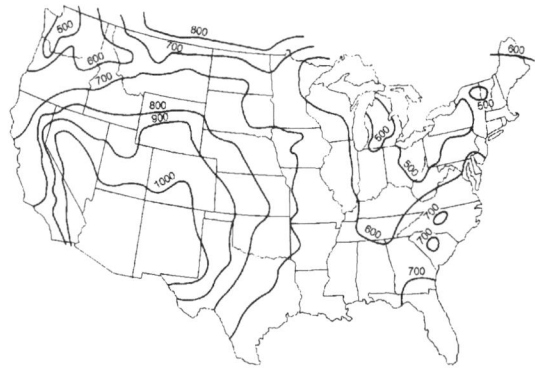

Fig. 2. Passive solar heating potential of south-facing windows (Btu/SF/day). Source: Dr. Douglas Balcomb, National Renewable Energy Laboratory.

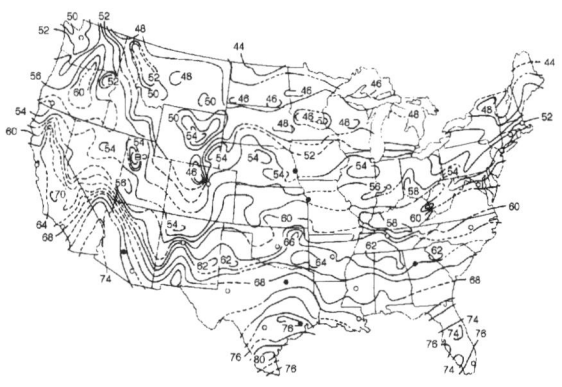

Fig. 3. Deep ground temperature (F) Source: National Well Water Association.

DEFINITIONS & PHYSICAL PRINCIPLES

the energy required to maintain comfort conditions.

■ **"Internal load dominated"** buildings—such as hospitals, offices, commercial kitchens, windowless stores—experience high internal gains imposed by the heat of occupancy, lights, and equipment.

• In such cases, the external climatic conditions may have a more complex influence on achieving comfort and low energy utilization.

• As internal loads are reduced through energy-efficient design—low-wattage equipment and lighting, occupancy scheduling and zoning—the effects of climate become more obvious and immediate.

■ All buildings can benefit from available daylighting, so that its related heating and cooling impacts and means of control are essential for all buildings.

There is a limited number of "pathways" by which heat is gained or lost between the interior and the external climate (Fig. 4). These can be understood in terms of the classic definitions of heating energy transfer mechanics, and from these, the resulting bioclimatic design strategies are defined:

• **Conduction**—from hotter object to cooler object by direct contact.

• **Convection**—from the air film next to a hotter object by exposure to cooler air currents.

• **Radiation**—from hotter object to cooler object within the direct view of each other regardless of the temperature of air between.

• **Evaporation**—the change of phase from liquid to gaseous state: The sensible heat (dry-bulb temperature) in the air is lowered by the latent heat absorbed into air when moisture is evaporated.

• **Thermal storage**—from heat charge and discharge both diurnally and seasonally, a function of its specific heat and weight. Although not usually listed alongside the four classic means of heat transport, this role of thermal storage is helpful in understanding the heat transfer physics of building climatology.

BIOCLIMATIC DESIGN STRATEGIES

In winter (or underheated periods), the objectives of bioclimatic design are to resist loss of heat from the building envelope and to promote gain of solar heat. In summer (or overheated periods), these objectives are the reverse, to resist solar gain and to promote loss of heat from the building interior. The strategies can be set forth as:

■ **Minimize conductive heat flow.** This strategy is achieved by using insulation. It is effective when the outdoor temperature is significantly different either lower or higher than the interior comfort range. In summer, this strategy should be considered whenever ambient temperatures are

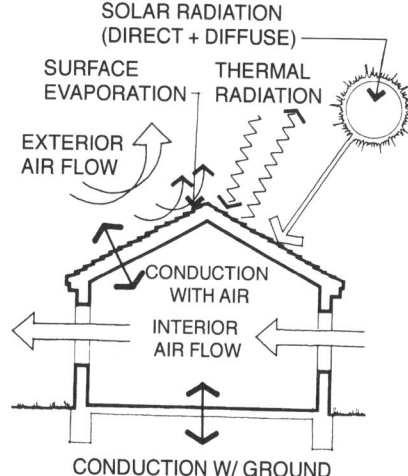

Fig. 4. Paths of energy exchange at the building microclimate (Watson and Labs 1993)

within or above the comfort range and where natural cooling strategies cannot be relied upon to achieve comfort (that is, mechanical air conditioning is necessary).

■ **Delay periodic heat flow.** While the insulation value of building materials is well understood, it is not as widely appreciated that building envelope materials also can delay heat flows that can be used to improve comfort and to lower energy costs. Time-lag through masonry walls, for example, can delay the day's thermal impact until evening and is a particularly valuable technique in hot arid climates with wide day-night temperature variations. Techniques of earth-sheltering and berming also exploit the long-term heat flow effect of subsurface construction.

■ **Minimize infiltration.** "Infiltration" refers to uncontrolled air leakage through joints, cracks, and faulty seals in construction and around doors and windows. Infiltration (and the resulting "exfiltration" of heated or cooled air) is considered the largest and potentially the most intractable source of energy loss in a building, once other practical insulation measures have been taken.

■ **Provide thermal storage.** Thermal mass inside of the insulated envelope is critical to dampening the swings in air temperature and in storing heat in winter and "coolth" in summer. (The term "coolth," coined by John Yellott, describes the heat storage capacity of a cooled thermal mass, that is, its capacity to serve as a heat sink for cooling).

■ **Promote solar gain.** The sun can provide a substantial portion of winter heating energy through elements such as equatorial-facing windows and greenhouses, and other passive solar techniques which utilize spaces to collect, store, and transfer solar heat.

DEFINITIONS & PHYSICAL PRINCIPLES

BIOCLIMATIC DESIGN (cont.)

DEFINITIONS & PHYSICAL PRINCIPLES

■ **Minimize external air flow.** Winter winds increase the rate of heat loss from a building by "washing away" heat and thus accelerating the cooling of the exterior envelope and also by increasing infiltration (or more properly, exfiltration) losses. Siting and shaping a building to minimize wind exposure or providing wind-breaks can reduce the impact of such winds.

■ **Promote ventilation.** Cooling by air flow through an interior may be propelled by two natural processes, cross-ventilation (wind driven) and stack-effect ventilation (driven by the buoyancy of heated air even in the absence of external wind pressure). A fan can be used to augment natural ventilation cooling in the absence of sufficient wind or stack-pressure differential.

■ **Minimize solar gain.** The best means for ensuring comfort from the heat of summer is to minimize the effects of the direct sun, the primary source of overheating, by shading the building from the sun, or otherwise minimizing the building surfaces exposed to summer sun, by use of radiant barriers, and by insulation.

■ **Promote radiant cooling.** A building can lose heat effectively if the mean radiant temperature of the materials at its outer surface is greater than that of its surroundings, principally the night sky. The mean radiant temperature of the building surface is determined by the intensity of solar irradiation, the material surface (film coefficient) and by the emissivity of its exterior surface (its ability to "emit" or re-radiate heat). This contributes little, however, if the building envelope is well insulated.

■ **Promote evaporative cooling.** Sensible cooling of a building interior can be achieved by evaporating moisture into the incoming air stream (or, if an existing roof has little insulation, by evaporatively cooling the exterior envelope, such as by a roof spray.) These are simple and traditional techniques and most useful in hot-dry climates if water is available for controlled usage. Modern evaporative cooling is achieved with an economizer-cycle evaporative cooling system, instead of, or in conjunction with, refrigerant air conditioning.

In regions of the world where extensive climatic data are not available and where, for example, data is limited only to monthly averages of temperature and humidity, weather data may not be coincident and must be interpreted with caution and only for "rough-cut" analysis. However, increasingly available throughout the world as in the United States, coincident climatic data are compiled from long-term readings and available on computer files, so that designers can obtain quite complete reference data.

BIOCLIMATIC DESIGN TECHNIQUES

Each locale has its own bioclimatic profile, sometimes evident in indigenous and long established building practices appropriate for different regions. Bioclimatic design techniques can be set forth as a set of design opportunities, which the designer may chose from for both region- and site-specific microclimatic response:

■ **Wind breaks (winter):** Two design techniques serve the function of minimizing winter wind exposure:

• Use neighboring landforms, structures, or vegetation for winter wind protection.

• Shape and orient the building shell to minimize winter wind turbulence.

■ **Thermal envelope (winter):** Isolating the interior space from the hot summer and cold winter climate, such as:

• Minimize the outside wall and roof areas (ratio of exterior surface to enclosed volume).

• Use attic space as buffer zone between interior and outside climate.

• Use basement or crawl space as buffer zone between interior and grounds.

• Centralize heat sources within building interior.

• Use vestibule or exterior "wind-shield" at entryways.

• Locate low-use spaces, storage, utility and garage areas to provide climatic buffers.

• Subdivide interior to create separate heating and cooling zones.

• Select insulating materials for resis-

tance to heat flow through building envelope.

• Apply vapor barriers to warm side to control moisture migration.

• Develop construction details to minimize air infiltration and exfiltration.

• Select high-capacitance materials to dampen heat flow through the building envelope.

• Provide insulating controls at glazing.

• Minimize window and door openings on north, east, and/or west walls.

• Detail window and door construction to prevent undesired air infiltration.

• Provide ventilation openings for air low to and from specific spaces and appliances.

• Use heat reflective (or radiant barriers) on (or below) surfaces oriented to summer sun.

■ **Solar windows and walls (winter):** Using the winter sun for heating a building through solar-oriented windows and walls is provided by a number of techniques:

• Maximize reflectivity of ground and building surfaces outside windows facing the winter sun.

• Shape and orient the building shell to maximize exposure to winter sun.

• Use high-capacitance thermal mass materials in the interior to store solar heat gain.

• Use solar wall and roof collectors on equatorial-oriented surfaces.

• Optimize the area of equatorial-facing glazing.

• Use clerestory skylights for winter solar gain and natural illumination.

■ **Indoor/outdoor rooms (winter and summer):** Courtyards, covered patios, seasonal screened and glassed-in porches, greenhouses, atriums and sun spaces can be located in the building plan for summer cooling and winter heating benefits, as in these three techniques:

• Provide outdoor semi-protected areas for year-round climate moderation.

• Provide solar-oriented interior zone for maximum solar heat gain.

• Plan specific rooms or functions to coincide with solar orientation.

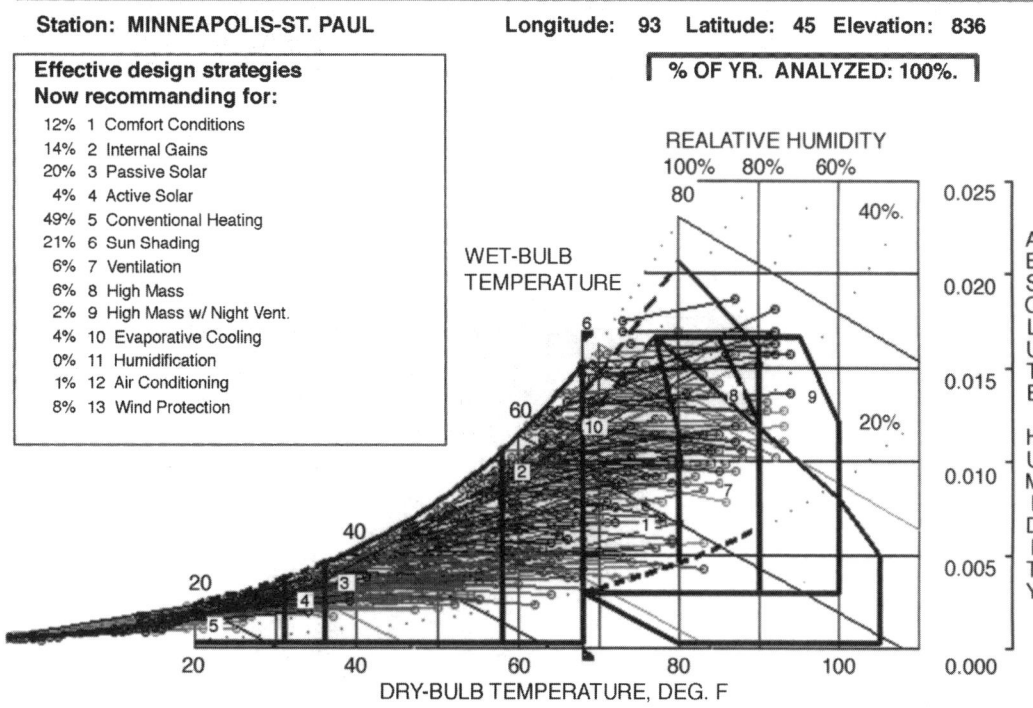

Station: MINNEAPOLIS-ST. PAUL   Longitude: 93 Latitude: 45 Elevation: 836

Effective design strategies
Now recommanding for:

%		Strategy
12%	1	Comfort Conditions
14%	2	Internal Gains
20%	3	Passive Solar
4%	4	Active Solar
49%	5	Conventional Heating
21%	6	Sun Shading
6%	7	Ventilation
6%	8	High Mass
2%	9	High Mass w/ Night Vent.
4%	10	Evaporative Cooling
0%	11	Humidification
1%	12	Air Conditioning
8%	13	Wind Protection

% OF YR. ANALYZED: 100%.

Fig. 20. Computer display (in numerous colors) of the Building Bioclimatic Chart for Minneapolis, MN. Tabulation on left of screen is similar to data in Table 1 above. (Milne and Li 1997).

■ **Earth-sheltering (winter and summer):** Techniques such as covering earth against the walls of a building or on the roof, or building a concrete floor on the ground, have a number of climatic advantages for thermal storage and damping temperature fluctuations (daily and seasonally), providing wind protection and reducing envelope heat loss (winter and summer). These techniques are often referred to as earth-contact or earth-sheltering design:

• Recess structure below grade or raise existing grade for earth-sheltering.

• Use slab-on-grade construction for ground temperature heat exchange.

• Use earth-covered or sod roofs.

■ **Thermally massive construction (summer and winter):** Particularly effective in hot arid zones, or in more temperate zones with cold clear winters. Thermally massive construction provides a "thermal fly wheel." absorbing heat during the day from solar radiation and convection from indoor air which can create comfort if it is cooled

at night, if necessary through nighttime ventilative cooling (if air temperatures fall within the comfort zone).

• Use high mass construction with outside insulation and nighttime ventilation techniques in summers.

■ **Sun shading (summer):** Because the sun angles are different in summer than in winter, it is possible to shade windows from the sun during the overheated summer period while allowing it to reach the window surfaces and spaces in winter. Thus the concept to provide sun shading does not need to conflict with winter solar design concepts.

• Minimize reflectivity of ground and building surfaces outside windows facing the summer sun.

• Use neighboring landforms, structures, or vegetation for summer sun.

• Shape and orient the building shell to minimize exposure to summer sun.

• Provide seasonally operable shading, including deciduous trees.

■ **Natural ventilation (summer and seasonal):** Natural ventilation is a simple concept by which to cool a building.

• Use neighboring landforms, structures, or vegetation to increase exposure to summer breezes.

• Shape and orient the building shell to maximize exposure to summer breezes.

• Use "open plan" interior to promote airflow.

• Provide vertical airshafts to promote "thermal chimney" or stack-effect air flow.

• Use double roof construction for ventilation within the building shell.

• Orient door and window openings to facilitate natural ventilation from prevailing summer breezes.

• Use wingwalls, overhangs, and louvers to direct summer wind flow into interior.

• Use louvered wall for maximum ventilation control.

• Use roof monitors for "stack effect" ventilation.

DEFINITIONS & PHYSICAL PRINCIPLES

BIOCLIMATIC DESIGN (cont.)

■ **Plants and water (summer):**
Several techniques provide cooling by the use of plants and water near building surfaces for shading and evaporative cooling.

• Use ground cover and planting for site cooling.

• Maximize on-site evaporative cooling.

• Use planting next to building skin.

• Use roof spray or roof ponds for evaporative cooling.

COMPUTER-AIDED CLIMATIC DESIGN

Recently developed energy design tools make it possible to utilize hourly weather data to accurately analyze climate. This enables the designer to apply sophisticated bioclimatic analysis to any location in the United States (and elsewhere where data are compiled in electronic files), providing a systematic basis to guide design judgment.

The majority of climatic analysis protocols create an "average" climate profile by selecting the most representative months from long-term data. The Typical Meteorological Year (TMY) contains simultaneous climatic data for 8,760 hours in a "typical" year. Electronic files of climatic data for most U. S. locations (major airports) are available through various sources on the World-Wide Web, NREL (1996) and Rutgers University (1994).

Climate Consultant: This software plots weather data, including temperature, wind velocity, sky cover, percent sunshine, beam and horizontal irradiation. It uses these data to create psychometric charts, timetables of bioclimatic needs, sun charts and sundials showing times of solar needs and shading requirements. It can be downloaded at no cost from the World Wide Web (Milne 1997). Fig. 5 indicates a typical bioclimatic chart generated by Climate Consultant, indicating an annual summary for Minneapolis and in the upper left, the percent that bioclimatic strategies are effective.

DEFINITIONS

Temperature is defined as the thermal state of matter with reference to its tendency to communicate heat to matter in contact with it. Temperature is an index of the thermal energy content of materials, disregarding energies stored in chemical bonds and in the atomic structure of matter.

■ **Fahrenheit temperature (F)** refers to temperature measured on a scale devised by G.D. Fahrenheit, the inventor of the alcohol and mercury thermometers in the early 18th century. On the Fahrenheit scale, the freezing point of water is 32F and its boiling point is 212F at normal atmospheric pressure. Fahrenheit reportedly chose the gradations he used because it divides into 100 units the range of temperatures most commonly found in nature.

■ **Celsius temperature (°C)** refers to temperatures measured on a scale devised in 1742 by Anders Celsius, a Swedish astronomer. The Celsius scale is graduated into 100 units between the freezing temperature of water (0°C) and its boiling point at normal atmospheric pressure (100°C) and is, consequently, commonly referred to as the Centigrade scale.

■ **Dry-bulb temperature (DBT)** is the temperature measured by an ordinary (dry-bulb) thermometer, and is independent of the moisture content of the air. It is also called "sensible heat temperature."

■ **Wet-bulb temperature (WBT)** is an indicator of the total heat content (or enthalpy) of the air, that is, of its combined sensible and latent heats. It is the temperature measured by a thermometer having a wetted sleeve over the bulb from which water can evaporate freely.

■ **Dew point temperature (DPT)** is the temperature of a surface upon which moisture contained in the air will condense. Stated differently, it is the temperature at which a given quantity of air will become saturated (reaching 100% relative humidity) if chilled at constant pressure. It is thus another indicator of the moisture content of the air. Dew point temperature is not easily measured directly; it is conveniently found on a psychrometric chart if dry-bulb and wet-bulb temperatures are known.

Humidity is a general term referring to the water vapor contained in the air. Like the word "temperature," however, the type of "humidity" must be defined.

■ **Absolute humidity is defined as the weight of water vapor contained in a unit volume of air;** typical units are pounds of water per pound of dry air or grains of water per cubic foot. Absolute humidity is also known as the water vapor density (Dv).

■ **Relative humidity (RH)** is defined as the (dimensionless) ratio of the amount of moisture contained in the air under specified conditions to the amount of moisture contained in the air at saturation at the same (dry bulb) temperature. Relative humidity can be computed as the ratio of existing vapor pressure to vapor pressure at saturation, or the ratio of absolute humidity to absolute humidity at saturation existing at the same temperature and barometric pressure.

■ **Water vapor pressure (Pv)** is that part of the atmospheric pressure ("partial pressure") which is exerted due to the amount of water vapor present in the air It is expressed in terms of absolute pressure as inches of mercury (in. Hg) or pounds per square inch (psi).

REFERENCES:

Donald Watson and Murray Milne "Bioclimatic Design" in Reference 1.

Milne, Murray (199) Energy Design Tools: Climate Consultant. http://www.aud.ucla.edu/energy-design-tools.

NREL. 1996. "TMY-2 Typical Meteorological Year Climate Data Files." National Renewable Energy Laboratory. http://rredc.nrel.gov:80/solar/old_data/nsrdb/tmy2/ (If this web address changes, e-mail: webmaster@nrel.gov).

Rutgers University. 1994. Department of Engineering. "TMY Typical Meteorological Year Climate Data Files."

http://oipea www.rutgers.edu/html_docs/TMY/tmy.html (If this web address changes, e-mail: webmaster@rutgers.edu).

ENVIRONMENTAL IMPACT OF CONSTRUCTION

ENVIRONMENTAL IMPACTS OF BUILDING

Architects and designers have available an increasing number of information sources to learn about the environmental impact of building materials and to select the materials to improve quality and reduce negative environmental impacts in the manufacture, construction and use of buildings.

To help improve positive environmental impacts, construction practices and/or products can be used to achieve one or more of the following:

■ Construction material or product has low or no emission of gases or outgassing or (such as Volatile Organic Compounds (VOCs) that are either a human health concern or are a risk to deterioration of archive materials. Specifying the recommended material aids in improving indoor air quality standards in the completed construction. Products exposed to outdoor air are also included, because outdoor emissions fall under the Clean Air Act standard for smog control.

■ Material has high-recycled content in its manufacture, either at the producing plant or at the construction site, thus reducing manufacturing and/or construction waste.

■ Material and/or product is longer lasting than its alternates, even its first cost is higher, requiring less replacement over its life cycle (with the likely result of achieving a lower life-cycle cost).

■ Material and/or product are produced by an environmentally responsible process, including a manufacturer/supplier plan to recycle the product when it requires replacement.

■ Wood supplier uses sustainable yield forestry practices that are well managed for ecological goals, and its entire production and processing ("chain of custody") is certified by recognized authorities (recommendation applies to all wood products).

■ Material manufacture uses environmentally benign processes and products that have less contribution to global ozone depletion or other negative environmental impacts (such as air, ground and water pollution) than do the commonly used materials.

■ Material and/or product result in energy savings, thus reducing energy cost and its indirect environmental impacts (pollution).

■ Material and/or product results in water savings, thus reducing water supply costs and waste stream volume and indirect environmental impacts (pollution).

■ Material and/or product is available from local sources, thus reducing transportation costs and its indirect environmental impacts (pollution).

PRODUCT LIFE-CYCLE

Any approach that considers materials from an environmental perspective must consider the entire product life cycle. A standard format for this type of study, which is termed environmental life-cycle assessment (LCA)—not to be confused with life-cycle cost analysis, which is strictly financial. For the purposes of the study, a product's life cycle is usually broken down into several stages, as illustrated in Fig. 1.

Most LCAs are based on an inventory of inputs and outputs. A researcher identifies all of the raw materials and energy consumed in the production, use, and disposal of the product, as well as pollutants and byproducts generated. Depending on the available data and resources, this inventory may be detailed and quantitative or cursory, looking only to the most significant inputs and outputs. A subset of this inventory is the energy required to extract, transport, and process a material. Called the embodied energy of the material, this estimate is often a good place to start because of the pollution associated with energy generation.

Following the inventory, the LCA examines the environmental impacts of each of these material and energy flows.

This involves the nearly impossible task of tracking ecological impacts as they ripple endlessly through the world's natural systems. LCAs done for specific products may include a final step—identifying areas for improvement.

RESOURCES FOR DESIGNERS

Designers who make a point of specifying environmentally preferable materials rely on a broad range of resources for their information. Representative sources are listed in the References.

Designers should learn about the environmental impacts of materials by querying manufacturers and suppliers directly. If sales representatives are knowledgeable about environmental issues, that is a fair indication that a company takes such issues seriously and is proactive in minimizing environmental impacts. It may be appropriate to speak with technical support personnel to find out specifics, such as where the raw materials come from and how they are processed.

Some design firms distribute questionnaires to their suppliers, asking for detailed information on the composition and environmental performance of their products. This information can then be compiled and utilized in a systematic way in the firm's specification process.

In some cases information is available in the form of a certification or seal of approval from an independent agency.

There are two common types of certification:

• those that establish the overall environmental performance of a product based on a predetermined set of criteria. In the U.S. the Washington, D.C.-based nonprofit organization Green Seal, performs this type of certification.

• those that simply verify a specific claim made by the manufacturer, such as a specific level of recycled content. For-profit companies, such as Scientific Certification Systems, Inc., Oakland, California, provide this type of certification.

DEFINITIONS & PHYSICAL PRINCIPLES

ENVIRONMENTAL IMPACT OF CONSTRUCTION

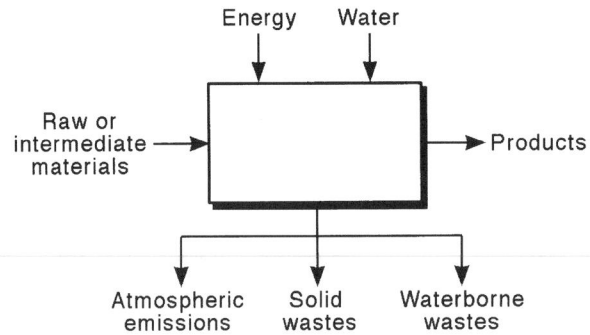

Fig. 1. Life cycle assessment template

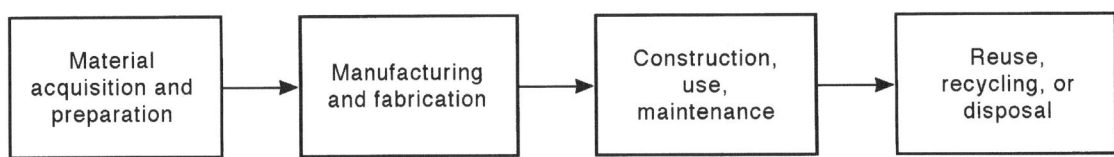

Fig. 2. Life cycle framework (after American Institute of Architects 1996)

Material Safety Data Sheets (MSDS) are available from manufacturers for almost all products. Obtaining an MSDS for a product is a relatively easy way to find out what it consists of, although some specifics may be left vague if considered proprietary. These data sheets also list potential health impacts of the ingredients in each product, so are particularly useful in assessing possible health impacts to construction workers and building occupants.

ASSESSMENT METHOD

Outlined below is a simplified methodology for choosing the most benign materials. These steps cannot take the place of a thorough understanding of the life cycles of the materials and their environmental impacts, but they offer a methodical way to apply that knowledge.

STEPS 1–3: USE PHASE

Two of the most significant sources of environmental impact from building materials are energy use in the building and possible impacts on occupant health. Considerations of impacts in the use phase depend not only on the

material in question, but also on the application for that material.

Step 1 – Energy use: Will the material in question (in the relevant application) have a measurable impact on building energy use?

If yes (as for materials such as glazing, insulation, mechanical systems), avoid options that do not minimize energy use. Also take care to design the application to minimize energy use. For materials that can be used in an energy-efficient manner only with the addition of other components, the impact of including those additional components must be factored in. Examples include glazing systems that require exterior shading systems for efficiency, and light-gauge steel framing that requires foam sheathing to prevent thermal bridging.

Step 2 – Occupant health: Might products in this application affect the health of building occupants?

If yes (interior furnishings, interior finishes, mechanical systems), avoid materials that are likely to adversely affect occupant health, and design systems to minimize any possible

adverse effects when sources of indoor pollution cannot be avoided.

Step 3 – Durability and maintenance: Are products in this application likely to need replacement, special treatment, or repair multiple times during the life of the structure?

If yes (roofing, coatings, sealants), avoid products with short expected lifespans (unless they are made from low-impact, renewable materials and are easily recycled), or products that require frequent, high impact maintenance procedures. Also, design the structure for flexibility so that materials that might become obsolete before they wear out (such as wiring) can be replaced with minimal disruption and cost.

STEPS 4–6: MANUFACTURING

The remaining steps pertain less to the application (how a material or product is used) and more to the material itself. They require knowledge of the raw materials that go into each product.

Step 4 – Hazardous by-products: Are significant toxic or hazardous

(cont.)

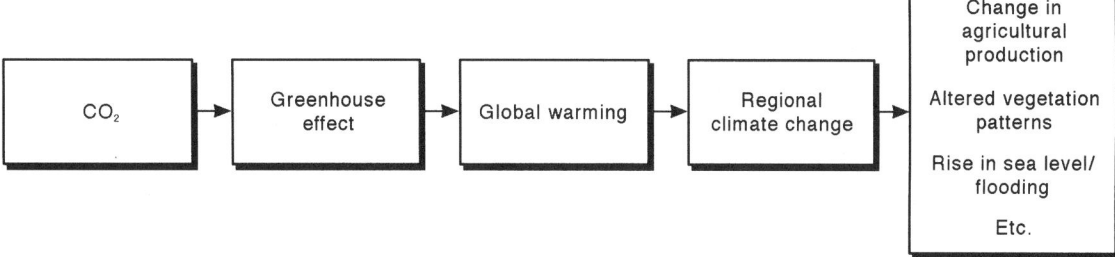

Fig. 3. Sample impact chain

intermediaries or by-products created during manufacture, and if so, how significant is the risk of their release to the environment or risk of hazard to worker health?

• Where toxic by-products are either generated in large quantities or in small but uncontrolled quantities (smelting of zinc, production of petrochemicals), the building material in question should be avoided if possible, or sourced from a company with strict environmental standards.

Step 5 – Energy use: How energy-intensive is the manufacturing process?

• If the manufacture of a building material is very energy-intensive compared to the alternatives (aluminum, plastics), its use should be minimized. It is not the energy use itself that is of concern, but the pollution from its generation and use; industries using clean-burning or renewable energy sources have lower burdens than those relying on coal or petroleum.

Step 6 – Waste from manufacturing: How much solid waste is generated in the manufacturing process?

• If significant amounts of solid waste are generated that are not readily usable for other purposes (tailings from mining of copper and other metals), seek alternative materials, or materials from companies with progressive recycling programs.

STEPS 7–9: RAW MATERIALS

Step 7 – Resource limitations: Are any of the component materials from rare or endangered resources?

• If yes (threatened tree species, old-growth timber), avoid these products, unless they can be sourced from recycled material.

Step 8 – Impacts of resource extraction: Are there significant ecological impacts from the process of mining or harvesting the raw materials?

• If yes (damage to rainforests from bauxite mining for aluminum, or timber harvesting on steep slopes with unstable soils), seek suppliers of material from recycled stock, or those with credible third-party verification of environmentally sound harvesting methods.

Step 9 – Transportation: Are the primary raw materials located a great distance from the construction site?

• If yes (Italian marble, tropical timber, New Zealand wool), seek appropriate alternative materials from more local sources.

FINAL STEPS; DISPOSAL OR REUSE

Step 10 – Demolition waste: Can the material be easily separated out for reuse or recycling after its useful life in the structure is over?

• While most materials used in large quantities in building construction (steel, concrete) can be at least partially recycled, others are less recyclable and may become a disposal problem in the future. Examples include products that combine different materials (such as fiberglass composites) or undergo a fundamental chemical change during manufacture (thermoset plastics such as polyurethane foams). Consider the future options to replace and recycle products chosen.

Step 11 – Hazardous materials from demolition: Might the material become a toxic or hazardous waste problem after the end of its useful life?

• If yes (preservative-treated wood), seek alternative products or construction systems that require less of the material in question.

Step 12 – Review the results: Go over any concerns that have been raised about the products under consideration, and look for other life-cycle impacts that might be specific to a particular material.

• For example, with drywall and spray-in open-cell polyurethane foam insulation, waste generated at the job site is a potential problem that should be considered.

REFERENCES:

Nadav Malin, Alex Wilson and Joel Ann Todd *"Environmental Life-Cycle Assessment"* in Reference 1.

DEFINITIONS & PHYSICAL PRINCIPLES

ENVIRONMENTAL IMPACT OF CONSTRUCTION

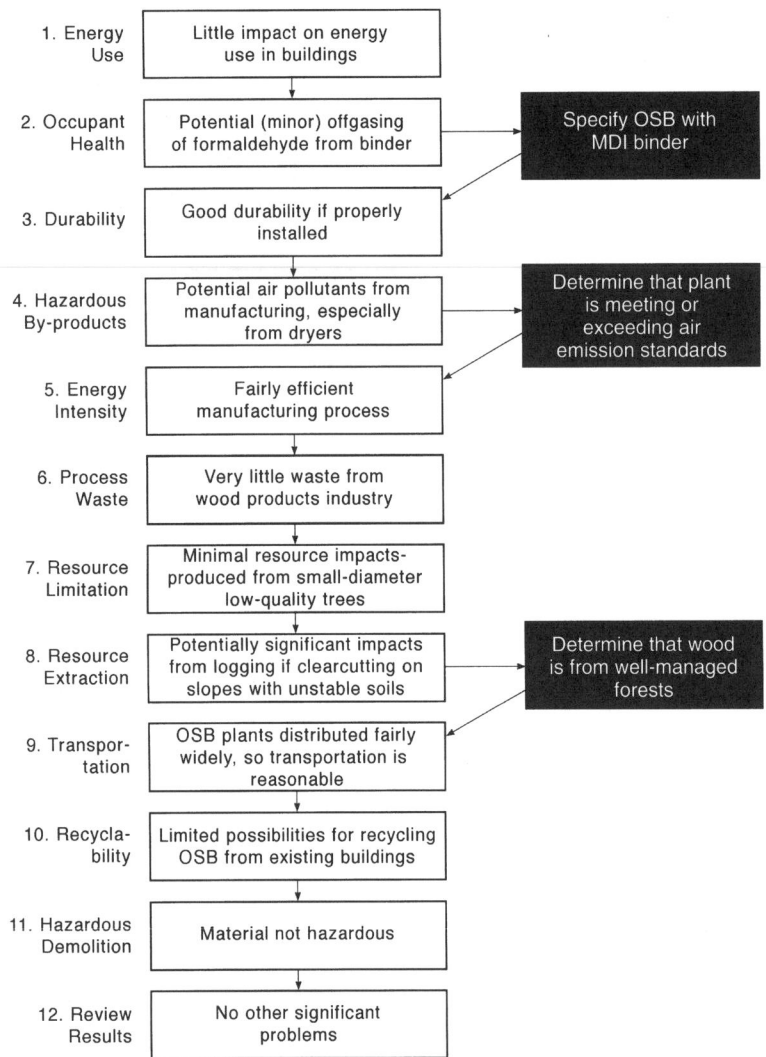

Fig. 4. Example: the simplified methodology applied to oriented-strand board sheathing

DEFINITIONS & PHYSICAL PRINCIPLES

Environmental Building News Product Directory. Brattleboro, VT. http://www.ebuild.com. http://www.usgbc.org

Forest Stewardship Council (FSC) Guideline Standard for Certification of Wood Products. FSC PRINCIPLES AND CRITERIA: http://www.fscoax.org/html/noframes/1-2.html

FSC ACCREDITED CERTIFIERS IN U.S.:

■ Rainforest Alliance Smart Wood Program, Ms Wendy Hall, #61 Millet Street, Goodwin Baker Building, Richmond, VT 05477, USA. E-mail: smartwood@ra.org.

Website:http://www.smartwood.org. Scope of accreditation: worldwide and chain of custody.

■ Scientific Certification Systems Forest Conservation Program, Park Plaza Building, 1939 Harrison Street, Suite 400, Oakland, CA 94612-3532. USA. E-mail: dhammel@scs1.com. Website: http://www.scs1.com. Scope of accreditation: worldwide and chain of custody.

LIST OF ALL FSC-ACCREDITED CERTIFIERS:

http://www.fscoax.org/html/noframes/5-3-1.html

Green Building Advisor. CD-ROM information resource on green building design. Center for Renewable Energy and Sustainable Technologies (CREST), 1200 18th Street Suite 900, Washington, DC 20036.

Leadership in Energy and Environmental Design (LEED) Green Building Performance Criteria. U.S. Green Building Council WEB SITE

WasteSpec: Model Specifications for Construction Waste Reduction, Reuse,and Recycling. Triangle J Council of Governments Publications. Contains recommended specifications of construction waste minimization. Triangle J Council of Governments. PO Box 12276, RTP, NC 27709.

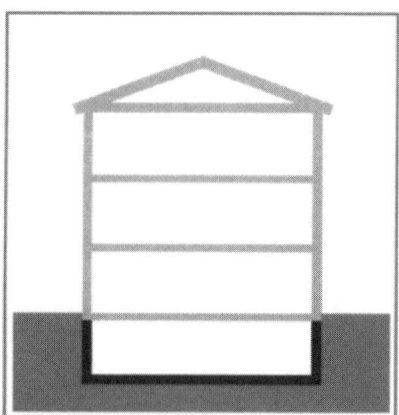

A • FOUNDATIONS

A1 • SUBSTRUCTURE

A1.1 SOILS & FOUNDATIONS

A2.1 SUBSURFACE MOISTURE
       CONDITIONS

A · FOUNDATIONS

A1. **1** SOILS & FOUNDATIONS

SOILS & FOUNDATION TYPES

A1.1 SOILS & FOUNDATIONS

EVALUATING THE BEARING CAPACITY OF SOIL

Subsurface investigation is most often done by borings, but test pits are also used. A typical boring rig consists of a tripod or frame with a pulley and a small winch.

• A hammer is raised by the winch and allowed to fall free, driving a pipe casing into the ground.

• The number of blows necessary to drive the spoon 1 ft. gives important information as to the compactness of the soil.

■ Many codes as well as good engineering practice dictate boring locations about 50 ft. (15 m) on center within the building outline.

• Abnormal ground conditions may require closer spacing.

• Depth of borings are typically 15 to 20 ft. (4.5 to 6 m) below foundation level, with one or more borings deeper to look for weak lower levels.

■ Test pits give a more immediate idea of the soil conditions but are limited to a depth of about 10 ft. (3 m).

• Dug with a backhoe, they give a method for economical and visually evident evaluation.

• Where rock is near the surface, a possible picture of the rock profile is obtained.

• Once the type and degree of compactness of soil has been established, its supporting ability must be evaluated.

SELECTING A FOUNDATION TYPE

The most common types of footings are the spread footings and wall footings. These are used where the soil bearing capacity is adequate for the applied loads.

• When good bearing material occurs directly under the building excavation, spread footings are designed for uniform bearing on the soil.

■ **Variations of spread footings include:**

• eccentric footings, where center of the superimposed load does not line up with the resultant center of the soil bearing pressure,

• combined footings, where two or

more columns must share one footing,

• matt footings, where the required superimposed loads require most of the building's footprint to transfer the accumulated loads to relatively weak soil bearing capacity.

■ **Pile foundations are required where poor surface and near surface soils are weak and column like shafts must be used to penetrate the weak soil and reach acceptable supporting stratum and greater depths below grade.**

• Piles are tied together with pile caps upon which the building's columns or walls are supported.

■ When large column loads exist, caissons are used as extensions to columns.

• **Caissons** typically are larger in diameter and longer. They rely on end bearing directly on earth with very high bearing capacity.

■ **Retaining walls** are used where a grade change occurs and the upper levels must be stabilized behind a wall. The wall portion of the foundation extends vertically cantilevered from a substantial and carefully designed footing.

The choice between walls and footings, piers and grade beams, or piles and grade beams is determined by soil conditions, by the requirements of the building's structural system, and cost. The requirement of many codes—that a pile be at least 10 ft. (3 m) long in order to provide adequate lateral stability—often determines the changeover depth between piers and short piles.

Mats can distribute loads to large areas, permitting light soil bearing loads on weak material. Hydraulic mats resist upward water pressure. Because of the various possible arrangements and loads, each mat becomes a specialized custom design.

FOUNDATIONS TO ROCK

Rock, having the highest bearing capacity, is often the only acceptable foundation available for heavy loads.

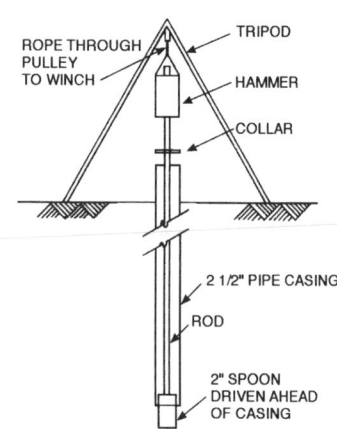

Fig. 1. Typical soil boring rig

Piers carry the loads directly to rock. On hard rock, piers require no footing, as the capacity of the rock is almost that of concrete. Typical column and grade beam construction is employed.

Where rock occurs more than 10 to 15 ft. (3 to 4.5 m) below the grade beam soffits, piers become too costly. Clusters of piles driven to rock and encased in a pile cap can support substantial loads. For heavier loads, caissons are used. Caissons are big holes drilled through the weak soil strata down to rock. The drilled voids are then filled with concrete. Piles or caissons may vary in length from 15 to over 100 ft. (4.5 to over 30 m).

PILES

Piles carry loads to strata below the ground surface either by end bearing, which are called bearing piles, or by surface friction along their sides which are called friction piles. The soft material through which the pile is driven provides lateral stability, but for structures over water the piles must be designed as columns.

Test load or driving resistance generally establishes pile capacity. Load tests are used to establish capacity. Driving resistance measurements are used to ensure that all piles are driven as hard as the test piles. Piles are generally grouped in clusters connected by pile caps.

Borings are essential for proper pile evaluation. Individual piles may test to a capacity greater than their contribution to the capacity of a cluster. A soft stratum underlying a hard one may not be able to support the total load delivered from the hard stratum even though the resistance of the hard stratum may indicate satisfactory pile support as indicated in Fig. 2. Different piles shown in Fig. 3 have evolved with certain characteristics, briefly described as follows:

■ **Types I and II are cast-in-place concrete piles.** A light-gage steel shell, driven on a mandrel, which is then withdrawn, is inspected and filled with concrete. Care must be taken to avoid collapsing of the shell when an adjacent pile is driven.

■ **Type III is similar to Types I and II except that the shell gage is heavier and no mandrel is required.**

■ **Type IV is an open-end steel pipe.** It is excavated, often by air jet, as it is advanced, and then filled with concrete after refusal has been reached. In lieu of reaching refusal, driving may stop while a concrete plug is placed and then redriving will seat it. The advantage is fewer disturbances to adjacent structures.

■ **Type V is a closed-end pile.** After driving, it is filled with concrete. Often it is used inside buildings with low headroom. Shorter lengths are simply spliced with steel collars.

■ **Type VI is a precast concrete pile.** It is good in marine structures but requires heavy handling equipment and accurate estimation of tip elevation as it is difficult to cut off in the field.

■ **Type VII is a wood pile-the least expensive.** Where the pile is partially exposed permanently above water level, it must be treated with a wood preservative.

■ **Type VIII, a composite wood and concrete pile,** is seldom used. The timber is kept below groundwater and a greater over-all length is achieved. A closed-end pipe pile may be used in place of the timber section.

■ **Type IX is a rolled steel H section.** It is the cheapest of the higher-capacity piles. Protection must be provided when driving through cinder fill or other rust-producing material.

■ **Type X is a drilled-in caisson.** A 24-in. (60 cm) round pipe is driven to rock and cleaned out. A rock socket is drilled and cleaned, a steel H-section core is set, and the shell is filled with concrete. This is good for very heavy loads.

Piles almost always are installed in groups of three or more. Table 3 is included to represent a few simple examples of pile cap sizes and shapes along with representative capacities of the cap and the column being supported. For heavier column loads the reader is referred to a structural engineer for analysis of specific foundation requirements of the building(s) under consideration.

Piles are located with a low degree of precision. They can easily be 6 inches or more from their desired location. If building columns, which are located with much greater precision, were to be located on single piles, the centerlines would rarely coincide. The resulting eccentric loads in both the column and the pile would generate unwanted moments in both members. A similar condition could exist around one axis for a column supported by a two-pile foundation.

Groupings of three or more piles provide a degree of safety and redundancy should one pile be driven slightly out of alignment. Lateral stability of the group increases with three piles as compared to fewer piles.

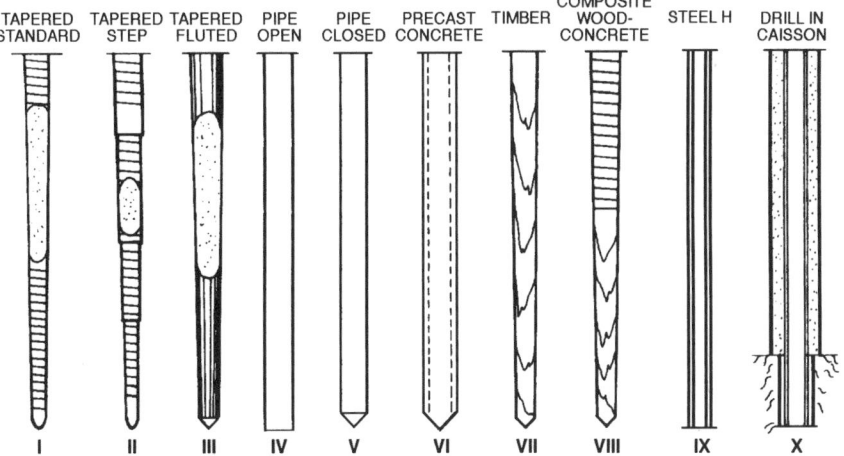

Fig. 2 Types of piles

A1. 2 SUBSURFACE MOISTURE CONDITIONS

SUBSURFACE MOISTURE PROTECTION

A1.2 SUBSURFACE MOISTURE CONDITIONS

SUBSURFACE MOISTURE PROTECTION

Because waterproofing of subsurface portions of a building is difficult to remedy, the waterproofing and moisture control strategies are critical. Methods of moisture control of substructures include:

■ **Dampproofing** which retards the passage of water in the absence of hydrostatic pressure.

■ **Waterproofing** that prevents the passage of water, under hydrostatic pressure, through subsurface foundation walls, slabs, or both.

■ **Subsurface drainage** that removes water from proximity to the foundations and subsurface slabs.

GROUNDWATER

In most regions, there is some dampness in the soil under and around buildings from both surface and underground water conditions.

• The dampness usually comes from rain water or local ground water near the surface, but in some desert regions the moisture movement is up from deep earth.

• Under the most severe conditions, there is standing water under hydrostatic pressure above or near the bottom of the foundations, either all the time or some of the time.

• More commonly, there is water in the ground during and after rain, and there is dampness which can penetrate the walls and slabs-on-grade by capillary action and through small cracks and voids.

■ Groundwater level tends to follow ground contour—deeper on hills, shallower in valleys.

■ Rainfall percolates through ground to recharge groundwater. Groundwater level varies with amount of rainfall.

■ Springs occur where local ground depressions place ground level below

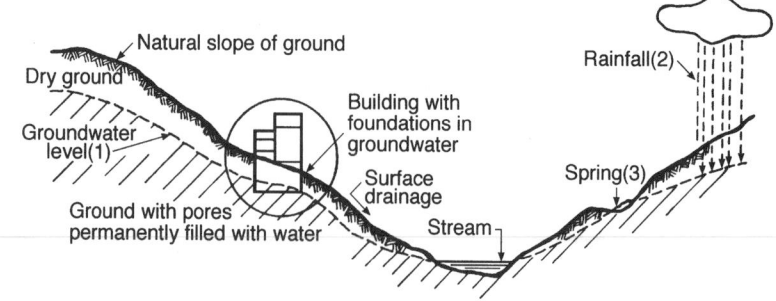

Fig 1. Basic Factors affecting groundwater flows

groundwater level.

Sources of information on the soil and water conditions, which prevail in the locality and at specific building sites include:

• all-season measuring of ground water in a test boring,

• consultation with a geotechnical engineer, and

• consultation with local building officials.

• long-term flooding records, as well as recent storm patterns (which in many localities are exceeding long-term records) provide equally critical reference data.

CONDITIONS REQUIRING SUBSURFACE MOISTURE PROTECTION

Dampproofing is generally adequate to retard passage of water into a basement, and subsurface drainage is provided by natural ground absorption and/or evaporation, under the following combination of conditions:

• If a building is built on very porous soil,

• If the standing ground water level is always well below the basement, and

• If moisture from ground runoff, roof drainage, and similar sources is directed away from the building (by swales, underground drainage pipes, and similar means).

Waterproofing—which is intended to exclude all water from a building under all foreseeable conditions—is the safe

choice if any combination of these factors exists:

• If the standing ground water is near or above the basement floor level,

• If water from other sources is not directed away from the building,

• If building contents and activities in the below ground spaces are valuable and critical.

• If discharge of water from a subsurface drainage system is not practical.

Subsurface drainage is an excellent method of avoiding water entry into the basement:

• If a building site has standing ground water which is sometimes above the basement floor, or

• If the soil is not sufficiently porous to act as a natural drainage bed.

A redundant combination of surface drainage, subsurface drainage, and dampproofing or waterproofing is a prudent design choice.

In critical or questionable situations, a good design decision might be to eliminate subsurface spaces or to make them noncritical:

• If there is no reliable outfall for subsurface drainage,

• If analysis shows that there may be troublesome ground water, and

• If the construction and maintenance budget won't permit waterproofing,

With any system of subsurface moisture protection, it is highly desirable to keep surface water away from the building. Slope the grade down away from the building, incorporating swales and area drains as needed. Do not discharge rainwater, parking lot drainage, and other surface water to areas near the foundations. Keep basement windows and hatches well above grade or in drained areaways.

If subsurface drainage is used to remove significant volumes of water, a civil engineer should be consulted to determine the size and slope of the pipe and the outfall.

• Many urban and suburban localities require that on-site storm water retainage tanks and/or on-site swales for percolation of surface runoff be provided. In most localities, surface runoff to adjacent properties is disallowed.

• Some surface runoff may contain harmful chemicals or pollutants.

• Discharging large volumes of water may also require approval by the Environmental Protection Agency, city engineer, and other officials.

Permeability of concrete and masonry foundations. If concrete is designed, formulated, and placed with sufficient care, it can be made waterproof. Formulation and mixing of waterproof concrete are specified in ACI 301, paragraph 3.4.2. Water stops are specified in ACI 301, paragraph 6.3.

Unless very special controls are applied to concrete foundation construction, and at masonry foundations, it is prudent to assume that there will be voids and cracks in the foundation materials, which will admit water. Water may wick through the foundation walls, and the ground may be temporarily saturated outside the walls during heavy rainfall.

In addition to water entry through basement slabs and foundation walls, water may "wick" slowly by capillary action upward in foundation walls, which are in contact with damp ground. This process is known as "rising damp". In new buildings the inclusion of a waterproof flashing course at the base of foundation walls is a good method of avoiding rising damp.

DAMPROOFING

Under those conditions listed above, when dampproofing is judged to be adequate, a brush or trowel coat of waterproofing material applied to the outside of the foundations is an inexpensive way to bridge over minute imperfections and cracks and to retard capillary infiltration. The surface should be cleaned and repaired first. A thick 1/8 in. (3.6 mm) coating with a non-asbestos fibrated trowel mastic will be more effective at filling voids and bridging small cracks than a thinner coating.

Waterproofing materials, as described below, may be used as dampproofing. They are generally more effective, and more expensive, than dampproof brush and trowel coatings.

SUBSLAB VAPOR RETARDERS

Subslab vapor retarders serve to retard the passage of water vapor from the earth up through the slab on grade and to retard the wicking of moisture from the earth into the slab. Subslab vapor retarders are not waterproofing; they are not intended to stop water under hydrostatic pressure. Granular fill under slabs on grade is more reliable than a vapor retarder in resisting capillary action. Factors determining whether or not to use a subslab vapor retarder include:

■ Based on an analysis of vapor flow, is the net vapor flow up from the earth or down to the earth? Under many conditions, a subslab vapor retarder will make the basement slab damper.

■ Will a subslab vapor retarder slow the initial drying of the concrete? The answer is often yes.

■ Do the requirements of manufacturers' associations and manufacturers require a vapor retarder? **Example:** The Resilient Floor Covering Institute.

A major cause of basement dampness is condensation of humid air on cool surfaces.
• A vapor retarder, waterproofing,

and dampproofing will have little effect on this process. In general, keeping the partial vapor pressure in the basement low (dry air) and keeping the surfaces in the basement warm can reduce condensation.

• Expanded, extruded polystyrene insulation or foamed glass insulation under the slab and outside the walls helps keep the basement surfaces warm.

• Designing the mechanical system to keep the basement warm in winter and dry in summer, or providing dehumidifiers, helps keep the partial vapor pressure low.

WATERPROOFING

First, and most important, determine the nature of the surface and subsurface water. This may require consultation with a geotechnical engineer and people familiar with the site. Determine whether there will be water under hydrostatic pressure under the basement slabs on grade.

Make sure that the structure is designed to resist the full displacement force of the water under all conditions.

Methods, materials, and details for waterproofing are included in NRCA 90. The following is a summary. In all cases the substrate should be clean, repaired, dry, and at the temperature recommended by the manufacturer.

■ Hot asphalt or coal tar bitumen built-up membranes (applied to earth side). These are similar to built-up roofing. The number of plies is recommended in NRCA 90.

■ Modified bitumen membranes, either hot applied or self-adhesive (applied to earth side). Hot-applied modified bitumen membranes are similar to modified bitumen roofing. Self-adhesive rubberized asphalt membranes are placed over patched, primed surfaces.

■ Butyl and EPDM rubber membranes (applied to earth side).

A1.2 SUBSURFACE MOISTURE CONDITIONS

SUBSURFACE MOISTURE PROTECTION

■ PVC membranes (applied to earth side).

■ Rubber and PVC membranes should be installed with water cutoffs dividing the waterproofed area into sections, since water which penetrates may travel between the foundation and the membrane.

■ Fluid-applied elastomeric membranes (applied to earth side). These materials achieve intimate bond to the surfaces, and thus water travel between the membrane and the wall is resisted.

■ Hot rubberized asphalt materials (applied to earth side). These are similar to fluid-applied elastomeric membranes.

■ Bentonite clay waterproofing (applied to earth side). This material swells greatly upon contact with water, and the gel thus produced waterproofs the surface. These materials can migrate and "heal" small voids and cracks, and they achieve intimate contact with the surface. They must be applied directly to the slab or wall. They are not suitable for above-ground use.

■ Metallic waterproofing (applied to earth side or interior side).

■ Cementitious waterproofing (applied to earth side or interior side).

■ Crystalline waterproofing (applied to earth side or interior side).

■ Metallic, cementitious, and crystalline waterproofing are rigid. Movement in the substrate may crack them. However, the substructure of a building is usually stable.

■ Other miscellaneous materials are listed in the NRCA Manual.

Waterproofing systems applied to the earth side have the advantage of being compressed between the foundations and the water. Systems applied to the interior side have the advantage of being applied after some

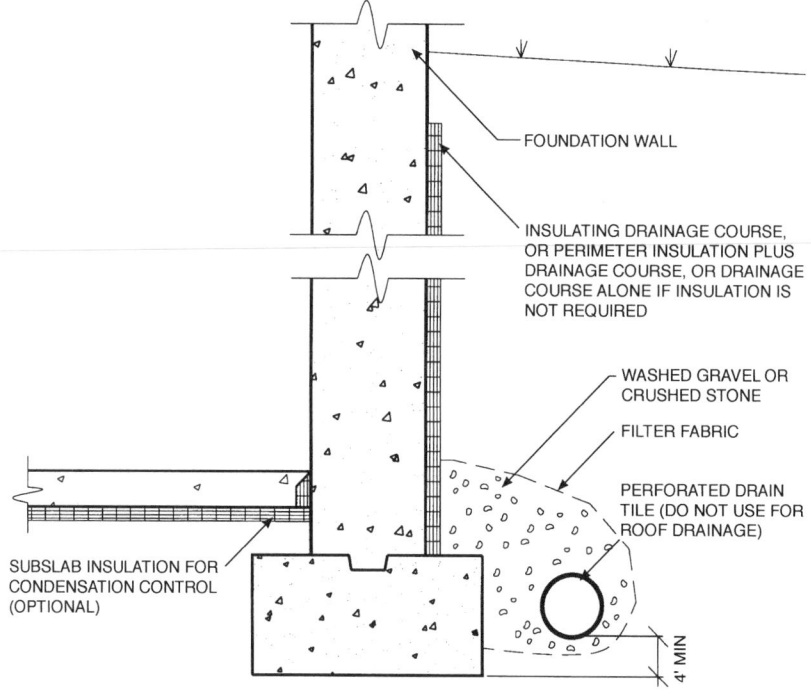

TYPICAL SUBSURFACE (FOOTING) DRAIN

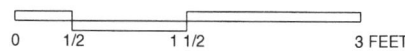

| 0 | 1/2 | 1 1/2 | 3 FEET |

Fig. 2. Typical subsurface (footing) drain

or all of the foundation shrinkage and settlement has occurred, and they may be inspected, maintained, and repaired while the building is in use, without disruptive, expensive excavation. It is good design to allow access to the basement surfaces, which are waterproofed by this method.

Application of a membrane waterproofing system under slabs on grade may require the placement of a subslab over which the waterproof membrane is installed. Protection board is then applied over the waterproofing, and the wearing slab is installed over that.

In all cases, the waterproofing must be protected against construction damage. If insulation is installed over the waterproofing, it may serve as protection. Otherwise, a special protection board is recommended. Full-time observation during backfilling is prudent.

REQUIREMENTS OF SUBSURFACE DRAINAGE

Subsurface drainage should, at best, drain to a fully reliable outfall such as a lower part of the site, a storm drain, or a drywell of adequate capacity. Although subsurface drainage can be directed to a sump pump, the same storm, which causes the heavy rain, may cause a power failure. If the outfall is a storm or combination sewer, there must be provisions against backflow during deluge conditions.

If the outfall is to grade or a natural waterway, there should be durable screening to keep animals out, and there should be rip-rap (fist-sized broken face rock) to prevent soil erosion.

If grade discharge, a storm sewer, or a reliable drywell system is not available, the storm water may drain to a sump pump. However, the sump pump and its power should be highly reliable and redundant.

(cont.)

OTHER TYPES OF SUBSURFACE DRAINAGE

If there is persistent or occasional water under hydrostatic pressure under the basement floor slab, especially if there is no effective waterproofing under the slab, an overall system of underfloor drainage may be used.

If there is persistent or occasional water under hydrostatic pressure outside the foundations or under the slab, waterproofing may be more appropriate than subsurface drainage, or it may be used in addition to subsurface drainage. For moisture prone sites and/or critical subsurface construction on sites sloping towards the building, the additional provision of swales, intercepting drains or curtain drains placed on the uphill sides offers a further prudent "first line" defense of water diversion and moisture control.

If the volume of water is great, its disposal may be a problem, and it may affect other parts of the project and neighboring sites. Also, subsurface drainage, like a well, tends to run more freely with time, as the silt clears from the soil.

Area ways sometimes become clogged with leaves and other debris and with silt, and they may cease functioning. Since areaways are seldom seen, they may not be maintained. During heavy rain, the areaways may overflow through doors or windows into the building.

REFERENCE:

Donald Baerman *"Subsurface Moisture Protection"* in Reference 1.

A1.2 SUBSURFACE MOISTURE CONDITIONS

B • BUILDING SHELL

B1 • STRUCTURE & MATERIALS

B2 • EXTERIOR WALL ASSEMBLIES

B3 • ROOFING ASSEMBLIES

B1 • STRUCTURE & MATERIALS

B1. ① STRUCTURE

STRUCTURE: introduction

FRAME

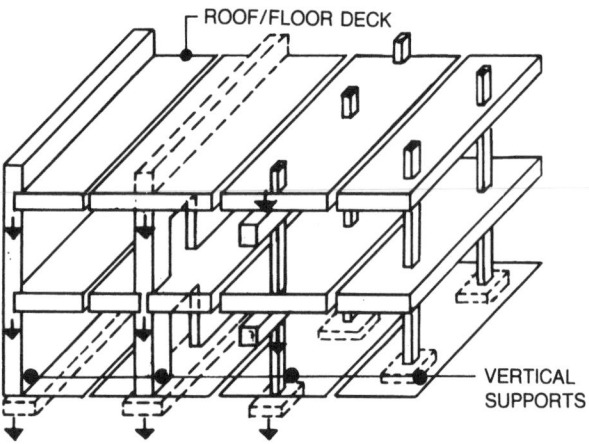

ROOF/FLOOR DECK

VERTICAL SUPPORTS

The structural frame of an enclosure should be selected to provide the most economical means of support for all loads and resistance to all forces that may be reasonably expected to be imposed upon the enclosure during its intended in-service life:
- without creating any hazard to its occupants or users.
- without excessive deformations and sideways and/or annoying vibrations.
- with proper provisions for possible or anticipated abnormal in-service conditions: such as fire, explosions, inadvertent over-loading.

The structural frame generally consists of:
- ■ Roof deck: either horizontal, pitched, or curved assemblies.
- ■ Floor decks: commonly flat horizontal assemblies:
- suspended above grade.
- supported above grade by piles driven into the ground.
- supported on the ground and independent of the structural frame.
- ■ Vertical supports or primary framing: to hold roof/floor decks in place and to carry all loads to the foundations.
- ■ Foundations: to transfer all loads to the ground.

VERTICAL SUPPORTS: TYPES

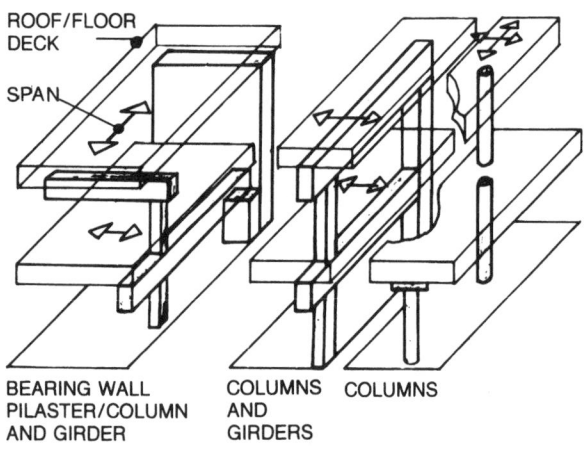

ROOF/FLOOR DECK

SPAN

BEARING WALL
PILASTER/COLUMN
AND GIRDER

COLUMNS
AND
GIRDERS

COLUMNS

Roof and floor decks may be supported by various means:
- ■ Bearing walls: which provide continuous support for the decks:
- bearing walls may be wood framed, of masonry, or of cast-in-place or precast concrete.
- ■ Pilasters: load bearing segments of nonbearing walls supporting girders, the horizontal component of a vertical support assembly, which in turn carry the roof/floor decks:
- pilasters are commonly tied into the nonbearing wall either of masonry or of concrete of which they are part of; when incorporated into a framed nonbearing wall they are also referred to as posts.
- ■ Column and girder assemblies: of wood, steel, or of reinforced concrete, either cast-in-place or precast:
- reinforced masonry columns are also used.
- ■ Columns only: which provide point support for decks, usually of monolithic reinforced concrete.
- columns are either of structural steel or of reinforced concrete.

ROOF/FLOOR DECK: TYPES

Roof/floor decks carry all loads and resist all forces: they are subjected to and transmit them to the vertical support assemblies between which they span.

The principal components of roof/floor deck assemblies are:
- ■ Decking: the structural top surface component of the deck.
- ■ Framing: structural components which support the decking. Framing and decking may be separate and distinct components or they may form a single element without any differentiation between them.

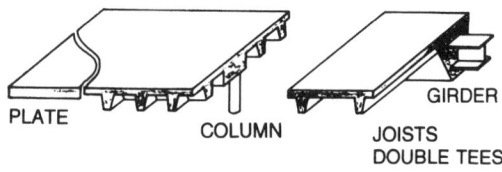

PLATE
COLUMN
GIRDER
JOISTS
DOUBLE TEES

FRAMING/DECKING: CAST-IN-PLACE/PRECAST

The assembly of framing and decking - the deck - may consist of:
- ■ Monolithic framing/decking: such as in cast-in-place reinforced concrete decks.
- ■ Fabricated components: which combined framing and decking into a single unit, such as precast concrete shapes, long span metal decks, stressed skin panels.
- ■ Framing and decking assembled at the site to function as roof/floor decks:
- framing may be prefabricated off-site to simplify site assembly, such as in pre-engineered space frames.

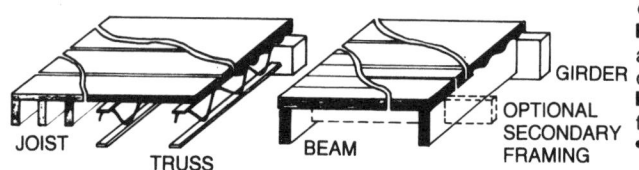

JOIST
TRUSS
BEAM
GIRDER
OPTIONAL
SECONDARY
FRAMING

FRAMING AND DECKING: SITE ASSEMBLED

B1.1 STRUCTURE

FRAMES: typical decks and supports

FRAMING/DECKING: CAST-IN-PLACE

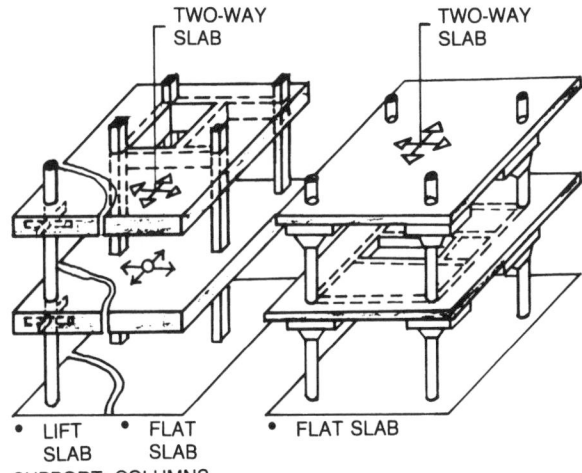

- LIFT SLAB
- FLAT SLAB
- FLAT SLAB

SUPPORT: COLUMNS

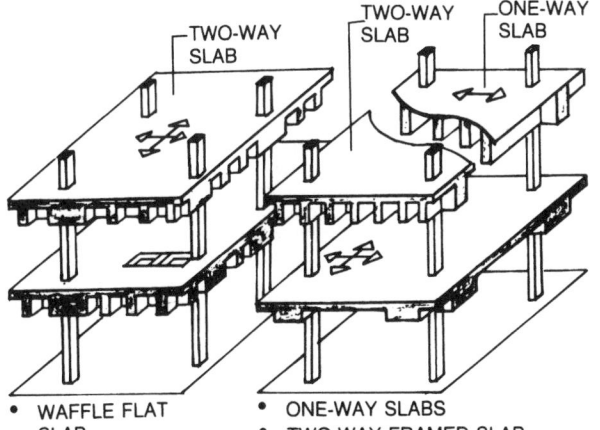

- WAFFLE FLAT SLAB
- ONE-WAY SLABS
- TWO-WAY FRAMED SLAB

SUPPORT: COLUMNS/COLUMNS AND GIRDERS

DECKS: cast-in-place reinforced concrete combining framing and decking into single element:

■ Two-Way Slabs: generally of uniform thickness, may be thickened at columns to increase resistance to shear thus increasing load carrying capacity:

- minimum of three continuous spans in each direction required for direct design of flat decks.
- generally limited to square or rectangular bays with ratios of width to length of less than two.
- relatively shallow depth of construction but extensive formwork generally required.
- when of uniform thickness throughout, slabs may be cast on the ground stacked, thus requiring minimal formwork, and then lifted into their final position.
- two-way slabs generally not recommended when numerous larger openings through decks are required: larger openings require special framing.
- conduits for electrical and communications wiring may be embedded in decks, but size of conduits generally limited.
- effects of deflection in decks and of cold flow, or creep, in concrete columns of multistory structures must be considered during selection and detailing of exterior walls, partitions, nonresilient flooring.

■ One-Way Slabs: thin sections functioning as decking cast-in-place monolithically with framing of uniformly spaced ribs of various depths:

- when closely spaced, the ribs are generally referred to as joists, when spaced further apart, as beams.
- the ribs are supported by girders spanning in one direction between columns.
- uniform depth construction may be attained by casting joists integrally with wide concrete girders of the same depth.

SUPPORTS: are generally concrete columns except for lift slabs where structural steel pipe columns are commonly used:

- point support of columns only for two-way flat and waffle slabs.
- columns and girders for two-way framed and one-way slabs.

■ Assemblies with concrete girders and, or columns may have fire resistance rating without the need for additional fireproofing.

FRAMING/DECKING: FABRICATED UNITS

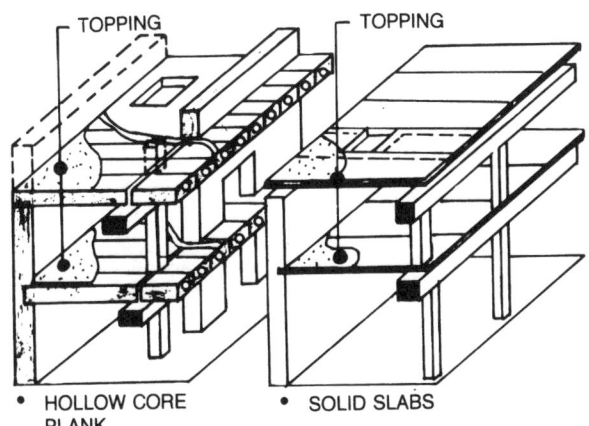

- HOLLOW CORE PLANK
- SOLID SLABS

SUPPORT: BEARING WALLS/COLUMNS AND GIRDERS

DECKS: of precast reinforced concrete components of essentially uniform overall depth, which combine decking and framing into a single unit and are capable of spanning between vertical supports:

- generally used for light to moderate loading conditions only.
- larger openings through decks require supplementary means of support.
- acceptable extent of deflection rather than strength of components may be the governing consideration during selection.
- when used for floor decks, addition of concrete topping is required to level the surface; topping may be required, and is often recommended, for roof decks.
- joints between units require grouting during installation.
- wiring may be run through cores of hollow-core plank.
- decks may have fire resistance rating without need for additional fireproofing.

SUPPORTS may be any combination of: bearing walls, either masonry or concrete; columns and girders of structural steel or concrete.

- spacing of supports: from about 12 up to 40 feet for hollow-core plank; 12 to 24 feet for solid slabs.

B1.1 STRUCTURE

FRAMES: typical decks and supports (cont.)

FRAMING/DECKING: FABRICATED UNITS

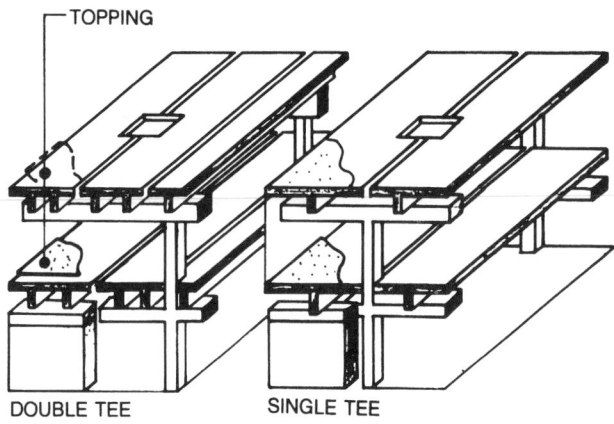

DOUBLE TEE SINGLE TEE
SUPPORT: BEARING WALLS/COLUMNS AND GIRDERS

DECKS: assemblies of precast reinforced concrete components in which framing and decking are cast monolithically:
* essentially precast sections of one-way slabs.
* generally used with widely spaced supports.
* smaller openings in decks may be made by cutting out decking between framing ribs; large openings require supplementary supports.
* acceptable extent of deflection rather than strength may govern selection, especially for upper ranges of allowable spans, camber usually provided.
* concrete topping required for floor decks, may be required for roof decks to provide level substrate for roofing.
* conduits for electrical/communications wiring may be embedded in topping, but size of conduit quite limited, may cause cracking in topping.
* decks may have fire resistance rating without need for additional fireproofing.

SUPPORTS may be any combination of: bearing walls of reinforced masonry or concrete, columns and girders of reinforced concrete or structural steel.
* spacing of supports from 40 to about 120 feet.

SOLID FRAMING AND DECKING: SITE ASSEMBLED

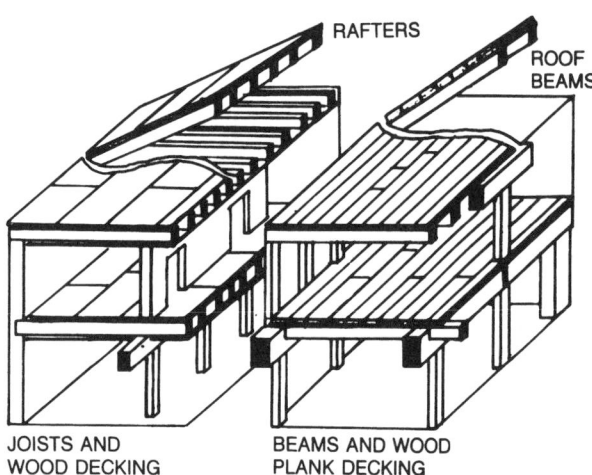

JOISTS AND   BEAMS AND WOOD
WOOD DECKING   PLANK DECKING
SUPPORT: BEARING WALLS/COLUMNS AND GIRDERS

DECKS: framing and decking as separate components assembled onsite in their final location.
■ Solid framing is commonly referred to as:
* joists: when horizontal and spaced 12 to 24 inches on centers, rafters or roof joists when pitched and part of a roof deck.
* beams: when spaced 4 to about 8 feet on centers and spanning between girders or bearing walls; also referred to as purlins when horizontal and spanning between pitched roof framing girders.
■ Spacing of framing is principally determined by properties of decking used:
* load carrying capacity of decking.
* extent of deflection allowable or acceptable.
* size of decking when joints between individual pieces have to fall over framing members for proper support.
* spacing may be reduced below the maximum allowable for specific type of decking to increase the load carrying capacity or span between supports of a section of a roof/floor-deck while maintaining the same overall depth of construction throughout.
■ Size of framing is generally controlled by allowable stresses in bending and/or shear for short spans, allowable deflection for long spans: especially when inelastic components of an enclosure are also supported by such framing, such as ceiling membranes of plaster or gypsum board, or inelastic flooring.
■ Framing may be of solid wood, laminated wood, light-gauge steel, structural steel:
* precast reinforced concrete beams may also be used with some types of decking, but such usage is not common.
■ Decking generally spans one-way between framing members and may be of: solid wood; laminated wood; wood composites; precast gypsum, or precast concrete of various densities; formed light-gauge steel with or without cementitious fill; composite of formboards, steel subpurlins, and cementitious fill.

SUPPORTS may be any combination of: framed, masonry or concrete bearing walls; columns and girders of solid wood, laminated wood, structural steel:
* reinforced concrete girders may also be used with some decks but such usage is not common.

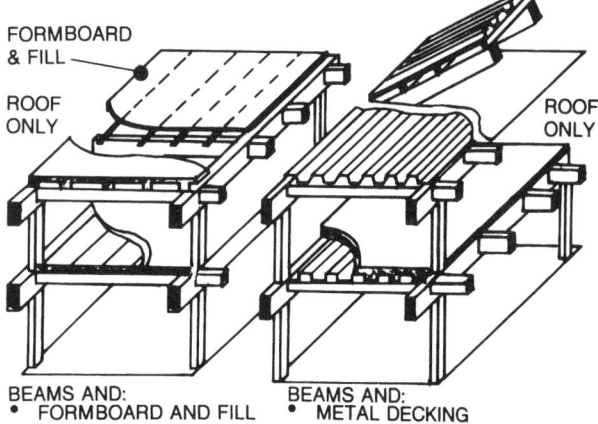

BEAMS AND:
* FORMBOARD AND FILL
* CHANNEL SLABS
* SOLID PLANKS
SUPPORT: COLUMNS AND GIRDERS

BEAMS AND:
* METAL DECKING
* METAL DECKING

B1.1 STRUCTURE

OPEN FRAMING AND DECKING: SITE ASSEMBLED

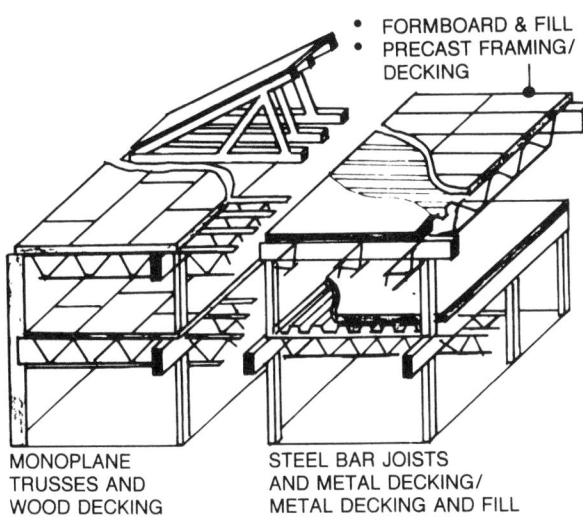

- FORMBOARD & FILL
- PRECAST FRAMING/ DECKING

MONOPLANE
TRUSSES AND
WOOD DECKING
SUPPORT: BEARING WALLS/COLUMNS AND GIRDERS

STEEL BAR JOISTS
AND METAL DECKING/
METAL DECKING AND FILL

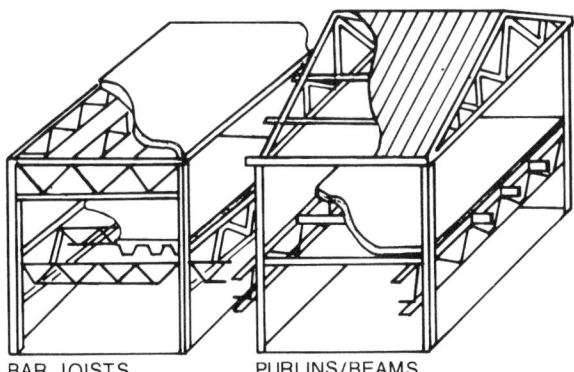

BAR JOISTS
AND VARIOUS DECKING
SUPPORT: COLUMN AND TRUSS GIRDERS

PURLINS/BEAMS
AND VARIOUS DECKING

DECKS: framing and decking as separate components assembled on site in their final location. Open framing may be:

■ Light trusses of solid wood or wood and steel bar composites: generally spaced 24 inches on centers and supporting solid or composite wood decking.

■ Short-span and long-span steel bar joists: commonly 24 or more inches on centers for floor decks, 4 to 6 feet or more on centers for roof decks, depending on properties of deck used:
- decking commonly used: formed light-gauge steel with or without cementitious fill; formboard, steel subpurlins and cementitious fill; precast cementitious slabs or planks; cementitious fill on metal lath; wood composites when nailing strips are attached to top flanges of steel bar joists.
- proprietary system of steel bar joists and cast-in-place concrete decking providing composite action under load in the deck assembly is available.
- steel bar joists may be used as rafters in pitched roof decks but such usage is not common.
- objectionable vibrations may occur in floor decks framed with short-span steel bar joist when their spans are in the upper range of those allowable.
- deflections in steel bar joists used in dead level roof-decks may result in ponding of rainwater unless drains are provided in all such low spots.

■ Purlins/beams in pitched roof-decks should be braced against rotation under eccentric load and lateral sag due to their own weight.

Open framing allows running electrical/communications wiring, small diameter piping, and small size ductwork within the depth of the deck assembly:
- more easily accomplished when deck assembly is supported on girders of open cross section.

SUPPORTS may be:
■ For light wood trusses: commonly bearing walls of wood frame, masonry, concrete; columns and girders of wood, structural steel, less often of concrete.
■ For steel bar joists: bearing walls of light-gauge steel frame, masonry, concrete; columns and girders of structural steel, less often of concrete.

OPEN FRAMING AND DECKING: PRE-ENGINEERED

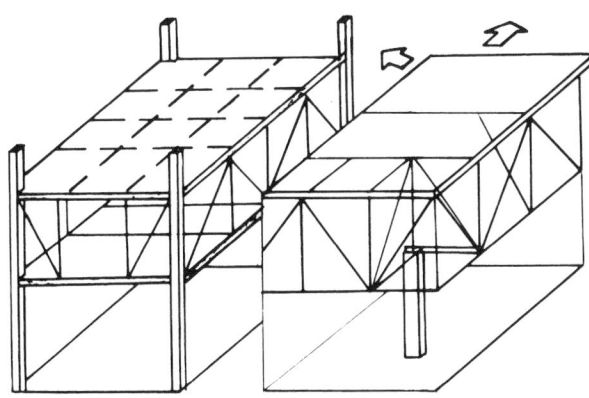

TWO WAY TRUSSES/SPACE FRAMES
SUPPORT: COLUMNS

DECKS: framing and decking as separate components, site assembled.
■ Framing:
- two-way interlocking braced truss system, in triangular, diagonal, hexagonal, or rectangular grid of structural steel or aluminum.
- horizontal or curved, used to roof over large open spaces.
- supported by columns, which may be randomly located, may be supported on bearing walls. System permits two-way overhangs.
- to simplify construction, the size of members is either the same throughout, or a limited number of sizes is used: the majority of members must be oversized so that the most heavily loaded would not be overstressed.
- may be assembled on the ground and lifted into place.
- ductwork, piping, conduits for electrical and telephone wiring may be run within space frame.
■ Decking may be: transparent or translucent, such as plastics, glass; formed light gauge metal; cementitious; wood or wood composites where permitted by building codes.
■ Most commonly used for roofs, but can be designed for floor loading also; full story-height space frames have been built to serve as mechanical equipment floors.

B1.1 STRUCTURE

DEFORMATION IN FRAMES

DEFORMATIONS DEFINED

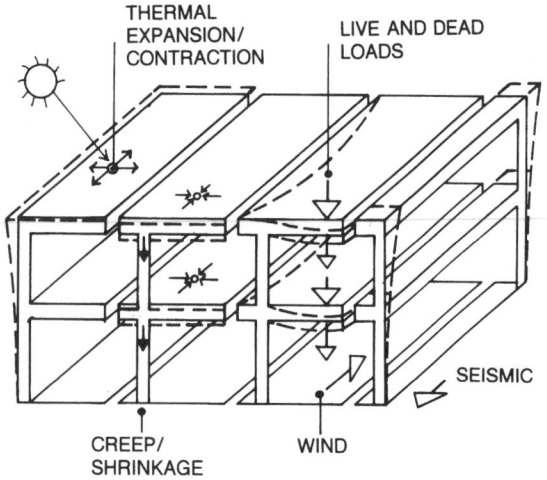

THERMAL EXPANSION/ CONTRACTION

LIVE AND DEAD LOADS

SEISMIC

CREEP/ SHRINKAGE

WIND

All structural frames are subject to deformations:

■ Deflection: the differential change in length between two op-posite faces of a horizontal or vertical assembly or component of a structural frame. Deflection may result from:
* bending loads: when one face shortens under compression while the other elongates under tension.
* temperature differential: when one face remains stable or con-tracts while the other expands.
* moisture differential: when one face remains stable or shrinks while the other swells.
■ Plastic flow: shortening of vertical components, such as columns, or deflection in horizontal assemblies and/or components, such as monolithic concrete decks, under long-term sustained loading; also commonly referred to as creep:
* concrete in particular and wood are subject to creep, while its effect on structural steel is insignificant.
■ Shrinkage: the overall volumetric change due to changes in moisture content.
■ Lateral displacement, also often referred to as sway or drift, of frames due to wind or seismic forces.

IN DECKS/GIRDERS

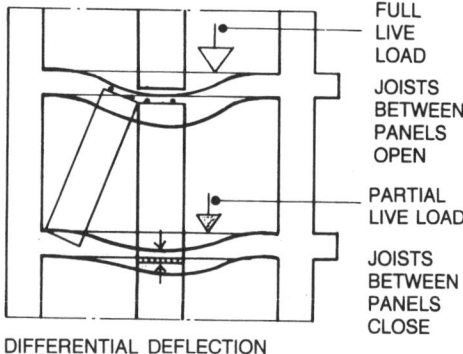

FULL LIVE LOAD

JOISTS BETWEEN PANELS OPEN

PARTIAL LIVE LOAD

JOISTS BETWEEN PANELS CLOSE

DIFFERENTIAL DEFLECTION

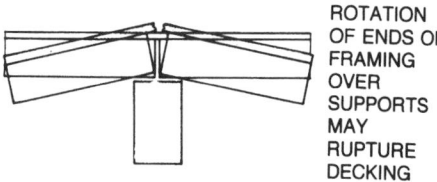

ROTATION OF ENDS OF FRAMING OVER SUPPORTS MAY RUPTURE DECKING

SIMPLY SUPPORTED DECK AT SUPPORTS

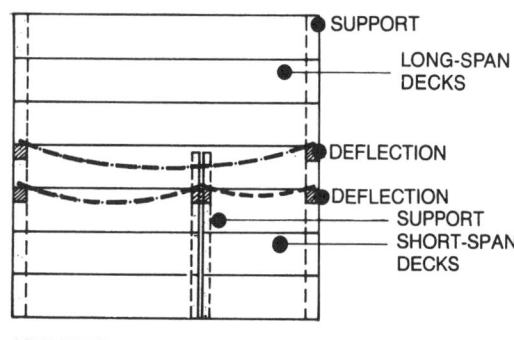

SUPPORT

LONG-SPAN DECKS

DEFLECTION

DEFLECTION

SUPPORT

SHORT-SPAN DECKS

ADJACENT UNEQUAL SPANS

Components of horizontal frames such as framing, decking, gir-ders are always subject to deflection due to bending and to varying extent the effects of lateral displacement:
* they may also be subject to: deflection due to temperature differential especially in roof decks, moisture diffential when of materials thus affected.
* plastic flow and shrinkage may further aggravate the effects of deflection.

Deformations to be expected in specific materials and their effects should be considered during preliminary selection of a structural frame:

■ Steel: is essentially elastic within allowable stresses, is not affected by moisture, and does not creep to any significant amount:
* deflection due to live load is the principal consideration with differential thermal expansion/contraction generally being less significant a factor.
* camber may be provided in girders to compensate for deflection generally for that due to dead load only which will also add to the cost of fabrication.
■ Concrete: is subject to deflection under load, creep, shrink-age, thermal expansion/contraction:
* deflection under permanent load continues to increase for several years due to shrinkage and creep: the total deflection to be provided for in design is the sum of creep deflection from per-manent or sustained long term loads which is largely irrever-sible deflection due to live loads, and the deflection effects of temperature and moisture differentials.
* creep may amount to as much as 2.5 to 3 times the load induced deflection, will result in loss of prestress, but will also re-lieve stress concentrations which otherwise might develop.
■ Wood: is affected by changes in overall moisture content, mois-ture and temperature gradients across a given section, deflection due to loads, and creep:
* wood is subject to continuous volumetric changes across the grain of about 3 percent due to changes in its moisture content under normal in-service conditions.
* creep will occur in wood when under sustained load, with the amount to be expected varying with different species.

IN COLUMNS/WALLS

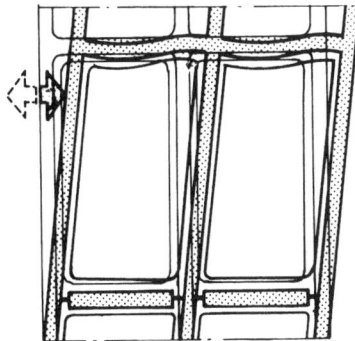

FRAME WILL DEFORM UNDER WIND OR SEISMIC FORCE AND MAY AFFECT EXTERIOR WALLS AND/OR FACINGS IN ADDITION TO POSITIVE OR NEGATIVE WIND PRESSURES ACTING ON SUCH WALLS

LATERAL DISPLACEMENT

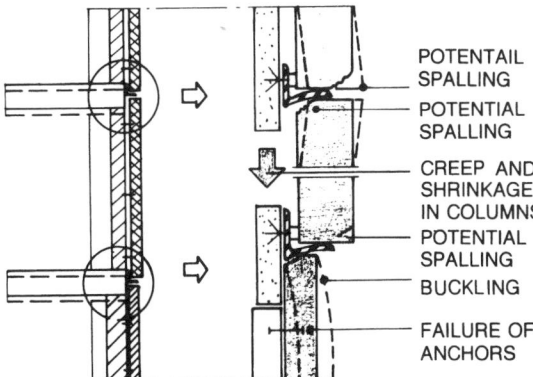

POTENTAIL SPALLING
POTENTIAL SPALLING
CREEP AND SHRINKAGE IN COLUMNS
POTENTIAL SPALLING
BUCKLING
FAILURE OF ANCHORS

SHORTENING OF FRAME

Vertical elements of a structural frame such as columns, bearing walls are always subject to deflection due to lateral loads, lateral displacement, and to varying extent effects of bending due to vertical loads:
* axial loads are seldom truly that, and any eccentricity will cause bending stresses to develop.
* lateral displacement or sway in tall structures may be well within safe limits structurally, but may be far in excess of maximum allowable values for a particular curtain wall system to function properly.
* columns and bearing walls may also be subject to: deflection due to temperature or moisture differential through their section and to plastic flow, depending on the properties of their constituent materials.

■ Steel: is generally affected by lateral forces, bending due to eccentric loading, and differential thermal expansion/contraction.

■ Concrete: is subject to creep and shrinkage in addition to bending, thermal and moisture differentials.
* shrinkage and creep in concrete during and after construction will result in shortening of the structural frame, which may amount to as little as 0.10 inch or more than 0.60 inch for a sixty foot high structure, depending on: at which stage of construction the frame is fully loaded, size of columns, reinforcing provided, differences in ambient relative humidity.
* reinforcing of concrete will tend to minimize creep but will not prevent it.
* when connections between a structural frame of concrete and a rigid wall assembly supported by it do not allow for creep related shortening of the frame, shearing action between the frame and the wall may develop, and may lead to damage or failure of the wall.

■ Wood columns will be affected by moisture and temperature differential across their section and bending:
* shrinkage along the grain is considerably less than across the grain and generally is not a significant factor.
* when vertical components of a multistory wood framed structure bear on horizontal framing at intermediate levels, the shrinkage and creep across the grain in such framing will result in shortening of the frame, with the extent varying constantly with changes in ambient relative humidity.

EFFECT ON PARTITIONS

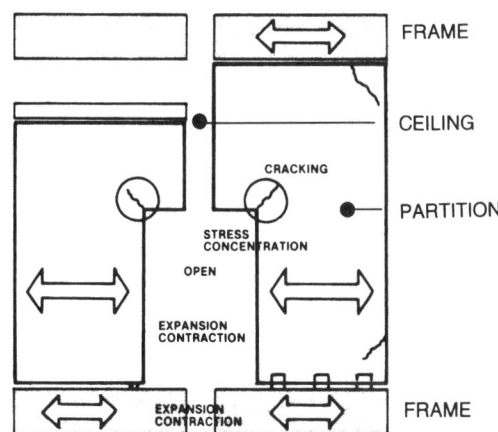

FRAME
CEILING
CRACKING
PARTITION
STRESS CONCENTRATION
OPEN
EXPANSION CONTRACTION
EXPANSION CONTRACTION
FRAME

Deformations in the structural frame will also affect interior elements of enclosures. Movement and subsequent cracking of partitions may be caused by:
* deflection in floor deck and/or girders which support the partition and/or foundation settlement, with either resulting in vertical cracking, commonly the full height of the partition.
* lateral displacement or distortion of the frame, with the resulting racking action commonly leading to corner cracking in partitions.
* thermal expansion/contraction in the frame, when at different rate than corresponding movement in the partition.
* shrinkage or moisture induced volumetric changes in the frame and/or partition.

Cracking in partitions may also be caused by factors not directly related to deformations in the structural frame:
* expansion/contraction in the partition itself.
* stress concentrations in abrupt changes in cross-sectional area, such as at openings.

B1.1 STRUCTURE

ROOF/FLOOR DECKS: typical decking

FRAMING & DECKING: site assembled

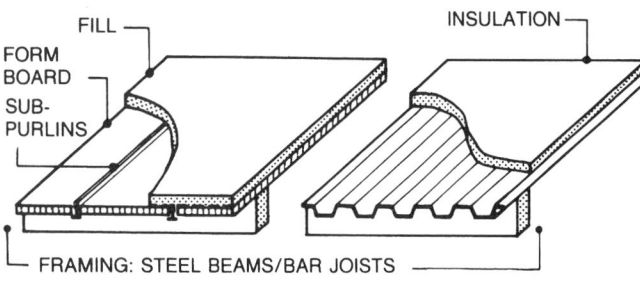

ROOF ONLY: PITCHED/FLAT

OPTIONAL INSULATION

OPTIONAL INSULATION
OPTIONAL FILL

FRAMING:
JOISTS/RAFTERS
COMPOSITION BOARD

FRAMING:
LIGHT BEAMS/BAR JOISTS
PRECAST PLANK/ CHANNEL SLAB

ROOF ONLY: FLAT

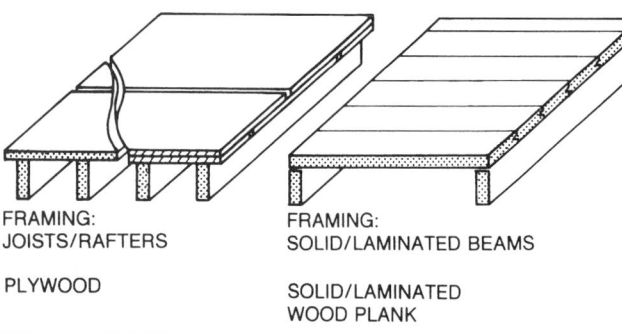

FILL

FORM BOARD

SUB-PURLINS

INSULATION

FRAMING: STEEL BEAMS/BAR JOISTS

FORMBOARD AND FILL   METAL DECK

ROOF: FLAT OR PITCHED/FLOOR

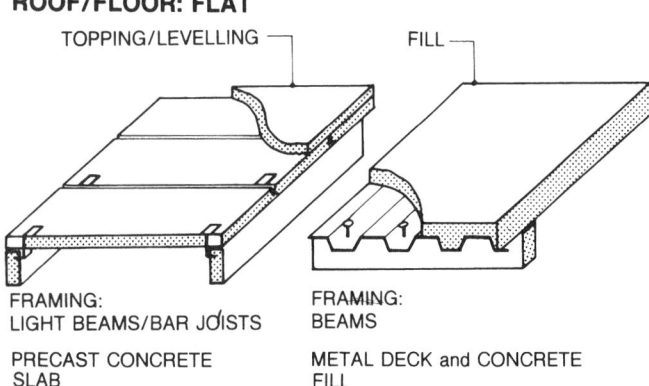

FRAMING:
JOISTS/RAFTERS

FRAMING:
SOLID/LAMINATED BEAMS

PLYWOOD

SOLID/LAMINATED
WOOD PLANK

ROOF/FLOOR: FLAT

TOPPING/LEVELLING   FILL

FRAMING:
LIGHT BEAMS/BAR JOISTS

FRAMING:
BEAMS

PRECAST CONCRETE
SLAB

METAL DECK and CONCRETE
FILL

Decking component of site assembled roof/floor decks may be:

■ Composition or wood particle board: generally used in framed structures as roof sheathing only:
• usually 4 feet wide, 8 to 12 feet long.
• strength in bending and dimensional stability under varying moisture conditions are primary considerations.
■ Plywood: for roof sheathing and floor decking in wood or metal framed structures:
• thickness varies from ⅜ inch for roof sheathing up to 1 ¼ inches with tongue and groove edges for floor decking.
• 4 feet wide, 8 to 12 feet long, with 8 feet being the most readily available length.
■ Wood plank: either solid or laminated
• solid wood boards of one inch or 1 ¼ inch nominal thickness may be used as roof sheathing or subfloor-decking, but such usage is no longer common.
• solid wood decking of 2 to 4 inch nominal thickness is available, but has largely been replaced by laminated decking.
• laminated decking is available either 3-ply or 5-ply with thickness ranging from 3 to 5 inches, in lengths of 6 feet or longer, in increments of 1 foot.
• spans for planks range from 5 to about 16 feet; lay-up generally random over 2 or more supports.
■ Planks of precast concrete or cement bound wood fiber; also precast concrete channel slabs:
• generally 2 to 4 inches thick, 16 to 48 inches wide, spanning 8 to 10 feet, plank generally tongue and groove, metal edged tongue and groove available.
• commonly secured to steel framing by metal clips; some may also be nailed.
• common usage is as roof decking only; has been used for lightly loaded floors.
■ Precast concrete slabs are similar to precast plank except that they are thicker, generally 4 to 8 inches, and used principally for floor decking:
• concrete topping of about 2 inches in thickness required for floors.
• spans range from 12 to 24 feet, may function as framing/decking combined.
■ Formboards of cement bound organic or mineral fibers, supported by steel subpurlins between framing, with site placed usually lightweight concrete fill:
• subpurlin spacing 24 to 33 inches: spans 6 to 10 feet.
■ Metal deck: usually of formed light-gauge steel either coated or galvanized; generally 28 to 20 gauge for depths of ½ to 1 ½ inches, 22 to 16 gauge for depths of 1 ½ to 3 inches commonly, but available up to 6 inches of various configurations.
• ½ to 1 ½ inch deep often used as centering or permanent formwork for cast-in-place concrete floor-decks over steel bar joist or light steel beam framing, spaced 2 to 8 feet on centers.
• metal decking for roofs may be used with site placed lightweight concrete or gypsum fill or more commonly with insulation only; types incorporating sound absorbing materials in ribs are available; spans for 1 ½ inch depth 4 to 8 feet, for 3 inch depth 8 to 12 feet.

B1.1 STRUCTURE

FRAMING & DECKING: site assembled

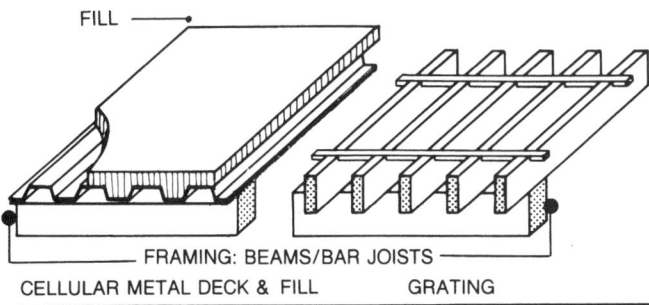

FILL

FRAMING: BEAMS/BAR JOISTS

CELLULAR METAL DECK & FILL   GRATING

■ Metal decking for floors: generally 22 to 16 gauge, 1 ½ to 3 inches deep:
• spans range from 6 to 14 feet depending on gauge and depth of deck, and thickness of concrete fill.
• available with closed cells, also referred to as cellular deck, to provide space for electrical/communications wiring.
• always filled with site placed normal weight or lightweight concrete, usually reinforced.
• decking may be formed to interlock with concrete fill for composite action; in addition metal studs may be welded through decking to top flanges of framing for their composite action with fill thus reducing their size.
■ Grating: of metal or glass fiber reinforced plastic.

FRAMING/DECKING COMBINED: fabricated

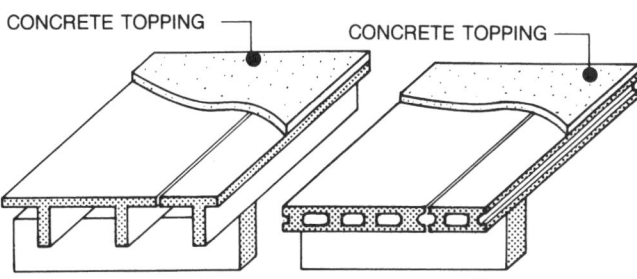

CONCRETE TOPPING   CONCRETE TOPPING

SUPPORT: BEARING WALLS/GIRDERS & COLUMNS

DOUBLE TEE'S   HOLLOW CORE PLANK
SINGLE TEE'S

■ Single or double tee's: precast of reinforced commonly prestressed concrete combining framing and decking into a single unit:
• double tee: generally 12 to 32 inches deep and 8 to 10 feet wide; normally spanning 20 to 80 feet, potentially up to 95 feet.
• single tee: Usually 24 to 48 inches deep and 8 to 10 feet wide; normally spanning 50 to 110 feet, potentially up to 120 feet.
• camber generally provided to minimize apparent deflection under load.
■ Hollow-core plank: precast of reinforced commonly prestressed concrete combining framing and decking into a single unit:
• depth varies from 4 to 12 inches; width, from 16 inches to 8 feet.
• spans generally from 12 to about 50 feet; potentially up to 55 feet.

FRAMING/DECKING COMBINED: cast-in-place

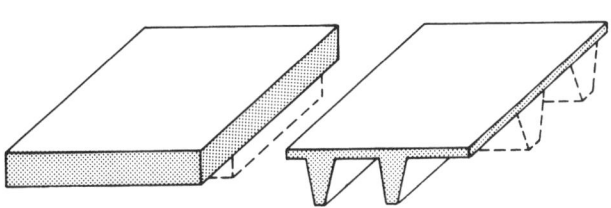

SUPPORT: COLUMNS/COLUMNS AND GIRDERS

FLAT PLATE/SLAB   ONE-WAY JOISTS
ONE- TWO-WAY SLAB   TWO-WAY JOISTS

Reinforced in directions monolithically placed concrete decks combining framing and decking:
• spans for light loading to about 30 feet; 20 to 25 feet for heavy loads.
■ Flat plate: a two-way slab of uniform thickness throughout.
■ Flat slab: generally thickened over columns by drop panels to increase resistance to shear; tops of columns may also be flared out for the same purpose.
■ Waffle flat slab: thin decking and a grid of joists cast-in-place monolithically:
• joists are omitted over columns to form solid panels, which may also be deeper than the joists to increase resistance to shear.
■ Two-way slab: solid relatively thin decking supported by girders along column center lines.

DECKING/ROOFING COMBINED

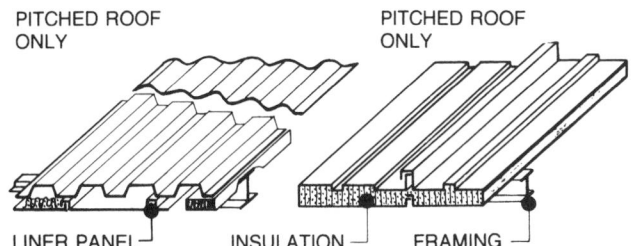

PITCHED ROOF   PITCHED ROOF
ONLY   ONLY

LINER PANEL   INSULATION   FRAMING

ROOF PANELS   ROOF PANELS
• SINGLE THICKNESS   • COMPOSITE
• SANDWICH

Formed metal panels which function as combined roof decking and roofing:
• available in: galvanized steel, either plain or prefinished, aluminum coated steel, aluminum, either plain or prefinished.
• may be single thickness or a built-up sandwich consisting of: exterior face panel, subgirts, insulation, and interior face panel.
• composite: with a rigid insulating core sandwiched between two face panels.
• spans generally 6 to 12 feet.
• supported by purlins which span between roof girders spaced about 20 to 30 feet on centers.
■ For additional information refer to ROOFING.

B1.1 STRUCTURE

ROOF/FLOOR DECKS: typical framing

JOISTS/RAFTERS

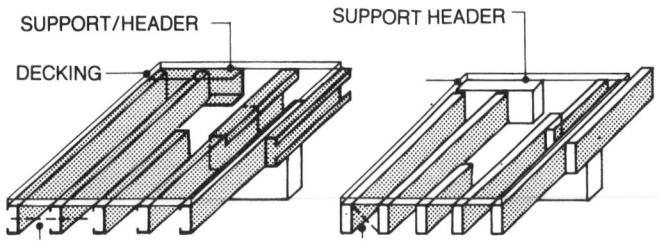

SUPPORT/HEADER   SUPPORT HEADER

DECKING

LATERAL SUPPORT FOR FLOOR FRAMING:
BRIDGING @ 8'0"O.C. MAX

STEEL: COLD ROLLED   WOOD: SOLID OR PLYWOOD/
LIGHT GAUGE   LUMBER COMPOSITES

■ Cold rolled light-gauge shapes used as roof/floor framing:
• gauge varies from 20 to 12; depth, from 6 to 12 inches.
• connections are made by welding and/or by self-drilling, self-tapping screws.
• allowable stresses and design in accordance with: Specification for the Design of Cold-Formed Steel Structural Members, of the American Iron and Steel Institute.
■ Dimensional lumber, 2 to 4 inches thick and 6 inches or more in width, used as roof/floor framing:
• moisture content should not exceed 19 percent.
• allowable stresses and design generally in accordance with: National Design Specification for Wood Construction, of the National Forest Products Association.
• for additional information refer to WOOD/WOOD PRODUCTS.

BEAMS

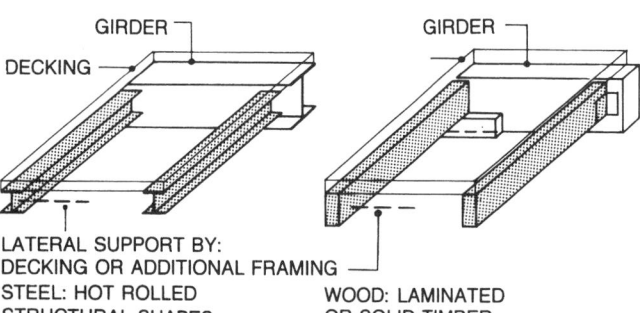

GIRDER   GIRDER

DECKING

LATERAL SUPPORT BY:
DECKING OR ADDITIONAL FRAMING
STEEL: HOT ROLLED   WOOD: LAMINATED
STRUCTURAL SHAPES   OR SOLID TIMBER

■ Hot-rolled structural shapes:
• spacing usually 6 to 14 feet when used as framing supported on girders.
• connections welded and/or bolted.
• rolling mill tolerances have been established under ASTM A6 Standard Specification for General Requirements for Rolled Steel Plates, Shapes, Sheet Piling, and Bars for Structural Use.
• decking usually used include: formed metal; precast concrete slabs, plank, channel slabs; cement bound organic fiber plank and boards.
■ Laminated dimensional lumber or solid timber:
• spacing depends on decking selected: up to 4 feet for plywood, 6 to 8 feet for 2 inch nominal plank, 9 to 16 for 3 inch plank, up to 20 feet for 5 inch plank.
• available treated for resistance to decay.

TRUSSES: flat top chord

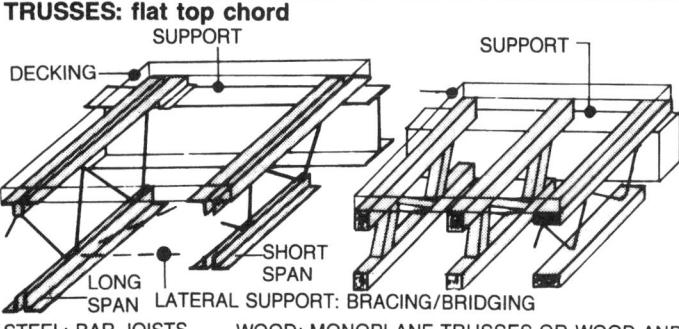

SUPPORT   SUPPORT

DECKING

SHORT SPAN
LONG SPAN   LATERAL SUPPORT: BRACING/BRIDGING
STEEL: BAR JOISTS   WOOD: MONOPLANE TRUSSES OR WOOD AND STEEL BAR JOISTS

■ Steel bar joists available as short-span and long-span framing:
• short-span bar joists usually 8 to 30 inches deep with spans from about 10 to 60 feet.
• long-span: 18 to 72 inches deep with spans from about 30 to 140 feet.
• connections to supports generally used: welding to steel girders or to steel bearing plates anchored in concrete or masonry; anchors to masonry.
• may be doubled, or trippled, to carry localized concentrated loads.

TRUSSES/BEAMS: pitched/curved

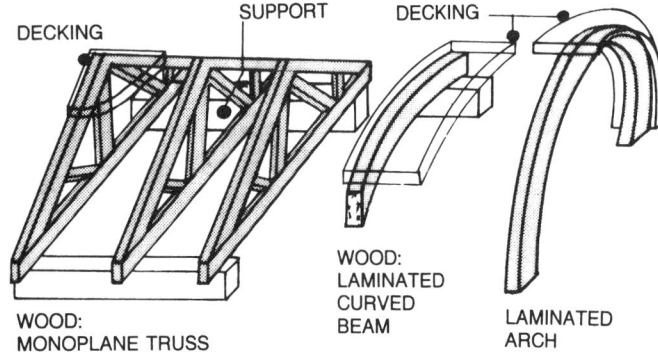

SUPPORT   DECKING

DECKING

WOOD:
LAMINATED
CURVED
BEAM   LAMINATED
WOOD: ARCH
MONOPLANE TRUSS

■ Monoplane trusses of dimensional lumber, or of top and bottom chords of wood with steel bar webs:
• spacing generally between 12 to 24 inches on centers.
• depth varies from 12 to 48 inches; spans for light loading, from 20 to 60 feet.
■ Monoplane trusses of dimensional lumber:
• spacing generally 2 feet on centers using ⅜ inch thick plywood decking.
• spans up to 60 feet based on allowable tension value in bottom chord.
• slope usually from 2 inches per foot to 6 inches per foot, allowable spans increase with increase in slope.
• monoplane trusses, either with pitched or flat chord, available of fire-retardant treated wood.
■ Curved beams and arches of laminated surfaced dimensional lumber in various shapes:
• nominal width commonly from 3 to 10 inches for beams, up to 16 inches for arches.
• available treated for resistance to decay, but not fire-retardant treated.
• commonly used with solid or laminated wood plank decking.

B1.1 STRUCTURE

PRELIMINARY CONSIDERATIONS

STRUCTURAL FRAME

The structural frame is an integral part of any site assembled enclosure and often a clear distinction cannot be made between what might be termed as a purely structural or a purely architectural component of such enclosure:

■ **Decking and framing of roof/floor decks may function:**

• structurally: by carrying all superimposed gravity loads between vertical supports, and commonly by transferring lateral loads to vertical supports or other elements of the structural frame.
• architecturally: by serving as the substrate for roofing or flooring, or as substrate and flooring combined.

■ **Bearing walls** are both a structural support for roof/floor decks and the vertical component of the envelope of such enclosure.

■ **Movement and/or deformations in the structural frame** will affect most or all architectural components; conversely, movement or deformation in such components may affect the structural frame.

Preliminary considerations of a structural frame should primarily include roof/floor-decks which shelter and support the activities for which an enclosure is being provided:

FRAMING/DECKING COMBINED:

Three basic types within the group are:

■ **Flat plate:** which has a completely flat underside.

■ **Flat slab:** with dropped panels at columns, columns generally round, with flared tops.

■ **Waffle slab:** with solid panels of the same depth as slab at columns, and the rest of slab waffled to reduce the dead load of construction.

■ **Flat plates and flat slabs** present few, or no obstructions to horizontal distribution of mechanical/electrical systems, but the vertical distribution must be carefully considered as openings in these assemblies are limited as to location and size.

Standard reusable forms available to be used in forming flat and waffle slabs include:

■ forms for waffle slabs.

■ **steel forms for round columns** with flared capitals, generally used for flat slab construction; and fiber tubes for forming round columns, often used with waffle slabs.

■ **Two-way slabs are:**

• generally economical for moderate spans and heavy loads.
• when minimum depth of construction is needed, edge beams may be made shallow and wide, in which case the columns need not line up as long as they fall within the width of the shallow beam.

■ **Two-way slabs with shallow** beams may be used in multistory apartment construction when the underside of the slab can remain exposed.

■ **Ceilings are generally not used with two-way slab assemblies,** nor are they required for a fire rated assembly.

FABRICATED FRAMING/DECKING:

■ **Precast concrete units** used for flooring usually require a concrete topping:

• to even out all irregularities between individual units.
• to improve the load carrying ability of the units.
• to improve fire resistance of the assembly.

■ **Concrete toppings** may also be required for roofs:

• to provide a smooth surface for the roofing.
• to improve the thermal resistance of the assembly when insulating concrete fill is used.

FRAMING AND DECKING SITE ASSEMBLED

■ **Joist framing** utilizes light, closely spaced members. A variety of materials, sizes, and shapes is available. Joist framing is:

• versatile, economical system for residential, commercial, and light industrial construction where: spans are short to moderate, loads are light.
• Prefabricated stressed skin panels consisting of plywood faces and wood joist ribs are available, widely used in prefabricated housing.

■ **Beam framing** is the most versatile

type with a wide choice of materials, sizes, and shapes. Beam framing is:
• most economical for moderate spans, and moderate loads, but
• can readily be adapted to carry heavy loads, with a corresponding increase in unit cost.
• not suitable for long spans.
• when beams frame into girders, the depth of girder usually determines the overall depth of construction.

■ **Framed assemblies,** with the exception of steel bar joists, will not permit ductwork, piping, and conduits to be run within the depth of the framing members, thus requiring additional depth between floors to accommodate such services.

Decking component selection will be influenced by the framing components:

■ **Wood planking or plywood may be used over metal framing:**

• should be secured using self-tapping metal screws or
• a wood nailer should be attached to the top flange of the member first.

■ **Roll-formed metal decking,** when in light gauges, is difficult to weld to supporting steel framing; the use of welding washers is recommended:

• decking spanning between widely spaced framing may require temporary shoring during placement of concrete fill.

■ **Precast planks,** whether of concrete, or fiberboard:

• are generally secured to metal framing using special clips supplied by plank manufacturer.
• diaphragm action is not provided in a roof assembly by precast planks thus attached.

■ **Wood planking,** either solid or laminated, is best suited for:

• post-and-beam or heavy timber framing.
• to span between laminated arches or rigid frames where it generally also serves as the finished interior surface.

SPANS AND STABILITY

Allowable spans of any structural systems are determined by:

■ **Combined dead and live loads.**

B1.1 STRUCTURE

B • BUILDING SHELL

PRELIMINARY CONSIDERATIONS (cont.)

■ **Stress capacity** of components of a given structural system.

■ **Allowable or acceptable deflection limits,** which may vary:

- allowable deflection is often given as a ratio of 1/360 of clear span, without regard to properties of the frame and/or components affected by it, and may not be sufficient to prevent deformations developing: inelastic materials may crack considerably before the limit of 1/360 of span is reached.

■ **Camber is normally provided in steel, glue-lam, and precast concrete members to compensate for live and dead load deflections.** In roof framing such camber may be increased to facilitate storm drainage.

Bracing between framing members is required in many structural systems as the forces on the beams, trusses or joists may cause lateral buckling.

■ **Diaphragm action** to resist lateral wind and seismic forces can be achieved in a number of ways to reduce isolated stresses and transmit them through the entire system:

- in wood framing assemblies, usually by diagonal planking or plywood used for decking.
- in steel framed assemblies, by diagonal bracing, or by roll-formed metal decking.
- in concrete assemblies by combined stress capacities of steel reinforcement and concrete coverage.
- to qualify as a diaphragm a floor and/or roof assembly must be capable of transmitting lateral forces to vertical components of the structural frame, such as shear walls, without deflecting to where such deflection could cause damage to a vertical component.
- the effects of lateral loads on the exposed components of roof and floor assemblies - the roofing membrane and flooring - should be investigated: structural frames adequate to resist all vertical and lateral loads may still deform too excessively for some types of flooring and/or roofing.

Other considerations may include:

■ **A column,** once installed, can, in most instances, no longer be removed.

■ **Each type of floor and roof assembly** has a certain range over which it

is most economical.

■ **The longer the span, the greater the overall depth of the assembly and the greater the cost per unit of area covered.**

TRUSSES AS PRIMARY SUPPORTS

Heavy trusses are used as girders, the primary supports, to carry roof/floor-decks:

■ **Typical configurations for trusses are:** crescent, also known as bowstring, with straight or curved bottom chord; double-pitched with straight bottom chord; double-pitched with straight bottom chord; double-pitched with pitched bottom chord, also known as scissors truss; single-pitch with straight bottom chord, referred to as sawtooth:

- lateral bracing for top chords must be provided; secondary framing selected will influence spacing of trusses.
- thermal expansion of long-span trusses must be considered in the design of supports.
- concentrated loads to be supported at panel points only.
- long-span, flat top chord trusses not economical in wood.
- pitched top chord trusses in wood not economical for spans over 60 feet; in long span trusses, wood is best suited for bowstring types.
- ceilings, light fixtures, or equipment may be suspended from the bottom chords of trusses.
- long-span trusses generally fabricated off-site. Shape and size of individual panels will be influenced by available transportation facilities, clearances at underpasses, and limitations imposed by applicable state laws.

■ **Floor to floor height trusses,** incorporated in partitions, have been used:

- in residential construction.
- in industrial buildings where mechanical equipment and services to the floor above could be located within the trusses.
- such trusses have also been used in multistory construction to carry suspended lower floors.

ARCHES

Arches are curved frames combining girders and columns into a single

unit. The configurations may be:

- radial, parabolic, tudor, gothic, A-frame, rigid frames.
- rigid frames are arches with straight rather than curved or sloping vertical components.
- arches may be two-hinged at supports, or three-hinged at supports and crown.
- waterproofing of hinged joint at crown may present a problem.
- outward thrust at supports must be resisted by foundations, buttresses or horizontal ties.

■ **Site assembled arches functioning as girders may be:**

- structural steel shapes, about 20 feet on centers.
- reinforced concrete, about 20 feet on centers.

ARCHES, DOMES, VAULTS consisting of curved monolithic decks, which combine means of support and envelope into one:

- of cast-in-place reinforced concrete.
- of essentially constant cross section throughout, or of heavier sections such as ribs, curved girders connected with thin diaphragms.

■ **Extensive formwork generally required:**

- small domes may be cast over flexible membranes inflated to resist the weight of fresh concrete.

■ **Arches and rigid frames** may be a structural frame supporting a separately placed concrete decking, but such usage is not common.

■ **Barrel vaults generally used in multiples, with adjacent shells bracing each other;**

- half-shells, with an opening at the crown to admit light and/or air are also used.

STRUCTURAL SPANS

The tables on the three pages that follow are a guide to preliminary selection of structural framing, based upon typical spans. The tables may be read directly, based upon typical spans (indicated by shaded bars, with the average span shown by a double line at the span mid-point).

B1.1 STRUCTURE

STRUCTURAL SPANS: timber

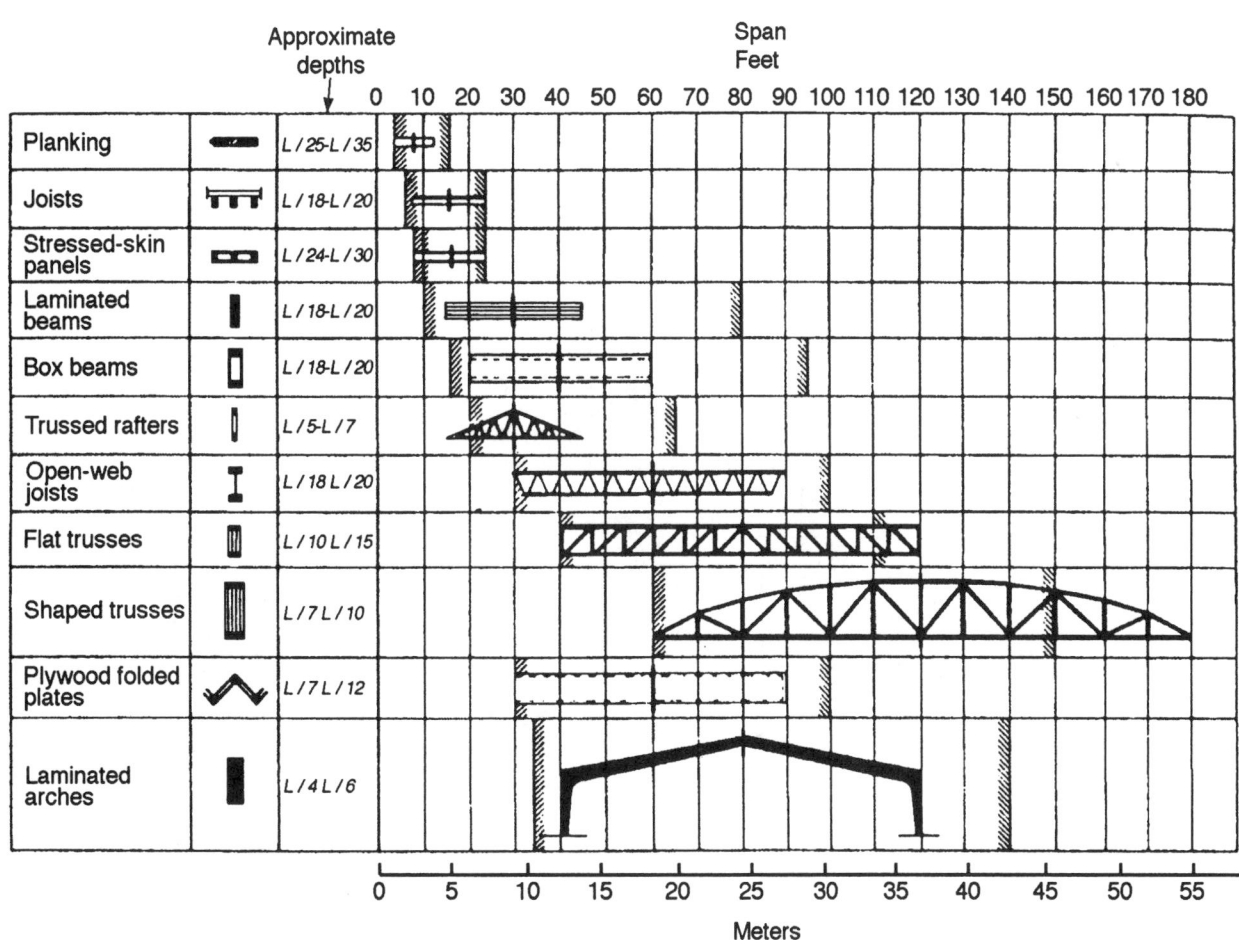

Approximate span ranges for timber systems. Source: Schodek, D., *Structures,* (Second Edition) Englewood Cliffs: Prentice Hall (1992) by kind permission of the publisher.

STRUCTURAL SPANS: steel

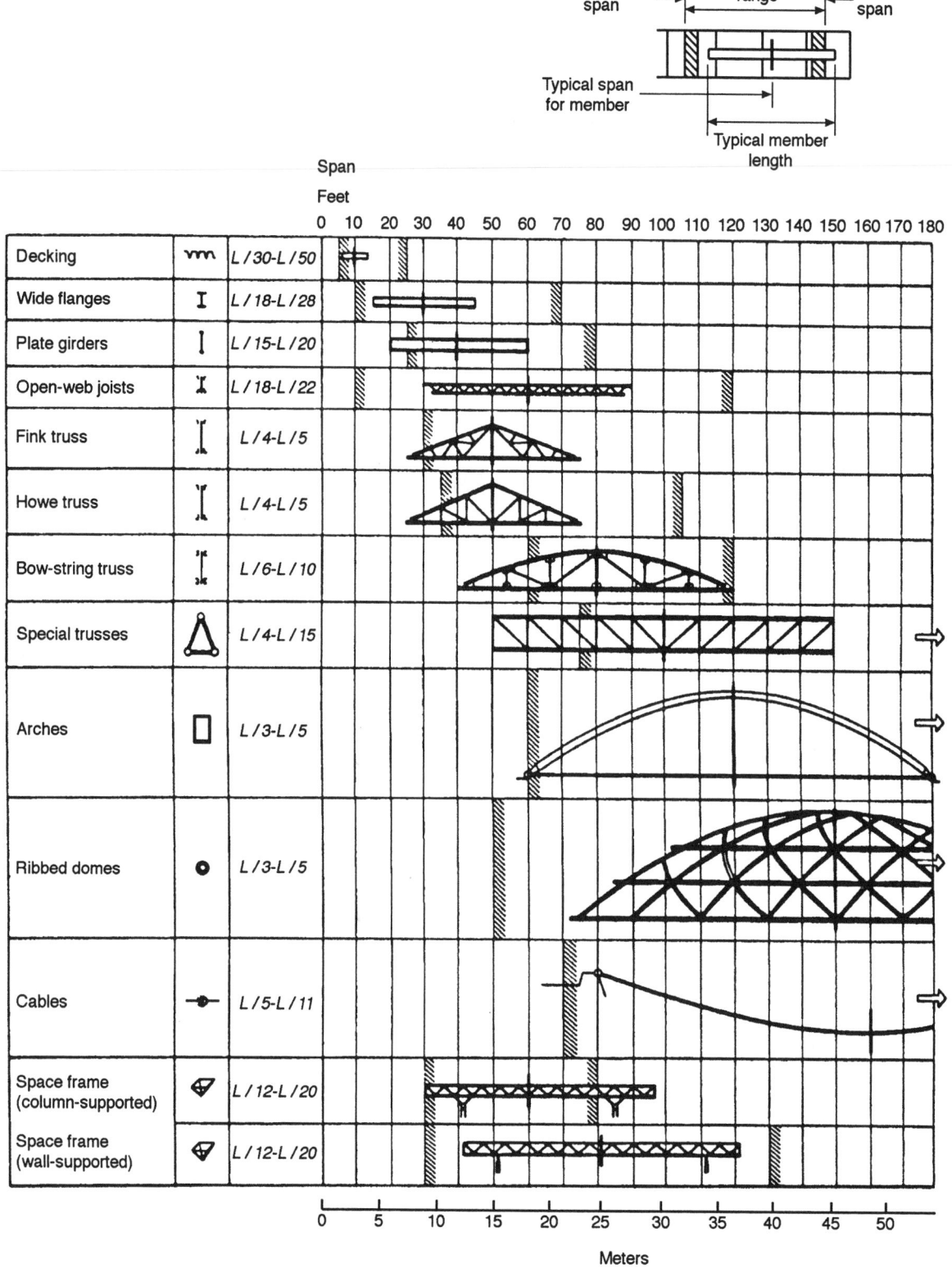

Approximate span ranges for steel systems. Source: Schodek (1992).

STRUCTURAL SPANS: reinforced concrete

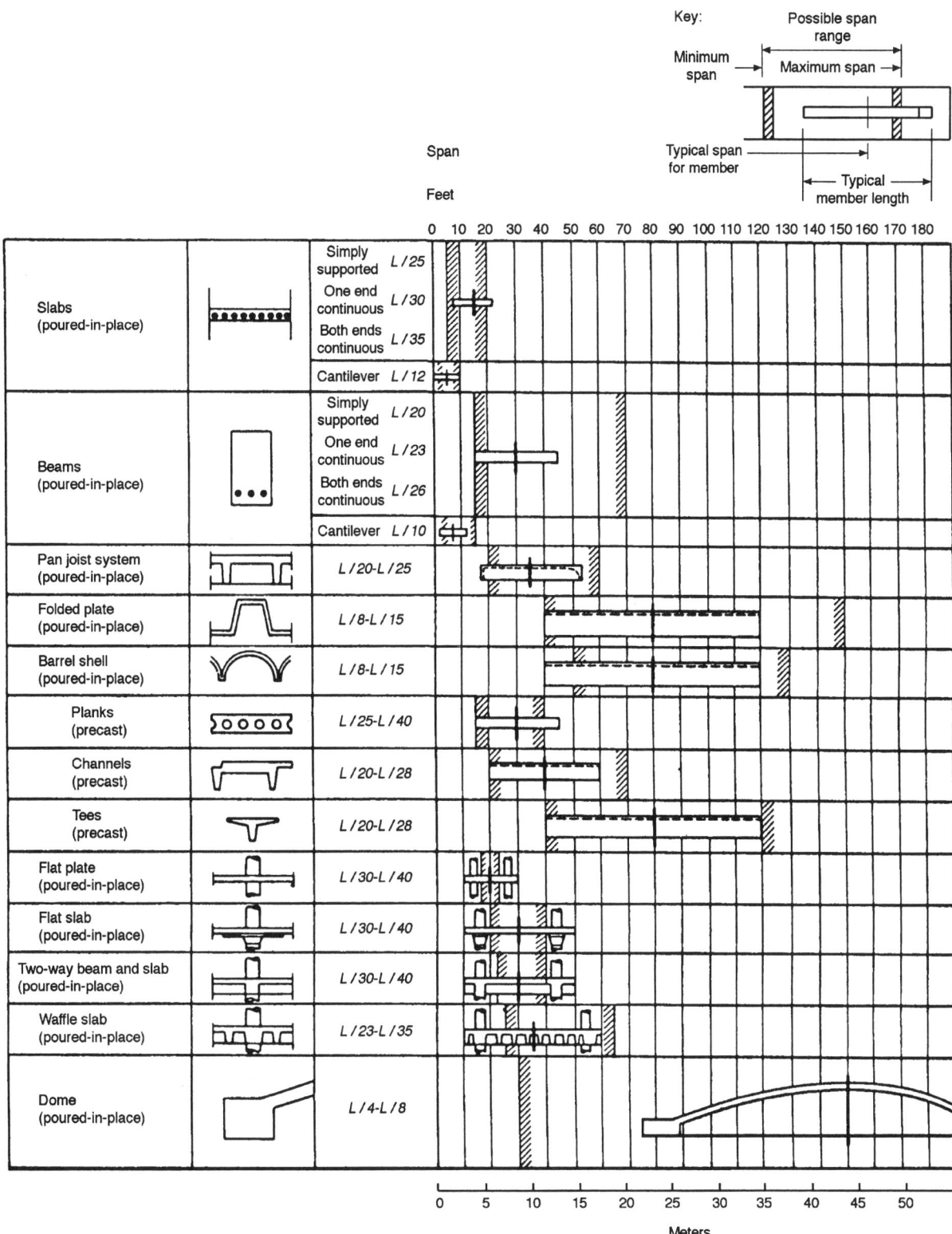

Approximate span for reinforced concrete systems. Source: Schodek (1992).

B1.1 STRUCTURE

SITE ASSEMBLED: pre-engineered

METAL BUILDING STRUCTURES

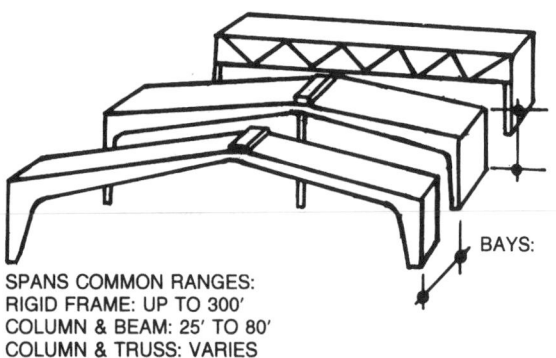

SPANS COMMON RANGES:
RIGID FRAME: UP TO 300'
COLUMN & BEAM: 25' TO 80'
COLUMN & TRUSS: VARIES

BAYS:

Metal buildings are commonly available as a complete envelope, i.e., a structural steel frame with a skin of roll-formed and prefinished metal roof and wall panels:
- secondary components of the envelope, such as doors, windows, louvers, roof curbs, gutters, canopies are generally available from the manufacturer of the system.
- roof and wall panels may incorporate insulation, or insulation is added during erection as backup to single thickness panels.
- framing systems commonly available: column and beam, column and truss, or rigid frame, with secondary framing between them supporting the panels.
- envelopes are pre-engineered for specific live and wind loads, which should be checked against requirements of local building codes.

GLASS AND PLASTIC STRUCTURES

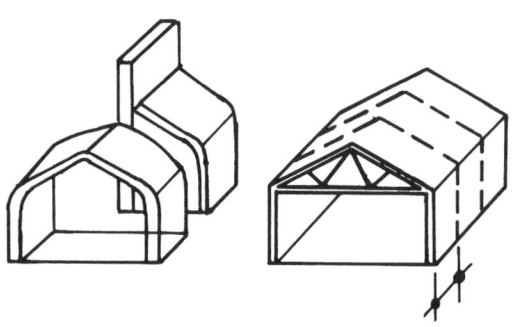

Greenhouses, sun rooms, framed skylights available as complete envelopes:
- framing: domes, arches, rigid frames, half-arches, trusses and columns; of steel, aluminum, laminated lumber.
- glazing: single light of tempered glass; single light of laminated glass; insulating panels of laminated glass, double or triple glazed; double-skinned plastic extruded or laminated; insulating panels of double-skinned plastic and laminated glass.
- secondary component options may include: doors, operable sash, louvers, shades, blinds, ventilators.
- envelopes are pre-engineered for specific live and wind loads which should be checked against requirements of local building codes.

DOME STRUCTURES

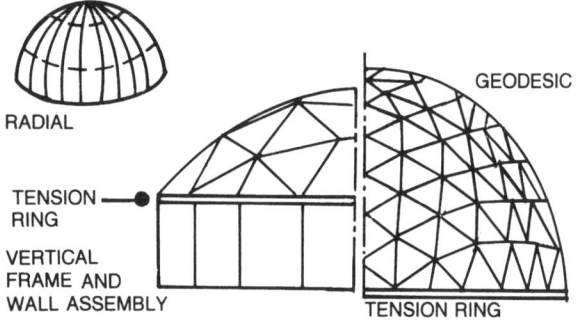

RADIAL

TENSION RING

VERTICAL FRAME AND WALL ASSEMBLY

GEODESIC

TENSION RING

Domed envelopes are available as:
- triangulated or geodetic of: dimensional lumber frames and plywood skin, laminated lumber frames with solid or laminated wood decking, aluminum space frames with metal faced panel or translucent plastic skin.
- sizes range from shallow skylights through multiple interconnecting units to envelopes enclosing stadia with independently supported structures for spectator seating.
- secondary component options may include skylights, doors, wall assemblies, insulation.
- conformance with requirements of local building codes must be ascertained.

MODULES

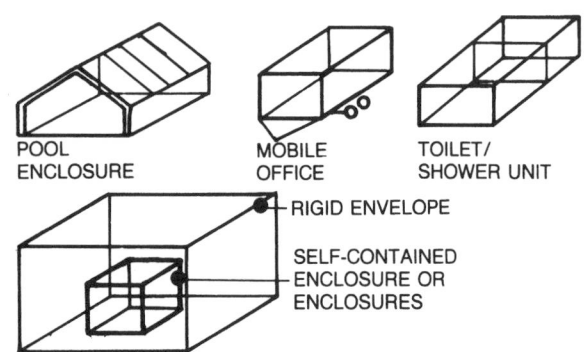

POOL ENCLOSURE

MOBILE OFFICE

TOILET/ SHOWER UNIT

RIGID ENVELOPE

SELF-CONTAINED ENCLOSURE OR ENCLOSURES

Modules are specific function enclosures: partially or completely preassembled; exposed to elements or protected within larger enclosures; fixed in place, relocatable, movable:
- complete office units, pool enclosures, guard houses, toll booths, toilet/shower units.
- complete units for: sound transmission resistant rooms, such as audiological, audiometric, music practice, recording; clean rooms; saunas; walk-in coolers/freezers; paint spray booths.
- mechanical/electrical services generally included or available as an option.
- conformance with requirements of local building codes should be ascertained.

FABRICATED ENCLOSURES: selection checklist

FOUNDATIONS

Fabricated envelopes are available as superstructures only, to be erected by manufacturer on prepared foundations: the manufacturer will generally only provide design loads to be supported by foundations, and anchor bolt location drawings.

SUSPENDED LOADS

Extent and/or location of suspended loads such as hoists, cranes, heavy mechanical equipment is generally limited:

- metal building systems such as column and truss frames are available specifically designed for heavy suspended loads.
- location of heavy suspended loads in dome structures generally limited to center or crown of dome.

MECHANICAL/ELECTRICAL SERVICES

Metal buildings, glass/plastic structures, and dome structures generally are supplied without any mechanical/ electrical services:

- required opening through the skin for such services such as roof curbs for exhaust fans will be provided when size and location is given to manufacturer prior to fabrication.
- design of structural frame generally includes an allowance for loads of suspended mechanical/ electrical service such as lighting fixtures, small diameter piping.

METAL BUILDINGS

Choice of framing system will primarily depend on functional requirements of enclosed space:

- when space is to be subdivided, column and beam framing should be considered: with spans kept relatively short it is the most economical system.
- for column free interior space, rigid frames may be selected; and when heavy suspended loads must also be supported, columns and truss systems may be the choice.
- bay spacing generally limited to manufacturer's standard.

Roof/wall panels are formed in various configurations, prefinished and/ or corrosion resistant or treated to resist corrosion:

- single thickness only when control of heat flow not required.
- single thickness with mineral fiber, generally glass fiber, insulation and interior plastic vapor retarder: economical system but vapor retarder exposed to potential damage.
- single thickness with exposed rigid insulation backup.
- site assembled sandwich with semi-rigid insulation between exterior and interior metal panels: interior panel may serve as finished wall.
- laminated panel of rigid foamed insulation faced with flat or formed metal panels: may be restricted by local building code when noncombustible materials only are permitted.
- when evaluating thermal resistance, the entire assembly, and not the panel only, should be considered: methods of securing panels in place may create thermal bridges thus minimizing effectiveness of insulation.

Other considerations may include:
- vehicular doors are generally not supplied by manufacturers of metal buildings: a framed opening only will be provided.
- NFPA requires lightning protection for metal-skinned buildings.

GLASS and PLASTIC STRUCTURES

Selection considerations should include:

- means to control daylight and/or solar heat gain/loss: insulating glazing, tinted glazing, shades, blinds.
- control of condensation: insulating glazing, vents either gravity or mechanical, operable sash, condensation gutters, corrosion resistant frames.
- protection against breakage: tempered glass, laminated glass, impact resistant plastic.
- resistance to rain penetration: glazing methods, flashing, runoff control.

DOME STRUCTURES

Types available include:

- ■ triangulated metal frame domes, since they work as diaphragms, are

ideally suited to withstand unusual loading conditions:

- the geodesic dome structure produces exceptional rigidity from its combination of many, relatively short structural members.
- the lamella dome, while it is also triangulated, is not as material efficient as the geodesic, since it utilizes longer, heavier structural members.

■ the radial arched dome is not triangulated, but rather a system of arches intersecting at midspan and reinforced with "hoops" around its circumference. Since its structural design is more conventional, consideration of lateral loading is more important.

These structural differences also affect their dimensional limitations:

■ the triangulated metal dome can be framed with members of varying lengths which are connected to form a uniform surface:

- the single layer dome utilizes structural elements solely at the dome surface and can span up to 150 feet.
- the double layer dome utilizes a framing layer within the sphere with web members connecting to the outer layer. The structure resembles that of a sphere frame and is used for long spans over 120 feet.
- geodesic domes on round, pentagonal and hexagonal bases with spans well over 200 feet are commercially available.

■ the radial arched dome is formed by intersecting arches of the same diameter and is more limited in clear span capability - 50 feet maximum. In general though, larger sizes do exist.

Fabricated envelopes should be evaluated as complete systems, with selection limited to standard options offered by the manufacturer of a specific system:

- modification to standards will affect cost-effectiveness.
- introduction of nonstandard components may affect performance of the total envelope.

B1.1 STRUCTURE

SITE ASSEMBLED: prefabricated

NONPRESTRESSED - SUSPENDED

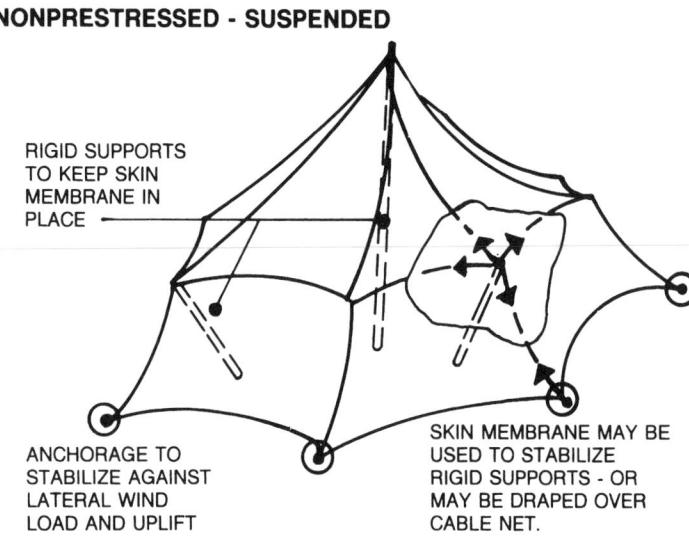

RIGID SUPPORTS
TO KEEP SKIN
MEMBRANE IN
PLACE

ANCHORAGE TO
STABILIZE AGAINST
LATERAL WIND
LOAD AND UPLIFT

SKIN MEMBRANE MAY BE
USED TO STABILIZE
RIGID SUPPORTS - OR
MAY BE DRAPED OVER
CABLE NET.

PRESTRESSED - PLANE

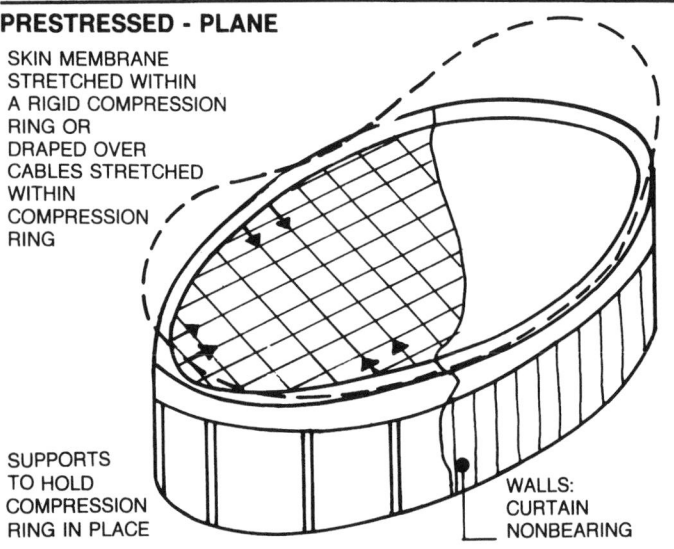

SKIN MEMBRANE
STRETCHED WITHIN
A RIGID COMPRESSION
RING OR
DRAPED OVER
CABLES STRETCHED
WITHIN
COMPRESSION
RING

SUPPORTS
TO HOLD
COMPRESSION
RING IN PLACE

WALLS:
CURTAIN
NONBEARING

PRESTRESSED - CUSHION

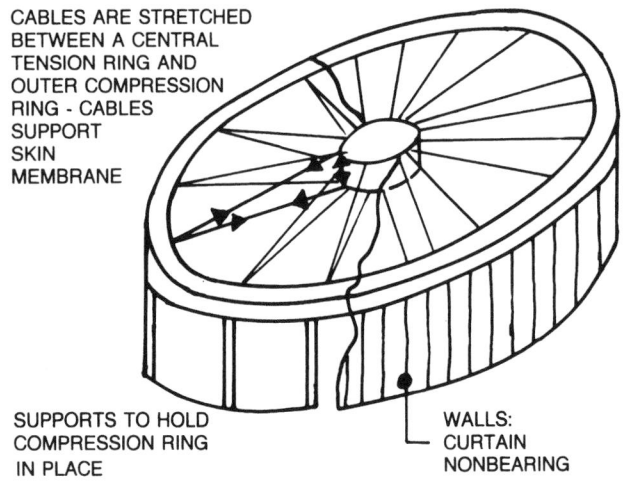

CABLES ARE STRETCHED
BETWEEN A CENTRAL
TENSION RING AND
OUTER COMPRESSION
RING - CABLES
SUPPORT
SKIN
MEMBRANE

SUPPORTS TO HOLD
COMPRESSION RING
IN PLACE

WALLS:
CURTAIN
NONBEARING

NONPRESTRESSED suspended enclosures generally consist of a flexible skin, commonly referred to as the membrane, suspended from rigid supports which hold it in the shape intended; the membrane may also be supported on a net of cables suspended from rigid supports.

* edge stresses in suspended membranes may be transmitted to points of suspension and/or anchorage by edge cables.

* membranes may be suspended using diverse methods, such as: central supports; between parallel rigid supports, either of individual posts or continuous; from a grid of individual posts of the same or of varying height with equal or varying spans between such posts; within compression rings. When a membrane suspended within a compression ring is further stretched to stabilize it against wind uplift, it then becomes a prestressed membrane. The basic difference between a nonprestressed tension-loaded membrane or cable and a prestressed one is that the former is given its shape only by the dead load, while the latter has the shape of a straight line even in weightless state, and only becomes catenary with dead load applied to it, or when deformed intentionally.

PRESTRESSED membranes may be constructed in a diversity of shapes, but all require:

* secure anchorage to allow for prestressing and to maintain the entire membrane in tension whether under gravity load alone or combined lateral/uplift forces and gravity loads.

A few of the more common shapes of prestressed membranes are:

* Plane surface which is akin to a drumhead, with the membrane stretched within a ring rigid in compression. The membrane may also be nonprestressed and suspended, or nonprestressed and air-supported.

* Curved saddle-shaped anticlastic surfaces which result when the compression ring of a plane surface membrane is raised at two points opposite each other. Since the membrane will tend to pull the high points of the ring together, bending stresses will develop in the ring unless additional supports are provided.

* Centrally supported membranes in a variety of configurations: the stresses in the membrane at the point of support generally control; cable nets may be required for larger enclosures.

* Cushion enclosures which result when two plane parallel membranes are stretched within a plane compression ring, and then are forced apart in the middle. Such self-contained structures require no force equilibration by the foundation: it can be used without anchoring in a horizontal, inclined, or vertical position. The spoked wheel of a bicycle is a well-known cushion structure.

B1.1 STRUCTURE

SITE ERECTED: fabricated

AIR SUPPORTED - UNREINFORCED

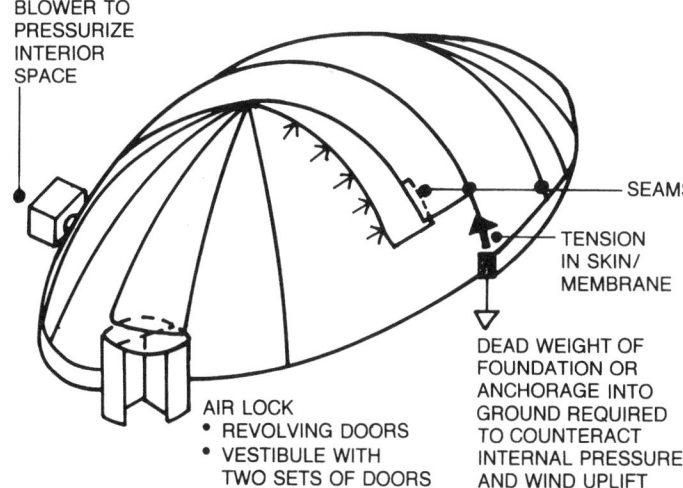

BLOWER TO PRESSURIZE INTERIOR SPACE

SEAMS

TENSION IN SKIN/ MEMBRANE

AIR LOCK
• REVOLVING DOORS
• VESTIBULE WITH TWO SETS OF DOORS

DEAD WEIGHT OF FOUNDATION OR ANCHORAGE INTO GROUND REQUIRED TO COUNTERACT INTERNAL PRESSURE AND WIND UPLIFT

AIR SUPPORTED - REINFORCED

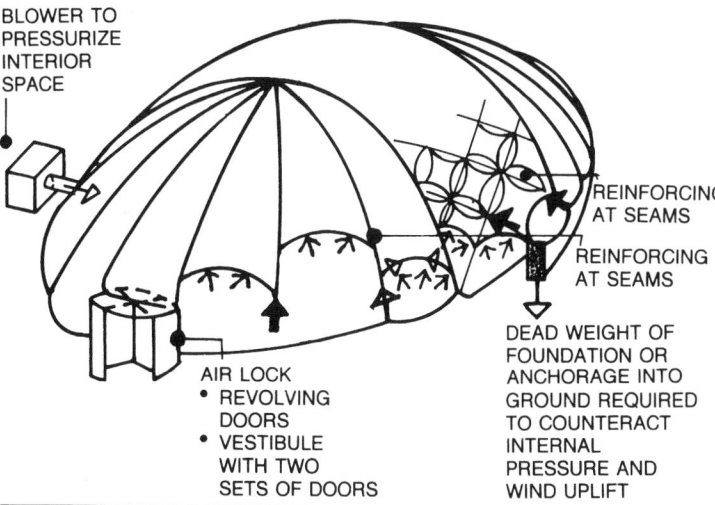

BLOWER TO PRESSURIZE INTERIOR SPACE

REINFORCING AT SEAMS

REINFORCING AT SEAMS

AIR LOCK
• REVOLVING DOORS
• VESTIBULE WITH TWO SETS OF DOORS

DEAD WEIGHT OF FOUNDATION OR ANCHORAGE INTO GROUND REQUIRED TO COUNTERACT INTERNAL PRESSURE AND WIND UPLIFT

AIR INFLATED

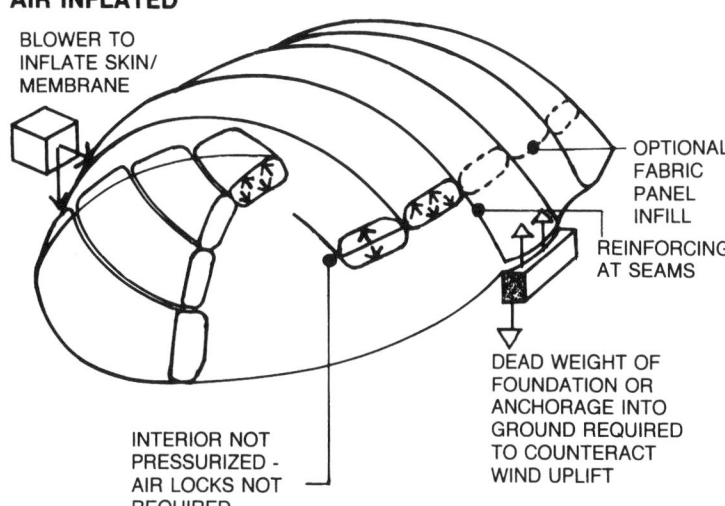

BLOWER TO INFLATE SKIN/ MEMBRANE

OPTIONAL FABRIC PANEL INFILL

REINFORCING AT SEAMS

INTERIOR NOT PRESSURIZED - AIR LOCKS NOT REQUIRED

DEAD WEIGHT OF FOUNDATION OR ANCHORAGE INTO GROUND REQUIRED TO COUNTERACT WIND UPLIFT

AIR-SUPPORTED ENCLOSURES: comprised of a flexible envelope which encloses a pressurized, occupied space. Two categories exist:
• unreinforced membrane structures in which the membrane is the primary structural component
• reinforced membrane structures, where a network of webbing or cables acts as the primary structural system and the membrane spans only between the cables.

In air-supported structures, in the absence of load the membrane is under internal pressure which causes stresses in it.

■ With uniform snow load, stress decreases and at one point theoretically vanishes.

■ With nonuniform moving loads, e.g. wind loads, membrane stresses are unavoidable.

■ Being generally very flexible, air-supported structures are subject to large displacements under vertical (snow) and lateral (wind) loads.

■ High-profile shapes have the following advantages:
• a more nearly uniform distribution of stresses in membrane.
• snow slides off the steep slopes which also helps prevent ponding of water.

■ On the other hand, high-profile shapes are:
• more subject to wind deformation.
• when snow slides off part, but remains on part, nonuniform loads are created which can lead to serious damage.
• are subject to positive wind stresses in part, whereas low-profile membranes are almost entirely in suction under wind action.

■ High winds are also countered by raising interior pressures to the design maximum and thereby reducing deformations and resulting stresses.

AIR-INFLATED ENVELOPES: in which self-enclosed membranes are inflated to form the structural system. The enclosed space is not pressurized. Two variations exist:
• low-pressure in which the inflated elements are continuous over the entire enclosure.
• high-pressure, where the inflated elements are spaced apart with the spaces between infilled by an enclosing membrane.

■ Some advantages are:
• neither airlocks nor a tight perimeter air seal are needed and normal ventilation prevails.
• a wider variety of shapes and sizes can be obtained; design flexibility is infinitely greater.

■ Some of the problems include:
• due to the higher pressure required to sustain the superimposed load, stresses in the membrane are much higher and so is required strength of fabric and seams. For this reason reinforced membranes are the rule.
• if damage to membrane occurs, or pressure fails, the structure may collapse considerably faster than an air-supported enclosure which may take days to deflate.
• heating within the structures to prevent snow accumulation is still needed.

HYBRID enclosures which combine elements of the air-supported and air-inflated types, also exist, e.g., there is a double-skin inflated membrane and the space is also pressurized.

B1.1 STRUCTURE

SITE ERECTED/FABRICATED: selection checklist

AIR-SUPPORTED ENCLOSURES are the type most widely used and commercially available. Air-supported enclosures might be considered when:

■ Coverage of very large spans at minimum installation cost and time is the problem, especially where:

- anticipated life span of enclosure is not very long, say up to 20-25 years.
- potential for easy relocation is important.
- interior environmental control requirements not too stringent.
- ultralight weight is an important asset - spans of 300 feet at .8 pound/square foot - for transportation over long distances.
- total factory prefabrication is desired, e.g., for locations where no local material, labor or equipment are available.

Principal characteristics of air-supported enclosures to be considered during selection and preliminary design:

SHAPE

■ The most popular shape of air-supported structures to date has been the **cylindrical segment.** Most of the "standard" enclosures are of this type, but the end treatment varies:

- ends are essentially portions of spheres.
- others are cylindrical or even rectangular.
- the fabric seams and the cables for reinforced membranes at ends are often radial, sometimes annular.

■ **Variants on these basic shapes** are many and are most commonly obtained by:

- combining two or more shapes of equal size or of varying sizes under a connecting membrane and reinforcing cable net.
- by combining air-supported membranes with rigid bases, endwalls or shapes.

■ **Custom-designed air-supported structures,** though not unlimited in the shapes feasible, are almost limitless in plan and massing variations.

■ In general, the most ecomonical form of air-supported structures for a given enclosed volume is the **dome.**

■ **Shapes should be selected** for distribution of stresses in the membrane under full pressure to be as uniform as possible.

■ All but the most basic shapes are generally constructed with the use of **reinforcing cable nets,** whether integral with the membrane or separate from it.

REINFORCING

Reinforcing webs, cables and cable nets are used in all larger enclosures with the membrane spanning only between cables.

■ **Cable nets** are used in various configurations, e.g., square, diamond and polygonal meshes and are especially effective when net is uniform mesh and intermediate membrane sections have similar radii of curvature.

■ **When membrane is pushed against cable net, shape of net is little affected** and cables form nearly straight lines between sections - an important factor in preventing accumulation of water

■ **Properties to look for in reinforcing cables,** in addition to tensile strength, include:

- resistance to abrasion.
- good fit with membrane in extensional stiffeners, expansion and contraction characteristics.
- environmental resistance, either inherent or through protective coatings or jackets.

■ **Reinforcing cables and cable nets may be:**

- independent of the membrane attached to it, e.g., inserted into sleeves.
- integrated into membrane seams, mostly webbing.

■ **Attached reinforcement offers advantages in erection and relocation,** but may not be always practicable, especially in very large structures.

MEMBRANE

■ **Presently available coated and laminated membrane materials provide:**

- life spans of 20 years and better service temperature ranges of -60° to 145°F.

- loss of tensile strength of less than 10 percent after 10 years.

■ **Materials commonly used are:**

- Glass fiber fabric coated with elastomeric resins; life expectancy up to 20 years and over depending on resin used; light transmission 5 to 25 percent.
- Polyester fabric coated with elastomeric resins: life expectancy over 10 years depending on type of resin used.
- Nylon fabric.
- Canvas.
- Multiple interior liner-layers for improved thermal and acoustic performance may be available.

■ **In considering the membrane,** it should be remembered that the seam strength and fire resistivity will determine membrane performance.

The methods in use include:

- sewing - double and triple seams with/without tape or web reinforcement used mostly in smaller enclosures.
- mechanical connectors - may have excellent strength, but present air-loss problems.
- cementing.
- heat sealing.

The last two methods are most frequently used on larger structures; a combination of methods may be also considered.

■ **Majority of failures are due to damage resulting from tearing of membrane.** Therefore it is important that:

- tear should not propagate if membrane is damaged while under maximum working stress
- this raises membrane strength requirements since tear-propagation resistance is generally estimated at 10-15 percent of tensile strength.

■ **Fire resistance of envelope membrane will be generally tested according to NFPA 701,** but it must be remembered that it is the fire resistance of the seam that may determine performance.

PRESSURIZATION

Interior pressures must be maintained higher than the outside air pressure:

■ The space being large in relation to envelope, pressure differential

B1.1 STRUCTURE

required is low: 0.65 to 1 inch water gauge for normal operation.

- continuous anchorage of membrane is necessary to minimize air infiltration.
- vents for natural ventilation may be provided.
- fire vents should be provided.

■ **Blowers are required to inflate structure and maintain required pressure levels in:**

- normal conditions.
- wind and snow conditions.
- olowers should be equipped with back-draft dampers to minimize air loss when not in operation.
- 100 percent blower overcapacity generally required.

■ **Controls commonly provided are:**

- integral pressure limits at the design pressure to prevent over-pressurization.
- automatic activation of auxiliary blowers with loss of pressure.
- centralized controls, manual or computerized available.

■ **Air locks are required at all points of entry and exit.**

- revolving doors used for general egress/ingress.
- emergency exits: - counterpressure balanced, self-closing with panic hardware.
- vehicular air locks consist of two overhead doors separated by a space and covered by canopy, normally kept closed, never opened simultaneously.

■ **Emergency power is generally required for all public assembly enclosures,** and all but smallest other enclosures.

- consists of:
auxiliary engine generator powering pressurization equipment
or
supplementary blower with independent on-site power source
or
auxiliary power line from supplementary power source.

■ **Safety:** if structure is damaged or air supply fails the enclosure deflates very slowly, may take days to collapse, thus posing no threat to life.

ENVIRONMENTAL

Environmental performance of an air-supported enclosure depends largely on the materials and construction of the enveloping membrane.

■ **A range of optical properties can be obtained even with a single monolithic membrane, including:**

- opaque membranes, best where cooling is important.
- opaque to solar but transparent to infrared to permit heat to escape.
- light-admitting membranes, ranging from 1 percent to 6 percent transmissibility, sufficient for daylighting in many cases to 98 percent transmissibility in some films where needed.

■ **Thermal and acoustic performance of single monolithic membrane is poor.**
U = approximately 1.2.

■ **Thermal performance can be improved up to U = .35 - .25 and better through**

- coated and/or laminated integrally reinforced membranes.
- multilayer construction which traps air between layers.
- low-emissivity outer layer materials, e.g., metallic aluminum foil.
- light-colored, reflective surfaces.

■ **Acoustic performance can be improved by:**

- convoluted shaping of enclosure or of inner liner.
- inner liners with absorptive coefficients of 0.1 to 0.65.

Weathering characteristics of enveloping membrane are critical to life span of enclosure. The major aspects are:

- shrinkage, embrittlement due to loss of plasticizer.
- resistance to:
ultra-violet radiation
fungal and bacterial attack
temperature extremes and also
color change, discoloration.

HEATING

■ Heating may be necessary to prevent snow accumulation and water ponding and/or cracking of fabric.

- snow detectors may be used to activate system to levels needed.
- system may be independent, combined with air conditioning and/or with pressurization.

- heaters for maintaining fuel and lubricant temperatures required available in case of power failure.

LIGHTING

■ **Lighting is generally independent of envelope:**

- placed on floor, straight or inclined poles.
- care is required to prevent fixtures from damaging envelope in case of deflation.

OTHER CONSIDERATIONS

In the selection of prefabricated, flexible enclosures the following should be kept in mind:

■ Standardized pre-engineered enclosures should be utilized as available without any modifications, or with the most minor ones when unavoidable.

■ Ignoring the above principle may result in either or both:

- loss of advantages in cost and quality.
- Problems with performance over time, and with its maintenance.

■ Code requirements should always be checked, since, depending on location, they may:

- preclude the use of a certain type of enclosure.
- require provisions that will render it uneconomical.

EARTHQUAKE RESISTANT DESIGN

EARTHQUAKE RESISTANT DESIGN

Earthquakes—causes and effects

The crust and upper mantle of the earth are made up of **rigid plates** (or segments of the lithosphere). The plates slide over the interior of the earth slowly, continuously and independently. Plate motion is thought to create earthquakes, volcanoes and other geologic phenomena.

■ **Geological faults** are vertical plane intersections along which plate motion takes place and are the source of the ground shaking characteristic of an earthquake. Fault types are illustrated on this page.

Four basic causes of earthquake-induced damage:

■ **Ground rupture in fault zones.** An earthquake may or may not produce ground rupture along the fault zone. A structure directly astride such a break will be severely damaged.

■ **Ground failure.** Earthquake-induced ground failure has been observed as landslides, settlement and liquefaction. Ground failures are particularly damaging to building infrastructure such as water lines, sewers, gas mains, communication lines, and transportation facilities.

■ **Tsunami** is a seismic sea wave produced by abrupt movement of landmasses on the ocean floor.

■ **Ground Shaking.** The effect of ground shaking on structures is the principal consideration in earthquake resistant design. Induced vibrations and displacements can destroy a structure unless designed and constructed to be earthquake resistant. Reasonable design practices can mitigate life safety hazards under earthquake conditions.

Ground shaking factors. Earthquake location and depth of focus are significant factors in ground shaking.

• **Short-period ground motions** tend to die out more rapidly with dis-

tance than do longer period motions.

• **Long-period vibrations** coincide with the longer natural periods of vibration of tall structures, causing *resonance*, which has a significant damaging effect on buildings.

EFFECTS OF EARTHQUAKES ON STRUCTURES.

Earthquake forces in structures result from the erratic omnidirectional motion of the ground.

■ **Ground motions** are described in terms of acceleration, velocity, and displacement of the ground at a particular location. These vary with time as the ground vibrates. The building should be able to undergo extended periods of ground shaking without failure.

• Structures that are fixed to the ground in a rigid manner respond to the ground motions. As the base of the structure moves, the upper portions tend to lag behind due to inertia. The resultant force (F) is equal to mass (M) times acceleration (A). The higher the acceleration, the greater the resultant forces on the structure.

• The building acts as a pendulum with respect to the ground. The rate and frequency of the swing (*i.e.*, the swaying) is a function of building height, mass, and cross-section.

■ **Relation of wave motion to structural behavior.** The rate of oscillation, or "natural period" of a structure, is an important factor because earthquakes do not result in ground movement in only one direction, as in the example above.

• **Complex deflections** may result as the building vibrates in all its modes of vibration in response to ground motion.

METHODS OF DEALING WITH THE EARTHQUAKE FORCES.

The way the structure absorbs or transfers energy of an earthquake will determine the success or failure

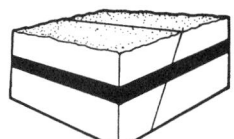
Fig. 1a. Quiescent fault

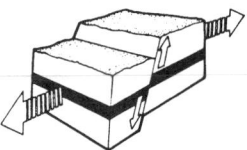
Fig. 1b. Normal fault

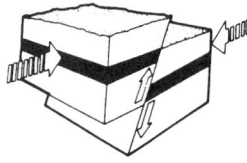
Fig. 1c. Thrust or reverse fault

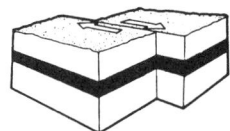
Fig. 1d. Lateral slip, strike slip or transform fault

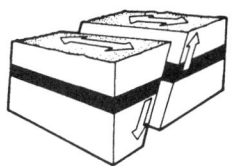
Fig. 1e. Normal and slip fault combination

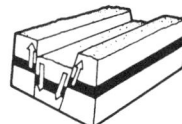
Fig. 1f. Graben

Fig. 1g. Horst

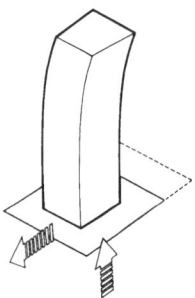
Fig. 2 The building should be able to undergo extended periods of ground shaking without failure.

B1.1 STRUCTURE

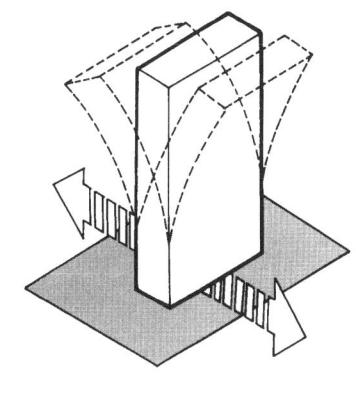

Pendulum action

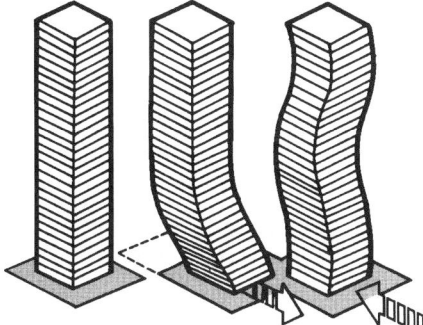

Fig. 3 Effects of cyclic reversals on ground acceleration. At the same time that the upper part of the structure begins to move to catch up with the initial displacement, the ground motion reverses itself.

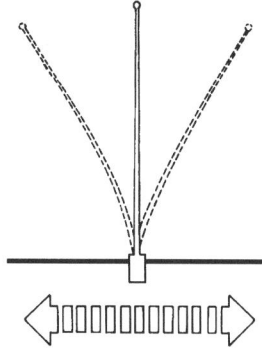

Fig. 4 A. Flexibility illustrated by a thin flagpole that can sway considerably without fracture or permanent displacement.

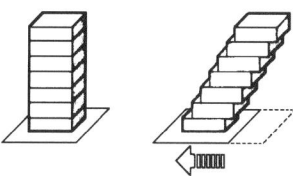

Fig. 4B. The opposite situation is represented by a stack of unreinforced bricks whose movement results in permanent displacement of each brick when a horizontal force is applied.

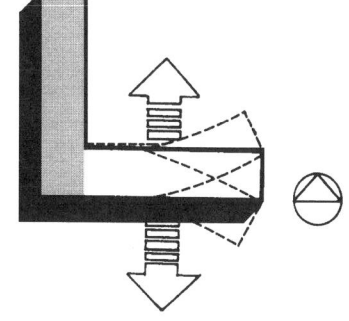

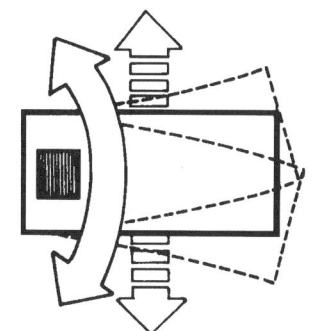

Fig. 5A Stiffness of structure related to building plan

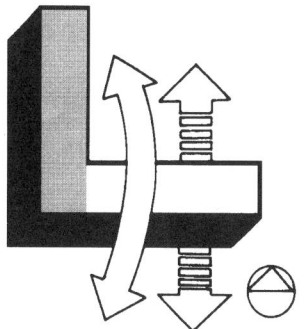

Fig. 5B Torsion effects on building plan.

of the building's seismic resistant design and construction.

■ **The principle of base isolation** is to reduce the lateral accelerations and velocities that a building's structural system will experience by lengthening the period (of the building) and allowing for increased displacements.

■ **Tuned mass dampeners** are designed to reduce the lateral displacement of a building moving in its first mode of vibration. This system has been used primarily to increase human comfort in tall buildings subjected to wind loads.

Impact of architectural form on stiffness and flexibility.

Nearly all buildings combine elements that are "flexible" with elements that are "stiff." The improper combination of such elements may create problems under earthquake loading.

Effect of building shape on response to seismic forces.

The choice of basic plan shape and configuration is a critical decision. Earthquake forces can come from any and all directions and act upon the building virtually simultaneously. The theoretical "best choice" is a building that is symmetrical in plan and elevation, equally capable of withstanding forces imposed from any direction.

• Given the practical constraints such as shape of site and functional requirements, the designer must understand how variations in plan and elevation can affect performance.

• Consider a building with an irregular plan shape, such as an "L" or "T" configuration. The wings might experience different movements depending upon their orientation relative to the direction of earthquake force. Unless designed with adequate capacity to absorb and dissipate the forces, the structure can suffer greater damage, particularly where wings connect.

■ **Torsion effects.** Torsion is the result of rotation of an eccentric or a less rigid mass about the base or the

EARTHQUAKE RESISTANT DESIGN (cont.)

more rigid mass of the building. Under earthquake motion, it can cause rotation of the mass of an E-W wing relative to the mass of a N-S wing.

■ **Effect on building systems.** Most buildings are designed with a mixture of stiff and flexible components. Some combinations if used unwisely may cause serious damage and collapse of structures, e.g.:

The "open first floor" concept commonly used today—placement of a rigid upper structure on a flexible column system—exemplifies this problem. The flexible columns are expected to resist exaggerated and concentrated forces, yet may not be designed to take these loads.

• A similar problem is created when the designer inadvertently weakens a stiff wall (shear wall) with many openings.

■ **Materials.** Different materials react differently with respect to inelastic behavior, as a function of their ductility.

• Ductility refers to the ability of a material to absorb energy while undergoing inelastic deformation without failure, particularly when the direction of the forces involved changes several times.

• Ductile building systems include steel frames, ductile concrete frames and wood diaphragm construction. Where the connections of the system used are ductile and numerous, the overall performance is improved considerably.

• Ductility can be thought of as providing a quality of toughness, which, to a large extent, determines a building's survival under seismic conditions.

■ **Architectural design concept.** The shape chosen by the designer for the structure will determine its response to seismic forces, including:

• The development of torsion effects as well as differential movements of parts of the building.

• The extent of glazing, the number of glazed facades, the size of spandrel elements, and the location of the exterior column line are among the architectural design factors which directly affect a building's seismic performance.

• Cantilevered balconies, parapets, railings, sunshades, statues, signs and planters must be structurally designed with sufficient capability to resist seismic forces.

■ **Critical need to tie the structural system together.** Since seismic forces affect all elements, the building must act as a unit to resist these forces. If the structure is not tied together to respond as a unit, separate elements or components of the building will respond individually and failure can occur beginning at the weakest element or component. Typical connection conditions which can fail, include:

• use of brittle rather than ductile connections,

• spacing of fasteners at too close intervals so that connecting members fail.

• reinforcement bars may not be adequately anchored or spliced to develop the full strength of the connection.

• in masonry construction, if the floor systems are not properly tied to the walls, under seismic forces the walls may move independent of the floors causing either the walls to fail or the floors to drop.

■ **Dissimilarity of wind and earthquake loads.** There is fundamental difference in the way in which lateral loads are transmitted to a building from earthquake and wind. In the case of earthquake, the load is transmitted to the building from its base. Thus, the entire building as well as the building contents will experience the force.

■ **Building drift.** The horizontal displacement of basic building elements is usually most critical to nonstructural

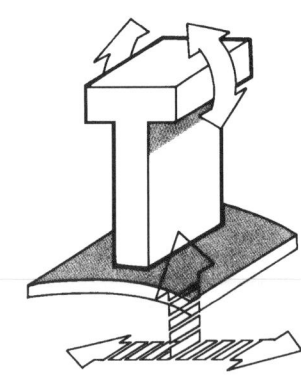

Fig. 6 Oblique view of vertical torsion effect.

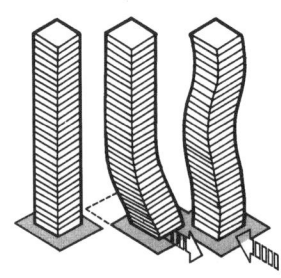

Fig. 7A Some floors of the building tend to move in one direction while floors above or below these tend to move in the opposite direction in a relatively tall building.

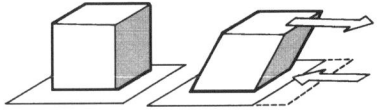

Fig. 7B Drift diagram showing lateral displacement and resulting foreshortening.

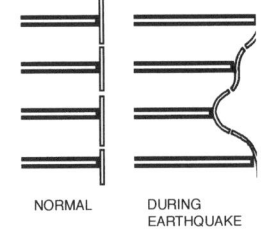

NORMAL   DURING EARTHQUAKE

SIMPLE SPAN CURTAIN WALL

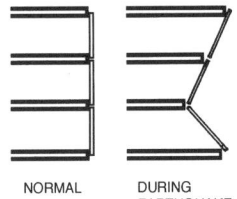

NORMAL   DURING EARTHQUAKE

Fig. 8 Effect of cantilevered exterior walls vs. simple span. The exterior curtain wall that is anchored at each floor slab and is cantilevered both up and down can be severely affected.

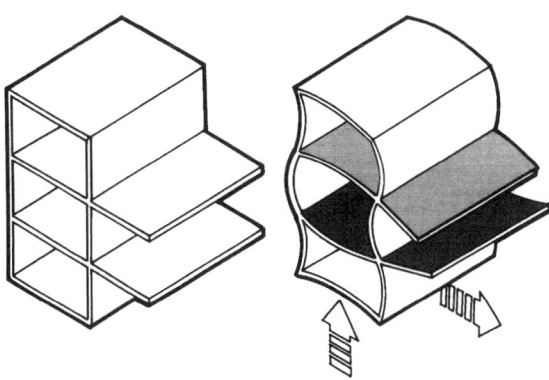
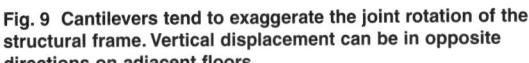

Fig. 9 Cantilevers tend to exaggerate the joint rotation of the structural frame. Vertical displacement can be in opposite directions on adjacent floors.

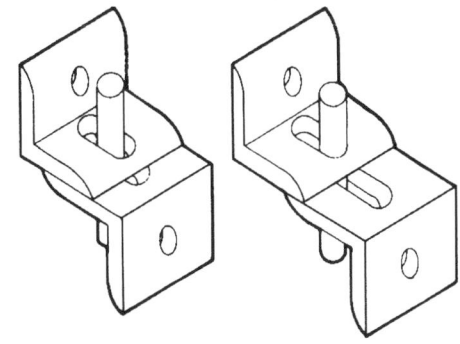

Fig. 10 Connection of double-clip angles. The extended position is the critical condition when subject to stresses.

components. All floors do not drift at the same rate or time, and this action causes a horizontal displacement between floors. This differential movement between floors can and does affect all full-floor height elements of a building

The accumulation of drift affects only those nonstructural components that are continuous over more than one floor. Even here, the effect is dependent upon the detailing of the component. Unless properly designed, the imposed racking of the elements can result in major failures of the wall system.

• **Effect of cantilevered exterior walls vs. simple span:** The exterior curtain wall that is anchored at each floor slab and is cantilevered both up and down can be severely affected.

• **Cantilevers** tend to exaggerate the joint rotation of the structural frame. Vertical displacement can be in opposite directions on adjacent floors.

• **Simple shearing or racking action** due to drift can be imposed on all floor-to-floor and some floor-to-ceiling components by the differential lateral movement between adjacent floor systems.

■ **Building torsion** usually brought about by the eccentric lateral resistance or mass of the basic structure causes the building to twist vertically. Torsion in a building sometimes results

from the stiffness of rigid or massive nonstructural components such as in-fill walls. The basic effects of torsion on components are similar to drift.

■ **Displacement of cantilevered members.** Due to their unique nature, cantilevers tend to exaggerate the joint rotation of the structural frame. Under seismic loading cantilevers must receive special consideration. The unrestrained end condition can result in vertical displacement of considerable magnitude and can create a significant hazard to life safety because of glass breakage and falling wall elements.

IMPORTANCE OF CONNECTIONS AND FASTENINGS

The importance of proper connection design cannot be overemphasized. Attention to this requirement of earthquake resistant design often can make the difference between success and failure under seismic loading. Commonly, connections are the weakest links in seismic design. Review of nonstructural component failures has shown that many occur at points of connection. Causes of these excessive stresses in nonstructural components include:

■ **Inadequate tolerances for seismic movement** will transmit impact loads to adjacent parts. Tolerances for siesmic movement must be provided in addition to normal construction tolerances.

■ **The design must take into account the limitations of bearing pressures on fastenings.** This is particularly true in threaded fastenings where the threads cause a sizable reduction in cross-section as well as bearing area of members.

■ **Another critical area is in light gauge material,** particularly aluminum. Excessive bearing pressure will cause yield in hole size and then "pull out." One example is the use of screws in extruded slots. These connections are extremely weak and should be avoided in critical elements.

Some connections for the attachment of components use various adjustable connections such as the double clip angle. In usual practice these are drawn in their normal position with construction tolerances not indicated. Often the connection is considered in its normal position and not in its extended position which is the critical condition when subjected to stresses.

Welding is used frequently in contemporary construction. In cases of on-site field welding, residual stresses remain in the material welded, requiring care of how the connection is detailed. Points of concern are:

• **Welding builds up local internal stresses,** particularly at end points. These residual stresses can increase the chance of failure when the

EARTHQUAKE RESISTANT DESIGN (cont.)

Summary of Seismic Performance of Structural Systems				
Structural System	EQ Performance	Test Data	Specific Bldg. Perf. & Energy Absorption	General Comments
Wood Frame	SF 1906, etc. ALA 1964 Variable to *Good*	1950's DFPA etc.	-SF Bldgs. performed reasonably well even though not detailed. -Energy Absorption is excellent.	-Connection details are critical. -Configuration is significant.
Unreinforced Masonry Wall	SF 1906 SB 1925 LB 1933 Variable to *Poor*	? Recent SEADSC	-Unreinf. masonry has performed poorly when not tied together. -Energy absorption is good if system integrity is maintained.	-Continuity and ties between walls and diaphragm is essential.
Steel Frame w/Mas Infill	SF 1906 Variable to *Good*	?	-SF Bldgs. performed very well. -Energy absorption is excellent.	-Bldg. form must be uniform, relatively small bay sizes.
R/C Wall	SF 1957 ALA 1964 JAPAN 1966 Variable to *Poor*	?	-Bldgs. in Alaska, SF and Japan performed poorly w/spandrel and pier failure -Brittle system.	-Proportion of spandrel and piers is critical, detail for ductility and shear.
Steel Brace	SF 1906 TAFT 1952 Variable to *Good*	Univ. of Mich. Japan UCB	-Major braced systems performed well. -Minor bracing and tension braces performed poorly.	-Details and proportions are critical.
Steel Moment Frame	LA 1971 JAPAN 1978 ? *Good*	Lehigh UCB	-LA and Japanese Bldgs. performed well. -Energy absorption is excellent	-Both conventional and D.F. have performed well if designed for drift.
Concrete S.W.	CARAC 1965 ALA 1964 LA 1971 ALG. 1980 *Variable*	PCA U. of Ill. UCB	-Poor performance w/ discontinuous walls. •San Francisco, Alaska, Algeria, Caracas. -Uneven energy absorp.	-*Configuration* is *critical*, soft story or L-shape w/torsion have produced failures.
P/C Concrete	ALA 1964 BULGARIA 1978 SF 1980 Variable to *Poor*	Japan ?	-Poor performance in 1964, 1978, 1980 -Brittle Failure	-Details for continuity are critical. -*Ductility* must be achieved.
R/C DMF	LA 1971 ? *Good*	PCA Texas Toronto UCB	-Good perf. in 1971, LA. -System will crack. -Energy absorption is good.	-Details *critical*

(after C. Arnold. "Buildings at Risk" Washington, DC: AIA/ACSA Council on Architectural Research, 1994).

B1.1 STRUCTURE

connection is further stressed due to seismic action.

• **Light gauge welding** often results in "burn through," particularly when light gauge material is connected to heavy structural shapes.

• **Welded steel moment frames** have been commonly used as a lateral bracing system for medium and high-rise construction. Prior to the Northridge Earthquake, it had been considered one of the most seismic-resistant structural systems. The wide-spread damage of these systems during the 1994 Northridge earthquake promoted intense research into the reasons for these unexpected failures. Preliminary conclusions are that certain welding electrodes are suspected of having insufficient "fracture toughness" to perform well during dynamic loading.

SUMMARY: Checklist of earthquake considerations in architectural design

Architects can best set meaningful design and review priorities by a deliberately defined strategy for earthquake resistant design:

■ Design and build to the expected standards of performance of the building as it affects life safety and property damage.

■ Establish basic planning and design parameters (form, shape) that will best meet the performance criteria.

■ Integrate the various building components within the planning and design parameters, giving full attention to appropriate life safety criteria.

To apply a strategy for earthquake resistant design, consider the following:

■ **Protection of building occupants** and the public adjacent to a building during an earthquake.

■ **Disaster control and emergency subsystems** must remain operable after an earthquake, which include

emergency and safety provision for:

• **Fire:** Fires can begin at a variety of locations during an earthquake, such as in mechanical rooms, kitchens, laboratories; that is, wherever fuel or electric lines rupture.

• **Electrical hazards:** Collapse of ceilings or partitions or dislocation of electrical appliances may leave wiring exposed which creates danger of shock, or results in sparking which can lead to fire or explosion.

• **Flooding:** Broken water pipes or sanitary lines may lead to flooding of various parts of the building.

■ **Occupants must be able to evacuate a building quickly and safely after an earthquake when it is safe to do so.**

■ **Rescue and emergency workers** must be able to enter the building immediately after an earthquake, encountering minimum interference and danger.

■ **The building must be returned to useful service** as quickly as possible, and include:

• Sewage disposal and potable water supply.
• Electric power
• Mechanical systems

■ **The building and personal property within the building should remain as secure as possible after the earthquake.**

REFERENCE:

Elmer F. Botsai, FAIA *"Earthquake Resistant Design"* (Reference 1).

B • BUILDING SHELL

B1 • STRUCTURE & MATERIALS

B1. 2 CEMENT/CONCRETE

CEMENT: introduction

Cement may be defined as a material which, when mixed with water or other liquid substance, will:

- Form temporarily a plastic paste; easily molded or deformed;
- After a short period of time will harden or set to a rigid mass.

By far the most important group of cementitious materials used in construction is the one known as Portland cement.

■ The common designation of Portland cement, as used in building construction, is **cement,** which designation will be used throughout this selection guide.

Portland cement is made by blending and processing several raw materials:

■ **Calcareous:** such as limestone, sea shells, or marl.

■ **Siliceous or silicous and aluminous:** such as sand, clay, shale, or slate.

■ **Iron compounds:** such as iron ore or mill scale.

■ **Color of cement** depends on the raw materials used and on the manufacturing process: a fine-ground cement is normally lighter in color than a coarse cement of the same chemical composition.

Raw materials generally also contain small amounts of impurities, such as alkalies, which may remain in the finished product:

■ **Alkalies present in cement,** even in small amounts, may lead to internal expansion and failure when such cement is used with reactive aggregates in a conrete mix.

- cement with low alkali content, generally less than 0.6 percent, is available.
- any type of cement can have a low or high alkali content, depending on raw materials used, and on the manufacturing process.

Portland cement is commonly manufactured to conform to ASTM C150 - Standard Specification for Portland Cement - which covers eight types:

■ **Type I:** for general use when special properties are not required, such as resistance to sulfate attack:

- commonly used for structural work, bridges, pavements, concrete masonry units.

■ **Type IA:** is Type I cement with air-entraining additives incorporated during the manufacturing process:

- air entrainment improves resistance to freeze-thaw cycles of concrete mix in which such cement is used.

■ **Type II:** generates less heat and at slower rate during the hydration process than Type I; it also provides moderate resistance to sulfate attack:

- used in structures of considerable size such as large piers, heavy abutments to minimize detrimental effects of heat of hydration.
- used in structures in contact with ground water when sulfate concentration in such water is higher than normal but not unusually severe.

■ **Type IIA:** is Type II cement with air-entraining additives incorporated during the manufacturing process.

■ **Type III:** for use where high strength is desired at early periods; usually within a week or less.

■ **Type IIIA:** air-entrained Type III cement.

■ **Type IV:** for use where low heat of hydration is desired; develops strength at a much slower rate than Type I:

- used in massive structures, such as large gravity dams.
- availability may be limited to specific localities.

■ **Type V:** is used when high resistance to sulfate attack is desired; gains strength at a slower rate than Type I:

- availability may be limited to specific localities.

Other Portland cements generally available include:

■ **White cement:** manufactured to conform to ASTM C150 using selected raw materials so that finished product will be white rather than gray:

- used for: decorative architectural concrete; stucco; cement paint; white or pigmented grout.

■ **Portland blast-furnace slag cement:** manufactured to conform to ASTM C595 - Standard Specification for Blended Hydraulic Cements - for use in general construction when special properties are not required:

- manufactured using granulated blast-furnace slag interground with Portland cement clinker, or by blending the two.
- has slower strength development at early age than Type I cement.

■ **Portland - pozzolan cement:** conforming to ASTM C595.

- manufactured by intergrinding - or blending - Portland cement or Portland blast-furnace slag cement with pozzolan, e.g., siliceous or siliceous and aluminous material:
- has slower strength development at early age than Type I cement.

■ **Pozzolan - modified Portland cement:** conforming to ASTM C595

- should not be used where special characteristics attributable to Portland-pozzolan cement are desired.

■ **Masonry Cement:** is a mixture of Portland cement, air-entraining additives, plasticizers, and other supplemental materials. Workability, strength, and color are closely controlled. Masonry cement is covered by requirements of ASTM C91.

Special types of Portland cement include:

■ **Block cement:** Similar to Type III cement, except that close control of uniform color is provided. Range of formulations available.

■ **Expansive Cement:** Increases significantly in volume during setting, and retains a portion of such increase in volume after hardening. It is used to inhibit shrinkage of concrete, thus minimizing cracking; and also as self-stressing cement in prestressed concrete work. Limited availability.

■ **Gun plastic cement:** Is similar to plastic cement, except that it was developed for application by compressed air guns, or pumps. A small amount of fine mineral fibers is generally blended into the plastic cement. Limited availability.

■ **Oil-well cement:** Slow-setting,

resistant to high temperatures and pressures, used for sealing oil wells. Limited availability.

■ **Pipe cement:** Specially made for use in centrifugally spun pipe.

■ **Plastic cement:** Is a mixture of Type I or Type II Portland cement and of plasticizing agents. Plastic cement is used in making mortar, cement plaster, and stucco. Not available from all manufacturers.

■ **Regulated set cement:** Proprietary formulations for early strength and fast setting: from a few hours to under one hour.

■ **Waterproofed cement:** Is a standard or white cement incorporating water-repellent additives. Limited availability.

Another group of cementitious materials used in construction include the special types and blends of cement, formulated to be used under specific conditions or for specific purposes:

■ **Flyash:**

• Can be used as a replacement for cements in amounts up to 20 percent to reduce the permeability.
• It is not effective with Type II and Type V cements: ASTM C618.

■ **High alumina cement:**

• also known as aluminous cement or calcium aluminate cement.
• resists hydroxylic compounds, such as phenols, glycerols, and sugars.
• not resistant to alkalies; durable in sea water.
• stable under high temperatures; used in manufacture of refractory products.
• may be used as accelerating admixture in Portland cement. Availability generally limited.

■ **Magnesium-oxychloride cement:**

• also known as sorel cement or magnesite cement
• has good resistance to: fire; abrasion; grease, oil.
• fairly resistant to: alkalies; organic solvents; common salts; sulphates.
• does not require wet curing.
• may be damaged by acids; is corrosive to metals such as aluminum and steel.
• unstable in water and loses strength on prolonged exposure to it.
• used for floors in industrial, commercial buildings as wearing surface, underlayment, or for resurfacing old floors.

■ **Magnesium-oxysulphate cement:** has good bindig properties, but is generally considered to be weaker than oxychloride cement.

• most properties similar to those of oxychloride cement, including instability in water; but is less sensitive to higher temperatures.
• limited availability.

■ **Natural cement:**

• used principally in masonry mortars, and as admixture in Portland cement concrete.
• composition not controlled, thus properties vary widely.
• gradually disappearing from the market.

■ **Phosphate cement:**

• several formulations are available
• quick-setting: generally used in patching; sprayable foamed insulation; flame-resistant coatings.

■ **Waterproof cement:**

• proprietary formulations; generally used in cementitious waterproofing membranes.

Cementitious materials used in building construction, which are not cements, but may be included under this definition, are:

■ **Limes:** which are slow setting and hardening. Their principal use is to plasticize harsh cements and to add resilience to mortars and stucco.

Several types are available:

• Quicklime: used primarily in masonry mortar. Must be hydrated, or slacked, before use.
• Mason's hydrated lime: is made from quicklimes in the plant. Used in mortars, base-coat plaster, and concrete.
• Finishing hydrated lime: is characterized by whiteness and plasticity. Used in finishing coat of plaster.

■ **Gypsum cement:** used in plaster, gypsum board, gypsum block: Solubility of gypsum limits its use to applications not exposed to weather nor to high humidity conditions.

Several types are available:

• Plaster of paris: used as molding plaster for preparing ornamental plaster objects; or as gaging plaster, used with finishing lime for white-coat finish on plaster walls.
• Unretarded plaster of paris is used for making gypsum board, gypsum lath, and gypsum blocks.
• Retarded plaster of paris, mixed with fibers and sand, is used for base-coat plastering.
• When mixed with the proper aggregate, lightweight plaster and acoustical plaster will be produced.

■ **Keene's Cement:** which is a gypsum derivative rather than a Portland cement, is much more resistant to water than plaster of paris; and also has good fire-retarding properties.

HYDRATION PROCESS:

Development of strength is the result of chemical reactions between the cement constituents and water and occurs by the formation of a rigid interlocking matrix of hydration products, which gradually replace the water between the cement grains and finally bind the composite cement mass together:

■ **The amount of water required in a normal mix is always much in excess of that required for the hydration of cement:** the final density of the hardened cement paste will be determined by the ratio of water to cement used in the original mix.

■ **Cement products undergo small changes in volume with changes in moisture content:**

• dimensional changes in cement paste may be as much as ten times those occuring in the aggregate or in the hardened concrete mix.
• concrete made with more cement paste than is required to fill the voids between the aggregate, will tend to take on more of the shrinkage characteristics of the cement paste, with resultant excessive cracking.

B1.2 CEMENT/CONCRETE

CEMENT: types, properties, uses

LEGEND:
- **I** primary property
- **II** secondary property
- **●** general usage
- **A** general availability
- **a** limited availability

	PROPERTIES														USES CONCRETE ①					
		HEAT OF HYDRATION				DEVELOP STRENGTH			RESISTANCE TO SULFATE ATTACK											
	GENERAL PURPOSE	HIGH	NORMAL	MODERATE	LOW	SLOW	NORMAL	RAPID	POOR	MODERATE	HIGH	RESISTANCE TO FREEZE/THAW	CONTROLLED COLOR	AVAILABILITY	THIN SECTIONS	HEAVY SECTIONS	MASSIVE SECTIONS	NORMAL WEIGHT	LIGHTWEIGHT	CELLULAR
CEMENT–STANDARD																				
TYPE I — GRAY	I		II				II		II					A	●			●	●	●
TYPE I — BUFF			II				II		II				I	a						
TYPE I — WHITE			II				II		II				I	A						
TYPE II				II			II			I				A	●	●		●	●	
TYPE III — GRAY		II						I	II					A	●			●	●	
TYPE III — WHITE		II						II	II				I	a	●			●	●	
TYPE IV					I	II			II					A			●			
TYPE V				II		II					I			A	●	●		●		
CEMENT–AIR ENTRAINING																				
TYPE IA — GRAY			II				II		II			I		A	●			●	●	
TYPE IA — WHITE			II				II		II			II	I	a	●			●	●	
TYPE IIA				II			II			II		I		a	●	●		●	●	
TYPE IIIA — GRAY		II						II	II			I		a	●			●	●	
TYPE IIIA — WHITE		II						II	II			II	I	a	●			●	●	
BLAST FURNACE SLAG																				
TYPE IS	I		II				II		II					a	●			●	●	●
TYPE IS-A			II				II		II			I		a	●			●	●	
POZZOLAN																				
TYPE IP	I		II				II		II					a	●			●	●	●
TYPE IP-A			II				II		II			I		a	●			●	●	
TYPE P	I		II				II		II					a	●			●	●	●
TYPE P-A			II				II		II			I		a	●			●	●	
SLAG CEMENT																				
TYPE S	I													a						
TYPE S-A	I													a						

MORTARS			FINISHES						SPECIAL APPLICATIONS				
MASONRY MORTAR [2]	GROUT	PATCHING MIXES	CEMENT BOUND AGGREGATE	CEMENT PLASTER	STUCCO	CEMENT PAINT	CEMENT-BOUND MINERAL FIBER	CONCRETE MASONRY [3]	CONCRETE PIPE	PRE-CAST SPECIALTY ITEMS	SHOTCRETE/GUNITE	FERROCEMENT	SOIL CEMENT

REMARKS

[1] the use of cement in concrete does not change whether the concrete is precast or placed-on-site, plain or reinforced

[2] for more information on masonry mortars, see STONE/MASONRY.

[3] for more information on concrete masonry units of all types, see STONE/MASONRY.

Matrix 1

MASONRY MORTAR	GROUT	PATCHING MIXES	CEMENT BOUND AGGREGATE	CEMENT PLASTER	STUCCO	CEMENT PAINT	CEMENT-BOUND MINERAL FIBER	CONCRETE MASONRY	CONCRETE PIPE	PRE-CAST SPECIALTY ITEMS	SHOTCRETE/GUNITE	FERROCEMENT	SOIL CEMENT
●	●	●	●	●			●	●	●	●	●	●	●
				●				●		●			
●			●		●	●				●	●		
							●		●		●	●	●
	●									●			
										●			
									●				

- heat of hydration maximums may be specified when used in heavy sections.
- cement for plaster and stucco needs addition of plasticizing agents.

Matrix 2

Matrix 3

MASONRY MORTAR	GROUT	PATCHING MIXES	CEMENT BOUND AGGREGATE	CEMENT PLASTER	STUCCO	CEMENT PAINT	CEMENT-BOUND MINERAL FIBER	CONCRETE MASONRY	CONCRETE PIPE	PRE-CAST SPECIALTY ITEMS	SHOTCRETE/GUNITE	FERROCEMENT	SOIL CEMENT
●	●	●	●	●			●	●	●	●	●	●	●
●	●	●	●	●			●	●	●	●	●	●	●
●	●	●	●	●			●	●	●	●	●	●	●

- moderate heat of hydration, moderate resistance to sulfate attack, or both, may be specified by adding the suffixes (MH) and (MS).
- low heat of hydration, moderate resistance to sulfate attack, or both, may be specified by adding the suffixes (LH) and (MS).
- Pozzolan Modified Portland Cement, Types I (PM), and I (PM)-A, air entrained, are also covered under ASTM C595. (MH) and (MS) properties may be specified.

Matrix 4

- both types used in combination with Portland cement for concrete, and in combination with hydrated lime for masonry mortar.

B1.2 CEMENT/CONCRETE

CEMENT: selection checklist

GENERAL CONSIDERATIONS

Cement is manufactured using locally-available raw materials to the greatest extent possible to save cost of transportation, and even though conformance to ASTM C150 is required, variations in the finished product will exist:

- to achieve close uniformity of color of cement to be used on a given project, all cement should be from the same batch.
- to ensure close similarity of properties, all cement should be purchased from the same manufacturer.
- to achieve economy in construction, cement with special properties should not be selected unless such properties are definitely required.
- most types may be used as general purpose cement but types with specific properties commonly cost more.

SPECIFIC CONSIDERATIONS

Selection of cement for a specific use should generally consider factors which affect - or are affected by - the cement paste component of a finished product incorporating it:

■ **Development of strength** when the finished product has to be put to use as soon as possible, or will be exposed to below freezing temperatures soon after being formed, or placed in its final location:

- high early strength cement - Types III or IIIA - should be considered as it will gain strength faster than general-purpose cement and will also generate more heat during hydration thus counteracting to some extent the effect of low temperature; or
- the amount of general-purpose cement in a mix may be increased, which will achieve essentially the same results-as far as early strength and heat generated are concerned-as the use of high early strength cement.
- admixtures known as accelerators are available to increase the setting rate and development of strength; the advisability of using such admixtures should be thoroughly investigated to ensure that no detrimental effects will develop.

■ **Heat of hydration:**
Heat is released during the chemical reaction between cement and water; the larger the mass of material and/or the higher the ambient temperatures, the higher the heat developed within the material and the more slowly it dissipates.

- if heat is not rapidly dissipated, thermal expansion may result and the subsequent cooling of hardened concrete may create undesirable stresses.
- a rise in temperature may be beneficial in cold weather to help maintain favorable curing temperatures.
- when heat of hydration is undesirable, Type IV cement or means of cooling the plastic concrete should be considered.
- cement with moderate or low heat of hydration will reduce expansion of cement paste during setting thus minimizing subsequent shrinkage cracking; but strength will be developed at a slower rate than when general purpose cement is used.

■ **Resistance to chemicals:**

- Portland cements which resist the attack of sulfates in the soil or in ground water are available; the type with moderate resistance is more readily available than the type with high resistance.
- cement with high resistance to sulfates should not be selected unless such property is definitely required: this type of cement is more expensive and not available in all localities.

■ **Cement may also react with some chemically unstable aggregates;** the process being known as the alkali-aggregate reaction:

- based on the type of aggregate the reaction may be alkali-silica or alkali-carbonate; both generally result in a pattern cracking on the exposed surface of concrete.
- when the use of reactive aggregates cannot be avoided, cement with low alkali content - generally less than 0.6 percent - should be selected.
- when the use of reactive aggregates and high alkali cement cannot be avoided, partial replacement of cement with pozzolan should be considered; but it should be realized that some poz-

zolans may contain substantial amounts of alkalies.
- addition of pozzolans may however not be effective in controlling alkali-carbonate reactions.
- use of air entrainment may also be beneficial in controlling the effects of alkali-aggregate reaction.

■ **Resistance to freeze thaw cycles:** will be improved by using air-entrained cement.

- resistance to attack by deicing salts is also improved through the use of air-entrainment.
- air-entraining cements generally provide an adequate amount of entrained air and avoid the problem of adding an extra ingredient at the mixer; adding air-entraining admixtures to cement allows for the adjustment of the amount of entrained air to meet specific job conditions.

Considerations when using pozzolan to replace a portion of Portland, Portland blast-furnace slag, or Portland-pozzolan cement in a mix:

- early strength will in most cases be reduced.
- may reduce resistance of concrete to the effects of freeze thaw cycles.
- may reduce entrained air content, and air-entraining admixtures may have to be added to compensate for such reduction.
- generally will improve resistance to seawater attack.

Cement and water pastes are rarely used in building construction by themselves, and are generally combined with:

■ Fine aggregate, such as sand, to form mortar.

■ Fine and coarse aggregate, such as sand and gravel, or sand and crushed stone to form concrete.

GENERAL DESCRIPTION

Concrete consists of inert, graded aggregate of controlled sizes embedded within a paste of Portland cement and water.

Since the cement paste ordinarily constitutes only 25 to 40 percent of the total volume of concrete, the properties of concrete are determined not only by the properties of the cement selected but also by:

■ Properties of coarse aggregate.

■ Properties of fine aggregate.

■ Effects of any chemical or organic compounds, or admixtures introduced.

■ Amount of water added in relation to the amount of cement used; commonly referred to as the water/cement ratio.

Strength and durability of concrete will not only depend on the properties of ingredients but also on their proportions in a given mix. The principal variations are:

■ Water/cement ratio.

■ Cement/aggregate ratio.

■ Size of coarse aggregate.

■ Ratio of fine aggregate to coarse aggregate.

■ Type of cement.

Depending on type, properties, and proportions of ingredients, strength of concrete may vary from a low of 100 psi for expanded plastic lightweight concrete to over 8000 psi for normal weight concrete; and durability from practically none unless protected to almost complete.

DESIGN MIX

There is no set formula for mixing concrete. For each mix, aggregate tests must be made; for each application, a concrete mix designed.

Thus in selecting materials for a mix to be used under particular placement and in-service conditions, consideration is to be given to:

■ Properties of each constituent material.

■ The compatibility and interaction of each with the other materials.

■ The effects of such interaction on the ultimate properties of the mix, as the enhancement of one property may reduce the effectiveness of others, with full understanding that:

■ **Composition of cement** may be varied to provide desirable properties.

■ **Properties of cement** may vary in different areas and between products of different manufacturers.

■ **Properties of aggregates** available will vary in different localities.

■ **Weather conditions,** such as high temperatures, low temperatures, high humidity, low humidity, high winds, etc., will also influence the ultimate product.

■ **The equipment** available and the skills of the local labor force may become a factor.

With all these variables to be considered, the ultimate mix can seldom, if ever, be achieved.

FRESH CONCRETE

For a relatively short period of time after the introduction of water into the mixture, concrete will remain plastic. During this period of time, concrete may be:

■ Sprayed

■ Formed

■ Molded

■ Levelled

■ Smoothed

■ Textured

■ Roughened

In its plastic state and for several days after hardening, concrete has to be **protected from the detrimental effects of:**

■ High temperatures

■ Below freezing temperatures

■ Rapid loss of mixing water

■ Stresses imposed by external loads

Properties of fresh concrete to be considered are:

■ **Workability:** the ability of fresh concrete to flow without segregation of its ingredients.

■ **Setting time:** the time required for the cement paste to harden since fresh concrete can only be worked before it sets.

It should be noted, that as long as there is moisture in the concrete, the hydration process, with resultant increase in strength, will continue, but at a progressively diminishing rate.

HARDENED CONCRETE

Hardened concrete is concrete in which the hydration process has developed to a point where the concrete has some strength.

■ As a measure of the quality of concrete the **compressive strength** is one of the most important standards; which is also often used to predict the durability of concrete:

• the prime factor affecting strength is water/cement ratio with all other ingredients of a mix being the same.

Concrete can be used by itself and often is, but being heterogeneous material, it is much more resistant to compressive stresses than to shearing and tensile stresses. To compensate for this deficiency, concrete frequently is reinforced with steel in various forms.

Reinforcement changes the basic strength properties of concrete and consequently its behavior as a structural material. Reinforcing is done to help concrete perform in two basic ways:

■ **To resist tensile and shear stresses** induced by superimposed loads. This is termed primary reinforcement.

■ **To resist stresses induced by changes in volume:**

• Contraction, or shrinkage, due to the loss of moisture.
• Contraction, or expansion, due to changes in temperature.

This is termed secondary reinforcement.

■ The high compressive strength of concrete may also be utilized by maintaining it in compression at all times through prestressing, e.g., by tensioning the steel reinforcing to eliminate or substantially reduce tensile stresses in concrete under load.

The information herein presented must of necessity be rather general in scope, and reflects prevalent practice. With a complex and versatile material such as concrete, there are not many rules that cannot be broken; and the unusual can be, and often is, done.

B1.2 CEMENT/CONCRETE

CONCRETE: types

*   Special type or method; not included in evaluation charts.
**   Special type or method; included in material evaluation chart only.

NORMAL WEIGHT
- ■ Weight of coarse aggregate used determines type.
  - • Weight of aggregate: 135 to 165 lbs/cf.
  - • All types of cement may be used.
  - • Compressive strength generally between 2000 to over 8000 psi.
  - • Typical uses: structural framing; pavements; floors.

LIGHTWEIGHT-STRUCTURAL
- ■ Weight of coarse aggregate used determines type.
  - • Weight of aggregate: 85 to 115 lbs/cf.
  - • Resistance to heat flow better than of normal weight concrete.
  - • Typical uses: load-bearing and exterior walls; prestressed concrete; floor fill.

LIGHTWEIGHT-INSULATING
- ■ Weight of coarse aggregate used determines type;
  - • Weight of aggregate: 15 to 90 lbs/cf.
  - • Aggregates used may be: perlite, vermiculite, expanded polystyrene, wood chips or fibers.
  - • Resistance to heat flow increases with decrease in density.
  - • Typical uses: fill over metal roof decks; partitions and panel walls.

HEAVYWEIGHT**
- ■ Weight of coarse aggregate determines type.
  - • Weight of aggregate: 130 to 290 lbs/cf.
  - • All types of cement may be used.
  - • Typical uses: walls of spaces containing radioactive materials; sometimes as counterweights in various applications.

CELLULAR
- ■ Air or gas bubbles, suspended in mortar, characterize type.
  - • Small amounts or no coarse aggregate provided.
  - • Typical uses: where high insulating properties are required.

GAP-GRADED
- ■ Omission of intermediate sizes of coarse aggregate charcterizes types.
  - • Typical uses: where aggregate is to be exposed; as inexpensive concrete for foundations.

SHOTCRETE, or GUNITE
- ■ Method of placement characterizes type.
  - • Pneumatic equipment, using dry or wet method.
  - • Typical uses: wherever construction without formwork is very desirable, as in complex forms such as in shells, domes, swimming pools.

PRE-PLACED*
- ■ Method of placement characterizes type.
  - • Coarse aggregate is placed dry, then mortar is pumped into it.
  - • Typical uses: special forms, surfaces, e.g., exposed aggregate finishes on cast-in-place concrete columns, walls.

PUMPED*
- ■ Method of conveying plastic concrete for placement characterizes type.
  - • Commonly normal, lightweight or lightweight-insulating types of concrete.
  - • Typical uses: when concrete is to be placed high above grade or in formwork of complex shape.

FERROCEMENT*
- ■ Mix, method of placement and reinforcement provided characterize type.
  - • Mortar with large amount of light-gauge wire reinforcing is used.
  - • Typical uses: containers, e.g., bins, boat hulls, other thin complex shapes.

FIBER**
- ■ Addition of short fibers to mix characterizes type.
  - • Fibers used may be steel, glass, polypropylene.

B1.2 CEMENT/CONCRETE

Con't.	• Fibers are used as secondary reinforcing to control plastic shrinkage and associated cracking; and will generally increase flexural and tensile strength of concrete. Used where prevention of cracking is important, e.g. slabs on grade, pavements, pneumatically placed concrete, precast concrete, overlays.
NAILING*	■ Nail-holding strength characterizes type.
	• Commonly lightweight-insulating types; often using wood fiber/chip aggregate.
	• Has high insulating value. Used in roof decks.
NO-SLUMP*	■ Very low water/cement ratio characterizes type.
	• Economical but requires vibration during placement. Has high strength and low shrinkage. Typically used for plant pre-cast items.
POROUS (no fines)**	■ Omission of fine aggregate from mix characterizes type.
	• Typical uses: mostly for porous drainage pipe; also cast walls, panels; possesses some insulating value.
TREMIE*	■ Method of placement characterizes type.
	• Used in placing concrete under water.
POLYMER-IMPREGNATED*	■ Additives and treatment characterize type.
	• Hydrated Portland cement concrete which has been impregnated with a monomer and subsequently polymerized after being placed.
POLYMER-PORTLAND CEMENT*	■ Additives used principally characterize type.
	• Monomer or polymer is added to fresh concrete mix and subsequently allowed to cure; and if needed polymerized in place.
POLYMER*	■ Type of binder characterizes type.
	• Monomer replaces cement; the polymerized monomer acts as the binder for the aggregate.
	• Used for curtain wall and wall facing panels.
SULPHUR*	■ Type of binder characterizes type.
	• Sulphur replaces cement; sulphur binder usually modified with plasticizers.
	• Solidifies rapidly - generally within one day - and has good hardening capabilities; strength similar to Portland cement concrete.
	• Used for industrial floors; bridge decks; leach tanks; sewer pipes.
REINFORCED*	■ Concrete in which steel has been incorporated in order to compensate for its lack of tensile and shear strength; also may be used for additional compressive strength.
	• Primary reinforcing may be bar or rod mats; plain or deformed bars; any other less-conventional shapes which will bond to concrete.
PRESTRESSED*	■ Reinforced concrete which is placed in compression prior to its integration into the structural system: before it starts interacting with the other members of the structure and before loads are superimposed.
	• Pretensioned - section is placed in compression using pretensioned steel wires when cast in plant.
	• Posttensioned - section is placed in compression using posttensioned cables after being cast in place.
	• Both methods reduce amount of concrete required, increase load carrying ability, reduce load-induced cracking.

B1.2 CEMENT/CONCRETE

CONCRETE: types, properties required, uses,

- ● denotes common usage
- ○ denotes possible usage
- ■ denotes principal properties required
- □ denotes secondary, optional or additional properties which may be required

USES

Column groups:
- FOUNDATIONS: PILES and fills for metal shells · FOOTINGS · GRADE BEAMS · WALLS - FOUNDATION · WALLS - RETAINING
- ROADWAY STRUCTURES: PAVEMENTS · WALKS
- BRIDGE STRUCTURES: ABUTMENTS - PIERS · FRAMING · DECKS · CURBS · RAILINGS · LIGHTING POLES, STANDARDS
- STRUCTURES IN CONTACT WITH WATER: DAMS · SPILLWAYS · PIERS · SEAWALLS · CANAL LININGS · MANHOLES · PIPE - WATER - SEWER · PIPE - DRAINAGE · SWIMMING POOLS

TYPE	PILES	FOOTINGS	GRADE BEAMS	WALLS-FOUND	WALLS-RET	PAVEMENTS	WALKS	ABUTMENTS-PIERS	FRAMING	DECKS	CURBS	RAILINGS	LIGHTING POLES	DAMS	SPILLWAYS	PIERS	SEAWALLS	CANAL LININGS	MANHOLES	PIPE-WATER-SEWER	PIPE-DRAINAGE	SWIMMING POOLS
CAST-IN-PLACE																						
NORMAL WEIGHT, plain/reinforced	○	●	○	●	●	●	●	○		○		●		●	●	○	●	●	●			○
LIGHTWEIGHT-STRUCTURAL, plain/reinforced																						
INSULATING, CELLULAR, NAILING																						
SHOTCRETE (Reinforced Only)															●			●				●
PRE-CAST, PLANT OR ON-SITE																						
NORMAL WEIGHT	●	○	○	○				●	●	●	○	●	●							●	●	
LIGHTWEIGHT, Structural			○	○	○			●				●	●							○	○	
INSULATING, CELLULAR, NAILING																						
POROUS																					●	
PROPERTIES REQUIRED																						
RAPID DEVELOPMENT OF STRENGTH	□	□	□	□		□	□	□	□	□	□	□							□	□	□	□
RESISTANCE TO: Freeze/Thaw Cycles						■	■	□	■	■	■	□	□	□	□	□	□	□				■
RESISTANCE TO: Sulfate Attack	□	□	□	□	□	□	□	□			□			□		□	□	□	□	□	□	□
RESISTANCE TO: Water Permeability																		□	■	■		■
RESISTANCE TO: Abrasion/Wear						■	□			■				■								
CONTROLLED HEAT OF HYDRATION				□				■						■	□	□	□					

RECOMMENDED MAX. WATER/CEMENT RATIO OR MIN. COMPRESSIVE STRENGTH AT 28 DAYS.

TYPE	PILES	FOOTINGS	GRADE BEAMS	WALLS-FOUND	WALLS-RET	PAVEMENTS	WALKS	ABUTMENTS-PIERS	FRAMING	DECKS	CURBS	RAILINGS	LIGHTING POLES	DAMS	SPILLWAYS	PIERS	SEAWALLS	CANAL LININGS	MANHOLES	PIPE-WATER-SEWER	PIPE-DRAINAGE	SWIMMING POOLS
NON AIR-ENTRAINED CONCRETE:																						
Water/Cement — 0.82 or 2000 psi		●												○								
Water/Cement — 0.68 or 3000 psi	●	○	●	●	●										○					●	●	
Water/Cement — 0.57 or 4000 psi																						●
Water/Cement — 0.48 or 5000 psi																						
AIR-ENTRAINED CONCRETE:																						
Water/Cement — 0.74 or 2000 psi														●	○	○						
Water/Cement — 0.59 or 3000 psi							●	●		●				●	●	●	●					
Water/Cement — 0.48 or 4000 psi						●			●	●		●	●	●			○					○
Water/Cement — 0.40 or 5000 psi																						

BUILDING ELEMENTS

FRAMING / **WALLS**

COLUMNS	PLANKS	SINGLE TEES	DOUBLE TEES	CHANNELS	JOISTS	LINTELS	BEAMS	GIRDERS	SLABS - FRAMED	SLABS - FLAT - WAFFLE	SHELLS - PARABOLOIDS	SLABS ON GRADE	EXTERIOR - BEARING	EXTERIOR - NON-BEARING	INTERIOR - BEARING	INTERIOR - NON-BEARING	MODULAR COMPONENTS	TOPPINGS/FILLS
●		●	●	●	●	●	●	●	●			●	●	●	●	●		●
○		●	○	●	●	●	●	●	●				●	●	●	●		●
																		●
											●							

Secondary reinforcing provided to:
- control cracking due to drying shrinkage. Reinforcing will prevent large cracks; cracking will be limited to numerous small tight cracks.

Primary reinforcing provided to:
- resist all stresses induced by dead and live loads. Reinforcing provided will also control cracking due to drying shrinkage.

COLUMNS	PLANKS	SINGLE TEES	DOUBLE TEES	CHANNELS	JOISTS	LINTELS	BEAMS	GIRDERS	SLABS - FRAMED	SLABS - FLAT - WAFFLE	SHELLS - PARABOLOIDS	SLABS ON GRADE	EXTERIOR - BEARING	EXTERIOR - NON-BEARING	INTERIOR - BEARING	INTERIOR - NON-BEARING	MODULAR COMPONENTS	TOPPINGS/FILLS
●	●	●	●	●	●	●	●	●			●		●	●	●	●	●	
●	○	○	○	○	○	○	○	○			○		●	●	●	●	●	
●																		

Primary reinforcing generally provided to:
- resist all stresses induced during erection only.
- resist all stresses induced by dead and live loads as well as erection stresses.
- in either case reinforcing will also control shrinkage cracking.
- plant-precast structural elements generally prestressed/ pretensioned.

COLUMNS	PLANKS	SINGLE TEES	DOUBLE TEES	CHANNELS	JOISTS	LINTELS	BEAMS	GIRDERS	SLABS - FRAMED	SLABS - FLAT - WAFFLE	SHELLS - PARABOLOIDS	SLABS ON GRADE	EXTERIOR - BEARING	EXTERIOR - NON-BEARING	INTERIOR - BEARING	INTERIOR - NON-BEARING	MODULAR COMPONENTS	TOPPINGS/FILLS
□	□	□	□	□	□	□	□	□	□	□	□	□	□	□	□	□	□	□

- may be required because of weather or placement conditions.

COLUMNS	PLANKS	SINGLE TEES	DOUBLE TEES	CHANNELS	JOISTS	LINTELS	BEAMS	GIRDERS	SLABS - FRAMED	SLABS - FLAT - WAFFLE	SHELLS - PARABOLOIDS	SLABS ON GRADE	EXTERIOR - BEARING	EXTERIOR - NON-BEARING	INTERIOR - BEARING	INTERIOR - NON-BEARING	MODULAR COMPONENTS	TOPPINGS/FILLS
												□	□	□			□	□
												□						
													□	□		□		
												□				□		

- resistance to freeze/thaw cycles also provides good resistance to weathering.
- high insulating value generally results in low strength.

COLUMNS	PLANKS	SINGLE TEES	DOUBLE TEES	CHANNELS	JOISTS	LINTELS	BEAMS	GIRDERS	SLABS - FRAMED	SLABS - FLAT - WAFFLE	SHELLS - PARABOLOIDS	SLABS ON GRADE	EXTERIOR - BEARING	EXTERIOR - NON-BEARING	INTERIOR - BEARING	INTERIOR - NON-BEARING	MODULAR COMPONENTS	TOPPINGS/FILLS
●												●	●	●	●	●		
●	●	●	●	●	●	●	●	●	●	●	●	○					●	●
○																		

- compressive strength of insulating fills generally low.
- special concrete with compressive strength greater than 5000 psi available for columns.
- compressive strength shown as recommended is subject to variations due to structural requirements, in-use conditions, severe exposure, etc. Under certain conditions a lower water/cement ratio may be required which would result in higher compressive strength.
- air-entraining mixture may be used in lieu of air-entrained cement.

COLUMNS	PLANKS	SINGLE TEES	DOUBLE TEES	CHANNELS	JOISTS	LINTELS	BEAMS	GIRDERS	SLABS - FRAMED	SLABS - FLAT - WAFFLE	SHELLS - PARABOLOIDS	SLABS ON GRADE	EXTERIOR - BEARING	EXTERIOR - NON-BEARING	INTERIOR - BEARING	INTERIOR - NON-BEARING	MODULAR COMPONENTS	TOPPINGS/FILLS
													○	○				
																○	○	

REMARKS:

B1.2 CEMENT/CONCRETE

B • BUILDING SHELL

CONCRETE: types, materials

● denotes material commonly used; one only within each group of materials.

□ denotes optional materials in addition to the commonly used materials.

S denotes material requiring special selection or processing.

	CEMENT ASTM C150 / ASTM C595						COARSE AGGREGATE ① NORMAL 135-160 LBS./CF ASTM C33				LIGHTWEIGHT 85-115 LBS./CF ASTM C330				LIGHTWEIGHT 15-90 LBS./CF ASTM C332					
	TYPE I, TYPE IS, TYPE IP, TYPE P	TYPE II, TYPE IIIA	TYPE II, TYPE IS (MH), TYPE IP (NH)	TYPE IV, TYPE P(LH)	TYPE IA, TYPE IS-A, TYPE IP-A, TYPE P-A	TYPE II, TYPE I, TYPE IS (MS), TYPE P (M), TYPE IP (M)	GRAVEL	CRUSHED GRAVEL	CRUSHED STONE	AIR COOLED BLAST FURNACE SLAG	EXPANDED SHALE	EXPANDED CLAY	EXPANDED SLATE	EXPANDED SLAG	DIATOMITE	FLYASH	PERLITE	PUMICE	SCORIA	VERMICULITE
STANDARD TYPES																				
NORMAL WEIGHT	●	●	●	●	●	●	●	●	●	●										
LIGHTWEIGHT, Structural	●	●		●							●	●	●	●						
LIGHTWEIGHT, Insulating	●	●													●	●	●	●	●	●
SPECIAL - BY AGGREGATE																				
EXPANDED PLASTIC	●		●																	
FIBER	●	●				●	S	S	S	S										
HEAVY WEIGHT	●		●	●																
NAILING	●		●																●	●
SPECIAL - BY PROCESS																				
CELLULAR	●		●																	
GAP - GRADED	●	●					S	S	S											
POROUS			●		●				S	S										
SHOTCRETE	●	●	●			●	S	S	S	S										

HEAVY WEIGHT 130-290 LBS./CF ASTM C637	SPECIAL TYPES NO ASTM SPECS	FINE AGGR ASTM C33	ADMIXTURES ASTM C260 ASTM C494 ASTM C618 ASTM C688	REINFORCING ASTM.A82-A.A184.A185. A416.A421.A496. A497.A615.A616. A704.	NOTES:

NOTES:

① Weight refers to concrete, not to aggregate alone.

Column headers (left to right):

HEAVY WEIGHT: BARITE · HEMATITE · ILMENITE · LIMONITE · MAGNETITE · STEEL PARTICLES

SPECIAL TYPES: WOOD FIBERS · WOOD CHIPS · SAW DUST · EXPANDED POLYSTYRENE

FINE AGGR: NATURAL SAND · MACHINE SAND

ADMIXTURES: AIR-ENTRAINING · WATER-REDUCING · SET-RETARDING · SET-ACCELERATING · FOAMING · GAS-FORMING

REINFORCING: SYNTHETIC FIBERS · GLASS FIBERS · STEEL FIBERS · LIGHT-GAUGE WIRE FABRIC · HEAVY-GAUGE WIRE FABRIC · DEFORMED-WIRE FABRIC · BAR OR ROD MATS · BARS PLAIN & DEFORMED (BILLET STEEL · AXLE STEEL · RAIL STEEL) · WIRE FOR PRETENSIONING · CABLE FOR POSTTENSIONING

WATER FOR HYDRATION

Heavy Weight	Special Types	Fine Aggr	Admixtures	Reinforcing	Water	Notes
	□ □ □ □	● ●	□ □ □ □	□ □ □ □ □ □ □ □ □ □ □ □	●	Polymer latex and water soluble polymers may be used
		● ●	□ □ □	□ □ ● ● ● ● ● ● ● ● ●	●	
	□ □ □ □	● ●	□ □ □ □	□ □ □	●	
	● ● S	● S		● □	●	Insulating
		● ●	□ □ □ □	● ● ● □ □ □	●	
● ● ● ● ● ●		● ●	□ □ □	□ □ □	●	
	● ● ● □		□ □	□	●	Insulating
			● ●	□ □ □	●	Insulating
		S S	□ □ □ □	□ □ □	●	
		□	□ □		●	
		● ●	□ □	□ □ □ □ □ □ □ □ □ □	●	

CONCRETE: materials, properties

- ● denotes material or method of primary importance for control of a given property
- □ denotes other materials or methods used or required for control of a given property
- ○ denotes substitute material which may be selected to control a given property

	PROPERTIES													
	DEVELOPMENT OF STRENGTH				RESISTANCE TO:							INSULATING VALUE		
	SLOW	NORMAL	FAST	HIGH COMPRESSIVE STRENGTH	FREEZE/THAW CYCLES	MOST DEICING CHEMICALS	ABRASION/WEAR	WATER PERMEABILITY	SULFATE ATTACK - MODERATE	SULFATE ATTACK - HIGH	ALKALI - AGGREGATE REACTION	LOW	MODERATE	HIGH
CEMENTS - TYPES														
TYPE I - TYPE IS - TYPE IP - TYPE P		●												
TYPE III - TYPE IIIA			●											
TYPE II - TYPE IS (MH) - TYPE IP (MH)	○													
TYPE IV - TYPE P (LH)	●													
TYPE IA - TYPE IS-A - TYPE IP-A - TYPE P-A		●			●	●								
TYPE II - TYPE IS (MS) - TYPE IP (MS) - TYPE P (MS)		●							●					
TYPE V	●									●				
AGGREGATES - PROPERTIES														
RESISTANCE TO ABRASION				□			●							
RESISTANCE TO FREEZE/THAW					□	□								
CHEMICAL STABILITY				□	□		□				●			
GRADING				□	□			□						
SHAPE AND SURFACE				□				□						
WEIGHT IN LBS. PER CUBIC FOOT:														
135-160				□				□				●		
85-115													●	
15- 90														●
ADMIXTURES - TYPES														
AIR-ENTRAINING					○	○		○						
WATER REDUCING	○		○											
SET RETARDING	●													
SET ACCELERATING			○											
DAMPPROOFING AGENTS								○						
GAS FORMING/FOAMING AGENTS														
POZZOLANS									□	□	□			
LOWEST PRACTICABLE WATER/CEMENT RATIO				●	□	□		●						

B1.2 CEMENT/CONCRETE

NOTES:

① Shrinkage at setting generally controlled through water/cement ratio. The less water used the lower the shrinkage. Proprietary cements are available which induce expansion in concrete during curing; generally only used in reinforced concrete floors.

HEAT OF HYDRATION			TIME OF SETTING			SHRINKAGE AT SETTING ①			
LOW	MODERATE	HIGH	SLOW	NORMAL	FAST	LOW	NORMAL	HIGH	
	●			●			●		• type I also available in buff or white.
		●			●			●	• type III also available in white.
	●			●			●		• type II will have moderate heat of hydration property when so specified; available in white.
●				●		□			• cooling of fresh concrete may be employed to control heat of hydration.
	●			●			●		• regular cement with air-entraining admixture may be substituted for these types.
	●			●			●		• also same types with air-entraining addition.
	●			●			●		

LOW	MODERATE	HIGH	SLOW	NORMAL	FAST	LOW	NORMAL	HIGH	
									• also used as a measure of overall quality of aggregate.
									• measure of volume stability; important in all structures subject to weathering.
									• affects strength and durability of all structures.
						□			• affects workability of plastic concrete. • affects porosity shrinkage and durability of hardened concrete.
						□			• affects workability and strength of concrete.
									• normal weight aggregate.
									• lightweight - structural aggregate.
									• lightweight - insulating aggregate.

LOW	MODERATE	HIGH	SLOW	NORMAL	FAST	LOW	NORMAL	HIGH	
									• ASTM C260. Decreases compressive strength at recommended amounts.
									• ASTM C494, type A May increase drying shrinkage.
○	○		○			○			• ASTM C494, type B for set retarding. • ASTM C494, type D for water reducing and set-retarding.
		○			○			○	• ASTM C494, type C for accelerating. • ASTM C494, type E for water reducing and accelerating.
									• investigate prior to selecting. Dense concrete will resist water penetration.
						□			• to induce expansion in concrete during setting. • will reduce strength of hardened concrete.
									• ASTM C618; used as substitutes, supplements or modifiers of cements and of other admixtures.
						●			• ratio of water, in pounds, to cement, in pounds.

B1.2 CEMENT/CONCRETE

CONCRETE: selection checklist

B1.2 CEMENT/CONCRETE

TYPE OF CONCRETE

Concrete in common usage for ordinary construction when no unusual conditions exist:

■ **Normal weight** aggregate and Type I cement:

• most economical assuming locally available aggregate can be used.

■ **Lightweight**-structural and Type I cement:

• generally more expensive than normal weight; used when reduction in weight of concrete is of importance, such as in site precast tilt-up panels; fill for floors above grade.

Special types commonly used:

■ **Polymer concrete** for curtain wall and wall facing panels.

• higher strength allows reduction in thickness.

■ **Fiber-reinforced concrete,** principally for wall facing panels but also for pavement overlays, floors.

• improved resistance to shrinkage cracking and/or higher strength results from incorporating various fibers in the mix.

When insulating value of concrete is a consideration - such as in roof fills - lightweight aggregates or cellular concrete may be used:

• the lower the density the higher the insulating value and the lower the strength.

NORMAL WEIGHT and LIGHTWEIGHT-STRUCTURAL CONCRETE

Considerations during the design of a mix and subsequent placement of concrete may include:

Strength and/or durability is generally of primary importance and will largely depend on the water/cement ratio in the mix:

■ Strength, durability, watertightness and other desirable properties decrease as the water/cement ratio increases.

• Theoretically, only enough water, as required by the hydration process of the cement paste, needs to be used:
• Practically such low water/cement ratio is impossible to achieve

and more water has to be added to allow the concrete to be mixed, placed, and consolidated.

• Admixtures - such as water reducers, plasticizers - may be used to improve workability while maintaining the same water/cement ratio.
• Low water/cement ratios result in stiff mixes which are difficult to place and consolidate.

Durability under exposure to freeze/thaw cycles will be improved by using air entrainment:

■ **Use of air-entrained cement,** or air-entrained admixture, will:

• Reduce strength for the same aggregate to cement ratio, but will also
• Improve overall resistance to weathering.
• Improve workability of plastic concrete.
• Reduce bleeding.

■ **Use of surface coatings,** such as penetrating oils, is highly recommended to protect the surface of fresh concrete pavements during the first winter season when subject to freeze/thaw cycles and deicing salts.

Resistance to abrasion/wear is of special importance in all traffic-supporting surfaces.

■ **In all cases,** concrete must be:

• Low slump.
• Well graded.
• Well consolidated.
• May require selection of special aggregate.
• May require special finish, such as steel trowelling.

■ **For floors subject to severe wear,** a topping with selected aggregate is recommended because:

• Placing and finishing a topping is easier to control;
• The more expensive select aggregate need not be used throughout the entire thickness of slab; but labor costs will increase.

Resistance to water permeability:

■ **Concrete with water/cement ratio of less than 0.49 by weight,** of low slump, properly placed and cured, will generally be watertight.

■ **The effectiveness of permeability-reducing agents,** and of dampproofing admixtures, is questionable: before selecting, suitability for a spe-

cific application should be investigated.

■ **In all instances,** effectiveness of concrete to resist passage of water will ultimately depend on:

• Strength of concrete;
• Reinforcing, if any;
• Proper detailing and treatment of joints;
• Proper placement of concrete;
• Proper curing;
• Density of concrete.

■ **Leaking joints or through-cracks** will render any treatment ineffective.

Setting time and its control:

■ **Setting time** of concrete depends mainly on ambient temperatures on the mass of concrete being placed. Control of setting time is needed:

■ **To minimize time** required for concrete to set during cold weather thus reducing length of time protection against freezing of fresh concrete is to be maintained:

Concrete placement in cold weather requires special precautions:

■ **Fresh concrete must be protected from freezing:**

• Water, and aggregates may have to be heated.
• Type III cement, set-accelerating admixture, or lower water/cement ratios may be necessary.

■ **The use of so-called antifreeze compounds** should not be permitted; the amount of such compounds required to lower the freeze point of concrete is so large that strength and other properties would be seriously affected.

■ **Insulation blankets,** bat insulation, or heated enclosures may be necessary. Dry straw, covered to keep it dry and in place, may be used for horizontal surfaces such as pavements and walks.

■ **When using heaters to protect fresh concrete:**

• Make certain concrete will not dry out too quickly.
• When using oil-burning heaters, keep them off the concrete slabs, and vent them properly; otherwise carbon dioxide will combine with calcium hydroxide in fresh concrete and will produce a weak surface layer.

Control of setting time is also required to increase time required for concrete to set during hot weather thus increasing time available to finish fresh concrete:

Concrete placement in hot weather requires special precautions and good planning:

■ **High temperatures** accelerate hardening of cement paste: set-retarding admixtures may be required.

■ **More mixing water is generally required for the same working consistency:**

• If water/cement ratio is not maintained, strength will be reduced.
• Larger water quantity will also result in higher drying shrinkages.

■ **Protection from too-rapid drying of concrete** is very important:

• Make certain concrete is wet at all times.
• Do not depend on formwork to prevent too rapid a loss of moisture.
• Aggregates, mixing water, conveying and placing equipment, as well as subgrade to receive concrete should be kept as cool as possible.

Control of setting time may also be required to increase time required for surface layer of concrete to set, when aggregate is to be exposed by removing such surface layer of cement paste; or to control:

■ **Bleeding:** the movement of water to the surface of freshly-placed concrete commonly needs to be controlled because

• Excessive bleeding will increase the water/cement ratio at the surface.
• Create a weak layer or laitance of poor durability.

Requirement for reinforcing:

■ **Evaluate need for reinforcing:**

• Reinforcing adds to material and placement costs.
• If to be selected for control of shrinkage cracking only, providing properly designed and spaced control joints may eliminate the need for reinforcing.

Fiber reinforcing:

The incorporation of short discontinuous fibers into cement matrix serves to increase the fracture

toughness of the composite through resultant crack-arresting process and results in an increase in the tensile and flexural strengths; the fibers act as secondary reinforcing for the concrete.

• High tensile strength of fibers is essential for a substantial reinforcing action.
• Fibers in a mix may affect workability depending on type of fiber used, and admixtures to improve workability, such as superplasticizers, may have to be added to concrete.

Some of the fibers available - and some typical uses of fiber-reinforced concrete include:

■ **Glass:** used for precast panels, facing panels, thin shell roofs, cement plaster and surface bonding mortars.

• glass fibers should be formulated to resist detrimental effects of alkalies in Portland cement; or admixtures to counteract such effects should be used in the concrete.

■ **Steel:** used in pavement overlays, bridge decks, runways, cellular concrete roofing units.

■ **Polypropylene:** for facing panels, floor slabs, pavements, pneumatically placed concrete, precast/extruded concrete, foundation piles, water containment structures, road patching/overlays.

Protection against corrosion of reinforcing steel:

■ **Corrosion can occur where:**

• Concrete is exposed to deicing salts, or seawater.
• Chloride admixtures are used.

■ **Mica flakes:** used as a replace, ment for mineral/ fibers in cement bound mineral fiberboards, facing panels, drainage pipes.

■ **Deterioration of concrete due to corrosion of steel manifests itself as:**

• Surface spalling.
• Cracks parallel and extending to the steel.

■ 2 percent calcium chloride addition by weight of cement is sufficient to cause corrosion

■ **Whether, and when, corrosion of reinforcing occurs** is determined by:

• Extent of concrete cover over steel.
• Rate of penetration of chloride to steel: the denser the concrete, the slower.

■ **Use of high quality concrete** can greatly slow down corrosion or even prevent it from starting.

■ **Measures to avoid corrosion include the use of**

• Cathodic protection.
• Polymer concrete.
• Coated reinforcing bars.

A Good and economical mix will have:

■ **The highest proportion of aggregate to cement possible,** consistent with the water/cement ratio required for strength and durability.

■ **The consistency necessary** for good workability or ease of placement.

■ **Reinforced** to resist all induced tensile, shearing, and compressive stresses.

■ **Properly placed,** consolidated, and finished.

■ **Properly cured.**

Field-placed concrete may be the most widely used and abused construction material:

■ **Quality control during blending of ingredients** - either at the plant or at the site - may be poor or non-existent.

• aggregate may be kept in open bins exposed to rain or snow and its moisture content may vary constantly thus affecting the ultimate water/cement ratio in a mix.
• aggregate may be stockpiled on the ground subject not only to rain or snow but also to contamination by dirt or ground moisture.
• admixtures may not be properly measured or not added at the proper time.

■ **Placement of concrete may not follow proper procedures:**

• concrete may be dropped into forms to cause separation of ingredients.
• concrete may be vibrated excessively causing separation; or undervibrated.
• excess water may be added in the field to improve workability at the expense of strength and durability.
• finishing may be started prematurely

CONCRETE: selection checklist (cont.)

which generally results in a poor wearing surface.

■ **Curing procedures** may be haphazard; or may not be used at all.

- curing compounds may not completely cover the surface.
- edges may not be protected.
- gaps may be left in coverings; or wet coverings may be allowed to dry out.

Stringent quality control measures should be adopted and enforced both at the plant and at the site. In addition, consideration should be given to measures which may minimize possibilities of detrimental practices in the field such as:

■ **Using plasticizers** to improve workability thus obviating the need to add excess water during placement;

■ **Selecting methods** of curing which may not be the most effective but simpler to follow and/or control.

■ **Vacuum dewatering** which is a special process, designed to remove excess mixing water from fresh concrete before final finishing is started; generally used for heavyduty industrial floors, it will:

- increase ultimate compressive strength
- provide higher initial strength
- provide control of finishing time
- which advantages must be considered versus higher costs.

POLYMER CONCRETE

Generally used in plant cast wall panels and wall facing panels:

- utilizes organic polymers such as polyester, methyl methacrylate, epoxy in lieu of cement as the binder for aggregate.
- properties of polymer concrete are largely dependent upon the properties of the polymer binder and its relative amount in the mix.
- generally compressive and tensile strength, durability, ductility, and chemical resistance is superior to that of Portland cement concrete; thermal expansion is high relative to Portland cement concrete; creep increases rapidly with increase in temperature.

POLYMER PORTLAND CEMENT CONCRETE

Organic material in liquid, powdery, or dispersed form is added to a fresh

concrete mixture. Considerations during selection may include:

■ **Polymer latexes:**

- materials used are the same as in regular concrete except for the addition of the latex.
- air-entrained cement should not be used; antifoaming agents may have to be used to control excessive air entrainment resulting from the use of latexes.
- latexes generally act as water reducing agents.
- curing characteristics differ from those of Portland cement concrete.

■ **Water-soluble polymers:**

- curing characteristics differ from those of polymer latexes.

Polymer Portland cement concrete generally has:

- higher tensile and compressive strength and improved durability than of regular concrete; also reduced permeability, water absorption, and water-vapor transmission.
- some formulations may lose strength when immersed in water or exposed to high humidities.
- resistance to chemicals depends on properties of polymer used.
- formulations using various polymers with differing properties may be available; generally as proprietary products.

AGGREGATES

The selection of aggregates should generally consider:

■ **Record of satisfactory performance** under conditions similar to those under which the aggregate to be selected will be used:

- if performace record is not available, tests to establish suitability of aggregate are recommended.

Specific properties of aggregate which may affect its performance;

■ **Poor resistance to abrasion** increases quantity of fines in aggregate, and may increase water requirements, resulting in lower strength of concrete.

■ **Fine aggregate should be checked for suitability for use in pavements,** because certain manufactured sands produce slippery pavement surfaces.

■ **Porous aggregate** will cause sur-

face pop-outs or spalls when exposed to freeze/thaw cycles.

■ **Alkali-silica reaction** of some reactive sands may cause surface pop-outs or spalling.

■ **Size of aggregate influences resistance to failure under freeze/ thaw cycles.**

- Larger size aggregates will generally minimize shrinkage and resultant surface cracking.
- Large size aggregate, properly graded, will generally require less water, consequently less cement for a given water/cement ratio.

■ **Size of aggregate is generally limited by conditions of placement.** Maximum size should not exceed:

- One-fifth the dimension of non-reinforced members.
- Three-fourths the clear spacing between reinforcing bars and forms.
- One-third the depth of slabs.

■ **Rough textured, or flat and elongated particles** will require more water as well as cement, if the required strength is to be maintained.

■ **Avoid having long silvery particles** or limit such to no more than 15 percent by weight.

■ **Rough, angular aggregate bonds** better with cement paste. Select when

- Flexural strength is important.
- High compressive strength is required.

■ **Use of low alkali cement, pozzolanic additives, or replacing approximately 30 percent of alkali-reactive aggregate with crushed limestone aggregate,** is often effective in preventing deterioration due to alkali-aggregate reaction.

Types of Aggregate: In addition to crushed stone, river pebbles and other natural aggregate used as available locally, there are special types of aggregate, manufactured and distributed nationally:

- Vermiculite, perlite: higher thermal and fire resistance; low strength, high shrinkage.
- Expanded slag, shale or clay—medium insulating value; low shrinkage, good strength.
- Volcanic: pumice, scoria, cinders — intermediate strength, shrinkage.

B1.2 CEMENT/CONCRETE

- Diatomite: low strength.

Other types include:

- Blast-furnace slag: air-cooled for use with regular concrete; foamed slag for lightweight insulating concrete.
- Reclaimed concrete: compressive strength lower than of normal aggregate.
- Various industrial wastes such as: by-products from power plants; metalurgical slags other than blast-furnace; mining and quarrying wastes; incinerator residue.

ADMIXTURES

Substances added to a concrete batch - immediately before or during its mixing - in order to alter or enhance specific property or properties of fresh and/or hardened concrete. Some of the more commonly used admixtures are:

■ **Air-entraining agents:** used with nonair-entraining cements when content of entrained air in concrete must be varied between batches or when air-entraining cement is not readily available.

■ **Accelerators:** used in cold weather to increase the rate of early strength development to reduce the curing and protection period necessary.

- Calcium chloride is the most widely used admixture of the many available chemical compounds: may reduce tensile strength of concrete; will reduce flexural strength; will reduce resistance to sulfate attack, and to alkali-aggregate reaction; will increase shrinkage and creep in concrete; should not cause significant corrosion in reinforcing steel if amount added does not exceed 1.5 percent.

■ **Reducers:** are used to reduce water requirement of concrete for a given consistency; with such reduction varying from 5 to 15 percent.

Higher water reductions - 25 to 35 percent - are possible with super plasticizers: high-range water reducers.

- Basic advantages of reducers are: high workability of concrete without reduction in cement content and strength; high strength concrete with normal workability but lower

water content; a concrete mix with less cement but normal strength and workability.
- Various formulations with differing properties available.

■ **Retarders:** used in hot weather to extend the time for placing and/or finishing operations by preventing early set of cement.

- Use will result in some loss of strength in concrete at early age; over dosage may increase such loss significantly.
- Some formulations may result in excessive shrinkage with resultant cracking.

■ **Gas-forming agents:** may be used in small quantities in grouts to cause slight expansion prior to hardening; when used in large quantities will produce cellular concrete.

■ **Compatibility of an admixture with cement,** aggregate and any other admixtures to be used, should be thoroughly investigated for effects, as well as any side effects of admixture on properties of plastic and of hardened concrete, including:

- Workability
- Time of setting
- Drying shrinkage
- Strength
- Permeability

■ **The possibility should be investigated of avoiding the use of admixtures by:**

- Modifying design of mix
- Selecting different type of cement, or aggregate
- Modifying cement content
- Modifying water/cement ratio

■ **The use of admixtures will never correct deficiencies in:**

- Concrete mix design
- Placement and consolidation
- Finishing
- Curing

POLYMER-IMPREGNATED CONCRETE:

Is a hydrated Portland cement concrete in its final form which is impregnated with a monomer and subsequently polymerized in place:

- monomers used may include: methyl

methacrylate; sulphur; styrene; vinyl acetate.
- polymerization may be through: heating above ambient temperatures; use of promoters or accelerators at ambient temperatures; ionizing radiation, such as by gamma rays emitted by cobalt-60 at room temperatures.

■ **Polymer-impregnated concrete** looks very much like ordinary concrete, but its properties are different:

- strength in compression, tension, and flexure may be increased as much as four times over that of ordinary concrete.
- creep is generally less than for ordinary concrete.
- durability under most forms of environmental attack is significantly improved.
- both normal weight and lightweight concrete may be polymer-impregnated.
- polymer-impregnated concrete tends to be more brittle than ordinary concrete: the addition of steel-fiber reinforcement will increase flexural strength as compared to plain polymer-impregnated concrete.
- fiber reinforced polymer-impregnated concrete will cost about double that of ordinary concrete.

■ **Polymer-impregnated concrete** has been used in: marine structures; bridge decks; pipe linings.

- common usage to date has been: in the construction and repair of concrete bridge decks to reduce the potential damage due to deicing salt penetration; for chemical resistant floors.

■ **Mineral Waste Admixtures**

- Using admixtures in cement saves energy, cuts CO_2 emissions, reduces the limestone that is mined and uses material that is otherwise landfilled.

- A leading candidate is fly ash, the particulated mineral impurities that remains after coal is burned. Fly ash improves workability and reduces the amount of water that's needed.

- Research studies are being conducted by the Civil Engineering Research Foundation (CERF) to determine the results of admixtures derived from fly ash and other industrial waste.

B1.2 CEMENT/CONCRETE

FINISHES: types, uses

TYPE	DESCRIPTION	EXPOSED FACE
INTEGRALLY PIGMENTED	**Coloring agents** - commonly proprietary formulations - are added to a batch of concrete during mixing. • available colors include various shades of: cream, buff, brown, red, green, blue, gray, black. • blending of several colors should not be attempted as uniform results are extremely difficult to obtain; even when using a single coloring agent variation between batches may be expected. • uniformity of all ingredients in a specific mix must be maintained. • use of white cement results in cleaner and purer colors; buff cement may be available in some localities. • compatibility with any other admixture to be used should be verified.	**Exposed face may be:** • smooth trowelled • lightly to heavily textured • sand blasted • tooled; stamped • form liners may be used • Coloring agents may be used in conjuction with select aggregate when to be exposed in the finished work.
SCREEDED TROWELLED	**Screeding** - or strike off - is a rough finish used on horizontal surfaces not exposed to weather, wear, or view; and for base course of floors to receive cementitious toppings. Steel trowelled: for horizontal surfaces exposed to wear: • may be used to embed hard, abrasion-resistant aggregate into the surface of fresh concrete during finishing operations for floors subject to severe wear. • aggregate for areas expected to be frequently wet should be slip and rust resistant. • non-sparking aggregate available for areas where danger of explosion exists.	• Steel trowelling produces dense, smooth surface. • When properly done no tooling marks will be noticeable.
SWIRL	**Moderately rough finish for horizontal surfaces where slip resistance is required and appearance is a consideration:** • aluminum and magnesium floats will produce a medium-textured surface. • wood and cork floats will produce coarser textures. • patterns are introduced by using the float in a series of uniform arcs or twists. • maintaining uniformity of finish with several workmen involved in the finishing operation may present a problem.	similar to broomed
BROOMED	**Light-to-heavy rough finish for horizontal surfaces where slip resistance is required and appearance is a consideration:** • brushes are used and the stiffness of bristles, as well as time before setting of concrete at which finishing takes place, will determine depth of texture. • when uniformity of texture is to be maintained, timing of finishing process becomes of primary importance. • pattern may be introduced by moving the brush in a series of uniform arcs or twists.	

B1.2 CEMENT/CONCRETE

B · BUILDING SHELL

TYPE	DESCRIPTION	EXPOSED FACE
BURLAP DRAG	**Medium-to-heavy texture for large surfaces, such as roadway pavements, where slip resistance is required and appearance is of secondary importance** • water-soaked burlap is dragged over the finished concrete surface. • economical method of producing slip-resistant surfaces; but uniformity is difficult if not impossible to achieve.	SIMILAR TO BROOMED
TRAVERTINE and ROCK SALT	**Travertine finish:** • finish coat of pigmented white cement is applied by throwing it vigorously over screeded - and broomed for better bonding - horizontal concrete slab to produce an uneven surface with ridges and depressions. • after the finish coat hardens sufficiently, it is steel trowelled to flatten the ridges. **Rock Salt Finish:** • is similar in appearance to travertine finish. • produced by scattering rock salt over a trowelled surface and then dissolving the salt after the concrete has hardened.	
STAMPED	**Proprietary aluminum tools are used to imprint various patterns in freshly placed horizontal concrete surfaces.** • used in lieu of scratching a design into fresh concrete with a tool. • patterns generally available are: brick; tile, cobblestone. • before imprinting, concrete is trowelled to the texture desired. • coloring agents are commonly used; generally as surface shake-on. • joints may be left open or filled with matching or contrasting mortar.	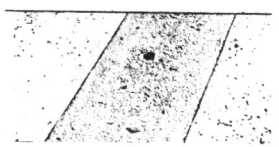 **FLAGSTONE PATTERN**
EXPOSED AGGREGATE horizontal surfaces - rough	**Aggregate of the mix** - or selected aggregate embedded in the fresh concrete after it has been placed and screeded - exposed by removing the surface layer of cement paste by brush and water. • aggregate should be spread evenly over the surface. • uniformity of size and color should be closely controlled. • when the surface to receive aggregate is large, a set-retarding admixture may have to be used to extend the available time for finishing. • curing method selected should be such that the surface will not be stained.	EXPOSED AGGREGATE - see following page, similar to vertical surfaces.

B • BUILDING SHELL

B1.2 CEMENT/CONCRETE

FINISHES: types, uses (con't)

TYPE	DESCRIPTION	EXPOSED FACE

EXPOSED AGGREGATE
horizontal surfaces - smooth

Crushed stone, most commonly marble aggregate in concrete toppings over concrete base slabs:

* surface is ground smooth after concrete has hardened to fully expose the aggregate and to provide a wearing surface.
* may be plant precast as: flooring tile; stair treads; shower receptors; and also as wall facing panels.
* when used as flooring, surface sealers are commonly applied for protection.
* cement used may be gray but generally white or white and pigmented cement is selected.

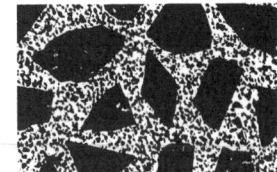

EXPOSED AGGREGATE
vertical surfaces

Aggregate may be embedded into precast concrete panels - during casting when in horizontal position - using the same procedures as for horizontal surfaces; or may be exposed by sandblasting. Aggregate may be incorporated into vertical surfaces cast-in-place by the aggregate transfer method; or may be exposed by sandblasting if integral.

* aggregates used may include: crushed marble; crushed rock; select gravel; enamelled stone; flagstone; plastic chips, such as polyester.
* aggregate added by the transfer method generally requires sandblasting to properly expose it.
* flagstones generally incorporated into site-cast, tilt-up panels by casting concrete over them.

FORM BOARDS

Form material - boards or panels such as plywood - may be used to impart a surface texture to concrete:

* form boards may be selected for their grain pattern; and such pattern may be enhanced by soaking the boards in water or by sandblasting their surface.
* plywood may be textured and/or patterned.
* joints between boards or plywood panels must be tight to prevent leakage of fines in the concrete mix and the treatment of joints may also be used to impart a texture to the concrete.
* differential water absorption of boards if not properly sealed may effect color of hardened concrete.

FORM LINERS
plastic

Thermoplastic or synthetic rubber sheets formed in a variety of patterns and/or textures are attached to the inside faces of formwork; concrete is then placed and, after it hardens, the formwork and liners are stripped.

* joints between liner panels should be watertight to prevent bleeding-through of cement paste and fines.
* form-release agents may have to be used on some form liners for ease in stripping.
* patterns which channel water runoff are recommended since streaking resulting from such runoff will tend to enhance the pattern.
* liners are expensive and means for their repeated use should be considered.

B1.2 CEMENT/CONCRETE

B • BUILDING SHELL

	DESCRIPTION	EXPOSED FACE

FORM LINERS
plaster of paris
glass fiber

Sculptured surfaces are obtained by placing concrete over or against formwork such as:

- molded plaster of paris; glass-fiber-reinforced plastic.
- wood, plywood, or fiberboard forms to which batten strips of various sizes have been attached.
- polystyrene panels built up in various forms and thicknesses; cut to various shapes and/or gauged out; heat shaped.
- sand molds.
- formwork is the most expensive part of the process, and multiple use of such formwork will reduce the cost; sand molds cannot be reused.

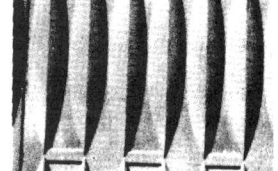

DIMPLED

Concrete cast over a bed of crushed stone or gravel aggregate against a polyethylene sheet between them to prevent bonding:

- texture may be fine to medium depending on the size of aggregate used.
- placing concrete over plastic film will produce slick, shiny surface.
- quality control during placing is very important: dirt may be tracked on to the surface of the plastic and become embedded in the concrete; reinforcing chair legs may puncture the film; color variations from one batch to another should be kept to a minimum.

SAND BLASTED
and
GROUND

Surfaces of hardened concrete may be sandblasted to:

- remove surface blemishes or surface layer of laitance for an even slightly-textured appearance.
- expose aggregate to varying degrees.
- create textured patterns through the use of templets.

Grinding is used to remove fines and form marks from hardened concrete for a smooth, relatively-uniform appearance; except that voids in the surface will remain.

- wet grinding, commonly referred to as wet rubbing, of hardened concrete uses a cement slurry to fill in voids in the surface prior to grinding.

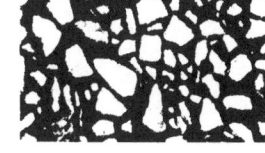

BUSHHAMMERED

Hardened concrete may be tooled - bushhammered or point tooled - to remove the smooth outer layer of hardened cement paste and reveal the rough textured aggregate below.

- joints in formwork may show after tooling, and their location should be considered at the design stage.
- tooling cannot be used when soft aggregate, such as sandstone, is used.
- appearance of finished surface depends on the person doing the work: adjacent surfaces tooled by different workmen may have somewhat different textures

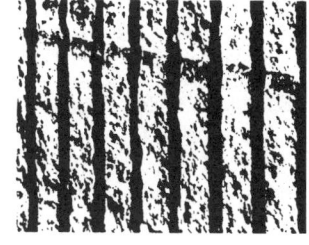

B1.2 CEMENT/CONCRETE

FINISHES/FORMS/CURING: selection checklist

The selection of concrete to remain exposed in the finished work should generally include considerations of:

■ **Ingredients of the concrete mix:**

- Coarse aggregate type: size, color, properties, proportional amount in the mix.
- Fine aggregate: type, color, properties, proportional amount in the mix.
- Cement: type, color.
- Admixtures, if any, and their effect.

■ **Formwork and reinforcing:**

- Materials to contain plastic concrete and their effect on the appearance of hardened concrete.
- Inserts, liners to impart pattern and/or texture to hardened concrete.
- Methods of consolidating plastic concrete during placement.

■ Methods of curing fresh concrete.

CONCRETE MIX AND ITS EFFECTS ON SURFACE FINISH

As-cast finish is one where concrete is left as it appears after the form is removed:

■ **Interior, face or faces of** forms may be:

- Smooth: such as metal, glass fiber, reinforced plastic, overlaid plywood; fiberboard.
- Textured: such as formboards, batten strips, rough-sawn veneered plywood; attached form liners.
- Sculptured, molded: using molded plastic or plaster forms; polystyrene laminations.

■ **Cement will be the primary factor in appearance;** with the influence of coarse aggregate minimal, and that of fine aggregate of some importance.

- Darker and more colorful fine aggregates influence the color of the concrete to a higher degree when white cement is used.

■ **Special attention should be given to the quality of formwork.**

- To minimize discoloration due to varying rates of absorption of form materials;
- And to surface variations which may result from leakage of fines through joints.

■ **Form joints are critical** since their location will be visible in the finished surface.

■ **Uniformly-graded concrete mixes may be successfully used for as-cast finishes,** but the consistency must be uniform from one batch to another.

- Slump should be held in the range of three inches plus or minus one-half inch, and never over four inches.
- Effective consolidation techniques and thorough blending of lifts are very important.
- Mix should not be oversanded nor should additional water be added even when pumping is used for placement.

■ **As-cast smooth uniform surface, using gray cement in the mix, is the most difficult finish to accomplish successfully.**

- Form texture will help considerably to achieve a visually-uniform surface.

Sandblasted finish is commonly classified as:

■ **Brush blast:** where the surface is lightly blasted to remove minor surface variations:

- Leaves surface slightly textured.
- Surface will seldom appear uniform at close range and should be inspected from a distance for evenness.
- Concrete mix may be evenly graded, as color is largely determined by cement used.
- Requirements for formwork are essentially the same as for as-cast finish; form butt joints should not be taped as they are likely to show in the finished surface.
- Proper placement of concrete is extremely important since any segregation or variation in the mix will remain visible.
- Finishing may generally be started seven days after concrete is placed which usually is the preferred time.

■ **Light blast:** removes the surface skin sufficiently to expose the larger particles of the fine aggregate as well as some of the coarse aggregate.

- Leaves surface relatively flat with little texture.
- Conventional concrete mixes may be used; except that it is recommended that at least ten percent

more of coarse aggregate be added.
- Color of finished surface will be primarily determined by color of fine aggregate.
- Will remove slight surface imperfections, such as those resulting from taping of joints in formwork.
- Finishing may be started right after forms are removed, but a delay of seven or more days after concrete is placed is commonly suggested.

■ **Medium blast:** exposes the coarse aggregate to the extent that it projects slightly from the hardened matrix of cement paste and fine aggregate.

- Coarse aggregate should be uniformly distributed over the surface to produce a good finish.
- Concrete mix should be designed with a higher proportion of coarse to fine aggregate to minimize the probability of uneven distribution of coarse aggregate at the surface.
- Coarse aggregate should be dense, hard material to avoid the possibility of it being eroded during the finishing process.
- Finishing should be done prior to seven days after concrete is placed.

■ **Heavy blast:** erodes the matrix of cement paste and fine aggregate to significantly expose the coarse aggregate.

- Lack of uniform distribution of coarse aggregate will result in unsatisfactory finish; approximately 80 percent of the exposed surface should contain coarse aggregate.
- Special-mix design - such as gap grading or preplacement of aggregate - should be selected: heavy blast finish will only be successful if there is sufficient aggregate to be exposed.
- Surface imperfections will be removed to a greater extent than with lighter blasting; but uneven absorption of moisture by the formwork or leakage through joints will result in hard, dark areas which may remain after finishing.
- Forms should be removed within 24 hours after concrete is placed and finishing should start right after removal of forms.
- When forms cannot be stripped at the proper time for finishing, the use of surface set retarders should be considered; but their use may

also create problems, such as over-, under- or uneven-retarding.

EXPOSED AGGREGATE

Principal considerations for exposed aggregate finish should include:

■ Methods of exposing aggregate:

- Sandblasting to expose aggregate in any hardened concrete surface; vertical, horizontal, curved, flat, patterned, molded, sculptured.
- Seeding aggregate over the horizontal surface of concrete right after placing and screeding is completed; forcing the aggregate into the cement matrix and then removing the surface layer of the matrix to expose the aggregate.
- Casting concrete over a horizontal bed or aggregate with sand to partially fill the voids between the aggregate: after the concrete cures and is removed from the form, the sand is brushed away to expose the aggregate: this method may be used to embed large-size aggregate, such as flagstone.
- Gluing aggregate to the inside face of vertical formwork which aggregate then remains embedded in the concrete after forms are stripped.

■ Properties of aggregate to be exposed:

- Color: Marble generally offers the widest choice among natural aggregates; ceramic aggregates and glass available in varied and brilliant colors.
- Hardness: As a measure of durability under weathering; to resist effects of sandblasting when such process used to expose it.
- Size: Will determine extent to which it can be exposed and appearance when viewed at a distance; exposure should be not more than one-third the average diameter of the aggregate particles.
- Shape: Will affect surface pattern and texture; will affect bonding of aggregate to cement matrix; rounded aggregate is practically self-cleaning, while angular aggregate tends to collect dirt with the dirt generally confined to the matrix.
- Gradation: should be closely controlled to ensure uniformity of surface texture.
- Chemical stability: Glass aggregate should be resistant to alkali-attack; when acid will be used to clean the finished surface, aggre-

gates with high calcium content - such as marble, limestone - should not be used as they will discolor and may dissolve in muriatic acid.

FORM MATERIAL and LINERS

Considerations during selection of form materials to impart a texture and/or pattern to vertically- or horizontally-placed concrete surfaces may include:

■ Formboards and plywood may be planed, rough sawn; grain pattern may be exaggerated by sandblasting the surface or by soaking the boards.

- Boards should have tongue and groove joints to minimize leakage of fines.
- Edges of boards may be: chamfered to produce a raised joint in concrete - which, however, may be damaged during stripping of forms - or covered by a batten strip which results in a recessed joint.
- Direction of boards will influence visual effects of streaking by water runoff over the surface: worst effects on relatively smooth surfaces will be minimized when the direction of the boards is vertical.

■ A range of patterns may be achieved by adding batten strips - or other materials - to the inside of ordinary plywood forms to produce special effects.

■ Form liners may be used to produce a wide variety of surface patterns and/or textures.

- Liners should be selected which will not bond to fresh concrete; or form release coatings should be used.
- Reuse of liners should be planned to minimize costs.
- Joints should be sealed to prevent leaking through of fines and discoloration of concrete.
- Deep patterns produced through use of form liners may be further tooled to impart a texture.

TOOLING

A satisfactory tooled finish can generally be achieved on any dense, properly graded and placed concrete. The aggregate selection should consider:

- Natural gravels tend to shatter when bushhammered.
- Some hard aggregates, such as quartz, granite, are difficult to busham-

mer uniformly.
- Coarse aggregate size commonly used for tooled surfaces is ⅜ inch or less; larger aggregate is preferred for jackhammered surfaces.
- Bushhammering should not be started before concrete reaches a compressive strength of a least 3750 psi.

COMBINATION FINISHES

More than one finishing process may be used to achieve different effects, such as: form liners and tooling; form liners or inserts and sandblasting.

SEALERS

Sealers may be used to:

- Minimize absorption of moisture by the concrete, which moisture will temporarily darken its surface, depending on amount absorbed.
- Protect finished surfaces from dirt, stains, concrete rundown during and/or after construction.

Sealers commonly used are based on silicones, stearates, and acrylics:

- Clear sealers found more effective than others by the Portland Cement Association, based on tests conducted both in the laboratory and outdoors, were methyl methacrylate forms of acrylic resin.
- Sealers used for exterior surfaces should be water vapor permeable; impermeable sealers or surface coatings are not recommended.
- Some sealers will permanently darken the surface of the cement matrix; more noticeably when white cement and light color fine aggregate is used.
- Sealers may have to be periodically renewed to remain effective; may be applied or may weather unevenly resulting in blotchy appearance due to differential absorption of moisture.

FORMWORK

Selection of forming material should generally include considerations of:

■ Water absorbency characteristics of materials to be used:

- Absorbent materials remove water from the plastic concrete and reduce water/cement ratio at its surface: the lower the water/cement ratio, the darker will be the color of hardened concrete.

B1.2 CEMENT/CONCRETE

FINISHES/FORMS/CURING:

- Nonuniform water absorption by forming materials will result in variations in color over the surface of concrete.

■ **Plywood forming materials** for architectural concrete work commonly used are:

- Plastic-coated birch plywood produced in Finland: may be the best material currently available for architectural finishes; has high average reuse factor; available in larger special sizes to reduce amount of butt joint treatment required.
- Plastic overlaid plywood, classified as medium or high density depending on thickness of overlay: problems sometimes encountered include embrittlement of overlay or development of pink discoloration in concrete, both due to reaction with cement matrix or admixtures when used; reuse for high density type may be 10 or more.
- Oil-impregnated construction-grade: generally does not produce acceptable finish due to lack of uniform absorbency; graining and patches may be reflected in the finished surface; wood sealers or special coatings should be considered for a moisture-impervious surface; wood sealers may last for an average of five reuses.

■ **Steel forms will produce high quality finishes:**

- Economical when multiple use is possible; generally may be reused 50 to 100 times when properly handled.
- May have to be epoxy-coated to eliminate problems of rusting.
- Jointing problems are minimized as compared to plywood forming systems.

■ **Glass-fiber reinforced plastic** forms are similar to steel forms but lighter in weight and have no problems with rusting.

- Reuse is commonly 20 to 30 times; up to 100 reuses have been achieved with proper construction and handling.

Considerations during detailing and construction of form work should generally include:

■ **Preventing loss of moisture** from the plastic concrete:

- Joints should be treated to: eliminate leakage through them; and/or differential moisture absorption into the forming material through exposed edges at such joints.
- Rustication strips at joints in plywood forms is the most common method used.
- When using medium or heavy exposed, aggregate finish joints is plywood forms may be taped with thin adhesive tape to prevent loss of moisture; tape marks may remain visible when brush or light sandblasted finish is used.
- Chamfer strips - wood or plastic, with plastic the preferred type - should be considered to prevent leakage through corner joints.

PLACING OF CONCRETE

Placement and compaction must be done properly the first time as most mistakes cannot be hidden from view in the finished surface:

■ **Concrete should be placed within the formwork without segregating due to being dropped an excessive distance,** and the use of a tremie may have to be considered to avoid such possibility.

■ **Splattering of concrete on the formwork must be minimized** to greatest extent possible:

- Steel forms when hot can cause a flash set of cement paste splatter which then will be clearly visible on as-cast or even lightly sandblasted surfaces.

■ **Proper vibration of concrete** to consolidate and also to blend lifts of concrete within the formwork is extremely important:

- If not properly performed, highly-objectionable lift lines will be visible on the finished surface.

■ **Lightweight concrete,** due to lack of particle weight, presents far greater problems in consolidation than normal weight concrete:

- Considerably more surface blowholes should be expected with lightweight aggregates.

■ **Accessibility to equipment:**

- The best and least expensive method of placement is from a stationary or mobile mixer directly to final location.
- Any other method will be more costly, as well as potentially detrimental to quality of concrete being placed, due to possible segregation of ingredients.
- Pumping concrete is generally the second best choice.

CURING

Desirable properties of concrete continue to improve with age as long as

■ **Moisture is present.**

■ **Temperatures are favorable** - the hydration process practically stops at near or below freezing temperatures. Desirable properties include:

- Resistance to freeze/thaw cycles
- Strength
- Watertightness
- Wear resistance
- Dimensional stability

Undesirable effects due to excessive evaporation of water from newly-placed concrete will:

■ Significantly retard the cement hydration process.

■ Induce plastic shrinkage cracking, especially during hot weather.

■ Induce shrinkage and resulting tensile stresses at the drying surface.

■ Result in surface and through-cracking if drying occurs before concrete has attained adequate strength.

CURING PERIOD

Concrete should be protected during the first few hours after being placed and finished until proper curing is started:

- A light fog spray is an effective means but should not be used during finishing; it might weaken the cement paste at the surface.

■ **The curing period** should in all cases be as long as practicable except where concrete is going to be

selection checklist (cont.)

exposed to deicing chemicals. In that case it should be allowed to:

* air-dry after curing for at least one month before anticipated exposure to such chemicals to enhance its resistance to scaling.

■ **Length of time the newly-placed concrete should be protected** from loss of moisture depends on:

* Type of cement used
* Proportions of mix
* Required strength
* Size, thickness, and shape
* Weather conditions
* Future exposure conditions.

METHODS OF CURING

Water ponding: used for curing of all horizontal surfaces, such as pavements, floors, horizontally cast panels, for containing water over the surface with sand or earth dams:

* Effective but expensive method:
* Sand or earth dams must be first constructed to properly contain the water and then completely removed after curing is completed.
* Earth used for dams may stain the surface of green concrete.
* Will help to maintain uniform temperature in concrete for proper curing.

Continuous sprinkling: may be used for horizontal and vertical surfaces, such as floors, walls, pavements as well as for stucco.

* Fine spray is continuously applied through nozzles or soil-soaker hoses.
* Constant supervision or sprinkling is required thus increasing labor costs.

Wet coverings - such as burlap, cotton mats - used on horizontal and curved surfaces:

* sand, earth, straw, hay may be used if kept constantly wet; but some may stain concrete.
* supervision required to ensure that coverings do not dry out with corresponding increase in labor costs.

Waterproof paper and plastic sheets: for horizontal surfaces

* Effectiveness depends on preventing moisture rising from concrete or

escaping through joints between sheets, and such sheets should be not only lapped but often also sealed at joints.
* When exposed to wind, coverings must be well secured in place to prevent blowoff.
* May stain concrete surfaces.
* Waterproof paper and plastic should conform to requirements of ASTM C171.

Forms left in place for vertical surfaces should be kept moist, especially in hot weather, to prevent premature drying out of concrete.

* Commonly spray or soil-soaker hoses are used.
* Constant supervision may be required.

Steam curing: almost exclusively used in precasting plants; very effective.

Curing compounds: are applied in liquid form by spray, brush, or roller on horizontal or vertical surfaces.

* Curing compounds retard loss of surface water and may also be formulated to:
* Harden and seal the surface of concrete;
* Act as bond breakers in site precast tilt-up panel construction.
* Available clear; clear with a fugitive dye which fades out after a few days; or white pigmented.
* May be used for further curing of vertical surfaces after forms are removed; or for additional curing after initial wet curing is completed.
* Curing compounds used should conform to requirements of ASTM C309.

■ **When selecting curing compounds,** consideration should also be given to potential drawbacks:

* Quality control to ensure proper application for complete coverage at the right time is very important.
* May prevent bond between hardened concrete and fresh concrete or plastering materials.
* Some formulations may not be compatible with adhesives or bonding agents to be used in the installation of facings, flooring.
* May prevent proper bonding of applied coatings.

■ Should not be used on surfaces placed in late fall that may be soon

be exposed to deicing chemicals:

* Compound remaining on surface may prevent the air-drying necessary for resistance to scaling.

B1.2 CEMENT/CONCRETE

REFERENCES: Cement/Concrete

STANDARDS

ASTM Specifications for:

C33	• Concrete Aggregates
C91	• Masonry Cement
C94	• Ready-Mixed Concrete
C150	• Portland Cement
C171	• Sheet Materials for Curing Concrete
C226	• Air-Entraining Additions for Use in the Manufacture of Air-Entraining Portland Cement
C260	• Air-Entraining Admixtures for Concrete
C309	• Liquid Membrane-Forming Compounds for Curing Concrete
C330	• Lightweight Aggregates for Structural Concrete
C332	• Lightweight Aggregates for Insulating Concrete
C387	• Packaged, Dry, Combined Materials for Mortar and Concrete
C470	• Molds for Forming Concrete Test Cylinders Vertically
C494	• Chemical Admixtures for Concrete
C595	• Blended Hydraulic Cements
C618	• Fly Ash and Raw or Calcined Natural Pozzolan for Use as a Mineral Admixture in Portland Cement Concrete
C685	• Concrete Made by Volumetric Batching and Continuous Mixing
C979	• Pigments for Integrally Colored Concrete
D994	• Preformed Expansion Joint Filler for Concrete (Bituminous Type)

ASTM Tests/Methods for:

C31	• Making and Curing Concrete Test Specimens in the Field
C143	• Slump of Portland Cement Concrete
C172	• Sampling Freshly Mixed Concrete
C666	• Resistance of Concrete to Rapid Freezing and Thawing
C672	• Scaling Resistance of Concrete Surfaces Exposed to Deicing Chemicals
C779	• Abrasion Resistance of Horizontal Concrete Surfaces

SELECTED REFERENCES

ACI	**American Concrete Institute**	
	• Admixtures for Concrete and Guide for Use of	
	• Admixtures in Concrete	
	• Polymers in Concrete	
	• Manual of Concrete Practice: Part I: Materials and Properties of Concrete Part II: Construction Practices and Inspection - Pavements Part III: Use of Concrete in Buildings - Design, Specifications and Related Topics Part V: Masonry; Precast Concrete; Special Processes	
ACI-12-01	• Formwork for Concrete	
ANSI	**American National Standards Institute**	
ANSI/ACI 211.1-81	• Normal Heavyweight, and Mass Concrete.	
ANSI/ACI 318-80	• Building Code Requirements for Reinforced Concrete.	
ANSI/ACI 104-71	• Preparation of Notation for Concrete.	
ANSI/ACI 301-79	• Structural Concrete for Buildings	
CRSI	**Concrete Reinforcing Steel Institute** • CRSI Handbook	
CSDA	**Concrete Sawing & Drilling Association** • Make Openings in Concrete a Better Way • Concrete Sawing and Drilling the Modern Way	

	• Concrete Sawing - What Choice Joints	
NPCA	**National Precast Concrete Association** • Fundamentals of Quality Precast Concrete, J. J. Waddell	
NSA	**National Stone Association** • Quality Concrete with Crushed Stone Aggregate • High Strength Concrete • Stone Sand for Portland Cement Concrete	
NRCC	**National Research Council of America**	
CBD 15	• Concrete	
22	• Concrete Floor Finishes	
93	• Precast Concrete Walls - Problems With Conventional Design	
94	• Precast Concrete Walls - A New Basis for Design	
103	• Admixtures in Portland Cement Concrete	
116	• Durability of Concrete Under Water Conditions	
119	• Volume Change and Creep in Concrete	
136	• Concrete in Sulphate Environments	
145	• Portland Cements in Building Construction	
153	• Removal of Stains from Concrete Surfaces	
165	• Calcium Chloride in Concrete	
187	• Non-Destructive Testing of Concrete	
203	• Superplasticizers in Concrete	
215	• Waste and By-Products As Concrete Aggregates	
223	• Fiber Reinforced Concrete	
PCA	**Portland Cement Association** • Effect of Various Substances on Concrete and Protective Treatments, Where Required • Corrosion of Nonferrous Metals in Contact with Concrete	

B1.2 CEMENT/CONCRETE

- Design of Shrinkage-Compensating Concrete Slabs
- Admixtures for Concrete
- Air-Entrained Concrete
- Design of Concrete Mixtures
- The Concrete Approach to Energy Conservation
- Heavy Building Envelopes and Dynamic Response
- Acoustics of Concrete in Buildings
- Fire Resistance of Lightweight Insulating Concretes

PCI   **Prestressed Concrete Institute**

MNL-120-78
- PCI Design Handbook— Precast Prestressed Concrete, Third Edition.

MNL-121-77
- PCI Manual for Structural Design of

MNL-122-73
- PCI Architectural Precast Concrete.

MNL-123-73
- PCI Manual on Design of Connections for Precast Prestressed Concrete.

MNL-124-77
- PCI Design for Fire Resistance of Precast Prestressed Concrete.

PCA-13-04
- Design and Control of Concrete Mixtures

PTI   **Post-Tensioning Institute**
- Post-Tensioning Manual
- Fire Resistance of Post-Tensioned Structures

- **Finish Hardened Concrete, CCP-01- 21-D, Concrete Construction**
- **Finishing Plastic Concrete, CCP-01-22-E, Concrete Construction**
- **Recommendations for the Production of High Quality Concrete Surfaces, L. S. Blake, Concrete Construction**
- **Vibration Concrete, CCP-01-40-X, Concrete Construction**
- **Visual Concrete: Design and Production, W. Monks, CCA-18-08, Concrete Construction**

B1.2 CEMENT/CONCRETE

B1. ❸ STONE/MASONRY

STONE: introduction

Stone is fabricated from rocks into individual panels, veneers, ornamental pieces, facing units or blocks, commonly referred to as dimensional stone; or, fragments of rocks designated as broken or crushed stone. Natural stone is generally selected for its aesthetic qualities and strength.

■ **There are three classifications of rock:**
- **Igneous:** Such as granite, basalt.
- **Sedimentary:** Such as limestone, marble, sandstone.
- **Metamorphic:** Derived from recrystalized igneous or sedimentary rocks such as slate.

■ **Important characteristics of rock include:**
- **Mineral composition:** Properties of rock depend to a large extent on the physical and chemical composition of its mineral constituents.
- **Fabric:** The spatial orientation of mineral components, whether crystals or fragments.
- **Texture:** The geometric aspect of mineral components, such as size, shape.
- **Structure:** Such as bedding, flow bonding which affect appearance and/or strength.
- **Color:** The visual/aesthetic aspect.

Dimension stone used in construction includes:

■ **Granite:** A visibly granular, crystalline rock of igneous origin:
- grain size varies from fine through medium to coarse;
- grain size and distribution should be uniform throughout when uniformity of color and texture is desired;
- very hard and durable surface finishes vary from very coarse to polished;
- gneiss, a metamorphic rock with a granite look, is often classified as commercial granite.

■ **Limestone:** Amorphous, semicrystalline or crystalline rock of sedimentary origin:
- oolitic limestone - calcite-cemented small rounded grains of lime carbonate (shell fragments) practically noncrystalline in character; quite uniform in composition, texture and structure; of high internal elasticity.
- dolomitic, or magnesian limestone - with more than 5 percent magnesium carbonate.

- travertine - a product of chemical precipitation from hot springs, characterized by the presence of numerous irregular cavities; some will take a fair polish and are then generally classified as marble; some varieties are quite resistant to wear of foot traffic.
- onyx - translucent layers of calcite precipitated from cold-water solutions in limestone caves; often called Mexican Onyx or onyx marble.
- oolite and dolomite limestone is classified by density into three catagories with Category I being low density and Category III, high density; but even the densest limestone may be easily scratched with a knife.

■ **Marble:** A crystalline metamorphic rock resulting from the recrystallization of limestone; commercial marble is commonly any calcareous or serpentine rock capable of taking a polish. The Marble Institute of America classifies marble as:
- Group A: Sound stone with uniform and favorable working qualities.
- Group B: Stone similar to, but with less favorable working qualities than Group A; may have natural faults.
- Groups C and D: Stone with variations in working qualities; containing geological flaws, voids, veins. Group D includes many of the highly colored decorative marbles.

■ **Sandstone:** Sedimentary rock usually of quartz, cemented with silica, iron oxide, or calcium carbonate.
- Quartzite: Quartz rock derived from sandstone; essentially homogeneous; low degree of porosity.
- Bluestone: Dark, bluish-gray sandstone; splits easily into thin slabs; commonly used as flagstone.

■ **Slate:** Microgranular metamorphic rock, characterized by excellent parallel cleavage entirely independent of original bedding; may easily be split into thin smooth slabs.
- classifications of slate for structural and roofing are established by end use rather than by properties.

■ **Obsidian:** Volcanic glass; generally black, although red, green and brown occur naturally.

■ **Serpentine:** Metamorphic rock, generally of very dark green color with markings of white, light green or black.
- One of the hardest varieties of natural building stone.
- Serpentine marble (verde antique) is classified commercially as marble but

is not truly a marble as it consists chiefly of serpentine.

■ **Soapstone:** A massive variety of talc with a soapy or greasy feel.
- Commonly divided into three groups; soft, regular and hard.
- Highly stain-resistant; heat and chemical resistant; easily worked.

■ **Lava:** Commonly pumice and volcanic scoria, lava is used as lightweight aggregate in concrete for structural rather than decorative purposes.
- The rocks are crushed, screened and washed to prepare them for use in concrete.
- Lava aggregate is of low density, strong and sound.
- Color varies between pink, purple and black.

■ **Broken stone includes:**
- **Rubble stone:** Stone of irregular size and shape laid up with unsquared or crudely-squared corners.
- **Chunk stone:** A large rubble-type stone that is usually shaped for laying in the field.
- **Ashlar:** Stone that has been squared by sawing or dressing - usually available in modular sizes.

■ **Crushed stone, most commonly marble, is processed through specialized machinery to produce a broad range of particle sizes for various uses:**
- Pulverized, granular or larger aggregate stone is often combined into panels of various thickness using cement or epoxy matrix binders. Thin panels are often reinforced with inorganic fibers.
- Coarsely-crushed marble is utilized in roofing chips, terrazzo chips and for landscaping as a ground cover.
- Finely crushed to resemble table salt, it is used as for swimming pool finishes, play sand and stucco.
- Marble dust, ground to a flour-like appearance, is combined with other elements to produce resilient tile, carpet backings, augmented limestone and a stainless molding clay.
- Ground or crushed material may be used as a filler for imitation marble.

■ **Simulated stone** is a material which may be formed from a variety of substances to resemble the appearance of stone.

B1.3 STONE/MASONRY

STONE: uses

● denotes common usage
○ denotes possible usage
1 for retaining walls, bridge piers, abutments, arch stones

	VERTICAL APPLICATION					HORIZONTAL APPLICATION									AGGREGATE FOR					
	FACING PANELS, EXTERIOR	FACING PANELS, INTERIOR	FACING UNITS, EXTERIOR	FACING UNITS, INTERIOR	LARGE BLOCKS 1	FLOORING, EXTERIOR	FLOORING, INTERIOR	PAVING BLOCKS	CURBING	TREADS/RISERS	SILLS/SADDLES	COPINGS	HEARTHS/MANTELS	ROOFING TILE	ROOF SURFACING	WALL SURFACING	PANELS, TILE: PRECAST	TERRAZZO	CONCRETE, LIGHTWEIGHT	LANDSCAPING
GRANITE																				
Building	●	●				●	●	●	●	●	●	●	●		○	●	●	●		○
Veneer	●	●																		
Masonry			●	●	●															
LIMESTONE																				
OOLITIC																				
Category I	●	●	●	●	○	●	●			●	●	●	●		●	●	●	●		●
Category II	●	●	●	●	○	●	●			●	●	●	●		●	●	●	●		●
Category III	●	●	●	●	●	●	●			●	●	●	●		●	●	●	●	●	●
DOLOMITIC																				
Category I	●	●	●	●	○	●	●			●	●	●	●		●	●	●	●		●
Category II	●	●	●	●	○	●	●			●	●	●	●		●	●	●	●		●
Category III	●	●	●	●	●	●	●			●	●	●	●		●	●	●	●	●	●
TRAVERTINE	●	●	●	●		●	●			●	●	●	●					●	●	
MARBLE																				
Group A	●	○			○	●	●			●	●		●		●	●	●	●		●
Group B, C & D		●		○	○		●			●	●		●		●	●	●	●		●
SANDSTONE																				
Standard			●	●	○	●	●			●	○		●							○
Quartzite			●	●	○	●	●			●	●		●							○
Bluestone			●			●	●			●	●		●							○
SLATE																				
Structural	●	●				●	●			●	●	●	●							
Roofing														●						
SPECIALTY STONES																				
Serpentine			●	●					○			○								○
Soapstone			○	○																○
Lava																			●	
AGGREGATE/SIMULATED	●	●					●			●	○		○					●		

STONE: sizes, properties

	Rubble, Chunk		Ashlar			Panels*		Flags, Pavers	
	Thickness, range, in inches	Face Dimensions, Average range in square feet	Thickness range, in inches	Face Dimensions, range	Rise (Height), range, in inches	Thickness range,* in inches	Face Dimensions, common limits* Larger sizes may be available	Thickness range*, in inches	Max. Face Area, range*, in square feet
GRANITE									
All Grades	¼–1½	½–2	4–12	½–10 sq. ft.	4–13 1½ incr.	1¼ to 4 and over	4′ × 10′	1¾–3	4–16
LIMESTONE									
Category I—Low density					2 ¼ 4 ⅞ 7 ½ 11 ½	⅞ 2¼ 3	5′ × 14′	1¾ 2½	1–6
Category II—Medium density	½–4	1–5	3½–6	Length: 10″–36″					
Category III—High density									
MARBLE									
Group A	¼–3	½–4	½–2		2½ 5 7¾	½–2	6′ × 7′	⅞ to 1	1–2
Groups B, C, D									
SANDSTONE									
Standard—Sandstone	1½–5		4	HEIGHT: ¾″–14″		2¼–4		1½–3	1–4
Quartzite—Quartzitic Sandstone		1–4		LENGTH: 8″–48″			4′ × 10′		
Bluestone—Quartzite	1–4		4–12		2 ¼ 5 7 ¾	1½–2		⅝–2	1–10
SLATE—Structural									
Exterior						1–1½	4′ × 8′	¼–2	OVER 4
Interior									

NOTES *Sizes and thicknesses shown are only indicative of some of the sizes and thicknesses generally used, and do not imply size and/or thickness limitations. Intended use and size will generally dictate minimum thickness. In all instances consult Industry Associations or individual manufacturers before making final decision.

Properties

Absorption by weight, maximum, percent; when tested in accordance with ASTM C 97 [1]	Density, minimum, pounds per cubic foot; range when tested in accordance with ASTM C 97	Compressive strength, minimum, psi; range when tested in accordance with ASTM C170 [2]	Compressive strength, psi, common range of available stones	Modulus of Rupture, minimum, psi; range when tested in accordance with ASTM C 99 [3]	Abrasion Resistance, minimum Ha, when tested in accordance with ASTM C241 [4]	Acid Resistance, maximum, inches; when tested in accordance with ASTM C217
.04	160	19,000	20,000 to 36,000	1,500	NA	NA
12	110	1,800	2,600	400		
7.5	135	4,000	to	500	10	NA
3	160	8,000	20,000	1,000		
0.75	144 to 175	7,500	8,000 to 23,000	1,000	10	NA
20	140	2,000	3,000	300		
3	150	10,000	to	1,000	8	NA
1	160	20,000	20,000	2,000		
0.25	NA		10,000 to 15,000	9,000 across grain	8	0.015
0.45				7,200 along grain		0.025

[1] Except for slate, which is tested in accordance with ASTM C121

[2] Load per unit area under which failure occurs by shear or splitting

[3] Except for slate, which is tested in accordance with ASTM C120

[4] Ha = The abrasive hardness value is the reciprocal of the volume of material abraded multiplied by ten. Minimum recommended value or 10 for flooring.

Slate ground-up and recast with cement binder available in panel form.

NA = Not available or not established in cited ASTM Standard Specifications.

B1.3 STONE/MASONRY

STONE: common finishes & color ranges

• denotes commonly-used finishes, generally available color ranges

	Honed	Polished	Rubbed	Sand Blasted	Sawed	Split Face	Textured	Thermal	Tooled	White	Cream	Light	Medium	Dark
FINISHES / **COLOR RANGES** (Gray = Light–Medium)														
GRANITE														
Building	•	•	•	•	•	•	•		•	•	•	•	•	
Veneer	•	•	•	•	•		•	•	•	•	•	•	•	
Masonry				•	•	•			•	•	•	•	•	
LIMESTONE														
OOLITIC														
Select	•		•	•	•	•	•		•	•	•	•		
Standard	•		•	•	•	•	•		•	•	•	•		
Rustic				•	•	•	•		•	•	•	•		
Variegated					•	•	•		•				•	•
DOLOMITIC	•		•	•	•	•	•		•	•	•	•		•
TRAVERTINE	•	•	•							•	•			
MARBLE														
Group A	•	•	•							•	•	•	•	•
Group B	•	•	•									•		
Group C		•	•											
Group D	•	•												
SANDSTONE														
Standard					•	•			•	•	•			
Quartzite			•		•	•				•	•	•	•	•
BLUESTONE	•		•		•	•		•						•
SLATE														
Structural	•	•	•			•							•	
Roofing					•	•							•	
SPECIALTY STONES														
SERPENTINE	•	•		•	•									•
SOAPSTONE	•				•	•						•	•	•
EXPANDED STONE											•	•	•	•
RECONSTITUTED STONE	•	•								•	•	•	•	•
SIMULATED STONE	• commonly imitates appearance of variegated rubble stone walls.													

B1.3 STONE/MASONRY

COLOR RANGES

REMARKS

— color classifications are generalized and may not be identical for all stone types.

— colors are dependent on mineral deposits at quarries — not all are available at all quarries.

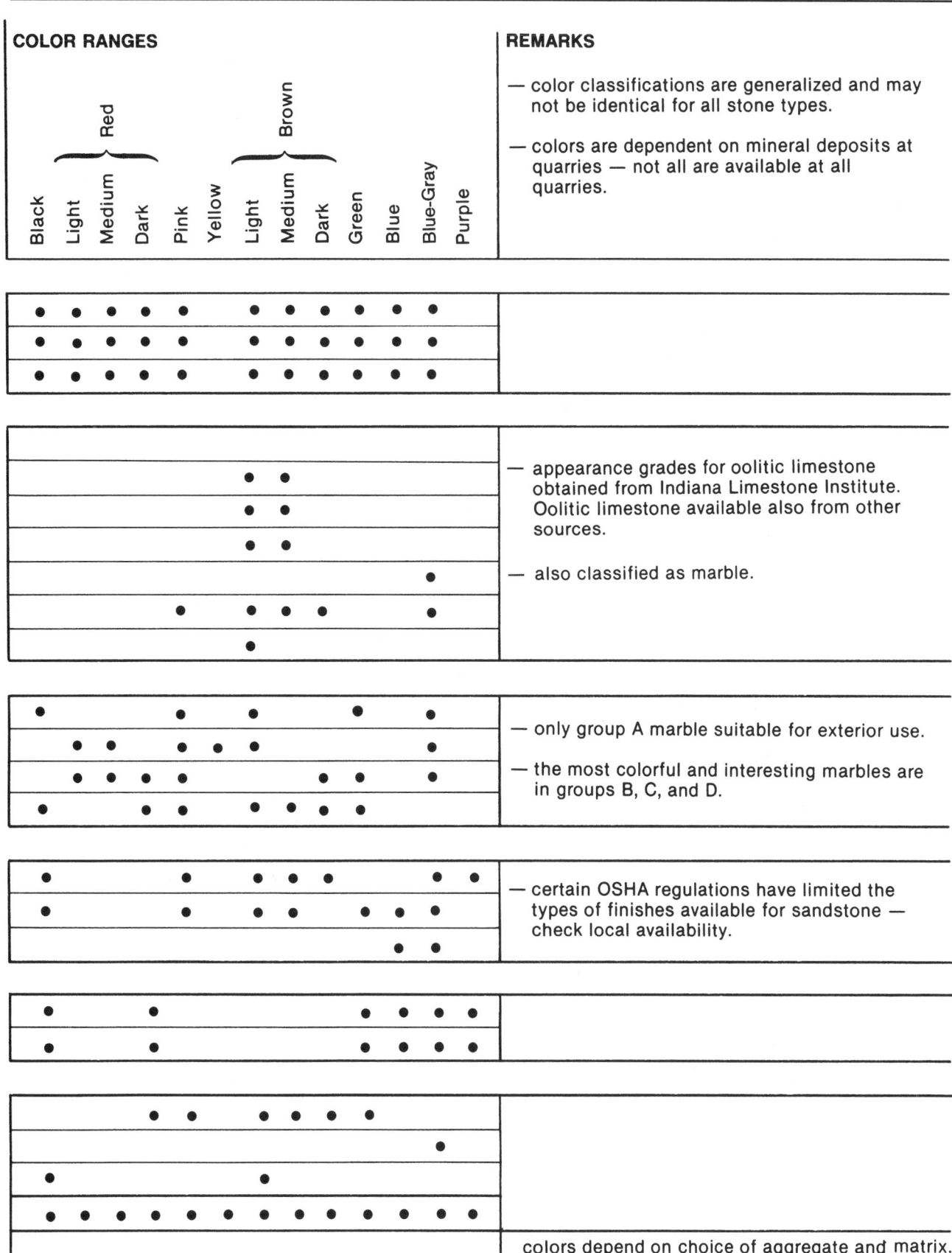

— appearance grades for oolitic limestone obtained from Indiana Limestone Institute. Oolitic limestone available also from other sources.

— also classified as marble.

— only group A marble suitable for exterior use.

— the most colorful and interesting marbles are in groups B, C, and D.

— certain OSHA regulations have limited the types of finishes available for sandstone — check local availability.

colors depend on choice of aggregate and matrix.

STONE: assemblies

HEAVY-STEEL-FRAME TRUSS

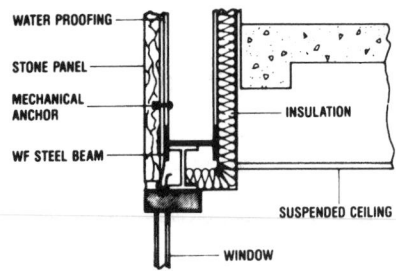

LIGHTGAGE-STEEL-FRAME TRUSS

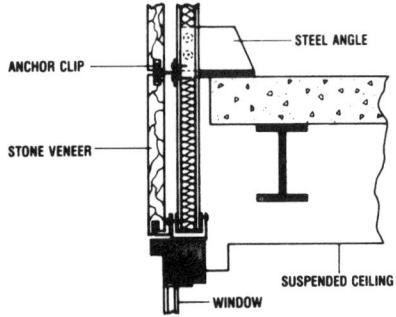

STONE-FACED CONCRETE

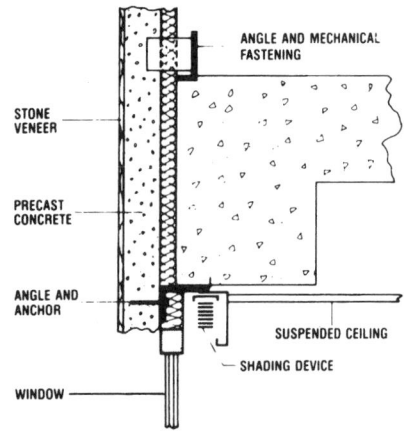

ALUMINUM GRID

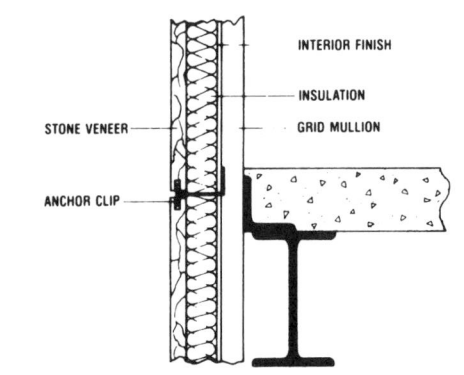

Dimension Stone Veneers

■ **Dimension stone, in the form of large slabs or panels, may be used for interior and exterior facings or veneers. Curtain-walls of stone can be economical selections (lower raw material and transportation costs) due to manufacturing techniques which produce thinner and lighter weight veneers.**

- Panels often range from over 4 inches thick to under 1/2 inch thick. Panel/slab strength will be affected by veining, physical impurities and crystalline structure.
- Marble and granite veneers may be very thin - 3/8 inches to 2 inches - while other stone types are often in the 3 to 4 inch range. At 2 inches thick or less, panels are generally more susceptible to staining, efflorescence and frost damage due to moisture absorption. This is due to the porosity found in all stone; the property which also allows stone to be highly polished.
- Ultrathin panels (1/4 inch) are available in standard-cut panel sizes up to 4 foot by 8 foot.
- Larger stone panels used in veneer designs are generally 1 1/4 to 1 1/2 inches thick, varying due to strength of the stone slab, the spacing and method of fastening, and imposed structural loading. Thicker panels are warranted when kerfing or drilling is required for anchoring.
- Panels of various sizes and thicknesses are frequently used to create ornate traffic-bearing surfaces, such as stairways and inlaid floors. Thinner stone panels and tiles may be 3/8 to 3/4 inches thick and are often limited in size to about a 2 foot square.

■ **Prefabricated stone panels can be an effective selection where facades are large, flat and repetitive. Several basic systems, often incorporating insulation and vapor barriers, are designed to support stone panels in curtain-wall applications.**

- **Steel-truss-framed:** Structural steel members, often heavy I- or S-shapes, are fabricated to form simple frames. Mechanical fasteners generally connect the veneer to the steel frame. The veneer and frame are then attached to the building structure at columns and/or along spandrel beams and slab edges.
- **Stone-faced concrete:** When contours are designed into the facade, or when the building structure is predominantly concrete, precast or glass-fiber-reinforced concrete panels may be appropriate. The stone veneer is placed, finished face down, into the mold and is prepared for connection to the concrete (drilling for and installation of mechanical fasteners; bond-breaker and/or back parging may be applied). Then the concrete is poured or sprayed into the molds to form a substrate that provides the structure for the stone veneer. **Glass-mesh mortar units** (in board form) may also be used as the substrate. The stone veneer may be adhered to the backer board using a mortar method.
- **Aluminum grid:** Similar to glass and metal curtain-wall systems, aluminum framing, as a concealed or exposed grid, supports stone panels by extending a continuous structural flange into a kerf cut in the four sides of the panel. Structural capacity will be limited by the overall height (unbraced) of the panel.
- **Diaphragm:** Lightweight steel framing is fastened to metal decking to form a structural diaphragm to which the stone veneer is attached using mechanical anchors, angles and structural (silicone) adhesive. The system boasts a durable, lightweight assembly which is capable of enduring greater amounts of deflection than many other systems.

DIAPHRAGM

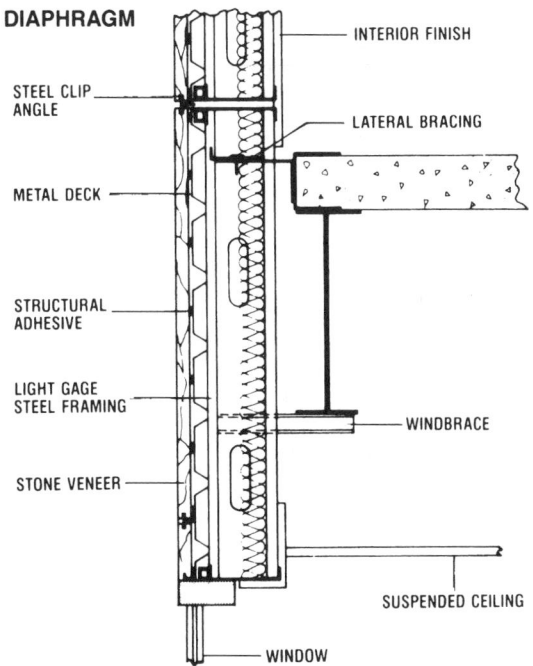

- INTERIOR FINISH
- STEEL CLIP ANGLE
- LATERAL BRACING
- METAL DECK
- STRUCTURAL ADHESIVE
- LIGHT GAGE STEEL FRAMING
- STONE VENEER
- WINDBRACE
- SUSPENDED CEILING
- WINDOW

RUBBLE VENEER

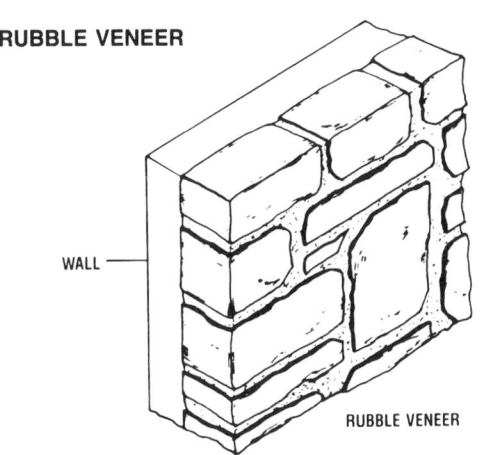

WALL

RUBBLE VENEER

ASHLAR VENEER

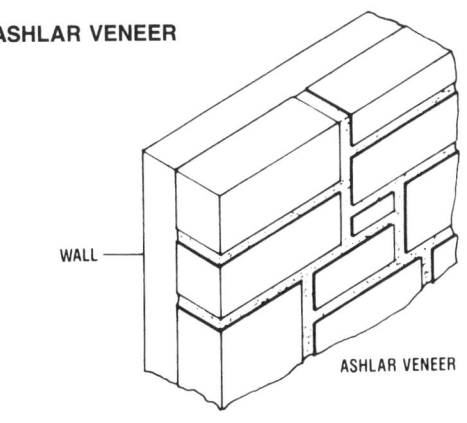

WALL

ASHLAR VENEER

■ When stone panels and veneers are to be used, it is highly recommended that the designer turn to the panel/ veneer manufacturer for assistance and expertise in developing the construction details and specifications.

■ Mechanical fastenings of veneer to steel frames can vary:
● Metal angles may be welded to the frame and then embedded and epoxied into slots formed in the stone panel.
● A system of dowels and bent anchors spaced intermittently across the back of the panel can provide lateral support for veneers which are set on an angle which is used to transfer the vertical load component to the frame.
● Masonry bolting systems are advantageous where tolerances are critical. Threaded bolts run through the steel frame members and are set into the veneer using a polyester resin adhesive or similar adhesive material.
● **In all cases where steel is in contact with masonry, non-corrosive materials must be used unless protective measures (waterproofing membranes, adhesive/epoxy isolation, etc.) are taken.**

Broken Stone Veneers

■ Stone veneers can also be constructed using broken stone. Installation is typically an on-site process, using similar techniques to that of standard masonry construction. Veneer construction types include:
● **Solid veneers:** Anchored directly to a load-bearing wall with masonry ties;
● **Cavity veneers:** Incorporate an airspace between the veneer and the supporting wall;
● **Bonded veneers:** Similar to solid veneers except that the veneer actually interlocks with ledges that are built into the supporting structure; and
● Thin stone veneers, which are 1/2 to 2 inches thick, are bonded directly to the support wall as a facing.

■ Veneers may be set in random, mosaic or coursed patterns. Rubble and ashlar veneers are usually set in courses similar to concrete masonry in pattern and construction.
● Standard one-height veneers appear similar to brick or block veneers;
● Two-height, three-height and four-height patterns use different shapes and sizes to fill in between large pieces of stone.

Crushed Stone Veneers

■ Veneers with the appearance of stone may be made from crushed stone, or aggregate.
● Some panels are available in a prefabricated form, with the aggregate set in a binder which laminates it to a rigid substrate.
● Some aggregate and binder mixtures are available in a premixed form and are applied to the installed substrate. Once applied to the substrate, the surface of the aggregate is molded, textured and otherwise prepared to meet the desired appearance.
● Similar appearances can be attained by using materials which simulate stone.

■ For further information on facing panel products with laminated aggregate surfacings, refer to chapter on EXTERIOR WALLS:

B1.3 STONE/MASONRY

STONE: selection checklist

Stone is a heterogeneous substance with a great diversity of chemical compositions, and its durability under in-service conditions varies widely. Properties affecting durability are:

■ Porosity of stone is generally the principal consideration for vertical and horizontal surfaces exposed to weathering, as moisture and its effects are the primary causes of decay of stone.

■ Porosity and hardness to resist abrasive wear are the important considerations for exposed horizontal surfaces subject to foot traffic.

Other important properties are:

■ Rupture or shear strength to minimize possibility of splitting when not fully bedded; when bridging over joints or cracks in substrate; or when subject to accidental heavy impact loads.

■ Acid resistance to minimize detrimental effects of pollutants, spills of drink or food, cleaning solutions.

Major factor affecting durability of stone is: MOISTURE, both as vehicle for transport of chemical substances and as disruptive agent. Sources of moisture are:

• water vapor condensing when temperature of the stone is below the dew point of air/water vapor mixture
• rainwater: both direct and runoff
• capillary travel of ground water.

Effects of moisture:

■ Moisture within the stone may be heated by solar radiation and may develop pressures far in excess of the tensile strength of the stone with resulting surface spalling and flaking.

■ Moisture will act as the vehicle for salts absorbed from ground water, derived from stone weathering, or from airborne pollutants.

• salt solutions will then be carried by diffusion, and by capillarity through the stone
• diffusion of salts may result in surface efflorescence, near surface subflorescence, salt crystallization, and salt hydration in stone, leading to considerable pressures which will cause rapid decay of the stone.
• rainwater may contain acids which will attack stone.

Other factors:

■ **Creep,** the time-dependent deformation, under dead load or sustained live load:

• marble is very susceptible to creep, and deformation of freely suspended slabs due to their own weight only is possible, especially in localities with high relative humidity.

■ Dissimilar stones, when installed next to each other, may react with one another:
• granite next to marble often disintegrates, possibly through infiltration of dissolved components of marble into the granite.

SELECTION CONSIDERATIONS should include:

■ Durability under in-service conditions:

• Exposure to moisture:

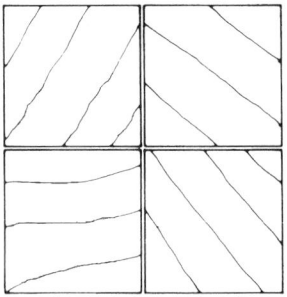

Dense stone should be selected for areas of most severe exposure.
• Surface depressions especially in horizontal units may hold water for extended periods of time; such trapped water may freeze with resultant spalling.

• Surface coatings which are impermeable to water vapor may trap moisture inside: low temperatures will cause such moisture to freeze, resulting in progressive surface spalling and deterioration.
• Dense, relatively smooth surfaces will trap and retain fewer airborne pollutants, but smooth horizontal surfaces are slippery when wet.
• Interior wall facing, or veneer, limitations are less stringent and allow widest range of stone selections to fit aesthetic requirements.

■ **Pattern:** Stone with distinctive texture, such as certain marbles, lends itself to specific pattern arrangements. The Marble Institute of America identifies common arrangements as:

• Blend pattern: panels of the same variety of stone but not necessarily from the same block arranged at random:

• Side-slip or end-slip pattern: panels from the same block are placed side by side or end to end in sequence to give a repetitive pattern and blended color in the horizontal and/or vertical.

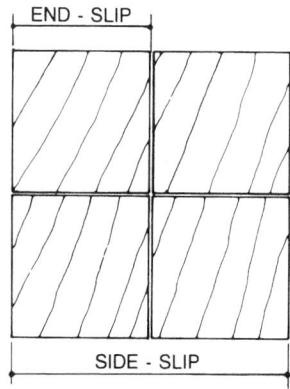

- End-match; book-match, or quarter-match patterns: in end-match the adjacent faces of panels A and B are finished and panel B is inverted above panel A; in book-match panel B is placed next to panel A; in quarter - or diamond - match panels A and B are book-matched and panels C and D are book-matched and then inverted over the top of panels A and B:

- Perfect, or nearly perfect, match is impossible to achieve due to intervening portions of stone lost during the sawing process and natural shifting of veining.

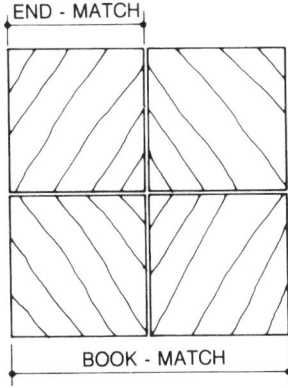

- Formal patterns require selectivity which increases installed cost.

■ **Sealers:** generally serve a dual purpose as impregnators and hardeners of stone:

- water-vapor tight surface sealers are not recommended as they may trap moisture inside the stone
- treatments shoud be applied after removal of water-soluble salts from the stone
- for information on sealers, refer to sections on COATINGS and to CEMENT/CONCRETE

■ **Fire Resistance:**
Exposure of stone to intense heating results in differential expansion of the minerals: stone rich in quartz, orthoclase, and sodium plagioclase is more susceptible to fire damage than stone without those minerals.

■ **FINISHES** available for exposed faces range from very rough to smooth; even though there is a great variety of finishing methods used, some may be limited to specific kinds of stone and the manufacturer should be consulted before making the final selection:

- **Bushhammered:** corrugated surface with interrupted parallel markings. National Building Granite Quarries Association designates fine texture as 8-cut; medium as 6-cut; and coarse as 4-cut.
- **Fine Rubbed** : smooth and free from scratches without gloss.
- **Gang-Sawed:** granular surface resulting from gang sawing alone.
- **Honed:** a satin smooth surface with little or no gloss.
- **Machine:** standard machine finish produced by the planer tool.
- **Nicked-Bit:** obtained by planing the stone with a planer tool in which irregular notches have been made in the cutting edge.
- **Plucked:** obtained by rough planing the surface of stone, breaking or plucking out small particles to give rough texture.
- **Polished:** the finest and smoothest glossy finish available in stone; generally only possible on hard, dense material such as granite, marble. Brings out the full color and character of stone. Polished finish is generally not recommended for marble intended to be used on building exterior.
- **Rubbed:** plane surface with occasional slight scratches.
- **Sand-Blasted:** a matte texture, coarse to fine, with no gloss. Recommended by the Marble Institute of America for marble intended to be used on building exterior.
- **Sand-Sawn:** surface left as the stone comes from the gang saw; moderately smooth, granular surface varying with the texture and grade of stone.
- **Sawed-Face:** a finish obtained from the process used in producing building stone; varies in texture from rough to smooth; such as diamond sawn, sand sawn, shot sawn.
- **Shot-Sawn:** a rough gang-saw finish produced by sawing with chilled-steel shot.
- **Split-Face:** usually sawed on the beds, is split either by hand or with machine so that the surface face of the stone exhibits the natural quarry texture.
- **Thermal:** plane surface with flame finish applied by mechanically controlled means to insure uniformity. Surface coarseness varies depending on grain structure of stone. Available for granite.
- **Wire-Sawn:** a method of cutting stone by passing a twisted multi-strand wire over the stone and immersing the wire in a slurry of

abrasive material; finish obtained similar to rubbed.

TYPICAL FORMS

- **Rubble Stone:** a rough stone as it comes from the quarry —laid up with unsquared or crudely-squared corners and of irregular size and shape.
- **Chunk Stone:** a large rubble-type stone that is usually shaped for laying in the field.
- **Ashlar:** a stone that has been squared by sawing or dressing—usually available in modular sizes.

Crushed or ground stone, most commonly marble, is processed through specialized machinery to produce a broad range of particle sizes for various uses:

- coarsely-crushed marble is utilized in exposed aggregate facings, roofing chips, terrazzo chips and for landscaping as a ground cover.
- crushed to a finer size resembling table salt, it becomes an aggregate for swimming-pool finishes, play sand and stucco.
- marble dust ground to a flour-like appearance is combined with other elements to produce resilient tile, carpet backings, augmented limestone and a stainless molding clay.
- ground or crushed material used as a filler for imitation marble.

MASONRY: types, uses

- denotes common usage
- ○ denotes possible or limited usage

	USE		VERTICAL APPLICATION									
					BEARING			NON-BEARING				
	INTERIOR/EXTERIOR	INTERIOR ONLY	RETAINING	FOUNDATION/BASEMENT	Single Wythe	Multi-Wythe	Cavity	Single Wythe	Multi-Wythe	Cavity	FACING	PREFABRICATED PANELS
CLAY												
BRICK												
Building/Common	●		○			●	●	●	●	●		
Facing	●				●	●	●	●	●	●	●	●
Facing, Glazed	●		○	○	●	●	●	●	●	●	●	●
Hollow	●				●	●	●	●	●	●		○
Industrial Floor	○	●										
Paving: Glazed, Unglazed	○	●										
Refractory		●										
TILE												
Structural, Facing [1]		●						●	●		●	
Structural, Load-Bearing		●				●		●	●		●	
Terra Cotta	●										●	
SAND-LIME												
Face Brick	●			○	●	●	●	●	●	●	●	○
CONCRETE												
BRICK												
Common	●		○	○								
Split Face/Slump	●		○		●	●	●	●	●	●	●	○
BLOCK [2]												
Solid, Load Bearing	●		●	●	●	●	●	●	●	●		
Hollow, Load Bearing [3]	●		●	●	●	●	●	●	●	●		
Hollow, Nonload Bearing	●							●	●	●		
PAVERS												
Solid	●											
Perforated/Grid	●											
MISCELLANEOUS												
Gypsum Block [4]		●						●				
Glass Block [5]	●							●				

HORIZONTAL APPLICATION:

FLOORING	TREADS/RISERS	SILLS, SADDLES	COPINGS	PAVEMENT	WALKS/TERRACES	EROSION CONTROL

WHEN RESISTANCE REQUIRED AGAINST

HEAVY FOOT TRAFFIC	STEAM/HOT WATER	EXPOSURE TO CHEMICALS	HIGH TEMPERATURE	CONTAMINANTS

REMARKS

[1] Structural clay facing tile and structural load-bearing available glazed.

[2] Perforated available; generally used as screen walls.

[3] Available glazed; resistances shown for glazed units.

[4] Gypsum block no longer in common use.

[5] Refer to section on GLASS/ PLASTICS for additional information.

- Brick may be molded or extruded; extruded brick has closer dimensional tolerances.

■ Generally, hollow brick may be substituted for solid brick as long as structural capacity and material properties are not changed.

FLOORING	TREADS/RISERS	SILLS, SADDLES	COPINGS	PAVEMENT	WALKS/TERRACES	EROSION CONTROL		HEAVY FOOT TRAFFIC	STEAM/HOT WATER	EXPOSURE TO CHEMICALS	HIGH TEMPERATURE	CONTAMINANTS
○	○	○										
○	○	●	●		○	○		●	○	○		●
○	○	●	●									
●								●	●	●		●
●								●	●	●		●
											●	
								●	●			●
								●	●			●

		SILLS, SADDLES	COPINGS									
		●	●									

FLOORING	TREADS/RISERS	SILLS, SADDLES	COPINGS	PAVEMENT	WALKS/TERRACES	EROSION CONTROL		HEAVY FOOT TRAFFIC	STEAM/HOT WATER	EXPOSURE TO CHEMICALS	HIGH TEMPERATURE	CONTAMINANTS
		○										
								●	●			●
								●	●			●
				●	●	○						
				○	●	●						

[3]

B1.3 STONE/MASONRY

MASONRY: properties

	PROPERTIES & TEST METHODS					NOTES
	① COMPRESSIVE STRENGTH, minimum, spi, ASTM C67 and C140	② WATER ABSORPTION, maximum, percent, ASTM C67	③ SATURATION COEFFICIENT, maximum ASTM C67	④ WATER ABSORPTION, maximum, pcf, range ASTM C140	⑤ MODULUS OF RUPTURE, minimum, p.s.i.	
BURNED CLAY BRICK						
BUILDING						
Grade SW	2500	20	.80	NT	NA	
Grade MW	2200	25	.90	NT	NA	
Grade NW	1250	NO LIMIT		NT	NA	
GLAZED FACING	2500–1500	NA	NA	NT	NA	
PAVING, Types SX, MX, NX	7000 to 2500	11 to no limit	80	NT	NA	
INDUSTRIAL FLOOR, Types T, H	NA	12	NA	NT	750	
Types M, L	NA	1.5	NA	NT	1500	
FACING						
Grade SW	2500	20	.80	NT	NA	
Grade MW	2200	25	.90	NT	NA	
HOLLOW						
Grade SW	2500	20	.80	NT	NA	
Grade MW	2200	25	.90	NT	NA	
BURNED CLAY TILE						
LOAD BEARING WALL TILE						
Type LBX	1000 [7] / 500	19 [8]	NA	NT	NA	
Type LB	700 [7] / 500	28 [8]	NA	NT	NA	
FACING TILE [9]						
Type FTX	1000 [10] / 500	11 [8]	9	NT	NA	
Type FTS	2000 [11] / 1000	19 [8]	16	NT	NA	
CONCRETE BRICK						
Grade N-1 and N-11	3000	NT	NT	15–10	NA	
Grade S-1 and S-11	2000	NT	NT	18–13	NA	
CONCRETE BLOCK						
HOLLOW LOAD BEARING						
Grade N-1 and N-11	800	NT	NT	18–13	NA	
Grade S-1 and S-11	600	NT	NT	20	NA	
HOLLOW NON-LOAD BEARING	500	NT	NT	NA	NA	
SOLID LOAD BEARING						
Grade N-1 and N-11	1500	NT	NT	18–13	NA	
Grade S-1 and S-11	1000	NT	NT	20	NA	
SAND-LIME BRICK						
Grade SW	3500	NA	NA	NT	400	
Grade MW	2500	NA	NA	NT	300	
GYPSUM BLOCK	75 [6]					

NOTES

① Gross area of individual unit. ASTM C67 for Brick; ASTM C140 for Concrete Brick and Block.

② Tested by submersion for 5 hours in boiling water; except for Paving, which is tested in cold water.

③ The saturation coefficient is the ratio of absorption by 24 hour submersion in cold water to that after 5 hour submersion in boiling water.

④ Average of 3 units tested. Weight range of units from less than 105 pcf to over 125 pcf.

⑤ Measured over average gross area of individual units.

⑥ Tested in accordance with ASTM C473.

⑦ Higher value for end construction tile; lower for side construction.

⑧ Water absorption by submersion for 1 hour in boiling water.

⑨ Available as Standard and Special Duty.

⑩ Standard Class; higher value for end construction.

⑪ Special Duty Class; higher value for end construction.

- For ASTM Standard Specifications for Masonry Units, see Evaluation Chart.

- NA = Not available or not established in respective ASTM Standard Specifications.

- NT = Not tested under cited ASTM Test Method.

- Range of densities, pcf:
  Clay Brick: 103–145
  Concrete Brick: 105–140
  Concrete Block:
    Pumice 75
    Expanded Shale 85
    Expanded Slag 95
    Air-Cooled Slag 120
    Crushed Stone 135 - 140

B1.3 STONE/MASONRY

• denotes common sizes

① size varies between manufacturers

BRICK TYPE	Thickness	Height	Length	Smooth	Wirecut	Sandstruck	Waterstruck	Brushed	Tapestry	Textured	Ribbed/Fluted	Scored	Flashed	Matte	Semi-Gloss	High-Gloss
NON MODULAR																
standard	3¾	2¼	8	•	•	•	•	•	•	•			•	•	•	•
jumbo	3¾	2¾	8	•	•	•	•	•	•	•			•	•	•	•
3-inch	3	2¾	9¾	•	•	•	•	•	•	•			•		•	•
MODULAR																
standard	4	2⅔	8	•	•	•	•	•	•	•			•	•	•	•
engineer	4	3⅕	8	•	•	•	•	•	•	•			•	•	•	•
jumbo closure	4	4	8	•	•	•	•	•	•	•			•	•	•	•
double	4	5⅓	8	•	•			•					•	•	•	
Roman	4	2	12	•	•	•	•	•	•	•	•		•			
Norman	4	2⅔	12	•	•	•	•	•	•	•	•		•			
Norwegian	4	3⅕	12	•	•	•	•	•	•	•	•		•	•		
6" norwegian	6	3⅕	12	•	•			•					•	•		
triple	4	5⅓	12	•	•			•					•	•		
SCR	6	2⅔	12	•	•			•					•			
jumbo utility	4	4	12	•	•	•	•	•	•	•			•	•	•	•
6" Jumbo	6	4	12	•	•	•	•	•	•	•			•	•		
8" Jumbo	8	4	12	•	•			•					•	•		
1" veneer	1	2⅔	8	•	•	•	•	•		•			•			
OVERSIZE																
8" square	3⅝	8	8	•	•			•		•	•		•	•	•	
12" square	3⅝	12	12	•	•			•		•	•		•	•	•	
high brick	4	8	16	•	•			•		•	•	•	•	•	•	
SQUARE AND RECTANGULAR																
¾" & 1⅛" thick		4	4	•	•	•							•	•	•	
¾" & 1⅛" thick		6	6	•	•	•							•	•	•	
¾" & 1⅛" thick		4	8	•	•								•	•	•	
¾" & 1⅛" thick		4	12	•	•								•	•	•	
¾" & 1⅛" thick		6	12	•	•								•	•	•	
1½" & 2¼" thick		4	4	•	•	•							•	•	•	
1½" & 2¼" thick		6	6	•	•	•							•	•	•	
1½" & 2¼" thick		8	8	•									•	•	•	
1½" & 2¼" thick		12	12	•	•	•							•	•	•	
1½" & 2¼" thick		4	8	•	•								•	•	•	
1½" & 2¼" thick		4	12	•	•								•	•	•	
1½" & 2¼" thick		6	12	•	•								•	•	•	
HEXAGONAL																
⅛" thick		6	6	•	•	•							•	•	•	
1½" & 2¼" thick		6	6	•	•	•							•	•	•	
1½" & 2¼" thick		8	8	•	•	•							•	•	•	
1½" & 2¼" thick		12	12	•	•	•							•	•	•	

REMARKS

• special shapes generally available for all sizes; for availability of shapes and sizes check local manufacturers

• masonry is generally coursed to nominal dimensions, which include thickness of joints - ⅜" to ½"

• dimensional tolerances of brick go up to ⅜" - see ASTM C9, C62, & C216

• sand-lime brick available in modular & non-modular sizes

B1.3 STONE/MASONRY

CONCRETE: block, brick

SHAPE

STANDARD BLOCK

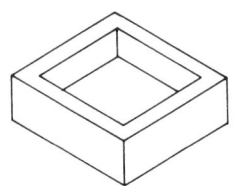

PLAIN END

CONCAVE END

THREE-CORE

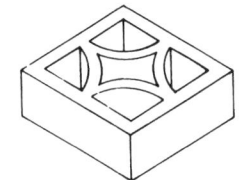

PLAIN END

TWO-CORE

- Classified as lightweight when made of concrete weighing less than 105 lb/cu ft; medium weight when 105 to 125 lb/cu ft; and normal weight when over 125 lb/cu ft.
- Number of cores varies with manufacturer; both types are commonly available. Two-core units are preferred for reinforced masonry as they provide more room for reinforcing and grout.
- Concave ends generally for 8-, 10-, and 12-inch thick units; 4- and 6-inch units commonly with plain ends.

SCREEN BLOCK

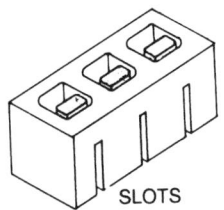

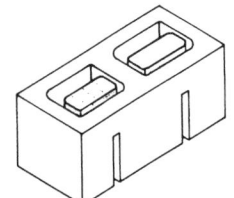

- Available in a wide variety of designs. Commonly 4 inches thick.
- Seldom used as load-bearing walls; some building codes may not permit such usage. When designed to support load, 6-inch thick units should be used.
- NCMA recommends that all units should have a compressive strength exceeding 1000 psi of gross area, with a minimum thickness of any section being not less than ¾ inch, and that they conform to ASTM C90.

SOUND ABSORBING BLOCK

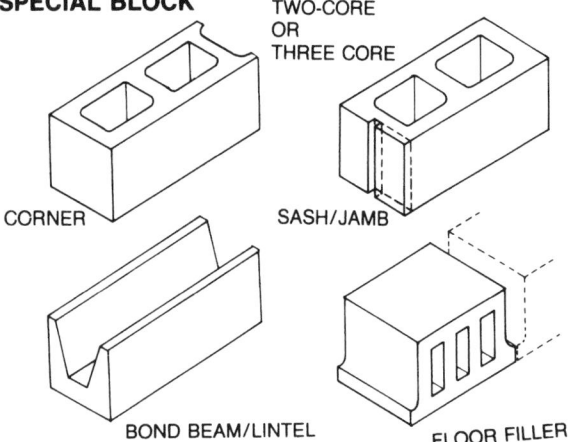

SLOTS

- Slotted opening in one face shell of block allow sound waves to enter the cores where their energy is dissipated.
- Sound absorbing inserts are available to partially fill the cores.
- Generally available as two-core types; three-core units may be limited to specific thickness of block.
- Top surface of unit either solid or open; ends commonly plain.
- Available in concrete block and in structural glazed Facing Tile.

SPECIAL BLOCK

TWO-CORE OR THREE CORE

CORNER

SASH/JAMB

BOND BEAM/LINTEL

FLOOR FILLER

- Other special shapes available: column/pilaster block; pilaster insert; chimney block; bullnose; double bullnose; header; soffit; manhole; control joint.
- Proprietary systems are available which use high strength blocks assembled into floor/roof plank: blocks have openings for reinforcing rods or prestressing wire to connect them; these are then grouted in place.
- Filler, or joist, blocks are used as a form for cast-in-place, one-way concrete joist floors; blocks are made 4 to 12 inches in depth with solid tops.
- Special shapes for use in reinforced masonry walls are also available.
- Custom-designed units may be available if the quantities required for a given project are sufficiently large.
- Not all shapes are available in all localities; availability should be ascertained before final selection.

BRICK/PAVERS

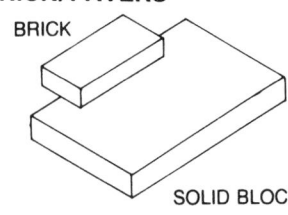

BRICK

SOLID BLOCK

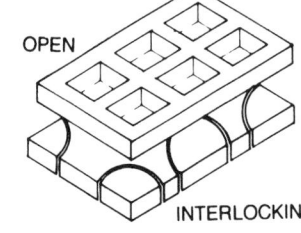

OPEN

INTERLOCKING

- Concrete brick is made essentially in the same sizes as clay or shale brick; also available as solid block of the same height and width as standard block.
- Concrete pavers are available as solid or open units in a variety of shapes, sizes, and thickness; may be integrally pigmented. Not all shapes and sizes available in all localities.

FACE FINISH

TEXTURED/SMOOTH/GLAZED

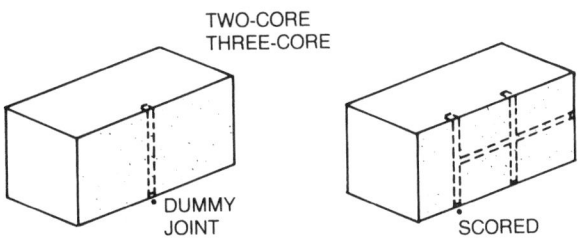

TWO-CORE
THREE-CORE

DUMMY JOINT

SCORED

- Texture may vary from fine to coarse depending on the aggregate used.
- Block with dummy joint is available to create the appearance of square units in stacked bond.
- Face may be ground smooth which will also bring out the color of aggregate.
- Face may be glazed in a variety of colors; may also be scored in addition to glazing; glazed, sound-absorbing block and structural glazed Facing Tile are available.

FLUTED/STRIATED/RIBBED

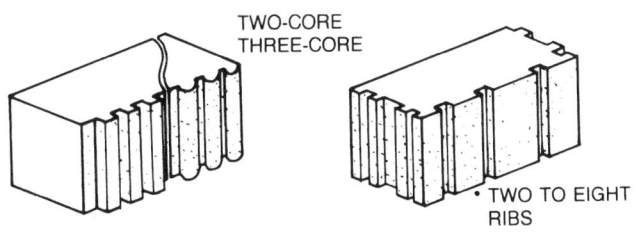

TWO-CORE
THREE-CORE

TWO TO EIGHT RIBS

- Fluted, striated and ribbed block are commonly manufactured to allow the vertical pattern to align when block is laid in running bond; some patterns may necessitate the laying of block in stacked bond to achieve the desired effect.
- Continuity of the pattern around corners should be investigated; with some patterns cutting the block may be necessary.

SCULPTURED/OFFSET

TWO-CONE COMMONLY

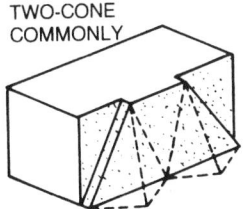 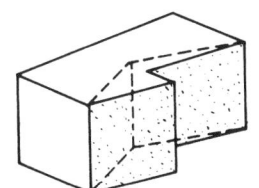

- Several different blocks may be required to develop the pattern.
- Pattern may also be changed by changing the bond, or arrangement, in which the blocks are laid.
- Shell and web thickness, and percent solid volume is generally the same as for standard block.
- Local availability of some specific patterns may be limited.

SPLIT/SLUMP

TWO-CORE COMMONLY

 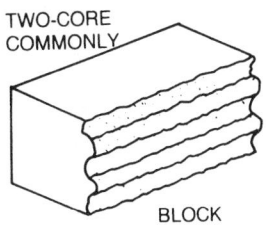 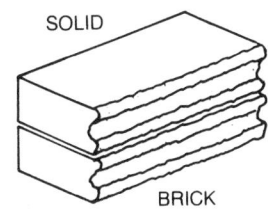

SOLID

BLOCK

BRICK

- Appearance of split units is similar to rough quarried stone, commonly produced by lengthwise splitting of solid units; special hollow block may also be available for splitting.
- Irregular or overhanging surfaces are produced by using concrete mixes which are slightly wetter than normal and which are then slumped right after they come out of the mold.

SIZE/THICKNESS OF HOLLOW CONCRETE BLOCK

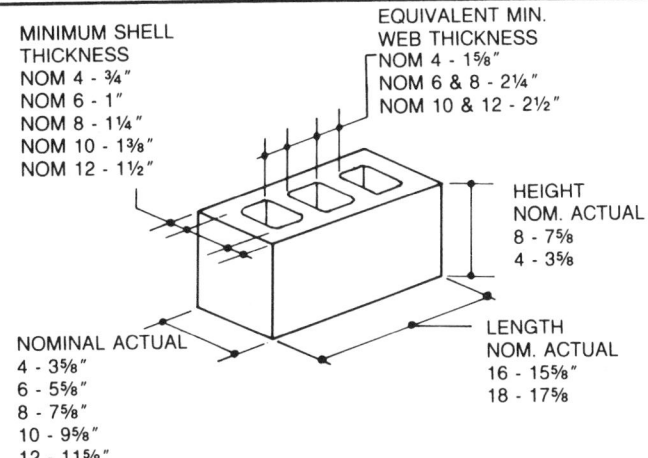

MINIMUM SHELL THICKNESS
NOM 4 - ¾"
NOM 6 - 1"
NOM 8 - 1¼"
NOM 10 - 1⅜"
NOM 12 - 1½"

EQUIVALENT MIN. WEB THICKNESS
NOM 4 - 1⅝"
NOM 6 & 8 - 2¼"
NOM 10 & 12 - 2½"

HEIGHT
NOM. ACTUAL
8 - 7⅝
4 - 3⅝

LENGTH
NOM. ACTUAL
16 - 15⅝
18 - 17⅝

NOMINAL ACTUAL
4 - 3⅝"
6 - 5⅝"
8 - 7⅝"
10 - 9⅝"
12 - 11⅝"

- Minimum values for shell and web thickness have been established for hollow concrete block under ASTM C90, and are the same for two- and three-core units.
- Equivalent web thickness is the sum of the measured thickness of all webs in the unit, multiplied by 12, and divided by the length of the unit.
- ASTM specifications permit a maximum variation in overall dimensions of plus or minus ⅛ inch.
- Resistance of hollow concrete block to fire is based on equivalent thickness: the solid thickness that would be obtained if the same amount of concrete contained in a hollow unit were recast without cores. It is obtained by multiplying the actual thickness of a unit by the percent solid volume. Solid volume of hollow block commonly ranges from 45 to 75 percent.
- Concrete block with 75 percent or more solid volume is classified as solid block and is covered by ASTM C145.

Time-Saver Standards for Building Materials and Systems

B1.3 STONE/MASONRY

MASONRY: selection checklist

TYPES AND GRADES

BRICK is generally classified as:

■ **Building brick** is made from clay or shale with three grades based on anticipated exposure to weather per ASTM C62:

- grade SW for exposure to severe weather conditions or when in contact with the ground.
- grade MW for exposure to moderate weather conditions.
- grade NW when not exposed to weather in walls only.

■ **Face brick** is made from clay or for exterior and interior use, with three types based on range of variations in color and in dimensional tolerances per ASTM C216:

- type FBS for general use where wider color ranges and greater variations in size are permitted.
- type FBX for general use where a high degree of mechanical perfection, narrow color range, and minimal variations in size are required.
- type FBA: selected to produce characteristic architectural effects resulting from nonuniformity in size, color, and texture.

■ **Ceramic glazed:** structural clay facing tile, facing brick - ASTM C126: made from clay or shale:

- grade S - select ٮ for use with narrow mortar joints
- grade SS - select sized - minimum variation in face dimensions
- type I: one finished face
- type II: two opposite finished faces

■ **Sand-Lime brick** - ASTM C73: Manufactured by the semidry method in presses and cured in autoclaves for 24 hours with steam; made from 5 to 10 percent hydrated lime and sand. Sand-lime brick is graded the same as common and face brick and has good frost resistance and better fire resistance than common or face brick. Its natural color is pearl-gray, but other colors are available:

- grade SW, to be used where exposed to temperatures below freezing in presence of moisture.
- grade MW used where exposed to temperatures below freezing but unlikely to be saturated with water.

■ **Concrete brick** - ASTM C55 made in the same sizes as common and face brick from controlled mixture of Portland cement and aggregates:

- type I: Moisture-controlled units
- type II: Non-moisture-controlled units.

- grade N: for use as architectural veneer and facing units in exterior walls; and where high strength and resistance to moisture penetration and to severe frost action are required; such as in pavements, walls, bearing walls.
- grade S: for general use where moderate strength and resistance to frost action and moisture penetration are required; such as veneer walls, interior walls.

■ **Industrial floor brick** - ASTM C410: made from clay or shale and suitable for surfacing industrial floors:

- type T: where high degree of resistance to thermal and mechanical shock is required, but low absorption is not a requirement.
- type H: where resistance to chemicals and thermal shock is important and low absorption is not.
- type M: where low absorption is required.
- type L: where minimal absorption and high degree of chemical resistance are required.
- types M and L generally have low thermal and impact resistance, but are highly resistant to abrasion.

■ **Pedestrian and light traffic paving brick** - ASTM C902: made from clay, shale, fire clay, or mixtures thereof for use in patios, walkways, floors, plazas, residential driveways.

Classified according to exposure to weather as:

- class SX: where it may be frozen while saturated with water
- class MX: for exterior use where resistance to freezing is not a factor
- class NX: for interior use and when effective surface sealer will be applied.

Classified in addition based on exposure to traffic:

- type I: for exposure to extensive abrasion, such as entrances to public or commercial buildings
- type II: exposed to intermediate

traffic, such as floors in commercial buildings, and exterior walkways
- type III: where exposed to light traffic, such as residential floors or walkways

■ **Building brick for paving:** for floors, low absorption, dense building such as residential or nonindustrial floors low absorption dense building brick meeting requirements of ASTM C62 for grade SW may be used: Brick Institute of America recommends that brick selected for flooring should meet or exceed the following:

- for extruded brick: 8000 psi minimum average compressive strength; 8 percent maximum average cold water absorption; 0.78 maximum saturation coefficient.
- corresponding values for molded brick are: 4500 psi compressive strength; 8 percent absorption; 0.78 saturation coefficient.
- whenever possible, selection should be based on actual performance data obtained for brick under similar conditions as anticipated.
- rough-textured units should be selected for exterior use to provide better resistance to slipping; smooth units are easier to maintain and are more suitable for interior use.

■ **Chemical-resistant brick** - ASTM C279: made predominantly of clay or shale:

- type H: where subject to thermal shock and minimum absorption is not required.
- type L: where minimum absorption is required and thermal shock is not a service factor.
- commonly used with chemical-resistant mortars.
- Red Shale Brick: most frequently used chemical-resistant brick; meets requirements for type L. This clay is also used to manufacture quarry tile employed in food processing plants and in light-traffic chemical areas.
- Fireclay: second most frequently employed acid brick; meets requirements for Type H. Also used in refractory exposure. Absorption generally higher than for red shale brick.
- Carbon Brick: primarily for exposure to strong alkalies and hydrofluoric acid. Has much higher rate of absorption than shale or

B1.3 STONE/MASONRY

fireclay and lower rate of expansion.
- Silicon carbide: high abrasion resistance; suitable for high temperature use.
- Silica: for very high concentrations of acid; particularly phosphoric acid.

■ **Fire brick:**

- For use in fireplaces, flues and other areas of high temperature extremes. See ASTM C155 for classification of insulating fire brick.

CONCRETE BLOCK is classified as:

■ **Hollow load-bearing** - ASTM C90: made with lightweight or normal weight aggregates:
- grade N: for use in exterior walls below and above grade that may or may not be exposed to moisture penetration or to weather, and for interior walls.
- grade S: limited to use above grade in exterior walls with weather-protective coatings, and in walls not exposed to weather.

Both grades are available as:
- type I: moisture-controlled units or
- type II: nonmoisture controlled.
- use of nonmoisture controlled block in walls greatly increases possibility of shrinkage cracking.
- control of shrinkage same as for brick, but more critical for block.

- **Non-load-bearing** - ASTM C129: either hollow or solid; made with lightweight or normal weight aggregates:

- type I: moisture-controlled
- type II: non-moisture-controlled.

■ **Solid load-bearing** - ASTM C145: made with lightweight or normal weight aggregates and of 75 percent or more net area:

- grades and types are the same as for hollow load-bearing units under ASTM C90.

■ **Block for mortarless application with surface bond:**
- special mortars needed
- special application methods and procedures
- block fabricated to full dimensions, not to nominal

STRUCTURAL CLAY TILE:

■ **Two basic size series available:**

- 8W series - nominal face dimension 8" x 16"
- 6T series - nominal face dimension 5 1/3" x 12"

■ **Standard thicknesses:** 2", 4", 6", 8"

■ Large variety of glaze colors and patterns available; glazed one or two sides.

■ Variety of preformed trim shapes available to meet most job conditions:

- cove base, corners, bullnose, wall and end caps, square corner, curved inside, etc.
- available either right hand or left hand with glazes on one or both sides.

■ **Structural clay tile is generally classified as:**

- standard - suitable for general use interior or exterior
- special duty - superior resistance to impact, moisture transmission; withstands larger lateral compressive loads.

■ **Structural clay load bearing wall tile** - ASTM C34:

- grade LBX - locations exposed to weather and frost action, provided they are burned to the normal maturity of the clay
- grade LB - locations where protected from frost.

■ **Structural clay non-load-bearing tile** - ASTM C56 - grade NB.

Typical uses:

- furring and partition
- back-up tile (non-load-bearing)
- fireproof tile

■ **Structural clay facing tile** - ASTM C212:

- type FTX - smooth-face tile suitable for general use in exposed exterior and interior masonry walls and partitions; adapted for use where tile low in absorption, easily cleaned, and resistant to staining is required, and where a high degree of mechanical perfection, only slight variations in color range, and minimum variation in face dimensions are desired.
- type FTS - smooth- or rough-tured-face tile, suitable for general use in exposed exterior and interior masonry walls and

partitions; adapted for use where tile of moderate absorption, moderate variation in face dimensions, and medium color range may be used, and where minor defects in surface finish, including small handling chips, are not objectionable.

DIMENSIONAL STABILITY

Masonry units expand and contract due to changes in temperature; and shrink or swell due to changes in their moisture content.

Thermal expansion/contraction:

■ Clay or shale brick will expand approximately 7/16 inch per 100 feet for 100°F increase in temperature.

- corresponding values for concrete masonry are:

dense aggregate	⅝ inch
expanded shale aggregate	½ inch
cinder aggregate	⅜ inch

- under the same conditions, normal weight concrete will expand ¾ inch; structural steel 13/16 inch; and aluminum 1 9/16 inch.

■ Effects of differential expansion between masonry units, which are inelastic and have low resistance to tensile stresses, and various other materials to which they are connected; or by which they are supported should always be considered.

Expansion/contraction due to changes in moisture content:

■ Burned clay products expand slowly upon contact with water or humid air, and such expansion is not reversible by drying at atmospheric temperatures: the use of a tentative coefficient of moisture expansion of 0.0002 is recommended by the Brick Institute of America.

■ Concrete shrinks as it dries after being cast, with the greatest amount of shrinkage occurring during the first two months, and then continuing indefinitely in decreasing amounts.
- coefficients of initial shrinkage commonly range from 0.0004 to 0.0006, but may be as high as 0.0010.

■ Shrinkage in concrete masonry units will largely depend on:
- aggregate used: units with sand and gravel have been found to have substantially less shrinkage than those produced with lightweight aggregates.

B1.3 STONE/MASONRY

MASONRY: selection checklist (cont.)

- method of curing: units which have been high pressure steam cured shrink less by one third or more than units cured at atmospheric pressures.

■ **Creep,** the gradual yielding in the direction of stress application in continuously loaded masonry units, is a consideration primarily for concrete:

- burned clay units are not subject to creep, and only the mortar in joints will be affected. The creep in joints may be of advantage as it will tend to compensate for the moisture expansion in the units themselves.
- creep in walls built of reinforced concrete masonry were found to be similar in magnitude to those in cast-in-place concrete. For additional information, refer to CEMENT/ CONCRETE chapter.

EFFLORESCENCE

Efflorescence is the deposit of water-soluble salts, usually white in color, upon the surface of masonry units. Efflorescence is most obvious in winter when rate of evaporation is slower or after heavy rains and drops in temperature.

- surface efflorescence is objectionable principally due to visual considerations.
- efflorescence may also occur within the pores of the masonry units and cause cracking and disintegration of the units.

■ Efflorescence results from several conditions occuring simultaneously within the masonry:

- presence of soluble salts within or in contact with the masonry
- source of moisture in contact with the salts for a sufficient period of time to dissolve them
- driving force - evaporation or hydro-static pressure - and pathways through the masonry, which cause and allow the dissolved salts to move to the surface or to other locations.
- if any one of above conditions is completely eliminated, efflorescence will not occur.

In conventional masonry exposed to the weather, such requirement is virtually impossible to meet and the practical approach is to reduce all contributing factors to a minimum:

- burned clay units which will not contribute to efflorescene are

readily available, and their potential for efflorescence may be determined by testing in accordance with ASTM C67 - Standard Methods of Sampling and Testing Brick.

- concrete masonry generally has greater potential for efflorescence than burned clay due to soluble salts which may be present in constituent materials, such as cement, aggregates, or as a result of hydration process; testing for efflorescence is done under ASTM C67.
- concrete masonry back-up of burned clay units may contain large quantities of soluble salts and will then contribute to efflorescene on the face of such composite wall, when sufficient moisture is present and pathways exist for the solution to migrate.

Sources of efflorescence other than masonry units themselves are:

■ **Mortar:** with the free-alkali or other chemical solutions present in mortar migrating into the masonry:

- free-alkali content of 0.1 percent in Portland cement may cause efflorescence, and such content should be kept as low as possible.
- sand used in mortar may contribute to efflorescence when contaminated with seawater, soil runoff, decomposed organic compounds.
- admixtures should be investigated for potential contribution to efflorescence.
- calcium chloride is generally not permitted in masonry containing metal anchors or reinforcing as its use creates potential for corrosion; when used in limited amounts it will not materially contribute to efflorescence.

■ **Groundwater:** generally contains high concentrations of salts, and when masonry is in contact with it, such solutions may be absorbed by capillary action.

■ **Atmospheric pollutants,** such as sulfurous gases, may contaminate masonry with soluble salts through soaking by acid rain.
Sources of free water to take any salts into solution in masonry are:

- rain water penetrating masonry.
- water vapor condensing within masonry.

- for additional information on water penetration and water vapor condensation, refer to EXTERIOR WALLS chapter.

COATINGS as means of minimizing water penetration through exposed faces of masonry units commonly used are:

- clear penetrating sealers, such as acrylics, silicones.
- opaque coatings, such as latex, rubber base, expoxies, alkyds.
- for additional information, refer to chapters on COATINGS and CEMENT/CONCRETE.

Concrete masonry is generally porous, and when a wall is constructed of a single wythe of such units, coatings may be required to minimize rain water penetration. Coatings used may be:

■ Clear water repellents in solvent-based or water-based solution, such as silicones, acrylics, will not materially change the appearance of the masonry:

- generally they will retard water absorption of masonry while allowing water vapor from within the units to escape to the outside.

■ clear or opaque coatings which cover the surface of masonry:

- coatings which prevent water vapor diffusion through them may trap moisture within the units causing possible peeling or flaking of the coating; mildew or efflorescence under the coating may develop.

Clear water repellent sealers for burned-clay masonry are often used in an attempt to remedy problems, such as water penetration and/or the occurence of efflorescence in poorly constructed walls:

■ Brick Institute of America lists some possible dangers inherent in indiscriminate application of clear sealers.

- they may trap moisture inside the wall thus causing or contributing to spalling and/or disintegration of the units through formation of crystalline deposits of salts within the units;

- they may not completely stop staining and efflorescence while covering such staining and efflorescence sufficiently to prevent their removal.

DEFLECTION OF SUPPORTS

Excessive deflection of structural supports for masonry may cause cracks to develop in such masonry.

The recommended maximum deflection in a short structural element - generally up to 8 feet in length - supporting masonry is 1/600 of span; and may have to be less when long unbroken panels of masonry are supported on a structural frame.

BRACING DURING CONSTRUCTION

Masonry walls should be temporarily braced during construction until design lateral strength of such walls is reached:

- in order to prevent collapse due to wind or other forces, such as unbalanced backfill;

- based on the assumption that bonding strength of mortar at this time is zero.

MORTAR

Mortar is the bonding agent for masonry units in an assembly thereof:

It must be strong, durable, capable of keeping the assembly intact, and it may have to create a water-resistant barrier.

Principal constituents of mortar are:

- cement, generally Portland cement, which provides durability and strength
- lime: which imparts workability, water retentivity, and elasticity

- sand: which acts as filler, providing the most economical mix and contributing to strength
- water: the mixing vehicle which creates plastic workability and initiates the cementing action.

■ Portland cement commonly used is Types I, II, and III. Use of air-entraining cement is generally not recommended, as it may tend to reduce bond between mortar and masonry units.

- Natural cements are also not recommended; while blended cements may be used when tested for such usage, or for use with nonstructural masonry.

- for additional information refer to CEMENT/CONCRETE chapter.

■ Lime for mortar should be hydrated, with Type S, as defined under ASTM C207, generally recommended when hydrated lime rather than quicklime is selected.

- Quicklime must be slacked before use; requirements are given by ASTM C5.

■ Masonry cement with the cement and lime premixed is available.

- Requirements for masonry cement are under ASTM C91.

- Masonry cements are proprietary, and their formulae are seldom disclosed by the manufacturer; since ASTM C91 does not impose limitations on additives, the constituents of proprietary mixes can vary widely among different brands.

- ASTM C91 also requires a minimum air content of 12 percent with a maximum of 22 percent, and air content in excess of 12 percent generally results in reduced bond between the mortar and masonry units.

- Performance data for masonry cement should be investigated prior to making final selection.

■ Aggregate used in either natural or manufactured sand with physical requirements and gradation limits given in ASTM C144.

- Brick Institute of America recommends gradation limits different than given under ASTM C144.

■ Water for mixing should be clean and free from acids, alkalies, or organic materials.

Other constituents of mortar may be:

■ Coloring agents, either incorporated into the mix at the site, or premixed in masonry cement.

■ Admixtures are available for air entrainment, water retentivity, workability, accelerated set.

- admixtures are proprietary products and their composition is generally not fully disclosed.

- National Concrete Masonry Association recommends that performance data in actual use as well as test results for a specific admixture be considered prior to selection, as some admixtures may do more harm than good.

Properties of plastic mortar:

■ Workability: the smooth plastic consistency which makes it easy to spread.

- workable mortar should hold the weight of masonry in place and make final alignment easy.
- workability is improved by: well-graded aggregate; air entrainment in cement; increase in lime content; amount of mixing water.

■ Water content, as recommended by the Brick Institute of America, should be the maximum amount consistent with the workability to provide maximum bond strength within the capacity of the mortar.

- retempering should be permitted but only to replace water lost by evaporation; with all mortar in a batch to be used within two hours after mixing.

- retempering of mortar within such constraints will reduce compressive strength slightly, but bond strength may be materially lowered if mortar is not retempered.

- moist curing of mortar is generally not required.

B1.3 STONE/MASONRY

MASONRY: selection checklist (cont.)

Properties of hardened mortar:

■ Compressive strength: of mortar depends largely on cement content and the cement/water ratio.

• compressive strength is a factor in walls carrying superimposed loads, but generally not the principal one.

■ Bond strength may be the most important single physical property of hardened mortar, but also the least predictable one.

• variables affecting bond strength include: air content of cement; elapsed time between spreading mortar and laying of masonry; suction of masonry units; water retentivity of mortar; pressure applied to masonry joint during forming; texture of masonry surfaces being bonded.

• bond strength is considerably less than compressive strength, and mortar joints subject to relatively small tensile stresses are likely to fail.

• lack of bond between masonry and mortar may lead to moisture penetration through the unbonded areas.

• according to National Concrete Masonry Association, with other factors being equal, bond strength of mortar will increase as compressive strength increases, but not in direct proportion.

Types of mortar:

Mortars for general use in masonry construction are classified under ASTM C270 as:

• Type M: with high - 2500 psi - compressive strength and somewhat greater durability than other mortar types. It is specifically recommended for unreinforced masory below grade and in contact with earth, such as in foundations, retaining walls, sewers, and manholes.

• Type S: with tensile bond strength approaching the maximum obtainable with cement-lime mortar; and reasonably high - 1800 psi average -

compressive strength. Recommended for reinforced masonry: unreinforced masonry where maximum flexural strength is required; and where mortar adhesion is the sole bonding agent between facing and backup wall.

• Type N: with medium - 750 psi average - strength suitable for general use in exposed masonry above grade. It is specifically recommended for parapet walls, chimneys, and exterior walls when subject to severe exposure.

• Type O: with low - 350 psi average - strength suitable for general use in interior non-load-bearing masonry. May be used in bearing walls of solid masonry where compressive stresses do not exceed 100 psi. Should not be used where it might be subject to freezing.

• Type K: with very low - 75 psi average - compressive strength for non-load bearing interior partitions. Use may be restricted by building codes.

Mortars for special use include:

• Non-staining: Portland cement mortar used for limestone, marble, terra cotta, glazed brick, cut stone.

• Non-staining, waterproof: for installation of caps, copings, sills, etc.

• Refractory: chimneys, incinerators, monolithic walls; types vary with different additives and heat requirements.

• Epoxy: structural & facing tile, glazed brick, stone veneers, bathrooms, kitchens.

• Surface Bonding: Portland cement with glass fibers; eliminates mortar between masonry units.

• Tuck-Pointing: prehydrated mix for stone and brick facing; should match flexibility and strength of existing mortar.

• High-Bond: with additives used to strengthen mortar; bond.

• Dirt-Resistant: additives used to increase stain resistance.

Chemical-Resistant Mortars

The chemical resistance of mortars may be evaluated by Standard Method of Test for Chemical Resistance of Mortars - ASTM C267.

The more commonly used chemical-resistant mortars may also be used alone, without masonry units, to form thick coatings usually applied by trowel. These mortars are:

• Asphaltic and bituminous mortars: for use over a limited range of low temperatures. Some are sand-filled and some are not. They may be applied either as mastics that depend upon evaporation of solvent or as hot-melt compounds.

• Epoxy resin mortars: two- or three-part systems with either amine or polyamide curing agents, they should conform to Standard Specifications for Chemical-Resistant Resin Mortars - ASTM C395 - or Specifications for Resin Chemical-Resistant Grouts - ASTM C658. For use, see the Recommended Practice for Use of Chemical-Resistant Resin Mortars - ASTM C399.

• Furan resin mortars: should conform to ASTM C395 or C658. They require a primer to ensure satisfactory adhesion to concrete. For use, see ASTM C399.

• Hydraulic cement mortars: for their use, see the Recommended Practice for Use of Hydraulic Cement Mortars in Chemical-Resistant Masonry - ASTM C398.

• Phenolic resin mortars: should conform to ASTM C395. For use, see ASTM 399.

• Polyester resin mortars: should conform to ASTM C395. Limited in resistance to strong chemicals, but will withstand mildly oxidizing solutions such as bleaches. For use, see ASTM C399.

• Silicate Mortars: should conform to Standard Specifications for Chemically Setting Silicate and Silica Chemical-Resistant Mortars - ASTM C466. For use, see the Recommended Practice for Use of

B1.3 STONE/MASONRY

Chemically Setting Chemical-Resistant Silicate and Silica Mortars - ASTM C397.

- Sulfur Mortars: should conform to Standard Specifications for Chemical Resistant Sulfur Mortar - ASTM C287. For use, see the Recommended Practice for Use of Chemical-Resistant Sulfur Mortars - ASTM C386.

Mortar for Surface Bonding

Concrete masonry units are first stacked without mortar and then both sides of the wall are coated with a thin layer of mortar that is reinforced with glass fibers:

- tensile strength of surface bonding mortar is approximately 1500 psi.

- mortar may be integrally pigmented.

- mortar should effectively resist rain penetration.

- surface bonded wall is less resistant to vertical loads than a conventionally constructed one; bearing capacity may be improved by using units with their bearing surfaces ground smooth.

- mortar available premixed

- equipment for spray application is available.

MASONRY REINFORCEMENT

TWO-WIRE

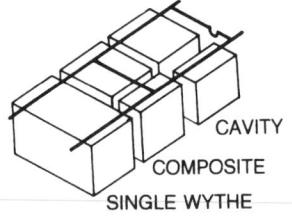

CAVITY

COMPOSITE

SINGLE WYTHE

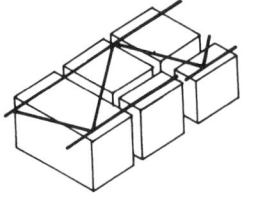

THREE-WIRE

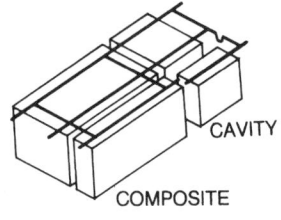

CAVITY

COMPOSITE

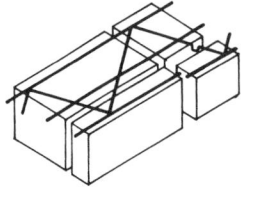

FOUR-WIRE

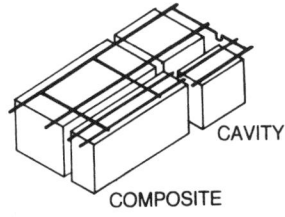

CAVITY

COMPOSITE

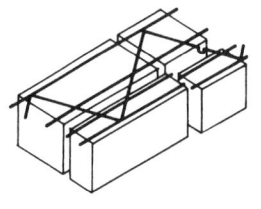

Joint reinforcement for masonry is a rigid factory fabricated welded-wire assembly in a ladder or truss design. The primary function of joint reinforcement is the control of cracking in a masonry wall due to thermal or moisture induced expansion or contraction; but it will also function to tie two wythes of masonry together.

■ **Joint reinforcement is commonly used in:**
- single wythe concrete masonry walls
- composite walls: with either two wythes of concrete masonry,
  or one of burned clay and the other a backup of concrete masonry;
- cavity walls: generally with one wythe of burned clay and the other of concrete masonry.

■ **Width generally available are:**
- for single wythe concrete masonry walls 4 to 12 inches in nominal thickness, at 2-inch increments
- for composite and cavity walls of up to 18 inches in overall nominal thickness

■ **Size of reinforcing wire:**
- light duty: 11-gauge longitudinal and cross wires
- standard: 9-gauge longitudinal and cross wires
- medium duty: 8-gauge longitudinal and 9-gauge cross wires
- heavy duty: 3/16-inch longitudinal and 9-gauge cross wire
- extra heavy: 3/16-inch longitudinal and cross wires.

■ **Finishes commonly available are:**
- plain, uncoated; also referred to as brite basic
- copper coated; generally 5 to 25 percent of copper
- mill galvanized; .40 or .80 oz. per square foot
- hot-dip galvanized after fabrication; 1.50 oz. per square foot.

TWO-WIRE WITH FIXED TIES

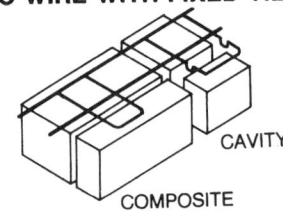

CAVITY

COMPOSITE

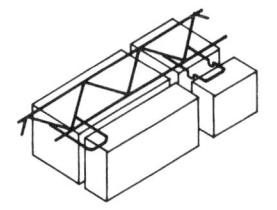

Joint reinforcement with fixed or adjustable ties is used for composite and cavity walls when one wythe only is to be reinforced and the other wythe is to be tied into the first one for lateral stability of the assembly.

- spacing of ties horizontally is commonly 16 inches on center; and spacing of reinforcement should be 24 inches maximum vertically for 9-gauge wire or heavier in solid composite walls, and in cavity walls with cavities not over 3½ inches in width.
- fixed ties generally used when joints in both wythes line up and both are built up at the same time.
- adjustable ties are commonly used when: adjacent wythes do not course out level; one wythe is built up before the other one, such as when rigid insulation in a cavity wall is to be mechanically attached to the backup wythe rather than inserted during construction of the wall.
- adequacy of flexible ties to provide the required lateral stability to the assembly when ties are under compression should be ascertained.

TWO-WIRE WITH ADJUSTABLE TIES

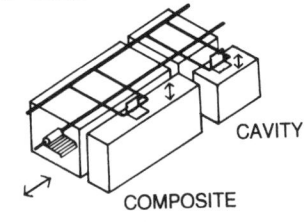

CAVITY

COMPOSITE

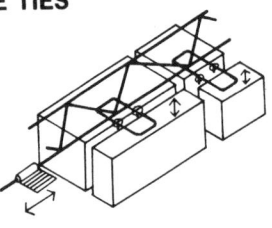

REINFORCING BARS

LINTEL BLOCK/
BOND BEAM
GROVED CORE

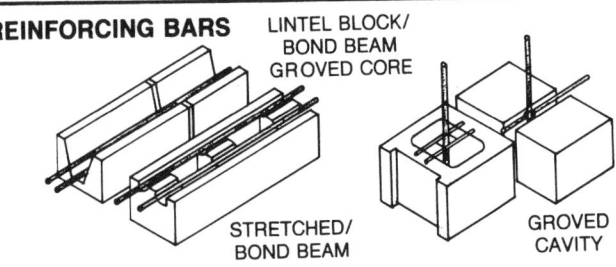

STRETCHED/
BOND BEAM

GROVED
CAVITY

Reinforcing bars commonly used in masonry walls are deformed bars No. 4 or heavier. Principal uses are:

- in lintels and bond beams
- in grouted cavities of reinforced masonry bearing walls
- in solidly grouted cavities of reinforced cavity walls Positioners and centering devices to hold reinforcing bars aligned prior to being grouted-in are available.

ANCHORS/TIES: selection checklist

ANCHOR

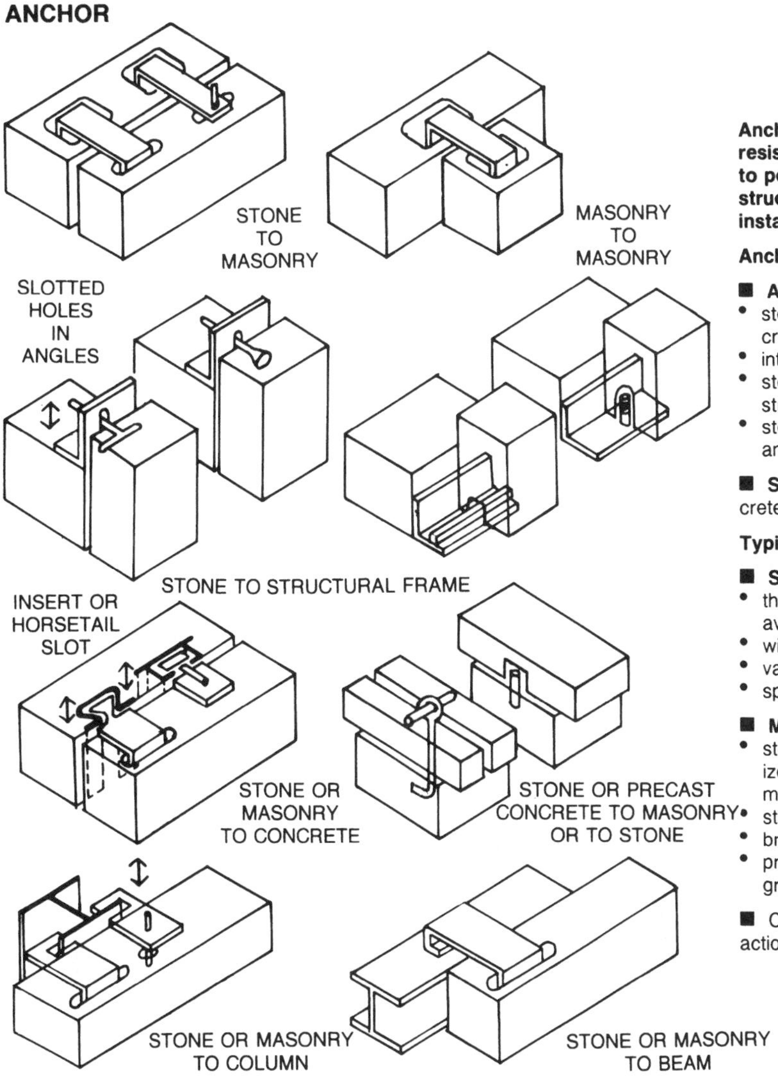

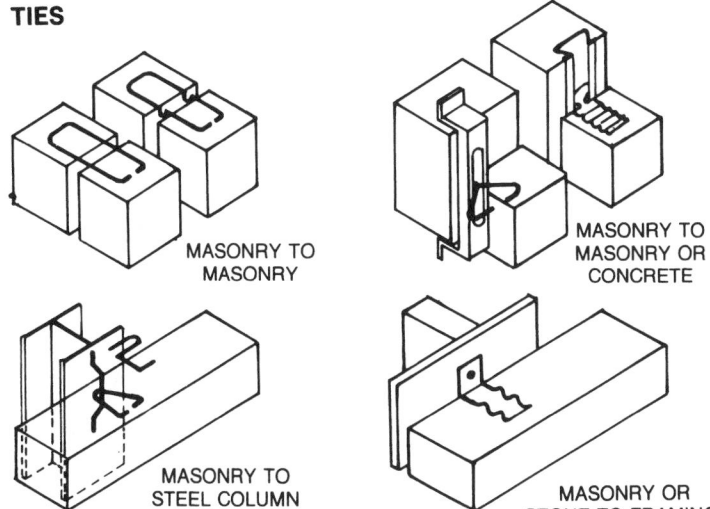

STONE
TO
MASONRY

MASONRY
TO
MASONRY

SLOTTED
HOLES
IN
ANGLES

STONE TO STRUCTURAL FRAME

INSERT OR
HORSETAIL
SLOT

STONE OR
MASONRY
TO CONCRETE

STONE OR PRECAST
CONCRETE TO MASONRY
OR TO STONE

STONE OR MASONRY
TO COLUMN

STONE OR MASONRY
TO BEAM

Anchors are strips or bars of metal sufficiently rigid to resist tension, compression and shear; and not intended to permit differential movement between dissimilar structural elements unless specifically designed and/or installed to accommodate such movement.

Anchors are commonly used to:

■ **Attach:**
* stone facing panels or masonry veneer to masonry or concrete backup walls.
* intersecting masonry walls to one another.
* stone or masonry to vertical or horizontal elements of the structural frame.
* stone or precast-concrete coping sections to one another and/or to supporting wall.

■ **Support:** and secure in place stone facing panels to concrete or steel structural frame.

Typical sizes and materials are:

■ **Size:**
* thickness of strips commonly used: ¼ inch; but also available in 3/16 and ⅛ inch.
* width generally ranges from one to two inches.
* various lengths are available.
* special sizes may be available from manufacturers.

■ **Materials:**
* steel: plain, uncoated; mill galvanized; hot-dip galvanized. Proprietary systems for attaching and supporting marble facing panels are available.
* stainless steel.
* brass, bronze.
* proprietary systems using aluminum to attach and support granite facing panels are available.

■ Corrosion of fasteners due to moisture and/or galvanic action should be considered during selection.

TIES

MASONRY TO
MASONRY

MASONRY TO
MASONRY OR
CONCRETE

MASONRY TO
STEEL COLUMN

MASONRY OR
STONE TO FRAMING

Ties are strips of lightgauge metal or wire intended to resist tension and compression but not shear; commonly used when differential movement between dissimilar structural elements is anticipated.
Ties are commonly used to:

■ **Bond or tie:**
* two wythes of masonry in lieu of masonry bonding.
* stone masonry to concrete or burned clay masonry.

■ **Attach:**
* stone, concrete, or burned clay masonry to structural frame or backup construction.

Commonly mill or hot-dip galvanized; wire may be copper clad or stainless steel:

* strips: generally 12 to 26 gauge.
* wire: commonly 3/16- to ¼-inch diameter.

REINFORCEMENT/TIES/ANCHORS:

JOINT REINFORCEMENT/TIES

Masonry walls may be of single or multiple wythes; each single wythe wall must be tied together longitudinally; and each multi-wythe wall must be tied together longitudinally and laterally.

Tying of walls longitudinally may be by means of masonry units or by metal reinforcement/ties:

■ **Longitudinal metal reinforcement or joint reinforcement** is required in walls where masonry units in the same wythe do not overlap units in adjacent courses by at least three inches.

- joint reinforcement for walls in running bond is principally used to control shrinkage cracking: reinforcement will not prevent cracking but its use will generally result in more but narrower cracks rather than fewer and larger ones.

- without joint reinforcement the strength of a stacked bond wall is approximately 40 percent that of a wall laid in running bond; when joint reinforcement is added at 16 inches on center, the strength of the two different bonding patterns is increased to approximately the same level.

- joint reinforcement may also be used as structural reinforcement for walls spanning horizontally between supports, but it will not improve strength in the horizontal span.

- only straight, uncoiled joint reinforcement is recommended; proper installation of coiled reinforcement is generally difficult to achieve.

■ **Vertical spacing of longitudinal reinforcement** for control of shrinkage depends on the length and height of wall and the moisture content of masonry units used:

- the larger the wall area between wall ends or control joints and the greater the moisture content of the masonry units, the closer should be the spacing of reinforcement:

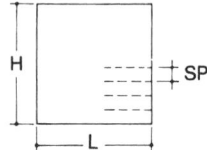

For moisture controlled Type I concrete masonry units and:
- L=45 feet maximum and L/H = 2 ½;

SP = 24 inches.
L=50 feet maximum and L/H = 3;
SP = 16 inches.
L=60 feet maximum and L/H = 4;
SP = 8 inches.

- spacing of reinforcement for nonmoisture controlled concrete masonry units should be decreased or length of wall between its ends, or between control joints, reduced for the same spacing of joint reinforcement.

For tying of multi-wythe composite walls and of intersecting walls, condensation on ties in cavity as thermal bridge may be by means of masonry units overlapping the wythes or by metal ties or anchors; tying of cavity walls is commonly by means of metal ties only:

■ **The use of metal ties rather than masonry bonding** increases the resistance of a masonry wall to rain water penetration.

- joint reinforced walls will maintain greater structural integrity after cracking.

- walls subjected to racking loads of sufficient magnitude to cause diagonal cracking may be protected from failure by the longitudinal wires of the joint reinforcement: horizontal steel is roughly three times as efficient as vertical steel in resisting racking shear loads.

■ **Wire ties only,** either incorporated into joint reinforcement or individual, are recommended for use in multi-wythe grouted walls or in cavity walls.

- adjustable wire ties may be used in multi-wythe walls where the collar joint, the longitudinal vertical joint between wythes, is solidly grouted.

- adjustable wire ties for cavity walls should be structurally designed for each specific configuration and in-service conditions; although the strength of adjustable ties in multi-wythe walls with grouted collar joints is nearly equivalent to the strength of fixed ties, the flexible connection of adjustable ties in cavity walls may result in structural problems.

■ **National Concrete Masonry Association recommends** that lateral ties in cavity walls not be crimped, since crimping reduces the compressive strength of the tie by about 50 percent, and such crimping does not contribute to the resistance of a cavity wall to water penetration.

- lateral ties in cavity walls act as columns when in compression, thus their strength is then inversely proportional to the square of the ratio of unsupported length of the ties to its radius of gyration.

- when cavity wall ties are placed at a 45 degree angle instead of perpendicular to the plane of the cavity, as would be the case with truss type reinforcing, their effective length is increased by 41 percent and the number of such ties must then be doubled to provide equivalent compressive strength.

- ladder-type reinforcing is recommended for long sections of cavity walls - 50 feet or more between ends of control joints - to allow for longitudinal differential movement between the two wythes.

- when 3/16-inch diameter longitudinal wires are used, the mortar should be Type S or Type M.

- commonly greater economy is achieved by using ties of larger cross sectional area at maximum allowable spacing.

MATERIALS FOR REINFORCEMENT/TIES

Metal joint reinforcement/ties are commonly made of:

- cold-drawn steel wire, either plain, meeting requirements of ASTM A82 or deformed, meeting requirements of ASTM A496.

- hard-drawn copper-clad steel wire, meeting requirements of ASTM B227, grade 30HS; this specification requires a 9 millimeter coating of copper on a 3/16-inch diameter wire; a copper sulphate bath gives the appearance of copper but does not provide a coating in accordance with above specification.

- hot rolled steel wire, ASTM A510, grade 1021, for mesh.

- hot rolled steel sheet, ASTM A570, grade E, for corrugated ties.

- galvanized coating: mill galvanized, ASTM A116, Class 1 or Class 3; hot-dip galvanized after fabrication, ASTM A153, Class as applicable.

- hot-dip galvanized after fabrication provides initial coverage of cuts and welds and a coating almost four times as heavy as under ASTM A116, Class 1; or nearly twice as heavy as under

selection checklist

ASTM A116, Class 3; at a cost which is proportionately less than the increase in thickness of coating.

- relative to hot-dip, zinc-coated steel wire, the 25 percent copper-coated wire costs about twice as much.

- the expense of stainless steel, nickel-copper alloys, brass and bronze for masonry wire ties is generally not warranted, except in instances where potential for corrosion is considerable.

- eliminating the possibility of corrosion in anchors for stone facing panels may be more critical than in masonry ties: anchors for stone are commonly spaced further apart than masonry ties, and failure of even one may result in a complete separation of a panel from supporting construction.

- local building codes may establish requirements for spacing, size, and materials to be used for metal ties and anchors.

REINFORCED MASONRY

The flexural strength of masonry walls may be increased by embedding steel reinforcing bars in cells, cores, or cavities filled with grout or concrete:

- if the area of steel reinforcement is more than 0.002 times the cross-sectional area of the wall; or the maximum spacing of the principal reinforcement is not more than six times the wall thickness or 48 inches; then the wall may be considered to be reinforced.

- when above requirements are not met, or when only portions of a wall are thus reinforced, the wall is considered to be partially reinforced.

- reinforced portions of partially reinforced walls are considered to be those which have an effective width not greater than six times the wall thickness in running bond; nor more than three times the wall thickness in stack bond.

BOND BEAMS

Bond beams are horizontal structural elements which integrate the components of a wall into a stuctural unit:

- where vertical steel reinforcement is not used, bond beams may be constructed of special channel shaped units.

- where vertical steel reinforcement must pass through a bond beam at intermittent locations, as in fully reinforced masonry walls, bond beams are constructed of concrete masonry units with recessed webs to accommodate the horizontal reinforcing bars.

- bond beams provide lateral stability to a wall when they are tied into vertical supports.

- as means of crack control, a bond beam is effective for approximately 24 inches above and below it thus bond beams located at four feet intervals serve to influence the entire height of a wall.

CONTROL JOINTS

Control joints are used to relieve horizontal tensile stresses in walls by reducing restraining and permitting movement to take place:

- ■ **Vertical separations** are provided at locations where stress concentrations may occur.

- during expansion, long walls with openings incorporated into them may develop shearing stresses in areas of minimum cross section; diagonal cracks often occur in sections between window or door openings unless control joints are provided at jambs of such openings.

- control joints may be omitted if adequate tensile reinforcement is placed above and below an opening in a wall.

- during contraction the tensile strength of masonry is not sufficient for it to move back thus causing cracks near corners unless control joints are provided at such locations.

- when concrete floor or roof slabs are poured directly on masonry walls, curling of the slab may occur due to shrinkage, deflection of plastic flow in concrete: if the slab is not separated from the wall and warps, it can cause ruptures in masonry, particularly at corners.

- vertical cracks are quite common at wall offsets or setbacks; at changes in height of wall; at construction joints in foundations, roofs, floors unless separated by control joints.

- control joints should extend through plaster applied directly to masonry units; plaster applied on lath which is

furred out from masonry may not require vertical separation at control joints but will still require its own control joints.

- control joints in a backup wall should also extend to any facing when such facing is rigidly bonded to the backup wall.

- horizontal slip planes should be provided at junctions of roof slab and load-bearing masonry walls occurring at a control joint.

- bond between rigid roof construction, such as concrete slabs, and bearing walls should be broken twelve to fifteen feet back from corners by providing a slip plane.

- ■ **Control joints must accommodate lateral loads** across the joint while permitting free lateral movement.

- reinforcement should not extend through a control joint.

- control joints may be formed in concrete block walls by filling the void between the concrete units with mortar, but preventing the mortar from bonding to block on one side of the joint.

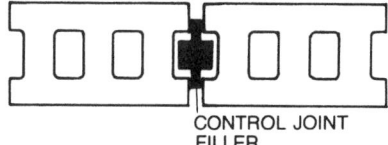

CONTROL JOINT
FILLER

- prefabricated cross-shaped control joint fillers of plastic are available; the section of the filler parallel to the plane of the wall should be sufficiently rigid to act as a shear key.

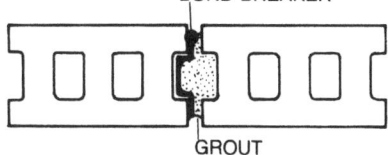

BOND BREAKER

GROUT

ACCESSORIES

Accessories for masonry may include

- wall plugs; inserts to allow wood to be nailed to masonry.

- anchors for hollow-metal frames to be built into masonry.

- plastic inserts to form weep holes in cavity walls.

B1.3 STONE/MASONRY

STAIN REMOVAL/CLEANING

• indicates commonly used applications

SURFACE TO BE CLEANED — POROUS, DENSE, ROUGH, SMOOTH/POLISHED, PATTERNED/CARVED, LIGHT COLOR, DARK COLOR

STAIN/PROBLEM — DIRT/POLLUTANTS, OIL/TAR, EFFLORESCENCE, WHITE BLOOM, GREEN STAIN, BROWN STAIN, RUST, GRAFFITI, ORGANIC STAIN

CLEANING METHOD	POROUS	DENSE	ROUGH	SMOOTH/POLISHED	PATTERNED/CARVED	LIGHT COLOR	DARK COLOR	DIRT/POLLUTANTS	OIL/TAR	EFFLORESCENCE	WHITE BLOOM	GREEN STAIN	BROWN STAIN	RUST	GRAFFITI	ORGANIC STAIN
WATER																
Low pressure	•		•	•		•	•	•		•		•				
Medium/High pressure	•		•	•		•	•	•			•					
Steam	•	•	•	•	•	•	•	•								
CHEMICAL																
Acids	•		•				•	•	•					•	•	•
Cleansers, Detergents	•		•	•		•	•	•	•	•	•	•			•	•
Steam and Chemicals	•	•	•	•	•	•		•	•						•	•
Proprietary Compounds	•	•	•	•	•	•	•	•	•	•	•	•	•	•	•	•
MECHANICAL																
Sand Blasting	•		•			•	•	•								•
Sanding Discs, Grinders	•		•					•								•
Wet Sand	•							•				•				•
Wet Aggregate	•	•	•	•		•	•	•								•

B1.3 STONE/MASONRY

APPLICATION NOTES

REMARKS

- gentle, safe to surface
- not effective for heavy staining

- pressure can cause damage
- generates large volume of water

- generally cleans most structures
- especially good on rough surfaces and buildings with varied surfaces

- generally safest and least expensive type of method
- do not use in cold weather
- chemical salts present in masonry may react with water and form stains or efflorescence
- provide for disposal of water

- do not use on high lime content stones and mortars, or on light-colored, glazed, or polished masonry

- many common household cleansers can be used

- takes off heavy applied stains that simple steam cleaning will not remove

- consult manufacturer's literature for specific product recommended for various masonry types

- generally contain water in base mixture and in rinse can create all of the problems of plain water as well as:
  discoloration
  damage to plant and animal life
  damage to adjacent materials, e.g., metals, glass
- brown stain may appear after cleaning with acid to remove efflorescence
- for efflorescence, dilute HCL acid generally used; not to be used on light-colored soft brick

---

- do not use on polished surfaces or detailed carvings
- choice available as to gradation of abrasive and amount of pressure

- removes outer surface, do not use on polished or carved surfaces

- proprietary method
- less abrasive than blasting or grinding; eliminates dust

- proprietary method
- contains a silica-free friable aggregate to lessen abrasion

- generally better for heavier staining potentially damaging due to: dust; surface erosion and need to preserve or weatherproof masonry; abrasion of adjacent surfaces.
- mechanical treatment may be required to remove old graffitti; steam and chemicals may also be effective.
- mortar droppings which have hardened may be removed by chisel and stiff brush.

B1.3 STONE/MASONRY

STAIN REMOVAL/CLEANING:

CLEANING masonry is, for the most part, a trial-and-error procedure:

- all cleaning procedures and chemical cleaning solutions should be tested under conditions of temperature and humidity approximating those under which the actual use is expected to occur.

CLEANING OF NEW MASONRY
Cleaning methods most commonly used are:

- bucket-and-brush hand cleaning
- pressurized water
- sandblasting, or other abrasive methods.

■ **Bucket-and-brush hand cleaning**

may be used for all types of burned clay and concrete masonry:

- hydrochloric acid solutions, proprietary compounds and emulsifying agents may be used.
- rough textured units generally require rinsing by pressurized water.
- muriatic acid should not be used on light-colored burned clay units as it may cause "acid-burn."
- manganese-colored units tend to react to muriatic acid solutions and stain.
- acid solutions for glazed units should only be used to remove very difficult mortar stains.
- acids should not be used on salt glazed or metallic-glazed burned clay masonry.
- chemical cleaning solutions are generally not recommended when colored mortar is used, as most acids tend to bleach colored mortars; mild detergent solutions should be considered.

■ **Pressurized water** may be used for most types of burned clay and concrete masonry; use for glazed units may not offer any significant advantages over hand cleaning.

■ **Sandblasting** is not recommended for sand-finished burned clay units; nor for glazed units.

■ **External stains** on new or existing masonry may be removed using a poultice:

- Poultice is a paste made with a solvent or reagent and an inert material applied over the stain.
- It acts by dissolving the stain and leeching the solution into the paste of the poultice.
- Principal advantage of poultices is

that they tend to prevent the stain from spreading during treatment
- Commonly used for removal of isolated, relatively small stains.

CLEANING OF EXISTING MASONRY
Cleaning methods most commonly used are:

■ **High-pressure steam:**
- smooth, hard and glazed units should generally be cleaned by this method;
- chemical solutions may have to to added to remove certain stains.

■ **Sandblasting:**
- should be used only when masonry will not be damaged through its abrasive action;
- and when high-pressure steam cannot be successfully used for cleaning.
- Check applicable state and local codes to determine acceptability of this method.

■ **Pressurized cold water:**
- generally gives satisfactory results
- involves the disposal of large volumes of water.

■ **Hand washing:**
- is more costly and slower.
- generally limited to buildings of smaller size.

GRAFFITI
Removal of graffiti from masonry surfaces, especially when such surfaces are relatively porous, presents a major problem in cleaning of masonry:
- Markings commonly consist of: spray paint; felt-tip pens; crayons; lipstick.

■ All such markings tend to be absorbed to varying extent into the surface of porous masonry thus making their complete removal difficult:
- traces of the markings may remain as shadows because pigments and binder resins which penetrate deep into the pores of masonry units cannot be easily removed.

■ **Fresh paint may generally be removed** from burned clay and concrete masonry surfaces using a commercial paint remover or a solution of trisodium phosphate in water.

■ **Proprietary chemical formulations for removing graffiti from masonry surfaces are available:**
- Graffiti removers should be quick

acting. Graffiti removers are best for removing stains on surface layers; slow acting strippers are best for removing deep-set stains.
- Strong chemical solutions may remove most of a marking but may also etch the masonry.

■ **Porous masonry surfaces** may be protected from graffiti by sealing them with a variety of commercially available coatings, such as acrylics, epoxies, polyesters, polyvinyl acetates, silicones.
- wide differences exist in performance characteristics of coatings within the same generic group, and the in-service record of a specific coating should be investigated prior to making a selection.

■ **Masonry units may be glazed** which will effectively prevent or minimize absorption of various marking substances but attention should be then given to protect the mortar joints between them.
- Specially formulated coatings, such as acrylics, urethanes, may be used to protect the mortar joints of glazed masonry surfaces.

STAINS

■ **Efflorescence:** salts are water-soluble and may be removed by dry brushing or with clear water.
- may disappear of their own accord with normal weathering.
- heavy accumulations may be removed using solution of muriatic acid and scrubbing; surface-active poultices and other cleaning compounds may be safer than acid processes; suction methods can also be employed.
- light deposits of efflorescence are often referred to as "white bloom" or "new building bloom."

■ **Greenstain:** green or yellow efflorescence resulting from vanadium salts in raw materials used to manufacture burned clay units:
- stains may result from washing such units with acid solutions.
- may be removed with sodium hydroxide; proprietary cleaning compounds are available.

■ **Brownstain:** may develop in mortar joints of manganese-colored burned clay units when such units are cleaned with hydrochloric acid or exposed to acid rain.
- proprietary cleaning compounds are available; their effectiveness should be investigated prior to selection.

B1.3 STONE/MASONRY

selection checklist

■ **Rust:** due to corrosion of iron, or of welding splatter, embedded in the masonry.
• may be readily removed from burned clay and concrete masonry by scrubbing with sodium or ammonium citrate and glycerine solutions.

■ **Oil and tar stains** may be effectively removed by commercial emulsifying agents.

■ **Dirt:** may be difficult to remove, especially from heavily textured units.
• High-pressure steam is generally the most effective method.
• Chemical cleaners are available which are formulated to remove heavy deposits of dirt.

■ **Organic stains** may generally be removed with household bleach or oxalic acid.
• organic stains may be identified by applying concentrated sulphuric acid which will turn an organic substance black.

■ **Detailed information on cleaning of masonry** may be found in Technical Notes No. 20 as published by the Brick Institute of America; and in TEK-45 as published by the National Concrete Masonry Association.

CLEANING FAILURES

Improper cleaning methods may result in irreparable damage to masonry walls:

■ **Failure to thoroughly saturate the masonry units** before and after application of chemical or detergent cleaning solutions:
• saturation of masonry units reduces their rate of absorption thus keeping the cleaning solution on the surface rather than being absorbed by the units.

■ **Solutions not properly mixed** or overly concentrated can etch out cementitious material from the mortar joints; and
• may discolor the masonry and/or promote the development of green or brown stains.

■ **Cleaning solutions,** particularly acid solutions, may have a corrosive effect on other components incorporated into a wall, such as metal, glass, limestone.
• Applying cleaning solutions with high pressure equipment will drive the cleaning compound deep into

the masonry such that residuals do not rinse away. As a result, scumming, efflorescence and metallic staining may occur. The use of high pressure equipment is appropriate for prewetting and rinsing of masonry only.

SURFACE CHARACTERISTICS

Important surface characteristics include:

■ **Finish and texture;** rough, smooth, polished;
• generally cleaning is easier on smoother surfaces.

■ **Color:** light, dark
• certain methods can cause discoloration of masonry.
• light colors are generally more susceptible.

■ **Abrasive hardness:** hard, soft
• soft masonry is more easily abraded and may rule out certain methods of cleaning.

■ **Absorption:**
• stains are more difficult to remove from porous masonry
• cleaning agents can also penetrate masonry and create new problems, e.g., efflorescence.

RESTORATION OF BUILDINGS

Some of the preliminary considerations may include:

■ **Use:**
• building is to continue functioning for its current end use.
• building is to be recycled for a new end use.
• building is of historical/symbolic importance and is to be restored to its original condition.

■ **Existing conditions:**
• cracks, extend and characteristics.
• spalling, crumbling of unit faces
• loose joints, loose and/or missing masonry units.
• failure of masonry-to-mortar bond.
• masonry departing from horizontal and vertical lines: bulging out, caving in.
• the need to preserve stone carvings, moldings still in good condition which could be destroyed by sandblasting.
• damp walls, staining, discoloration.
• deterioration of masonry faces.
• damage to floors, ceilings.
• damage to utilities/services network, e.g., to plumbing pipes, to electrical conduit, etc.

■ **Type of structural failure**
• foundation settlement.
• inadequate structural design.
• omission or inadequacy of expansion, control joints provided.

■ **Moisture penetration caused by:**
• inadequate original design - caulking, flashing, roofing, weeps, parapets, below grade conditions.
• failure of mortar bond.
• failure of masonry units themselves.
• deterioration of accessory materials: metal ties, anchors, lintels, etc.

■ **Adverse environmental conditions:**
• temperature extremes.
• advent of air pollution and corrosive atmospheres.

■ **Possible environmental or ecological problems such as:**
• damage to adjacent property or materials.
• damage to plant or animal life.

■ **Remedial action:**
• repointing of mortar joints should include matching composition of existing mortar for: color; texture; strength.

■ **Replacement of masonry units using:**
• units from other parts of structure.
• new units: trying to match existing.
• methods of reproducing decorative masonry elements: glass fiber; reinforced plastic or concrete; precast concrete; cut stone.

■ **Refinishing masonry**
• filling in cracks and chips: epoxy, grout
• resurfacing: stucco, mortars

■ **Options for preservations after restoration/cleaning:**

■ **Waterproof treatments: opaque coatings**

■ **Water-repellent treatments:** clear coatings like silicone

■ **While considering:**
• application characteristics
• manufacturers guarantees and recommendations
• possible problems: trapping moisture within masonry; further cracking and spalling
• cost-effectiveness
Before any final decisions are made, the services of competent restoration experts should be considered.

B1.3 STONE/MASONRY

REFERENCES: Stone/Masonry

STANDARDS

ASTM Specifications for:

C34	• Structural Clay Load-Bearing Wall Tile.
C55	• Concrete Building Brick
C56	• Structural Clay Non-Load Bearing Tile.
C62	• Building Brick
C73	• Calcium Silicate Face Brick (Sand-Lime Brick).
C90	• Hollow Load-Bearing Concrete Masonry Units
C126	• Ceramic Glazed Structural Clay Facing Tile, Facing Brick, and Solid Masonry Units
C129	• Non-Load-Bearing Concrete Masonry Units
C144	• Aggregate for Masonry Mortar
C145	• Solid Load-Bearing Concrete Masonry Units
C212	• Structural Clay Facing Tile.
C216	• Facing Brick (Solid Masonry Units Made from Clay or Shale)
C270	• Mortar for Unit Masonry
C279	• Chemical-Resistant Masonry Units
C287	• Chemical-Resistant Sulfur Mortar
C331	• Lightweight Aggregates for Concrete Masonry Units
C404	• Aggregates for Masonry Grout
C406	• Roofing Slate
C410	• Industrial Floor Brick
C503	• Exterior Marble Building Stone
C568	• Limestone Building Stone
C615	• Granite Building Stone
C616	• Sandstone Building Stone
C629	• Slate Building Stone
C652	• Hollow Brick (Hollow Masonry Units Made from Clay or Shale)
C744	• Prefaced Concrete and Calcium Silicate Masonry Units.
C902	• Pedestrian and Light Traffic Paving Brick

ASTM Definitions of Terms Relating to:

C119	• Natural Building Stones

ASTM Classification of:

C27	• Fireclay and High Alumina Refractory Brick.
C155	• Insulating Firebrick.

SELECTED REFERENCES

• Chemically Resistant Masonry, Vol. 7, W. L. Sheppard, Marcel Dekker, NY.

ACI   **American Concrete Institute**
- title 76-23 - Concrete Masonry, ACI Manual of Practice
- title 531-1/81 - Specification

B/A   **Building Stone Institute**
- Brick Architectural Details
- Building Code Requirements for Engineering Brick Masonry
- Recommended Practice for Engineered Brick Masonry
- 1 Rev. Cold Weather Masonry Construction
- 1A Construction and Protection Requirements
- 24A Rev. Building Code Requirements
- 26 Six-Inch Brick & Tile Walls
- 8 Rev. Portland Cement-Lime Mortars for Brick Masonry
- 8A Standard Specification for Portland Cement-Lime Mortar for Brick Masonry
- 8B Mortar for Brick Masonry; Selection & Controls
- 11E Guide Specs. for Brick Masonry Part V - Mortar
- 17 Reinforced Brick Masonry Part 1
- 17C Reinforced Brick Masonry Part IV
- 17D Rev. High-Life Grouted Reinforced Brick Masonry
- 20 Rev. Cleaning Brick Masonry
- 23 Re. Efflorescence, Causes
- 23A Efflorescence, Prevention & Control

FTI   **Facing Tile Institute**
- Specifications for Glazed and Natural Finish Structural Facing Tile
- Specifications for Ceramic Glazed Facing Brick
- Design Data for Structural Facing Tile, Ceramic Glazed Brick

ILI   **Indiana Limestone Institute**
- The Indiana Limestone Handbook
- Textures in Indiana Limestone
- New Developments in Indiana Limestone
- Fire! and Indiana Limestone
- Preassembled Systems in Indiaana Limestone

MIA   **Marble Institute of America**
- Marble Engineering Handbook, Appendix, E, Definitions, A1A File No. 8 B-1, 1962

NBGQA   **National Building Granite Quarries Association**
- 4.1 New Specifications for Architectural Granite

NCMA   **National Concrete Masonry Assoc.**
- 12 Estimating U Factors for Concrete Masonry Construction
- 67 Tables of U Values for Concrete Masonry Walls
- 68 New Findings on Energy Conservation with C/M
- 82 Energy Conscious Design for Buildings
- 44 Maintenance of Concrete Masonry Walls XXX45 Removal of Stains from Concrete Masonry Walls
- 92 Control & Removal of Efflorescence

NLA   **National Limestone Association**
- Efflorescence of Masonry
- Lime - The Oldest Proven Mortar Material for

Durable Watertight
Construction
XXXStrength
Considerations in Mortar
& Masonry
- Durability or Mortar &
  Masonry
- Bond of Mortar to
  Masonry Units

NLI   **National Limestone
             Institute**
             - Limestone, Published
               Quarterly by Institute

NRCC   **National Research Council
             of Canada**

CBD2   - Efflorescence
6   - Rain Penetration of Wall
               sof Unit Masonry
123   - Cold Weather Masonry
               Construction
138   - One Using Old Bricks in
               New Buildings
140   - Thermal Performance of
               Concrete Masonry Walls
               in Fire
163   - Masonry Mortar
169   - Bricks
194   - Cleaning of Brickwork

PCA   **Portland Cement
             Association**
             - CM Handbook for
               Architects, Engineers,
               Builders - EB008M
             - How to Calculate Heat
               Transmission Coefficients
               and Vapor Condensation
               Temperatures of CM
               Walls - IS015M
             - Load Tests of Patterned
               CM Walls, Trowel Talk
               No. 3 - ISI02M
             - ST6-2 Efflorescence
             - 100 Removal of Graffiti
             - IS040M Mortars for
               Masonry Walls

B1.3 STONE/MASONRY

B1. 4 METALS

METALS: introduction

Evaluation of metals in construction may be broadly divided into two categories:

■ **When the properties of the metal used for a component are of primary importance,** irrespective of the form the component takes or of the amount of fabrication involved:

* seamed flat sheet roofing.
* formed sheet siding, horizontal or vertical; roofing panels.
* castings, forgings.
* extrusions: such as curtain wall framing, mullions, column covers.

■ **When the function of a component fabricated fully or partially of metal is the primary** concern, with the properties of the metal used of secondary importance:

* finish hardware: such as locks, closers.
* electrical fixtures, motors, wiring.
* information on the basic materials in a product within this category becomes progressively less and less important as the product becomes more complex: in an elevator cab the only material decision often left to the architect is the selection of finishes from the options provided by manufacturer.

Metals which fall under the category where the properties are of primary importance include:

FERROUS METALS:

Principal ferrous metals used in construction are:

■ **Iron:** containing no substantial amounts of carbon, soft and ductile, easily worked, relatively resistant to corrosion.

■ **Malleable iron:** iron cast at low carbon content which has been re-heated and slowly cooled again, or annealed, to improve its working properties:

* has good resistance to shock.

■ **Gray cast iron:** generally used in castings:

* hard, brittle, cannot be hammered or bent.
* has good resistance to corrosion.

■ **Steel:** iron with low to medium carbon content, relatively mild, relatively easy to work by cutting, punching, welding.

■ **Steel alloys:** plain carbon steel to which other elements are added to modify its properties:

* aluminum for surface hardening.
* chromium for corrosion resistance.
* copper for resistance to atmospheric corrosion.
* manganese: in low concentrations for hardenability, in high concentrations for wear resistance.
* molybdenum: generally combined with other alloying metals such as chromium and nickel, increases resistance to corrosion and raises tensile strength without reducing ductility.
* nickel increases tensile strength without reduction in ductility; in high concentrations, improves corrosion resistance.
* silicon: strengthens low-alloy steels, improves oxidation resistance, in high concentrations provides hard, brittle castings resistant to corrosive chemicals.
* sulfur promotes free machining, especially in mild steels.
* titanium prevents intergranular corrosion of stainless steels.
* tungsten, vanadium, and cobalt: for hardness and resistance to abrasion, generally for high-speed tool steels.

■ **Stainless steel:** steel containing a minimum of 11 ½ percent chromium:

* other metals such as nickel or molybdenum may be added.
* where maximum resistance to corrosion is required, such as resistance to pitting by seawater and chemicals, molybdenum containing types are best.
* for resistance to ordinary atmospheric corrosion, steels containing chromium and nickel are commonly used.
* straight chromium alloys generally used for interior applications.

NONFERROUS METALS

Principal nonferrous metals used in construction are:

■ **Aluminum:**

* high purity aluminum is soft and ductile, highly resistant to corrosion, but of low strength.
* when unfinished aluminum is exposed to the atmosphere, a thin protective oxide film begins immediately to form on exposed surfaces; this film retards corrosion and heals quickly when damaged, but upon prolonged exposure to weathering it will be penetrated resulting in spotty or chalky appearance.
* aluminum alloys are generally harder and stronger than the pure metal.
* alloying elements used are copper, manganese, silicon, magnesium, zinc.
* aluminum alloys may be clad with either pure aluminum or other alloys which are anodic, thus sacrificially protective, to the core.

Aluminum alloys commonly used in construction include:

■ **Sheets and plates:**

* 1100: for sheet metal applications where high strength is not required, can be porcelain enameled, has excellent resistance to corrosion and high degree of forming or drawing.
* 3003: general purpose alloy, commonly used for sheet metal applications, high strength than 1100 alloy, has good resistance to corrosion.
* 3004: for formed curtain wall panels, moderate strength.
* 5005: is comparable to 3003 alloy in strength, workability, and resistance to corrosion, when anodized will have close color match with anodized alloy 6063 extrusions.
* 5052: has higher strength than 3003 or 5005 alloys, good resistance to corrosion, when anodized will have fair color match with anodized alloy 6063 extrusions.
* 6061: high strength heat-treatable alloy, frequently used for formed structural shapes.

■ **Extrusions:**

* 6061: commonly for extruded structural shapes, pipe and tubing, or other applications requiring high strength; heat-treatable.
* 6063: most commonly used alloy for extruded building and ornamental metal shapes, may be welded readily and has good resistance to corrosion.

■ **Castings:**

• 43: used for hardware and ornamental metal castings where low strength is acceptable.
• 214: higher strength than 43 alloy, best appearance match with 6063 extrusions when anodized.
• 356: high strength, turns gray when anodized.

■ **Copper:** is resistant to corrosion, ductile, resistant to impact and fatigue:

• will oxidize on continuous exposure to the atmosphere resulting in the formation of green patina.
• when arsenic is added to copper, it improves resistance to pitting corrosion.
• copper-zinc alloys are commonly referred to as brass, available in various compositions, such as commercial bronze, red brass, muntz metal.
• leaded brass: with approximately two percent added to copper-zinc alloy to improve machinability, with several compositions available, such as architectural bronze, forging brass.
• copper-nickle-silver alloys are generally referred to as nickel silver, highly resistant to corrosion and tarnish, malleable and ductile.

CORROSION

The principal types of corrosion are:

■ **Galvanic action,** or corrosion, between dissimilar metals or metals and other materials when sufficient moisture is present to carry an electric current.

■ **Pitting corrosion** is a nonuniform type of corrosion which occurs as the result of local cell action which is produced when particles are deposited on a metal surface either as flakes, solids, or bubbles of gas, such as may occur in water system piping:

• Oxygen deficiency under a deposit sets up anodic areas to cause pitting.
• Oxygen may also cause pitting corrosion.

■ **Concentration cell corrosion:** is similar to galvanic action, except that the potential arises from the differences in electrolyte:

• Potential may be produced by differences in ion concentration, oxygen concentration, or foreign matter adhering to the surface.

GALVANIC ACTION

The susceptibility of metals to galvanic action varies:

■ **Below is a table of Galvanic Series** which is a good indicator of susceptibility to corrosion due to galvanic action:

Galvanic Series
Zinc
Galvanized steel
Aluminum alloys: 5052, 3004, 3003, 1100, 6053
Aluminum alloys: 2117, 2017, 2024, in this order
Low-carbon steel Wrought iron Cast iron
Type 410 stainless steel, active Type 304 stainless steel, active Type 316 stainless steel, active
Lead Tin
Copper alloy 280 Copper alloy 675 Copper alloys 404, 465, 466, 467 Nickel 200
Copper alloy 270 Copper alloys 443, 444, 445 Copper alloys 708, 614 Copper alloys 230 Copper alloy 110 Copper alloy 651, 655 Copper alloy 715 Copper alloy 923, cast
Nickel 200 Monel alloy 400
Type 410 stainless steel, passive Type 304 stainless steel, passive Type 316 stainless steel, passive
Silver

■ **Roughly speaking, the further apart the metals are from each other in the table, the greater the possibility for corrosion,** with the metals on top being the most susceptible to corrosion and those at the bottom the least susceptible:

• when zinc and iron are joined, the zinc is corroded and the iron is not affected.

• when copper and steel are joined, corrosion is accelerated on the steel but not on the copper: the greater the copper area relative to the area of steel, the faster the rate of corrosion.
• when metals from the same group are joined, they are relatively safe.

CORROSIVE ENVIRONMENTS: are commonly defined as:

■ **Marine:** Air containing high salt concentrations or salty fog and mist due to the proximity of a body of salt water.

■ **Mild Atmosphere:** Air which does not contain the high concentrations of acids or bases normally associated with heavily industrial or marine environments.

■ **Mild Chemical:** An environment containing concentrations of acids, bases, or salts.

■ **Oxidizing:** An environment containing high concentrations of acids which tend to return a metal to the unrefined ore state.

■ **Reducing:** An environment containing high concentrations of basic, or alkaline, substances causing deoxidation with a noncommitant plating of new metal upon the old.

RELATED CONSIDERATIONS

■ **Exposure to:**

• high pressure or lateral loads, heavy usage, possibility of accidental impact which may require high strength.
• forced entry: which may require high strength and hardness to resist bending and cutting.

■ **Contact with dissimilar materials:**
• galvanic action between metals in an assembly and dissimilar metal fasteners may cause corrosion especially when moisture is present.
• metals in contact with chemically active materials may corrode, especially when moisture is present, e.g., aluminum directly in contact with concrete or mortar, steel in contact with certain types of treated wood.

■ **Cut edges of galvanized and coated steel panels** generally will form protective film by galvanic action, and may not pose a problem.

■ **Cut edges of aluminum coated steel** should be touched up to prevent corrosion development.

ARCHITECTURAL USES

● denotes common uses
○ denotes possible or limited use, or limited availability

MAJOR METALS, ALLOYS	BUILDING ROOFS					WALLS							CEILING	
	SEAMED SHEETS	PANELS, FORMED, LAPPED	FLASHING	FASCIAS, COPINGS	GUTTERS, LEADERS	PANELS, FORMED	SIDING, HORIZONTAL	FACINGS, COLUMN COVERS	STUDS, JOISTS	FOIL INSULATION	PREFABRICATED PARTITIONS	GRILLES AND SCREENS	PANS, PANELS, CLAD TILES	SUSPENSION GRIDS
ALUMINUM ①	○	●	●	●	●	●	●	●	●	●	●	●	●	●
COPPER ②	●		●	●	●			●				●		
STAINLESS STEEL	●	○	●	●	○	○		●			●	●	○	
CARBON STEEL AND ALLOYS ③	○	●	●	●	○	●	●		●		●	●	●	●
IRON ④												○		
OTHER METALS														
LEAD ⑤	○		○											
MONEL ⑥	●		○	○	○									
TERNEPLATE ⑦ ⑧	●		●	●	○									
ZINC ⑨	●		●	○	○									
CHROMIUM, NICKEL														
PROPRIETARY ALLOYS	●		●	●	○	○	○							

B1.4 METALS

	MISCELLANEOUS													SITE	
	STAIRS, FIRE ESCAPES	FLOOR PLATES, GRATINGS	RAILINGS, HANDRAILS	LADDERS	ACCESS FLOORS SYSTEMS	DOORS, DOOR FRAMES	WINDOWS, WINDOW WALL FRAMING	HARDWARE: ROUGH, FINISH	SUN CONTROL DEVICES-EXTERIOR	LOUVERS: EXTERIOR, INTERIOR	ACCESS DOORS, DAMPERS, DUCTS	BATHROOM ACCESSORIES	FASTENERS, ANCHORS	FENCING, GATES	DRAINS, CATCH-BASINS
1	○	●	●	●	●	●	●	●	●	●			●	●	○
2		●			○	○	●	●	○			●	●		
3			●	●	○	●	○	●	●	●	○	●	●	○	
4	●	●	●	●	●	●	●	●	●	●	●	○	●	●	
5	○		●									●	●	○	●
								●							
												●			
														⑩	
										○					

REMARKS

① Aluminum is subject to attack by alkalies and should be protected from wet concrete, mortar, and plaster: protective coatings used are methacrylate lacquers, bituments, other moisture and alkali resistant coatings.

• Electrolythic action between aluminum and other metals should avoided: e.g., steel bolts should be isolated, drainage from copper alloy surfaces onto aluminum must be avoided.

• Galvanizing or cadmium coating anodic to aluminum and helps to protect it.

② Includes bronze and brass.

③ Generally galvanized and/or coated.

④ Other metals are often used for coatings or plating e.g., chromium, lead, nickel, zinc, can be applied to aluminum, copper, iron and steel.

⑤ Lead often used for flashing of roof drains, sound isolation, radiation shielding, vibration control.

⑥ Trade name.

⑦ Terneplate – sheet iron or steel plated with an alloy of three or four parts of lead to one part of tin.

⑧ Primarily used for plating.

⑨ Includes special combinations such as zinc-copper-titanium.

⑩ Generally cast.

B1.4 METALS

ALUMINUM: properties, uses

FORM / ALLOY [1]	TENSILE STRENGTH [2] psi, minimum	FABRICATION CHARACTERISTICS [3]		
		COLD FORMING	MACHINA–BILITY	WELDA–BILITY [4]
SHEET/PLATE				
PURE ALUMINUM ALLOY 1100 • AT LEAST 99 PERCENT • AVAILABLE CLAD • NON-HEAT-TREATABLE	11,000 through 22,000 depending on temper.	VERY GOOD to FAIR	FAIR to POOR	VERY GOOD to GOOD
COPPER BASED ALLOY 2024 • AVAILABLE CLAD • HEAT-TREATABLE	30,000 to 32,000 maximum for 0 temper; 57,000 to 71,000 for other tempers	FAIR	GOOD	GOOD to FAIR
MANGANESE BASED ALLOYS 3003 3004 • AVAILABLE CLAD • NON-HEAT-TREATABLE	• 13,000 to 27,000 for alloy 3003 • 21,000 to 38,000 for alloy 3004	VERY GOOD to FAIR	FAIR	VERY GOOD to GOOD
MAGNESIUM BASED ALLOYS 5005 5050 5052 • NON-HEAT-TREATABLE	• 15,000 to 27,000 for alloy 5005 • 18,000 to 29,000 for alloy 5050 • 25,000 to 39,000 for alloy 5052	VERY GOOD to FAIR	FAIR	VERY GOOD
MAGNESIUM SILICON BASED ALLOY 6061 • AVAILABLE CLAD • HEAT-TREATABLE	22,000 maximum for 0 temper; 27,000 to 42,000 for other tempers	VERY GOOD to FAIR	FAIR	VERY GOOD
ZINC BASED ALLOY 7075 • AVAILABLE CLAD • HEAT-TREATABLE	40,000 maximum for 0 temper 67,000 to 78,000 for other tempers	FAIR	GOOD to FAIR	GOOD to FAIR
EXTRUSIONS				
MAGNESIUM SILICON BASED ALLOYS 6061 6063	• Alloy 6061: 22,000 maximum for 0 temper; 27,000 to 42,000 for other tempers • Alloy 6063: 19,000 maximum for 0 temper; 22,000 to 41,000 for other tempers	GOOD to FAIR	FAIR to GOOD	VERY GOOD

RESISTANCE TO CORROSION		TYPICAL USES	REMARKS
GENERAL	STRESS [5]		
VERY GOOD	VERY GOOD	• Sheet metal work. • Low strength rivets and washers. • May be used as substrate for porcelain enamel.	
FAIR	FAIR	• Bolts, screws. • Should not be used in corrosive industrial and marine environments.	
VERY GOOD	VERY GOOD	• Sheet metal work. • Formed curtain wall panels. • Builders hardware. • Roofing/Siding. • Storage tanks • Conduit.	
VERY GOOD	VERY GOOD	• Sheet metal work. • Builders hardware. • Wall facing panels. • Heat exchanger tubing. • Conduit.	
GOOD	VERY GOOD to GOOD	• Formed structural shapes. • Heavy duty structures requiring good corrosion resistance. • Furniture. • High strength rivets and bolts.	
FAIR	GOOD to FAIR	• Aircraft and other structures requiring very high strength. • One of the highest strength alloys available.	
VERY GOOD to GOOD	VERY GOOD to GOOD	• Extruded structural shapes, pipe, tubing. • Railings, mullions, ornamental products.	

[1] Alloys and forms more commonly used in construction only, for information on other alloys refer to Aluminum Standards & Data, a publication of The Aluminum Association.

[2] Ultimate tensile strength based on requirements of ASTM B209.

[3] • Very good: generally weldable by all commercial procedures and methods.
   • Good: weldable with special techniques.
   • Fair: limited weldability because of crack sensitivity or loss in resistance to corrosion and mechanical properties.

[4] Commonly used gas welding methods may not be available for all tempers.

[5] Stress-corrosion cracking: very good denotes no known failure in service or in laboratory tests.

■ Galvanic corrosion of aluminum in contact with dissimilar materials will not occur indoors under dry conditions. Aluminum used outdoors or indoors where moisture is present or condensation may occur may require protection. Galvanic corrosion between aluminum and:
   • iron and steel is slow and can be prevented by painting the iron or steel with a primer and a bituminous coating.
   • zinc is significant and even tends to protect the aluminum.
   • galvanized and cadmium plated steel is significant, but when the coating over steel is consumed, the steel will rust which may cause staining. Protection is recommended.
   • stainless steel, lead, and monel is insignificant.
   • copper, brass and bronze will occur and direct contact of aluminum with these materials should be prevented.

■ • Non-heat-treatable alloys may be strengthened by various degrees of cold working.
   • Heat-treatable alloys are generally strengthened in two steps: solution heat treated and precipitation hardened.
   • Heat-treatable alloys in which copper or zinc are major alloying constituents are less resistant to corrosion than the majority of non-heat-treatable alloys. To increase the resistance to corrosion resistance of these alloys in sheet and plate form, they are often clad with high purity aluminum, or other alloys. Cladding is generally 2.5 to 5 percent of the total thickness on each side. Cladding also exerts a galvanic effect which further protects the core material.

■ • Aluminum alloys in both sheet and extrusions may be porcelain enameled. Where strength after application is important heat-treatable alloys are commonly used since non-heat-treatable alloys are annealed by the firing temperatures during fabrication.

■ • Coefficient of thermal expansion: 12.9 to 13.4 x 10 -6/°F.

B1.4 METALS

COPPER/COPPER ALLOYS: properties

BEST COLOR SHEET/PLATE: ALLOY	MATCH EXTRUSIONS: ALLOY	TENSILE STRENGTH psi, range	FABRICATION: COLD/HOT FORGING	MACHINABILITY	EASE OF BRAZING/ SOLDERING
COPPER					
110	110	36,000 to 40,000	VERY GOOD	FAIR	GOOD/ VERY GOOD
122		40,000	NOT COMMONLY USED	POOR	VERY GOOD
ARCHITECTURAL BRONZE/COMMON BRASS					
220 COMMERCIAL BRONZE	314 LEADED COMMERCIAL BRONZE	ALLOY 220: 37,000 to 61,000 52,000 for half hard sheet	ALLOY 220: VERY GOOD	ALLOY 220: POOR	ALLOY 220: VERY GOOD
230 RED BRASS	385 ARCHITECTURAL BRONZE	ALLOY 230: 57,000 for half hard sheet ALLOY 385: 60,000	ALLOY 230: VERY GOOD to GOOD ALLOY 385: VERY POOR for cold forging VERY GOOD for hot forging	ALLOY 230: POOR ALLOY 385: GOOD	ALLOY 230: VERY GOOD ALLOY 385: GOOD/ VERY GOOD
260 CARTRIDGE BRASS NOTE: Primarily recommended for interior use	260 CARTRIDGE BRASS	62,000 for half hard sheet	VERY GOOD to GOOD	FAIR	VERY GOOD
280 MUNTZ METAL	385 ARCHITECTURAL BRONZE	ALLOY 280: 70,000 for half hard sheet	ALLOY 280: FAIR for cold forging VERY GOOD for hot forging	ALLOY 280: GOOD	ALLOY 280: VERY GOOD
NICKEL-SILVER/SILICON BRONZE					
655 HIGH SILICON BRONZE	655 HIGH SILICON BRONZE	78,000	VERY GOOD to GOOD	FAIR	VERY GOOD/ GOOD NOTE: May be welded
745 NICKLE-SILVER	796 LEADED NICKEL-SILVER	ALLOY 745: 73,000 ALLOY 796: 60,000	ALLOY 745: VERY GOOD for cold forging	ALLOY 745: FAIR	ALLOY 745: VERY GOOD

STAINLESS STEEL: properties

ALLOY	TENSILE STRENGTH psi, range	RESISTANCE TO CORROSION	REMARKS
CHROMIUM-NICKEL-MANGANESE			
201 PLATE/SHEET/ STRIP 202 PLATE/SHEET/ STRIP	ALLOY 201: 95,000 to 185,000 ALLOY 202: 90,000 and 125,000	GOOD	• Alloy 201 available annealed, ¼ hard, ½ hard, ¾ hard, and full hard. Similar to 301 but higher tensile strength. • Alloy 202 available annealed and half hard. Similar to alloy 302 but higher tensile strength. • Requirements given under ASTM A412 and ASTM A480.
CHROMIUM-NICKEL			
301 PLATE/SHEET/ STRIP	75,000 for annealed to 185,000 for full-hard	GOOD	• Available annealed, ¼ hard, ½ hard, ¾ hard, and full hard. • Requirements given under ASTM A167, ASTM A177, and ASTM A480.
302 • PLATE/SHEET/ STRIP • BARS/SHAPES	75,000 for annealed to 125,000 for hardened bar	GOOD	• Readily fabricated for decorative uses, very good drawing and stamping properties, very good welding properties. • Also referred to as 18-8 steel: chromium content 17 to 19 percent, nickel content 8 to 10 percent. • Alloy 302B has silicon added to increase resistance to saling at high temperatures.
304 • PLATE/SHEET/ STRIP • BARS/SHAPES	75,000 for annealed to 125,000 for hardened bar	GOOD	• Low carbon 18-8 steel, weldable with less danger of intercrystalline corrosion. • Readily fabricated for decorative uses by drawing or stamping. • Extensively used for architectural work.
308 • PLATE/SHEET/ STRIP • BARS/SHAPES		VERY GOOD	• Used when resistance to corrosion greater than that of 18-8 steels is required.
316 • PLATE/SHEET/ STRIP • BARS/SHAPES	75,000 for annealed	VERY GOOD	• Has superior resistance to corrosion by seawater and chemicals. • Very good welding properties. • Good drawing and stamping properties. • Alloy 317 has similar properties.
321 • PLATE/SHEET/ STRIP • BARS/SHAPES		GOOD	• Essentially 18-8 steel stabilized against intercrystalline corrosion at elevated temperatures, such as due to welding. • Very good welding properties. • Good drawing and stamping properties. • Alloy 347 has similar properties.
CHROMIUM			
430 PLATE/SHEET/ STRIP	70,000	FAIR	• Will resist ordinary atmospheric corrosion. • Commonly for interior use, lowest in cost. • High chromium steels are resistant to oxidizing corrosion and are useful in chemical plants.

B1.4 METALS

OTHER METALS

PROPERTIES	SHAPES/APPLICATIONS	REMARKS
CARBON STEELS		
• Categorized according to carbon content • Increased carbon content provides greater hardness and strength and reduced weldability and ductility. • Other alloys are also added–within allowable maximums–to improve properties such as toughness, machinability, weldability, corrosion resistance. • Properties of carbon steel vary greatly depending on chemical composition, thermal treatment and fabrication methods.	• Plate; sheet; strip; bar; rod; tube; pipe; wire; structural shapes and castings. • Extrusions are used mostly for fasteners and small structural shapes.	• Resistance to corrosion is increased by applying a finish i.e. galvanizing, organic coatings, etc. • Given variances of properties in carbon steels a simple designation system is not available.
ALLOY STEELS		
• Alloying elements at levels higher than those allowed for carbon steels permit better and more varied performance properties. • The properties that can be altered include: strength, hardness, toughness, resistance to wear and abrasion, machinability, weldability, surface quality, resistance to corrosion, etc. • High Strength Low Alloy Steels (HSLA) are a class of alloy steels made usually to attain improved strength and corrosion resistance. Some of these are sometimes called "weathering steels." These steels perform well when welded, cut, formed, etc. Precautionary measures are needed in high humidity, industrial or marine environments.	• Non HSLA alloy steels are not prevalently used in architectural applications. • HSLA Steels are available in the forms of plate, sheet, strip, bar, tube, pipe, wire, and structural shapes. • HSLA steels are suitable for uses where weight must be held down. • Weathering steels are used where unpainted/uncoated steel components are exposed to view & to weather—use requires specific weathering conditions and special detailing.	• HSLA are marketed as trade name steels. • Given variances of properties in carbon steels a simple designation system is not available. • Water run off from weathering steel can stain adjacent materials. • HSLA corrosion resistance is twice that of carbon steel.
IRON		

• Pure iron is malleable, ductile, tough, easy to work; oxidizes rapidly in air, and is attacked by most acids.
• Property improvement is possible by changing chemical composition, thermal treatments and forming methods.
• Cast iron is tough, brittle, has high compressive strength and capacity to absorb vibration.
  • Grey Cast Iron is soft, tough and easily machined.
  • Malleable Cast Iron is strong, tough and easily machined.
• Wrought Iron is soft, resistant to corrosion and fatigue, easily worked and has good machinability.
	• Iron used primarily as main element of steels.
	• Cast Iron: castings for gratings, manhole covers, stair components, ornamental work, hardware, etc.
	• Wrought Iron: Plate, sheet, bars, pipes, special shapes. Applications include railings, fences, grilles and screens, ornamental work, etc.
		• Use of wrought iron has been decreasing for many years and is quite limited today.

PROPERTIES	SHAPES/APPLICATIONS	REMARKS
CHROMIUM		
• Is hard, brittle and corrosion resistant. • Can take a bright polish and does not tarnish in air.	• Plating on aluminum, copper iron, nickel, zinc & others • As alloying element in stainless steel and other metals.	• Sufficient thickness of plating is needed to reduce corrosion of plated metal.
LEAD		
• Is malleable, easily worked, resistant to corrosion. Has high density and ability to shield radiation. • Hardness and strength can be increased by alloying with antimony. Other alloys are used to improve properties. • Lead coatings provide corrosion resistance and shielding properties combined with those of the base metal.	• Waterproofing; isolation from sound, vibration; shielding from radiation • Sheet, strip, foil, bar, rod, pipe, castings and other. • Lead coatings. Most common is terne-plated (lead-tin on steel). • As alloying element.	• Lead in vapor or dust form is toxic if ingested. • Lead-coated copper is used for flashing.
MONEL		
• Is a nickel-copper alloy • Is highly resistant to corrosion, strong, ductile; has good fatigue strength, low coefficient of expansion, is machinable, workable, weldable and solderable.	• Plate, sheet, strip, bar, rod, tube and wire. • Major applications include roofing, siding, flashing, downspouts, copings.	• Marketed under the trade name Monel. • Will take a bright polish.
NICKEL		
• Is strong, tough, hard and corrosion resistant. • Is workable, weldable, brazeable and solderable. • Imparts control of expansion when alloyed to non ferrous metals.	• Major use as alloying element in aluminum, copper, steels and iron • Also used for plating.	• Can take a high polish.
ZINC		
• Is brittle, low in strength, corrosion resistant to water and air. • Is machinable, workable, weldable and solderable.	• Major use is galvanizing. • Sheet/strip for roofing/flashing • Die castings for hardware, etc. • As alloying element.	• Galvanizing is achieved by hot dipping iron or steel in molten zinc.

B1.4 METALS

FABRICATING/FORMING: definitions

FABRICATION is the primary processing applied to metal to obtain the basic shape/form needed. Secondary forming processes are also applied when more complex shapes and special dimensions are needed:

■ Rolling:

Rolling is used to produce plate, sheet, foil, rod, bar, wire and complex shapes. By passing the metal between rollers under pressure, the desired shapes, sizes, and surface qualities can be achieved. Metal can be rolled hot or cold, or in a sequence combining the two. The choice of rolling temperature will determine properties such as:

- surface qualities: grain, smoothness.
- mechanical properties: tensile strength, hardness, ductility.
- rolling is applicable to most metals except iron.

■ Extruding:

This is the process of pushing metal under pressure through a die orifice. Size or capability of die is the main limiting factor:

- Greater widths are obtainable by extruding desired shape in bent or curved form and then straightening it out. Another way of obtaining larger size is by joining different extrusions either mechanically or by welding, brazing, or soldering.
- Sizes available in industry can accommodate most architectural needs. Specialized die designs are worth considering when large orders are involved.
- Extruding is applicable to all metals except iron.

■ Casting:

This is a method by which molten metal is poured into molds and is allowed to solidify into the desired shape. Virtually all metals are cast into basic shapes and forms - ingots, billets, blooms, etc. - before they are further processed, i.e., rolled, extruded, etc.

- Special shapes for architectural uses also are achieved with castings, such as posts, balusters, hardware

as well as drains, valves, piping and similar specialties. Usually special casting alloys are utilized for this purpose, i.e., copper alloy series 800 and 900.
- Casting is applicable to aluminum, copper, iron, steel, and others.
- Choice of casting treatment will determine properties affecting surface quality and physical characteristics, i.e., dimensional tolerances.

■ Drawing:

This is the process of pulling solid metal through dies. It is used to alter or reduce cross sectional shape, to attain three-dimensional shapes for sheets and to alter physical properties and finishes:

- drawing hot or cold improves hardenability of metal and affects its surface qualities.
- shapes that can be drawn include sheets, tubes, pipes, rods, bars, wires, and complex shapes.
- drawing is applicable to all metals except iron.

■ Forging:

This is a process by which metal is worked to a particular shape through hammering and pressing. It can be done in either hot or cold form:

- Forging usually enhances the strength characteristics of the metal.
- Surface quality properties also are affected by forging processes: grain fineness, hardness, etc.
- Forging is applicable to aluminum, copper and steel.

■ Welding: Fusing of metals above molten point with or without the addition of metal filler:

- Welding is usually performed at temperatures higher than in brazing or soldering. The following types of welding can be identified:
- Arc Welding: Where heat for fusion is derived from an electric arc, suitable electrodes, and a suitable gas. There are various types of arc welding, such as: Oxyacetylene, gas shielded, where a gas shields the molten metal, and other. Arc welding equipment is costly and not as portable as gas welding equipment. Arc welding in general provides a high quality weld.

- Gas Welding: Where heat for fusion is provided by a torch with a suitable gas. Most economic and portable of welding methods.
- Resistance Welding - Spot and Seam: Where metals are fused by being butted together in close contact with electrodes which feed strong alternating current. Good for high heat-treatable/high strength alloys. Limited to shop application.
- Cold Welding: A method of joining metals such as aluminum, by subjecting the thoroughly cleaned joining surfaces to pressures in specially shaped dies. When the combined thickness of the surfaces are reduced by a specific percentage, a weld occurs at normal temperatures.

FORMING is a process by which metals are shaped by mechanical operations, excluding machining. Forming includes:

■ Bending:

This method is used to achieve curved shapes, generally applied to tubes, rods and extrusions.

■ Brake forming:

This is a process of successive pressing to achieve specific shapes:
- Usually applied to plate, sheets, and strips.

■ Spinning:

This is a method of shaping by smooth tool while metal is being rotated on an axis.

■ Embossing and Coining:

This is a method to achieve textured patterns, raised, usually on flat shapes.

■ Blanking:

This entails cutting by punch press, sawing, shearing; it is performed to achieve outline:

- Usually applied to sheets.

■ Perforating:

This is punching or drilling of holes in metal, usually flat shapes.

■ Piercing:

This means piercing a hole through

B1.4 METALS

metal without removing any of the metal.

■ **Forming processes are in general applicable to all metals,** though some may have size limitations; effects on physical and surface properties are determined by alloy and primary fabrication processes used.

THERMAL TREATMENTS

These treatments have an important impact upon the performance of metal, especially on strength, hardness, machinability, cold forming characteristics and ductility. The basic types are annealing, quenching, and tempering:

■ **Annealing:** A heat treatment which reduces the hardness of a metal and which can be controlled to increase machinability or cold forming characteristics.

■ **Quenching:** A thermal treatment to achieve desired properties, strengthened workability being the most common, in which steel is heated to a specified critical temperature, held there for a specified length of time, rapidly cooled to a specified temperature by submersion in a liquid.

■ **Tempering:** A thermal treatment to achieve desired properties, strength and workability being the most common, essentially the same as quenching, but which does not involve rapid cooling by submersion in a liquid.

DEFINITION OF SOME PROPERTIES

■ **Ductility:** The capability of certain metals, when cold, to undergo large permanent deformations without rupture. Opposite of brittleness.

■ **Workability:** Ease with which a material can be rolled, extruded, forged, bent, pressed, etc.

■ **Machinability:** A measure of the ease with which a metal can be cut to form, as on a lathe, while retaining acceptable surface quality and mechanical properties.

■ **Brazeability:** Capacity for two metals to be fused at an intermediate temperature, using a nonferrous filler metal with a melt-

ing point above 800°F, but generally lower than that of the base metals.

■ **Weldability:** Capacity of metals to be joined through fusion of the base metals, carried out at high temperature and/or pressure, with or without the addition of a filler metal:

• The filler metal either has a melting point approximately the same as the base metals or has a melting point below that of base metals, but above 800°F.

■ **Solderability:** Capacity for two metals to be fused at a relatively low temperature, lower than either brazing or welding, using lead-base or tin-base alloy solder filler metals which melt below 500°F.

■ **Modulus of Elasticity:** The ratio of direct stress to corresponding strain throughout the range where the stress and strain are proportional:

• Applicable for tension test.

■ **Modulus of Rigidity:** The ratio of direct shear stress to corresponding shear strain throughout the range where they are proportional. Often referred to as shear modulus of elasticity.
• Applicable for torsion test.

■ **Hardness:** The measure of the resistance of a material to wear, abrasion, or indentation:

• Two types of tests are used to measure hardness: Brinell and Rockwell:
• Both involve dropping an indenter ball with a predetermined force upon the surface of material and measuring the resulting indentation.

■ Hardenability: The measure of the propensity of a metal to harden through heat treating and/or cold working.

■ **Toughness:** The measure to resist fracture upon impact.

■ **Allowable or Working Stress:** Certain fraction of the yield point or of the ultimate strength which can be safely used for a given metal:

• for ductile metals the yield point is usually chosen as the criterion for determining allowable stress:

• for brittle metals which have no yield point, the ultimate strength is used.

■ **Yield Strength:** The point at which an increase in stress in a material is not necessary to increase strain, i.e., no addition in increase of force is needed to change size or shape per unit length.

■ **Ultimate Strength:** The maximum stress which a material can sustain under a gradual and uniformly applied load.

■ **Tensile Strength:** The maximum stress which a material can sustain when tension is applied.

■ **Compressive Strength:** The maximum stress which a material is capable of sustaining under compressive forces.

■ **Shear Strength:** The maxiumum stress which a material can sustain when a force is applied tangentially to the section upon which it acts.

■ **Fatique Strength:** The maximum stress caused by repeated or fluctuating application of stresses, which a material is capable of sustaining indefinitely.

■ **Strain:** Change in size or shape in a material unit length

■ **Elastic Limit:** The maximum unit stress which can be applied to metal without causing permanent deformation.

■ **Coefficient of Thermal Expansion:** The change in dimension of a material due to temperature change, per unit of dimension, per degree of change in temperature.

■ **Oxidation:** A combination of oxygen with metal, forming rust on ferrous metal, and patinas, tarnishes, or coating of varying depths and color on non-ferrous metal.

■ **Electrolysis:** Chemical decomposition by galvanic action caused by contact of dissimilar metals in the presence of moisture.

B1.4 METALS

FINISHES

● commonly used
○ rarely used

TYPE	ALUMINUM	COPPER ALLOYS	STAINLESS STEELS	CARBON STEEL	REMARKS
MECHANICAL					
MILL OR "AS FABRICATED"	●	●	●	●	• may be cast, extruded, hot or cold-rolled
BRIGHT ROLLED	●	●	●		• produced by cold-rolling only
DIRECTIONAL, GRIT TEXTURED	●	●	●	○	• may be polished, buffed, hand-rubbed, brushed, or cold-rolled
NON-DIRECTIONAL, MATTE	●	●	●	○	• sand or shot-blasted
BRIGHT POLISHED	●	●	●		• polished or buffed
PATTERENED	●	●	●	●	• available in light gage sheets of all metals
CHEMICAL					
NON-ETCH CLEANING	●	●	●	●	
MATTE FINISH	●	○	○	○	• etched finish common in aluminum
BRIGHT FINISH	○	○			
CONVERSION COATINGS	●	●		●	• provide color variations in copper
COATINGS					
ANODIC	●				
ORGANIC	●	●x	○	●	
VITREOUS	●	○		●	• intermediate & top coats must be compatible.
METALLIC	○	○	○	●	• organic coatings are the most important finish for carbon steels.
LAMINATED	●	○		●	• clear coatings commonly used for copper only.

B1.4 METALS

TYPE	DURABILITY/RESISTANCE TO:		RESISTANCE TO SOLVENTS:
ORGANIC COATINGS: backed-on			
FLUOROCARBON • THERMOPLASTIC • COMPARATIVE UNIT COST: VERY HIGH	• FADING • CHALKING • MILDEW • SALT SPRAY • ABRASION • MARRING	EXCELLENT EXCELLENT EXCELLENT EXCELLENT EXCELLENT FAIR	• ALIPHATIC HYDROCARBONS: EXCELLENT • AROMATIC HYDROCARBONS: EXCELLENT • KETONE OR OXYGENATED: EXCELLENT • GENERAL CHEMICAL RESISTANCE: EXCELLENT
SILICONIZED ACRYLIC • THERMOSETTING • COMPARATIVE UNIT COST: HIGH	• FADING • CHALKING • MILDEW • SALT SPRAY • ABRASION • MARRING	EXCELLENT EXCELLENT GOOD GOOD GOOD GOOD	• ALIPHATIC HYDROCARBONS: GOOD • AROMATIC HYDROCARBONS: GOOD • KETONE OR OXYGENATED: GOOD • GENERAL CHEMICAL RESISTANCE: GOOD
SILICONIZED POLYESTER • THERMOSETTING • COMPARATIVE UNIT COST: HIGH	• FADING • CHALKING • MILDEW • SALT SPRAY • ABRASION • MARRING	EXCELLENT GOOD GOOD EXCELLENT TO GOOD GOOD GOOD	• ALIPHATIC HYDROCARBONS: EXCELLENT TO GOOD • AROMATIC HYDROCARBONS: EXCELLENT TO GOOD • KETONE OR OXYGENATED: GOOD • GENERAL CHEMICAL RESISTANCE: GOOD
PLASTISOL • THERMOPLASTIC • COMPARATIVE UNIT COST: LOW	• FADING • CHALKING • MILDEW • SALT SPRAY • ABRASION • MARRING	GOOD EXCELLENT GOOD EXCELLENT EXCELLENT FAIR	• ALIPHATIC HYDROCARBONS: EXCELLENT • AROMATIC HYDROCARBONS: FAIR • KETONE OR OXYGENATED: POOR • GENERAL CHEMICAL RESISTANCE: EXCELLENT

B1.4 METALS

FINISHES: aluminum

TYPE	APPEARANCE	LIMITATIONS	REMARKS
CHEMICAL			
NON-ETCHED	Natural mill appearance.	Surface blemishes resulting from fabrication will not be removed.	Commonly used to clean in preparation for other finishes.
ETCHED	Rough surface: • Fine matte • Medium matte • Coarse matte.	Not recommended for large assemblies or those consisting of different alloys.	• Often used for surfaces to be anodized • Etching may be followed by protective coating of clear lacquer.
BRIGHT	Ranges from highly specular mirror-bright to diffuse bright.	• Limited architectural use • Not recommended for large surfaces or for assemblies consisting of different alloys.	• Uniformity difficult to maintain • Highly specular finish often used for reflectors in lighting fixtures.
CONVERSION COATED	Various methods produce different appearance: • clear to yellow colors • clear or greenish colors • gray colors.	• Not recommended for assemblies consisting of different alloys if used as final finish.	• Generally used to improve adhesion of surfaces for painting, and some may also be used as final finish. • Most methods are proprietary.
MECHANICAL [1]			
MILL OR AS FABRICATED	Hot rolled and heat treated surfaces are relatively dull; cold rolled surfaces are brighter with more metallic appearance.	Should not be selected where uniformity of surface finish is desired.	Variations generally available are: • Unspecified: no special methods used. • Specular as fabricated: not available for extrusions, forgings, castings • Nonspecular as fabricated.
BUFFED	Smooth, highly reflective surface.	Should not be selected for wide flat surfaces as it may highlight surface irregularities.	Variations are: • Smooth specular: mirror-like; brightest finish obtainable • Specular: may show some evidence of scratches or surface defects.
TEXTURED: DIRECTIONAL	Smooth surface with satiny sheen of limited reflectivity.	May tend to collect and retain dirt.	Variation commonly available: • Fine satin • Medium satin • Coarse satin • Brushed • Hand rubbed.
TEXTURED: NON-DIRECTIONAL	Matte surface of varying degrees of roughness.	Will pick up and hold dirt; will show fingerprints. Uniform appearance difficult to achieve.	• Principally used for castings. • Produced by blasting using sand, glass beads, metal shot, dust, or slurry of very fine abrasive and water. [2]
PATTERNED	Patterns/textures impressed either on both or one side of a thin sheet.	Generally limited to light gauge sheets only.	Patterns/textures may be highlighted by buffing.

B1.4 METALS

TYPE	APPEARANCE	LIMITATIONS	REMARKS
COATINGS			
ANODIC ☐3 CLASS I	• Clear: showing the ☐4 natural color of aluminum • Integral color: light-to-dark gray; pale gold through dark bronze; and dead black. • Electrolitically deposited: medium bronze to black.	No significant limitations: may be used for interior and exterior components.	• Thickness of coating: at least 0.7 mil • Aluminum Association Designations: A41: clear A42: integral color A43: impregnated color A44: electrolytically deposited.
ANODIC ☐3 CLASS II		For interior components not subject to excessive wear or abrasion and for exterior components regularly cleaned and maintained.	• Thickness of coating: 0.4 mil to 0.7 mil • Aluminum Association Designations: A31: clear A32: integral color A33: impregnated color A34: electrolytically deposited.
ORGANIC: factory applied	• Wide choice of glosss and/or color.	Thorough cleaning of surfaces to receive coatings of critical importance.	Commonly used organic coatings include: • Fluorocarbons • Siliconized polymers • Plastisols.
VITREOUS: porcelain enamel	• Wide range of colors • Full matte to high gloss • May be textured.	• Limited impact resistance • Brittle and may crack under deformations • Sharp corners not practicable.	High firing temperatures required during manufacture may cause warping.
LAMINATED ☐5	• Clear or pigmented plastic films • May be embossed.	No significant limitations.	Plastic generally used: • Polyvinyl chloride • Polyvinyl flouride • Acrylic.

NOTES:

☐1 Mechanical finishes should not be selected for clad sheets because the cladding may be penetrated by the finishing process.

☐2 • Sandblasted finish should not be selected for light gauge sheet to avoid possibility of distortion.
• Thickness of metal should be at least ¼ inch.
• Coatings of clear lacquer commonly used to prevent dirt pick up.

☐3 • Anodic coatings can only be restored by refinishing at the factory: components subject to heavy wear or abrasion should therefore either receive a sufficiently heavy coating or should have a type of finish which can be restored in the field.
• All surface protective coverings such as strippable plastic films or pressure sensitive adhesive paper should be removed as soon as possible after installation: exposure to sunlight and weather tends to make them brittle and unpeelable.
• Quality control is generally performed in accordance with the following:
ASTM B244: Coating Thickness
B137: Coating Weight
B136: Resistance to Staining.

☐4 • Clear anodic finishes are generally preceded by either a chemical etch or some form of mechanical finish; both treatments may also be combined with etching then following mechanical finishing.
• Each alloy will produce its own characteristic shade of color.
• Joints between large anodized panels should be detailed to provide visual interruption since slight variations in shade may be expected.
• Exposed welds will never exactly match adjacent metal areas.

☐5 • Highly durable, excellent color retention when properly formulated.

FINISHES: carbon steel/iron

TYPE	APPEARANCE	LIMITATIONS	REMARKS
CHEMICAL			
CLEANED	• As fabricated surfaces with scale and oxide coatings removed • As fabricated surfaces with mineral and animal fats and oils removed.	Not a surface finish: surface defects are not removed.	Methods commonly used: • Pickling: to remove mill scale, rust, dirt • Alkaline cleaning: to remove oil and grease • Vapor degreasing: to remove oil and grease.
CONVERSION COATED	• Fine matte surface texture • Usually slate gray in color.	Not a surface finish: surface defects are neither removed nor masked.	• Used on cabon and galvanized steel to "convert" the chemical nature of surface, thus improving bond for applied coatings • Methods commonly used are acid phosphate solutions.
MECHANICAL ☐1			
MILL or AS FABRICATED	• Hot rolled: tight mill scale and rust powder • Cold rolled: smooth, greasy surface.	• Surfaces not suitable for applied coatings. • Will continue to oxidize, or corrode, until the metal reverts back to its natural state.	High-strength low-alloy steels are available which form a protective surface film when exposed to weathering, thus inhibiting further corrosion. Rate of film formation varies from 18 months to 3 years; but in dry climates the formation will be much slower.
CLEANED	• As fabricated surfaces with mill scale and rust removed • Surface defects may be removed when grinding or blasting is used.	Surfaces may also have to be cleaned of oil and/or grease to make them suitable to receive applied coatings.	Methods used include: • Shot or sand blasting: best method for obtaining a clean even surface, but relatively expensive • Power tool cleaning • Flame cleaning: for heavy components only • Hand cleaning: for small areas only.
COATINGS: metallic ☐2			
ZINC ☐3	• Bright spangle through dull gray for hot dip galvanized • Gray, crystalline, matte surface for electrogalvanized.	Light coatings, such as electrogalvanized, of approximately 0.10 oz per square foot require additional protection for satisfactory durability.	Two types of zinc-coated sheets are commonly available: • Hot-dip galvanized: grade G90, with minimum of 0.90 oz per square foot of zinc on both sides of sheet, is the grade most widely used in architectural applications • Electrolytically galvanized.
ALUMINUM	Looks and weather same as pure aluminum.	• Subject to galvanic corrosion same as aluminum. • Does not provide protection to base metal through anodic action, and base metal at cut edges may rust.	• Coatings are applied by hot-dip process • General resistance to corrosion similar to that of aluminum alloy 1100. • Type 2, with commercially pure aluminum coating, is generally used in architectural applications.

B1.4 METALS

TYPE	APPEARANCE	LIMITATIONS	REMARKS
COATINGS: organic, vitreous, laminated			
ORGANIC: factory applied [4]	Wide choice of gloss and/or color.	• Thorough cleaning of surface to receive coatings is of critical importance • Special surface treatments may be necessary	Organic coatings commonly used for architectural applications include: • Fluorocarbons • Siliconized polymers: acrylic, polyester • Plastisols.
VITREOUS: porcelain enamel	• Wide range of colors available • Full matte to high gloss • May be textured	• Brittle and may crack under impact or when subject to deformations • Sharp corners should not be used: minimum recommended radius is 3/16 inch.	• Selection of proper base material and surface preparation are critical considerations • High firing temperatures required during manufacture may cause warping, especially in light gauge sheets.
LAMINATED	• Clear or pigmented plastic films • May be embossed or grained • Several colors may be used for the same film.	• No significant limitations • Resistance to weathering should be ascertained.	Plastic films commonly used include: • Polyvinyl chloride: may be embossed or grained • Polyvinyl fluoride: widely used for residential siding • Acrylic: highly resistant to ultraviolet radiation.

NOTES

[1]
• Directional and nondirectional finishes may be, but rarely are, used on carbon steel.
• Patterned finishes, such as embossing, are used for light gauge sheets.

[2]
• Other metals also often used for the coating of carbon steel include: cadmium; chromium; lead; nickel; and terne, which consists of 80 to 90 percent of lead and 20 to 10 percent of tin.
• Coating processes include:
Hot-Dip: used to apply zinc, aluminum, and terne.
Thermal Spraying: extensively used for zinc, aluminum, and most other metals. Thermal spraying is the only method of metallic coating which can be used on-site.
Electroplating: used for applying most metallic coatings.
Cladding: produces bimetallic sheets consisting of a steel core covered with a

thin sheet of the coating metal.
Cementation: by which the coating metal is alloyed into the surface of the base metal.
Fusion Welding: fusing the surface of the coating metal to the surface of the base metal.

[3]
• Two variations of the hot-dip process are used: one for continuous coating of sheets and strips; the other, for fabricated shapes.
• After-Fabrication galvanizing by the hot-dip process provides an average minimum coating of 2.0 oz per square foot for each side of surface.
• Untreated commercial galvanized finishes may require special primers or preparation when to be coated with some organic coatings: zinc dust, zinc oxide, and Portland cement-in-oil coatings

will generally adhere well with minimum of preparation.
• The standard G90 coating, ASTM A525, is usually adequate for most conditions, but for severe exposure, such as in aggressive industrial atmospheres, a coating weight of 2.0 oz per square foot is recommended by the Zinc Institute.
• Zinc coatings will be burned off by ordinary welding operations: the coating can be restored by painting with zinc dust, zinc oxide, or zinc rich coating, or by the use of thermal spraying.

[4]
• For information on field-applied coatings for steel, and on surface preparation, refer to the chapter on COATINGS.

B1.4 METALS

FINISHES: copper

TYPE	APPEARANCE	REMARKS
CHEMICALS		
NON-ETCH CLEANING [1]	CLEAN AS FABRICATED SURFACE FREE FROM OIL, GREASE, FINGERPRINTS	[1] • Cleaning only before applying other finishes, does not alter surface of metal. • Two commonly used methods are: vapor degreasing and cleaning by spray or immersion, usually at elevated temperatures.
CONVERSION COATED: PATINATED [2]	DUPLICATE COLOR CHANGES DUE TO EFFECTS OF WEATHERING	[2] • Patinated finishes, also referred to as verde antique, are dependent upon a number of variables such as temperature, humidity, surface conditions during application, and uniformity of color over large surface areas cannot be readily controlled. • May excessively stain adjacent surfaces. • Clear organic coatings may be applied to improve resistance to wear, but such coatings may alter the color of the patina.
CONVERSION COATED: STATUARY [3]	COLOR VARIES FROM LIGHT GOLDEN TO BLACK	[3] • Natural color and surface texture of the metal and the thickness of the conversion film will affect the ultimate appearance. • May be highlighted by cutting back the conversion coating with an abrasive. • Uniformity of color cannot be readily controlled.
MECHANICAL		
AS FABRICATED	• DULL TO BRIGHT [4] • MAY SHOW STAINS, DARKENING OR DISCOLORATION	[4] • Finishes are imparted through normal production processes such as hot or cold rolling, extrusion, or casting. • Typical variations are: Unspecified: as extruded, cast, or rolled with unpolished rolls; for rolled products it may vary from dull to rather bright, and may show stains. Specular: a mirror-like cold rolled surface produced on one or both sides of a sheet by final pass through highly polised rolls. Matte: a dull surface resulting from hot rolling, extruding, casting, or cold rolling followed by annealing. • As Fabricated rolled finishes, except for specular, are seldom acceptable as the final finish. • On extrusions the As Fabricated finish may be acceptable. • Unless As Fabricated finish is carefully protected during subsequent fabrication, it is likely to be discolored and/or marred and cannot be restored.
BUFFED	VARIES FROM BRIGHT [5] WITH SOME EVIDENCE OF SCRATCHES OR IMPERFECTIONS TO MIRROR-LIKE	[5] • Highly reflective and will highlight surface irregularities, not recommended for wide flat surfaces. • Generally available as smooth specular, which is the brightest mechanical finish obtainable, and specular, which may show some evidence of scratches and/or other surface imperfections. • Refinishing on-site of smooth specular is impossible and that of specular quite costly.
TEXTURED: DIRECTIONAL	SMOOTH SATIN [6] SHEEN OF LIMITED REFLECTIVITY	[6] • Most popular finish for architectural copper alloys. • Standard types are: fine satin, medium satin, coarse satin, uniform, brushed, and hand rubbed.
TEXTURED: NON-DIRECTIONAL	MATTE SURFACE OF [7] VARYING DEGREES OF ROUGHNESS	[7] • Produced by sand or metal shot blasting, and limited to metal at least ¼ inch thick to minimize the possibility of distortion in the metal. • Rough surfaces will pick up and hold dirt.
PATTERNED	WIDE VARIETY OF [8] PATTERNS OR TEXTURES ON ONE OR BOTH SIDES OF SHEET	[8] • Produced by passion a light gauge As Fabricated sheet between engraved design rolls thus impressing patterns either on both sides or on one side of the sheet only. • Impressing a pattern on both sides of the sheet, or embossing, will generally increase its stiffness.
COATINGS		
ORGANIC	COMMONLY CLEAR [9] FILMS TO PROTECT THE SURFACE FROM STAINS, WEAR, AND ABRASION	[9] • Copper will develop a superficial discoloring tarnish when exposed to weathering and handling, and clear organic coatings are often used to preserve the natural color. • Coatings will not perform satisfactorily unless the surface is properly prepared and the coating is applied immediately after the finishing process is completed. • Clear coatings often used include: acrylic, alkyd, cellulose acetate butyrate, epoxy, nitrocellulose, silicone, urethane. • Lead coated copper is used in roofing and flashing applications where water runoff may subject stone, concrete, or other materials to staining from copper. • Chrome plated copper is extensively used for hardware and toilet accessories.
OIL/WAX	TO BRING OUT THE LUSTER AND TO ENHANCE DEPTH OF COLOR	

B1.4 METALS

TYPE	APPEARANCE	REMARKS
CHEMICALS		
CONVERSION COATED [1]	• LIGHT TO DARK BRONZE BY CONTROLLED OXIDATION • BLUISH TO DARK BROWN OR BROWN-BLACK BY HEAT TREATMENT	[1] • Surface color on stainless steel ma be produced by controlled oxidation; a modified ultrahard silicate is then applied to the oxidized surface, providing excellent resistance to abrasion and weathering. • Proprietary hot-dip processes are used to produce colors ranging from black through green, gold, bronze, blue, and red. • Color uniformity is generally good. • Surface finish will affect the color: matte finishes will result in matte colors; satin finish, in satin sheen colors; polished surfaces will exhibit their degree of brightness in the color. • Different effects may be achieved by selective polishing prior to the color treatment. • Color treated stainless steel may be fabricated, bent, notched, and embossed generally without damage to the surface color.
MECHANICAL		
MILL [2]	• RELATIVELY ROUGH DULL FOR HOT ROLLED • DULL TO BRIGHT FOR COLD ROLLED	[2] • Hot-rolled mill finish available only on heavy gauge sheets, bars, and shapes. • Dull finish for cold-rolled is available as Sheet Finish or Strip Finish. • Bright finish for cold-rolled is also available as Sheet Finish or Strip Finish, and is obtained by a final pass through highly polished rolls. • Bright annealed finish is a cold-rolled mill finish retained by final annealing of the metal, and may be used without further fabrication, or may be buffed.
POLISHED [3] **POLISHED and BUFFED**	• DULL TO BRIGHT, MODERATELY REFLECTIVE FOR POLISHED • BRIGHT, HIGHLY REFLECTIVE FOR POLISHED AND BUFFED	• Dull and bright sheet finishes cannot be matched by other means, and should not be selected when subsequent fabrication requires extensive bending or welding and final appearance is critical. • For welded assemblies in which fusion welds must be finished to blend inconspicuously with adjacent surfaces, a directional polished finish or a special blended finish is recommended. [3] • Polished finishes are directional, produced by wheel or belt polishing, used generally for, but not limited to, sheet stainless steel, with the direction of the finish usually parallel to the long dimension of the sheet.
PATTERNED [4]	WIDE CHOICE OF EMBOSSED OR COILLED PATTERNS AND TEXTURES	• Directional finishes are classified as: No. 3 Sheet Finish: coarsest of the polished finishes. No. 4 Sheet Finish: most widely used architectural finish. No. 6 Sheet Finish: obtained by additional brushing with abra sive and oil, which results in a dull satin surface. No. 7 Sheet Finish: is a bright highly reflective finish resulting from additional buffing of the polished surface. No. 8 Sheet Finish: bright mirror-like finish, obtained by further buffing. • Highly polished/buffed finishes are not recommended for large flat surfaces, since they will tend to emphasize any surface irregularities or even slight damages sustained during and after installation.
COATINGS		• Stainless steel with directional textured finish may be fabricated, such as brake or roll formed, punched, sheared, or welded, after finishing.
METALLIC [5]	• COPPER COATED: NATURAL COLOR OF COPPER • TERNE COATED: DULL GRAY	[4] • Available for light gauge sheet, may be differentially polished. • When embossed, may add stiffness to the sheet. • Will reduce optical distortion and will hide surface irregularities. • May be conversion color coated before fabrication and then polished, resulting in colored and natural polished surfaces. [5] • Terne coated stainless steel sheets available as proprietary product for roofing applications. [6] • Applied organic coatings and organic coatings modified with chromium complexes, both factory applied, are available: generally they are sufficiently flexible to withstand normal forming operations.
ORGANIC VITREOUS [6]	WIDE CHOICE OR COLORS AND/OR GLOSS	• Porcelain enamel coatings may also be used over stainless steel, but such usage is not common.

B1.4 METALS

FINISHES: selection checklist

METAL FINISHES are used to protect the metal surface from corrosive elements and/or to improve its appearance, initial and weathered. Organic coatings for metals may be either shop or field-applied. Other than that today most finishes for metals are shop-applied. Finishes may be applied to the base materials, or to the finished product after fabrication, or both:

APPLIED ORGANIC COATINGS

Applied finishes are those which involve the application of some added material as a surface covering over metal. They may be inorganic, such as porcelain enamel, or organic, either field-applied or factory-applied baked-on synthetic coatings.

Most commonly used factory-applied baked-on organic coatings are:

■ **Fluorocarbons:**

- various proprietary formulations are available, generally based on resins produced by Pennwalt Corporation under the trade name of Kynar 500.
- may be applied over primers for added resistance to corrosion and for improved adhesion, such as over epoxy-zinc chromate primers.
- available in a wide range of colors, but only in the low to medium range of gloss.
- air-dry formulations generally available for touch-up of scratches and/or minor damage.

■ **Siliconized Polymers:** are based on a combination of a high quality organic polymer and silicone intermediates, which impart to the unmodified polymer: improved weathering characteristics, greater resistance to color change and chalking, extended gloss retention:

- formulations, thus properties, may vary widely, time-proven performance record for a specific coating being considered should be investigated prior to making a selection.
- applied as either top coat only or as primer plus top coat, the two-coat system is preferred for maximum performance as the primer-top coat combination insures improved uniformity of coating and more complete film continuity.
- available in a wide range of colors: low to high gloss.
- air-dry formulations generally available for touch-up of scratches and/or minor damage.

Principal types of siliconized polymers commonly available are:

- Siliconized Acrylic.
- Siliconized Polyester.

■ **Plastisols:** are dispersion coatings based on polyvinyl chloride homopolymer resins dispersed in a plasti-sizer:

- plastisols are the topcoat of a two-coat system, requiring special primers to achieve proper bond to the substrate.
- film thickness of the topcoat varies from 3 mils to as much as 15 mils, with the greater thicknesses being used when the coating is to be embossed.
- physical properties of the coating can be varied considerably during the formulation process; time-proven performance record for a specific coating being considered should be investigated prior to making a selection.
- choice of colors is available in the low to medium gloss range.

SURFACE PREPARATION FOR APPLIED ORGANIC COATINGS

The ultimate performance of any applied coating largely depends on the preparation and treatment of the metal surface to receive it. The primary objectives of pretreatment of metal surfaces are:

- removal of residual organic and inorganic surface contaminants.
- provision of a nonmetallic chemical film integral with the metal surface for bonding the coating, generally referred to as conversion coating.
- preventing under-coating corrosion.

■ **Conversion coating** generally recommended for aluminum used in architectural applications is an amorphous chrome phosphate coating:

- for additional information refer to Voluntary Standard AAMA 605.1 as promulgated by Architectural Aluminum Manufacturers Association.

■ Conversion coatings for steel generally employ phosphate or various strong salt solutions:

- for additional information on surface preparation of steel, refer to section on COATINGS.

VITREOUS COATING: A coating finish composed of inorganic glassy materials, e.g., porcelain enamel, that are bonded to the metallic substrate by fusion:

■ **Porcelain enamel** is the hardest and most durable architectural metal finish available.

- available in very wide range of permanent colors and suface textures, ranging from full matte to high gloss with excellent color uniformity.
- textures are obtained either by embossing the metal prior to the application of the coating or by stripping or graining the coating itself prior to firing.
- porcelain enamel is brittle and almost any deformation in the substrate may lead to cracking and spalling.
- impact resistance is limited.

ANODIC FINISHES

The anodizing process for aluminum consists of producting under controlled conditions an oxide coating which is much harder and substantially thicker than the natural oxide film which would form on aluminum upon exposure to the atmosphere.

■ **The final appearance** produced by any anodic coating depends largely on the method of surface treatment before anodizing:

- Surface treatments commonly consist of: mechanical finishing and pre-anodic treatment.
- Mechanical finishing will change the properties of the surface and alter its response to the anodizing process: if one part only of an item to be anodized received a mechanical finish, the color of such part may be drastically different from the rest.
- Preanodizing treatments are used to remove all foreign matter from the surface and to establish to some extent optical characteristics desired in the finished product: a matte surface in an etch bath, a mirror-like appearance in a brightening bath.

■ **The choice of the proper alloy is of primary importance:**

- Alloys must be selected not only for their mechanical properties but also for their suitability for anodizing.

- Fabrication processes, such as rolling or extrusion, and thermal treatments, including welding, influence the response of aluminum to anodizing.
- Color variations may occur from one element to another, especially between sheets and extrusions; welds will always show when anodized.
- Alloy 6063 for extrusions and alloys 5005 and 1100 for sheets and plates will properly accept Class I and Class II anodic finishes.
- Alloy 6061 for extrusions will give a good color match with alloy 6061 for sheets and plates; but neither will provide a good color match with other alloys when anodized.

OTHER FINISHES:

■ **As Fabricated:** A mechanically produced finish, the texture and surface appearance of which is originally given to the metal by the extrusion press, rolling mill, or casting mold which shapes it, also known as "mill finish."

■ **Brightened:** A variety of chemically induced, bright finishes obtained by electrolytic brightening or by dipping the metal in specific acid solutions.

■ **Buffed:** A mechanically produced finish generated by successive operations of polishing and buffing with power-operated soft fabric wheels using fine abrasives with a lubricant.

■ **Chemical Finish:** A finish which is chemically induced, and which may or may not impart a physical effect on the metallic surface.

■ **Directional Textured:** A mechanically produced, process finish of soft texture and smooth satiny sheen of limited reflectivity produced by imparting tiny, almost parallel, scratches in the metal surface by wheel or belt polishing with fine abrasive, by handrubbing with stainless wool, or by brushing with abrasive wheels.

■ **Etched:** Treatment consisting of acid, alkaline solutions to produce a matte finish and frosted surface, imparts a satin/dull sheet and varying degrees of roughness.

■ **Laminated Coating:** A coating finish of nonmetallic, precast films that are bonded to the metal surface by adhesives.

■ **Metallic Coating:** A coating composed of one metal applied to another metal by electrolytic action, hot dipping, electroplating, or other techniques.

■ **Nondirectional Textured:** A mechanically produced, matte finish of varying degrees of roughness, generated by blasting sand, glass beads or metal shot against the metal under controlled conditions.

■ **Patterned:** A mechanically produced finish, available in a wide variety of textures and designs, produced by passing "as fabricated" sheet either between two machined matched-design rolls, impressing patterns on both sides of the sheet, embossing, or between a smooth roll and a design roll impressing a pattern, coining, one side of the sheet only.

OTHER CONSIDERATIONS

■ **Color uniformity** may vary due to the process of baking on the coatings:

- **heavy sections may not attain the same temperature in the baking oven as light sections with resultant slight variations in the color of the finished products.**

■ **Compatability and adhesion** of sealants to be used in contact with coated surfaces should be considered during the selection of the sealant as well as of the coating itself.

■ **Touch-up of organic coatings** is possible, but should be held to absolute minimum:

- air-drying touch-up coatings do not weather at the same rate as the baked-on coatings, neither are they as durable:
- some difference in color and gloss should be expected.
- organic touch-up coatings for porcelain enamel are available.

■ **Installation of finshed product** into building:

- precautions required in storage and during installation.
- dissimilar metals in close contact.
- contact with corrosion-triggering

nonmetallic materials.
- cleaning and touch-up.

■ **Appearance requirements,** both initial and after weathering, including:

- color, texture, pattern, gloss.
- finish durability required, taking into account years of service and maintenance program anticipated.
- control of color variations from alloy to alloy, from form to form.

■ **Degree of exposure,** as defined by:

- location in the building: orientation, contact with soil, etc.
- corrosive and marine atmospheres, pollution.
- temperature range and extremes.

■ **Usage, type and degrees:**

- probability of surface abrasion, scratching.
- impact.

In the case of field-applied finishes and coatings, in addition to all of the above considerations, it is important to consider also:

■ **Surface preparation** required for a successful field application of finish/coating:

- old finishes to be removed.
- surface smoothness, cleanness, soundness.
- need for primer and intermediate coats.
- need for all coats to be compatible with each other.

■ **Application methods,** conditions:

- timing of coating application within sequence of trades in construction schedule.
- application type of applied coating in field, spray, roller, brush, etc., and special tolls, skills needed.
- shelf life of coating.
- temperature and humidity requirements during application.
- number of coats and treatment between coats.
- drying and curing time.
- recoat time, full use time.
- advisability of providing a temporary clear finish to protect permanent finish surface from damage during installation.

B1.4 METALS

REFERENCES: Metals

STANDARDS

ASTM Specifications for:

A167   • Stainless and Heat-Resisting Chromium-Nickel Steel Plate, Sheet, and Strip.

A176   • Stainless and Heat-Resisting Chromium Steel Plate, Sheet, and Strip.

A177   • High-Strength Stainless and Heat-Resisting Chromium-Nickel Steel Sheet and Strip.

A276   • Stainless and Heat-Resisting Steel Bars and Shapes.

A368   • Stainless and Heat-Resisting Steel Wire Strand.

A412   • Stainless and Heat-Resisting Chromium-Nickel-Manganese Steel Plate, Sheet and Strip.

A480   • General Requirements for Flat-Rolled Stainless and Heat-Resisting Steel Plate, Sheet and Strip.

A484   • General Requirements for Stainless and Heat-Resisting Wrought Steel Products (Except Wire).

A580   • Stainless an Heat-Resisting Steel Wire.

A666   • Austenitic Stainless Steel Sheets, Strip, Plate, and Flat Bar for Structural Appli-

B26   • Aluminum-Alloy Sand Casting.

B36   • Brass Plate, Sheet, Strip and Rolled Bar.

B43   • Seamless Red Brass Pipe, Standard Size.

B75   • Seamless Copper Tube.

B88   • Seamless Copper Water Tube.

B98   • Copper-Silicon Alloy Rod, Bar and Shapes.

B108   • Aluminum-Alloy Permanent Mold Casting. Copper-Nickel-Tin Alloy

B122   • Copper-Nickel-Zinc Alloy (Nickel Silver) and Copper-Nickel Alloy Plate, Sheet, Strip, and Rolled Bar.

B133   • Copper Rod, Bar, and Shapes.

B135   • Seamless Brass Tube.

B152   • Copper Sheet, Strip, Plate and Rolled Bar.

B209   • Aluminum and Aluminum-Alloy Sheet and Plate.

B211   • Aluminum-Alloy Bars, Rods and Bars.

B221   • Aluminum-Alloy Extruded Bars, Rods, Wire, Shapes, and Tubes.

B241   • Aluminum-Alloy Seamless Pipe and Seamless Extruded Tube.

B308   • Aluminum-Alloy Standard Structural Shapes, Rolled or Extruded.

B316   • Aluminum-Alloy Rivet and Cold Heading Wire and Rods.

B447   • Welded Copper Tube.

B455   • Copper-Zinc-Lead Alloy (Leaded Brass) Extruded Shapes.

B584   • Copper Alloy Sand Casting for General Applications.

SELECTED REFERENCES

AA   **Aluminum Association**
  • Aluminum Extrusion Application Guide.
  • Specifications for Aluminum Structures.
  • Aluminum Standards and Data.
  • Designation System for Aluminum Finishes.
  • Specification for Aluminum Sheet Metal Work in Building Construction.
  • Standards for Aluminum Sand and Permanent Mold Castings.

AAMA   **Architectural Aluminum Manufacturers Association**
  • Voluntary Specification and Inspection Methods for Clear Anodic Finishes for Architectural Aluminum.
  • Voluntary Guide Specification and Inspection Methods for Electrolytically Deposited Color Anodic Finishes for Architectural Aluminum.
  • Voluntary Guide Specification and Inspection Methods for Integral Color Anodic Finishes for Architectural Aluminum.
  • Voluntary Specification for High Performance Organic Coatings for Architectural Extrusions.

AISI   **American Iron and Steel Institute**
  • Stainless Steel Uses in Architecture.
  • Steel Products Manual (very technical and metallurgically oriented).

AWS   **American Welding Society**
  • Welding Handbook
  • Various Publications.

CDA   **Copper Development Association**
  • Sheet Copper Applications.
  • Copper, Brass and Bronze Design Handbook: Architectural Applications.
  • Standards Handbook for Wrought and Cast Copper and Copper Alloy Products:
   - Wrought Products
    Part 1 - Tolerances
    Part 2 - Alloy Data
    Part 3 - Terminology
    Part 4 - Engineering Data
    Part 5 - Sources
    Part 6 - Specifications/ Cross Index
   - Cast Products

LIA   **Lead Industries Association**
  • Lead as a Modern Design Material.
  • Modern Uses of Lead in the Construction Industry.

NAAMM   **National Association of Architectural Metal Manufacturers**
  • Metal Finishes Manual.

NRCC   **National Research Council of Canada**
CBD170   • Atmospheric Corrosion of Metals.

B1. 5 WOOD/WOOD PRODUCTS

WOOD/WOOD PRODUCTS: introduction

SOLID WOOD

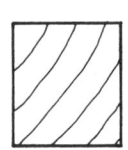

HARDWOOD: most commonly used for flooring. Standard thickness for strip flooring is 25/32 to ⅜ inch, with face width of 1½ to 3¼ inches and random lengths of 2 feet or more. Thin, narrow strips are available preassembled into parquet tile, commonly ⅜ inch think by 12 inches square.
* Hardwood is also used for stair treads, moldings, trim.

SOFTWOOD: used in construction, may be broadly classified into:
■ **Stress-graded lumber:**
* timber: posts and beams. Availability may be limited, especially of seasoned stock. Has been largely replaced by parallel laminated wood in heavy timber type of construction.
* joists, rafters, studs, decking: for structural framing. Grades of all species for high allowable stresses may not always be readily available; installation should be inspected to ascertain that the proper grade lumber is being used.
■ **Appearance graded lumber:** for trim, paneling, siding, cabinets, shelving.
■ **Nonstress-graded lumber:** generally boards, planks for sheathing, underflooring, formwork, utility shelving.
Softwood end-grain blocks are also used as heavy duty flooring.

TREATED WOOD

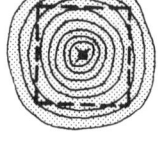

DECAY-RESISTANT TREATED: to minimize the effects of decay, fungi, termites, marine borers. Creosote, creosote solutions, pentachlorophenol, and waterborne preservatives are commonly used.

* decay resistant treatments have little effect on the strength of the treated wood.
* some preservatives may react with other materials used in construction, such as metals, roofing bitumens and compatibility should be ascertained.
* arsenic-free/ chromium-free preservatives are available for environmentally sensitive applications.

FIRE-RETARDANT TREATED: to protect against effects of exposure to fire, especially wood of small dimensions, such as studs, joints, rafters, plywood roof sheathing.

* two types available: exterior for exposed locations which resists leeching out of treating chemicals, and interior for used in protected locations.
* some treatments used for interior types will absorb moisture under high relative humidity conditions, which may corrode fasteners used. Treatments which are negligibly affected by exposure to sustained high relative humidity, up to 95 percent, are available.
* allowable stresses are reduced in lumber and plywood due to fire retardant treatment.

LAMINATED WOOD, parallel laminated

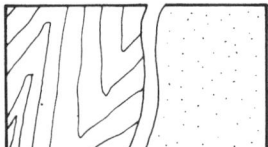

LAMINATED LUMBER: Parallel laminates have properties similar to those of sawn wood except that: load bearing properties are generally better; they are less subject to checking, and dimensional stability is improved.
* butt, finger, and scarf joints are used to connect the laminae in longer beams, with the latter two more common. Butt joints are inexpensive but also inefficient; scarf joints work well but are wasteful of material and are being replaced with finger joints.
* adhesives are available for either interior use only, or for exterior exposure.

LAMINATED VENEER LUMBER:
Laminae of wood veneer, with the direction of grain parallel, rather than lumber are used, thus allowing for more efficient material utilitzation. Generally manufactured in width of up to four feet, and then cut into required sizes.
* used for truss members, as outer laminae for laminated lumber, or for wood I beams.
Laminated lumber available in various forms, generally manufactured for each specific project.

LAMINATED WOOD, cross-laminated

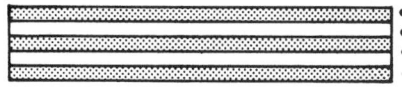

GENERALLY OF
ODD NUMBER OF
PLIES: 3-5-7

• GRAIN DIRECTION
  IN ALTERNATE
  PLIES PARALLEL

• TYPES WITH EVEN
  NUMBERS OF PLIES
  AVAILABLE

• FACES
• CROSSBAND
• CORE
• CROSSBAND
• BACK

Plywood: is manufactured using several thin sheets or plies of wood peeled from logs. As the geometry of grain orientation in plywood is clearly defined, reasonably accurate calculations of the properties of the composite board can be made:
• the strength in tension or compression is largely determined by the percentage of long grain in the cross section being loaded and on the long grain properties of the wood used.
• in bending not only the direction of the grain but also the position of the plies in relation to the surface of the panel should be considered.
  shear properties in the plane of the panel are always considerably improved because of the reinforcement given by adjacent plies against tension failure across the grain.
  improvement in dimensional stability is also considerable, but the construction of the panel must be balanced: each ply on one side of a sheet must be balanced by another ply of equal properties and orientation on the other side.
• adhesive used is either for interior dry locations only, or for exposed exterior use.
• available in appearance and utility/construction grades, with a variety of face, crossband, and core plies.

PARTICLE BOARD

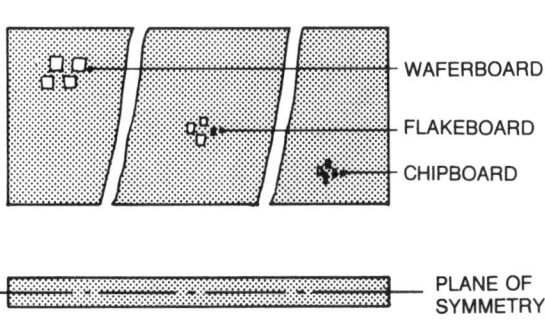

— WAFERBOARD

— FLAKEBOARD

— CHIPBOARD

— PLANE OF
  SYMMETRY

Particle boards are fabricated using relatively small fragments of hardwoods or softwoods produced by cutting or mechanical fracture. Particles are bound with an adhesive, predominantly of synthetic origin.
Additives such as paraffin are used to impart desired properties, such as dimensional stability.
• particles commonly used are: wafers, relatively large flakes. produced by cutting; flakes, which are smaller than wafers, also by cutting chips, which are produced by mechanically fracturing wood. into small fragments.
• in the finished board particles generally lie with their principal plane nearly parallel with the large plane of the board which results in considerable differences between the physical properties of the board perpendicular or parallel to its plane.
• made by either extrusion or mat-forming: extruded boards due to manufacturing process have a plane of weakness across the length as extruded. They are generally used a cores in veneered by overlaid construction; mat-formed panels are used both as cores and panel stock. Balanced construction in laminates using particle board core is important to prevent warping or twisting.

FIBERBOARD

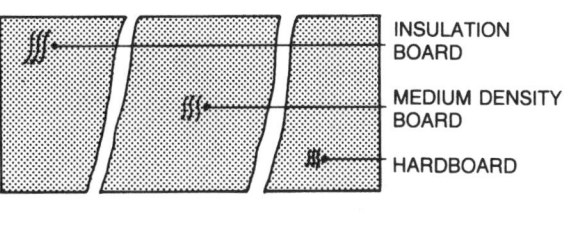

— INSULATION
  BOARD

— MEDIUM DENSITY
  BOARD

— HARDBOARD

— PLANE OF
  SYMMETRY

Fiberboard is essentially similar to particle board, the difference being that fibers are used in lieu of flakes or chips.
Fiberboards are manufactured to include a wide range of densities.

■ Insulation board with density range from 10 to 30 pounds per cubic foot. Used as sheathing, horizontal form board, sound-deadening, or low cost insulation/levelling in reroofing.

■ Medium-density fiberboard generally used as siding, either as panel or horizontal strip siding; generally factory prefinished.
• also available as prefinished interior paneling.

■ Hardboard: both are produced by hot-pressing, with the tempered hardboard further impregnated with oils or synthetics and then baked.
• tempering improves water resistance, hardness, and strength, but embrittles the board.
• hardboard is used as: backing for metal faces of wall panels; interior facing with a plastic film laminated to its exposed surface, or with various patterns impressed during manufacture.

B1.5 WOOD/WOOD PRODUCTS

WOOD/WOOD PRODUCTS: (cont.)

SANDWICH PANELS

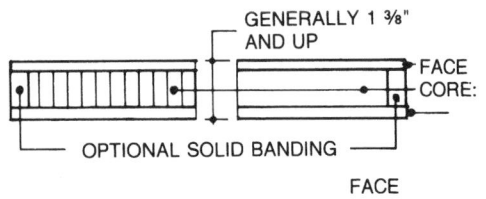

Sandwich panels are a special group of laminates which commonly consist of three laminae of which the core is much thicker and lower in both stiffness and rigidity than the faces.

- Faces of plywood, hardboard, particle board; cores of fiber honeycomb or foamed insulation.
- Cores of low stiffness contribute little to total transformed moment of inertia or area moment, but do contribute significantly to deflection of the panel.
- Dimension stability of panels depends principally on balanced construction, i.e., both faces of the same light material: differential movement between faces of dissimilar materials may cause bowing, warping in the panel.
- Failure under load may occur through: buckling, core crushing and/or shear crimping of faces under compressive load.
- Faces are generally thin and subject to damage through impact, especially in panels with honeycomb cores.
- Honeycomb cores may be filled with loose insulation, such as perlite, to minimize heat flow.
- Sandwich panels edged with solid lumber are used as low-cost doors, prefabricated unitized wall and roof panels.

SIDING

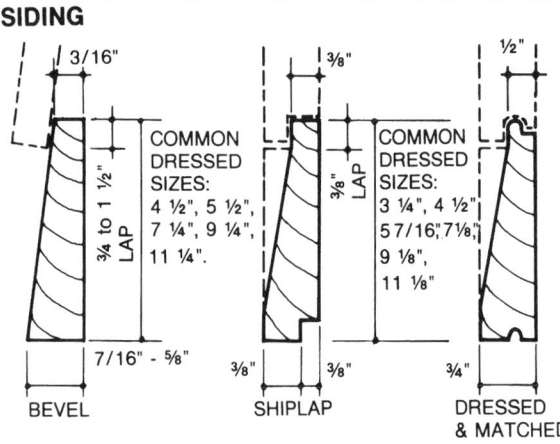

HORIZONTAL SIDING: horizontal strips of solid wood, plywood, or fiberboard: bevel, either smooth or rough resawn; shiplap, either smooth or rough resawn; drop siding with tongue and groove edges in several patterns.

- Siding with tongue and groove edges has better resistance to rain penetrations. Sheathing generally required to attach siding.
- Species of wood best suited and generally used: cedars, eastern white pine, suger pine, western white pine, cypress, and redwood.
- Medium density overlaid plywood, and medium density fiberboard, either prefinished or ready for site applied finishes are available.

VERTICAL SIDING: boards with shiplap, tongue and groove, or batten vertical joints; panels, generally grooved vertically, with various surface finishes and either butt or shiplap joints.

- Panels may be used directly over studs without additional sheathing.
- Species of wood best suited are the same as for horizontal siding.
- Plywood is available with rough sanded, rough sawn, brushed, or striated face finishes; medium or heavy density overlaid; either prefinished or for finishing at the site.
- Fiberboard is available with various embossed surface patterns.
- For additional information on siding and paneling refer to WALL FACINGS.

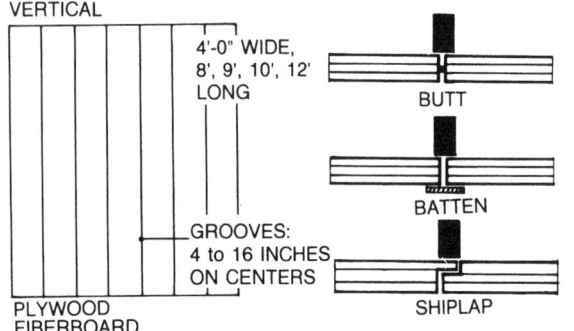

SHINGLES AND SHAKES: Generally available is standard lengths of 16,18 and 24 inches.

- Specie of wood most commonly used in western red cedar, northern white cedar, bald cypress, and redwood are also used. All have decay-resistant hardwood which is desirable when shingles are left unfinished.
- Available fire retardant treated, which may be a building code requirement for roofing shingles. COLUMNS, CORNICES, GABLE ORNAMENTS, PEDIMENTS and other items of architectural woodwork are available in a variety of designs. Except for ornamental columns, which use wood, most of the other trim items are made of molded plastic, both for exterior and interior use.

B1.5 WOOD/WOOD PRODUCTS

CONNECTORS: types, uses

JOIST AND PURLIN HANGERS

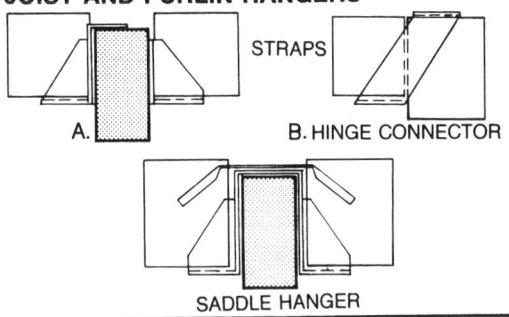

STRAPS

A.

B. HINGE CONNECTOR

SADDLE HANGER

JOIST AND PURLIN HANGERS:

- Type A used for light loads; available in galvanized sheet metal, 16 to 18 gauge.
- Type B used for heavier loads; available in galvanized sheet metal, 7 to 12 gauge; and 1/8 to 3/16 inch steel plate, generally prime painted.
- Saddle hangers available in 12 to 18 gauge galvanized steel or primed steel plate.
- Hangers, seats, hinge connectors for gluelam are available generally of prefinished steel plate.
- Straps used to anchor joists or purlins to ledger, or to tie them across intervening framing.

FRAMING ANCHORS

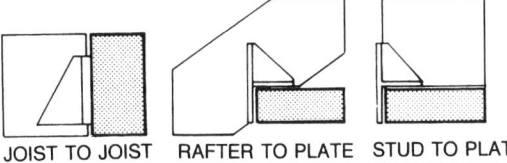

JOIST TO JOIST   RAFTER TO PLATE   STUD TO PLATE

FRAMING ANCHORS

Used as tie-downs to provide positive connection. Generally required in high wind and seismic zones: available in 16 or 18 gauge galvanized steel.
- Special sill plate anchors are available for casting into concrete foundation in lieu of anchor bolts.

POST CAPS AND BASES

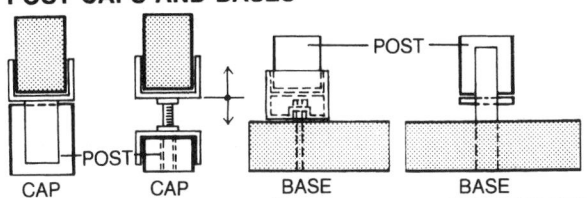

POST

POST

CAP   CAP   BASE   BASE

POST CAPS AND BASES

Caps available in galvanized steel, 16 to 12 gauge; or in steel plate, generally 1/8 to 3/16 inch, prime painted, galvanized, or plated.
- Caps to provide for adjustment after installation are available.
Bases available of galvanized steel, prefinished steel plate, and cast aluminum.

JOIST AND PURLIN ANCHORS

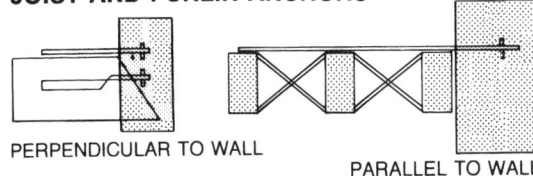

PERPENDICULAR TO WALL

PARALLEL TO WALL

JOINT AND PURLIN ANCHORS

Commonly every fourth joist perpendicular to wall, and 8 foot on center for joists parallel to wall.
- Available in prime painted steel plate, 3/16 to 1/4 inch. Other finishes generally available.

ACCESSORIES

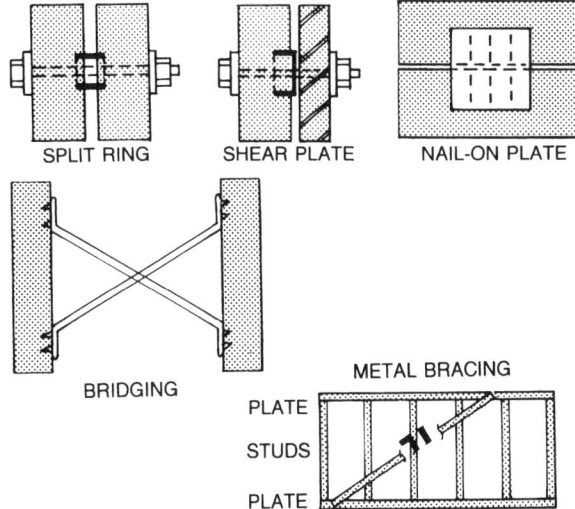

SPLIT RING   SHEAR PLATE   NAIL-ON PLATE

BRIDGING

METAL BRACING

PLATE

STUDS

PLATE

ACCESSORIES

■ Split ring connectors are commonly used in truss member connections to distribute shearing stresses over a larger area. They are installed in prepared grooves, and expand upon tightening to assure positive simultaneous bearing in members being joined.
- Available in 2½ and 4 inch diameters.

■ Shear plates are used in wood to steel connections for the same purpose as the split ring connectors.
- Available in 2½ and 4 inch diameters.
- May also be used in wood-to-wood connections using two plates back to back; generally when connection must be demountable.

■ Nail-on plates are commonly used for single plane assembly of light roof trusses, as well as for other wood-to-wood connections.
- Generally in 20 gauge galvanized steel.
  Other connectors available are: spike grids for pole connections; plain and flanged clamping plates.

■ Bridging of 18 gauge galvanized steel used in lieu of wood bridging; ends of metal bridging are nailed directly into joists.

■ Bracing of 16 or 18 gauge galvanized steel, either flat or angle used in lieu of let-in wood bracing when sheathing used is nonstructural.

B1.5 WOOD/WOOD PRODUCTS

WOOD

SHRINKAGE ACROSS GRAIN

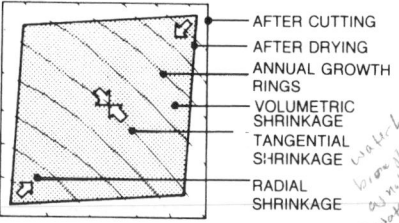

AFTER CUTTING
AFTER DRYING
ANNUAL GROWTH RINGS
VOLUMETRIC SHRINKAGE
TANGENTIAL SHRINKAGE
RADIAL SHRINKAGE

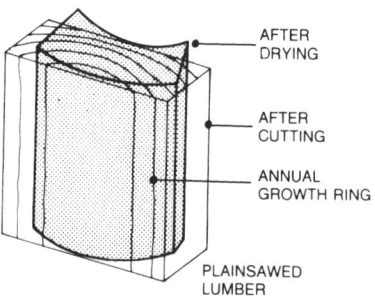

AFTER DRYING

AFTER CUTTING

ANNUAL GROWTH RING

PLAINSAWED LUMBER

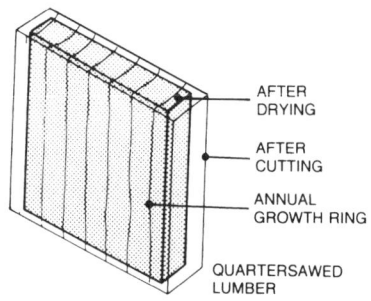

AFTER DRYING

AFTER CUTTING

ANNUAL GROWTH RING

QUARTERSAWED LUMBER

SHRINKAGE ALONG GRAIN

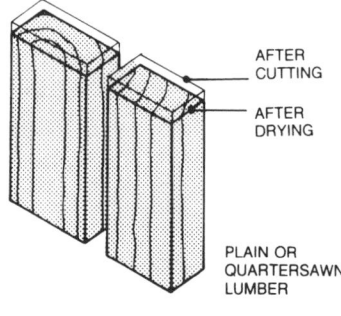

AFTER CUTTING

AFTER DRYING

PLAIN OR QUARTERSAWN LUMBER

MOISTURE IN WOOD

■ **Freshly cut lumber begins to lose moisture immediately if the surrounding air is at less than 100 percent relative humidity. Given the relative humidity and temperature of its environment, wood will lose water by evaporation until equilibrium is reached between the moisture content (MC) in the wood and that in the air.**

- Water is held by wood cells in two ways; free water in the body of the cell and water bonded amidst the structure of the cell walls. When the free water has evaporated leaving only the bonded water, the wood has reached the condition known as the **fiber saturation point.**
- Fiber saturation is typically around 30 percent MC in the sapwood and in the area of 20 percent MC in the heartwood.
- Green wood typically refers to that with moisture content at or above fiber saturation; air-dry lumber averages 20 to 25 percent MC for hardwoods and 15 to 20 percent for softwoods, with a minimum attainable level of 12 to 15 percent MC; surface dry refers to lumber at 19 percent MC or less at the time of milling; kiln dry equates to lumber at 15 percent or less MC, often about 12 percent.
- Dimensional lumber, 4 inches or less in thickness, is regarded to be stronger at 19 percent MC than when green (e.g., 25 percent greater in bending and tension). Lumber is even stronger at 15 percent MC.
- The lower the moisture content, the less attractive the wood is to attack by insects and fungi.

MOISTURE CONTENT UNDER IN-SERVICE CONDITIONS

■ **The nature of the structure of wood is to continually seek equilibrium with its changing environment causing the moisture content to rise and fall with relative humidity, even after the wood has been thoroughly dried.**

- Moisture content varies inversely with temperature; that is, with constant humidity, a rise in temperature will reduce moisture content.
- Generally in North America, wood products for interior use are recommended to be at 6 to 10 percent MC at time of installation; 9 to 14 percent for exterior uses. In damp, warm coastal areas, increase to 8 to 13 percent for interior use; in warm, dry areas, use 4 to 9 percent MC for interior uses and 7 to 12 percent for exterior uses.
- **The relation of moisture content at installation to that in use may be critical.** Surface finishes may crack and buckle due to the shrinkage of horizontal framing if the finish is applied to wood which has a moisture content over 5 percent above its in-service percentage. Finish flooring will crack if not allowed to reach equilibrium before installation as it absorbs moisture and dries during the heating season. Designs should accommodate differential shrinkage in structures using heavy timber and/or green lumber for framing. Connections between wood members and especially between wood and other materials must account for the effect of shrinkage; tops of joists framed into a steel girder need to be kept above the top of the steel enough to allow for expected wood shrinkage.
- During the winter heating season when indoor humidity is low, moisture content may be as low as 5 percent; as temperature and humidity increase in spring, moisture content may rise to 15 percent.
- During the summer cooling season when indoor humidity is about 40 to 50 percent, moisture content will be between 7 and 10 percent.
- The evaporation and absorption of moisture in wood can be inhibited by applying water repellants, deeply penetrating paints, and other such sealants and coatings.

DIMENSIONAL CHANGES

■ **Dimensional lumber shrinks and swells with its environment's relative humidity. These changes in dimension occur due to the loss or gain of moisture at different rates and in different places in the piece. As wood dries to below fiber saturation, the cell structure tightens where water had been.**

- The amount and effect of shrinkage will vary with the drying factors (faster rates cause greater movement and checking; and extent - MC level attained) and the structural characteristics in the wood (presence of resin; density - directly proportional to shrinkage; direction of grain and abnormal growth - compression wood, rubber wood, etc. - may cause distortion in shape). Uneven drying results in checking and warping (twisting, cupping, bowing).

SHRINKAGE IN TRIM

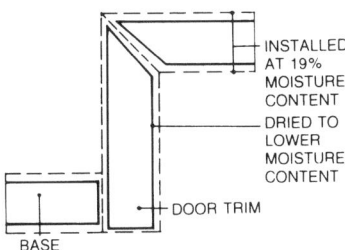

INSTALLED AT 19% MOISTURE CONTENT

DRIED TO LOWER MOISTURE CONTENT

DOOR TRIM

BASE

LOADING CONDITIONS

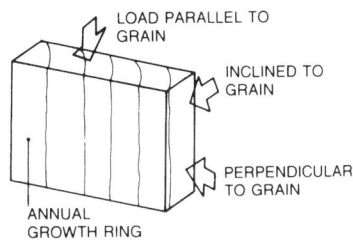

LOAD PARALLEL TO GRAIN

INCLINED TO GRAIN

PERPENDICULAR TO GRAIN

ANNUAL GROWTH RING

SUSTAINED LOADING

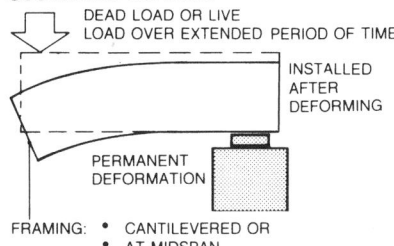

DEAD LOAD OR LIVE LOAD OVER EXTENDED PERIOD OF TIME

INSTALLED AFTER DEFORMING

PERMANENT DEFORMATION

FRAMING: •   CANTILEVERED OR
             •   AT MIDSPAN

SHRINKAGE IN FRAMING

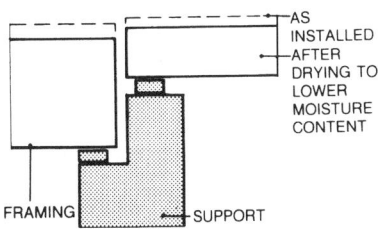

AS INSTALLED

AFTER DRYING TO LOWER MOISTURE CONTENT

FRAMING   SUPPORT

■   In most softwoods, the average shrinkage in overall width when taken from green to dry is 5 to 8 percent (0.185 to 0.3 percent per one percent change in MC). In thickness, the average shrinkage is 2.4 to 5.5 percent (0.089 to 0.205 percent per one percent change in MC). The average shrinkage in length is about 0.1 to 0.2 percent, but it is affected more by deformation (warp).

● Every dimension across grain in wood and of any wood structure which includes cross-grain wood is subject to change with variations in moisture content.

● Softwood shrinkage values range from 2.1 to 5.4 percent in radial directions, 4.4 to 9.2 percent tangentially, and 6.8 to 14 in volumetric measurement. Hardwood shrinkage values range from 3 to 7.9 percent in radial directions, 6.2 to 12.7 percent tangentially, and 10.2 to 19.2 in volumetric measurement.

PLAINSAWED AND QUARTERSAWED WOOD

A harvested tree, or bole, is milled into dimensional lumber using one of two distinct sawing techniques.

■   Lumber sawn at a tangent to the annual rings is called flat- or slash-grained for softwood and plainsawed for hardwood. The cut typically results in exposure of annual rings at angles of 45 to 90 degrees with the wide dimension of the piece.

● Growth characteristics which tend to be radial (knots, rays) will be minimized as they are exposed at or near perpendicular cross-section. Those that are tangential (shake, pitch pockets) may be maximized in a particular piece but will extend through fewer pieces.

● Figure patterns (elliptical/parabolic growth rings - faint or distinct, curly, wavy; burl) are exposed more conspicuously by plainsawing.

● Pieces shrink and swell less in thickness but are susceptible to cupping.

■   Quartersawed (hardwood) or vertical-/edge-grained (softwood) lumber is created by sawing radially. The cut typically results in exposure of annual rings at angles of 0 to 45 degrees with the wide dimension of the piece.

● Radially-oriented growth characteristics have greater impact on appearance and strength. Shake may actually appear as a split. Tangentially oriented defects will occur in more pieces.

● Figure is marginal - growth rings appear as stripes or streaks.

● Dimensional change in the width is less than in plainsawed. Improper drying is likely to cause a piece to bow lengthwise, usually in the direction of the outermost, narrow face and in the direction of the grain.

STRENGTH OF WOOD

■   The mechanical properties of wood have been determined for most North American species and many imported ones. The established stress values are based on tests described by ASTM Standard D2555 Method for Establishing Clear-Wood Strength Values. Appropriate safety and design factors responding to size variation (versus the 2 inch test sample), growth characteristics (slope of grain, knots, and other defects), moisture content, intended use, and duration of load are applied to these values.

● Wood has a phenomenal capacity to withstand bending stresses (important for horizontal members - joists, rafters, beams) as well as compression parallel to the grain (typically lengthwise in the piece - important for posts and studs).

● Resistance to shear (in the direction of the grain) and compression perpendicular to the grain is considerably lower. For example, failure (crushing and splitting) may occur at the end of a heavily loaded beam near the edge of its bearing point. This possibility, coupled with the nature of wood fiber around knots is the reason that knots are limited in the end one-third of beams and stringers. Also, a wood column safely carrying a concentrated axial load may crush a wood sill plate on which it is resting.

● In cases where large tension loads must be transferred at an angle to one member, intermediate elements (such as gusset plates) are usually necessary to distribute the load so as not to exceed the allowable shear stress. Connectors, such as plates, bolts or nails, are located a safe distance from the ends of the members.

● Duration of load should be considered since appreciable deformation, or creep, may result with long-term loading (continuous loading for more than ten years). Ultimately, failure could occur at lower loads when resisted continuously over the long term. Similarly, higher loads, such as snow loads, may safely be carried over short periods of time.

B1.5 WOOD/WOOD PRODUCTS

WOOD SPECIES: uses & properties

● denotes common uses & properties, placed on a scale of 1 to 10, where 1 is the lower value & 10 the greater

NA denotes information not available

■ treated wood only

○ possible or limited usage

	COMMON FORMS				PROPERTIES								
	Veneers	Boards/Planks	Dimension	Strips/Blocks	Shrinkage (Volumetric)	Bending Strength	Compression Strength, Parallel	Compression Strength, Perpendicular	Hardness, Side	Impact Bending	Resistance to Decay	Weathering Properties	Coating Retention
HARDWOODS													
ASH, WHITE, BLACK	●			○	5	6	6	6	6	6	4	5	5
BEECH	●			●	8	5	5	5	5	5	5	5	6
BIRCH, YELLOW	●			●	7	5	4	4	5	6	5	5	6
CHERRY	●			●	3	4	5	3	4	3	6	5	5
ELM, AMERICAN	●				6	3	3	3	4	4	5	5	5
LOCUST, BLACK					2	8	8	8	8	5	8	5	5
MAHOGANY	●			●	NA								
MAPLE (HARD), SUGAR	●			●	6	6	6	6	6	5	5	5	6
OAK, RED	●			●	7	3	3	5	5	3	5	5	5
POPLAR, YELLOW	●				4	3	3	2	3	3	5	6	7
ROSEWOOD	●				NA								
TEAK	●			●									
WALNUT, BLACK	●	○			4	6	6	4	6	4	7	6	5
SOFTWOODS													
CEDAR, WESTERN RED	○	●	●		2	4	5	4	3	4	8	7	7
CYPRESS, BALD					5	6	6	6	6	6	8	7	7
FIR, DOUGLAS (COAST)		●	●	●	7	7	7	6	7	6	6	5	4
HEMLOCK, WESTERN		●	●		7	6	6	5	6	7	5	5	5
LARCH, WESTERN					8	3	7	6	7	6	6	5	4
PINE—LODGEPOLE		●	●		5	4	5	4	4	5	5	5	5
—PONDEROSA		●	●		4	3	4	5	4	5	5	5	6
—RED		●	●		5	5	5	5	5	6	5	5	4
—SOUTHERN (WHITE)		●	●	●	7	7	7	6	7	6	5	5	5
—SUGAR					3	3	4	3	3	4	5	5	6
REDWOOD—OLD GROWTH	○	●	●		2	5	6	5	5	4	8	7	7
SPRUCE—BLACK					5	4	5	3	5	5	5	5	5
SPRUCE—ENGELMANN		●	●		5	3	3	3	3	4	5	5	5
—RED					6	5	5	5	5	4	5	5	5
—SITKA		●	●		6	5	5	5	5	5	5	5	5

B1.5 WOOD/WOOD PRODUCTS

BUILDING				PARTITIONS		FLOORS			ROOFS		FOUNDATION & OUTDOOR								EQUIPMENT	
Posts, Columns	Framing	Sheathing	Siding, Facing	Framing	Paneling	Joists, Beams	Rough	Finish	Rafters, Purlins	Decking Planking	Piles	Other Foundations	Retaining Walls	Bulkheads	Fence Posts	Boats	Decks, Patios, Walks	Furniture, Outdoor	Cabinets, Casework	Furniture, Indoor
					○		○													●
					●														●	●
					●		○												●	●
					●		○												●	●
					●															●
○					○							○	○	○	○					
					●			●								●			○	○
					●			●											○	●
					●			●											○	○
			○		●														○	○
					●														●	●
					●		○			○							●	○	●	●
					●		○												●	●
●	●	●	●	●	●	○			●	●							●	●	○	○
●	●	●	●	●	●	○			●	●							●	●	○	○
●	●	●	●	●	○	●	●	●	●	●	■●	■●	■●	■●	■●	■●	■●			
●	●	●	●	●		●			●	●			■●	■●	■●	■●	■●			
●	●	●	●	●		●			●	●	■○						●			
○	●	○	○	○		○			●								■●			
○	●	○	●	●	●	○			●	●		■●	■●	■●			■●		○	○
○	●	○	●	●	●	○			●	●			■●				●		○	○
○	●	●	●	●	●	○	●	●	●	●	■●	■●	■●	■●	■●	■●	■●			
●	●	●	●	●	●	○			●	●							■●		○	○
●	○	○	●	○	●	○			○	●						●	●	●	○	○
○	○	○	○	○		○			○	○							■○			
○	○	○	○	○		○			○	○							■○			
○	●	○	○	○		○			○	○							■○			
○	●	○	○	○		○			●	●							■○			

B1.5 WOOD/WOOD PRODUCTS

WOOD: properties

SPECIES	SHRINKAGE, percent from 20% moisture content to oven dry, range			SPECIFIC GRAVITY, average, range	MODULUS OF RUPTURE, psi, average, range	COMPRESSION PARALLEL TO GRAIN psi, average, range	COMPRESSION PERPENDICULAR TO GRAIN, psi, average, range	SHEAR STRENGTH, psi, average, range	MODULUS OF ELASTICITY, psi, average, range
	RADIAL	TANGENTIAL	VOLUMETRIC						
CEDAR: ALASKA, INCENCE, PORT OR FORD, ATLANTIC WHITE, NORTHERN WHITE, EASTER RED, WESTERN RED, CYPRESS	1.5 to 3.1	3.3 to 4.6	4.8 to 6.7	.30 to .46	3800 to 7000	1900 to 3600	200 to 750	660 to 1000	520,000 to 1,400,000
FIR: ALPINE, BALSAM, DOUGLAS, SUBAL-PINE	1.9 to 3.3	4.6 to 5.2	6.0 to 7.8	.31 to .45	4900 to 7700	2300 to 3600	190 to 460	700 to 980	1,000,000 to 1,600,000
HEMLOCK: EASTERN, MOUNTAIN, WESTERN	2.0 to 2.8	4.5 to 5.4	6.5 to 9.8	.39 to .41	6300 to 7000	2900 to 3600	360 to 400	750 to 920	1,000,000 to 1,400,000
PINE: JACK, EASTERN, LODGEPOLE, MONTEREY, PONDEROSA, RED, SUGAR, WESTERN	1.6 to 3.1	3.4 to 4.8	3.2 to 7.7	.34 to .46	4700 to 6600	2400 to 3300	190 to 440	670 to 870	1,000,000 to 1,400,000
PINE, SOUTHERN: LOBLOLLY, LONGLEAF, SHORTLEAF, SLASH	3.0	5.0	8.0	.47 to .54	7300 to 8700	3500 to 4300	350 to 530	870 to 1000	1,400,000 to 1,600,000
REDWOOD: OLD GROWTH, SECOND GROWTH	1.5	2.9 to 3.3	4.5 to 4.7	.34 to .39	5900 to 7500	3100 to 4200	270 to 420	900 to 900	950,000 to 1,200,000
SPRUCE: BLACK, ENGELMANN, RED, SITKA, WHITE	2.3 to 2.9	4.4 to 5.0	6.9 to 7.7	.35 to .41	4700 to 6000	2200 to 2800	190 to 300	630 to 800	1,000,000 to 1,400,000
BIRCH: PAPER, SWEET, YELLOW	4.2 to 4.3	5.7	10.4 to 10.8	.48 to .60	6400 to 9400	2400 to 3800	270 to 480	840 to 1200	1,200,000 to 1,600,000
MAPLE: BIG LEAF, BLACK, SUGAR, RED, SILVER	2.5 to 3.3	4.7 to 6.3	7.7 to 9.9	.44 to .57	5800 to 9400	2500 to 4000	370 to 650	1000 to 1500	950,000 to 1,500,000
OAK, WHITE: CHESTNUT, LIVE, POST, SWAMP CHEST-NUT, WHITE, BUR, OVERCUP, SWAMP WHITE	2.7 to 3.7	4.8 to 6.5	9.2 to 11.1	.56 to .81	8000 to 12000	3300 to 5400	530 to 2000	1200 to 2200	800,000 to 1,600,000

NOTES:

- Strength values shown are average range for species within group, and are based on rounded off, average values given in ASTM D2555 for woods grown in the United States and Canada; taken from Table 2: Clear wood strength values unadjusted for end use and measures of variation for commercial species of wood in the unseasoned condition - Method B.
- Modulus of rupture: nominal failure stress in bending, or the unit stress calculated from Mc/I when the bending moment used is the maximum before rupture.
- Modulus of elasticity: measure of stiffness, or the ability to resist deformation.
- Compression parallel to grain: or the crushing strength.
- Compression perpendicular to grain: stress at proportional limit, or at a point where stress is no longer proportional to strain.
- Values for tension parallel to grain are not given: for clear wood specimen values for modulus of rupture may be used.
- Wood increases in strength, and modulus of elasticity also increases as wood dries below the fiber saturation point. Allowable unit stresses are generally established at a moisture content of 19 percent, and take into account such increases in strength, modified for strength-reducing effects, such as checks, knots, duration of load, and a safety factor is also applied; generally in accordance with ASTM D245 - Standard Methods for Establishing Structural Grades and Related Allowable Properties for Visually Graded Lumber.

WOOD PRODUCTS: properties

NA denotes information not available

	LINEAR EXPANSION, percent, from 50 to 90 percent of relative humidity	DENSITY, pounds per cubic foot, range	SPECIFIC GRAVITY	MODULUS OF RUPTURE, psi, range	ULTIMATE TENSILE STRESS, parallel to surface, psi	ULTIMATE COMPRESSION parallel to surface, psi	MODULUS OF ELASTICITY, psi, range	THERMAL CONDUCTIVITY; Btu/in/ft²/hr/°F
PARTICLEBOARD								
CHIPS — **SPLINTERS**	.30 AVERAGE FOR LOW DENSITY	37 AND UNDER- LOW DENSITY	0.50 to 0.80	1500 to 5000	500 to 4000	1400 to 5200	150,000 to 450,000	.55 to 1.25
CHIPS — **PLANER SHAVING**		37 to 50 MEDIUM DENSITY		1500 to 4000			150,000 to 400,000	
FLAKE	.55 AVERAGE FOR HIGH DENSITY			2000 to 6300			300,000 to 650,000	
WAFER		50 AND OVER- HIGHER DENSITY	0/60 to 0.70	2400 to 3300			500,000 to 700,000	
FIBERBOARD								
INSULATING BOARD	.20 to .50	20 to 26	0.16 to 0.42	200 to 800	200 to 500	1 to 10	25,000 to 125,000	.27 to .45
LAMINATED PAPERBOARD	.20 to 1.50	31 to 37	0.50 to 0.59	900 to 2000	650 to 2000	500 to 950	100,000 to 700,000	.50 to .60
MEDIUM DENSITY FIBERBOARD	.20	37	0.59	1900 to 2300	NA			.55
HIGH DENSITY HARDBOARD	.15 to .45	50 to 80	0.80 to 1.28	3000 to 7000	3000 to 6000	1800 to 6000	400,000 to 800,000	.80 to 1.40
TEMPERED HARDBOARD		60 to 80		5600 to 10000	3800 to 7800	3700 to 6000	640,000 to 1,100,000	1.10 to 1.50

NOTES:
- Information principally based on values obtained from tests conducted in accordance with ASTM D1037-Standard Methods of Evaluating the Properties of Wood-Base Fiber and Particle Panel Materials. Values have generally been rounded off.
- Particle board available with additives to improve dimensional stability, fire retardance, fungus and termite resistance.
- Particle size, moisture content of particles, temperature and pressing time will influence the ultimate strength properties of particle boards. Research to date suggests that using a range rather than a single size of chip will improve modulus of elasticity and strength.
- Tests conducted by Forest Products Laboratory suggest that steam posttreatment of particle board improves its strength retention and thickness swell during extended term weathering.
- Fiberboard properties shown are considered minimum requirements and should be considered only as indicative of properties to be expected.
- Principal additives used in the manufacture of fiberboard are: resin, wax, and oil. Resin improves internal bonding thus modulus of rupture; wax additives, used to impart moisture resistance, when in large amounts tend to reduce many mechanical properties; oil tempering is used exclusively on hardboard and can double the strength and dimensional stability.

B1.5 WOOD/WOOD PRODUCTS

PLYWOOD: uses & properties

● denotes common usage

	VENEER Strength Group	GLUE LINE Interior	GLUE LINE Exterior	EDGE Square	EDGE Tongue & Groove	EDGE LAP	1/4	5/16	3/8	1/2	5/8	3/4	4×8	4×9	4×10
SOFTWOOD PLYWOOD															
ENGINEERING GRADE															
EXTERIOR — STRUCTURAL I C-C	1		●	●				●	●	●	●	●	●		
STRUCTURAL II C-C	1,2,3		●	●				●	●	●	●	●	●		
FDPA FINPLY	1		●	●	●		●		●	●	●	●	●	●	●
FDPA COMBI	1 & 3		●	●	●		●		●	●	●	●	●	●	●
C-C PLYWOOD	ALL		●	●	●		●		●	●	●	●	●		
B-B PLYFORM	ALL		●	●							●	●	●		
INTERIOR — STRUCTURAL I C-D	1	●		●				●	●	●	●	●	●		
STRUCTURAL II C-D	1,2,3	●		●				●	●	●	●	●	●		
C-D PLUGGED	1,2,3	●	●	●	●		●		●	●	●	●	●		
UNDERLAYMENT	1,2,3	●	●	●	●		●		●	●	●	●	●		
2-4-1	1,2,3	●	●	●	●								●		
FDPA FIN PLY	1	●	●	●	●		●		●	●	●	●	●	●	●
FDPA COMBI	1 & 3	●	●	●	●										
CRA PREMIUM CUSTOM															
APPEARANCE GRADE															
EXTERIOR — APA A-A	ALL		●	●			●		●	●	●	●	●		
APA A-B	ALL		●	●			●		●	●	●	●	●		
APA A-C	ALL		●	●			●		●	●	●	●	●		
APA B-B	ALL		●	●			●		●	●	●	●	●		
APA B-C	ALL		●	●			●		●				●		
APA HDO	ALL		●	●					●	●	●	●	●		
APA MDO	ALL		●	●					●	●	●	●	●		
303 SIDING	ALL		●	●		●		●	●	●				●	●	●
303 T1-11	ALL		●	●		●						●		●	●	●
PLYRON	ALL		●	●							●	●	●	●	●	
MARINE	1		●	●			●		●	●	●	●	●		
CRA PREMIUM CUSTOM	ALL														
FDPA FINPLY	1		●	●	●		●		●	●	●	●	●	●	●
FDPA COMBI	1,3		●	●	●		●		●	●	●	●	●	●	●
INTERIOR — APA N-B	ALL	●		●							●	●	●		
APA N-D	ALL	●		●			●						●		
APA A-A	ALL	●		●			●		●	●	●	●	●		
APA A-B	ALL	●		●			●		●	●	●	●	●		
APA A-D	ALL	●		●			●		●	●	●	●	●		
APA B-B	ALL	●		●			●		●	●	●	●	●		
APA B-D	ALL	●		●			●		●	●	●	●	●		
DECORATIVE PANELS	1	●		●				●	●				●	●	
PLYRON		●		●							●	●	●		
CRA PREMIUM CUSTOM	ALL														
FDPA FINPLY	1	●	●	●	●		●		●	●	●	●	●	●	
FDPA COMBI TWIN	1,3	●	●	●	●		●		●	●	●	●	●	●	

4 x 12	SURFACE TREATMENTS				STRUCTURE						ENCLOSURE					INTERIOR			EQUIPMENT & FURNISHINGS				
	Sanded	Mill Oiled	Grooved	Textured	Concrete Formwork	Foundations	Structural Members	Plywood Web Joist	Built-up Arch Forms	Roof Systems	Stressed Skin Panels	Decking/Flooring	Sheathing/Roofing	Siding	Fascias & Soffits	Subfloor/Underlay	Tile Backing	Paneling	Casework Cabinets	Counter Tops	Stairs/Platforms	Doors	Base for Surfacings
						•	•	•	•	•	•	•	•								•		
						•	•	•	•	•	•	•									•		
•						•	•					•				•	•						
•					•							•											
	•						•	•	•	•	•		•			•	•				•		•
	•	•			•																		
						•	•	•	•	•	•		•										
						•	•	•	•	•	•		•										
													•		•	•	•		•				•
	•															•	•						•
	•															•	•						•
•																•							
	•					•	•	•	•	•	•				•				•		•	•	
	•					•	•	•	•						•				•		•	•	
	•						•	•	•	•	•				•				•		•	•	
	•						•	•	•										•		•	•	•
	•						•												•	•			•
			•				•							•	•			•	•				
		•	•				•							•	•			•	•				
							•					•						•	•	•		•	
	•						•												•				
•																							
•																							
	•																	•	•		•		
	•																	•					
	•																•	•	•		•	•	•
	•					•	•		•		•						•	•	•		•	•	•
	•											•					•		•		•	•	•
	•											•					•		•		•	•	•
	•														•		•		•		•	•	•
			•															•	•				
																			•	•	•	•	•

WOOD/WOOD PRODUCTS: selection checklist

B1.5 WOOD/WOOD PRODUCTS

FRAMING:

Wood used for both horizontal and vertical framing is commonly soft-wood lumber: dressed solid for light construction; parallel laminated, usually referred to as gluelam, for heavy framing. Timber is also used for posts and beams, principally in heavy construction, but generally not readily available in all areas: availability of particular sizes and/or species should be ascertained before making selection.

SOLID WOOD:

■ **Soft wood used in light construction is graded under several grading rules promulgated by regional associations of producers, such as:**
- Western Wood Products Association: which produce the major portion of all lumber used.
- Redwood Inspection Bureau.
- Southern Forest Products Inspection Bureau.

Wood may also be graded under National Grading Rule of the American Lumber Standard.

■ **Grading generally separates softwood lumber into major categories, principally based on intended end use:**
- Select or appearance grade: for wood where appearance is the primary consideration: paneling, siding, exterior and interior trim, flooring.
- Common, or standard, or utility or light framing grade: for wood used in general construction and light framing when neither appearance nor high structural strength is a primary consideration.
- Stress, or structural, or dimension grade: for wood used in light framing when high structural strength is required. Lumber may be either visually or machine stress rated.

Selection of framing lumber should consider:
- Using lumber with higher allowable stresses in bending to increase span will also result in larger deflections in the framing.
- Using nonstress graded lumber, or stress graded with lowest allowable stresses, even when it necessitates increase in size of

the framing members may generally be more economical.
- Building codes may require on-site inspection to ensure that high stress graded lumber is being used when thus specified.
- Stress rated lumber may allow reduction in total depth of assembly, or maintaining the same depth when adjacent spans vary.

PARALLEL LAMINATED WOOD

Two basic principles guide the design of gluelams: to maximize performance in use while minimizing material and manufacturing costs:
- solid wood beams may not be the most efficient product for a particular load-carrying purpose: knots, checks and other imperfections may limit its load-carrying capacity. By sawing the pieces into thinner layers or laminae and regluing with the imperfections randomly distributed horizontally, a stronger but not necessarily stiffer beam is produced.
- next level of improvement would be the removal of all imperfections, such as knots, and then finger jointing the clear material; but as the load carrying efficiency is increased, so is the cost of manufacturing.

■ **Advantages of gluelams are:**
- ease of manufacturing large structural components from standard commercial sizes of lumber.
- designing on the basis of the strength of seasoned wood as the laminae can be dried individually to achieve the same moisture content throughout the finished component.
- structural components may vary in cross section along their length in accordance with strength requirements, may be bent to form arches or rigid frames; but
- cost is higher than for solid wood of comparable load carrying capacity.

■ **Gluelams may be preservative treated to minimize decay when used in exposed locations:**
- treatment may take place after fabrication is completed; or the laminae may be treated before gluing.
- treatment after fabrication may be limited by the size of equipment used for treating when large com-

ponents are to be treated.
- not all preservative treated wood can be glued with all glues.
- allowable stresses and design criteria for gluelams are established by the American Institute of Timber Construction.

FIRE-RETARDANT TREATED WOOD

The use of fire-retardant treated wood should be considered when:

■ **Building Codes prohibit the use of combustible materials in a particular type of construction, generally in those classified as fire-resistive:**
- fire-retardant treated wood is generally allowed by the Model Building Codes - BOCA, Standard, and Uniform - to be used in partitions of fire-resistive types of construction.
- fire-retardant treated wood may be allowed to be used in roof construction of fire-resistive types of construction.
- may be allowed by some codes in roof decks of combustible types of construction, when fire-retardant treated decking is used over a party or fire wall terminating below it, in lieu of extending such wall above the roof to form a parapet.
- may be allowed in roof structures such as spires, cupolas, cooling towers in buildings of fire-resistive construction.
- fire-retardant treated wood may be required or may ease limitations imposed by code in exterior finish and/or trim for buildings in fire districts.
- fire-retardant treated wood paneling may be used for interior finishes when such finishes are required to have a flame spread rating of 25 or less when tested in accordance with ASTM-E84, 30 minute duration.
- use of fire-retardant treated wood may offer benefits of greater safety and reduced fire insurance rates.

■ In all instances the specific provisions of the local building code should be ascertained.

OTHER CONSIDERATIONS

Shrinkage and swelling due to changes in moisture content to be expected may influence selection:

■ **Amount of shrinkage varies between species of wood:** same species or species with similar shrinkage characteristics only should be used in the same assembly.

■ **Moisture content at time of installation should be the same as that to be expected under in-service conditions.**
• when lumber with higher moisture content is used, it should be allowed to dry before finishing materials are installed.
• detailing should take into account shrinkage after installation when wood with higher moisture content than that to be expected later is used:

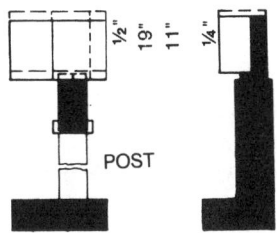

INSTALLED AT 19 PERCENT MOISTURE CONTENT AFTER DRYING TO 8 PERCENT MOISTURE CONTENT

• framing joists into center girder would eliminate effect of cumulative strinkage.
• when exposed to the weather and/or water vapor condensation lumber (such as trim, window frames, siding) should be backprimed and plywood backprimed and edgesealed before installation when to be coated in the finished work.

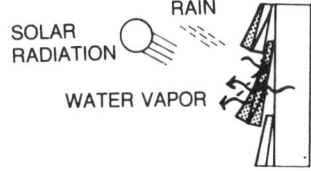

WHEN HEATED BY SOLAR RADIATION, MOISTURE IN SIDING WILL VAPORIZE AND MAY LIFT APPLIED COATING.

• wood flooring must be protected from any moisture migrating to it through or from the substrate it is supported by such as a concrete slab on grade and should be installed at the same moisture content as that to be expected later.

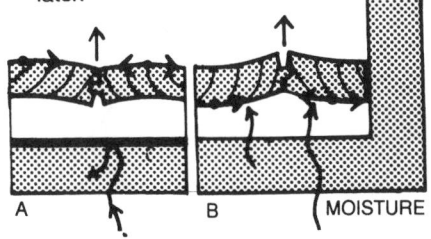

A. FLOORING IS INSTALLED AT LOWER MOISTURE CONTENT AND SWELLS AFTER INSTALLATION

B FLOORING PICKS UP MOISTURE FROM SUBSTRATE

■ **Thermal expansion in wood across the grain radially and tangentially varies directly with specific gravity of the species and is relatively small.**
• Longitudinal thermal expansion, parallel to grain, is independent of specific gravity and varies from 1.7×10^{-6} to 2.5×10^{-6} per °F, which is lower than coefficients of expansion for most other materials.
• Thermal expansion is often offset by concurrent drying shrinkage thus seldom a factor in design and/or selection.

■ **Thermal conductivity of softwoods averages k=.80 Btu/hr/sq.ft/in/°F.**
• Corresponding values for some other building materials are: k=4.8 for brick; k=5.5 for glass; k=12.6 for concrete.

IMPREGNATED WOOD

Wood may be impregnated with a polymer to improve some of its properties: frequently both strength and stiffness are improved, but other mechanical properties may be reduced:
• the use of phenolic polymer produces a wood-polymer composite which is relatively brittle and generally does not perform well when subjected to tensile stresses or impact bending: stress at proportional limit range from 7500 to 7900 psi; modulus of rupture from 8250 to 9300 psi; crushing

strength parallel to grain from 6990 to 7900 psi.
• wood impregnated with methyl methacrylate will exhibit increased hardness, compression strength, and impact bending strength. Species of hardwoods and softwoods such as birch, pine, are available impregnated with methyl methracrylate.
• impregnated flakeboard is also available to limited extent.
• uses of impregnated wood in building construction are quite limited.

PLYWOOD

Basic principles which guide the design of parallel laminates are applicable to plywood, except that instead of being parallel, the adjacent laminae are oriented through 90 degrees:
• properties are contingent on properties of species of wood used in the plies; number of plies used; type of glue.
• grades and mechanical properties for various types of softwood plywood established by the American Plywood Association.
• grades and mechanical properties for hardwood plywood are established by the Hardwood Plywood Manufacturers Association.

PARTICLE BOARD

Particle board is an intermediate product between fiberboard and plywood: because of the larger size of particles, the properties of particle boards are more nearly like those of plywood; and such properties are strongly influenced by particle properties, size, and quantity of adhesive used.

FIBERBOARD

Specific gravity of fiberboard influences its physical properties: modulus of rupture is most sensitive to changes in specific gravity, and tensile strength perpendicular to the plane of the board the least.

B1.5 WOOD/WOOD PRODUCTS

TREATED WOOD: types, uses

- NA denotes Not Applicable
- AWPA denotes the American Wood Preservers Association
- AWPB denotes the American Wood Preservers Bureau

1 AWPA Standard CI applies to all other standards listed
2 Dual treatment = creosote + water-borne salts

	AWPA STANDARD	TYPE OF TREATMENT AND AWPB MARK						
		[1] WATER BORNE	OIL-BORNE: PENTACHLOROPHENOL			CREOSOTE		[2]
			LIGHT HYDROCARBON-PENTA SOLUTION	HEAVY HYDROCARBON-PENTA SOLUTION	VOLATILE HYDROCARBON-PENTA SOLUTION	CREOSOTE	CREOSOTE or CREOSOTE/COAL TAR SOLUTION	DUAL TREATMENT
LUMBER, TIMBER FOR GENERAL USE								
ABOVE GROUND, NOT IN CONTACT WITH SOIL OR FRESH WATER	C2	LP-2	LP-3	LP-7	LP-4	NA	LP-5	NA
IN CONTACT WITH GROUND OR FRESH WATER		LP-22	LP-33	LP-77	LP-44		LP-55	
PLYWOOD FOR GENERAL USE								
ABOVE GROUND, NOT IN CONTACT WITH SOIL OR FRESH WATER	C9	LP-2	LP-3	LP-7	LP-4	NA	LP-5	NA
IN CONTACT WITH GROUND OR FRESH WATER		LP-22	LP-33	LP-77	LP-44		LP-55	
LUMBER AND PLYWOOD FOR SPECIAL USE								
ALL-WEATHER WOOD FOUNDATIONS	C2, C9	FDN	NA					
LUMBER AND PLYWOOD FOR MARINE USE								
SUBJECT TO ATTACK BY PHOLAD BUT NOT LIMNORIA TRIPUNCTATA	C18	NA	NA				MLP	NA
SUBJECT TO ATTACK BY PHOLAD AND LIMNORIA TRIPUNCTATA							NA	MLP
SUBJECT TO ATTACK BY LIMNORIA TRIPUNCTATA BUT NOT PHOLAD		MLP						NA
POLES, POSTS FOR CONSTRUCTION								
WHEN RELACEMENT NOT DIFFICULT	C16	LP-22	LP-3	LP-77	LP-44	NA	LP-55	NA
WHEN REPLACEMENT DIFFICULT OR IMPOSSIBLE	C23	CP	CP			CP		NA
PILES FOR CONSTRUCTION								
FOUNDATION LAND AND FRESH WATER	C3	CP	CP			CP		NA
PILES FOR MARINE USE								
SUBJECT TO ATTACK BY PHOLAD BUT NOT LIMNORIA TRIPUNCTATA	C18	NA	NA			MP-2	NA	NA
SUBJECT TO ATTACK BY PHOLAD AND LIMNORIA TRIPUNCTATA						NA		MP-1
SUBJECT TO ATTACK BY LIMNORIA TRIPUNCTATA BUT NOT PHOLAD		MP-4						NA
LUMBER AND PLYWOOD - FIRE RETARDANT								
INTERIOR/PROTECTED	C20 LUMBER	PROPRIETARY-TO MEET REQUIREMENTS OF ASTM E-84, 30 MINUTE DURATION						
EXTERIOR/EXPOSED	C27 PLYWOOD							

B1.5 WOOD/WOOD PRODUCTS

TREATED WOOD: selection checklist

GENERAL

- Wood when kept constantly dry does not decay; when constantly and fully submerged in fresh water will not decay significantly over long periods of time.
- Heartwoods of some species of wood such as cedars, black locust, oaks, redwood have varying degrees of natural decay resistance; sapwoods of substantially all species of wood have low resistance to decay.
- Wood which is either moderately decay resistant, or not resistant at all should generally be preservative treated when exposed to wetting and drying; in contact with ground; partially submerged in fresh water; fully or partially submerged in salt water; exposed to prolonged or constant high temperatures and/or high humidity.

WOOD TREATED FOR RESISTANCE TO DECAY

Degree of protection depends on: type of preservative; treatment process; penetration and retention of preservative. Wood can only be well protected when the preservative substantially penetrates it, and is retained for a prolonged period of time under in-service conditions.

Wood preservatives are broadly classified as:
- Creosote
- Oil-borne
- Water-borne

■ **Some preservatives may react with metals or other materials, such as bitumens in roofing:** compatibility should always be investigated.

OIL-BORNE preservatives for pressure treatment include:
■ Creosote; creosote-coal tar solutions; creosote-petroleum solution; creosote-pentachlorophenol solution.
- Advantages are: high toxicity to wood-destroying organisms; relative insolubility in water and low volatility; relative low cost.
- Disadvantages are: cannot be painted; odor may be unpleasant; vapors harmful to plants and to foodstuffs which absorb it; will burn skin on contact.
- When creosote is fortified with pentachlorophenol, retention requirements may be reduced, thus minimizing bleeding of preservative.

- Creosote-coal tar solutions, compared to straight creosote, tend to reduce weathering and checking of the treated wood; but they are less toxic to wood-destroying organisms
■ Pentachlorophenol in either light hydrocarbon solvent; heavy hydrocarbon solvent, or volatile hydrocarbon solvent.
- Wood treated with pentachlorophenol in light or volatile solvent can be painted, but "paintable" should be specified.
- Pentachlorophenol is ineffective against marine borers and should not be used for marine piling or for timbers in coastal waters.
■ **Solubilized Copper - 8-Quinolinolate:**
- approved as a preservative for wood used in harvesting, storage, and transportation of foodstuffs.

WATER-BORNE preservatives for pressure treatment include: acid copper chromate; ammoniacal copper arsenate; chromated copper arsenate; chromated zinc chloride; fluor chrome arsenate phenol.
- Water-borne preservatives leave the wood surface comparatively clean and paintable.
- Since water is added during treatment, wood must be dried after treatment to moisture content required for intended use.
- Ammoniacal copper arsenate and three types of chromated copper arsenate provide good protection to wood in marine environment, provided pholad type borers are not present. They are also used as part of dual treatment of creosote and water-borne salts for marine piles when subject to attack by pholad and limnoria tripunctata types of marine borers.

FIRE-RETARDANT TREATED WOOD

Fire-retardant treated wood is lumber or plywood pressure impregnated with fire-retardant chemicals. Two types of fire-retardant treated wood are available:
■ **Interior:** with a number of proprietary chemical treatments available.
- wood treated for interior use must not be directly exposed to weather.
- some proprietary treatments, especially when mineral salts are used, result in high moisture absorption at higher relative humidities, generally over 80 percent, and should not be used in areas where sustained high humidity conditions

are expected: the free moisture primarily - and in some instances the salts themselves - will corrode metals in contact with the fire-retardant treated wood.
- finishes for wood treated with compositions which result in high hygroscopicity are generally limited, as the excessive surface moisture absorption may cause coatings to fail. Varnishes and acrylic latex coatings generally give the best results.
- proprietary treatments are available which leave the treated wood with moisture absorption characteristics approximately equal to those of untreated wood: such fire-retardant treated wood may be used effectively under conditions where relative humidity reaches 95 percent; finishing characteristics are also improved, being similar to those of untreated wood.
- appearance and color of wood will be affected to varying degrees by treatment, depending principally on species used.
- when to be left exposed, light sanding or wire brushing may be used to remove surface defects, such as raised grain or chemical deposits due to treatment.
- special handling may be specified to avoid sticker marks, generally on lumber to one inch in thickness, and on plywood to ⅝ inches thick.
- All treated lumber should be dried after treatment to a maximum moisture content of 19 percent or less, and plywood to 15 percent or less.
- allowable stresses are reduced by treatment for both fire-retardant treated lumber and plywood.
- fire retardant lumber may be field cut to length without the necessity of applying a treating solution to end cuts. Fire-retardant treated plywood may be field cut at any plane.
■ **Exterior type fire retardant treated lumber and plywood has similar characteristics to the interior type,** except that treatment resists leaching of chemicals when exposed to the weather.
- tested after treatment in accordance with ASTM D2898; the UL label then states that: "There is no increase in the listed classification when subject to the standard rain test."
- may also be used in protected or interior applications.

B1.5 WOOD/WOOD PRODUCTS

REFERENCES: Wood/Wood Products

B1.5 WOOD/WOOD PRODUCTS

STANDARDS

ASTM Specifications for:

D25 • Round Timber Piles.

D390 • Coal-Tar Creosote for the Preservative Treatment of Piles, Poles, and Timbers for Marine, Land and Fresh Water use.

D1032 • Chromated Zinc Chloride.

D1034 • Fluor-Chrome-Arsenate-Phenol.

D1272 • Pentachlorophenol.

D1324 • Modified Wood.

D1624 • Acid Copper Chromate.

D1625 • Chromated Copper Arsenate.

D1760 • Pressure Treatment of Timber Products

D2277 • Fiberboard Nail-Base Sheathing.

D2559 • Adhesives for Structural Laminated Wood Products for Use Under Exterior (Wet Use) Exposure Conditions.

D3024 • Protein-Base Adhesives for Structural Laminated Wood Products for Use Under Interior (Dry Use) Exposure Conditions.

ASTM Tests/Methods for:

D245 • Establishing Structural Grades and Related Allowable Properties for Visually Graded Lumber.

D1037 • Evaluating the Properties of Wood-Base Fiber and Particle Panel Materials.

D2016 • Moisture Content of Wood.

D2898 • Accelerated Weathering of Fire-Retardent-Treated Wood for Fire Testing.

D2899 • Establishing Design Stresses for Round Timber Piles.

D2915 • Evaluating Allowable Properties for Grades of Structural Lumber.

D3200 • Establishing Recommended Design Stresses for Round Timber Construction Poles.

D3737 • Establishing Stresses for Structural Glued Laminated Timber (Glulam).

ANSI Specifications:

ANSI 05.1-1979 • Wood Poles.

ANSI 05.2-1983 • Structural Glued Laminated Timber for Utility Structures.

ANSI A208.1-1979 • Particleboard, Mat-Formed Wood.

ANSI Vol. Prod. • Structural Glued Laminated Timber.

STD PS 56-73

ANSI Vol. Prod. • Construction and Industrial Plywood.

STD PS 1-74

ANSI/AHA A135.4-1982 • Basic Hardboard

ANSI/AHA A135.5-1982 • Panels

SELECTED REFERENCES

AITC **American Institute of Timber**
• Timber Construction Manual

APA **American Plywood Association**
• Effect of Fire-Retardant Treatment on the Strength of Plywood (W370).
• Plywood Diaphragm (E315)
• AFG-01 Adhesives for Field- Gluing Plywood to Wood Framing.
• Grades & Specifications (C20).
• Plywood Design Specification (Y510).
• Fire-Rated Systems (W305)
• Product Standard PSI-74 for Construction Industrial Plywood with typical APA Grade marks (Y800).

AWI **Architectural Woodwork Institute**
• Architectural Woodwork Institute Quality Standards Guide Specifications, Certification Program.

AWPB **American Wood Preservers Bureau**
• Standards for Softwood Lumber, Timber and Plywood Pressure-treated with Water-Borne Preservatives LP-2 LP-22.
• Pressure Treated with Light Petroleum Solvent-Penta Solution LP-3 LP-33.
• Pressure Treated with Volatile Petroleum Solvent (LPG) - Penta Solution LP-4 LP-44.
• Pressure Treated with Creosote or Creosote Coal Tar Solution LP-5 LP-55
• Pressure Treated with Heavy Petroleum Solvent-Penta Solution LP-7 LP-77.
• Pressure Treated for Marine Exposure MLP.
• Pressure Treated with Water- Borne Preservatives and Creosote for Use in Marine Waters MP-1.
• Pressure Treated with Creosote for Use in Marine Waters MP-2.
• Pressure Treated with Water- Borne Preservatives for Use in Marine Waters. MP-4.

AWPI **American Wood Preservers Institute**
• TD Pressure Treated Wood- Unlimited Versatility in Design.
• FRG Fire-Retardant Treated Lumber and Plywood in Western States.

CRA **California Redwood Association**
• 2B1-2 Redwood Lumber Grades and Uses
• 2D1-2 Redwood Properties and Uses
• 2D2-1 Durability of Redwood
• 2D2-3 Shrinkage of Lumber

- 2D2-4 Lumber Density, Weight, Specific Gravity
- 2D2-5 Lumber Stiffness, Hard-ness, Shock Resistance
- 2D2-7L Fire Hazard Classification - Lumber

HPMA   Hardwood Plywood Mfg. Assoc.

- The Story of Hardwood Plywood
- Structural Design Guide for Hardwood Plywood, HP-SG-80
- Hardwood Plywood Features Design Expansion and Sturdy Construction

NPA   National Particleboard Assoc.

- Technical Bulletins 1-13
- Thermal Conductivity of Particleboard
- Joints Used with Particleboard

NRCC   National Research Council of Canada

- **CBD85** Some Basic Characteristics of Wood.
- **86** Some Implications of the Properties of Wood.
- **88** Use of Wood in Construction.
- **111** Decay of Wood.
- **112** Designing Wood Roofs to Prevent Decay.

NFoPA   National Forest Products Assoc.

- National Design Specification for Wood Construction
- Fire Insurance Rates and Wood Construction
- Lumber and Wood Products Literature
- Heavy Timber Construction Details — Wood construction Data No. 5
- Comparative Fire Test on Wood and Steel Joists — Tech. Report No. 1

SFPA   Southern Forest Products Assoc.

- Wood Engineering — Textbook
- Technical Note on Construction Poles

WWPA   Western Wood Products Assoc.

- Western Wood Species Book — Vol. 1 buildings
- Western Woods Use Book
- G-16 "Standard Patterns"
- TG-2 "Technical Guide — Properties of Sections (S4S)"

USFPL   U. S. Forest Products Lab.

- Effect of Fire-Retardant Treatment on Bending Strength of wood FPL 145, 1970
- Evaluation of Fire-Retardant Treatments for Wood Shingles, FPL 158, 1981
- Differences Between Heartwood and Sapwood, FPL 0147, 1966
- Comparative Decay Resistance of Heartwood of Native Species, FPL 0153, 1967
- Factors Influencing Decay of Untreated Wood, FPL 0154, 1967

B1.5 WOOD/WOOD PRODUCTS

B • BUILDING SHELL

B2• EXTERIOR WALL ASSEMBLIES

B2. 1 INSULATION

INSULATION: introduction

All building materials have some resistance to the flow of heat, which is directly proportional to the thickness of the material and inversely proportional to its density.

■ **Economic and structural** considerations preclude the use of most materials in a thickness sufficient not only to serve as the structural element but also to provide the desired resistance to heat flow.

■ **Better control of the flow of heat** is generally achieved more economically and efficiently by incorporating a separate non-structural element into the assemblies of the structures for that specific purpose: **insulation.**

INSULATION

Insulation may be defined as those materials or features of construction that are provided in order to reduce the flow of heat between the spaces separated. **By reducing heat flow, insulation will:**

■ **Minimize consumption** of energy required to maintain the desired interior environment.

■ **Reduce temperature** fluctuations within the enclosed space.

The effect of insulation on the flow of heat through an assembly may be **compared to the flow of water** through two containers, one with a large orifice, the other with a restricted one:

■ **In order to maintain a constant level** of water in both containers, the container with the large orifice requires a much larger supply of water than the one where the outflow is restricted:
* water flowing into the containers represents the supply of heat to an enclosed space.
* the constant water level represents a constant temperature in such a space.
* water flowing out represents heat being lost to colder outside air.

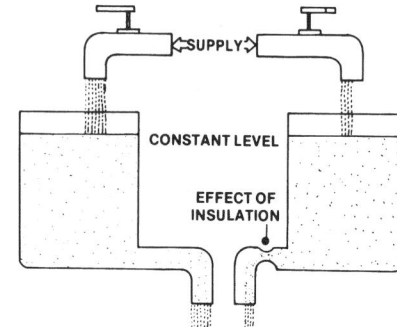

■ **In addition** to reducing heat flow between spaces, **insulation may be used to:**
* control surface temperatures of building components (such as piping, ductwork, equipment) for economy in operation, comfort of occupants, or safety.
* prevent water vapor condensation on cold surfaces.
* resist water vapor transmission.
* provide support for a surface finish or facing.

HEAT FLOW AND COMFORT

Body comfort in an enclosed space largely depends on the balance between heat produced internally in the body and:

■ The **temperature** and humidity of the surrounding air.

■ **Surface temperatures** of the surrounding envelope.

Any changes in the ambient or surrounding surface temperature, when the factor of humidity is disregarded, will change the comfort level as the following example illustrates:

■ **There are two people** in a space with air temperature at 75°F and wall surface at 70°F.

■ **The first person, who is at rest,** will lose about 400 Btu, of which:
* 160 Btu will be radiated to the wall
* 140 Btu by convection to surrounding air
* 100 Btu by evaporation of perspiration.

■ **The second person, engaged in strenuous exercise,** will lose about 1200 Btu, of which:
* only 160 Btu will be radiated to the wall
* 400 Btu would be lost by convection
* 640 Btu lost by evaporation of perspiration
* an imbalance between heat produced and heat readily dissipated into surrounding air will result in a rise of body temperature, and the active person will feel uncomfortable.

■ Now let's assume that the **surface temperature of the wall rises** to 80°F while the **air temperature decreases** to 60°F.

■ This change would **not significantly alter** the comfort level of the person at rest.

■ **The active person** will feel more comfortable:
* the slight reduction in heat loss by radiation to the wall surface would be offset considerably by the increase in heat loss by convection, assuming that such convection is not restricted by clothing.
* the amount of heat to be lost by evaporation would then be reduced from about 640 Btu under the first set of conditions to about 140 Btu under the second.

Since insulation incorporated into an assembly reduces the heat flow through it, **the temperature of the interior surface of the assembly may be controlled** by increasing or decreasing the thickness of the insulation.

■ **Uncontrolled heat flow from heated spaces outward in cold weather** may lower the interior surface temperature well below that of the heated interior air and may result in:
* drafts, when heated air loses heat at a colder surface and sinks, setting up convective currents: the greater the temperature differential between the heated air and the interior surfaces, the greater and more uncomfortable will such convective currents be.
* radiation of body heat to colder surfaces: the greater the temperature differential, the larger and more uncomfortable will be the loss of body heat.
* stratification of hot air at the ceiling, and of cold air at the floor, will be more pronounced.

■ **Uncontrolled heat flow inward into cooled spaces in hot weather** may raise the temperature of the interior surfaces (of exterior walls, or roof) to above the temperature of the inside air and may result in:
* increase in the temperature of inside air.
* decrease or reversal of radiation of body heat: surfaces with temperatures above body temperature will radiate heat to the body and cause discomfort even at moderate air temperatures.

Another significant contribution of most types of insulation to comfortable interior conditions is in **reducing levels of sound transmitted** through walls, partitions, floors, and ceilings.

EXTERIOR, vertical or horizontal

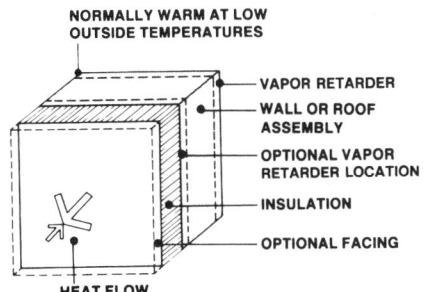

NORMALLY WARM AT LOW
OUTSIDE TEMPERATURES

VAPOR RETARDER
WALL OR ROOF
ASSEMBLY
OPTIONAL VAPOR
RETARDER LOCATION
INSULATION
OPTIONAL FACING

HEAT FLOW

Rigid boards or blocks; also foamed-in-place, sprayed-on and poured-in-place.

■ **Boards or blocks, installed vertically** over:
• foundation walls as perimeter insulation: when below grade should be resistant to moisture and to exposure to chemicals dissolved in ground water.
• exterior masonry and framed walls; cold storage room walls: generally with an applied protective/decorative facing.

■ **Boards or blocks, installed horizontally:**
• under concrete slabs on grade as perimeter insulation: when in contact with ground, should be resistant to moisture and groundwater.
• over reinforced concrete, insulating concrete, gypsum, wood, or metal roof decks; may also be over existing roofing. May be installed over the roofing membrane, or be faced by the roofing membrane. When installed over roofing membrane, should be resistant to the elements.

■ **Poured-in-place, sprayed-on, foamed-in-place installed horizontally:**
• Over metal roof decks, wood-fiber structural formboards: faced with the roofing membrane. Some foamed-in-place and poured-in-place types may also be installed over existing roofing.

■ **Locating insulation on the exterior** will minimize thermal movements in the assembly and/or framing, but will generally require **protective facings** or coverings over the insulation.

INNER COMPONENT, vertical or horizontal

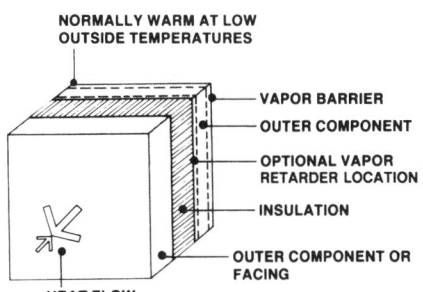

NORMALLY WARM AT LOW
OUTSIDE TEMPERATURES

VAPOR BARRIER
OUTER COMPONENT
OPTIONAL VAPOR
RETARDER LOCATION
INSULATION
OUTER COMPONENT OR
FACING

HEAT FLOW

■ **Loose fill, fibrous or granular; fibrous batts; rigid foamed boards; foamed-in-place; reflective.**

■ **Granular loose fill is used:**
• In cavities of hollow masonry units; or to fill air spaces in cavity walls. May also be used as fill between attic joists.

■ **Fibrous loose fill and fibrous batts are used:**
• Between studs and floor and ceiling joists in framed construction.
• In metal sheet sandwich and metal sheet assembly walls and roofing.

■ **Rigid foamed boards are used in:**
• Laminated metal sheet sandwich wall panels.
• Masonry cavity wall air spaces; precast concrete sandwich panels.

■ **Foamed-in-place is used in:**
• Cavities of hollow masonry units, or air spaces of masonry cavity walls.
• Between studs of framed walls.

■ **Reflective is used in:**
• Air spaces of vertical or horizontal structural assemblies.

■ **Differential thermal movements** may result between the outer and inner components of the assembly due to higher differentials in temperature.

INTERIOR, vertical

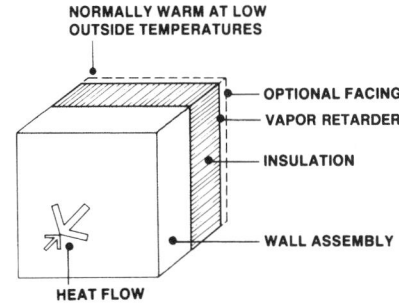

NORMALLY WARM AT LOW
OUTSIDE TEMPERATURES

OPTIONAL FACING
VAPOR RETARDER
INSULATION
WALL ASSEMBLY

HEAT FLOW

Rigid foamed boards, blocks; also sprayed-on.

■ **Rigid foamed boards used over:**
• exterior concrete or masonry walls, either laminated/attached directly to such walls, or over metal or wood furring strips. Generally faced with a decorative/protective facing, such as a factory finished gypsum wallboard, or prefinished paneling.
• foamed blocks generally bonded over exterior concrete or masonry walls and/or masonry partitions in cold storage rooms.

■ **Sprayed-on used over:**
• Exterior or interior concrete or masonry walls, especially over molded surfaces; generally in protected locations only; for insulation and/or sound absorption.
• Exposed or concealed structural elements for fire protection.

■ **Locating the insulation on the warm side** of the assembly may result in large differential thermal movements between the exterior wall assemblies and the structural frame due to higher differences in temperature between them.

B2.1 INSULATION

INSULATION: types, materials, uses

TYPE	DESCRIPTION	MATERIALS
LOOSE		
FIBROUS	• **Fibrous materials** have the appearance of loose wool or cotton, and are differentiated principally by the raw materials from which they are made. • **Available as** loose wool for hand packing, modulated for pneumatic application; or pre-formed in batts or blankets, unfaced & faced; facing generally serves as vapor retarder.	• Silica sand is the principal ingredient for **glass fibers.** • Slag from iron, copper, and lead used for **rock wool**; natural rock may be added, or used as the sole ingredient. • Waste paper or virgin wood fiber used for **cellulose.** Chemicals are added to cellulose to provide resistance to fire, water absorption and fungal growth.
GRANULAR	• **Perlite:** Lightweight inorganic granules or pellets containing tiny sealed air cells. Available loose or formed with binder into rigid boards. • **Vermiculite:** Mica, exfoliated, or expanded into threads with a vermicular foamed motion. Available as loose fill only. • **Plastic:** Pellets of foamed plastic, containing tiny closed cells.	• **Perlite:** Glassy volcanic rock expanded by heating. May be treated with silicone to increase resistance to water penetration. • **Vermiculite:** Hydrated magnesium aluminum iron silicate with foliated structure. Exfoliated or expanded by heating. May be treated for water repellency. • **Foamed plastic:** polystyrene, polyurethane, polyisocyanurate.
RIGID		
FIBROUS	• **Organic or inorganic fibers** compressed into rigid boards of varying densities. • Surfaces may be coated to resist moisture; prefinished when exposed to view, or faced to protect the material and/or facilitate installation.	• **Glass fibers,** felted, treated with resinous binder, and compressed. Faced for use as roof insulation, unfaced for walls. • **Organic fibers** such as wood or cane, felted and compressed; generally impregnated with a binder to resist moisture absorption. • **Wood fibers** bound with Portland cement. Will absorb moisture. Generally available factory primed.
GRANULAR	• **Inorganic granules** combined with organic fibers and binders, and formed into rigid boards. Surfaced with moisture resistant coatings. • **Organic granules** combined into boards under heat and pressure.	• **Perlite:** Glassy volcanic rock expanded by heating. Thermal resistance provided by tiny sealed air cells within the granules. • **Cork:** Granules made from the bark of cork oak.
FOAMED	• **Plastic substances or glass** expanded into foam by mixing with air, carbon dioxide, fluorocarbons, or other gaseous media and molded into rigid boards. • Gaseous media are trapped in closed cells, with the cells comprising up to 90% of the material.	• **Polystyrene,** using air or air and fluorocarbon. • **Polyurethane** and polyisocyanurate, using fluorocarbon. • **Glass,** ground and re-melted, using air as foaming agent. • Phenolic base thermosetting foam.

B2.1 INSULATION

DENSITY lbs/cu ft range	Fire Resistance	Water Absorption	Uses	REMARKS
0.6 to 3.0	rock wool non-combustible cellulose: combustible	2% to 15% by weight	• Thermal insulation in framed walls, floors and attic spaces: generally as faced or unfaced batts, but can be packed in by hand or blown in. • Blankets used over suspended ceilings. • Rock wool may be used as high temperature insulation. • Easily installed; available in various thicknesses, and several densities.	• Fibrous loose fill insulation is most widely used in retrofit applications: generally blown into walls and attic space. Also used to resist sound transmission. • Insulating existing buildings may create condensation problems unless migration of water vapor can be controlled: interior surfaces may have to be specially painted to minimize vapor transmission: exterior surfaces should not trap vapor. • Shredded waste of foamed insulation in large nodules is also available. Because of large size of nodules, generally hard to install in effective densities. • Foamed insulation inserts for concrete block are available, to be inserted during installation of block.
2 to 10	perlite and vermiculite: non-combustible	high if not treated	• Used as loose fill thermal insulation in voids or cavities of masonry wall assemblies, or between attic joists. • Also used as aggregate with or without sand in insulating concrete for densities from 20 to 60 lbs/cu ft. • Other uses include: aggregate for gypsum plasters.	
10 to 30	organic: combustible inorganic: non-combustible except for binder.	5% to 20% by weight	• Glass fiber: Thermal insulation in roof and wall assemblies, such as in metal sheet sandwich wall assemblies. • Organic fiber: Thermal insulation in roof assemblies. • Wood fiber: generally used as formboard in poured gypsum or insulating concrete roof deck installations. Also as aggregate for nailable concrete.	• Organic fiber boards poor in resisting moisture. Insulating value relatively low as compared to foamed insulation. • Glass fiber boards will not deteriorate due to absorbed moisture, but insulating value will decrease. If moisture is trapped within the insulation, it may be removed by venting to the outside, as water vapor will migrate through the felted fibers. • Wood fibers bound with cement have relatively low thermal resistance. May be exposed to view in finished work, and also are sound absorbing. • Perlite will degrade due to absorbed moisture; and freezing of trapped moisture may cause deterioration. • Foamed plastics with trapped fluorocarbons have higher thermal resistances than those using air. Size of cells will affect thermal resistance. • Closed cell foamed insulations generally have good resistance to moisture absorption and water vapor permeability. • Foamed polyurethane and polyisocyanurate boards are available with aluminum foil facings to minimize vapor permeability and effects of aging. • Foam/concrete composite types, 2'x4' in various thicknesses available. * When permitted by Building Codes.
5 to 8	perlite: non-combustible except for binder.	high if not treated	• Perlite: Thermal insulation in roof construction over nailable or non-nailable decks. • Perlite: May be combined with foamed insulation in composition boards. • Cork: Principally as cold storage insulation. No longer in wide use.	
1.5 to 9.5	combustible except for foamed glass	negligible to 5% by weight	• Thermal insulation in roofs; exterior non-structural sheathing in framed construction. • Exposed under roof decks; or suspended as ceilings.* • Incorporated into cavity of multi-wythe masonry walls or concrete sandwich panels. • As perimeter insulation for foundation walls and slabs on grade.	

B2.1 INSULATION

B • BUILDING SHELL

INSULATION: types, materials, uses (cont.)

TYPE	DESCRIPTION	MATERIALS
RIGID		
COMPOSITION	• **Two or more layers** of rigid materials selected for specific properties of each and laminated together into a rigid board, e.g., perlite board with lower thermal resistance, but better resistance to exposure to fire, laminated to urethane board with high thermal resistance but poor resistance to fire.	• **Rigid foamed boards,** such as polyurethane, polyisocyanurate, polystyrene laminated to one or several rigid boards of perlite, mineral fiber or cement bound wood fiber.
MISCELLANEOUS		
FOAMED-IN-PLACE	• **Plastic materials:** foaming and hardening agents, combined while being forced under air pressure into voids or cavities in wall or roof construction, or sprayed over exposed wall or roof surfaces. • Foam will expand and cure in place. • **Cellular concrete:** blend of foaming agent and cement paste (slurry). • Forced under air pressure into voids and cavities in wall.	• **Polyurethane, polyisocyanurate, urea-formaldehyde*,** and other urea-base foams. • Water vapor permeability and water absorption resistance of ureaformaldehyde* and urea-based foams are relatively poor compared to other foams; vapor retarders generally recommended. • Cellular concrete: Portland cement foaming agent and water.
SPRAYED-ON	• **Built-up coatings** over formed or molded surfaces, such as underside of metal decks, structural framing shapes. • Applied with pneumatic equipment, or by trowelling. • Surfaces generally rough; may be levelled during application by rolling.	• **Organic,** such as cellulose, and inorganic fibers with binders for porous coatings. • **Cementitious materials,** such as magnesium oxychloride for hard, dense coatings. • Generally supplied pre-mixed as proprietary products.
POURED-IN-PLACE	• **Monolithic slabs of cementitious material;** foamed or with lightweight aggregate to reduce density. • Materials are combined at the site, or pre-mixed at a plant and delivered by special trucks to site ready for pouring. May be pumped to areas not accessible to direct pouring from truck. • Surface finishes vary from rough to smooth trowelled.	• **Foamed:** Portland cement, and fine aggregate; foaming agent. Coarse aggregate may or may not be used. Foamed gypsum has also been used on an experimental basis. • **Lightweight aggregate:** perlite, and vermiculite, but also flyash, pumice, scoria, cinders, wood chips, sawdust, expanded polystyrene. • **Binder:** Portland cement, gypsum, asphalt.
REFLECTIVE	• **Reflective surface films or foils,** either laminated to components of wall or roof assemblies, or assembled in multiple layers at the factory or on site. • Reflective films or foils **require an air space** of at least ¾ inch adjacent to at least one surface to be effective as thermal insulation. • Generally impermeable to water vapor, and will also serve as vapor retarder.	• Usually **bright aluminum foil.** • Thicknesses used for multilayer application: 0.008 to 0.02 inch. • Thickness when laminated to backing, such as paper, or to rigid components, such as gypsum wallboard: approximately 0.005 inch. May vary with manufacturer. • Available faced with plastic film.

B2.1 INSULATION

DENSITY lbs/cu ft range	Fire Resistance	Water Absorption	Uses	REMARKS
about 2 for urethane	combustible	<1% for urethane	• Generally as thermal insulation over metal roof decks. • When laminated to cement-bound wood fiber boards: as formboards in poured gypsum or insulating concrete roof deck installations.	• Rigid boards used in flat roof assemblies should be capable of supporting light foot traffic without deforming or breaking. • Composition, foamed, and granular type boards available tapered to provide positive drainage on flat roofs. • Composition with nailable surface available.
.7 to 3 for plastic 15 to 90 for cellular concrete	combustible except for cellular concrete	negligible to 32% by weight	• Thermal insulation in wall, floor and roof assemblies. • Re-roofing applications: over existing roofing, generally then surfaced with liquid applied membrane. May be installed to provide slope to drains. • Retrofit applications: pressure injected into existing walls.	• Foamed-in-place insulation prevents air circulation around electrical wiring and outlets to dissipate heat generated by flow of electric current. Instances of over-load on circuit breakers and wiring related fires have been reported. • Cellular concrete has density of 15 to 90 lbs/cu ft. R-value of 2 for 20 lbs/cu ft. and .43 for 90 lbs/cu ft. Has good insulating properties, but its strength depends on density.
2.5 to 70	cementitious: non-combustible organic: combustible		• Porous coatings for insulation of exposed surfaces. May be applied over complex shapes; will provide various degrees of sound absorp-tion depending on applied thick-ness and density. • Cementitious coatings generally for fireproofing structural components, such as girders, beams, columns.	• Fibrous sprayed-on coatings easily dam-aged by impact, and may flake off, or be eroded by high velocity air flow. Use of asbestos fibers severely restricted be-cause of potential health hazard. • Poured-in-place insulating concrete of densities less than 30 lbs/cu ft may be damaged by frost and should be pro-tected when used in exterior applications.
12 to 88	cement and gypsum: non-combustible	varies with density	• Roof deck and thermal insulation combined: poured over wood fiber cement bound boards, paper backed metal lath, metal roof decks. • Over structural reinforced concrete slabs for additional insulating value, and/or to provide slope to roof drains. • Asphalt bound aggregate also used for re-roofing.	• Poured-in-place insulating concrete and gypsum roof fill should be allowed to dry before roofing membrane is installed. Ex-cessive moisture remaining may damage the roofing membrane through expansion of trapped water vapor. • Reflective foils no longer in wide usage as thermal insulation. Largely restricted to low temperature applications. * Urea-formaldehyde has been banned by Consumer Product Safety Commission from use in residences and schools. Ban has been overturned by court action 1983.
not applicable	non-combustible	negligible	• In multiple layers with air spaces as thermal insulation in wall and roof assemblies; mostly in cold storage applications. • Primarily as vapor retarder when laminated to batts or blankets of loose fibrous insulation; or to gypsum wallboard.	

B2.1 INSULATION

INSULATION: types, properties

O denotes possible usage
● denotes common usage
G denotes good resistance
F denotes fair resistance
P denotes poor resistance
NA not applicable

PROPERTIES

		THICKNESS, range, in inches	DENSITY, range, in lbs./cu.ft	THERMAL RESISTIVITY r = 1/k at 75°F, range ①	WORKING TEMPERATURE °F, range	THERMAL EXPANSION COEFFICIENT, range	LINEAR EXPANSION, percent, from 50% to 90% relative humidity, range	VAPOR PERMEANCE, range ②	WATER ABSORPTION, by volume, maximum percent
LOOSE, BLOWN OR POURED-IN									
FIBROUS, CELLULOSE	⑤	varies see note ④	2 to 3	2.78 to 3.7	−50 to 180	not a factor	not a factor	100	15
MINERAL OR ROCK WOOL			1.5 to 2.5	2.9 to 3.7	up to 600				Adsorption 2% max. by weight
GLASS FIBER			.6 to 1	2.2	−50 to 180				
GRANULAR, PERLITE	⑦	varies with width of cavity	5 to 8	2.63	−50 to 1400				not available
VERMICULITE			3.5 to 10	2.0					
LOOSE, BATTS OR BLANKETS									
FIBROUS, MINERAL FIBER		1 to 12	1.5 to 2.5	3.2 to 3.7	up to 600	not a factor	not a factor	⑨ ⑩ 100	not available
GLASS FIBER		1 to 9	1.5 to 4	3.16	−50 to 180 *				
RIGID BOARDS									
FIBROUS, ORGANIC FIBER	⑫	½ to 3	10 to 30	2.27 to 2.63	not available	3 x 10⁻⁶	.5 to .6	5	7 to 15
GLASS FIBER	⑭	¾ to 4	4 to 9	3.85 to 4.76	−150 to 180	not available	<.05	⑩ 100	10
GRANULAR, PERLITE	⑮	1 to 3	5 to 8	2.77	up to 200	negligible	.1	25 to 35	1.5
POLYSTYRENE, EXPANDED	⑯	1 to 6	1 to 1.5	3.85 to 4.35	㉛ up to 165	35 x 10⁻⁶	not available	1.2 to 5	< 2 to 3+
EXTRUDED		¾ to 4	1.6 to 3	4 to 6				.3 to .9	< 1 to 3
POLYURETHANE	⑰	1 to 4	<1.7 to 4.0	⑰a 5.6 - 6.2	−100 to 250	40 x 10⁻⁶	<1 to 4	⑩ ⑱	<1.0
POLYISOCYANURATE	⑰	½ to 4¼		⑰a 5.6 - 6.2 to 7-8	−100 to 250			2.5 to 3	
CELLULAR GLASS		1½ to 5	7 to 9.5		−450 to 800	5 x 10⁻⁶	0	.0005	0.5
PRESSED, CORK		1 to 4	6 to 8	3.22 to 3.57	up to 180	not available	not available	2.6	not available- high
COMPOSITION: URETHANE BACKED BY FIBROUS OR GRANULAR BOARD ㉒		1 to 4½	composite materials	R = 10 to12 for 2″ thick	composite materials			<1	not available

B2.1 INSULATION

STRENGTH, range, in psi (3)	FLAME SPREAD, per ASTM E84	ASTM STANDARD SPECIFICATIONS	RESISTANT TO: MOISTURE, CONDENSATION	MILDEW	VERMIN	FIRE	ULTRA-VIOLET RADIATION	SECURED BY: BITUMEN	ORGANIC ADHESIVE	METAL FASTENERS
not a factor	Type I: to 25 Type II: 26–50	C739	F	F	F	F	G		NA	
	15 to 25 max.	C764	F	G	G	G	G			
		(6)	F	G	G	G	G			
	0	C549	F	G	G	G	G			
	0	C516	F	G	G	G	G			
not a factor	15	(11)	F	G	G	G	G			●
	25 max.	C665	F	G	G	F	G			●
a, 40 to 400	200 max. for interior grade.	C208 (13)	P	G	G	P	G	●		○
e, 12	25 max.	C726	F	G	G	F	G	●		○
c, 20	25	C728	P	G	G	G	G	●		○
b, 8 to 30	25	C578	F to G	G	G	P to F	P	●	●	●
b, 20 to 100		C578	G	G	G	P to F	P	●	●	●
d, 15 to 30	25 to 75	(19)	F	G	G	F to P	P	●	●	●
	25	C591	F	G	G	F	F (20)	●	●	●
b, 75	0	C552 (21)	G	G	G	G	G	●	○	○
75	not available	C640	F	G	G	P	P	●	○	●
b, 20 to 25	not available	NA	F to G	G	G	G	F	●		●

REMARKS

1. k = Btu • in/h • ft² • deg F.
2. Grains/h • ft² • in Hg pressure differential. Values given are for insulation not faced to reduce permeance.
3. Strength:
   a. Modulus of Rupture, minimum, psi.
   b. Compressive, psi.
   c. Compressive, psi, at 5% consolidation.
   d. Compressive, psi, at 10% deformation.
   e. Compressive, psi, at 25% deformation.
   - Thickness as recommended by manufacturer to achieve desired thermal resistance value (R). R values generally 13 to 40.
4. - Use of vapor barriers added to assembly recommended with all types of loose fibrous insulation.
   - Fibrous fill will settle after installation with resultant decrease in insulating value.
   - Granulated polystyrene and cork also available as loose fill insulation.
5. - Cellulose not recommended in locations subject to prolonged exposure to heat and humidity.
   - Resistance of cellulose to exposure to fire and bacterial growth is improved by chemical treatment, which generally is required, and provided, for commercially available products. Resistances shown are for treated cellulose.
   - Chemicals used in treating of cellulose may be detrimental to metals in contact with it.
6. Generally Federal Specification HH-1-1030B.
7. Perlite and vermiculite should be water repellent treated to resist effects of condensation or rain penetration. Moisture collecting within the insulation will decrease thermal resistance.
8. - Available as faced or unfaced batts or blankets.
   - Unfaced type friction fitted between supports. Supports should be spaced not more than 24 inches on centers to prevent sagging of insulation.
   - Semi-rigid types, either faced or unfaced, available in boards for horizontal and vertical applications.
9. - Facings of kraft paper or kraft paper backed with aluminum foil are available to serve as vapor barriers to reduce permeance.
   - Recommended permeance of facing; not more than one perm.
10. For selection of vapor barriers for thermal insulation, see ANSI/ASTM C755.
11. Generally Federal Specifications:
    - HH-1-521F for faced and unfaced batts and blankets.
    - HH-1-558B for faced and unfaced semi-rigid boards.
    - Also ASTM C665—Mineral Fiber Blanket Thermal Insulation for Wood Frame and Light Construction Buildings.
12. - Organic fiber boards are asphalt impregnated and generally surface-coated to minimize moisture penetration.
13. ASTM C532 for Structural Insulating Formboard. ASTM D2277 for Fiberboard Nail-Base Sheathing.
14. - Glass fiber boards of several densities available foil faced as sheathing, and impregnated kraft paper faced as roof insulation.
    - Mineral fiber insulating planks available.
    - Flame spread: 25 maximum on unfaced side, when specified.
15. Binder may be detrimentally affected by moisture penetrating into boards. Surfaces but not edges usually sealed to minimize moisture penetration.
16. - Hot asphalt should not be applied directly to polystyrene; coal tar pitch should not be used.
    - Premolded polystyrene inserts are available to insulate hollow cove masonry units.
17. Application of membrane directly over insulation not recommended. Use of interply perlite board, wood fiberboard, glass fiber or venting-type felt is recommended method.
17a. - Aging will degrade insulating value. Protective facings will minimize effects of aging, and those having a high resistance to gas permeation will give best results. Thermal resistivity values shown are stabilized values, the higher one for insulation faced with facings having high resistance to gas permeation. Manufacturers' literature should be consulted for specific product performance. ASTM C591 gives "r" value as 5.88.
18. - Commercially available with aluminum foil facings, which give the product permeability of less than one perm.
19. ASTM Specification cited is for rigid preformed cellular urethane thermal insulation, intended for use on pipes and flat surfaces operating within temperature range of −100°F to +230°F. Specifications generally used by industry are: HH-1-530A for faced boards, and HH-1-530B for unfaced boards, with testing of the products generally in accordance with various ASTM Methods.

B2.1 INSULATION

INSULATION: types, properties (cont.)

O denotes possible usage
● denotes common usage
G denotes good resistance
F denotes fair resistance
P denotes poor resistance
NA not applicable

PROPERTIES

	THICKNESS, range, in inches	DENSITY, range, in lbs./cu.ft.	THERMAL RESISTIVITY, r=l/k at 75°F, range ①	WORKING TEMPERATURE °F, range	THERMAL EXPANSION COEFFICIENT, range	LINEAR EXPANSION, percent, from 50% to 90% relative humidity, range	VAPOR PERMEANCE, range ②	WATER ABSORPTION, by volume, maximum percent
FOAMED-IN-PLACE ㉓								
UREA-FORMALDEHYDE ㉔	½ to 4	.7 to 3	4.2 to 4.5	up to 400	30×10^{-6}	25	4.5 to 100	high
POLYURETHANE ⑰		1.9 to 3	5.6 to 7.14	up to 230	40×10^{-6}	not available	2 to 3	not available
POLYISOCYANURATE ⑰								
EPOXY, TWO PART ㉕		1.8 to 2.3	6.6 to 9.0	up to 200	30×10^{-6}	low	.9 to 1.2	not available
PHENOLIC ㉖		2.5 to 4	4.5 to 8.0	up to 350	5×10^{-6}	< 1.0 for rigid	1.0 for rigid	1 to 2 for rigid
POURED-IN-PLACE ㉘								
CONCRETE, PERLITE AGGREGATE	1 to 3 or more	25 to 35	1.08 to 1.5	40 to 1000	4.3 to 6.1×10^{-6}	Less than one percent after initial shrinkage	varies with density-low	27
VERMICULITE AGGREGATE			.98 to 1.0		4.6 to 7.9×10^{-6}			44
CELLULAR		20 to 40	.85 to 2	40 to 1800	varies with density			varies with density
ASPHALTIC, PERLITE AGGREGATE	1½ to 5	20 to 22	2.5	−20 to 250	not available	not available		4.5
GYPSUM, PERLITE AGGREGATE	2 to 3	40 to 50	.57 to .66	up to 1800	8×10^{-6}	not available		not available
SPRAY-APPLIED ㉙								
FIBERS, ORGANIC	2 to 4	2 to 3	.95 to 4.5	to 150	not available	not available	varies-high	
MINERAL	¾ to 3	2.5	3.0 to 4.5	32 to 212	not available	less than 1%	varies-high	high
CEMENTITIOUS	½ to 2	17 to 70	1.6 to 3.4	up to 1800	0.12×10^{-6}		varies-low	
REFLECTIVE								
SPACED ALUMINUM FOIL ㉚	.008 to .02	not a factor	not a factor	not a factor	not a factor	not a factor	0.1 to 0.3	0

B2.1 INSULATION

③ STRENGTH, range, in psi	FLAME SPREAD, per ASTM E84	ASTM STANDARD SPECIFICATIONS	RESISTANT TO: MOISTURE DAMAGE	MILDEW	VERMIN	FIRE	ULTRA-VIOLET RADIATION	SECURED BY: SELF-BONDING	ORGANIC ADHESIVE	METAL FASTENERS
25 to 40 b	25	NA	F to G	G	G	P	F	●		
20 to 40 b	25	⑲	F to G	G	G	F to P	F	●		
			G	G	G	F to P	F	●		
20 to 30 b	not available	NA	F	G	G	P	P	●		
20 to 50 d	not available	㉗	G	G	G	G	F	●		●
150 to 300 b	0	C332 and C150 for cement	G	G	G	G	G	●		
			G	G	G	G	G	●		
100 to 250 b		NA	F	G	G	G	G	●		
40 d	not available	C332 and D312-binder	G	G	G	G	G	●		
500 minimum	0	C317	P	G	G	G	G	●		
low	10 to 25	NA	P	F	G	F	G	●		
200	0 to 15	C195	F	G	G	G	G	●		
150 to 1000	5 to 15	C196	G	G	G	G	G	●		
not a factor	5	NA	G	G	G	G	G			●

REMARKS, cont'd.

⑳ • Commercially available roof insulating boards generally cannot be applied directly over metal roof decks when Factory Mutual Class I approval is required. Insulation specially developed for direct application to metal roof decks as a component of FM Class I Insulated Steel Deck Roofs is available. Consult manufacturers' literature.

㉑ • Four types are covered under ASTM C552: Type I—Flat block; Type II—Pipe and Tubing insulation; Type III—Special shapes; Type IV—Roof board.

㉒ Urethane over fibrous or granular insulating board generally for roofs. With the insulating board placed between metal roof deck and urethane, this type can generally be used in FM Class I installations.

㉓ • Foamed insulation, whether foamed-in-place or rigid boards, may not be left exposed to interior spaces unless specifically permitted; Building Codes generally require a protective facing, such as gypsum wall board, to provide at least 15 minute fire resistance.
• Foamed insulation will soften and start melting after maximum working temperatures are exceeded.
• Polystyrene may break down under freeze-thaw cycles.

㉔ • Consumer Product Safety Commission has banned the use of urea-formaldehyde in residences and schools. Ban has been overturned by court action in 1983.
• Urea-formaldehyde is more permeable than other foams: water vapor retarder may be required.
• Urea-formaldehyde releases moisture and exudes formaldehyde during the curing period.

㉕ Not in wide usage; special applications only.

㉖ Phenolic also available in rigid boards for roof and cavity wall insulation.

㉗ Generally under Federal Specification HH-1-530A.

㉘ • Poured-in-place generally used as structural roof decks over formed metal roof decks, or over prefabricated form boards.
• Insulating concrete may be poured over structural reinforced concrete deck.
• Formboards used are: gypsum board, cement-bound mineral fiber, glass fiberboards, cement-bound wood fiber, and mineral fiberboards; all must be strong enough to support the wet mix between structural supports; some will contribute to the overall thermal resistance of the assembly; some may not be used in high humidity areas.
• Insulating concrete with wood chips may not be classified as incombustible; consult Building Codes and manufacturers' literature.
• Precast planks of perlite and cellular concrete, and of gypsum are available.

㉙ • Mineral fibers will provide insulating value, sound absorption and may provide fire resistance.
• Cementitious of low insulating value, used primarily as fireproofing of structural components.
• When spray-applied materials are used as fireproofing, such assemblies should be tested, generally in accordance with ASTM E119.

㉚ • Reflective foil used in multiple layers with air spaces between them; may be installed in wall and roof assemblies; mostly used in cold storage installations.

㉛ • Polystyrene may be subjected to temperatures up to 250°F for short-term or one-face exposure without damage to insulation.

B2.1 INSULATION

INSULATION: selection checklist

■ The transfer of heat within and through a building envelope at low exterior temperatures is diagramatically illustrated below:

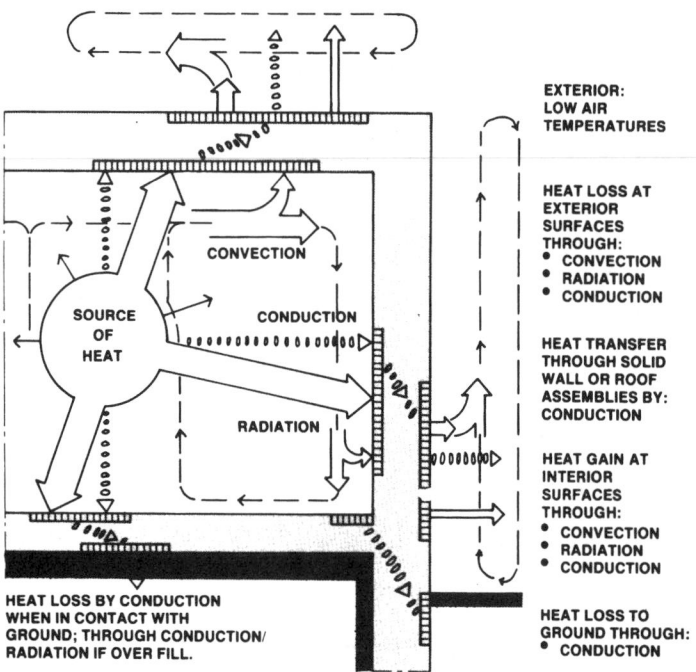

EXTERIOR:
LOW AIR
TEMPERATURES

HEAT LOSS AT
EXTERIOR
SURFACES
THROUGH:
• CONVECTION
• RADIATION
• CONDUCTION

HEAT TRANSFER
THROUGH SOLID
WALL OR ROOF
ASSEMBLIES BY:
CONDUCTION

HEAT GAIN AT
INTERIOR
SURFACES
THROUGH:
• CONVECTION
• RADIATION
• CONDUCTION

HEAT LOSS TO
GROUND THROUGH:
• CONDUCTION

CONVECTION

CONDUCTION

SOURCE
OF
HEAT

RADIATION

HEAT LOSS BY CONDUCTION
WHEN IN CONTACT WITH
GROUND; THROUGH CONDUCTION/
RADIATION IF OVER FILL.

TYPES OF INSULATION

The best insulation is a vacuum or air when kept completely motionless in a space separating two solid components.

■ Air, however, cannot be kept motionless even in a narrow vertical cavity (as in a wall assembly): **convective currents develop,** which transfer heat from the warm side of the cavity to the colder one.

■ **Radiation from the warm side** to the colder one takes place whether the air moves or is still.

■ In an air space **broken up horizontally** into tiny compartments convective currents can be effectively minimized, and the excellent insulating properties of still air utilized.

■ This is the basic principle in the development of **mass or bulk type insulation.**

Mass type insulation reduces the flow of heat by preventing convection in entrapped air and also by forming a barrier to radiation.

■ Some types (such as foamed plastics, or cellular glass) **trap small quantities of air** or other gaseous substances in closed cells. The heat flow through the cells is greatly reduced because convection currents are virtually eliminated in small cells.

■ **Size of the cells is critical:**

• if they are too large, convective heat flow within them may become significant.

• if they are too small, or there are too few of them, conduction through the solid material surrounding the cells increases, offsetting the insulating value of the cells.

■ **Granular materials** (such as perlite, vermiculite, granulated foam) trap air in relatively large voids and consequently may have poorer insulating properties than materials with numerous small cells.

■ **Fibrous materials** (such as glass fibers or cellulose) depend for performance on the air's characteristic to cling to all exposed surfaces in thin films, thus reducing the heat flow.

■ Fibrous materials will perform best at **a specified optimum density:**
• if compressed to higher than optimum density, heat flow will increase since fiber will touch fiber and some of the surface air film will be lost.
• if fluffed up too much, more heat may be transmitted by convection or radiation through the large voids.

■ **Bulk density** of insulating materials is a very good indication of their insulating properties.

Reflective insulations reduce the transfer of heat through air spaces by minimizing radiation of energy from the warmer, or emitting, surface of one of the components which enclose an air space to a colder, or receiving, surface of the other component:

■ **Emissivities of various building materials at the same surface temperature vary:** radiation across an air space between two polished aluminum surfaces will be only about 3 percent of that between two black surfaces.

■ **Reflective materials** act as insulation:

• because of their low surface emissivity

• by reflecting incident radiant energy: in a cavity wall up to 60 percent of heat transfer is estimated to be by radiation.

INSULATION & CONDENSATION

Air almost always contains a certain amount of water vapor; the maximum amount of vapor that can be contained at constant pressure is directly proportional to the temperature of the air/vapor mixture:

■ When air at a given temperature, saturated with water vapor, is cooled, or comes into contact with a colder surface, **water vapor will continuously condense** as long as the temperature of the air/vapor mixture drops.

Insulation incorporated into assemblies of an enclosure changes the temperature gradients through them thereby **increasing the likelihood of condensation** within the assemblies:

■ **Condensation may occur** within the insulation, if it is permeable, and increase its density, thereby lowering its thermal resistance.

B2.1 INSULATION

INSULATION: thermal transmittance of assemblies

MASS, vertical or horizontal

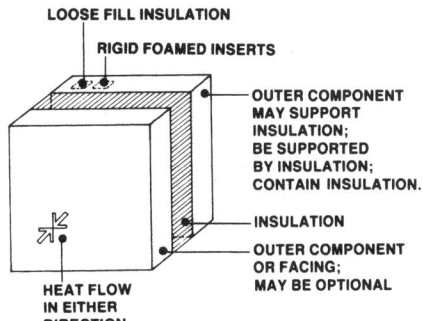

LOOSE FILL INSULATION

RIGID FOAMED INSERTS

OUTER COMPONENT MAY SUPPORT INSULATION; BE SUPPORTED BY INSULATION; CONTAIN INSULATION.

INSULATION

OUTER COMPONENT OR FACING; MAY BE OPTIONAL

HEAT FLOW IN EITHER DIRECTION

Thermal transmission coefficients (or the rate of heat flow through any assembly) **may be modified by replacing components** of high conductivity with components of low conductivity; or by **adding insulating materials** to the assembly:

■ **Changing components** of an assembly is not in general a practical solution: components of low conductivity will also have low structural strength and density.

■ **Adding insulation** is the usual approach, but each successive layer of the same thickness is progressively less effective than the first:
- adding one inch of fibrous insulation to an assembly with U-.38 will reduce the U value to .16, or about 57 percent;
- adding another inch of insulation will result in U-.10, or a 73 percent reduction;
- three inches will give U-.07, or a reduction of 81 percent over the coefficient of transmission of the uninsulated assembly.

■ The cost of each inch of insulation being the same, the **pay back period for each successive** layer of added insulation will be longer unless there is a sharp rise in the cost of energy over the expected in-service life of the assembly.

REFLECTIVE, vertical

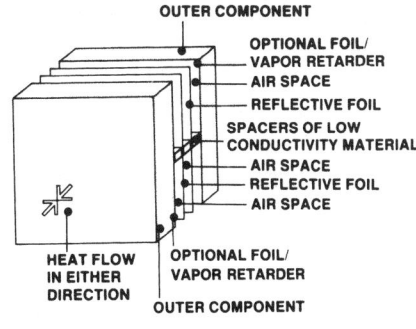

OUTER COMPONENT

OPTIONAL FOIL/ VAPOR RETARDER

AIR SPACE

REFLECTIVE FOIL

SPACERS OF LOW CONDUCTIVITY MATERIAL

AIR SPACE

REFLECTIVE FOIL

AIR SPACE

HEAT FLOW IN EITHER DIRECTION

OPTIONAL FOIL/ VAPOR RETARDER

OUTER COMPONENT

In assemblies enclosing an air space, thermal transmission coefficients may be modified, when the components facing the air space have surfaces of high emissivity, by covering them with a material of low emissivity, such as bright aluminum foil.

■ **Additional sheets of foil** may be used to divide the air space into two or more sections: that is to replace the single air space with multiple air spaces bounded by surfaces of the same or different emissivities.

■ **When reflective sheets are added** to one or both sides of an air space at least ¾ inch wide & enclosed within an assembly with a U value of .38, the coefficient of transmission will be reduced to U = .25, or by about 34 percent.

■ **Two reflective sheets** positioned within the cavity of the same assembly will reduce its coefficient of transmission to U = .09, or by about 76 percent.

■ **Higher heat transmission rates** may occur at spacers/supports, depending upon their coefficients of conductance.

■ **Surface condensation greatly reduces the effectiveness of foil insulation:** it increases surface emissivity from about 5 percent when foil is dry, to a value of over 90 percent when foil is wet.

REFLECTIVE, horizontal

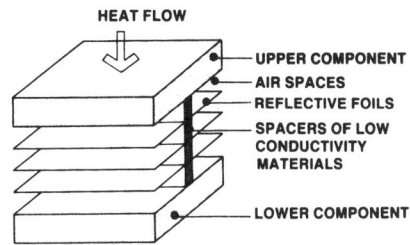

HEAT FLOW

UPPER COMPONENT

AIR SPACES

REFLECTIVE FOILS

SPACERS OF LOW CONDUCTIVITY MATERIALS

LOWER COMPONENT

In horizontal assemblies with an air space between the components, thermal transmission coefficients differ from those for vertical assemblies:

■ **Air in vertical spaces will always be in motion** as long as a temperature differential exists between the two surfaces facing the air space, and so will air in a horizontal space when the flow of heat is upward.

■ **When the flow of heat is down,** the air will stratify, and heat transfer by convection will be negligible.

■ **Adding reflective sheets** to a horizontal assembly will therefore result in **appreciable differences** in thermal transmission coefficients for the same assembly under **different directions of heat flow:**
- two reflective sheets added to an assembly with a U value of .38 with the flow of heat up will reduce the transmission coefficient to U = .12, or 68 percent.
- the same assembly with the flow of heat down will have a U = .04, resulting in a reduction of about 89 percent.

B2.1 INSULATION

RADIANT BARRIER SYSTEMS

RADIANT BARRIER SYSTEMS

Background: In warm climates, a number of strategies are depended upon to keep heat out of buildings. Mostly, these affect heat gains by conduction or convection. In the average building, insulating walls and ceilings primarily restrict conduction. Double-glazed windows restrict both conductive and convective heat gain. However, **radiation**—the third means of heat transfer—is largely ignored except in window treatments and coating applications that reflect, absorb or shade from solar energy. Research and recent developments point to the potential for reducing heat gain in buildings by controlling radiation through the use of radiant barrier systems (RBS).

■ **To help understand why RBS are effective, consider that:**

• **Radiation travels only in a strait line.** On earth, regions of different temperatures that "see" each other exchange energy via far infrared radiation in the 4 to 40 micron wavelength band. (A micron is a millionth of a meter.)

• **Sunlight,** on the other hand, consists of much shorter wavelengths in the 0.2 to 2.6 microns band. Unlike the visible portion of the solar spectrum (0.4 to 0.7 microns), the "near infrared" portion of the solar spectrum (0.7 to 2.6 microns) is invisible. "Far-infrared" radiation is also invisible.

• **Near infrared radiation** is generated by the sun and far infrared radiation is generated by all bodies on earth.

• **Far infrared radiation** is sometimes called "thermal" or "long-wave" radiation. The effect of both is heat and, in air-conditioned buildings, this heat is unwanted.

■ **Radiant barrier systems are a method of stopping far-infrared radiation from getting to building interiors and increasing air conditioning loads.**

A radiant barrier system is defined by the American Society for Testing and Materials (ASTM 1990) in Standard C 1158-90 as "a building construction consisting of a low emittance (normally 0.1 or less) surface (usually aluminum foil) bounded by an open air space." The definition given by

this Standard goes on to stipulate that, "a RBS is used for the sole purpose of limiting heat transfer by radiation . . ."

■ **Radiant barrier systems comprise an air space** with one or more of its boundaries functioning as a radiant barrier. Radiant barriers are materials that restrict the transfer of far-infrared radiation across an air space. They do this by not emitting radiant energy. A material with this capability is said to have a very low emissivity. The lower the emissivity, the better the radiant barrier.

■ **Emissivity values range from 0 to 1.** The laws of optics stipulate that for any given wavelength, a material's emissivity plus its transmissivity plus its reflectivity must equal one. Opaque materials have a transmissivity of zero, so their emissivity plus their reflectivity must equal one. It follows, therefore, that their emissivity must equal one minus their reflectivity.

■ **Materials that radiate very well have high emissivities** and those that radiate very poorly have low emissivities. Most common building materials, including glass and paints of all colors, have high emissivities of 0.9 or greater. Such materials are capable of transferring far infrared radiation at 90% or more of their temperature potential. These materials are ineffective barriers to radiant energy transfer. On the other hand, aluminum foil is an excellent radiant barrier. It has a low emissivity (0.05), therefore, it eliminates 95% of the far infrared radiation energy transfer potential.

■ **Aluminum foil, however, is a very good thermal conductor.** Consequently, it has an extremely low R-value. However, if it is placed between materials that are attempting to transfer energy by radiation (rather than conduction) and if it is separated from these materials by an open air space, the foil effectively eliminates the normal radiant energy exchange across the air space. (If the air space is evacuated, the result is a Dewar's flask, or "thermos bottle"—one of the most effective heat transfer reduction systems known.)

This is the operating principle of a radiant barrier system. It can be used to significantly reduce the flow of heat through building components and systems.

SUNLIGHT AND HEAT

A material's response to far-infrared radiation can be quite different from its response to sunlight. Since a large percentage of sunlight is in the visible range, we characterize materials by color and clarity. White paint reflects far more solar radiation than does black paint. But in the far-infrared band, white paint absorbs slightly more radiation than does black paint. This surprising fact indicates that a material's far-infrared properties cannot be judged by sight. The accompanying table compares the solar and far infrared characteristics of some common opaque building materials.

The table shows only opaque materials. Transparent materials also respond differently to solar and far-infrared radiation. Common window glass, for example, transmits more than 85% of incident sunlight but absorbs more than 85% of the far-infrared radiation that strikes it. The "solar greenhouse effect" results in part from this phenomenon. Solar energy readily passes through the glass and is absorbed by the opaque surfaces within the space. When these heated surfaces begin to radiate to cooler surfaces, the glass absorbs most of this far-infrared radiation, trapping much of the original solar gains inside the space as heat.

■ **Roof systems.** A house attic offers excellent potential for use of radiant barrier systems: first, because the roof is the surface most exposed to solar radiation, and second, because most of the solar gain absorbed by the roof is transmitted down to the attic floor by far infrared radiation. Since the attic airspace separates the hot roof surface from the ceiling, no heat will move down by conduction, and the heat will not convect down from the hot roof to the ceiling because heated air rises.

■ **If one places a radiant barrier (layer of foil) in the airspace** between the hot roof deck and the cooler attic floor (insulation), almost all radiant heat transfer can be eliminated.

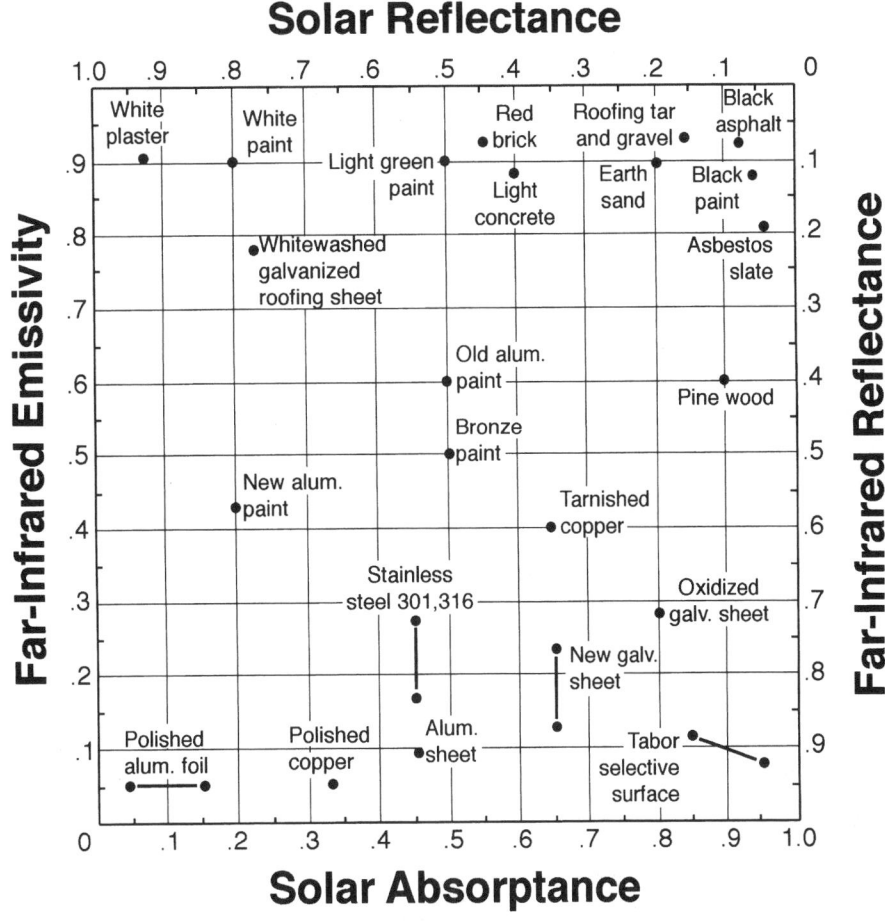

Solar Reflectance

Far-Infrared Emissivity

Far-Infrared Reflectance

Solar Absorptance

Fig. 1 Emissivity values of common roofing and radiant barrier materials

Studies at the Florida Solar Energy Center (FSEC) indicate that, under peak day conditions, total heat transfer down through ceilings can be reduced by more than 40% in this way. These results occur because the radiant barrier significantly reduces the top surface temperature of the ceiling insulation.

■ **Heat transferred upward through attics (winter heat loss) won't be affected as much because a greater part of total upward heat transfer occurs by convection (heated air rising).** That is why radiant barriers in roof systems are a more effective as a cooling strategy in overheated periods rather than as a heating strategy and thus why they may be of great benefit in warm- and hot-climate applications. In a typical home in southern U.S. for example, an attic radiant barrier could cut annual cooling costs by 6–12% and peak cooling loads by

15%. An important component of an effective attic radiant barrier system is effective attic ventilation, which can normally be achieved by continuous soffit and ridge vents.

■ **Most roof types already contain some kind of attic or airspace that can accommodate an effective radiant barrier system.** In new construction, it should be easy to install radiant barrier systems regardless of roof pitch. The accompanying figure Alternatives for radiant barrier placement illustrates three alternate locations for radiant barriers in attics. When first installed, there will be no significant difference in the effectiveness of these locations. But in time, location 3 will suffer because of dust accumulation, which decreases performance. Dust can't collect on the underside of the radiant barriers at locations 1 or 2.

ALTERNATIVES FOR RADIANT BARRIER PLACEMENT

Location 2 is often considered best because it offers the potential for separately ventilating the space between the radiant barrier and hot roof deck and the attic space itself. This results in an attic air temperature somewhat closer to the conditioned space temperature in both winter and summer. As with location 3, dust may collect on the top of location 2, but a radiant barrier surface facing downward will perform as well as one facing upward. Therefore, for reasons of dust accumulation, use location 1 or 2 and depend on the down side for radiation control.

In new construction, another alternative may offer the advantages of location 2 and the construction ease of location 1. This construction places the radiant barrier on top of the roof rafters (or trusses) before the roof decking is applied. It is installed so that it droops

B2.1 INSULATION

RADIANT BARRIER SYSTEMS (cont.)

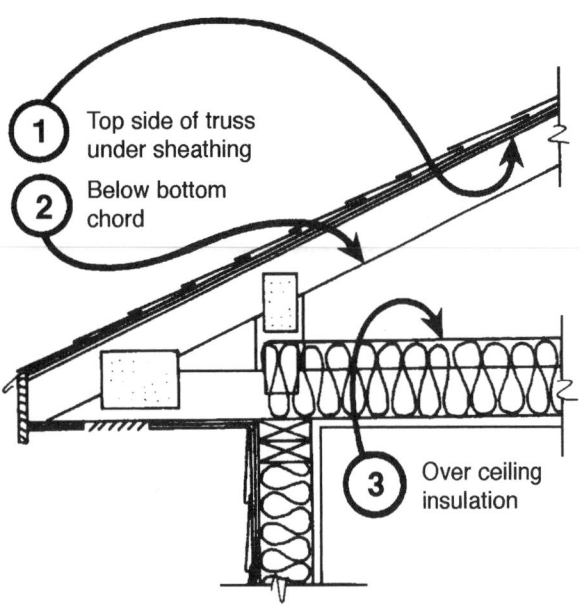

Fig. 2 Alternatives for radiant barrier placement

1. Top side of truss under sheathing
2. Below bottom chord
3. Over ceiling insulation

approximately 2 in. (5 cm) below the upper surface of the roof structure. When the roof decking is applied, an airspace separates it from the radiant barrier in a way similar to that of location 2. This airspace also can be vented separately from the attic. As with location 2 the most reflective radiant barrier surface should face downward toward the attic airspace.

■ **Multiple layers.** Economics bode against more than one radiant barrier in attics. The first barrier surface eliminates about 95% of the radiant heat transfer across the attic. Adding more layers can affect only 95% of the remaining 5%. (This is not necessarily true in wall systems, where heat transfer by air convection can account for a greater percentage of total heat transfer).

■ **Tightness.** It is not necessary to form airtight seals with radiant barriers; radiant energy travels in a straight line through the air but not in the air. If you choose location 3, you should use a perforated foil product that will allow the free passage of vapor out of the insulation during winter. This may also apply to location 1 in some cases, because the barrier is in contact with the roof decking. Location 2 should not have moisture condensation problems because it has an airspace on both sides of the radiant barrier.

RADIANT BARRIER MATERIALS

There are many types of radiant barrier materials on the market, and more are being developed as radiant barriers become more widely used. Five generic types are most common:

■ **Single-sided foil** (one foil side) with another material backing such as Kraft paper or polypropylene. Fiber web-

bing sandwiched between foil and backing further strengthens some products. The strength of the backing materials is important since unreinforced foil tears very easily.

■ **Double-sided foil** with reinforcement between the foil layers. Reinforcement may be cardboard, Kraft paper, Mylar or fiber webbing.

■ **Foil-faced insulation.** The insulating material may be polyisocyanurate, polyethylene "air-bubble" packing or other materials that impede heat conduction.

■ **Multi-layered foil systems.** When fully extended and installed so that the foil layers do not touch, these products also form insulating air spaces.

Some of these products have R-values, which may be properly claimed as a representation only if the product was tested according to Federal Trade Commission regulations for insulation.

Although it is not by definition a radiant barrier, there is a low-emissivity paint available that can be applied directly to the underside of the roof decking.

Characteristics of radiant barriers provide a guideline to material

selection:
- Emissivity (the lower the better)
- Fire rating (as required by building codes)
- Ease of handling
- Strength of reinforcement
- Width appropriate for installation
- Low cost

REFERENCES:

Fairey *"Radiant Barrier Systems"* in Ref. 1; also: ASTM. 1990. Standard C 1158-90, *"Standard Practice for Use and Installation of Radiant Barrier Systems in Building Construction."* Philadelphia, PA: American Society for Testing and Materials.

B2.1 INSULATION

REFERENCES: Insulation

STANDARDS

ASTM Specifications for:

C208 • Insulating Board (Cellulosic Fiber), Structural and Decorative.

C332 • Lightweight Aggregates for Insulating Concrete.

C516 • Vermiculite Loose Fill Insulation.

C532 • Structural Insulating Formboard (Cellulose Fiber).

C549 • Perlite Loose Fill Insulation.

C552 • Cellular Glass Block and Pipe Thermal Insulation.

C553 • Mineral Fiber Blanket and Felt Insulation Industrial Type.

C578 • Preformed, Cellular Polystyrene Thermal Insulation.

C591 • Unfaced Preformed Rigid Cellular Polyurethane Thermal Insulation.

C592 • Mineral Fiber Blanket Insulation and Blanket-Type Pipe Insulation (Metal Mesh Covered).

C610 • Expanded Perlite Block and Pipe Thermal Insulation.

C612 • Mineral Fiber Block and Board Thermal Insulation.

C665 • Mineral Fiber Blanket Thermal Insulation for LIght Frame Construction and Manufactured Housing.

C726 • Mineral Fiber and Rigid Cellular, Polyurethane Composite Roof Insulation Board.

C728 • Perlite Thermal Insulation Board.

C739 • Cellulosic Fiber (Wood-Base) Loose-Fill Thermal Insulation.

C764 • Mineral Fiber Loose-Fill Thermal Insulation.

C984 • Perlite Board, Rigid Cellular Polyurethane Composite Roof Insulation.

C1013 • Membrane Faced Rigid Cellular Polyurethane Roof Insulation.

C1014 • Spray-Applied Mineral Fiber Thermal or Acoustical Insulation.

C1029 • Spray-Applied Rigid Cellular Polyurethane Thermal Insulation.

C1050 • Rigid Cellular Polystyrene Cellulosic Fiber Composite Roof Insulation.

D2341 • Rigid Urethane Foam.

ASTM Definition of Terms Relating to:

C168 • Thermal Insulating Materials.

ASTM Recommended Practice for:

C653 • Determination of the Thermal Resistance of Low-Density Blanket-Type Mineral Fiber Insulation.

C687 • Determination of the Thermal Resistance of Loose-Fill Building Insulation.

C727 • Use of Reflective Insulation in Building Constructions.

Federal Specifications:

HH-I-521F • Mineral Fiber Blankets.

HH-I-558B(3) • Industrial-Type Mineral Fiber.

HH-I-1030A • Mineral Fiber Loose-Fill.

H-H-C-561 • Cork; compressed (Corkboard) (for Thermal Insulation).

H-H-I-515D • Insulating Blanket, Thermal-Acoustical, and insulating Thermal, Cellulose or Wood fiber.

H-H-I-574B • Insulation, Loose-Fill (Perlite).

H-H-I-585C • Insulation, Thermal (Vermiculite).

H-H-I-1252B • Insulation, Thermal, Reflective (Aluminum Foil).

H-H-I-524B • Insulation Board, Thermal (Polystyrene).

H-H-I-530A • Insulation Board, Thermal (Urethane).

H-H-I-551 • Insulation Block and Boards, Thermal (Cellular Glass).

SELECTED REFERENCES

AIA **American Institute of Architects**
• AIA Energy Notebook (Information on Energy and Built Environment)
• Architect's Handbook of Energy Practice, The Building Envelope

AAMA **Architectural Aluminum Manufacturers Association**
• Design for Energy Conservation in Aluminum Curtain Walls

CSI **Construction Specifications Institute**
• Insulation: Ally or Enemy.

DOE **U.S. Department of Energy**
• An Assessment of Thermal Insulation Materials and Systems for Building Applications.

NAHB **National Association of Home Builders of the U.S.**
• Thermal Performance Guidelines.

NBS **National Bureau of Standards**
• Retrofitting an Existing Wood Frame Residence for Energy Conservation—An Experimental Study.
• National Bureau of Standards Heating and Cooling Load Determination Program.

NPA **National Particleboard Assn.**
• Thermal Conductivity of Particleboard.

NRCC **National Research Council of Canada**
CBD 16 • Thermal Insulation in Dwellings.
70 • Thermal Considerations in Roof Design.
102 • Thermal Environment and Human Comfort.
144 • Tonic Gases and Vapours Produced by Fires.
149 • Thermal Resistance of Building Insulation.
166 • Plastic Foams.
167 • Rigid Thermoplastic

B2.1 INSULATION

REFERENCES: Insulation (cont.)

Foams.

168 • Rigid Thermosetting
Plastic Foams.

178 • Fire and Plastic Foam
Insulation Materials.

218 • Effects of Insulation on
Fire Safety.

OCE   **Owens/Corning Fiberglas.,**
• Design Guide for
Insulated Buildings

PCA   **Portland Cement Assn.**
• Simplified Thermal Design
of Building Envelopes for
Use with ASHRAE
Standard 90-75.

PI   **Perlite Institute**
• Perlite Technical Data
Sheet, No. 2-4.

• Fire Resistance of
Lightweight Insulating
Concretes.

SPI   **Society of the Plastics
Industry.**

U-100R • Fire Safety guidelines for
Use of Rigid Polyurethane
Foam Insulation in
Building Construction.

U-102R • An Update Report on
Findings of Fire Study of
Rigid Cellular Plastic
Materials for Wall and
Rood Ceiling Insulation.

U-103 • Large-Scale Corner Wall
Fire Tests of Spray-On
Coatings Over Rigid
Polyurethane Foam
Insulation.

U-107 • Room-Scale Compartment
Tests of Spray-On
Coatings Over Rigid
Polyurethane Foam
Insulation.

U-108 • An assessment of the
Thermal Performance of
Rigid Polyurethane and
Polyisocyanurate Foam
Insulations for Use in
Building
Construction.

B2.1 INSULATION

B2. 2 WATERPROOFING/DAMPROOFING

CONTROL OF PASSAGE OF MOISTURE:
introduction

SOURCES OF MOISTURE

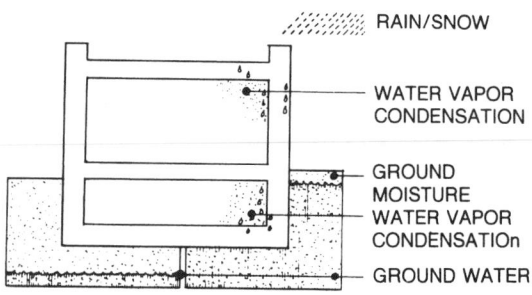

RAIN/SNOW

WATER VAPOR
CONDENSATION

GROUND
MOISTURE
WATER VAPOR
CONDENSATIOn

GROUND WATER

External sources:

* rain, melting snow accumulating on horizontal surfaces above or at grade.
* rain striking and running over vertical or inclined surfaces.
* rain or melting snow seeping into the ground and temporarily trapped by partially impervious soils.
* rain or melting snow seeping into the ground and permanently contained in water bearing strata.

External/internal sources:

* water vapor suspended in air/vapor mixture when condensing, upon reaching its dew point.

MODES OF PASSAGE

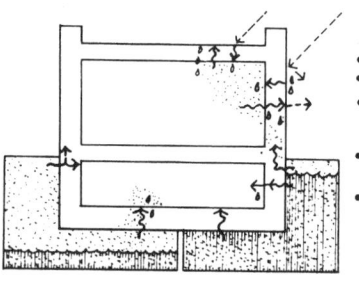

WATER THROUGH:
* CAPILLARIES
* VOIDS/CRACKS
* JOINTS

WATER VAPOR:
* BY DIFFUSION
THROUGH PERMEABLE MATERIALS
* JOINTS/CRACKS
AS PART OF
AIR/VAPOR
MIXTURE

Water passage may occur through:

* capillaries, especially when they are relatively large; small capillaries such as in hard-burned brick or dense concrete draw and hold water with high suction and seldom contribute significantly to passage of water.
* cracks which may develop due to: shrinkage in concrete or masonry; differential thermal movement between components.
* joints which fail due to: loss of bond between mortar and masonry units; loss of adhesion in sealants; splits through sealants; poorly installed sealants.

DRIVING FORCES: water passage

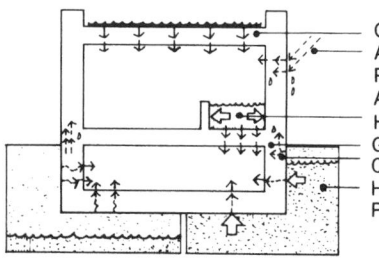

GRAVITY
AIR MOVEMENT/
PRESSURE CAPILLARY
ATTRACTION
HYDROSTATIC PRESSURE
GRAVITY
CAPILLARY ATTRACTION
HYDROSTATIC
PRESSURE

* Gravity: weight of accumulated/contained water.
* Kinetic energy of wind driven rain.
* Capillary attraction; especially when in conjunction with the additional force of wind pressure on the outside generally coupled with lower inside pressure.
* Hydrostatic pressure of ground water acting: laterally on vertical surfaces; upward on submerged horizontal surfaces.
* For additional information on water penetration under pressure differentials, refer to WATER VAPOR MIGRATION, and WATERTIGHT EXTERIOR WALLS.

DRIVING FORCES: water vapor passage

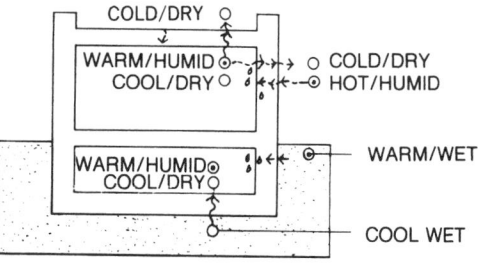

COLD/DRY

WARM/HUMID
COOL/DRY   COLD/DRY
                HOT/HUMID

WARM/HUMID
COOL/DRY   WARM/WET

COOL WET

Water vapor will migrate from one area to another whenever a vapor pressure differential exists between them:

* from a warm interior space to a cold exterior air.
* from hot and humid exterior air to a cooled interior space, especially when the relative humidity of interior air is lower than that of exterior air.
* from cool wet ground to cool but dry interior space; or the vapor in such space may condense on surfaces cooled through contact with the ground.

B2.2 WATERPROOFING/DAMPROOFING

MEANS OF CONTROL

INTEGRAL: admixture in concrete

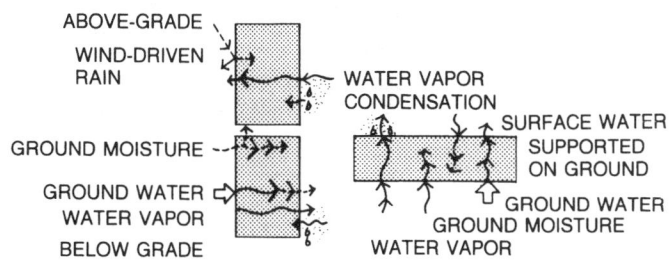

Concrete, when properly designed, placed and cured, will be essentially watertight; but it will still absorb moisture through capillary attraction:

- rain water penetration through dense concrete is seldom a problem.
- concrete continuously exposed to ground moisture may resist passage of water but may remain damp throughout.

Admixtures to reduce capillary attraction may be added to concrete during mixing; such additives generally function as water repellents only, and their effectiveness under specific conditions should be investigated prior to selection.

SURFACE SEALERS: penetrating

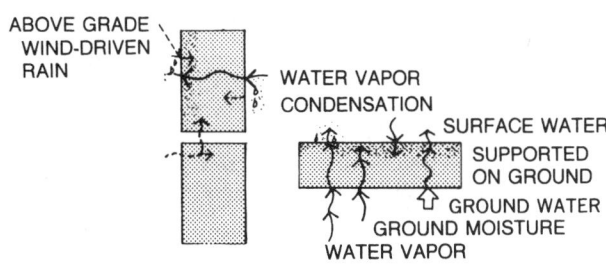

Chemical solutions applied to vertical masonry or concrete surfaces above grade to function as water repellents:

- will minimize rain water absorption and dirt retention.
- generally will allow water vapor migration through sealed surface; may prevent bonding of subsequent cementitious coatings.

Chemical solutions are applied to horizontal concrete surfaces - commonly when such surfaces are of inferior quality - to improve their resistance to wear and to minimize dusting of surface layer. Surfaces treated with chemical solutions will:

- resist absorption/penetration of surface water not under hydrostatic pressure; may resist some chemical solutions.
- generally remain permeable to water vapor.

For additional information refer to CEMENT/CONCRETE.

SURFACE COATING
- vapor permeable
- vapor impermeable

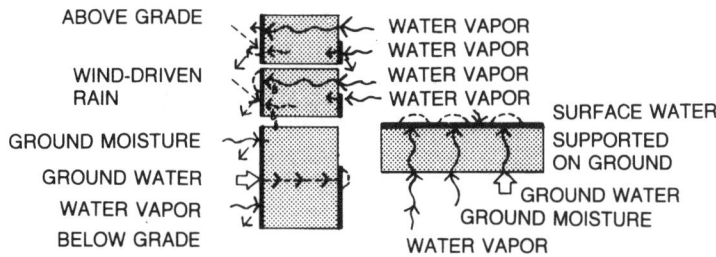

Water vapor permeable coatings:

- resist rain water penetration and dirt retention; commonly also function as decorative finishes.
- available clear, tinted, or opaque.

Water vapor impermeable coatings:

- resist rain water/ground moisture penetration.
- when such coatings block escape of water vapor through a surface, condensation may occur and destroy their bond with such surface leading to failure of the coating.

For additional information refer to COATINGS.

MEMBRANES

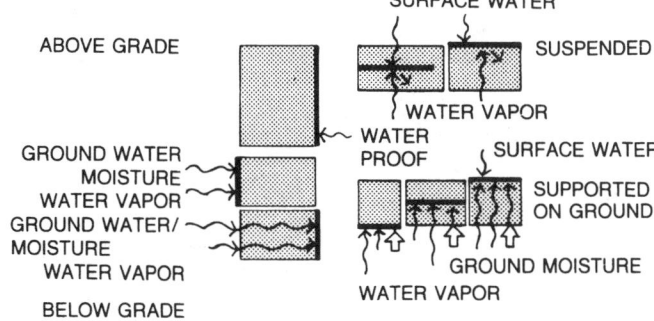

Membranes are commonly available with widely differing properties and are used to:

- prevent moisture absorption from the ground.
- minimize water vapor migration.
- prevent passage of surface water and water vapor diffusion.
- resist passage of water under hydrostatic pressure.
- protect substrate from effects of: chemical solutions being contained; chemicals spilled either routinely, or accidentally, or in cleaning solutions.

B2.2 WATERPROOFING/DAMPROOFING

MEANS OF CONTROL: types, uses

INTEGRAL: Admixtures in concrete

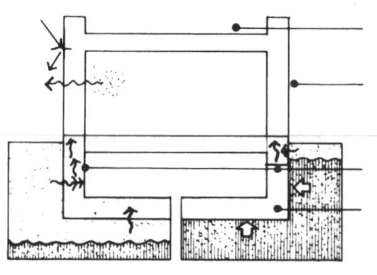

WILL NOT PROVIDE POSITIVE PROTECTION TO ENCLOSED SPACE

MAY BE USED BUT SUCH USAGE NOT COMMON

MOST FREQUENT USAGE

WILL NOT RESIST HYDROSTATIC PRESSURE- NOT USED TO CONTROL PASSAGE OF WATER

Admixtures commonly used include stearates, oleic acids, tallow; alkalies liberated during hydration of cement may cause such additives to deteriorate with eventual loss of water repellency:

- Integral water repellents are not effective in preventing passage of water when under hydrostatic pressure.
- Permeability of concrete may also be reduced without the use of water repellents by using: low porosity aggregates, air-entraining admixtures, minimum water-cement ratios, proper placement and curing.

For additional information refer to CEMENT/CONCRETE.

SURFACE SEALERS: penetrating

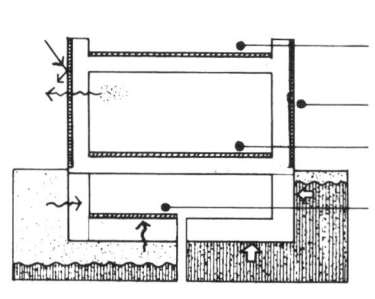

WILL NOT PROVIDE POSITIVE PROTECTION. MAY BE USED TO MINIMIZE SURFACE WATER PENETRATION

FUNCTION AS WATER REPELLENTS

USED AS PROTECTION AGAINST OCCASIONAL SPILLS AND WEAR

USED AS PROTECTION AGAINST OCCASIONAL SPILLS, WEAR AND TO MINIMIZE GROUND MOISTURE PASSAGE

Horizontal surfaces of concrete:

Sealers harden the treated surface and increase its resistance to chemical attack, freeze-thaw cycles, and to passage of water. Generic products used may include: sodium silicate; silicofluorides of zinc, sodium, and manganese; linseed or soybean oil.

Vertical surfaces of masonry and concrete above grade: Clear penetrating sealers such as silicones, stearates, acrylics, resins are used as water repellents to change the capillary angle of the pores at the surface from positive to negative, thus minimizing absorption of water.

COATINGS: water vapor permeable

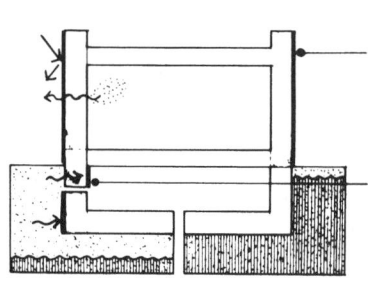

PROTECTION AGAINST RAIN PENETRATION ALSO AS DECORATIVE FINISH: ORGANIC AND CEMENTITIOUS

PROTECTION AGAINST GROUND MOISTURE COMMONLY CEMENTITIOUS ONLY

Coatings commonly used as protection against rain water absorption on vertical masonry and concrete surfaces above grade:

- organic water-thinned emulsions: styrene butadiene, vinyl, acrylic, alkyd, and multicolored lacquers.
- cementitious: thin coatings generally referred to as cement paints, white or pigmented stucco, shotcrete.
- organic solvent-thinned special purpose, such as chlorinated rubber, epoxy.

Coatings for below grade masonry and inferior quality concrete: commonly cementitious; may contain integral water-repellent or expansive metallic admixtures.

COATINGS: water vapor impermeable

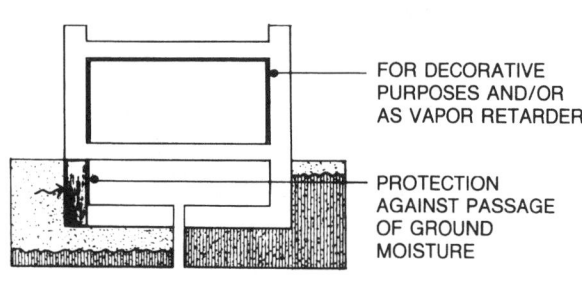

FOR DECORATIVE PURPOSES AND/OR AS VAPOR RETARDER

PROTECTION AGAINST PASSAGE OF GROUND MOISTURE

Vapor impermeable coatings when used on exterior or cold side of walls above grade may fail due to moisture trapped inside the wall; if thus used the warm side of the wall should have a vapor retarder less permeable than the exterior coating.

Coatings commonly used on masonry and concrete walls below grade where exposed to occasional water build-up: bituminous emulsions, cutbacks; roofing cement; hot-applied asphalt

- roofing cement and hot-applied asphalt is also used in built-up membranes.

B2.2 WATERPROOFING/DAMPROOFING

PROTECTIVE BARRIERS

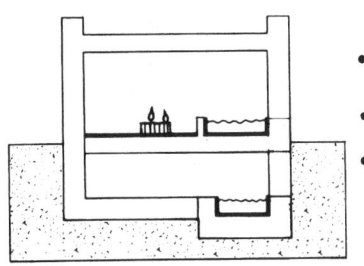

- TO PROTECT AGAINST SPILLED CHEMICALS
- TO CONTAIN CHEMICAL SOLUTIONS
- TO PROTECT CONTAINED FLUIDS AGAINST CONTAMINATION

Protective barriers are coatings or single-ply elastomeric membranes specifically selected to protect the substrate from degradation by chemicals; or conversely, the contained solutions from contamination by the substrate.

- coatings commonly used as barriers include: asphalt, coal tar, chlorinated rubber, epoxy resins, polyester resins, urethane resins, polyvinyl butyral.
- single-ply membranes include: polychloroprene - also referred to as neoprene, plasticized polyvinyl chloride (PVC).
- for additional information on single-ply elastomeric membrane materials refer to ROOFING.

MEMBRANES: protected/concealed

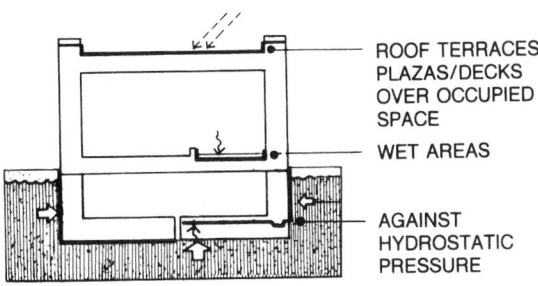

ROOF TERRACES/ PLAZAS/DECKS OVER OCCUPIED SPACE

WET AREAS

AGAINST HYDROSTATIC PRESSURE

Membranes are used to prevent passage of water, whether under hydrostatic pressure or not, and of water vapor.

- Membranes of materials which are easily damaged - such as liquid applied or single-ply elastomerics, built-up bituminous - must be protected from damage during and after installation.
- Cementitious membranes being resistant to traffic, abrasion, and erosion may be used over structural slabs without the need for protection.

Membranes may be installed on the positive or negative side:

- The positive side is the side in direct contact with the hydrostatic pressure; generally all liquid applied, single-ply, and built-up membranes are installed on the positive side.
- The negative side is the side opposite to the expected hydrostatic pressure; generally cementitious membranes may be installed on the negative, as well as on the positive side.

MEMBRANES: exposed

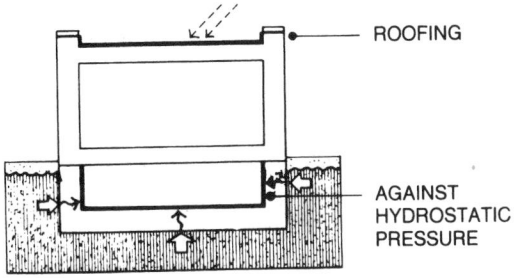

ROOFING

AGAINST HYDROSTATIC PRESSURE

Most membranes require a stable, rigid, and level substrate for their application:

- Generally a subslab is used when the membrane is below the structural slab; or when placed on the structural slab, a protective cover, such as another concrete slab, is provided.
- Bentonite clay panels for waterproofing slabs on grade may be placed on level and dense compacted fill when dry, and the concrete placed over them.

VAPOR RETARDERS

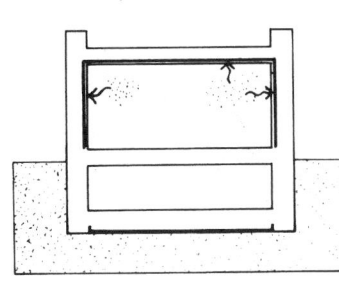

- WARM SIDE OF WALLS AND ROOF TO MINIMIZE VAPOR MIGRATION TO EXTERIOR
- UNDER SLABS ON GRADE TO MINIMIZE VAPOR MIGRATION INTO SPACE

Vapor retarders are used to minimize the migration of water vapor by diffusion.

- Impermeable coatings on the warm side or exterior walls when used for decorative purposes may also function as vapor retarders.
- Plastic sheets, either single thickness of laminated to kraft paper for added resistance to puncturing, or single-ply of 55 lbs. asphalt rollroofing are commonly used under slabs on grade to minimize passage of moisture and water vapor from the ground to the space above.

B2.2 WATERPROOFING/DAMPROOFING

MEMBRANES

● denotes common usage
○ denotes possible or limited usage

B2.2 WATERPROOFING/DAMPROOFING

USE	SINGLE PLY ELASTOMERIC BUTYL	ETHYLENE PROPYLENE (EPOM)	POLYCHLOROPRENE (NEOPRENE)	POLYVINYL CHLORIDE (PVC)	CHLORINATED POLYETHYLENE (CPE)	MODIFIED BITUMEN	CEMENTITIOUS MEMBRANE	BENTONITE CLAY PANELS	BUILT-UP BITUMINOUS HOT APPLIED ASPHALT & FELTS	COAL TAR PITCH & FELTS	COLD APPLIED [4] FIBRATED & GLASS FABRIC	FIBRATED & COTTON FABRIC [4]	CUTBACKS & COATED FELTS	EMULSION & COATED FELTS
WATER / BACKFILL / MEMBRANE / SUBSTRATE	●	●	●	●	●	●	●	●	●	●	●	●	●	
TRAFFIC WATER OPTIONAL COATING / PROTECTIVE COVER ① / MEMBRANE / SUBSTRATE	●	●	●	●	●	●	①		●	●	●	●	●	●
TRAFFIC / OPTIONAL COATING / SUBSTRATE / MEMBRANE / WATER OR VAPOR — FILL							●							
MEMBRANE / SUBSTRATE / STRUCTURAL SLAB / MEMBRANE / FILL (BACKFILL, GROUND WATER)							●							
BACK FILL / GROUND WATER ② MEMBRANE / SUBSTRATE / PROTECTIVE COVER OR STRUCTURAL / MEMBRANE / SUBSTRATE	●	●	●	●	●	●	①		●	●	●	●	●	④
BACK FILL / GROUND WATER / MEMBRANE / SUBSTRATE / DRAINAGE — FILL							①							
BACKFILL / GROUND MOISTURE / DRAINAGE — COATING OR MEMBRANE / OPTIMAL PARGING SUBSTRATE / OPTIMAL MEMBRANE / FILL	○	○	○	○	○	○	①		○	○	●	●	●	④

LIQUID APPLIED ELASTOMERIC				VAPOR RETARDERS [3]			SUBSTRATE			
POLYCHLOROPRENE	EPOXY	POLYESTER, glass reinforced	URETHANE	POLYETHYLENE, reinforced [1]	POLYETHYLENE/vinyl	POLL ROOFING [2]	MASONRY: clay; concrete	CONCRETE	CEMENT PLASTER	COMPACTED FILL
●	●	●	●					●	②	
●	●	●	●					●		
				③	③	③				●
										●
●	●	●	●				●	●	⑤	
								●		
○	○	○	○	⑦	⑦	⑦	●	●	⑤	

REMARKS

- For additional information on single-ply, built-up and liquid-applied membranes refer to ROOFING.
- [1] Polyethylene laminated to kraft paper for added resistance to tearing.
- [2] Generally single-ply with joints lapped and sealed.
- [3] Horizontal surfaces on or below grade only.
- [4] Fibrated bitumen generally referred to as roofing cement.

① cementitious membranes may be used on both positive/negative side without need of protective cover.

② Optional as parge coat.

③ For protection against water vapor only.

- When backfill is placed over membranes, protection for membrane is recommended. Membranes under cast concrete slabs are subject to damage during placing of the protective slab.
- Protective cover over membranes should preferably be readily removable for repair of damage sustained by the membrane during and after the installation of the protective cover.
- Neoprene, ehtylene propylene, polyvinyl chloride membranes may be used in locations partially or fully exposed to weathering.
- Bentonite clay may react with chemicals dissolved in ground water; consult manufacturers' literature.
- Membranes under concrete slabs on grade generally are used as vapor retarders.
- Optional coatings generally used when protective slabs are subject to heavy traffic, deicing salts, oil spillage.

④ May reemulsify in contact with water if not fully cured.

⑤ Coatings/parge coats on vertical surfaces.

⑥ Single-ply; built-up; liquid-applied may be used to contain liquids where not exposed to wear and/or abrasion. Exposure to sunlight may be a limiting factor for some.

⑦ Under slab on grade.

⑧ Coatings generally bituminous.

- Continuous membranes used when walls and floor slabs are below ground water table, or subject to frequent hydrostatic pressure.
- Bentonite panels may be placed over level, well-compacted fill and should be protected using polyethylene film from premature release by wet concrete placed over them.
- Footing drains recommended: when ground water level may rise above top of floor slab, or when subject to hydrostatic pressure after heavy rain.
- Bituminous coatings over masonry or rough concrete surfaces generally require a cement plaster parge coat; bituminous coatings not recommended when possibility of hydrostatic pressure exists; glass fiber mesh reinforced heavy coatings generally limited to three feet of water head, or less.

B2.2 WATERPROOFING/DAMPROOFING

WATERPROOFING/DAMPROOFING:

Components of an enclosure when exposed to moisture may be:

■ **Water resistant:** passage of water may occur through capillaries only, but not in significant amounts.

■ **Water repellent:** passage of water is prevented through capillary attraction alone but may occur when water is under pressure; either kinetic such as wind driven rain or hydrostatic.

■ **Waterproof:** when no openings exist to permit passage of water, whether under pressure or not.

Means of controlling passage of water are commonly classified as:

■ **Dampproofing:** means of protection against the passage of water present in small quantities and not under pressure; commonly against penetration through capillary attraction only.

■ **Waterproofing:** means of protection against the passage of water present in large quantities and under pressure.

WATERPROOFING

■ **Waterproofing is used to prevent water penetration under hydrostatic pressure:**

• to the surfaces of components, such as structural steel.
• into and through assemblies such as exterior masonry walls, foundation walls, roofs.

Waterproofing of horizontal assemblies exposed to the weather - such as roof terraces, plazas over below grade spaces, or parking decks - is essentially roofing with a protective traffic-bearing cover, and as such subject to most of the same in-service conditions as roofing:

■ **Differential thermal movements occur between the substrate and the protective cover, which the waterproofing separates; e.g.:**

• the substrate over a heated space may be in thermal equilibrium, or may be expanding, while the protective cover contracts under low outside temperatures.
• the cover may be expanding due to heat gain through solar radiation while the substrate is stable.
• water may penetrate through faulty

joints or cracks into the protective cover, freeze and cause localized expansion while the cover is contracting; this may rupture the waterproofing membrane if it is bonded to the underside of the cover.

■ **To allow thermal movements to take place, one of two methods can be adopted:**

• the waterproofing should be kept free of both the substrate and the protective cover.
• the waterproofing should have sufficient resistance to such movements.

■ **Structural movements in the substrate and in the protective cover should also be considered during the selection of the type of waterproofing;** some types will resist them better than others; these movements include:

• deflection under moving loads.
• variations in live loads.
• deformations under dead loads.

■ **Solar radiation and wind uplift generally do not affect waterproofing, unless it is exposed in some parts of the assembly it protects.**

■ **Waterproofing of interior horizontal surfaces** - e.g., floors in shower rooms or interior swimming pools - generally is not subject to wide temperature fluctuations; in these cases in addition to protection from damage during construction, structural movements only need be considered.

■ **Waterproofing is generally an elastomeric or built-up membrane that is easily punctured, ruptured, or abraded;** it must be protected from damage during and after installation, especially when used in horizontal assemblies subject to any type of traffic:

• walkways may be used for light localized foot traffic.
• the entire membrane may be covered with dense, wear resistant materials such as ceramic tile, stone, paving blocks, or cast concrete slabs, a protective cover.

Waterproofing must of necessity be installed before the protective cover is placed and therefore is subject to accidental damage during its placement.

■ Even when properly constructed,

the protective cover is unlikely to prevent water penetration through it at all times: positive drainage at the waterproofing level should be provided.

■ Since damage to the waterproofing may occur during the installation of the protective covering, **the use of small, easily removed units rather than large monolithic sections offers the advantage of making subsequent repairs easier.**

■ Protective covers made of small units with open joints which allow water to drain to the waterproofing level may offer advantages provided that: joints can be kept open to prevent the intrusion of rigid particles which may cause spalling or cracks in the units when they expand.

Waterproofing under slabs on grade may be:

■ **A single sheet of impermeable plastic material,** essentially a vapor barrier rather than true waterproofing:

• water impermeable membrane under a concrete slab on grade does affect curing conditions of concrete and often leads to extensive shrinkage cracking of the concrete surface.
• advisability of using water impermeable membranes or water vapor barriers directly under concrete slabs should be considered when the concrete is to remain exposed in the finished work.
• a sand bed between an impervious membrane and a concrete slab may improve curing conditions and minimize shrinkage cracking.

Waterproofing for vertical surfaces under hydrostatic pressure is essentially the same as for horizontal surfaces subject to the same in-service conditions; the horizontal membrane is extended up to protect the walls; the requirements for providing for structural movements or protecting the membrane during construction remain similar.
Waterproofing for walls of shallow foundations subject to occasional moderate hydrostatic pressures of short duration may consist of an impermeable coating over a level dense substrate, such as concrete or a parge coat of Portland cement plaster over concrete masonry.

B2.2 WATERPROOFING/DAMPROOFING

selection checklist

Considerations during the selection of specific means for control of the passage of water through a component of the enclosure may include:

WATER REPELLENTS: either integral or surface sealing are generally used for dampproofing only:

* above grade water repellents such as silicones have to be periodically reapplied to maintain their effectiveness.
* they will not stop moisture penetration through cracks and faulty or incompletely sealed joints.
* solvent based surface water repellents are generally more effective than water based ones; solids content should be at least five percent, with over seven percent recommended.
* use of integral or surface water repellents may prevent proper bonding of subsequent cementitious coatings, such as cement paint, stucco.
* surface water repellents may impede removal of efflorescence from exterior surfaces of masonry walls, or may result in buildup of crystalline salts near the exterior surface and the expansion of such salts may then cause spalling.
* for additional information on causes and effects of effloerescence, refer to STONE/MASONRY.

COATINGS

Considerations during selection of coatings to serve either as dampproofing or as protection barriers should generally include:

* water may be absorbed by a component of the enclosure through cracks, faulty joints, flashings, capillary attraction from the ground; when coatings do not allow such water to evaporate, the trapped moisture may cause peeling and cracking of the coating; or such moisture may freeze within the component with resultant spalling and cracking of the component itself.
* moisture within the substrate such as water vapor diffusing through it may produce voids in a coating and lead to blistering and peeling after the coating has cured: when moisture is present or anticipated at the surface during application, the ability of the material to displace such mois-

ture for proper bonding should be investigated.

■ **Coatings for dampproofing** should be almost impermeable to water but should allow water vapor to diffuse through them:

* coatings for dampproofing are not effective when subjected to even intermittent hydrostatic pressure; generally they will not bridge cracks; therefore, if cracks are present or may develop later, they should not be used. Some single component polymer coatings such as chlorinated rubber, if applied in sufficient thickness, may be capable of maintaining their integrity if minor cracks develop in the substrate after their application.
* when to be applied to surfaces with high initial moisture content, or when moisture is liable to enter the reverse side of a coated component, permeable coatings such as cement or latex base should be used. In some cases specially formulated two component polymer coatings could be used when they are capable of resisting blistering under pressure of moisture and have a sufficiently high water vapor transmission rate.
* when coated surfaces may be exposed to high humidity or moisture on the coated side, less permeable coatings should be considered.

■ **Coatings,** some of which may also be considered as liquid applied membranes, when used as protective barriers, should:

* resist effects of chemicals to be contained without dissolving, cracking, or losing adhesion with the substrate.
* have adequate resistance to wear and abrasion during and after installation.
* have adequate bond strength to the substrate: at least equal to the tensile strength of the substrate at their interface. Stresses in coatings may result from shrinkage during curing or from differential volume change in the substrate and the coating.

Protective coatings are commonly used over dense concrete surfaces against:

■ **Mild chemical solutions;** deicing salts; freeze-thaw cycles; staining; or to prevent contamination of high purity water by the substrate:

* some of the more commonly used coatings are: polyvinyl butyral, acrylics, epoxy, polyurethane, chlorinated rubber, asphalt, coal tar.

■ **Abrasion and concurrent intermittent exposure to dilute acids** in chemical, dairy and food processing plants:

* some of the coatings used include: sand-filled epoxy, sand-filled polyester, sand-filled polyurethanes, bituminous formulations.

Considerations during the selection of a specific protective coating:

* **Asphalt:** provides good protection against acids and salt solutions, but resistance to solvents is poor. Asphalt is hot applied; for cold application, solvent base type, or cutback, and emulsions are available.
* **Bituminous emulsions:** are made using either asphalt or coal-tar base; they are more water vapor permeable than cutbacks; excellent atmospheric exposure characteristics.
* **Coal-tar:** provides excellent resistance to water; moderate to good resistance to acids and alkalis. Coal-tar is applied hot; for cold application, solvent base type, or cutback, and emulsion type are available. Some coal-tar solvents may impart objectionable taste and odor to potable water supplies.
* **Chlorinated rubber:** has excellent resistance to alkalies, moisture, and abrasion; fair resistance to a wide range of common acids and aliphatic hydrocarbons but not to aromatic hydrocarbons, fatty acids, and animal and vegetable oils. When exposed to direct sunlight, only the pigmented types with ultraviolet absorbers should be selected.
* **Epoxy:** commonly two part with a curing agent or hardener, provides excellent resistance to caustics, acids, and solvents. Some formulations will adhere well to damp concrete surfaces.
* Sand filled epoxy resin is commonly used as trowelled-on floor

B2.2 WATERPROOFING/DAMPPROOFING

WATERPROOF/DAMPROOFING: (checklist cont.)

topping where such floors are subject to intermittent exposure to chemicals; may be sealed with low viscosity epoxy resin.

- Glass fiber reinforced epoxy resin toppings are used to protect concrete floors from acids and other aggressive chemicals which might cause rapid disintegration of the concrete.

- Polyester resins are used in more severe chemical environments; may be sand-filled or glass fiber or glass flake reinforced. Water will inhibit hardening of polyesters and a primer should be used first.

- **Polyvinyl butyral:** has excellent resistance to weathering.

- **Urethane:** provides protection against deicing salts and other chemicals; impact and abrasion; will prevent fungus growth. May be either two-component chemical cure or single-component moisture cure type. Should not be applied to damp surfaces.

■ When selecting a protective coating its **resistance to specific in-service conditions and time-proven performance record should be investigated,** since a wide variety of formulations for each type are commonly available.

MEMBRANES

Membrane waterproofing is the most reliable means to prevent water under hydrostatic pressure from entering an underground space, but the membrane must be continuous, and all penetrations such as drains, joints must be properly sealed:

- good adhesion to the substrate is essential to prevent water migration if small cracks develop in the membrane.

- fully bonded membranes should be considered if possibility of cracking or tearing in the membrane exists; leaks in unbonded membranes are difficult to trace.

- protective coverings over membranes should not be sealed in a way which would trap moisture between two impermeable barriers, as blistering and/or delamination of the top barrier may result. An exception would be if an unbonded and vented membrane is used over another membrane.

- liquid applied membranes will not cover, hide or level substrate surface irregularities.

- joint design should include means of transferring loads from one side to the other without differential movements which could damage the membrane.

- insulating materials, if used over membranes, should not be left exposed, should be chemically compatible with the membrane system, and should be dimensionally stable.

- continuous support for membranes must be provided.

- when exposed, membranes must be resistant to ultraviolet radiation and ozone.

Considerations during selection of specific membranes may include:

■ **Single-ply membranes** are commonly precured elastomeric sheets which offer the advantage of uniform thickness and relative freedom of imperfections over liquid applied elastomeric materials. Sheet applied materials generally available are:

- **Butyl**, or polyisobutylene, best suited for below grade concealed application. Resistance to ozone, soil acids, fungi, and ultraviolet radiation generally good.

- **Polychloroprene,** also referred to as neoprene, has good resistance to ultraviolet radiation, ozone, acids, and to intermittent exposure to oil. May be used exposed.

- **EPDM,** or ethylene propylene, and **PVC,** or polyvinyl chloride, provide good resistance to weathering; remain elastic over a wide range of temperatures. Resistance to specific in-service conditions should be ascertained prior to making selection.

- for additional information on elastomeric and modified bitumen single-ply membranes refer to ROOFING.

■ **Built-up bituminous may be either hot or cold applied:**

- hot-applied membranes employ either asphalt or coal-tar pitch with either felt or fabric reinforcement.

- fabric reinforcement has two advantages over felt: it is stronger and more pliable; it can absorb movement and vibration better.

- glass fiber fabric is more resistant to deterioration than cotton or jute fabric.

- felt reinforcement is commonly less permeable than fabric containing cotton or glass fibers.

- cold-applied membranes employ bituminous cutbacks or emulsions and either coated felt or fabric reinforcement.

- for additional information on felts and bituminous materials refer to ROOFING.

■ **Liquid-applied membranes** are commonly elastomeric materials, such as polychloroprene, urethane, which cure in place to form a continuous film; may be applied over irregular complex substrates.

■ **Cementitious membranes** are proprietary formulations using waterproof cement:

- may be installed on either the positive side, the side in direct contact with water under hydrostatic pressure, or on the opposite, the negative, side.

- may be used as wearing surface without the need for protective covering.

- may receive decorative coatings when exposed in the finished work.

- should be nonshrinking and high bonding; may offer resistance to salt water, sulfates, certain chemicals.

- should not require moist curing to be effective.

- generally have excellent weathering characteristics.

- Cementitious membranes may be installed over wet, dry, smooth, rough and/or uneven surfaces.

- thermal movement and proper joint design are the main considerations.

- unstable, nonrigid substrates subject to deflection and/or constant movement are not suitable for cementitious membranes.

- resistance to and suitability for specific in-service conditions of a particular formulation should be ascertained prior to making final selection.

■ **Bentonite clay panels** depend on kraft paper to hold the clay in place: after installation the biodegradable paper is dissolved by ground moisture and the clay expands upon absorption of water to form an impermeable liquid membrane, or gel, over the surface to be waterproofed.

VAPOR RETARDERS

■ **Water vapor retarders—or vapor barriers—are vapor impermeable materials** or coatings installed to resist the diffusion of vapor through roof or wall assemblies:
- vapor retarders should be located as close to the warm side of an assembly as practicable.
- the temperature at the vapor retarder should always be above the dew point of the actual air/vapor mixture under the most severe in-service conditions: that to prevent the possibility of condensation occuring on the vapor retarder itself.

■ **Vapor retarders are generally thin materials,** easily damaged during or after installation; the most impermeable, most durable barrier may be rendered virtually ineffective by the presence of accidental or intentional but improperly sealed openings.

VAPOR RETARDERS & AIR LEAKAGE

Properly installed vapor retarders will also prevent air leakage. The prevention of vapor migration through air leakage is considered **more critical** than preventing vapor migration by diffusion. Some of the reasons are outlined below:

■ **the amount of water vapor carried by air leakage** through a crack in the interior component of an assembly – with a vapor retarder which is likely to crack along with the component - may be far greater than the amount that would migrate through the assembly by diffusion if there were no vapor retarder.

■ **For example,** permeance of 100 s.f. of a 6 inch thick concrete wall with paper backed aluminum foil cemented to one face is 250 times less than the permeance of a crack 12 inches long by 1/32 inches wide.

■ Vapor carried by air leakage may be **concentrated** in larger amounts **within a relatively small area,** and may:
- condense within the assembly if free and direct passage to the outdoors is impeded or blocked, in amounts too large to be readily reevaporated.
- freeze and expand;
- accumulate as frost during periods of low outdoor temperatures, then thaw out and run to the outside when temperatures rise briefly, and freeze again on the outside forming icicles.
- accumulate as frost and, upon thawing, run to the inside, damaging facings or furnishings;
- penetrate into roof insulation, be

trapped there between the vapor barrier and roofing membrane and reevaporate when heated by solar radiation.

Providing a continuous airtight vapor retarder will minimize condensation problems:

■ Air tightness of the exterior envelope will also significantly **contribute to energy conservation** by minimizing heat losses or gains through infiltration/exfiltration.

■ **But airtightness will also:**
- reduce the rate of water vapor removal from the interior space thus increasing humidity;
- increase concentration of pollutants in the air of interior spaces: excessive levels of pollutants generated by cooking, smoking, propellants in spray products, cleaning products, including pollutants of plastic foams in furniture or insulation have been found in tightly sealed residential buildings.

WHEN TO USE VAPOR RETARDERS

The analysis of an assembly for the potential problem of water vapor condensation should consider:

■ **Temperature and partial water vapor pressure differentials,** or relative humidities:
- between the outdoor and indoor spaces
- between indoor spaces with different controlled environments.

■ **Temperature gradients** through the assembly, and the dew point location under design temperatures and relative humidities.

■ **Resistances to vapor** transmission of the components of the assembly.

Design outdoor temperatures and relative humidities are fixed values, and the only controllable variables are:

■ **Indoor temperatures,** which generally vary over a narrow comfort zone;

■ **Indoor relative humidities,** which will vary depending on the occupancy:
- from under 20 percent for dry occupancies, such as warehouses and factories, to
- over 60 percent for high moisture occupancies, such as commercial kitchens, public bath and shower rooms, or where high humidity conditions are artificially maintained.

■ **Higher indoor relative humidity** will shift the dew point closer to the warm side of the assembly, increasing

the amount of condensation to be expected within it.

When the dew point occurs within an assembly, consideration should be given to several possible methods for remedying the situation:

■ **Adding a vapor retarder** on the warm side: this reduces the amount of water vapor migration by diffusion, but changes neither the temperature gradients nor the dew point location.

■ **Increasing permeance of materials** between the dew point and the outside in order to maintain saturation vapor pressure above vapor pressure for continuity of flow past the dew point. This may require changing some components of the assembly, or the assembly itself.

■ **Changing temperature gradients** to shift the dew point to the cold-side of the assembly:
- relocating insulation from the inside face to the outside face may be sufficient to accomplish this.
- adding insulation on the outside; but insulation over the exterior face may require a protective facing, which then should be highly permeable, yet resistant to rain penetration.

MATERIALS

To classify materials as to their resistance to vapor migration, a standard has been established by which materials can be rated in terms of their permeability or permeance to vapor.

■ **Permeance coefficients** have been derived to indicate the rate at which vapor will diffuse through a given material; such coefficients are expressed in units called "perms":
- one perm represents a transfer rate of one grain per square foot per hour under a vapor pressure difference of one inch of mercury.
- the lower the perm rating, the greater the resistance of a given material to vapor migration.

■ **The considerations in selecting a material to serve as vapor retarder should include** not only its permeance, but, depending on in-service or installation conditions, also:
- fire resistance
- tensile strength, pliability
- durability under bacterial action
- freezing, thawing, wetting, drying.

B2.2 WATERPROOFING/DAMPROOFING

VAPOR RETARDERS: types, permeance

Additional information may be found in ASTM C755,
Standard Recommended Practice for selection of vapor
barriers for thermal insulations.

TYPE	PERMEANCE: range in perms
FOILS AND FILMS	
ALUMINUM FOIL, 1 mil thick	0.0
ALUMINUM FOIL, applied to gypsum lath	0.085 to 0.385
POLYETHYLENE, 4 mils thick	0.08
POLYETHYLENE, 6 mils thick	0.06
POLYVINYLCHLORIDE, unplasticized, 2 mils thick	0.68
POLYVINYLCHLORIDE, plasticized, 4 mils thick	0.8 to 1.4
BUILDING PAPERS, FELTS	
DUPLEX SHEET, asphalt impregnated, aluminum foil one side	0.176
ROLL ROOFING, asphalt saturated, coated	0.213 to 0.77
VAPOR-BARRIER PAPER, asphalt saturated, coated	0.36 to 0.64
ROOFING FELT, 15 lb, asphalt saturated	5.6
ROOFING FELT, 15 lb, coal tar pitch saturated	18.2
SHEATHING PAPER, asphalt saturated, uncoated	20.2
KRAFT PAPER, single, double infused	42.0
LIQUID APPLIED, paint, two coats	
ASPHALTIC, on plywood	0.4
ALUMINUM IN VARNISH, on wood	0.3 to 0.5
ENAMEL, on smooth gypsum plaster	0.5 to 1.5
PRIMER plus ONE COAT FLAT OIL BASE, on plaster	1.6 to 3.0
LIQUID APPLIED, paint, three coats	
WHITE LEAD AND OIL, exterior, on wood siding	0.3 to 1.0
STYRENE-BUTADIENE LATEX COATING, at 2 ounces per square foot	11.0
POLYVINYL ACETATE LATEX COATING, at 4 ounces per square foot	5.5
ASPHALT CUTBACK, 1/16 inch thickness, dry	0.14
ASPHALT CUTBACK, 3/16 inch thickness, dry	0.0
ASPHALT, hot applied at 2 ounces per square foot	0.5
ASPHALT, hot applied at 3.5 ounces per square foot	0.1
BUILDING MATERIALS	
SHEATHING, 25/32 inch thick, uncoated	23.26 to 62.50
SHEATHING, 25/32 inch thick, coated	0.44 to 10.92
PLYWOOD, douglas fir, exterior glue, 1/4 inch thick	0.7
HARDBOARD, tempered, 1/8 inch thick	4.78 to 5.0

B2.2 WATERPROOFING/DAMPROOFING

REFERENCES: Waterproofing/Damproofing

STANDARDS

ASTM Specifications for:

C836
- High Solids Content, Cold Liqued-Applied Elastomeric Waterproofing Membrane for Use With Separate Wearing Course.

D41
- Asphalt Primer Used in Roofing and Waterproofing.

D43
- Creosote Primer Used in Roofing, Dampproofing and Waterproofing.

D173
- Bitumen-Saturated Cotton Fabrics Used in Roofing and Waterproofing.

D226
- Asphalt-Saturated Organic Felt Used in Roofing and Waterproofing.

D227
- Coal-Tar Saturated Organic Felt Used in Roofing and Waterproofing.

D250
- Asphalt-Saturated Asbestos Felt Used in Roofing and Waterproofing.

D449
- Asphalt Used in Dampproofing and Waterproofing.

D450
- Coal-Tar Pitch for Roofing, Dampproofing and Waterproofing.

D491
- Asphalt Mastic Used in Waterproofing.

D1327
- Bitumen-Saturated Woven Burlap Fabrics Used in Roofing and Waterproofing.

D1668
- Glass Fabrics (Woven and Treated) for Waterproofing and Roofing.

D3468
- Liquid-Applied Neoprene and Chlorosulfonated Polyethylene Used in Roofing andWaterproofing.

SELECTED REFERENCES

ACI — **American Concrete Institute**
- Guide to the Use of Waterproofing, Dampproofing, Protective, and Decorative Barrier Systems for Concrete.

BIA — **Brick Institute of America**

7
- Dampproofing & Waterproofing Masonry Walls.

CSI — **Construction Specifications Institute**

07175
- Specifying: Water-Repellent Coatings

07M115
- Bentonite Clay Waterproofing

NCMA — **Natinal Concrete Masonry Association**

10A
- Decorative Waterproofing of Concrete Masonry Walls

55
- Waterproofing Coatings for Masonry

121
- Waterproofing Concrete Masonry Basements and Earth-Sheltered Structures.

NRCA — **National Roofing Contractors Association**
- The NRCA Roofing & Waterproofing Manual

NRCC — **National Research Council of Canada**

CBD 13
- House Basements

38
- Bituminous Materials

74
- Properties of Bituminous Mem branes

150
- Protected-Membrane Roofs

B2.2 WATERPROOFING/DAMPROOFING

B2. 3 EXTERIOR WALLS

EXTERIOR WALLS: introduction

FUNCTION

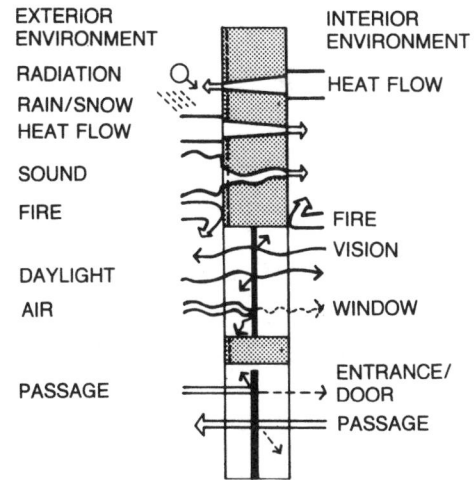

Walls separate the external environment from the internal one while functioning as a barrier and/or selective filter to:
* Control heat flow between the two environments by reducing the rate of its transfer through them.
* Minimize warer vapor migration from one environment to another while preventing or minimizing its condensation on or within them.
* Admit daylight to the interior environment or control its transmittance.
* Allow for controlled movement of air from one environment to another, minimize air infiltration/exfiltration through them.
* Minimize the transmission of sound through them.
* Control passage between different environments or enclosed spaces.
* Resist penetration of rain, effects of external or internal fire.

While remaining in place and intact during the entire intended life span of the enclosure.

STABILITY

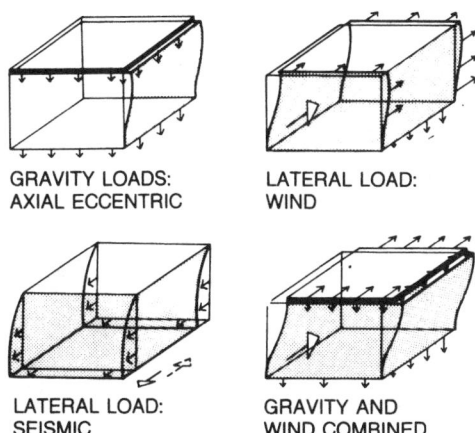

GRAVITY LOADS:
AXIAL ECCENTRIC

LATERAL LOAD:
WIND

LATERAL LOAD:
SEISMIC

GRAVITY AND
WIND COMBINED

Exterior walls will always be subject to lateral wind forces and may also be subject to seismic laods.
■ **Wind pressure may be:**
* positive, pushing the wall against the supporting horizontal frame and causing it to deform, or deflect, inward.
* negative, pulling the wall away from supporting frame and causing outward deflection.
* parallel to wall, which may result in shearing stresses in the wall due to interaction between it and the supporting horizontal frame.

■ **Walls which do not carry superimposed loads of the structural frame must only resist wind loads between horizontal and/or vertical supports, and transmit such loads to the supporting frame.**
■ **Walls which carry superimposed loas of the structural frame may also be subject to eccentric loading which will result in further deflection in them, either adding to that induced by lateral loas or counteracting it.**
■ **Effect of seismic forces on walls and provisions required to resist them are similar to those of wind pressures.**

MOVEMENT

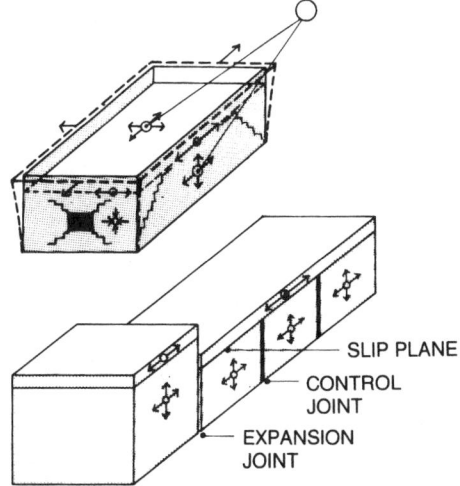

SLIP PLANE

CONTROL
JOINT

EXPANSION
JOINT

Walls will be affected by thermal expansion/contraction with-in them as well as that in the structural frame to which they are connected:
* solar radiation striking a wall or parts thereof will cause expansion, and such expansion may be differential.
* juncture of walls with different exposures to solar radiation should provide for differential movement between the walls joined.
* long walls may require control joints to limit extent of cumulative movement thus magnitude of resultant stresses; expansion joints, when the structural frame itself is divided into sections to limit expansion/contraction.
* differential movement may occur between walls and horizontal assemblies: flat roof assembly may be expanding while walls which are continuously connected to it or are supporting it may be expanding at different rates, or may be contracting depending on their exposure.

Monolithic walls of concrete and walls of masonry units may also be subject to shrinkage induced stresses in addition to thermal stresses, and may require control joints to minimize the possibility of cracking.

B2.3 EXTERIOR WALLS

EXTERIOR WALLS: types

BEARING/NONBEARING: STACKED, MONOLITHIC

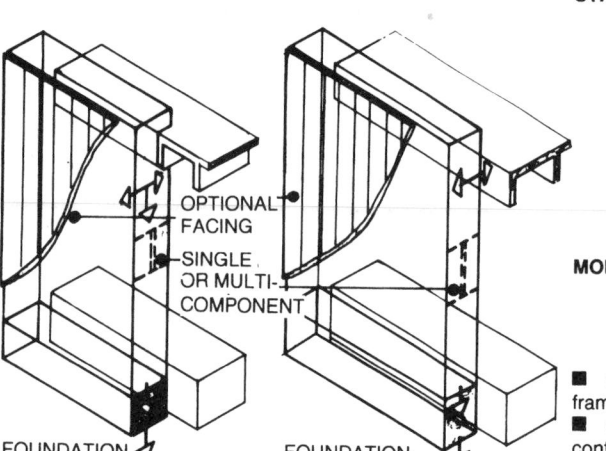

OPTIONAL FACING

SINGLE OR MULTI-COMPONENT

FOUNDATION

FOUNDATION

STACKED UNIT: • Single or double wythe of concrete block, generally with reinforcing in horizontal joints to control shrinkage induced cracking; bearing or nonbearing. Outer wythe may be left exposed, may receive applied coating, or may be faced. • Single wythe of reinforced brick masonry, generally nonbearing; or outer wythe of brick bonded to an inner wythe of concrete block or structural tile, bearing or nonbearing. • Cavity: two wythes separated by an airspace, generally two inches wide. Outer wythe usually of brick, inner wythe either brick or concrete block, bonded with metal ties.

MONOLITHIC: • Concrete, poured-in-place, reinforced as a minimum to control shrinkage stresses when nonbearing. • Concrete, site precast large panels, reinforced to control erection and shrinkage and/or superimposed load induced stresses.

■ **Compatible structural frame:** steel, concrete, light-gauge metal or wood framed.

■ **Careful consideration** must be given to location and spacing of joints to control temperature/moisture induced expansion and contraction.

BEARNG/NONBEARING: FRAMED

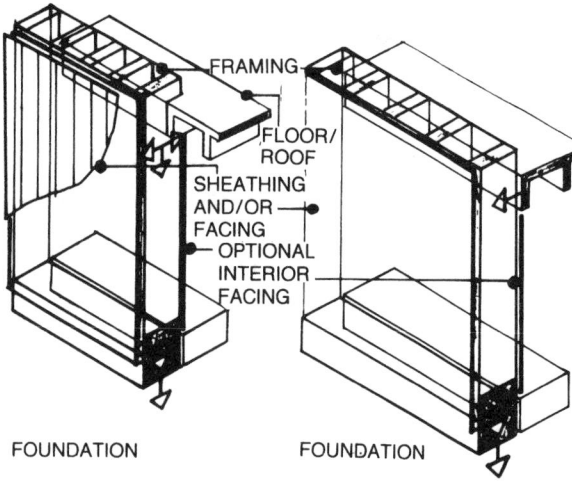

FRAMING

FLOOR/ROOF

SHEATHING AND/OR FACING

OPTIONAL INTERIOR FACING

FOUNDATION

FOUNDATION

Closely spaced light-gauge metal or wood studs, with exterior faces secured against lateral displacement in the plane of the wall by structural sheathing connected to them, or by diagonal braces when sheathing used in nonstructural. Generally capable of supporting superimposed loads, whether thus used or not.:

• posts of support concentrated loads of the structural frame may be incorporated into the framing.
• studs may be faced both sides with structral sheathing, generally plywood glued or glued and nailed to studs, to function as stressed skin panels.
• exterior generally faced over sheathing; some facings may act as combined facing and/or structural sheathing.
• interior generally faced, with the facing often also functioning as secondary bracing to the framing.
• insulation generally placed between studs, wiring and small diameter piping may be run through studs.

■ **Generally used with framed floor/roof assemblies only.**

CURTAIN: GRID

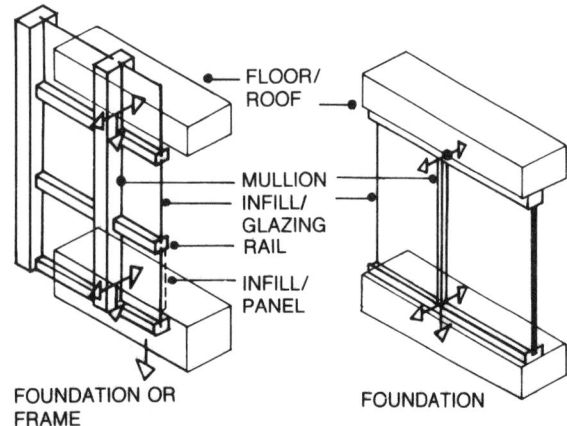

FLOOR/ROOF

MULLION

INFILL/GLAZING

RAIL

INFILL/PANEL

FOUNDATION OR FRAME

FOUNDATION

■ **Mullions spanning between floor, or floor and roof framing, supported and laterally braced by such framing.**

• Rails connected to, supported by, and laterally braced by mullions. • Mullions and rails generally of aluminum. Other corrosion resistant metals may also be used. • Subtypes: stick, unitized stick, unitized.

■ Infill panels, also referred to as glazing panels, generally with edges of the same thickness whether transparent or opaque, held in place by mullions and rails. Windows and/or doors may be used in lieu of panels. Infill panels may be:

• monolithic of single thickness; glazing of tinted, patterned, opaque glass; panels of cement bound mineral fiber, patterned, pigmented, coated. • monolithic, spaced: glazing of two or more lights of glass or of plastic, or combinations thereof, assembled into units with air spaces between them to reduce flow of heat through them. • composite, laminated: glazing of glass and plastic, panels of prefinished metal faces with cores of rigid insulation or rigid boards. • formed: panels of metal faces with formed edges, generally with insulation between them.

■ **Compatible structural frames:** steel, concrete, may be also used in conjuction with bearing walls.

B · BUILDING SHELL

B2.3 EXTERIOR WALLS

CURTAIN: WALL PANELS

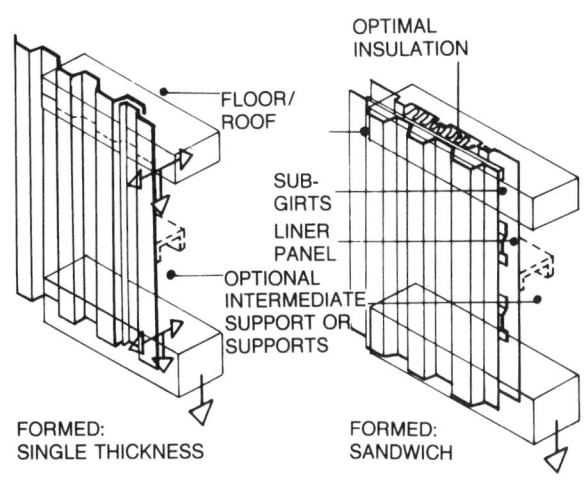

FORMED:
SINGLE THICKNESS

OPTIMAL
INSULATION

FLOOR/
ROOF

SUB-
GIRTS

LINER
PANEL

OPTIONAL
INTERMEDIATE
SUPPORT OR
SUPPORTS

FORMED:
SANDWICH

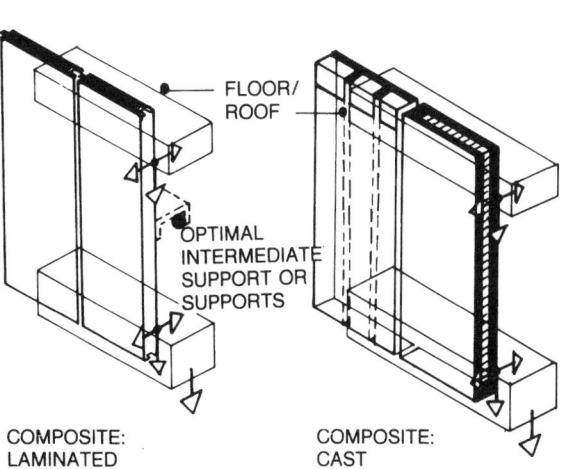

COMPOSITE:
LAMINATED

FLOOR/
ROOF

OPTIMAL
INTERMEDIATE
SUPPORT OR
SUPPORTS

COMPOSITE:
CAST

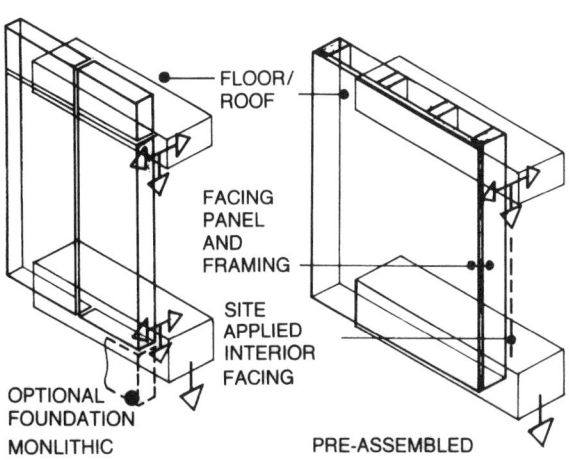

MONLITHIC

FLOOR/
ROOF

FACING
PANEL
AND
FRAMING

SITE
APPLIED
INTERIOR
FACING

OPTIONAL
FOUNDATION

PRE-ASSEMBLED

CURTAIN: wall panels
Panels, generally intended to function as the entire wall assembly; spanning between floor/roof assemblies; supported and laterally braced by such assemblies:
* intermediate supports between the floor/roof assemblies may be used to reduce the span of thin panels for more economical installation.

Length of panels may be limited by manufacturing processes and/or shipping constraints.

FORMED:
■ **Single thickness:** of metal, either nonferrous or anticorrosion treated ferrous, formed in various configurations to impart rigidity to the section; of plastic in various configurations; of cement-bound mineral fiber, generally corrugated.

■ **Sandwick:** single thickness metal panel outer face, connected to and supported through subgirts by a formed metal inner face, generally with insulation between the two faces.
* inner face attached to and supported by continuous structural steel framing • outer face connected to subgirts using exposed fasteners or concealed clips • generally site assembled.
* sandwich panels available as fire-resistant rated assemblies.

■ **Metal panels may be unfinished when nonferrous, surface treated or coated.**

■ **Compatible structural frame:** steel. May be installed on concrete frame if structural steel supports are added.

COMPOSITE:
■ **Composite, laminated:** Outer face of: metal either flat, formed, or stamped, generally with applied coating; flat cement-bound mineral fiber, either textured or coated.
* cores of rigid boards, rigid insulation, fiber or metal honeycomb. • interior face may be exposed core or similar to outer face. • secured to steel supports with concealed fasteners or formed metal clips.
* Compatible structural frame: steel. May be used with concrete frame if steel supports are added.

■ **Composite, cast:** of polymer or regular concrete with insulation between the outer and inner faces.
* outer face may be lightly to heavily textured or of exposed aggreaget • inner face generally smooth • secured to framing with structural steel clips • intermediate supports not commonly used.
* Compatible structural frame: steel, concrete.

MONOLITHIC:
Solid panels generally of regular concrete, with either normal or lightweight aggregate: • outer face either flat, textured, or molded/sculptured • inner face generally smooth
* secured to framing with structural steel clips • intermediate supports seldom used • when one or two stories in height may be supported directly on foundation and laterally braced by the structural frame.
* Compatible structural frame: steel, concrete.

PRE-ASSEMBLED:
Prefinished facing panels attached to preassembled framing and shipped to the site as ready-to-be-installed units.
* Compatible structural frame: steel, concrete.

B2.3 EXTERIOR WALLS

WALL FACINGS: types

PANELS

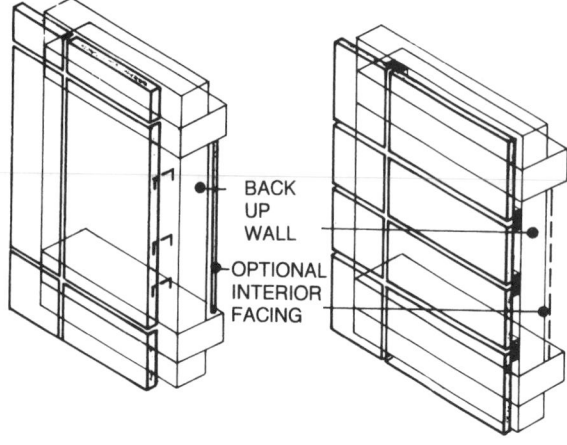

BACK UP WALL

OPTIONAL INTERIOR FACING

All walls, whether exterior or interior, consist of at least two elements: the outer faces and the core or body between them. The wall may be single component, such as the fabric of a tent, or it may be an assembly of multiple layers of different components, but the two basic elements will still be present:

■ The outer surfaces of a solid stone or brick wall may be finished for appearance or durability, but that will not constitute a facing. But when stone is combined with a concrete wall to protect and enhance it, it becomes a facing, whether dependent on the concrete for stability or not.

■ A sheet of metal such as aluminum will always constitute a facing, whether single thickness or laminated to a backing, formed or flat, when used as the outer surface of a solid core wall assembly or attached to open framing.

■ Facings then are surface components that are integrated, added, bonded, or attached to the core of a wall assembly, and specifically selected to provide:

- protection for the wall assembly from external or internal environmental factors.
- durability under in-service conditions.
- an aesthetically pleasing appearance.
- economy, by allowing for differentiation of the functional requirements of each component of the wall assembly.

FACING PANELS: applied over a back-up wall and secured to the wall and/or structural frame:

- back-up may be of stacked units, generally concrete block, or framed; facing panels are not commonly used over concrete walls, principally because of economic considerations.
- facing panels may require subframing to be installed over the back-up wall for proper attachment and/or alignment; light, thin panels may be attached to solid plumb surfaces using adhesives, such as high modulus silicone, but such methods of attachment is not common.

UNITS/STRIPS: ATTACHED

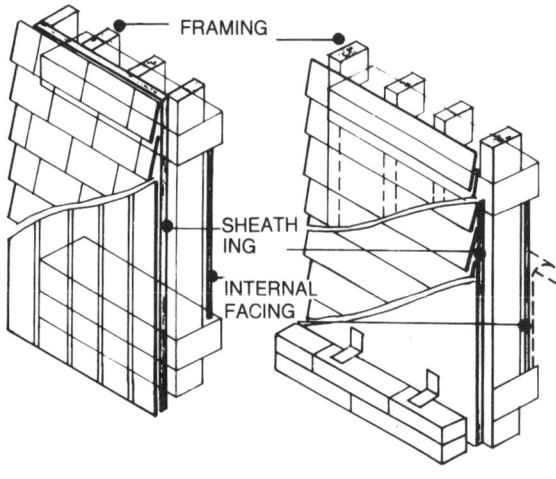

FRAMING

SHEATHING

INTERNAL FACING

UNITS: such as shingles, stone or brick masonry; and
STRIPS: such as vertical or horizontal siding; applied over a back-up wall:

- back-up generally framed with nailable sheathing, may also be stacked units, such as concrete block, but nailable furring strips have to be then installed over the wall.
- horizontal siding, when sufficiently rigid, may be attached directly to the studs of a framed wall, and sheathing omitted.

UNITS: such as ceramic tile, face brick, glazed tile; and
SURFACING: such as stucco, select aggregate, bonded to a back-up wall:

- back-up may be: stacked units, generally concrete block, monolithic concrete, either cast-in-place or precast, sheathed framing, or rigid insulation secured to back-up wall.
- stucco applied over framed back-up wall may be over sheathing, or the sheathing may be omitted.

INTERIOR face of a back-up wall may also require a facing, such as over the studs of a framed wall, or the use of an interior facing may be optional, such as over masonry or concrete:

- interior facings may be: panels, such as veneered plywood, fabric faced gypsumboard; strips, such as solid wood panelling; units, such as ceramic tile; surfacing, such as plaster; fabrics, either woven or felted; semirigid or flexible plastics.

UNITS/SURFACING: BONDED

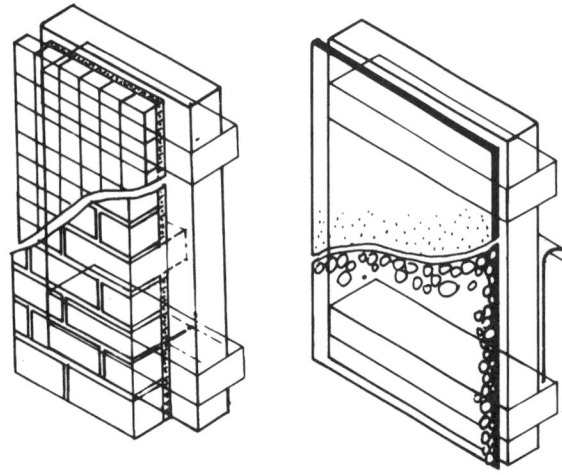

B2.3 EXTERIOR WALLS

WALLS: preliminary selection

PRINCIPAL CONSIDERATIONS during selection of an exterior wall assembly should generally include:

■ **Function of the building and requirements it imposes:**
* extensive areas of glazing required and their distribution over the plane of the walls may narrow the choice of wall type, or preclude the efficient use of a bearing walls.

■ **Form of the building, whether low- or high-rise, may influence the choice of a particular type:**
* bearing walls are generally only economical up to about ten stories in height.

■ **Structural frame, whether short or long span, and the spacing of horizontal framing members may affect the choice:**
* short spans and closely spaced horizontal framing members would allow for efficient use of bearing walls, as would uniform compartmentilization of interior space.

■ **Ground conditions may impose limitations on the entire building:**
* soils with poor bearing capacity may require that all components be as light as practicable.
* differential settlement to be expected may preclude the use of rigid wall assemblies.

■ **Structural stability and integrity under all loads the wall assembly may be subjected to:**
* Walls may interact with the structural frame to contribute to its strength or rigidity, or the structural frame may impose loads on walls they were not intended to resist through dimensional changes in the frame.

■ **Durability under all environmental factors, or service life as a measure of the time until some loss of function occurs:**
* The environment at any plane of the wall assembly is determined by the arrangement and properties of its components, and durability is reflected in the ultimate cost of maintaining all the required functions of the wall assembly over its intended service life.

■ **Economy in initial and maintenance costs:**
* Initial cost may be reduced by selecting lower quality components, but generally only at the expense of

increased maintenance cost or reduced service life.

■ **Aesthetic quality:**
* Form, overall pattern of components, color, texture may be varied over a wide range without affecting other considerations, but inadequacies in design which allow problems, such as cracks, runoff staining, etc., to develop may severely affect the aesthetic quality of a building.

CURTAIN WALLS are classified in a variety of ways, none of them fully representative of all types available, nor universally accepted. Classifications used, and specific considerations for each type, are:

■ **Grid:** a pressure equalized system of vertical and horizontal framing members attached to the structural frame and supporting transparent, translucent, or opaque infill panels. **Selection of the grid curtain wall may be influenced by:**
* **Functional considerations:** when occupancy activities require a large amount of daylighting and/or natural ventialtion, resulting in a high opening to wall ratio. Also secondary components such as windows, doors may easily be incorporated into the grid.
* **Erection considerations:** grid members may be selected to be erected and glazed from the interior, eliminating the need for scaffolding.
* **Aesthetic considerations:** a variety of arrangements is possible, such as mullions projecting from the plane of the walls, or recessed inside for a virtually flush exterior face.
* **Weight:** grid curtain walls are relatively light thus minimizing loads imposed on the structural frames and foundations.

The grid curtain wall is generally further classified as:
* **Stick system:** with the vertical members, or mullions, and horizontal members, or rails, assembled at the site using sleeves to connect mullions and splines to connect the rails. Movements in the wall assembly and/or the structural frame are generally accommodated at splices of framing-members and in the glazing pockets of rails.

The stick system is the least expensive of the grid systems, but has a limited capacity for accom-

modating thermal expansion and contraction.
* **Unitized stick system:** with prefabricated vertical mullions and two-piece interlocking rails, which are connected to mullions with splines. Movements in wall and/or the structural frame are taken at joints in mullions and in the interlocking rails. This system is generally more expensive than the stick system, but is able to better accommodate thermal expansion and contraction.
* **Unit system:** with shop prefabricated interlocking mullions and rails, often preglazed and shipped to the site as units. Movements in the wall assembly and/or structural frame are accommodated in the interlocking mullions and rails. This system has the advantage of closer tolerances and greater capacity for accommodating movement. The disadvantages are increased material and fabrication costs, and necessity of maintaining closer tolerances in the structural frame.

Additional considerations applicable to all grid systems are:
* Amount of adjustment available within the system to accommodate field conditions, such as misalignments in the structural frame.
* Maintaining integrity of the system under in-service conditions, such as horizontal displacement, or sway, in the structural frame under lateral loads.
* indow washing equipment may be required for high-rise buildings with fixed glazing, and the additional load imposed thereby on mullions becomes a factor.

■ **Wall Panel:** a system of an array of panels capable of functioning as a complete wall assembly. Preliminary considerations of the wall panel system may be influenced by:
* **Functional considerations:** when occupancy activities require either a virtually opaque wall, or when a limited amount only of openings, such as windows and/or doors is required. Openings in most wall panels require secondary framing to support the frame of the opening and the free edges of adjacent panels.
* **Erection considerations:** wall panels are always erected from the outside, and connections to the structural frame are generally also made from the outside, except for some types which may be backfastened.

B · BUILDING SHELL

B2.3 EXTERIOR WALLS

WALLS: preliminary selection (cont.)

- **Aesthetic considerations:** a variety of applied coatings, patterns and textures are available: from flat and smooth to sculptured.
- **Weight:** ranges from very light, such as single thickness of formed metal, to very heavy, such as sculptured of precast concrete.

Additional considerations of specific types of wall panels are:

- **Formed:** whether of single thickness, or field assembled sandwich, the method of attaching panels to structural supports results in thermal bridges in the system. Erection of panels with exposed fasteners is generally nonsequential, thus a panel already installed and then damaged during or after construction may be replaced with minimal effect on adjacent panels; panels with concealed fasteners are erected sequentially, and replacement of a damaged panel is rather difficult.
- **Composite-laminated:** thermal breaks are available in some to minimize heat flow.
  Erection is generally sequential, using concealed fasteners, and replacement of a damaged panel will then affect adjacent panels. Prefabricated corners for some are available, and if such are used, maintaining close tolerances in the structural frame is a necessity.
- **Preassembled:** essentially a faced framed wall except that panels are assembled in a shop. Interior face generally open, to be finished after erection is completed. Thermal breaks generally not provided, and fasteners may create themal bridges. Erection generally nonsequential with butt joints between individual panels.
- **Composite-cast:** generally of regular or polymer concrete. Polymer concrete has the advantage of reduced weight and of greater strength. Available prefabricated only. Regular concrete panels are either plant or site precast. Plant precasting generally used for molded/sculptured panels and allows maintaining closer tolerances, but shipping considerations may limit size of panels. Site cast panels generally flat, with various surface textures or patterns, with sizes limited by thermal or moisture induced expansion/-contraction considerations, or by capacity of equipment required for erection.

BEARING WALLS act simultaneously as structure and enclosure to:
- support superimposed loads from floor, roof and other wall assemblies and transfer the resultant forces downward to the foundations.
- withstand the flexural moment and shears caused by lateral and vertical loads.
- serve as bracing to other parts of structure.
- below grade withstand lateral soil and sometimes also hydrostatic pressure.
- resist seismic tremors in some areas.

Bearing walls may be:
- identical to a curtain wall in construction and a nonbearing wall, or a partition when one story or less in height.
- may become a curtain wall without any change if its bearing capacity is not utilitzed.
- may be bearing for a certain portion of it and nonbearing, or curtain wall, for remaining portions.
- When numberous large openings have to be provided, even with uniformly applied floor/roof framing loads, the wall will become a series of piers or posts; the loading on the footing will be uneven and the wall assembly will no longer function as a true bearing wall.

Under such conditions, a portion or the entire wall may have to be replaced with a structural frame.

Bearing walls are generally classified as:

- ■ **Stacked unit:** walls made up of relatively small units stacked upon each other:
- in various patterns or bonds.
- in one or more wythes, contiguous or separate.
- in various forms: brick, block, ashlar, rubble.
- of various materials: stone, burned clay, concrete.
- bonded by mortar of various types or laid dry.
- reinforced or unreinforced.

- ■ **Monolithic:** of reinforced concrete. There are three major types: cast-in-place, tilt-up and precast:
- Cast-in-place thin concrete walls are usually integrally connected to the floor and roof slabs and to each other, thus forming a crate-like form with excellent lateral stiffness. Buildings of 16 stories and more have

been erected by this method.
- An often more convenient method is the so-called "tilt-up," where walls are cast flat on the ground or on a platform next to their final position, then picked up by a crane and tilted into place. Connecting pours to surrounding structure are then generally made.
- Precast concrete bearing walls are generally ribbed panels, single or double tees. Connections are usually made by preformed fasteners with/without additional concrete placed.

- ■ **Framed,** made of small, closely spaced vertical members, connected to plates top and bottom and covered by a skin.
- wood framed walls of which balloon and platform frames are best known.
- metal framed walls, where metal studs replace wood.

Secondary components of bearing walls for all types may be any/all of the following:
- exterior facings, surfacings, coatings.
- interior facings, surfacing, coverings, coatings.
- insulation of all types.
- waterproofing and vapor retarders.
- openings and perforations: windows and doors.

DEFORMATIONS IN WALLS

Walls must perform under conditions subject to frequent and often sudden changes:

Variations in the magnitude and/or distribution of lateral and vertical loads acting upon them, variations in external and internal temperatures and/or moisture content.

Changes in loading conditions will alter internal stresses thus the magnitude of deformation caused by them in a wall assembly, and may even lead to cyclic stress reversals and potential premature failure of components subjected to them.

CURTAIN WALL:

Negative air pressure on a given wall of a tall building, especially at corners, may significantly exceed positive air pressures the same wall could be subjected to if direction of wind reverses. Negative air pressure may also

B2.3 EXTERIOR WALLS

be augmented when buildings are maintained under positive pressure:
- Air pressure, either negative or positive, acting on a grid curtain wall will cause bending in the framing members and infill panels held by them, with each component then deflecting proportionately to its relative rigidity.
- Mullions being the principal framing members are relatively rigid, and can easily be reinforced, thus bending stresses and resulting deflections can be kept within safe limits.
- Rails are generally short members and relatively rigid.
- Infill panels, especially transparent ones, may when subjected to the same air pressures deform, fail in tension, or be pulled out of the framing members.

■ **In grid curtain walls** excessive deformation or deflection of the framing members under lateral loads may affect the infill panels or their connections to framing member, or may be visually unacceptable.
- generally deflections should not exceed 1/180 of the span or a ¾ inch maximum deviation from a straight line between supports.

■ **Deflections in panels of curtain wall** panel systems may generally be as high as 1/120 of span if not visually objectionable:
- or allowable stresses in bending will determine maximum spans.

BEARING/NONBEARING WALLS

■ **Lateral stability of walls is always a consideration:**
- For bearing and nonbearing walls, the allowable stresses in bending under lateral load, or under combined lateral and gravity loads will determine maximum unsupported height or length.

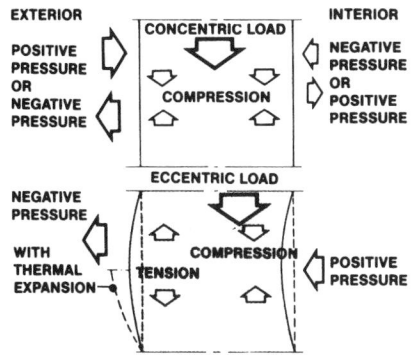

- Empirically established ratios of unbraced height/length to thickness of wall, or slenderness ratio, may be used for some types in lieu of calculating actual stresses.

■ **Concentrically loaded,** single-component wall assembly will be in compression thoughout its thickness:
- horizontal loads will cause bending in the assembly and change the magnitude of compressive stresses; in slender wall assemblies tension may occur under heavy wind loads.
- effect on facings is seldom if ever significant.

■ **Eccentrically loaded** wall assemblies may develop tensile stresses due to bending depending on the magnitude and/or eccentricity of the load:
- facings may be affected when eccentric loading, horizontal forces and thermal expansion combine to increase bending stresses in the wall assembly.

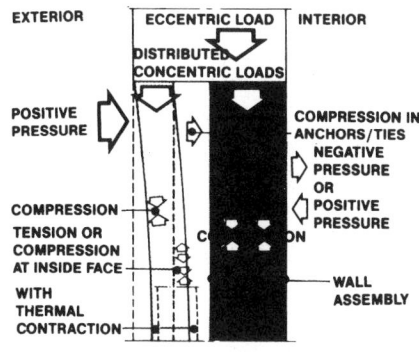

■ **Multi-component bearing wall assemblies,** such as a bearing cavity wall, may not develop bending stresses even when carrying an eccentric load, provided the load is distributed between the two wythes:
- bending may develop in slender exterior facing wythes under extreme horizontal load and the influence of thermal expansion/constraction.
- thermal and moisture induced movements rather than horizontal loads will generally be the more important considerations.
- vertical loads will affect wall assembly only when safe working stresses in component materials are exceeded.

■ **When walls are faced to enhance their resistance to detrimental effects of external environment,** or because of aesthetic considerations, the physical

properties of such facings may dictate the structural requirements of the wall:
- deflection in walls faced with stucco, plaster, or other inelastic materials my be limited to minimize the possibility of cracks developing in such facings.

■ **Walls when supported by or connected to the structural frame will generally be affected by deformations in such frame:**
- the entire envelope must there fore be considered during the selection of its constituent parts.

INSULATION

Insulation may be installed to reduce heat flow between all spaces with different environments. It may be:

■ **Incorporated into the roof construction** separating the exterior from a controlled interior environment.

■ **Incorporated into the ceiling assembly** separating controlled interior environment from attic space above.

■ **Incorporated into a floor construction,** separating one controlled interior environment from a space below it with a different environment.

■ **Attached to or incorporated into a wall assembly separating:**
- the interior from the exterior.
- two interior spaces with different environments.

■ Insulation may serve the following additional functions:
- Provide support for a surface finish.
- Resist water vapor transmission.
- Reduce noise.

B2.3 EXTERIOR WALLS

WALLS: rain penetration

PRESSURE EQUALIZATION, general

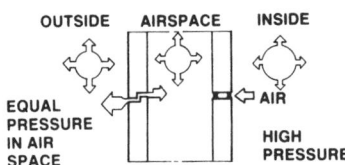

OUTSIDE　AIRSPACE　INSIDE

EQUAL PRESSURE IN AIR SPACE　　AIR　　HIGH PRESSURE

AIR/WATER VAPOR EXFILTRATION UNDER CHIMNEY EFFECT ONLY. WIND FORCES MAY REVERSE PRESSURES.

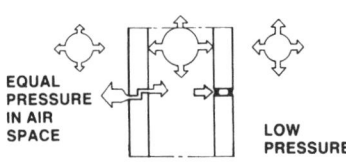

EQUAL PRESSURE IN AIR SPACE　　LOW PRESSURE

AIR/WATER INFILTRATION UNDER WIND AND/OR CHIMNEY EFFECT.

To prevent water penetration through an exterior wall assembly:
• the components should be waterproof and without cracks
• all joints in such assembly should be air and water tight.
Such a condition is very difficult if not impossible to achieve: sealants may fail due to movements in the components, the assembly, the structure itself, or because of faulty application.

Water penetration will not occur if the driving forces can be effectively controlled:
• gravity and capillarity by proper joint design
• differences in pressure by their equalization.

Pressure equalization is also referred to as the rain screen principle or two-stage weather-proofing.
It basically consists of:
• **first stage** is to provide a confined air space behind a waterproof exterior baffle, either within a joint or between joints; this air space is connected to the outside to equalize its pressure and to drain any water which may penetrate into it.
• **second stage** consists of an air-tight seal between the air space and the interior; the air seal must be effectively protected from direct effects of rain, extreme variations in temperature and from weathering caused by solar radiation.

PRESSURE EQUALIZATION: WALLS

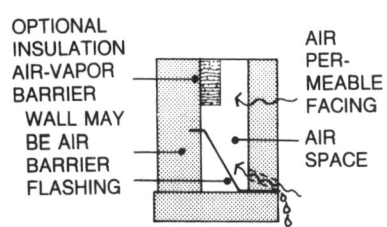

CURTAIN: GRID　　　　PANEL

MULLION
JOINT
AIR SPACE AROUND PERI-PHERAL
RAIL
PRESSURE EQUALIZED WINDOW FRAME

PRESSURE EQUALIZED AIR SPACE　　INTERIOR AIR SEAL

AIR SEAL

FILLER
AIR SPACE

DETERRENT SEAL

WITH RAIN BAFFLE

MULLION　　　　JOINT

CAVITY WALL

OPTIONAL INSULATION
AIR-VAPOR BARRIER
WALL MAY BE AIR BARRIER
FLASHING

AIR PER-MEABLE FACING

AIR SPACE

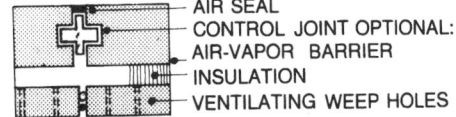

AIR SEAL
CONTROL JOINT OPTIONAL
AIR-VAPOR BARRIER
INSULATION
VENTILATING WEEP HOLES

GENERAL:
Providing a confined airspace within or behind the exterior component of a wall assembly and connecting such airspace to the exterior would satisfy the general requirements of pressure equalization, but such airspace may not perform as expected unless consideration is also given to: unequal intensities and distribution of air pressures over a given wall of a building, the effects of projecting elements of a wall which will locally modify the pattern of air pressures, the reversal of air pressures from positive to negative at edges of walls.

Differential air pressures at the open ends of a wide or long airspace may result in airflow through such space: this airflow may move significant amounts of water or snow into the airspace, with the risk of water penetration to the interior.

The airspace therefore should be: interrupted at intervals to minimize lateral or vertical air movement; the spacing of closures should be such that the expected variations in air pressure outside the compartmentalized space are at a minimum; closures must be provided in airspace at edges or corners of a wall to prevent air being drawn from the space around to an area of negative air pressure.

The effectiveness of the airspace will be negated if significant air infiltration from it to the interior occurs: interior, or air, seals of a double sealing system should effectively prevent air leakage through them.

CURTAIN WALL, GRID:
Confined air spaces, or air pressure equlaization chambers, are incorporated into mullions and rails, with openings to the outside generally located in soffit areas of rails to protect them from heavy wetting. Air pressure equalization chambers are compartmentalized to prevent differential air pressures of developing within each chamber.

Double seals are provided at connections of infill panels to mullions and rails: the outer seal acts as a deterent seal to water penetration but is not relied upon to completely prevent it; the inner seal acts as an air seal to substantially reduce air from the interior to the air chamber.

Pressures in air chambers will not be effectively equalized unless the aggregate area of all openings to the outside is considerably larger than the total area of all openings to the inside. The ratio of ten to one is considered minimal by the Architectural Aluminum Manufacturers Association.

FACED WALL:
Joints between facing panels are, fully or partially open to the airspace behind the panels; airspace may have to be compartmentalized: joints in backup wall should be sealed airtight, and the backup walls should effectively resist exfiltration of air, or an air-water barrier should be used.

BEARING, NONBEARING WALLS:
Pressure equalization may be inherent in the method of construction itself: vented airspace in a cavity wall, when the outer component allows for air infiltration and the inner one does not, will partially equalize air pressures; open joints between wood shingles or clapboard siding will effectively equalize pressures between the outside and the wall sheathing, when such sheathing is sealed airtight.

MOISTURE CONTROL & BUILDING ENVELOPE

Moisture mitigation can be summarized as control of rain, ground water and moisture vapor in building design. Rain control and moisture vapor control is discussed in this section. Control of ground water is described in SUBSURFACE MOISTURE PROTECTION.

RAIN CONTROL

Rain control strategies are varied based on the frequency and severity of rain.

• The amount of annual rainfall determines the type of approach necessary to control rain.

• Rainwater flow over the building surface will be determined by gravity, wind flow across the surface, and wall-surface features such as overhangs, flashings, sills, copings, and mullions.

Rain penetration into and through building surfaces is governed by capillary, momentum, surface tension, gravity, and wind (air pressure) forces. Capillary forces draw rainwater into pores and tiny cracks, while the remaining forces direct rainwater into larger openings.

• **Capillarity** can be controlled by capillary breaks, capillary resistant materials or by providing a receptor for capillary moisture (Fig. 1).

• **Momentum** can be controlled by eliminating openings that go straight through the wall assembly (Fig. 2).

• **Rain entry by surface tension** can be controlled by the use of drip edges and kerfs (Fig. 3).

• **Using flashings, and layering the wall assembly elements to drain water to the exterior** (providing a "drainage plane") can be used to control rain water from entering by gravity flow (Fig. 4), along with simultaneously satisfying the requirements for control of momentum and surface tension forces.

• Sufficiently **overlapping the wall assembly** elements or layers comprising the drainage plane can also control entry of rain water by air pressure differences.

• Finally, **locating a pressure equalized air space** immediately behind the exterior cladding can be used to control entry of rain water by air pressure differences by reducing those air pressure differences (Fig. 5).

Combining a pressure equalized air

space with a capillary resistant drainage plane—which typify state-of-the-art for Norwegian and Canadian rain control practices—addresses all of the driving forces responsible for rain penetration into and through building surfaces under the severest exposures.

■ **The first level approach to rain control involves overall architectural design decisions including:**

• locating buildings so that they are sheltered from prevailing winds,

• providing roof overhangs and massing features to shelter exterior walls and reduce wind flow over building surfaces,

• providing architectural detailing to shed rainwater from building faces.

■ **The second level approach to rain control involves details that deal with capillary, momentum, surface tension, gravity and air pressure forces acting on rainwater deposited on building surfaces.**

This second level detailing approach employs two general design principles:

■ **FACE-SEALED/BARRIER APPROACHES**

• **Storage/reservoir systems** (Fig. 6), appropriate for all rain exposures.

• **Non-storage/non-reservoir systems** (Fig. 7), appropriate for locations with less than 30 inches average annual precipitation.

■ **WATER MANAGED APPROACHES**

• **Drain-screen systems** (Fig. 8), appropriate for locations with less than 50 inches average annual precipitation.

• **Rain-screen systems** (Fig. 9), appropriate for locations with less than 60 inches average annual precipitation.

• **Pressure equalized rain-screen (PER) systems** (Fig. 10), appropriate for all rain exposures.

Rain is permitted to enter through the cladding skin in the three systems listed above as the water-managed approach: drain-screen, rain-screen or pressure equalized rain-screen (PER) systems. **"Drain the rain"** is the cornerstone of water managed systems. In the three water-managed systems, drainage of

water is provided by a capillary resistant drainage plane or a capillary resistant drainage plane coupled with an air space behind the cladding. If the air space has sufficient venting to the exterior to equalize the pressure difference between the exterior and the cavity, the system is classified as a PER design.

In the face-sealed barrier approach, the exterior face is the only means to control rain entry. In storage/reservoir systems, some rain is permitted to enter and is stored in the mass of the wall assembly until drying occurs to either the exterior or interior. In non-storage/non-reservoir systems, no rain can be permitted to enter.

Frequency of rain, severity of rain, system designs, selection of materials, workmanship, and maintenance determine the performance of a specific system.

• In general, water-managed systems outperform face-sealed/barrier systems due to their more forgiving nature.

• However, face-sealed/barrier systems constructed from water resistant materials that employ significant storage have a long historical track-record of exemplary performance even in the most severe rain exposures.

The least forgiving and least water-resistant assembly is a face-sealed/barrier wall constructed from water sensitive materials that does not have storage capacity. Most external insulation finish systems (EIFS) are of this type and, from the point of view of moisture control, should be considered as best limited to climate zones which see little rain, less than 30 in. (76 cm.) average annual precipitation.

The most forgiving and most water-resistant assembly is a pressure equalized rain screen wall constructed from water-resistant materials. These types of assembles perform well in the most severe rain exposures (more than 60 in. (152 cm.) average annual precipitation).

Face-sealed/barrier strategies should be carefully considered.

Non-storage/non-reservoir systems constructed out of water sensitive mate-

B2.3 EXTERIOR WALLS

MOISTURE CONTROL & BUILDING ENVELOPE

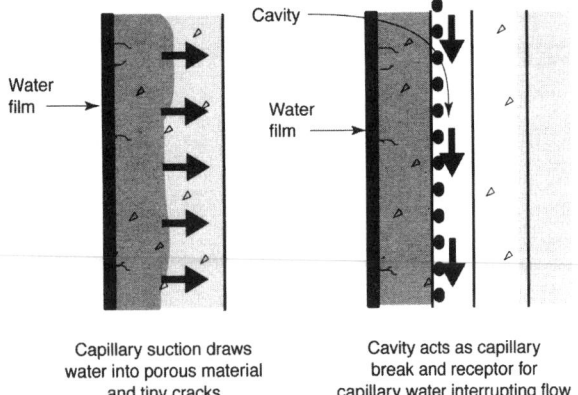

Capillary suction draws
water into porous material
and tiny cracks.

Cavity acts as capillary
break and receptor for
capillary water interrupting flow

Fig. 1. Capillarity as a driving force for rain entry
- Capillary suction draws water into porous material and tiny cracks.
- Cavity acts as capillary break and receptor for capillary water interrupting flow.

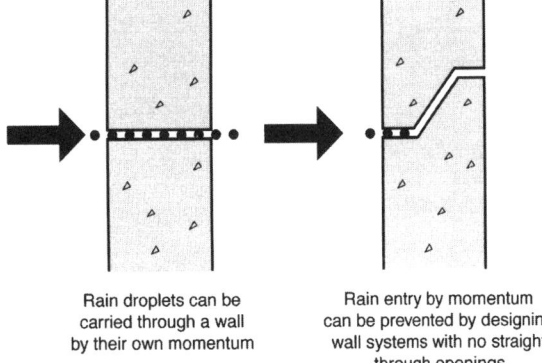

Rain droplets can be
carried through a wall
by their own momentum

Rain entry by momentum
can be prevented by designing
wall systems with no straight
through openings

Fig. 2. Momentum as a driving force for rain entry
- Rain droplets can be carried through a wall by their own momentum.
- Rain entry by momentum can be prevented by designing wall systems with no straight through openings.

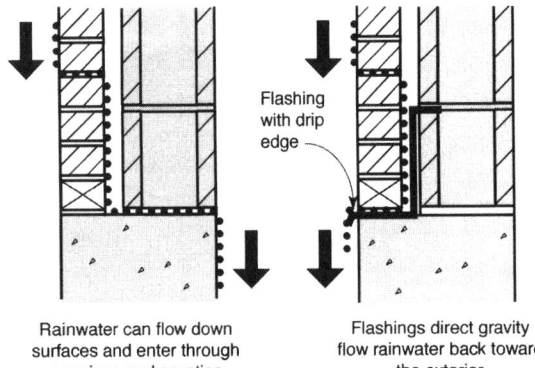

Rainwater can flow down
surfaces and enter through
openings and cavaties

Flashings direct gravity
flow rainwater back toward
the exterior

Fig. 4. Gravity as a driving force for rain entry
- Rainwater can flow down surfaces and enter through openings and cavities.
- Flashings direct gravity flow rainwater back toward the exterior.

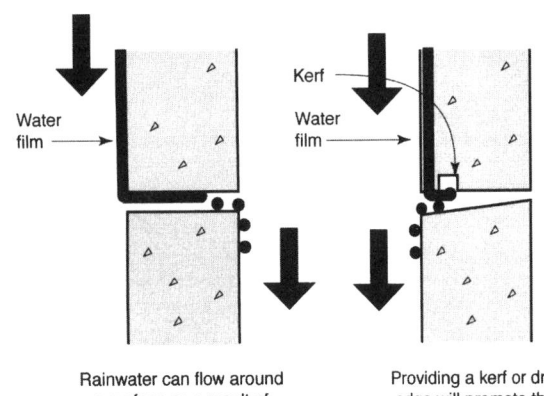

Rainwater can flow around
a surface as a result of
surface tension

Providing a kerf or drip
edge will promote the
formation of a water
droplet and interrupt flow

Fig. 3. Surface tension as a driving force for rain entry
- Rainwater can flow around a surface as a result of surface tension.
- Providing a kerf or drip edge will promote the formation of a water droplet and interrupt flow.

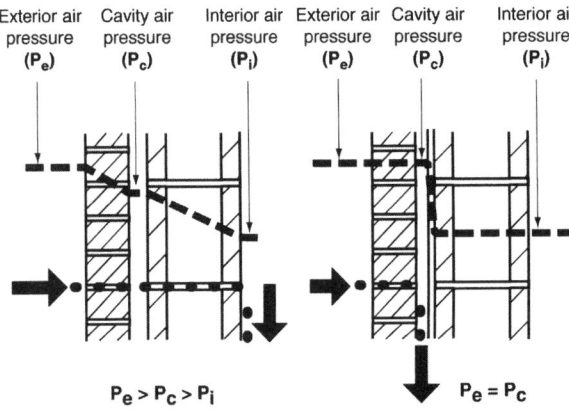

$P_e > P_c > P_i$

$P_e = P_c$

Driven by air pressure differences,
rain droplets are drawn through
wall openings from the exterior
to the interior

By creating pressure equalization
between the exterior and cavity air,
air pressure is diminished as a
driving force for rain entry.

Fig. 5. Air Pressure Difference as a Driving Force for Rain Entry
- Driven by air pressure differences, rain droplets are drawn through wall openings from the exterior to the interior.
- By creating pressure equalization between the exterior and cavity air, air pressure is diminished as a driving force for rain entry.

B2.3 EXTERIOR WALLS

(cont.)

rials should be limited to regions where average annual rainfall is less than 30 in. (76 cm.). Storage/reservoir systems constructed with water-resistant materials can be built anywhere. However, their performance is design, workmanship, and materials dependent. In general, these systems should be limited to regions or to designs with high drying potentials to the exterior, interior or, better still, to both.

Drainage plane continuity. The most common approach to rain control is the use of a drainage plane. This drainage plane is typically a "tar paper" or building paper. More recently, the term "housewrap" has been introduced to describe building papers that are not asphalt-impregnated felts ("tar papers"). Drainage planes can also be created by sealing or layering water-resistant sheathings such as a rigid insulation or a foil covered structural sheathing.

In order to effectively "drain the rain," the drainage plane must provide drainage plane continuity especially at "punched openings" such as windows and doors. Other critical areas for drainage plane continuity are where roofs and decks intersect walls.

MOISTURE VAPOR CONTROL

Two seemingly innocuous requirements for building envelope assemblies challenge designers and builders almost endlessly:

■ Keep moisture vapor out.

■ Let the moisture vapor out, if it gets in.

It gets complicated because, sometimes the best strategies to keep moisture vapor out also trap moisture vapor in. This can be a problem if the assemblies start out wet because of rain and the use of wet materials (wet framing, concrete, masonry or damp spray cellulose, fiberglass or rock wool cavity insulation).

It gets more complicated because of climate. In general, moisture vapor moves from the warm side of building assemblies to the cold side of building assemblies. This means that different strategies are needed for different climates. The designer also has to take into account differences between summer and winter.

Water vapor moves in two ways, by

vapor diffusion and by air transport. If the designer understands the two ways, and knows the climate zone, the problem can be addressed and solved. However, techniques that are effective at controlling vapor diffusion can be ineffective at controlling air-transported moisture, and vice versa.

Building assemblies, regardless of climate zone, need to control the migration of moisture that results from both vapor diffusion and air transport. Techniques which are effective in controlling vapor diffusion can be very different from those which control air transported moisture.

VAPOR DIFFUSION AND AIR TRANSPORT OF VAPOR

Vapor diffusion is the movement of moisture in the vapor state through a material as a result of a vapor pressure difference (concentration gradient) or a temperature difference (thermal gradient).

• **Vapor diffusion** moves moisture from an area of higher vapor pressure to an area of lower vapor pressure, as well as from the warm side of an assembly to the cold side.

• **Air transport of moisture** will move moisture from an area of higher air pressure to an area of lower air pressure if moisture is contained in the moving air (Fig. 12).

Vapor pressure is a term used to describe the concentration of moisture at a specific location. It refers to the density of water molecules in air. For example, a cubic foot of air containing 2 trillion molecules of water in the vapor state has a higher vapor pressure (or higher water vapor density) than a cubic foot of air containing 1 trillion molecules of water in the vapor state. Moisture will migrate by diffusion from where there is more moisture to where there is less. Hence, moisture in the vapor state migrates by diffusion from areas of higher vapor pressure to areas of lower vapor pressure.

Moisture in the vapor state also moves from the warm side of an assembly to the cold side of an assembly. This type of moisture transport is called "thermally driven diffusion." Transported mois-

ture vapor condenses on cold surfaces. These cold surfaces act as "dehumidifiers" pulling more moisture towards them.

Vapor diffusion and air transport of water vapor act independently of one another.

• Vapor diffusion will transport moisture through materials and assemblies in the absence of an air pressure difference if a vapor pressure or temperature difference exists.

• Furthermore, vapor diffusion will transport moisture in the opposite direction of small air-pressure differences, if an opposing vapor pressure or temperature difference exists. For example, in a hot, humid climate, the exterior is typically at a high vapor pressure and high temperature during the summer.

• In addition, the interior air-conditioned space is maintained at a cold temperature and at a low vapor pressure through the dehumidification characteristics of the air conditioning system. This causes vapor diffusion to move water vapor from the exterior towards the interior. This will occur even if the interior conditioned space is maintained at a higher air pressure (a pressurized enclosure) relative to the exterior (Fig. 13).

VAPOR DIFFUSION RETARDERS

The function of a **vapor diffusion retarder** is to control the entry of water vapor into building assemblies that would occur by the mechanism of vapor diffusion. The vapor diffusion retarder may be required to control the diffusion entry of water vapor into building assemblies from the interior of a building, from the exterior of a building or from both the interior and exterior.

Vapor diffusion retarders should not be confused with airflow retarders whose function is to control the movement of air through building assemblies. In some instances, airflow retarder systems may also have specific material properties, which also allow them to perform as vapor diffusion retarders. For example, a rubber membrane on the exterior of a masonry wall installed in a continuous manner is a very effective airflow retarder. The physical properties of rubber also give it the characteristics of a vapor diffusion retarder. Similarly, a continuous, sealed polyethylene ground

B2.3 EXTERIOR WALLS

MOISTURE CONTROL & BUILDING ENVELOPE

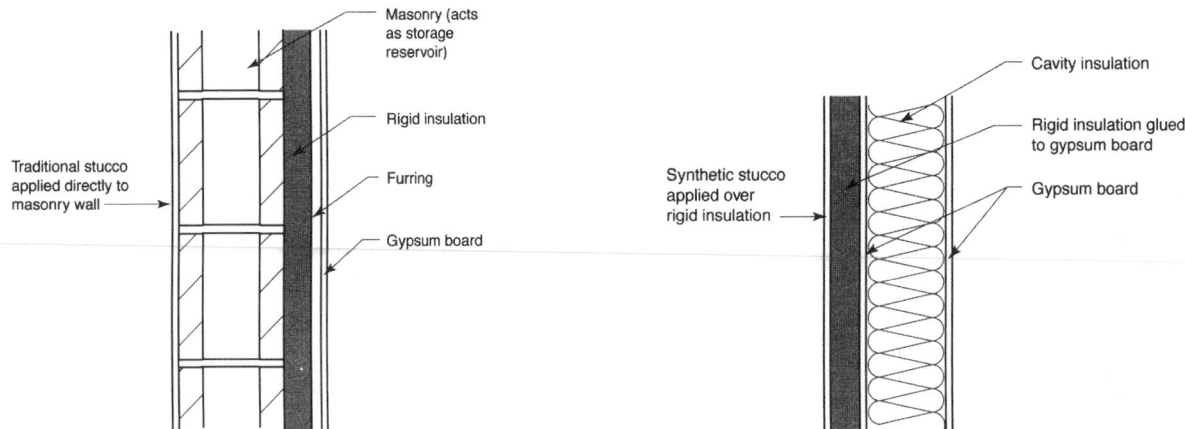

Fig. 6. Face-sealed barrier wall storage reservoir system
- Some rain entry past exterior face permitted.
- Penetrating rain stored in mass of wall until drying occurs to interior or exterior.

Fig. 7. Face-sealed barrier wall non-storage non-reservoir system
- No rain entry past exterior face permitted.

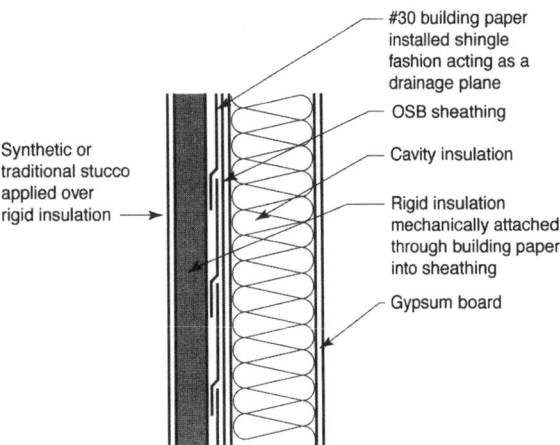

Fig. 8. Water managed wall drain-screen system (drainage plane)
- Should not be used in regions where the average annual precipitation exceeds 50 inches.
- Should not be used in regions where the average annual precipitation exceeds 30 inches.

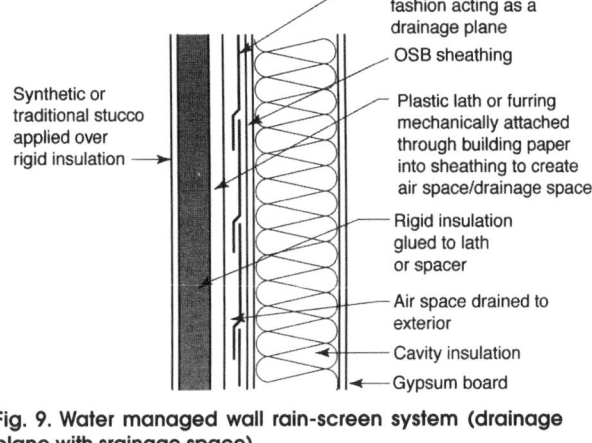

Fig. 9. Water managed wall rain-screen system (drainage plane with srainage space)
- Should not be used in regions where the average annual precipitation exceeds 60 inches.

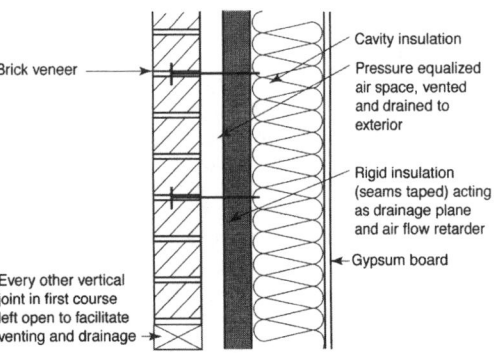

Fig. 10. Water managed wall pressure equalized rain-screen system (drainage plane with pressure equalized drainage space)
- Should be used in regions where the average annual precipitation exceeds 60 inches.

B2.3 EXTERIOR WALLS

(cont.)

cover installed in an unvented, conditioned crawl space acts as both an airflow retarder and a vapor diffusion retarder. The opposite situation is also common. For example, a building paper or a house wrap installed in a continuous manner can be a very effective airflow retarder. However, the physical properties of most building papers and house wraps (they are vapor permeable, that is, they "breathe") do not allow them to act as effective vapor diffusion retarders.

WATER VAPOR PERMEABILITY

The key physical property which distinguishes vapor diffusion retarders from other materials, is permeability to water vapor. Materials, which retard water vapor flow, are said to be impermeable. Materials, which allow water vapor to pass through them, are said to be permeable.

The unit of measurement typically used in characterizing permeability is a "perm." Many building codes define a vapor diffusion retarder as a material, which has a permeability of one perm or less.

Materials which are generally classed as impermeable to water vapor are: rubber membranes, polyethylene film, glass, aluminum foil, sheet metal, oil based paints, bitumen impregnated kraft paper, almost all wall coverings and their adhesives, foil faced insulating and non-insulating sheathings.

Materials which are generally classed as semi-permeable to water vapor are: plywood, OSB, expanded polystyrene (EPS), extruded polystyrene (XPS), fiber-faced insocyanurate, heavy asphalt impregnated building papers (30# building paper) and most latex-based paints. Depending on the specific assembly design, construction and climate, all of these materials may or may not be considered to act as vapor diffusion retarders. Typically, these materials are considered to be more vapor permeable than vapor impermeable.

Materials which are generally classed as permeable to water vapor are: unpainted gypsum board and plaster, fiberglass insulation, cellulose insulation, dimensional lumber and board lumber, unpainted stucco, some latex-based paints, masonry, brick, lightweight asphalt-impregnated building papers (15# building paper), asphalt-impregnated fiberboard sheathings, and "house wraps."

AIR FLOW RETARDERS

The key physical properties, which distinguish airflow retarders from other materials, are continuity and the ability to resist air pressure differences. Continuity refers to absence of holes, openings and penetrations. Large quantities of moisture can be transported through relatively small openings by air transport if the moving air contains moisture and if an air pressure differential also exists. For this reason, airflow retarders must be installed in such a manner that even small holes, openings and penetrations are eliminated.

Airflow retarders must also resist the air pressure differences, which can act across them. These air pressure differences occur as a combination of wind, stack and mechanical system effects. Rigid materials such as interior gypsum board, exterior sheathing and rigid draft stopping materials are effective air retarders due to their ability to resist air pressure differences.

MAGNITUDE OF VAPOR DIFFUSION AND AIR TRANSPORT OF VAPOR

The differences in the significance and magnitude of vapor diffusion and air-transported moisture are typically misunderstood. Air movement as a moisture transport mechanism is typically far more important than vapor diffusion in many (not all) conditions. The movement of water vapor through a 3/4" (19 mm) square hole as a result of a 10 Pascal air pressure differential is 100 times greater than the movement of water vapor as a result of vapor diffusion through 32 sq. foot (2.9 sq. meter) sheet of gypsum board under normal heating or cooling conditions (see Fig. 14).

• In most climates, if the movement of moisture-laden air into a wall or building assembly is eliminated, movement of moisture by vapor diffusion is not likely to be significant. The notable exceptions are hot, humid climates or rain wetted walls experiencing solar heating.

• Furthermore, the amount of vapor which diffuses through a building component is a direct function of area. That is, if 90 percent of the building envelope area is covered with a vapor diffusion retarder, then that vapor diffusion retarder is 90 percent effective. In other words, continuity of the vapor diffusion retarder is not as significant as the continuity of the airflow retarder.

• It is possible and often practical to use one material as the airflow retarder and a different material as the vapor diffusion retarder. However, the airflow retarder must be continuous and free from holes, whereas the vapor diffusion retarder need not be.

In practice, it is not possible to eliminate all holes and install a "perfect" airflow retarder. Most strategies to control air transported moisture depend on the combination of an airflow retarder, air pressure differential control and interior/exterior moisture condition control in order to be effective. Airflow retarders are often utilized to eliminate the major openings in building envelopes in order to allow the practical control of air pressure differentials. It is easier to pressurize or depressurize a building envelope made tight through the installation of an airflow retarder than a leaky building envelope. The interior moisture levels in a tight building envelope are also much easier to control by ventilation and dehumidification than those in a leaky building envelope.

COMBINING APPROACHES

In most building assemblies, various combinations of materials and approaches are often incorporated to provide for both vapor diffusion control and air transported moisture control.

• For example, controlling air-transported moisture can be accomplished by controlling the air pressure acting across a building assembly. The air pressure control is facilitated by installing an airflow retarder such as glued (or gasketed) interior gypsum board in conjunction with draft stopping.

• For example, in cold climates during heating periods, maintaining a slight negative air pressure within the conditioned space will control the exfiltration of interior moisture-laden air.

• However, this control of air-transported moisture will not control the migration of

B2.3 EXTERIOR WALLS

MOISTURE CONTROL & BUILDING ENVELOPE

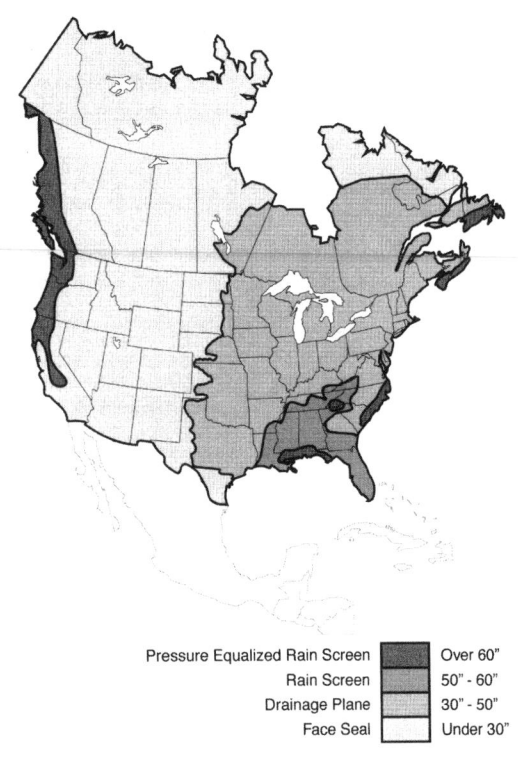

Pressure Equalized Rain Screen	Over 60"
Rain Screen	50" - 60"
Drainage Plane	30" - 50"
Face Seal	Under 30"

Fig. 11. Annual rainfall map

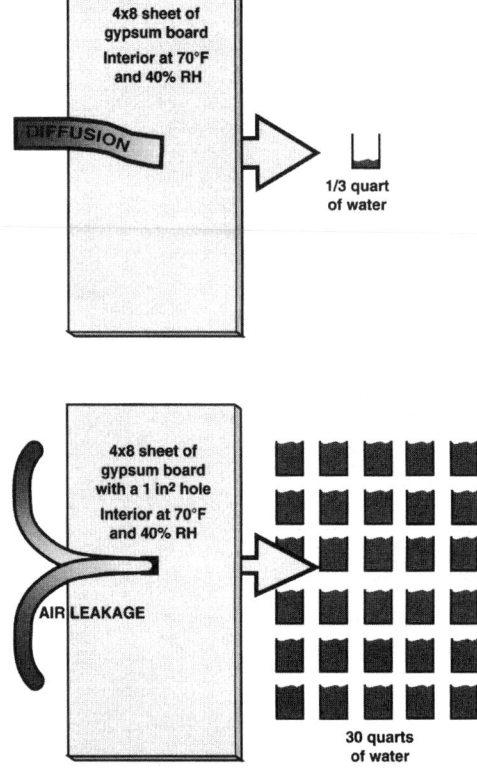

Diffusion vs air leakage

In most cold climates, 1/3 of a quart of water can be collected by diffusion through gypsum board without a vapor diffusion retarder; 30 quarts of water can be collected through air leakage.

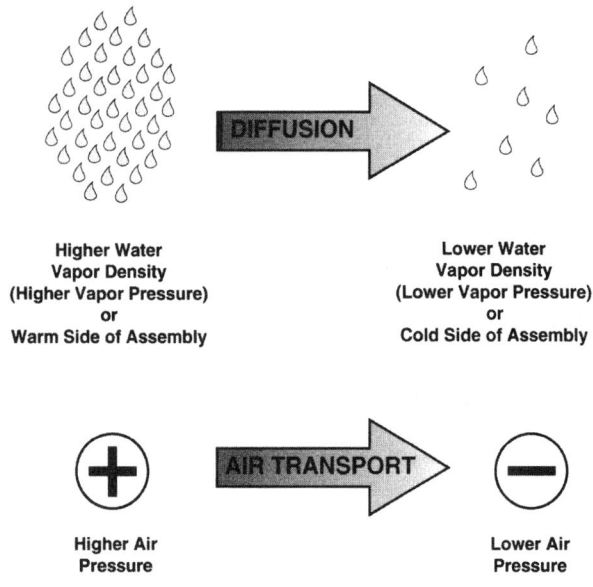

Higher Water Vapor Density (Higher Vapor Pressure) or Warm Side of Assembly	Lower Water Vapor Density (Lower Vapor Pressure) or Cold Side of Assembly

Higher Air Pressure	Lower Air Pressure

Fig. 12. Water Vapor Movement

water vapor as a result of vapor diffusion.

• Accordingly, installing a vapor diffusion retarder towards the interior of the building assembly, such as the **kraft paper backing** on fiberglass batts is also typically necessary. –

• Alternatives to the kraft paper backing are **low permeability paint** on the interior gypsum board surfaces, the foil backing on foil-backed gypsum board, sheet polyethylene installed between the interior gypsum board and the wall framing, or almost any interior wall covering.

In the above example, control of both **vapor diffusion and air transported moisture** in cold climates during heating periods can be enhanced by maintaining the interior conditioned space at relatively low moisture levels through the use of controlled ventilation and source control. Also, in the above example, control of air transported moisture during air-conditioning periods (when moisture flow is typically from the exterior towards the interior) can be facilitated by maintaining a slight positive air pressure across the building envelope, thereby preventing the infiltration of exterior, hot, humid air.

REFERENCES:
Joseph Lstiburek, P. Eng. *"Moisture Control"* in Reference 1.

Lstiburek, Joseph. 1997. *"Builder's Guide: Cold Climates"*. Westford, MA: Building Science Corporation.

Lstiburek, Joseph and John Carmody. 1994. *"Moisture Control Handbook"*. New York: Van Nostrand Reinhold.

B2.3 EXTERIOR WALLS

WEATHERTIGHT EXTERIORS

WATERTIGHT EXTERIOR WALLS

Two fundamental approaches to waterproofing exterior walls are considered:

■ barrier wall construction that utilizes the exterior surfacing as the sole waterproofing barrier.

■ cavity wall construction that provides a waterproof barrier behind the exterior surfacing to collect and drain water that penetrates the veneer back to the exterior.

BARRIER WALLS

Barrier wall designs require that the exterior wall materials and joinery are able to block passage of all water at the exterior face of the wall. These systems typically have no waterproofing redundancy and little tolerance for construction variations and defects.

The basic cladding element must be relatively impermeable and cannot develop through cracks in the course of weathering and reacting to thermal or moisture cycles.

Design considerations include:

• **Joints in the wall system** at openings or between cladding elements must be sealed with materials that do not split or debond from the cladding.

• The **cladding** must be continuous and uncracked along the sealant bond line. Typically, cladding joints are sealed during construction (in the field) with liquid-applied sealants. It is unreasonable to expect the application of these materials to be perfect.

• **Substrate surfaces** must be sound and uncracked and then cleaned and prepared for sealing. Joint back-up materials need to be positioned properly, and sealant materials must be mixed in some cases and then applied.

• The **sealant materials** must withstand joint movement and weathering without deterioration.

Given these variables, some deficiencies in the joint seals are likely to occur

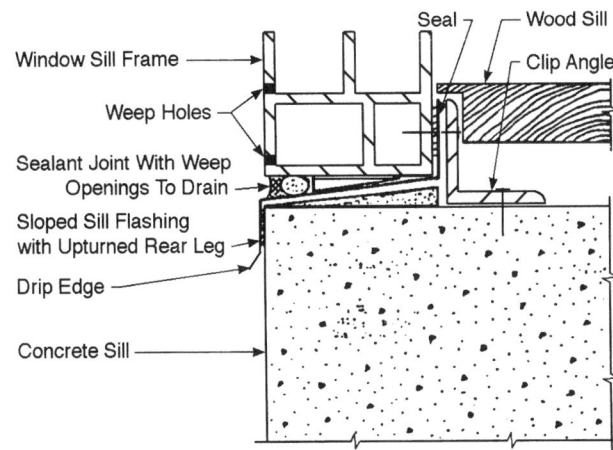

Fig. 1. Schematic cross-section of window sill flashing. The flashing collects water that penetrates the window, such as at corners, and drains it back to the exterior through weep holes. Window frame also has drainage ability.

both upon initial installation and as the sealant ages. Under the best circumstances, the number of deficiencies is small and leakage is not widespread. but maintenance of the seals is necessary to avoid increased leakage. A common method to improve the watertightness of sealant joints is to provide two seals in one joint.

Incorporating shielding elements within the cladding to protect the joints can improve barrier wall performance significantly. Overlapping the wall elements at joints, recessing the seals and windows from the face of the wall, and providing overhangs and drip edges are examples. Unfortunately, recent trends in wall design eliminate such features and, instead, set the glazing and joint seals flush with the exterior surface, providing little or no shielding from rainwater and from the deteriorating effects of ultraviolet (UV) radiation on organic sealants.

Elements within wall openings, such as windows, must be watertight and cannot leak from frame corners or face joints. Windows typically contain joints between the horizontal and vertical framing members that are sealed with gaskets or liquid-applied sealants. Corner seals that are constructed with liquid-applied sealants are not likely to be watertight. In addition, handling and installation of the window frame can

disturb or break these seals. For these and other reasons, it is prudent to install a flashing, such as a sheet metal pan, along the bottom of the window to collect leakage through the window glazing or frame joints and direct it back to the outside (Fig. 1).

Wall openings interrupt the cladding. Some barrier walls, such as **multi-wythe brick masonry**, may absorb and contain some water within the cladding. As this water seeps down within these materials, it can leak to the inside at the top of the wall openings, unless a flashing is installed in this location to collect water and drain it to the exterior.

Field experience is that barrier walls generally are problematic because of imperfect average workmanship and degradation of materials by weathering in the barrier that results in some water leakage. The extent of leakage problems depends on the materials used, quality of workmanship, and frequency of maintenance.

CAVITY WALLS

The cavity wall concept differs fundamentally from the barrier wall concept, in that the exterior surfacing screens the rain from the waterproofing layer that is placed behind it, rather

B2.3 EXTERIOR WALLS

WEATHERTIGHT EXTERIORS (cont.)

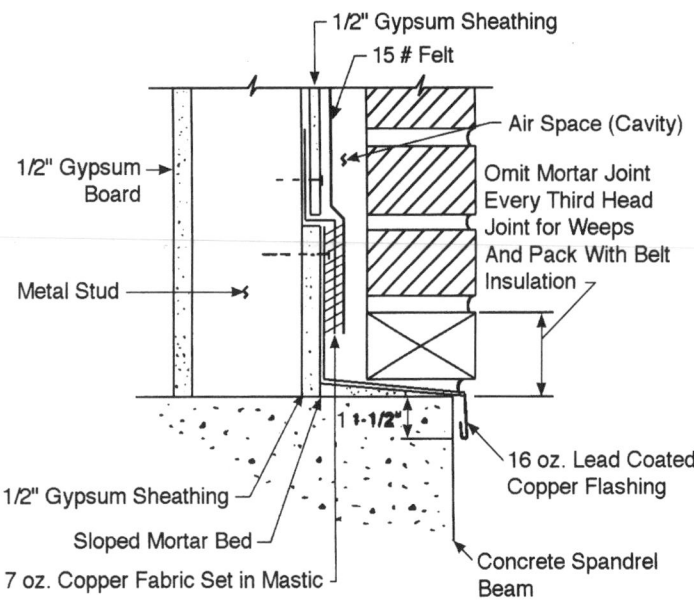

Fig. 2. Through-wall flashing of brick veneer/steel stud wall.

than acting as the sole barrier to water entry. This concept acknowledges, and accounts for, the inevitable penetration of some water through the exterior veneer and joinery. As such, it avoids some of the primary drawbacks of the barrier wall approach and can possess a high degree of reliability and durability. The details of cavity wall construction can take different forms depending upon the veneer type and back-up construction. Its fundamental design elements include:

■ The **exterior veneer** provides the initial barrier to water penetration. While the veneer is not expected to prohibit all water entry, it should not contain significant cracks, openings, or unsealed joints. Differential air pressure acts across this veneer and drives water through it.

■ An **air space** isolates the inner, or back-up wall, from the exterior veneer. Water that penetrates the veneer flows downward in this cavity, minimizing any contact with the back-up wall construction. The width of the air space varies depending upon the veneer materials and the likelihood of creating obstructions during construction of the veneer, but generally ranges from 1 to 2 in. (2.5 to 5 cm).

■ **A continuous waterproofing layer** should cover the back-up wall to shed any small amounts of water that inevitably cross the air space by splashing or by direct flow at cavity obstructions or at veneer anchor ties that span the cavity. Because the veneer and cavity control much of the water and the veneer shields the cavity from wind-driven rain, the requirements for this waterproofing layer are much less severe than if it were exposed on the face of a building. The combination of a protective screen and a waterproofing layer provide significant redundancy in these systems with resultant long-term reliability.

■ **Horizontal runs of through-wall flashings must be located at regular vertical intervals** to collect the water that flows downward within the veneer and cavity space. The inboard end of the flashing should turn upward at the back-up wall and the wall waterproofing layer should shingle over it. The flashing should extend from the back-up wall, across the cavity, through the veneer, and terminate with an exposed drip edge at the front of the veneer to prevent water from running back underneath the flashing (Fig. 2). Providing slight outward slope to the horizontal part of the flashing to pro-

mote drainage and avoid ponding on the flashing enhances reliability and durability. Sloped quick-set mortar beds or closely spaced tapered shims beneath the flashing can provide such slope.

■ Along the length of the wall, the flashing needs to be continuous and seamed watertight at joints and corners. Expansion joints should be incorporated in continuous flashings that are made with rigid materials, such as sheet metal, to accommodate thermally induced movement of the flashing and cladding. At terminations, the flashing should turn up and the corner should be sealed watertight to prevent water from draining off the end of the flashing and into the building. Weep openings are needed in the veneer at the flashing level to permit drainage of water from the flashing to the exterior. Size and spacing of these weep holes varies with veneer materials.

PRESSURE EQUILIZED DESIGN

An approach that is related to the cavity wall concept is pressure-equalized design, which provides an air barrier inboard of the veneer, instead of, or in addition to, a waterproofing layer. By preventing air penetration through the back-up wall and by sufficiently venting the cavity (air chamber) to the outside air, the pressure differential across the exterior veneer is reduced, or eliminated, during wind-driven rains, thus removing a primary driving force for water penetration. Essential elements for pressure-equalized systems include:

■ **The air barrier must be continuous** and properly sealed to all wall openings such as windows and doors. The air chamber is not simply a ventilated space. Because wind pressures vary considerably over the face of the wall, the air chamber should be compartmentalized to avoid air flow, and accompanying water flow, from high pressure to low pressure regions.

■ The air barrier and its supporting wall, typically the back-up wall, must have adequate strength to resist wind loads on the building.

■ The **exterior veneer** serves as the primary rain screen or barrier to water penetration. However, the joints between veneer elements are left open to some degree to allow efficient pressurization of the air chamber behind the veneer. Wind-driven rain inevitably penetrates the open joint areas due to momentum of the raindrops.

■ **Back-up waterproofing layers** are needed at the joints or the joints should be configured to control this form of penetration, e.g., shiplap geometry.

■ **Internal drainage devices,** such as through-wall flashings, are required at regular vertical Intervals to collect water that penetrates the cladding and direct it back to the exterior.

SUMMARY

■ Exterior wall systems that incorporate cavity-wall waterproofing principles are the most reliable in preventing water leakage to the building interior.

• The key component for these systems is the through-wall flashing which should be durable and have an expected service life equivalent to that of the entire wall system.

• Proper attention to the detailing and installation of these flashings is crucial to the success of a cavity wall system.

• Lower durability flashings with limited track records should be avoided due to the high cost of future replacement of failed flashings.

■ Barrier wall systems with modifications to incorporate some degree of secondary drainage capability, particularly at vulnerable joints, can provide levels of watertightness acceptable to some building owners, if sound, durable materials are used to form the barrier.

• Barrier walls that rely solely on surface seals and which use components that deteriorate readily from water that penetrates flaws in those seals do not provide a level of waterproofing reliability acceptable to most building owners.

■ All wall systems, and in particular barrier walls, can benefit from shielding provided by proper articulation of the wall surface to promote water drainage away from vulnerable joints.

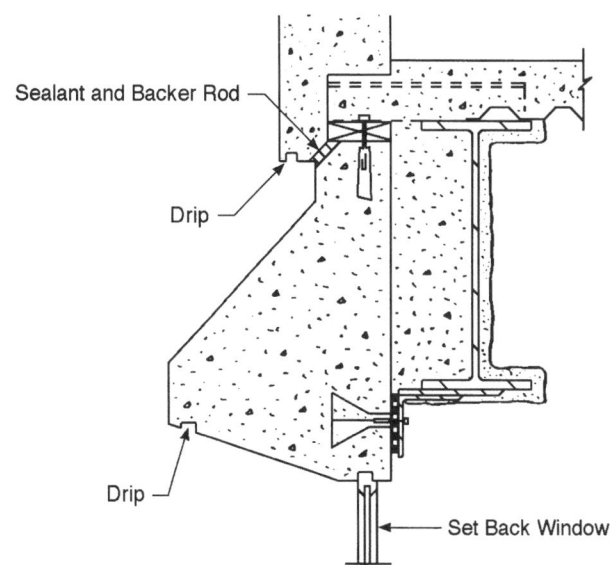

Fig. 3. Vertical section showing horizontal joinery in precast concrete panels. Note ship-lap geometry and recessed sealant to shield the joint from the weather.

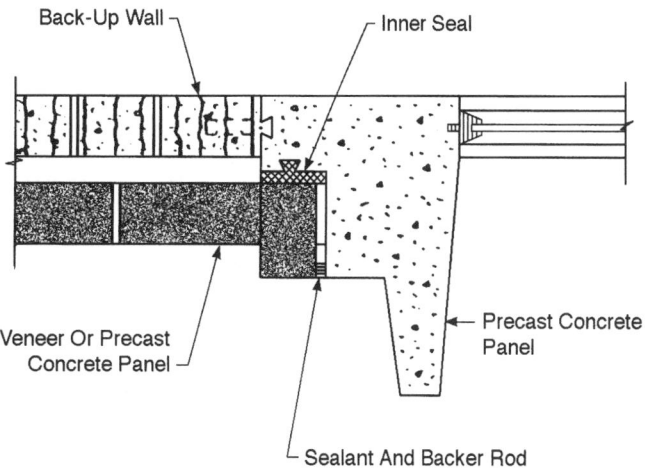

Fig. 4. Plan section showing vertical joinery in precast concrete panels. Note that panel geometry shields vertical joint from weather. Water that penetrates outer seal does not have a direct path to the interior.

REFERENCES:

The following references provide additinal details of watertight design principles applied to common wall systems, including construction, diagnostic evaluation and remediation.

Stephen S. Ruggiero and James C. Myers *"Design and Construction of Watertight Exterior Walls"* in Reference 1.

Ruggiero, S. S., and Myers, J. C. 1991. *"Design and Construction of Watertight Exterior Building Walls."* Water in Exterior Building Walls: Problems and Solutions. ASTM STP 1107. Thomas A. Schwartz, editor. Philadelphia: American Society for Testing and Materials.

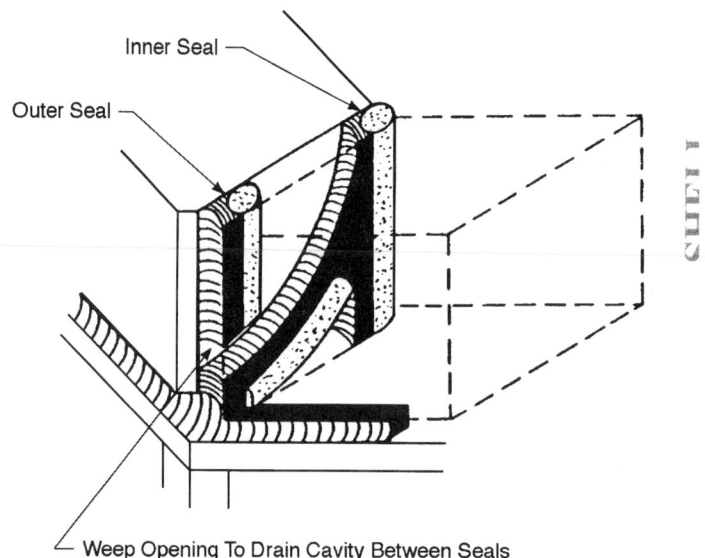

Fig. 5. A two-stage remedial vertical seal with weep openings at the base of the joint.

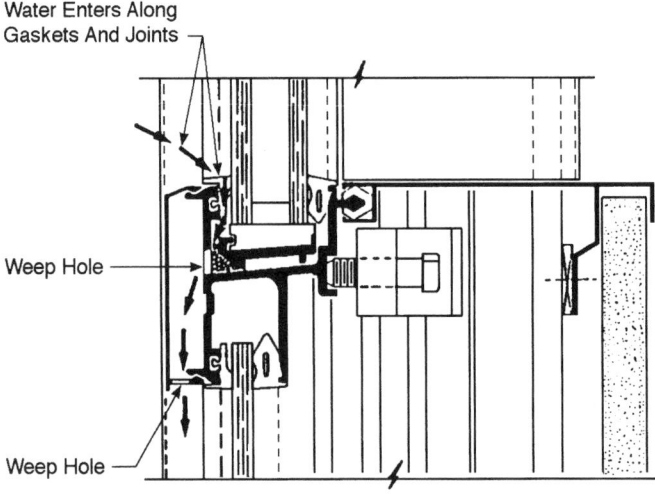

Fig. 6. Cross-section of a curtain wall. Note secondary drainage capability of the pocket below glass. Base of the pocket is sloped outward to promote drainage through weep holes.

B2.3 EXTERIOR WALLS

B2. 4 BEARING WALLS

BEARING WALLS: introduction

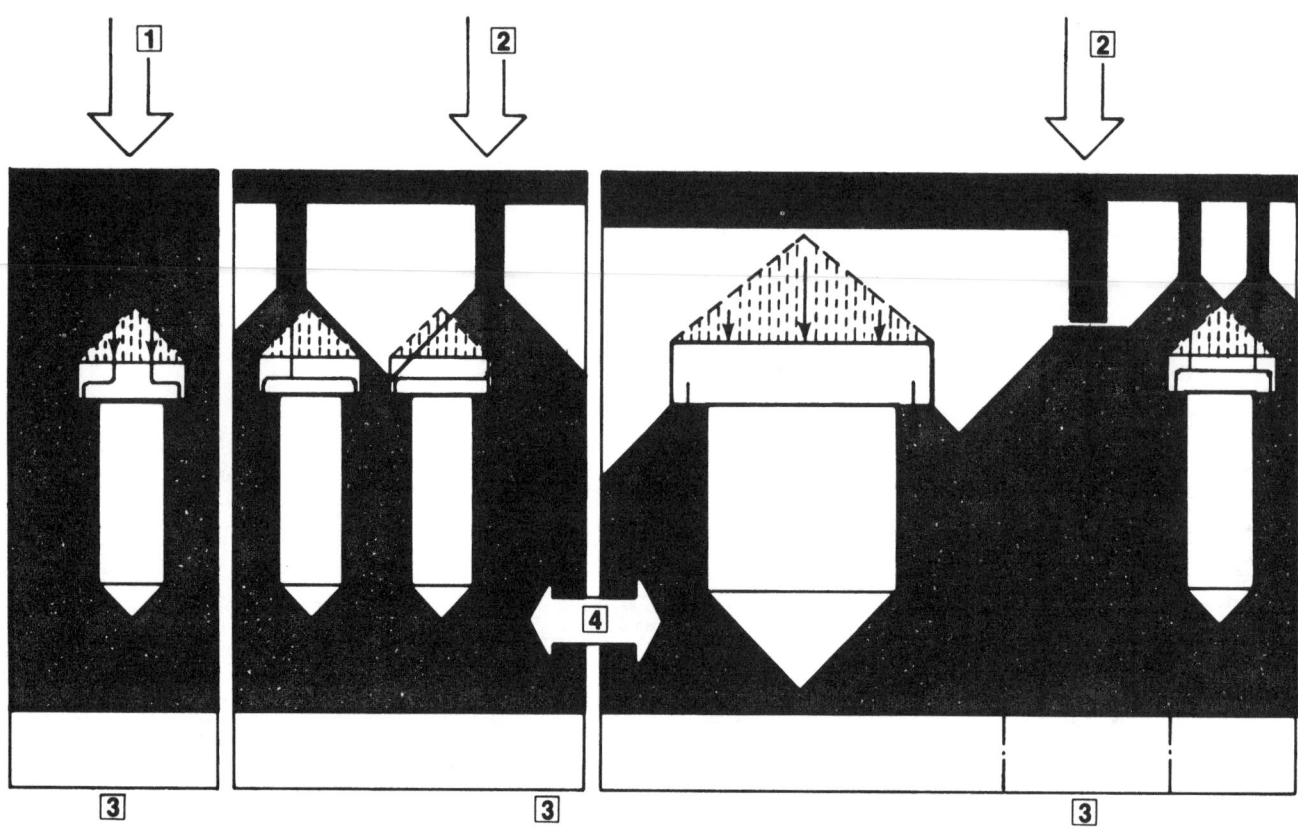

B2.4 BEARING WALLS

1
- evenly distributed load.
- when bearing surface for load is not over full depth of wall, bending stresses in addition to compressive stresses will be induced in wall.
- superimposed load is distributed over entire wall in stacked or monolithic types; in framed type, each stud acts as a post in carrying the superimposed load.
- lintels must carry weight of wall above plus a portion of the superimposed load.

2
- concentrated load.
- when bearing surface for load is not concentrically located on wall, bending stresses in addition to compressive stresses will be induced in wall.
- allowable bearing capacity of material used for wall construction must not be exceeded under concentrated loads; bearing plates or other means may be required to minimize unit compressive stresses at such points.
- superimposed loads may not be evenly distributed over the entire wall, and parts of wall between concentrated loads will then support their own weight, and resist horizontal forces only, resulting in inefficient utilization of material.

- openings should not be located under concentrated loads to avoid excessive loads on, and stress concentrations at, lintels.
- heavy concentrated loads may exceed bearing capacity of wall, and pilasters, piers, or posts may have to be provided to carry them. In such cases, the wall on either side of the support may become non-bearing.

3
- load on footing is uniform for evenly distributed loads, and is the recommended condition.
- load on footing may not be uniform for concentrated loads, and will require modification to carry pilasters or posts.

4
- walls, especially the stacked and monolithic types, will be subject to thermal expansion and contraction; and to differential expansion and contraction between the wall and the floor/roof assemblies. Provisions must be made to control such movements.

DESIGN CONCEPT OF BEARING WALL framing system assumes the combined structural action of the floor and roof assemblies with the walls:

• the floor and roof assemblies transmit vertical live and dead loads and also, acting as diaphragms, lateral loads to the walls.
• the walls then transfer all vertical and lateral loads to the foundation.
• primary stresses in the walls will be compression and shear resulting from both lateral and gravity loads.
• bending may be induced at floor/roof to wall connections due to one-sided or unequal loading.
• walls which are parallel to the direction of lateral loads at any given time will act as shear walls at such time when tied into the floor and roof assemblies.

DISTRIBUTION OF LATERAL FORCES:

Horizontal distribution of lateral forces of wind or earthquake to shear walls through floor or roof assemblies assumes that:

• such assemblies will transmit lateral forces without deflecting to a degree which would cause distress to any wall.
• assemblies are properly tied into the shear walls.

DISTRIBUTION OF AXIAL LOADS:

When a superimposed axial load is applied to a masonry bearing wall, it is generally assumed to be distributed uniformly through a triangular section of the wall. There is no complete agreement as to the angle of distribution, with values from 30 degrees to not more than 45 degrees from the direction of the axial load commonly used. Generally, the angle of distribution for concentrated loads superimposed on a masonry bearing wall should not exceed 30 degrees.

■ **Superimposed loads** may be eccentrically applied and will then cause bending in the wall

• even when the framing of a floor/roof assembly bears on the full cross section of a wall, some eccentricity may result due to deformation in such framing.

MOVEMENT CONSIDERATIONS

All building elements, including walls, are under almost constant movement.

■ **Main causes of movement are:**

• variations in temperature and exposure to sun;
• differentials in temperature between interior and exterior faces;
• variations in moisture content within the wall assembly.

■ **Other causes may be:**

• settlement of foundations

■ **Variations in temperature and exposure will result in:**

• expansion and contraction of the complete wall assembly.
• differential expansion and contraction between interior and exterior faces of the wall assembly.
• differential movement between the wall assembly and floor/roof assemblies framing into it.

■ **Variations in moisture content will affect the components of the wall assembly:**

• brick will expand when in contact with water. This initial expansion will not be reversed by drying at atmospheric temperatures.
• concrete will expand when allowed to absorb water but will tend over a long period of time to shrink, if exposed to the weather.
• wood, when seasoned, will absorb or shrink depending upon the surrounding environment. The change in movement will continue indefinitely depending upon the moisture differential.

■ **Framed wall assemblies,** when of limited size, are generally flexible enough to accommodate movement:

• due to variations in moisture content
• thermal induced expansion and contraction
• without any special provisions

■ **Stacked wall assemblies** require careful consideration to prevent or minimize cracking caused by shrinkage and thermal induced tensile stresses:

• moisture content of masonry units at time of construction must be controlled; it should be kept to a minimum in relation to expected in-service exposure.

• walls may be reinforced to control tensile stresses by the use of:
• horizontal joint reinforcing.
• bond beams located approximately four feet on centers;
• horizontal and vertical reinforcing bars in reinforced masonry walls.

Control joints are the principal method by which movement is managed in stacked walls:

■ **Control joints should be provided:**

• to limit cumulative effect of expansion and contraction.
• to allow movement to take place.

■ **Location of control joints should include:**

• at all changes in wall height or thickness.
• at construction joints in foundations and floor (roof assemblies).
• at jambs of openings.
• at pilasters, columns or piers.
• near corner.
• at planes of weakness, such as piers, near corners; at chases in walls.

■ **Detailing should accommodate lateral forces** across the joint by the use of a shear key while providing free movement longitudinally.

■ **Control joints should divide the wall** into a series of isolated panels, and horizontal joint reinforcing should not be continuous through the joints.

■ **Spacing of control joints** must be determined in each individual case based on:

• characteristics of locally available materials
• climatic conditions
• length and height of walls
• size and location of openings
• location and amount of reinforcing provided
• characteristics of floor/roof assemblies being supported

■ **When stacked assemblies support a flat concrete roof slab or similar construction,** horizontal slip planes to isolate differential movement in each should be provided for 12 to 15 feet back from all corners; such slip planes should terminate at a control joint in the wall assembly.

■ **Large, long, or irregularly shaped structures** may require expansion joints to allow for free thermal movement; expansion joints will completely separate all the component parts of a structure.

B2.4 BEARING WALLS

BEARING WALLS: types, properties

● denotes common usage
○ denotes possible usage

TYPE		DESCRIPTION	PROPERTIES			
			THICKNESS, nominal, average, minimum and maximum	SLENDERNESS RATIO, h/t [1]	SOUND TRANSMISSION COEFFICIENT (STC)	FIRE RESISTANCE, range in hours
FRAMED						
	WOOD	• wood studs: 2 x 4 inches nominal; 2 x 6 inches used for heavier loads/wider air space; spaced 16 to 24 inches on centers. • exterior face diagonally braced or structural sheathing used to provide lateral stability in plane of wall. • platform framing, with studs one story in height between horizontal framing generally used.	[2] 5" to 7"	[4] l/d $\leq$ 50 or Deflection $\leq$ 1/360 [5]	35	Combustible, 1 to 2 if protected
	METAL	• galvanized cold rolled steel, 18 to 20 gauge; 16 or 14 gauge for heavier loads. May be punched for wiring and/or horizontal bracing. Diagonal bracing or structural sheathing required. • connections made using self-tapping sheet metal screws, welding, or bolts.	[2] 5" to 7"	[5] Deflection $\leq$ 1/360	42	1 to 2
STACKED-MASONRY						
	SINGLE COMPONENT	• burned clay brick, concrete brick, hollow concrete block stacked in various patterns; held in place by mortar in joints, or by surface bonding; horizontal reinforcing commonly used.	6" to 12"	Brick $\leq$ 20 to 40 Block $\leq$ 18 to 20	57	1 to 4
	MULTI-COMPONENT	• materials as for single component but a variety of combinations possible, such as burned clay brick and hollow concrete block. Bonding of the separate wythes of masonry is required.	8" to 12"	$\leq$ 20 Average	52	2 to 4
	CAVITY	• materials as for single component. Air space between two wythes generally 2 inches. Wythes tied with horizontal reinforcing and/or ties. Cavity may be insulated. Weep holes at bottom of cavity required.	[3] 10" to 14"	$\leq$ 20 Average	54	2 to 4
CAST-CONCRETE						
	IN-PLACE	• normal or lightweight aggregate and Portland cement binder; generally reinforced with deformed bars, ½ inch in diameter or larger, horizontally and vertically. Height and thickness established by structural analysis.	6" to Unlimited	Varies	55	2 to 4 and more
	TILT-UP	• normal or lightweight concrete. Size of tilt-up panel up to 800 square feet; may be limited by capacity of erection equipment. Size of plant precast: 8- to 10-feet wide, up to 40 feet long; based on shipping limitations. Generally reinforced as a minimum to resist erection stresses; ½ inch deformed bars or larger. Plant precast panels may also be prestressed. • May have boards of rigid insulation sandwiched between two layers of concrete.	5" to 8"	Varies	55	2 to 4
	PRECAST		4" to 8"	Varies		2 to 4

B · BUILDING SHELL

B2.4 BEARING WALLS

COMPATIBLE FLOOR AND ROOF FRAMING

CONCRETE					STEEL				METAL/WOOD				
FLAT PLATE/FLAT SLAB	TWO-WAY SLAB	GIRDERS, BEAMS	PAN JOISTS/PRECAST TEE'S	HOLLOW CORE PLANK	GIRDERS, BEAMS	BAR JOISTS/JOIST GIRDERS	TRUSSES	SPACE FRAMES	JOISTS, RAFTERS	TRUSSED RAFTERS	TRUSSES, SHORT SPAN	BEAMS, SOLID/LAMINATED	TRUSSES, LONG SPAN
									●	●	●	○	
									●	●	○		
		●	●		○	●	○		●	●	●	●	○
	○	●	●		○	●	○		●	●	●	●	○
		●	●			○			●	●	●	○	
○	○	●	●	●	●	●	○		○	○	○	○	○
		●	●	●	○	○	○		●	●	●	●	○
		●	●	●	○	○	○		●	●	●	●	○

REMARKS

[1] Ratio of the effective height of a wall (h) to its effective thickness (t).
[2] With ½-inch sheathing, and ½-inch interior facing.
[3] With 2-inch air space.
[4] Ratio of unbraced length (1) to effective depth of member (d).
[5] Deflection of 1/360 when rigid gypsum base finishes attached directly to wall framing. For semirigid facings deflection may be greater.

- provisions must be made in wall framing at points of transfer of concentrated loads from floor/roof assemblies to safely support such loads.
- bracing in the plane of the wall framing must be provided either through sheathing, or by diagonal braces.
- in areas subject to high winds or to earthquakes, roof framing and walls should be tied into foundations.
- wood and metal framing may be used in the same structure; such use, however, is not recommended because of potential problems during erection.

- provisions must be made at points of transfer of concentrated loads from roof/floor assemblies to safely support and distribute such loads.
- in areas subject to high winds or to earthquakes, walls should be tied into, or anchored to, foundations.
- floor and roof framing should be anchored to walls at bearing points; and when parallel to wall to provide lateral bracing to walls.
- walls should be braced during construction.
- control joints empirically located 4-12 feet from corners, and about every 20 feet of run. Locations of control joints should be established by analysis.

- provisions to support and distribute concentrated loads from floor/roof assemblies may be required.
- walls generally tied into, or anchored to, foundations. In areas subject to high winds or earthquakes, tilt-up and pre-cast types may need special anchorage.
- for tilt-up and pre-cast types, floor and roof framing should be anchored to walls to provide lateral bracing.
- tilt-up and pre-cast types must be braced during erection.
- wood framing seldom used, especially with cast-in-place type, but could be used with all three types.

B2.4 BEARING WALLS

BEARING WALLS: load bearing capacities

NOTE: values given are for typical
assemblies based on the assumptions
listed below:

- effective height = 96"
- all mortar types for masonry

$$\text{allowable load} = \frac{\text{load-bearing capacity}}{\text{factor of safety}}$$

ASSEMBLY TYPE		WEIGHT	ALLOWABLE LOAD	REMARKS
		lbs./sq. ft.	lbs./lin. ft.	
FRAMED				
WOOD:	2"x4" STUDS (pine), 16" O.C. 3/4" EXT. PLYWOOD SHEATHING	7.0	2100	• ½" interior gypsum board reflected in weight of assembly. • compressive strength assumed: pine-f_C = 500 psi; fir-f_C = 600 psi. • strength of sheathing defines assembly resistance to lateral and shear forces.
	2"x6" STUDS (fir), 24" O.C. 5/8" EXT. GYPBOARD SHEATHING	7.5	2700	
METAL:	3⅝" ST'L STUDS, 18 GA, 16" O.C. 7/8" EST. CEMENT-LIME PLASTER on M.L.	12.0	4900	• ½" interior gypsumboard reflected in weight of assembly. • C type, punched studs assumed (1⅝" flanges). • allowable compressive strength f_S = 30,000 psi.
	3⅝" ST'L STUDS, 16 GA, 16" O.C. 5/8" EXT. GYPBOARD SHEATHING	6.0	5380	
STACKED MASONRY, SINGLE COMPONENT				
BRICK:	4" SOLID	40.0	21000	• strength of brick varies widely, from 2000 psi to 14,000 psi. • assemblies shown without grouting of cores or loose-fill insulation.
	6" SCR	65.0	33000	
	8" HOLLOW	50.0	32800	
	8" SOLID	80.0	45000	
CONCRETE BLOCK:	6" TWO-CORE	40.0	13340	• assemblies shown without grouting or loose-fill insulation. • grouting of voids in block could significantly increase strength.
	8" THREE-CORE	47.0	15300	
	8" SOLID	80.0	26860	
	12" THREE-CORE	66.0	27000	
STRUCTURAL CLAY TILE:	6"	41.0	5180	• 4" thick structural clay tile is non-loadbearing.
	8"	60.0	7900	

B2.4 BEARING WALLS

ASSEMBLY TYPE	WEIGHT lbs./sq. ft.	ALLOWABLE LOAD lbs./ft.	REMARKS
STACKED MASONRY, MULTI-COMPONENT			
8″ BRICK - 4″ SOLID+4″ HOLLOW	65.0	34000	• dense units on outside, hollow on inside in all assemblies.
8″ COMPOSITE - 4″ BRICK+4″ BLOCK	72.0	29500	
8″ BLOCK - 4″ BLOCK+4″ BLOCK	60.0	17200	
12″ COMPOSITE - 4″ BRICK+8″ BLOCK	87.0	36300	
STACKED MASONRY, CAVITY			
10″ BRICK/BLOCK, BOTH WYTHES LOADED	72.0	19650	• 66%-75% of similar composite assemblies. • 2 inch wide cavity
12″ BRICK/BLOCK, BOTH WYTHES LOADED	80.0	23600	
12″ BRICK/BLOCK, ONE WYTHE LOADED	80.0	14000	• capacity of loaded wythe only; 10″ assemblies not usually built.
STACKED UNIT, REINFORCED			
SINGLE WYTHE, 8″ BRICK	80.0	55000	• for 1/h<15, reinforcement has negligible influence on capacity for vertical axial loading. • with min. reinforcement as recommended by BIA.
SINGLE WYTHE, 8″ CONC. BLOCK (3-CORE)	48.0	19000	
MONOLITHIC CONCRETE, cast-in-place and precast*			
8″ WALL, STONE AGGREGATE	94.0	57600	• with nominal reinforcement in both directions. • concrete strength, cast-in $f_c = 3000$ psi.
8″ WALL, LIGHTWEIGHT AGGREGATE	74.0	57600	* pre-cast concrete similar, but higher allowable stresses, of $f_c = 5000$ psi and up more easily attained
8″ WALL, 2″ INTEGRAL INSULATION	55.0	43200	

B2.4 BEARING WALLS

BEARING WALLS: component options

LEGEND:
- ● denotes common usage
- ○ denotes possible usage or unlimited availability
- □ integral with wall

TYPE		EXPOSED	SURFACING				PANELS					UNIT FACINGS			
			STUCCO	STONE CHIPS IN MATRIX	CLEAR COATINGS	PIGMENTED COATINGS	STONE	SIDING, WOOD	METAL	GLASS	PORCELAIN ENAMEL [1]	STONE	MASONRY, BRICK	TEXTURE BLOCK	TERRA COTTA
FRAMED															
	WOOD		●	○				●	●			●	●	○	○
	METAL		●	○				●	●			●	●	○	○
STACKED-MASONRY															
	SINGLE COMPONENT	●	●	●	●	●	●			○	○	●	□	□	○
	MULTI-COMPONENT	●				●						□	□	□	□
	CAVITY	●				●							□	□	
CAST CONCRETE															
	IN-PLACE	●	○	●	●	●	○			○	○	○	○		○
	TILT-UP	●		○	●	○						○			
	PRECAST	●		○	●	○									

B · BUILDING SHELL

B2.4 BEARING WALLS

SIDING, WOOD, HARDBOARD	METAL	PLASTIC	SHINGLES, WOOD	MINERAL FIBER	TILE, GLAZED BLOCK	CERAMIC [2]	EXPOSED	STONE	GYPSUM BOARD	PANELING	TILE, GLAZED BLOCK	CERAMIC [2]	TERRAZZO, PRECAST PANELS	VINYL FABRIC [2]	WALLPAPER [2]	PLASTER	CLEAR COATING	PIGMENTED COATING	REMARKS
								INTERIOR FACING											
								FACINGS						COVERINGS					
●	●	●	●	○				○	●	●		●		●	●	●	●	●	• Unit facings used as veneers over sheathing.
●	●	●	●	○				○	●	●		●		●	●	●	●	●	• Stone chips may be applied over sheathing, or separate backing required. • Exterior and interior coatings applied over facing material. • Backing, such as gypsum board, recommended for paneling. • Coatings used over interior facings.
					○	●	●	●	●	●	●	●	●	●	●	●	○	●	• Exterior facings may be integral with all.
					○	○	●	●	●	●	●	●	●	●	●	●	○	●	• Exterior brick generally left exposed. • Exterior coatings may be for waterproofing only; or also for appearance. • Interior clear coatings seldom used.
					○	○	●	●	●	●	●	●	●	●	●	●	○	●	• Backing or furring generally required for paneling.
					○	●	●		●	●	○	●	○	●	●	●	○	●	• Exterior face of tilt-up and pre-cast types usually left exposed. Various finishes are available; refer to "CEMENT/CONCRETE"
						●	○	○	●	●	○	●	○	●	●	●	○	●	• Stone chips in matrix generally used for cast-in-place. • Unit stone facing may be cast integral with wall. • Gypsum board generally installed over furring as interior facing.
						●	●	○	●	●	○	●	○	●	●	●	○	●	• Backing or furring required for paneling. • Interior clear coatings seldom used.

REMARKS

[1] Furring or separate framing required.

[2] Smooth surface or backing required.

B2.4 BEARING WALLS

BEARING WALLS: selection checklist

FRAMED ASSEMBLIES

Consider framed construction:

- in low- and medium-rise projects, up to three floors or so;
- where superimposed loads are not excessive.

■ **Wood framing should be considered:**

- for economy.
- where termites and rot are not an extreme hazard.
- where fire resistance rating requirements can be met by protection from facing materials.

■ **Added protection against fire hazard may be achieved through the use of fire retardant treated framing and/or sheathing:**

- fire-retardant treated wood may also provide protection against decay
- type of treatment selected should have low hygroscopicity to minimize possibility of corrosion in fasteners

■ **Metal framing should be considered:**

- when it is competitive with wood.
- for added ease of installing electrical conduit and plumbing.
- for somewhat more resistivity to fire.
- but not in corrosive atmospheres unless extremely well protected.
- care to avoid thermal bridges should be exercised during design and erection.

STACKED MASONRY ASSEMBLY

Consider stacked masonry assemblies:

- for high-rise projects.
- where superimposed dead and live loads are greater.
- where superior fire resistance is required.
- where climatic conditions are such that the heat storage capabilities of the wall become of value to total thermal performance.
- when interior and/or exterior wall surfaces can be left exposed, without any additional facings.

■ **Choice of the particular stacked assembly will depend on a combination of many factors, including:**

- strength requirements;
- degree of protection from water/

moisture needed, cavity walls generally provide maximum impermeability.
- thermal and acoustic performance needed.
- type of exterior/interior finish needed or anticipated.
- availability of skilled labor and degree of inspection anticipated;
- local availability and cost of materials.

■ **Single wythe construction is simpler, faster, and lower in cost. On the other hand:**

- thickness is limited to 12 inches and thereby also bearing capacity and allowable height.
- it is likely to prove less water/ moisture and air tight.
- it is less likely to perform adequately as its own exterior/interior facing.

■ **Multi-wythe construction can be:**

- brick plus brick or
- face brick plus concrete block.

■ **It provides:**

- additional bearing capacity;
- more impermeability if laid with a full bed of mortar between the wythes.
- opportunity for placing reinforcement between wythes.
- opportunity for placing insulation between wythes.
- possibility of using exterior and/or interior wythes as self-facing, e.g., face brick on outside, glazed concrete block on inside.

■ **For both the single wythe and multi-wythe construction:**

- floor/roof framing should bear on the full depth of the wall; when the bearing area is less than the full depth of the wall, such eccentricity in bearing will induce tensile stresses in the wall.
- horizontal joint reinforcing to control shrinkage stresses should be provided. Joint reinforcing will not prevent cracking, but the cracks should be more numerous and narrower than would be the case without reinforcement.

Cavity walls are more difficult and costly to construct well but offer additional advantages:

■ **Any moisture which penetrates the outer wythe will be wept right out,** thus keeping the inner wythe and the interior surface dry.

■ **Air space also acts as sound and thermal barrier.**

■ **Insulation can be placed within the cavity.**

■ **The two wythes are connected with flexible ties and act more independently;** therefore, any cracks which occur in one wythe need not be transmitted to the other wythe.

■ **A cavity wall will work well only if:**

- ties and flashing are properly and carefully installed.
- weep holes are properly spaced and constructed.
- the cavity is kept open and free of mortar droppings and any other debris which may block the weep holes.

■ **In computing the bearing capacity of a cavity wall, consider:**

- if floors bear only on the inner wythe, only the cross section of that wythe is seen as the bearing surface.
- if floors bear on both wythes, only 50 percent of the outer wythe's cross section is accepted as the bearing surface.
- in all cases, only the cross section of the wythes is counted and the air space area has to be deducted.

REINFORCED MASONRY

Reinforced masonry is increasingly used today:

- for high-rise buildings.
- where heavy lateral wind and/or seismic loads prevail.
- where heavy superimposed and/or dynamic loadings occur.
- where solid conditions cause high risk of differential settlement.

■ **Vertical reinforcement is placed in several ways:**

- masonry units are placed with voids lined up and rebars dropped into position and grouted. When reinforcement is dropped after the wall is built-up to one story height, pressure grouting will be required to ensure the entire void is filled.
- rebars are placed first and open-ended units are built around them and grouted as the wall is laid-up.
- rebars are placed between the two wythes of multiwythe masonry; pressure grouting is generally required.

■ **Horizontal reinforcement is placed as the walls are laid up. It may con-**

sist of:

- plain No. 2 bars in joints.
- deformed rebars in bond beam shapes.
- truss or ladder type welded-wire joint reinforcement.
- a combination of both bond beam and joint reinforcement.

■ **Masonry can be deemed reinforced if:**

- the area of steel equals or exceeds 0.002 times the cross-sectional area of the wall (not more than two thirds of which is used in either direction).
- if the maximum spacing is less than six times the wall thickness or 48 inches.

■ **Reinforced masonry which does not meet the above requirements is defined as "partially reinforced" and can be used:**

- where spans between lateral supports exceed those permitted for unreinforced masonry;
- as shear walls against overturning moments in less active seismic regions (zones 0 and 1).
- reinforcement in partially reinforced walls is not counted toward support of superimposed loads.

■ **Control joints must be provided in all long walls of stacked masonry units to allow expansion and contraction, due to drying and wetting of wall, and to allow variation in temperature to take place:**

- The spacing of control joints will depend on the horizontal reinforcing provided; generally the more reinforcing is provided, the wider the joint spacing may be.

CAST CONCRETE ASSEMBLIES

Cast concrete bearing walls may be considered:

■ **When this is the local building tradition and skilled labor is available**

■ **Where structural requirements are severe; e.g.,**

- floors have to bear heavy machinery.
- withstand vibration.

■ **In high-rise buildings** where walls are to act as shear walls and withstand heavy wind and/or seismic forces: in such buildings the core walls may be structural concrete while the outer perimeter may be load-bearing masonry for medium height buildings or curtain walls for tall buildings.

■ **When very fast erection, especially**

in cold weather, is needed, walls may be cast integrally with floors and heated from inside to allow erection to proceed continuously.

■ **Cast concrete walls are always reinforced,** even when carrying light loads, to control shrinkage stresses:

- reinforcing may not prevent shrinkage cracks from developing, but such cracks will be numerous hairline cracks rather than few wide cracks which would develop if reinforcing were not provided.

Precast concrete bearing walls may be considered where:

■ **Large elements are available within transportable distance.**

■ **When a decorative integral exterior finish,** a three-dimensional molded exterior face, and/or integral insulation is required.

■ **Minimum on-site work and speed of erection are important.**

■ **Note that in high-rise structures,** care must be taken in design to prevent progressive collapse of all vertically stacked units when one of them is destroyed.

■ **Precast concrete bearing walls** are always reinforced to resist handling and erection stresses, and to control shrinkage:

- generally, the amount of reinforcing provided to resist handling and erection stresses is more than adequate to control shrinkage.
- reinforcing may be incorporated to tie individual panels at joints to resist high wind and/or seismic forces.

Tilt-up concrete bearing walls are essentially site precast bearing walls and may be considered where:

■ **Economical walls for low-rise structures are required.**

■ **Walls may be used as deep girders** to support floor and roof framing on the same unit.

■ **Sufficient space** is available at the site to cast panels.

■ **Three-dimensional molded shapes,** or special finishes are not required.

■ **Special methods,** such as site-cast walls and floors with hinged wall and floor/roof panels to speed erection may be employed in low- and medium-rise structures.

■ **As the wall panels are generally cast in the open,** weather conditions

will greatly affect the construction schedule during that phase of the work.

■ **Reinforcing for tilt-up walls is** provided in the same way and for the same reasons as for precast walls.

■ **Tilt-up walls** may be cast on the ground floor slab, or on specially prepared casting beds, and the wall panels may be stacked.

■ **Erection of precast and tilt-up walls** requires the walls to be erected first and braced before floor/roof assemblies are constructed. Such bracing is an important part of the erection process, and has to be considered during the design phase.

■ **The tilt-up walls are cast at the site and therefore:**

- their size is not limited by transportation limitations.
- the capacity of the equipment used for erection will generally determine the maximum size, with six to eight hundred square feet being a good average.

■ **Floor/roof framing** must be tied into both precast and tilt-up walls for lateral stability; wall panels must be connected across joints.

■ **Various finishes:** exposed aggregate; patterns using liners or mats; cast-in large stones are available.

■ **Preassembled masonry panels** may be seen as another type of monolithic load-bearing construction.

In considering these, remember;

- size limitations for ease of erection.
- additional stresses at connections to floors and other bearing members and the possible resulting requirements for special connections.

ADDITIONAL CONSIDERATIONS

■ **Metal stud framing** is more dependent upon wind-load restrictions than wood framing when bearing capacity is reviewed. Since metal studs are formed from sheet material, their load capacity will be regulated by crippling and buckling which normally occurs well below the bearing stress limits.

■ **The strength of surface bonded concrete masonry bearing walls:**

- depends on the proper application of surface bonding mortar.
- when ground masonry units are used, the strength will equal that of a conventionally constructed concrete masonry-bearing wall.

B2.4 BEARING WALLS

B • BUILDING SHELL

BEARING WALLS: insulating values,

- concrete masonry values acc. to NCMA
- brick masonry values acc. to BIA

U-value at 40°F
Exterior surface film   R = 0.17
Interior surface film   R = 0.65
3/4" air space   R = 0.95
1/2" gypsum board   R = 0.45

Note:
- For thermal resistances of various insulating materials refer to INSULATION
- For information on heat flow through various assemblies refer to ENVIRONMENT.

TYPE OF INSULATION	SINGLE WYTHE								
	BRICK, CLAY			CONCRETE			BLOCK		
	4"	6"	8"	6"			8"		
	REINF. HOLLOW	HOLLOW	HOLLOW	80 LBS. DENSITY	100 LBS.	120 LBS.	80 LBS.	100 LBS.	120 LBS.
UNINSULATED									
• exposed both faces	.66	.58	.54	.37	.42	.47	.34	.38	.43
• 1/2" gypsum① on 3/4" furring	.34	.32	.31	.24	.26	.28	.23	.25	.27
• 1/2" foil-backed gypsum on 3/4" furring	.21	.20	.20	.16	.17	.18	.15	.16	.17
LOOSE-FILL (in cores)									
• exposed both faces (mineral)	.53	.43	.37	.18	.22	.26	.14	.18	.21
• 1/2" gypsum on 3/4" furring in	.30	.27	.24	.15	.17	.19	.12	.14	.17
• 1/2" foil-backed gypsum board on furring	.19	.17	.16	.11	.12	.14	.10	.11	.12
BOARD									
• 1" glass fiber — direct to	.20	.18	.17	.14	.15	.15	.14	.15	.15
• 1" polystyrene foam — surface with	.17	.16	.16	.12	.13	.13	.12	.13	.13
• 1" polyurethane — 1/2" gypsum board	.12	.12	.12	.11	.11	.11	.10	.11	.11
• R-7 blanket insul.	.11	.11	.11	.09	.10	.10	.09	.10	.10
• double-furred system with 1" polystyrene, two air spaces and 1/2" gypsum board	.13	.12	.12	.08	.08	.08	.08	.08	.08
BOARD & LOOSE-FILL									
• 1" glass fiber — perlite/vermiculite	.17	.16	.15	.10	.11	.12	.09	.10	.11
• 1" polystyrene — in cores	.16	.15	.14	.09	.10	.11	.08	.09	.10
• 1" polyurethane — + 1/2" gypsum board	.12	.11	.11	.08	.09	.10	.07	.08	.09
FOAM									
• in cores (urethane)	.44	.34	.28	.15	.17	.22	.10	.13	.16
• on inside, mineral fiber — w/1/2" gypsum	.17	.16	.12	.12	.13	.13	.12	.13	.13
• on inside, urethanes — board over	.12	.12	.12	.11	.11	.11	.10	.11	.11

B2.4 BEARING WALLS

B • BUILDING SHELL

stacked assemblies

COMPOSITE															REMARKS
12"			BRICK+BLOCK 4"+4"			4"+8"			BRICK+BLOCK 4"+4" (dense on outside, hollow on inside)			4"+8"			U-value of framed assemblies depends largely on resistance and placement of insulation. U-value of cavity walls and of monolithic assemblies calculated similarly to assemblies shown.
80 LBS.	100 LBS.	120 LBS.	80 LBS.	100 LBS.	120 LBS.	80 LBS.	100 LBS.	120 LBS.	80 LBS.	100 LBS.	120 LBS.	80 LBS.	100LBS.	120 LBS.	

12" 80	12" 100	12" 120	4+4 80	4+4 100	4+4 120	4+8 80	4+8 100	4+8 120	d4+4 80	d4+4 100	d4+4 120	d4+8 80	d4+8 100	d4+8 120	REMARKS
.29	.34	.38	.31	.34	.37	.27	.30	.34	.34	.37	.41	.29	.33	.36	① for paneling assume
.20	.23	.25	.22	.24	.25	.20	.22	.23	.23	.24	.26	.21	.22	.24	½" plywood .63
.14	.15	.16	.15	.16	.16	.14	.15	.15	.15	.16	.17	.14	.15	.16	¾" hardwood .78 / ¾" softwood .60
.10	.13	.17	.28*	.30*	.33*	.13**	.15**	.18**	N/A			.14	.16	.20	*loose-fill in brick only
.09	.11	.13	.21*	.22*	.23*	.11**	.13**	.14**				.11	.13	.15	**loose-fill in brick and in block
.08	.09	.10	.14*	.15*	.15*	.10**	.12**	.13**				.09	.10	.12	
.13	.13	.14	.14	.15	.15	.15	.16	.16	.14	.14	.14	.13	.13	.14	any number of variations exist, including:
.11	.12	.12	.13	.14	.14	.14	.14	.15	.12	.12	.13	.11	.12	.12	• double air spaces with foil-backed sheathing
.10	.10	.11	.10	.10	.11	.10	.11	.11	.10	.11	.11	.10	.10	.11	• patented thermal-break furred systems
.09	.10	.10	.09	.10	.10	.10	.10	.10	.10	.10	.10	.09	.10	.10	• board applications of 3" and more
.08	.08	.08	.08	.09	.09	.08	.08	.08	.09	.09	.09	.08	.08	.08	• clip-on PVC furring strips
.07	.08	.10	.13*	.14*	.14*	.08**	.09**	.10**	N/A			.08	.09	.10	*loose-fill in brick only
.07	.08	.09	.13*	.13*	.13*	.08**	.08**	.10**				.08	.09	.10	**loose-fill in brick and in block
.06	.07	.08	.10*	.10*	.10*	.07**	.07**	.08**				.07	.08	.09	
.07	.09	.12	.28	.31	.33	.08	.10	.12	N/A			.09	.11	.14	• spray-on insulation used where fireproofing and acoustic performance needed and/or over irregular/exposed surfaces
.11	.12	.12	.13	.14	.14	.14	.14	.15				.11	.12	.12	
.10	.10	.11	.10	.10	.11	.10	.11	.11				.10	.10	.10	

B2.4 BEARING WALLS

B • BUILDING SHELL

WALL INSULATION: selection checklist

INSULATION AND VAPOR RETARDERS

Selection of the type of insulation to be used should also include consideration of the method of its installation within the wall assembly. Framed bearing-wall assemblies may be insulated by:

■ **Batt insulation,** between studs, generally with an integral vapor retarder.

- insulation should be packed behind outlet and switch boxes, and any piping or wiring run through the studs.
- take care to fill the space between studs completely from top to bottom to avoid the possibility of convective air currents developing with resultant high heat losses.
- insulation which is not the full thickness of the stud generally is installed on the warm side of the wall.

■ **Foamed-in-place insulation between studs.**

- place insulation after all piping and wiring work has been completed within the walls.
- take care to fill all spaces behind outlet and switch boxes with insulation.
- Foamed-in-place insulation does not allow heat buildup in wiring, when such wiring is completely embedded in it, to dissipate; this heat buildup, if not prevented, may lead to premature degradation of insulation and a potential fire hazard.

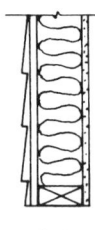 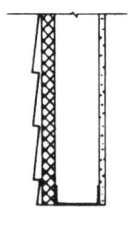

BATT   RIGID BOARD

■ Rigid-board insulation sheathing secured to the exterior face of studs:

- when more insulation than can be provided within the thickness of the studs is required
- the rigid board insulation sheath-

ing is not structural and diagonal bracing has to be provided for the wall framing.
- For additional information on insulating materials, refer to INSULATION.

Single wythe and multiwythe stacked assemblies may be insulated by:

■ **Loose insulation fill within the voids of concrete masonry units:**

- such insulation must be moisture-resistant as water vapor may condense within the voids.
- weep holes may have to be provided to drain the voids.
- after top of wall is covered, no openings may be punched in the faces of the masonry units or the insulation within may all run out.

■ **Shredded rigid insulation is also used to fill voids within concrete masonry units;** care must be taken to ensure that the voids are completely filled.

■ **Foamed-in-place insulation** may be placed within the voids of masonry units.

■ **Concrete masonry units** are available with rigid insulation inserts cast-in during fabrication.

■ **Rigid board insulation laminated,** or clip attached, inside face of wall. Furring strips may have to be provided between the boards to facilitate attachment of interior facing materials.

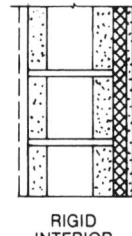 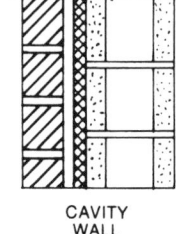

RIGID   CAVITY
INTERIOR   WALL

Cavity wall stacked assemblies may be insulated in all the ways listed above and also:

■ **Rigid board insulation secured to exterior face of interior wythe (within the cavity):**

- unless the exterior wythe is built separately, or is kept back during construction installation of rigid board, insulation becomes very difficult.

- if the insulation fills the entire cavity, it no longer acts as an air space and may not function to prevent moisture penetration.

■ **Loose insulation fill within the cavity.**

■ **Foamed-in-place insulation within the cavity.**

Monolithic in-place concrete assemblies may be insulated with rigid board insulation laminated or clip attached to interior face.

Precast and tilt-up concrete assemblies may be insulated by:

■ **Rigid board insulation between interior and exterior courses of a sandwich panel.**

- thickness of insulation is generally limited to about two inches, and fabrication is relatively expensive.
- interior and exterior faces have to be tied with metal ties through the insulation.
- the perimeter of panel is of solid concrete to tie the two courses together, creating thermal bridges resulting in higher heat loss and somewhat uneven temperatures across the surface.

■ **Rigid board insulation laminated or clip attached to interior face.**

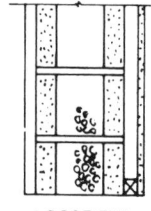 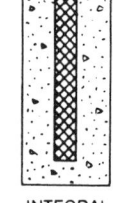

LOOSE FILL   INTEGRAL

All stacked and monolithic assemblies may be:

■ **Furred out to provide an air space or**

■ **The space between furring and interior facing may be filled with insulation,** either rigid board or batt.

Water vapor, if allowed to penetrate into the wall assembly, may condense on surfaces with temperatures below the dew point; such condensation may

B2.4 BEARING WALLS

B • BUILDING SHELL

cause a rapid deterioration of the entire wall assembly.

■ **Water vapor migration** is either by diffusion, or by air leakage; and is generally controlled by providing a vapor retarder on the warm side of the wall:

• vapor retarders consist of materials that resist the diffusion of vapor through them under the action of a difference in pressure, such as plastic film, metallic foil, coated paper, and, to a certain extent, applied coatings.
• vapor retarders should be continuous over entire wall surface, and properly sealed around all penetrations, such as outlet or switch boxes.
• properly installed vapor retarders will also prevent or minimize air leakage, which often is the major source of harmful condensation in wall assemblies.

■ **Walls in existing buildings** may be insulated with blown-in or foamed-in insulation, however, consideration should be given to:

• ensuring that all voids in wall assembly are completely filled.
• possible settlement of blown-in insulation.
• water vapor migrating by diffusion or by air leakage into the wall assembly, condensing within the wall and causing rapid deterioration of exterior facing or even the wall assembly itself.

UTILIZATION OF HEAT STORAGE CAPACITY

■ Heat storage capacity of walls can be used to significant advantage:

• to reduce peak heat gain and thereby reduce cooling loads on mechanical equipment
• to reduce heat losses through time lag.
• to store heat absorbed through solar energy and release it when needed.

The calculation of heat gain/loss of a structure based on specific design outdoor temperatures for winter and summer, the "steady state" method, does not accurately determine the maximum heating and especially not the maximum cooling loads over a 24-hour period. As temperatures rise on one side of

the wall exposed to sunlight during a specific period of time, heat starts flowing towards the cool side.

Before heat transfer through the wall can be achieved, the temperature of the wall assembly itself must rise above temperatures on the cool side of the wall.

■ **if the period of time the wall is exposed to sunlight is of limited duration and**

■ **the amount of heat required to raise the temperature of the wall assembly itself is large, no heat transfer, or heat gain, to the interior may occur.**

Time lag is a factor of considerable importance, especially in calculating heat gain where temperature variations are such that the direction of heat flow may be reversed one or more times in a 24-hour period.

The "dynamic analysis" of heat gain/loss takes into account the hourly changes in weather conditions as well as the thermal storage capacity of the structure, and closely predicts the peak loads required to determine the size of equipment needed to control the interior environment of a structure:

• In tests performed by the National Bureau of Standards, reported in Building Science Series 45, "Dynamic Thermal Performance of an Experimental Masonry Building", July 1973, it was found that "steady state" calculations gave heat flow rates about 52 percent higher than those actually measured.

Based on above tests, the National Concrete Masonry Association investigated the effects of building weight on the capacity of equipment for cooling the building. Cooling loads are selected because:

■ cooling loads generally consume more energy than heating loads.

■ "steady state" is typically more nearly approached in winter than in summer. The design day criteria used in the dynamic analysis were: 65°F to 90°F for a 24-hour period.

The results of this investigation indicate that substituting concrete masonry walls with a U-value of 0.10 for insulated wood frame walls with the same U-value, results in reducing the peak cooling load for

the particular test building from 40,500 BTU/hr to 33,900 BTU/hr. The effect of the weight of building elements on the cooling load for the test building is illustrated in the table below:

Weight of Roof, Walls and Floors, lbs. per sq. ft.	Calculated Maximum Cooling Load	
	BTU/hr.	Percent
10	35300	100
15	34100	97
20	32700	93
25	31400	89
30	30200	86
35	29200	83
40	28300	80
45	27500	78
50	26800	76
55	26200	74
60	25600	72
65	25100	71
70	24600	70

Comparison of Peak Load for Cooling When Weight of Structure Varies but U-Value Remains Constant

B2.4 BEARING WALLS

BEARING WALLS: anchorage, connections

FRAMED

FOR LOW-PITCHED ROOFS PROVIDE 1 OR 2 LAYERS OF **ROOFING FELT** OVER SHEATHING. **CONTINUOUS FLASHING** TO EXTEND PAST WALL/CEILING INTERSECTION, TO PREVENT WATER PENETRATION IF ICE DAMS FORM AND MELT AT EAVES.

GUTTERS IN AREAS OF HIGH PRECIPITATION.

FASCIA BOARD: 1X BACKING RECOMMENDED TO FACILITATE INSTALLATION OF FASCIA BOARD. WIDE BOARDS SUBJECT TO WARPING AND SPLITTING. SHOULD BE BACK PAINTED.

CONTINUOUS **SCREENED VENTS,** TO PREVENT MOISTURE BUILD-UP IN ATTIC SPACE. INSULATION SHOULD NOT BLOCK AIR FLOW TO ATTIC SPACE.

SOLID **BLOCKING** IN ALL CASES TO PROVIDE FOR NAILING.

RIGID BOARDS OF FOAMED INSULATION, SUCH AS EPS OR ISOCYANURATE MAY BE USED OVER SHEATHING FOR ADDED THERMAL RESISTANCE OF WALL.

HORIZONTAL **SIDING**/VERTICAL SIDING OF A VARIETY OF MATERIALS (WOOD, ALUMINUM, ETC.) SHINGLES, BRICK OR STONE VENEER.

GROUT OR FIBERBOARD TO LEVEL TOP OF FOUNDATION WALL FOR FULL BEARING

TERMITE SHIELD WHERE NECESSARY.

SLOPE GRADE AWAY FROM BUILDING.

WATERPROOFING, DAMPPROOFING AS REQUIRED.

PROVIDE **LEDGE** FOR BRICK OR STONE VENEER MIN. 2 COURSES BELOW FIN. GRADE AS REQUIRED.

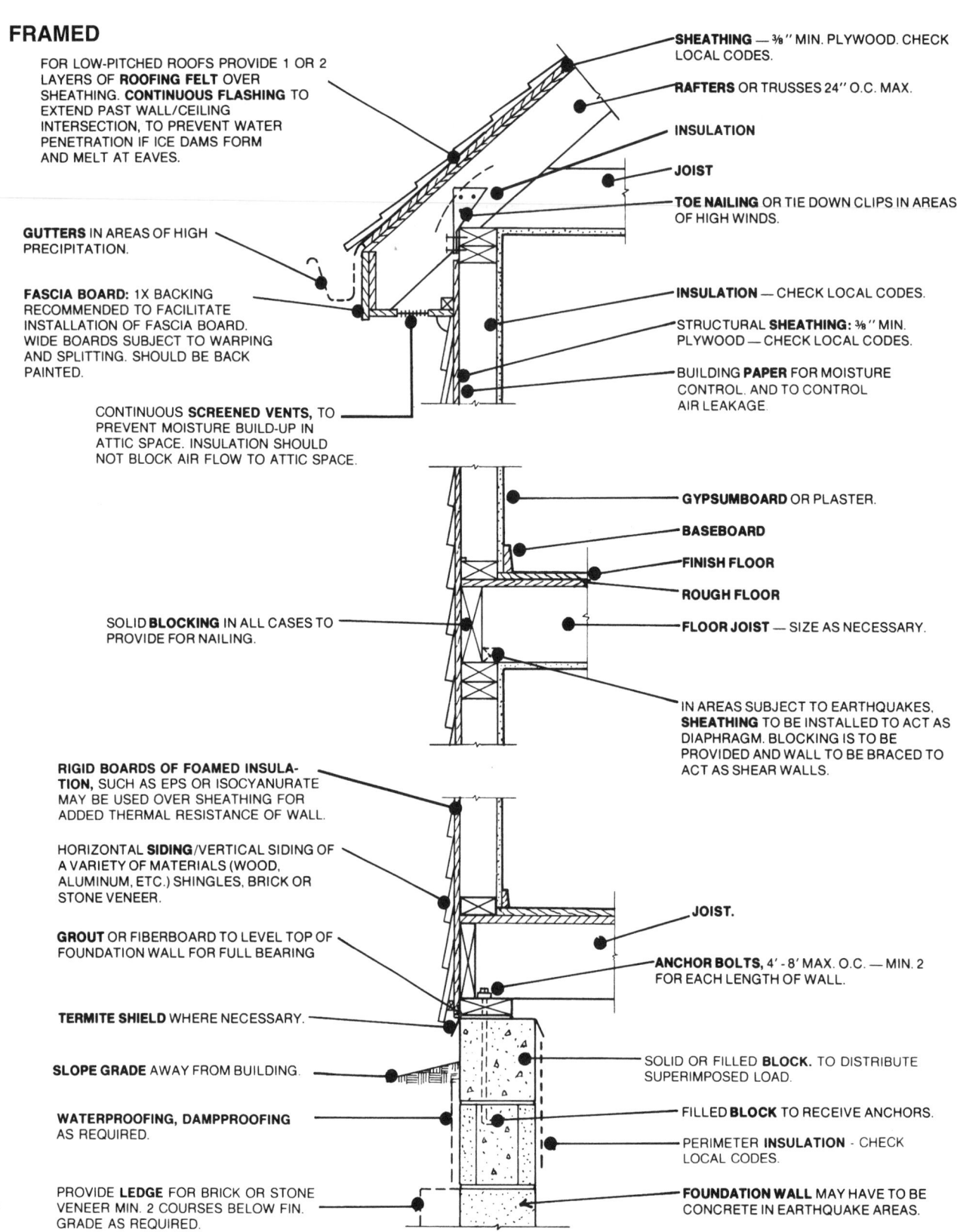

SHEATHING — ⅜″ MIN. PLYWOOD. CHECK LOCAL CODES.

RAFTERS OR TRUSSES 24″ O.C. MAX.

INSULATION

JOIST

TOE NAILING OR TIE DOWN CLIPS IN AREAS OF HIGH WINDS.

INSULATION — CHECK LOCAL CODES.

STRUCTURAL **SHEATHING:** ⅜″ MIN. PLYWOOD — CHECK LOCAL CODES.

BUILDING **PAPER** FOR MOISTURE CONTROL. AND TO CONTROL AIR LEAKAGE.

GYPSUMBOARD OR PLASTER.

BASEBOARD

FINISH FLOOR

ROUGH FLOOR

FLOOR JOIST — SIZE AS NECESSARY.

IN AREAS SUBJECT TO EARTHQUAKES, **SHEATHING** TO BE INSTALLED TO ACT AS DIAPHRAGM. BLOCKING IS TO BE PROVIDED AND WALL TO BE BRACED TO ACT AS SHEAR WALLS.

JOIST.

ANCHOR BOLTS, 4′-8′ MAX. O.C. — MIN. 2 FOR EACH LENGTH OF WALL.

SOLID OR FILLED **BLOCK.** TO DISTRIBUTE SUPERIMPOSED LOAD.

FILLED **BLOCK** TO RECEIVE ANCHORS.

PERIMETER **INSULATION** - CHECK LOCAL CODES.

FOUNDATION WALL MAY HAVE TO BE CONCRETE IN EARTHQUAKE AREAS.

B2.4 BEARING WALLS

B • BUILDING SHELL

MASONRY: SINGLE WYTHE

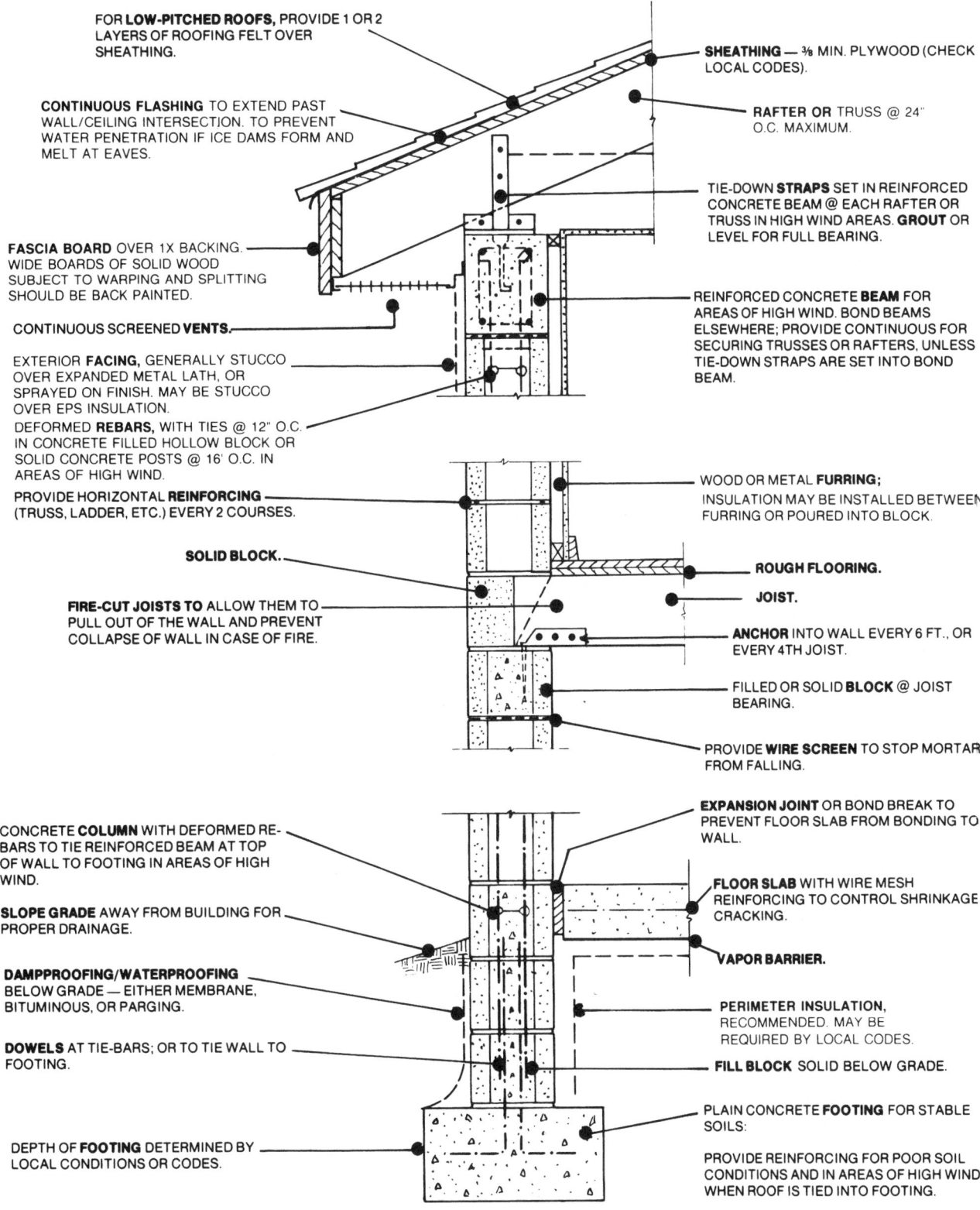

FOR **LOW-PITCHED ROOFS,** PROVIDE 1 OR 2 LAYERS OF ROOFING FELT OVER SHEATHING.

CONTINUOUS FLASHING TO EXTEND PAST WALL/CEILING INTERSECTION. TO PREVENT WATER PENETRATION IF ICE DAMS FORM AND MELT AT EAVES.

FASCIA BOARD OVER 1X BACKING. WIDE BOARDS OF SOLID WOOD SUBJECT TO WARPING AND SPLITTING SHOULD BE BACK PAINTED.

CONTINUOUS SCREENED **VENTS.**

EXTERIOR **FACING,** GENERALLY STUCCO OVER EXPANDED METAL LATH, OR SPRAYED ON FINISH. MAY BE STUCCO OVER EPS INSULATION.

DEFORMED **REBARS,** WITH TIES @ 12" O.C. IN CONCRETE FILLED HOLLOW BLOCK OR SOLID CONCRETE POSTS @ 16' O.C. IN AREAS OF HIGH WIND.

PROVIDE HORIZONTAL **REINFORCING** (TRUSS, LADDER, ETC.) EVERY 2 COURSES.

SOLID BLOCK.

FIRE-CUT JOISTS TO ALLOW THEM TO PULL OUT OF THE WALL AND PREVENT COLLAPSE OF WALL IN CASE OF FIRE.

CONCRETE **COLUMN** WITH DEFORMED RE-BARS TO TIE REINFORCED BEAM AT TOP OF WALL TO FOOTING IN AREAS OF HIGH WIND.

SLOPE GRADE AWAY FROM BUILDING FOR PROPER DRAINAGE.

DAMPPROOFING/WATERPROOFING BELOW GRADE — EITHER MEMBRANE, BITUMINOUS, OR PARGING.

DOWELS AT TIE-BARS; OR TO TIE WALL TO FOOTING.

DEPTH OF **FOOTING** DETERMINED BY LOCAL CONDITIONS OR CODES.

SHEATHING — ⅜ MIN. PLYWOOD (CHECK LOCAL CODES).

RAFTER OR TRUSS @ 24" O.C. MAXIMUM.

TIE-DOWN **STRAPS** SET IN REINFORCED CONCRETE BEAM @ EACH RAFTER OR TRUSS IN HIGH WIND AREAS. **GROUT** OR LEVEL FOR FULL BEARING.

REINFORCED CONCRETE **BEAM** FOR AREAS OF HIGH WIND. BOND BEAMS ELSEWHERE; PROVIDE CONTINUOUS FOR SECURING TRUSSES OR RAFTERS, UNLESS TIE-DOWN STRAPS ARE SET INTO BOND BEAM.

WOOD OR METAL **FURRING;** INSULATION MAY BE INSTALLED BETWEEN FURRING OR POURED INTO BLOCK.

ROUGH FLOORING.

JOIST.

ANCHOR INTO WALL EVERY 6 FT., OR EVERY 4TH JOIST.

FILLED OR SOLID **BLOCK** @ JOIST BEARING.

PROVIDE **WIRE SCREEN** TO STOP MORTAR FROM FALLING.

EXPANSION JOINT OR BOND BREAK TO PREVENT FLOOR SLAB FROM BONDING TO WALL.

FLOOR SLAB WITH WIRE MESH REINFORCING TO CONTROL SHRINKAGE CRACKING.

VAPOR BARRIER.

PERIMETER INSULATION, RECOMMENDED. MAY BE REQUIRED BY LOCAL CODES.

FILL BLOCK SOLID BELOW GRADE.

PLAIN CONCRETE **FOOTING** FOR STABLE SOILS:

PROVIDE REINFORCING FOR POOR SOIL CONDITIONS AND IN AREAS OF HIGH WIND WHEN ROOF IS TIED INTO FOOTING.

B2.4 BEARING WALLS

BEARING WALLS: anchorage, connections (cont.)

STACKED: MULTI-WYTHE

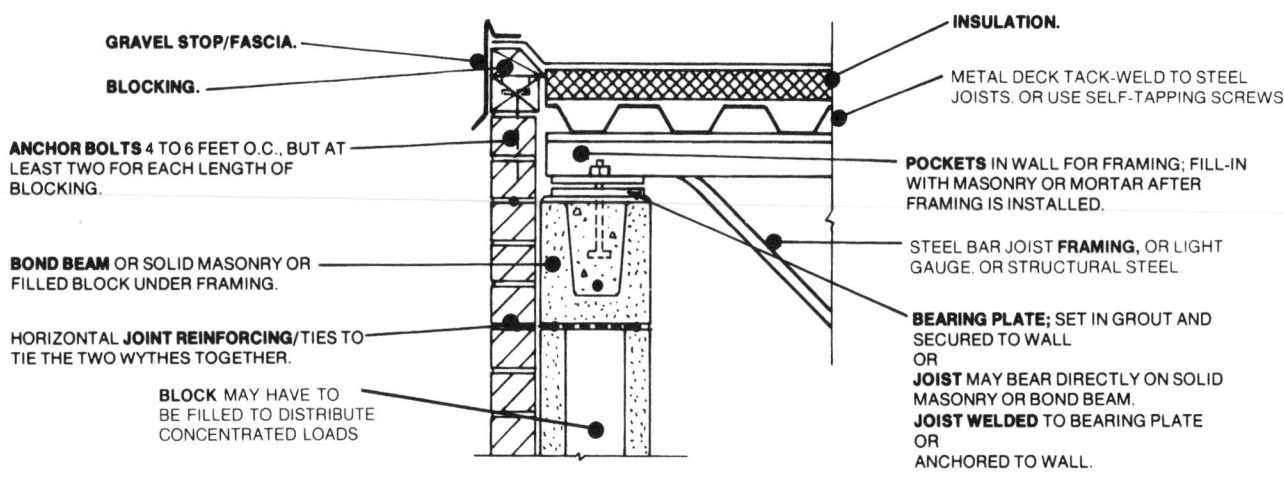

GRAVEL STOP/FASCIA.

BLOCKING.

ANCHOR BOLTS 4 TO 6 FEET O.C., BUT AT LEAST TWO FOR EACH LENGTH OF BLOCKING.

BOND BEAM OR SOLID MASONRY OR FILLED BLOCK UNDER FRAMING.

HORIZONTAL **JOINT REINFORCING**/TIES TO TIE THE TWO WYTHES TOGETHER.

BLOCK MAY HAVE TO BE FILLED TO DISTRIBUTE CONCENTRATED LOADS

INSULATION.

METAL DECK TACK-WELD TO STEEL JOISTS. OR USE SELF-TAPPING SCREWS.

POCKETS IN WALL FOR FRAMING; FILL-IN WITH MASONRY OR MORTAR AFTER FRAMING IS INSTALLED.

STEEL BAR JOIST **FRAMING,** OR LIGHT GAUGE. OR STRUCTURAL STEEL.

BEARING PLATE; SET IN GROUT AND SECURED TO WALL
OR
JOIST MAY BEAR DIRECTLY ON SOLID MASONRY OR BOND BEAM.
JOIST WELDED TO BEARING PLATE
OR
ANCHORED TO WALL.

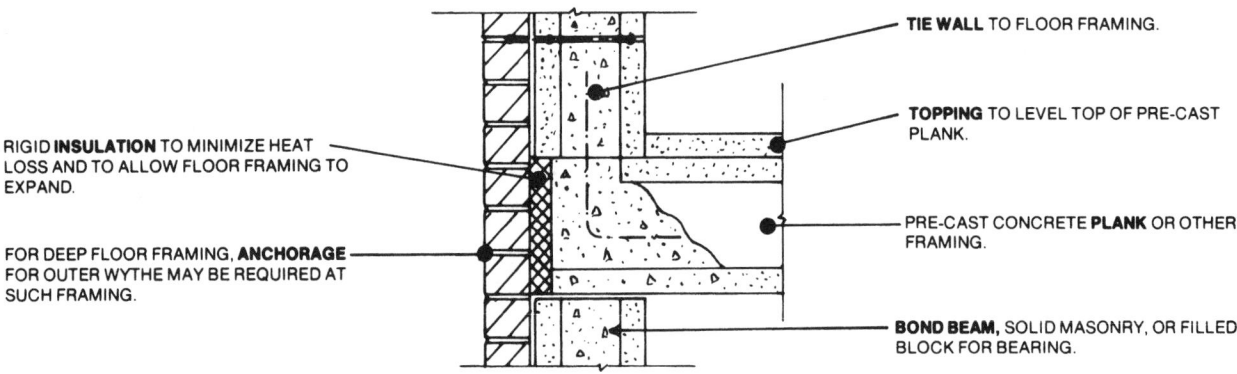

RIGID **INSULATION** TO MINIMIZE HEAT LOSS AND TO ALLOW FLOOR FRAMING TO EXPAND.

FOR DEEP FLOOR FRAMING, **ANCHORAGE** FOR OUTER WYTHE MAY BE REQUIRED AT SUCH FRAMING.

TIE WALL TO FLOOR FRAMING.

TOPPING TO LEVEL TOP OF PRE-CAST PLANK.

PRE-CAST CONCRETE **PLANK** OR OTHER FRAMING.

BOND BEAM, SOLID MASONRY, OR FILLED BLOCK FOR BEARING.

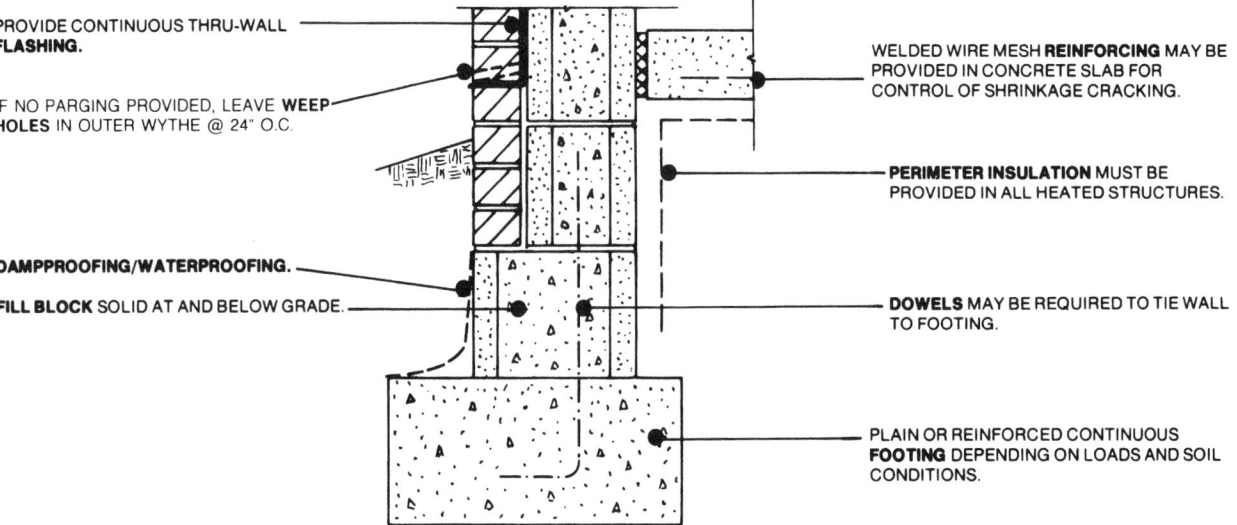

PROVIDE CONTINUOUS THRU-WALL **FLASHING.**

IF NO PARGING PROVIDED, LEAVE **WEEP HOLES** IN OUTER WYTHE @ 24" O.C.

DAMPPROOFING/WATERPROOFING.

FILL BLOCK SOLID AT AND BELOW GRADE.

WELDED WIRE MESH **REINFORCING** MAY BE PROVIDED IN CONCRETE SLAB FOR CONTROL OF SHRINKAGE CRACKING.

PERIMETER INSULATION MUST BE PROVIDED IN ALL HEATED STRUCTURES.

DOWELS MAY BE REQUIRED TO TIE WALL TO FOOTING.

PLAIN OR REINFORCED CONTINUOUS **FOOTING** DEPENDING ON LOADS AND SOIL CONDITIONS.

B2.4 BEARING WALLS

CONCRETE: PRE-CAST AND TILT-UP

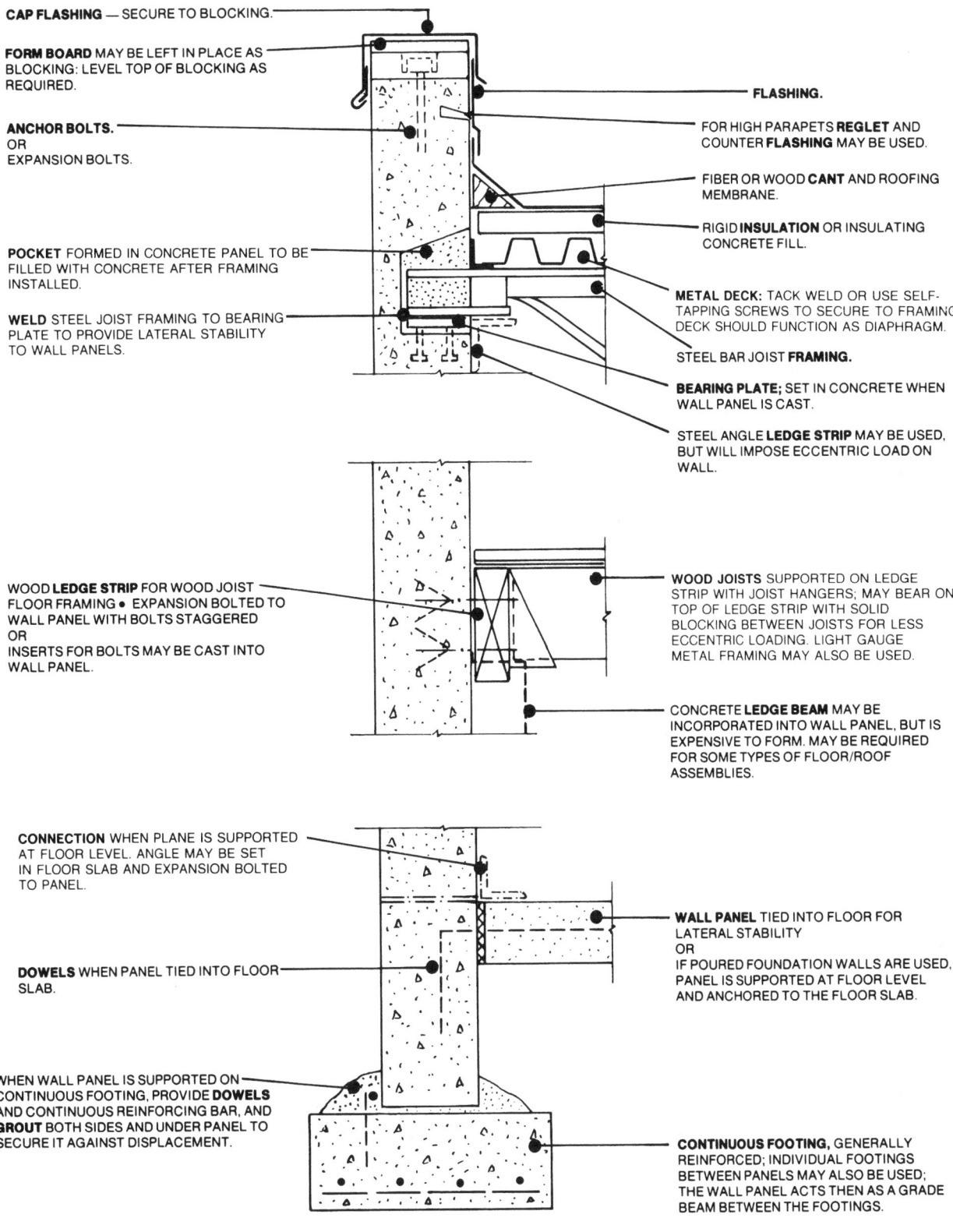

CAP FLASHING — SECURE TO BLOCKING.

FORM BOARD MAY BE LEFT IN PLACE AS BLOCKING: LEVEL TOP OF BLOCKING AS REQUIRED.

ANCHOR BOLTS.
OR
EXPANSION BOLTS.

POCKET FORMED IN CONCRETE PANEL TO BE FILLED WITH CONCRETE AFTER FRAMING INSTALLED.

WELD STEEL JOIST FRAMING TO BEARING PLATE TO PROVIDE LATERAL STABILITY TO WALL PANELS.

FLASHING.

FOR HIGH PARAPETS **REGLET** AND COUNTER **FLASHING** MAY BE USED.

FIBER OR WOOD **CANT** AND ROOFING MEMBRANE.

RIGID **INSULATION** OR INSULATING CONCRETE FILL.

METAL DECK: TACK WELD OR USE SELF-TAPPING SCREWS TO SECURE TO FRAMING DECK SHOULD FUNCTION AS DIAPHRAGM.

STEEL BAR JOIST **FRAMING.**

BEARING PLATE; SET IN CONCRETE WHEN WALL PANEL IS CAST.

STEEL ANGLE **LEDGE STRIP** MAY BE USED, BUT WILL IMPOSE ECCENTRIC LOAD ON WALL.

WOOD **LEDGE STRIP** FOR WOOD JOIST FLOOR FRAMING • EXPANSION BOLTED TO WALL PANEL WITH BOLTS STAGGERED
OR
INSERTS FOR BOLTS MAY BE CAST INTO WALL PANEL.

WOOD JOISTS SUPPORTED ON LEDGE STRIP WITH JOIST HANGERS; MAY BEAR ON TOP OF LEDGE STRIP WITH SOLID BLOCKING BETWEEN JOISTS FOR LESS ECCENTRIC LOADING. LIGHT GAUGE METAL FRAMING MAY ALSO BE USED.

CONCRETE **LEDGE BEAM** MAY BE INCORPORATED INTO WALL PANEL, BUT IS EXPENSIVE TO FORM. MAY BE REQUIRED FOR SOME TYPES OF FLOOR/ROOF ASSEMBLIES.

CONNECTION WHEN PLANE IS SUPPORTED AT FLOOR LEVEL. ANGLE MAY BE SET IN FLOOR SLAB AND EXPANSION BOLTED TO PANEL.

DOWELS WHEN PANEL TIED INTO FLOOR SLAB.

WHEN WALL PANEL IS SUPPORTED ON CONTINUOUS FOOTING, PROVIDE **DOWELS** AND CONTINUOUS REINFORCING BAR, AND **GROUT** BOTH SIDES AND UNDER PANEL TO SECURE IT AGAINST DISPLACEMENT.

WALL PANEL TIED INTO FLOOR FOR LATERAL STABILITY
OR
IF POURED FOUNDATION WALLS ARE USED, PANEL IS SUPPORTED AT FLOOR LEVEL AND ANCHORED TO THE FLOOR SLAB.

CONTINUOUS FOOTING, GENERALLY REINFORCED; INDIVIDUAL FOOTINGS BETWEEN PANELS MAY ALSO BE USED; THE WALL PANEL ACTS THEN AS A GRADE BEAM BETWEEN THE FOOTINGS.

B2.4 BEARING WALLS

BEARING WALLS: used as curtain walls

FRAMED

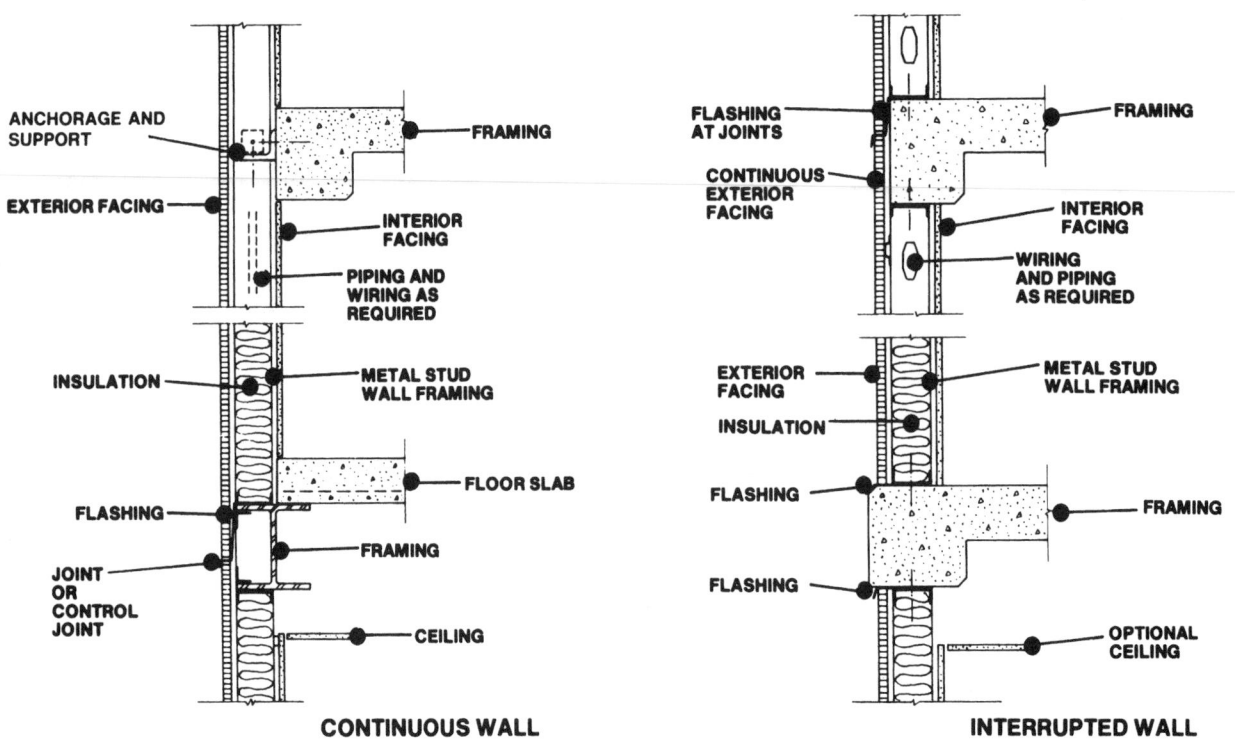

ANCHORAGE AND SUPPORT
EXTERIOR FACING
INTERIOR FACING
PIPING AND WIRING AS REQUIRED
INSULATION
METAL STUD WALL FRAMING
FLASHING
FLOOR SLAB
JOINT OR CONTROL JOINT
FRAMING
CEILING
FRAMING

CONTINUOUS WALL

FLASHING AT JOINTS
CONTINUOUS EXTERIOR FACING
FRAMING
INTERIOR FACING
WIRING AND PIPING AS REQUIRED
EXTERIOR FACING
INSULATION
METAL STUD WALL FRAMING
FLASHING
FRAMING
FLASHING
OPTIONAL CEILING

INTERRUPTED WALL

STACKED

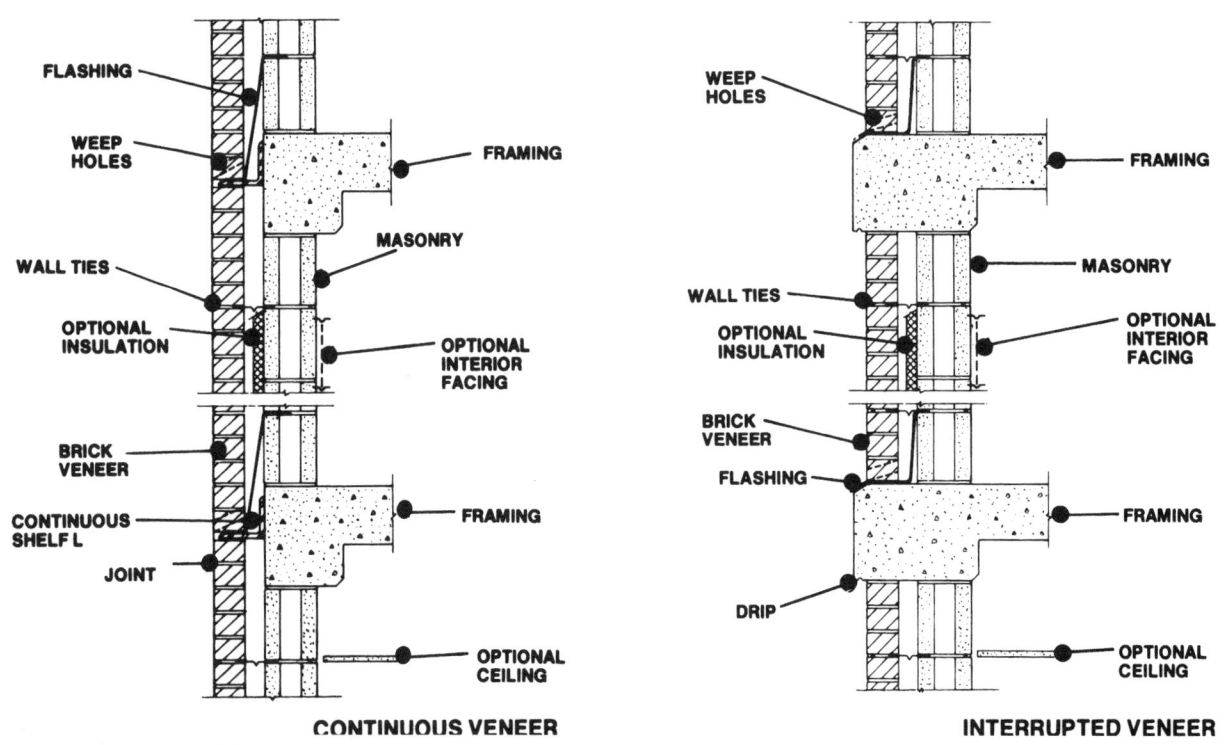

FLASHING
WEEP HOLES
FRAMING
WALL TIES
MASONRY
OPTIONAL INSULATION
OPTIONAL INTERIOR FACING
BRICK VENEER
CONTINUOUS SHELF L
FRAMING
JOINT
OPTIONAL CEILING

CONTINUOUS VENEER

WEEP HOLES
FRAMING
MASONRY
WALL TIES
OPTIONAL INSULATION
OPTIONAL INTERIOR FACING
BRICK VENEER
FLASHING
FRAMING
DRIP
OPTIONAL CEILING

INTERRUPTED VENEER

B2.4 BEARING WALLS

description:
- **metal framed** bearing wall used as a non-bearing curtain wall; generally when a limited number of openings in wall is provided.
- not practical when large, closely spaced openings are required.
- wall is **site assembled** using standard readily available components; no special equipment is required.
- brick veneer may be used as continuous **exterior facing** for the interrupted wall type.
- exposed edge of framing in the interrupted wall type provides a thermal bridge, resulting in high heat loss.
- insulation is generally provided between the metal studs.

installation:
- in the continuous wall, metal studs are installed outside the structural frame; supported from the frame by clip angles at each stud, or supported on the structural steel spandrel beams.
- in the interrupted wall, metal studs are installed between framing assemblies, and secured top and bottom to such assemblies; the exterior facing may either be continuous over the outside face of structural framing, or the edge of the structural framing may be exposed.
- continuous flashings should be provided to drain any moisture which might penetrate the exterior facing, or condense within the wall to the outside.

limitations:
- when stucco is used for exterior facing, vertical and horizontal control joints should be provided not more than 10 feet apart, and/or whenever the framing is discontinuous.
- exterior facing, especially in the interrupted wall type, will be subjected to differential movement in relation to the frame; provisions should be made to accommodate such movement.
- wall assembly will be subject to stresses imposed by deflection, shrinkage, and creep in reinforced concrete framed structures; deflection in steel framed structures; and deformations due to vertical forces, especially in tall structures.

■ fire-retardant treated wood studs may be used when permitted by building code.

description:
- cavity wall used as a non-bearing curtain wall; generally when a limited number of openings in wall is provided.
- not practical when large, closely spaced openings are required.
- may be used as spandrel panel if lateral bracing is provided.
- stone or pre-cast concrete veneers may also be used.
- wall is built using standard components, and standard construction methods.
- more often used in reinforced concrete framed structures than in steel framed structures.
- exposed edge of framing in the interrupted wall type provides a thermal bridge, resulting in high heat loss.
- insulation may be provided in the cavity.
- interior face of inner wythe may be painted; or plastered and painted; or faced.

installation:
- construction of wall is similar to that for bearing cavity wall.
- sufficient ties must be provided below, at, and above spandrels to prevent rotation, or bulging of outer wythe.
- inner wythe is always supported on the structural frame; outer wythe generally supported on continuous shelf angles outside the structural frame, but may be supported on the structural frame.
- continuous flashings and weep holes should be provided to drain any moisture which will penetrate the wall, or might condense within the cavity to drain to the outside.

limitations:
- wall assembly will be subject to horizontal movement due to thermal expansion and contraction; and to differential expansion and contraction of exterior facing relative to frame. Control joints should be provided to accommodate such movement.
- wall assemblies, especially in reinforced concrete framed structures, will be subject to vertical movement and resultant stresses due to shortening of columns caused by elastic, thermal, shrinkage and creep strains, and to deflections in horizontal framing members due to dead and live loads, shrinkage, and creep; and to deformations due to vertical forces, especially in tall structures.
- due to vertical movements, stress concentrations may develop at the shelf angles, where the joint is pointed only. Use of caulking compound to seal the joint at shelf angles should be considered.

Time-Saver Standards for Building Materials and Systems

B · BUILDING SHELL

B2.4 BEARING WALLS

WALL FLASHING: materials, uses

• denotes common availability or use
○ denotes limited use or availability

MATERIALS

DESCRIPTION		CONCRETE/STONE/TILE	STEEL, GALVANIZED	STEEL, GALV., COATED	STEEL, STAINLESS	ZINC ALLOY/TERNE	COPPER	COPPER, COMPOSITE	ALUMINUM, MILL FINISH	ALUMINUM, COMPOSITE	ALUM., COATED/ANODIZED	POLYVINYL CHLORIDE, RIGID	POLYVINYL CHLORIDE, FLEXIBLE	FABRIC, BITUMEN SATURATED
COPINGS Copings protect the top of a parapet wall from water penetration • Precast concrete of stone copings should be flashed at least at joints in coping; anchorage to wall may be required under severe exposures • Metal copings may also serve as counterflashing, but such use is not recommended; metal copings are secured to the wall by splice plates.	A		○	●	○	●	○		●			●			
	B	●												
REGLETS Reglets are installed in walls to secure counterflashing or base flashing of roofing membranes: type A is cast into concrete, or built into masonry joints; type B is attached to wall, either concrete, masonry, or wood, and the top is field sealed with bulk sealants.	A		●		●	●	●		●		●	●		
	B		●		●	●	●		●		●			
FLASHING: SPANDREL, SILL Wall flashing or through-wall flashings function to divert any moisture penetrating the exterior face of the wall, or condensing within the wall from migrating through the wall to the interior. • Flashings are commonly used at changes in wall assembly and/or between different wall components: under concrete or stone copings; at intersections of masonry walls and roof or floor assemblies; over window and door openings; under window sills; at shelf angles, at bases of hollow masonry and cavity walls, • When flashings extend through a wall, they should be turned up inside to prevent water penetration. Water trapped at wall flashings must be diverted to the outside: weep holes, 24 to 36 inches on centers should be provided in masonry joints immediately above all flashings.	A		●				●	●	○		●			○
	B		●				●	●	○		●			○
FLASHING: BASE, SPANDREL	A		●				●	●	○		●			○
	B		●				●	●	○		●			○

B2.4 BEARING WALLS

MATERIALS

COPPER: is durable and an excellent moisture barrier:

- When exposed, copper may tend to stain adjacent surfaces but it is not materially affected by caustic alkalies present in masonry mortar and concrete.
- Copper can safely be embedded in fresh mortar, and will not deteriorate in continuously saturated hardened concrete unless excessive amounts of chloride additives are present in the mortar or in concrete.
- Copper is also available as thin sheets, two- to seven-ounce weight, laminated to various flexible coverings, such as bitumen-saturated glass fiber fabric, polyethylene, kraft paper; or coated with bitumen. The coverings add protection and/or stiffness to the thin sheet of metal; the metal provides strength, and resistance to puncture and cold flow under dead load of the assembly.
- Lock-slip joints to prevent water penetration can readily be formed in plain copper.
- Joints in copper composites generally bonded with bituminous cement/adhesive.

STAINLESS STEEL: is as durable as copper and will not stain adjacent surfaces:

- Sealing of joints and field modification of preshaped sections are more difficult than with other flashing materials.
- Generally, Type 304 with minimum thickness of 0.010 inches is used.

GALVANIZED STEEL is subject to corrosion in fresh mortar:

- Although the corrosive products apparently form a very compact film around zinc and bond well with the mortar, the extent of corrosion, thus durability, cannot be accurately predicted.
- Proprietary zinc alloys with improved durability are available.

ALUMINUM will be attacked by caustic alkalies in fresh unhardened mortar and concrete.

- Dry, cured mortar and concrete will not materially affect aluminum but corrosion may again occur when such mortar or concrete becomes wet.

- Uncoated aluminum is not recommended for flashing masonry and concrete walls.
- Aluminum composites of .004- or .005-inch thick aluminum sheets faced on both sides with bitumen saturated cotton fabric are available.
- Bitumen-coated thin sheets for flashing are also available.
- Joints in coated aluminum and aluminum composites are generally bonded using bituminous cement/adhesive.

ELASTOMERIC flashing, generally polyvinyl chloride, is one of the most widely used flashing materials:

- Formulations vary from one manufacturer to another and may affect the long term performance of the material. Performance record of material being selected and compatibility with alkaline masonry mortar should be investigated.
- Elastomeric flashings are more prone to be torn and/or punctured during installation.
- Joints between individual lengths must be sealed to prevent water penetration· but proper sealing is more difficult to achieve than in metal based flashings due to the flexibility of the elastomeric material, and may require constant field inspection.

BITUMINOUS flashing of bitumen saturated organic or mineral fabrics is used as low-cost flashing/dampproofing over openings, at spandrels, and at window sills:

- They can be effective if properly installed, but are not as durable as elastomerics or metals.

LEAD flashing is seldom used and not commonly available:

- Lead, like uncoated aluminum, is susceptible to corrosion in fresh mortar and concrete.
- Lead, when only partially embedded in mortar or concrete, will develop, in the presence of moisture, a differential electrical action gradually disintegrating the embedded lead.

INSTALLATION

- Reglets embedded in plant precast concrete panels are difficult to align in field; and almost impossible when used in site-cast tilt-up concrete panels.
- Flashings in masonry walls should

be set into thin bed of mortar with another thin bed or mortar over the flashing to minimize the possibility of masonry puncturing the flashing material and also to improve the strength of the wall assembly.

- When flashing is placed directly on masonry with no mortar under it, the flexural strength of wall at that point is zero, and resistance to shearing under lateral load depends on friction between masonry and flashing.
- Strength of walls under axial loads is generally not affected by through-wall flashing.
- Lock-slip joints should be provided in long length of metal wall flashings to accommodate expansion and contraction while preventing water penetration at such joints.
- Differential movement between wall and concrete or stone copings may also develop due to different exposure to solar radiation. When joints between sections of solid coping are fully grouted, expansion may be cummulative and push copings out at corners.

COST

Replacement cost of flashings that fail will generally exceed the original cost of material and installation; thus initial cost should always be considered in the context of ultimate durability.

- Average comparative installed costs per square foot for commonly used flashings, with the basis being elastomeric flashing, are:

17 oz fabric	90
.30" elastomeric	100
.004" aluminum, mastic-coated	110
.32" aluminum	200
5 oz. copper, mastic-coated	200
7 oz. total weight lead-coated copper	200
5 oz. copper, fabric-backed	220
16 oz copper	380

B2.4 BEARING WALLS

B · BUILDING SHELL

WALL EXPANSION JOINT COVERS

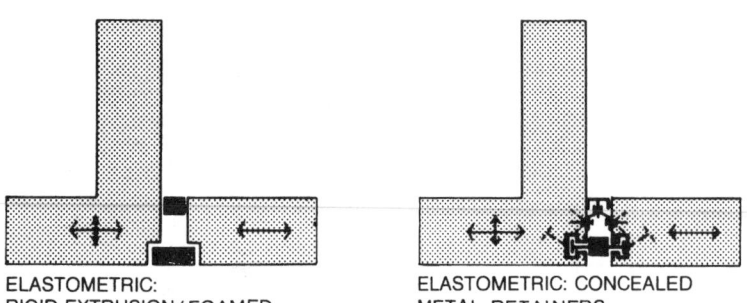

ELASTOMETRIC:
RIGID EXTRUSION/FOAMED

ELASTOMETRIC: CONCEALED
METAL RETAINERS

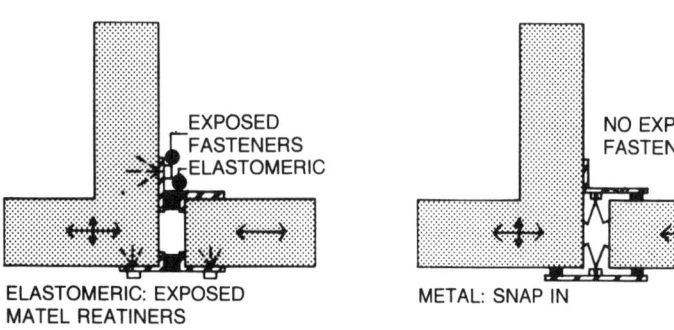

EXPOSED
FASTENERS
ELASTOMERIC

ELASTOMERIC: EXPOSED
MATEL REATINERS

NO EXPOSED
FASTENERS

METAL: SNAP IN

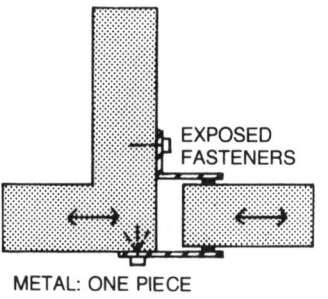

EXPOSED
FASTENERS

METAL: ONE PIECE

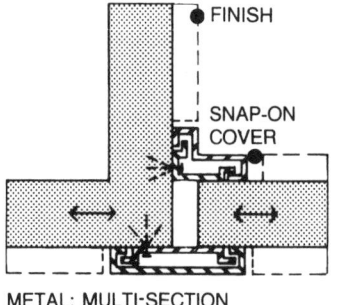

FINISH

SNAP-ON
COVER

METAL: MULTI-SECTION

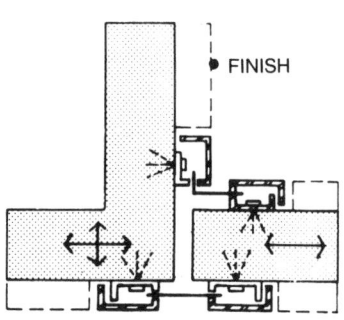

FINISH

METAL: MULTI-SECTION

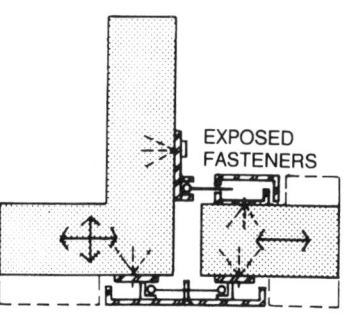

EXPOSED
FASTENERS

METAL: MULTI SECTION

B2.4 BEARING WALLS

Wall expansion joints isolate sections of
enclosures to limit cumulative thermal ex-
pansion/contraction. Covers for expansion
joints must accommodate such movements and
may also be subject to seismic forces acting upon the enclosure. Expansion joint
covers may be used to:

■ Close the joint primarily because of
visual considerations, and/or to prevent
debris being lodged in the joint.

■ Seal the joint against air leakage to or
from the outdoors - or between spaces with
different environments - and because of
visual considerations.

■ Seal the joint against fire penetration
as well as for air leakage and visual consideration

• means to prevent fire penetration may
be incorporated into the assembly of the
joint cover or may be a separate added
component.

Considerations during selection may include:

■ Elastomeric: either rigid extrusions or
closed-cell foams.

• foams are generally relatively soft and
may be subject to damage.
• all have to be compressed in the joint
- generally about half of uncompressed
width - and rigid extrusions may damage
the wall when closing of joint compresses
them excessively.
• difficult to maintain plumb during installation unless provisions are made to
limit depth to which they can be inserted
into the joint.
• will not accommodate irregularities in
the width of joint unless such irregularities are small and gradual.
• when provided with retaining frames of
metal to maintain them in place will perform better, but will be more costly and
more difficult to install.
• generally will accommodate movement in
compression and shear
• for additional information on rigid extrusions and foams, refer to SEALANTS.

■ Metal covers: commonly of extruded aluminum

• snap-in type: easy to install but cannot
be removed later without damaging the
cover and/or wall.
• multi-section types: commonly provide
widest choice of functions and/or finishes.
for information on materials and finishes,
refer to description chart on floor and
ceiling expansion joint covers under
FLOORING.

REFERENCES: Bearing walls

STANDARDS

ASTM Specifications for:

C34
- Structural Clay Load-Bearing Wall Tile.

C55
- Concrete Building Brick.

C62
- Building Brick (Solid Masonry Units Made from Clay or Shale).

C90
- Hollow Load-Bearing Concrete Masonry Units.

C126
- Ceramic Glazed Structural Clay Facing Tile, Facing Brick, and Solid Masonry Units.

C145
- Solid Load-Bearing Concrete Masonry Units.

C216
- Facing Brick (Solid Masonry Units Made from Clay or Shale).

ASTM Tests/Methods for:

C271
- Density of Core Materials for Structural Sandwich Constructions.

E72
- Conducting Strength Tests of Panels for Building Construction.

E119
- Fire Tests of Building Construction and Materials.

E176
- Terminology Relating to Fire Standards.

E564
- Static Load Test for Shear Resistance of Framed Walls for Buildings.

ANSI Specifications

ANSI/ACI 318-80
- Building Code Requirements for Reinforced Concrete.

ANSI/ACI 315-80
- Details and Detailing of Concrete Reinforcement.

ANSI/ACI 301-79
- Structural Concrete for Buildings.

ANSI/ACI 531-79
- Building Code Requirements for Concrete Masonry.

SELECTED REFERENCES

ACI   **American Concrete Institute**
- Pre-cast Concrete: Handling and Erection
- ACI Manual of Concrete Practice, Part 3; Products and Processes
- Pre-cast Concrete Wall Panels

AITC   **American Institute of Timber Construction**
- Timber Construction Manual

BIA   **Brick Institute of America**
- **7 Series** • Moisture Control
- **16 Series** • Fire Resistance
- **17 Series** • Reinforced Brick Masonry
- **18 Series** • Differential Movements
- **21 Series** • Cavity Walls
- **24 Series** • Bearing Walls
- **28 Series** • Brick Veneer
- **31 Series** • Structural Design
- **35 Series** • Strength
- **39 Series** • Engineered Brick Masonry

IMI   **International Masonry Institute**
- The New and Modern Capabilities of Engineered Load-bearing walls.

ML/SFA   Metal Lath/Steel Framing
- Steel Framing Systems Manual

NCMA   **National Concrete Masonry Assoc.**
- **3** • Control of Wall Movement with Concrete Masonry
- **4B** • Concrete Masonry Retaining Walls
- **5** • Concrete Masonry Screen Walls
- **7B** • Load-Bearing Block in HighRise Buildings
- **12** • Estimation U-Factors for Concrete Masonry Construction
- **13** • Details for Building Dry Concrete Masonry Walls
- **15** • Compressive Strength of Concrete Masonry
- **24** • Engineered Concrete Masonry-Wind Loads
- **25** • Concrete Masonry Lintels
- **27** • Flexural Design of Non-reinforced Engineered Concrete Masonry
- **29** • Concrete Masonry Walls in Multi-Family Housing
- **31** • Eccentric Loading of Non-reinforced Concrete Masonry
- **34** • Combined Loads on Concrete Masonry Walls
- **35A** • Fire Safety with Concrete Masonry
- **43** • Concrete Masonry Foundation Walls
- **48** • Design Composite Masonry Walls
- **53** • Design of Concrete Masonry for Crack Control
- **56A** • Concrete Masonry Basement Walls
- **59** • Reinforced Concrete Masonry Construction
- **61A** • Concrete Masonry Load Bearing Walls - Lateral Load Distribution
- **62** • Concrete Masonry Cavity Walls
- **63** • Partially Reinforced Concrete Masonry Walls
- **66A** • Design for Shear Resistance of Concrete Masonry Walls
- **74** • Surface Bonded Concrete Masonry
- **78** • Concrete Masonry Beam Walls
- **79** • Concrete Masonry Veneers
- **85** • Building Weathertight Concrete Masonry Walls
- **89** • Engineered Concrete Masonry Warehouse Walls
- **95** • Design Details for Concrete Masonry Fire Walls
- **103** • Tall Concrete Masonry Walls

NFPA   **National Fire Protection Assoc.**
- Fire Protection Handbook

NPCA   **National Precast Concrete Assoc.**
- Fundamentals of Quality Precast Concrete

B2.4 BEARING WALLS

REFERENCES: Bearing walls (cont.)

NRCC   **National Research Council of Canada**

CBD 48   • Requirements for Exterior Walls

50   • Principle Applied to an Insulated Masonry Wall

93   • Pre-cast Concrete Walls - Problems With Conventional Design

94   • Pre-cast Concrete Walls - A New Basis for Design

PCA   **Portland Cement Association**

             • The Concrete Approach to Energy Conservation

USFPL   **U. S. Forest & Products Laboratory**

             • Longtime Performance of Sandwich Panels in FPL Experimental Unit

             • Diaphragm Action of Diagonally Sheathed Board Panels

             • Testing of A Full-Scale Hour Under Similated Snowloads and Windloads

             • Houses Can Resist Hurricanes

B2 • EXTERIOR WALL ASSEMBLIES

B2. 5 CURTAIN WALLS

GRID TYPES: introduction

STICK

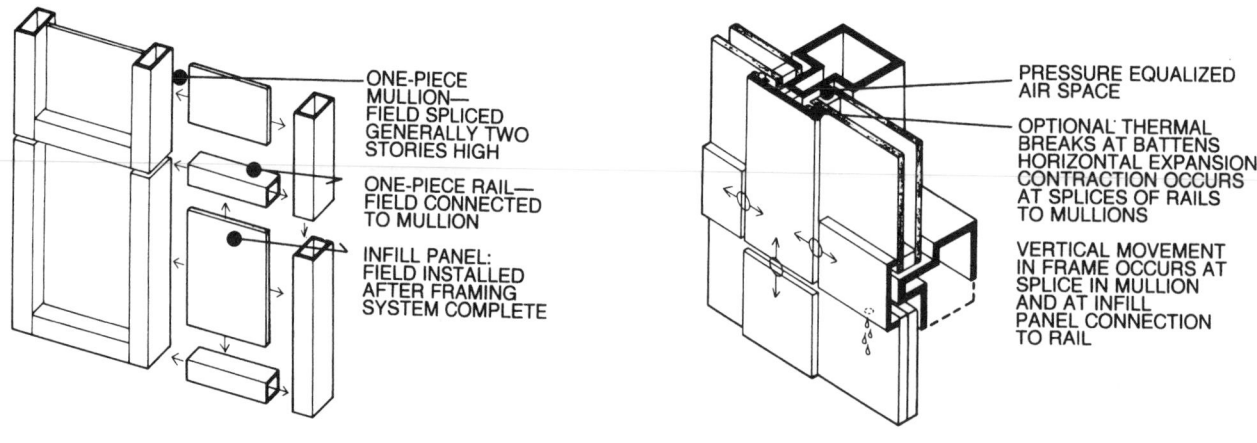

ONE-PIECE MULLION— FIELD SPLICED GENERALLY TWO STORIES HIGH

ONE-PIECE RAIL— FIELD CONNECTED TO MULLION

INFILL PANEL: FIELD INSTALLED AFTER FRAMING SYSTEM COMPLETE

PRESSURE EQUALIZED AIR SPACE

OPTIONAL THERMAL BREAKS AT BATTENS HORIZONTAL EXPANSION CONTRACTION OCCURS AT SPLICES OF RAILS TO MULLIONS

VERTICAL MOVEMENT IN FRAME OCCURS AT SPLICE IN MULLION AND AT INFILL PANEL CONNECTION TO RAIL

UNITIZED STICK

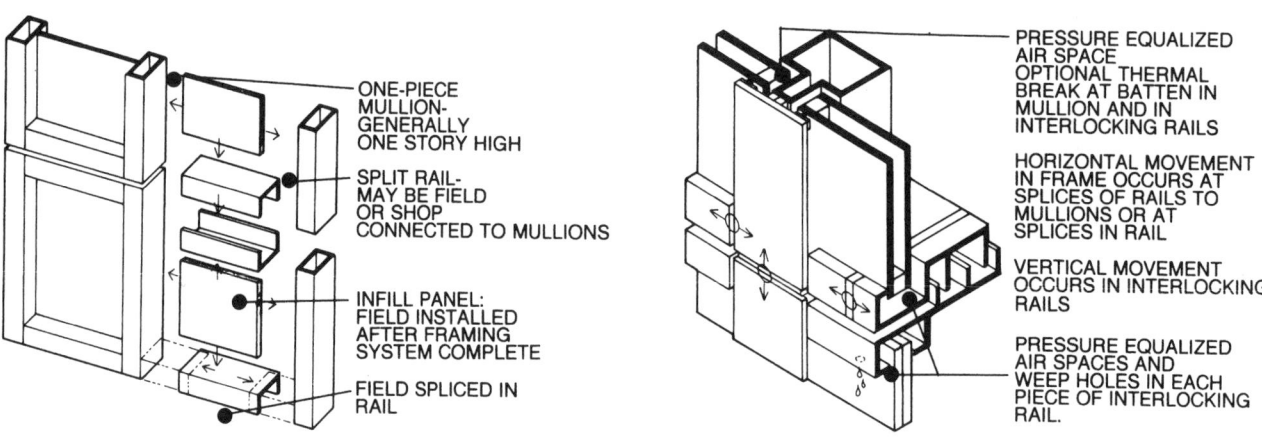

ONE-PIECE MULLION- GENERALLY ONE STORY HIGH

SPLIT RAIL- MAY BE FIELD OR SHOP CONNECTED TO MULLIONS

INFILL PANEL: FIELD INSTALLED AFTER FRAMING SYSTEM COMPLETE

FIELD SPLICED IN RAIL

PRESSURE EQUALIZED AIR SPACE OPTIONAL THERMAL BREAK AT BATTEN IN MULLION AND IN INTERLOCKING RAILS

HORIZONTAL MOVEMENT IN FRAME OCCURS AT SPLICES OF RAILS TO MULLIONS OR AT SPLICES IN RAIL

VERTICAL MOVEMENT OCCURS IN INTERLOCKING RAILS

PRESSURE EQUALIZED AIR SPACES AND WEEP HOLES IN EACH PIECE OF INTERLOCKING RAIL.

UNITIZED

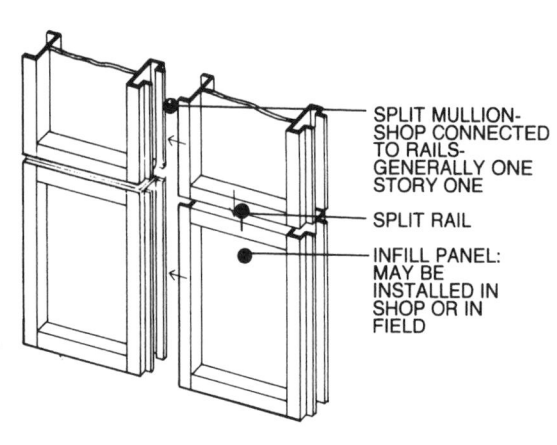

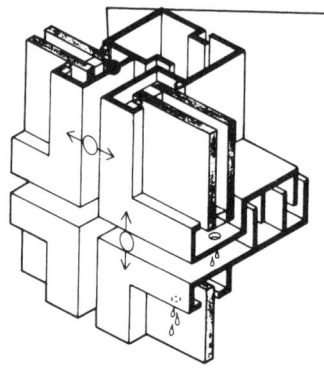

SPLIT MULLION- SHOP CONNECTED TO RAILS- GENERALLY ONE STORY ONE

SPLIT RAIL

INFILL PANEL: MAY BE INSTALLED IN SHOP OR IN FIELD

PRESSURE EQUALIZED AIR SPACE OPTIONAL THERMAL BREAKS IN EACH PIECE OF SPLIT MULLION AND IN INTERLOCKING RAILS

HORIZONTAL AND VERTICAL MOVEMENTS OCCUR AT INTERCONNECTING MULLIONS AND RAILS

PRESSURE AND EQUALIZED AIR SPACES AND WEEP HOLES IN EACH PIECE OF INTERLOCKING RAIL

B2.5 CURTAIN WALLS

MULLIONS: types

BATTEN GLAZING

WALL PANELS

EXPOSED FASTENERS
NO THERMAL BREAK

OPTIONAL SNAP-ON COVER

INFILL PANELS

NO THERMAL BREAK

OPTIONAL SNAP-ON
COVER
FIN FOR ADDED STRENGTH

INFILL PANELS

OPTIONAL SNAP-ON COVER

NO THERMAL BREAK

OPTIONAL SNAP-ON COVER

INFILL PANELS

WITH THERMAL BREAK

STOP GLAZING

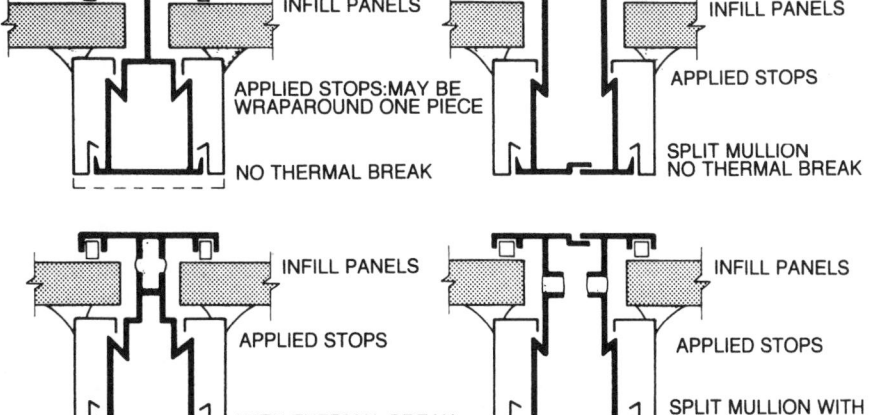

INFILL PANELS

APPLIED STOPS:MAY BE
WRAPAROUND ONE PIECE

NO THERMAL BREAK

INFILL PANELS

APPLIED STOPS

SPLIT MULLION
NO THERMAL BREAK

INFILL PANELS

APPLIED STOPS

WITH THERMAL BREAK

INFILL PANELS

APPLIED STOPS

SPLIT MULLION WITH
THERMAL BREAK

GASKET AND BUTT GLAZING

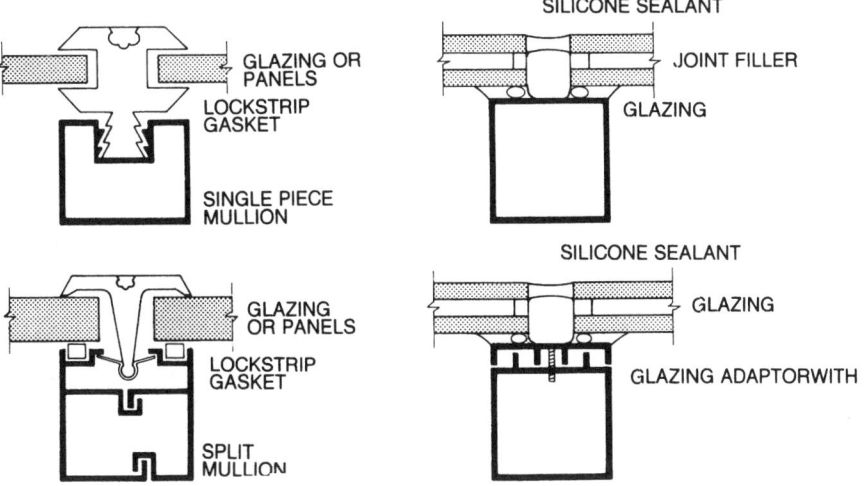

GLAZING OR
PANELS
LOCKSTRIP
GASKET

SINGLE PIECE
MULLION

GLAZING
OR PANELS
LOCKSTRIP
GASKET

SPLIT
MULLION

SILICONE SEALANT

JOINT FILLER

GLAZING

SILICONE SEALANT

GLAZING

GLAZING ADAPTORWITH

REMARKS

Materials: Aluminum extrusions commonly used. Alloys for extrusions: 6063 generally, or 6061 when higher strength required.

Finishes commonly used, per Aluminum Association designations: Clear Anodized: AA-A41; Color Anodized: either integral AA-A42, or electrolytically deposited AA-A44. Applied organic coatings may also be used, such as fluorocarbon, acrylic, polyester, or plastisol. Stainless steel is available, but generally on special order basis only. Carbon steel is seldom used unless fully protected from the weather.

Bronze has also been used in custom designed installations.

Movement: Thermal expansion and contraction generally accommodated in splices in or between mullions and rails; or in split mullions and rails. Mullions may also be designed to deform when wall assembly expands, such as a bellows mullion, but care must be taken not to overstress the material, and such practice is not widely employed.

Thermal movement of infill panels in the lockstrip gasket systems is accommodated within the gasket, but provisions are still necessary for thermal movement in the mullions and rails.

Deflection: deflection limits gen-generally accepted are 1/175 of span or not more than ¾ inch between supports; except that when a plastered surface may be affected by bending in the mullions and/or rails, deflection should not exceed 1/360 of span.

Control of heat flow: Mullions without thermal breaks act as continuous thermal bridges in the wall assembly: heat loss and the possibility of condensation forming on interior surfaces in winter and inward heat flow into cooled interior space in summer will result. When no thermal breaks are provided in the mullions/rails, their exposure to the outside should be kept to a minimum. Even with thermal breaks, thermal bridges may be formed if fasteners connecting one part of a mullion to another penetrate such thermal breaks.

General: The cost of a die for a new extrusion is relatively low, and mullions may be designed for a particular project. Most systems allow for glazing/installation of infill panels from either inside or outside. Mullions only are illustrated; rails are generally similar to mullions but different configurations are possible and often used. Pressure equalization, as defined in WALL ASSEMBLIES AND COMPONENTS should be provided in any system.

B2.5 CURTAIN WALLS

GRID TYPES: infill panels

• denotes common usage or availability
○ denotes possible usage or availability

REMARK
- INFILL PANELS ARE ALSO REFERRED TO AS GLAZING PANELS. MOST OPAQUE PANELS MAY BE ALSO USED AS WALL FACING PANELS.
- THICKNESS GENERALLY FROM ¼ TO 2 INCHES.
- [1] APPROXIMATE ONLY FOR WINTER CONDITIONS. ACTUAL VALUE DEPENDS ON THICKNESS AND PROPERTIES OF MATERIALS SELECTED.
- [2] SAME OR SIMILIAR: FINISHED OR PRIMED ONLY.

TYPE	THERMAL RESISTANCE: R=°F/SF/BTU/HR, range [1]	SMOOTH	PATTERNED	METAL	GLASS	PLASTIC	CEMENT-BOUND FIBER	STONE	INTEGRAL	APPLIED	RIGID BOARD	HONEYCOMB	INSULATION	CORE/BACKING	SAME AS EXTERIOR [2]	REMARK
MONOLITHIC: single thickness	.80 to 1.20	•	•		•	•	•	•	•	•						Cement bound mineral fiber; ¼ to ⅝ inches thick; sizes up to 60 x 120 inches; larger sizes may require stiffening. Available with applied organic coatings, integrally colored, or uncoated. Amorphous: glass, either clear, tinted, reflective coated, patterned, or opaque. May be heat strengthened, tempered. Acrylic, ⅛ to ¼ inches thick is available.
MONOLITHIC: spaced (DESICCATED AIR SPACE)	1.5 to 2.5	•			•	•			•							Generally two lights of glass with insulating air space between. Three lights also available. Each light may be different and various combinations are available, such as tinted, reflective coated, opaque, or laminated outside lights usually with clear inside lights. Insulating units of two spaced sheets of plastic, or of plastic and glass are available.
COMPOSITE-LAMINATED (EXTERIOR FACE / BACKING / OPTIONAL INSULATION)	.90 to 1.00	•	•	•	•	•	•		•	•	•		○	•		Exterior face of: aluminum, either mill finished, color anodized, or with applied coating; galvanized carbon steel, with applied coating; opaque glass, single thickness; or laminated glass. Applied coatings generally used: porcelain enamel, fluorocarbon, acrylic, polyester, or plastisol. Exposed surface may be flat or with light embossing. Other metals, such as stainless steel, copper, are also available, but generally on special order basis only.
COMPOSITE-LAMINATED (EXTERIOR FACE / CORE:INSULATION / OPTIONAL BACKING)	3.0 to 12.0	•	•	•			•		•	•	•		•	•	•	Outside faces generally laminated to rigid-board backing to stiffen the light-gauge metal against thermal deformations. Backing: usually cement-bound mineral fiber, hardboard, fiberboard, plywood. Core: generally foamed urethane isocyanurate, polystyrene; cellular glass; perlite board; or semirigid glass fiber.
FORMED (EXTERIOR FACE / INTERIOR FACE / OPTIONAL INSULATION)	3.0 to 12.0	•	•	•					•	•	○	○	•		•	Two sheets of metal with edges connected around perimeter of panel. Thermal break, generally used between exterior and interior faces, and venting of core, should be provided. Exterior face may be flat; stamped in various configurations; with integral of applied finishes. Core generally insulated with semirigid fibrous insulation. Panels available on special order basis only.

B2.5 CURTAIN WALLS

GRID TYPES: selection checklist

PRINCIPAL CONSIDERATIONS in selecting a grid curtain wall assembly should always include:

- effects of thermal expansion and contraction on the various components which might have different coefficients of expansion, and/or different heat storage capacities.
- effective means to minimize rain and air penetration under varying air pressures, both positive and negative.
- provisions to accommodate deformations in the structural frame of the building.

■ **In-service record and/or laboratory test results** for an assembly being considered should be thoroughly investigated.

- if the assembly is being designed for a specific application, laboratory testing of a mock-up is highly recommended.

MULLION selection should generally include:

■ Size and shape of mullions for appearance and strength:
- tubular mullions may be internally reinforced with standard structural shapes.

■ **Configuration of mullions:**
- for inside or outside installation of infill panels and windows.
- for thermal breaks to prevent through-metal conduction of heat.

■ **Without thermal breaks** the location of the mullion in relation to the plane of the wall will affect the fin heat loss: the less the mullion is exposed to the outside, the lower the heat loss.

■ **Finish for the material selected:**
- aluminum comes in anodized finishes, either clear or in colors,
- stainless steel, is available in several finishes, ranging from bright polished to patterned.
- bronze is generally used without any applied finishes, except for clear protective coatings.
- alloy steel, sometimes called weathering, steel which develops an integral protective coating upon exposure to weather, is available; precautions must be taken in the design and detailing to: avoid prolonged contact with moisture; to expose all surfaces directly and equally to the atmosphere; and to prevent water runoff from staining adjacent surfaces during the time the protective coating is forming.

■ **Texture and color of finish will also affect thermal performance**
- the darker the color, the greater the heat absorption.
- the smoother the surface, the lesser the cooling convection effect of wind.

INFILL PANELS are generally selected on aesthetic as well as technical considerations:

■ Appearance of opaque panels is of primary importance as it will largely determine the overall appearance of the structure:

- the exterior facing of the panel can be made of just about any durable material and is generally not dependent on the core material of the panel.

■ **Strength and stiffness of opaque panels** to resist lateral loads and stresses induced by thermal movement:

- infill panels must transmit wind loads to supporting mullions without excessive deformation in the panels.
- metal-faced laminated panels may deform with changes in temperature; bowing in during cold weather and bulging out when heated by the sun; such deformation should be controlled primarily because of visual considerations.
- hollow metal panels and unstiffened metal sheets, when restricted at their edges will tend to bow, or "oil-can;" this tendency may be prevented, or at least minimized, by allowing for free movement in all directions in the plane of the panel.

■ **Panels may also be preassembled horizontally** with a secondary back-up framing system to span between vertical components of the structural frame. This system is generally referred to as column cover and spandrel system. Other components of the system are horizontal strips of glass infill panels or of window units.

■ **Water/vapor penetration and condensation** on, or within, infill panels can be controlled by:

- providing a vapor retarder on or near the warm side of the wall.
- providing sufficient insulation to keep temperatures at critical surfaces higher than the dew point of inside air.
- providing for release of vapor on the cold side of panel.
- avoiding the use of materials which may act as vapor retarders on both sides of the wall thus creating a vapor trap.

■ **Double glazing** should be considered for transparent panels and for fixed or operable windows:

- to prevent condensation
- to minimize heat loss
- to reduce heat gain for air-conditioned spaces.

■ **Sound transmission** ratings of curtain walls are generally similar to those of most other types of wall assemblies having equivalent fenestration; the fenestration usually being the weakest link in resisting airborne sound transmission. Without any fenestration a light curtain wall will not perform as well as a heavy bearing wall, since the transmission of sound through any barrier is inversely proportional to the mass of the barrier.

THERMAL PERFORMANCE

■ **Air infiltration** through joints and cracks must be minimized.

■ **All thermal bridges** must be prevented through proper thermal break design of:

- connections of curtain wall to structure.
- connections of panels to framing members.
- within the panels: including panel bracing, ties and reinforcement.

■ **Before sophisticated measures are decided upon,** it must be remembered that a thermally highly-resistant wall assembly will be of real value only when:

- heat loss/gain through exterior walls is a significant factor in the energy budget of the building.
- thermal resistance of floor, roof and other wall assemblies in the building is comparable to it.
- thermal resistance of operable openings within the assembly does not constitute an excessive heat loss.

B2.5 CURTAIN WALLS

WALL PANELS: types, properties

● denotes common usage
  or availability
○ denotes possible usage
  or limited availability

WALL PANELS DEFINED:
Rigid panels at least one story in height and capable of:
● functioning as a complete wall assembly in separating exterior and interior environments
● supporting own weight
● resisting lateral loads between supports and transferring such loads to the structural frame.

TYPE	DESCRIPTION	Width, range in inches	Length, maximum in inches	Thickness, range in inches
FORMED: single thickness CORRUGATED THICKNESS = DEPTH OF SECTION STRUCTURAL SUPPORT RIBBED	● Corrugated panels of cement-bound mineral fiber - also referred to as transite; ● Corrugated or ribbed - in various configurations - panels of plastic, commonly of glass-fiber-reinforced polyester. ● Exterior face panels of sandwich systems may also be used as single-thickness wall panels.	12 to 40	30	½ to 1½
FORMED: sandwich OPTIONAL INSULATION INTERIOR LINER PANEL STRUCTURAL SUPPORT SUBGIRT FACE PANEL	Interior liner panel of metal attached to structural frame: subgirts attached to liner panel with exterior face panel of metal, attached to and supported by subgirts: ● Exterior face panel ribbed in various configurations or corrugated ● Optional semirigid insulation between panels ● May be split to enclose structural supports.	12 to 36	40	2¾ to 9¼
COMPOSITE: laminated INTERIOR FACE CORE EXTERIOR FACE PANEL	● Flat face panels laminated to honeycomb core of fiber-board or aluminum. Loose-fill insulation, such as perlite, may be added to the core ● Flat or formed panels with a core of rigid insulation between them ● Edges commonly closed; thermal breaks between exterior and interior faces generally provided.	24 to 72	30	2 to 4
INTERIOR FACE BACKING CORE BACKING FACE PANEL	Exterior face of flat panels laminated to rigid backing for stiffness, or of single, flat, rigid panel; in turn laminated to insulating core ● Interior face may be: similar to exterior face; exposed core; or of rigid boards, such as hardboard, gypsum board. ● Edges commonly open ● Also often used as glazing or facing panels.	48 or less to 60	12	2 to 4
COMPOSITE: cast INTERIOR FACE CORE: INSULATION EXTERIOR FACE	Precast concrete - Portland cement or polymer - with insulating core ● Polymer concrete has the advantage of reduced weight and increased strength ● Insulation completely encased in concrete with resultant thermal bridges at perimeter of panel ● Interior face commonly flat and smooth finished.	48 or less to 96 or more	20 or more	2½ to 4
MONOLITHIC INTERIOR FACE EXTERIOR FACE	Solid panels of precast concrete with either normal weight or lightweight aggregate ● Precasting may be done at plant or at site ● May be: molded or sculptured in various configurations; cast over form liners of various designs to impart a pattern or texture to exterior face.	96 or less to 240 or more	40 or more	4 to 6 or more
PREASSEMBLED FIELD INSTALLED INTERIOR FACE INTERIOR FACE CORE: FRAMING EXTERIOR FACE	● Exterior face panels attached to framing preassembled at plant and shipped to site as units. ● Optional insulation and interior facing installed after panels are erected ● Exterior and interior face panels attached to preassembled framing at plant and shipped to site as complete unit.	48 to 96	30	4½ to 7

SIZE ①

B2.5 CURTAIN WALLS

JOINTS ②					COMPATIBLE STRUCTURAL FRAME						③	④			
BUTT	LAP	TONGUE & GROOVE	INTERLOCKING	MULLIONS/FRAMES	STEEL, ROLLED	STEEL, LIGHT GAUGE	CONCRETE, CAST	CONCRETE, PRECAST	WOOD FRAMING	USE AS ROOFING	SPACING OF SUPPORTS, range in feet	EXPOSED FASTENERS	CONCEALED FASTENERS	STRUCTURAL CONNECTIONS	'U' VALUE, range WINTER CONDITIONS
●			○		●	○	⑤	⑤	○	●	3 to 10	●			1.50
●	●				●	○	⑤	⑤		●	8 to 25	●	●		1.15 to 0.90 or 0.14 to 0.86
●	●		○		●	●	⑤	⑤		●	8 to 16	●			.55 to .04
				●	●	●	●	●			8 to 12		○		.08 to .04
●		●		●	●		●	●			8 to 16			●	.20 to .04
●		●		●	●		●	●			10 to 30 or more			●	2.50 0.80
●				●	●	●	●	●	●	⑧	12 to 20			●	.47 to .15 or .08 to .05

REMARKS

① Length of fabricated panels generally limited by manufacturing processes and/or shipping restrictions; site fabricating equipment may be available when panels longer than stock length are required.

① Panels with tonque-and-groove joints or with interlocking joints and concealed fasteners have to be erected sequentially: erection can only start at specific points, such as at corners or expansion joints in the wall assembly.

③ Spacing of supports based on approximately 20 psf wind load and a deflection of 1/180 of span; range shown is average for different panels of the same type.

④ Values shown are approximate only; when insulation is optional, the lower value is for a panel with insulation added.

⑤ Continuous supports of steel, such as sheet metal subgirts or light structural steel shapes, should generally be added to concrete frame since fasteners commonly used to attach panels are not suitable for concrete.

⑥ Face panels may also be used as single-thickness panels; and may be insulated with plastic-faced glass-fiber batts, which are then exposed on the inside.

⑦ Panels may also be used as wall facing panels; when thus used, may be available of lighter gauge metal and/or thinner cores. Prefabricated corners or bent panels may be available.

⑧ Plastic-faced panels only.

■ Thermal expansion/contraction may be a factor when several materials with different coefficients of expansion are combined.
Coefficients of linear expansion for some commonly used materials, in in/°F:

aluminum .... 13×10^{-6}
steel, carbon .... 7×10^{-6}
steel, stainless .... 10×10^{-6}
cement-bound mineral
fiber .... 5×10^{-6}
hardboard/plywood .... 3×10^{-6}
plastic .... $12 to 49 \times 10^{-6}$
glass .... 5×10^{-6}
stone .... $8 to 13 \times 10^{-6}$

B2.5 CURTAIN WALLS

WALL TYPES: materials

● denotes common availability or usage
○ denotes limited availability or usage

EXTERIOR FACE FINISH

TYPE / EXTERIOR FACE: BASE MATERIAL OR SUBSTRATE	INTEGRAL/NO FINISH	TREATED	TEXTURED	PIGMENTED	CLEAR/STAIN	PORCELAIN ENAMEL	FLOUROCARBON	ACRYLIC	POLYESTER	PLASTISOL	ANODIZED: clear/color	COATED: with aluminum	AGGREGATE: in matrix
FORMED: single thickness — cement-bound mineral fiber	●												
PLASTIC: glass fiber reinforced polyester or acrylic modifed polyester	●												
FORMED: sandwich — ALUMINUM	○	●			○	●	●	●	●		●		
STEEL	○	●				●	●	●	●			○	
COMPOSITE: laminated — ALUMINUM	○	●			○	●	●	●	●		●		
STEEL		●				●	●	●	●				
ALUMINUM/STEEL		●				●	●	●	●	●	●		
CEMENT-BOUND MINERAL FIBER	○	●	●		●			●	●	●			●
COMPOSITE: cast — CONCRETE-PORTLAND CEMENT	●	●	●	●	●								○
CONCRETE-POLYMER	○	○											●
MONOLITHIC — CONCRETE-PLANT PRECAST	●	●	●	●	○								○
CONCRETE-SITE PRECAST	●	●	●	●	○								○
PREASSEMBLED — CEMENT BOUND MINERAL FIBER	●		●		○								●
PLASTIC: glass fiber reinforced	●												

Notes under headers: INTEGRAL COATING ① / APPLIED FINISH OPTIONS / BAKED-ON ② / ③ (AGGREGATE: in matrix)

Diagram labels:
- FORMED: single thickness — CORRUGATED, THICKNESS = DEPTH OF SECTION, STRUCTURAL SUPPORT, RIBBED
- FORMED: sandwich — OPTIONAL INSULATION, INTERIOR LINER PANEL, STRUCTURAL SUPPORT, SUBGIRT, FACE PANEL
- COMPOSITE: laminated — INTERIOR FACE, CORE, EXTERIOR FACE PANEL; INTERIOR FACE BACKING, CORE, BACKING, FACE PANEL
- COMPOSITE: cast — INTERIOR FACE, CORE: INSULATION, EXTERIOR FACE
- MONOLITHIC — INTERIOR FACE, EXTERIOR FACE
- PREASSEMBLED — FIELD INSTALLED INTERIOR FACE, INTERIOR FACE, CORE: FRAMING, EXTERIOR FACE

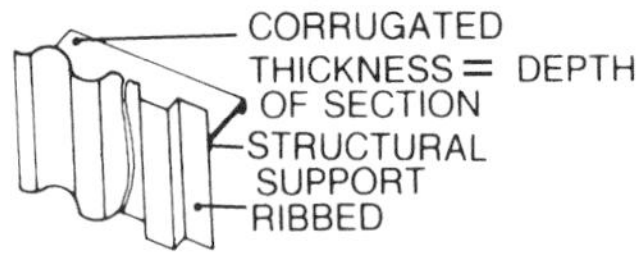

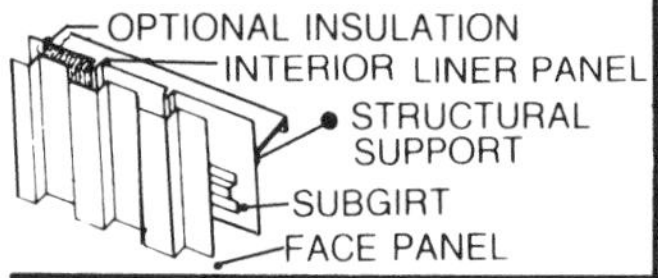

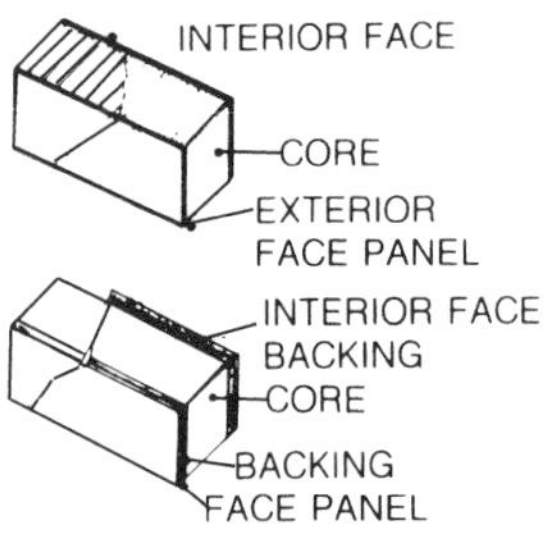

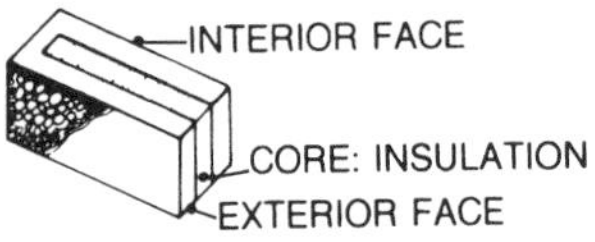

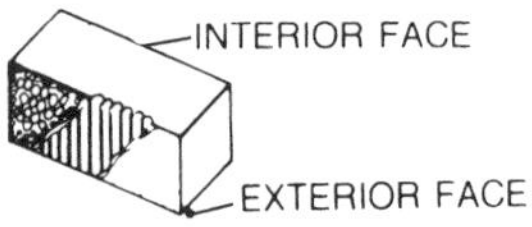

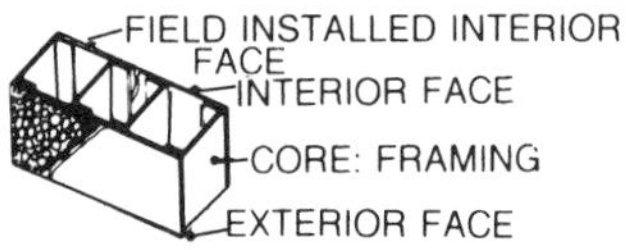

B2.5 CURTAIN WALLS

INTEGRAL WITH FACE	HARDBOARD/PLYWOOD CEMENT-BOUND FIBER	HONEY COMB, fiber	HONEY COMB, aluminum	FRAME, aluminum	FRAME, metal studs	SUBGIRTS	FOAMED, rigid	FIBER, semi-rigid	FIBER, batt	EXPOSED CORE	STEEL/HARDBOARD	PLYWOOD/GYPSUMBOARD	SAME AS EXTERIOR FACE	NONE/SITE APPLIED	PRIMER	SAME AS EXTERIOR FACE	USED AS FACING PANEL
•														•			
•														•			○
						•		•	•	○			•	•	•	•	
						•		•	•	•				•	•	•	
		•	•										•	○	○	•	•
		•	•										•		○	•	•
	•	•					•			•	•	•		•	•		•
	•	•					•			•	•	•	•	•	•	•	•
							•						•	•			
							•						•	•			○
•													•	•			
•													•	•			
					•		•						•	•			
				•			○						•		•		

REMARKS

① • Chemical treatment: acid wash, retarding agents for concrete.
• Mechanical treatment: wire brushing, sandblasting.
• Integral texturing: embossing, mold casting for metals; form liners for concrete.
• Mechanical texturing: stamping, brake forming, for metals; bush-hammering for concrete.

② For specific information, refer to Selection Checklist. Metals may be also embossed after coating. Built-up coatings are available but not commonly used.

③ Selected stone aggregate embedded in surface-applied matrix.

④ Light-gauge metals may require rigid board backup for stiffness.

⑤ Rigid insulation generally isocyanurate or urethane. Extruded polystyrene, perlite, and foamed cellular glass also used. Semirigid and batts generally glass fiber, but rockwool may also be used, especially in fire-resistance rated assemblies.

⑥ Steel generally galvanized: 18 to 20 gauge for exterior panel; 18 to 24 gauge for interior liner panel.

⑦ Steel generally galvanized: 20 to 26 gauge.

⑧ Steel generally galvanized: 24 to 28 gauge.

⑨ Differential expansion/contraction may occur between exterior and interior faces.

⑩ May be cast over various form liners; may have aggregate exposed; may be cast over flagstone, stone rubble. Size of plant precast panels generally limited due to shipping constraints; site-cast panels generally up to 800 square feet with capacity of cranes used for erection being the limiting factor.

⑪ Plastic generally polyester or acrylic modified polyester; available with surface-applied protective film but generally for smooth surface panels only.

B2.5 CURTAIN WALLS

WALL PANELS: typical assemblies

● denotes common usage or characteristic

○ denotes possible or limited usage or characteristic

ASSEMBLY TYPE	COMPARATIVE COST, range Basis = 100	STEEL, ROLLED STEEL, LIGHT GAUGE	CONCRETE, CAST CONCRETE, PRECAST	WOOD FRAMING	MASONRY BEARING	TYPICAL WALL JOINTS
PANEL FORMED — WALL ASSEMBLY — SINGLE THICKNESS	100 to 230	● 2	2 2		2	**THICKNESS RANGE: ¾" to 4"** — SHEET METAL SCREWS — CRIMPED, BUTTON PUNCHED — LAP: METAL, PLASTIC, MINERAL FIBER — INTERLOCKING: METAL ONLY
PANEL FORMED: SANDWICH — INTERIOR FACE — WALL ASSEMBLY	200 to 400	●				SNAP-ON TRIM — SUB-GIRT — FACING — CONCEALED CLIP AT JOINTS — THICKNESS RANGE: 3¼" to 9¼" — LINER — STRUCTURAL SUPPORT — INTERIOR FACING/LINER
PANEL FORMED: SPLIT SANDWICH — WALL ASSEMBLY — INTERIOR FACE	250 to 300	●				CONCEALED FASTENERS — EXPOSED FASTENERS — MUST BE INSTALLED BEFORE EXTERIOR FACING. — MAY BE INSTALLED TOGETHER WITH EXTERIOR FACING. — STRUCTURAL SUPPORT
PANEL COMPOSITE LAMINATED — WALL ASSEMBLY — PANEL		● 2	2 2	○	2	THICKNESS RANGE: 1½" TO 6" — DOUBLE TONGUE AND GROOVE — BATTEN — SEALANTS MAY BE USED IN JOINTS.
PANEL MONOLITHIC — WALL ASSEMBLY	650 to 900	●	● ●			THICKNESS RANGE 4" TO 7" — PLASTIC CONCRETE — SEALANT AND JOINT FILLER — OPTIONAL SEALANT — BUTT: MOST COMMONLY USED — OPTIONAL SEALANT — LAP: HIGHER COST — ALSO T & G FOR PLASTIC CONCRETE

Column header note: 1 | COMPATIBLE STRUCTURAL FRAME

B2.5 CURTAIN WALLS

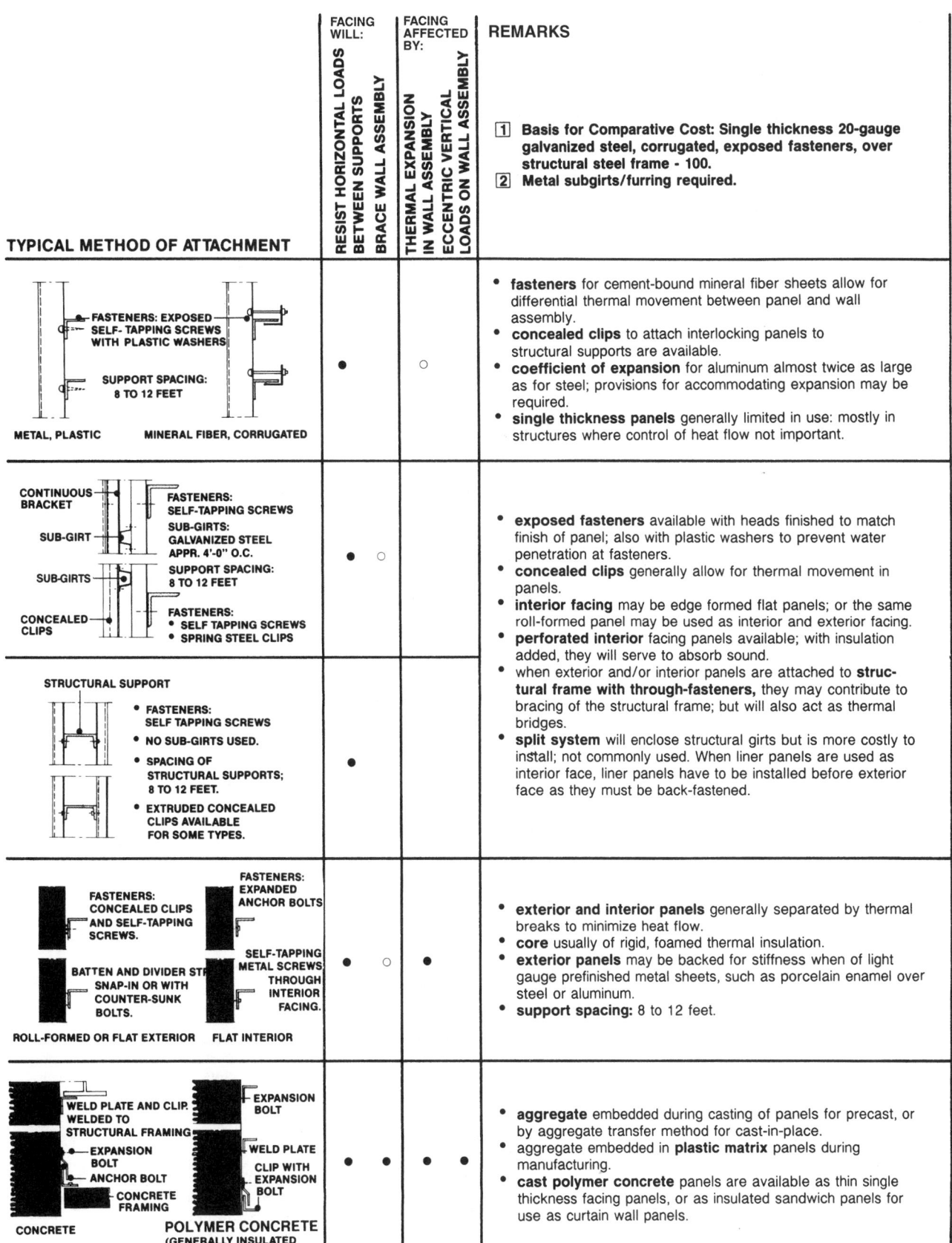

TYPICAL METHOD OF ATTACHMENT	FACING WILL: RESIST HORIZONTAL LOADS BETWEEN SUPPORTS	FACING WILL: BRACE WALL ASSEMBLY	FACING AFFECTED BY: THERMAL EXPANSION IN WALL ASSEMBLY	FACING AFFECTED BY: ECCENTRIC VERTICAL LOADS ON WALL ASSEMBLY	REMARKS
					1 Basis for Comparative Cost: Single thickness 20-gauge galvanized steel, corrugated, exposed fasteners, over structural steel frame - 100. 2 Metal subgirts/furring required.

FASTENERS: EXPOSED SELF-TAPPING SCREWS WITH PLASTIC WASHERS **SUPPORT SPACING: 8 TO 12 FEET** METAL, PLASTIC — MINERAL FIBER, CORRUGATED	●		○		• **fasteners** for cement-bound mineral fiber sheets allow for differential thermal movement between panel and wall assembly. • **concealed clips** to attach interlocking panels to structural supports are available. • **coefficient of expansion** for aluminum almost twice as large as for steel; provisions for accommodating expansion may be required. • **single thickness panels** generally limited in use: mostly in structures where control of heat flow not important.
CONTINUOUS BRACKET **FASTENERS: SELF-TAPPING SCREWS** **SUB-GIRT** **SUB-GIRTS: GALVANIZED STEEL APPR. 4'-0" O.C.** **SUB-GIRTS** **SUPPORT SPACING: 8 TO 12 FEET** **CONCEALED CLIPS** **FASTENERS: • SELF TAPPING SCREWS • SPRING STEEL CLIPS**	●	○			• **exposed fasteners** available with heads finished to match finish of panel; also with plastic washers to prevent water penetration at fasteners. • **concealed clips** generally allow for thermal movement in panels. • **interior facing** may be edge formed flat panels; or the same roll-formed panel may be used as interior and exterior facing. • **perforated interior** facing panels available; with insulation added, they will serve to absorb sound. • when exterior and/or interior panels are attached to **structural frame with through-fasteners,** they may contribute to bracing of the structural frame; but will also act as thermal bridges. • **split system** will enclose structural girts but is more costly to install; not commonly used. When liner panels are used as interior face, liner panels have to be installed before exterior face as they must be back-fastened.
STRUCTURAL SUPPORT • **FASTENERS: SELF TAPPING SCREWS** • **NO SUB-GIRTS USED.** • **SPACING OF STRUCTURAL SUPPORTS; 8 TO 12 FEET.** • **EXTRUDED CONCEALED CLIPS AVAILABLE FOR SOME TYPES.**	●				
FASTENERS: CONCEALED CLIPS AND SELF-TAPPING SCREWS. **FASTENERS: EXPANDED ANCHOR BOLTS** **BATTEN AND DIVIDER STR SNAP-IN OR WITH COUNTER-SUNK BOLTS.** **SELF-TAPPING METAL SCREWS THROUGH INTERIOR FACING.** ROLL-FORMED OR FLAT EXTERIOR — FLAT INTERIOR	●	○	●		• **exterior and interior panels** generally separated by thermal breaks to minimize heat flow. • **core** usually of rigid, foamed thermal insulation. • **exterior panels** may be backed for stiffness when of light gauge prefinished metal sheets, such as porcelain enamel over steel or aluminum. • **support spacing:** 8 to 12 feet.
WELD PLATE AND CLIP WELDED TO STRUCTURAL FRAMING **EXPANSION BOLT** **EXPANSION BOLT** **ANCHOR BOLT** **CONCRETE FRAMING** **WELD PLATE CLIP WITH EXPANSION BOLT** CONCRETE — POLYMER CONCRETE (GENERALLY INSULATED	●	●	●	●	• **aggregate** embedded during casting of panels for precast, or by aggregate transfer method for cast-in-place. • **aggregate** embedded in **plastic matrix** panels during manufacturing. • **cast polymer concrete** panels are available as thin single thickness facing panels, or as insulated sandwich panels for use as curtain wall panels.

B2.5 CURTAIN WALLS

WALL PANELS: selection checklist

Panel selection will depend primarily on visual considerations, the finished panel being the principal component of the wall assembly:

■ **Metal or metal-faced panels may be:**

* flat, stamped
* roll-formed in different configurations
* finished with different coatings in a wide choice of colors.

Finishes for metal generally available are:

■ **Aluminum may be clear or color anodized.** Aluminum Association has established two classes of finishes, the difference being primarily the thickness of the coating: Class I coatings should be used for all exterior applications and Class II for interior.

■ **Carbon steel is generally hot dip galvanized** to minimize corrosion; may be galvanized and coated with aluminum by a hot dip process; or galvanizing may be omitted and a heavier coating of aluminum provided. Generally the heavier the applied coating, the better the resistance to corrosion.

■ Aluminum and galvanized steel may also receive baked-on protective/decorative coatings:

* porcelain enamel: a vitreous inorganic coating bonded to substrate by fusion at temperatures of 800°F or higher. Hardest and most durable applied finish, with a variety of colors available. Surface may be from full matte to high gloss. Embossing is available. Color uniformity from one part to another can be excellent.
* fluorocarbon: considered to be among the most stable resins known, highly resistant to degradation from effects of ultraviolet radiation and weathering. Wide choice of colors, but generally only in the low to medium gloss range. Cost is high.
* siliconized polymers, either acrylic or polyester: based on organic polymers and silicone intermediates, to improve weathering characteristics and resistance to fading. Wide

choice of colors in a full low-to-high gloss range. Cost is lower than of fluorocarbons.
* plastisols: have high resistance to chemical attack and corrosion, good weatherability and durability. Wide choice of colors in low- to medium-gloss range. Physical properties of coating may be varied considerably in the formulating process, and the extended in-service record of a particular formulation should be checked before making a selection. Cost is moderate.

■ **Concrete panels may be:**

* flat, ribbed
* patterned, or sculptured
* ground smooth, textured, pigmented
* Selected aggregate may be embedded into freshly placed concrete: appearance of finished surface depends on the skill of workmen.
* Large size aggregate or flat stones may be placed in a sand bed and the concrete placed over: appearance of finished surface depends on the skill of the workmen.
* refer to CEMENT/CONCRETE for additional available finishes.

■ **Availability of certain types of panels may be limited:**

* stamped metal panels will generally require special production facilities; the cost of design and fabrication may preclude their use in small- or medium-size projects.
* sculptured precast concrete panels require special formwork, the cost of which must be distributed over a large number of panels to keep the total cost of panels competitive.

■ Hollow core precast concrete plank may also be used as wall panels.

THERMAL TRANSMISSION OF COMPONENTS

The evaluation of the resistance to thermal transmission of a wall assembly should include all components of the assembly to arrive at a true value of the U-coefficient of heat transmission.
* Generally, any metallic part

extending through the thickness of the wall, such as tabs, stiffeners, subgirts, even fasteners, will provide a path for heat flow, contributing to heat loss, and possibly even causing water vapor to condense at such points.

INSULATION

■ The preferred location of insulation is outside the structural frame, in order to:
* minimize heat loss through uninsulated edges of floor/roof framing.
* minimize thermal movement in the structural frame and the resultant effect on the curtain wall panels.

■ Concrete sandwich unit panels require solid edges to hold the two concrete faces together.
* the interior surfaces of panels will then have uneven surface temperatures as more heat will be lost through solid concrete than through the sandwich part.
* such panels may nevertheless be economical, where the interior face remains exposed, and durable interior surfaces are required.

■ **Solid concrete panels may**

* have rigid insulation boards laminated, or clip-attached, with the finish surfacing, such as gypsum board, in turn laminated to the insulation.
* may be furred out with metal, or wood strips to which the finish surfacing is attached.

■ **Single thickness** of panels will result in high heat losses from heated interiors unless insulated; when insulation is used with this type:

* it is then exposed to the interior and subject to damage;
* its appearance may be objectionable.

JOINTS

Joints between components of a curtain wall are critical to its proper functioning and must be detailed to accommodate movement in the wall due to thermal expansion and contraction as well as to lateral loads:

■ Larger panels require fewer joints; joint design then becomes more critical since thermal movement across the joint increases in proportion to the size of the panel:

■ **Jointing precast concrete panels generally presents the greatest problems:**

- close dimensional tolerances are hard to achieve.
- joints other than straight butt joint are difficult and costly.

■ **Generally, joint design should be as simple as possible:**

- with the smallest number of loose parts.
- of the same design, or with a minimum of variations throughout the entire structure.

OTHER CONSIDERATIONS

■ **Material isolation may have** to be considered as a variety of materials may be incorporated into a curtain wall assembly:

- aluminum alloys are highly reactive with some other material types.
- cement products react with aluminum alloys. Provide coatings or physical separation.
- water runoff over metal surfaces can produce ion flow which causes staining or corrosion.
- water runoff from the exterior skin can permanently stain adjacent work; sealants may discolor; glass may be stained or etched.
- ferrous metal products incorporated into masonry or concrete should be galvanized and/or protective-coated to provide added protection.

■ **Corrosion in panels** of edges where panels are field cut and not protective-coated during erection.

■ **Thermal expansion/contraction** in metal panels:

- metal panels are stiffer in the direction of corrugations/ribs, and crushing of metal at through fasteners may occur.
- concealed clip fasteners will generally allow for thermal movements without damage to the facing.
- thermal movement perpendicular to corrugations/ribs is taken up in such corrugations/ribs between closey spaced fasteners.

■ **Formed plastic and cement-bound,** mineral-fiber panels are generally stiffer than metal sheets in both directions and some crushing of material at fasteners may occur unless provisions are made to:

- provide for flexibility at fasteners - standard practice for mineral-fiber panels.
- reduce the spacing of fasteners and thus minimize extent of movement at each.

■ **Formed panels** should not be restrained at ends perpendicular to corrugations/ribs, since permanent deformations may then result.

■ **Vertical expansion joints** in facings are generally required only at expansion joints in the structure.

■ **Gauge or thickness,** and depth/configuration of corrugations/ribs determines resistance to bending and extent of deflection under horizontal loads:

- recommended maximum deflection for metals is generally 180th of span.
- visual considerations or building codes may require lower limits.

■ **Formed panels** have poor resistance to impact and/or indentation: they generally cannot be readily repaired but can be easily removed and replaced.

■ **Formed single thickness panels** may be limited by span and/or deflection, and may not be sufficiently strong to be used between floors or between floor and roof; structural girts may then be used to reduce the span:

- such girts will add to the cost of the structure;
- being exposed to the interior will also collect dust and/or moisture.

■ **Formed sandwich panels** are commonly insulated with the interior liner enclosing the insulation and adding to structural strength of assembly:

- spans of up to 30 feet or more are possible without the use of intermediate structural girts;
- sheet assemblies generally are field assembled with some types available factory preassembled;
- factory preassembly usually does not offer any significant benefits.

Formed panels are most suitable for wall assemblies with a limited number of openings.:

- additional structural framing and trim pieces for facing are

required at each opening and increase the overall cost of the assembly.
- when multiple openings in the wall assembly are required, curtain walls of the grid type should be considered.

Special type of curtain wall panel is the all-glass wall:

- floor height panels of glass suspended from structural floor or roof framing.
- glass stabilizing fins may be cemented perpendicular to glass sheets for added resistance to lateral loads.
- sliding connection is provided at floor level to allow for deflection in structural framing, and to secure lower edge of glass against lateral displacement.
- thermal movement vertically is provided for by sliding connection at floor level; and horizontal expansion is provided for by resilient seals between panels.
- sizes limited by manufacturing and shipping restrictions.
- expensive system, requiring special considerations to minimize deflection of supporting structural frame, and for the transfer of lateral loads to structure.
- generally used for one-story buildings, or for ground floors of multistory buildings.

B2.5 CURTAIN WALLS

CURTAIN WALLS: energy notes

Air Leakage
The uncontrolled movement of air out, or into a building through the fabric of the exterior wall, or air leakage, is an important consideration for any type of exterior wall, but becomes even more critical for the grid type curtain wall, because of the numerous joints required in the finished wall assembly.

Controlling air leakage is important to:
- Reduce heat loss in cold weather.
- Reduce heat gain in air-conditioned buildings.
- Prevent exfiltrated air from condensing and freezing within the wall assembly in cold weather, with resulting rapid deterioration of the wall assembly.
- Prevent dust from infiltrating the building.

Air leakage is caused by:
- Pressure differentials due to positive, or negative, wind pressures on the walls.
- Chimney effect, when air on opposite sides of a wall is at different temperatures.
- Operation of HVAC systems, which generally are designed to provide a slight positive pressure inside the building.

Air leakage may be controlled by:
- Sealing interior surfaces of exterior walls air-tight, and controlling any rain penetration through exterior face by the rain screen principle.
- Sealing all joints at the interior face to maintain air-tightness.
- Providing control joints in the backup materials to control any possible cracking, and sealing such joints to make them air-tight.
- Sealing the perimeter of all fixed frames within the wall, as well as the perimeter of operating panels within such frames.
- Using vapor barriers, provided to control migration of water vapor by diffusion, also in order to prevent, or minimize air leakage.
- Sealing the interior surfaces of all porous materials.
- Restricting air flow between floors to reduce pressure differences resulting from chimney effect.

Thermal Conductors
Elements of a wall assembly with lower resistance to heat flow than the main construction result in thermal conductors, or thermal bridges, which may seriously interfere with the total thermal performance of the wall assembly, in the following ways:
- The temperature of the interior face of a thermal bridge will be lower than the temperature of adjacent surfaces, thus it may not be possible to maintain the desired inside relative humidity without having to contend with condensation over the interior surface of the thermal bridge.
- The difference in temperatures may induce additional stresses due to differential movements.
- Heat loss through the wall assembly will be increased.
- Moisture will condense at thermal bridges within the wall in cold weather.

Thermal bridges are formed by:
- Single-section mullions.
- Single-section window and door frames.
- Single glazing; or single sheet panels.
- Metal-faced sandwich panels without thermal breaks between exterior and interior facing sheets.
- Any material with low resistance to heat flow extending through the thickness of the exterior wall assembly, such as structural framing; anchors supporting the wall assembly; or ties connecting the exterior and interior faces of panels.

Theoretical calculations indicate that if a thermal bridge, consisting of an aluminum tab of ½ inch cross-sectional area, is introduced into an insulated aluminum faced panel 16 square feet in area, **the insulating value of the entire panel may be reduced by up to 85 percent.**

The effects of thermal bridges may be controlled by:
- Avoiding, or at least keeping to a minimum, all through-metal elements.
- Providing thermal breaks in all elements with low resistance to heat flow.
- Locating insulation to cover the entire exterior of the building, with only the exterior facing placed outside of such envelope.

REFERENCES: Curtain walls

STANDARDS

ASTM Specifications for:

A525 • Steel Sheet, Zinc-Coated (Galv.) by the Hot-Dip Process.

C33 • Concrete Aggregates.

C56 • Structural Clay Non-Load-Bearing Tile.

C126 • Ceramic Glazed Structural Clay Facing Tile, Facing Brick, and Solid Masonry Units.

C212 • Structural Clay Facing Tile.

C220 • Flat Asbestos-Cement Sheets.

C221 • Corrugated Asbestos-Cement Sheets.

C223 • Asbestos-Cement Siding.

C542 • Lock-Strip Gaskets.

C551 • Asbestos-Cement Fiberboard Insulating Panels.

C659 • Asbestos-Cement Plastic-Foam Core Insulating Panels.

C716 • Installing Lock-Strip Gaskets and Infill Glazing Materials.

D3679 • Rigid Poly (Vinyl Chloride) (PVC) siding.

D4216 • Rigid Poly (Vinyl Chloride) (PVC) and Related Plastic Building Products Compounds.

ASTM Tests/Methods for:

C272 • Water Absorption of Core Materials for Structural Sandwich Constructions.

D1037 • Evaluating the Properties of Wood-Base Fiber and Particle Panel Materials.

D1151 • Effect of Moisture and Temperature on Adhesive Bonds.

D4226 • Impact Resistance of Rigid Poly (Vinyl Chloride) (PVC) Building Products.

E72 • Conducting Strength Tests of Panels for Building Construction.

E84 • Surface-Burning Characteristics of Building Materials.

E96 • Water Vapor Transmission of Materials.

E119 • Fire Tests of Building Construction and Materials.

E283 • Rate of Air Leakage Through Exterior Windows, Curtain Walls, and Doors.

E330 • Structural Performance of Exterior Windows, Curtain Walls, and Doors by Uniform Static Air Pressure Difference.

E331 • Water Penetration of Exterior Windows, Curtain Walls and Doors by Uniform Static Air Pressure Difference.

E546 • Frost Point of Sealed Insulating Glass Units.

E547 • Water Penetration of Exterior Windows, Curtain Walls, and Doors by Cyclic Static Air Pressure Differential.

ASTM Definitions of Terms Relating to:

C274 • Structural Sandwich Constructions.

C717 • Building Seals.

E241 • Recommended Practice for Increasing Durability of Building Constructions Against Water-Induced Damage.

E413 • Classification for Determination of Sound Transmission Class.

ANSI Specifications:

A108.1 1976 • A Collection - Includes A108.1, A108.4, A108.5, A108.6, A108.7, A118.1, A118.2, A118.3, A118.4, A136.1.

SELECTED REFERENCES

• The Performance of a Curtain Wall Panel Faced with Ceramic Tile, test performed by Department of Engineering Research, Penn. State University.

• A Reflective Method for Testing Flatness and Thermal Buckling of Metal Panels (Study No. 4), John H. Callender, School of Architectural, Princeton University, N.J.

AA **Aluminum Association**

• Aluminum Standards and Data

• Illustrative Examples of Design

• Specifications for Aluminum Structures

• Aluminum Siding Application Manual

AAMA **Architectural Aluminum Mfg. Association**

• Methods of Test for Curtain Walls

• Voluntary Specifications for Aluminum Siding

• Curtain Wall Design Manual, Contents: Testing, Types & Systems Rain Screen Principle Pressure-Equalized Wall Design Primary Design Concerns Wall Detail Guidelines Lock-Strip Gasket Application Curtain Wall History High-Rise Fire Safety Sealants, Glass, & Glazing Curtain Wall Installation Finishes Design for Energy Conservation

• Sound Control for Aluminum Curtain Walls and Windows

• Design Wind Loads for Aluminum Curtain Walls

• Fire-Resistive Design Guidelines for Curtain-Wall Assemblies

• Metal Curtain Wall, Window, Store Front and Entrances Guide Specifications Manual

ACI **American Concrete Institute**

• Precast Concrete: Handling and Erection

• ACI Manual of Concrete Practice, Part 3; Products and Processes

• Precast Concrete Wall Panels

REFERENCES: Curtain walls (cont.)

- Design of Wall Panels
- Fabrication, Handling and Erection of Wall Panels
- Materials for Wall Panels
- Quality Standards and Tests for Wall Panels
- Tilt-Up Construction Compilation Panels

AHA — American Hardboard Association
- Field Finishing Hardboard Siding
- Hardboard Paneling
- Hardboard Siding

APA — American Plywood Association
- E300 — APA Product Guide: 303 Plywood Siding
- F800 — APA Product Guide: Panel Care and Installation
- F405 — APA Product Guide: Performance Rated Panels
- C465 — Plywood Siding Over Rigid-Foam Insulation Sheathing
- D481 — Buckling of Plywood Sheathing
- D485 — Corrosion-Resistant Fasteners for Plywood Construction
- F410 — Buckling of Plywood Panel Siding
- X330 — Joint Details for Exterior Plywood Wall Systems

BIA — Brick Institute of America
- 13 — Ceramic Glazed-Brick Facing for Exterior Walls
- 17L Rev. CDA — Four-Inch RBM Curtain and Panel Walls Copper Development Association
- Copper, Brass & Bronze Building Products Sources

CRA — California Redwood Association
- 3A4-1 — Redwood Siding Applications
- 3A4-3 — Redwood Channel Rustic and Shiplag V Siding.
- 3A4-3 — Redwood Bevel Siding
- 3A4-6 — Redwood T&G Siding
- 3A4-2 — Redwood Board and Batten Siding
- 4B1-1 — Redwood Exterior Finishes
- 4B1-3 — Painting Exterior Redwood

- 3A9 — Redwood Plywood Guide (Sweet's).
- 2B1-2 — Redwood Lumber Grades and Uses
- 4A1-1 — Nails and Nailing

FTI — Facing Tile Institute
- Design Data for Structural Facing Tile Ceramic Glazed Brick
- Standard Units and Shapes

HPMA — Hardwood Plywood Mfg. Association
- Structural Design Guide for Hardwood Plywood, Wall Panels
- Design Guide for Wood Composition Board Panels
- Installation of Hardwood Plywood

ML/SFA — Metal Lath/Steel Framing Association
- Fire-rated Metal/Steel Stud Curtainwalls
- Curtainwalls

NAAMM — National Assoc. of Architectural Metal Mfg.
- Metal Curtain Wall Manual
- Design Wind Loads for Walls of Rectangular Buildings
- Field Check for Water Leakage of Metal Curtain Walls
- Metal Finishes Manual; For Architectural Metals and Metal Products

NCMA — National Concrete Masonry Association
- Concrete Masonry Faces and Finishes
- Prefabricated Concrete Masonry Wall Panels
- Concrete Masonry Veneers
- Curtain and Panel Walls of Concrete Mansonry
- Structural Backup Systems for Concrete Masonry Veneers

NPA — National Particleboard Association
- Technical Bulletin: Control of Warping of Wood Panels Faced with Plastic Laminates

NPCA — National Precast Concrete Association
- Fundamentals of Quality Precast Concrete

NRCC — National Research Council of Canada
- CBD 48 — Requirements for Exterior Walls
- 93 — Precast Concrete Walls - Problems With Conventional Design
- 94 — Precast Concrete Walls - A New Basis for Design
- 97 — Look at Joint Performance
- 125 — Cladding Problems Due to Frame Movements
- 126 — Influence of Orientation on Exterior Cladding
- 185 — Failure of Brick Facing on High-Rise Buildings

NTMA — National Terrazzo and Mosaic Association
- Architects' Guide to Terrazzo

PCA — Portland Cement Association
- Finishing Concrete Slabs - Exposed Aggregate, Patterns and Color
- Color and Texture in Architectural Concrete
- Design and Construction of Large-Panel Concrete Structures
- Jobsite Precast Concrete Panels - Textures, Patterns, and Designs
- Special Considerations for the Selection of Tilt-Up Concrete Sandwich Panel
- Tilt-Up Concrete Buildings - A Value Decision
- Tilt-Up Concrete Walls

PCI — Prestressed Concrete Institute
- PCI Handbook
- PCI Architectural Precast Concrete
- Design for Weathering of Buildings Using Architectural Precast Concrete
- Architectural Precast Concrete Joint Details

B2.5 CURTAIN WALLS

TCA   **Tile Council of America, Inc.**
- Recommended Standard Specifications for Ceramic Tile

USDA   **U. S. Department of Agriculture Forest Service**
- Basic Properties of Three Medium-Density Hardboards

VSI   **Vinyl Siding Institute**
- The Cleaning of Vinyl Siding
- What Builders Want to Know About Vinyl Siding and Vinyl Windows and Doors
- Rigid Vinyl Siding Application Manual

WWPA   **Western Wood Products Association**
- Lumber Paneling
- Siding Installation Information
- Wood Siding

B2. 6 WALL FACINGS

FACING PANELS: types, materials

● denotes common usage or availability
○ denotes possible or limited usage and/or limited availability

TYPE	DESCRIPTION	EXTERIOR FACE: BASE MATERIALS	USE — EXTERIOR/INTERIOR	USE — INTERIOR ONLY	WALL TYPE — MASONRY	WALL TYPE — FRAMED: metal	WALL TYPE — FRAMED: wood
FORMED: single thickness THICKNESS & DEPTH OF SECTION	**Panels formed in various configurations;** may also be of stiffened flat or curved sheets ● formed panels more commonly used for deep fascias, mansards, equipment screens; flat or curved sheets generally as trim, such as column facing.	ALUMINUM	●		③	③	○
		STEEL: galvanized	●		③		○
COMPOSITE: single thickness	**Panels of single thickness after manufacturing:** laminated, molded, or cast ● either prefinished or site finished after installation ● Plywood and hardboard used as vertical siding when exterior type; as paneling when interior-only type.	PLYWOOD/ HARDBOARD	●	●	③	●	●
		CEMENT-BOUND MINERAL FIBER	●		③	③	○
COMPOSITE: laminated INSULATION, INTERIOR FACE, CORE, EXTERIOR FACE	**Thin sheets of metal laminated to a rigid board or a rigid core material** ● inside face sheet generally used to stabilize the assembly against warpage; may be finished when exposed ● Also commonly used as glazing panels ● Edges open or formed.	ALUMINUM	●		③	③	
		STEEL: galvanized	●		③	③	
INTERIOR FACE, CORE: insulation, EXTERIOR FACE	**Rigid boards or heavier gauge metal sheets laminated to an insulating core of various thicknesses** ● may be made sufficiently rigid to function as wall panels; or sufficiently thin to be used as glazing panels ● Edges may be formed or open.	ALUMINUM	●		③	③	
		STEEL: galvanized	●		③	③	
		CEMENT-BOUND MINERAL FIBER	●		③	③	○
INTERIOR FACE, BACKING, CORE: insulation, BACKING, EXTERIOR FACE	**Exterior face of thin sheets of metal** laminated to rigid boards for rigidity and then laminated to insulating core ● **inside face of various materials** ● may be made sufficiently rigid to function as wall panels ● edges may be open closed, and/or formed.	ALUMINUM	●		③	③	
		STEEL: galvanized	●		③	③	
COMPOSITE: surfaced BACKING/SUBSTRATE, EXTERIOR FACE	**Surfacing laminated to a substrate of various rigid boards:** plywood, hardboard, cement-bound mineral fiber, polystyrene; or embedded in polymer concrete ● may be used as wall panels when preassembled with framing.	AGGREGATE: in matrix	●		③	●	●
		BRICK: thin face or CERAMIC TILE	●		③	●	●
MONOLITHIC OPTIONAL BACKING	**Solid panels of stone;** such as marble, granite ● may also be backed for rigidity ● Opaque or nearly opaque glass, either integrally pigmented or coated; commonly heat strengthened and/or laminated.	STONE	●		●		
		GLASS	●		●		

B2.6 WALL FACINGS

B · BUILDING SHELL

SIZE WIDTH x LENGTH x THICKNESS, average in inches	EXTERIOR FACE: FINISH OPTIONS — INTEGRAL/NO FINISH	APPLIED — CLEAR/TINTED	APPLIED OPAQUE — PORCELAIN ENAMEL	SILICONIZED POLYMERS	ANODIZED, COLOR	PLASTIC SHEET	BACKING/CORE — INTEGRAL WITH FACE	RIGID BOARD ①	INSULATION ②	THERMOPLASTIC	JOINTS — BUTT/GASKETED	LAP/TONGUE & GROOVE	FRAME/MOLDINGS	METHOD OF ATTACHMENT — FACE: SCREWS/NAILS	BACK: CLIPS/SCREWS	ADHESIVE BONDED	ANCHORS-SHELF ANGLES	REMARKS
12 to 34x up to 480x 1 ½	○			●	●		●					●	○	●	●			**Formed panels** essentially the same as wall panels ● May also be flat with standing or batten seams for use as mansard roofing ● Generally available as complete systems with trim pieces and subgirts and/or framing.
			●				●					●		●	●			
48x 120x to ¾		●				●	●				●	○		●	○			**Prefinished gypsum board** available for interior use ● Plywood/hardboard panels for interior use only commonly prefinished ● Face nailing often used with adhesive ● Holes for fasteners generally plugged in cement-bound mineral fiber panels.
60x 120x to ½	●		●				●				●	○		●	○			
48x to 360x ½	○	●	●	●	●			●	●	●	○	●				○	●	**Proprietary systems of aluminum sheets** with thermoplastic cores and of insulating cores with exterior face only of aluminum sheet are available ● Panels of both systems available bent to various radii.
				●	●					●	○	●				○	●	
48 x 144 x ½	○			●	●				●		●	●		●				**Edges of cement-bound mineral fiber board** faced panels commonly open ● Edges of metal faced panels generally closed and formed to: interlock; to receive gaskets; or to fit a curtain wall grid system.
			●						●		●	●			●		
	●		●						●		○	●				○	○	
36x 96x 2	○		●	●	●			●	●		○	●					○	**Available with interior faces of:** prefinished gypsum board; rigid boards, such as plywood, cement-bound mineral fiber, hardboard; or metal-faced rigid boards ● Edges may be formed to fit curtain-wall grid system.
				●	●			●	●		○	●					○	
48x 96x ½	●	●						●	●		●	●		●	○	●		● Clear sealers may be used over aggregate ● Panels with polystyrene substrate generally attached using adhesive.
48x 96x ½	●								●		●			●	○	○		
48x 120x ½	●						●				●						●	**For additional information on stone** refer to STONE/MASONRY ● For additional information on glass - commonly referred to as spandrel glass - see GLASS/PLASTICS
72x 144x x ¼ to 2	●						●				●	●					●	

REMARKS

① **Rigid boards:** plywood, hardboard, cement-bound mineral fiber.

② **Insulation commonly used:** urethane, isocyanurate, polystyrene; perlite.

③ **Metal or wood furring or metal subgirts commonly required;** some panels may also be adhesive attached to a solid level substrate.

B2.6 WALL FACINGS

FACINGS: types, materials

- denotes common usage or availability
- ○ denotes possible usage or limited availability

B2.6 WALL FACINGS

TYPE	DESCRIPTION	EXTERIOR FACE: BASE MATERIAL	USE: EXTERIOR/INTERIOR INTERIOR ONLY	WALL: MASONRY/CONCRETE FRAMED	SIZE: WIDTHxLENGTHxTHICKNESS average, in inches [1]
UNITS: attached FACING • MASONRY	Stone or brick masonry units bonded together with mortar and attached to a bearing or nonbearing wall assembly for lateral support • weight of facing supported by wall assembly, or by structural frame.	**STONE**	•	• •	VARIES
		BRICK **●GLAZED TILE**	•	• •	4x8 x2
FACING • SHINGLES	Thin units of various widths and lengths butted horizontally and overlapping vertically • may be preassembled into, or manufactured as horizontal panels 4 to 8 feet long • attached directly to wall assembly for support.	**WOOD**	•	[5] •	6x18 x½
		METAL	•	[5] •	12x48 x.02
STRIPS: attached FACING •HORIZONTAL SIDING	• Horizontal siding: narrow strips butted or overlapping at ends, and overlapping vertically • attached to sheathing or directly to studs in framed construction • may be manufactured as panels, generally two strips wide, with a center horizontal joint imitating a lapped joint. • Vertical siding: narrow strips with butted or overlapping vertical joints • available as facing panels imitating vertical siding.	**WOOD**	•	[5] •	8 to 12x x144x x½ or x.03
		●PLYWOOD **●HARDBOARD**	•	[5] •	
FACING • VERTICAL SIDING		**METAL**	•	[5] •	
		PLASTIC	•	[5] •	
UNITS: bonded FACING • TILE	Burned clay units continuously bonded to wall assembly • with grouted joints between units • wide choice of sizes available • choice of colors when glazed; choice of shades when without glaze.	**●CERAMIC TILE** **●TERRA-COTTA** **●GLAZED TILE**	•	• ○	1x1 to 12x12 x½
		QUARRY TILE	•	• ○	6x6 x½
SURFACING: bonded FACING • STUCCO • PLASTER • AGGREGATE	• Stucco: of white cement; may be internally pigmented • Cement plaster: of gray Portland cement • Plaster: of gypsum, for dry interior locations only • Select stone aggregate: hand embedded in adhesive matrix.	**●STUCCO** **●PLASTER**	• •	• •	NO LIMIT
		STONE AGGREGATE	•	• •	NO LIMIT
FACING • FABRIC	Organic or synthetic fibers woven or felted and continuously bonded to wall assembly • lightweight carpeting for use as wall facing is available. • Semi-flexible plastic sheets; flexible plastic without or with fabric reinforcing • continuously bonded to wall.	**FABRIC:** **● WOVEN** **● FELTED**	•	[6] •	27 or 54 WIDE in LONG LENGTH
		PLASTIC: **●HOMOGENEOUS** **●FABRIC REINFORCED**	•	[6] •	

B • BUILDING SHELL

REMARKS (notes)

[1] Sizes shown are common average; various sizes generally available as stock items.

[2] Pattern denotes the appearance of the facing after installation. Stucco and plaster may be trowelled, tooled or stamped into various patterns.

[3] Refer to COATINGS for information on available field-applied coatings.

[4] Field-formed joints to control expansion/contration and/or for decorative purposes.

[5] Usage not common. Nailable furring strips required for installation.

[6] Smooth and level substrate required; generally installed over gypsum board or plaster.

SMOOTH/TEXTURE: light	TEXTURE: medium to heavy	PATTERN: regular [2]	PATTERN: random	INTEGRAL COATING: preapplied	COATING: field applied [3]	GLAZE: preapplied	BUTT	LAPPED	FORMED: in field [4]	NAILS/SCREWS: face	METAL TIES	MORTAR BONDED	ADHESIVE BONDED	SELF-BONDING	EXTENSIVE	AVERAGE	REMARKS
●	●	●	●	●			●			●						●	Facing supports own weight between supports • In multistory construction, shelf angles commonly provided at each floor level to transfer load to structural frame • When one story high may be supported on foundation or by wall assembly.
●	○	●		●	●		●			●						●	
●	●	●	●	●	●		●	●		●						●	Preassembled wood shingles are attached to plywood backing which then serves as sheathing over wall framing • Wood commonly of decay-resistant species; may be field coated • Metal generally aluminum, prefinished • Shingles of cement-bound mineral fiber are available.
○	●	●		●			●	●								●	
●		●			●		●	●		●						●	• Joints between individual lengths of wood, plywood, and hardboard strips generally butted; joints between lengths of metal and plastic strips lapped to keep them aligned and to provide for expansion/contraction. • Strips nailed directly to wall framing must be sufficiently rigid to resist lateral movement in the plane of the wall. • Metal and plastic sheets may have insulation incorporated within the formed section or may be installed over rigid insulation attached to the wall assembly.
●	●	●		●	●		●	●		●						●	
●		●						●		●						●	
●		●					●			●						●	
●		●	●	●		●	●					●	●			●	Units for exterior use should be frostproof; not all are available in such grade. Large units of terra-cotta when glazed may tend to develop cracks or crazing in the glaze.
●		●		●			●					●	●			●	
●	●	○	○	●	●				●					●	●		For occasionally and moderately wet indoor locations gypsum, known as Keene's cement, should be used • Appearance of stone aggregate facing depends on the skill of workmen
	●	●	○						●				●		●		
●		●		●			●						●			●	Plastic fabrics are generally of vinyl and fabric reinforced; wide choice of colors and printed patterns available. Vinyl-faced paper is also available; use generally limited to residential occupancies.
●		●		●			●						●			●	

B2.6 WALL FACINGS

FACINGS: typical assemblies

[1] Basis for Comparative Cost: corrugated single thickness 26 gauge galvanized steel formed panel; exposed fasteners; over structural steel frame = 100.

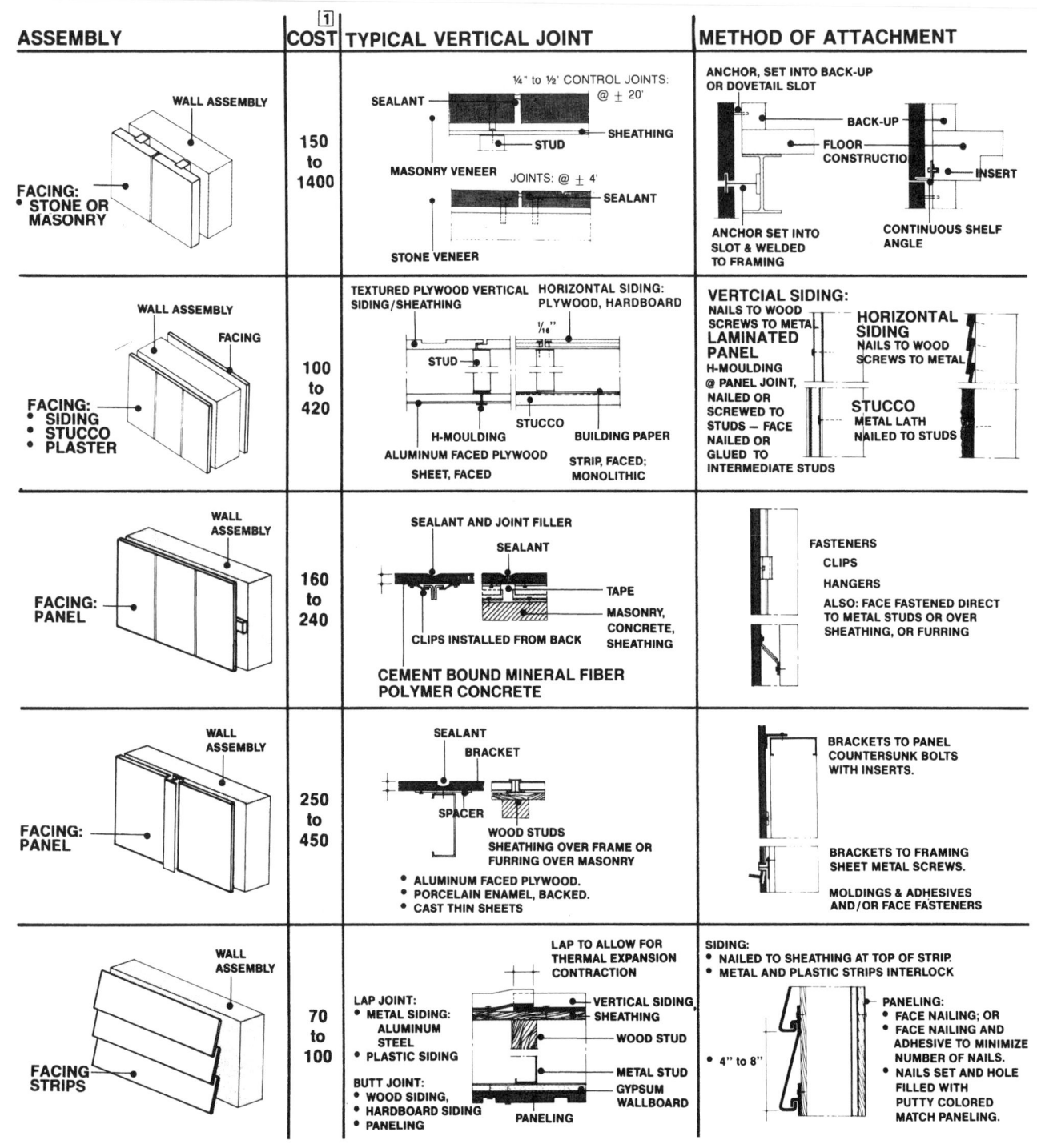

ASSEMBLY	COST[1]	TYPICAL VERTICAL JOINT	METHOD OF ATTACHMENT
WALL ASSEMBLY — FACING: • STONE OR MASONRY	150 to 1400	SEALANT, ¼" to ½' CONTROL JOINTS: @ ± 20', SHEATHING, STUD, MASONRY VENEER, JOINTS: @ ± 4', SEALANT, STONE VENEER	ANCHOR, SET INTO BACK-UP OR DOVETAIL SLOT, BACK-UP, FLOOR CONSTRUCTION, INSERT, ANCHOR SET INTO SLOT & WELDED TO FRAMING, CONTINUOUS SHELF ANGLE
WALL ASSEMBLY — FACING — FACING: • SIDING • STUCCO • PLASTER	100 to 420	TEXTURED PLYWOOD VERTICAL SIDING/SHEATHING, HORIZONTAL SIDING: PLYWOOD, HARDBOARD, ¹/₁₆", STUD, STUCCO, H-MOULDING, BUILDING PAPER, ALUMINUM FACED PLYWOOD, STRIP, FACED; SHEET, FACED, MONOLITHIC	VERTCIAL SIDING: NAILS TO WOOD SCREWS TO METAL, LAMINATED PANEL H-MOULDING @ PANEL JOINT, NAILED OR SCREWED TO STUDS — FACE NAILED OR GLUED TO INTERMEDIATE STUDS, HORIZONTAL SIDING NAILS TO WOOD SCREWS TO METAL, STUCCO METAL LATH NAILED TO STUDS
WALL ASSEMBLY — FACING: PANEL	160 to 240	SEALANT AND JOINT FILLER, SEALANT, TAPE, MASONRY, CONCRETE, SHEATHING, CLIPS INSTALLED FROM BACK, CEMENT BOUND MINERAL FIBER POLYMER CONCRETE	FASTENERS, CLIPS, HANGERS, ALSO: FACE FASTENED DIRECT TO METAL STUDS OR OVER SHEATHING, OR FURRING
WALL ASSEMBLY — FACING: PANEL	250 to 450	SEALANT, BRACKET, SPACER, WOOD STUDS SHEATHING OVER FRAME OR FURRING OVER MASONRY, • ALUMINUM FACED PLYWOOD. • PORCELAIN ENAMEL, BACKED. • CAST THIN SHEETS	BRACKETS TO PANEL COUNTERSUNK BOLTS WITH INSERTS. BRACKETS TO FRAMING SHEET METAL SCREWS. MOLDINGS & ADHESIVES AND/OR FACE FASTENERS
WALL ASSEMBLY — FACING STRIPS	70 to 100	LAP TO ALLOW FOR THERMAL EXPANSION CONTRACTION, LAP JOINT: • METAL SIDING: ALUMINUM STEEL • PLASTIC SIDING, BUTT JOINT: • WOOD SIDING, • HARDBOARD SIDING • PANELING, VERTICAL SIDING, SHEATHING, WOOD STUD, METAL STUD, GYPSUM WALLBOARD, PANELING	SIDING: • NAILED TO SHEATHING AT TOP OF STRIP. • METAL AND PLASTIC STRIPS INTERLOCK, • 4" to 8", PANELING: • FACE NAILING; OR • FACE NAILING AND ADHESIVE TO MINIMIZE NUMBER OF NAILS. • NAILS SET AND HOLE FILLED WITH PUTTY COLORED MATCH PANELING.

B2.6 WALL FACINGS

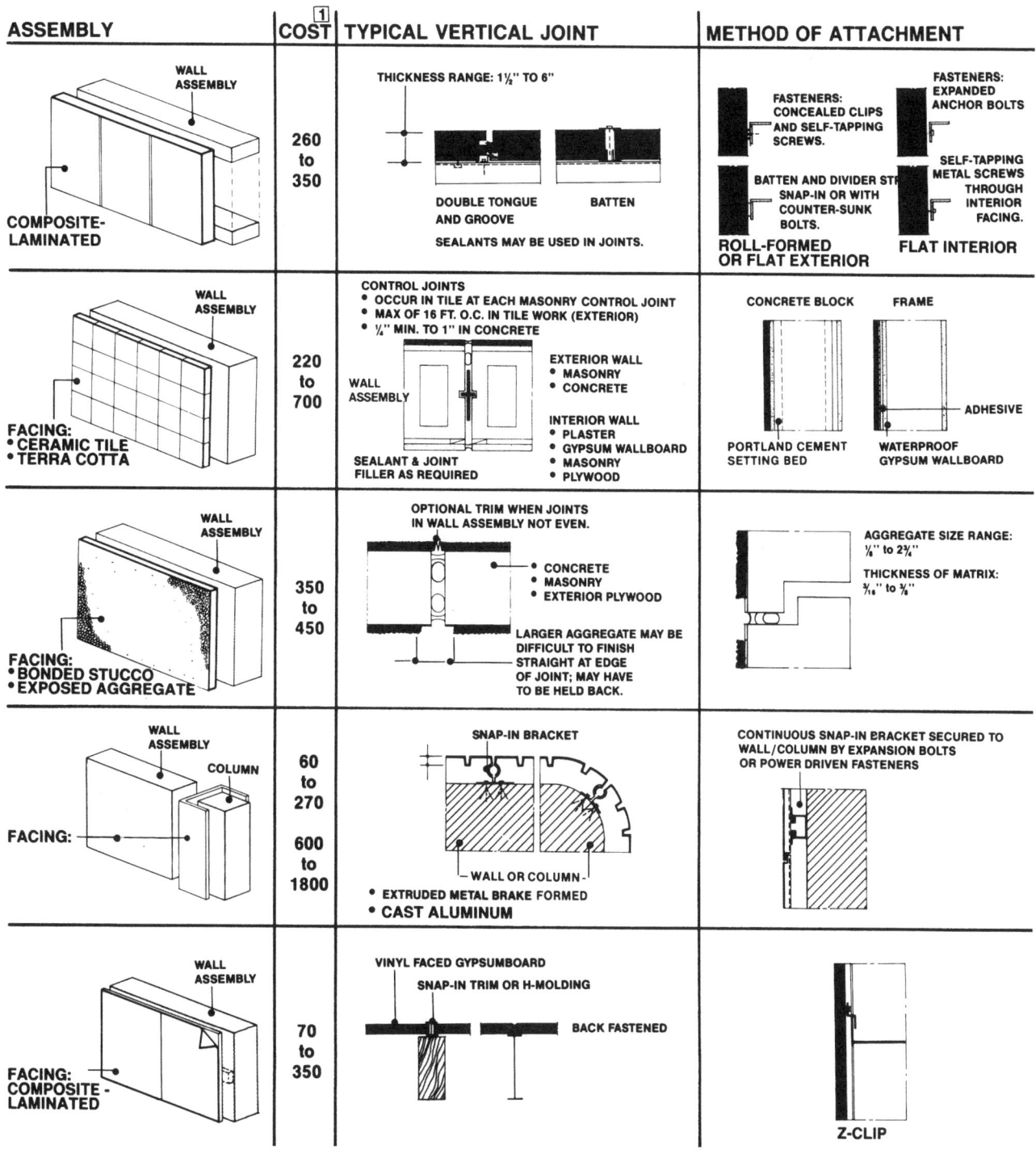

ASSEMBLY	COST [1]	TYPICAL VERTICAL JOINT	METHOD OF ATTACHMENT
COMPOSITE-LAMINATED (WALL ASSEMBLY)	260 to 350	THICKNESS RANGE: 1½" TO 6" — DOUBLE TONGUE AND GROOVE, BATTEN. SEALANTS MAY BE USED IN JOINTS.	FASTENERS: CONCEALED CLIPS AND SELF-TAPPING SCREWS. FASTENERS: EXPANDED ANCHOR BOLTS. BATTEN AND DIVIDER STR SNAP-IN OR WITH COUNTER-SUNK BOLTS. SELF-TAPPING METAL SCREWS THROUGH INTERIOR FACING. ROLL-FORMED OR FLAT EXTERIOR / FLAT INTERIOR
FACING: • CERAMIC TILE • TERRA COTTA (WALL ASSEMBLY)	220 to 700	CONTROL JOINTS • OCCUR IN TILE AT EACH MASONRY CONTROL JOINT • MAX OF 16 FT. O.C. IN TILE WORK (EXTERIOR) • ¼" MIN. TO 1" IN CONCRETE. WALL ASSEMBLY. EXTERIOR WALL • MASONRY • CONCRETE. INTERIOR WALL • PLASTER • GYPSUM WALLBOARD • MASONRY • PLYWOOD. SEALANT & JOINT FILLER AS REQUIRED	CONCRETE BLOCK / FRAME. ADHESIVE. PORTLAND CEMENT SETTING BED / WATERPROOF GYPSUM WALLBOARD
FACING: • BONDED STUCCO • EXPOSED AGGREGATE (WALL ASSEMBLY)	350 to 450	OPTIONAL TRIM WHEN JOINTS IN WALL ASSEMBLY NOT EVEN. • CONCRETE • MASONRY • EXTERIOR PLYWOOD. LARGER AGGREGATE MAY BE DIFFICULT TO FINISH STRAIGHT AT EDGE OF JOINT; MAY HAVE TO BE HELD BACK.	AGGREGATE SIZE RANGE: ⅛" to 2¾". THICKNESS OF MATRIX: ³⁄₁₆" to ⅞"
FACING: (WALL ASSEMBLY / COLUMN)	60 to 270 / 600 to 1800	SNAP-IN BRACKET. WALL OR COLUMN. • EXTRUDED METAL BRAKE FORMED • CAST ALUMINUM	CONTINUOUS SNAP-IN BRACKET SECURED TO WALL/COLUMN BY EXPANSION BOLTS OR POWER DRIVEN FASTENERS
FACING: COMPOSITE-LAMINATED (WALL ASSEMBLY)	70 to 350	VINYL FACED GYPSUMBOARD. SNAP-IN TRIM OR H-MOLDING. BACK FASTENED	Z-CLIP

B2.6 WALL FACINGS

FACING PANELS/FACINGS: selection checklist

Facings, as the outer components of a wall assembly, are generally subject to:

■ **Expansion/contraction** due to variations in temperature, particularly when exposed to solar radiation.

■ **Shrinking/swelling** due to changes in internal moisture content.

■ **Tensile and/or compressive** stresses due to their position within the wall assembly:

■ **Facing component** may be restrained by the wall assembly and deform even though the wall assembly itself is not restrained. Deformation in facing may be caused by:

• excessive differences in temperature between facing and wall assembly.

• rapid heat gain in a thin facing incorporated into a massive wall assembly, which will gain heat and expand much more slowly than the facing.

• different rates of expansion between the facing and the wall assembly.

• ties or fasteners, if incorporated in the field of the facing, will be in tension and may cause localized deformation.

• deformations may occur at restrained ends in the facing, in restraining elements, or both.

• tensile stresses at the surface may cause cracking in monolithic inelastic materials, such as stucco; or grouted joints between units, such as burned clay tile, may open up.

• fasteners in the field of the facing/wall component may be in tension, may cause localized stress concentrations, may pull-out or fail.

• restraint may be provided by any element stiffer than the facing such as: structural framing, mullions, frames, closely spaced & tightly drawn through-fasteners.

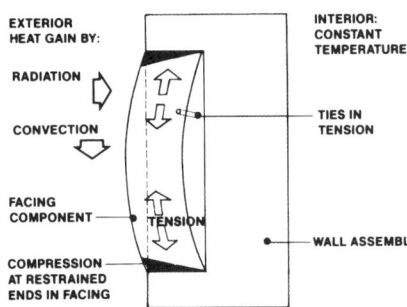

■ **Facing component** and wall assembly may be free to expand/contract under heat gain/loss or changes in internal moisture content but thermal movements may still be differential:

• facing component will gain/lose heat before wall assembly does, and may be subject to fluctuations or reversals when sunlight is intermittently shaded.

• large differences in heat storage capacity may exist between facing and wall assembly.

• coefficients of thermal expansion of facing and wall assembly materials may be different.

• differential movement between facings and wall assembly will affect the bonding agent or the fasteners which have to accommodate it: tension may develop in the bonding agent and, if elastic limits are exceeded, rupture may result.

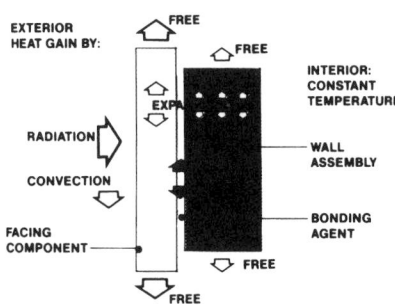

■ **Deformations** within the wall assembly which, having been caused by thermal or moisture induced movements, are then transmitted to the facing.

■ **Movements in the structure,** such as deflection, creep, and settlement may affect the facing as well as the entire wall assembly.

Moisture

Shrinking and swelling due to changes in internal moisture content in generally not an important consideration for most facing materials, the exceptions being wood or wood products and concrete, if fully saturated, where provisions should be made to accommodate such movements.

■ **Moisture penetrating into a facing,** either as rain from the exterior, or as condensing water vapor from the exterior or interior:

• may be trapped inside the facing material and cause bond failure in laminated facing.

• may freeze and expand resulting in rapid deterioration of the facings.

■ **Facings should either be impermeable, sealed to prevent penetration of moisture or porous to** allow absorbed moisture to evaporate quickly.

■ **Moisture absorbed by porous facings** may carry dust and/or airborne pollutants with it which then become trapped inside the facing.

■ **Rain water runoff over facings,** especially when localized, is likely to result in dirt streaking even on smooth impermeable surfaces.

Heat Flow

Facings generally do not contribute to any significant extent to the control of heat flow through the total wall assembly but their color, reflectivity, and surface pattern may affect the thermal performance of the assembly:

■ **Light colored and/or reflective surfaces** will minimize radiant heat gain and thus also reduce thermal movements.

Other considerations during selection of facings should include :

■ **Durability, or the suitability** of the facing to function under the expected in-service requirements, including those that might be imposed upon it by the wall assembly selected.

■ **Appearance:** the choices available in color and/or texture.

■ **Ease of maintenance:**

• are periodic cleaning and/or refinishing required?

• would damaged portions be easy to repair or replace?

• would the materials be readily available after a longer period of time if replacement of damaged parts became necessary, and how closely would they match the adjacent original installation?

■ **Cost of the facing in place.**

Facing Panels

Flat panels, whether single thickness or laminated, depend on the stiffness of the material(s) to resist normal thermal movements:

■ **Expansion** is accommodated by providing edge clearances, slip joints, or elastic connectors.

■ **When edges are restrained,** internal stresses will increase and deformation may result.

■ **When intentionally restrained,** flat sheets may need special stiffening to reduce internal stresses and to minimize deformation.

■ **Bending** due to horizontal loads is seldom a problem; bending must be considered when it may be combined

B2.6 WALL FACINGS

with concurrent thermal movement.

■ **Deflection** in single thickness panels may have to be limited either for visual considerations or to minimize the possibility of cracking of an applied surface finish.

■ Some flat panels may serve a **double purpose** as combined facing and bracing in framed assemblies.

In considering stone veneer, note:

■ **Periodic cleaning** is generally required; most stones will stain and may decay rapidly in polluted atmospheres.

■ **Clear surface sealers** may be used to protect the exterior faces, but sealers have to be periodically renewed and some may trap soot and dirt.

■ **Impervious surface sealers** should never be used over porous stone: moisture may become trapped inside the stone, freeze, and damage the stone.

Facings

■ **Formed strips of light gauge metal:**
• dent easily.
• may oil-can if some fasteners are not properly installed and locally restrain thermal movement.
• ends have to be lapped to allow for thermal movement and will show in the finished work.

■ **Formed strips of plastic:**
• may deform if locally restrained at fasteners.
• will generally have larger thermal movements than the supporting substrate.
• ends have to be lapped for thermal movement and will show in the finished work.

■ **Shingles of weathering steel** are available but while the protective coating is forming, runoff will stain adjacent surfaces unless protected.

Tile: glazed, ceramic, mosaic and quarry:

■ **Tile requires** dimensionally stable and level substrate.

■ **Expansion joints in tile have to be provided:**
• overall expansion/control joints in the substrate.
• where tile abuts other materials and 12 to 16 feet on centers horizontally and vertically for exterior tile.
• 24 to 36 feet on centers and where tile abuts other materials for interior tile.
• expansion/control joints are also

required at any location, exterior or interior, where materials of the substrate change.

■ Tile for exterior use should be frostproof.

Glazed block, glazed brick, terracotta:

■ **Not all bricks are equally resistant to frost:** some will decay rapidly when wet and subject to freezing temperatures.

■ **Thermal movements in masonry are small,** but may become a factor if a dark colored veneer is exposed to intense solar radiation with resulting differential movement between it and the wall assembly, e.g., in a cavity wall.

In considering stone chips in matrix as a facing, note:

■ **Joints have to be** provided at all expansion/control joints in substrate.

■ **In general,** stone chips are manually applied in the field:
• finished appearance will depend upon the skill of workmen;
• uniform application over large unbroken areas is difficult to achieve.

■ **Cracks developing in substrate** after application will usually also rupture the matrix.

In considering stucco, note:

■ **Joints over expansion/control joints** in solid substrate are required; large areas should have expansion joints 12 to 16 feet on center both horizontally and vertically.

■ **In framed wall assemblies,** stucco may be applied either directly to framing, or over sheathing first covered with building paper; when applied directly to framing stucco will generally have better resistance to cracking but walls must be independently braced.

■ **Proprietary systems for exterior** and/or interior application of stucco over thermal insulation are available.

INSTALLATION OF FACINGS

Facings are either bonded or attached during the construction of a wall assembly.

Bonded Application

Bonded application essentially integrates the facing into the wall assembly which may then act as a composite material rather than as several joined materials. Important considerations are:

■ **Compatibility** of bonding agent with both facing and substrate.

■ **Dimesional stability** of substrate and facing.

■ **Coefficients of expansion** of facing and substrate.

■ **Ability of bonding** agent to accommodate differential movement between the facing and the substrate whether thermal, moisture, or superimposed load induced.

■ **Performance characteristics** of the bonding agent under in-service conditions to be expected.

Attached Application

Attached application employs **resilient mounting or mechanical fasteners;** in general, there is no composite action with the core of the wall assembly.

Mechanical attachment is either by through-fasteners (such as nails, screws or bolts), or by concealed clips, interlocking strips, ties or anchors.

■ **Through-fasteners should**
• have adequate pull-out strength.
• be compatible with the materials of the facing and substrate.
• be non-corroding,
• match the color of the facing when exposed to view.

■ **When covered in the finished work, fasteners such as nails, screw, ties, and anchors should be:**
• compatible with the materials being jointed.
• should be noncorroding when the possibility of exposure to moisture exists; rust may stain facing or finish, may cause expansion in the facing or substrate and reduce the strength of the fastener enough to cause it to fail.

■ **Method of attachment may have to provide for:**
• inaccuracies in the substrate.
• thermal- or moisture-induced differential movements between the facing and the supporting wall assembly.
• stresses under superimposed loads in the facing, the wall assembly and/ or the structural frame.

■ **Two- or three-way adjustments** at connections of large-size facing panels may be necessary to compensate for:
• discrepancies in structural frame dimensions.
• tolerances within the panels.

■ **Method of attachment** should be economical: simplicity of design contributes to low cost.

B2.6 WALL FACINGS

B2. 7 GLASS/PLASTICS

GLASS/PLASTICS: introduction

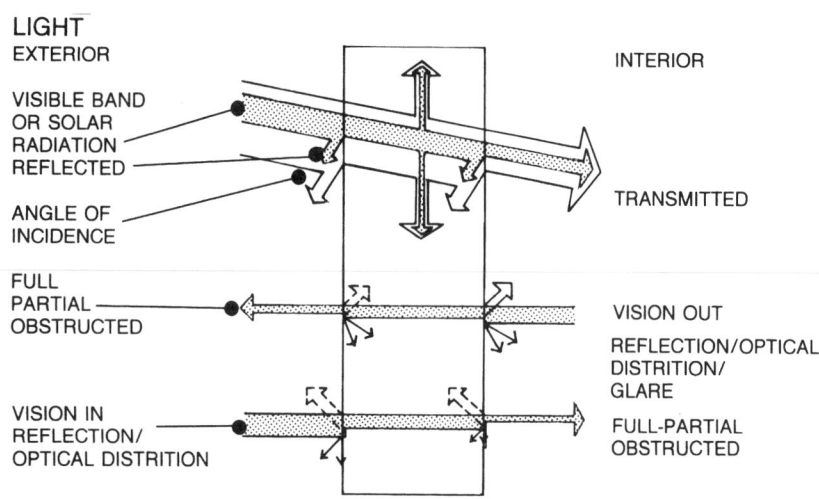

LIGHT
EXTERIOR

VISIBLE BAND
OR SOLAR
RADIATION
REFLECTED

ANGLE OF
INCIDENCE

FULL
PARTIAL
OBSTRUCTED

VISION IN
REFLECTION/
OPTICAL DISTRITION

INTERIOR

TRANSMITTED

VISION OUT

REFLECTION/OPTICAL
DISTRITION/
GLARE

FULL-PARTIAL
OBSTRUCTED

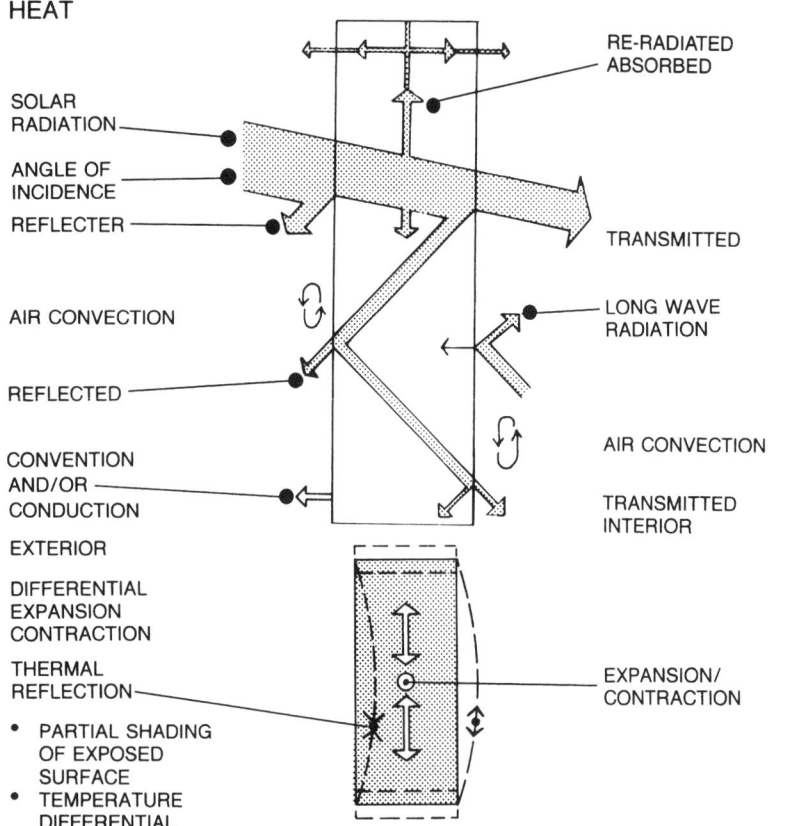

HEAT

SOLAR
RADIATION

ANGLE OF
INCIDENCE

REFLECTER

AIR CONVECTION

REFLECTED

CONVENTION
AND/OR
CONDUCTION

EXTERIOR

DIFFERENTIAL
EXPANSION
CONTRACTION

THERMAL
REFLECTION

• PARTIAL SHADING
  OF EXPOSED
  SURFACE
• TEMPERATURE
  DIFFERENTIAL

RE-RADIATED
ABSORBED

TRANSMITTED

LONG WAVE
RADIATION

AIR CONVECTION

TRANSMITTED
INTERIOR

EXPANSION/
CONTRACTION

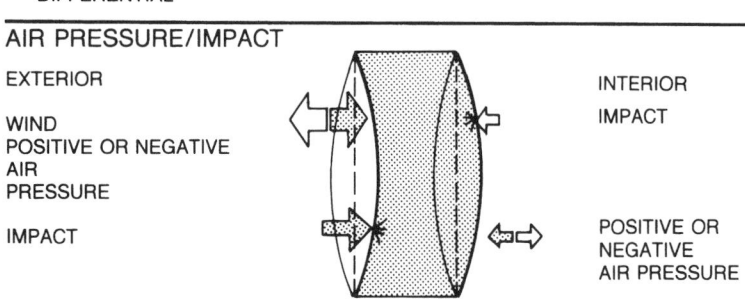

AIR PRESSURE/IMPACT

EXTERIOR

WIND
POSITIVE OR NEGATIVE
AIR
PRESSURE

IMPACT

INTERIOR

IMPACT

POSITIVE OR
NEGATIVE
AIR PRESSURE

The transparency of glass and plastic to visible radiation is the primary reason for their use:

• Transparent or translucent sheets of glass or plastic are incorporated into the exterior envelope of a building as window or skylight glazing, or as in-fill panels in grid curtain wall systems, to admit daylight into the interior spaces.

• Transparent sheets of glass or plastic also allow for visual contact between the interior/exterior environment to be maintained.

• Glass may also serve as practically the entire vertical component of a building envelope, such as an all-glass curtain wall; but whether used as a component of the building envelope or as the envelope itself, glass and plastic are still subject to the same factors as any building envelope:

HEAT FLOW resulting from:
Radiant energy striking glass or plastic which may be:

• **Short-wave, high-intensity solar radiation,** with wavelengths from 0.3 to 4.5 microns, which includes ultraviolet, visible, and infrared bands. Short-wave radiation will degrade glass and plastics over an extended period of time.

• **Long-wave, low intensity infrared radiation** emitted by ground and building surfaces with wavelengths from 5 to 50 microns.

Short-wave radiant energy striking glass or plastic will be:

• **Partially reflected,** with the amount reflected dependent on the angle of incidence at which radiation strikes the surface. Maximum transmission occurs at zero angle of incidence.

• **Partially absorbed and converted to heat** within the thickness of the glass or plastic.

• **Partially transmitted,** with the amounts absorbed and transmitted dependent on characteristics of the glass or plastic.

Heat transfer through glass or plastic glazing due to air temperature differentials on either side will be by:

■ **Conduction:** the flow of heat through a single thickness of glazing from the warmer surface to the colder one.

■ **Convection:** the transfer of heat from the surfaces of glazing to a fluid medium, such as air or water when such medium is at a lower temperature than the surface or surfaces of glazing:

- when air temperature at one surface of glazing is higher than at the opposite surface, heat will be transferred from the warm air to glazing by convection, flow through the glazing by conduction, and will be transferred at the opposite surface to the cold air by convection.

- resistance to heat transfer at surfaces is affected by a layer of stagnant air clinging to the surfaces of glazing or the surface film.

- thermal conductivity of soda-lime glass is between 5 and 7 Btu/hour square foot °F/inch, and offers negligible resistance to heat transfer between the inside and outside of a building, with the thermal resistance of glass being predominantly due to surface films.

■ **Radiation:** the transfer of heat from a warmer body to colder ones in line of sight of it by radiant energy through space without the need for a medium of transfer. All radiant energy, both visible and invisible, produces heat when absorbed.

Heat buildup in glass or plastic will result in:

THERMAL EXPANSION AND CONTRACTION.

■ **Thermal stress** resulting from differences in temperature within a single unit of glazing may be a factor:

- sudden exposure to solar radiation, other heat sources, or heat trapped at one surface of the glass may quickly raise temperature in the central portion of the glass while the edge temperature may rise more slowly, partly because the edge is shaded from radiation, or protected from the heat source, partly because of the relatively large heat storage capacity of the frame:

- these temperature differentials may produce high tensile stresses in the glass near the edges and compressive stresses in the center.

- the strength of glass is generally sufficient to withstand the stresses resulting from temperature differentials alone, but if the glass is already stressed from other causes, the added thermal stresses may cause breakage.

- coefficients of expansion for most plastics are considerably higher than for glass, and provisions to allow for such larger movements have to be incorporated into the design of framing systems.

AIR PRESSURE/IMPACT

Glass is a brittle material and does not deform plastically before failure: it fails in tension regardless of the nature of loading:

- The potential tensile strength of glass may be as high as 1,000,000 psi, but failure occurs at average stresses far below this value due to stress concentrations at surface imperfections both inherent in the glass and mechanically created.

- Glass is most vulnerable at its edges as surface imperfections from cutting and handling add to the possibility of failure.

  —Breakage tends to originate at microscopic flaws at or near the edge regardless of whether the cause is thermal or uniform pressure.

- Because the effect of stress concentrations is indeterminate, the allowable tensile strength of glass is established statistically and a sizable safety factor is included:

■ **Design factors** are used to select glass thickness and maximum size for a design wind pressure condition. with the design factor of 2.5 - the statistical probability of 8 out of 1000 units of given thickness and size breaking under given uniform static loading most often used.

■ All other conditions being equal, a glass panel will break more easily under long-term, steady loads:

- forces of short duration such as gusts of wind can double over forces of long duration without causing breakage.

- typical breaking stresses in pounds per square inch for a relatively large light of annealed float glass might be:

- 6,000 psi for a sonic boom.

- 5,500 for a 10 second wind gust.

- 4,000 for a 1 minute wind load.

- 3,000 for pressure lasting for several hours.

- Glass thicknesses sufficient for uniform wind loading may not be adequate for concentrated point loads, e.g., those caused by roof gravel, other flying debris, or human impact.

■ **Plastics are lighter in weight** and generally have better shatter resistance than glass; also they can be formed into diverse shapes:

- may discolor upon weathering.

- are combustible and cannot be used in fire resistance rated assemblies.

B2.7 GLASS/PLASTICS

GLASS/PLASTICS: types, properties

FLOAT GLASS

CLEAR
- HEAT STRENGTHENED
- TEMPERED

TINTED
- HEAT STRENGTHENED
- TEMPERED

REFLECTIVE
- CLEAR
- TINTED
- HEAT STRENGTHENED
- TEMPERED

REFLECTIVE OR OPAQUE COATING

ROLLED GLASS

CLEAR

FIGURED/ PATTERNED
- LIMITED TEMPERING
- OPTIONAL FIGURED/ PATTERNED

LIMITED TINTING

FIGURED/ PATTERNED
- OPTIONAL FIGURED/ PATTERNED

WIRED
- TINTED NOT AVAILABLE

SMOOTH/ ROUGH/ PATTERNED
- IMPACT
- FIRE TEMPERING NOT AVAILABLE

SMOOTH/ROUGH PATTERNED WIRE MESH

LAMINATED GLASS

LIGHTS:
- CLEAR
- TINTED
- REFLECTIVE

PLASTIC:
- CLEAR
- TINTED

- HEAT STRENGTHENED
- TEMPERED

OPTIONAL PLASTIC AND LIGHTS
- RADIATION
- IMPACT
- SOUND

PLASTICS

- CLEAR
- TINTED
- SMOOTH
- FIGURED/ PATTERNED

- ACRYLICS
- POLY- CARBONATE
- POLYVINYL CHLORIDE
- GLASS FIBER REINFORCED POLYESTER

INSULATING

OUTER LIGHT:
- CLEAR
- TINTED
- REFLECTIVE
- PLASTIC

INNER LIGHT:
- CLEAR
- TINTED
- REFLECTIVE
- PLASTIC

OPTIONS FOR GLASS: HEAT- STRENGTHENED - TEMPERED

GLASS

■ Glass is made from silica sand with added alkaline salts and soda. It is produced by heating the mixture to a molten state and then allowed to cool to a rigid condition without crystallization.
- While molten, glass may be formed into a smooth flat sheet (float), rolled into patterned/textured sheet, or pressed or cast into a variety of shapes.
- Molten glass production is also referred to as annealed glass.
- Ordinarily transparent or translucent, it may be clear or tinted depending on the contents of the mixture. Tinted glass is made by mixing amorphous or crystalline materials into a mixture-batch.

FLOAT GLASS

■ Float glass is the most common glass used for glazing.
- It is produced by pouring the molten glass mixture continuously from a furnace onto a large pool of molten tin, spreading as it seeks its own level for the required thickness.
- It may be clear, tinted or coated. The tints and coatings are generally designed to improve thermal and solar transmission qualities.
- Where tinted or coated glazings are used, the glass will be exposed to higher temperatures due to reflected solar radiation. To reduce the possibility of glass failure, heat strengthening or tempering may be required.

■ **Clear float glass** is colorless with 80 to 90 percent transmission of visible light, depending on thickness. Transmission quality will vary depending on the type and quantity of impurities (minerals) in the batch of glass.
- Clear glass will reflect ultraviolet radiation, but will allow most of the visible and infrared bands to pass through.
- A clear glass window, totally exposed to the sun, has a shading coefficient of 1.0. When the window is shaded completely, it's shading coefficient is 0. This is also true when a total reflective surface is covering the inside of the window.
- Low-iron glass has the highest transmission qualities of clear glasses; double glazing transmittance can be improved 20 percent by using it.

■ **Tinted glass, also known as heat absorbing glass,** is used for the purposes of reflecting and absorbing solar radiation. When multiple layers are used, the outermost layer is the one which should be tinted.
- Heat absorbing glass will reduce the amount of infrared radiation transmitted, but will also decrease transmittance of visible light.
- Reduction in visible light transmission ranges from 7 to 83 percent, dependent on color and thickness. Higher transmission values are desirable; values below 25 percent may be objectionable.
- Lower shading coefficients are desirable; values typically range from .15 to .84.
- Increased iron content produces a green glass with the best transmittance of daylight.
- Gray glass has the lowest transmittance of visible and solar light.
- Plan applications carefully so that winter solar gain is not reduced during the heating season.

■ **Coated glass, reflective glass and low-emissivity (low-e) glass** are all names for a special type of glazing that has a very thin, transparent metallic coating applied to one surface. It is used for its transmission and reflectance properties:
- Visible light is permitted to pass;
- Solar gain is controlled;
- Thermally insulative values and the potential for condensation are enhanced by reducing the temperature difference between the inside and outside temperature of the glazing;
- Fabric-fading ultraviolet light is filtered.
- **When the glass must be heat strengthened or tempered, the coating must be applied after the heat treating is completed.**

■ **Coatings are available in different colors or clear, each color providing different solar and thermal characteristics. The glazing layer to be coated may be clear or tinted and characteristics of the glazing will combine with that of the coating. For coated glazings:**
- Visible light transmittance is in the range of 5 to 27 percent.
- Shading coefficients range from 0.06 to 0.33.
- A broad range of colors are available for coatings - silver, pewter, blue, grey, earth or bronze, gold, copper and transparent (low-e).

■ **Low-e glass is a coated glass which does not appear reflective.** It may be soft-coated or hard-coated. The thickness of the coating determines the insulating properties of the glazing, but obscures vision as thickness increases.

- The performance of **soft-coated** low-e glass is usually superior, but it is quite delicate. It requires special handling and cutting, care in shipment and storage, and protection from moisture.
- Soft-coatings may be applied directly to the glass or to a thin sheet of transparent plastic. In either case, the coated surface must be protected, such as between double glazing layers. A coated plastic sheet would be stretched across the sash opening between two layers of standard glazing; a 1/2 to 5/8 inch airspace is optimal.
- Soft-coatings are applied in a separate operation, off the standard glass production line. Methods for applications include magnetic process in a vacuum chamber called sputtering, by wet chemical spray deposition (requires protection from mechanical and chemical degradation) or electron-beam evaporation (interior side only). When silver is used in the coating, exposure to moisture and air must be minimized to keep the silver from oxidizing.
- Durable, **hard-coatings** are tin-oxide solutions sprayed on as a part of a standard float glass production line. These pyrolytic coatings are deposited on the surface of a hot sheet of glass, essentially baking it into the surface.
- Hard-coated glass is often guaranteed for 5 years.

■ **Low-e glazing provides** considerable thermal benefit, having an emittance of 0.40 or less (range = 0.35 to 0.05), compared to that of standard glass (0.84). Low-e surfaces should face the airspace of insulated glazings; on the inner panel for cooler climates.

- Application to double glazed windows will improve thermal performance to that equal to triple glazing.
- Comparison of low-e glazings can be made on the basis of R-value.
- R-values of double glazed windows can be doubled, while visible light transmission is marginally reduced (less than 10 percent).
- Most infrared radiation will be reflected back into the space for the heating season; outside infrared

radiation is reflected to help during the cooling season.

■ Newer glazing technologies include:

- Thermochromic glazings sandwich a thermally-sensitive gel or liquid between glazing layers to change transmission properties relative to variation in temperature.
- Photochromic glazings are coated to change their performance to match ambient lighting levels.
- Electrochromic glazings will change their reflectance or absorption in response to an electric current.

ROLLED GLASS

■ Molten glass is fed through rollers to produce the required thickness, with the rollers imparting various patterns to one or both surfaces of the glass.

- The glass is usually clear, but tinted glass is available.
- Deep or intricate patterns may "wash out" on tempered glass due to the required reheating of the glass.

■ **Wired glass** is rolled glass with a welded wire mesh incorporated during manufacture.

- Cannot be tempered and is not available in tinted glass.
- May be rough, patterned, ground or polished.

■ **Art/opalescent/cathedral glass** is produced by the rolling process in small batches, commonly with variegated colors within each sheet. It is used to produce stained glass windows.

LAMINATED GLASS

■ Laminated glass is made from two or more layers of glass which are laminated to an interlayer of clear or tinted polyvinyl butyral (PVB) sheet. Use dictates the thickness and type of glass and interlayer.

- Lamination minimizes the hazard of shattering glass as the pieces will stay in place; effectively resists forced entry.
- All types of glazing can be laminated - tinted, coated, tempered, etc.

HEAT TREATED GLASS

■ Regular annealed glass is reheated and then rapidly cooled. The cooling process causes shrinkage in the glass, thus placing the outer surfaces into high compression with the core of the glass in compensating tension.

- Heat treated glass cannot be worked (cut, drilled) after treatment.
- It is subject to possible breakage after installation due to subsequent transformation of impurities in the batch (compounds of nickel or sulphur).
- Where the possibility of breakage can cause bodily harm, the glass should be specially treated or laminated.

■ **Heat treating methods are either vertical or horizontal.** Each method produces a degree of bow and warp evident as optical distortion patterns in the treated glass.

Tempered Glass

■ **Tempered glass (FT = fully tempered) is treated at a temperature of about 1300°F, which changes the physical characteristics of the glass:**

- Upon failure, tempered glass will break into a multitude of small fragments which are generally less than one square inch in size.
- Tempering increases resistance to impact; resistance to wind (deflection) loads and thermal stresses is 4 to 5 times that of regular annealed glass.
- It is typically required for glazed areas which are within 12 inches of a door or 18 inches from the floor.

Heat Strengthened Glass

■ **Heat strengthened glass (HS) is treated at lower temperatures than tempered.**

- The breaking pattern is similar to that for regular annealed glass.
- Resistance to wind loads and thermal stresses is twice that of regular annealed glass.
- Usage includes all general glazing applications where more strength is needed.

Spandrel Glass

■ **Spandrel glass is heat treated glass, either tempered or heat strengthened, with a ceramic coating of various colors permanently fused to its interior surface.**

■ **Ceramic enamelled glass** is a heat-treated, opaque glass that has a durable and stable, ceramic enamel coating applied to one surface for color.

- Used as a wall finish and in curtainwalls.
- Available in a range of geometric patterns, in an unlimited range of colors, metallic colors and as simulated stone.

B2.7 GLASS/PLASTICS

GLASS/PLASTICS: types, properties (cont.)

PLASTICS

■ Properties common to plastics used as glazing include:
- Lightweight.
- Optical properties may vary from highly transparent to opaque.
- Solar energy transmission may be up to 90 percent; up to 93 percent for visible light.
- Some compositions will allow ultraviolet transmission; short-wave infrared radiation may be absorbed or transmitted.
- Long-wave infrared radiation will be reflected, similar to glass.
- Plastics are easily and economically manufactured into any thickness and shape. They can be cold-formed on site to provide curved surfaces and can be easily cut.
- Plastics may be used where resistance to shattering or vibration is required. However, even though plastics have good resistance to impact, they are breakable.
- Plastics are typically water-permeable. For this reason, plastic glazing systems should employ adequate venting of moisture from air spaces created by multiple glazing layers.
- Compared to glass, plastic glazing is less rigid, lighter in weight and not as subject to breakage.

Acrylic Glazing

■ Acrylic glazing is most commonly made from polymethyl methacrylate (PMMA).
- It is produced by the cell casting method, which is considered to provide the best overall properties, especially optical clarity.
- Although easily scratched, acrylics have good resistance to weathering, salt spray, and corrosive atmospheres; degradation occurs from contact with aromatic and chlorinated hydrocarbons, esters, ketones, etc.
- The coefficient of expansion is over 8 times that of plate glass.
- Plastic modifiers can be added for improved impact resistance, but acrylics are still susceptible to breakage.
- Acrylic glazings may be limited in surface area due to their flammability and burning characteristics.

Polycarbonate Glazing

■ Polycarbonate (PC) glazing has considerably higher impact resistance than acrylics. Light transmittance of 82 to 90 percent is combined with low infrared transmittance.
- It should be stabilized against the effects of ultraviolet light when used for outdoor applications. Prolonged exposure will result in yellowing, cracking and reduced strength.
- As glazing, it may prove to be nearly unbreakable.
- Its coefficient of expansion is greater than that of acrylic glazing, but its melting point and water absorption is low.
- Maximum glazing area can be up to 50 percent of wall area when the glazing has a CC-1 rating (per ASTM D635).
- Bowing is common in large panels because of its high coefficient of expansion.

Fiberglass Reinforced Polyester Glazing

■ Fiberglass Reinforced Polyester (FRP) glazing is made from a thermosetting polyester resin reinforced with glass fiber. It is low in cost with light transmission up to 85 percent.
- It has good impact resistance, but prolonged ultraviolet (outdoor) exposure will result to discoloration, fiber pop-outs, surface cracking and reduced light transmittance. It is stronger than acrylic, but not as strong as PC.
- Surface pop-out, or fiber bloom, requires restoration about every 10 years, which is accomplished by recoating the glazing in place.
- High coefficient of thermal expansion and wavy appearance may limit usage.
- Surface treatment or protective films may be available to improve weathering characteristics.

Twin Wall Glazing

■ Twin wall glazing was developed to provide good thermal performance with an plastic glazing panel. Materials include acrylic, polycarbonate and FRP. Triple wall panels are also available in polycarbonate.
- Twin wall panels have good structural and impact-resistant properties.

- They are easy to work with in the field, can be fabricated in curved shapes and may be cold-formed for bending.
- Acoustical properties and resistance to abrasion are not as good as other plastic glazing materials.
- Vision is limited by support ribs.
- Thicknesses range from 5/32 to 13/32 inches; widths are typically 4 or 6 feet; lengths are up to 39 feet.

Other Plastic Glazings

■ Several other plastic glazing materials are available for more specific use:
- Cellulose acetate butyrate:
  - widths of 3 to 6 feet;
  - lengths of 6 to 10 feet;
  - thicknesses of 1/16 to 1/4 inch;
  - transparent, moderate impact resistance, and 10 year life;
  - solar transmittance of 90 percent;
  - sensitive to heat, softens and has high coefficient of expansion;
  - useful for most outer glazing applications and curved designs.
- Fluorinated ethylene propylene (FEP):
  - widths up to 63 inches;
  - thicknesses up to 3/32 inch;
  - transparent, noncombustible, low impact resistance, and 20 year service;
  - solar transmittance of 96 percent;
  - maximum unsupported span is 48 inches;
  - difficult to install; wrinkles easily;
  - limited to interior applications.
- Polyvinyl fluoride (PVF):
  - widths up to 50 inches;
  - lengths up to 100 feet;
  - thickness of .004 inch;
  - nearly transparent with low impact resistance and 5 to 10 year life;
  - maximum unsupported span is 48 inches;
  - difficult to install; wrinkles easily;
  - solar transmittance of 90 percent; 58 percent infrared transmittance;
  - use limited to outer applications; for outer layer of solar collectors.
- Polyethylene:
  - used for short term (8 mo. to 3 year life) glazing of greenhouses;
  - milky appearance; allows 87 to 90 percent solar transmittance.
- Polyvinyl chloride (PVC)-
  - economical, but less suitable to glazing needs.

B2.7 GLASS/PLASTICS

B • BUILDING SHELL

Coatings for Plastic Glazing

■ **Polymer topcoatings are available to improve performance and/or durability of plastic glazing.**

- Coatings are desirable for durability (often increasing warranties to 5 or 10 years) but may distort optic qualities.
- Silicone-based coatings are available for acrylic and polycarbonate glazing. They provide resistance to abrasion, moisture and chemical attack; they enhance overall durability. These are applied using dip, flow or spread coating techniques.
- Acrylic-based, protective surfacings are available to improve weathering characteristics, resistance to ultraviolet radiation, and reduction of solar transmittance (tinting) for polycarbonate glazing.
- Ultraviolet absorbers can be added to the above coatings.
- Effective reflective coatings not available.

GLAZING FILMS

■ Like tinted and coated glass, the reflective and absorptive qualities of glazing can be enhanced by adhering thin, metallic-coated polyester films to the glazing surface. The benefits of glazing films include:

- Infrared and ultraviolet solar radiation is reflected, while visible light is transmitted. Anti-reflective coatings are best for blocking ultraviolet radiation (up to 98 percent), while minimizing the reflectance of desired, visible light.
- Clear films with special, high-strength adhesives improve shatter resistance.
- Mirror-like appearance provides privacy from the outside.
- Some coatings are scratch-resistant.
- For optimum energy savings during the heating season, films with a high shading coefficient are most effective on north windows; for cooling, films with lower shading values should be put on east and west window glazing.

■ Glazing films are frequently used for retrofit situations.

- Color is added via a second layer of polyester which is tinted. This is not a preferred treatment as this layer is heat absorptive and may transfer heat into the space.
- Shading coefficients can range from 0.30 to 0.40 on single, clear glass. Solar heat and glare is often reduced by 60 to 80 percent.

- Reflective films may be located on the surface of the outer or the inner glazing layer.
- When on the inside surface of the outside glazing layer, reflective films will reduce the transmission of heat, but the outer glazing will get quite hot under direct radiation, increasing the chance of thermal breakage.
- Locating the film on the outer surface of the inner light will reduce the possibility of thermal breakage.
- Films are usually warranted for five years. Silver-coated films maintains its reflective properties for long periods of time.

INSULATING GLASS:

■ **Insulating glass is an assembly of two or more layers of glazing, each layer separated by an air space. The air space is sealed to maintain a moisture-free and dust-free separation of glazing.**

- Moisture is relieved through weep holes or vents or is absorbed by dessicant materials held in spacer bars. It should be noted that dessicant materials have a limit to absorption, at which point they must be recharged (adding heat will release the moisture) or replaced.
- A breather tube may be used to equalize air pressures when the assembly is manufactured at an elevation substantially different from the place of its installation.
- Aluminum spacers are soldered or welded. The glazing is sealed to the spacer by an extruded shape made of one of many resilient materials (polysulfide, polyisobutylene, silicone, urethane, butyl). Butyl is the most water resistant whereas silicone is stronger.
- The air space functions to reduce the difference in temperature between the inside air and the outside air, thereby reducing the conductive heat flow through the glazing assembly.
- Air spaces are 1/4 to 1 inch wide; wider spaces result in increased convection in the air space, which causes increased thermal transfer/flow.

■ For the purposes of improving the thermal resistance through the glazed area, researchers have developed three new technologies to replace air in the space between glazing layers.

- Evacuated glazing refers to a system that uses a vacuum between layers and a low-e layer to attain R-values up to 10. Improvements in viewing quality and seal reliability are being developed.
- Aerogel glazing encapsules the highly transparent, pure silica aerogel material between glazing layers to attain R-values up to 20. The high insulative value is derived by presence of trapped air within the aerogel. The technology is still developing.
- Gas-filled glazing encapsules gases that are heavier than air to reduce the internal convective motion of air in order to reduce heat loss. An R-value of up to 5 can be achieved using argon gas. Research into improved seal reliability is in process.

■ Insulating glass assemblies may consist of:

- Two or more layers of clear glass;
- An outer layer of heat absorbing glass with an inner layer of clear glass;
- Two layers of float glass with a third layer made of coated plastic;
- One layer of heat reflective glass and another of clear glass; or
- Various other assemblies are available, such as incorporating laminated glass, single thickness plastics, single and double skin plastics, glass and plastic, etc.
- Insulating glass may be used for sloped glazing applications. The angle and degree of the glazing should be in conformance to manufacturer recommended specifications.

B2.7 GLASS/PLASTICS

GLASS/PLASTICS: types, sizes, uses

● denotes common usage or availability
○ denotes possible usage or limited availability
NA denotes non availability

| | OPTION | | | | USES | | | | | | | |
| | | | | | EXTERIOR CURTAIN WALL | | | INTERIOR PARTITIONS | | | | [2] |
	CLEAR	TINTED	REFLECTIVE	SIZE & THICKNESS, approximate maximum availability in inches	WINDOW GLAZING	GLAZING INFILL	SPANDREL/INFILL	ALL GLASS	FRAMED, GLAZED	FIRE SEPERATION	GLASS SLIDING PANELS	CEILING PANELS
FLOAT GLASS: FLAT												
ANNEALED	●	●	●	120 x 200 x x½	●	●			●			
HEAT-STRENGTHENED	●	●	●	80 x 144 x x¼	●	●	●		●			
TEMPERED	●	●	●	80 x 144 x x½	●	●	●	●	●		●	
ROLLED GLASS: FLAT, PATTERNED												
ANNEALED	●	○		60 x 132 x x 7/32	●	○			●			
HEAT TEMPERED	●			60 x 120 x x ⅜	●	○		●	○		●	
ANNEALED, WIRED	●			60 x 144 x x ⅜	●	○		○	○	●	○	
PLASTICS: FLAT, MOLDED, PATTERNED												
ACRYLIC (single or double wall)	●	●		108 x 156 x x ⅝	○				○		○	●
POLYCARBONATE (single, double or triple wall)	●	●		96 x 468 x ¾	○				○		○	○
POLYVINYL CHLORIDE	●	●		NA	○				○		○	○
FIBERGLASS REINFORCED POLYESTER	●	●		60 x 600 x x ¼	○				○		○	○
ASSEMBLIES												
LAMINATED, GLASS	●	●	●	84 x 180 x x 1	○	○	○	●	●		○	
LAMINATED, PLASTICS	●	●		48 x 96 x x ¼	○			●	●		○	
INSULATING, GLASS	●	●	●	84 x 144 x ¼ x ¼	●	●					○	
INSULATING, GLASS AND PLASTICS	●	○	○	NA	●							
INSULATING, PLASTIC	●	●		NA	○							
UNITS												
GLASS BLOCK	REFER TO PAGE 5-46 AND 5-47											

B2.7 GLASS/PLASTICS

B • BUILDING SHELL

DOORS			SKYLIGHTS				Special conditions when resistance is required against:					
ALL GLASS/NARROW STYLE	SIDE-LIGHTS [3]	VISION PANELS [4]	UNITS: FLAT	UNITS: MOLDED	FRAMED: FLAT, SLOPING	FRAMED: CURVED	HIGH WIND LOAD	HIGH HEAT BUILDUP	FORCED ENTRY	BULLETS	EFFECTS OF FIRE [5]	SOUND TRANSMISSION
	○	●										
	○	○					●	●				
●	●	○	●		●		●	●	○			
	○											
	○											
●	●				●				○		●	
	○	○	●	●	●				○			
	○	○	●	●	●				●			
				●	○	○						
					○							
	○		●		●		○		○	●		●
	○		●		●		○			●		●
●			●		●						○	
			○		○							
	○		●		●							

REMARKS

[1] Sizes shown are representative only; particular usage may impose limitations on size: larger sizes may be available from specific manufacturers. Flat glass 1½ inches thick and up to 86 inches wide is available.

[2] Ceiling panels generally used in luminous ceilings and plastic used is commonly patterned to diffuse light.

[3] Building codes generally require safety glazing in full height sidelights used with all glass doors.

[4] Vision panels for fire-resistant rated door assemblies are limited as to size and type.

[5] For use in fire-resistant rated assemblies.

FLOAT GLASS:
- custom bent shapes are available.
- sizes of tinted and/or reflective coated annealed glass may be restricted: heat strengthened or tempered glass may have to be used for larger sizes.
- tempered glass for skylight glazing may require backing or safety screens to prevent pieces from falling on occupants in case of breakage.
- STC ratings range from 28 to 36 for ¾ inch thick.
- choice of tints for all glass doors limited for thicknesses over ½ inch.

ROLLED GLASS:
- deep patterns may not be available heat treated; choice of tints limited.
- wired glass: not available tinted; cannot be heat-treated; available polished one or two sides.

PLASTICS:
- acrylics and polycarbonate available in double-skinned panels.
- protective films for PVC to protect it from weathering in skylights available.
- acrylics may be available in several types.
- flat glass fiber reinforced polyester available with added steel mesh reinforcing.

LAMINATED:
- various combinations of different types of glass available.
- available to restrict vision in one direction.
- laminated plastic, generally using polycarbonate, commonly for forced entry and bullet-resistant applications.
- STC ratings range from 30 to 40.

INSULATING:
- interior light generally clear.
- laminated or wired glass can be used in most insulating units except those which are tinted as thermal stresses may be too high.
- sizes of units using laminated or wired glass are usually smaller by 30-50 percent. Consult manufacturers.
- different glass thicknesses can usually be used if the difference does not exceed 1/16 inch.
- if units are to comply with safety glazing codes, both lights must be tempered or else be of some other approved safety glazing material.
- reflective coatings can be applied on almost any of the four surfaces of an insulating unit, on clear or tinted glass. Sizes offered may vary accordingly.
- reflective coated clear glass unit over 30-40 square feet must usually be heat strengthened.
- almost all reflective units which are also tinted must be heat strengthened to withstand the higher thermal stresses.
- gas-filled (argon) options are available to increase insulative value.

B2.7 GLASS/PLASTICS

GLAZING MATERIALS: solar-optical properties

- Values shown indicate the range of commonly available materials.
  Some variance is due to thickness variances in glass and air space.
- Percent transmitted shown is for zero angle of incidence.
- NA denotes information not available.

			Light = Visible Band		Solar Radiation		U-Value Btu/hr.sq.ft.°F	
			PERCENT: Transmitted	Reflected	PERCENT: Transmitted	Reflected	Winter [1]	Summer [2]
SINGLE GLAZED								
FLOAT,	CLEAR		79–91	8	53–89	7–8	1.00–1.16	0.98–1.05
	TINTED, GREEN		77–86	7–8	49–70	6–7	1.12–1.16	1.00–1.10
	BLUE		57–71	6–7	48–65	5–6	1.12–1.16	1.00–1.10
	BRONZE		28–69	5–7	24–68	6	1.08–1.16	1.00–1.10
	GRAY		19–62	4–6	22–65	5–6	1.06–1.16	1.00–1.11
	REFLECTIVE COATED		4–77	6–44	4–37	8–64	0.43–1.15	0.42–1.12
	LOW-IRON		91					
PLASTIC,	CLEAR, ACRYLIC		73–92	4	83–92	7	0.56–1.09	0.55–1.02
	FRP		75–82		72–88		0.42–1.00	
	POLYCARBONATE		67–86		67–86		0.36–1.00	
	TINTED/TRANSLUCENT		21–80	4	31–85	varies	0.80–1.06	0.75–1.02
	TWIN WALL				74–86		0.36–0.58	
DOUBLE GLAZED								
CLEAR +	CLEAR		78–82	14–15	60–76	11–15	0.42–0.61	0.51–0.62
CLEAR +	LOW-e		49–86	12–15	17–56	17–25	0.23–0.52	0.28–0.54
CLEAR +	TINTED	GREEN	62–78	12–14	36–62	8–12	0.49–0.58	0.52–0.62
		GRAY	13–56	5–13	22–56	7–9	0.49–0.60	0.51–0.62
		BRONZE	19–62	8–13	26–57	8–9	0.49–0.60	0.50–0.64
		BLUE	50–64	8–13	38–56	7–9	0.49–0.58	0.52–0.63
CLEAR +	COATED	SILVER	7–19	22–41	5–14	18–34	0.39–0.48	0.43–0.52
		BLUE	12–27	16–32	12–18	15–20	0.42–0.46	0.44–0.54
		BRONZE	7–18	14–54	5–16	13–24	0.39–0.50	0.40–0.54
		COPPER	25	30–31	12	45	0.29–0.30	0.29–0.30
		GOLD	7–13	42–59	3–6	42–49	0.32–0.39	0.32–0.35
GREEN +	COATED	SILVER	6–16	17–41	3–8	13–19	0.41–0.45	0.43–0.53
		BLUE	15–24	13–32	8–12	11–13	0.42–0.46	0.44–0.54
BLUE +	COATED	SILVER	5–11	12–38	3–7	10–17	0.41–0.45	0.43–0.53
		BLUE	5–17	9–36	3–11	6–16	0.42–0.47	0.46–0.55
BRONZE +	COATED	SILVER	5–11	11–41	3–8	11–15	0.41–0.45	0.43–0.53
+	LOW-e		35–56	7–26	23–43	8–18	0.31–0.43	0.30–0.47
GRAY +	COATED	SILVER	4–9	9–41	3–7	10–14	0.41–0.45	0.43–0.53
TRIPLE GLAZED								
CLEAR			70–79	20	46–61	19–20	0.31–0.46	0.40–0.49
TINTED			11–56	8–17	25–46	8–16	0.31–0.38	0.40–0.46
COATED			16–70	8–44	5–66	8–44	0.23–0.32	0.25–0.41
QUAD GLAZED			63–73		50–63		0.25–0.26	0.34
GLASS BLOCK		SEE GLASS BLOCK EVALUATION CHART PAGE 5-46						

B2.7 GLASS/PLASTICS

Relative Heat Gain BTU/Hr. sq. ft. Summer [3]	Shading Coefficients (SC) [4]				
	No Shading	Dark Drapes	Light Drapes	Light Blinds	White Shades
186–217	0.73–1.00	0.56–0.70	0.46–0.56	0.53–0.55	0.25
154–189	0.69–0.87				
155–183	0.70–0.84				
115–185	0.51–0.85	0.41–0.63	0.37–0.50	0.41–0.54	0.30
120–185	0.49–0.84	0.39–0.62	0.35–0.50	0.39–0.54	0.30
56–112	0.22–0.80	0.21–0.52	0.20–0.44	0.20–0.45	
	0.90–1.00				
		Use SC of glass with similar solar transmittance			
	0.44–0.90				
169–192	0.79–0.92	0.51–0.53	0.47–0.50	0.49–0.51	0.25
133–157	0.23–0.78				
116–161	0.54–0.76	0.41–0.49	0.37–0.44	0.35–0.45	0.22
74–152	0.33–0.71				
76–152	0.25–0.73				
120–154	0.56–0.73				
36–59	0.14–0.36	0.15–0.40	0.15–0.40	0.15–0.40	
58–73	0.25–0.33				
38–70	0.15–0.31				
44	0.20				
30	0.13				
37–51	0.15–0.22				
50–62	0.21–0.27				
37–51	0.15–0.22				
39–60	0.16–0.26				
37–52	0.16–0.22				
74–101	0.35–0.47				
37–50	0.15–0.21				
148–164	0.78–0.86	0.56–0.59	0.46–0.49	0.48–0.51	
92–132	0.43–0.68	0.40–0.50	0.34–0.43	0.33–0.43	
74–80	0.34–0.67	0.13–0.39	0.12–0.35	0.14–0.35	
	0.74				

REMARKS

[1] Winter U-Value:
Outside = 0°F
15 mph wind
inside = 70°F

[2] Summer U-Value:
Outside = 90°F
7.5 mph wind
inside = 75°F

[3] Relative heat gain = heat gain due to conduction plus heat gain due to solar radiation when outside = 80°F with a 7.5 mph wind and inside = 75°F with a solar intensity of 200 BTU/HR. SQ. FT.

[4] Shading Coefficient,

$$S.C. = \frac{\text{Solar Heat Gain of Fenestration}}{\text{Solar Heat Gain of DS Sheet Glass}}$$

Shading Coefficient is the ratio of the total solar heat gain through it to the total solar gain through a standard sheet of clear glass under exactly the same conditions. It is a dimensionless number with a value from zero to one: the smaller the value of the shading coefficient the better the glazing is at stopping the entry of solar radiation.

B2.7 GLASS/PLASTICS

GLAZING ASSEMBLIES: general considerations

THERMAL MOVEMENT, in frame and glass

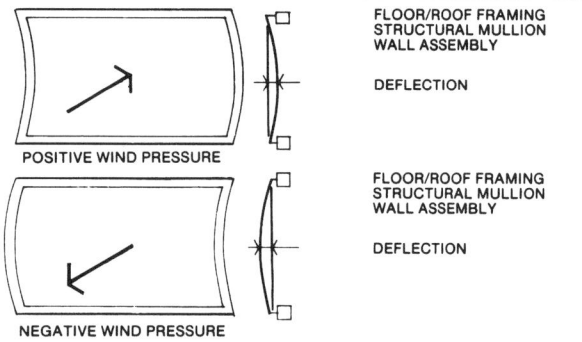

RADIATION•SHORT-WAVE

CONVECTION•HOT AIR EXPANSION IN GLASS AND FRAME

HEAT LOSS TO ADJACENT MATERIALS

HEAT GAIN

RADIATION•SHORT-WAVE

CONVECTION•COLD AIR

RADIATION•LONG-WAVE

DIFFERENTIAL CONTRACTION IN GLASS AND FRAME

HEAT LOSS TO ADJACENT MATERIALS

HEAT LOSS

DEFLECTION, vertical framing members

FLOOR/ROOF FRAMING
STRUCTURAL MULLION
WALL ASSEMBLY

DEFLECTION

POSITIVE WIND PRESSURE

FLOOR/ROOF FRAMING
STRUCTURAL MULLION
WALL ASSEMBLY

DEFLECTION

NEGATIVE WIND PRESSURE

DEFLECTION, horizontal framing members

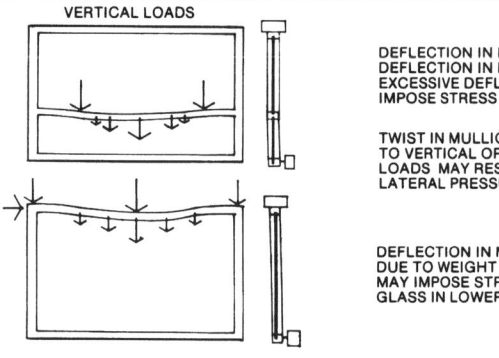

VERTICAL LOADS

DEFLECTION IN LINTEL
DEFLECTION IN MULLION
EXCESSIVE DEFLECTION MAY
IMPOSE STRESS ON GLASS

TWIST IN MULLION DUE
TO VERTICAL OR LATERAL
LOADS MAY RESULT IN
LATERAL PRESSURE ON GLASS

DEFLECTION IN MULLION
DUE TO WEIGHT OF GLASS
MAY IMPOSE STRESSES ON
GLASS IN LOWER FRAME

CLEARANCES, SHIMS, DRAINAGE

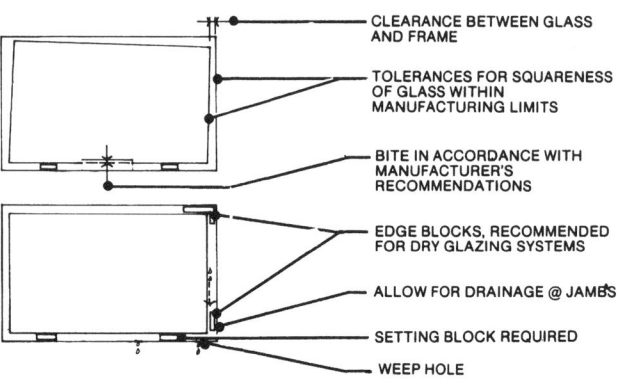

CLEARANCE BETWEEN GLASS
AND FRAME

TOLERANCES FOR SQUARENESS
OF GLASS WITHIN
MANUFACTURING LIMITS

BITE IN ACCORDANCE WITH
MANUFACTURER'S
RECOMMENDATIONS

EDGE BLOCKS, RECOMMENDED
FOR DRY GLAZING SYSTEMS

ALLOW FOR DRAINAGE @ JAMBS

SETTING BLOCK REQUIRED

WEEP HOLE

- Heat Gain: Glass, especially when heat absorbing or reflective, will generally absorb more heat than surrounding frame when exposed to solar radiation.

- Heat Loss: Glass may be expanding under solar radiation, while frame may be contracting. Coefficient of expansion for frame may be higher than for glass, aggravating the condition.

- Heat absorption or heat loss of frame will depend on its surface characteristics, such as reflectivity, emissivity. Frames may also lose heat to surrounding materials, thus lowering the temperature of glass adjacent ot them, leading to differential thermal stresses in glass.

- Draperies and venetian blinds should be kept at least 2 inches from the glass, and should allow air to circulate over and under them to prevent heat buildup.

- Deflection under positive and negative wind pressures in vertical members of framing systems for glass, such as in curtain walls, should be limited to 1/175 of the span of such member between supports, or to a maximum of ¾ inch to avoid imposing additional stresses on the glass.

- Strength of glass under wind loads is a function of glass area and thickness when supported on all four edges up to about a dimensional ratio of 5:1, for glass supported on two opposite edges only, the strength is a function of thickness and unsupported span.

- Deflection may also occur in long framing members with end restraint as a result of thermal expansion, especially in members with high coefficient of expansion such as aluminum.

Deflection in horizontal members may result from:
- Deflection in structural frame being transmitted to the framing member. Deflection in structural frame may result from excessive loading on structural steel, cold flow, or creep, in reinforced concrete frames, excessive spans.

- Lateral movement of structural frame due to wind loads on the structure, seismic forces, creep in reinforced concrete frame.

Vertical loads on vertical framing members may deform horizontal members, and if such deformation is excessive, bending stresses may be imposed on the glass.

- Water condensing within or penetrating into the frame should be drained to prevent damage to the glass. Weep holes must be provided for insulating glass with organic seals, laminated and wire glass; recommended for all others to prevent moisture collecting and freezing.

- Tolerances for frame and bite for glass edge must be considered since field adjustments in glass size should be avoided in annealed glass, and are not possible for tempered and insulating glass.

B2.7 GLASS/PLASTICS

The first parameter to establish will probably be the expansion and contraction of the glazing material and the resulting movement and stresses the glazing system has to cope with. This will be determined by:

■ Size of light to be glazed:

- the larger the light and the frame, the more it expands and contracts as temperature rises and falls . . .
- thus, the larger the movement the glazing system must cope with.
- soda-lime glass has a coefficient of expansion of about 4.5×10^{-6}, while the coefficient for aluminum is about 13×10^{-6} and an aluminum frame contracting under low temperatures may impose stresses on glass when edge clearance is insufficient to accommodate differential thermal movement.

■ Maximum temperatures likely to be reached by the glazing materials selected as a result of:

- exterior conditions - ambient temperature, incident solar radiation and shading devices.
- interior conditions - shading devices which may trap heat, air flow conditions due to HVAC design.
- tinted and/or coated materials which get much hotter than clear.
- thicker materials which get hotter than thinner ones.
- plastic materials which have an inherent thermal expansion and contraction coefficient that is up to eight times that of glass.

■ Sealed insulating units get hotter due to air trapped in air space.

In addition to the above considerations, the following dimensions must be examined:

- face clearance for channel glazing joints: larger face clearances can accommodate larger movements with less stress on sealants.
- edge clearance for all glazing methods, including lock-strip gaskets, must be large enough to permit expansion and contraction of glazing material and frame.
- channel depth for all methods: must be sufficient to provide proper support for glazing material and prevent "blowout" of

light when glazing material deflects, shortens, and slips out of frame.

Transfer of wind loads to surrounding structure will probably be checked out next. Factors to consider are:

■ Proportion as well as size of opening, span between supports, deflection.

■ Method of support for the glass pane:

- supported on all four sides
- top and bottom
- suspended or cantilevered one side only.

■ Movement of the surrounding structure

- short-term, due to wind loads, seismic.
- long-term, due to creep of framing members, foundation settlement.
- cyclic loads, wind conditions especially around complexes of tall buildings create alternating positive and negative lateral loads which complicate load transfer and create a "pumping action."

The basic concept in glazing systems to deal with the requirements cited is floating the glazing material within its frame by means of:

■ Setting blocks placed under bottom edge of glass. Their performance varies in:

- positioning for even distribution of loads, usually at quarter points from each end.
- shape - L-shaped used when "pumping out" of sealants due to cyclic lateral loading occurs; rectangular for normal use.

■ Space shims - to assure proper clearances between face of glazing material and framing channels. These vary in:

- form - intermittent blocks or continuous perimeter strips which avoid pressure points on glazing material.
- hardness - softer where resilient polymer- based sealants which rely on adhesion for sealing are used; harder where preformed tapes and gaskets which rely on pressure to seal are used.

■ Squareness and flatness tolerances of surrounding channel need to be close enough to avoid creating stress points in glazing materials.

EXISTING ASSEMBLIES
to minimize heat transfer through existing clear glazing and/or to reduce glare, tinted plastic films are available for laminating to the surface of the existing glazing:

- different transmission/reflection characteristics are available.
- films may hold glazing in place when broken.
- consideration should be given to resultant higher thermal stresses in the glazing, especially when films are laminated to the interior surface of the glazing.

Framing systems to allow the addition of another light of glazing to existing assemblies are available.

B2.7 GLASS/PLASTICS

GLASS BLOCK: types, properties

TYPE	Actual Dimensions height x width x thickness	Approx. Weight, in lbs.	% Light Trans- mission	Shading Coef- ficient	U-Value BTU/Hr. Sq.ft.°F
CLEAR, SMOOTH • Hollow	5¾ x 5¾ x 3⅞ 7½ x 7½ x 3¼ & 4 7¾ x 7¾ x 3⅞ 9½ x 9½ x 3⅛ 11¾ x 11¾ x 3⅞ & 4	4 to 18	75	.65	.60 to .52
CLEAR, SMOOTH • Hollow • With Reflective Coating	7¾ x 5¾ x 3⅛ & 3⅞ 7¾ x 7¾ x 3⅛ & 3⅞ 11¾ x 11¾ x 3⅞	6 to 16	5 to 20	.20 to .44	.56 to .52
CLEAR, PATTERNED • Hollow • Light Diffusing or • Decorative	3¾ x 7¾ x 3⅛ & 3⅞ 4½ x 4½ & 9½ x 3⅛ 5¾ x 5¾ x 3⅞ 5¾ x 7¾ x 3⅛ & 3⅞ 7½ x 7½ x 3⅛ & 4 9½ x 9½ x 3⅛ 11¾ x 11¾ x 3⅞ x 4	2 to 18	75	.65	.60 to .52
CLEAR, PATTERNED • Hollow • Light Diffusing With Fibrous Insert	5¾ x 5¾ x 3⅞ 7¾ x 7¾ x 3⅞ 11¾ x 11¾ x 3⅞	4 to 16	43	.44	.48 to .44
TINTED, PATTERNED • Hollow • Decorative	4½ x 4½ & 9½ x 3⅛ 5¾ x 7¾ x 3⅛ & 3⅞ 7½ x 7½ x 3⅛ 7¾ x 7¾ x 3⅛ & 3⅞	2 to 6	60	.52	.60 to .56
CLEAR, SMOOTH • Solid	7⅝ x 7⅝ x 3	15	80	NOT AVAILABLE	.87
• Paver Units	5¾ x 5¾ x 1 7⅝ x 7⅝ x 1½				

B2.7 GLASS/PLASTICS

MAXIMUM RECOMMENDED SIZE

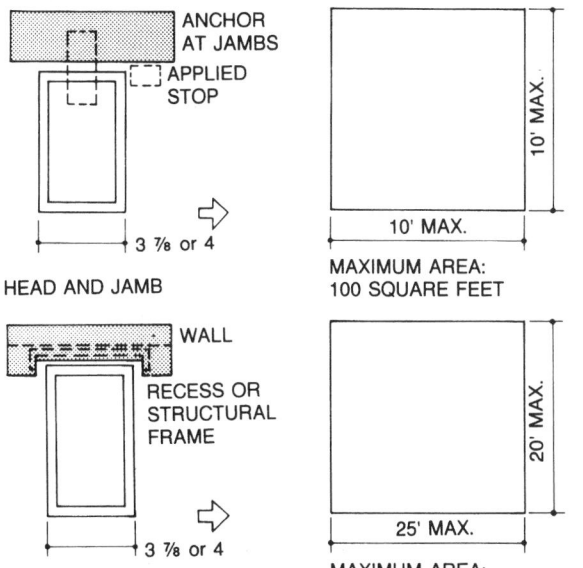

HEAD AND JAMB

3 ⅞ or 4 — ANCHOR AT JAMBS — APPLIED STOP

MAXIMUM AREA: 100 SQUARE FEET — 10' MAX. — 10' MAX.

HEAD AND JAMB

3 ⅞ or 4 — WALL — RECESS OR STRUCTURAL FRAME

MAXIMUM AREA: 144 SQUARE FEET — 25' MAX. — 20' MAX.

OPERABLE SASH

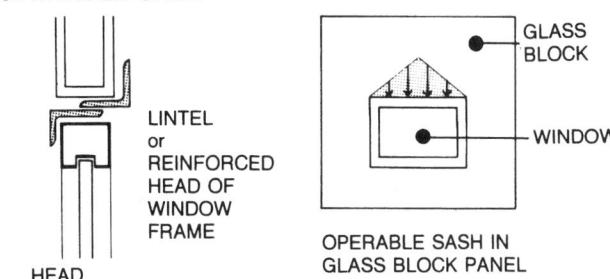

HEAD

LINTEL or REINFORCED HEAD OF WINDOW FRAME

GLASS BLOCK — WINDOW

OPERABLE SASH IN GLASS BLOCK PANEL

MINIMUM RADII

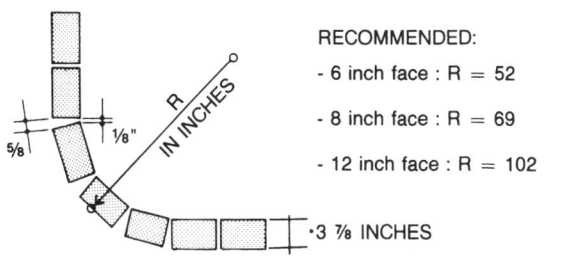

⅝ — ⅛" — R IN INCHES — 3 ⅞ INCHES

RECOMMENDED:

- 6 inch face : R = 52

- 8 inch face : R = 69

- 12 inch face : R = 102

THERMAL

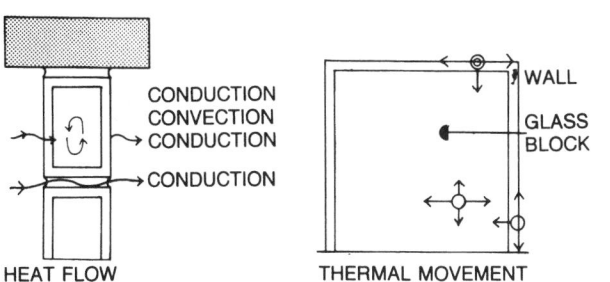

HEAT FLOW

CONDUCTION
CONVECTION
CONDUCTION
CONDUCTION

WALL — GLASS BLOCK

THERMAL MOVEMENT

GENERAL:

Glass block provides light transmission, controls thermal transmission and reduces noise transmission. Installation can be vertical or horizontal. Blocks are available in hollow or solid form in a variety of clear and textured patterns.

- installation can be the same as for masonry using mortar and welded wire reinforcing in joints; a mortarless frame system using neoprene gaskets; or a rebar-reinforced concrete or steel grid system with mortar joints of high density grout.
- glass block units can be installed on site or furnished in prefabricated panels ready for placement.
- for horizontal traffic applications the supporting structure must be capable of supporting its own weight, the weight of floor (dead load) including the glass block, and the live load limit as determined by the application and building code requirements.
- in horizontal traffic applications sandblasting is recommended for added slip resistance.
- in vertical installations windows may be incorporated into glass block, but lintels for large openings should not bear on glass block either side of the opening: structural channels should be used to transmit such concentrated loads directly to supporting structural components.
- for small windows the head of the window may be reinforced to act as a lintel: reinforcing must be strong enough not to allow deflection in the window frame.

USES:

Some of the uses for glass block are:
- translucent/partially transparent infill panels within structural frame or within openings in bearing walls.
- interior partitions; screens.
- floors, walkways and stairways.
- skylights.
- security and vandalism protection generally using solid glass block.

THERMAL MOVEMENT: between the glass block panel and surrounding construction must be allowed for in the perimeter joint by compressible fillers and caulking.
- thermal movement within the panel itself is controlled by horizontal joint reinforcing.

HEAT FLOW: will be differential through body of the glass block, its edges, and the mortar in joints between the blocks.
- for information on heat transfer through solids and air spaces refer to HEAT FLOW.

Resistance to air and water penetration through joints similar to that for masonry.

Sound transmission class for hollow block, 37 to 40; about 45 for solid block.

REFERENCES: Glass/Plastics

STANDARDS

ASTM Specifications for:

D702
- Cast Methacrylic Plastic Sheets, Rods, Tubes, and Shapes

D1547
- Extruded Acrylic Plastic Sheet

ASTM Tests/Methods for:

E546
- Frost Point of Sealed Insulating Glass Units

E576
- Dew/Frost Point of Sealed Insulating Glass Units in Vertical Position

E972
- Solar Photometric Transmittance of Sheet Materials Using Sunlight

E1084
- Solar Transmittance (Terrestial) of Sheet Materials Using Sunlight

ANSI Specifications:

297.1-1975 American National Standard Performance Specifications and Methods of Test for Safety Glazing Material Used in Buildings.

Federal Specifications:

DD-G-451D Specifications for flat glass for glazing, mirrors, and other used.

DD-G-001403B(1) Specifications for plate, float, sheet, patterned and spandrel glass - heat strengthened and tempered.

SELECTED REFERENCES

AAMA **Architectural Aluminum Manufacturers Association**
- Design Guide, Aluminum Curtain Wall Design Guide Manual
- Metal Curtain Wall Window Store Front and Entrance Guide Specification Manual

GTA **Glass Tempering Association**
64-3-16
- Specification for Fully-Tempered Glass for General Construction Use.

Also: Engineering Standards Manual

FGMA **Flat Glass Tempering Association**
- Architects' Guide to Glass, Metal and Glazing
- Glazing Sealing Systems Manual

LIA **Lead Industries Association**
- The Stained-Glass Renaissance

LOF **Libby-Owens-Ford— Technical Information**
ATS-109
- Strength of Glass Under Wind Loads
ATS-110
- Heat-Treated Glasses
- Tuf-flex Tempered Safety Glass

NRCC **National Research Council of Canada**
CBD 46
- Factory-Sealed Double-Glazing Units
60
- Characteristics of Window Glass
101
- Reflective Glazing Units
129
- Potential for Thermal Breakage of Sealed Double-Glazing Units
132
- Glass Thickness for Windows
205
- Glass Fibre-Reinforced Polyester Composites
213
- Plastics in Glazing and Lighting Applications

PPG **PPG Ind. Glass-Div. Technical Service Reports**
101A
- Wind Load Performance
104C
- Insulating Glass
104D
- Tinted Glass
103
- Installation Recommendations, Tinted & Reflective Glass
PDS-fl
- Glass

B2. 8 WINDOWS

WINDOWS: introduction

COMPONENTS OF ASSEMBLY

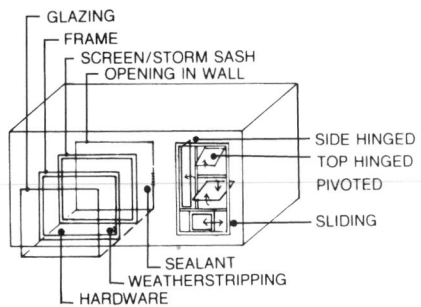

EXTERNAL FACTORS

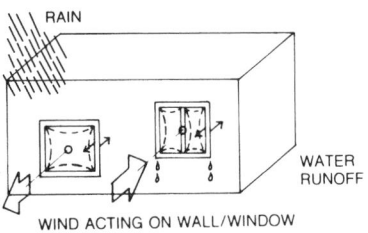

WIND ACTING ON WALL/WINDOW

SOLAR RADIATION

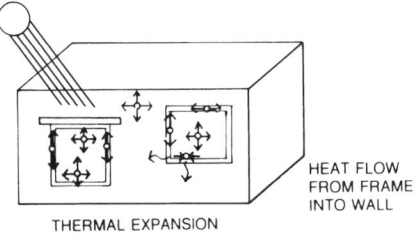

THERMAL EXPANSION

THERMAL/ACOUSTIC FACTORS

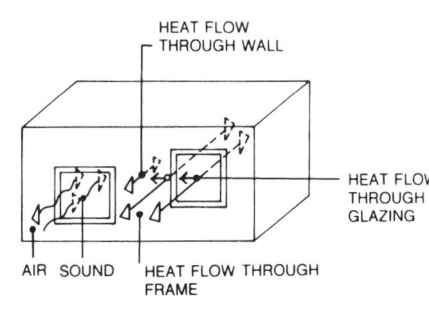

- Windows are assemblies which are used to allow vision, light, ventilation and emergency egress through an opening in an exterior wall while excluding outside elements.
- Vision is allowed via a transparent or translucent material referred to as glazing -- glass or plastic. When the glazing is surrounded by a rigid material to protect the edges, it becomes the glazing panel, or sash. When the sash has to be opened and closed, a frame has to be provided.
- The window assembly consists of the frame and one or more sashes. The frame consists of the jambs, sash stops, sill and head. An exterior trim or nailing flange may be incorporated for installation. The sash may pivot, slide, swing on a hinge mechanism, or be fixed in place. When fixed in an opening without provision to operate it, the sash may constitute the entire window assembly. Hardware must be installed when the sash is operable.
- Thermal performance may be improved when incremental layers of glazing are added to sashes. Screened sashes allow ventilation while preventing entry of insects.

GENERAL: Windows are installed into openings in wall assemblies and function as an integral part of the wall, resisting all external factors the wall may be exposed to:

- **AIR PRESSURES:** Both positive and negative air pressures will cause deformations in the glazing; deformations in the frame when free to deflect; shear at anchorage of perimeter frame to wall; and pressure differentials through the frame, tending to drive rain water to the interior.
- For information on pressure equalizations as a means to minimize water penetration, refer to WALL ASSEMBLIES AND COMPONENTS.

- **WATER RUNOFF:** Runoff of water over the plane of a wall is never equally distributed and is affected by differential air pressure and projections from the plane of the wall (such as window frames, reveals and window sills).
- Differential runoff may result in excessive localized staining.
- Water penetration into the wall is possible wherever the plane is broken, such as at panel seams, window framing, etc. Since water can cause considerable damage when inside the wall, these seams require careful sealing.

- **THERMAL EXPANSION/CONTRACTION:** Changes in dimension due to variation in thermal factors must be allowed within the assembly by providing adequate tolerances and flexible sealing methods. This is because different materials will expand and contact at different rates; i.e., an aluminum window frame in a masonry wall can expand at a rate twice that of the masonry. The changes may be compounded by differential stresses resulting from localized cool spots (in the shade) and hot spots (in the sun) existing within the same thermally sensitive material; taken to an extreme, this can lead to a thermal crack/break in glazing.

- **HEAT FLOW:** The transfer of heat occurs through all materials in a window assembly, but typically has the highest coefficient of thermal transfer through a single layer of glazing (about 1.10 Btu/hr/sf/°F). This coefficient is reduced to less than half by using two glazing layers separated by a 1/2 inch air space. Framing members may be made from low-conductance materials with lower coefficients or interrupted between the exterior and interior surfaces by low-conductance materials (thermal breaks).
- An issue which exists parallel to thermal conductance is condensation. Materials used in the construction of the window which have higher coefficients of thermal transfer will tend to have a greater occurrence of condensation. Lower rates of heat loss allow indoor relative humidity to rise to higher levels before condensation occurs.

- **AIR INFILTRATION/EXFILTRATION:** The movement of air from outside to inside (and vice versa) results in unwanted drafts, burdens on the heating/cooling load, and introduces outside air pollutants to the inside. Air leakage through the window assembly is minimized by using weatherstripping; between the frame and the wall leakage is controlled through the use of sealants.

- **SOUND ATTENUATION:** The transfer of outside sounds through a closed or fixed window may be minimized by isolating pockets of air within the assembly and by using materials which do not readily transmit sound waves. Multiple layers of glazing, storm windows and hollow frames help in sound attenuation.

■ **To determine the objectives** as to the extent, configuration and design of windows in each part of the building, the foremost considerations should be:
• **Temperature** - range and extremes.
• **Humidity** - annual rainfall and direction of prevailing winds for water and air infiltration requirements.
• **Air quality** - special environmental exposures to marine or industrial atmosphere to determine materials and/or finishes.
• Surrounding conditions and possible need for **sound attenuation**.
• **Building configuration** - height; positive and negative air pressures on the exterior and interior; stack effects within the building.
• Relationship to surrounding buildings - possible "canyon" effect of wind; surrounding environment for direction and configuration of air currents; natural physical formations; open rural area; proximity to large bodies of water.

Structural Performance

■ **The structural performance of the window must be compatible with the specifications and requirements applied to the wall.**
• While the wall is designed to transfer vertical loads around the window opening, the window must be capable of sustaining lateral loading. It is not designed to withstand vertical loading other than its own weight.
• The pressure exerted by the wind on a window is assumed to be uniformly distributed across its face.
• All certified windows are tested in accordance with ASTM Standard E330 whereby both the exterior and the interior faces of the window unit are subjected to a uniform static air pressure. Acceptable performance of this testing is evidenced by sustaining minimum pressures (per performance grade) without damage and without exceeded given deflection criteria. The deflection test is met when no exterior member deflects more than 1/175 of the span when stated uniform loads are applied.
• The performance grades are often set to the design pressures (ranging from 15 to 60 psi). Test pressures include a 150 percent safety factor over the design pressures (15 psi design = 22.5 psi min. test).
• **It is the designer's responsibility to stipulate the required design**

pressure in the specifications for each project. In the absence of this information, window manufacturers will be unable to properly bid in accordance with the desires or needs of the architect or owner.
• Check applicable building codes for requirements of windows to accommodate **movements due to seismic forces** for resistance to torsion, in order to preclude failures at or near connections and joints. See ANSI A58.1.

■ **Clearances and tolerances are required to accommodate movement in materials and to facilitate installation.** These allowances are usually considered by manufacturers when recommending rough opening sizes (frame size plus).
• Generally, the rough opening provides additional space around the jambs and head of the window unit to the wall.
• Clearances between the frame and sash are important for proper sash operation.
• Tolerances around glazing will allow the glazing to expand and contract while maintaining its weatherproof relation to the sash assembly.

Thermal Performance

■ **Heat loss through windows is caused by:**
• air infiltration;
• heat flow.

■ **Air infiltration is the leakage of unwanted outside air into the interior, conditioned space.** It occurs wherever a seam or crack in the exterior skin occurs. Since windows can account for 10 percent or more of the area of an otherwise solid wall, they represent the potential of considerable infiltration. Openings for windows and window assemblies themselves present the type of crack which is conducive to air leakage.
• Air infiltration is usually expressed as flow rate per foot of sash crack (operable sashes) or flow per unit area (non-operative windows).
• Certified window units are tested (under a uniform static air pressure difference of 1.57 psf) for air leakage according to ASTM E283. Maximum levels of infiltration vary with performance grade.
• Air infiltration is controlled through weatherstripping of operable sashes, gasketing and sealing of glazing, and

thorough caulking and sealing of compressible/expandable joints between frame and wall.

■ **Heat loss occurs both through the window itself and through its frame at a rate proportional to the temperature differential between the inside and the outside and to the nature of the material. It is controlled by the following insulative methods:**
• **increasing the number of glazing layers installed within the sash;**
• **installation of fixed or operable storm windows;**
• **introduction of nonconductive material (a thermal break)** into design of window frames and sashes made of conductive materials.
• Some **heat-absorbing glazing and glazing films** can also be used to improve heat flow characteristics.
• Testing and energy-use modeling has shown that additional layers of glazing are most beneficial when the air space is kept in the range of 1/4 inch to 4 inches. Outside of this range, the insulative value of the air space drops off rapidly.
• Any method which reduces the heat flow from outside to inside will reduce the occurrence of condensation on the interior surface during the winter heating season.

■ **Heat gain** is another element of window performance that must be considered. During the period that solar radiation passes through a window, the radiant heat gain may actually exceed the window's heat loss. While this is good for cooler climates, it is not for warmer ones.
• Reflective glazings and coatings are often combined with insulative techniques to keep solar radiation out and cool air inside.

■ **Water infiltration** is tested according to ASTM E331 and is a critical element of window performance: No water shall pass beyond the inner window plane or through the frame, when tested at air pressures so stated, with water applied at 5 gallons/square foot/hour (equal to a rainfall of 8 inches per hour).

B2.8 WINDOWS

WINDOWS: hinged operation

SIDE HINGED: casement (C)

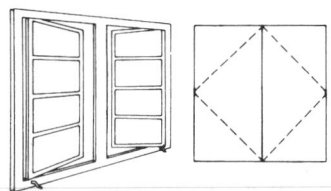

BOTTOM HINGED: hopper (H)

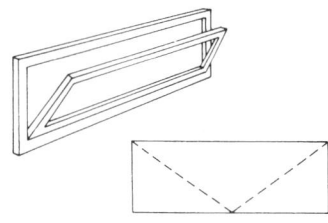

TOP HINGED: awning (A)

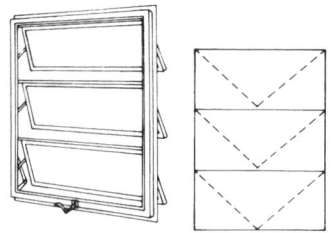

PROJECTED (P)

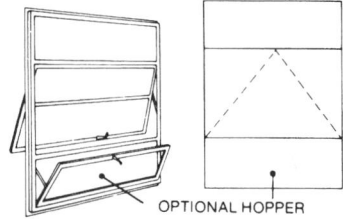

OPTIONAL HOPPER

TILT & TURN

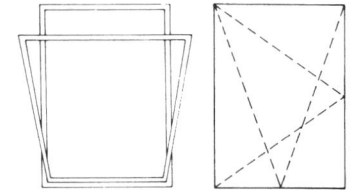

■ **Casement windows are single sash assemblies which are hinged or pivoted on one side (vertical axis), typically swinging outward. Maximum (100 percent) ventilation is achieved when the window is fully open (90° or more). The hinge may occur on either side jamb, requiring specification to denote left-hand or right-hand swing.**
● Hardware typically consists of a locking handle on the jamb or mullion and a rotary crank installed at the sill. Lever operators may also be used. Hinges may vary between extension type, butt (close-up) type or four-bar friction type.
● Windows of equal height can be assembled in any combination of fixed, right-hand or left-hand units or in any combination of widths.
● These units are often used in residential design and can be sized for emergency egress; speed and ease of operation may be a selection factor.
● Outswinging operation allows for easy cleaning of both glazing surfaces, but also places the sash and hardware under stress when the wind blows. Screens and storms are typically applied to the inside.

■ **Hopper windows swing single sash assemblies inward from hinges on the bottom, providing a maximum opening position of 30 to 50 percent for ventilation.**
● Hardware typically consists of a combination lock/handle on the head and elbow hinges on the sides. To open, turn the handle to unlock and pull inward.
● Windows can be assembled in any combination of vertically stacked units, side-by-side units or fixed/operable combinations. These units are also combined below casement or projecting windows.
● Hoppers should only be used where the sash will not interfere with interior conditions. Commonly used in residences (lower cost/quality units in basements) and apartments, these units are not appropriate for use in egress situations.
● Inswinging sashes allow for easy cleaning of both glazing surfaces. Screens and storms are typically installed to the outside.

■ **Awning windows are single sash assemblies which swing outward, hinged on the top. Maximum ventilation area of about 50 percent is provided. These windows are one of the few types that effectively protect against rain penetration.**
● Windows can be assembled in any combination of vertically stacked units, side-by-side units or fixed/operable combinations. Where the view is important, they are commonly used as the lower, ventilating element of picture windows.
● Hardware is typically mounted on the sill and consists of a rotary crank or lever operator which is capable of operating all sashes, securely closing them without the need for locks. When in multiple, vertical stacks, the mechanical operation will allow for the bottom vent to open before the other vents, which will then open in unison.
● These units are seldom used for emergency egress.
● Outside glazing surfaces of sashes are not easily cleaned from the inside. Screens and storms are typically installed to the inside.

■ **Projected windows swing their sash assemblies outward. The sash may be hinged or pivoted on the top and provides a maximum ventilating area of about 30 percent.** These windows are one of the few types that effectively protect against rain penetration.
● The operating sash may be assembled in vertically stacked units, combined with a fixed lite above and a hopper unit below; often used with lower hopper sashes.
● Commonly used in commercial, institutional and industrial applications.
● Hardware typically consists of a combination lock-handle on the sill and balance arms on the sides. To open, turn the handle to unlock and push outward.
● These units are generally not appropriate for use where egress is an issue.
● Outside glazing surfaces of sashes are not easily cleaned from the inside. Screens and storms are applied to the inside when lever or rotary crank operators are used.

■ **Tilt & turn windows, also known as casement-hopper windows, are designed to allow the sash to operate as a casement or a hopper window. The sash is normally operated (inward) by hinges near the bottom of the jambs, but the sash can be converted by use of secondary hinges into an inswinging casement.**
● Conversion to casement operation allows for ease of cleaning.
● These units can be opened to provide 100 percent ventilation area.
● When insulated glass is used, maximum glazing area is limited to 35 square feet.

WINDOWS: sliding, pivot operation

VERTICAL SLIDERS: double hung, single hung (DH, SH or VS)

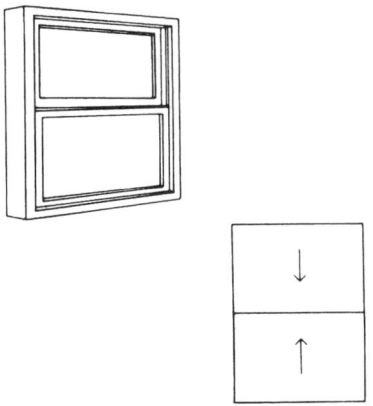

■ **Vertical sliding windows operate one or more sash assemblies, often in combination with a fixed sash, in the same frame.** The number of sashes may affect sash and frame thickness, or extent of operation. Operating units are held in place securely and in level by counterbalance systems such as weights and pulleys, tension springs or friction devices.
- **Double hung windows** have two sash assemblies which slide the full height of the jamb in separate tracks. Top and bottom openings optimize natural stratification ventilation.
- **Single hung windows** have a fixed upper sash and an operable lower sash. The operable sash assembly generally slides to the full height of the frame.
- **Triple hung windows** are available. Also, some double hung units actually contain three sashes with one being fixed. Check manufacturer literature for ventilation area.
- Ventilation area is 50 percent of light area in single and double hung units.
- Hardware consists of a lock on the sashes and finger grips on operating sashes.
- Windows can be assembled in any combination of operable and fixed units of varying widths and uniform height. Many sizes are available which comply with emergency egress requirements; egress operation is excellent.
- Methods for sash removal, sash tilting or other techniques are employed to improve ease of cleaning of outside glazing surfaces of sashes.
- Self-storing storm and screen units are typically installed on the outside, but some storm panels are designed to be applied to the inside.

HORIZONTAL SLIDING (HS, GL or SL)

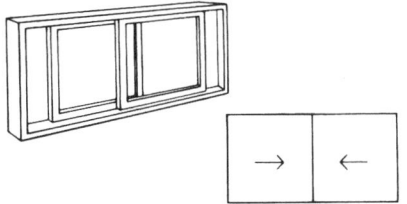

■ **Horizontal sliding windows operate one or more sash assemblies in the same frame. Sash height to width ratio should not exceed 1:2 for good operation.** The number of sashes may affect sash and frame thickness, or extent of operation.
- Ventilation area is 50 percent of light area, typically.
- Hardware consists of a sash lock and finger pulls on the operable sash.
- Windows can be assembled in any combination of operable and fixed units of varying widths. The fixed sash can be on the right, left or in the center (with two operating sashes) of the unit. Window height will be uniform.
- Ease of operation and size availability make these units excellent for egress.
- Methods for sash removal, sash tilting or other techniques are employed to improve ease of cleaning of outside glazing surfaces of sashes.
- Operable sashes generally employ metal or plastic roller and track systems; tracks generally require periodic cleaning.
- Self-storing storm and screen units are typically installed on the outside, but some storm panels are designed to be applied to the inside.

DUAL VENT (DW)

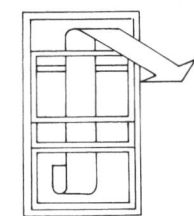

■ **Dual vent windows are essentially two sets of sliding windows set in one frame. There are generally four or more sashes working in tandem with an air space between; air circulates through the bottom outer sash and then through the top inner sash.**
- Operation may be vertical or horizontal. Vertical operations may be single or double hung. Fixed sashes are set in opposite positions.
- Units can be effective for egress and ventilation (50 percent of area).
- Units provide protection from rain and drafts. Screens and storms are applied in the same manner as with other sliding units. Cleaning from inside is difficult without removable sashes, tilting sashes or other technique.

PIVOT (VP, HP)

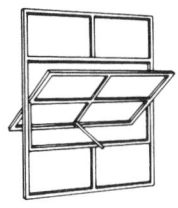

■ **Pivot windows are single sash assemblies which turn 180° or 360° within the frame. The axis of rotation may be vertical or horizontal and is centered on the respective frame members. The primary purpose of the pivot window is to accommodate cleaning; they are not primarily designed for ventilation.**
- Pivot windows are generally limited to an opening of about 4 inches for ventilation when special hardware is installed. They are seldom used in emergency egress situations.
- When insulating glass is used, limit the maximum glazing area to 50 square feet.
- Windows can be assembled in any combination of horizontal or vertical groupings to form entire walls. Operating sashes can be combined with fixed sashes in the same frame.
- Units mounted high on walls are often operated by a pull-chain (pulling on the chain withdraws the spring-catch and then draws the top of the sash inward), rotary crank with extension hardware or pole-hook (using a pole to unlock and pull open the window).

B2.8 WINDOWS

WINDOWS: special operations

BOW AND BAY WINDOWS

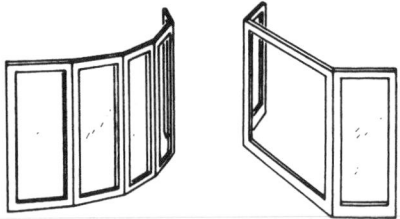

DETENTION WINDOWS

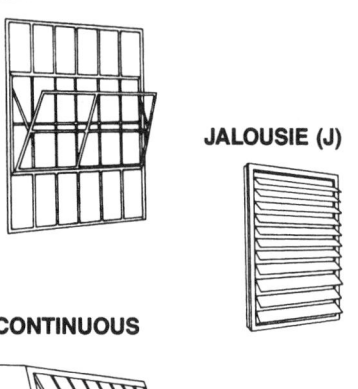

JALOUSIE (J)

CONTINUOUS

AUSTRAL

REVERSIBLE

GREENHOUSE

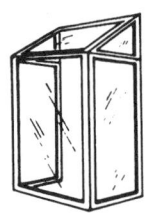

GLAZED DOOR ASSEMBLIES (SGD)

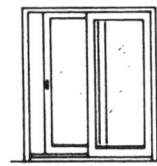

■ **Bow windows and bay windows are often used to create an interior feature, such as a window seat, and provide an expanded exposure to the outside.**
- Bow windows curve out, combining several narrow windows. Most often, bow windows are created with casement windows on the ends and fixed units in the center.
- Bay windows have operating windows slanting away from the wall plane and meeting at either side of a picture unit which is parallel to the wall plane.
- Assemblies may include head or seat boards, often included to maintain structural integrity during transport.
- Both window types require consideration of protection from weather at the top and bottom of the unit.

■ **Decorative windows are operative and non-operative windows with frames that do not conform to conventional rectangular shapes. These units are often custom made, and they may consist of one or more window assemblies.**
- Typical shapes include oval, arch, ellipse, octagon and circle.
- Glazing may also be decorative; stained, etched, beveled, diffusing, etc.

■ **Windows used for detention purposes are typically made of metal (usually steel) with reinforced frames. Venting sections are usually of the hopper type.**
- Use may be in commercial and industrial buildings where resistance to forced entry is of major importance. These units are also used to prevent forcible exit, such as in housing for mental patients where the appearance of restraint is minimized.
- Operable sashes are generally limited to a clear opening of about 6 inches or are divided by muntins to limit glazed areas to 88 square inches. When larger sashes are desired, muntins are installed in the frame opening.

■ **Jalousie windows are primarily used for sunrooms, porches, and the like where protection from the weather is desired with maximum ventilation. Due to the length of crack between glazing panels, jalousie windows will seldom pass air infiltration testing for prime window usage.**
- Multiple vents combine good vision quality with controlled ventilation.
- Louvered glazing panels operate from rotary crank type hardware.
- To improve thermal qualities, interior storm sashes may be installed.

■ **Continuous windows are used primarily for top lighting and ventilation (20 to 40 percent of opening) in monitor and sawtooth roof construction.** Units are hinged at the top and swing outward. Window cleaning from the inside is difficult. Mechanical operators may be either manual or motorpowered.
■ **Austral windows have two operating sashes which pivot in the frame.** Upper and lower sashes are counterbalanced and operate simultaneously. These windows are difficult to screen, shade or curtain; they are not effective for egress.

■ **Reversible windows are similar to vertical sliders in appearance, but may be tilted for better control of ventilation or reversed for cleaning.** They are not appropriate for emergency egress.
■ **Greenhouse windows (GH) are units which are designed to fit in a normal window opening but extend beyond the exterior plane of the wall.** These units typically have an insulated opaque bottom, glass sides and top, and a picture sash parallel to the wall's plane.
- Various types of ventilation systems are used to release unwanted trapped heat during the summer. Venting sashes are usually screened.

■ **Glazed door assemblies are designed to allow circulation from one area to another as well as function in all other aspects as a window.**
- The sliding glass (patio) door unit is designed and assembled with the same basic considerations as a horizontal sliding window with the exception that components are larger and stronger to accommodate the weight of the large "sash." Typically, one sash will operate and one will be fixed, but additional fixed or operable sashes may be added.
- An alternative to the sliding door is the swinging door. Unlike side-hinged windows, these units will have normal door hardware attached to the sash and frame. One or more operating sashes, or an operating and fixed sash(es) may be assembled together.

HARDWARE

CAM LOCKING HANDLE

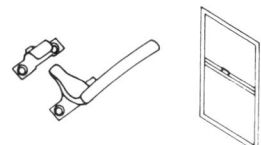

PIVOT LOCKING HANDLE

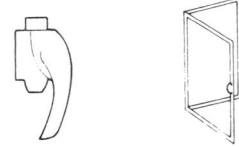

ROTO (CRANK) OPERATOR

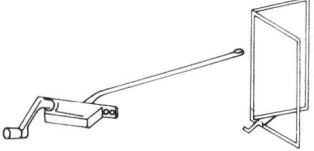

EXTENSION HINGES

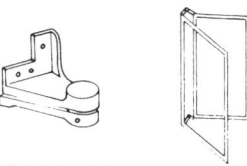

ANDERBER ARM

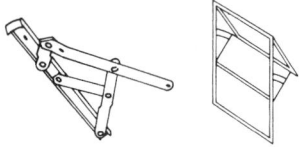

LIFE SAFETY: HOLD OPEN BAR

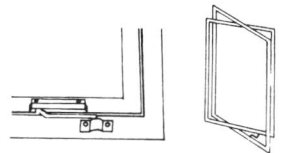

LIFE SAFETY: DUO-FOLD HOLD OPEN DEVICE

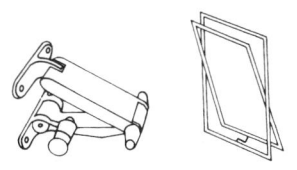

CUSTODIAN LOCK

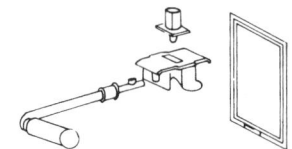

■ **The functional life of a window will depend to a great extent on the quality and durability of its operational hardware.** The operating life of these devices will vary with the type of occupancy and the frequency and amount of use (cycles). All operating devices are **performance tested** for number of operating cycles without deterioration in function. Check manufacturer literature for performance or number of cycles, guarantees and replacement parts. Most guarantees cover 20 year replacement of defective hardware.

Locking and Operating Devices

■ Cam locking handles are commonly used as locks and push-pull operators on hinged/pivot windows, but are used solely for locking on sliding units.
● These locks, often used without lever-type handles, use a cam shape on the moveable section to provide pressure to the sashes against movement.
● These locks are hard to use unless the window sashes are in perfect alignment.

■ Pivot locking handles are appropriate for side-swinging vents only; casement, tilt-and-turn, and pivot window types. The handle on the vertical member of the sash drops, raising a bolt out of its recess in the frame. Depending on sash size, they may be arranged such that a vertical bar operates top and bottom latches.

■ Rotary crank (roto) operators can be manually operated or motorized. Motorized operators are usually hard to reach hinge/pivot type windows or are multiple unit assemblies. A revolving handle connects to an extension bar which slides along a bar which is attached to the edge of the window sash, gradually pushing the window open or pulling it closed. For high, inaccessible windows, control systems may be surface mounted or concealed within walls.
● The sash is held steady in all positions by the extension bar.
● Removable handles may be appropriate to manually operated, single hinged units.
● Heavy-duty electric powered operators are commonly used for multiple hinged window assemblies, continuous/ribbon windows and some jalousie and awning units.

■ Common pole and ring operation is used for hinged and pivot windows which are out of normal reach.

■ Extension hinges, which permit cleaning of exterior surfaces from the inside, can be applied to the individual sash of any casement or hopper type window.
● Extension hinges are offset pivot hinges which create a space between the open ventilator and the window frame.
■ The "Anderberg" arm is a particular product used for top hinged and pivot type windows.
● It is a four-bar hinge sufficiently strong to hold the sash in an open position despite strong winds.
● They are mostly used on top-hinged windows.

Safety Devices

■ Life safety bars are available on tilt-and-turn, pivot and other such window types. Standard hold-open bars limit opening of the window to a size that will prevent passage through the opening (maximum opening of 4 to 6 inches) while allowing ventilation. These devices are often required by building codes.
■ Swing-tilt assemblies are available for tilt-and-turn windows.
● Duo-fold devices are similar to hold-open bars in purpose.
● These devices also serve as push-pull locking handles.

■ Custodian (keyed) locks are available for any unit where security or safety are an issue. These are typically used on tilt-and-turn units and on pivot windows, where operation is primarily for cleaning.
● Master-keyed locks are openable only by specially shaped keys.
● These locks prevent casual window operation.
■ Vertical sliding windows use spring balance or counterbalance systems to hold sashes in place.

WINDOWS: grades and performances

NOTE: The grades and performance criteria shown here have been developed and set by industry associations in compliance with ASTM Standards. Not all manufacturers will comply with these criteria, and others will exceed them. Refer to manufacturer literature and association certification programs.

MATERIAL: ASSOCIATION & STANDARD	GRADE	PERFORMANCE CLASS/ DESIGN PRESSURE	MAXIMUM OPERATING FORCED	TESTS AIR INFILTRATION ASTM E283	WATER LEAKAGE ASTM E331 (1)	UNIFORM WIND LOAD STRUCTURAL ASTM E330 (2)	UNIFORM LOAD DEFLECTION ASTM E330 (3)
ALUMINUM ANSI/AAMA 101 (4)	Residential (R)	15 psi min. (5)	20, 35 lb.	0.37 cfm per lineal foot of crack (6)	2.86 psf 3.00 psf	22.5 psf	15 psf
	Commercial (C)	20 psi min. (5)	25, 45 lb.		3.00 psf	30 psf	20 psf
VINYL ANSI/AAMA 101V		25 psi			3.75 psf	37.5 psf	25 psf
		30 psi			4.50 psf	45 psf	30 psf
		35 psi			5.25 psf	52.5 psf	35 psf
	Heavy Commercial	40 psi min. (5)	45 lb.		6.00 psf	60 psf	40 psf
		45 psi			6.75 psf	67.5 psf	45 psf
		50 psi			7.50 psf	75 psf	50 psf
STEEL SWI Specifications	Residential Std. Intermediate (7)			0.5 cfm per lineal foot of crack at 1.56 psf (8)	NOT ESTABLISHED		
	Heavy Intermediate Heavy Custom						
WOOD NWWDA I.S.2	Residential	20 psi	25 lb.	0.34 cfm/lf/crack	2.86 psf	20 psf	
	Light Commercial	40 psi	30 lb.	0.25 cfm/lf/crack (9)	4.43 psf	40 psf	
	Heavy Commercial	60 psi	35 lb.	0.10 cfm/lf/crack	6.24 psf	60 psf	

NOTES:
(1) ● Resistance to water penetration is defined as no leakage to the inside at a specific air pressure difference.
   ● ASTM E331 is used for Heavy Commercial grades of aluminum and vinyl windows; ASTM E547 test methods are used for wood windows, and for Residential and Commercial grades of aluminum and vinyl windows.
   ● Test pressure is generally 15 percent of design wind pressure.

(2) ● Uniform load structural test pressure for aluminum and vinyl windows is generally 150 percent of design wind pressure; and 100% of design wind pressure for wood windows.
   ● Wood windows are also subject to a Preliminary Loading test conducted at about 2/3 of design wind pressure.

(3) ● Uniform load deflection test pressure for aluminum and vinyl windows generally is the design wind pressure. The maximum deflection generally acceptable is 1/175 of span; most glass manufacturers recommend this.
   ● Wood windows are subjected to Physical Load Testing at design wind pressure; residual deflection is limited to 0.4 percent of span.

(4) ● Aluminum windows are also available graded under AAMA GS-001-Voluntary Guide Specifications for Aluminum Architectural Windows. GS-001 includes the same performance requirements as ANSI/AAMA 101, but requires better resistance to air and water penetration than for Heavy Commercial grade under AAMA 101.
   ● Resistance to wind load under GS-001 is to be selected by user, and additional testing for uniform load deflection is required.

(5) ● Minimum Performance Class – which has the same value as the design wind pressure – for grade. Windows in each grade are available meeting requirements of all higher Performance Classes e.g. Residential grade window may be selected meeting requirements of Performance Class 50.

(6) ● All operable styles except jalousies: 0.37 cfm per foot of operating crack length at 1.57 psf; except Heavy Commercial grade casement, projected, top hinged and pivoted types which are tested at 6.24 psf.
   ● All fixed windows: 0.15 cfm per square foot of area at 1.57 psf; except Heavy Commercial grades at 6.24 psf.
   ● Jalousies: 1.5 cmf per square foot at 1.57 psf.

(7) ● Windows defined by minimum combined weight of outside frame and vent members in pounds per lineal foot: Standard Intermediate = 3.0; Heavy Intermediate = 3.5; Heavy Custom = 4.2.
   ● Weight for Residential not established by SWI.

(8) ● Maximum air infiltration for Architectural Projected type 1.0 cfm at 1.56 psf; rates of infiltration not established for all types under SWI Specifications.

(9) ● Test pressure for air infiltration at 1.57 psf.

WINDOWS: restoration/replacement

■ One of the major factors in the restoration of exterior facades is fenestration. On a case by case basis, the decision will be to repair or replace.

■ When a building is to be restored with the intent of maintaining historical integrity, an assessment of the windows must consider the following:

- Are the windows original to the building?
- Do they make up a large area of the facade ?
- How visible is the facade from the street? From neighboring buildings
- What is the exposure of that facade to the elements ?
- What kind of aesthetic do the windows, themselves, present? Both exterior and interior
- In what condition are the windows ?

■ Factors which respond to aesthetics involve window details such as:

- Muntin patterns and widths.
- Mullion profiles.
- Reveal and installation details.
- Interior and exterior trim.
- Reflective qualities of the glass and frame.
- Color of sash and frame.

■ The condition of the windows may answer the repair/replace question:

- Can the hardware be reconditioned? Concealed hardware should be updated when possible; exposed vintage hardware should be recovered.
- Does weatherstripping need to be replaced or added?
- How effective is sash operation - sash ropes, pulleys, stops, etc.? Do sash joints need tightening? Is the opening still square ?
- What is the condition of the sash rails and sill ?

■ Improper sash operation may be affected be a function of an opening which is no longer square.

- Window frames commonly experience racking as the building structure settles, causing deformation of the original opening and resulting in binding sashes.

- Measure the frame opening for the operable sash carefully before replacing the sash.

■ Replacement windows are used for renovation and historic preservation where performance, operation and economy are primary considerations. Aesthetics may be a concern where original materials and window lines are to be matched.

- Steel windows are generally replaced with aluminum units, but the frames, mullions and meeting rails of the aluminum windows are wider than those of the steel units.
- Wood windows with single glazing and sashes divided by muntins are difficult to replicate with insulated glazing. Muntin grids are often not desirable because of the flat reflection of the glazing layer which is on one side or the other of the grid. Metal grids may also need to be considerably wider than the original muntin.
- Replacement windows may have particular glazing systems which are designed to give the appearance of molded wood sash and muntin pieces. These systems eliminate the need for glazing putty and its subsequent maintenance.
- Replacement of steel windows with wood, aluminum or vinyl replacement windows will require careful planning because the strength of the steel window can support a larger glass area.

■ When energy issues are the paramount concern, the following techniques can be effective in restoring the thermal performance of existing windows:

- Air infiltration may be reduced through basic maintenance programs which **examine caulking around frames, weatherstripping, glazing putty, the fit of the sash in the frame, and replace cracked panes.**
- **Storm windows** provide an air film between windows, increasing both thermal and acoustical performance considerably over that of the individual storm and primary windows. This is generally the preferred method of increasing the number of glazing layers because most existing windows will not have been designed to accept dual-glazing.

- Wood windows may provide enough sash and muntin material to allow a glazing rabbet to be routed into the sash stile. An additional glazing layer can then be installed to the rabbet, providing a 1/2 to 3/4 inch air space. A vent hole should be drilled so that moisture can escape.

■ Storm windows can play an important role in the restoration of historic windows. Reflections, shadows, details, and other features of the historic window must be considered as to whether an exterior or interior storm is most desirable.

- Exterior storm windows can protect the outer surfaces of the prime window from weathering, lowering maintenance requirements.
- Interior storms are generally used when prime window details are esthetically significant.

■ The condition of the sash rails and sill are a good indicator of the type and amount of moisture in contact with the window. While steel windows may have rusted glazing channels, wood windows show the worst deterioration at the sill. This area is a major source of air and water penetration, with the latter capable of causing serious damage. For wood sills, typical repairs include:

- Applying coatings - durable paints are good for sealing sill cracks. Water repellants should be applied to the entire window unit, and preservatives are recommended for warm, moist climates.
- Installing metal pans - be sure installation protects the wood sill against further degradation, such as decay.
- Filling bad or missing wood with fillers - epoxy can be used to reconstitute a wood member/sash.
- Replacing the sill altogether - preferably, use treated wood.
- The design of insulative double glazing eliminates glazing putty and its general maintenance.

B2.8 WINDOWS

WINDOWS: materials

Aluminum Windows

■ **Aluminum windows are widely used in all types of construction. They offer a variety of colors and broad availability.**
● Regarded as a well priced unit, aluminum windows are often found to offer more size options (but less standard sizes), easier installation and maintenance and a better manufacturer guarantee.
● Aluminum windows still face the metal window problems of corrosion, difficulty in refinishing, general maintenance issues and poorer reputation for thermal integrity.
● Except for casement and awning, most window styles are commonly available.

■ **Characteristics and properties of aluminum windows include:**
● size limitations are placed on operable windows due to the inherent strength of the frame;
● for double glazed units, muntins can be sandwiched between glazing layers or grid assemblies can be installed on the interior or exterior of the unit;
● thermal breaks are often required to separate the inside and outside surfaces of frames and sashes;
● framing members tend to expand and contract rapidly, potentially causing stress in thermal barriers;
● installation generally is via fastening through a fin which runs along the top and sides of the frame; exterior finish runs over the fin; interior finish butts up to the frame;
● aluminum windows can be left in their natural mill finish (silvery), can have anodic coatings or can be painted, enamelled or lacquered.

■ **Compatibility of materials is a key element of design and installation of aluminum windows.**
● When welding is used for fastening or joining, corrosion can result if the completed connection is not thoroughly cleaned of all flux.
● Anchoring devices must be made of aluminum, non-magnetic stainless steel, zinc or cadmium plated, or other such corrosion-resistent materials which are compatible with aluminum. Non-compatible devices can be used if they are well insulated from any aluminum surfaces.
● Aluminum surfaces should be insulated from all non-compatible materials and all surfaces which are absorptive or repeatedly wet (including drainage).
● Insulation from non-compatible surfaces is accomplished by painting, caulking, or gasketing; for masonry use alkali resistant coatings; for steel use galvanized coatings; no paints or coatings should contain lead pigmentation.

■ According to ANSI/AAMA 101, aluminum used in window construction has the following minimum properties:
● tensile strength - 22,000 psi;
● yield strength - 16,000 psi.
● The coefficient of thermal expansion for aluminum is about 13×10^{-6} in./in./°F.

■ The American Architectural Manufacturers Association (AAMA) provides a certification and labelling program to ensure that aluminum window products conform to ANSI/AAMA 101-85.

Steel Windows

■ **Steel windows are often used in institutional, industrial and commercial types of construction.**
Industrial applications often do not require attention to thermal issues and manufacturers specializing in this area may not provide thermal information/testing.
● Windows are available in most types and styles, however casement, continuous/ribbon, pivot, projected and reversible are most common.
● Specialty windows for detention or security are often made of steel. Steel windows are naturally vandal resistant.

■ **Window shapes are hot rolled (from 12 gauge steel) to tolerances of 1/4 inch. Windows are classified by the minimum combined weight of outside frame and vent members:**
● **Residential** - minimum 2.0 pounds per lineal foot (plf), with a maximum of 1 inch from front to back of ventilator and supporting frame; may have a maximum dimension of 6 1/2 feet; for combined units, maximum spacing of mullion is 3 1/2 feet.
● **Standard Intermediate** - minimum 3.0 plf, with a maximum of 1 1/4 inches from front to back; a 3/4 inch vertical muntin in projected vents over 4 feet, 6 inches wide; may have maximum glazed areas of 60 sf, and a maximum dimension of 10 feet; for combined units, maximum spacing of mullion is 6 1/2 feet.
● **Heavy Intermediate** - minimum 3.5 plf, with a maximum of 1 5/16 inches from front to back; a 3/4 inch vertical muntin in projected vents over 5 feet wide; may have maximum glazed areas of 84 sf, and a maximum dimension of 12 feet; for combined units, maximum spacing of mullion is 7 feet.
● **Heavy Custom** - minimum 4.2 plf, with a maximum of 1 1/2 inches from front to back of ventilator and supporting frame.

■ **Characteristics and properties of steel windows include:**
● the great strength of steel allows glazing areas to be larger than with other frame materials and allows the use of narrow framing members, reducing sight lines and resisting deflection and distortion of shape;
● tendency for frame and glazing to exist on one plane, presenting little depth or shadow;
● thermal expansion is minimal, roughly half that of aluminum;
● installation often includes an interior steel pan;
● thermal breaks may be required to separate the inside and outside surfaces of frames and sashes.

■ **Steel windows require maintenance to ensure proper operation and durability.**
● A variety of protective coatings, such as rust-inhibitive primers, paint and PVC coatings, bonderizing or hot-dip galvanizing, are available that can increase the life of steel windows. These coatings minimize the corrosive effect of moisture and various environmental factors.
● Some windows are chemically treated, with a baked on prime coat so that a final coat of any color can be applied in the field. Others are galvanized with shop or field applied finishes, including phosphate treating, prime coating and finish coating. Still others have special factory-applied finish coatings.
● Caulking and glazing putty will require maintenance.

Plastic & Vinyl Windows

■ **Vinyl windows are extruded from the plastic polymer, polyvinyl chloride (PVC). For windows, rigid PVC is used because it is a hard, strong and tough material which can be produced at reasonable cost.**
● According to ASTM D4216, the minimum properties of PVC used in building products has impact resistance of 0.65 ft.lbf./in. of notch; tensile strength of 5000 psi; modulus of elasticity in tension of .29 x 10^6; deflection temperature under load of 140°F; and a coefficient of linear expansion of less than 2.2 x 10^{-5} in./in./°F.

■ **Vinyl windows were originally developed in Europe and became established in the replacement window market in the United States. Since then, they have entered the prime window market in several areas of construction.**
● Vinyl units are available in all styles, shapes and sizes.
● They are designed to give the appearance of wood windows, but are as easy to install as metal windows.
● Some disadvantages of vinyl windows are that they are difficult to refinish, can be damaged due to extreme cold, may not age well and, like wood, are vulnerable to fire. Some dark-colored vinyl products have been known to distort under the heat of the sun; stabilizers are often added to reduce damage from the sun.

■ **Characteristics and properties of vinyl windows include:**
● frames and sashes are hollow extrusions, similar to metal windows but with better sound and thermal qualities; encapsuled air spaces act as an insulator and the vinyl has excellent thermal properties (1/8 inch thick sheet vinyl has a u-value of 0.38 to 0.27), reducing the occurrence of condensation and eliminating chances of a "thermal wick";
● flexible vinyl weatherstripping is integral (coextruded) to the frame and sash or may be applied after window assembly;
● resistant to denting, abrasions, condensation/moisture, and corrosion due to weather, salt air or fumes;
● exhibit little to no deformation due to heat and cold - no warping or sticking;
● low maintenance is due to finish quality of PVC, which does not require painting or other protective coating;
● limited color selection - white, gray, tan, and brown;
● when glazing is of substantial weight (large panes, triple glazing), metal frame and sash reinforcement is necessary.
● Care must be taken that materials used in and around the window units are compatible with PVC. Fasteners, weatherstripping, hardware, mullions, installation fins, etc., should be of stainless steel, aluminum or other corrosion-resistant material.

■ The Society of the Plastics Industry, Inc. (SPI) and the American Architectural Manufacturers Association (AAMA) provide certification and labelling programs to ensure that vinyl window products conform to appropriate standards. SPI requirements are based on ASTM D4099, while AAMA requirements are based on AAMA 101V-86.

■ **Another plastic is becoming available for the manufacture of window frames and sashes - fiberglass.** Slated for the housing market, fiberglass windows offer good structural properties and a wide range of colors.
● Fiberglass strands are bound in a polyester matrix and then pulled through a heated die in order to set the form. This is called a pulltruded form.

Wood Windows

■ **Wood windows are often preferred because of the beauty and aesthetic warmth they offer. They also have a reputation for good thermal qualities.**
● Wood windows are predominantly used in residential and light commercial applications.
● Problems that might occur to wood windows are due to weather exposure (deterioration), moisture and temperature changes (sticking).

■ **Due to the nature of wood, specific assembly criteria is desirable.**
● The wood parts of the window are typically milled from softwoods (ponderosa pine or similar). The wood is kiln dried to a moisture content of 6 to 12 percent before milling.
● Wood quality is important, thus predominantly clear (no knots, stains or structural defects) wood is used.
● Wood parts are fingerjointed or edge bonded with wet adhesives.
● All wood parts which may be exposed to any type of moisture are to be treated with preservative and water repellent.

■ **Characteristics and properties of wood windows include:**
● thick framing members set on various planes, give depth and shadow, but take up glazing area;
● for double glazed units, muntins can be sandwiched between glazing layers, or grid assemblies can be installed on the interior or exterior of the unit;
● available in a wide range of standard styles, shapes and sizes;
● often, windows are virtually held together by factory-installed exterior trim which should not be removed until they are set in place.

■ **Wood windows are available in a variety of finishes in order to protect the wood from weathering elements. Water seepage can result in wood decay, and ultraviolet light from the sun can degrade wood structure.**
● Natural wood finishes are often provided on interior surfaces, but are less frequently offered for exteriors. The advantages of natural finishes are that the wood can be stained or painted to match any decor. Additionally, color can be changed by repainting/refinishing.
● Windows may have a natural finish on the interior yet have a protective finish coating on the exterior. The coating can be as simple as a paint primer or may be the attachment of a vinyl or metal cladding. For properties of vinyl cladding, see Plastic & Vinyl Windows.
● Some wood windows will be completely finished; painted or clad. Cladding is recommended where exposure to moist salt air is possible.
● Vinyl clad windows are easily sealed at seams, eliminating chances for moisture penetration.
● Metal (aluminum) clad windows should be sealed at butt joints by welding, fusing or finish-coating.

B2.8 WINDOWS

WINDOWS: selection checklist

Overall Design

■ **Functional and aesthetic considerations are typically the most important specification criteria. They should include:**
- configuration of windows -
  - punched openings;
  - horizontal, vertical ribbons;
  - large combination assemblies.
- exterior/interior design features -
  - window surrounds, awnings;
  - louvers, grilles, fixed and operating sashes;
  - interior treatments, shades, blinds;
  - colors and finishes as related to other building materials, weather resistance and painting requirements;
  - flush or indented planes of buildings; shadow casting, reflectiveness of materials;
  - location of windows within interior spaces and relationships to use of space.
- type of glazing -
  - tinted, reflective;
  - tempered, vandalproof;
  - single or multiple layers;
  - plastic or glass.
  - Refer to SD.5/Gl, Glass/Plastics for more information.
- environmental factors -
  - amount and type of daylight;
  - amount and type of ventilation;
  - exterior noise;
  - visual perspective.
- building codes -
  - requirements to provide natural light to a space on the basis of a minimum percent of the floor area (residential buildings are 8 or 10 percent);
  - unless mechanical ventilation is adequately provided, natural ventilation is required at a rate which is usually half that of natural lighting (4 or 5 percent of floor area);
  - safety glazing/safety bars are required in windows which extend down near the floor;
  - energy codes will impact window quality and quantity; options for determining compliance may include maximum u-values for glazing and maximum percentage of glazing based on the insulative values of other envelope assemblies.

■ **Although glass curtainwalls have dominated commercial construction in recent years, operable windows may still be desirable where natural ventilation can benefit indoor environmental conditions. Window ventilation must also consider:**
- Optimum distribution of windows relative to prevailing breezes so as to create cross-ventilation in warm weather, but minimal drafts in cold.
- Optimum distribution of windows in height to encourage favorable movement of warm and cold air for comfort of occupants.
- Optimum interface of windows with mechanical heating, cooling and ventilating systems.
- Potential for excessive accumulation of dust, soot and corrosive elements in polluted environments.
- Protection from intruders, whether insect or human.

■ **Building orientation also affects window use and placement.**
- Solar penetration angles in brightness and glare in various seasons and various times of the day.
- Views in each direction may be desirable and/or objectionable; studies have shown that office workers generally prefer a view of trees and natural landscape over other buildings.
- Occupancy and activity patterns of immediate interior spaces are affected in terms of:
  - replacement of artificial light;
  - pattern and quantity (area) of light admitted;
  - heat gain/radiation contributed to the space;
  - transmission of disturbing, even debilitating noise from outside;
  - and the degree of visibility toward the outside.

■ **Sound transmitting properties of windows** may be important when exterior noise is commonly loud. Typically, windows with good air/water infiltration qualities will have good sound attenuation.
- Windows may be tested in accordance with ASTM E413 to determine a particular window's sound transmission class (STC).
- Test results are usually reported at a number of frequencies in the range between 125 and 4000 Hertz.

Safety

■ **Window safety covers issues of basic operation and of maintenance/cleaning.**
- Windows that have ventilators projecting into pedestrian walkways are hazardous because a person could walk into them or children could be playing beneath them and rise up, hitting projecting corners, rails and points.
- Open vents should be balanced or held open so that they cannot close on parts of the body which may be within the window opening. The hardware which provides this measure must be designed and tested to endure other factors, such as wind loads.
- Window operation should also consider tolerances between moving and stationary parts to ensure that knuckles are not scraped and fingers are not pinched.
- Window cleaning becomes a safety issue in high-rise construction. Are window-washing bolts required? Can the windows be cleaned from the inside? Do pivot windows lock in place for washing so that the washer does not lean through

Size Ranges and Limitations

■ Window units are designed and produced by individual manufacturers and fabricators. They vary widely in design, layout, size, and method of manufacture. They do not lend themselves to rigid standardization.
- Individual manufacturers literature should be consulted for this information.
- The Steel Window Institute produces a specifications manual which delineates standard sizes on the basis of window style and frame weight grade.

Window Construction

■ **An important aspect of window design is the relation of the sash to the frame.**
- The performance of the window in terms of air infiltration is based on the lineal footage of crack between the operating sashes and between the sash and frame. Fixed sashes are reated on the basis of area rather than crack.

Window designs which minimize the movement of sash against frame generally provide a tighter seal and longer lasting weatherstripping. Hinged windows (casement, hopper, awning) typically outperform their sliding-sash counterparts (single-hung, double-hung, horizontal sliders). Hinged sashes usually employ interlocking weatherstripping techniques which can last almost indefinitely. Sliding sashes can gradually wear down weatherstripping to the point where performance is questionable. Locked casement and awning units typically have 1/3 to 1/6 the infiltration rate (in cfm/foot of crack) of vertical or horizontal sliding units; when considered in an enclosure such as a house, this may equate to 10 times greater air leakage.

- The construction of the jamb/frame is often determined by sash operation. Sliding sashes will require a wider jamb, as two tracts are provided for sash movement.

- Also, the jamb finish may be determined by sash operation. For instance, to hold double hung sashes open, white vinyl tracks with a concealed spring system may be used to apply pressure. If the sashes and frame are to be finished in white, the tracks are acceptable; for natural finished wood windows and for others of different color, the white tracks may be objectionable.

■ Windows with metal sashes and frames require special consideration to eliminate the possibility of a "thermal wick" from the outside to the inside. In response, the **metal windows incorporate a nonconductive separation of inside and outside surfaces known as a thermal break.** Several methods are available:

Integral Extrusion

Socket Extrusion

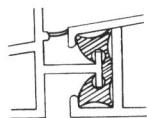

Wrap-around

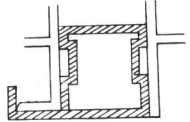

■ **The primary method for reducing air infiltration is to use integrated weatherstripping techniques.** To be an effective barrier to air and moisture, weatherstripping needs to be durable. In order to comply with tests requirements and specifications for prime windows, weatherstripping will be installed by the manufacturer.

- Interlocking vinyl or metal extrusions may be used for sashes that close into one another, such as on hinged and pivot units.
- Spring metal strips are most durable for vertically sliding sashes.
- Some vinyl and rubber extrusions are extruded in tube form with an integrated fastening flange. Some are left hollow and some are fiber filled. The tubing is applied to the sashes so that it is compressed as the window is closed to fill any gaps.
- Pile, set into metal, vinyl or rubber extrusions, is also effective when installed on window frames such that the sash closes down on it or slides over it.
- Similar in application to pile types, closed-cell foam tapes can be used. The foam is generally by adhesive; for replacement, foam attached to a peel-away paper backing protects the adhesive surface until application.

■ **The following characteristics for weatherstripping should be considered:**
- Resistance to air, airborne particles and water;
- Ability to withstand atmospheric conditions;

- Compatibility with frame and sash materials;
- Mechanical and physical properties;
- Resistance of plastic and rubber materials (especially foams) to ultraviolet light;
- Ease and continuity of application;
- Resistance to thermal degradation.

Compatibility of Materials

■ **Consider that concrete, mortars and cements in contact with aluminum have a corrosive effect:**
- Separate all aluminum from cementitious materials with bituminous, zinc chromate or alkali-resistant paints.
- Provide spacers or anchors chemically inert to aluminum.

■ **Whenever metal windows are used, there is a chance of galvanic corrosion due to moisture penetrating protective surfaces or from the reaction of two different connecting metals which are in contact with moisture.**
- Check exterior finishes.
- Check fasteners.

■ In selecting materials and finishes, also consider the environment to which the windows will be exposed.

Emergency Egress

■ **Fire safety concerns may affect the construction, placement and size of windows. Fire code requirements often determine:**
- Window size, such that firemen with backpacks can enter;
- Minimum clear opening area, height and width of egress windows;
- Maximum sill height above floor;
- Ease of operation for rapid sash movement.
- An emergency exit sign may be required to denote an egress window.

■ **Windows may also be tested and rated for particular fire issues:**
- Fire rating may be required for openings in corridor and room partitions or for windows in exterior walls in special locations.
- Fire-resistance rated windows are normally glazed with wired glass 1/4 inch thick. The exposed area of glass and maximum glass dimensions are often restricted.

B2.8 WINDOWS

WINDOWS: selection checklist (cont.)

Detention & Security

■ **Physical security of window units is tested according to ASTM Standards and methods established by the National Institute of Law Enforcement and Criminal Justice.** Window designs are available with different degrees of detention:

- Maximum detention windows will have tool-resistant steel bars at a maximum spacing of 6 inches on center.
- Minimum and moderate detention units can have either mild steel bars or detention screens.
- Steel/aluminum screens are often used to improve security. Screens may be of woven rod mesh or wire cloth.
- Frames are usually metal and incorporate heavy duty security lock and operating hardware as well as concealed hinges and brackets.

■ **If the building being designed requires resistance to forced entry, consider how to classify the different types.** The units are tested for:

- Ability to resist force applied to the locking device by moving sash.
- Resistance to unlocking motion.
- Strength of locking device to resist static load.
- Impact resistance of sash frame.
- Impact resistance of security bars.
- Impact resistance of glazing.

Accessibility for the Disabled

■ **Considerations of barrier-free design for the physically disabled will include several questions:**

- Does window operation present a problem to a wheelchair bound person? - Vertical sliding windows are most difficult to open or close from a sitting position.
- How much force does it take to operate the window? - ANSI A117.1 states that it must be less than 5 pounds.
- Do operating mechanisms require finger dexterity and coordination?
- Are locking devices in reach of a wheelchair bound person? - The height of the operating hardware should not be greater than 54 inches above the finished floor.

Storm Windows

■ **Storm windows may be a single, solid pane of glass or may be a combination of panes and screen.**

- Single pane units may be side or top hinged.
- Combination units commonly have double or triple track frames.
- Mounting is usually to the window jamb.
- Some interior storms are large, single glazed panels which are held in place with turn-clips.

■ **A problem encountered with storm windows is condensation.** This is a result of warmer, moister inside air coming in contact with a cold surface.

- This may occur more often with interior storms.
- Weep holes are required for drainage of potential condensate.

■ **Special features can make the window more attractive and/or more functional. A list of options would include:**

- removable, tilting or other method of adjustment to make window sashes easier to clean;
- storm windows with or without screens;
- multiple locks, remote controls, environmental sensing and response systems and other hardware options;
- leaded, beveled or stained glass;
- decorative grilles, muntins or true divided lights;
- awnings, shades, shutters, canopies;
- safety, tinted and reflective glass;
- insect screens, made of aluminum or fiberglass, to cover the opening of the operating sash; in the case of sliding windows, the screen may be half or whole frame;
- exterior louvered sun screens;
- mini-blinds between glazing layers.

WINDOW GLAZING

NOTE: FRAMING SHOWN IS DIAGRAMATIC ONLY.

SINGLE or DOUBLE GLASS: wet; wet/dry; dry systems

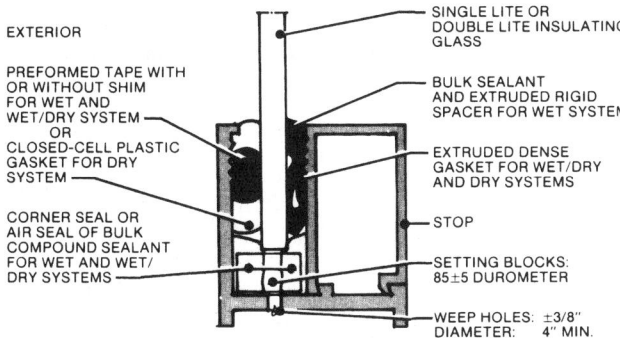

- EXTERIOR
- PREFORMED TAPE WITH OR WITHOUT SHIM FOR WET AND WET/DRY SYSTEM OR CLOSED-CELL PLASTIC GASKET FOR DRY SYSTEM
- CORNER SEAL OR AIR SEAL OF BULK COMPOUND SEALANT FOR WET AND WET/DRY SYSTEMS
- SINGLE LITE OR DOUBLE LITE INSULATING GLASS
- BULK SEALANT AND EXTRUDED RIGID SPACER FOR WET SYSTEM
- EXTRUDED DENSE GASKET FOR WET/DRY AND DRY SYSTEMS
- STOP
- SETTING BLOCKS: 85±5 DUROMETER
- WEEP HOLES: ±3/8" DIAMETER: 4" MIN.

SINGLE GLASS: stopless system

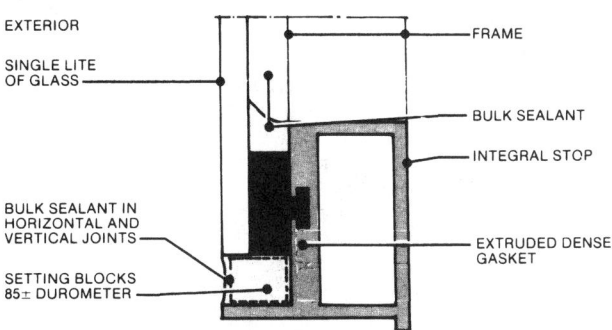

- EXTERIOR
- SINGLE LITE OF GLASS
- BULK SEALANT IN HORIZONTAL AND VERTICAL JOINTS
- SETTING BLOCKS 85± DUROMETER
- FRAME
- BULK SEALANT
- INTEGRAL STOP
- EXTRUDED DENSE GASKET

SINGLE or DOUBLE GLASS: lock-strip system

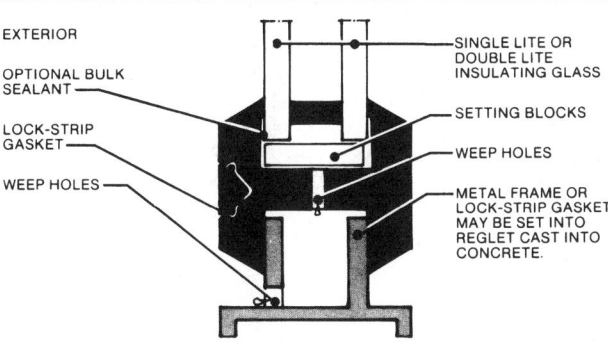

- EXTERIOR
- OPTIONAL BULK SEALANT
- LOCK-STRIP GASKET
- WEEP HOLES
- SINGLE LITE OR DOUBLE LITE INSULATING GLASS
- SETTING BLOCKS
- WEEP HOLES
- METAL FRAME OR LOCK-STRIP GASKET MAY BE SET INTO REGLET CAST INTO CONCRETE.

SINGLE PLASTIC: wet/dry; dry systems

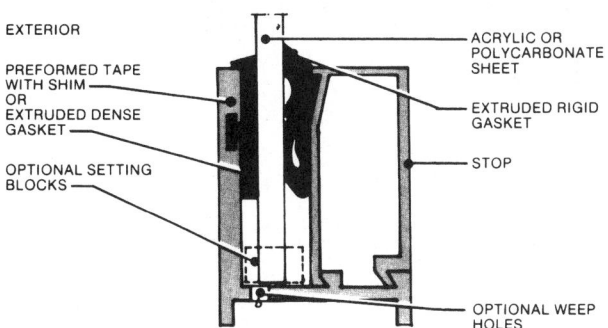

- EXTERIOR
- PREFORMED TAPE WITH SHIM OR EXTRUDED DENSE GASKET
- OPTIONAL SETTING BLOCKS
- ACRYLIC OR POLYCARBONATE SHEET
- EXTRUDED RIGID GASKET
- STOP
- OPTIONAL WEEP HOLES

- ■ Common considerations to all glazing/sealing systems are:
- ● Tapes and gaskets must be placed in compression to ensure watertight seals. Constant uniform pressures of 4 to 10 (maximum) lbs. per inch are recommended.
- ● Outer surfaces of gaskets, tapes and sealant beads should be sloped to provide for water runoff.
- ● Exterior shims, in addition to interior spacers, are recommended for larger glass sizes; follow glass manufacturers' recommendations.
- ● Sealing systems must accommodate tolerances for the squareness of the glass/glazing as well as the clearance between glazing and frame for expansion and contraction.
- ● Weep holes are required for insulating units, laminated and wire glass. They are recommended for all glazing applications.

- ■ There are three basic types of glazing systems:
- ● **Wet systems offer the advantage of adhesion of sealant to glass using gunned-in-place sealants.** Gunable sealant beads should be used at the stop, or air seal, in all pressure-equalized glazing systems.
- ● **Dry systems use preformed rubber (neoprene and EPDM) gaskets which wrap around the glazing layer(s).** They offer installation cost savings, but are often of higher material cost. Edge blocks are recommended while weep holes and drainage allowance should be typically specified.
- ● **Wet/dry systems** are available which combine the advantages of both; i.e. the dry system provides the inner moisture protection and a silicone outer seal provides strength.

- ■ **Stopless systems, also known as structural glazing, use silicone sealants to install glazing to interior structural framing members from the outside.** A high-modulus silicone bulk sealant is used. It must be capable of transferring all loads from the glazing to perimeter framing.
- ● **All glazing materials must be compatible to avoid the possibility of staining.**
- ● High-modulus silicone is also used in butt-glazing, where glazing is retained at the head and sill by the metal frame. The sealant is used in vertical joints and as exterior cap bead at head and sill. Frames at vertical joints may be omitted, or glass stiffeners may be used.

- ■ **Lock-strip systems provide pressure against the glazing by inserting a continuous wedge-like gasket into the main gasket which separates the glazing from the supporting frame.** They are typically molded or extruded rubber shapes which are designed to accommodate single glazing or double glazing with various air space dimensions.
- ● **H-type** gaskets engage both the glass and the surrounding frame. Metal frames should be properly aligned, of constant thickness, and have smooth surfaces to ensure a tight seal.
- ● **Reglet type** gaskets are inserted into a voids which have been cast into the surrounding construction, generally concrete. Alignment of the reglet is critical.
- ● Proprietary types are made for specific extruded framing sections used by metal and vinyl window manufacturers.

- ■ Glazing methods for plastics should:
- ● Allow for free expansion and contraction in plastic sheets as their coefficient of thermal expansion is typically about eight times that of glass and about three times that of aluminum.
- ● Allow sufficient rabbet depth to avoid the possibility of withdrawl of the sheet from frame due to thermal contraction and concurrent wind pressures.
- ● Setting blocks may be required in pressure-equalized systems to prevent blockage of weep holes.

B2.8 WINDOWS

LOUVERS

PERFORMANCE CONSIDERATIONS

First, review the basic performance required of glazing methods — glazing materials must be held in place without being excessively stressed.

More specifically, the glazing system must:

- transfer wind and impact loads to surrounding structure while cushioning glass.
- accommodate thermal expansion and contraction of glazing materials and of frame.
- accommodate temperature differentials caused by various adjoining materials, shading patterns, shading devices.
- prevent water penetration; prevent or minimize air infiltration or ex-filtration.
- create thermal barriers to prevent heat loss through frame and condensation on frame.
- present appearance consistent with design goals.
- retain appearance and function over life span of building anticipated, given the maintenance program planned.
- match any special performance required of the rest of the glazing components.

Consideration which may influence choice of "wet" vs "dry" systems may include:

- initial and replacement costs
- workmanship available since wet systems require better workmanship
- location of glazing joint; locations from least to most demanding are:
- completely interior
- interior joint of exterior light
- protected exterior
- exposed exterior

Glazing material used:

- tinted and reflective-coated glass can heat sealants to high temperatures.
- reflective also increases intensity of ultraviolet radiation on exterior cap beads.

Sealants to be used:

- polysulfides, polyurethanes, and silicones are generally well suited for severe conditions; acrylics and butyls can be used in less demanding locations.

- degree of movement to be accommodated; polysulfides, polyurethanes, and silicones can withstand more joint movement than acrylic or butyl based sealants.
- For additional information, refer to SEALANTS.

Location of sealant bead in glazing joint:

- cap bead - expected to provide primary weather seal; therefore, its adhesive strength, weather resistance, and movement capability will be major properties to consider.
- toe bead - applied on exterior side of channel before glazing material is installed; provides secondary seal under cap bead or preformed tape; weather resistance less critical but adhesive and movement capabilities remain important.
- heel bead - applied on interior side after glazing material is installed; provides seal at removable stop; same considerations as those for toe beads.
  Solvent-release butyls and acrylics should not be used for insulating units as the sealant may not be compatible with the sealing material used for insulating units.

INSTALLATION CONSIDERATIONS

Application temperature range:

- most sealants cannot be installed below 40°F; silicone can be applied at very low temperatures (-20°F) with proper precautions; acrylics may require warming if temperature is below 60°F.
- if sealant bead is installed on hottest day, it will be in constant tension during cold weather; if installed on coldest day, bead will be in constant compression during hot weather. Ideally, sealant should be installed at a temperature about halfway between the extremes.
- tack-free time — affects time available for tooling and clean-up.
- setting or full-cure time; if new sealant bead must attain properties quickly to withstand wind loads or thermal movements, fast curing sealants should be considered.

The basic concept in glazing systems to deal with the requirements cited is floating the glazing material within its frame by means of:

- setting blocks placed under bottom edge of glass.
  material - neoprene rubber for general use; lead for very thick monolithic glass
- spacer shims - to assure proper clearances between face of glazing material and framing channels.

For additional information refer to GLASS/PLASTICS.

COMMON CAUSES OF GLAZING FAILURES:

- Installation done at temperatures below 40°F. (4°C.).
- Failure to properly seal miter and butt joints.
- Sash rabbet not clean and free of contamination.
- Lateral shifting or "walking" of glass.
- Failure to properly bed, cushion, or center the glass.
- Improper glazing system used - not suited for sash design and building conditions.
- Setting blocks used incorrectly or not at all.
- Frames or surrounds out of plane, improperly anchored, and/or out of square.
- Lack of or improper positioning of spacers or edge blocks.
- Damage to sash, rabbet, or stops.

B2.8 WINDOWS

B · BUILDING SHELL

SHADING DEVICES

STATIONARY

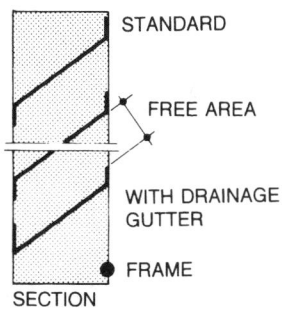

STANDARD

FREE AREA

WITH DRAINAGE GUTTER

FRAME

SECTION

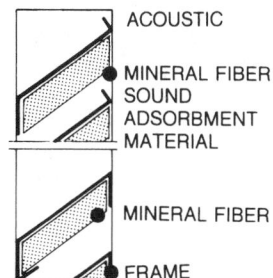

ACOUSTIC

MINERAL FIBER SOUND ADSORBMENT MATERIAL

MINERAL FIBER

FRAME

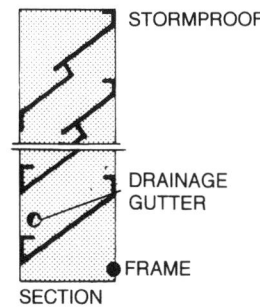

STORMPROOF

DRAINAGE GUTTER

FRAME

SECTION

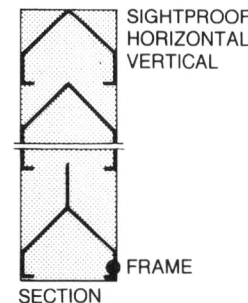

SIGHTPROOF: HORIZONTAL VERTICAL

FRAME

SECTION

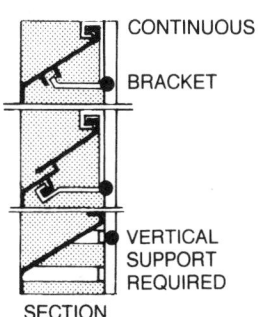

CONTINUOUS

BRACKET

VERTICAL SUPPORT REQUIRED

SECTION

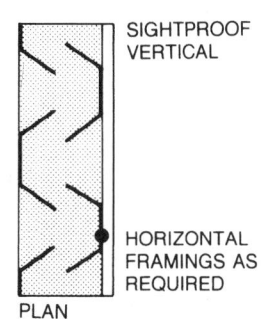

SIGHTPROOF VERTICAL

HORIZONTAL FRAMINGS AS REQUIRED

PLAN

Horizontal louvers may be of unlimited height; depths commonly vary from 4 to 6 inches; deeper louvers have better resistance to rain penetration.

■ **Standard louvers** commonly used where exposure to wind driven rain is not severe; or when a certain amount of leakage is acceptable; will provide maximum free area for passage of air.

• different configurations of blades available.

■ **Stormproof louvers** used for protection against severe weather; but free area is reduced.

• different configurations of blades available.

■ **Continuous:** for horizontal effect; vertical supports hidded behind louvers, with louvers mounted on brackets

• available with standard or stormproof blades

■ **Acoustic:** with sound absorbent materials to reduce ambient and transmitted noise; depth generally 12 inches.

■ **Louvers are generally available** in: extruded aluminum; galvanized steel, sheet aluminum; glass-fiber-reinforced polyester.

• other metals, such as stainless steel, copper, Monel Metal may be available on special order
• finishes available are: mill; coatings; baked-on coatings; anodized.

■ **Method of assembly** is either by welding, or mechanical; welded assemblies generally give better service.

■ **Screens** to protect against insects, birds, vandalism are generally used.

OPERABLE/AUTOMATIC

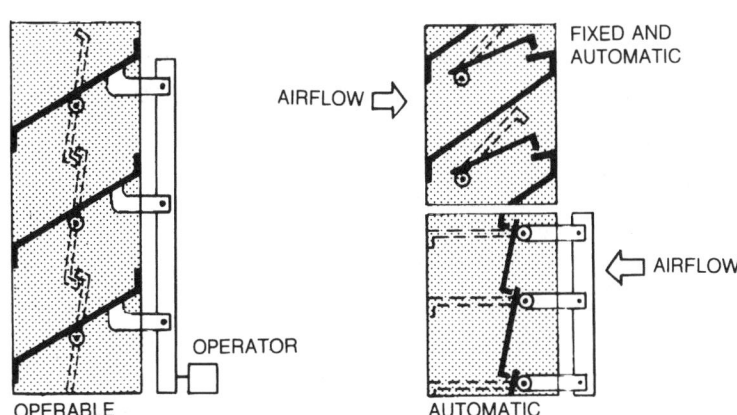

AIRFLOW ⟹

FIXED AND AUTOMATIC

⟸ AIRFLOW

OPERATOR

OPERABLE

AUTOMATIC

■ **Operable louvers** are used to control flow of air through them:

• may be hand, chain, crank, pneumatic, or electric motor operated.
• hand and chain operation commonly for small louvers only.
• when air flow has to be modulated, pneumatic or electric operation should be considered.
• screens are generally available.

■ **Automatic louvers** are used with exhaust or supply fans, or for automatic pressure relief/equalization: commonly in spaces to be kept under specific positive pressure

• louvers may be kept open by means of fusible links, and then automatically closed in case of fire.

B2.8 WINDOWS

SHADING DEVICES (cont.)

TYPE   DESCRIPTION

OVERHANGS, AWNINGS

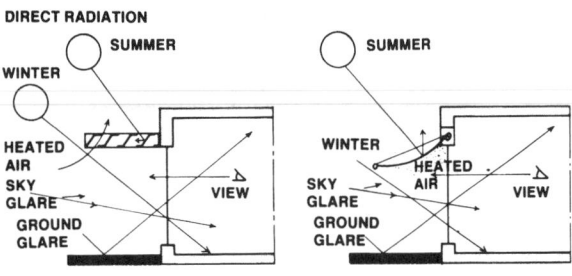

Overhangs may be solid and opaque; solid and translucent to admit diffuse light; or may be louvered:

■ **Solid overhangs** will trap heated air which may increase heat gain into the space;

■ **Louvered overhangs** may have reflective surfaces to admit diffuse light into the interior while excluding direct radiation;

■ **Eggcrate** type louvers may be used to control oblique radiation.

■ **Awnings allow:**
• better control of extent of shading at lower angles of incidence,
• may be retracted to allow entry of winter sun

■ **Shading by fixed overhangs follows the seasons of sun,** not climatic seasons: overhangs intended to shade during mid August, when temperatures are high, will also shade in early May, when solar heat might be welcome.

LOUVERS, horizontal, vertical

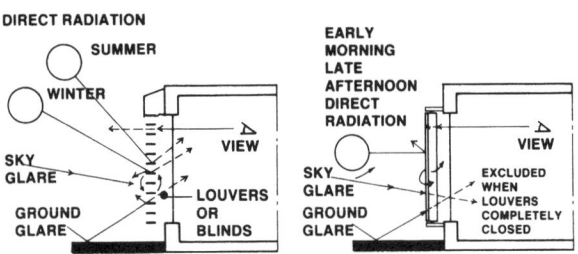

Horizontal louvers may be fixed or adjustable; attached to the structure, or incorporated into window assemblies.

■ **Operable louvers** are manually adjustable from the interior, or power operated.

■ **Power operated** louvers may be heliostat controlled for automatic adjustment.

■ **Operable louvers are effective** in controlling direct radiation, sky glare, and ground glare:
• most effective on south facing openings, but may also be used on east and west facing openings.
• view to the outside may be partially or fully blocked, depending on angle of adjustment.
• may be reflectively coated to admit diffuse light to interior while excluding direct radiation.

■ **Exterior blinds** are similar to manually adjustable louvers in operation, except that they can be retracted when not needed for control.

■ **Vertical louvers** are effective on east and west facing openings; on south facing openings when combined with overhang only. May be fixed or adjustable.

SCREENS, GRILLES

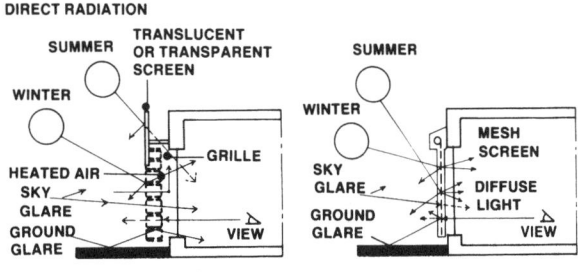

■ **Vertical fixed glass screens** are generally translucent or transparent to reduce heat gain while admitting maximum amount of light. Glass screens may require periodic maintenance to remove accumulated dust.

■ **Wire mesh exterior screens** are intended to absorb some of the heat before it reaches the window and to diffuse visible light in direct radiation:
• may be retractable, but usually available in removable panels.
• essentially similar to insect screens, except that wire used is heavier gauge.
• view to outside partially obstructed at all times.
• cannot be repaired when damaged.

■ **Grilles, or screen walls,** may be:
• prefabricated of metal in various configurations and shading patterns;
• site assembled of hollow clay units, concrete masonry, clay tile;
• are principally used to reduce heat gain through openings, but may also be used to shade solid walls in hot, arid climates.
• grilles will partially obstruct view to the outside.

TYPE

DESCRIPTION

INTERPANE BLINDS, as components of windows

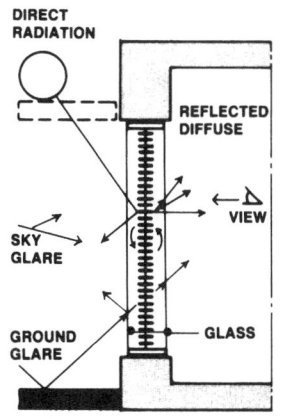
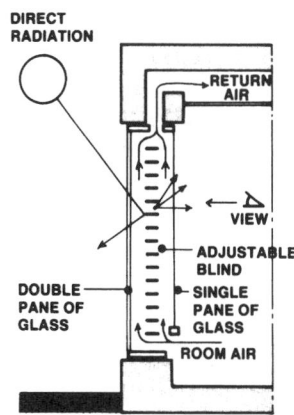

Slat type blinds, fixed or adjustable, may be installed between the panes of a double glazed window:

■ **Direct radiation** transmitted through the outer pane will be absorbed in the blind, with some of the heat then transferred to the enclosed space and the balance to the outside;

■ **Interpane blinds** are a compromise between exterior blinds, which transfer a relatively small amount of heat to the space and interior blinds, which transfer all of it.

■ Size is generally limited, some of the view to outside will be blocked at all times. Maintenance is difficult.

■ **A variation** of interpane blinds used in Europe has the room air flowing between the panes to pick up heat absorbed by the blind. Heated air is then either returned to the central heating system or exhausted to the outside.
• the flow of room air will also deposit dust inside the window assembly and frequent maintenance is generally required.

BLINDS, horizontal, vertical

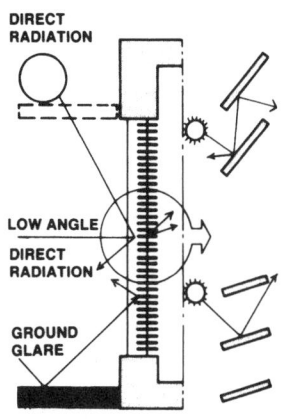
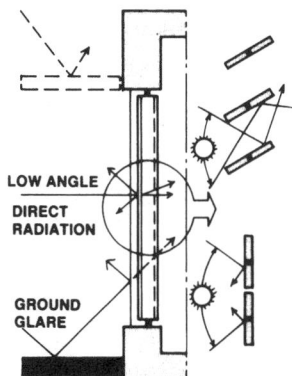

Horizontal slat type blinds can transmit up to 15 percent of direct solar radiation reaching them even when there is no direct sunshine passing between the slats.

■ **The actual amount** will depend on the color of their surface, the angle at which they are set, and the shading mask for the opening.

■ Slightly more than 50 percent of the light transmitted enters the space **in an upward** sloping direction and **is reflected** from the ceiling, adding to the general illumination of the space.

■ **The remainder** enters the space as though the blind itself were a source of diffuse light.

■ **The brightness of the blind** may be controlled by changing reflectivity of slats: top surface may be reflective to increase amount of diffuse light on ceiling surface, while the underside may be dark to reduce brightness.

■ **Light transmission** characteristics of **vertical** blinds are similar to those of horizontal ones.

■ **The differences are:**
• when reflective surfaces are used resultant glare may cause visual discomfort to some of the occupants,
• light distribution is not as good as for horizontal blinds.

B2.8 WINDOWS

SHADING DEVICES (cont.)

TYPE   DESCRIPTION

SHADES, SHUTTERS, DRAPES, SCREENS

Shades are usually opaque, made of flexible materials:

■ **Translucent, partially translucent, and insulating** types are available, such as laminated film and fabric, insulation faced with fabric, multiple layers of fabric, inflatable with reflective outer surface.

■ **Shades are easily adjusted** for different angles of insolation, but will completely block view to outside when in down position. Periodic replacement may be required.

■ **Drapes,** like shades, do not reflect light into the space & therefore do not increase the level of illumination as blinds do. Drapes can be easily adjusted by the occupants to suit outside conditions.

■ **Louvered screens** consist of narrow fixed metal slats held in position with wires. The fabric is flexible and requires a frame to hold it in place. The frames may be stationary, sliding, or swinging. View to outside partially obstructed at all times.

■ **Shutters may be** of rigid plastic, either solid, or with fixed or adjustable louvers. Louvered interior types are similar to blinds in function and use. Sizes generally limited. Insulating types are available.

■ **Shutters** are also available for **exterior use,** solid and louvered.

GLASS BLOCKS, INSULATING WINDOW

■ **Prismatic glass blocks may be used** to admit and diffuse light to the interior. Their use is limited, because:
• interior surfaces of blocks tend to get quite hot when exposed to direct radiation,
• visual contact with outside is obstructed,
• heat flow cannot be controlled.

■ **Assemblies using heat absorbing** or heat reflecting glazing as well as glass blocks have the disadvantage of reducing heat flow into the interior also when such heat may be wanted. For additional information, refer to GLASS/PLASTICS.

■ **An insulating window is** a double glazed assembly in which the air space can be filled with loose-fill polystyrene beads to reduce heat flow through the assembly:
• the beads are blown into the assembly, and sucked back into a storage container through a reversible suction pump.
• assembly may be filled to different levels.
• view is completely obstructed when assembly is filled.
• may be installed at angles down to 45 degree.
• maintenance of mechanical system is required;
• in-service record has not yet been fully established.

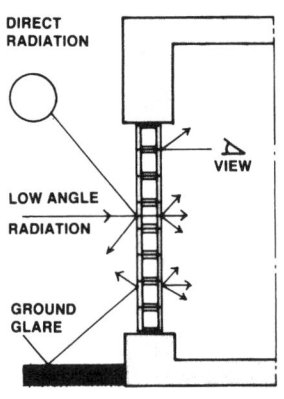

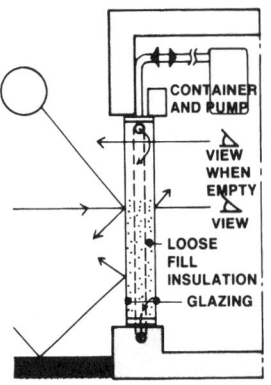

B2.8 WINDOWS

B • BUILDING SHELL

TYPE	DESCRIPTION

GLASS, heat absorbing

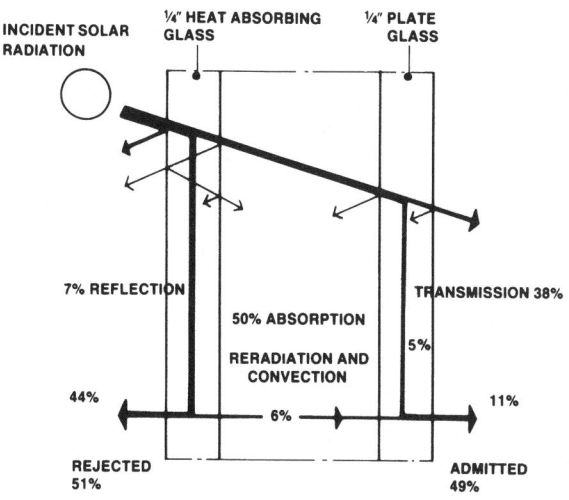

SHADING COEFFICIENT: 0.55
COEFFICIENT OF LIGHT TRANSMITTANCE: 0.45

Shading devices outside, within or inside glazed openings function to control solar radiation either before or after it is transmitted through the glazing.

■ **They are useful when** the flow of radiation into a space is to be controlled, but offer little or **no benefit in reducing heat flow out** of the space during the heating season.

■ Transmission of solar radiation into a space **may also be controlled by the glazing itself,** and the glazing may be arranged to minimize heat flow into and out of the space.

■ Transmission of solar radiation **through glass can be modified by:**
• increasing or decreasing surface reflectivity
• changing the chemical composition of glass to vary its absorptance characteristics.

■ **One approach commonly used** is to increase the absorptance of infrared radiation to the highest degree, with smallest decrease in transmittance of the visible band.

■ **A second pane of glass** may be incorporated into the assembly to reduce heat flow.

GLASS, heat reflecting

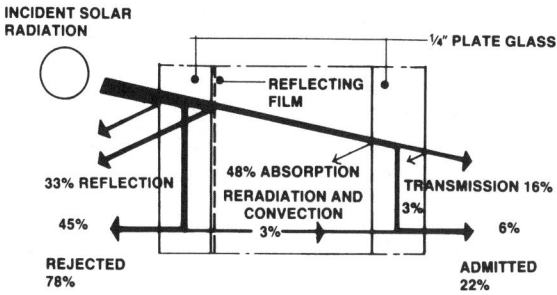

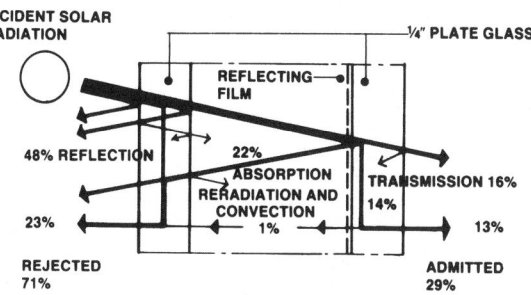

SHADING COEFFICIENT: 0.25
COEFFICIENT OF LIGHT TRANSMITTANCE: 0.35

Another approach also commonly used is to increase surface reflectivity by the addition of a thin, highly reflective metallic film on the surface of a glazing unit in a single or double pane assembly.

■ **Glass thus treated is referred to as heat reflecting:** it is more transparent to the visible than to the infrared band.

■ **In a double pane arrangement,** the film may be located on the surface of the outer or inner pane.

■ **Reflective film on the inside** surface of the outer pane will result in maximum reduction of total transmittance of heat, but the outer pane gets quite hot under direct radiation, increasing chances of thermal breakage.

■ **Locating the film on the outer** surface of the inner pane will reduce the possibility of such breakage.

■ **In double pane assemblies,** the air space functions to reduce heat transfer through them. For additional information on heat flow through air spaces, refer to ENVIRONMENT.

■ For additional information on insulating, heat absorbing, and reflective glass, refer to GLASS/PLASTICS.

B2.8 WINDOWS

RESIDENTIAL WINDOWS

	Description	Heating Climate	Mixed Climate	Cooling Climate
Heat Flow (U-value) 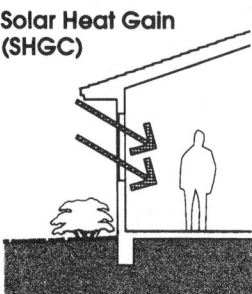	The rate of heat transfer is indicated in terms of the U-value (U-factor) of a window assembly. The insulating value is indicated by the R-value, which is the inverse of the U-value. The lower the U-value, the greater a window's resistance to heat flow and the better its insulating value.	PRIMARY FACTOR: A low U-value is the most important window property in cold climates.	PRIMARY FACTOR: A low U-value is one important window property in mixed climates.	A low U-value is helpful during hot days or whenever heating is needed, but it is less important than SHGC in warm climates.
Solar Heat Gain (SHGC) 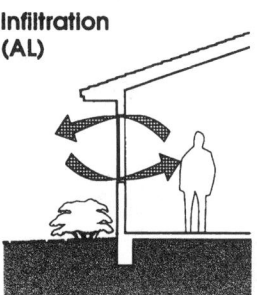	The SHGC is the fraction of incident solar radiation admitted through a window, both directly transmitted, and absorbed and subsequently released inward. SHGC is expressed as a number between 0 and 1. The lower a window's SHGC, the less solar heat it transmits.	A high SHGC increases passive solar gain for heating, but reduces cooling season performance. A low SHGC improves cooling season performance, but reduces passive solar heating.	PRIMARY FACTOR: A low SHGC is one important window property in mixed climates.	PRIMARY FACTOR: A low SHGC is the most important window property in warm climates.
Infiltration (AL) 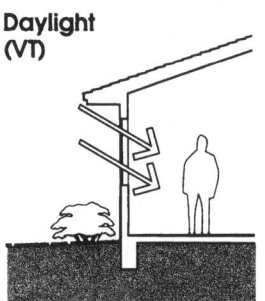	Heat loss and gain occur by infiltration through leaks in the window assembly. The air leakage rating (AL) is expressed as cubic feet of air passing through an equivalent square foot of window area. The lower the AL, the less air will pass through leaks in the window assembly.	The air leakage rating (AL) is an important window property in cold climates. Air leakage should not exceed 0.56 cfm/sq ft.	The air leakage rating (AL) is an important window property in mixed climates. Air leakage should not exceed 0.56 cfm/sq ft.	The air leakage rating (AL) is generally less important in warm climates. However, infiltration can contribute to excessive summer cooling loads by introducing humid outdoor air.
Daylight (VT)	The visible transmittance (VT) is an optical property that indicates the amount of visible light transmitted through the glass. VT is expressed as a number between 0 and 1. The higher the VT, the more daylight is transmitted.	A high VT is desirable to maximize daylight and view.	A high VT is desirable to maximize daylight and view.	A high VT is desirable to maximize daylight and view, but this must be balanced against the need to control solar gain and glare in hot climates.

Fig. 3. Energy-related properties of windows. Source: Carmody, Selkowitz and Heschong (1996)

B2.8 WINDOWS

B • BUILDING SHELL

RESIDENTIAL WINDOWS

Window units are one of the most important components affecting energy performance and comfort in residential buildings. New window technologies provide improved performance and an array of design options. This article provides an introduction to the energy-related aspects of windows and provides guidelines for window selection in different U. S. climates.

Choosing a window involves many considerations related to aesthetics, function, energy performance, and cost. This article focuses primarily on the energy performance considerations in comparing windows. First, recent technological advances are identified, followed by descriptions of the energy-related properties of windows, condensation potential, and window rating systems. Then, window selection based on energy performance is discussed in more detail for different U.S. climate zones.

Technological improvements

Some technological innovations appearing in today's window products are described briefly below.

■ Glazing unit structure.

Multiple layers of glass or plastic films improve thermal resistance and reduce the heat loss attributed to convection between window layers. Additional layers also provide more surfaces for low-E or solar control coatings.

■ Low-emittance coatings.

Low-emittance or low-E coatings are highly transparent and virtually invisible, but have a high reflectance (low emittance) to long-wavelength infrared radiation. This reduces long-wavelength radiative heat transfer between glazing layers by a factor of 5 to 10, thereby reducing total heat transfer between two glazing layers. Low-emittance coatings may be applied directly to glass surfaces, or to thin sheets of plastic (films) which are suspended in the air cavity between the interior and exterior glazing layers. In effect, a window with a low-E coating can transmit a significant amount of daylight as well as passive solar heat gain, while signif-

icantly reducing heat loss.

■ Low-conductance gas fills.

With the use of a low-emittance coating, heat transfer across a gap is dominated by conduction and natural convection. While air is a relatively good insulator, there are other gases (such as argon, krypton, and carbon dioxide) with lower thermal conductivity. Using one of these nontoxic gases in an insulating glass unit can reduce heat transfer between the glazing layers.

■ Solar control glazing and coatings.

To reduce cooling loads, new types of tinted glass and new coatings can be specified that reduce the impact of the sun's heat without sacrificing view. Spectrally selective glazing and coatings absorb and reflect the infrared portion of sunlight while transmitting visible daylight, thus reducing solar heat gain coefficients and the resulting cooling loads. These solar control coatings can also have low-emittance characteristics. In effect, a window with a spectrally selective coating or tint can significantly reduce solar heat gain while providing more daylight than traditional reflective or tinted glazing.

■ Warm edge spacers.

Heat transfer through the metal spacers that are used to separate glazing layers can increase heat loss and cause condensation to form at the edge of the window. "Warm edge" spacers use new materials and better design to reduce this effect.

■ Thermally improved sash and frame.

Traditional sash and frame designs contribute to heat loss and can represent a large fraction of the total loss when high-performance glass is used. New materials and improved designs can reduce this loss.

■ Improved weather-stripping.

Better weather-strips are now available to reduce air leakage, and most are of more durable materials that will provide improved performance over a longer time period.

ENERGY RELATED PROPERTIES OF WINDOWS

Heat flows through a window assembly in three ways: conduction, convection, and radiation. When these basic mechanisms of heat transfer are applied to the performance of windows, they interact in complex ways. Three energy performance characteristics of windows are used to portray how energy is transferred and a fourth indicates the amount of daylight transmitted

■ Heat flow

When there is a temperature difference between inside and outside, heat is transferred through the window frame and glazing by the combined effects of conduction, convection, and radiation. This is indicated in terms of the U-factor of a window assembly. It is expressed in units of Btu/hr-sq. ft-F (W/sq. m-°C). The U-factor may be expressed for the glass alone or the entire window, which includes the effect of the frame and the spacer materials. The lower the U-factor, the greater a window's resistance to heat flow. A window's insulating value is indicated in terms of its R-value, which is the reciprocal of U-value.

■ Heat gain from solar radiation

Regardless of outside temperature, heat can be gained through windows by direct or indirect solar radiation. The ability to control this heat gain through windows is indicated in terms of the solar heat gain coefficient (SHGC). The SHGC is the fraction of incident solar radiation admitted through a window, both directly transmitted, and absorbed and subsequently released inward. The solar heat gain coefficient has replaced the shading coefficient as the standard indicator of a window's shading ability. It is expressed as a number between 0 and 1. The lower a window's solar heat gain coefficient, the less solar heat it transmits, and the greater its shading ability.

■ Infiltration

Heat loss and gain also occur by infiltration through cracks in the window assembly. This effect is measured in terms of the amount of air (cubic feet or meters per minute) that passes

B2.8 WINDOWS

RESIDENTIAL WINDOWS (cont.)

through a unit area of window (square foot or meter) or window perimeter length (foot or meter) under given pressure conditions. It is indicated by an air leakage rating (AL). In reality, infiltration varies with wind-driven and temperature-driven pressure changes. Infiltration also contributes to summer cooling loads in some climates by raising the interior humidity level.

■ Visible transmittance

Visible transmittance (VT) is an optical property that indicates the amount of visible light transmitted through the glass. Although VT does not directly affect heating and cooling energy use, it is used in the evaluation of energy-efficient windows. For example, two windows may have similar solar heat gain control properties, however one may transmit more daylight as indicated by the visible transmittance. The visible transmittance may then be the basis for choosing one window over another. Specifically, VT is the percentage or fraction of the visible spectrum (380 to 720 nanometers) weighted by the sensitivity of the eye that is transmitted through the glazing. The higher the VT, the more daylight is transmitted.

■ Condensation potential

Reducing the risk of condensation on windows is an important aspect of selecting a window. Fig. 2 shows condensation potential on glazing (center of glass) at various outdoor temperature and indoor relative humidity conditions. Condensation can occur at any points that fall on or above the curves. (Note: All air spaces are 1/2 inch; all coatings are e = 0.10).

WINDOW RATING SYSTEMS

The National Fenestration Rating Council (NFRC) was established in 1989 to develop a fair, accurate, and credible rating system for fenestration products. This was in response to the technological advances and increasing complexity of these products, which manufacturers wanted to take credit for but which cannot be easily visually verified. NFRC has developed a window energy rating system based on whole product performance. This accurately accounts for the energy-related effects of all the products' component parts, and prevents information about a single com-

ponent from being compared in a misleading way to other whole product properties. At this time, NFRC labels on window units give ratings for U-value, solar heat gain coefficient, and visible light transmittance. Soon labels will include air infiltration rates and an annual heating and cooling rating. NFRC procedures started to be incorporated in state energy codes in 1992. The 1992 National Energy Policy Act provided for the development of a national rating system. The U. S. Department of Energy has selected the NFRC program and certified it as the national rating system. In addition, the NFRC procedures are now referenced in and being incorporated into the Model Energy Code and ASHRAE Standards 90.1 and 90.2.

SELECTING AN ENERGY EFFICIENT WINDOW

One important practical reason to select energy efficient windows is to reduce the annual cost of heating and cooling your home. This makes good economic sense for most building owners and it also contributes to national and global efforts to reduce the environmental impacts of non-renewable energy use. It can be a relatively painless and even profitable way for every family to help improve the environment in which we live. In order to select a window, which will lower heating and cooling costs, you first need to estimate how much energy the furnace and air conditioner will consume. This is influenced not only by the window properties as you would expect, but by a series of other factors including the house location and microclimate, house characteristics, occupant use patterns, and cost of energy.

To evaluate windows with respect to energy performance, various types of information and tools are available:

• Evaluate the window based on its energy-related properties applied to your climate.

• Use an annual energy performance rating system to evaluate heating and cooling energy use.

• Use a computer program to compare energy use and utility costs.

Selecting a window based on energy performance may involve two additional considerations—the impact on peak heating and cooling loads, and the long-term ability of the window unit to maintain its energy performance characteristics.

USING THE BASIC ENERGY-RELATED PROPERTIES

The three key properties are U-factor, solar heat gain coefficient (SHGC), and air leakage rating. Visible transmittance (VT) is another property used in comparing windows. These are the first properties to appear on NFRC window labels (U-factor was introduced first, followed by SHGC and VT; air leakage ratings will be added soon).

The accompanying table Energy-related properties of windows indicate guidelines for using the basic energy properties in choosing a window. Note that these guidelines are different for distinct climate regions. Until there is a reliable annual performance rating in place or unless one is using a computer program, these properties are the main basis for making energy performance decisions.

WINDOW SELECTION IN AN UNDERHEATED CLIMATE

Using a computer simulation program, four possible window choices are compared for a typical house in Madison, Wisconsin. In addition, the energy use for the same house with poor windows is shown in the first bar of the table Comparing windows in an underheated climate. to provide a comparison to an older, existing structure with single-glazed windows. **Window A** in the table is a typical clear, double-glazed unit—the most common cold-climate window installed in the U. S. during the period from 1970 to about 1985. **Window B** has a high-transmission low-E coating, while **Window C** has a spectrally selective low-E coating. Window B is designed to reduce winter heat loss (low U-factor) and provide winter solar heat gain (high SHGC). Window C also reduces winter heat loss (low U-factor) but it reduces solar heat gain as well (low SHGC). **Window D,** with triple glazing and two low-E coatings, is representative of the most efficient window on the market today with respect to winter heat loss (very low U-factor).

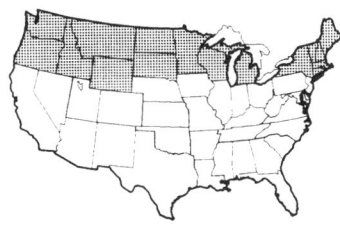

U.S. Cooling Climate Zone.
(Boundaries of the climate zones are approximate.)

A. Clear double glazing

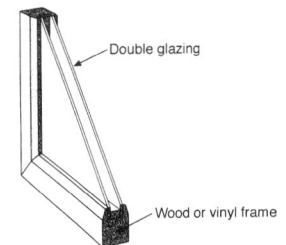

Double glazing

Wood or vinyl frame

B. High-transmission low-E coating

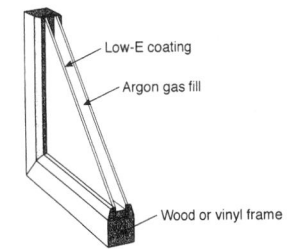

Low-E coating

Argon gas fill

Wood or vinyl frame

C. Spectrally selective low-E coating

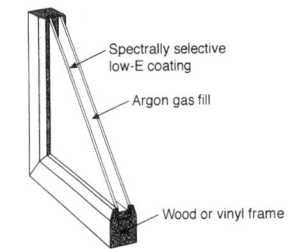

Spectrally selective low-E coating

Argon gas fill

Wood or vinyl frame

D. Triple glazing with low-E coating

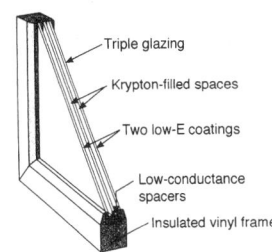

Triple glazing

Krypton-filled spaces

Two low-E coatings

Low-conductance spacers

Insulated vinyl frame

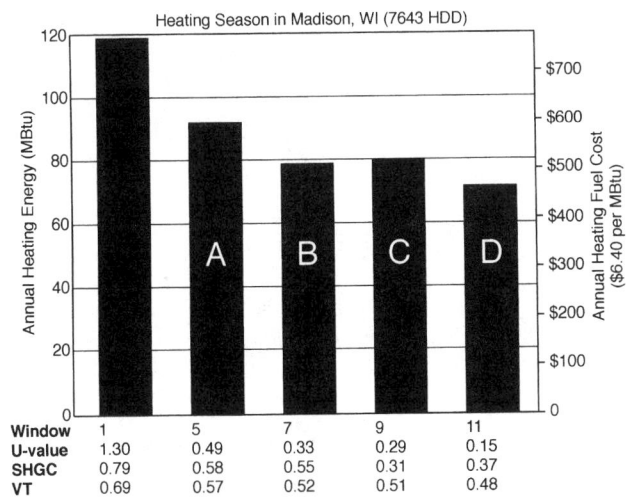

Heating Season in Madison, WI (7643 HDD)

Window	1	5	7	9	11
U-value	1.30	0.49	0.33	0.29	0.15
SHGC	0.79	0.58	0.55	0.31	0.37
VT	0.69	0.57	0.52	0.51	0.48

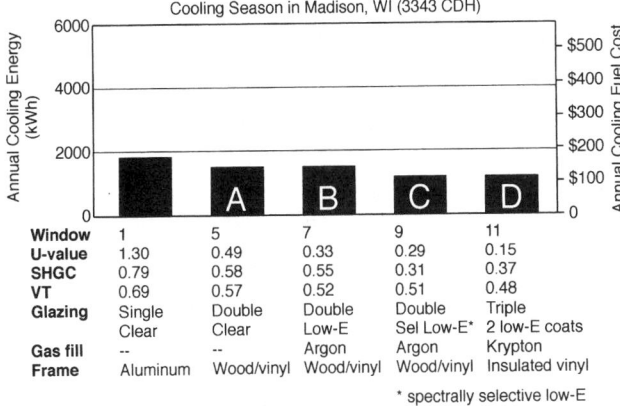

Cooling Season in Madison, WI (3343 CDH)

Window	1	5	7	9	11
U-value	1.30	0.49	0.33	0.29	0.15
SHGC	0.79	0.58	0.55	0.31	0.37
VT	0.69	0.57	0.52	0.51	0.48
Glazing	Single Clear	Double Clear	Double Low-E	Double Sel Low-E*	Triple 2 low-E coats
Gas fill	--	--	Argon	Argon	Krypton
Frame	Aluminum	Wood/vinyl	Wood/vinyl	Wood/vinyl	Insulated vinyl

* spectrally selective low-E

Note: The annual energy performance figures shown here are for a typical 1540 sq ft house with 231 sq ft of window area (15% of floor area). The windows are equally distributed on all four sides of the house and are unshaded. U-factor, SHGC, and VT are for the total window including frames. HDD=heating degree days. CDH=cooling degree hours. **kWh-kilowatt hours. Mbtu=millions of Btu. The fuel and electricity prices represent national averages.**

Fig. 4. Comparing windows in an underheated climate. Source: Carmody, Selkowitz and Heschong (1996)

B2.8 WINDOWS

RESIDENTIAL WINDOWS (cont.)

U.S. Cooling Climate Zone.
(Boundaries of the climate zones are approximate.)

A. Clear double glazing

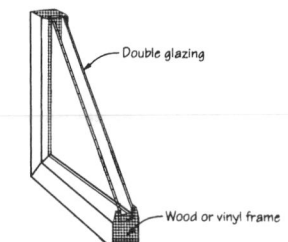

Double glazing

Wood or vinyl frame

B. Double glazing with bronze tint

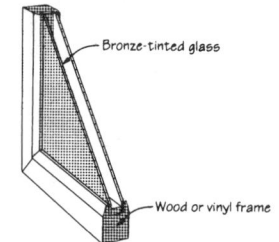

Bronze-tinted glass

Wood or vinyl frame

C. Spectrally selective low-E coating

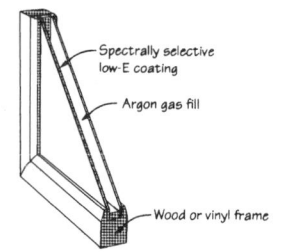

Spectrally selective low-E coating

Argon gas fill

Wood or vinyl frame

D. Spectrally selective low-E coating with tinted glass

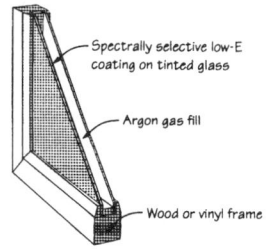

Spectrally selective low-E coating on tinted glass

Argon gas fill

Wood or vinyl frame

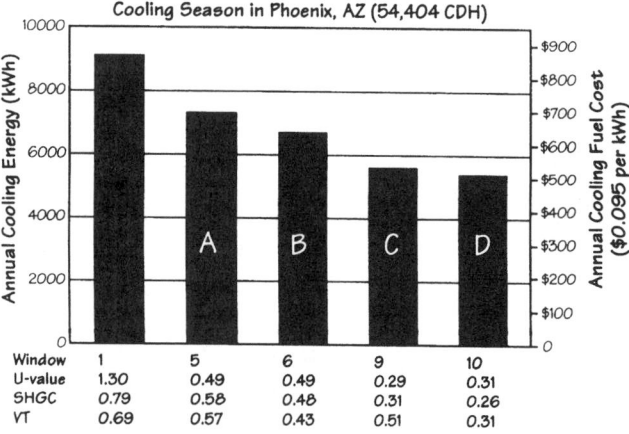

Cooling Season in Phoenix, AZ (54,404 CDH)

Window	1	5	6	9	10
U-value	1.30	0.49	0.49	0.29	0.31
SHGC	0.79	0.58	0.48	0.31	0.26
VT	0.69	0.57	0.43	0.51	0.31

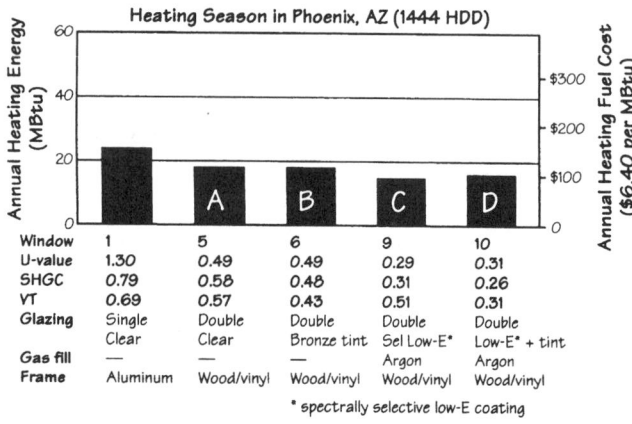

Heating Season in Phoenix, AZ (1444 HDD)

Window	1	5	6	9	10
U-value	1.30	0.49	0.49	0.29	0.31
SHGC	0.79	0.58	0.48	0.31	0.26
VT	0.69	0.57	0.43	0.51	0.31
Glazing	Single Clear	Double Clear	Double Bronze tint	Double Sel Low-E*	Double Low-E* + tint
Gas fill	—	—	—	Argon	Argon
Frame	Aluminum	Wood/vinyl	Wood/vinyl	Wood/vinyl	Wood/vinyl

* spectrally selective low-E coating

Note: The annual energy performance figures shown here are for a typical 1540 sq ft house with 231 sq ft of window area (15% of floor area). The windows are equally distributed on all four sides of the house and are unshaded. U-factor, SHGC, and VT are for the total window including frames. HDD=heating degree days. CDH=cooling degree hours. kWh=kilowatt hours. Mbtu=millions of Btu. The fuel and electricity prices **represent national averages.**

Fig. 5. Comparing windows in an overheated climate. Source: Carmody, Selkowitz and Heschong (1996)

WINDOW SELECTION IN AN OVERHEATED CLIMATE

Similar to the heating climate example above, a computer simulation program is used to compare four possible window choices for a typical house in Phoenix, Arizona. Again, the energy use for the same house with poor windows is shown in the first bar of Fig. 5 to provide a comparison to an older, existing structure with single-glazed windows. **Window A** in Fig. 5 is a typical clear, double-glazed unit. **Window B,** with bronze-tinted glass, represents a traditional approach to reducing solar heat gain (note the somewhat reduced SHGC accompanied by a significant reduction in daylight—lower VT). **Window C** represents the relatively _new technology of using a spectrally selective low-E coating (a low SHGC combined with a relatively high VT). **Window D,** which combines a spectrally selective low-E coating with tinted glass, represents further reduction in summer heat gain (very low SHGC), but at the cost of losing daylight as well (low VT).

The accompanying table, **Comparing windows in an overheated climate,** illustrates that there are significant savings in annual cooling costs by using windows with low solar heat gain coefficients (Windows C and D) instead of double-glazed, clear units or traditional bronze-tinted glass (Windows A and B). Savings are even greater when compared to the single-glazed case, which is common in many existing homes of warmer regions. Windows C and D, with their low U-values, also reduce heating costs in a warm climate where there is some heating required.

As noted for the overheated climate example, consider that this is for a relatively small house (1500 sq. ft) with an average amount of window area (231 sq. ft). The fuel and electricity rates shown on the figures are national averages.

WINDOW SELECTION IN A MIXED CLIMATE

The previous two examples have focused on the regions of more extreme climate in the United States. In terms of analyzing energy performance, these climates are easier to address because one season clearly predominates, so decisions are clearly weighted in favor of winter heating in the north and summer cooling in the south. The great area in between is often referred to as a mixed heating and cooling or temperate climate zone. In these cases, the relative importance of the heating and cooling season performance will vary with location, and utility costs.

The comments for the previous heating and cooling climate examples all apply to some degree in a mixed climate.

DEFINITIONS

ABSORPTANCE. The ratio of radiant energy absorbed to total incident radiant energy in a glazing system.

EMITTANCE. The ratio of the radiant flux emitted by a specimen to that emitted by a blackbody at the same temperature and under the same conditions.

AIR LEAKAGE RATING. A measure of the rate of infiltration around a window or skylight in the presence of a specific pressure difference. It is expressed in units of cubic feet per minute per square foot of window area (cfm/sq. ft) or cubic feet per minute per foot of window perimeter length (cfm/ft). The lower a window's air leakage rating, the better its airtightness.

COMPOSITE FRAME. A frame consisting of two or more materials—for example, an interior wood element with an exterior fiberglass element.

DOUBLE GLAZING. In general, two thicknesses of glass separated by an air space within an opening to improve insulation against heat transfer and/or sound transmission. In factory-made double glazing units, the air between the glass sheets is thoroughly dried and the space is sealed airtight, eliminating possible condensation and providing superior insulating properties.

GAS FILL. A gas other than air, usually argon or krypton, placed between window or skylight glazing panes to reduce the U-factor by suppressing conduction and convection.

LIGHT-TO-SOLAR-GAIN-RATIO. A measure of the ability of a glazing to provide light without excessive solar heat gain. It is the ratio between the visible transmittance of a glazing and its solar heat gain coefficient. Abbreviated LSG.

LOW CONDUCTANCE SPACERS. An assembly of materials designed to reduce heat transfer at the edge of an insulating window. Spacers are placed between the panes of glass in a double- or triple-glazed window.

LOW-EMITTANCE (LOW-E) COATING. Microscopically thin, virtually invisible, metal or metallic oxide layers deposited on a window or skylight glazing surface primarily to reduce the U-factor by suppressing radiative heat flow. A typical type of low-e coating is transparent to the solar spectrum (visible light and short-wave infrared radiation) and reflective of long-wave infrared radiation.

R-VALUE. A measure of the resistance of a glazing material or fenestration assembly to heat flow. It is the inverse of the U-factor ($R = 1/U$) and is expressed in units of hr-sq ft-F/Btu. A high-R-value window has a greater resistance to heat flow and a higher insulating value than one with a low R-value.

REFLECTANCE. The ratio of reflected radiant energy to incident radiant energy.

SHADING COEFFICIENT (SC). A measure of the ability of a window or skylight to transmit solar heat, relative to that ability for 1/8-inch clear, double-strength, single glass. It is being phased out in favor of the solar heat gain coefficient, and is approximately equal to the SHGC multiplied by 1.15. It is expressed as a number without units between 0 and 1. The lower a window's solar heat gain coefficient or shading coefficient, the less solar heat it transmits, and the greater is its shading ability.

SOLAR HEAT GAIN COEFFICIENT (SHGC). The fraction of incident solar radiation admitted through a window or

B2.8 WINDOWS

RESIDENTIAL WINDOWS (cont.)

skylight, both directly transmitted, and absorbed and subsequently released inward. The solar heat gain coefficient has replaced the shading coefficient as the standard indicator of a window's shading ability. It is expressed as a number between 0 and 1. The lower a window's solar heat gain coefficient, the less solar heat it transmits, and the greater its shading ability. SHGC can be expressed in terms of the glass alone or can refer to the entire window assembly.

SPECTRALLY SELECTIVE GLAZING. A coated or tinted glazing with optical properties that are transparent to some wavelengths of energy and reflective to others. Typical spectrally selective coatings are transparent to visible light and reflect short-wave and long-wave infrared radiation. Usually the term spectrally selective is applied to glazing that reduces heat gain while providing substantial daylight.

SUPERWINDOW. A window with a very low U-factor, typically less than 0.15, achieved through the use of multiple glazing, LOW-E coatings, and gas fills.

TRANSMITTANCE. The percentage of radiation that can pass through glazing. Transmittance can be defined for different types of light or energy, that is, visible light transmittance, UV transmittance, or total solar energy transmittance.

U-FACTOR (U-VALUE). A measure of the rate of non-solar heat loss or gain through a material or assembly. It is expressed in units of Btu/hr-sq ft-F (W/sq m-°C). Values are normally given for NFRC/ASHRAE winter conditions of 0F (18° C) outdoor temperature, 70F (21° C) indoor temperature, 15 mph wind, and no solar load. The U-factor may be expressed for the glass alone or the entire window, which includes the effect of the frame and the spacer materials. The lower the U-factor, the greater a window's resistance to heat flow and the better its insulating value.

VISIBLE TRANSMITTANCE (VT). The percentage or fraction of the visible spectrum (380 to 720 nanometers) weighted by the sensitivity of the eye that is transmitted through the glazing.

REFERENCES:

John Carmody and Stephen Selkowitz *"Residential Windows"*: in Reference 1

Carmody, John, Stephen Selkowitz, and Lisa Heschong. 1996. *"Residential Windows: New Technologies and Energy Performance"*. New York: W.W. Norton & Company.

B2.8 WINDOWS

REFERENCES: Windows

STANDARDS

ASTM Specifications for:

D4099
- Poly (Vinyl Chloride) (PVC) Prime Windows.

D4216
- Rigid Poly (Vinyl Chloride) (PVC) and Related Plastic Building Products Compounds.

ASTM Tests/Methods for:

D4226
- Impact Resistance of Rigid Poly (Vinyl Chloride) (PVC) Building Products.

E163
- Fire Test of Window Assemblies.

E283
- Rate of Air Leakage through Exterior Windows, Curtain Walls and Doors.

E330
- Structural Performance of Exterior Windows, Curtain Walls and Doors by Uniform Static Air Pressure Difference.

E331
- Water Penetration of Exterior Windows, Curtain Walls and Doors by Uniform Static Air Pressure Difference.

E413
- Determination of Sound Transmission Class

E1017
- Performance Specification for Exterior Residential Window Assemblies.

F588
- Resistance of Window Assemblies to Forced-Entry, "Excluding-Glazing"

ANSI Specifications:

A39.1
- Safety Requirements for Window Cleaning

A134.1
- Specifications for Aluminum Windows (AAMA 302.9)

ANSI/NWMA 1.S.2
- Standard for Wood Window Units

ANSI/NFPA 80
- Standard for Fire Doors and Windows

ANSI/ASTM E163
NFPA 257/ UL 9
- Methods of Fire Tests of Window Assemblies

ANSI A58.1
- Design Loads for Buildings and Other Structures.

Federal Specifications:

08520
- Corps of Engineers - Guide Specifications: Aluminum Windows, Nov 1980

GSH-8.6a
- Department of Defense - Guide Specifications for Military Family Housing: Aluminum Windows and Jalousies, Jul. 1970

TS-08520
- Department of Navy, Naval Facilities Engineering Command - (NAVFAC) Type Specifications: Aluminum Windows, Nov. 1971

08-1
- Federal Construction Guide Specifications: Windows, Nov. 1971

8G3
- Production Systems for Architects & Engineers, Inc. - MASTERSPEC: Aluminum Windows, Nov. 1978 Veterans Administration - Office of Construction - Master Construction (Guide) Specifications: Aluminum Windows (Double Glazed) Aluminum Windows Aluminum Storm Windows

RR-W-365(1) Wire Fabric (Insect Screening)

FS L-S-125B Federal Specification Screening, Insect, Non-Metallic

SELECTED REFERENCES

AA — **Aluminum Association**
- Aluminum Curtain Wall Series: Anodic and Applied Exterior Finsihes; Guide Lines for Specifying Aluminum Work.
- Aluminum Standards and Data
- Designation System for Aluminum Finishes
- Drafting Standards for Aluminum Extruded.

SAA-46-78
- Standards for Anodized Architectural Aluminum

AAMA — **Architectural Aluminum Mfg. Assoc.**
- Aluminum Windows for Every Home
- Combination Aluminum Insulating
- Thermalized Aluminum Windows for Every Home
- Energy Efficient Windows

1302
- Voluntary Specifications for Forced Entry Resistant Aluminum Prime Windows

1502
- Voluntary Test Method for Condensation Resistance of Windows, Doors and Glazed Wall Sections

1503
- Voluntary Test Method of Thermal Transmittance of Windows, Doors and Glazed Wall Sections

1504
- Voluntary Standards for Thermal Performance of Residential Windows and Sliding Glass Doors
- Curtain Wall Design Guide Manual
- Architectural Aluminum Window Selection Guide Manual
- Metal Curtain Wall, Window, Store Front and Entrance Guide Specifications Manual

TIR-A1
- Sound Control for Aluminum Curtain Walls and Windows

CSI — **Construction Specifications Institute**

08520
- Aluminum Windows

NBS — **National Bureau of Standards**
- "A Guide for Window Design" Environmental Design Research Division
- "Door and Window Security Demonstration" Environmental Design Research Division by J. Stroik
- A New Look at Windows
- Performance Evaluation of Window Strategies
- Economic Evaluation of

REFERENCES: Windows (cont.)

Windows in Buildings -
Ruegg & Chapman
NBS-BSS-119

NRCC   **National Research Council of Canada**

CBD-25 • Window Air Leakage

110   • Ventilation and air quality

129   • Potential for thermal breakage of sealed double glazing units

132   • Glass thickness for windows

NWMA   **National Woodwork Manufacturers**

• Advantages of Wood Windows

• Care & Finishing of Wood Windows

NWMA/ANSI
1-S-2-80 • Wood Window Units Standard

NWMA/ANSI
I.S.3-70 • Wood Sliding Patio Door Standard

NWMA
1.5.4   • Water-Repellent Preservative Non-Pressure Treatment for Millwork

SWI   **Steel Window Institute**

• Recommended Specifications for Steel Windows

VWD   **Vinyl Window and Door Institute**

• What Builders Want to Know About Vinyl Siding and Vinyl Windows and Doors.

• Recommended Procedures for Installation of Vinyl Windows.

• A Profile of Vinyl Windows and Doors.

• Getting at the Source of Window Condensation.

• Vinyl Window Certification: A Quality Assurance Program.

• Vinyl Windows Questions and Answers

• An Introduction to the Vinyl Windows and Door Institute.

B2.8 WINDOWS

B2. 9 ENTRANCES/DOORS

ENTRANCES/DOORS: introduction

OPENING

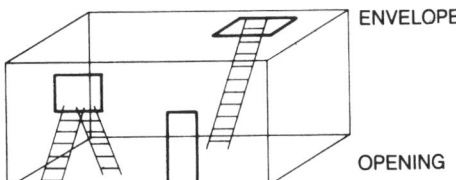

ENVELOPE

OPENING

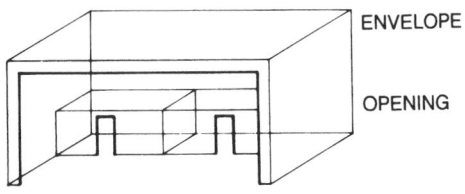

ENVELOPE

OPENING

- Spaces are always enclosed to serve a specific purpose, and therefore, have to be accessible: a portion of the envelope may be left out during construction, or a segment of the envelope may be cut out to provide an opening. Such opening, or openings, can be of any size and at any location within the envelope as long as they provide the required degree of accessibility to an enclosed space, either at, or above, grade; but openings in an envelope may negate the original intent of enclosing a given space since the internal environment can no longer be fully separated from the external one, and accessibility to the space cannot also be readily controlled.
- While some provisions may be made for at least partial control of accessibility to the enclosed space, the control of internal environment is only possible if the segment of the envelope, cut out or left out to provide access to the space, can be put back into the opening during times when access to space is not required.
- Openings in an envelope are therefore seldom left without provisions to close them.

DOOR

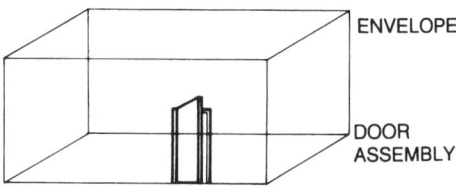

ENVELOPE

DOOR ASSEMBLY

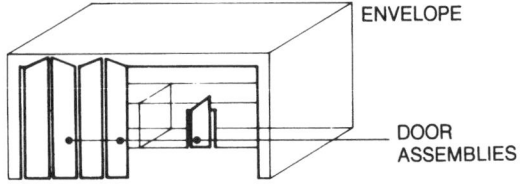

ENVELOPE

DOOR ASSEMBLIES

- Closures for openings essentially are movable segments of the envelope of a space made to open and close quickly and easily whenever passage to and from an enclosed space through its envelope is required; such movable closures are the doors.
- Functions of a door may be: separating different environments while controlling passage; controlling passage only; providing visual privacy; providing visual control through it; isolating or resisting the effects of external/internal factors, such as sound, light, air, fire, explosion, radiation, forced entry; or any combination of the above. To function properly, doors have to: move freely when passage is required; be held securely within the opening when passage is to be prevented; seal the opening completely when environments have to be separated, or internal/externals factors isolated.
- devices used to hold doors in place while, at the same time allowing for controlled opening and closing, are the hardware.
- devices used to separate environments are seals, often referred to as weatherstripping.
- frames to hold the components of a door assembly within an opening may also be required.

ENTRANCE

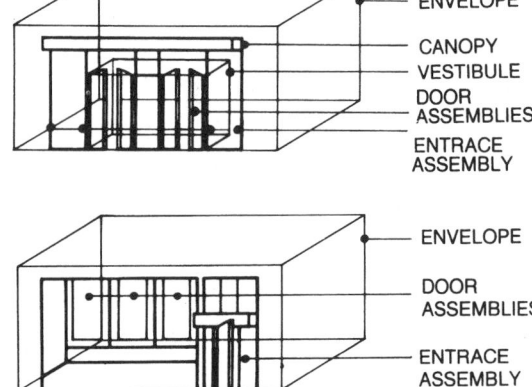

ENVELOPE
CANOPY
VESTIBULE
DOOR ASSEMBLIES
ENTRACE ASSEMBLY

ENVELOPE
DOOR ASSEMBLIES
ENTRACE ASSEMBLY

Door assemblies may also become components of a larger assembly: the entrance, either pedestrian or vehicular/service, to a building.
- Pedestrian assembly may consist of: a series of doors, of one or more functional types; fixed panels - transparent, translucent, or opaque; on one or both sides of the door, or above the door; the entrance doors may be separated with an air lock, or a vestibule, which may also include fixed panels on the sides or above the doors; canopies or awnings to provide shelter from the weather may also become a component of the entrance assembly; a continuous stream of air or an air curtain, to keep out insects and dust, or to prevent cold or hot outside air from mixing with interior air, may also be a major component of such an assembly.
- ■ In the case of vehicular/service entrances, a range of components and accessories may be included which are designed to help the convenient and sheltered loading and unloading of goods, including:
- canopies; loading docks, with movable platforms, bumpers guards, etc.; loading-dock shelters.
- SAFETY and UNIVERSAL DESIGN criteria apply to all entry system selections.

B2.9 ENTRANCES/DOORS

THE TOTAL ASSEMBLY

Entrance Size & Configuration

The first consideration should be the size and configuration of the assembly and the major components contained within it:

■ Number, distribution, and type by operation of the entrance doors.

■ **Number and distribution of the fixed dividing components** - sidelights and transoms.

■ **Design and location** of sheltering components: awnings, canopies, vestibules, air curtains.

■ **Design and structure** of the assembly encompassing frame. The frame system can also develop into a curtain-wall system.

Building type and the resulting type of occupancy will be the most important single determinant:

■ **Total capacity,** or the number of exit units needed, will be determined by the applicable Building Codes which will establish maximum number of occupants allowed. This determination will hold unless the toal number of building occupants is known in advance and can be demonstrated to remain unchanged to authorities having jurisdiction.

■ **The configuration** will be basically determined by the pattern of the building occupants coming and leaving:
- Maximum number of people entering/leaving the building at peak load time.
- Number of people entering/leaving at other times of the day. Number of hours in twenty-four when building remains open.

■ **Security requirements** or the barrier integrity of the building envelope at the entrance assembly should be considered:
- What security requirements are created by the design program?
- Is vandalism a concern?
- Is visual recognition through the entrance required?

Handicapped accessibility should also be considered:

■ **Entrance selection for handicapped** usage must be concerned with movement and dimensional limitations of the individual. Generally, accessibility is determined based on requirements included in prevailing codes.

■ **Some items for consideration are:**
- Minimum separation between sets of doors in vestibules.
- External entrance space.
- Minimum door width/clear opening.
- Maximum force to open door.
- Level change at entrance.
- Hardware requirements and location.
- Vision panels; size and location.

■ **Revolving doors are not acceptable.**

■ **Swinging doors** are acceptable as long as the force required to open the door is within certain stated limits; or power operation must be provided

■ **Power actuated sliding door** entrances are considered the best solution for handicapped.

ENVIRONMENTAL INFLUENCES

Climatic conditions which have a direct influence on the design of the entrance assembly include:

■ **Temperature range and extremes.**

■ **Prevailing winds.**

■ **Precipitation.**

■ **Atmospheric conditions:**
- Humidity
- Presence of salt spray, corrosive agents.
- Presence of pollutants.

Aspects of entrance design most influenced by the climatic considerations above should include:

■ **Incorporation of sun** and rain protecting devices, wind screens.

■ **Use of vestibules** and of revolving doors to minimize drafts and for better separation of outside and inside environments.

■ **Concern for sufficient structural strength** of entrance frame.

■ **Selection of types of door operation** and of hardward for efficient opening in all weather conditions and for adequate durability.

■ **Selection of component construction,** glazing and weatherstripping to minimize air and water infiltration and heat loss.

DOOR TYPES

The closure panel, or door, is generally classified by the method used to allow its opening and closing.

■ **There are three major types of pedestrian entrance doors:**
- Swinging, where the door panel is attached to a supporting frame by hinges or is pivoted top and bottom.
- Sliding, where the door panel slides to one side either top hung from a supporting frame or bottom supported; with either bottom or top guides.
- Revolving, where door panels are attached to a center rotating post; operating within a self-supporting enclosure.

■ **The three major functional types of vehicular/service entrace doors are:**
- swinging
- sliding - to one or both sides - in large or small segments
- vertical rise - in large or small segments

WEATHERTIGHTNESS

Exterior doors are subjected to all the effects of natural forces: solar heat, rain, and wind.

■ **For ordinary installations,** closed doors cannot be expected to exclude water or stop air movement completely under all conditions. One important reason for this is that clearances must be provided around each door. These are necessary to permit ease of operation, thermal expansion, and construction tolerances. See ASTM E331.

■ **In mechanically ventilated buildings,** there is likely to be a difference in air pressure between the inside and outside at entrances.

■ **This air leakage can be controlled in several ways:**
- Entrance vestibules.
- Revolving door entrances.
- Weatherstripping.

■ **For doors whose operation is impaired by air pressure;** consider:
- Balanced pivots.
- Power actuators.

■ **Entrances may need to isolate the interior from the exterior acoustically:**
- Weatherstripping is effective in sound isolation.

■ **When selecting weatherstripping, consider:**
- Do the weather sealing components being reviewed comply with energy code restrictions?
- Tighter fit due to seals means more effort to operate.

B2.9 ENTRANCES/DOORS

ENTRANCES: assemblies

COMPONENTS OF ASSEMBLY

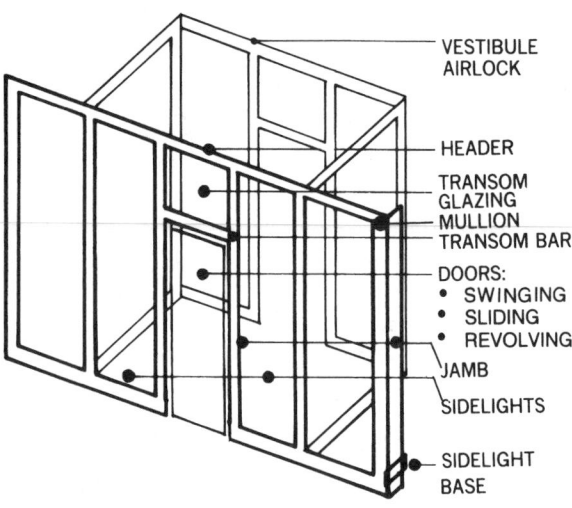

- VESTIBULE AIRLOCK
- HEADER
- TRANSOM GLAZING
- MULLION
- TRANSOM BAR
- DOORS:
  - SWINGING
  - SLIDING
  - REVOLVING
- JAMB
- SIDELIGHTS
- SIDELIGHT BASE

- Jamb/header: vertical and horizontal frame members forming the sides of an entrance or door assembly. In a door assembly, the hinge jamb is the frame member at which the hinges or pivots are mounted.
- Mullion: vertical framing member holding and supporting fixed glazing or opaque infill panels. May be single piece or split; with or without thermal breaks.
- Transom bar: horizontal framing member which separates the door opening from the transom above. Transom bars may contain operating hardware for doors, such as closers or automatic operators.
- Transom bracket: a bracket to support all-glass transom over an all-glass door when no transom bar is used.
- Sidelight: fixed light or lights of glass located adjacent to a door opening. Wet or dry glazed, with dry glazing more commonly used.
- Sidelight base: may be a single piece, built up of several framing members, or a masonry or concrete curb may be used for support of the bottom frame. In either case, provisions to seal the gap between the bottom of the entrance assembly and supporting construction should be incorporated in the design of the assembly.

DESIGN COMPONENTS

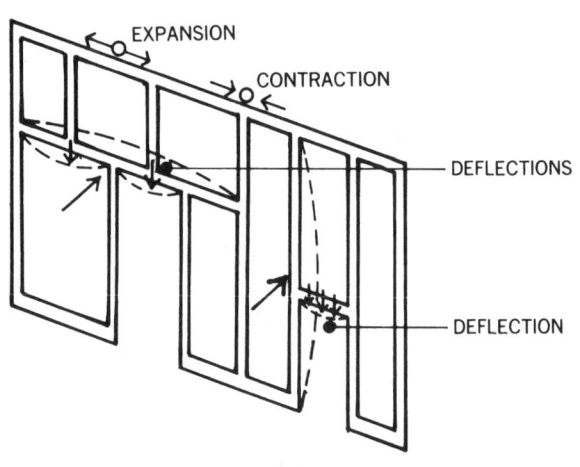

- EXPANSION
- CONTRACTION
- DEFLECTIONS
- DEFLECTION

- Expansion/contraction: in the assembly must not be restricted within the opening and proper clearances in the opening and at connections must be incorporated in the design. Movement within the assembly may be accommodated by providing for slippage at joints or in split mullions.
- Door frames, especially in large assemblies, should be independent of other framing to minimize effects of thermal movement.
- Deflection in horizontal members may impose loads on glass and may cause breakage and/or prevent proper operation of doors. When staggered concentrated loads have to be carried by a horizontal framing member, such member may have to be reinforced to limit deflection, or the assembly should be redesigned. Lateral loads may result in excessive deflection in long horiztonal members and reinforcing may be required.
- Deflection in vertical members may affect operation of doors, and may have to be less than the acceptable maximum of approximately 1/180 of span for other framing members.
- Split mullions supporting doors hung from them may require reinforcement since they generally are not stiff enough to hold doors securely and in proper alignment.

FRAMING MEMBERS

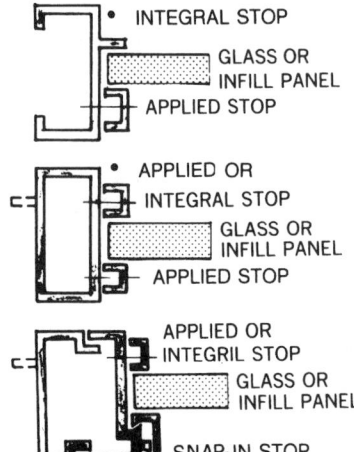

- INTEGRAL STOP
- GLASS OR INFILL PANEL
- APPLIED STOP

- APPLIED OR INTEGRAL STOP
- GLASS OR INFILL PANEL
- APPLIED STOP

- APPLIED OR INTEGRIL STOP
- GLASS OR INFILL PANEL
- SNAP-IN STOP

CHANNEL
- HEADER
- JAMB

TUBE
- MULLION
- TRANSOM BAR
- HEADER
- JAMB

SPLITUBE
- MULLION
- TRANSOM BAR
- HEADER
- JAMB

Framing members for entrance assemblies, commonly used and available as stock items, are of extruded aluminum, generally clear or color anodized. Baked-on finishes, such as fluorocarbons and siliconized polymers, are also available.

CHANNELS are generally used as perimeter framing when the assembly is to be installed in a masonry or framed wall. Options available in stock shapes are: integral stop, recess for flush glazing. Modified shapes available with thermal breaks, and for lockstrip gasket glazing.

TUBES are used as perimeter framing, vertical and horizontal framing members and as covers for structural steel channel or tub reinforcement. Options available in stock shapes are: integral stop, recesses for flush glazing. Modified shapes incorporating thermal breaks and/or provisions for lockstrip gasket glazing are available.

SPLIT TUBES may be used for perimeter framing and as vertical/horizontal framing. Integral stops and recesses for flush glazing available in stock shapes. Modified shapes with thermal breaks and/or for lockstrip glazing available.

Also refer to CURTAIN WALLS for additional information on mullions.

SAFETY CONSIDERATIONS

- Codes may require safety glazing to reduce accidents due to broken glass.
- Door swings, location of stops and other applied hardware must be considered to avoid interference with traffic flow through door.
- The arc of a door swing should exceed 90° to allow the full width of the door opening to be unobstructed.
- Hinge jambs should be at least 6 inches from a wall perpendicular to a building face to prevent the user's hands from being pinched between the door and the wall.
- If hinged jambs for two door have to be adjacent, there should be enough distance between them to permit the doors to swing through an arc of 110°.
- If doors hung on center pivots are hung in pairs, they should be hinged at the side jamb and not at the center mullion.
- Doors hinged off a common mullion may create problems: hardware to prevent one door from swinging into the path of the other could subject the door to excessive forces with resulting severe damage.
- All-glass doors and sidelights should be clearly identified to prevent the possibility of people walking into them accidentally.
- Swinging doors should be located to clear passing pedestrian traffic without interference.

ENTRANCE DOORS

In general, operating panels, or doors, are selected on the basis of desired type of operation:

■ **Swinging, either single-action or double-action.** This type of operation:
- Provides versatility in permitting large or small passage capacity.
- Lends itself to manual or power actuation, but:
- May require considerable strength to open in certain weather conditions.
- Does not provide the best protection against drafts and heat loss.

■ **Sliding:** These are:
- Safest, especially where people with objects pass; as in air terminals, shopping centers, etc.
- Present least obstruction inside and outside and will act in any weather. but:
- Must be power-actuated when serving as entrances of any size.
- Will permit penetration of outside air inside in harsher climates unless a vestibule design is used.

■ **Revolving:** These are most effective for:
- Sealing the outside environment from the interior without use of a vestibule.
- Handling an orderly trickle of pedestrians in and out of buildings, but:
- Have limited capacity and cannot handle peak loads.
- Are difficult for people with large objects and for the handicapped, and inaccessible to those in wheelchairs.

■ **Other general considerations in selecting type of door operation will include :**
- Capacity as a means of egress to satisfy code requirement.
- When considered as an exit, direction of swing must be in direction of flow of traffic to a place of refuge.
- Restriction on door operation type: can sliding door or revolving be used?
- Ease of opening; maximum force to open door needed.
- Self-closing operation.

In selecting a swinging door, consider:

■ **Single-action doors** can swing 90° or more in one direction only.

■ **Double-action doors** can swing 90° or more in each direction.

■ **Far more styles, types, materials, and accessories** are available for swinging entrances than for any other type.

■ Since all components which comprise swinging entrances can be manufactured independently, **the selection of the proper components for a specific application is as important as the selection of the basic entrance type.**

■ **Horizontal sliding doors are advantageous for :**
- Unusally wide openings.
- Where clearances do not permit use of swinging doors.

■ **Operation of sliding doors** is not hindered by windy conditions or differences in air pressure between indoors and outdoors.

■ **For certain usage, horizontal sliding doors:**
- May have to be motorized to increase traffic flow through the entrance.
- Panic exit features may need to be incorporated to let people move

through the door panels without injury during emergency egress.

■ **The basic door** and frame components of swinging doors apply to horizontal sliding doors.

■ **Sliding entrances are manufactured as assemblies.** Since a single manufacturer provides all components, the selections to be made will deal more with the style, materials, finishes, and accessories available than with the range of components and combinations available.

■ **Sliding entrances are normally selected for view and illumination as well as ingress/egress.** Unless the doors are power-operated, sliding entrances are normally used for infrequent or low traffic entrances.

When selecting revolving doors, consider:

■ **Revolving door entrances are generally selected for:**
- Entries which carry a continuous flow of traffic without very high peaks.
- They keep interchange of inside and outside air to a relatively small amount compared to other types of doors.
- They are usually used in combination with swinging doors because of revolving doors' inability to handle large volumes of people in short periods of time.

■ **Building codes prohibit the use of revolving doors** for some types of occupancy because of the limited traffic flow in emergencies:
- Where permitted as exits, they have limitations imposed by local building codes.
- May not, in some instances, provide more than 50 percent of the required exit capacity at any location. The remaining capacity must be supplied by swinging doors within close proximity.

■ **Revolving doors may be power assisted** to facilitate traffic through them.

■ **Revolving-door entrances normally provide a panic-releasing device** which automatically releases a door panel in case of:
- Entrapment of the user.
- Accidental jamming or impact on the door leaves.

■ **Revolving doors are provided with speed governors to limit the maximum speed** at which the door leaves will travel.

■ **Revolving door entrances are manufactured as assemblies.** The manufacturer will provide all components required for installation.

B2.9 ENTRANCES/DOORS

DOOR ASSEMBLIES: swinging

COMPONENTS OF DOOR ASSEMBLY

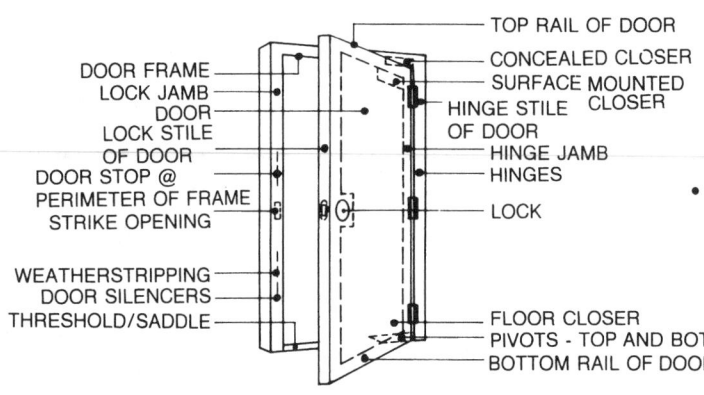

DOOR FRAME
LOCK JAMB
DOOR
LOCK STILE
OF DOOR
DOOR STOP @
PERIMETER OF FRAME
STRIKE OPENING

WEATHERSTRIPPING
DOOR SILENCERS
THRESHOLD/SADDLE

TOP RAIL OF DOOR
CONCEALED CLOSER
SURFACE MOUNTED CLOSER
HINGE STILE OF DOOR
HINGE JAMB
HINGES
LOCK

FLOOR CLOSER
PIVOTS - TOP AND BOTTOM
BOTTOM RAIL OF DOOR

- Doors may be mounted either on butt hinges or on pivots. Hinges provide maximum free opening width, but impose strain on jamb. Commonly three hinges for standard 7 foot high door. Pivots may be center, offset, or swinging. Center pivots required for double-acting doors, and generally required for automatically-operated doors. Pivots preferred to hinges for heavy doors or for unusually severe service.
- Closers may be mounted concealed in the top rail of the frame, exposed on the top rail, or surface mounted on the door. Floor closers are recessed in the floor construction.

FLUSH DOORS

- HINGED OR PIVOTED

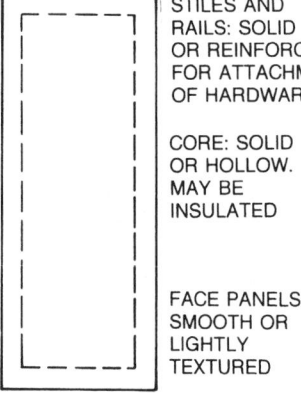

STILES AND
RAILS: SOLID
OR REINFORCED
FOR ATTACHMENT
OF HARDWARE

CORE: SOLID
OR HOLLOW.
MAY BE
INSULATED

FACE PANELS:
SMOOTH OR
LIGHTLY
TEXTURED

- METAL
- WOOD

- HINGED OR PIVOTED

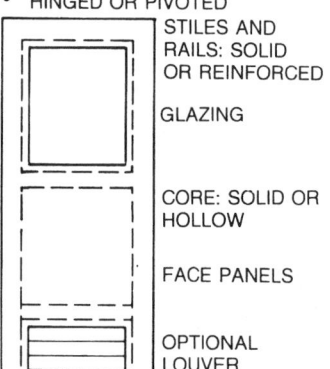

STILES AND
RAILS: SOLID
OR REINFORCED

GLAZING

CORE: SOLID OR
HOLLOW

FACE PANELS

OPTIONAL
LOUVER

- METAL
- WOOD

- PIVOTED ONLY

METAL FITTING

TEMPERED
GLASS

METAL FITTING

- ALL-GLASS

STILE AND RAIL DOORS

- PIVOTED OR HINGED

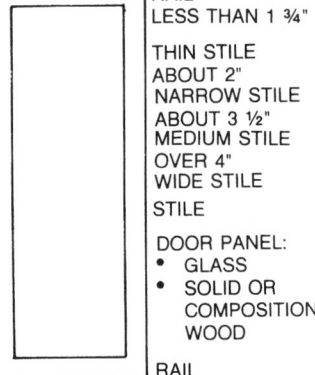

RAIL
LESS THAN 1 ¾"

THIN STILE
ABOUT 2"
NARROW STILE
ABOUT 3 ½"
MEDIUM STILE
OVER 4"
WIDE STILE
STILE

DOOR PANEL:
- GLASS
- SOLID OR
  COMPOSITION
  WOOD

RAIL

- METAL FOR ALL WIDTHS OF STILES
- WOOD FOR MEDIUM
  AND WIDE STILE ONLY

- PIVOTED OR HINGED

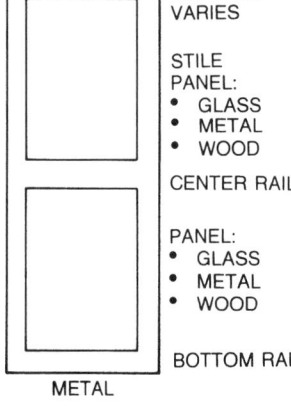

TOP RAIL
VARIES

STILE
PANEL:
- GLASS
- METAL
- WOOD

CENTER RAIL

PANEL:
- GLASS
- METAL
- WOOD

BOTTOM RAIL

METAL

WOOD

TOP RAIL

PANELS:
- WOOD
- GLASS

STILE

PANEL:
- WOOD

PANEL:
- WOOD

BOTTOM RAIL

- WOOD
  MAY BE METAL
  CLAD

METHOD OF OPERATION

TYPE	DESCRIPTION
SINGLE ACTING 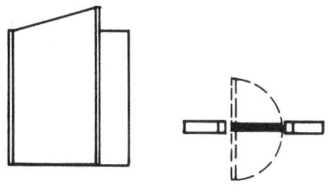	
	• Single-panel swinging type most commonly used when high winds, heat loss or gain through air exfiltration/infiltration during the time when door is opened are not critical considerations. • Difficult to operate manually against wind. Operation may be impaired when pressure differentials exist between interior and exterior environments, such as might be the case in pressurized air conditioned buildings. • Generally installed within the frame; single-acting types available to overlap the frame, in which case hinged only. • Hinged doors may be effectively sealed to prevent air leakage; pivoted doors difficult to weatherstrip completely. • Operation easiest when balanced pivot mounted, but effective clear opening will be reduced more than when hinges, center, or offset pivots are used. • Automatic operators, either hydraulic, pneumatic, or electric, are available. • Time to open when motor-operated: one-half to three seconds; may be adjustable up to five seconds. • Hinged and offset pivoted doors may open up to 180 degrees; center-pivoted doors to a maximum of 13 degrees. • For additional information on double-acting doors, refer to Pair-Single or Double-Acting, below.
DOUBLE ACTING 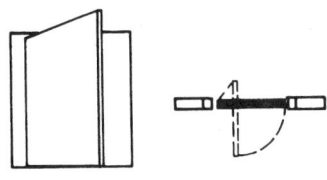	
SINGLE ACTING - BALANCED 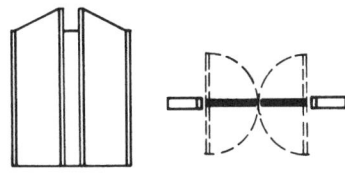	
PAIR - SINGLE OR DOUBLE ACTING 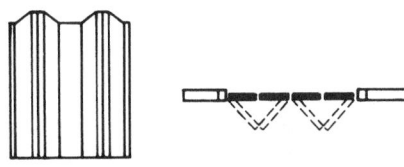	Pair of doors in common frame. Single-action or double-action; generally installed within frame. • Double-acting pair of doors, just as a single door, provide less traffic control, increasing the likelihood of interference between in-traffic and out-traffic; cannot be made as weathertight as single-acting doors; may not remain closed when air pressure is different on one side of the door from the other side; not appropriate for panic exit hardware; local codes may not permit double acting doors to be used as required exits. • Single-acting doors may be automatically operated. Time to open when motor operated: one-half to three seconds; adjustable up to five seconds.
BI-FOLDING 	Four or more doors, hinged in pairs in a common frame. May be hinged to jambs or pivoted jambs. Guides for leading edges required, either in head, at floor, or in both locations. • Special provisions, such as surface bolts at hinged mating edges, required to secure doors in place when closed. • Single-acting only.

B2.9 ENTRANCES/DOORS

DOOR ASSEMBLIES: types, operation

TYPE	DESCRIPTION

SLIDING

TRACK

- Single door installed to overlap opening. May also be within opening with a sidelight. Generally hung from overhead track with bottom guides. May also ride in a bottom track with top guided only.
- When fire-resistance rated, inclined top track may be used for automatic closing or door may be counterweighted; fusible link then used to hold door in open position.
- Opening width generally up to 16 feet. Side clearance required for stacking door when open; additional space required when counterweight is used.
- Opening speed when motor-operated: 4 to 8 inches per second.

TRACK

- Pair of doors installed to either overlap the opening or within the opening with side panels. Generally glazed when within the opening. Hung from top track with bottom guides; or ride in bottom track. When used as entrance, operation is automatic, and doors are designed to swing out when pressure is applied from inside the building during panic exit. • Inclined tracks available for automatic closing during fire, or doors may be counterweighted; generally for industrial application only.
- Opening speed when motor-operated: 4 to 8 inches per second.

GUIDE / TRACK

- Multiple doors installed within opening. Generally ride in bottom tracks with top guides when large; small doors may be hung from top tracks with bottom guides only. Opening width and height not limited; often used as aircraft hangar doors. May slide to one side, or be bi-parting. Clearance required for stacking doors alongside the opening. When large, constructed of rolled structural shapes with various wall panels available for facing. Generally motor-operated: large doors may have individual motors. Opening speed for large doors: about 60 feet per minute.

REVOLVING

- Revolving entrance doors are only available as a complete package from manufacturer. Diameter varies from 6.5 to 7.5 feet, with 7 feet being the standard height. Doors are available up to 8 feet in diameter and to 9 feet in height on special order. Use of wider doors generally recommended for easier traffic flow, especially in buildings where hand-carried luggage is a factor. Revolving doors should be equipped with emergency-release mechanism to book-fold wings and should have manual or power assisted speed controller. Effective speed range is 8 to 12 revolutions per minute.

- Special type to prevent light penetration into a space when door is operated. Generally used as door to a darkroom.
- Available only as a complete package from manufacturer. Sizes are limited, and once installed will not allow any large equipment to be taken into or out of the space.
- Manual operation only.

VERTICAL RISE, SLIDING

- Single panel installed to overlap opening. Side guides and counterweight required. Clearance required over opening for stacking panel, and at sides of opening for guides and counterweight. Generally used to protect openings in fire-rated walls; automatic closing with fusible link releasing counterweight.
- Cannot be used as required exit unless a swinging door incorporated into door panel but such use not common.
- May be motor-operated. Opening speed when motorized: 45 to 60 feet per minute.

- Multiple panels installed within opening.
- Overhead clearance for stacking panels required. Guides within opening; counterweight required but may be remotely located if space near door is not available. Opening width generally up to 24 feet; wider doors also available.
- Generally used for high openings when headroom for stacking panels limited, or when speed in opening and closing is important.
- Generally motor-operated, with opening speeds from 45 to 60 feet per minute.

VERTICAL RISE, TELESCOPING

- Horizontal panels nesting one into the other.
- Installed to overlap opening. Headroom required over opening for header box and nesting of door panels; clearance at sides for guides.
- Opening size 20 by 20 maximum; door limited to 400 square feet.
- Availability limited. Motorized operation only with opening speed of 40 feet per minute.

B2.9 ENTRANCES/DOORS

TYPE	DESCRIPTION

VERTICAL RISE, SECTIONAL

OVERHEAD CLEARANCE HORIZONTAL PROJECTION	• Vertical-lift type. Installed to overlap opening. • Simplest to install. Minimum projection into inside space, but overhead clearance the height of opening required. Vertical tracks at jambs; torsion springs are used to hold door in open position. • Opening width to 36 feet; opening height to 20 feet. Maximum opening size about 600 square feet. • Opening speeds when motor-operated: 40 to 60 feet per minute.
	• High-lift type. Installed to overlap opening. • Limited projection into inside space. Overhead clearance required is less than for vertical-lift type. • Clearance required at jambs for tracks; tracks bent on top to partially project into inside of space. • Door held open with torsion springs. • Opening width to 36 feet; height to 20 feet. Maximum opening size about 600 s.f. • Opening speeds when motor-operated: 40 to 60 feet per minute.
	• Standard-lift type. Installed to overlap opening. Full projection into inside of space. Vertical tracks at jambs bent at top to project into the space the full height of door to support door in open position. • Torsion springs used to operate door. Door may also be motor-operated, with opening speeds from 40 to 60 feet per minute. • Opening width to 36 feet; height to 20 feet. Maximum size of opening limited to about 600 square feet.
	• Low-headroom type. Similar to standard-lift type, except that maximum size limited to about 240 square feet. GENERALLY: • interior use of all types of sectional doors is limited, principally because of overhead clearance requirements and the fact that fire-resistance rated types are not available. • pedestrian pass doors may be incorporated into sectional doors to provide access to space with door in closed position.

VERTICAL RISE, ROLL UP

	• Interlocking metal slats. May be insulated with rigid foamed insulation and also weatherstripped. May be fire-resistance rated but only up to 120 square feet in size. May be installed within opening; generally mounted to overlap opening. Guides required at jambs and torsion springs for opening. • Opening size up to 30 feet wide by 30 feet high. Larger sizes available on special order, but width may be limited. • May be motor-operated, with speeds from 30 to 60 feet per minute.

VERTICAL RISE, CANOPY

	• Single panel installed within opening only. Top tracks and counterweight are required. • Opening width generally to 60 feet; up to 130 feet wide available on special order. Height generally up to 30 feet. • Clearance required: approximately 2/3 of door height in front of opening; 1/3 in back plus space for overhead tracks and counterweight. • Smaller doors may be crank and gear operated; larger sizes generally motorized with opening speed of about 30 feet per minute.
	• Bi-folding, installed within opening only. Side guides, top tracks, and counterweights required. • Opening widths generally to 60 feet; sizes to 130 feet wide available on special order. Height to 50 feet. • Clearance required in front of opening less than for single panel: clearance in back for top tracks and counterweight similar to that required for single panel. • Generally motor-operated only, with opening speed of about 30 feet per minute.

PLUG

	• For specific applications only, such as isolating high-level radiation. Will continue to shield outside space against radiation even when opened. Also used where resistance against explosion required. • Generally of reinforced and/or lead-shielded concrete. • track mounted and motorized for operation • Always custom designed.

B2.9 ENTRANCES/DOORS

DOORS: types, materials

● Denotes common usage or availability
○ Denotes possible usage or limited availability

FACE PANEL
CORE
INSET PANELS
STILES AND RAILS

		SIZE [1] WIDTH x HEIGHT maximum, in feet	FLUSH/COMPOSITE/LAMINATED FACE PANELS							CORE					HOMOGENEOUS FACE AND CORE			STILE and RAIL FRAME				
			ALUMINUM	STEEL/TIN	STEEL, STAINLESS/BRONZE	PLASTIC LAMINATE	PLYWOOD VENEER	HARDBOARD	SOLID WOOD/CARVED	LUMBER STAVED	PARTICLE BOARD/PLYWOOD	INSULATION/MINERAL BOARD	METAL, STIFFENERS/HONEYCOMB	FIBER BOARD/HONEYCOMB	GLASS, TEMPERED	STEEL PLATE	RUBBER/PLASTIC	ALUMINUM, COATED/ANODIZED	BRONZE	STEEL, STAINLESS	STEEL, GALVANIZED, COATED	WOOD, SOLID
FLUSH																						
	SWINGING	4 x 10	●	●	●	●	●	●	●	●	●	●	●	●	●							
	SLIDING	No Limit	●	●	●	●	●	●	●	●	●	●	●	●	●							
	SWINGING	4 x 10	●	●	●	●	●	●	●	●	●	●	●	●								
	SLIDING	No Limit	●	●	●	●	●	●	●	●	●	●	●	●								
	SWINGING ONLY	4 x 8													●	○	●					
STILE AND RAIL																						
	SWINGING	4 x 10																●	●	●	●	●
	SLIDING	7 x 10																●	●	●	●	●
	SWINGING	4 x 10														●		●	●	●	●	●
	SLIDING	7 x 10														●		●	●	●	●	●
	SWINGING	4 x 10																●	●	●	●	●
	SLIDING	7 x 10																●	●	●	●	●

B2.9 ENTRANCES/DOORS

| [2] INSET PANELS | | | | | | | | | | | [3] | | | | | [4] | [5] |
Wood, metal/plastic clad	Plastic	Glass: clear/wire	Glass, tempered	Metal, solid	Wood: solid/plywood	Metal slats/mesh	Wood slats	Louvers: metal/wood	Automatic operation	Fire rated type available	Heat flow	Forced entry	Bullets	Explosion	Radiation	Air/gas pressure	Comparative cost, range for door and frame
		●							●	●	●	●	●	●	●	●	[6] 100 to 250 for wood
		●							●	●	○	●	●	●	●		
		●				●			●	○							[6] 250 to 450 for metal
		●				●			●	○							
			●				●		●								[6] 130 to 570 for folding
●		●	●	●	○	○	●			○							[7] 460 to 700 sliding
●		●	●	●	○	○											[8] 600 to 1300 sliding
●	●	●	●	●	●	●	●			○							
●	●	●		●	●	●	●										
●	●	●		●	●					○							
●	●	●		●	●												

[1] Generally, larger sizes are available for sliding doors, and on special order for swinging doors. Doors in fire-rated assemblies all have size limitations.

[2] One-piece molded plastic units are available; use generally limited to residential closed doors.

[3] Heavily insulated doors, such as for coolers, freezers. Freezer door·assembly will generally include heater cables for efficient operation.

[4] Generally available as single doors, single-acting only.

[5] Comparative average cost per square foot of door. Basis for cost: 1⅜-inch thick hollowcore wood door, hardboard faced, in wood frame = 100. Cost of finish hardware not included.

SWINGING DOORS:

[6] Cost of single panel swinging doors:
* comparative cost is for standard construction doors, 3'-0" wide by 6'-8" high.
* fire-rated doors will be at the high range for wood and metal doors.
* cost of special-purpose doors will in all cases be higher.
* comparative cost range for bifolding closet doors: 70 to 80.

■ **General:**
* fire-rated doors are rated for specific fire resistance and are limited in size and/or design. Consult NFPA 80 or manufacturer's literature for specific requirements.
* size of vision panels in fire-rated doors limited based on rating of door required.
* air/gas or lightproof doors require weatherstripping or seals to seal perimeter of door against frame;
* bi-folding doors may be motor-operated.
* double-panel flexible doors are used to minimize movement of air between two spaces in opening subject to frequent passage of material-handling equipment.
* horizontally hinged type available for crane entranceways.
* four-fold type, motor-operated, also available.
* hollow core metal doors may be insulated to deaden sound and/or minimize heat flow through door.
* bullet-resistant doors may have laminated glass vision panels.

[7] **SLIDING DOORS:**
Single-panel, wood-faced, commercial/industrial-use assemblies. Double-panel, tin clad, commercial/industrial-use assemblies

[8] **GENERAL:**
* available for very large openings, such a aircraft hangars; constructed using structural steel rolled shapes and formed-metal wall panels.
* cannot be used as emergency exits unless special provisions made.
* generally motor-operated when used to isolate controlled environments such as cold-storage rooms.

B2.9 ENTRANCES/DOORS

DOORS: types, materials (cont.)

- • denotes common usage or availability
- ○ denotes possible usage or limited availability

FACE PANEL
CORE
INSET PANELS
STILES AND RAILS

	TYPE		SIZE [1]	FACE PANEL						FRAME		INSET PANEL			
	FLUSH	STILE AND RAIL	WIDTH x HEIGHT, average maximum, in feet	WOOD/PLYWOOD	HARDBOARD	STEEL, GALV./COATED	STEEL, STAINLESS	TIN CLADDING	ALUMINUM	WOOD, SOLID	METAL	METAL	PLASTIC	HARDBOARD	PLYWOOD
VERTICAL RISE															
SLIDING, SINGLE SECTION MULTI-SECTION	•		24 x 27	•	•	•	•		•						
SECTIONAL SECTIONS HINGED	•	•	36 x 20	•	•	•				•	•	•	•	•	•
ROLL UP	•		30 x 30	•		•	•		•						
ROLLING GRILLE	•		Varies	•		•	•		•						
TELESCOPING	•		20 x 20	[4]											
CANOPY	•		60 x 30	[5]											

TEMPERED GLASS PANELS	VISION PANELS	WEATHERSTRIPPING	INSULATION	FIRE RATED AVAILABLE	MANUAL	CRANK AND GEAR	MOTORIZED	AUTO-REVERSE	COUNTERWEIGHT	COMPARATIVE COST, range, per square foot [3]	REMARKS
	OPTIONS				OPERATION OPTIONS [2]						
•	•	•		•	•		•	•	•	400 to 600	• Single-panel types generally used to protect opening in fire rated walls when such opening must be at all times to allow for continuous traffic. • Single-panel types may be insulated when normally closed, but difficult to weatherstrip effectively. • When overhead clearance is restricted, multisectional, sectional, roll-up, or telescoping types should be considered. • When quick opening and closing is important, multisectional types rather than single panel types should be considered. counterweight used.
•	•	•		•	•	•	•	•		120 to 1200	• Four basic types available: vertical-lift, high-lift, standard, and low headroom. • Flush-type sectional panels available with added insulation for all four basic types. • Pedestrian doors may be incorporated into all types, but are not recommended. • Torsion springs rather than counterweights commonly used.
•	•	•		•	•	•	•	•		300 to 600	• Available with solid-metal slats, either flush or curved. Vision strips, generally of transparent plastic may be provided in solid slats. Also available as grilles with interlocking open slats. • Horizontal ceiling types, either with solid slats or open grille type, are also available. • Available with solid insulated slats; but not in fire-resistance rated types. • Side coiling types available.
•						•	•	•	•	custom	• Clear polycarbonate panels available. • Side-coiling or side-stacking types available. • Various designs of grilles available. • Generally used to close openings for security and allows for visibility.
	•						•	•		custom	[4] • Limited availability. Insulated and fire-resistance rated types not available. • Minimum headroom and quick opening. • Side guides will reduce effective width of opening and should be protected from damage.
•	•				○	○	•	○	•	custom	[5] • Generally the larger sizes are custom-designed for each specific application. • Will form a canopy over opening when open. Should be pitched for drainage, and clearances for both sides of opening are required. • Unless center-pivoted, counterweights are required for ease in opening.

REMARKS

[1] Larger sizes are generally available on special order. Canopy doors available on special order only.

[2] Counterweights generally required when door is used as opening protective. Counterweight holds door in open position and is released by fusible link, in case of fire on either side of opening.

[3] Comparative cost is average per square foot of door based on the cost of 1⅜-inch thick hollow-core wood door, hardboard faced, in wood frame = 100.

[4] Face panel and core homogeneous since panels are formed of light-gauge galvanized and coated steel.

[5] Face panels may be of formed prefinished metal. Refer to CURTAIN WALLS.

B2.9 ENTRANCES/DOORS

B · BUILDING SHELL

DOORS: selection checklist

PEDESTRIAN DOORS

General considerations should include:

- Vision panels for all swinging doors unless locked at all times, to prevent the leaf from swinging in the face of a person approaching entrance from the opposite side.
- All-glass doors should be marked to prevent injury to persons from walking into the glass.
- Package-type entrance should be reviewed for the number of cycles the assembly can be used without deterioration in function.

■ **Corrosion may be an important factor** in the selection of entrances. When corrosion resistance is reviewed, it should concern both the base materials and finishes of the components. Consider the building microclimate:

- Humidity and salt spray will affect certain finishes.
- Sunlight may degrade certain materials.
- Environmental pollution will affect certain materials and finishes.
- Materials which are vulnerable to corrosion can be upgraded by applied coatings or finishes.
- Aluminum alloys are highly reactive with some metal types.
- Cement products react with aluminum alloys. Provide coatings or physical separation.
- Water runoff over metal surfaces can produce ion flow which causes staining or corrosion. Review all materials used in the construction of the building envelope.

■ **Sound transmission** through the door and along its perimeter should fit the STC of the entire assembly; accordingly, doors selected may be:

- hollow doors with sound deadening insulation.
- solid core doors.
- solid core doors faced with sound absorbing material.
- doors in partitions with high STC requirements have to be completely sealed arount the entire perimeter: even a small gap will result in sound leakage enough to drastically lower the STC of the entire assembly.

■ **Doors located in a wall or a partition** required to be fire rated:

- size and construction of door panel, the frame construction and anchorage and all rough and finish hardware should conform to all requirements

for a rated assembly, and be made by a manufacturer who had an identical assembly tested and approved.
- generally, doors in fire-rated walls and partitions should have the same rating as the assembly, but building code requirements differ, and should be checked prior to making a selection.
- refer also to NFPA 80 - Standard for Fire Doors and Windows.

When pressure differentials exist between the two sides of the door.

■ **Air/gas leakage under pressure should be prevented:**

- compressible or mechanically inflatable seals along the entire perimeter are required.
- seals will also effectively prevent light passage; simple weatherstripping may be adequate when resistance to light leakage only is required.
- in certain continuous traffic locations flexible doors may be selected which minimize leakage of air - conditioned, polluted, etc. - when opened.

■ **Doors located under impact of constant traffic:**

- solid-core metal-clad wood doors are good in resisting frequent impact but high in cost.
- hollow-core wood doors are easily broken and hollow metal doors are easily dented.
- doors may be reinforced to minimize damage when accidentally hit.
- flexible doors, which bend when hit, then quickly return to a closed position, may sometimes be used.

■ **In doors used for cooler or freezer assemblies:** heat flow is critical:

- freezer doors are generally provided with perimeter heating cables to prevent freeze-ups due to water vapor condensation and freezing.

■ **Doors may be located in radiation-resistant assemblies:** range of requirements is:

- from a low level such as of an X-ray machine in a physician's office
- to that of an atomic reactor.
- for low-level radiation protection, lead-lined doors and frames are available.
- most other applications require special design.

In selecting construction of door

operating panel for overall strength, consider:

■ **Resistance to impact:**

- Solid core doors have high impact resistance.
- Under severe conditions, metal-clad solid-corewood doors, or kalamein (metal-clad wood) doors, are recommended.
- Flush glass doors and hollow-core wood or metal doors are least resistant.

■ **Flush doors are generally stronger than stile and rail doors.**

■ **Face width, and thickness of stile-and-rail frame members will affect overall strength.**

■ **Hollow-core wood doors are not recommended for exterior exposure.**

■ **Oversized doors and lead-lined doors,** because of their weight:

- Require more rugged hinges, frames, closers, etc.
- Power actuators may be needed if operating effort becomes excessive.
- Due to the stresses incurred by oversized doors, the choice of hardware may be limited.

■ **Vision panels, louvers required;** and any special configurations, such as dutch doors.

■ **Fire-resistance rating requirements:** Not all door panels can be fire rated.

■ **Standards used in the construction of door panels** are in general as follows:

- For wood panels - NWMA and AWI.
- For metal and glass doors - NAAMM and AAMA.
- For all-glass doors - FGMA and SIGMA, but safety standard ANSI as enforced by the Consumer Product Safety Commission must be complied with.

In selecting panel materials, note:

■ **Wood components** are available in a wide variety of wood species and finish grades:

- Flush doors come in architectural, premium, custom and standard grades.
- Stile and rail doors are fabricated from solid wood sections, either hardwood or softwood, and in various grades.
- Wood components will normally not withstand exterior exposure without some type of protective coating.

SERVICE/VEHICULAR ENTRANCES AND DOORS

General considerations should include:

Anticipated Traffic:

■ **Pedestrian,** with some light, manually-operated materials-handling equipment:

• Select swinging doors, either single leaf or double leaf.
• Single panel and multipanel sliding types should also be considered if power operators are used.

■ **Power-operated materials-handling** equipment, such as fork lift trucks:

• Consider biparting sliding, multipanel vertical sliding, sectional, roll-up, and telescoping types.

■ **Vehicular traffic,** such as automobiles and trucks:

• Consider multipanel vertical sliding, sectional, roll-up and telescoping types.
• For low frequency use, where available clearances between entrances must be kept to a minimum, consider swinging bifolding, or four-folding types.

■ **Large equipment,** such as aircraft:

• For openings up to 30' high x 130' wide: single-panel canopy type.
• For openings up to 50' high x 130' wide: bifolding canopy type.
• For higher and wider openings: multipanel sliding. Multipanel types require room for stacking panels beyond the clear opening provided.

Frequency of Use:

■ **High frequency use may:**

• Preclude selection of certain types of entrances.
• Require power operators, special controls, or other options.

■ **In all instances the need for power operators should be investigated;** power operators add to:

• Cost of the installation.
• Maintenance requirements.

■ **Ease of operation under adverse conditions,** such as high winds, especially for large doors, may limit selection to horizontal sliding, vertical sliding, sectional, and roll-up types.

■ **Time required to open and close** the door panels may also be a factor, especially for high frequency use entrances.

■ **Insulation within the door panel** and possibility of efficient weatherstripping may be a consideration for low frequency use entrances.

ENVIRONMENTAL FACTORS:

■ **Wind loads** to which large entrances will be subjected will affect the choice of:

• Type of entrance: to avoid damage to entrance, and to allow operation under high wind conditions.
• Type of core material or frame construction: to resist horizontal wind forces without excessive deflection.
• Canopy-type doors will, in addition, be subject to vertical wind force or uplift when in open position.

■ **Thermal movement** should be considered for very high or very wide door panels.

■ **Impact resistance** may be a consideration for high frequency use vehicular entrances.

• Generally sectional and roll-up types are easily damaged, but also are simpler to repair.

CLEARANCES

Clearance required for the proper operation should be considered.

SWINGING TYPE ENTRANCES

■ **Width** of panels generally limited by allowable loads on hinges and frames.

■ **Width and height** limited for panels requiring fire-resistance ratings.

■ **Select surface** mounted hinges when appearance is not a consideration.

■ **Use mortise or unit locks/latches** for durability and/or security.

■ **Power operation** should be considered:

• For high frequency use entrances.
• For oversized door panels.
• When opening speed is important.

■ **Bifolding, and four-folding types** generally

• Require top and bottom guides.
• Bottom guides subject to snow, ice, and dirt accumulation.

■ **When speed of opening is important,** consider using:

• Bifolding instead of double panel type or
• Even four-folding which will provide faster opening than the bifolding type.

SLIDING ENTRANCES

■ **Width and height of panels** generally not limited except for panels requiring fire-resistance ratings.

■ **Top track limited to** smaller sizes of single panel and biparting types.

■ **Inclined top track** with fusible link catch available for normally open entrances required to be self-closing in case of fire.

■ **Counterweights** may be added:

• For ease in opening and closing.
• To make door with horizontal track self-closing in case of fire.
• Clear space should be provided near door for counterweights.

■ **Multipanel type** generally power operated using endless chain, or motor-driven bottom rollers.

■ **For multipanel type in wide openings,** excessive deflection of framing, which supports the top guides, will prevent opening or closing of door.

■ **Side-coiling** type may also be used with curving tracks. Only top track with bottom guides can be used for this type.

VERTICAL RISE ENTRANCES

■ Width and height of panels generally not limited except for panels requiring fire-resistance ratings.

■ Multipanel vertical slide and canopy types generally used for special conditions, and are individually designed for each application.

■ Sectional and roll-up types most widely used; standard sizes are available.

■ Roll-up types are limited in width by structural properties of slats, which have to transmit wind forces to jamb guides.

■ Multipanel vertical sliding type may also be used with curving tracks.

B2.9 ENTRANCES/DOORS

FRAMES: types, uses, materials

● Denotes common usage or availability
☐ Denotes limited availability

USES

PANEL TYPE

TYPE	DESCRIPTION	SWINGING	SLIDING	REVOLVING	FIXED [1]
BUILT-UP					
SECTION / PLAN	• Rabbeted frame, or rectangular frame, with applied stop. • Exterior casing, drip cap, and interior trim generally required. • Frame secured to rough buck in framed walls.	●			●
	• Rectangular frame with extruded metal sub-frame secured to it. • Exterior casing, drip cap, and interior trim generally required.			●	
	• Structural shape, or heavy bent plate with applied stop. • Metal strap anchors are welded to frame at jambs to secure frame to wall.	●			●
BRAKE or ROLL FORMED					
SECTION / PLAN	• Brake formed sheet metal, either single rabbeted, or double rabbeted. • Anchors provided at jambs to secure frame to wall: • Masonry and dry wall types. • Thermal-break type available. • Usual gauge of metal: 14, 16 and 18.	●			
	• Brake formed sheet metal sticks, assembled in various configurations either single rabbeted, or double rabbeted. • Applied stops used to secure fixed panels in frame. • Section at walls similar to standard brake formed frame. • Usual gauge of metal: 14 and 16.	●			●
EXTRUDED • FORMED • TUBULAR					
SECTION / PLAN	• Extruded shapes, with applied or integral stops. • Thermal-break types available. • Adjustable types available. • Usual thickness of metal: ⅛ inch.	●	●	●	●
	• One or two section tubular. • Available with integral or applied stops. • May be internally reinforced. • May be assembled in various configurations to receive operating and fixed panels. • Usual thickness of metal: ⅛ inch.	●			●

B2.9 ENTRANCES/DOORS

B • BUILDING SHELL

| MATERIALS | REMARKS |

MATERIALS

WOOD
CARBON STEEL, heavy
CARBON STEEL, sheet
ALUMINUM
BRONZE
STAINLESS STEEL

REMARKS

1 Frame can be used to hold a fixed light of glass, such as sidelights or transom; opaque infill panels may also be used.

- Assembled frames with pre-hung doors available.
- Two piece, adjustable frames are available.
- When wood frames are used in masonry walls, a sub-frame is recommended.
- Structural shape or bent plate frames generally limited to industrial type construction; hinges are generally surface mounted, recess for latch is drilled in the field.
- Drip cap at head recommended.
- Closers cannot be concealed in frame; may be surface mounted on frame or on door.
- Wood and structural shape frames generally prepared to receive hardware in the field.
- Wood frames and trim for sliding doors available clad in aluminum or plastic.

- Usually available bonderized and prime painted; galvanized metal available when specified.
- Various sizes and shapes available; wrap-around shapes generally used for drywall construction.
- When installed in masonry walls, jambs and head generally filled solid with mortar.
- Prefabricated shapes in standard lengths — the stick system — available for local fabrication of entrance assemblies.
- Drip cap recommended when face of frame is flush with outside face of wall.
- Frames should be factory prepared to receive hardware.

- Aluminum extrusions available clear or color anodized, as well as painted.
- Various shapes and sizes available.
- Extruded sections with curved glass or metal fixed panels are used for revolving door enclosures.
- Drip caps, either attached to frame, or installed in wall at head of frame, recommended when face of frame is flush with outside face of wall.
- Frames should be factory prepared to receive hardware.

B2.9 ENTRANCES/DOORS

DOOR CLOSERS: types, uses

● Denotes common usage or availability

TYPE			USES DOOR TYPE		
			SINGLE ACTING, hinged	SINGLE ACTING, pivoted	DOUBLE ACTING
OVERHEAD					
	SURFACE MOUNTED	● **Location:** mounted on hinge face of door, or on stop face of door. ● **Some types** may be bracket mounted on door frame.	●	●	
	CONCEALED IN FRAME	● **Location:** installed in prepared recess in door frame, or in transom bar. ● **Available** with concealed arm, or exposed arm.	●	●	
	CONCEALED IN DOOR	● **Location:** installed in prepared recess in door panel. ● **Available** with exposed arm, or concealed arm.	●	●	
FLOOR					
	OFFSET PIVOT	● **Location:** installed in a recess in the floor slab. ● **Used with** offset pivoted doors.	●	●	
	CENTER PIVOT	● **Location:** installed in a recess in the floor slab. ● **Used with** center pivoted doors.	●	●	●
	HINGED	● **Location:** installed in a recess in the floor slab. ● **Used with** hinged doors.	●		

B2.9 ENTRANCES/DOORS

SERVICE CONDITIONS			OPTIONS								REMARKS
			Operators			Controls			Holders		
LIGHT DUTY	HEAVY DUTY	POWER OPERATED	ELECTRIC	PNEUMATIC	HYDRAULIC	SENSOR	PHOTOELECTRIC	ACTUATOR IN MAT	MECHANICAL	ELECTRO-MAGNET	• Power operators are also available for sliding and revolving doors. • All closers offer closing speed adjustment. • Some closers offer one or more of the following features: latching speed adjustment; back check; dead stop or selective hold-open.
•	•	•	•	•	•	•	•	•	•	•	• For outswinging exterior doors use special brackets to keep closer inside. • Closer size will vary with door size and weight. • Available with power assist feature for easier opening. • Available normally disengaged; will automatically engage and close door when sensor activated by fire or smoke. • Should be mounted on interior side of entrance.
•	•	•	•	•	•	•	•	•	•	•	• Head frame must be of sufficient depth to accept closer. • Concealed arm type not recommended for moderate or severe exposure. • Door and/or frame must be notched at arm. • Factory preparation of frame to receive closer required. • Rigidity of header should be checked. • Pivoted type available.
•									•		• Not generally available for fire rated assemblies. • Not recommended for stile and rail doors unless rail is of required depth. • Door and/or frame must be notched at arm. • Factory preparation of door to receive closer required.

LIGHT DUTY	HEAVY DUTY	POWER OPERATED	ELECTRIC	PNEUMATIC	HYDRAULIC	SENSOR	PHOTOELECTRIC	ACTUATOR IN MAT	MECHANICAL	ELECTRO-MAGNET	
•	•		•	•		•	•	•			• Floor closers are normally more durable than overhead closers. • Recommended for high frequency use entrances. • Floor closers must be coordinated with threshold selection. • Hold-open and stops available. • Recess in floor required. Before selecting a floor closer, it should be ascertained that such recess can properly be provided in the floor assembly. • Floor closers most commonly used with pivoted doors.
•	•		•	•		•	•	•			
•											

B2.9 ENTRANCES/DOORS

DOOR HINGES: types, uses

● Denotes common usage
□ Denotes possible usage

TYPE	DESCRIPTION	FLUSH	STILE AND RAIL	SINGLE ACTION	DOUBLE ACTION
BUTT					
	Full Mortise ● Two equal square-edged leaves. ● Location: One leaf mortised into edge of door panel, the other into rabbet in frame. ● Standard weight, and heavy weight. Extra heavy weight available. ● Usual height: 4½", 5", 6".	●	●	●	
	Half Surface ● Two equal leaves; one square-edged, the other bevel-edged. ● Location: Square-edged leaf mortised into rabbet in frame; bevel-edged leaf mounted on face of door. ● Standard weight, and heavy weight. ● Usual height: 4½", 5", 6".	●		●	
	Half Mortise ● Two equal leaves: one square-edged, the other bevel edged. ● Location: Square-edged leaf mortised into edge of door; bevel-edged leaf mounted on face of frame. ● Standard weight, and heavy weight. ● Usual height: 4½", 5", 6".	●	□	●	
	Full Surface ● Two bevel-edged leaves of unequal size. ● Location: Both leaves surface mounted; one on face of frame, the other on the face of door. ● Standard weight, and heavy weight. ● Usual height: 4½", 5", 6".	●		●	
BUTT • SPECIAL					
	Swing-Clear — Full Mortise **Also: Half Surface; Half Mortise; Full Surface.** ● Location: same as for respective type of Butt hinges. ● All types heavy weight; only some types in standard weight. ● Provides unobstructed clear frame opening when door is opened 90° ● Usual height: 5" for heavy weight.	●		●	
	Spring Hinge ● Double acting. Single acting. ● Mortised into door and frame; or mortised into frame, and attached to both faces of door with clamp flanges.	●	●	●	●

B2.9 ENTRANCES/DOORS

FRAME TYPE					SERVICE CONDITIONS			REMARKS
BUILT-UP, wood	BUILT-UP steel shape	BRAKE FORMED	EXTRUDED	EXTRUDED, tubular	HEAVY USAGE	OVERSIZE DOOR	LINED DOOR	
								• Materials: plated steel, brass/bronze, stainless steel. • Standard weight and heavy weight types with anti-friction or ball bearings. • Light weight type generally without bearings.

BUILT-UP, wood	BUILT-UP steel shape	BRAKE FORMED	EXTRUDED	EXTRUDED, tubular	HEAVY USAGE	OVERSIZE DOOR	LINED DOOR	REMARKS
●	●	●	□		●	●	□	• Heavy weight recommended for high frequency use doors, oversize doors, and heavy doors, such as lead-lined doors. • Three hinges should generally be used, except for light, low frequency use doors. • Light weight hinges generally used for light, low frequency use doors only. • Half mortise and full surface types generally used with metal clad or hollow metal doors in built-up steel shape frames. If full mortise type is used with such frames, door must be double-mortised.
●	●	●			●	●	●	• Non-removable pins available for added security. • Cam-lift types available. • Excess force imposed by top closers, holders, or stops, oversized or heavy doors, traffic abuse, wind conditons, all have to be considered in the selection of hinges. • Accurate installation critical for proper functioning of all other hardware. • Minimum recommended size: 4½ inches high by 4 inches wide.
●					●	●		
●					●	●	●	

BUILT-UP, wood	BUILT-UP steel shape	BRAKE FORMED	EXTRUDED	EXTRUDED, tubular	HEAVY USAGE	OVERSIZE DOOR	LINED DOOR	REMARKS
●	●	●	□	□	●	●	●	• Use, installation, and limitations similar to those for regular butt hinges.
●		●	□					• Single acting type available with U.L. rating for use in fire-rated assemblies. • Single acting types similar in appearance to full mortise butt hinges.

B2.9 ENTRANCES/DOORS

DOOR HINGES: types, uses (cont.)

● Denotes common usage
☐ Denotes availability on special order only

TYPE	DESCRIPTION	USES — DOOR TYPE			
		FLUSH	STRIKE AND RAIL	SINGLE ACTION	DOUBLE ACTION
PIVOTS					
	Pivot Reinforced ● Location: Hinge leaves mortised into edge of door panel, and rabbets in frame at jamb and head. ● Heavy weight only.	●		●	
	Offset Pivot ● Location: Mortised into top and bottom edges of door panel, and into jamb of frame at bottom, and head of frame at the top. ● Floor mounted bottom pivots available. ● Jamb mounted intermediate pivots available.	●	●	●	
	Center Pivot ● Location: Mortised or attached to top and bottom edges of door panel, and into jamb of frame at bottom, and head of frame at the top. ● Floor mounted bottom pivots available.	●	●		●
SPECIAL					
	Invisible Hinge ● Location: Full mortised into edge of door and jamb of frame.	●	☐	●	
	Paumelle Hinge **also Olive Knuckle Hinge** ● Location: Full mortised into edge of door and jamb of frame.	●	●	●	

B2.9 ENTRANCES/DOORS

FRAME TYPE					SERVICE CONDITIONS			REMARKS
BUILT-UP, wood	BUILT-UP, steel shape	BRAKE FORMED	EXTRUDED	EXTRUDED, tubular	HEAVY USAGE	OVERSIZE DOOR	LINED DOOR	

BUILT-UP, wood	BUILT-UP, steel shape	BRAKE FORMED	EXTRUDED	EXTRUDED, tubular	HEAVY USAGE	OVERSIZE DOOR	LINED DOOR	REMARKS
●		●	□		●	●	●	• Materials: plated steel, brass/bronze, stainless steel, aluminum. • Recommended for high frequency use doors with overhead closers, and for any exterior door subject to severe wind conditions and sudden shock openings or stops. • Pivots are stronger and more durable than hinges and more resistant to racking stresses to which doors are subject. • Pivot knuckles of offset type are visible when door is closed. • Pivots of center type are completely invisible when door is closed.
□	□	●	●		●	●	□	• Double acting pivots available to allow single action door to be swung in opposite direction in emergencies. • Center pivots always required on double-acting doors. • Center pivots generally required on automatically-operated doors. • Some types allow for both vertical and horizontal adjustment. • Offset pivots: two for doors up to 7½ feet in height; three for higher doors.
□	□	●	●		●	●	●	

BUILT-UP, wood	BUILT-UP, steel shape	BRAKE FORMED	EXTRUDED	EXTRUDED, tubular				REMARKS
●		●	□	□				• Hinge not visible when door is closed. • May be used for high security rooms. • Installation more difficult than for other types.
●		●	●	●				• Only knuckle is visible when door is closed.

B2.9 ENTRANCES/DOORS

LOCKS: types, uses

● Denotes common usage

TYPE		DESCRIPTION	SINGLE ACTING	DOUBLE ACTING	SLIDING
LATCH/LOCK					
	BORED	• Lock uses the key-in-the-knob principle. • Installed through holes bored in the door panel faces and edge. • Also known as a cylindrical lock. • In locations not requiring lock action the unit can be provided with a latch only.	●		
	MORTISED	• The lock assembly is installed through the edge of the door in a prepared recess, or mortise. • All parts are contained in a rectangular case, with holes for cylinder and knob spindle. • Lever handles may be used instead of knobs.	●		
	PREASSEMBLED	• Preassembled, or unit locks are completely assembled at the factory. • All parts are contained within a cast or extruded metal frame, usually brass. • The unit is installed in a notch cut in the door edge.	●		
DEADBOLT					
	BORED	• The construction of a bored dead bolt is similar to that of a bored lock. • The deadbolt is different in that the lock bolt is not beveled or spring loaded. • The lock bolt is controlled by a key or thumb turn. • Mortise type, similar to mortise lock in construction is available.	●	●	
	MORTISED	• Similar in construction to mortise lock; hook bolt instead of straight bolt used for sliding panels.			●
	THREE POINT **Three point latch —** **panic hardware**	• Mortised type deadlock which also simultaneously operates two flush bolts in either leaf of a double door, or within a single door panel. • Two point type available. • Three-point and two-point latches available for fire-rated doors.	●	●	

SERVICE CONDITIONS				OPTIONS				REMARKS
LIGHT DUTY	HEAVY DUTY	HIGH SECURITY	ADDED FOR SECURITY	ELECTRIC RELEASE	TIME DELAY	CARD CONTROL	ROTATING LATCH	
•	•			•			•	• Bored locks come in light, medium, and heavy duty as well as security grades. • A wide variety of functions are available. • Comply with Fed. Spec. 160 or 161 standards. • Good range of finishes. • Some types are rated for fire assemblies. • Not recommended for security "type" conditions. • Security properties are improved by use of separate deadbolt. • Flush and surface bolts added for double doors.
•	•			•	•	•	•	• Mortise locks come in heavy duty and extra heavy duty. • Best standard lock mechanism for security. Wide range of armoured/vandalproof accessories. • Generally more durable than the other lock types. • Wide variety of functions are available. • Comply with Fed. Spec. 106, series 85, 86, 87. • Wide variety of trims and knob sets.
•	•							• Preassembled locks come in various duty grades. • Stile and rail doors need sufficient stile width to accept lock notch. • Good range of finishes. • Some types are rated for fire assemblies.

SERVICE CONDITIONS				OPTIONS				REMARKS
LIGHT DUTY	HEAVY DUTY	HIGH SECURITY	ADDED FOR SECURITY	ELECTRIC RELEASE	TIME DELAY	CARD CONTROL	ROTATING LATCH	
		•	•	•	•	•	•	• Give good security features when used with either bored or preassembled locks. • Normally a bored deadbolt will fit the same cut out of a bored lock. • Stile-and-rail door member sizes may not provide enough recess depth; check specific manufacturers. • Can be used on doors having push/pull plates for locking. • Provide added security on doors with other locks. • Surface mounted deadbolts available. • Deadbolt with less than ¾-inch throw does not provide adequate security on aluminum entrance doors. • For maximum security, hardened steel inserts in bolt are required to prevent forced entry by cutting. • Dead locks may be operated by single cylinder or thumb turn, a pair of cylinders, or a combination of both. For maximum security, double cylinder operation is recommended. • Some types available with indicators to show whether door is locked or unlocked.
•	•							
•	•							

B2.9 ENTRANCES/DOORS

WEATHER-STRIPPING: jambs, head, meeting stiles

● Denotes common usage
☐ Denotes availability on special order only

TYPE	DESCRIPTION	FLUSH	STILE AND RAIL	SINGLE ACTION	DOUBLE ACTION
METAL SEAL					
	● Material: Spring type metal. ● Location: Between frame and door panel at jambs and head; or between stiles. ● Performance: Long wearing; low maintenance.		●	●	●
	● Material: Interlocking elements of spring type or extruded metal. ● Location: Surface mounted or mortised into door panel or frame. ● Performance: Long wearing; low maintenance.		●	●	●
COMPRESSIBLE SEAL					
	● Material: Compressible filler with or without metal trim. ● Location: Surface applied to door panel; or to jambs and head of frame. ● Performance: Medium wearing; medium maintenance. ● Magnetic type available for metal doors and/or frames.		●	●	●
	● Material: Compressible filler held in metal trim. ● Location: Half mortised or surface mounted at meeting stiles. ● Performance: Medium wearing; medium maintenance. ● Spring loaded swing-away, or magnetic type astragals available.		●	●	●
	● Material: Compressible filler held in metal trim. ● Location: Surface mounted or mortised into edge of door panel. ● Performance: Medium wearing; medium maintenance.	●	●	●	●
SPECIAL					
	● Material: Resilient material, interlocking elements of spring type metal, or extruded metal. ● Location: Surface mounted to edges or sides of sliding or revolving door panels. ● Performance: Medium to long wearing; medium to low maintenance.	●	●		

B2.9 ENTRANCES/DOORS

MATERIALS

SEAL					SPRING TYPE				TRIM			REMARKS
WOOD PILE	NEOPRENE, sponge	NEOPRENE, extruded	WATERPROOF FELT	VINYL, extruded	ALUMINUM	BRONZE	STAINLESS	ZINC	ALUMINUM	BRONZE	STAINLESS	
												• Dissimilar materials must be isolated when used on metal frames or door panels.
					•	•	•	•				• Will not compensate for warpage or thermal movement of entrance components. • Very limited adjustment. • Also available in plastic.
					•	•	•	•				
												• Will not compensate for warpage or moderate thermal movement of entrance components. • Limited adjustment available. • Trim suitable for hinge side only; cannot be used at head. • Extruded shapes available for heavy duty doors. • Generally mortised for wood doors and frames; surface mounted for metal doors and frames.
•	•	•	•	•					•	•	□	• Sponge neoprene without metal trim subject to high wear and maintenance. • Not recommended for extremely low temperatures. • Spring loaded swing-away astragals used for security. • Astragals required for fire rated double doors. • Some types cannot be adjusted. • ASTRAGAL is a vertical molding secured to the meeting edge of one leaf of a double door as protection from weather; or to minimize passage of light; or to retard passage of smoke, or other products of combustion in fire rated assemblies. • Flap seals of rigid/flexible vinyl for swinging doors available.
•	•	•							•	•	□	
•	•	•							•	•	□	
	•	•	•	•	•	•	•		•	•	•	• Various other types available for sliding doors. • Extruded neoprene seal generally used for service or industrial type doors. • Flap seals of rigid/flexible vinyl for swinging doors available.

B2.9 ENTRANCES/DOORS

WEATHER-STRIPPING: saddles, seals

● Denotes common usage
□ Denotes availability on special order only

USES
DOOR TYPE

TYPE	Description	FLUSH	STILE AND RAIL	SINGLE ACTION	DOUBLE ACTION

METAL SEAL

| | • Seal: Spring type or extruded metal.
• Location: Surface mounted on face of door panel.
• Performance: Long wearing; low maintenance.
• Special type saddle required. | ● | ● | ● | |

| | • Seal: Spring type or extruded metal.
• Location: Mortised into bottom edge of door panel.
• Performance: Long wearing; low maintenance.
• Special type saddle required. | ● | ● | ● | |

COMPRESSIBLE SEAL

| | • Seal: Compressible.
• Location: Mortised or surface applied to bottom edge of door panel; or attached to saddle.
• Performance: Medium wearing and maintenance.
• When attached to saddle, special type of saddle required. | ● | ● | ● | ● |

| | • Seal: Compressible seal in special housing torsion bar activated. Resilient seal with metal trim.
• Location: Mortised into bottom edge of door panel; semi-mortised; surface mounted.
• Performance: Medium wearing and maintenance. | ● | ● | ● | |

| | • Seal: Compressible. Integral or applied metal stop.
• Location: Compressible seal attached to or inserted in saddle. Metal stop integral or attached to saddle.
• Performance: Long wearing, low maintenance for metal stop. Medium wearing and maintenance for compressible seal.
• Special type of saddle may be required. | ● | ● | ● | |

B2.9 ENTRANCES/DOORS

MATERIALS

SEAL COMPRESSIBLE					SPRING TYPE OR EXTRUDED				SADDLE		REMARKS
WOOL PILE	SPONGE NEOPRENE	EXTRUDED NEOPRENE	EXTRUDED VINYL	WATERPROOF FELT	ALUMINUM	BRONZE	ZINC	STAINLESS	ALUMINUM	BRONZE	
											• Dissimilar materials must be isolated. • Saddles for exterior doors should be set in a full bed of caulking. • Some types of saddles available with a thermal break. • Some types of saddles available with non-slip surface treatment. Some means to provide a non-slip surface highly recommended.
					●	●	●	□	●	●	• Use generally restricted to light traffic doors. • For in-opening doors a rain drip is recommended. • Mortise type seal generally used for wood doors only. • Available with sponge neoprene inserts for better sealing. • Very limited adjustment. • Saddles should project above finished floor as little as possible. • Attachment of saddle generally by machine screws and expansion shields. • Saddles should not be used to cover inaccuracies in floor construction. • Saddles available in: aluminum, brass, or bronze.
					●	●	●	□	●	●	
●		●	●						●	●	• Seals attached to saddle subject to wear and damage due to traffic. • Rain drip recommended. • Spring type metal, or plastic, attached to bottom edge of door available for use with flat saddles. • Limited adjustment.
●	●	●	●						●	●	• Torsion bar operated drop seal available for surface mounting. • Door panel must be specifically prepared to receive recessed drop seal. • Can be adjusted. • Rain drip recommended. • Also used on sound and light proof interior doors.
●		●	●						●	●	• Generally for light traffic doors, and doors with panic hardware. • Dirt accumulation may prevent proper sealing. • Stops available with compressible insert. • Rain drip recommended. • Very limited adjustment.

B2.9 ENTRANCES/DOORS

DOOR HARDWARE: miscellaneous

AUTOMATIC DOOR OPERATORS

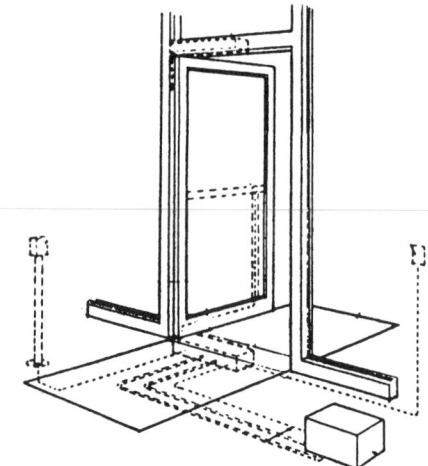

ULTRASONIC DEVICE
OPERATOR-OPTIONAL
LOCATIONS.

CONTACT SWITCH

PHOTOELECTRONIC
CELL

DEMOTE SWITCH

FLOOR MAT

POWER UNIT

Components of automatic door operators are similar for both swinging and sliding doors:

- Modes of operation available are: electrohydraulic; electropneumatic; and electromechanical.
- Center-pivoted doors are normally used for automatic operation, with the pivots, either top or bottom, being part of the door operator. Operators may be recessed in the floor or mounted in transom bar of frame.
- Overhead surface mounted operators may be added to existing swinging doors.
- Activators most commonly used are: floor mats; ultrasonic devices; photoelectric cells; contact switches on push bar of door. Operators may also be activated by remote switches; kick plate activated switches; pull cords. Special provisions may be required if a swinging door has to accommodate two way traffic.
- Panic breakaway and automatic reset is generally provided.
- Opening speeds are adjustable.
- Provisions for doors to close automatically when power fails, and to be operated manually may be included.

PANIC EXIT DEVICES

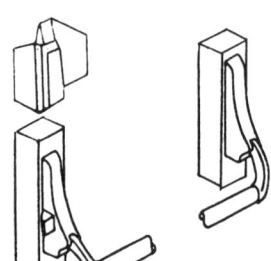

LATCH FOR
CONCEALED ROD
TYPE-LATCHES
TOP AND BOTTOM

LATCH FOR CENTER
RIM LATCH-BOLT TYPE

- Panic exit devices function to permit safe, instant emergency exit without need for keys, tools, or special knowledge, through a door normally locked against entry. Panic exit devices also serve as a push bar for normal exit traffic.

Two basic types are available:

- Center latch-bolt type: either rim or mortise. Used for single door; active door of a pair of doors; both doors of pair with center mullion.
- Vertical rod type: either exposed rod, or concealed rod. Used for: single door; inactive door of a pair of doors; both doors of pair.

OVERHEAD HOLDER

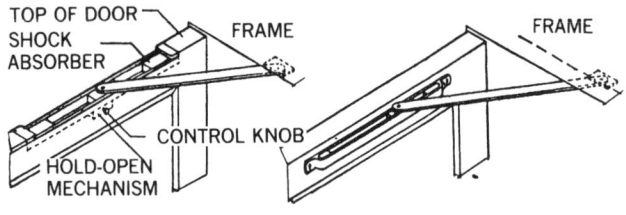

TOP OF DOOR

SHOCK
ABSORBER

FRAME

FRAME

CONTROL KNOB

HOLD-OPEN
MECHANISM

- Closers may incorporate hold-open feature, but severe load conditions, such as winds in exposed location, may necessitate a separate device. In some instances a selective hold-open device may be desirable.
- Whenever possible, the minumum hold-open angle should be 105 degrees. Care should be taken that holders do not interfere in the operation of other hardware in the same general location.
- Holders with fusible links or detectors to release open doors in case of fire are available.

DEADLATHES

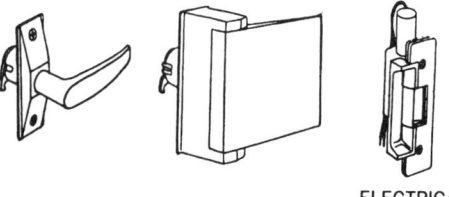

ELECTRICAL

- Deadlatches function to automatically secure door upon closing.
- Unlatching may be accomplished by: key only; outside key-inside lever; electrical operation through remotely operated strike.
- When used on pair of doors, inactive leaf must be secured, and active leaf should have a header stop.
- Due to short bolt travel, deadlatches should not be used as security devices. An additional deadlock should be provided for security.

FLUSHBOLTS

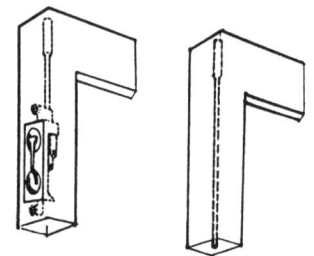

Flush bolts function to secure inactive leaf of a pair of doors, usually at both top and bottom of door.

- Automatic flush bolts are available, which hold the inactive leaf in closed position until the active leaf is opened.
- Building codes may restrict or prohibit the use of flush bolts for specific occupancy groups.
- COORDINATOR: is a device used on a pair of swinging doors to prevent active leaf closing before inactive leaf closes.

B2.9 ENTRANCES/DOORS

FRAMES, HARDWARE: selection checklist

The entrance frame can be a simple frame surrounding a door panel, or a series of members holding fixed panels as well as the door panels.

■ Entrance frame must be strong enough to **resist wind load** without excessive deflection, to avoid the possibility of deformation of the frame with resultant:
• Cracking of fixed glass panels.
• Binding of door panels.
• Opening of joints in frame.

■ The frame and the trim may be integral or made of separate pieces.

■ Door frames in **wood construction:**
• Require rough bucks.
• Joints between frame and wall are covered by trim.
• Finished wood frames are generally field fitted to exact conditions.
• Many standard profiles can be made for a specific design usually at additional cost.

■ Door frames in **metal construction:**
• Are often set in before the wall is filled in and serve as framing members.
• Frames and doors may be pre-hung and pre-assembled and come as a package.
• Protection of metal frames and pre-hung doors during installation and remainder of construction often requires considerable care.

■ Metal entrance frames **may be of:**
• Built-up rolled sections.
• Brake formed metal sections.
• Roll formed metal sections.
• Extrusions.

CLOSERS
The next major entrance component to be considered are closers, manual and powered:

■ **Closers** for doors are either overhead or floor-type and either fully concealed, semi-concealed, or surface mounted.

■ **All manually operated** closers require a certain amount of force to open the door panel.

■ Therefore, depending on the type of entrance desired, **power-actuation** may be required:
• Oversize doors may become too heavy to be opened manually.
• All entrance types are generally available with power-actuation.

When selecting power-actuation, consider safety:

■ People must be protected from power-actuated door leaves. Review the **type of sensors** to be used.

■ Swinging doors normally require **guard rails** or other architectural barriers to protect people from their swing.

■ Sliding doors need **a pocket or other barrier** to prevent contact with people during operation.

■ Revolving doors must have a **speed control** to limit number of revolutions.

HINGES, PIVOTS
The other entrance components which have to be considered simultaneously with the selection of closers are hinges, or devices on which doors turn or swing, to open and close:

■ **Hinges** may be concealed or exposed.

■ **Butts** are the most common type of hinge used today. They are usually mortised into the edge of the door.
• Are generally mounted on a door 5" from the head and 10" from the floor.
• When a third butt is required to minimize warping of door, it is mounted equidistant between top and bottom hinges.

■ **Pivots** are stronger and more durable than hinges and are better able to withstand the racking stresses to which doors are subjected. Their use is generally recommended for:
• Oversize doors, heavy doors.
• Entrance doors of high frequency use.

LOCKS, LATCHES
Locks and latches are used to hold doors in closed position:

■ **A deadbolt** is often used in conjunction with a latch, in which case the unit is known as a lock.
• For doors which need not be latched or locked during the normal work day, Push Pull plates are normally used in lieu of latch sets or locks.
• Doors with push pull plates may be provided with a dead bolt if there is a need to secure them at certain times.
Hardware is a factor in determining the **ultimate security** of the entrance:

■ **Mortise** locks are the most secure type of lock.
• Deadbolts provide superior protection to latch bolts.

■ The proper location of door **silencers** on swinging entrances can prevent latch lock tampering.

■ Special armor plates are available for protecting lock cylinders.

■ Electronic latches, hinges, card readers, and other devices are available for specific security requirements.

■ **Panic devices** for mass exit in emergencies, are installed on exterior doors which serve as legal exits from a building.

Entrance, exit and door hardware is a very diverse and highly intricate subject. Architects rarely specify such hardware without the benefit of the expertise of a hardware consultant.

Installation of some entrance types and component types create special requirements:

■ **One piece** hollow metal frames are normally installed before the wall construction is complete.

■ **Recessed** floor closures need openings in floor slabs and adequate slab depth.

■ **Revolving** door entrances require special care in the installation of the floor within the enclosure for smoothness and flatness.

■ Hardware should be reviewed to determine its **operating life.**

SOME DEFINITIONS:
• **Automatic closing device:** causes the door to close when activated by detector through rate of temperature rise, smoke, or other products of combustion.
• **Automatic closing door:** is normally in open position, and is closed by automatic closing device in case of fire.
• **Center latch:** is used to hold two leaves of bi-parting doors together.
• **Self-closing door:** will return to the closed position after having been opened and released.

B2.9 ENTRANCES/DOORS

REFERENCES: Entrances/Doors

STANDARDS

ASTM Tests/Methods for:

E283
- Rate of Air Leakage Through Exterior Windows, Curtain Walls and Doors

E330
- Structural Performance of Exterior Windows, Curtain Walls and Doors by Uniform Static Air Pressure Difference

E331
- Water Penetration of Exterior Windows, Curtain Walls and Doors by Uniform Static Air Pressure Difference

ANSI Specifications:

ANSI/AAMA 402.9-1977
- Aluminum Sliding Glass Doors

A151.1-1980
- Performance Test for Standard Steel Doors, frames, anchors, hinge reinforcings, and exit device reinforcings

A123.1-1982
- Nomenclature for Steel Doors and Steel Door Frames

A117.1-1980
- Specifications for Making Building and Facilities Accessible to and Usable by Physically Handicapped People

**Series
A115.1-1982**
A115.17
- Covers the Preparation of Doors and Frames for Bolts, Closers, Latches, Locks, Pivots and Strikes

A156.3-1978
- Exit Devices (BHMA701)

A156.4-1980
- Door Controls - Closers (BHAM301)

A156.6-1979
- Architectural Door Trim (BHMA1001)

A156.1-1981
- Butts and Hinges (BHMA)

FS   **Federal Specifications**

RR-W-365(1)
- Wire Fabric (Insect Screening)

RR-D-575B • Door, Metal, Sliding and Swinging: Door Frame, Metal (Flush and Semi-flush)

FF-H-106C/Gen
- Hardware, Builders'; Locks and Door Trim: General Specification for

FF-H-106/1
- Hardware, Builders'; Locks and Door-Trim: Cylinder Entrance Door Type 121A and 122A

FF-H-106/2
- Hardware, Builders'; Locks and Door Trim: Cylinder Entrance Door Type 123A and 123B

FF-H-106/3
- Hardware, Builders'; Locks and Door Trim: Letter Box Plates

FF-H-00116D
- Hinges, Hardware, Builders'

FF-H-121D • Hardware, Builders'; Door Closers

FF-H-1819 • Hardware, Builders'; Auxiliary Devices

FF-H-1820 • Hardware, Builders'; Exit Devices

AA-C-30D • Cabinet, Key (Boxes and Racks, Metal, and Identification Systems)

FF-H-111C • Hardware, Builders'; Shelf and Miscellaneous

SELECTED REFERENCES

AA   **The Aluminum Association**
ASD-1
- Aluminum Standards and Data

DAF-45
- Designation System for Aluminum Finishes

SAA-46
- Standards for Anodized Architectural Aluminum
- Specifications for Aluminum Structures, 1971

ED-33
- Engineering Data for Aluminum Structures
- Drafting Standards for Aluminum Extruded and Tubular Products

---

AAMA   **Architectural Aluminum Manufacturers Association**

402.9
- Voluntary Specifications for Aluminum Sliding Glass Doors

1102.7
- Voluntary Specification for Aluminum Storm Doors

1303
- Voluntary Specifications for Forced-Entry-Resistant Aluminum Sliding Glass Doors

AWI   **Architectural Woodwork Insititute**
- Architectural Woodwork Quality Standards, Guide Specification and Quality Certification Program Section 1300-Architectural Flush doors

BHMA   **Building Hardware Manufacturer's Association**

BHMA-1601
- Power Operated Pedestrian Door Standard

BHMA-301 • Door Controls (Closers) (ANSI A156.4-1980)
- Product Standards and Performance Specifi cations
- Hardware for Labeled Fire Doors
- Standardization for Terms and Nomenclature of Keying
- Materials and Finishes Standard
- Sliding and Folding Door Hardware
- Locks and Lock Trim

CS   **Construction Specifier**
- Scheduling of Doors, Frames and Hardware

DHI   **Door & Hardware Institute**
Society of Architectural Hardware Consultants
- Tech Talk Bulletins SP-1 through SP-4- Specifications
- Tech Talk Bulletin EH-1 - Electronic Hardware
- Tech Talk Bulletin HH-1 - Hospital Hardware Problems
- Tech Talk Bulletin HTL-1 - Hotel Keying

- Tech Talk Bulletin HTL-2 - Hotel/Motel Hardware
- Architectural Hardware Schedule Sequence and Scheduling Format, January, 1974
- Recommended Procedure for Processing Hardware Schedules and Templates, December, 1974
- Keying Procedure and Nomenclature, February, 1975
- Hardware Reinforcements on Steel Doors and Frames

NAAMM   **National Association of Architectural Metal Manufacturers**
- Hollow Metal Technical & Design Manual
- Fire-Rated Custom Doors & Frames, 1974
- The Entrance Manual
- Custom Hollow Metal Doors and Frames
- Hollow Metal Technical Design Manual, 1977
- Metal Finishes Manual, 1976

NFPA   **National Fire Protection Association**

NFPA 80
- Standard for Fire Doors and Windows (ANSI A2.7-73)

NFPA 252
- Standard Methods of Fire Tests of Door Assemblies

NWMA   **National Woodwork Manufacturers Association**

NWMA 1.S.3-70
- Wood Sliding Patio Doors

NWMA 1.S.1-80
- Wood Flush Doors

NWMA/ANSI 1.S.5-73
- Ponderosa Pine Doors Standard

SMA   **Screen Manufacturers Association**

SMS-2003
- Specifications for Aluminum Sliding Screen Doors

SDI   **Steel Door Institute**
- Standard Steel Doors & Frames
- Specifications
- Erection Instructions

- Standard Nomenclature
- Selection & Usage Guide
- Standard Details
- Standard Minimum Acceptance Values
- Standard Thermal, Acoustical, Water Resistance
- Rate of Air Flow Tests
- Manufacturing Tolerances

107
- Hardware on Steel Doors

109
- Hardware on Standard Steel Doors & Frames

UL   **Underwriters Laboratories, Inc.**

UL 63
- Safety Standard for Fire Doors and Frames (ANSI 155.1-70)

UL 10A
- Fire Test of Door Assemblies

UL 325
- Door, Drapery, Gate, Louver and Window Operators and Systems
- Fire Protection Equipment List
- Building Materials Directory

B2.9 ENTRANCES/DOORS

B2. 10 COATINGS

COATINGS: introduction

COATINGS DEFINED

Coatings are substances applied in liquid form which then solidify and become thin facings over surfaces of components. They are selected:

■ To protect the components from:
- moisture: rain, ground or sea water
- chemicals: in solution, or airborne
- corrosion: rust, galvanic action
- solar radiation
- soiling: by dirt, dust, graffiti
- abrasion: foot traffic, scrubbing
- heat: whether controlled or accidental such as fire.

■ To provide an aesthetically pleasing appearance.

■ Coatings should be viewed as an entity, consisting of:
- the surface, or substrate, prepared to receive the coating.
- the prime coats or undercoats, when needed.
- the finish coatings, or topcoats.

The entire coating system must be compatible through each layer, from the substrate to the topcoat. Any incompatibility between coating and substrate will reduce adhesion and accelerate deterioration. Incompatibility may result from:

■ Incompatible preservatives in wood;
■ Incompatible primers and top-coats: the solvent in the topcoat may act as a paint remover on the primer.
■ Applying a relatively hard topcoat over a relatively soft primer or undercoat; this may cause alligatoring in the topcoat;
■ Applying a coating over a surface which it fails to wet; e.g., a water-based coating over a glossy solvent-based coating may cause crawling or peeling;
■ Poor surface preparation:
- with oil, grease, chemicals remaining
- moisture collecting on the surface
- excessive moisture in the substrate.
■ Coatings which deteriorate and fail because of substrate defects or incompatibility may no longer be recoatable and must then be removed completely at considerable inconvenience and expense.

Coatings may be classified by their appearance as:
■ **Clear:** the surface of the component they are applied over is only slightly obscured.

■ Semi-transparent: the surface of the component they cover is partially obscured.
■ **Opaque:** the surface of the component covered is completely obscured and its texture may also be changed.

Within the constraints imposed by the protective properties available with each group, **the visual considerations for selection of coatings are:**

■ **Clear coatings: are used** when it is important to preserve the appearance of the surface of the substrate, e.g. the color and graining of wood, or the color of selected aggregate facing of concrete.

■ **Semi-transparent coatings:** are used when the surface of the substrate requires modification, e.g. when the color of wood is to be changed without hiding the grain.

■ **Opaque coatings:** are used when the color and/or texture of the surface of the substrate is to be changed for:
- aesthetic considerations
- control of light diffusion or reflection
- visual organization or identification through color coding.

COMPOSITION

Coatings are composed of a vehicle, alone when they are clear, or of a vehicle and pigments when opaque.

■ **The vehicle is** the liquid portion of a coating and generally consists of:
- a nonvolatile part, or the binder
- and a volatile part, the solvent.

■ **Binders form the film** of a coating, and bind or hold pigments, when such are used.

■ **Solvents dissolve** the binder to adjust the viscosity during application:

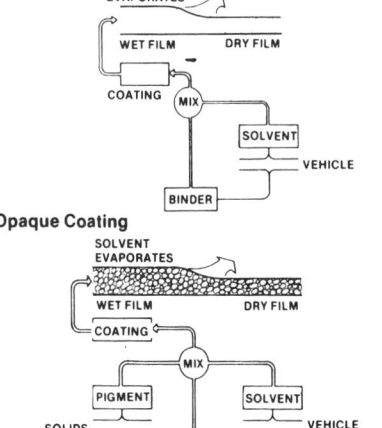

Clear Coating

Opaque Coating

Coatings are classified by com-composition into:
■ **Solvent based:** with binders containing or dissolved in organic solvents.
■ **Water-based:** with binders either soluble or dispersed in water.

pigments combined with the binder constitute the **solids** of a coating:
■ **Clear coatings** generally contain the vehicle only;
■ **Semi-transparent** coatings have a limited amount of pigment added to a vehicle; and are an intermediate step toward:
■ **Opaque coatings** which generally contain sufficient amounts of pigment in a vehicle to fully obscure the surface of the substrate.

BINDER

Binder is the fundamental constituent of all film-forming coatings:

■ **It bonds itself** to the substrate it is applied over, while incorporating all other ingredients used to impart to the coating any special properties required to meet in-service conditions, such as:
- driers to increase the rate of curing; plasticizers to impart flexibility or moisture resistance;
- stabilizers to control effects of heat and solar radiation;
- thinners to adjust viscosity for proper application.

■ **The properties of a coating** will largely depend on the properties of the binder used, or on intermediate properties when two binders are combined:

Coatings are quite complex, composed of many ingredients which may be combined in various proportions, with the properties of the film varying markedly in the final formulation. It is not possible to have all of the best properties in a single material, and compromises are inevitable:

■ **Binders composed of small molecules,** e.g., drying oils:
- penetrate rough surfaces and adhere well
- but are slow to dry and do not resist chemicals.

■ **Binders with large molecules,** built-up - or polymerized - of a large number of small molecules:
- yield strong, chemically resistant films

B2.10 COATINGS

- but when they are formulated to dry by solvent evaporation only are resoluble in the same solvent at any time.

■ **Large molecules in a coating may be attained by:**
- reaction between small molecules to form large ones after having been applied in a thin layer; e.g., in linseed oil.
- the polymer may be made before application and dissolved in a solvent to lower its viscosity for the purpose of application; the film is formed upon evaporation of the solvent.
- a combination of the two methods in various proportions.

■ **Coatings with good resistance to chemicals** usually have poor adhesion; conversely most coatings that adhere well are, because of their molecular structure, most easily attacked by chemicals. Also: the binder must remain liquid in bulk, but must solidify into a film when applied over a surface. For the film to solidify, large molecules are required, since small molecules may evaporate or remain in a liquid state when spread in a thin film:

PIGMENTS

Pigments add opacity and/or color to the coating film:

■ Depending on their type and concentration also **may increase the durability and protective characteristics** of the coating by:
- screening out ultraviolet radiation
- controlling transmission of moisture and gases
- inhibiting corrosion of substrate
- preventing degradation of substrate by air or water borne chemicals, or high temperatures.

The primary purpose of a pigment is to hide the substrate by preventing the transmission of light to the substrate, and its reflection back through the coating:

■ Colored pigments accomplish this primarily by absorbing some of the light rays and reflecting others.

■ White pigments absorb relatively little, thus their hiding power depends on their ability to scatter and reflect the incident light; this in turn depends on: the size of the particles, their distribution and their refractive index.

EFFECT OF PIGMENT CONTENT ON GLOSS OF COATING

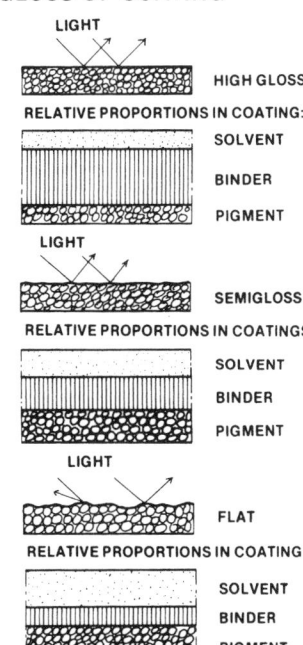

Principal Properties

Desirable properties of coatings may be selectively controlled by adjusting the formulation, except that when one property is improved, another may suffer and compromises must be made.

Properties which generally should be considered are:

■ **Flow:** or the ease with which a coating can be applied.
- too much ease may cause the coating to run, especially on vertical surfaces.
- poor flow may result in brush or roller marks.

■ **Leveling:** or the ability of a coating to smooth out after application: coatings should spread evenly, since their protective value is reduced by surface imperfections.

■ **Film thickness:** is directly related to the degree of protection a coating will provide. When selecting a specific coating, the film thickness to which it must be applied should also be selected.

■ **Drying time:** indicates the period of time the coating may be subject to surface contamination. **Two stages of drying should be considered:**
- **set-to-touch,** or surface drying, at which point the surface will resist contaminants.

- **through-dry,** when drying is complete to the substrate, and film may be recoated.

■ **Permeability:** indicates the degree to which water vapor can migrate through the coating to an area of lower vapor pressure to:
- impermeable coatings on exterior surfaces of a porous substrate may prevent water vapor migration to the cold low vapor pressure outside zone in winter.
- this may cause the vapor to condense within the substrate, with resultant blistering or peeling of the coating.

■ **Wetting:** the maximum distance or penetration the vehicle is capable of delivering the coating on a specific substrate. The lower the wetting ability of a coating, the more thorough must be the surface preparation to ensure adequate adhesion.

■ **Adhesion:** to the substrate must be good to avoid the possibility of cracking and peeling.

■ **Other important properties are:**
- **flexibility,** or the ability to accommodate thermal and/or moisture movements in the substrate;
- **abrasion** and impact resistance;
- **stain** resistance;
- **cleanability.**

EFFECTS OF ENVIRONMENTAL EXPOSURE

The exposure, at the time or after the application of the coating, or the environment, may also affect the coating and/or the substrate.

■ **Mildew:** may attack coatings in humid climates or in warm damp rooms, such as shower rooms.

■ **Atmospheric contamination,** such as sulfur-containing gases:
- may discolor coatings
- also will accelerate chalking and deterioration
- salt laden air of marine environments will have a similar effect.

■ **Sudden drop in temperature,** especially right after the application of the coating, may cause dew or even frost deposits on the coating with resultant flatting.

■ Sudden rise in temperature may cause air entrapped in porous substrate to vaporize partially and cause dry blisters in the coating film.

B2.10 COATINGS

FIELD APPLIED: types, uses, properties

PRINCIPAL BINDER	BASE/ CURE	TYPICAL USES	COMPARATIVE COST, range
CLEAR			
ACRYLIC, methyl methacrylate copolymer	SOLVENT WATER	Waterproofing and surface sealer against dirt retention, graffiti; for vertical surfaces of concrete, masonry, stucco; may be pigmented.	MODERATE TO HIGH
ALKYD, spar varnish	SOLVENT	For interior and protected exterior wood surfaces. Also as vehicle for aluminum pigmented coatings.	MODERATE
PHENOLIC, spar varnish	SOLVENT	Exterior wood surfaces subject to moisture. May be used in marine environments. Also vehicle for aluminum pigment.	MODERATE TO HIGH
SILICONE	SOLVENT	Waterproofing and surface sealer against dirt retention for vertical surfaces of concrete, masonry, stucco.	MODERATE
URETHANE, one-part	MOIST ① CURE	Surfaces subject to chemical attack; abrasion, graffiti, heavy or concentrated traffic, such as gymnasium floors.	MODERATE TO HIGH
STAIN			
ACRYLIC	SOLVENT WATER	Pigmented translucent or semiopaque exterior surface sealers; solvent based for masonry, concrete; water-based for wood.	MODERATE TO LOW
ALKYD	SOLVENT WATER	Pigmented exterior or interior surface sealer for wood surfaces, such as shingles. Does not impart sheen to surface.	MODERATE
OIL	SOLVENT	Pigmented exterior or interior surface sealer for wood, such as shingles, trim. Opaque or semitransparent.	MODERATE
OPAQUE			
ACRYLIC	WATER	For exterior/interior vertical surfaces of wood, masonry, plaster, gypsum board, metals. Good color retention. Permeable to vapor.	MODERATE TO LOW
ACRYLIC, epoxy modified, two-part	WATER	High performance coating for interior vertical surfaces subject to graffiti, stains, heavy scrubbing. May be used in food preparation areas.	HIGH
ALKYD	SOLVENT WATER	For exterior/interior vertical and horizontal surfaces, such as wood, metals, masonry. Poor permeability to vapor.	MODERATE
CHLORINATED RUBBER	SOLVENT	Swimming pool coatings. Corrosion protection; isolating dissimilar metals.	HIGH TO VERY HIGH

B2.10 COATINGS

B • BUILDING SHELL

IN-SERVICE LIFE, range in years	GLOSS RETENTION	STAIN RESISTANCE	WEATHER RESISTANCE	ABRASION IMPACT RESISTANCE	FLEXIBILITY	REMARKS
5 to 10	excellent to good	fair	excellent to good	good	good	• solvent-based acrylic impermeable to water vapor, high gloss. • water-based acrylic is semigloss, water vapor permeable.
up to 1 exterior	fair to good	poor	poor	fair	good	• phenolic varnish has a dark tint; will darken with age; may be top-coated with clear alkyd. • clear varnishes not recommended for exterior wood because of limited durability.
up to 2 exterior	fair to good	fair	good	good	good	• urethane may be formulated to yield hard, glossy surface so that graffiti can be removed with strong solvents.
5 to 7	flat	fair	good	penetrating coating		• fillers may be required when using clear coatings over hard-wood, such as oak; abraded wood may limit choice; consult manufacturer's literature.
up to 15	excellent to good	good to excellent	good to excellent	good to excellent	excellent	• stains may be used as surface sealers to change color of wood and then be top coated with clear coatings.
3 to 5	flat finish		good to fair			• stains over exterior wood surfaces generally will provide better protection than clear coatings, but usually will not last as long as opaque coatings.
3 to 5	flat finish	not a factor	fair	penetrating coatings – resistance same as for substrate		• alkyd may be modified with silicone for better color retention.
3 to 5	fair		fair			① solvent base, oil-modified urethane is also available; for use on interior/exterior vertical and horizontal wood surfaces. Cost is moderate.
5 to 8	good to fair	fair	good	good to fair	good to excellent	
10 to 15	good	good	good to excellent	good to excellent	good to excellent	
5 to 8	good to excellent	fair	fair to good	fair to good	fair to good	
up to 10	fair	fair	good	fair to good	good	

B2.10 COATINGS

FIELD APPLIED: types, uses, properties (cont.)

PRINCIPAL BINDER	BASE/ CURE	TYPICAL USES	COMPARATIVE COST, range
OPAQUE, continued			
CHLOROSULFONATED POLYETHYLENE	SOLVENT	Protective coating for tanks, piping, valves, elastomeric roofing membranes.	VERY HIGH
EPOXY, two-part EPOXY ESTER, one part	SOLVENT CURE SOLVENT	Moisture/alkali resistant. Two-part for nondecorative interior uses highly resistant to chemicals. Esters in wide choice of colors.	HIGH TO VERY HIGH
PHENOLIC	SOLVENT	Chemical and moisture resistant coatings. May be used over alkaline surfaces.	MODERATE TO HIGH
POLYCHLOROPRENE	SOLVENT ①	Marketed as "Neoprene"; resistant to chemicals, moisture, ultraviolet radiation. Also used as roofing membrane; generally covered with Hypalon.	VERY HIGH
POLYESTER	SOLVENT	Limited application in field: over cementitious surfaces, metal, plywood for exterior exposures.	HIGH
SILICONE	SOLVENT	Surfaces with temperatures up to 1200°F. Often with aluminum pigments. Corrosion and solvent resistant.	VERY HIGH
SILICONE – modified acrylic, alkyd, epoxy	SOLVENT	High performance exterior coatings. Industrial siding, curtain walls, when shop applied baked-on.	HIGH TO VERY HIGH
STYRENE- butadiene	WATER	Interior coating for gypsum board, plaster, masonry. Limited exterior use over cementitious substrate, as filler over rough porous surfaces.	MODERATE TO LOW
URETHANE, one- or two-part	MOIST OR CHEMICAL CURE ②	Heavy duty wall and floor coatings. Resistance to stains, chemicals, graffiti, scrubbing, solvents, impact, abrasion.	HIGH TO VERY HIGH
VINYL, polyvinyl chloride-acetate	SOLVENT	Residential metal siding and trim, gutters, leaders, baseboard heating covers, when shop applied, baked on.	HIGH
VINYL, polyvinylidiene chloride	WATER	Metal and concrete surfaces in contact with dry and wet food, potable water, waste water, jet and diesel fuels.	HIGH
VINYL, polyvinyl acetate	WATER	Exterior and interior vertical surfaces, such as masonry, concrete, wood, plaster, gypsum board, metals. Permeable to vapor.	MODERATE TO LOW
BITUMINOUS, coal tar pitch, asphalt: emulsions, cut-backs	SOLVENT	Waterproofing of metals, concrete, masonry, Portland cement plaster, piping when below grade or immersed.	LOW
CEMENT	WATER	Leveling coat over porous masonry or concrete not subject to abrasion or scrubbing. Cement and oil used as primers for metal surfaces.	LOW

B · BUILDING SHELL

IN-SERVICE LIFE, range in years	GLOSS RETENTION	STAIN RESISTANCE	WEATHER RESISTANCE	ABRASION IMPACT RESISTANCE	FLEXIBILITY	REMARKS
up to 15	not applicable	fair	excellent	fair to good	excellent	• epoxy-esters have intermediate properties between two-part epoxies and alkyds and phenolics.
15 to 20 up to 10	poor to good	excellent for two-part	good to excellent	excellent	good to excellent	• bitumen-epoxy formulations are available for use as heavy-duty waterproofing of underground piping, structural members.
up to 10	fair	fair	good to excellent	good to excellent	good	• phenolic may chalk upon exterior exposure; high degree of resistance to acids, alkalies, and solvents; some formulations available for surface temperatures of up to 300-350°F.
up to 25	not applicable	good	excellent	excellent	good	
up to 15	good to excellent	good to excellent	good to excellent	good	good to excellent	• polyesters available glass fiber reinforced; also used widely for baked-on factory applied finishes for formed metal wall panels.
varies	not applicable, special purpose coating			good	good	• silicone for high temperature applications generally with aluminum pigment.
15 to 20	good to excellent	good	good to excellent	good to excellent	good	• polyvinyl chloride film used for factory applied finishes for formed metal wall panels.
4 to 6	poor to fair	fair	poor	fair	good	• cement paint will absorb rain water and will darken until water evaporates; requires moist curing after application: if not properly cured will tend to dust and rub off.
15 to 20	excellent	good to excellent	good to excellent	good to excellent	excellent	• for high performance coatings under severe conditions, life expectancy may be less.
up to 15	good	fair	good	good	good to excellent	• for additional information on special coatings, such as baked-on finishes, fire retardant coatings, reflective coatings, refer to Selection Checklist.
up to 10	good	fair	good	good	good	① may be obtained as water-reducible coating; use as field-applied coating very limited; generally used as tank linings.
5 to 8	good to fair	fair	good	good to fair	good	② solvent base, oil-modified urethane is also available; for use on vertical and horizontal surfaces. Cost is moderate; but durability lower than other types.
10 to 15 protected	not a factor		good	poor	fair	
varies	flat finish	poor	poor for color	good	poor	

B2.10 COATINGS

B • BUILDING SHELL

EXTERNAL FACTORS

SUBSTRATE COATING	EXTERNAL FACTOR	DESCRIPTION AND/OR EFFECT

SOLAR RADIATION

- **ultra-violet radiation band** in solar radiation may cause fading in colored pigments, chemical reaction in some binders or solvents; may degrade the substrate when coating is not opaque to such radiation.
- **visible light radiation band** may have to be reflected, scattered, or absorbed by the coating.

TEMPERATURE

- **solar radiation raises the temperature** of the coating, causing thermal expansion and accelerated evaporation of solvent.
- **heat may also be gained** through convection of hot air or other gaseous substances, or through conduction through the substrate; both may occur simultaneously, e.g., during accidental exposure to fire.
- **freezing temperatures** may affect some vehicles and therefore the proper curing of coating.

RAIN

- **rain may cause thermal shock** in heated coatings; may be absorbed and cause swelling of the coating; pigments may be washed out of the coating.
- **rain may penetrate** through checks and cracks into coating and freeze, thus accelerating deterioration of coating and/or substrate.

WATER VAPOR

- **water vapor may be required** for the curing of some coatings; it may have to be allowed to permeate the coating from the substrate to prevent condensation at their interface from affecting the bond between the coating and the substrate.
- in other cases **vapor must be prevented** from permeating the coating from the outside in order to protect the substrate.

CHEMICAL FUMES

- **generated by chemical processes,** or the burning of fossil fuels may be deposited on the coating; they may react there directly with the coating, or when dissolved by rain or by condensing water vapor.

DUST, DIRT

- **may penetrate porous coatings** and be lodged within them; may also trap airborne pollutants when leeched out and/or reacting with rain water; may cause staining and/or degradation of the coating; marring of coating may also be intentional, as in the case of graffiti.

ABRASION, IMPACT

- **abrasion of coatings** on vertical and horizontal surfaces may be caused by air-borne dust; high velocity flow of gaseous or liquid substances; traffic, vandalism.
- **impact** may be through natural agents such as hailstones, accidental, or intentional, such as in vandalism.

SURFACE WATER

- surface water external to the coating **may be fresh or sea water;** its level may vary thus exposing parts of the coating normally submerged to: effects of **solar radiation and oxygen** in the air and to **differential thermal movements** between the exposed and submerged parts.

CHEMICAL SOLUTIONS

- coatings may be submerged in chemical solutions, such as sea water, sewage, oils, lubricants, solvents; some of these may **chemically react with** specific constituent parts of the coating and may **degrade or dissolve it.**

B2.10 COATINGS

B • BUILDING SHELL

SELECTION CONSIDERATIONS

The selection of a coating, or of the principal binder, should start with a definition of the external factors, or the environment under which the coating will have to perform:

■ **Type and degree of exposure:** to environmental factors (whether exterior or interior):
• solar radiation
• rain, wind
• temperature extremes
• salt spray, polluted atmospheres
• chemicals.

■ **Degree to which substrate** is likely to deteriorate rapidly if coating fails.

In evaluating in-service conditions, remember that:

■ Conditions **may vary** over time.

■ Conditions may **be accidentally changed** sufficiently to affect coating performance, as the following examples show:

A change in substrate conditions occurring long after coating has been applied may affect coating performance:

■ **A porous alkaline substrate** such as concrete may be dry and protected from moisture at the time a coating is applied.

■ If the **substrate remains dry,** an impermeable coating not resistant to alkali may perform well.

■ If the same substrate is **temporarily or accidentally subject to** rain penetration or extensive water vapor condensation:
• the moisture will be trapped within the substrate by the impermeable coating.
• the moisture will dissolve the alkali salts locked in the substrate.
• the coating will fail.

In another example **a wet substrate may be treated or sealed** and a coating not resistant to substrate moisture applied.

■ If the **treatment holds,** all will be well.

■ But such treatment may be **improperly done and fail.** Then the coating which is not resistant to substrate moisture may also fail rapidly.

Conditions may vary from place to place on the same surface, affecting coating:

■ **Coating over a horizontal surface** subject to light foot traffic may perform well for an extended period of time over most of the surface except in a few localized areas, such as at doors, where all of the traffic becomes concentrated.

■ **A value judgment** may have to be made in such cases as to whether:
• a higher performance coating should be used throughout to guard against the possibility of failure.
• several different coatings should be used with different properties, and possibly different appearance, in specific portions of a contiguous surface; or

• the possibility that a coating may fail should be accepted.

In arriving at such decisions it is well to remember:

■ **In-place cost** of a coating includes not only the cost of the coating itself, but also the cost of surface preparation and of the actual application.

■ Coatings which fail prematurely may have to be **removed completely,** sometimes at considerable expense, before the substrate can be recoated.

■ Failure of a coating may also result in **permanent damage to the substrate.**

The selection of a coating should begin with the properties of the principal binder, and then proceed to the evaluation of different formulations and modifications through additions of other binders, within the basic type:

■ **The principal binders** and more important modifications only are listed, as the evaluation of all available modifications and formulations is beyond the scope of this section.

■ **The fact** that a specific binder is included in a given formulation is no guarantee that the coating will have all the desirable properties of the binder:
• the binder may not be properly formulated
• the proportion of binder to solvent may not be sufficient to perform as expected.

SAFETY AND HEALTH CONSIDERATIONS

Safety and health hazards which may be encountered **during surface preparation** to receive a coating, or during its **application** should also be considered:

■ **Surface preparation** may include sandblasting, grinding, fire, strong solvents:
• toxic fumes, toxic dust may result from surface preparation, especially in confined spaces.

■ **Solvents** used in formulating the coating **may be toxic** and may be absorbed through the skin or by inhalation.

■ The use of coatings formulated with photochemically reactive solvents may be restricted or limited by **air pollution controlling ordinances.**

CHART ORGANIZATION

In the charts in this section, coatings have been **evaluated on the basis of two different sets of in-service conditions:**

■ Coatings applied to protect the substrate from external factors (such as solar radiation, rain, air pollution) which **will not rapidly** cause deterioration of the substrate if the coating fails:

■ Coatings applied to protect the substrate from exterior or interior factors

which **may cause rapid deterioration** of the substrate if the coating fails.

When danger of substrate damage in case of coating failure is not of prime importance:

■ **First consideration** is the most durable binder for a particular substrate listed, assuming that there **may be** temporary or accidental **changes in the conditions of the substrate** e.g.:
• a cementitious substrate, even when dry, may still contain alkali which may be released at any time if moisture penetrates into the substrate.
• porous substrates in general may be subject to water vapor migration from a warm and humid interior space to cold outdoors unless completely sealed: impermeable exterior coatings may trap such vapor within the substrate and result in condensation and moisture build-up.

■ **Coating first considered may not:**
• have the surface sheen desired
• may not offer choice of colors
• may be more costly
• may require more extensive surface preparation than coatings listed as alternate considerations.

■ **Alternate considerations given** include coatings which should provide durable protection assuming that the **stated conditions of the substrate will not change.**

■ **They may be:**
• less durable than the coating under first consideration
• may offer wider choice of surface sheen or colors
• may be lower in cost.

■ **When there is no coating** which may be expected to provide long lasting protection, **none will be** listed under first consideration, for example:
• clear exterior coatings over wood will seldom last for more than two years and are not recommended.

The second set of conditions listed above exists when:

■ **Substrate** from either exterior or interior external factors may deteriorate rapidly if coating fails.

■ **The coating must** have a high degree of resistance to external factors.

In these cases:

■ **The specific extent** of exposure should be considered prior to the final selection:
• a higher performance coating may be required
• a lower performance coating may be suitable.

■ **The in-service record** of a particular coating as reported by reputable manufacturers should generally be part of the selection process.

B2.10 COATINGS

INTERIOR: general purposes, types, uses

● denotes property, or requirement

EX excellent
G   good
F   fair
P   poor
(S) solvent
(W) water

EXTERNAL FACTORS

■ **LIGHT TO MODERATE USAGE AND MAINTENANCE**

■ **MILD DETERGENTS**

■ **NOT EXPOSED TO SPILLS OR SPLASHES OF CHEMICALS; CHEMICAL FUMES**

■ **SPECIFIC CONDITIONS AS LISTED**

SUBSTRATE	COATING SYSTEM	TOPCOAT: TYPE and BASE	PRINCIPAL BINDER FIRST CONSIDERATION [1] / [2]
GYPSUM BOARD WALLS, CEILINGS			
■ Subject to: • light scrubbing • mild detergents	opaque; water or solvent	topcoat	ALKYD (S)
		primer	VINYL, polyvinyl acetate; water
■ Subject to: • periodic scrubbing • occasional spatter of grease, food stains	opaque; water	topcoat	ACRYLIC
		primer	self-priming
WOOD, DRY, VERTICAL, HORIZONTAL			
■ Doors, wood veneered; trim; paneling	clear; solvent	topcoat	ALKYD; may be over stain
■ Doors, hardboard veneer. ■ Doors, wood veneer; ■ Trim	opaque; water or solvent	topcoat	ALKYD (S)
		primer	ALKYD (S)
■ Floors, light to moderate use	clear; solvent	topcoat	ALKYD self-priming
■ Floors, heavy use	clear; solvent	topcoat	URETHANE, one-part moisture-cure
■ Floors, light to moderate use	opaque; solvent	topcoat	ALKYD, self-priming
CONCRETE, MASONRY, PORTLAND CEMENT PLASTER			
■ Dry, not exposed to moisture penetration, such as ground moisture ■ Typical components: • concrete and concrete masonry walls and partitions	opaque; water ②	topcoat	VINYL, polyvinyl acetate
		primer	self-priming or STYRENE-BUTADIENE
	opaque; solvent	topcoat	ALKYD
		primer	STYRENE-BUTADIENE
CONCRETE FLOORS			
■ Dry, not exposed to ground moisture penetration ■ Light to moderate usage	opaque; solvent	topcoat	URETHANE, one-part moist-cure
		primer	self-priming
GYPSUM PLASTER			
■ Dry, fully cured, no signs of efflorescence protected from moisture penetration	opaque; water or solvent	topcoat	ACRYLIC (W)
		primer	self-priming
METAL, shop primed			
Doors and frames	opaque; water or solvent	topcoat	ALKYD (S)
Electrical cabinets, equipment		topcoat	ACRYLIC, epoxy modified, two-part
Structural steel framing		topcoat	VINYL, polyvinyl acetate (W)

B2.10 COATINGS

B • BUILDING SHELL

GLOSS	SEMIGLOSS	FLAT	COLOR RETENTION	AVE. PREPARATION	EXT. PREPARATION	DRY ONLY	MAY BE DAMP	ALTERNATE CONSIDERATION	GLOSS	SEMIGLOSS	FLAT	COLOR RETENTION	AVE. PREPARATION	EXT. PREPARATION	DRY ONLY	MAY BE DAMP
•	•	•	G			•		ACRYLIC (W)	•	•		EX				•
				•		•		self-priming						•		•
	•	•	EX			•		ACRYLIC, epoxy modified, two-part	•	•		EX			•	
				•			•	VINYL, polyvinyl acetate							•	•
	•		G	•		•		URETHANE, one-part oil-modified	•	•		G	•		•	
•	•		G			•		ACRYLIC (W)		•		EX				•
		•		•			•	ALKYD (S)						•	•	
•	•		G	•		•		URETHANE, one-part self-priming	•	•		G	•		•	
•	•		G	•		•		EPOXY, two-part self-priming	•	•		G	•		•	
•			G	•		•		URETHANE, one-part self-priming	•	•		G	•		•	
	•	•	G			•		ACRYLIC; may be self-priming	•	•		EX				•
		•		•			•	STYRENE-BUTADIENE filler for porous block						•		•
•	•	•	G			•		ACRYLIC, epoxy modified, two-part	•			EX			•	
		•		•			•	STYRENE-BUTADIENE filler for porous block						•		•
•	•		G			•		ALKYD-CHLORINATED RUBBER	•	•		G			•	
				•		•		self-priming						•	•	
•	•		EX			•		ALKYD (S)	•	•	•	EX			•	
		•				•		CHLORINATED RUBBER base						•	•	
•	•		G	•		•		ACRYLIC, epoxy modified, two-part (W)	•			EX	•		•	
•			EX	•		•		ALKYD (S)	•	•		G	•		•	
	•	•	G	•		•		ALKYD (S)	•	•	•	G	•		•	

REMARKS

[1] Other properties may limit or prevent use.

[2] Principal binder/resin and some major modifications listed. Modifications generally are available.

■ Federal consumer safety regulations prohibit use of products containing lead in or around areas where children may congregate.

• acrylic, epoxy modified for severe exposure in food preparation areas. Available USDA approved when required.
• solvent-base coatings should not be used directly over gypsum board as they tend to raise the nap of the paper facing.
• joints in gypsum board should be taped and spackled; absorption over spackled areas may differ from that of paper facing.
(1) when resistance to fire is required, intumescent coatings, either solvent or water based, may be selected. Also see note (1), page 4-67.

• urethane: single component, may be applied over stain.
• abraded or rough surfaces may restrict use of some coatings; consult manufacturers' literature.
• fillers for open grain wood, such as oak, are recommended to smooth out the surface; stain may be added to filler, when required.
• edges of doors should be sealed to prevent absorption of moisture.
• particleboard generally finished with opaque coatings; for clear use filler, and stain; absorption may be uneven.
• hardboard generally finished with opaque coatings only; primers required.
• alkyd for wood veneer and trim may be self-priming.

(2) cement-water paints may be used in damp areas, such as basement walls; colors generally limited to light tints; moisture required during curing period; usually 24 to 48 hours.
• coatings not recommended over alkaline substrate; coating of fresh concrete, masonry, or plaster should be delayed for as long as possible.
• heavy bodied primers/fillers recommended for porous rough surfaces.

• coatings for floors subject to water, chemicals, heavy traffic may be found under evaluation chart for contaminated, industrial environment.
• dusting surfaces should be sealed first.

• substrate may be alkaline, and primers/coatings should be alkali resistant.
• coating of plaster should be delayed for as long as possible to allow it to dry out.

• coatings listed are to be used over a properly applied rust inhibiting primer; for primers refer to evaluation chart for rural/suburban environment.
• primer should be touched up where abraded or chipped prior to applying topcoat.
• for coatings under severe exposure refer to evaluation chart for contaminated/industrial environment.

B2.10 COATINGS

EXTERIOR: low pollution environment

● denotes property, or requirement

EX excellent
G   good
F   fair
P   poor
(S) solvent
(W) water

		COATING		
		SYSTEM		PRINCIPAL BINDER
EXTERNAL FACTORS	**SUBSTRATE**	**TOPCOAT: TYPE and BASE**		**FIRST CONSIDERATION** 1 2

EXTERNAL FACTORS

■ SOLAR RADIATION
● RAIN
● FREEZING RAIN
● MODERATE DAILY TEMPERATURES
● LOW POLLUTION LEVEL
● AVERAGE HUMIDITY

WOOD, DRY, VERTICAL

SUBSTRATE	TOPCOAT: TYPE and BASE		PRINCIPAL BINDER
SIDING, vertical/horizontal ● recommended moisture content not over 12 percent. ● protected from moisture; or limited occasional exposure to water. ■ Typical components: ● veneered plywood siding ● MDO plywood siding ● hardboard siding ● redwood siding ● cedar siding, shingles and shakes	clear; solvent	topcoat	PHENOLIC, ① tung-oil
		primer	self-priming, top coat; or SHELLAC
	stain; water, or solvent	topcoat	ALKYD, oil-base, self-priming (S)
	opaque; solvent	topcoat	ALKYD
		primer	ALKYD, oil base
	opaque; water	topcoat	ACRYLIC
		primer	ALKYD, oil-base ACRYLIC, emulsion
TRIM ● recommended moisture content not over 12 percent ● occasional exposure to moisture or water ■ Typical components: ● shutters ● doors ● accent areas of limited size ● railings	clear; solvent	topcoat	URETHANE, one part oil-modified
		primer	self-priming
	stain; solvent	topcoat	none recommended 3
	opaque; water or solvent	topcoat	ALKYD, oil-base (S)
		primer	ALKYD, oil-base

WOOD FLOORS, DRY

SUBSTRATE	TOPCOAT: TYPE and BASE		PRINCIPAL BINDER
● recommended moisture content not over 12 percent ● exposed to moisture, rain, snow ● subject to light to moderate traffic ■ Typical components ● porch decking ● exterior stairs	clear; solvent	topcoat	none recommended 3 ②
		primer	not applicable
	clear; solvent	topcoat	URETHANE, one part moist cure
		primer	self-priming; follow for hardwood recommended
	opaque; solvent	topcoat	URETHANE, one-part, moisture-cure
		primer	self-priming

B2.10 COATINGS

REMARKS

[1] Other properties may limit or prevent use.

[2] Principal binder/resin and some major modifications listed. Modifications generally are available.

[3] Refer to "Alternate Considerations": because of limitations, requirements for this particular system should be further investigated.

■ Federal consumer safety regulations prohibit use of products containing lead in or around areas where children congregate.

COATING SHEEN			SUBSTRATE SURFACE CONDITION					ALTERNATE CONSIDERATION [2]	COATING SHEEN			SUBSTRATE SURFACE CONDITION				
GLOSS	SEMIGLOSS	FLAT	COLOR RETENTION	AVERAGE PREPARATION	EXTENSIVE PREPARATION	DRY ONLY	MAY BE DAMP		GLOSS	SEMIGLOSS	FLAT	COLOR RETENTION	AVERAGE PREPARATION	EXTENSIVE PREPARATION	DRY ONLY	MAY BE DAMP
•	•		P			•		ALKYD ①	•		•	F			•	
				•	•			over clear primers or stains					•	•		
•		F	•			•		ACRYLIC, water-base not for redwood or cedar (W)		•		EX	•		•	
•	•		G			•		OIL-base, low cost, lower durability	•	•		P			•	
		•				•		self-priming						•	•	
	•		EX			•		VINYL, polyvinyl acetate	•	•		G			•	
		•				•		ALKYD, also ACRYLIC, but not for redwood, cedar.					•	•		
•	•		F			•		PHENOLIC, tung oil	•	•		P			•	
		•				•		self-priming top coat or SHELLAC					•	•		
								ALKYD, oil-base, may rub off, self-priming		•		F	•		•	
•	•		G			•		ACRYLIC, water-base ALKYD, emulsion (W)	•	•		EX			•	
		•				•		ALKYD, oil-base					•	•		

GLOSS	SEMIGLOSS	FLAT	COLOR RETENTION	AVERAGE PREPARATION	EXTENSIVE PREPARATION	DRY ONLY	MAY BE DAMP		GLOSS	SEMIGLOSS	FLAT	COLOR RETENTION	AVERAGE PREPARATION	EXTENSIVE PREPARATION	DRY ONLY	MAY BE DAMP
								PHENOLIC, tung oil	•	•		P			•	
								self-priming, filler for hardwood recommended					•	•		
•	•		F			•		URETHANE, one-part, oil-modified	•	•		F	•		•	
		•				•		self-priming, filler for hardwood recommended					•	•		
•	•		F			•		ALKYD	•	•		G			•	
				•	•			self-priming					•		•	

① clear coatings for plywood not recommended.
- light color stains have shorter durability than heavily pigmented ones.
- PVA used on yellow pine and red cedar.
- acrylic resistant to ultraviolet rays, thus no embrittlement or yellowing.
- no coating for wet wood has been recommended; wood should be dry before any coating is applied.
- opaque stains hide surface imperfections and will last longer, but will also hide grain of wood.
- wood requires primer to equalize absorption; hardboards require filler to smooth out grain.
- on cedar and redwood always use oil based primer under any coating.
- wood and locations subject to occasional moisture penetration, or to water vapor migration and/or condensation should be backprimed and edge sealed; unless properly sealed, permeable coatings only, such as acrylic, should be used; even then they may peel.
- clear phenolic coatings may be protected with alkyd type clear coatings for better color retention.
- all knots and pitch streaks should be sealed with shellac; all nails set and nail holes filled-in.
- even galvanized, ferrous metal nails may corrode and stain water based coatings, because such coatings allow water vapor to penetrate to the nails increasing the possibility of rusting.
- alkyds may react with chemicals in previous coatings; refer to Selection Checklist.
- clear finishes for trim and doors may be pigmented to stain the wood, or a staining primer may be used.
- extensive surface preparation, when required, applies to both previously coated and uncoated surfaces, but principally to previously coated ones.

② clear coatings for exterior floors are not recommended: ultraviolet radiation may degrade not only the coating but the substrate as well; once the substrate fails, it has to be completely refinished in order to receive another coating; clear coatings may last one to two years and may require yearly maintenance.
- pigmented coatings only are recommended when exposed to sunlight; pigments used should block penetration of ultraviolet radiation to the substrate.
- water-based coatings generally are porous and not sufficiently abrasion resistant for use on surfaces subject to abrasion.
- urethane has excellent resistance to abrasion, alkali, acids, solvents, strong detergents and fuels; clear type not recommended for exterior exposure due to possible degradation of substrate by ultraviolet radiation.

B2.10 COATINGS

EXTERIOR: low pollution environment (cont.)

● denotes property, or requirement

EX excellent
G good
F fair
P poor

COATING SYSTEM

TOPCOAT: TYPE and BASE

PRINCIPAL BINDER
FIRST CONSIDERATION [1]

[2]

EXTERNAL FACTORS

■ SOLAR RADIATION
● RAIN
● FREEZING RAIN
● MODERATE DAILY TEMPERATURES
● LOW POLLUTION LEVEL
● AVERAGE HUMIDITY

SUBSTRATE

CONCRETE, MASONRY, STUCCO, DRY, VERTICAL			
● aged over 90 days. ● no visible signs of efflorescence. ● protected from moisture entry ● limited water vapor diffusion. ■ Typical components: ● precast concrete panels. ● concrete, clay masonry. ● stucco, cement-bound mineral fiber.	clear; solvent	topcoat	SILICONE min. 5% solution
	clear; water	topcoat	ACRYLIC, methyl methacrylate
	opaque; solvent	topcoat	ALKYD
		primer	STYRENE-BUTADIENE
	opaque; water	topcoat	ACRYLIC
		primer	self-priming

CONCRETE, FLOORS, DRY			
● aged over 90 days. ● light to moderate traffic. ● surface intact, dusty.	clear; solvent	topcoat	EPOXY ESTER
		primer	self-priming
● aged over 90 days. ● moderate to high traffic. ● surface worn, dusty.	opaque; solvent	topcoat	URETHANE, one-part moisture cure
		primer	self-priming. substrate to be patched

CONCRETE, WET			
● aged under 30 days or when subject to water penetration or water vapor condensation. ● surfaces cleaned to remove efflorescence. ■ Typical components: ● concrete walls. ● concrete floors.	clear, or opaque; solvent	topcoat	HARDENING-SEALING COMPOUNDS ②
		primer	
	opaque; water	topcoat	none recommended [3]
		primer	not applicable

METAL			
Bare, ferrous ■ Typical components: ● structural steel not exposed to abrasion. ● miscellaneous metal		topcoat	ALKYD, oil-base, choice of colors.
		primer	RED LEAD, alkyd-base
Base, nonferrous ■ Typical components: ● aluminum trim, flashings	opaque; water or solvent	topcoat	ALKYD, oil-base, choice of colors.
		primer	ZINC-CHROMATE, alkyd-base
Zinc coated, ferrous ● generally does not require protective coating.		topcoat	ALKYD, oil-base choice of colors.
		primer	ZINC DUST-ZINC CHROMATE

B2.10 COATINGS

REMARKS

[1] Other properties may limit or prevent use.

[2] Principal binder/resin and some major modifications listed. Modifications generally are available.

[3] Because of limitations, requirements for this particular system should be further investigated.

■ Federal consumer safety regulations prohibit use of products containing lead in or around areas where children congregate.

GLOSS	SEMIGLOSS	FLAT	COLOR RETENTION	AVERAGE PREPARATION	EXTENSIVE PREPARATION	DRY ONLY	MAY BE DAMP	ALTERNATE CONSIDERATION [2]	GLOSS	SEMIGLOSS	FLAT	COLOR RETENTION	AVERAGE PREPARATION	EXTENSIVE PREPARATION	DRY ONLY	MAY BE DAMP
	•		NA			•		ACRYLIC, methyl methacrylate ①	•			G	•		•	
•			G	•		•		none available [3]								
•	•	•	G			•		OIL-base; low cost low durability	•	•		P			•	
		•		•		•		STYRENE-BUTADIENE					•			•
•	•		EX	•		•		VINYL, polyvinyl acetate	•	•		G			•	•
		•		•		•		self-priming					•			•
•			F	•		•		ACRYLIC	•			G			•	
					•	•		self-priming. sealer-hardener recommended					•		•	
•	•		F	•		•		ALKYD, oil base; or oil and rubber base	•	•		G			•	
					•	•		self-priming. substrate to be patched.							•	•
								HARDENING-SEALING COMPOUNDS ②								
								ACRYLIC or VINYL, polyvinyl acetate	•	•		G				•
								self-priming					•			•
•	•	•	G			•		ACRYLIC, water-base.	•	•		EX				•
	•					•		RED LEAD, alkyd-base.					•		•	
•	•	•	G			•		ACRYLIC, water-base.	•	•		EX				•
					•	•		ZINC-CHROMATE, alkyd-base							•	•
•	•	•	G			•		ZINC DUST-ZINC CHROMATE, gray only		•		F			•	
	•					•		self-priming					•		•	

① solvent based coatings are not recommended as first consideration over exterior concrete or masonry surfaces as such coatings form an impermeable film preventing the escape of any moisture which may still be present in or may later penetrate into, the substrate. Condensation at the interface of coating and substrate may also contain soluble alkaline salts; either one, or both may cause blistering or peeling. Solvent-based coatings should only be used when the substrate is completely dry, and there is no possibility of substantial moisture penetration into it.

• water-based coatings generally allow water vapor to escape to the outside, and do not present the same problem as do solvent based coatings.

• silicone should be considered more as a water repellent than as a coating film.

• heavy bodied water-based coatings are available as fillers for rough surface masonry units.

② sealers-hardeners preferably applied to fresh concrete. For additional information refer to CEMENT/CONCRETE

• epoxy is a high performance, high cost coating: suitable for floors exposed to heavy wear, chemical spills, moisture. May be used to resurface worn floors after proper patching.

• no coating will perform well over a poor quality substrate.

• compatibility of coating with bond breakers, curing agents and hardeners which may have been used over the substrate should be checked.

• epoxy may be used over damp surfaces. High performance, high cost coating.

• application of coating should be delayed as long as possible to allow substrate to dry out. Refer to remarks for aged concrete, masonry above.

• coatings considered should be water vapor permeable.

• bleeding of alkaline salts to surface may result in brown spots over permeable coatings. If impermeable coatings are used over permeable primer; such coatings are likely to blister and peel.

• generally one coat of primer is used in rural/suburban environment.

• rusty, galvanized steel requires two coats of primer. Second coat may be the finish coat if color is acceptable.

• alkyd darkens with age on exterior exposure.

• primers listed for ferrous metal may be applied over lightly rusted surfaces.

• silicone-alkyd coatings may be selected for superior durability and color retention, but cost is higher and surface preparation required more extensively.

• bituminous coatings generally used in concealed locations to isolate aluminum from dissimilar metals and materials. For additional information, refer to Metals.

B2.10 COATINGS

EXTERIOR: marine/humid environment

● denotes property, or requirement

EX excellent
G   good
F   fair
P   poor

EXTERNAL FACTORS	SUBSTRATE	COATING SYSTEM TOPCOAT: TYPE and BASE		PRINCIPAL BINDER FIRST CONSIDERATION [1] [2]
• HIGH CONCENTRATION OF SALT LADEN MOISTURE • SALT WATER SPRAY AND SPLASHES • HIGH HUMIDITY • MILDLY CORROSIVE ATMOSPHERE • SOLAR RADIATION	**WOOD, DRY, VERTICAL**			
	SIDING, vertical, horizontal • moisture content not over 12 percent • back-primed and edge sealed against moisture absorption. ■ Typical components: • veneered plywood siding. • MDO plywood siding. • hardboard siding. • redwood siding. • cedar siding, shingles, and shakes.	clear; solvent	topcoat	URETHANE, one-part ① oil-modified ②
			primer	self-priming
		opaque; solvent	topcoat	VINYL URETHANE
			primer	self-priming
		opaque; water	topcoat	ACRYLIC
			primer	ALKYD, oil-base ACRYLIC, emulsion
	• wood veneered plywood. solid wood siding.	stain; solvent	topcoat	ALKYD, oil base self-priming
	• MDO plywood siding. • hardboard siding.	stain; water	topcoat	ACRYLIC, self-priming
	TRIM • recommended moisture content not over 12 percent. • back primed and edge sealed against moisture absorption.	clear; solvent	topcoat	none ③ recommended ②
			primer	not applicable
	■ Typical components: • shutters • doors • accent areas	opaque; solvent	topcoat	PHENOLIC, pigmented
			primer	self-priming
	WOOD FLOORS, DRY			
	• recommended moisture content not over 12 percent • exposed to salt water and moisture. • subject to moderate to heavy traffic. ■ Typical components: • porch decking • platforms • stairs	clear; solvent	topcoat	none ③ recommended ③
			primer	not applicable
		opaque; solvent	topcoat	URETHANE, one-part, moisture-cure
			primer	self-priming
			topcoat	see above
			primer	
	CONCRETE, MASONRY, VERTICAL			
	• aged over 90 days. • no signs of efflorescence. • subject to high humidity, salt spray, splashes. ■ Typical components: • precast concrete panels. • concrete, clay masonry. • stucco	clear; solvent	topcoat	SILICONE ④ min. 5% solution
		clear; water	topcoat	ACRYLIC, methyl ⑤ methacrylate
		opaque; solvent	topcoat	CHLORINATED RUBBER
			primer	self-priming
		opaque; water	topcoat	ACRYLIC ⑥
			primer	self-priming

B2.10 COATINGS

B · BUILDING SHELL

GLOSS	SEMI-GLOSS	FLAT	COLOR RETENTION	AVE. PREPARATION	EXT. PREPARATION	DRY ONLY	MAY BE DAMP	ALTERNATE CONSIDERATION	GLOSS	SEMI-GLOSS	FLAT	COLOR RETENTION	AVE. PREPARATION	EXT. PREPARATION	DRY ONLY	MAY BE DAMP
•		F				•		PHENOLIC, tung-oil	•	•		P			•	
				•		•		self-priming topcoat						•	•	
•			EX			•		ALKYD, choice of colors	•	•		G				
•					•	•		oil-base						•		
			EX				•	VINYL, polyvinyl acetate	•		•	G				•
				•		•		ALKYD, oil-base / ACRYLIC, emulsion						•	•	
	•		G	•		•		oil-base, self-priming			•	F			•	
	•		G				•	none available								
								PHENOLIC, tung-oil	•	•		P			•	
								self-priming topcoat						•	•	
•			F			•		ALKYD, choice of colors	•	•		G			•	
				•		•		ALKYD, oil-base						•	•	

GLOSS	SEMI-GLOSS	FLAT	COLOR RETENTION	AVE. PREPARATION	EXT. PREPARATION	DRY ONLY	MAY BE DAMP	ALTERNATE CONSIDERATION	GLOSS	SEMI-GLOSS	FLAT	COLOR RETENTION	AVE. PREPARATION	EXT. PREPARATION	DRY ONLY	MAY BE DAMP
								URETHANE, one-part moisture-cure	•	•		F			•	
								self-priming						•	•	
•	•		F			•		EPOXY, two part, high performance	•	•		G				•
					•	•		self-priming						•		•
								ALKYD-PHENOLIC	•	•		G			•	
								self-priming						•	•	

GLOSS	SEMI-GLOSS	FLAT	COLOR RETENTION	AVE. PREPARATION	EXT. PREPARATION	DRY ONLY	MAY BE DAMP	ALTERNATE CONSIDERATION	GLOSS	SEMI-GLOSS	FLAT	COLOR RETENTION	AVE. PREPARATION	EXT. PREPARATION	DRY ONLY	MAY BE DAMP
	•		NA	•		•		ACRYLIC, methyl methacrilate ⑤	•			G	•		•	
•			G	•			•	none available ③								
•			F				•	ALKYD	•	•	•	G				
					•		•	STYRENE-BUTADIENE					•			•
•	•		EX				•	VINYL, polyvinyl acetate	•	•						•
				•			•	self-priming						•		•

REMARKS

☐1 Other properties may limit or prevent use.

☐2 Principal binder/resin and some major modifications listed. Modifications generally are available.

☐3 Because of limitations, requirements for this particular system should be further investigated.

■ Federal consumer safety regulations prohibit use of products containing lead in or around areas where children may congregate.

① clear coatings not recommended for exterior plywood facing.

② clear coatings for wood not recommended as life expectancy when exposed to sunlight is relatively short and ultraviolet radiation penetrating through coating may degrade substrate.

- alkyds for porous surfaces should not be used directly over previous coatings containing lead or zinc. Primers should be applied first, or the previous coating removed.
- fillers to smooth out grain in hardwoods are recommended.
- all coatings used in hot, humid environments should be resistant to mildew and fungus growth.
- wood should be backprimed and edge sealed to prevent moisture absorption. Impermeable coatings over wood subject to moisture penetration may fail by blistering and peeling.
- stains may last up to three years, but generally can be readily recoated.
- water base coatings not recommended for trim and doors because they tend to pick up and hold dirt more readily than solvent base coatings.
- epoxy ester and acrylic epoxy may be used. They are high performance, higher cost coatings and should be considered when resistance to severe exposure to salt spray or fog and mild chemicals is required.

③ clear coatings not recommended when exposed to sunlight as ultraviolet radiation penetrating through the coating may degrade substrate.

- opaque coatings, mineral spirit solvent-based, are not included under first choice, as they generally have poor resistance to mildly corrosive atmosphere.
- epoxy is a high performance coating and should be considered when resistance to severe exposure to salt spray, heavy use, and chemicals is required.

④ clear coatings, mineral spirit solvent-based, are not recommended as they may trap moisture within the substrate.

⑤ clear coatings water-based, are not recommended due to the possibility of excessive moisture build-up in the substrate. If the substrate will only be subject to occasional moderate moisture build-up, they might be considered.

- use of clear coatings will generally result in uneven absorption of moisture during rain, and will show up as splotches on concrete or masonry surfaces until they dry out completely.

⑥ cement-water paints should be considered for high humidity locations.

B2.10 COATINGS

EXTERIOR: marine/humid environment (cont.)

● denotes property, or requirement

EX excellent
G   good
F   fair
P   poor

	COATING SYSTEM TOPCOAT: TYPE and BASE		PRINCIPAL BINDER FIRST CONSIDERATION [1] [2]

EXTERNAL FACTORS

• HIGH CONCENTRATION OF SALT LADEN MOISTURE

• SALT WATER SPRAY AND SPLASHES

• HIGH HUMIDITY

• MILDLY CORROSIVE ATMOSPHERE

• SOLAR RADIATION

SUBSTRATE

CONCRETE FLOORING

Aged over 90 days; protected from ground or other sources of moisture: Subject to: • water, mild chemicals. • moderate to heavy traffic. • salt water spashes.	clear solvent	topcoat	URETHANE, one-part moisture cure
		primer	self-priming patching as req'd.
	opaque solvent	topcoat	URETHANE, one-part moisture cure
		primer	self-priming

CONCRETE, MASONRY, STUCCO, WET, VERTICAL

• aged under 30 days or when subject to: • moisture penetration • vapor condensation • salt, fog or spray ■ Typical components • precast or site-cast concrete • masonry walls • stucco	opaque solvent	topcoat	none [3] recommended
		primer	not applicable
	opaque water	topcoat	ACRYLIC
		primer	self-priming or STYRENE-BUTADIENE

CONCRETE, submerged in water

• filter basins, reservoirs • water plants • swimming pools	opaque solvent	topcoat	CHLORINATED RUBBER
		primer	self-priming

METAL, above ground

Bare, ferrous ■ Subject to salt, fog or spray ■ Typical components: • structural steel not exposed to abrasion • miscellaneous metal	opaque solvent	topcoat	CHLORINATED RUBBER choice of colors
		primer	CHLORINATED RUBBER red lead
Bare, nonferrous ■ Subject to salt, fog or spray ■ Typical components • aluminum trim, flashings		topcoat	CHLORINATED RUBBER choice of colors
		primer	ZINC CHROMATE and ACID WASH
Zinc Coated, ferrous ■ Subject to salt, fog or spray • generally does not require protective coating.		topcoat	CHLORINATED RUBBER choice of colors
		primer	ZINC DUST-ZINC OXIDE

METAL, buried in ground

Bare, ferrous Typical components: • structural • piping	opaque solvent	topcoat	COAL TAR-EPOXY
		primer	self-priming

METAL, immersed in salt water

Bare, ferrous Typical components: • structural • piping	opaque solvent	topcoat	VINYL spray application only
		primer	VINYL or POLYSTYRENE, zinc dust

B2.10 COATINGS

B • BUILDING SHELL

REMARKS

1. Other properties may limit or prevent use.
2. Principal binder/resin and some major modifications listed. Modifications generally are available.
3. Because of limitations, requirements for this particular system should be further investigated.
- Federal consumer safety regulations prohibit use of products containing lead in or around areas where children may congregate.

GLOSS	SEMIFLAT	FLAT	COLOR RETENTION	AVERAGE PREPARATION	EXTENSIVE PREPARATION	DRY ONLY	MAY BE DAMP	ALTERNATE CONSIDERATION	GLOSS	SEMIGLOSS	FLAT	COLOR RETENTION	AVERAGE PREPARATION	EXTENSIVE PREPARATION	DRY ONLY	MAY BE DAMP
•	•		F			•		EPOXY ESTER	•			F			•	
					•	•		self-priming						•	•	
•	•		F			•		EPOXY, two part chemical cure	•	•		G				•
					•	•		self-priming, sealer may be required.							•	•
								none recommended [3]								
								not applicable								
•	•		G				•	VINYL, polyvinyl acetate	•	•		F				•
						•		self-priming or STYRENE-BUTADIENE						•		•
•			F			•		POLYCHLOROPRENE (NEOPRENE)	•			G				•
					•	•		self-priming							•	•
•			F			•		VINYL, limited colors; spray application only			•	EX			•	
					•	•		VINYL or POLYSTYRENE, zinc dust					•	•		
•			F			•		EPOXY ESTER, choice of colors	•			F			•	
					•	•		self-priming					•	•		
•			F			•		POLYCHLOROPRENE limits choice of colors	•			G				•
					•	•		self-priming							•	•
•			F			•		BITUMINOUS	•		•	F			•	
					•		•	self-priming						•		•
	•		EX			•		CHLORINATED RUBBER brush, spray applied	•			F			•	
					•	•		CHLORINATED RUBBER red lead						•	•	

REMARKS

- substrate should be protected from moisture: no coatings should be used if substrate is subject to moisture penetration.
- epoxy is a high performance, high cost coating.
- substrate must be non-dusting, and all laitance must be removed: no coating will perform well over a poor quality substrate.
- compatibility of coating with bond breakers, curing compounds, admixtures used for substrate should be checked.
- urethane may yellow on exterior exposure.

(1) application of coating should be delayed for as long as possible to allow substrate to dry out. Need for coating should be investigated.
- if coating is necessary, water vapor permeable and alkali resistant coatings only should be considered; fresh concrete, masonry and stucco are highly alkaline and moisture dissolves such salts and brings them to the surface as efflorescence. All efflorescence must be removed prior to applying coatings. Highly alkaline surfaces should not be coated.

- for bulkheads and piling use coal tar-epoxy.
- polychloroprene available in dark colors only, very good resistance to most chemicals.
- epoxy based coatings may also be used.

- surface preparation required for different primers varies: consult manufacturer's literature.
- the performance of the topcoat will largely depend on the performance of primer selected and on proper surface preparation: salt water and salt spray are highly corrosive and consideration should also be given to other high performance coatings, such as polychloroprene (neoprene) and two-component epoxy. Top coats listed for one substrate generally may be used over other primers. Consult manufacturers' literature.
- aluminum generally will pit in marine environment and should be coated.

- zinc coating, in addition to protective coating, should be considered for critical items, such as piping.
- allowance may be made in structural members to allow for the possibility of corrosion developing in the undercoating.

- proper surface preparation and application of coating is critical.
- coal tar-epoxy and polychloroprene may also be considered.

B2.10 COATINGS

EXTERIOR/INTERIOR:

● denotes property, or requirement

EX excellent
G   good
F   fair
P   poor
(S) solvent
(W) water

COATING SYSTEM

TOPCOAT: TYPE and BASE

PRINCIPAL BINDER FIRST CONSIDERATION ☐1

☐2

EXTERNAL FACTORS

■ FOR EXTERIOR/INTERIOR EXPOSURE

■ SPECIFIC CONDITIONS LISTED AND:

■ EXTERIOR FACTORS:
• SULPHUR CONTAINING FUMES
• RAIN BORNE POLLUTANTS
• SOLAR RADIATION

SUBSTRATE

WOOD, EXTERIOR, INTERIOR

■ End-grain wood block flooring ■ Subject to: Heavy traffic water, dilute chemicals	opaque; solvent	topcoat	EPOXY, two-part
		primer	self-priming
■ Vertical surfaces ■ Subject to: Graffiti and stains (such as ink, mustard), impact, abrasion, frequent scrubbing	opaque; water or solvent	topcoat	URETHANE, two-part (S)
		primer	self-priming
■ Vertical, overhead horizontal surfaces, interior ■ For protection against fire	intemescent:① clear, or opaque; solvent or water	topcoat	proprietary systems should have fire resistance rating
		primer	

CONCRETE AND MASONRY, EXTERIOR, INTERIOR

■ Walls, exterior ■ Subject to: chemical fumes, dust, dirt, scrubbing	opaque; solvent	topcoat	EPOXY ESTERS
		primer	self-priming
■ Walls, exterior ■ Subject to: Dilute acids, solvents, water, impact		topcoat	POLYCHLOROPRENE (NEOPRENE)
		primer	self-priming
■ Walls, interior ■ Subject to: Graffiti, stains, strong detergents, splashes of synthetic oils	opaque; water solvent	topcoat	URETHANE, two-part (S)
		primer	self-priming
■ Walls, interior ■ Subject to: Fatty acids, strong detergents, scrubbing, low temperatures, such as in food storage preparation		topcoat	ACRYLIC, epoxy modified, two-part (W)
		primer	self-priming

CONCRETE FLOORS, EXTERIOR, INTERIOR

■ Subject to: Heavy traffic water, dilute chemicals		topcoat	EPOXY, two-part
		primer	self-priming
■ Subject to: Steel wheel traffic, cutting oil, salts, acids, alcohol, weathering	opaque; solvent	topcoat	POLYCHLOROPRENE (NEOPRENE)
		primer	self-priming
■ Subject to: Fatty acids, strong detergents, such as in food storage preparation areas		topcoat	URETHANE, two-part
		primer	self-priming
■ Subject to: Petroleum solvents, jet or diesel fuel, synthetic oils, strong detergents		topcoat	URETHANE, two-part
		primer	self-priming

B2.10 COATINGS

B • BUILDING SHELL

contaminated/industrial environment

REMARKS

1. Other properties may limit or prevent use.
2. Principal binder/resin and some major modifications listed. Modifications generally are available.
3. Because of limitations, requirements for this particular system should be further investigated.
- ■ Federal consumer safety regulations prohibit use of products containing lead in or around areas where children congregate.

The following table reads, left to right: COATING SHEEN (GLOSS, SEMIGLOSS, FLAT), COLOR RETENTION, SUBSTRATE SURFACE CONDITION (AVERAGE PREPARATION, EXTENSIVE PREPARATION), DRY ONLY, MAY BE DAMP | ALTERNATE CONSIDERATION | COATING SHEEN (GLOSS, SEMIGLOSS, FLAT), COLOR RETENTION, SUBSTRATE SURFACE CONDITION (AVERAGE PREPARATION, EXTENSIVE PREPARATION), DRY ONLY, MAY BE DAMP

GLOSS	SEMIGL	FLAT	COL RET	AVG PREP	EXT PREP	DRY ONLY	MAY BE DAMP	ALTERNATE CONSIDERATION	GLOSS	SEMIGL	FLAT	COL RET	AVG PREP	EXT PREP	DRY ONLY	MAY BE DAMP
•	•		G				•	BITUMINOUS	•		•	F				•
					•		•	self-priming						•		•
•	•		F			•		ACRYLIC, epoxy modified two-part (W)	•	•		G				•
					•	•		self-priming							•	•
•	•	•	F			•		not applicable								
					•	•		not applicable								
•			F			•		ACRYLIC, solvent	•	•		G				•
					•	•		self-priming						•		•
	•		G				•	URETHANE, one-part, moisture-cure	•	•		F				•
					•	•		self-priming							•	•
•	•		G			•		ACRYLIC, epoxy modified, two-part (W)	•			G				•
					•	•		self-priming						•		•
•			G			•		URETHANE, one-part, moisture-cure (S)	•	•		F				•
				•		•		self-priming							•	•
•	•		G				•	URETHANE, one-part, moisture-cure	•	•		G				•
					•		•	self-priming							•	•
	•		G				•	EPOXY, two-part, high performance.	•	•		G				•
					•		•	self-priming						•		•
•	•		F			•		EPOXY, two-part	•	•		G				•
				•		•		self-priming						•		•
•	•		F			•		EPOXY, two-part	•	•		G				•
				•		•		self-priming						•		•

Remarks (right column):
- wood should generally be at 12 percent or less moisture content; backprimed and edge sealed to prevent absorption of moisture.
- epoxy may be applied over damp surfaces.
- bituminous coating for floors are generally heavy build, trowel applied. Also refer to FLOORING.
- high performance coatings to protect vertical surfaces from graffiti, stains generally may be applied over less resistant coatings to reduce costs. Check for compatibility.
- 1 fire protective coatings are formulated to bubble up when exposed to intense heat to form an insulating layer over the surface to be protected. Thickness of coating will determine degree of protection.
- coatings may have to be selected to limit flame spread. Consult manufacturers' literature for specific ratings.

- substrate should be dry and protected from moisture penetration.
- concrete should be allowed to dry for at least six months or more, depending on external conditions.
- coatings impermeable to water vapor may trap moisture at the interface of substrate and coating leading to blistering and peeling of coating.
- porous, rough substrate should be levelled using either cement based slurry, or water base coatings to provide a smooth surface for high performance coatings.
- high performance coatings used to protect vertical surfaces from graffiti, stains, chemicals may be applied over less resistant coatings to reduce cost. Check for compatibility.
- glassy surfaces of cast concrete should be sand or water blasted to improve adhesion of coatings. Form oils and waxes must be removed prior to application of coatings.
- urethane may yellow on exterior exposure.

- epoxy, two-part loses gloss and chalks on prolonged exterior exposure.
- abrasive aggregate chips should be added to minimize possibility of skidding.
- clear urethane available. Urethanes may yellow on exterior exposure.
- coatings in food preparation or storage areas should be USDA approved.
- glassy surfaces should be acid etched prior to application of coating to improve adhesion.
- epoxy may be applied to damp surfaces.
- concrete on grade should be protected from ground moisture absorption.
- curing compounds, bond breakers, admixtures used for concrete may prevent proper bonding of coating.
- laitance must be completely removed prior to application of coating. Coatings will not improve a poorly placed concrete floor.
- for additional information on floor coatings, refer to seamless flooring in FLOORING.

B2.10 COATINGS

B • BUILDING SHELL

EXTERIOR/INTERIOR:

● denotes property, or requirement

EX excellent
G   good
F   fair
P   poor
(S) solvent
(W) water

		COATING SYSTEM TOPCOAT: TYPE and BASE		PRINCIPAL BINDER FIRST CONSIDERATION [1] [2]

EXTERNAL FACTORS

■ FOR EXTERIOR/INTERIOR EXPOSURE

■ SPECIFIC CONDITIONS LISTED AND:

■ EXTERIOR FACTORS:
 • SULPHUR CONTAINING FUMES
 • RAIN BORNE POLLUTANTS
 • SOLAR RADIATION

SUBSTRATE

CONCRETE, IMMERSED, HORIZONTAL, VERTICAL

Substrate	Base	Coat	Binder
■ Walls, floors of fresh or potable water treatment, filtration plants or reservoirs	opaque; water, solvent	topcoat	VINYL, polyvinylidiene chloride (W)
		primer	self-priming
■ Walls, floors of sewage treatment reservoirs, treatment plants	opaque; solvent	topcoat	COAL TAR – EPOXY
		primer	self-priming
■ Walls, floors of aviation fuel; diesel fuel, oil storage tanks	opaque; water, solvent	topcoat	VINYL, polyvinylidiene chloride (W)
		primer	self-priming
■ Walls, floors subject to spills in Plating Shops, Aircraft maintenance shops, Paper Mills	opaque; water, solvent	topcoat	URETHANE, two-part walls/floors (S)
		primer	self-priming

METAL, FERROUS, EXTERIOR, INTERIOR

Substrate	Base	Coat	Binder
■ Subject to: Surface temperature to 350°F • equipment, piping	opaque; solvent	topcoat	PHENOLIC
		primer	self-priming
■ Subject to: Surface temperature to 500°F • equipment, piping	opaque; solvent	topcoat	SILICONE, alkyd
		primer	ZINC SILICATE
■ Subject to: Surface temperature to 1200°F • equipment, piping	opaque; solvent	topcoat	SILICONE, aluminum pigmented
		primer	ZINC SILICATE
■ Containing or subject to flow of Potable water, chlorine, as in bleaching, water treatment	opaque; water solvent	topcoat	VINYL, polyvinylidiene chloride (W)
		primer	self-priming
■ Containing or subject to flow of hot water • equipment, inside of pipes	opaque; solvent	topcoat	CHLOROSULFONATED POLYETHYLENE (Hypalon)
		primer	self-priming
■ Containing or subject to: Aviation fuel, diesel fuel, synthetic oils.	opaque; water solvent	topcoat	VINYL, polyvinylidiene chloride (W)
		primer	self-priming
■ Containing or subject to: Crude Oil • storage tanks	opaque; solvent	topcoat	BITUMEN-EPOXY
		primer	self-priming
■ Subject to: Fire • exposed structural framing; walls, ceilings	intumescent: clear, or opaque; solvent or water	topcoat	PROPRIETARY SYSTEMS SHOULD HAVE FIRE RESISTANCE RATING
		primer	

B2.10 COATINGS

contaminated/industrial environment (cont.)

REMARKS

1. Other properties may limit or prevent use.
2. Principal binder/resin and some major modifications listed. Modifications generally are available.
3. Because of limitations, requirements for this particular system should be further investigated.
■ Federal consumer safety regulations prohibit use of products containing lead in or around areas where children congregate.

COATING SHEEN			SUBSTRATE SURFACE CONDITION					ALTERNATE CONSIDERATION	COATING SHEEN			SUBSTRATE SURFACE CONDITION				
GLOSS	SEMIGLOSS	FLAT	COLOR RETENTION	AVE. PREPARATION	EXT. PREPARATION	DRY ONLY	MAY BE DAMP		GLOSS	SEMIGLOSS	FLAT	COLOR RETENTION	AVE. PREPARATION	EXT. PREPARATION	DRY ONLY	MAY BE DAMP
●			EX			●		BITUMINOUS (S)	●		●	F				●
				●		●		self-priming						●		●
●			F		●			BITUMINOUS	●		●	F				●
				●		●		self-priming						●		●
●			EX			●		URETHANE, two-part (S)	●	●		F				
				●		●		self-priming							●	●
●	●		F		●			vinyl, polyvinylidiene chloride, walls only (W)	●			EX				●
						●	●	self-priming						●		

COATING SHEEN			SUBSTRATE SURFACE CONDITION					ALTERNATE CONSIDERATION	COATING SHEEN			SUBSTRATE SURFACE CONDITION				
GLOSS	SEMIGLOSS	FLAT	COLOR RETENTION	AVE. PREPARATION	EXT. PREPARATION	DRY ONLY	MAY BE DAMP		GLOSS	SEMIGLOSS	FLAT	COLOR RETENTION	AVE. PREPARATION	EXT. PREPARATION	DRY ONLY	MAY BE DAMP
	●		G			●		none recommended [3]								
					●	●		not applicable								
	●		G			●		none recommended [3]								
					●	●		not applicable								
	●		F			●		none recommended [3]								
					●	●		not applicable								
●			EX			●		CHLORINATED RUBBER (S)		●		F				●
				●		●		CHLORINATED RUBBER red lead							●	●
	●		EX			●		EPOXY, phenolic	●			F				●
					●	●		self-priming						●	●	
●			EX			●		URETHANE, two-part (S)	●	●		G				
				●		●		self-priming						●	●	
●			F			●		EPOXY, two-part	●	●		G				●
					●	●		self-priming						●	●	
●	●	●	F			●		not applicable								
					●	●		not applicable								

Remarks (first group):

- hypalon and Chlorinated Rubber may also be used on surfaces exposed to potable water.
- bitumens oxidize and become brittle upon weathering. Periodic maintenance required.
- the need for a coating over dense concrete substrate should always be carefully evaluated: concrete is best left uncoated, as coatings once applied have to be maintained and periodically renewed. Protection usually is required when concrete is exposed to substances containing acids; alkali resistant cement is available, refer to CEMENT/CONCRETE.
- concrete subject to rapid water flow usually erodes over a period of time and generally requires protection.
- joint sealants should be compatible with coating used. Consult manufacturers' literature.
- for horizontal surfaces subject to traffic refer to evaluation chart for concrete floors in contaminated industrial environment.

Remarks (second group):

- surface preparation is critical for proper performance of coating: blast cleaning to white or near white metal usually required.
- consult manufacturer's literature for recommended surface preparation.
- two coats of primer usually recommended for severe exposure. Thickness of total film of coating should always be selected: consult manufacturer's literature for film thickness recommended for specific conditions and exposure.
- formulations may vary, and the mere fact that a particular binder is included in a formulation should not be taken as an indication of the suitability of a coating to perform under specific in-service conditions.
- for surfaces subject to traffic consult evaluation chart for concrete floors and Flooring some coatings listed may not be suitable for surfaces subject to sustained heavy traffic.
- for coatings over metal subject to high humidity, ground moisture, salt water, refer to evaluation chart for exterior coatings in marine/humid environment.
1. structural steel framing will lose strength rapidly when exposed to temperatures over 500°F, and protection against accidental exposure to fire may be required by building codes. Fire retardant-insulative coatings bubble-up when exposed to fire, providing an insulating layer or heat barrier over structural steel framing and briefly delaying the rise in temperature within such framing. Fire retardant coatings should not be applied over previous coatings, as heat may cause them to blister and spall with resultant failure of the fire retardant coating as well.
- flame spread rating of coating materials should be in accordance with ASTM E-84.

B2.10 COATINGS

TYPES/USES: selection checklist

B2.10 COATINGS

SOLVENT-BASED COATINGS

Solvent-based coatings are available in a variety of vehicles to form transparent, translucent or opaque films.

■ **Binders used** in the vehicle determine the characteristics of the cured film, whether clear or opaque.

■ Though the following descriptions are sectioned off into clear and opaque coatings, **many binders are available in both formulations and both sections should be referred to for information.**

Clear, Solvent-Based

The binders generally used for clear coatings are: drying oils combined with a resin, drier, and solvent.

Clear coatings are generally referred to as Varnishes.
Resins often used include:

■ **Phenolic:**
• has good resistance to water and weathering.
• spar varnish made with phenolic resin and tung oil is considered the most durable for exterior exposures, including marine environment, but its color is relatively dark, and it tends to further darken with age.
• Phenolic resin is also used in opaque coatings.

■ **Shellac resin:** or other resins or gums dissolved in a volatile solvent, referred to as spirit varnish. Generally used as sealer for porous surfaces under more durable top coats.

■ **Cellulose derivations:** in volatile spirits, referred to as lacquers. Have limited use in building construction.

■ **Silicone resin** in a solvent solution of mineral spirits, containing about five percent of silicone by weight is available as a clear coating to dampproof masonry and concrete walls.
• it wets the surface of the masonry and penetrates into the pores to seal it.
• life expectancy of silicone dampproofing is generally five to ten years.

■ **Clear finishes over wood that are exposed to ultraviolet radiation** generally will not last more than two years: they may deteriorate even earlier through either film, or substrate failure.

■ **Clear coatings transmit** ultraviolet radiation to the substrate which in combination with moisture in the wood will degrade the top surface of the wood causing delamination over large areas.

■ If a clear coating is **allowed to deteriorate** to a point where weathering in the wood can occur, any subsequent clear coating will fail.

Urethane, one-component moisture-cure and solvent-base formulations, are commonly used as a clear floor coating when superior wear resistance is required, such as in gymnasiums, bowling alleys:
• have outstanding resistance to abrasion good resistance to alcohol, water, grease, fuels, detergents.
• some formulations tend to yellow and chalk with age under exterior exposure.

Stains are pigmented clear coatings generally used over wood substrates:

■ **Pigmented type only** is recommended for exterior exposure; dyes are not sufficiently light fast and will transmit ultraviolet radiation with resulting degradation of the wood.

■ **Stains generally fail** by erosion, leaving a good surface for refinishing.

■ **Stains should not** be expected to last as long as opaque coatings.

Opaque, Solvent-Based

Opaque coatings offer the widest choice of properties.
■ **Alkyds are the principal binders used** in solvent based opaque, or pigmented, coatings.

■ **Alkyds:** are a designation derived from polyhydric alcohols and polybasic acids used in their manufacture.

■ **Available in** water dispersion as well as solvent-based.

■ **Alkyd-oil vehicle** may be formulated in flat, semigloss, gloss, and clear over a wide compositional range to provide:
• fast drying
• hardness flexibility, durability
• chalk resistance
• good gloss and color retention.

■ **Alkyds are:**
• not resistant to strong alkali
• and are not compatible with previous coatings containing zinc or lead: blistering or peeling may result as alkyds may react with them to combine into an impermeable layer trapping water vapor when the substrate is porous.
• conversely, since an impermeable layer is desirable over ferrous metals, alkyd-zinc formulations are used as primers for such surfaces.

■ **Alkyds are available modified with other binders** for intermediate properties of both, such as:
• with phenolics for improved resistance to water and alkali salts
• with vinyl-chloride acetate to improve gloss and resistance to chalking
• with silicone for very good color and gloss retention.

■ **Alkyds are also used** in shop applied baked-on coatings.

■ **Chlorinated rubber has:**
• good resistance to attack by microorganisms
• resistance to alkali and acids
• low permeability to water and water vapor
• good adhesion to wood and concrete surfaces, but surfaces must be dry or blistering will result.

■ **Chlorosulfonated polyethylene:** is used under severe corrosive conditions:
• is resistant to halogens, such as chlorine, bromine
• is resistant to action of hot water, oxygen, ozone, and ultraviolet radiation
• available clear or pigmented.

■ **Epoxy-ester:** is an epoxy resin reacted with drying oil; has similar properties to alkyd resins and phenolic varnishes, except that:
• gloss retention is poorer
• resistance to chemical fumes and marine environment better.
• generally can be applied over previous coatings without causing lifting or blistering.

■ **Epoxy:** specifically the polyamide cured type:
• has excellent resistance to corrosion and chemicals such as solvents, oils, acids, alkalies.

- will resist abrasion, strong detergents and frequent scrubbing, such as in hospitals, dairies, chemical plants.
- excellent adhesion to concrete, wood and metal substrates; 100% solid types will adhere and cure when applied under water.
- Bitumen-epoxy, both coal tar and asphalt, are available: generally used for heavy duty immersion service, such as buried structural steel, underground piping, oil storage tanks.

■ **Oil, or oleoresinous:** pigments in drying oil, varnish, or oleoresinous vehicle:
- oil degrades under sunlight and water by surface erosion, or chalking.
- oil is permeable to water vapor and will minimize blistering over porous surfaces, such as wood.
- oil should not be used in corrosive, especially in alkaline environments.
- widely used as a primer, because it wets the surface and adheres well to steel, even when some rust is present on the surface.

■ **Phenolic: polymerized** with formaldehyde reactant through polyamide catalysts:
- used when high degree of resistance to acids, alkalies, solvents is required.
- will withstand continuous immersion in hot distilled water, and is therefore widely used in process equipment.

■ **Polyester:** is available in a wide range of durability and flexibility:
- also available modified with amine and silicone
- more widely used in glass fiber reinforced structural plastics.

■ **Silicone:**
- is principally used in heat resistant coatings for surface temperatures of up to 1200°F.
- silicone is also used with alkyds, acrylics, polyesters, epoxies, and urethanes to improve their weathering and color retention properties.
- also see clear coatings.

■ **Urethane or polyurethane:** is available in:
- **one component moisture-cure** formulation which is generally not pigmented, since any moisture associated with the pigment may cause jelling in the can.

- **two-component chemical cure** formulation, clear or pigmented.
- is also available as **one component oil-modified** formulation with better gloss retention than other formulations, but lower resistance to chemicals.
- **two-component** types may range from rubber-like to glass-hard; adhesion to concrete and steel surfaces generally poor.
- **for general properties,** see clear coatings, preceding.

■ **Vinyl:** polyvinyl chloride copolymerized with polyvinyl acetate:
- **excellent** durability and resistance to acids, alkalies, salt water, oils, fats.
- **adherence** is poor and special primers must be used.
- **vinyl-mastics** are available for high-build coatings; they dry by solvent evaporation and tend to be porous.
- **polyvinyl chloride,** is commonly used in water-based coatings.

WATER-BASED COATINGS

Water-based coatings were first developed as interior coatings for plaster and gypsum board surfaces, and are commonly, but incorrectly, known as latex paints.

■ **The term "latex"** applies only to styrene, or styrene butadiene, the binder first used:
- **generally used** as interior coating or as primer for porous masonry surfaces only.
- it will **oxidize** on exterior leading to yellowing and brittleness.

Other binders include:

■ **Acrylic, or acrylic ester resin:** is best suited for exterior exposure, but also is the highest in cost within the group:
- **acrylic** is available pigmented and clear, and the clear acrylic based on methyl methacrylate has been found to offer the best protection for concrete against weathering.
- acrylics are produced in **many formulations,** and the mere fact that a particular formulation is an acrylic emulsion does not ensure the best durability.

■ **Alkyds, water-dispersed:** for properties, see solvent-based coatings.

■ **Vinyl, or polyvinyl acetate:** available as:
- **homopolymer** for economical interior application
- **copolymer** for exterior use
- **for other vinyl** formulations, see solvent-based coatings.

■ **All water-based opaque coatings have high permeability to water vapor:**
- this makes them **suitable** for use over **moist porous** surfaces such as concrete, masonry, or wood where coatings of low permeability might blister or peel.
- **should not be used** directly over chalking previous coatings.

■ **Cement: is a water-soluble** rather than a water-dispersed coating, composed of Portland cement, some hydrated lime, and silica aggregate:
- mixed on-site with water for application
- requires moist curing like any other cement-based product
- resists alkali, moisture
- may be pigmented but color retention is poor.

SPECIAL PURPOSE COATINGS

Fire-retardant insulative: or intumescent coatings are formulated to bubble and swell on heating to form an insulating cover over the substrate.

■ **Degree of protection** afforded depends on the film thickness and the nature of the substrate:
- a heat barrier is provided for incombustible surfaces
- on combustible substrates the coating provides an additional, more fire resistant surface layer.

■ **Should not be applied** over previous coatings: these will blister and spall.

Reflective coatings: will absorb the ultraviolet band of solar radiation and reflect it as visible light:

■ **Life expectancy** of reflective coatings is about one year when subject to sunlight.

Bituminous coatings are water emulsions or solvent cut-backs of coal tar pitch or asphalt:

■ Are used as low-cost coatings under corrosive and high humidity conditions, such as for buried structural steel, piping, below-grade concrete.

B2.10 COATINGS

SUBSTRATE: effect on coatings

■ Coatings are in liquid form before and during application, then cure to form a non-self-supporting film; they cannot exist without a solid, gener-ally rigid substrate to receive and support them. Since a coating bonds itself firmly and continuously to it, the exposure, condition and properties of the **substrate,** as well as its surface characteristics and defects, **will directly affect the coating during and after application.**

MOVEMENT

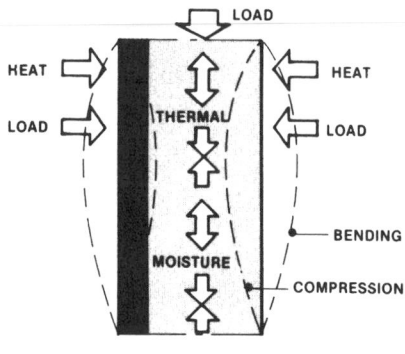

VERTICAL OR HORIZONTAL ELEMENT

Movement in the substrate may result from:

■ **Thermal expansion/contraction** due to exposure to:
• solar radiation
• sources of heat external to the substrate, such as heat gener-ating equipment adjacent to it
• heat generating processes con-tained by the substrate.

■ Thermal **movement may be dif-ferential** between the substrate and the coating it supports due to variations in exposure, e.g.:
• rain suddenly cooling a coating over a hot substrate
• a coating being heated by solar radiation over a cold substrate.

■ **Shrinking/swelling** due to changes in internal moisture con-tent of the substrate: variations may be as much as five percent or more across the grain in wood under extreme conditions.

■ **Deflection under load,** vertical or horizontal, will induce tensile and compressive stresses in the sub-strate which may affect the coating.

■ **Restrained end conditions** of the substrate may also cause bending stresses; vibration in the substrate will result in cyclical stress rever-sals.

■ Movements in substrate may **result in cracking of coating.**

MOISTURE

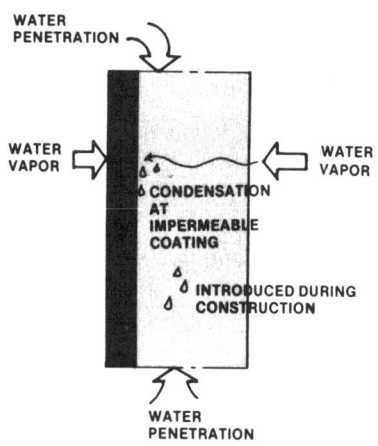

Moisture may penetrate into a porous substrate as:

■ **Water vapor migrating** from high vapor pressure areas, such as warm, humid interior spaces to low vapor pressure areas, such as cold, dry outdoors. If the **dew point occurs within the substrate,** vapor will condense, especially if the coating is of low permeability and blocks its free passage to the outdoors.

■ **Water vapor penetrating** a per-meable coating to condense on a cold impermeable substrate, such as metal.

■ **Rain penetrating** into the sub-strate through faulty joints, damaged or faulty flashings, cracks in coating.

■ **Water that was absorbed** when substrate was exposed to rain or ground moisture while improperly stored before or during con-struction.

■ **Water that was a constituent part** of the substrate during construction and has not yet evaporated, generally a slow process; this condition often occurs in concrete, mortar, or gypsum plaster, since generally more water is used than is required by the hydration process.

■ Moisture in the substrate during application of coating may **prevent proper adhesion.**

■ Moisture penetrating substrate after application of coating may **destroy the bond between substrate and coating.**

CHEMICALS

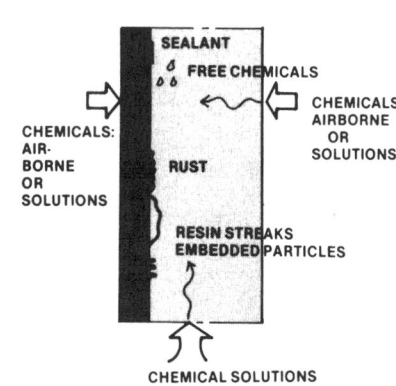

Chemicals may be contained within the substrate, or may be absorbed by it:

■ **Soluble alkaline salts** in concrete or mortar may be dissolved and crystallize on the surface.

■ **Resin streaks** in wood may react with coating and bleed through. Old coatings may chemically react with new.

■ **Rust deposits** may stain some coatings and also impair adhesion.

■ **Sealants used** in joints may stain substrate and/or coating.

■ **Fasteners may corrode;** loose particles of metal may lodge themselves in the substrate and corrode after coating is applied.

■ **Preservative treatments** used for wood may affect absorption, or adhesion of coating; or may react with coating.

■ **Admixtures, form oils,** curing agents, antifreeze solutions used during casting of concrete may prevent proper adhesion of coating, or may react with coating.

■ **Cleaning solutions** used during surface preparation, if not com-pletely removed prior to application of coating, may react with coating, or impair absorption and/or adhesion.

ABSORPTION

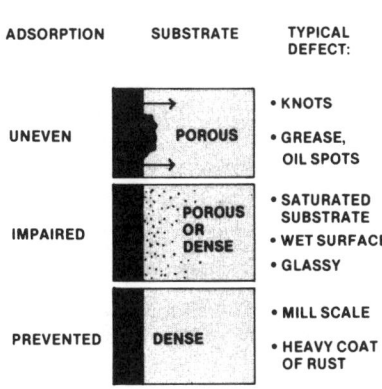

SURFACE	SUBSTRATE	TYPICAL MATERIAL
DENSE: GLASSY SMOOTH	DENSE OR POROUS	• METALS • GLAZED SURFACES • CONCRETE: PLASTIC FORMS
POROUS: SMOOTH	POROUS	• PLASTER • GYPSUM BOARD • WOOD
POROUS: ROUGH	POROUS	• MASONRY • CONCRETE

Absorption at the surface of the substrate, and that of the substrate itself may vary and affect the choice of coating or its performance characteristics:

■ **Glassy dense surfaces** even over a porous substrate will prevent absorption of a coating. Such surfaces may require roughening, e.g. by sandblasting or acid etching, to ensure proper adhesion of most coatings.

■ **Porous substrate** may have varying degrees of absorption within a continuous surface, e.g., different rates of absorption in bands of spring and summer growth of wood. Different rates of absorption will result in different degrees of adhesion and

may cause cracking of coating along junction lines between such bands.

■ **Rough surface of a porous substrate** may have varying degrees of absorption even though the absorption of a substrate does not vary significantly, as when the roughness of the surface prevents applying a film of uniform thickness over it. Different rates of absorption may cause changes in gloss of coating, overpigmentation in areas of excessive absorption of the vehicle, varying degrees of adhesion.

ADHESION

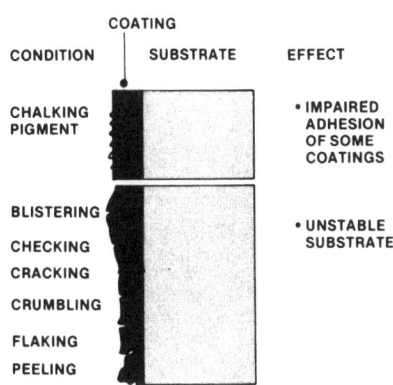

ADSORPTION	SUBSTRATE	TYPICAL DEFECT:
UNEVEN	POROUS	• KNOTS • GREASE, OIL SPOTS
IMPAIRED	POROUS OR DENSE	• SATURATED SUBSTRATE • WET SURFACE • GLASSY
PREVENTED	DENSE	• MILL SCALE • HEAVY COAT OF RUST

Adhesion of coating to the surface of a substrate will be affected by:

■ **Surface defects,** such as knots, resin streaks in wood, or surface contaminants, such as oil, grease, salt deposits over any type of substrate. All will impair bonding of coating to the substrate creating weak spots: vapor may condense in such areas and upon expanding break or lift the film.

■ **Moist or wet surfaces** may impair adhesion of certain coatings to such surfaces. Moisture collecting at the bonding surface of coatings may cause loss of adhesion with resulting blistering or lifting of the film.

■ **Adhesion to glassy surfaces** may be incomplete resulting in flaking and peeling of the film.

■ Adhesion will be prevented by **deposits of mill scale, heavy coats of rust, salts.**

■ All surface defects should be corrected, and all surface contaminants removed prior to the application of coating: **uneven or impaired adhesion will result in coating failure.**

COATINGS AS SUBSTRATE

CONDITION	COATING SUBSTRATE	EFFECT
CHALKING PIGMENT		• IMPAIRED ADHESION OF SOME COATINGS
BLISTERING CHECKING CRACKING CRUMBLING FLAKING PEELING		• UNSTABLE SUBSTRATE

New coatings may be applied over previously coated substrate, and the surface stability of the old coating may affect the performance of the new coating. All coatings deteriorate over a period of time, and deterioration of a coating may result in:

■ **Chalking:** when the vehicle is broken down by weathering, particularly solar radiation, leaving behind loose pigment. Chalking surfaces may be recoated with suitable coatings with little or no surface preparation; loose chalk should always be removed as a minimum measure.

■ **Checking and cracking:** breaks developing in a coating upon loss of flexibility by aging of the coating.

Both result from stresses imposed on the coating by thermal or moisture movements in the substrate, when such stresses exceed the strength of the coating.

■ **Crumbling, flaking, peeling:** may follow checking and cracking and are generally caused by moisture and airborne pollutants penetrating into the coating.

■ Coatings which have deteriorated to a point where **extensive checking is evident** should generally not be used as a receiving surface for new coatings; **cracked, crumbling, flaking or peeling surfaces** should never be used and the old coating must be completely removed from the substrate.

B2.10 COATINGS

B • BUILDING SHELL

SUBSTRATE/PRIMERS: selection checklist

The preparation of a surface to receive a coating, be it a primer or a top coat, is essential:

■ Unless the surface is in proper condition coatings will not:
• adhere well
• provide the required protection to the substrate
• have the desired appearance.

■ The selection of surface preparation systems depends on:
• nature of the substrate
• condition of the substrate
• type of exposure
• type of coating to be applied
• physical and/or environmental conditions present.

Primers are generally used to improve adhesion of the finish coating to a substrate.

■ Primers may also be used to impart other desirable properties to a coating system, such as resistance of corrosion in the substrate, while the finish coating provides resistance to impact and abrasion.

■ Primers are critical for the proper performance of a coating, especially for exterior exposure, where the substrate and the coating may be subject to heat, chemicals and to changes in, or high levels of humidity.

The condition of any substrate is an important consideration with any type of coating:

■ Surface contaminants and defects reduce adhesion of coating, and may cause blistering, peeling, flaking, rusting:
• all surface contaminants have to be removed prior to applying a coating.
• Conditioners may be required to seal a chalky masonry surface; fillers may be used to produce a smooth finish on rough masonry, or on open grain wood; knots have to be sealed to prevent resin oxidation; nail holes, crevices should be patched or filled; all particles clinging to the surface must be removed.

The nature of the substrate will impose specific requirements for its preparation and subsequent coating application:

Wood Substrates & Primers

Wood substrate presents the following problems:

■ It may exhibit different degrees of absorption, such as between spring and summer growth.

■ It may contain water soluble dyes, such as in redwood and cedar, which may be released by moisture penetrating the substrate.

Preparation of wood substrate should include the following:

■ Wood must be dry before coating is applied; any damp surface should be allowed to dry.

■ Wood should be sanded smooth; knots and resin streaks should be sealed.

■ Redwood and cedar must be primed with an oil primer to keep soluble dyes within them from bleeding through an opaque coating.

■ Any sealants used in joints should be compatible with the coating selected. Refer to SEALANTS.

■ Wood in contact with the ground or exposed to moisture or high humidity generally is treated with preservatives to prevent decay. Improperly selected preservatives may adversely affect adhesion of coating, or may bleed through it.

Primers for wood should:

■ Seal the surface against moisture penetration;

■ Have controlled penetration to obtain good adhesion, but not so much as to lose the binder, and thus adversely affect flexibility;

■ Flexibility and adhesion should be sufficient to resist movement (especially across the grain) caused by:
• variations in temperature,
• internal moisture content of the wood.

■ Primers generally used over wood substrate are:
• for uncoated surfaces such as vertical siding, smooth surface shingles, trim: oil-based with good penetration, adhesion, flexibility.
• for rough siding and shakes: alkyd-oil; thin bodied exterior coating, self-priming; may also be used as top coat.
• filler for open grain wood to minimize absorption of top coat: drying oil and/or varnish; used over stain and shellac-type wash coat.
• sealer for surface knots: shellac; used to prevent resin bleeding.

Cementitious Substrates & Primers

Concrete, masonry, stucco and other substrates which incorporate Portland cement:

■ Tend to be alkaline, especially when fresh.

■ Generally are relatively porous.

■ Coatings used on such surfaces should be alkali-resistant and have low water sensitivity, as the surfaces may also be damp.

In evaluating surface preparation, note:

■ Surfaces should be damp for cement-water coatings.

■ Surfaces may be damp when water-based coatings, such as acrylic, are applied.

■ Surfaces must be dry for solvent-based coatings, such as alkyd.

■ Smooth-trowelled surfaces should be acid etched for better adhesion of the coating.

■ Salts contained within the substrate may also be dissolved after the coating is applied:
• they may penetrate through a permeable primer or coating, such as water-based acrylic, and discolor the surface.
• if an impermeable coating, such as an alkyd, is used over the primer, or directly over the substrate, localized loss of adhesion may result.

■ Curing agents, bond breakers, anti-freeze compounds may remain on the surface, or within a cementitious substrate and:
• will impair proper adhesion of a coating
• may react with some coatings.

In evaluating use of primers for cementitious substrates, note:

■ Generally, coatings for alkaline surfaces should be alkali resistant and are thus self-priming.

■ Primers are required for coatings that are not resistant to alkali:
• when there is moisture in the substrate
• if there is a possibility of moisture penetrating into a dry substrate.

■ Coatings which are not resistant to alkali depend for their performance on the integrity of the primer being maintained at all times.

■ Primers generally used for concrete and masonry are:
• primer for uncoated concrete and masonry: acrylic or polyvinyl acetate, water-based. May also be used as top coat.
• primer for high humidity or wet conditions: chlorinated rubber. May also be used as top coat.
• primer for corrosive conditions: vinyl resin. Generally used as top coat also.

■ Plaster:
• should be allowed to age for at least 30 days if an oil-based coating is to be applied.
• water-based coatings may be applied sooner.
• like concrete or stucco, plaster also contains lime and other soluble salts: most of the conditions applicable to concrete also apply to plaster.

■ Gypsumboard: is quite porous:
• nonpenetrating primers or sealers only should be used.
• oil-based primers or coatings such as alkyd, not only are penetrating, but tend to raise the nap of the facing.

- water-based primers or coatings are generally recommended.

Metal Substrates & Primers

Metals may have oil, grease, or chemicals remaining on their surfaces, and when ferrous, also rust and mill scale.

■ The effects of the **substrate** itself on coatings **generally is not a factor.**

■ **Surfaces,** especially those of ferrous metals, usually **require more preparation** than those of other substrates.

■ **The degree of surface preparation required and the methods used vary** for different primers and top coats, and may range:
- from a simple manual wire brushing to remove loose rust and mill scale
- to extensive chemical treatments, grinding, or blast cleaning.

■ **The more extensive surface preparations** are more commonly done at the shop:
- field conditions may restrict the use of some of the methods
- the proper equipment may not be available.

Cleaning methods for ferrous metals include mechanical and chemical methods. Mechanical methods are:

■ **Hand cleaning:** generally restricted to surfaces where minimal preparation is acceptable;

■ **Power tool cleaning:** generally preceded by solvent or chemical treatment. Coatings should be applied as soon as possible after cleaning is completed;

■ **Flame cleaning:** followed by wire brushing to remove all loose particles. Oil and grease should be removed prior to flame cleaning. Coating is applied while the surface is still warm.

■ **Blast cleaning:** is accomplished through high velocity impact of sand, metal shot, metal or synthetic grit, or other abrasive particles.
- may be used in the field, most often employing sand.

■ **Four degrees of blast cleaning** have been established by Steel Structures Painting Council:
- white metal blast
- near-white metal blast
- commercial blast
- brush-off blasting, this being a relatively low cost method of cleaning to remove loose rust, mill scale and old finishes in poor condition.
- for additional information also refer to ASTM D2200.

Chemical treatments are:

■ **Solvent wiping** and degreasing:
- which may present health and safety hazards.

■ **Alkali cleaning:** with alkaline cleaners dissolved in water at relatively high temperatures of 150° – 200°F to remove oil, grease, dirt, old coatings, mildews.

■ **Steam cleaning:** to remove oil and grease.

■ **Acid cleaning to:**
- remove oil, grease, slight rust
- also to etch the surface lightly for better adhesion.

■ **Pickling** through immersion in an acid bath to remove rust, rust scale, mill scale and other surface contaminants.

Ferrous metals may also be pretreated after cleaning to improve adhesion and effectiveness of the coating to be applied.

■ **Hot phosphate treatment** utilizes zinc or iron phosphate to increase bond and adhesion and to minimize corrosion.

■ **Cold phosphate treatment:** which is similar but not as effective as the hot process.

■ **Wash primers** which are similar to the cold phosphate treatment.

■ **Zinc coating,** or galvanizing to prevent corrosion: this results in a surface different from the substrate it is deposited upon, and requires different primers and/or top coatings. Effectiveness depends on type of galvanizing process and thickness of resulting surface.

Primers for ferrous metals substrate, which are subject to rapid corrosion in the presence of moisture, must be corrosion inhibiting. This property is imparted to the primer by the **inclusion of corrosion inhibiting pigments in the formulation:**

■ **Red lead:**
- is reactive and chemically combines with some vehicles to form resistive films.
- particularly suited for protecting surfaces that cannot be thoroughly cleaned of rust.
- used with binders such as: oil, alkyd.

■ **Basic lead** – Silico-Chromate:
- has excellent anticorrosive and weather resistant properties.
- may be tinted to match some top-coats.
- the topcoats may also contain the same pigment.

■ **Zinc dust:**
- provides galvanic protection as the zinc is slowly sacrificed as a cathode in the presence of moisture, while the ferrous surface serves as the anode and remains intact.
- used for service under severe conditions with organic binders, such as acrylics, epoxies, alkyds, chlorinated rubber.

■ **Zinc chromate or zinc yellow:**
- of low sodium salt content

- imparts inhibitive chromate ions to the formulation.

■ **Zinc dust - zinc oxide:** usually 80 percent zinc dust:
- imparts excellent rust-inhibitive properties as well as adhesion, elasticity, and abrasion resistance.
- primers containing zinc dust-zinc oxide may be applied over moderately cleaned and over galvanized surfaces.

Binders or vehicles commonly used for the primers described above include:

■ **Oil:** often added to other vehicles to penetrate rust and promote adhesion.
- linseed oil, often combined with alkyd, is used with red lead.
- when the surface is poorly prepared, oil and straight red lead are used.

■ **Alkyd:**
- is usable with most corrosion inhibiting pigments.
- it serves as binder for moderately priced primers of intermediate durability where meticulous surface preparation is not practicable.

■ **Phenolic:**
- used with zinc chromate or basic silica lead chromate.
- provides good protection from excessive dampness or from immersion in water.
- may also be used with red lead.

■ **Epoxy:** has excellent adhesion to carefully prepared surfaces.
- may be combined with polyamide resins and zinc dust or lead pigments for high performance primers.
- epoxy esters are also used under less demanding conditions.

■ **Chlorinated rubber:** has excellent resistance to water vapor transmission, acids, alkalies, and various salts.

■ **Vinyl:** formulated with zinc chromate or iron oxide provide excellent resistance to chemicals.

■ **Silicate:** with zinc dust offers excellent resistance to weather, chemicals, and salt spray.

■ **Polyesters and acrylics** are used mostly with factory applied coatings.

Non-ferrous metals do not present serious problems with corrosion. When to be coated, aluminum and other non-ferrous metals (such as tin and copper), generally are pre-treated with a wash primer followed by a zinc chromate primer.

B2.10 COATINGS

REFERENCES: Coatings

STANDARDS

ASTM Specifications for:

B580 • Anodic Oxide Coatings on Aluminum.

D13 • Spirits of Turpentine.

D234 • Raw Linseed Oil.

D235 • Mineral Spirits (Petroleum Spirits).

ASTM Tests/Methods for:

C633 • Adhesion or Cohesive Strength of Flame-Sprayed Coatings.

D185 • Coarse Particles in Pigments, Pastes, and Paints.

D562 • Consistency of Paints Using the Stormer Viscometer.

D658 • Abrasion Resistance of Organic Coatings by the Air Blast Abrasion Test

D659 • Evaluating Degree of Chalking of Exterior Paints.

D662 • Evaluating Degree of Erosion of Exterior Paints.

D772 • Evaluating Degree of Flaking (Scaling) of Exterior Paints.

D870 • Water Immersion Test of Organic Coatings on Steel.

D968 • Abrasion Resistance of Organic Coatings by the Falling Abrasive Tester.

D1212 • Measurement of Wet Film Thickness of Organic Coatings.

D1360 • Fire Retardancy of Paints (Cabinet Method).

D1395 • Abrasion Resistance of Clear Floor Coatings.

D1546 • Performance Tests of Clear Floor Sealers.

D1654 • Evaluation of Painted or Coated Specimens Subjected to Corrosive Environments.

D2245 • Identification of Oils and Oil Acids in Solvent-Type Paints.

D2348 • Arsenic in Paint.

D2366 • Accelerated Testing of Moisture Blister Resistance of Exterior House Paints on Wood.

D2394 • Simulated Service Testing of Wood and Wood-Base Finish Flooring.

D2486 • Scrub Resistance of Interior Latex Flat Wall Paints.

D2932 • Exterior Solvent-Based House and Trim Coatings.

D3260 • Resistance to Acid and Mortar of Factory Applied Clear Coatings on Extruded Aluminum Products.

D3273 • Resistance to Growth of Mold on the Surface on Interior Coatings in an Environmental Chamber.

D3281 • Formability of Attached Organic Coatings with Impact-Wedge Bend Apparatus.

D3335 • Low Concentrations of Lead, Cadmium, and Cobalt in Paint by Atomic Absorption Spectroscopy.

D3624 • Low Concentrations of Mercury in Paint by Atomic Absorption Spectroscopy.

ASTM Definition of Terms Relating to:

ASTM	STANDARD DEFINITION OF TERMS:
D16	• Paint, Varnish, Lacquer, and Related Products.

Federal Specifications:

SS-W-110C • Water Repellant, Colorless Silicone Resin Base.

TT-C-535B(2) • Coating, Epoxy, Two Component, For Interior or Exterior Use on Metal, Wood, Concrete and Masonry.

TT-C-542E • Coating, Polyurethane, Oil-Free, Moisture Curing.

TT-E-4870(1) • Enamel, Floor and Deck.

TT-E-489F • Enamel, Alkyd, Gloss (For Exterior and Interior Surfaces).

TT-E-490E(3) • Enamel, Silicone, Alkyd Copolymer, Semigloss, Exterior.

TT-E-496B(2) • Enamel, Heat Resisting, Black.

TT-E-505A(3) • Enamel, Odorless, Alkyd Interior High Gloss, White and Tints.

TT-E-506K • Enamel, Tints and White, Gloss, Interior.

TT-E-509B(2) • Enamel, Odorless, Alkyd, Interior, Semigloss, White and Tints.

TT-E-522A(1) • Enamel, Phenolic Lustreless, Outside.

TT-E-529C(2) • Enamel, Alkyd, Semigloss.

TT-E-543A(1) • Enamel, Interior, Undercoater, Tints and White.

TT-P-26C(1) • Paint, Interior, White and Tints, Fire Retardant.

TT-P-30E • Paint, Alkyd, Odorless, Interior, Flat, White and Tints.

TT-P-34B • Paint, Exterior, Fire Retardant, White and Light Tints.

TT-P-35(1) • Paint, Cementitious, Powder, White and Colors (For Interior and Exterior Use).

TT-P-52D(2) • Paint, Oil (Alkyd-Oil) Wood Shakes and Rough Siding.

TT-P-55B(2) • Paint, Polyvinyl Acetate Emulsion, Exterior.

TT-P-81E • Paint, Oil: Ready-Mixed, Exterior, Medium Shades on a Lead-Zinc Base.

TT-P-86G • Paint, Red-Lead-Base, Ready-Mixed.

TT-P-91D(1) • Paint, Rubber-Base, For Concrete Floors (interior).

TT-P-95C(1) • Paint, Rubber: For Swimming Pools and other Concrete and Masonry Surfaces.

TT-P-650C(1)

B2.10 COATINGS

- Primer Coating, Latex Base, Interior, White (For Gypsum Wallboard).

TT-P-659C(1)
- Primer-Coating and Surfacer, Synthetic, Tints and White (For Metal and Wood Surfaces).

TT-S-711C • Stain, Oil Type, Wood Interior.

TT-V-714 • Varnish, Interior, Floor and Trim.

SELECTED REFERENCES

- **Applications Manual for Paint and Protective Coatings,** W. F. Gross, McGraw-Hill, New York.

AA   **Aluminum Association**
- Designation System for Aluminum Finishes.
- Organic Coatings.
- Aluminum Finishes for Architecture.
- Finishes for Aluminum in Building.

ACI   **American Concrete Institute**
- Guide for the Protection of Concrete Against Chemical Attack by Means of Coatings.
- Guide for Use of Epoxy Compounds with Concrete.
- Guide for Painting Concrete.

APA   **American Plywood Assn.**
V307 • Finishing for Exterior Exposure.

B407 • Stains & Paints on Plywood.

M60 • Finishing Softwood Plywoods.

S910 • Qualified Coatings for Plywood.

CRA   **California Redwood Assn.**
431-4 • Redwood Deck Finishes.
431-1 • Redwood Exterior Finishes.
431-3 • Painting Exterior Redwood.
431-2 • Brand List of Redwood Exterior Finishes.

NAAMM   **National Association of Architectural Metal Manufacturers**
- Metal Finishes Manual.

NCMA   **National Concrete Masonry Assn.**
55 • Waterproof Coatings for Concrete Masonry.

NOFMA   **National Oak Flooring Manufacturers Assn., Inc.**
- Hardwood Flooring Finishing Manual.

NPA   **National Particleboard Assn.**
- Direct Painting & Finishing Particleboard.

NPVLA   **National Paint, Varnish and Lacquer Assn.,Inc.**
796 • The Selection of Paint. Raw Materials Index.

NRCC   **National Research Council of Canada**
CBD 22 • Concrete Floor Finishes.
76 • Paint – What Is It?
78 • Paints and Other Coatings.
79 • New Organic Coatings.
90 • Coatings for Interior Walls.
91 • Exterior Coatings for Wood.
98 • Coatings for Exterior Metal.
172 • General Recommendations for Painting Buildings.

NWMA   **National Woodwork Manufacturers Assn.**
- Care and Finishing of Woodwork.
- Care and Finishing of Wood Windows.

PCA   **Portland Cement Assn.**
- Clear Coatings of Exposed Architectural Concrete.

1S001T • Effect of Substances on Concrete and Guide to Protective Treatment.

1S134T • Painting Concrete.

USFPL   **U.S. Forest Products Laboratories**
FPL0123 • Wood Finishing: Painting Outside Wood Surfaces.

FPL0125 • Blistering, Peeling and Cracking of House Paints from Moisture.

FPL0126 • Temperature Blistering of House Paints.

FPLD126 • Intercoat Peeling of House Paints.

FPL0128 • Mildew on House Paints.

FPL0133 • Wood Finishing: Finishing Exterior Plywood.

FPL0135 • Weathering of Wood.

B2.10 COATINGS

B2. 11 SEALANTS

SEALANTS: introduction

The construction process is one of assembling and joining diverse constituent parts into an integrated whole. It may be simple, such as sewing a few skins together and stretching them over a pole to provide shelter from the elements, or it may be complex, but the process remains essentially the same: whether a structure consists of a few or many component parts, the common denominator is still the requirement for joining them together in such a way that the **integrity of the whole is maintained under all in-service conditions.**

Sealants are materials introduced into joints in the fabric of a building in order to:
* **protect** the joint against the intrusion of foreign matter.
* **prevent** passage of water and/or air through the joint.
* absorb expansion or contraction of a moving joint.

This being their primary function, the **evaluation and selection of sealants** is rooted in an understanding of the various types of joints, the way they function and the resulting performance characteristics required of the sealants.

JOINTS

The process of joining building components **may be categorized based on conditions at the joints.** The components may be connected to provide:

■ **Continuity through the joint:** when joints between several homogeneous components are effectively eliminated, as when two pieces of steel are welded together.

■ **Composite action:** when the components and the joints between them form a single non-homogeneous assembly, e.g. when two pieces of steel are bolted together and, though separate, act as one; though there is a joint, there is no measurable movement within it.

■ **Transfer of stresses:** when components are discontinuous but connected at the joints; e.g. when two preformed panels interlock, or sheets are connected by a standing seam; a limited amount of movement within the joint occurs.

■ **Discontinuity:** when the components are simply supported without any restraint at their edges; movement is unrestrained.

Some of the characteristics of the methods described above are:

■ Discontinuity through the joints is **generally the simplest solution,** but it effectively destroys the integrity of a multi-component assembly by introducing planes of weakness.

* **Providing continuity** between components would restore the integrity of the assembly to meet performance requirements the components have been selected for, but:
* full continuity between field assembled components is generally difficult and costly to achieve.
* a large monolithic assembly might create problems in accommodating thermal expansion and contraction between it and other assemblies, or within the assembly itself; this might necessitate the re-introduction of joints into it.

■ Thus the **joint, its function, shape and location** are an essential part in the selection and evaluation of components, assemblies and of the structure itself.

Joints are commonly classified by function into working and non-working joints.

Working Joints

Working joints are provided to allow expansion or contraction in a component to occur without affecting adjacent components. Expansion or contraction in a component **may be caused by:**

■ **Thermal movement due to variations in temperature:**
* all solids are subject to thermal expansion or contraction, the rate being different for each.
* inelastic components will expand or contract at their free edges, and may deform or fail if the movement is restrained.
* thermal movement in large homogeneous or composite components may result in excessive stress concentration causing deformation or failure.
* thermal movement will also be affected by the heat storage capacity of the component; thin sections respond much more quickly to sudden changes in temperature than do heavy sections of the same material.

■ **Moisture movement, or shrinkage and swelling, due to variations in internal moisture content:**
* porous materials only are subject to moisture movement; the extent varies widely, with wood being the most affected, especially across the grain;
* moisture movement may be cyclical, as in wood; or irreversible, such as the initial shrinkage in fresh concrete during the evaporation of the excess water added to facilitate handling.

■ Working - or moving - joints may also have to **accommodate differential movement between components,** e.g., due to deflection and/or creep in supporting structural frame, or to foundation settlement.

Non-Working Joints

Non-working joints may be:

■ **Joints with limited movement** as in glazing joints or construction joints in a cast concrete slab.

■ **Joints with virtually no movement** at all, as between two small metal sections of similar thickness.

Joints may be further subdivided into:
■ Control joints
■ Construction joints
■ Isolation joints

■ **Control joints:** are used to introduce planes of weakness into monolithic or composite inelastic assemblies, (such as masonry or concrete walls, concrete slabs):
* to divide the assembly into smaller sections to prevent or minimize cracking.
* to induce cracking in pre-selected locations when cracking cannot be avoided.

■ Though considered non-working joints, **control joints can be working joints** in certain conditions:
* when they divide a masonry wall into separate sections and confine thermal and/or moisture movements within each section.
* in concrete subject to thermal movement due to wide temperature variations, joints in concrete not subject to significant thermal movement may be classified as non-working.

■ **Construction joints:** generally joints between monolithically poured sections of an assembly, such as concrete slabs or walls. Construction joints are normally non-working joints.

■ **Isolation joints:** when movement in one section, such as shrinkage, is to be isolated in order not to affect adjacent assemblies, e.g.:
• at the juncture of a masonry wall and a concrete floor slab when partial bonding of the concrete to the masonry is to be prevented.

Joint Protection
The need to protect the joint varies with:

■ **The degree of movement** within the joint; as well as with

■ **Additional performance** required to prevent:
• rain penetration
• air movement
• deterioration in appearance.

■ Some joints may **require no protection;** these will generally be:
• joints with no, or very limited, movement.
• vertical joints.

■ **All working joints** have to remain working and therefore must be kept free of intrusion by solid matter:
• such intrusion can freeze the joint and cause localized stress concentrations which may deform or crack the components.
• as an example, a small piece of stone wedged in a horizontal joint between two expanding concrete slabs may cause spalling or even cracking in such slabs.

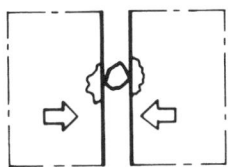

■ **Joints in exterior assemblies,** even where movement is limited, generally require protection from moisture penetration and air leakage.

Two basic methods of protecting joints are generally used:

■ **Prefabricated covers** of metal: generally for joints which have to accommodate large movements, such as expansion joints between sections of large buildings.

■ **Sealants** to protect the joint against the intrusion of foreign matter, and/or to prevent passage of water and/or air through it: the subject of this section.

SEALANTS

Sealants Defined

Open joints between components may be considered as voids created by the removal of a thin strip from each of the adjoining edges of such components to allow them to expand and contract freely.

■ **The sealant is then a more efficient substitute for the removed material:**
• it will absorb expansion and contraction in the original material; while
• meeting all other requirements applicable to the material which it replaces.

■ **A sealant should** maintain the integrity of the assembly of which the components are a part without restricting the movement between them.

■ **Performance requirements** for components and for the joints between them vary widely, and so do properties and use of sealants.

Types of Sealants

Sealants generally are classified based on their characteristics and on the state in which they are supplied to the installer, as follows:

■ **Bulk compounds,** which depend on their adhesion and cohesion to seal a joint.

■ **Formed resilient shapes,** which must be held constantly under pressure for an effective seal.

Bulk compounds are further classified based on the performance characteristics of specific compounds as:

■ **Low movement:** generally referred to as caulks and mastics, these are:
• low cost, easy to install, life expectancy 4 to 7 years.
• used in joints with limited movement, up to ±5 percent, such as glazing beads.

■ **Intermediate movement sealants:**
• cost approximately twice that of caulks or mastics, life expectancy also double: 7 - 14 years.
• use is similar to that of low performance sealants, for larger movements in joints, up to ±12.5 percent, used for joints around doors and windows.

■ **High movement sealants:**
• cost two to four times, or more than intermediate performance sealants.
• life expectancy 20 to 30 years.
• used in joints with large movements, up to ±25 percent in elongation or compression.
• installation generally more difficult than of low or intermediate movement sealants.

Bulk compounds may be available in several forms:

■ **Pourable,** self-leveling.
■ **Knife or tool** grade, non-sagging consistency:
• generally used for glazing and for pointing masonry.
■ **Gunnable, non-sagging consistency:** to be forced into the joint by a manually or pneumatically operated caulking gun and then to be consolidated and shaped by tooling.

■ **Pre-formed tapes:** bulk compounds extruded into various shapes either cured or not:
• used when a continuous uniform bead of sealant is required, and when it can be placed prior to joining the parts.

Performance Requirements

The type of sealant to be used will principally be determined by:

■ Percent elongation it can safely sustain, that is **the amount of movement through a joint relative to the width of the joint.**
• The cured sealants in moving joints are under a continuous contraction or expansion. The ideal sealant should return to its original state from the stressed or compressed state in the moving joint, when stress is removed. That means it should be 100 percent elastic.
• Some silicone or urethane sealants have close to 100 percent elasticity or recovery.
• Some sealants, which are too visco-elastic, do not return completely to their original state after removal of the extension or compression stress. The residual irreversible difference is called elongation set or compression set.
• The sealants with a high elongation set or compression set in a series of irreversible dimensional changes can buckle, crack, and foil.

B · BUILDING SHELL

B2.11 SEALANTS

SEALANTS: introduction (cont.)

- A high elastic recovery is a desirable property of sealants in moving joints.

■ When it is possible to increase the width of a joint with the extent of movement through it remaining the same, the performance requirements for the sealant will be lowered.

Other considerations generally are: Flow or rheology - of uncured and cured sealants will affect their performance.

■ **The flow of viscous uncured sealants can be pseudoplastic or dilatant:**
- in the case of pseudoplastic sealant, its viscosity with the increasing stirring increases less than proportionally.
- in the case of dilatant sealant its viscosity with the increasing stirring increases more than proportionally.
- Soft, easily extrudable, easily stirrable, although thick, sealant putties are pseudoplastic: Quick sand is a typical example of a dilatant flow.
- Polymeric formulations heavily pigmented with uniform particles of fillers are dilatant too.
- dilatancy is undesirable in sealants as well as in coatings.
- Pseudoplasticity is a beneficial property of sealants.

■ **Cohesiveness and workability over a wide range of temperatures:**
- the cohesive strength of a sealant should not be greater than its adhesive strength, but the cohesion should be strong enough to resist the constant expansion and contraction through joints.
- the low modulus of a high resiliency sealant reduces the excessive stresses in an expanding joint.

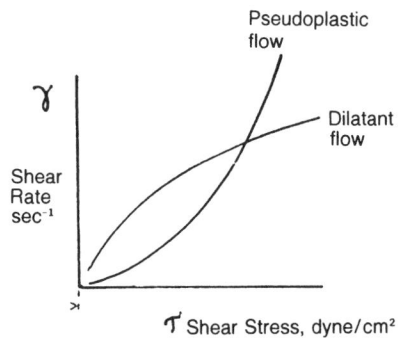

■ **Non-sagging sealants:**
- Non-sagging sealants are thixotropic: the property of changing by touch or becoming fluid when disturbed.
- their structure can be easily broken by stirring, and it restores quickly, when the agitation is stopped.
- the thixotropic sealants thin down upon gunning, brushing, knifing, and when placed in vertical joints, they immediately regain their structure and stay there without sagging.
- the degree of thixotropy of the system is determined by the area of the thixotropic hysteresis loop, enclosed between the ascending and descending curves in the shear stress and shear rate diagram.

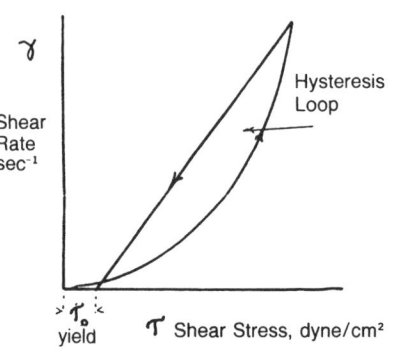

Thixotropic System with Hystersis Loop

- the additives imparting thixotropy are combinations of silica with amines; carbon black pigments with microcrystalline waxes; and others.
- the non-sagging efficiency is measured by a slump test method in accordance with ASTM D 2202.

■ **Chemical stability to avoid staining of adjacent surfaces and/or to provide a suitable surface for an applied coating;**

■ **Skinning properties:** the ability to form a non-tacky elastic surface skin to prevent pick-up of dirt and the intrusion of solid objects into the joint.

WATER STOPS

Water stops are formed seals, rather than sealants and are used in construction joints of monolithic assemblies subject to hydrostatic pressure.

■ Water stops are cast into adjacent, separately poured sections of such assemblies, to prevent water penetration through a plane of discontinuity: the construction joint.

■ Some types of water stops include provisions to accommodate movement in the joint, either perpendicular or parallel to the plane of the joint.

■ Water stops are generally formed of plastic materials, but copper ones are also available.

B2.11 SEALANTS

JOINTS: types, uses

● denotes condition expected or common usage of material

○ denotes possible or partial condition or limited use of material

TYPE	TYPICAL USE	MOVEMENT EXPECTED		MATERIALS MAY BE:		SEALANT TYPE			SEALANT MAY BE IN:				EDGE CONDITION MAY BE:	
		UNDER 10 PERCENT	OVER 10 PERCENT	SAME or with SIMILAR COEFFICIENT of EXPANSION	DISSIMILAR, with DIFFERENT COEFFICIENTS of EXPANSION	BULK COMPOUNDS	PREFORMED TAPES	FORMED SEALS/GASKETS	COMPRESSION or TENSION only	TENSION and COMPRESSION	SHEAR only	TENSION, COMPRESSION, and SHEAR	FREE to MOVE	RESTRAINED
[diagram]	CONCRETE: • PRECAST PANELS • SITE CAST • PAVEMENTS	○	●	●		●	○	○	●	●		○	●	
[diagram]	CONCRETE: • PRECAST PANELS METAL: • SANDWICH PANELS	○	●	●	○	●		○	●	●		○	●	
[diagram]	METAL: • FORMED FACING PANELS, e.g. prefinished steel, aluminum	○	●	●	○	●		○	●	●		○	●	
[diagram]	METAL OR WOOD FRAME, IN-FILL PANELS OF: • METAL • WOOD • PLASTIC	●	●	●	●	●		●	●	●		●	●	●
[diagram]	CONCRETE: • PRECAST PANELS METAL: • FLAT PANELS, single or sandwich	●	●	●	●	○	●				●	○		
[diagram]	METAL • FORMED PANELS, single or sandwich; connected at lap joint	●		●				●	●					○
[diagram]	METAL: • FORMED PANELS, single or sandwich, crimped at joint	●		●			●	●						●
[diagram]	CONCRETE: • SITE CAST: control, construction, expansion, isolation joints	●		●		●		●	●	●			●	○
[diagram]	CERAMIC TILE TO: MASONRY OR CONCRETE, Portland cement grout	●		●		●			●	●			●	
[diagram]	MASONRY: • BRICK OR • CONCRETE BLOCK; control joint	●	●	●		●		●	●	●			●	
[diagram]	WOOD OR METAL TO: • MASONRY • CONCRETE door, window frames	●			●	●					●			
[diagram]	WOOD OR COMPOSITE PANELS TO: • METAL FRAMES steel, aluminum	○	●		●	●			●	●		○	○	●
[diagram]	WOOD, COMPOSITE, METAL SANDWICH PANELS TO: • METAL FRAMES	○	●		●		●	●	○	●	●		●	○
[diagram]	WOOD TO WOOD: • SIDING PANELS or STRIPS; vertical, horizontal	●	●	●		●	○		●	●				●

SEALANTS: types, properties, uses

TYPE   **DESCRIPTION**

DEFORMABLE

* COMPONENTS CONTRACT AFTER
  SEALANT IS INSTALLED.

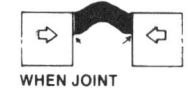

SEALANT RELAXES   WHEN JOINT
INTO STRETCHED   NARROWS NEW
SHAPE   SHAPE IS DEFORMED

* COMPONENTS EXPAND
  AFTER SEALANT IS INSTALLED.

SEALANT RELAXES   WHEN JOINT WIDENS
INTO COMPRESSED   TENSION DEVELOPS
SHAPE   IN NEW SHAPE

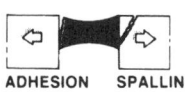

ADHESION   SPALLING   COHESION FAILURE:
FAILURE   FAILURE   MAY BE CAUSED BY
                        PUNCTURE OR NICK.

■ **Principal characteristics:** though they possess a degree of instantaneous elasticity (recovery) under loads of short duration:
* these sealants **creep (cold flow) when held** in a stressed (deformed) shape for a longer time and assume a new shape that relieves the stresses.
* also **tend to wrinkle** at the surface and become unsightly when subject to relatively large movements between components.

■ **Failure within a deformable sealant may occur in tension or in compression;** it is due to movements taking place after cold flow has occurred in the sealant, and after it has relaxed into either a new compressed shape, or into an extended one:
* when the new compressed shape is subject to **tensile stresses**, it may **yield at its minimum cross-section.**
* when the new extended shape is subject to **compressive stresses**, it may **buckle up** resulting in high peel stresses at lower corners, which may lead to **adhesion failure.**

■ **Common types of failure** (for both deformable and elastic sealants):
* **adhesion failure**, or the loss of bond between the sealant and the edges of the components; in most cases it may be prevented by using appropriate primers.
* **cohesion failure**, or a failure within the body of the sealant
* **spalling failure**, when cohesive strength of the sealant is greater than the cohesive strength of the component to which it is adhered.

ELASTIC/ELASTOMERIC

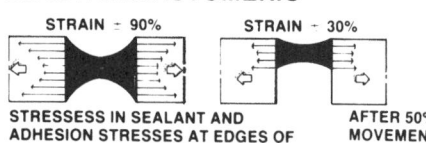

STRAIN ≈ 90%   STRAIN ≈ 30%

STRESSESS IN SEALANT AND   AFTER 50%
ADHESION STRESSES AT EDGES OF   MOVEMENT
COMPONENTS FOR DEEP AND
SHALLOW BEADS

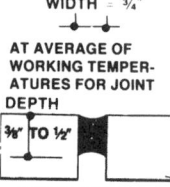

WIDTH = ¾″   WIDTH   1 ½″

AT AVERAGE OF   JOINT WIDENED TO
WORKING TEMPER-   IMPROVE SHAPE
ATURES FOR JOINT   OF SEALANT
DEPTH   DEPTH

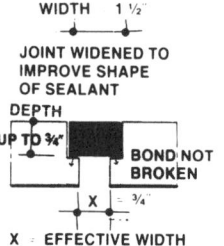

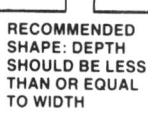

¾″ TO ½″   UP TO ¾″

                    BOND NOT
                    BROKEN

RECOMMENDED
SHAPE: DEPTH   X   ¾″
SHOULD BE LESS
THAN OR EQUAL   X   EFFECTIVE WIDTH
TO WIDTH   OF JOINT IF SEALANT
                    BONDS TO BOTTOM

■ **Principal characteristic:** ability to **return to the shape** given at the time of installation, after having stretched because of movements between components:

■ **Failure:** though the shape of a bead of cured elastomeric sealant can be changed by force, the volume will remain constant; therefore, **when stretched** through movement between components, the bead's **cross-section will "neck-in"** and extend its outer surfaces (which are free to deform) more than the material in the middle of the bead. This results in considerably **increased adhesion stresses** at the outer edges of the bead.

■ **Form of bead:** in order to accommodate the same degree of extension, the free surfaces of a deep bead of sealant must deform considerably more than those of a shallow one; therefore the **depth and shape** given the bead at the time of installation are **critical to the proper performance** of the sealant:
* the most **efficient shape** is a very thin bead slightly thinner in the middle than at the edges.
* theoretically, the width of a bead should be twice its depth, but such proportions are difficult to achieve and maintain during actual installation, especially in narrow joints:
* the proportions for the depth to width ratio of sealant beads should not exceed 1.0, e.g., the depth should not exceed the width.

■ **Common types of failure:** are similar to those encountered in deformable sealants: failures in **adhesion, cohesion and spalling** (see above).

FORMED

SEAL COMPRESSED   WATER/AIR
50% FOR INSERTION.

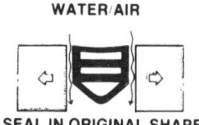

SEAL COMPRESSED   SEAL IN ORIGINAL SHAPE

EXTRUDED: MUST EXERT PRESSURE
AGAINST SIDES OF JOINT TO SEAL
THE JOINT AT ALL TIMES.

SEALANT INSTALLED   JOINT WIDTH AT
AT MINIMUM 25%   MAXIMUM OPENING
COMPRESSION

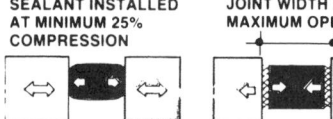

                                ADHESIVE

FOAMS: MUST EXERT PRESSURE
AGAINST SIDES OF JOINT AT ALL
TIMES. USING ADHESIVE MAY HELP
HOLDING SEAL IN JOINT AT MAXIMUM
OPENING OF JOINT.

■ **Formed sealants (or seals) are compression seals** and may be:
* **rigid extrusion seals,** shaped with internal voids which allow them to be compressed in the joint through the full range of movement within it.
* **foams made of high recovery elastomers** which depend on the properties of the material for compressibility.

■ **Failure occurs** when the joint opens to a point when the seal no longer exerts pressure against the sides of the joint; the effectiveness of the seal is then lost and it may even **slip out** of the joint.

■ **Form and installation:**
* **rigid extrusion** seals are compressed to about half their initial width during installation into a joint.
* **foams** are similar to extrusions as far as use and installation requirements are concerned: they seal a joint by exerting pressure against its sides.
* **adhesives** may be used with foams to improve their performance when sealing pressure is reduced at maximum opening in the joint; they are not effective with rigid extrusions.
* **extrusions and foams** require joints of constant width along their entire length; they also should be fused or bonded at all splices and changes in direction.

B2.11 SEALANTS

TYPE	DESCRIPTION

DEFORMABLE

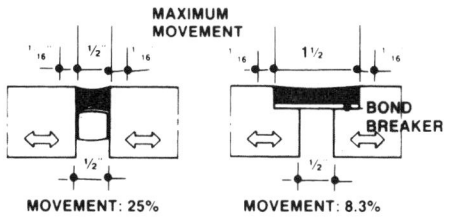

MAXIMUM MOVEMENT

MOVEMENT: 25%   MOVEMENT: 8.3%

TYPICAL NON-WORKING JOINTS

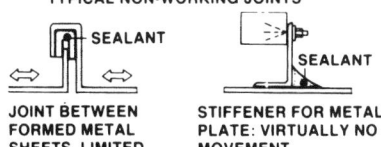

JOINT BETWEEN FORMED METAL SHEETS. LIMITED MOVEMENT

STIFFENER FOR METAL PLATE: VIRTUALLY NO MOVEMENT

■ **Deformable sealants generally are classified as:**
• **low performance,** used when the percentage of movement in a joint is limited to not more than ±5 percent; or
• **intermediate performance,** generally for movements of 5 to 15 percent, with some types suitable for use in joints with movements of up to ±12.5 percent.

■ **The movement in the joint is the principal consideration** in selecting a sealant, whether deformable or elastic, because failure is likely to occur if limits of elongation for the sealant are exceeded. Since the percentage of movement through a joint is the ratio of movement to width, **joints may be modified** to allow the use of a sealant with lower limits of elongation if the sealant offers other advantages, for example:
• a joint may be **partially widened** to reduce the percentage of movement through it, but the wider bead of sealant used should be free to move along its entire width.

■ **Elongation limits for sealants used in non-working joints,** (that is joints with little or virtually no movement through them) may become secondary to other considerations, such as workability; cohesiveness; retention of flexibility; resistance to shrinking, oxidation, ultraviolet radiation; hardness if exposed to indentation; paintability.

ELASTIC/ELASTOMERIC

RECOMMENDED BEAD SHAPE:

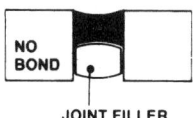

JOINT FILLER

• WIDTH TO DEPTH: NOT MORE THAN ONE-TO-ONE, BETTER IF LESS.
• SEALANT TOOLED CONCAVE AND TIGHT TO FILLER
• FILLER INSTALLED TO PROVIDE CONCAVE BOTTOM SURFACE.

BOND BREAKER

THREE-SIDED JOINT

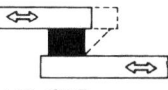

LAP-JOINT: SEALANT IN SHEAR

■ **Elastomeric sealants are classified as high performance,** with most types suitable for use in joints with movements of ±25 percent, and some types for movements of up to 50 percent in pure elongation.
• **depth to width** varies from half-to-one to one-to-one, and should be **closely controlled.**
• the sealant, whether deformable or elastic, generally is **forced** into the joint; therefore **fillers are required** in open joints to limit the depth of the bead:

■ **Fillers** should be of materials **compatible with the sealant** used:
• in order to **prevent bonding** of sealant to them so that free movement of the bead through the joint is not restricted; and
• fillers may be also used to **give the bead the proper shape** for best performance.

■ **Shallow joints with one end closed** may not require a filler:
• **bond breakers** over the closed end will be needed to prevent the sealant from bonding to the bottom of the joint and
• **adhesion to three sides** of a joint prevents free movement in the bead thus overstressing it and resulting in cohesive failure.

■ **Sealants in lap joints,** and joints subject to differential movement, will be subject to **shearing stresses;** a sufficient volume of sealant must be provided to resist them: generally the depth should be equal to the extent of movement expected.

FORMED

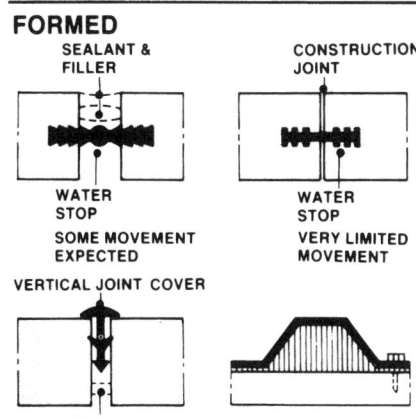

SEALANT & FILLER   CONSTRUCTION JOINT

WATER STOP   WATER STOP

SOME MOVEMENT EXPECTED   VERY LIMITED MOVEMENT

VERTICAL JOINT COVER

SEALANT AND FILLER

JOINT COVER AS RAIN BAFFLE

CLOSURE FOR FORMED METAL OR PLASTIC PANEL

■ **Formed seals** are available in various configurations:
• for use as **primary seals**
• as **secondary seals, joint fillers,** in conjunction with a sealant at the primary seal.

■ **Extrusions are extensively used** as the primary seal for horizontal joints in concrete, such as in pavements, where they offer the advantage of easy installation and durability. Their disadvantage is that in joints of **unequal width,** often the case with field-formed joints, they may **not be watertight.**

■ **Extruded gaskets** are also extensively used in glazing applications;
• in curtain wall construction as **thermal breaks** in mullions;
• as **seals** in batten joints;
• as **closures** for corrugated metal or plastic panels.

■ **Foams** are more often used as **the secondary seal** or joint filler, but may also be used as primary seals.

■ **Water stops** are used as an integral part of concrete construction. They may accommodate movements of up to 15 percent in either tension or shear. Water stops may be used as the **primary seal,** or in conjunction **with another sealant.** They should be fused or fully bonded at splices to maintain continuity.

B2.11 SEALANTS

SEALANTS, SEALS: uses

● common usage
○ some properties may limit or restrict usage

USE	JOINT WIDTH range in inches	LOW up to ±5 percent				INTERMEDIATE ±12.5 percent						HIGH ±25 percent						
		OIL-BASED	BUTYL, unvulcanized	POLYBUTENE	BITUMEN, rubberized	BUTYL, vulcanized, non-skinning	BUTYL, vulcanized, skinning	ACRYLIC-LATEX	POLYVINYL ACETATE LATEX	CHLOROSULFONATED POLYETHYLENE (HYPALON)	POLYCHLOROPRENE (NEOPRENE)	ACRYLIC, solvent-based	POLYSULFIDE, two component	POLYSULFIDE, one component	POLYURETHANE, two component	POLYURETHANE, one component	SILICONE, low, medium modulus	POLYMERCAPTAN*
BETWEEN SIMILAR MATERIALS, exposed location																		
CONCRETE	¼ to ¾	●	●				●	●	●	●	○	●	●	●	●	●	●	●
	¾ and over											○	●	○	●		○	
MASONRY, brick, concrete block	¼ to ¾	●	●				●	●	●	●	○	●	●	●	●	●	●	●
	¾ to 1½											○	●	○	●		○	
METAL	¼ to ¾	●	●				●	○	○	●	○	●	●	●	●	●	●	●
	¾ to 1½											○	●	○	●		○	
WOOD [1]	¼ to ¾	●	●				●	●	●	●	○	●	●	●	●	●	●	●
	¾ to 1½											○	●	○	●		○	
DISSIMILAR MATERIALS, exposed location, differential movement																		
METAL to METAL	¼ to ¾	○	●				○	○	○	●	○	●	●	●	●	●	●	○
	¾ to 1½											○	●	○	●		○	
METAL to WOOD [1]	¼ to ¾	○	●				○	○	○	●	○	●	●	●	●	●	●	○
	¾ to 1½											○	●	○	●		○	
METAL or WOOD to [1] MASONRY/CONCRETE	¼ to ¾	○	●				○	○	○	●	○	●	●	●	●	●	●	○
	¾ to 1½											○	●	○	●		○	
METAL or WOOD to [1] GLASS/PLASTIC	¼ to ¾	○	●				○	○	○	●	○	●	●	●	●	●	●	●
	¾ to 1½											○	●	○	●		○	
SPECIAL CONDITIONS																		
CONCEALED LOCATION	¼ to ¾	●	●	●		●	●	●	●		●	●	●	●	●	●	●	●
	¾ to 1½			●		●						○	●	○	●	○	○	
SWIMMING POOLS reservoirs, aquariums	¼ to ¾									●	●				●			●
	¾ to 1½														●			
ROADWAYS, WALKS	¼ to ¾				●										●			
	¾ to 1½				●										●			

PRE-FORMED TAPES:		FORMED FOAMS			SEALS: RIGID EXTRUSIONS			WATER STOPS		
POLYBUTENE, uncured	POLYISOBUTYLENE, cured/uncured	POLYCHLOROPRENE (NEOPRENE) cellular	POLYVINYL CHLORIDE (PVC)	POLYURETHANE, open cell	POLYCHLOROPRENE (NEOPRENE) non-cellular	POLYVINYL CHLORIDE (PVC)	STYRENE BUTADIENE RUBBER	POLYCHLOROPRENE (NEOPRENE)	POLYVINYL CHLORIDE (PVC)	STYRENE BUTADIENE RUBBER
		●	●	●	●	●	●	●	●	●
					●	●	●			
		●	●	●	●	●	●			
					●	●	●			
○	○	○	○	○	○	○	○			
○										
○		○	○	○						
○										
●		○	○	○	○	○	○			
●										
●		○	○	○						
●										
		○	○	○	○	○	○			
●		○	○	○						
●										
●		○	○	○	○	○	○	●	●	●
					○	○	○	●	●	●
					●	●	●			
								○	○	○
					●	●	●	●	●	●

REMARKS:

[1] sealants for use with wood treated for decay or fire resistance should be investigated for compatibility with specific type of chemicals used during a given treatment.

- water-release latex may bleed and stain concrete.
- components of dissimilar materials may expand/contract at different rates when their coefficients of expansion are different. Conditions assumed are for differential movements; if coefficients of expansion are similar, information given for similar materials applies.
- butyl, unvulcanized, should have a skinning agent added when in exposed locations.
- butyl use depends on a particular formulation.
- joint preparation for most high performance sealants is critical for proper performance.
- high modulus silicone should not be used on cementitious substrate.
- polyvinyl acetate, interior use only.
- polymercaptan in-service record not yet fully established.
- non-skinning butyl and polybutene will pick up dirt.
- polybutene, uncured, found to be satisfactory for concealed locations; partially cured type is available with superior adhesion.
- Hypalon may be installed over neoprene when color is required.
- acrylics may be used in wider joints, but wrinkling of the surface may result.
- movement in joints between metal and wood may be differential due to thermal movement in metal and to moisture movement in wood.
- curtain walls, stone veneers on high rise buildings are not readily accessible after installation: in-service life of sealant should be considered to minimize maintenance requirements.
- poor resistance to indentation will restrict use of low and moderate modulus silicone in wide joints subject to traffic.
- silicones tend to pick up and hold dirt.
- two-part silicone available for joints up to 1½ inches for water immersion.
- for structural glazing of aquariums, high modulus silicone should be used.
- tapes may be limited in use by conditions of installation: seldom used in concrete and masonry joints, except in bedding joints; may be pressed into joints.
- tapes used in bed joints of sound-proofed partitions, in laps between formed metal or plastic panels.
- formed seals for masonry: foams to seal against air/water penetration; rigid extrusions as shear keys in control joints.
- formed rigid extrusions extensively used in pavements.
- formed rigid extrusions generally not used with metal, glass, plastic and wood because of insufficient depth of joint.
- formed foam seals often used with additional bead of sealant thus acting as seal/joint filler; also used to isolate structural steel when built into masonry.
- polyurethane should not be exposed to ultra-violet radiation.
- water stops should be selected based on extent of movement to be expected in joint.

B2.11 SEALANTS

BULK COMPOUNDS: types, properties

● denotes common usage
* primer required in working joints
EX–excellent
G –good
F –fair
P –poor

PROPERTIES

TYPE	SERVICE LIFE, in years	JOINT MOVEMENT, percent	ULTIMATE TENSILE STRENGTH, range, psi [2]	SHORE A HARDNESS, range at 75°F [1]	ELONGATION AT RUPTURE percent, range [2]	SHRINKAGE, percent range	DEGREE OF COMPRESSION RECOVERY, percent, range	SERVICE TEMPERATURE, range °F	CURING TIME, in days	MAXIMUM JOINT WIDTH, inches
LOW PERFORMANCE										
OIL-BASED	3 to 5	up to ±5	2 to 4	10 to 30	100	5 to 10	10 to 15	-40 to 250	120	¾
BUTYL, unvulcanized	5 to 10		10 to 20	10 to 15	60 to 70	10 to 35	10 to 15	-10 to 200	120	¾
POLYBUTENE, noncuring	3 to 10		0 to 1	not available	50	0 to 5	15 to 20	-60 to 150	120	¾
BITUMINOUS, rubberized	5 to 10		200	not available	50	0 to 5	15 to 20	-60 to 135	120	2
INTERMEDIATE PERFORMANCE										
BUTYL, vulcanized	to 10	±7.5	250 to 400	20 to 30	75 to 125	10 to 35	10 to 30	-20 to 200	120	¾
ACRYLIC-LATEX	to 7	±7.5	40 to 60	30 to 35	60	5 to 10	5 to 10	-20 to 180	5	⅜
POLYVINYL ACETATE LATEX	to 7	±7.5	40 to 60	20 to 40	60	5 to 10	5 to 10	-20 to 150	5	⅜
CHLOROSULFONATED POLYETHYLENE (HYPALON)	to 10	±10	40 to 100	12 to 25	150	15	10 to 30	-10 to 220	30 to 180	¾
POLYCHLOROPRENE (NEOPRENE)	to 20	±10	100 to 400	20 to 40	300 to 600	35	10 to 30	-20 to 200	30	¾
HIGH PERFORMANCE										
ACRYLIC, solvent-base	to 20	±12.5	40 to 60	10 to 50	60 to 250	Negligible	10 to 30	-30 to 200	14 to 21	¾
POLYSULFIDE, two component	to 20	±25	120 to 770	20 to 40	450 to 650	Negligible	30 to 60	-60 to 250	4 to 7	2
one component	to 20	±25	85 to 120	20 to 40	800 to 1000	Negligible	30 to 60	-40 to 250	14 to 60	1
POLYURETHANE, two component	to 20	±20	150 to 400	20 to 40	350 to 800	Negligible	60 to 100	-65 to 250	3 to 6	1½
one component	to 20	±20	250 to 500	15 to 50	100 to 600	Negligible	60 to 100	-40 to 250	14 to 30	¾
SILICONE	to 20	±25	250 to 450	15 to 30	400 to 1000	Negligible	100	-65 to 350	2 to 5	1
POLYMERCAPTAN	Not Est.	±25	400 to 500	30 to 40	400 to 1000	Negligible	30 to 60	-40 to 250	2 to 3	⅝

B2.11 SEALANTS

REMARKS (legend)

1. Hardness on a scale ranging from 0 to 100, very soft to very hard.
2. Not primary consideration

Column groups: **ADHESION Self-Bonding To:** (Metal, Wood, Masonry, Plastic, Glass) · **RESISTANCE TO:** (Ultraviolet Rays, Oxidation on Exposure, Water Immersion, Alkalies, Acids, Oil/Grease, Abrasion/Indentation, Puncture/Tear, Fire)

Section 1

Self-Lev.	Non-Sag	May Stain	Metal	Wood	Masonry	Plastic	Glass	Primer Req.	UV	Oxid.	Water	Alk.	Acids	Oil/Grease	Abr./Ind.	Punct./Tear	Fire	ASTM / Fed. Spec.
		●	●	*	*		●	●	F	F	F	P	P	F	P	P	P	TT-C-598
	●		●	●	●	●	●		F to G	F to G	G	G	G	F	F	F	F	TT-C-598
	●		●	●	●		●		F to G	F	P	P	F	F	P	F	P	C920
	●				●				P	G	F to G	P	P	P	G	F	P	D1190

Remarks (Section 1):
- recovery of oil-based sealants is average to low.
- life expectancy and resistance to water exposure of oil-based sealants can be prolonged with paint.
- bituminous sealants not recommended for swimming pools, interior floors or vertical sloping joints because they soften under high temps.
- skin forming sealants will self-heal after a cohesive rupture.
- unvulcanized butyl generally used in residential and light commercial construction.
- polybutene remains permanently tacky; some dirt pickup will result if exposed; should be "topped off" with another elastomeric sealant or used in concealed locations.

Section 2

Self-Lev.	Non-Sag	May Stain	Metal	Wood	Masonry	Plastic	Glass	Primer Req.	UV	Oxid.	Water	Alk.	Acids	Oil/Grease	Abr./Ind.	Punct./Tear	Fire	ASTM / Fed. Spec.
	●		●	●	●	●	●		F to G	F	F to G	G	G	F to P	F	F	F	TT-S-1657
	●		●	●	●	●	●		G	G	P	F	F	F	P to F	F to P	F	C834
			●	●	●	●	●		F	F	P	F	F	F	F	P	P	C834
	●		●		●				Ex	Ex	G	G	G	G	Ex	F	G	C920
●	●		●		●				G	G	G	G	G	Ex	G	F	G	C920

Remarks (Section 2):
- skinning and non-skinning types in butyl available.
- butyls should be used in concealed joints, due to their tendency to pick up dirt.
- latex may be applied to damp, absorptive surfaces.
- polyvinyl acetate latex sealants are generally recommended for interior use only.
- hypalon has excellent colorability and color retention.
- neoprene has good adhesion to asphaltic cement.
- two-component neoprene is used only in specialized areas.

Section 3

Self-Lev.	Non-Sag	May Stain	Metal	Wood	Masonry	Plastic	Glass	Primer Req.	UV	Oxid.	Water	Alk.	Acids	Oil/Grease	Abr./Ind.	Punct./Tear	Fire	ASTM / Fed. Spec.
●	●		●	●	●	●	●		G	G	P	G	G	G	F	P	P	C920
●	●		●	*	*		●	●	F to G	G	G	G	F	G	P	G	P	C920
●	●		●	*	*		●	●	F	G	G	G	F	G	P	G	P	C920
●	●	●	●	*	*		●	●	P to F	G	P	G	F	G	G	G	P	C920
●	●		●	*	*		●	●	P to F	G	P	G	F	G	G	G	P	C920
●	●		●	*	*	●	●	●	G	G	G	F	G	F	P	F to P	G	C920
●	●		●		●		●		G	G	F	G	F	G	G	G	P	C920

Remarks (Section 3):
- offensive odor is released by acrylics and they should not be used in food storage or preparation areas.
- acrylic will self-heal after cohesive failure.
- polysulfide has unpleasant odor until it cures.
- avoid applying polysulfide to joints contaminated with bituminous materials.
- urethane may be applied to bituminous or asphaltic compounds.
- three-part polyurethane available for joints up to 5 inches.
- use of primer in all working joints recommended.
- for wider joints, two-part silicone to function properly.
- most silicones affect paint adhesion and cannot be painted.
- due to strong smell of silicone, ventilation recommended during installation.
- resistances of polymercaptans has yet to be established on long-term basis.
- two-part medium density, foamed-in-place silicone with fire-resistant properties available; not compatible with sulfur-containing materials.

B2.11 SEALANTS

TAPES, FORMED SEALS: types, properties

● denotes common usage
Ex – excellent
G – good
F – fair
P – poor

PROPERTIES

TYPE	SERVICE LIFE, in years	JOINT MOVEMENT, in percent, average	ULTIMATE TENSILE STRENGTH, range, psi	SHORE A HARDNESS, range at 75°F [1]	ELONGATION AT RUPTURE, percent, range	COMPRESSION IN JOINT, required for performance, percent [2]	DEGREE OF COMPRESSION RECOVERY, percent	SERVICE TEMPERATURE, range °F
PRE-FORMED SEALING TAPES								
NON RESILIENT POLYBUTENE, uncured	10 to 15		0 to 1	not available	50	not applicable	0	– 60 to 150
POLYISOBUTYLENE, uncured	15 to 20	to ± 10	4 to 10					– 45 to 200
POLYISOBUTYLENE, cured			10 to 25	10 to 20	250		10 to 30	– 20 to 200
FORMED SEALS, FOAMS								
POLYCHLOROPRENE (NEOPRENE) cellular	10 to 30		2,000	24 to 40	200 to 450	up to 40	30 to 60	– 40 to 150
POLYVINYL CHLORIDE (PVC)	10 to 20	to ± 7.5	50 to 150	15 to 40	100 to 250	15 to 25	10 to 30	– 20 to 130
POLYURETHANE, open cell			250 to 500	60 to 90	200 to 300	up to 50		– 20 to 160
FORMED SEALS, RIGID EXTRUSIONS								
POLYCHLOROPRENE (NEOPRENE) non-cellular	20 to 30 +		2,000 to 2,500	40 to 80	200 to 300	25 to 50	60 to 100	– 20 to 160
ETHYLENE PROPYLENE (EPDM)	5 to 15	to ± 7.5	1,400	40 to 60		up to 25	10 to 30	– 75 to 200
POLYVINYL CHLORIDE (PVC)	5 to 20		1,800	60 to 90		25 to 50		– 20 to 130
STYRENE BUTADIENE RUBBER	5 to 10		700 to 2,000	40 to 80	450	up to 25	30 to 60	– 10 to 160
WATERSTOPS, CLOSURE STRIPS								
POLYCHLOROPRENE (NEOPRENE)	20 to 30	to ± 7.5	2,500	60 to 70	450	not applicable	not applicable	– 20 to 160
POLYVINYL CHLORIDE (PVC)			1,800	75	300			– 30 to 175
STYRENE BUTADIENE RUBBER			2,500	60 to 70	450			– 10 to 160

B2.11 SEALANTS

REMARKS:

[1] Hardness on a scale from 0 to 100, very soft to very hard.

[2] Formed seals, foams and extrusions have to be pre-compressed prior to being inserted into a joint.

MAXIMUM JOINT WIDTH, inches	CHOICE OF COLOR	MAY STAIN ADJACENT SURFACE	ADHESION Self-Bonding To: METAL/WOOD	MASONRY	PLASTIC/GLASS	ADHESIVE MAY BE USED	PRE-COMPRESSION REQUIRED FOR INSERTION	RESISTANCE TO: ULTRAVIOLET RAYS	OXIDATION ON EXPOSURE	WATER IMMERSION	ALKALIES	ACIDS	OIL/GREASE	ABRASION/INDENTATION	PUNCTURE/TEAR	FIRE	APPLICABLE ASTM or FEDERAL SPECIFICATIONS
no * limit		•	•		•			G	F to G	P	P	F to G	P	P	P	P	
	•	•	•		•			G	F to G	P	G	P	P	P	P	P	D2581
	•	•	•		•			G	G	P	G	P	P	P	P	P	
1½						•	•	Ex	Ex	G	G	G	G	G	G	G	C509
½	•					•	•	G	G	G	F to G	G	G	F	F	F	D1565
						•	•	F to G	G	P	G	F	G	G	G	P	D3574
1½ up to 5" avail.							•	Ex	Ex	G	G	G	G	G	G	G	D2628
	•						•	F to G	G	G				G	G	P	C509
	•						•	G	G	G	F to G	G	G	F	F	F	C542
	•						•	P	F to P	G	P	F	P	G	G	F to P	D2000
up to 2								Ex	Ex	G	G	G	Ex	G	G	G	C509
								G	G	G	G	G	G	G	G	F to P	D412
								P	F to P	G	G	G	P	G	G	F to P	D412

Section 1 (no * limit) remarks:
- polybutene has a high degree of tack, but little strength.
- non resilient tapes should not be used in exposed locations; tend to pick up dirt and appear discolored.
- * joint, width is not limited, tapes available up to ½".

Section 2 (1½ / ½) remarks:
- foam seals are more suitable for use in joints of stone, concrete and brick than dense extrusions.

Section 3 (1½ up to 5" avail.) remarks:
- neoprene can be bonded with Hypalon to give it color.
- cellular neoprene can successfully be used with slightly irregular joints.
- cellular neoprene has excellent recovery; used extensively in highway construction.
- EDPM has limited resistance to stress or compression set.
- PVC has the tendency to flow under stress.
- SBR has poor color retention, high shrinkage; not suitable for exposed areas.

Section 4 (up to 2) remarks:
- waterstops generally selected on amount of movement required.
- metal waterstops available, no longer in wide use.
- metal and EPDM closures available; used for metal and asbestos roofing and siding.

B2.11 SEALANTS

BULK COMPOUNDS/PREFORMED TAPES:

Preliminary considerations in selecting a sealant for specific in-service conditions should include:

■ Type of joint to be sealed.

■ Materials to be joined.

■ Movement expected through the joint.

This then is viewed against sealant characteristics:

■ Sealant capability to **accommodate movement** expected.

■ In-service **life expectancy** of the sealant; including:

• retention of cohesiveness

• continuity of adhesion.

■ Suitability of sealant to perform satisfactorily under **other in-service conditions,** such as:

• exposure to weathering

• water immersion

• traffic

• alkalies, acids, oil, grease

• abrasion, indentation, puncture.

■ **Ease of installation,** clean-up.

■ **In-place cost.**

■ **Sealants are categorized based on their performance under movement through the joint,** but different formulations may be available for each type to meet other performance requirements, such as:

• easier installation; simpler clean-up; longer time available for tooling.

• resistance to abrasion, indentation, intentional abuse.

• better weatherability; color fastness; paintability.

• better adhesion; lower adhesion stresses.

• faster or slower curing; skinning or non-skinning.

■ **Sealants for joints with slight movement** through them will generally be selected for properties other than their ability to accommodate movement, such as:

• in-service life to be expected; durability.

• ease of installation and clean-up; paintability; choice of color.

• in-place cost.

When larger movements through the joint are to be expected, the ability of

the sealant to accommodate them should be the primary, but not necessarily the only, consideration, since other measures can be taken as well, including the following:

■ **Joint width may be increased** to reduce the percentage of movement and a lower performance sealant can then be used.

■ **Joint width may be decreased,** to increase the percentage of movement, when a higher performance sealant than the one meeting original requirements, combined with a narrower joint offer additional advantages, e.g.:

• simplified installation.

• a wider choice of sealants: some types may be restricted to narrow joints only.

• reduce the in-place cost.

Movement Through Joint

The degree of movement in components through a joint in relation to the width of the joint - the percentage of movement - is the first consideration in selecting the type of sealant to be used:

■ **The percentage of movement is not dependent on the width of the joint:**

• slight linear or differential movements in a narrow joint may exceed the capability of even high performance sealants to resist them.

• wide joints may move perceptibly due to seasonal changes and yet the percentage of movement may be low.

■ Use of a **low performance sealant** in a joint with a high percentage of movement may lead to failure of sealant.

■ Use of a **high performance sealant** in joints with low percentage of movement generally offers no advantages.

The properties of a sealant which indicate its performance characteristics under movement are:

■ **Elongation at rupture,** or ultimate elongation:

• this is a measure of stretchability of the sealant.

• it is only an indication of how much a specimen can elongate before it fails, not a measure of the amount of extension the sealant can safely resist under in-service conditions.

■ Sealants generally are **subject to elongation and compression;** their ability to resist such cyclic stresses is given by:

• the percentage of joint movement which will take place after the sealant is installed.

• if the sealant is installed at maximum extension of the components, the sealant will later be subject to tensile stresses only.

• if at maximum contraction to compressive stresses only.

• degree of compression recovery indicates the ability of the sealant to return to its original shape after being kept under compressive load for a period of time.

■ **Hardness of a sealant** is not directly an indication of the performance characteristics, but can be used as a rough measure of its modulus of elasticity which indicates resistance to deformation expressed as the ratio of stress to strain under load:

• hardness is measured on Shore A scale from 0 to 100: an automobile tire has a hardness of about 70, while a pink eraser has a hardness of about 15.

• mastic sealants generally are in the 10–30 Shore A range.

• sealants softer than 10 Shore A may be gouged out of joints or may be penetrated by stones or spiked heels in horizontal joints.

TYPES OF SEALANTS

Based on properties of elasticity and recovery, sealants may be subdivided approximately into:

■ High performance group which includes:

• high recovery sealants: silicones, urethanes.

• moderate recovery sealants: polymercaptans, polysulfides, polychloroprene.

selection checklist

■ **Intermediate performance** or the low recovery group: chlorosulfonated polyethylenes, acrylics, vulcanized butyl rubber.

■ **Low performance,** or the very low recovery group: oil-based caulks and glazing compounds, butyl mastic and tapes, polyisobutylenes.

Selection of a sealant within one of the groups may also include other considerations not directly related to the principal performance characteristics of the sealant.

Low Performance Group

■ **Oil-based caulks** are available in different formulations; their properties vary considerably between low and high grade:

* life expectancy varies from 3 to 5 years for low grade to 5 to 10 years for high. Generally all should be painted regularly to extend in-service life.

* all apply and tool easily. Special joint preparation or cleaning is not required.

* slow to skin and cure. May bleed and stain substrate or adjacent surfaces.

* lowest cost sealants.

■ **Butyl mastics** are available in a variety of formulations of unvulcanized rubber, solvents, and various plasticizers, such as polybutene; they cure by solvent release:

* application is fairly simple; no primers are required for any type of surface; joint preparation not critical.

* excellent resistance to immersion in water.

* slow curing, high shrinkage. Should not be used in traffic-bearing joints, such as in floors or sidewalks.

■ **Polybutene is used as** a gun grade sealant, in preformed tapes, or as plasticizer/tackifier in other sealants:

* has good adhesion, but poor elongation and recovery;

* remains soft and tacky; may stain porous surfaces; generally used in concealed joints only.

* low cohesive strength, but is self-healing.

■ **Bituminous sealants range** from oxidized asphalt through asphalt or coal tar compounded with synthetic rubber, such as neoprene or styrene butadiene:

* applied either hot, or at room temperature when solvent is added;

* generally used to seal joints in pavements and sidewalks, have good resistance to de-icing chemicals.

* very low recovery: may be permanently extruded from joint during maximum expansion of components.

Intermediate Performance Group

■ **Butyl rubber compounds in this group are the vulcanized type butyls** with better recovery properties than unvulcanized butyl compounds:

* skinning types are available, compounded with oxidizing agents to prevent dirt pick-up;

* other properties similar to butyl mastics.

■ **Vinyl acrylic and polyvinyl acetate latex** are one-component water-base gun grade sealants with good flexibility, but little recovery:

* installation and clean-up is simple: spills and spots can be cleaned up with water; generally non-staining execpt for concrete.

* will not bleed through water-base coatings as oil-base sealants would.

* joints should be cleaned; primer generally not required.

* fast skinning; can be painted over shortly after installation.

■ **Chlorosulfonated polyethylene (Hypalon):**

* excellent colorability and color retention.

* good adhesion to most substrate; primer not required.

* non-staining, non-corrosive, and self-extinguishing when cured.

* slow curing may limit application.

■ **Polychloroprene (Neoprene)** when properly compounded has excellent resistance to flexure cracking, deformation, and temperature extremes:

* excellent adhesion;

* will not support combustion.

* limited to dark colors only.

High Performance Group

■ **Acrylics are solvent based,** one component-gun grade, and cure by solvent evaporation:

* available in colors; good color stability.

* self-healing if cohesive failure occurs due to temporary excessive movement through joints.

* slow curing and remain tacky, which may lead to dirt pick-up and discoloration.

* require heating during installation.

* strong odor is offensive; may contaminate food products.

Polysulfide is available in one-component pourable grade, and non-sag one- or two-component formulations:

■ **Installation of two-component formulations** involves careful blending of the two components.

* working time after blending is short.

* sealant will fail in service if improperly mixed or contaminated.

* low temperatures will slow down curing process.

■ **One-component formulations** cure slowly, especially at low temperatures.

■ **Both one- and two-component formulations:**
* have a specific odor when curing which may limit their use.

* adhesion is good, but joints have to be carefully prepared; primers generally required, especially for porous surfaces.

* good resistance to deterioration when immersed in water, such as in swimming pools, reservoirs.

Polyurethane is available in one-component pourable grade, and in one- or two-component non-sag formulations, with a wide range of properties:

■ **Installation requirements** for two-component formulations are as stringent, or even more so, as for polysulfides:

* mixing must be slow and uniform.

B2.11 SEALANTS

BULK COMPOUNDS/PREFORMED TAPES:

- joints must be very clean and may require priming.

■ **Recovery from deformation** is superior to polysulfides and polyurethanes may be used in wide joints subject to large movements:
- adhesion is very good for all formulations; one-part polyurethanes cure slowly and excess moisture during curing may impair adhesion.
- good abrasion resistance and tear strength: may be used in joints subject to traffic, such as in parking decks, patios.
- polyurethane foam caulks available. Generally used around door and window frames and in large cracks in retrofit construction.

Silicones are available as one-part formulations in three basic types: low, medium, and high modulus:

■ All have very **good resistance to** ultraviolet radiation, ozone, and oxidation.

■ Silicones **may be applied** over a wider temperature range than other sealants and have a wider in-service temperature range.

■ **Recovery from deformation** is also superior, approaching 100 percent for low modulus types.

■ **Low modulus formulations** generally are used where a high degree of movement is expected:
- low modulus of the material reduces stresses at the sealant/substrate interface: a force of only 20 to 40 psi is required to extend a test specimen by 150 percent, as compared to 110 psi for a high modulus formulation.
- should not be used below grade or in traffic bearing joints.

■ **Medium modulus formulations** are used where a higher strength (95 psi at 150 percent elongation) and good recovery after deformation are required:
- they can be characterized as high performance, general purpose sealants.
- not recommended for traffic bearing joints.

■ **Both low and medium modulus formulations** cure tack free in about two hours, and allow sufficient time from 15 to 45 minutes for tooling.

■ **High modulus formulation,** the acid cure type, may be used as a sealant as well as an adhesive, and is the

only type recommended for structural glazing:
- should not be used on masonry substrates, because the acid released during curing may weaken the substrate leading to adhesion failure.
- curing is very fast, becoming tack-free in about 30 minutes, and tooling must be done very soon after application.
- may be used in joints with moderate movement and when immersed in water. Silicones are the highest cost sealants.

Polymercaptan is based on mercaptan terminated polymers, similar to those of polysulfides, anchored to polymer chains similar to those of urethane.

■ **Polymercaptan is available in one- and two-part formulations:**
- modulus similar to silicones and urethanes.
- recovery after compression between that for polysulfides and polyurethanes.
- compatible with coal tars.
- self-leveling and non-sagging formulations available.

■ **In-service record** not yet fully established.

Epoxy is a reaction product of bis-phenol A and epichlorhydrin:

■ **It is available in one- and two-part formulations:**
- high adhesion, good flexibility, tough and low shrinkage on cure.
- good resistance to alkali, heat and chemicals.

■ **Non-sticking, non-sagging** formulation may be used for vertical and horizontal applications as a patching material for filling holes, small cracks and joints of concrete, wood or steel.

■ **Diverse formulations** available and selection depends on the requirements of the end product.

■ **May be pressure injected** into cracks to bond them to restore structural integrity.

CONDITIONS AT INSTALLATION

In selecting a sealant for a given percentage of movement through a joint of a given width, consider the **conditions**

to be expected during the installation of the sealant:

■ **Seasonal and daily temperatures** which will affect thermal movements of the components the sealant is to join.

■ **Moisture conditions,** which may affect the components to be joined and/or the sealant itself.

Temperatures to be expected during the time the sealant is to be installed as well as exposure to heat gain of components through solar radiation may require the selection of a high performance sealant where under a different set of circumstances an intermediate, or even a low performance sealant could have been used:

■ As the example given in this section for cumulative movement demonstrates, **any sealant might fail under the extreme movement of 62.5 percent.**

■ If the sealant is installed either:
- after the initial shrinkage in the concrete has taken place, or
- when air temperatures are lower, **the percentage of movement can be lowered and the sealant can perform well.**

■ **If the installation takes place after shrinkage in concrete** and at an average temperature, an intermediate performance sealant could be used.

■ **Even a sealant installed at high outdoor temperature** after most of the initial shrinkage in concrete has taken place can perform well:
- the sealant will generally remain in tension throughout the seasonal movements in the joint; this is not detrimental to high performance sealants, some of which are capable of sustaining movement of up to 50 percent in a pure extension cycle.

■ **Widening the joint** is another method to remember for reducing the percentage of movement and thus reducing the performance requirements for the sealant.

Daily variations in temperature and exposure to solar radiation during installation may also affect the ultimate performance of a sealant:

■ A sealant installed **during the lower morning temperatures** may be com-

selection checklist (cont.)

pressed through expansion in components under solar radiation before the sealant cures, thus reducing the cured width of the bead.

■ When, subsequently, **low winter temperatures cause contraction** in the components, the expansion capability of the narrower bead may be exceeded and failure may result.

Conditions existing during installation will also affect the performance of formed seals:

■ **Formed seals, whether rigid or foamed,** should continuously exert pressure against the sides of a joint to be effective.

■ **A rigid extrusion seal,** compressed to 50 percent for insertion and installed in the joint between the panels described in this section for cumulative movement, would slip out of the joint even before maximum contraction in the panels had occured.

■ **Foamed seals** may not be suitable for use even under thermal movement conditions only - without shrinkage - since they are generally compressed about 25 percent for insertion.

Construction tolerances should always be part of the considerations during the detailing of joints and of the selection of sealants for them:

■ **Manufactured components,** such as metal panels, mullions, structural steel, are all subject to certain manufacturing tolerances; these tolerances may be offsetting or cumulative during installation, which generally takes place under rather difficult conditions, not conducive to maintaining close tolerances in the assemblies.

■ **Settlement, deflections, slippage** at connections after installation is completed may cause further loss of uniformity in the joints.

■ Formed seals require maintaining **a constant width of joint throughout:**
• rigid extrusions will not conform to localized or abrupt changes in the width of a joint.
• foams generally will conform, but the degree of compression will change at such locations and may lead to loss of seal upon further movement.

Buildings are never static: they move with changes in temperature, moisture and wind forces.

■ Allowance for such movements must be made by the joints and within them.

■ Unless this is done, the natural forces will create their own joints:
• cracks in masonry or concrete
• deformations such as buckling in metals.

COST OF SEALANTS

Evaluating the real cost of a sealant involves more than the selection of the lowest cost product capable of accommodating the expected movement in the joint. Sealants differ from each other in many other performance characteristics, besides their capability to accommodate movement. These other **characteristics may affect or even determine the in-place cost of sealants, as the following examples show:**

■ Reducing the width of a joint may require a higher performance sealant, but will also **reduce the amount used per unit length of joint:**
• assuming the shape factor to be one-to-one –not necessarily the case in wider joints–, the amount of sealant required per linear foot in joints of various widths would be:

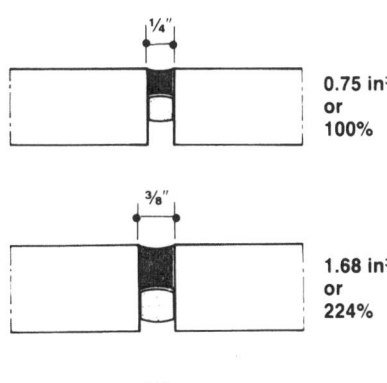

¼"	0.75 in³ or 100%
⅜"	1.68 in³ or 224%
½"	3.00 in³ or 400%

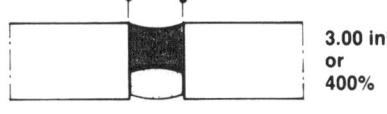

■ Increasing the width of a joint will increase the amount of sealant required,

but will **decrease performance requirements** for sealant used thus reducing the cost per unit of volume:
• the differential may be large enough to justify the change.
• there may be other offsetting advantages, such as easier installation when no mixing of two-component formulations is required.

■ Some types of sealants require **extensive and careful preparation of joints** for proper adhesion and the cost of such work is also a factor in the overall in-place cost of a sealant; changing to a different type of sealant may also change requirements for joint preparation.

■ Joints that accommodate little movement **may be in an inaccessible location** where their long term performance becomes critical: once maintenance and/or replacement costs have been added to the initial cost of a low performance sealant, the use of a high performance sealant of superior durability may prove more economical.

■ **The approximate range of comparative cost per unit volume of different sealants is:**
• Oil-based: 100 to 125
• Polybutane: 125 to 200
• Butyl: 150 to 200
• Acrylic: 150 to 200
• Polychloroprene (Neoprene): 200 to 350
• Polysulfide: 200 to 400
• Polyurethane: 200 to 400
• Silicone: 400 to 700 or more.
• the range in cost reflects the different formulations, such as one-part, two-part, or self-leveling, that are available within each type.

JOINTS: movement

| CONDITION | DESCRIPTION |

EFFECT OF WIDTH OF JOINT

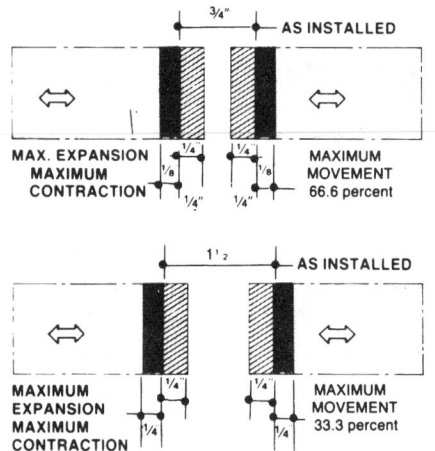

MAX. EXPANSION
MAXIMUM
CONTRACTION

AS INSTALLED

MAXIMUM
MOVEMENT
66.6 percent

MAXIMUM
EXPANSION
MAXIMUM
CONTRACTION

AS INSTALLED

MAXIMUM
MOVEMENT
33.3 percent

The extent of movement in components, to be accommodated in the joints between them, **may be the only criterion for establishing the required width of the joint** when it is to remain open; that is when no sealant is used in the joint.

■ **In a sealed joint, the ability of the sealant to return to the shape given to it** at the time of installation depends on its resistance to deformation both in tension and compression, unless the sealant is installed under conditions of maximum elongation or maximum contraction in the components.

■ **Different types of sealants have different rates of resistance to deformation;** the amount of movement through a joint relative to its width - the percentage of movement in a joint - will determine the type of sealant to be used:
* changing the width of a joint under the same set of conditions will change the percentage of movement in that joint and will either increase or lower performance requirements on the sealant.

■ **Percentages of rates of resistance to deformation** shown for sealants are given for both tension and compression:
* sealants capable of resisting ± 25 percent movement will resist 25 percent elongation, or 25 percent compression after installed in a joint.

CUMULATIVE MOVEMENT

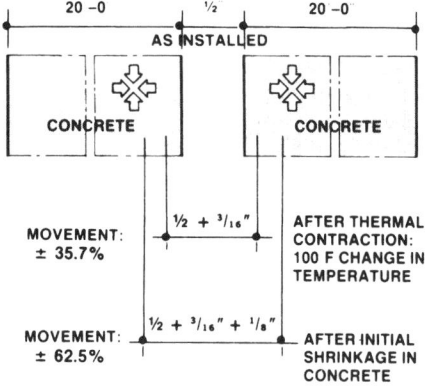

20 -0   ½   20 -0
AS INSTALLED

CONCRETE   CONCRETE

MOVEMENT:
± 35.7%

½ + ³/₁₆"

AFTER THERMAL
CONTRACTION:
100 F CHANGE IN
TEMPERATURE

MOVEMENT:
± 62.5%

½ + ³/₁₆" + ¹/₈"

AFTER INITIAL
SHRINKAGE IN
CONCRETE

Cumulative movements in components may affect the ultimate percentage of movement in a joint; for example:

■ **Tilt-up concrete panels** are erected as part of a north facing wall two weeks after being cast. Temperature of the panels is assumed at 90°F, and the joint between them has been selected to be ½ inch. At the design low temperature of − 10°F, the contraction in the panels would be approximately ³/₁₆ inch for a 37.5% movement through the joint. But if the effect of initial irreversible shrinkage in the concrete is included, the movement in the joint at maximum contraction will be 62.5%. Subsequent movements in the concrete due to changes in moisture content will be much lower than the initial shrinkage, and the **sealant is likely to remain in tension at all times:**
* when percentage of movement in a joint exceeds the capability of a sealant, **the width of the joint should be increased to lower such percentage.**

■ **In wood panels, shrinkage and swelling** will occur with each change in their internal moisture content; it would either counteract or aggravate effects of thermal movement in the panels.

■ **The best time to fill up the joints with sealant is when the width of the joints is at minimum:** then mostly tensile stresses will develop, and the compression stresses and strains will be kept at acceptable limits.

RESTRAINED CONDITION

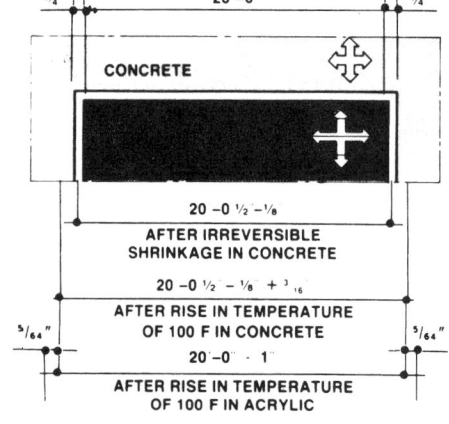

20 -0

CONCRETE

20 -0 ½ -¹/₈

AFTER IRREVERSIBLE
SHRINKAGE IN CONCRETE

20 -0 ½ - ¹/₈ + ³/₁₆

AFTER RISE IN TEMPERATURE
OF 100 F IN CONCRETE

20 -0 - 1

AFTER RISE IN TEMPERATURE
OF 100 F IN ACRYLIC

Restrained movement may occur when a component with a high coefficient of expansion: is installed between two other components, or when it is set into a component with a lower coefficient of expansion. For example:

■ **A corrugated acrylic panel** in a concrete frame may be subject to end (or perimeter) restraint due to:
* initial shrinkage of concrete which will reduce perimeter clearance.
* expansion in the acrylic panel which may exceed the end clearances provided.

■ **Restraint may result in:**
* cracking of the panel if its resistance to deformation is exceeded and/or
* spalling of concrete.

■ **The same acrylic panel in a wooden frame may be:**
* subjected to even greater restraint if rise in temperature results in simultaneous expansion and drying shrinkage in the wood; or
* end restraint may not occur if expansion due to rise in temperature is augmented by expansion in the frame due to increase in internal moisture content.

■ **At 100% movement,** a sealant would be completely squeezed out of a joint.

JOINT FILLERS: types, uses

TYPE	DESCRIPTION	COMPATIBILITY	USES
FOAMS, closed cell • POLYCHLOROPRENE (NEOPRENE) • POLYVINYL CHLORIDE (PVC) • POLYETHYLENE • POLYURETHANE	• high resistance to oils, chemicals, water and weathering. • non-absorptive • will not migrate or stain adjacent material • 90 percent recovery • available in flat sheets, strips, round and rectangular rods to fit most joints.	• compatible with: polysulfide, urethane and most other elastomerics. • not compatible with: hot poured and solvent-based sealants, solvents may attack filler. • polychloroprene not compatible with silicone.	■ in deep joints where the depth of joint to receive sealant must be limited and/or resilience is required, such as: • between precast concrete panels; • working joints between metal components; • joints in masonry.
BUTYL SPONGE RUBBER	• readily compressible, resilient and with low moisture absorption. • generally, colored cement gray to blend with concrete. • 95 percent recovery. • width 36"; up to 1 inch in thickness.	• compatible with: elastomeric sealants. • not compatible with: hot poured and solvent-based sealants, solvents may attack filler.	■ for visually exposed concrete panels, masonry and other structures when the filler may be exposed to view and resilience is required.
CORKBOARD, RESILIENT	• high resistance to acids, alkalies, decay. • waterproof. • light in color and flexible. • 95 percent recovery • standard widths 2" to 36"; thickness ¼" to 2".	• compatible with: all types of sealants.	■ joint filler generally used as bond breaker/full depth filler between individual sections of poured concrete assemblies such as: • flood walls, • sewage treatment ponds, • filtration plants.
CORKBOARD, RESILIENT, SELF-EXPANDING	• resistant to decay. • contains no asphalt or tar. • expands 50 percent by absorbing moisture after installation. • 95 percent recovery. • sizes similar to resilient corkboard.	• compatible with: all types of single and two-component sealants	■ assemblies of concrete, brick and stone including: • flood walls, spillways, canal linings; • sewage treatment plants; • may be used as primary seal, or in conjunction with bulk sealants.
BITUMINOUS IMPREGNATED FIBERBOARD	• easily compressible, waterproof, durable, self-healing and most economical. • not to be compressed more than 50 percent. • moderate recovery. • sizes similar to corkboard.	• compatible with: hot poured and solvent-based sealants. • not compatible with: elastomeric sealants and most non-asphaltic sealants. • polyurethane can be modified to be compatible.	• driveways, walks and curbs; • perimeter of concrete slab around columns through concrete slabs; • highways, bridges and tunnels; • generally exposed to view when black color is not objectionable.
BITUMINOUS IMPREGNATED CORKBOARD	• readily compressible, resilient, low moisture absorption. • 80 percent recovery. • sizes similar to corkboard.	• compatible with: hot poured and solvent-based sealants. • not compatible with: elastomeric sealants and non-asphaltic sealants.	• concrete pavements, highway work, driveways, parking areas; • in building construction where a more compressible material than fiberboard is needed.
OAKUM, JUTE, COTTON ROPE AND YARN	• should not be treated with extrudable oils.	• compatible with: all types of elastomeric and bituminous sealants.	• should not bond to sealant and allow for movement of sealant.

B2.11 SEALANTS

SEALANTS: a case history example

A CASE HISTORY

reprinted from CBD-97,
published by
Division of Building Research
National Research Council
Ottawa, Canada

Study No. 3. Through-wall joints are employed in many contemporary buildings, and a number of studies of individual cases have been made. A particularly interesting and informative experience was gained during construction of a tall building in Eastern Canada. During installation of the precast concrete wall panels, a bead of caulking compound was placed in the half-lapped joints (Figure 2)

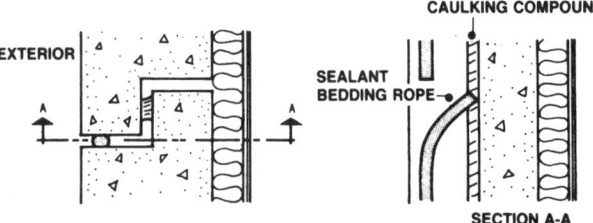

Figure 2 Vertical joint between concrete wall panels. Note: Although rain penetration can be controlled in the joints, the use of this wall design is not recommended

The fact that it did not provide an airtight seal was not considered important because a bead of elastomeric sealant was to be installed at the outside face of the joint. The installation of this sealant, however, had to be delayed because of cold weather, and only the sealant bedding rope was placed in the joint. Because the delay was expected to be more than six months, the sealant bedding rope was installed with the upper end of each length turned inward to flash the joint space, incidentally leaving a 2- to 3-in. gap at frequent intervals. No rain penetration was noticed during the delay. Only after installation of the elastomeric sealant at the outside of the joint was leakage observed, and after a few months it became a serious problem.

Study of the situation that existed during the delay indicated that although the caulking had not provided a complete air seal, it was considerably more airtight than that provided by the sealant bedding rope. The air pressure on both sides of the sealant bedding rope could equalize and the air pressure difference required to move water past it could not develop (CBD 96). When the elastomeric sealant was finally installed, however, it was far more airtight than the caulking bead, with the result that the air pressure difference between inside and outside the building was borne by it. With both air and water leakage resisted at the same point, air penetration occurred at even minor failures of the seal. As the outer seal was better than the inner seal, any water that entered the joint could only drain to the building interior. The same conditions and performance were experienced at the window joints.

REFERENCES: Sealants

STANDARDS

ASTM Specifications for:

C509 • Cellular Elastomeric Preformed Gasket and Sealing Material.

C542 • Lock-Strip Gaskets.

C834 • Latex Sealing Compounds.

C920 • Elastomeric Joint Sealants.

D994 • Preformed Expansion Joint Filler for Concrete (Bituminous Type).

D1056 • Flexible Cellular Materials—Sponge or Expanded Rubber.

D1190 • Concrete Joint Sealer, Hot-Poured Elastic Type.

D1565 • Flexible Cellular Materials - Vinyl Chloride Polymers and Copolymers (Open-Cell Foam).

D1667 • Flexible Cellular Materials-Vinyl Chloride Polymers and Copolymers, (Closed-Cell Vinyl).

D1751 • Preformed Expansion Joint Fillers for Concrete Paving and Structural Construction (Nonextruding and Resilient Bituminous Types).

D1752 • Preformed Sponge Rubber and Cork Expansion Joint Fillers for Concrete Paving and Structural Construction.

D1850 • Concrete Joint Sealer, Cold-Application Type.

D2287 • Nonrigid Vinyl Chloride Polymer and Copolymer Molding and Extrusion Compounds.

D2581 • Polybutylene (PB) Plastics Moulding and Extrusion Materials.

D2628 • Preformed Polychloroprene Elastomeric Joint Seals for Concrete Pavements.

D3542 • Preformed Polychloroprene Elastomeric Joint Seals for Bridges.

ASTM Tests/Methods for:

C510 • Staining and Color Change of Single or Multicomponent Joint Sealants.

C603 • Extrusion Rate and Application Life of Elastomeric Sealants.

C639 • Rheological (Flow) Properties of Elastomeric Sealants.

C661 • Indentation Hardness of Elastomeric-Type Sealants by Means of a Durometer.

C679 • Tack-Free Time of Elastomeric-Type Joint Sealants.

C711 • Low-Temperature Flexibility and Tenacity of One-Part, Elastomeric, Solvent-Release Type Sealants.

C712 • Bubbling of One-Part, Elastomeric Solvent-Release Type Sealants.

C719 • Adhesion and Cohesion of Elastomeric Joint Sealants Under Cyclic Movement.

C732 • Aging Effects of Artificial Weathering on Latex Sealing Compounds.

C792 • Effects of Heat Aging on Weight Loss, Cracking, and Chalking of Elastomeric Sealants.

D1851 • Concrete Joint Sealers, Cold-Application Type.

Federal Specifications:

SS-S 159b • Sealing Compound Cold-Application Mastic Multiple Component Type - For Joints in Concrete.

SS-S 1401B • Sealing Compound, Hot Applied, For Portland Cement and Asphalt Concrete Pavement.

SS-S-1614(1) • Sealing Compound, Jet-Fuel Resistant, Hot-Applied, Concrete Paving, One Component.

SS-S-169B(1) • Sealer, Joint, Sewer, Mineral-Filled, Hot-Pour.

SS-S-00195 a • Sealing Compound, Two-Component, Elastomeric, Polymer Type, Cold-Applied Concrete Paving.

SS-S-00200D (3) • Sealing Compound, Two-Component, Elastomeric, Polymer Type, Jet-Fuel Resistant, Cold-Applied, Concrete Paving.

TT-S-00227E (3) • Sealing Compound, Elastomeric Type. Two Component (For Caulking, Sealing, and Glazing in Buildings and Other Structures).

TT-S-00230C (2) • Sealing Compound, Elastomeric Type, Single Component, (For Caulking, Sealing, and Glazing in Buildings and Other Structures).

TT-S-001543 A • Sealing Compound: Silicone Rubber Base (For Caulking, Sealing and Glazing in Buildings and Other Structures).

TT-P-791A(3) • Putty; Pure Linseed Oil (For Wood-Sash Glazing).

TT-C-598B(2) • Caulking Compound, Oil and Resin Base Type (For Masonry and Others).

TT-G-00410 E(1) • Glazing Compound, Sash (Metal) for Bedding and face Glazing.

TT-S-001657 • Sealing Compound-Single Component, Butyl

B2.11 SEALANTS

REFERENCES: Sealants (cont.)

Rubber Based, Solvent
Release Type.

SELECTED REFERENCES

AAMA   **Architectural Aluminum**
           Manufacturers
           Assn.
801-809 • Voluntary Standard for
              Sealants.
810 • • Voluntary Specification for
              Expanded Cellular
              Glazing Tape
820   • Voluntary Test Methods
              for Sealants

ACI   **American Concrete**
           Institute
504   • Guide to Joint Sealants
              for Concrete

NRCA   **National Paint & Coatings**
           Assn., Inc.
            • Selection & Application of
              Caulks, Sealants,
              Putties & Glazing
              Compounds Structures.

NRCC   **National Research Council**
           of Canada
CBD 19 • Caulking Compounds
96   • Use of Sealants
97   • Look at Joint
              Performance
155   • Joint Movement of
              Sealant Selection

PCA   **Portland Cement**
           Association
D137   • Clear Coatings for
              Exposed Architectural
              Concrete

B2.11 SEALANTS

B3. ❶ ROOFING

ROOFING: introduction

■ **Roof assemblies consist of two principal elements:**
● **Substrate:** spaced framing only; framing + deck; framing + deck + insulation or fill;
● **Roofing:** the barrier formed to protect the substrate against the elements. Roofing may be a continuous membrane or individual overlapping units.

■ **Continuous roofing membranes are generally used on flat and low sloped roofs.** They may be categorized as:
● **Built-up membranes -** the alternate application of bitumen over successive layers of felt; often surfaced with aggregate such as: gravel, crushed stone, slag, mineral granules.
● **Liquid-applied membranes -** roll-on or spray-on elastomeric coatings.
● **Single-ply membranes -** include sheets of modified bitumen or of polymeric materials.

■ **Preformed panels and overlapping units though made of a variety of materials and shapes such as shakes, shingles, tiles, panels are installed and perform essentially alike:**
● They are mechanically fastened to the substrate.
● They are always installed with a pitch to shed water quickly.

ROOFING: BITUMINOUS MEMBRANES

■ **Premature failures** of roofing membranes are more often related to improper application or installation of materials than to the materials used. Failures are most commonly tied to:
● **Water penetration into membranes.**
● **Fastening and securing methods.**
● **Substrate considerations.**

Water Penetration into Membrane

■ **Moisture damage in a roof membrane is most often found as wrinkling or blistering. Moisture permeates the membrane in a number of ways:**

● **Through absorption and retention** by felts or insulation before installation.

● **Through migration of water vapor** through open joints or cracks in the substrate, condensing at lower temperatures in the felts.

● **Through leakage** problems of penetrations and interruptions of roofing layers.

● **Through condensation** caused by migration of water vapor due to temperature or pressure differential.

● **Through splitting, cracking, etc.** on the exposed side of protective coatings or the top ply of felts due to weathering, solar radiation, water ponding, freeze-thaw cycling, flow-off, alligatoring, or traffic damage due to foot-traffic, falling objects and temporary placement of equipment, tools, etc.

● **Through trapping of moist air** between plies of felts due to incomplete coatings of bitumen or embedment of felts into bitumen. Blistering occurs when moisture is trapped between or within the plies of felt or under the membrane itself.

● **Through fishmouths** or edge wrinkles in felts where they overlap other felts resulting from improper installation of the membrane.

■ **Provide adequate drainage to minimize ponding or standing water, which adversely affects adhesion and stress resistance.** Inadequate drainage is a primary cause of roof failures.
● Water runoff is promoted by sloping the roof surface toward roof drains.
● Provide at least a minimal amount of slope, typically about 1/4 inch in 12 inches, but not less than 1/8 inch in 12 inches.
● Maintain positive slope to roof drains; adjust for differential settlement and roof deflection when drains are near structural columns; locate mechanical equipment over structural members to minimize deflection potential.

■ **Vents may be provided to allow moisture trapped below a membrane to escape to the outside.** Their use must be carefully evaluated because lateral migration of water vapor cannot be readily controlled or predicted with any degree of certainty.

■ **Premature failure of built-up membrane may also occur due to:**
● **Splitting due to tensile stresses -** resistance depends on the strength of the membrane and the method of fastening.
● **Slippage -** the relative lateral movement between plies in built-up roofing usually appear in the first few years of the installation as a series of random wrinkles perpendicular to the slope. Slippage is caused by:
  —inadequate anchorage.
  —using bitumen with too low a softening point.
  —high roof surface temperatures.
  —the excessive coating of bitumen between plies or layering of aggregate surfacing.
● **Uplift resulting from air movement at high velocities over the roof surface can move and separate roofing layers.** Uplift is resisted by:
  —Adequate anchorage of the membrane to the substrate, which in turn must be properly anchored and rigid enough to prevent excessive deflections.
  —Aggregate surfacing which ballasts the membrane and also tends to break up the laminar flow of air, thus reducing negative pressure.

Fastening/Securing Methods

■ **Methods for securing built-up membranes may be:**
● **Nailing to substrate,** such as wood deck or nailable concrete.
● **Full adhesion using hot bitumen;** or a bituminous cutback for cold-applied membranes.
● **Spot** adhesion using bitumen or adhesive.

B3.1 ROOFING

■ **Common methods for fastening/ securing single ply membrane roofing are:**
- **Full adhesion -** contact adhesives are used; Self-adhesion is air tight, lightweight and easily inspected/ maintained, but expensive; application errors are difficult to correct.
- **Torching -** appropriate for modified bitumen in locations that are not convenient for the transportation of hot asphalt.
- **Spot anchoring -** mechanical fastening of loose-laid single-ply membranes using individual discs or plates; wind can cause uplift at edges and billowing. Beware of:
  - fastener and/or deck corrosion.
  - fasteners acting as thermal bridges, allowing condensation to occur on or around the fasteners.
  - fasteners puncturing the membrane when insulation around the stress plate is compressed.

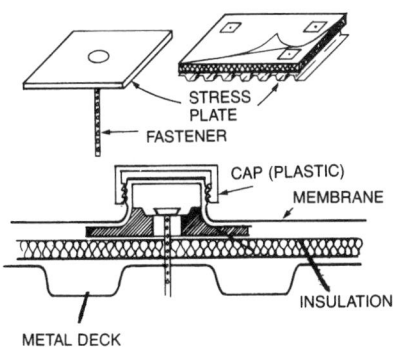

STRESS PLATE
FASTENER
CAP (PLASTIC)
MEMBRANE
INSULATION
METAL DECK

- **Spaced anchoring -** continuous bar or strip fasteners may be used for single-ply membranes.

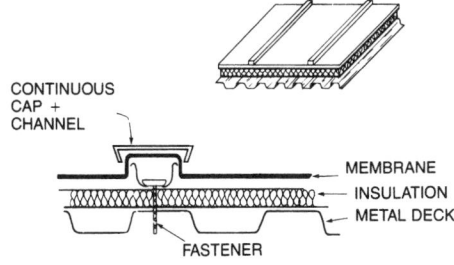

CONTINUOUS CAP + CHANNEL
MEMBRANE
INSULATION
METAL DECK
FASTENER

- **Loose-laid/ballasted -** placing smooth, round rock or other aggregate surfacing, or using prefabricated pavers/blocks over a single-ply membrane which is not attached to the substrate.
  - Minimize the use of slivered rock since it can puncture membranes.
  - Ballasting allows membrane elongation, but stress may be excessive at joints in membrane.
  - Low cost and easy installation.
  - Loose gravel can be moved by winds of 80 to 85 mph; pavers may be shifted by winds of 110 mph.

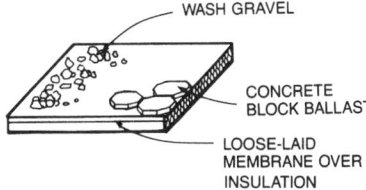

WASH GRAVEL
CONCRETE BLOCK BALLAST
LOOSE-LAID MEMBRANE OVER INSULATION

- **Protected/inverted membranes -** The waterproofing membrane is installed below the insulation. The insulation is generally protected by applying ballast. This method minimizes thermal movement in the membrane; it is susceptible to wind damage but good in resisting damage due to foot traffic.

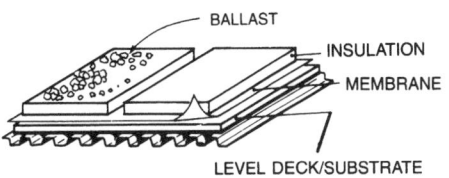

BALLAST
INSULATION
MEMBRANE
LEVEL DECK/SUBSTRATE

■ **Selection of placement and sealing methods should be evaluated for:**
- **Condition of materials:** Single-ply membrane surfaces may be coated with talc during manufacture to prevent sticking. If not completely removed prior to installation, the talc may prevent proper bonding/adhesion of seams/membrane.
- **Weather, temperatures and site conditions:** High winds may present problems in laying large single-ply sheets on the substrate. Low temperatures may impede proper sealing of joints.
- **Availability of skilled trades** who are familiar with the materials.

Substrate Considerations

■ The condition of the substrate supporting the roofing membrane may determine the quality of the roofing. Roofing delamination and fatigue can occur if thermal expansion and contraction rates between adjacent materials are not matched.
- **Cracks** and non-working joints should be covered with tape since liquid-applied membranes will not bridge or fill voids.
- **Working joints -** joints subject to continuous movement should be sealed to allow for movement in the joints without rupturing/splitting the membrane. Movement is common:
  - between roof and substrate
  - between roofing/flashing and parapet walls,
  - at through-roof penetrations, cant strips, gravel stops and fascias.

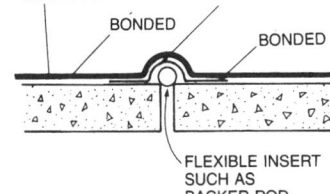

NO BOND
MEMBRANE
BONDED
BONDED

MEMBRANE
NO BOND
BONDED
BONDED
FLEXIBLE INSERT SUCH AS BACKER ROD

- **Cracks that develop** in the substrate after the membrane has been installed may split the membrane when degree of elongation required exceeds the limits of elasticity.
- **Substrate deflection -** Excessive deck deflection due to ponding or other loading is likely to cause splitting.
- **Liquid applied or fully adhered** polymeric materials have no significant strength and will not perform better than the substrate to which they are applied or bonded.

B3.1 ROOFING

ROOF/FLOOR: assemblies

ROOF ASSEMBLY:

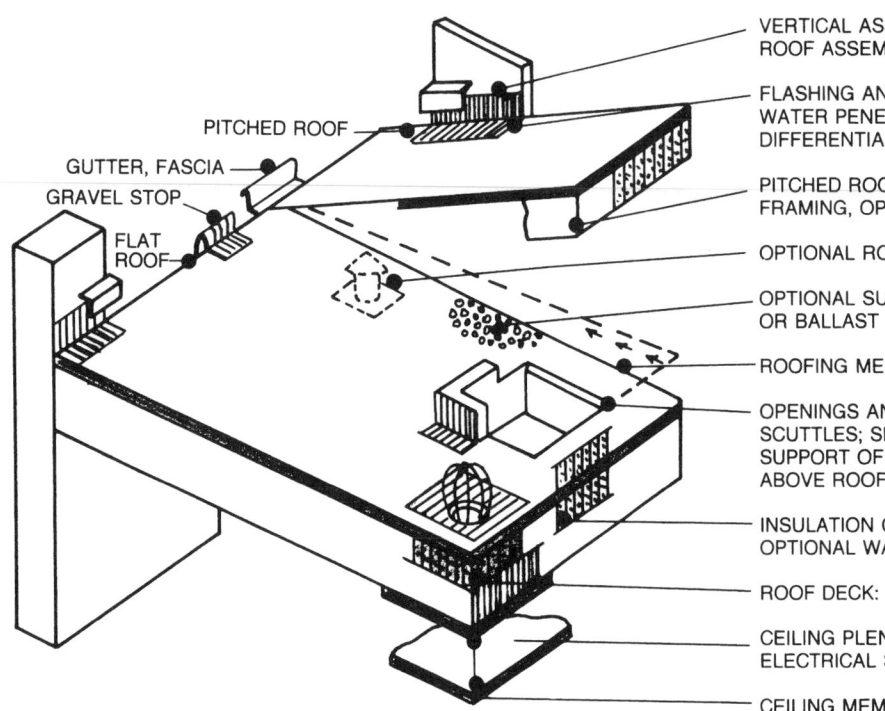

VERTICAL ASSEMBLIES/PROJECTIONS THROUGH ROOF ASSEMBLY

FLASHING AND COUNTERFLASHING TO PREVENT WATER PENETRATION AND/OR TO ACCOMMODATE DIFFERENTIAL MOVEMENT

PITCHED ROOF ASSEMBLY: ROOFING, DECKING, FRAMING, OPTIONAL INSULATION

OPTIONAL ROOFING VENTS

OPTIONAL SURFACING OVER ROOFING MEMBRANE OR BALLAST OVER INSULATION

ROOFING MEMBRANE

OPENINGS AND/OR CURBS FOR: SKYLIGHTS; SCUTTLES; SERVICE LINE PENETRATIONS; SUPPORT OF EQUIPMENT OR SERVICE LINES ABOVE ROOFING

INSULATION OVER OR UNDER ROOFING MEMBRANE; OPTIONAL WATER VAPOR RETARDER

ROOF DECK: FRAMING, DECKING

CEILING PLENUM SPACE - MECHANICAL/ ELECTRICAL SERVICES

CEILING MEMBRANE: ATTACHED TO OR SUSPENDED FROM STRUCTURAL FRAME

PITCHED ROOF

GUTTER, FASCIA

GRAVEL STOP

FLAT ROOF

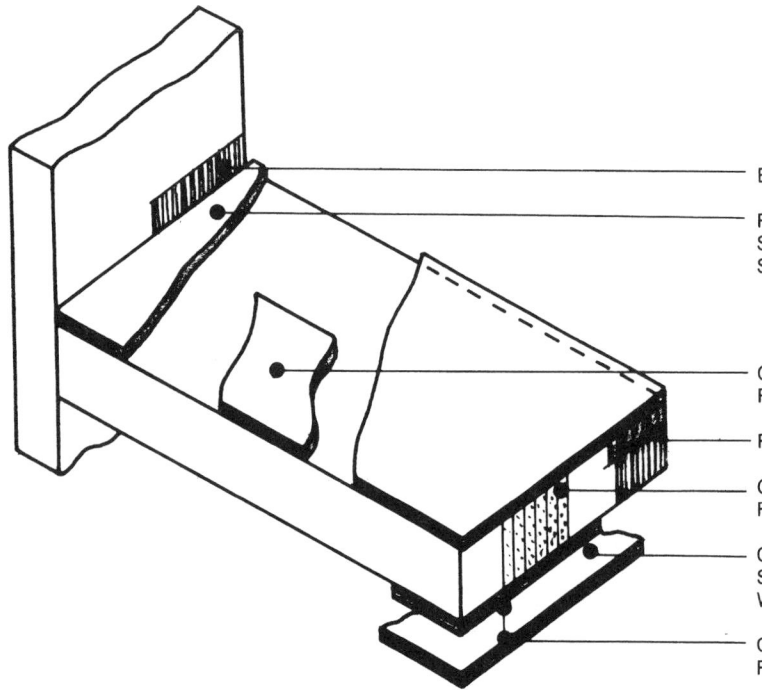

BASE AT JUNCTURE WITH VERTICAL ASSEMBLIES

FLOORING: INTEGRAL WITH SUBSTRATE; OR SEPARATE COMPONENT ATTACHED OR BONDED TO SUBSTRATE

OPTIONAL TOPPING AS FLOORING OR AS LEVELLING/ FURRING TO RECEIVE AND SUPPORT FLOORING

FLOOR DECK: FRAMING, DECKING

OPTIONAL INSULATION WITHIN THE STRUCTURAL FRAME

CEILING PLENUM: OPTIONAL SPACE FOR UTILITIES, SUCH AS PIPING: DUCT WORK, ELECTRIC/TELEPHONE WIRING

CEILING MEMBRANE: ATTACHED TO OR SUSPENDED FROM STRUCTURAL FRAME

FLOOR ASSEMBLY

B3.1 ROOFING

FORCES/DEFORMATIONS

PRESSURE/GRAVITY LOADS

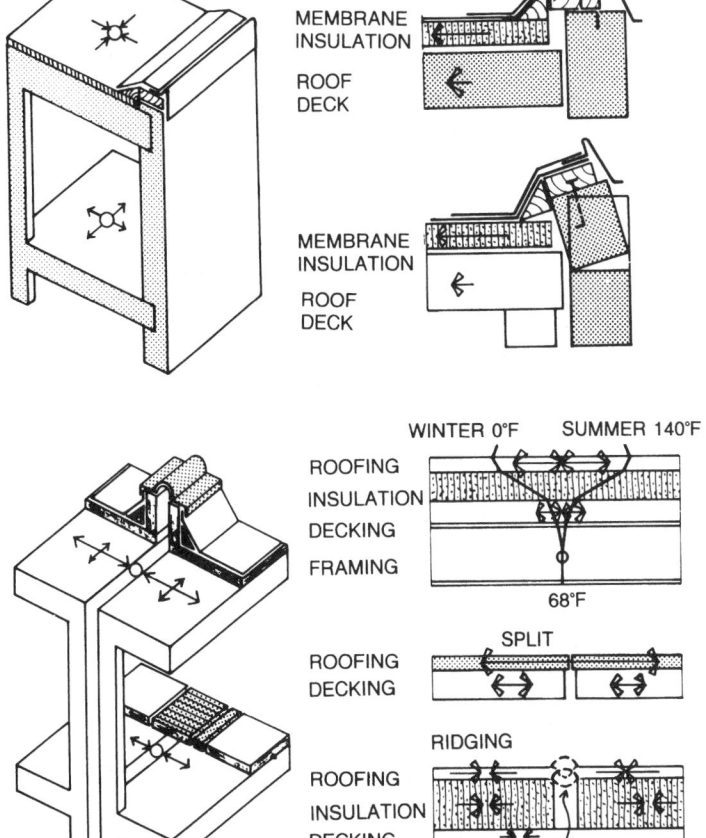

Floors are subject to deflections due to a gravity loads both live and dead and to shear when functioning as diaphragms under wind pressure or seismic forces.

Roofs are subject to shear when functioning as diaphragms and to deflection due to:

■ Dead load of the assembly, rooftop equipment.

■ Wind pressure:
• negative from flat to about 30 degree slope.
• positive when over about 30 degree slope.

■ Live loads due to rain/snow accumulation:
• rain water and/or meltwater from snow accumulation may collect in low spots resulting from deflection in flat roof assemblies.
• water may freeze in low spots and under extreme conditions may lead to splitting of the roofing membrane.
• snow accumulation on roofs will generally be unequal due to wind action during or after a snowfall with resulting unequal load distribution, which condition might be further aggravated by unequal melting of accumulated snow.

EXPANSION/CONTRACTION

Movement in roofs/floors results from:

■ Thermal expansion/contraction which affects all materials.

■ Shrinkage and creep in cementitious decks.
■ Shrinkage or swelling in wood due to changes in its moisture content.

Thermal movement in floors other than at building expansion joints is seldom a factor, unless the interior environments they separate are maintained at significantly different temperatures.

Thermal movement in roofs is generally an important factor:

■ Differential movement will generally occur due to temperature gradients through the roof assembly:

• the roofing membrane may contract when exposed to low outdoor temperatures and when the bond between the insulation and decking is insufficient to resist such movement with the contraction most noticeable and damaging along the perimeter of the membrane: it may carry blocking and flashing along with it unless such components are properly secured in place; subsequent thermal expansion will not bring the membrane nor the components back to the original position.
• roofing membrane may expand or contract considerably while thermal movement in the decking and framing could remain insignificant.

Moisture related movement in a roof assembly may also be a major factor:
• moisture such as condensing water vapor may penetrate through open edges into certain types of insulation resulting in differential expansion, or warpage, of the insulation and ridging of the membrane at joints.
• ridging of membrane at joints may also be caused by water vapor migrating to the roofing membrane through open joints between insulation boards.
• the cyclic changes in temperature and slippage between felt layers is also commonly considered to be one of the main contributing factor to ridging over joints in insulation.

EXPANSION JOINTS/MOVEMENT IN ROOFING

B3.1 ROOFING

ROOFS/FLOORS: roofing, ceilng, flooring

ROOFING

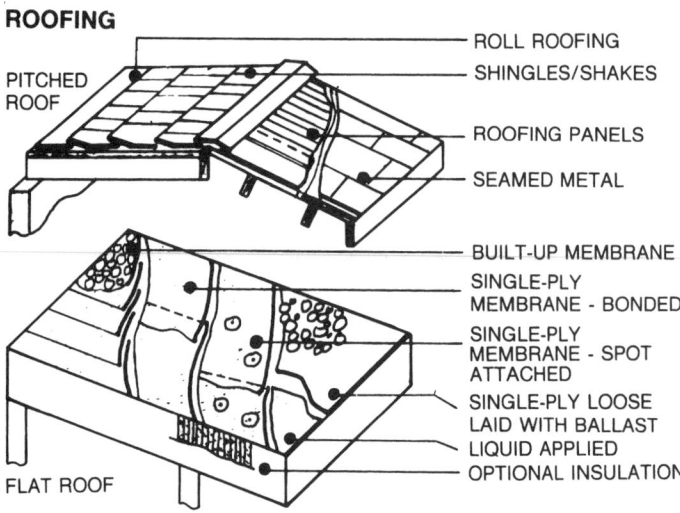

PITCHED ROOF — ROLL ROOFING / SHINGLES/SHAKES / ROOFING PANELS / SEAMED METAL

FLAT ROOF — BUILT-UP MEMBRANE / SINGLE-PLY MEMBRANE - BONDED / SINGLE-PLY MEMBRANE - SPOT ATTACHED / SINGLE-PLY LOOSE LAID WITH BALLAST / LIQUID APPLIED / OPTIONAL INSULATION

Preliminary considerations should include:
- **Pitch of the roof:** while membranes may be used on pitched roofs, overlapping units, such as shingles, and roof panels cannot be used on flat roofs.
- **Shape of roof assembly:** complex shapes may limit the choice of roofing to single-ply membranes or liquid applied types.
- **Durability:** not all roofing materials are equally durable.
- **Resistance to exposure to fire** may be a requirement of the local building code: not all roofing types are equally resistant.
- **Aesthetic considerations** may also influence the selection of a particular type of roofing.
- **Economy:** based not only on initial cost but also including durability and expected maintenance costs over the expected in-service life of roofing.

CEILINGS

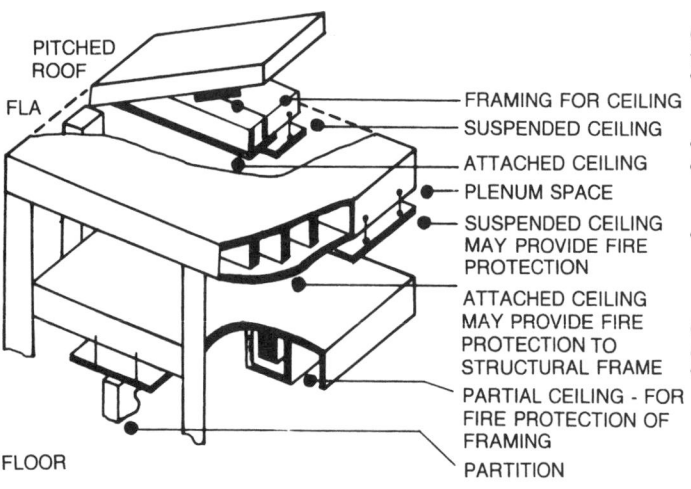

PITCHED ROOF / FLA

FLOOR — FRAMING FOR CEILING / SUSPENDED CEILING / ATTACHED CEILING / PLENUM SPACE / SUSPENDED CEILING MAY PROVIDE FIRE PROTECTION / ATTACHED CEILING MAY PROVIDE FIRE PROTECTION TO STRUCTURAL FRAME / PARTIAL CEILING - FOR FIRE PROTECTION OF FRAMING / PARTITION

Ceiling membranes are nonstructural components supported by the structural frame of a roof or floor assembly to function as:
- **Visual screen or visual/maintenance separation** between an inhabited space and the underside of the structural frame above.
- **Visual separation and sound absorbing surface.**
- **Baffle between inhabited space and space above membrane** or plenum space used for supply or return of conditioned air.
- Integral component of a fire-resistant rated roof or floor assembly, when the structural frame is not fire-resistive of itself.

Ceiling membranes may be attached directly to components of the structural frame or suspended from them:
- When attached, the deformations in such structural frames will directly affect the ceiling membrane, especially when the membrane is relatively rigid, such as plaster or gypsum board.

FLOORING

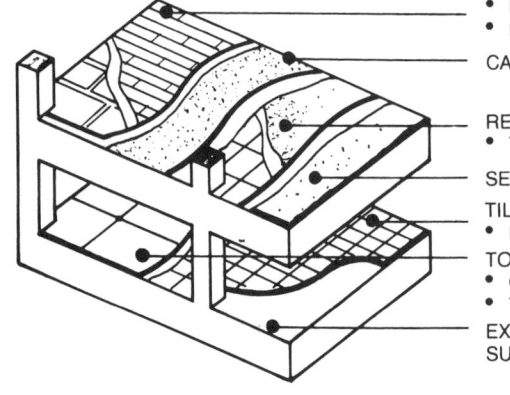

WOOD: STRIP / • PARQUET / • BLOCK / CARPETING / RESILIENT: SHEET / • TILE / SEAMLESS / TILE: STONE / • BURNED CLAY / TOPPING: / • CONCRETE / • TERRAZZO / EXPOSED SUBSTRATE

Preliminary considerations should include:
- **Traffic the flooring will be subjected to:** light foot, heavy foot and rolling loads, concentrated at specific locations.
- **Abrasion and scratches the flooring will be exposed to:** through wet or dry dust/dirt particles being ground into its surface by foot traffic, which will affect most types of flooring.
- **Moisture:** such as water spills; ground water or water vapor migration/condensation, which will affect wood flooring in particular, and may be detrimental to resilient, seamless, and carpeting.
- **Impact generated shearing stresses, and tensile/shearing stresses** due to deformation and/or shrinkage in the substrate when rigid flooring, such as stone, burned clay, terrazzo is continuously bonded to the substrate. Carpeting is least affected by shrinkage cracking and deflections in the substrate.

B3.1 ROOFING

ROOFS/FLOORS: preliminary considerations

ROOF DECKS AS SUBSTRATE FOR ROOFING

Choice of roofing and substrate will have a bearing on each other;

■ **Substrate must be compatible with the roofing materials** and/or with the materials used in attaching or bonding the roofing to it:
- certain types of roofing such as shingles require a nailable substrate.
- built-up membranes cannot be installed directly over formed metal decking.

■ **Deflection in decks may have to be limited:**
- large deflections, such as 1/180 of span sometimes allowed for roofs without any finish at the underside of decking, may cause damage to roofing materials attached directly to the decking: adhesive bond may fail in shear.
- maximum deflection of 1/360 of span is often recommended; in any event deflection in decks and its effect on the total assembly should be investigated.
- light-gauge steel decks though structurally adequate may be too flexible for a successful roofing application.

■ **Dimensional stability of a roof deck** is influenced primarily by its thermal and moisture characteristics and method of attachment:
- excessive movement in decking may cause warping and cracking in it with subsequent damage to materials directly attached to it.
- precast or preformed decking units are often secured to framing with metal clips which will resist wind uplift while presenting little or no resistance to lateral movement along the framing, such as under thermal contraction.
- high dimensional stability of substrate may be required to limit expansion/contraction when roofing is to be continuously bonded to it, especially where the substrate is exposed to high interior temperatures, high relative humidity, or both.

■ **Surface condition of the roof deck** is of considerable importance to the success of a roofing membrane installed over it:
- smoothness and dryness will influence how well a membrane can be adhered to the decking.
- some types of decking such as long span precast tee's may require a topping to achieve a level and smooth surface or insulation and/or other components of a roofing system

that do not need a high degree of smoothness and levelness should be selected.
- joints in substrate may have to be minimized and may require special treatment.

ROOF DRAINAGE

Roof drains should be:

■ **Located at low points of the roof to prevent water ponding on the roofing:**
- drains located near columns in a level roof may turn out to be at high points of the roof, since there will be no deflection in framing at such locations.
- drains at midspan of framing for a level roof may also be at high points if framing members have been fabricated with an excessive camber.
- creep in reinforced concrete framing members and decks may over a period of time change the runoff pattern of an initially level roof.
- drains should be sized and spaced to ensure rapid runoff of water; smaller closer spaced drains are preferable to larger but fewer drains.

Other considerations during selection of drains might be:

■ **Drains may be also selected for delayed drainage,** when capacity of storm sewers is insufficient to take full flow. Care must then be taken to prevent excessive accumulation of water on the roof.

■ **Roofs with parapets should,** in addition to drains, have overflow scuppers in the parapet walls, this in order to prevent water accumulating on the roof to excessive depth, should roof drains be clogged, or frozen.

■ **Determining factors in the sizing of roof drains are:**
- area of roof to be drained.
- intensity and duration of heaviest rainfall recorded.

■ **Connection to leaders** may have to be flexible to allow for movement in drains due to live load-induced deflections of the roof assembly.

■ **It is recommended that positive drainage to drains be provided;** water ponding on roofing may damage or destroy the roofing membrane.

■ **Dead level roofs** are not recommended, unless specifically designed and installed to retain and hold water; dead level roofs are almost never dead level; deflections in the roof

assembly, shrinkage due to changes in moisture content, settlement or creep in columns may result in low spots which allow water to pond.
- built-up membranes over dead level roof assemblies generally not bondable by most membrane manufacturers.

OTHER CONSIDERATIONS:

Requirements for the integration of mechanical/electrical services into a roof or floor assembly may influence the ultimate choice of a specific structural frame:

■ **Open framing,** such as steel bar joists or trusses may be considered when some or all of the service lines can readily be accommodated within the structural frame: the greater depth of the framing members thereby required may be offset by reduced clearance requirements below them.

■ **Service lines** suspended below solid framing members may be easier, thus less costly, to install.

■ **Framing may function as a service floor** when service lines must be easily accessible and readily relocatable, such as in a research facility.

■ **Certain structural frames,** such as flat slabs, cannot readily accommodate large openings for service line penetrations without special provisions, such as in a reseach facility.

■ **Roofing membranes** over flat or nearly flat roofs will be subject to solar radiation which will raise their temperature and result in thermal expansion. The substrate/decking is protected from direct solar radiation and may have lower internal temperature, while framing members exposed to the interior may remain at a constant temperature, the differences in temperature then resulting in differential expansion within the roof assembly.
- The effects of solar radiation on flooring are seldom significant enough to warrant special considerations, except in buildings employing principles of passive solar heating, when the floor assembly is utilized as a heat sink.

B3.1 ROOFING

ROOFS/FLOORS: integration of services

Type and extent of mechanical/electrical services to be provided will influence the selection of the entire assembly:

■ **Space must be provided** within the plenum for feeder lines serving the enclosed space or a group of enclosed spaces; feeder lines may need to run above the ceiling membrane of one space to serve adjacent space(s).

■ **The usual arrangement** and order of installation of mechanical/electrical service lines is:
• wiring/piping first;
• ductwork, next
• lights, diffusers, return grilles, public address systems, etc., last.

■ Of all services, **only the automatic sprinkler system** must be located in a rigid prescribed pattern; all other systems allow for varying degrees of flexibility.

■ **Depth of plenum space** above a level membrane will be determined primarily:
• at points where service lines cross
• where service lines clear the deepest member of floor/roof framing.

OPEN FRAMING

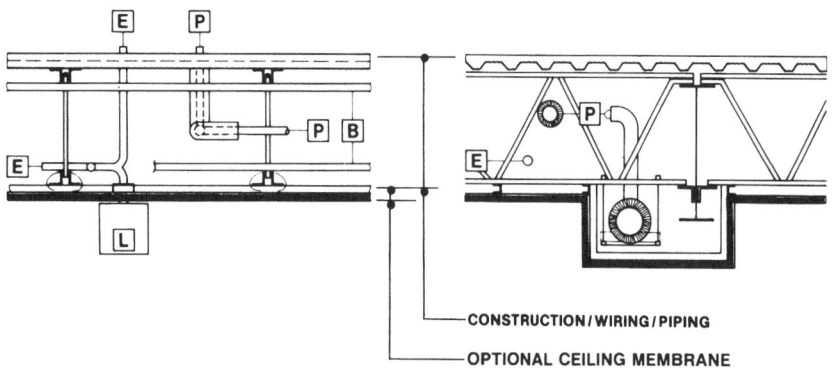

CONSTRUCTION/WIRING/PIPING

OPTIONAL CEILING MEMBRANE

■ **Framing:** open web steel bar joists and rolled steel or open web joist girders; short span wood trusses; generally 2' o.c. for floors.

■ **Wiring and piping:** may be run between or through joists and through joist girders, if provided:
• **web members** of joists should be lined up during installation.
• **clearance** between web members should be carefully checked, especially when flanged piping is used.
• **installation of piping** through joists is generally more costly since shorter lengths of pipe may have to be used and more labor is usually required.
• **running rigid ductwork** through open web joists is not practical; generally only flexible small-diameter circular ducts can be used.

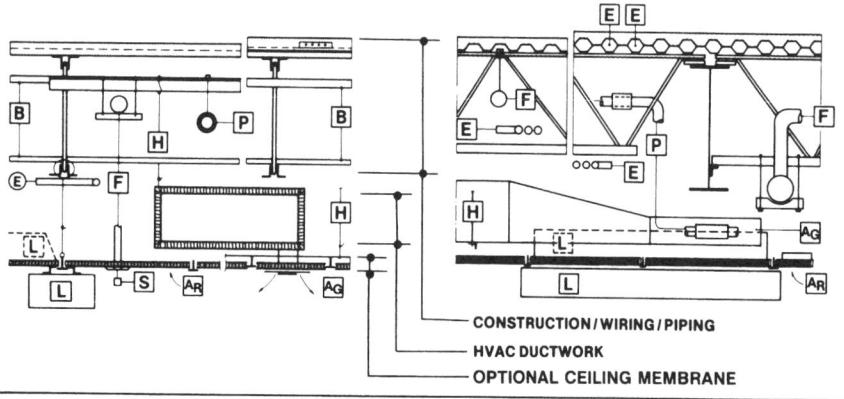

CONSTRUCTION/WIRING/PIPING

HVAC DUCTWORK

OPTIONAL CEILING MEMBRANE

■ **Framing:** open web steel bar joists and rolled steel or open web joist girders; steel or wood trusses.

■ **Wiring and ductwork:** may generally be suspended from metal deck, heavier piping should be supported from framing at panel points.
• **metal deck** with tabs for connecting hangers is available.
• **ducts** may be flattened to clear girders or where two ducts or duct and piping cross; flattening ducts will increase resistance to air flow.
• layout of **fire protection piping** should allow for draining entire system.
• **light fixtures** may be recessed, surface mounted, or suspended.

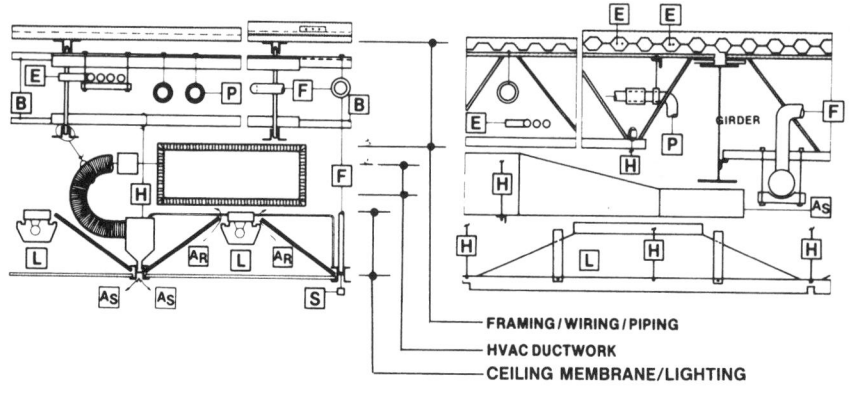

FRAMING/WIRING/PIPING

HVAC DUCTWORK

CEILING MEMBRANE/LIGHTING

■ **Framing:** open web bar joists and rolled steel or open web joist girders; or trusses.

■ **Wiring, piping and ductwork** location and supports similar to those for penetrated membranes:
• **modular components,** such as diffusers, air return grilles, and lighting fixtures to fit the suspension grid, are used;
• **pre-assembled modules** with mechanical/electrical/sound control components incorporated may be installed into a suspension grid.
• **clearances** must be provided for tilting tile, fixtures, or modules into place.
• larger diameter piping is difficult to install through joists.

B3.1 ROOFING

B • BUILDING SHELL

■ **The symbols used are:**
[As] Conditioned air supply ductwork and outlets.
[AR] Inlets for plenum or ducted air return to heating/cooling system.
[P] Plumbing and/or heating/cooling piping: such as domestic hot/cold water, steam, hot/chilled water for heating/cooling, roof drain leaders.

[S] Sprinkler heads.
[F] Fire protection piping: main feeders, branches.
[E] Electrical power and lighting wiring: communication systems wiring.
[L] Lighting fixtures.
[H] Hangers: either wire or strap.
[B] Bracing for framing members.

■ For information on structural frame and ceiling membranes, refer to respective sections.

SOLID FRAMING

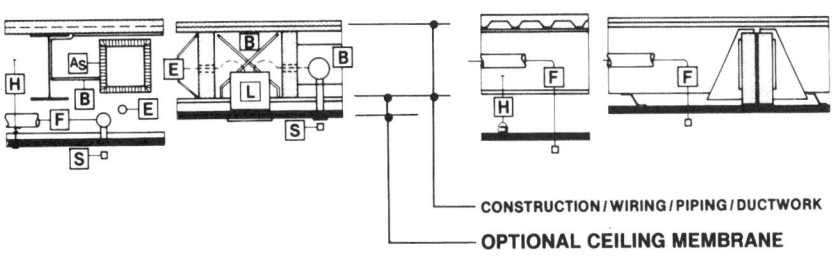

CONSTRUCTION/WIRING/PIPING/DUCTWORK
OPTIONAL CEILING MEMBRANE

■ **Framing:** rolled steel beams and girders; wood or metal joists with flush or dropped girders.

■ **Wiring and/or small diameter piping** may be run through field-drilled holes in wood joists, or through pre-punched holes in metal joists:
• installing **long straight runs** of piping through wood or metal joists is impractical; flexible tubing only can be installed.
• **steel beams/girders** can be drilled to allow wiring or small diameter piping, but drilling is costly and larger holes may require reinforcing the web.
• **ductwork** may be run parallel to rolled steel framing, but clearing lateral bracing between framing and girders may present a problem.
• **ductwork between** wood or metal joists would have to clear bracing: sizes are limited.

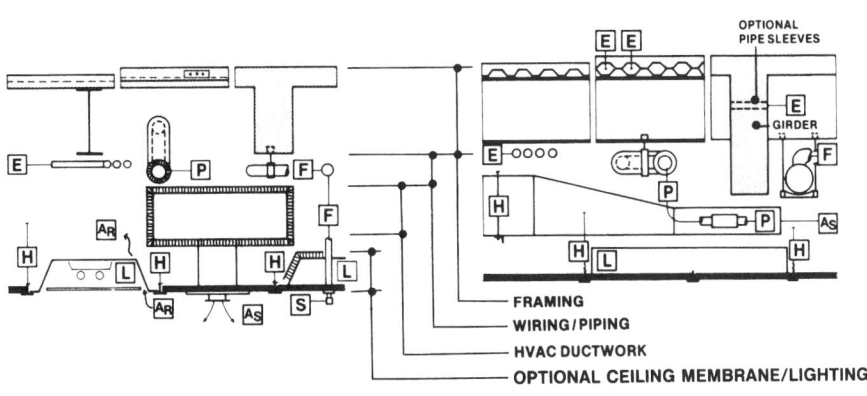

FRAMING
WIRING/PIPING
HVAC DUCTWORK
OPTIONAL CEILING MEMBRANE/LIGHTING

■ **Framing:** rolled steel beams and girders; reinforced concrete joists and beams; reinforced concrete beams and girders.

■ **Wiring and piping:** may be run between framing members, but has to clear lateral bracing members in steel framing; it is generally dropped to clear girders.
• **pipe sleeves** for wiring and small diameter piping may be cast into reinforced concrete girders.
• for fire-resistance rated assemblies, recessed **lighting fixtures are boxed-in** above the ceiling membrane to maintain its continuity.
• **in exposed grid** suspension systems, clearance for tilting tile, fixtures or panels into place must be provided.

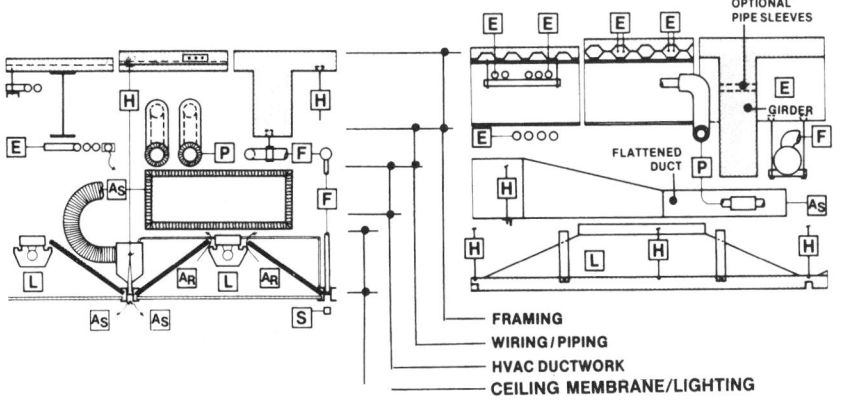

FRAMING
WIRING/PIPING
HVAC DUCTWORK
CEILING MEMBRANE/LIGHTING

■ **Framing:** rolled steel beams and girders; reinforced concrete beams and girders.

■ **Wiring, piping and ductwork:** location and supports similar to those for penetrated membranes:
• **modular components** to fit the suspension grid are generally used;
• **preassembled modules** with mechanical/electrical/sound control components may be installed into a suspended grid.
• **piping** may be run between beams, but is generally dropped to clear girders.
• **pipe sleeves** may be cast into reinforced concrete girders, but diameter is limited.
• **steel girders** may be cut to allow service lines to pass, but openings generally have to be reinforced.

B3.1 ROOFING

ROOFING: preliminary selection

ROOFING

Though all roofing types perform essentially the same function, the choice of roofing may be limited by:

■ **Pitch, or slope, of the roof assembly:**
* some types of roofing cannot be used on steep pitches.
* other types cannot be used below certain minimum pitches.

■ **Type of decking to which roofing is to be attached,** or by which it is to be supported.

■ **Type of framing when roofing is functioning as roofing and decking combined.**

MEMBRANE ROOFING

Membrane, including single-ply types, are defined as a continuous waterproof covering. Membranes are generally used for flat, or low pitch roofs in nonresidential construction. Several types are available.

HOT-APPLIED BUILT-UP

■ **In Hot-Applied Built-Up the membrane consists of coatings of bitumen, asphalt or coal tar pitch, separated by layers of felt:**

* bitumen is used as the waterproofing component and must be heated to lower its viscosity to workable range for application.
* felts are primarily used to reinforce the bitumen.
* three or five layers, or plies, of felt are generally used: the more coatings of bitumen and plies of felt, the longer the expected service life should be.

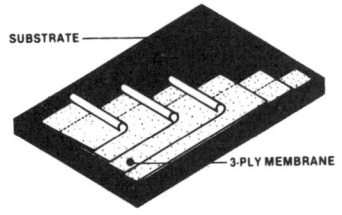

SUBSTRATE
3-PLY MEMBRANE

■ **Heating of bitumen during installation results in air pollution, fire hazard, and may present a health hazard:**
* The toxicity of asphalt is generally lower than of coal-tar pitch.

■ **Built-up is an economical roofing system and may be used on most flat, or low pitched roofs** when no unusual requirements or conditions exist:

■ **Counterindications could include:**
* excessive movments in the structure or substrate.
* ponding of water on the surface of roofing.
* heavy foot traffic over the roof.
* very low temperatures during installation.

■ **For slopes of two inches per foot and over:**
* nailing of the membrane to the substrate is generally required, or
* treated wood nailers have to be provided on nonnailable decks.

■ **For roofs with slopes up to two inches per foot,** a surfacing of aggregate embedded in a heavy flood coat of bitumen is a recommended method because it:
* allows the use of a heavier top coating of bitumen for better resistance to moisture penetration.
* protects the membrane from weathering, especially ultraviolet radiation.
* minimizes surface temperatures.
* serves as ballast for membrane.

■ **For steeper slopes,** where it becomes difficult to keep aggregate surfacing in place, roofs may be smooth surfaced, using:
* asphalt coatings, such as emulsion, cutbacks, or hot asphalt moppings.
* cap sheets of coated felt, factory finished with mineral granules.

■ **Smooth surfaced membranes are:**
* easier to repair, easier to maintain and resaturate, but
* do not generally last as long as aggregate-surfaced membranes without periodic maintenance.

■ **For slopes over six inches in twelve, or even less in hot climates,** the use of built-up hot-applied membranes must be carefully evaluated:
* danger of slippage between felts or of the entire membrane in relation to the substrate exists when temperature of the membrane approaches or reaches the softening point of bitumen used, and this danger increases with increase in slope of roof.

■ **Built-up membranes are also used as waterproofing for:**
* traffic carrying decks.
* roof terraces.
* building walls below grade.

COLD-APPLIED BUILT-UP

Cold-Applied Built-Up membranes employ coatings of bitumen cutback, which is bitumen dissolved in a solvent, to cement together two to three layers of coated organic or inorganic felts.

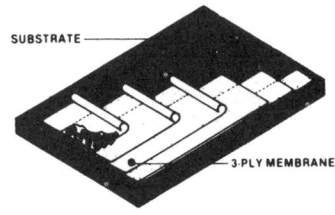

SUBSTRATE
3-PLY MEMBRANE

■ **Limitations for use of cold-applied membranes are essentially the same as for hot-applied except that they:**
* are more often used on small roofs and in reroofing than hot-applied.
* surfacing is limited to asphalt coatings, such as emulsions, or a cap sheet of coated felt factory or field finished with mineral granules may be used.
* large size aggregate surfacing is generally not used, as it would be impractical to provide the required depth of a top coat of bitumen for proper embedding of the aggregate.
* mineral surfacing, factory or field applied generally available.
* cutbacks solidify by evaporation of solvent, and emulsions by evaporation of water; as heating of bitumen is not required during installation, air pollution is negligible, and fire hazard is eliminated for emulsions.

■ **Cold applied membranes are generally built-up with fewer plies than hot-applied membranes,** thus reducing labor costs; but
* the coated felts used are relatively stiff and have to be flattened out before installation.

LIQUID-APPLIED MEMBRANE

Liquid applied membrane roofing consists of elastomeric fluid continuously applied and bonded directly to a solid, dimensionally stable substrate and cured in-place.

B · BUILDING SHELL

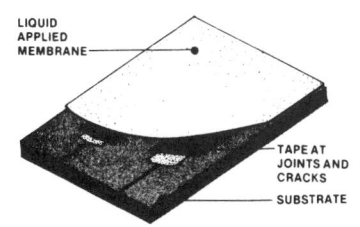

■ **It should be considered especially for complex, free-form roofs where:**
• built-up membranes would be difficult to apply.
• pitch varies from level to steep.
• appearance is important.

■ **Surfacing may be integral, or a compatible material,** more resistant to weathering, may be applied as a second coat.

■ **Solvents will be released during curing of the film and may present a fire and health hazard.**

■ **Cracks developing in substrate after liquid membrane is applied are likely to split the membrane.**

■ **Liquid-applied membranes are also used as waterproofing:**
• below traffic carrying decks.
• on building walls below grade.

SINGLE-PLY MEMBRANE

Sheet membrane roofing consists of:
• polymeric materials precured into large sheets; or
• sheets of modified bitumen reinforced and/or faced.
• permanently jointed at their edges during installation to form a continuous membrane.

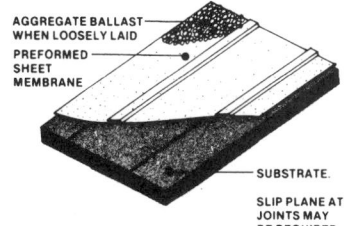

■ **Three methods of installation are used:**
• sheets are laid loose over the substrate to prevent movement in it from affecting the membrane and are then ballasted with large size stone against wind uplift.

• sheets are secured to the substrate using spot anchorage to minimize effects of movement in substrate on the membrane.
• sheets are continuously bonded to the substrate, moving together with it; when so bonded they may accommodate normal movement in substrate better than built-up types, but

■ **When fully bonded to substrate, cracks developing in the substrate/decking after installation of membrane may split the membrane.**

■ **Surfacing:**
• spot anchored and bonded elastomeric sheets have integral surfacing.
• modified bitumen single-ply sheets have a top surfacing resistant to weathering.

■ **Membranes are also used as waterproofing:**
• under traffic carrying decks.
• under floors below grade subject to hydrostatic pressure.
• on building walls below grade.

SHINGLES/SHAKES/TILE

In this type of roofing overlapping units of small size are installed to shed water rather than to form a continuous watertight covering.

■ **Pitch, or slope,** at which roofing is installed is the limiting factor for this type: if pitch is reduced below the minimum, water penetration may occur under high wind conditions:
• water may be pushed under and between the units.
• water may be pulled inward due to differences in pressure.

■ **Generally used in residential, and light commercial construction.**

■ **Aesthetic considerations** are an important factor in the selection of the specific type of overlapping unit to be used.

■ **Nailable substrate is generally required for installation.**

■ **Thermal movement** is provided for in the open joints between units and by spot anchorage to the substrate.

SHEET AND PANEL ROOFING:

SEAMED FLAT METAL

Roofing is composed of flat metal sheets, such as terne, copper, of light gauge, with interlocking edges to prevent water penetration:
• Aesthetic considerations are likely to be the determining factor in the selection of this type.

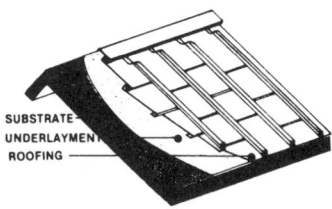

■ **Pitch should in all cases be provided, but:**
• roofing will perform satisfactorily from a low pitch to vertical.
• changes in pitch, or in shape can quite readily be accommodated.

■ **Surfacing is predominantly integral:**
• generally only metals subject to corrosion receive an applied protective coating.
• surface coating may also be provided for appearance.

■ **Installation** is spot anchoring to substrate to allow for thermal expansion and constraction; a nailable substrate, or separate nailers are required.

FORMED METAL PANELS

Lapped formed metal panel roofing is available in various shapes, such as ribbed, fluted, corrugated. It is commonly installed with interlocking side and lapped end joints to prevent water penetration:

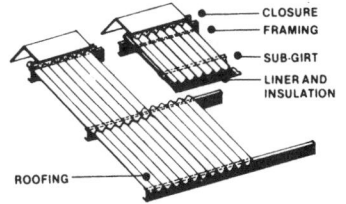

B3.1 ROOFING

ROOFING: preliminary selection (cont.)

■ **Generally used in industrial and commercial construction.**

■ **Pitch may be the limiting factor for this type:**
- when joints perpendicular to slope are necessary between panels and if pitch is reduced below the allowable minimum, water may penetrate through end joints under high wind conditions.

■ **Aesthetic considerations** may also influence the selection process.

■ **Thermal movement is accommodated:**
- through the shape of the panels in metals.
- by flexible connections for panels of rigid materials, such as cement-bound mineral fiber.

■ **Lapped panels:**
- do not require a continuous level substrate.
- are generally attached to and supported by structural framing, spaced approximately six to twelve feet apart.

■ **The panels act as roofing and decking combined,** but provide very limited bracing, or diaphragm action, to the roof assembly.

■ **Surfacing is:**
- integral: or
- factory applied protective coatings are provided for metals subject to corrosion or galvanic action.

ROOF TERRACES

Roof terraces are roofs designed to meet several functional requirements, while still excluding the weather from the space below:
- serve as play areas.
- allow parking of motor vehicles.
- allow the growing of plants.
- or any combination of the above.

When selecting the roofing for a roof terrace, note:

■ **Protective covering** for the membrane is required.

■ **Once installed, the membrane is difficult, if not impossible to repair;** materials should be selected with this consideration in mind.
- surfacing may consist of individual

units loose laid over the membrane to allow for their removal in the event repairs to the membrane become necessary.

■ **Drainage should be provided for the protection surfacing** and also for the membrane: any water penetrating to the membrane through the surfacing must not be allowed to accumulate over it.

■ **Types of roofing membranes, suitable for terraces include:**
- hot-applied built-up membranes.
- elastomers either liquid applied, or single-ply.

THERMAL INSULATION

In roof assemblies with thermal insulation roofing membrane may be located:

■ **Over thermal insulation,** exposed directly to weathering.

■ **Below thermal insulation:**
- to minimize the effects of thermal expansion/contraction on the membrane.
- to protect the membrane from weathering.

When membrane is installed under the insulation:

■ **Thermal insulation must be:**
- resistant to moisture penetration;
- protected from the effects of solar radiation; and
- must be properly secured against wind uplift.

■ **Wind uplift failures are generally caused by:**
- inadequate anchorage of roofing to substrate.
- excessive deflection in the substrate.
- failure in connections of substrate/ decking to framing members.
- faulty installation, such as application of too cold bitumen.
- edge flashings of too light a gauge, or insufficiently anchored.

MOISTURE

Moisture generated from within a building or brought in by infiltration of outside air can permeate through a roof assembly:

■ **If enough moisture hits a cold area,** it will condense inside the roof potentially leading to the deterioration of the assembly.

■ **Means of combatting this problem consist of:**
- Installation of a vapor retarder in the area of the roof adjacent to the occupied space; i.e., on the warm side of the insulation layer.

FLOORING/CEILINGS: preliminary selection

FLOORING

Flooring is the wearing surface component of a floor assembly which should be:

■ **Durable and safe.**

■ **Easily maintained.**

■ **Aesthetically pleasing.**
And remain so during the in-service life of the assembly of which it is part, at acceptable installed initial and in-service cost.

Flooring is subject to the following in-service conditions:

■ **Traffic:** both uniformly distributed over its surface and concentrated localized such as at:
• entrances to buildings, doors to individual spaces, traffic aisles in offices, at stairs, on stair treads.
• fixed work stations: work counters in kitchens, desks in an office, around movable furniture, dining room tables.

■ **Moisture:** either intermittent or frequent, such as:
• water spills, wet cleaning.
• ground water or water vapor migrating through the substrate.
• condensation of water vapor under hot and humid interior conditions.
• rain, snow, ice when exposed to the elements.

■ **Moisture within the substrate during installation** may prevent proper bonding of most types of flooring.

■ **Moisture migrating through the substrate** will affect most types of flooring:
• **wood:** must be protected from moisture entering it from, or through, the substrate.
• **resilient:** some types are not recommended for use in damp locations, such as over concrete slabs below or on grade, unless migration of moisture through the slab can be effectively prevented.
• **seamless:** moisture in the concrete slab, either trapped during installation of flooring, or migrating through the slab after installation, may loosen or destroy the bond between flooring and slab.
• **carpeting:** most types must be protected from moisture present within, or migrating through the

substrate: some backing materials may quickly deteriorate when exposed to alkalies leeching out of concrete slabs.

■ **Abrasion:** by wet or dry dust/dirt being rubbed over the surface through foot traffic:

■ **Scratches:** by hard sharp objects, such as cinders, brought in by foot traffic.

■ **Indentation:** from spiked heels, furniture, stored, moving, or dropped heavy objects.

■ **Impact-generated shearing/-compressive stresses.**

■ **Tensile/shearing stresses developing in the substrate and/or flooring due to:**
• thermal expansion/contraction, which may differ between flooring and substrate when temperature differentials and/or differences in thermal expansion of the two materials exist.
• shrinkage or swelling resulting from changes in moisture content in substrate and/or flooring.
• movement through joints in a substrate supporting a monolithic flooring.
• bending stresses when flooring has to bridge voids in substrate.
• excessive deflections in substrate.

■ **Inelastic flooring materials** are weak in resisting tensile stresses; any movement in the substrate may also crack the flooring, especially when self-bonding.

• to minimize the possibility of cracking cast-in-place terrazzo, flooring may be placed over a slip plane as an independent slab of a thickness sufficient to resist all superimposed liveloads.

■ **Inelastic materials** may be used over wood or steel-framed assemblies, if deflections are kept within safe limits.

■ **Wood will tolerate relatively large deflections,** especially when nailed.

■ **Resilient flooring requires a firm, level, smooth and dimensionally stable substrate;** however, hairline shrinkage cracks, developing before the installation of flooring, will generally not affect it.

■ **Seamless flooring is directly bonded to the substrate;** though able to accommodate larger deflections than the inelastic types, it may split if cracks develop in the substrate after flooring is installed.

■ Carpeting is least affected by shrinkage, cracking and deflections in the substrate

CEILINGS

Ceiling types range from the simplest to assemblies highly complex in their own right. Preliminary selection may include:

Appearance and design considerations:

■ Underside of concrete slab in monolithic flat plate or flat slab assemblies may:
• receive directly applied plaster or acoustic plaster.
• may be painted only.
• may be left exposed as a ceiling.

■ Suspended ceilings will hide from view:
• all horizontal runs of mechanical/electrical services.
• plumbing.
• unsightly maze of structural members.
• sprinkler systems.

■ Suspended ceilings may be used as a means to unify a number of disparate design elements into a single entity:
• lighting.
• heating and cooling supply/return.
• fire and smoke detection and sprinkler systems.
• audio and audiovisual communications systems.

Fire resistance ratings of the roof/floor assemblies may depend on the ceilings. For example:

■ **Ceilings may have to be provided in truss assemblies** when a fire resistance rating for the assembly is required, as fireproofing of all members of the trusses may become difficult and expensive.

Ceilings may become an active part of the mechanical system when:

■ **Used to enclose:**
• a supply plenum for conditioned air in conjuction with ventilating type acoustic tile.
• a return air plenum with ducted conditioned air supply.
• when ceiling or lighting fixture framing members act as air supply units.

B3.1 ROOFING

MEMBRANES: types, uses, materials

● denotes common usage
○ denotes possible usage

note: further detail on built-up
membrame roofing can be
found in chart "Built-up
Membrane: types, materials"

TYPE	DESCRIPTION	SERVICE RANGE (in years)	COMPARATIVE COST (100 = base) [1]	USES & MATERIALS					
				USES			SLOPE [2]		
				NEW ROOFING	RE-ROOFING	WATERPROOFING	MINIMUM, inches per foot	MAXIMUM, inches per foot	NO LIMIT
MEMBRANE, BUILT-UP									
SUBSTRATE — 3-PLY MEMBRANE	● **Hot Applied:** a continuous covering of 3 to 5 alternating layers of hot bitumen and felt plies, with a bitumen top coat; four layers are common. Optional mineral aggregate surfacing may be used to protect the membrane from weathering and ultraviolet radiation. ● **Installation:** felts are continuously bonded or spot anchored to substrate; laid in shingle fashion or single course.	15 to 20	100 to 160	●	●	●	0	[3] 6	
	● **Cold Applied:** a continuous covering of 2 or 3 layers, alternating coatings of cold bitumen with felt plies. ● **Installation:** felts, laid in shingle fashion or single course, are continuously bonded or spot anchored to substrate; solvent-based materials are applied with spray gun and squeegee; emulsions are applied with double-nozzled spray gun (films need more time to dry/cure); aluminum pigmented asphalt, mineral surfacing and colored coatings may be used.	10 to 15	90 to 130	●	●	●	½	4	
MEMBRANE, FLUID APPLIED OR SINGLE-PLY									
LIQUID APPLIED MEMBRANE — TAPE AT JOINTS AND CRACKS — SUBSTRATE	● **Liquid Applied - Single Coat:** an elastic substance capable of expanding or contracting with the substrate to which it is applied. ● **Installation:** built-up to a single coat by spraying, rolling or brushing.	5 to 10	140 to 200	●	●	●			●
	● **Liquid Applied - Two Coats:** essentially the same as single coat, except that the top coat is built-up of material with weathering characteristics superior to those of the base coat. ● **Installation:** applied in 2 coats, each built-up by using materials of different properties, rolled, sprayed, or brushed.	5 to 10	210 to 300	●	●	●			●
AGGREGATE BALLAST WHEN LOOSELY LAID — PREFORMED SHEET MEMBRANE — SUBSTRATE — SLIP PLANE AT JOINTS MAY BE REQUIRED	● **Single-ply, preformed polymeric sheets:** an elastic substance precured either in large sheets or narrow strips approximately 3 feet in width, joined by heat fusion or adhesion. ● **Installation:** the membrane is either bonded to substrate, spot anchored or loose laid and ballasted with aggregate.	10 to 20	340 to 450	●	●	●	0	6 and over	○
	● **Single-ply, Modified Bitumen:** a laminated membrane composed of flexible core of modified bituminous material, and layers of plastic or metallic film and surfacing and/or reinforcing. ● **Installation:** the membrane is either bonded to substrate or loose laid and ballasted with aggregate.	10 to 20	170 to 270	●	●	●	0	6 and over	●

B3.1 ROOFING

REMARKS

1. base = 3 ply hot applied gravel surfaced built-up membrane; including labor and materials.

2. slope for positive drainage should be provided to avoid possibility of water ponding.

3. when nailed:
- Roof quality relies on keeping materials free of moisture and executing flashing details carefully. Continuous jobsite visual inspection is often recommended.

SUBSTRATE — SOLID						PRINCIPAL MATERIALS								SURFACING				REMARKS
NAILABLE, FLAT	NAILABLE, PITCHED	NON-NAILABLE, FLAT	NON-NAILABLE, PITCHED	FRAMING ONLY, WOOD	FRAMING ONLY, STEEL	BITUMEN, HOT	BITUMEN EMULSION	BITUMEN CUTBACK	COATED/IMPREGNATED FELTS	SATURATED FELTS	ELASTOMERIC FLUID	ELASTOMERIC SHEET	MODIFIED BITUMEN	INTEGRAL	BITUMINOUS COATING	MINERAL, GRANULES	MINERAL, AGGREGATE	
●	●	●	●			●			○	●				●	○	●		• minimum slope of ¼ inch in 12 recommended. • on nonnailable substrates with slopes over 1 inch in 12, nailing strips should be provided. • coated, mineral surfaced cap sheet may be used as surfacing.
●		●					●	●	●	○				●	●	○		• min. slope of ¼ inch in 12, ½:12 recommended; slopes over 1:12 require nailing to substrate. • cutbacks used for interply mopping; emulsions for surfacing; freshly applied emulsions are vulnerable to rain and impact; application in cold weather not recommended. • coatings/adhesives may be spray applied; solvents may be flammable; primer may be required for proper adhesion.
●	●	●	●								●			●				• dimensionally stable substrate required; cannot be applied over rigid insulation boards, but can be used with foamed-in-place insulation. • joints and cracks in the substrate must be taped prior to application using glass fiber, nylon or cotton tape. • concrete substrate should be water cured, as curing agents may prevent bonding of the film. • glass fiber reinforcing mats embedded in the first coating may be used as reinforcing. • a minimum mil thickness should be specified for each coating system and application.
●	●	●	●								●			●				
●	●	●										●		●			●	• elastomeric sheets may be solidly bonded to substrate; spot anchored, or laid loose with aggregate surfacing used as ballast against wind uplift. • not all sheets can be bonded to substrate, consult manufacturer's literature. • modified bitumen sheets are prelaminated in two or more layers and installed as single sheet by either bonding/torching to the substrate, or laid loose with aggregate ballast. Joints are heat sealed in some types, solvent or adhesive attached in others.
●	●	●	●			●	●	●					●		○	●	●	

B3.1 ROOFING

BUILT-UP MEMBRANES: types, materials

● denotes common usage
○ denotes possible usage

MATERIALS

TYPE	APPROXIMATE WEIGHT (in pounds per 100 s.f.)	NUMBER OF PLIES, RANGE	BITUMEN							FELTS SATURATED				
			ASPHALT, TYPE I	ASPHALT, TYPE II	ASPHALT, TYPE III	ASPHALT, TYPE IV	ASPHALT CUTBACK	ASPHALT EMULSION	COAL TAR PITCH	ASPHALT, ORGANIC	ASPHALT, ASBESTOS	ASPHALT, GLASS FIBER	COAL TAR PITCH, ORGANIC	NONWOVEN POLYESTER MAT
HOT APPLIED, SMOOTH SURFACED														
SLOPE of SUBSTRATE: 0 to 1 INCH per FOOT			●							●	●	●	●	○
1 to 3 INCHES per FOOT	130	3		●						●	●	●		○
3 to 6 INCHES per FOOT	to 175	to 5			●					●	●	●		○
6 INCHES per FOOT						●				●	●	●		○
HOT APPLIED, AGGREGATE SURFACED														
SLOPE of SUBSTRATE: 0 to 1 INCH per FOOT	400	3	●						●	●	●	●	●	○
1 to 3 INCHES per FOOT	to 600	to 5			●					●	●			○
COLD APPLIED														
SLOPE of SUBSTRATE ¼ to ½ INCH per FOOT	90	2					●	●		○	○	○		●
½ to 3 INCHES per FOOT	to 130	to 3					●	●		○	○	○		●

B3.1 ROOFING

REMARKS

■ Field testing of the roof membrane may include:
- tension test on lap joints (90 percent of the membrane's tensile strength).
- low-temperature flexibility.
- dimensional change, assessed by a travelling microscope.
- embedment of ballast granules.
- puncturing tests: standing object (static) and moving object (dynamic).

COATED				SURFACING AGGREGATE				SUBSTRATE								REMARKS
ASPHALT, ORGANIC	ASPHALT, GLASS FIBER	CAP SHEET, SMOOTH	CAP SHEET, MINERAL	SLAG	GRAVEL	CRUSHED STONE	BITUMINOUS COATING	WOOD PLANK	PLYWOOD / PARTICLE BOARD	FIBERBOARD	GYPSUM PRECAST OR POURED	INSULATING CONCRETE	REGULAR CONCRETE	RIGID INSULATING BOARD	NAILING REQUIRED	
			○				●	○	●	●	●	●	●	●		• **thermal expansion coefficients** for bituminous membranes for temperature range of -30° to +30°F: parallel to fiber grain: —asphalt/organic felts = 11×10^{-6} —coal tar/organic felts = 19×10^{-6} perpendicular to fiber grain: —asphalt/organic felts = 21×10^{-6} —coal tar/organic felts = 29×10^{-6} • **coated base** sheet recommended, especially when organic felts are used; may not be required with glass fiber felts. • **smooth surface cap** sheet should be coated. • **walkways** should be provided to protect membrane from foot traffic. • **placing membrane** over wood planking not recommended unless surface is smooth. • **concrete** surfaces have to be primed.
			●				●	○	●	●	●	●	●	●	●	
○	○	○	●				●	○	●	●	●	●	●	●	●	
○	○	○	●				●	○	●	●	●	●	●	●	●	
				●	●	●		○	●	●	●	●	●	●	○	• **marble chips** may not resist ultraviolet radiation. • **base** sheet recommended but generally not required with glass fiber felts. • **thermal expansion** same as for smooth surfaced.
				●	●	●		○	●	●	●	●	●	●	●	
●	●	●	●				●	○	●	●	●	●	●	●		• **random-oriented glass fiber mat** with resinous binder recommended for emulsion coatings. • **polyester** is the most satisfactory reinforcing product; glass fiber may tend towards bridging and fishmouthing. • **interply** adhesive may be spray applied. • **mineral granule** surfaced cap sheet is recommended.
●	●	●	●				●	○	●	●	●	●	●	●	●	

BUILT-UP MEMBRANES: bitumen

● denotes property
  or usage
○ denotes possible usage

MATERIAL		TYPE OR GROUP	WEIGHT lbs/100sf MOPPINGS/COATINGS	WEIGHT lbs/100sf FLOOD COAT	SOFTENING POINT – °F 135-151	158-176	180-205	205-225	± 400	EVT (equiviscous temp., °F)
BITUMEN–ASPHALT BASE										
ASPHALT	DEAD LEVEL - slope up to 1:12	I			●					475
	FLAT - slope up to 3:12	II	25	[1] 60		●				500
	STEEP - slope up to 6:12	III					●			525
	SPECIAL STEEP	IV						●		525
	EMULSION	I	15							
		II								
	CUTBACK	I	10		●					
		II				●	●			
	ALUMINUM COATING	I	varies							
		II								
	ROOFING CEMENT	I	varies		●					
		II				●	●			
	PRIMER									
BITUMINOUS GROUT			varies						●	
BITUMEN–COAL TAR PITCH BASE										
	COAL TAR PITCH - slope up to 1:12	I	27	[1] 75		[2]				
	COAL TAR BITUMEN - slope up to 1:12	III				[2]				
	PRIMER									
BITUMINOUS GROUT			varies						●	

PROPERTIES

B3.1 ROOFING

B • BUILDING SHELL

PRINCIPAL USE / APPLIED BY / REMARKS

General remarks:

- when not hot enough, asphalt is difficult to apply, adhesion is poor, and it may be over-applied, resulting in splitting and interply slippage in the membrane.
- overheating the asphalt will make it flow easier but may result in under-application, incomplete coverage, voids and poor waterproofing qualities.
- asphalt base and coal tar pitch base resaturants are available for weathered membranes.
- asphalt emulsion may be used to secure rigid insulation to metal decks.

[1] for aggregate surfacing.

Hot Applied Membrane	Cold Applied Membrane	Roll Roofing	Surfacing	Flashings	Dampproofing	Water Proofing	Mopping	Brushing / Spraying	Trowelling	Remarks
●			●		●	●	●			recommended minimum air temperature at installation = 45°F; brittle point = 20 to 30°F.
●			●		●	●	●			incompatible with coal tar pitch.
●			●	○	●	●	●			three types (A, B, C) available for dampproofing and/or waterproofing. • poor bonding to damp or wet surfaces.
●			●	●			●			thickness of application is hard to control. • oxidizes in the sun.
	○	○	●					●		emulsifying agents: Type I = mineral clay colloid; Type II = chemical/soap.
	○	○	●					●		easy handling (heat not required); bonds well to most substrates.
	●	●	○	○	○	○		●		brushing consistency: Type I = light; Type II = heavy.
	●	●	○	○	○	○		●		Type I reinforced with inorganic fibers; Type II includes a filler and fibers.
			●		○	○		●		Type I non-fibrated; Type II fibrated with asbestos fiber.
			●		○	○		●		
○	●			●	○	○			●	Type I will self-heal. • includes mineral stabilizer and asbestos fibers.
○	●			●	○	○			●	
								●		Type III cutback is used as primer with no additives; for hot applications.
					●	●				either asphalt or pitch base.
●					●	●	●			brittle point = about 32°F; [2] softening point = 126-140°F for Type I, 133-147°F for Type III.
					●	●	●			
									●	with hot application.
					●	●				either asphalt or pitch base.

B3.1 ROOFING

BUILT-UP MEMBRANES: felts, surfacing

● denotes common usage
○ denotes possible usage
NA - not applicable

		NOMINAL WEIGHT OF ONE PLY TO COVER ONE SQUARE 100 square feet of roof, in pounds = No.	WEIGHT, pounds per 100 s.f.	PROPERTIES		RESISTANCE TO MOISTURE PENETRATION	RESISTANCE TO WEATHERING
				Breaking Strength psi @ 77°F minimum			
				WITH FIBER GRAIN	ACROSS FIBER GRAIN		
FELTS–SATURATED, IMPREGATED							
ORGANIC, ASPHALT SATURATED	TYPE I	15	13	30	15	poor	poor
	TYPE II	30	26	40	20		
	TYPE III	20	17	35	17		
INORGANIC, ASPHALT OR COAL TAR SATURATED	TYPE I	15	13	20	10	fair	poor
	TYPE II	30	28	40	20		
GLASS FIBER, ASPHALT IMPREGNATED	TYPE I	8	7.3	15	15	fair	poor
	TYPE III	10	9.7	22	22		
	TYPE IV	8	7.0	44	44		
	TYPE V	15	14.6	30	30		
NONWOVEN POLYESTER		25	25			good	good
ORGANIC, COAL TAR PITCH SATURATED		15	13	30	15	poor	poor
FELTS–ASPHALT SATURATED AND COATED							
FELT/BASE SHEET	ORGANIC	30	29	35	20	good	fair
	ORGANIC TYPES I & II	40	37; 39	35; 45	20	good	fair
	INORGANIC TYPES I & II	40	37; 39	30; 40	15; 20		
	VENTING, INORGANIC TYPES I & II		60; 50	30; 25	15; 20		
ROLL ROOFING, SMOOTH SURFACE	SPECIAL BASIS	50	46	Not Given in ASTM		good	fair
	TYPE II	60	56				
	MINERAL SURFACED, ORGANIC	90	83			good	good
	MINERAL SURFACED	45	42			good	good
	WIDE SELVAGE	55	52				
SURFACING							
MINERAL GRANULES		NA	NA	NA		NA	good
AGGREGATE		NA	40-50	NA		NA	good

B3.1 ROOFING

B • BUILDING SHELL

REMARKS

[1]
- saturated felts should be protected with bituminous surfacing or bituminous flood coat and aggregate surfacing.
- crushed dolomite surfacing is not resistant to ultra-violet radiation.
- emulsion may be used instead of cutbacks for cold applied membranes, but cost more.
- felts must be strong enough to resist stresses in the membrane caused by: **Temperature variations** in the membrane and supporting substrate, resulting in differential expansion/contraction of the two elements; **movement** through joints, deflection and cracks in the substrate, resulting in localized stress; **shrinkage** due to moisture content and weathering/aging changes.

SYSTEM		APPLICATION						SECURED TO SUBSTRATE			REMARKS
Hot Applied Membrane	Cold Applied Membrane	Roll Roofing	Cap Sheet	Ply Sheet	Base Sheet	Flashing	Dampproofing	Hot Bitumen	Cutback	Nailing	
●		●		●			○	●		●	• perforated felts are used to reduce blistering.
●					●		○	●		●	• unperforated used as underlayment. • weight gauged per 480 sq. ft. of area before saturation. • least expensive; subject to rotting from moisture.
●				●		●	○	●		●	• available perforated.
●				●	○		○	●		○	
●			○	●	○		○	●		●	• uniform application of interply bitumen required to prevent water leakage.
			○	●			○	●		●	• available perforated or unperforated.
●			○	●							
●	●	○	●	●	○		○	●	●	●	• may be grid-grooved and used as venting base sheets.
●				●	○		○	●		●	
○	●			●				○	●	●	• unperforated sheets only.
●	●				●			●	●	●	• may be used as vapor retarder base sheet.
●	●				●			●	●	●	• emulsion used for surfacing. • venting type available.
○	●	●	○		●	○	○	○	○	●	• joints sealed with lap cement.
○	●	●	●			○		○	○	●	• narrow selvage.
○	●	○	○			○		○	○	●	• polyester mat available. • inorganic.
○	○	○	●			○		○	●	●	• plant surfaced with granules.
○	○	○	●			○		○	●	●	• cap sheet used for hot applied instead of surfacing.
	++							++			++ may be field applied to emulsion.
●								●			weight in place.

SINGLE-PLY MEMBRANES: materials, properties

● denotes common usage or availability
○ denotes limited usage or availability

	CHOICE OF SURFACING	CHOICE OF COLOR	REINFORCED	ULTRAVIOLET RESISTANCE	LOOSE LAID	SPOT ADHESION	SPOT ATTACHED	FULLY ADHERED	HEAT FUSED	INVERTED	A. SUBSTRATE REQUIREMENTS / B. RESTRICTIONS	THICKNESS, range in mils
POLYMERIC (ELASTOMERIC/PLASTOMERIC)												
POLYCHLOROPRENE (NEOPRENE)	●	●	○		●		●	●		●	A. Loose laid not suitable for slopes over 2:12. B. Aromatic solvents, strong oxidizing chemicals.	39 min. 45 to 60
Liquid-applied, 1-part		○		P								
EPDM ● Ethylene Propylene Diene Monomer	○	●	●	G	●	●	●	●		●	B. Petroleum-base products, most oils, hydrocarbons, plastic roof cement.	39 min. 45 to 60
CSPE ● Chlorosulfonated Polyethylene (Hypalon)	●	○		G	○		○	●		○	B. Chlorinated, petroleum, aromatic solvents.	39 min. 35 to 45
Liquid-applied, 1-part	●			G								
CPE ● Chlorinated Polyethylene	●	●		G	●		●	●			B. Petroleum distillates, strong oxidizers concentrated acids and aromatic hydrocarbons; not compatible with flexible PVC.	39 min. 40 to 48
PVC ● Polyvinyl Chloride	●	●		P	●		●	●		○	A. Some partially adhered restricted to min. slope 6:12. B. Petroleum products, oils, asphalt, coal-tar pitch, solvents.	39 min. 34 to 60
PIB ● Polyisobutylene	○	●	●	F	●	●	●	○		●	B. Organic solvents, strong oxidizing chemicals, tar.	39 min. 57 to 120
NBR ● reinforced acrylonitrile butadiene polymer			●		●		●	●		●	B. Aromatic hydrocarbons, coal tar pitch.	30 to 45
ECB ● ethylene copolymer bitumen	○	○		G	●			●	●		B. Petroleum products; aromatic hydrocarbons.	35 to 50
EIP ● ethylene interpolymer alloy	○	●			●		●		●			32
URETHANE				P				●				15 to 30
MODIFIED BITUMEN												
MB ● Modified Bitumen	●	●	●		○			●	●	○	A. Some require ¼″ to ½″ slope. B. Petroleum distillates, concentrated acids, hydrocarbons, aromatic solvents, some chemicals.	60 to 80
MB/R ● Modified Bitumen ● Reinforced	●	●	●		○			●	●	○		40 to 160

B3.1 ROOFING

B · BUILDING SHELL

[5] WEIGHT, without ballast, range in lbs. per sq. ft.	[6] WORKABLE TEMPERATURE range, °F	[7] STRENGTH, TENSILE; in psi or BREAKING, in lbf	[8] ELONGATION AT BREAK, percent	[9] WATER ABSORPTION or RESISTANCE, maximum	[10] WARRANTY AVAILABLE, range, in years
.28 to .46	-25 to 180	1400 psi to 2000 psi / 1000 to 1200	300 min. 450 350 / 150 min.	5% to 12.2% / .2-.3 perms	10 max.
0.245 to .45	-69 to 300	1300 psi to 1755 psi	300 min. 300 to 500	5% to 2.1%	5 to 30
.25 to .33	10 to 158	800 psi to 1458 psi / 400 to 600	300 min. 400 to 560 / 150 min. 350	5% to 0% / .2-.3 perms	5 to 15
.28 to .33	0 to 120	2000 psi to 2500 psi	300 min. 210	5% to .8%	10 max.
.23 to .49	-15 to 140	1400 psi to 2850 psi	250 min. 300	2% to 1.33%	10 max
.57 to .60	20 to 160	380 psi to 550 psi	300 min. 445	5% to .52%	10 max.
.22	none	7500	28	.75	15
.25 to .32	-10 to 120	130 to 250 lbf			5 to 20
.25	0 to 120	8500 psi			5 to 10
72 to 91% solids by wt.	40-120	400 to 2500	150 to 500	0.5 to 4% by wt.	10 max.

.30 to 1.47	-30 to 150	Grade I 44 to 676 lbf / Grade II 124 to 324 lbf	4 min. 100	1.0g max. .32 g	5 to 12

REMARKS

[1] **Surfacing for** polychloroprene: CSPE is generally used. Surfacing for MB/R: Thin embossed metal (such as aluminum, copper, stainless steel), mineral aggregate and special coatings are used.
- **Choice of color:** Other than black - white, dark to pale grey, beige, various other light colors.
- **Reinforcing or backing:** Inorganic fibers, such as polyester, glass; woven fabric.

[2] **Generally using metal disks attached to substrate and membrane adhered to disks.** Resistance to wind uplift should be investigated. Proprietary system of mechanical fasteners without adhesives is available for spot attaching membranes.

[3] A. **Substrate may be dead level unless noted.** Slope for positive drainage may be required and is recommended. All may be installed over various types of insulation, unless noted.
B. **Restrictions refer to exposure to contaminants;** may vary with different manufacturers.

[4] **First value denotes required minimum cited under applicable standards;** second value denotes commonly available range.

[5] **Ballast generally of large round gravel;** 10 to 13 pounds per square foot.

[6] **Temperature during installation.**

[7] **Indicates the ability of the membrane to resist stresses induced by thermal changes, deformations and movements in substrate.** Minimum values for tensile strength is 870 psi; MB minimum breaking strength is 66 lbf for Grade I, 176 lbf for Grade II.

[8] **Indicates the ability of the membrane to stretch without rupture.** First value denotes required minimum; second value denotes commonly available average range. Values vary based on test method. A minimum 150 percent elongation is required to ensure that foam surface movement does not rupture liquid-applied coatings.

[9] **Indicates absorption characteristics when exposed to standing water** or resistance to absorb moisture and resulting dimensional change.

[10] **Warranties on materials vary with manufacturer.**

[11] **Ultraviolet resistance** is indicated as good (G), fair (F) or poor (P).

B3.1 ROOFING

MEMBRANES: selection checklist

BUILT-UP MEMBRANES

■ **Built-up membranes consist of:**
● **Bitumen (coal tar pitch or asphalt)** - used to saturate, coat or impregnate felts so as to resist moisture penetration.
● **Felts** - used to strengthen and dimensionally stabilize the bitumen.
● **Base flashing** - of heavy felt, cotton fabric, glass fiber mesh, polymeric sheet or modified bitumen.

■ **Built-up membranes are secured to the substrate by fully embedding felts in a continuous coating of bitumen, or by nailing or spot cementing with bitumen.**
● Full embedment allows membranes to move with the substrate, but cracks in the substrate or excessive differential movement in substrate or membrane may cause splitting.
● Nailing/spot cementing minimizes the detrimental effects of differential movement, but concentrates stresses at points of attachment.

Bitumens

■ **Bitumens are thermoplastic materials which become more fluid when heated and firm when cooled.**
They are refined from coal (coal tar and pitch) or crude oil (asphalt).
● Bitumens tend to stiffen at low temperature and become brittle with age.
● Low temperature flexibility may be important for winter application.
● Outdoor temperatures at the time of application of bitumen may be an important factor during cold weather. Improper applications can also result in poor adhesion and premature cracking.

■ **Characteristics required of bitumens used in roofing applications include:**
● **Resistance to water penetration and absorption.**
● **Good range of working temperatures** defined by the softening point at very high temperatures, and the brittle condition at very low temperatures (at which it loses its viscous properties and behaves as a brittle elastic solid).
● **Chemical stability** - when heated to working temperatures, bitumen should have the least volatile content and be

free of the tendency of its oily components to separate during application or subsequent weathering.
● **A high flash point** will diminish the possibility of fire during installation.
● **Compatibility with other roofing materials** - asphalt and coal tar pitch are not chemically compatible.
—If asphalt is applied over pitch, it may soften and flow off, allowing exposed pitch to rapidly weather.
—If pitch is applied over asphalt, the pitch may harden and crack.
—The bond between the pitch and asphalt will degrade as a film of oil generally develops.
● **Good adhesion to a solid surface** - bonding to felts or the substrate depends on:
—the viscosity of the bitumen.
—the condition and material of the solid surface.
—the presence of dust and moisture on/within the solid surface.
—the permeability of water into the solid surface after application.
—the need for a primer coat on some surfaces (metals, concrete) before hot-applied bitumen is applied.

■ **Asphalt bitumens are available in three types - straight-run, cutback, and emulsion.** These types are based on the means used to lower the highly viscous nature of asphalt to make it workable and to ensure adhesion.
● **Straight-run asphalt** is heated asphalt selected depending on the slope of the substrate (the greater the slope, the higher the softening point). As the softening point lowers, flow resistance at high outdoor temperatures lessens; crack resistance (shrinkage) improves at low outdoor temperatures; and waterproofing properties increase.
● **Cutback** is asphalt dissolved in solvent (usually kerosene) to lower viscosity. Generally used to bond the plies of felt in a cold-applied membrane, it can be used as a top coating though it does not weather as well as emulsion top coats. Bonding is poor on wet surfaces but improved when only damp.
● **Emulsions of asphalt** lower viscosity by using three types of water-based mixtures - soap type, with a soap emulsifier; clay-modified type, with a soap and clay emulsifier; clay-base type, with a mineral clay emulsifier (usually bentonite). They are used as

top coats in smooth-surfaced, hot- or cold-applied membranes and are less apt to deteriorate due to weathering than chemical types.

■ **The temperature of asphalt during application is critical to the quality of the built-up membrane.**
Asphalt should be applied according to its equiviscous temperature (EVT) and flash point.
● The EVT is the temperature needed at the time of application to achieve the proper application viscosity for best end results for a given asphalt.
● In general, the properties of asphalt begin to degrade when heated to above 325°F for a duration of more than 4 hours. Maximum heating temperatures in the Equiviscous Temperature (EVT) range of 475°-525°F can be tolerated only over a short time.

■ **Coal-based bitumens are heated or dissolved in a solvent (generally tolene) to make them workable and ensure good adhesion.** Coal tar characteristics are:
● High resistance to water penetration, leaching or alteration.
● Self-healing capacity: sealing small cracks when exposed to normal temperatures after installation provided by cold flow properties.
● Minimum volatility at high temperatures.
● Requires aggregate surfacing.

Felts

■ **Felts are made from organic or inorganic fibers that are saturated, coated or impregnated for waterproofing purposes.**
● **Saturated felts** - Organic/inorganic felts saturated with bitumen.
● **Coated felts** - Asphalt-saturated felts coated on both sides with a factory-applied film of asphalt; may contribute to interply slippage.
● **Impregnated felts** - Inorganic felts impregnated with asphalt; they provide an open, porous membrane for venting water vapor.
● **Mineral-surfaced felts** - Granules of rock or colored slate embedded into the weather-exposed surface of coated felts.
● Felts should be stored in rolls, on end and protected from moisture. Saturated felts will absorb moisture and allow water penetration.

B3.1 ROOFING

■ **Organic felts** are composed of wood fibers; primarily softwoods, but some hardwoods as supplement; paper waste or recycled; bond, corrugated, cartons, etc.; rags, untreated cotton textiles; or any combination of the above.
● They are susceptible to edge curling which can be minimized by applying a glaze coating at the end of each day of installation.

■ **Inorganic felts** are made of nonabsorbent fibers such as glass or polyester, making them more resistant to the effects of moisture.
● **Glass fiber** - made of matted glass fibers. The porosity of the felt allows air/gas pockets to be replaced by subsequent applications of bitumen without the need for brooming.
● **Polyester** - made of nonwoven (spunbonded, spunlaced, wet-laid, needlepunched) polyester fibers. Uses include reinforcement of modified bitumen and single-ply membranes, single-ply slip/base sheets, and stone separator mats in ballasted one-coat systems. It is more flexible, durable, puncture resistant, pliable, lighter in weight, tougher, and tear resistant than some organic and glass fiber felts. 20 to 50 percent elongation allows movement due to temperature changes; excellent recovery; negligible shrinkage.

■ **Workmanship is an important element in the quality of built-up roofs because the membrane is literally manufactured on the roof.** Inspect to ensure that:
● Corrosion-resistant metal fasteners are used per fastening schedule to affix the first layer of insulation.
● The vapor retarder is thoroughly mopped with asphalt.
● The vapor retarder if used, is present.
● Overspacing the felt rolls during application, resulting in less than the specified number of layers, does not occur.
● Inferior materials in flashing, such as untreated wood nailers or flimsy sheet metal, are not used.
● Asphalt is properly heated.
● Insulation boards are butted and firmly seated into the asphalt.
● Insulation is properly protected from the weather when construction is phased such that more insulation is installed than can be roofed in one day.

● Flashing bends are never greater than 45 degrees and that cant strips are used to support flashing materials.
● Traffic is kept off freshly hot-applied bitumen until it has cooled so that bitumen is not displaced.

Surfacings

■ **Surfacings are provided to protect the membrane from the effects of solar radiation and minimize surface temperatures.** Types include:
● **Coatings** - with/without reflective pigments not recommended where ponding of water is a possibility. Bituminous coatings, or resaturants, may be added to weathered membranes to restore their waterproofing characteristics provided that the bitumen base is compatible with the roofing bitumen. Acrylic elastomeric mastics may be useful for reflective coatings and substrate primers; rapid curing is an advantage.
● **Mineral granules** - factory applied to coated felts or spray applied over a bituminous top coat.
● **Aggregate ballast** - 1/4 to 5/8 inch size gravel, crushed stone, or slag applied over a heavy top coat of bitumen is used for roofs with a slope of 2 inches per foot or less. It protects against damage from hail, reduces fire hazard from flying brands, and protects the bitumen from ultraviolet radiation. Crushed marble may not be opaque to ultraviolet radiation.
● **Roof walkways** - required to protect membranes where frequent traffic occurs. Pavers, when used, may be set on pedestals to allow drainage beneath.

SINGLE-PLY MEMBRANES: POLYMERICS

■ **Polymeric:** Those materials made from synthetic polymers, including elastomerics which may be vulcanized or nonvulcanized, and thermoplastics.
● Overlaps between sheets are sealed by heat fusion, solvent welding, tacky tapes, or adhesives. Exposed seam lines are caulked for additional protection and endurance.
● Membranes are generally delivered in rolls of up to 200 feet in length. Widths vary from 3 feet to 50 feet; thicknesses range from 30 to 60 mil.

■ Single-ply roofing problems may include blistering, oxidation, roll back, corrosion of metal fasteners, rupture of adhered seams, difficulty in repairing installation errors, and migration of water under an entire ballasted roof membrane from one penetration.
● Defective laps and flashing account for nearly 40 percent of single-ply membrane failures.
● Single-ply membranes are specified by thickness, breaking and tearing strength, elongation, and resistance to ultraviolet rays, ozone, etc.

■ **Methods of securing polymeric single-ply membranes include:**
● **Continuous bonding** - using specially formulated adhesives; good for sloped roofs.
● **Spot anchoring to the substrate** - adhesives/mechanical fasteners are used to minimize effects of movements in the substrate on the membrane.
● **Ballasted weighting of loose-laid membranes.**

Vulcanized Elastomers

■ **Vulcanized elastomers** are thermosetting materials which have great stability at high temperatures. These rubberlike synthetic polymer sheets will stretch when pulled and will return quickly to their original shape when released. Membrane seams are usually sealed with contact adhesive.
● Material specifications include tensile strength and set, heat aging, tear resistance, water absorption, factory seam strength, brittleness point, ozone resistance, fabric adhesion, and ply adhesion.

■ **EPDM (ethylene propylene diene monomer):** A highly elastic, flexible (at low temperatures), but tough membrane which is ultraviolet, ozone, weathering and abrasion resistant; performance is largely determined by formulation. It does not require protective coating.
● Performance is optimal with a minimal number of seams. EDPM splice adhesive systems include two-part butyl tape, neoprene, neoprene/primer, neoprene primer wash, and one-part butyl.
● Formulations are available which can be heat welded.

B3.1 ROOFING

MEMBRANES: selection checklist (cont.)

- Field-formed lap seams are joined with contact adhesive (neoprene - phenolic) or 2-inch wide adhesive tape (cured or uncured synthetic rubber compounds); adhesive bond is improved by using primer on the lapping seam surfaces. Cleaning solvents and lap cements used during installation are flammable.
- Factory-formed seams minimize lap seam failure; sheets are available up to sizes of 100 by 150 feet. Fully bonded sheets are less prone to wind damage and are better at restricting water migration from punctures.
- Potential problems may occur from:
  - Substituting inferior insulation or membrane materials.
  - Using too few fasteners, inferior or non-corrosive-resistant fasteners.
  - Not butting insulation boards.
  - Lapping adjoining sheets less than 3 inches; using too little or too much bonding adhesive or splicing cement; not letting the adhesive/cement dry sufficiently before adhering the membrane or joining the lap.
  - Not cleaning laps thoroughly or with the specified product; talc or mica dust used for separation in shipping and handling must be cleaned off (with solvents - chlorinated hydrocarbons, white gas) before sealing.

■ **Neoprene, or polychloroprene (CR):** Synthetic rubber based on polymers of chloroprene; often used for flashing and roofing in chemical environments.

- Good resistance to abrasion, petroleum oils, solvents, greases, heat and weathering; more chemically resistant than EPDM.
- Cleaning solvents and lap cements used during installation are flammable.
- May require protective coating of chlorosulfonated polyethylene (Hypalon) to retard weathering or to provide lighter colors.
- May have UL Class A rating.

Nonvulcanized Elastomers

■ **Nonvulcanized elastomers** are not resilient during installation, but cure to a rubberlike substance. Membrane seams are usually sealed using heat or solvent welding while in the uncured state.

- Material specifications include low temperature bend, linear dimensional change, ozone resistance, and hydrostatic resistance.

■ **CSPE (chlorosulfonated polyethylene - Hypalon):** Sheet membranes are reinforced with polyester or laminated to a felt backing; self-curing, liquid-applied form available.

- Does not support combustion.
- Excellent ozone and weather resistance; also resistant to fats, oils and greases, but not tolerant of other hydrocarbons. Superior to neoprene in resisting ultraviolet radiation, heat and abrasion; may be liquid applied as top coat over neoprene in a two-coat system.
- Sheet seams are made by heat or solvent welding.

■ **CPE (chlorinated polyethylene):** Sheet membrane, generally backed with inorganic fibers, applied directly over an asphalt or coal tar roof without requiring a separator.

- Compatible with bitumen and foamed polystyrene, polyurethane and polyethylene.
- Resists hydrocarbons (except aromatic), oils, weather, ozone and airborne pollutants. Resists ultraviolet light and chemicals better than PVC.

■ **NBR (reinforced acrylonitrile butadiene polymer blends):**

- Material specifications include heat aging, low temperature bend, linear dimensional change, water absorption, ozone resistance, hydrostatic resistance, low vapor permeability.
- Resists tears, punctures and weathering; not resistant to aromatic hydrocarbons.

■ **PIB (polyisobutylene):** Sheet membrane laminated to a nonwoven reinforcement except at side laps.

- Resistant to weather, ultraviolet radiation and radiant heat.
- Loose-laid with ballast, partially adhered with hot asphalt or adhesive, mechanically fastened.

Thermoplastics

■ **Thermoplastics** (thermosetting plastics) are polymer materials that soften when heated (or exposed to solvents), harden when cooled and, when cured, stretch but do not recover fully. Membrane seams are usually sealed by heat or solvent/chemical welding. Like elastomeric sheets, these plastomeric sheets should be stored flat on site.

- Material specifications include tensile strength, heat aging, tear resistance, low temperature bend, linear dimensional change, water absorption, and seam strength.

■ **PVC (polyvinyl chloride):** Sheet membrane with good fire resistance and water/chemical resistive properties. The biggest disadvantage is membrane shrinkage.

- PVC membranes are available reinforced (glass fiber or polyester) and nonreinforced. Nonreinforced membranes exhibit greater shrinkage and expansion.
- Strong seams and patches are heat- or solvent-welded in the field.
- Good tensile strength and abrasion resistance, but generally lower weather-resistance than for neoprene or EPDM.

■ **ECB (ethylene copolymer bitumen):**

- Contains no plasticizers.

■ **EIP (ethylene interpolymer alloy):** Generally a polyester reinforced membrane with resistance to fire, chemicals, oils and tears; usually 32 mil thick.

Modified Bitumen

■ **Modified bitumen** are factory-made composite sheets consisting of bitumen, chemical additives - SBS, styrene butadiene styrene or APP, atactic polypropylene - which increase the softening point, and reinforcement fabric. Filler and polymer additives are used to modify the bitumen to improve plastic and elastic properties, durability, ultraviolet resistance and weatherability.

- APP is a thermoplastic that causes brittleness below 20°F. APP modified bitumens are applied using a torch to melt the back coating which provides an adhesive bond.
- SBS is a thermoplastic copolymer which widens the temperature range of asphalt, has good elongation properties, and becomes brittle below -10°F. SBS modified bitumens are applied with hot asphalt or torching; less heat is required for SBS modified bitumens.

- Reinforcement is often a polyester (better elongation and puncture resistance) or glass fiber mat (tensile strength) or both.
- The membrane sheets may be supplied with some type of film, foil and/or mat for ease in handling, improved weatherability and to screen out ultraviolet radiation. Field-applied surfacing options include loose gravel and aluminum coating.
- Compatibility with asphalt and asphalt-based products is a major advantage for re-roofing; excellent as base flashing for built-up roofing.
- Storage should be in roll form, on end, protected from moisture and at a temperature of at least 40°F before installation.
- Sheet laps are usually 3 to 4 inches on the side; 4 to 6 inches at ends.
- Installation may be full adhesion using hot asphalt, torching (softening of bitumen in successive layers to create adhesion and forming strong, monolithic seams) or ballasting. Spot anchoring is not used since it involves nailing through the membrane. Spot adhesion is difficult to accomplish and control.

■ **Multi-layer configurations may be used to reduce the probabilities of substrate or workmanship problems.** Base sheets are often used to provide a surface for modified bitumen membranes.
- **Glass fiber felts over insulation -** installed in hot asphalt, side lapped a minimum of 2 inches and end lapped a minimum of 4 inches.
- **Glass fiber felts over concrete deck -** may or may not be required; cold process applied; may require priming.
- **Venting-type felts over wet fill decks -** mechanically fastened to lightweight insulating concrete or poured gypsum.
- **Organic felts over wood deck -** nailed per given pattern and spacing to wood deck; glass base sheets may wrinkle in these applications.
- **Punched fiberglass base sheets -** 1-inch diameter holes, 6 to 8 inches on center throughout the sheet, allow asphalt mopping to attach the base sheet to urethane and isocyanurate insulation.

■ **MB, modified bitumen self-adhering membrane:**
- Peel-and-stick products are available which are applied directly to the base sheet or insulation. Solvents may also be used for fastening.
- Cold applied, priming of substrate may be required to receive membrane.
- Top surfaced generally with polyethylene film to serve as a protection barrier, or to receive optional liquid protective coating, or to serve as substrate for gravel or mineral granules.
- Joints between sheets are lapped and pressure sealed.
- Difficult to apply under low outdoor temperatures.

■ **MB/R, modified bitumen reinforced membrane:**
- Modified bitumen and reinforcing sheets, generally located at mid-depth of the composite sheet. Reinforcing may be plastic sheeting, polyethylene film, heavy polyester mat, glass fiber felt or fabric, or a combination of the above.

■ Flashing for modified bitumen must be carefully designed and applied. Leakage at flashed joints/openings can be minimized by:
- Using 24-gauge galv. steel or .032-inch aluminum with neoprene washers on fasteners.
- Using plenty of sealant on the back of the flashing and including a U-tube - a strip of flashing that seals the face of the opening.

■ Other problems noted from application of modified bitumen membranes include:
- The use of inferior grade/quality lumber used in place of specified, pressure-treated lumber.
- Misaligned or inadequately lapped seams, T-joints not sealed, failure to fully adhere membrane.
- Unrestricted construction traffic, including equipment, on the newly completed roof.

LIQUID-APPLIED MEMBRANES

■ **Liquid-applied elastomeric membranes have similar properties to their sheet-form counterparts. They may be used as exposed** roofing, often over foamed-in-place insulation, or as waterproofing for floors, either exposed or protected by solid surfacing, such as concrete, paving, tile, brick and the like.
- As coatings over foamed-in-place insulation, they are effective protection from ultraviolet radiation, vapor permeability, shear and impact loads, and thermal shock.
- Coatings high in volume solids are more expensive per unit but may be more cost effective for greater mil thicknesses since fewer gallons are required to attain that thickness.
- As primers, liquid applied elastomerics provide a better surface to which foam insulation can bond.

■ **Urethane foam** may be sprayed on the roof surface to add insulation, resurface an old roof, and/or cover irregular shapes and surfaces.
- A protective coating (liquid applied membrane) is required to eliminate deterioration from ultraviolet radiation and must be capable of resisting impact loads.

■ Materials commonly used for liquid applied membranes are:
- **Acrylic:** Copolymers or terpolymers of acrylic monomers are water-based products that may be used for high-pitch applications. Colors are available.
- **Butyl:** Two-part butadiene polymers are vapor barriers with excellent cold temperature flexibility. They may require protection from traffic and ultraviolet radiation. Colors are limited to black, tan or grey.
- **Chlorosulfonated polyethylenes:** Similar but not as strong as sheet membranes.
- **Polychloroprene (neoprene):** Similar properties as sheet membranes.
- **Polyurethane:** Some ultraviolet resistant formulations are available.
- **Silicone:** One or two-part elastomers with good ultraviolet resistance but poorer physical qualities; not recommended for flat roof applications.

B3.1 ROOFING

ROOFING: types, uses, materials

● denotes common usage
○ denotes possible usage

				USES & MATERIALS					
			[1]	USES			SLOPE		
TYPE	DESCRIPTION	SERVICE RANGE (in years)	COMPARATIVE COST (100 = base)	NEW ROOFING	RE-ROOFING	WATERPROOFING	MINIMUM, inches per foot	MAXIMUM, inches per foot	NO LIMIT

OVERLAPPING UNITS

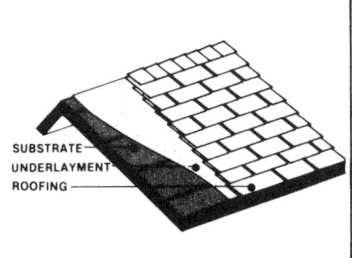

SUBSTRATE
UNDERLAYMENT
ROOFING

	DESCRIPTION								
	● **Shingles or Shakes:** small rectangular smooth or textured units; installed on pitched substrate to shed water rather than to form a continuous watertight cover. ● **installation:** shingles and shakes are nailed to substrate using the overlapping method either in single or double courses.	20	160 to 470	●	●		2		●
	● **Tiles:** small, thin, flat or molded smooth surfaced units; installed on pitched substrates to shed water rather than to form continuous watertight cover. ● **installation:** tiles are nailed to substrate using the overlapping method either end lap only or end and side laps in single course.	20	210 to 510	●	●		3 to 8		●

SEAM FLAT METAL SHEETS

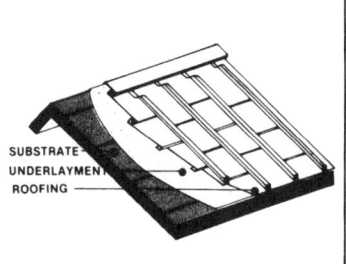

SUBSTRATE
UNDERLAYMENT
ROOFING

	DESCRIPTION									
	● **Seamed Flat Metal Sheets:** comparatively light gauge thin flat rolled sheet of metal with interlocking edges to prevent water penetration. Prefabricated units with standing or batten seams and flat or stiffened pans available. ● **installation:** sheets are spot anchored to substrate using cleats, nails or snap locks.	flat	+ 20		●	●		½	4	○
		standing	+ 20	440 to 1300	●	●		2½		●
		batten	+ 20		●	●		3		●

PREFORMED PANELS

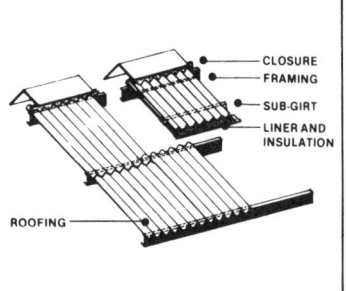

CLOSURE
FRAMING
SUB-GIRT
LINER AND
INSULATION
ROOFING

	DESCRIPTION								
	● **Lapped Panels:** a one piece unit that is factory prefabricated in various shapes with lapped or interlocking sides and lapped and sealed ends to prevent water penetration. Trim pieces generally of matching material. ● **installation:** panels are spot anchored to spaced framing using concealed or exposed fasteners.	single	10	150 to 300	●	●		1½		●
		sandwich	20	325 to 625	●	●		1½		●

B3.1 ROOFING

B • BUILDING SHELL

	SUBSTRATE – SOLID						PRINCIPAL MATERIALS										SURFACING		
	NAILABLE, FLAT	NAILABLE, PITCHED	NON-NAILABLE, FLAT	NON-NAILABLE, PITCHED	FRAMING ONLY, WOOD	FRAMING ONLY, STEEL	WOOD	ASPHALT SATURATED MAT	BURNED CLAY/SLATE	CONCRETE	METAL, FLAT	METAL, BRAKE-FORMED	METAL, ROLL-FORMED	CEMENT-ASBESTOS, FLAT	CEMENT-ASBESTOS, CORRUGATED	PLASTIC, RIBBED/CORRUGATED	INTEGRAL	APPLIED COATING	AGGREGATE
	●	○	●	○	●	●	●	●				●		●			●	●	
	●	○	●	○					●	●	●						●	●	
	●										●	●					●	●	
	●										●	●					●	●	
	●										●	●					●	●	
			○	○	●	●							●		●	●	●	●	
			○	○	●	●							●		●	●	●	●	

REMARKS

[1] **base** = 3 ply hot applied gravel built-up membrane, including labor and material.

- **wood shingles:** 4" slopes require No. 15 roofing felt laid over top portion of each course; two layers for 2" slope nonnailable substrate and steel framing require nailing strips.
- **aluminum shingles:** 4" slope per foot recommended.
- **cost for shingles:** low for asphalt; high for porcelain enamel.
- **cost for tiles:** low for burned clay; high for slate.

- **No. 30 asphalt** saturated felt over substrate recommended for 3" slope per foot; protection against galvanic action, corrosion, chemical agents required.
- **cost:** low range for terne over galvanized steel; high range for copper.

- **minimum** ½ inch per foot slope possible if installed in single lengths.
- **lapped panels** seldom used over solid substrate, except in re-roofing; metal, or wood sub-girts must be provided for attachment of panels.
- **cost:** varies depending on gauge or thickness of metal and finish, high range for heavy gauge steel with high performance finish.

B3.1 ROOFING

SHINGLES & SHAKES: tile, metal sheets & panels

● denotes common usage
○ denotes possible usage

MATERIAL	WEIGHT INSTALLED in pounds per 100 s.f.	TENSILE STRENGTH (psi)	LINEAR EXPANSION for 100°F RISE	RESISTANCE TO MOISTURE PENETRATION	RESISTANCE TO WEATHERING
PROPERTIES					
OVERLAPPING UNITS–SHINGLES OR SHAKES					
ASPHALT ORGANIC MAT	130-380		.021*	fair	good
GLASS FIBER MAT	215-300		.026*	good	good
WOOD RED CEDAR	220-350	tensile strength not a factor	average .002 per-pendicular to fibers	poor	fair to good
REDWOOD	200-250				
CYPRESS	180-200				
ALUMINUM, COATED, PREFORMED	50-100		.0012	good	good
COPPER, PREFORMED	150		.0009	very good	
STEEL, GALVANIZED, PREFORMED	250		.0007	good	poor
PORCELAIN ENAMEL ON METAL	Abt. 225		varies	good	good
CEMENT BOUND MINERAL FIBER	258-585		.0007	good	good
OVERLAPPING UNITS–TILES					
BURNED CLAY	900-1000	tensile strength not a factor	.0003	fair to good	good to very good
CONCRETE, FLAT	650-700		.0005		
SLATE	700-900		.0004		
SEAMED FLAT METAL SHEETS					
ALUMINUM, SHEET AND PREFORMED	55-90	22,000	.0012	very good	good
COPPER, SHEET	125	36,000	.0009		very good
COPPER, LEAD COATED, SHEET	125	36,000	.0009		
ZINC ALLOY, PREFORMED	187	45,000	.0017		very good
MONEL, SHEET	80-90	70,000	.0009		very good
STAINLESS STEEL, SHEET	100	110,000	.0009		
STEEL, COATED, PREFORMED	80	60,000	.0007		good
TERNE, COATED STEEL, STAINLESS STEEL, SHEET	89	45,000	.0008		fair
LAPPED PANELS					
ALUMINUM, PLAIN OR COATED	35-90	22,000	.0012	very	good
STEEL, GALVANIZED AND/OR COATED	120-240	60,000	.0007	good	good

B3.1 ROOFING

ALSO USED FOR:					SECURED TO SUBSTRATE:					REMARKS
ROOF FLASHING	WALL FLASHING	FASCIAS	EQUIPMENT SCREENS	WATERPROOFING	BONDING	DIRECT NAILING	CLEATS/CONCEALED	SELF-TAPPING SCREWS	AGGREGATE BALLAST	
						•				[1] linear expansion for asphalt shingles given across fiber grain, for temp. range -30°F to +30°F, values approximate.
						•				• highest UL classification: A, except B for wood.
						•				• red cedar: fire retardant treated;
						•				• 30 lbs. underlayment may be required under certain conditions.
						•				• cypress: shingles not readily available.
		○	○			•				• porcelain: may be on aluminum or steel.
		○	○				○	•		• steel shakes: 30 lbs. organic felt underlayment recommended.
		○	○			•				• steel shakes available galvanized only; or galvanized and prefinished.
						•				• Copper used as water stop in joints. 30 lbs. underlayment may be required under certain conditions.
						•				
						•				• concrete tiles: nailing strips recommended.
						•				• choice of colors and shapes available for glazed clay tiles.
						•				
•	○	•	•				•			• usual weight for copper: 16 oz.
•	•	•	○				•			• usual thickness for zinc alloy: .030" to .018".
•	•	•	○				•			• usual gauge for monel: 26 to 22.
•	•	○	○				•			• usual gauge for stainless steel: 28 to 26.
○	○	○	○				•			• usual gauge for terne: 30 to 26.
•	•	○	○				•			• lead also available but seldom used.
○		•	•				•			• terne over galvanized steel needs painting.
•	○	•	•				•			• aluminum, when used as flashing, needs protection.
		•	•					○	•	• usual thickness for aluminum: .050 to .019.
		•	•					○	•	• usual gauge for steel: 24 to 18.

B3.1 ROOFING

B • BUILDING SHELL

SHINGLES & SHAKES: selection checklist

OVERLAPPING UNITS:

■ **Shingles, shakes, tile -** either flat or formed - are generally used in residential or light commercial construction. Their selection is commonly based on:
● appearance,
● durability,
● economy.

■ **Asphalt shingles** are most widely used and are the most economic choice for pitched roofs with slopes of 3 1/2 inches in 12 inches or more.
● Square-tab strip shingles may be used for slopes as low as 1 inch per foot but double underlayment of No. 15 asphalt saturated felt in steep asphalt or mastic should be used for all low slope, below 3 1/2 inches per foot, applications; and exposure should be 4 inches, rather than the normal 5 inch exposure.
● The durability of asphalt shingles varies with
  —weight of shingle - the heavier, the more durable.
  —core material - either saturated and coated organic felts or coated mineral fiber (glass fiber). Glass fibers do not absorb moisture and have higher tensile strength.

■ **Wood shingles and shakes** are made from naturally decay resistant species.
● Visual weathering characteristics may be the primary consideration during selection.
● Durability largely depends on grade and thickness available.
● Minimum recommended slope is 4 inches per foot.
● Sheathing to receive shingles and shakes should be at least 1/2 inch thick plywood.
● Shingles and shakes may also be applied over spaced sheathing.
● Underlayment of No. 15 felt is required for shakes but may be optional for shingles depending on slope.
● For information on materials, refer to WOOD/PRODUCTS.

■ **Slate roofing** is a durable, dense and sound rock. It is a time-tested, weather and fire-resistant material.
● Generally selected for appearance.

● Choice of colors, textures and patterns available; thickness ranges from a minimum of 3/16 inch to 2 inches.
● Underlayment of No. 30 felt is recommended for standard slate roofing, two layers of felt for graduated slate roofing.
● Zinc coated, copper weld or copper nails are used for fastening through machine punched holes in the slate. Nail penetrations should be protected with sealant as well as by overlapping of slate.
● Minimum recommended slope is 4 inches per foot.

■ **Metal shingles, shakes and tile** generally imitate their wood or burned clay counterparts.
● Common finishes are plastisols, siliconized polymers, and porcelain enamel, with porcelain enamel being the most durable.

■ **Clay tile** is available either molded into several shapes or flat.
● Vitrified clay tile has water absorption of 3 percent or less.
  —Fast water runoff is promoted by the hard-fired clay.
  —It can withstand constant freeze/ thaw cycling.
● Minimum recommended slope is 4 inches per foot; double layer of felt set in mastic or hot asphalt should be used at low slopes.
● Wood plank or plywood substrate should be covered with No. 40 or 2 layers of No. 30 asphalt saturated felt with double layers on rough surfaces, hips, valleys and ridges.
● Concrete substrate requires 1x2 wood nailing strips to receive No. 30 minimum felt underlayment, and horizontal battens for securing tile.
● Copper flashing (16 oz.) is commonly used for valleys, fastened with cleats and not soldered. Eave flashing is often an extra layer of 40 pound felt installed with plastic cement. Flashing should extend 4 to 6 inches under tiles, with edges folded at least 1/2 inch to prevent water from getting under the tiles.
● Tiles are fastened with 1 to 3 large-headed copper nails.
● Care must be taken during installation:
  —ensure proper blending of colors.
  —use average width and length dimensions.

■ **Concrete tile** is available either as roll or flat, in shapes similar to clay tile.
● Exposed surfaces generally finished with synthetic oxide pigmented cementitious material.
● Moisture absorption should be investigated; if tiles absorb moisture, roof problems may develop.

SEAMED FLAT METAL SHEETS

■ **Seamed metal sheet roofing is typically selected when appearance is the primary consideration.**
● Though time-proven in performance, the high cost in labor and material often determine the specific metal selected.

■ **Three types of seams are used and are selected basically for appearance - flat, standing, and/or batten.**
● **All seams are detailed** to allow expansion/contraction due to variations in temperature within the material. Standing seam roofs, for example, all use clips to attach panels without penetrating the exposed panel surfaces while moving freely in the longitudinal direction. Transverse movement is taken up in joint configuration.
● Flat seam systems may be applied with the seam running parallel or perpendicular to the roof pitch. However, standing and batten seams must be installed in the direction of the slope so that water runoff is not blocked.
● **Designs which allow uneven or concentrated water runoff should be avoided** because it may result in:
  —differential weathering.
  —localized corrosion in uncoated metals.
  —stains or streaks on coated metals.

■ **Flat metal is available in rolled sheets for field forming only, in preformed panels and/or in prefinished panels.** The choice of metals used includes:
● aluminum (generally preformed and prefinished);
● copper and lead coated copper;
● lead;
● monel;
● terne (coated after installation); and

- steel - painted, stainless, aluminum coated, aluminum-zinc coated (Galvalume), galvanized (prefinished and generally preformed), and coil-coated - paints (polyesters, siliconized polyesters, fluorocarbons) and laminates (acrylic, fluorocarbon) applied to the entire coil in the manufacturing process.
- Materials and finishes should be examined for various performance factors such as:
  - resistance to fading and chalking.
  - adhesion capacity of coatings.
  - gloss.
  - hardness.
  - flexibility.
  - corrosion protection.
  - mar resistance.
  - humidity resistance.

■ Hard lead may be used for batten or standing seam roofs. Applications should have 15 pound asphalt impregnated felt underlayment (lapped 2 inches), with an optional layer of rosin sized paper.

- Concealed fastening is accomplished by using cleats (lead or 16 oz. copper) lead-burned to the back of the lead sheet; countersinking brass screws with washers and filling the countersink with solder or lead burned filling; hard copper wire nails.
- Sheet size is generally 16 square feet (8 feet long by 2 feet wide).
- Sheet thicknesses range from 1/32 to 1/4 inch with respective weights from 2 to 16 psf. 3 pound (3/64 inch thick) lead sheet is usually used for roofing, flashing, gutter linings, ridges, hips, valleys, copings, base flashing, etc.
- Asphalt coatings or felts can be used to separate lead sheet from dissimilar metals, cement or concrete. Aluminum will corrode severely when in contact with lead.

■ **Galvanic action between dissimilar metals must be prevented.** Precautions to be taken during design as well as during installation include:
- All fasteners should be of the same metal or alloy as the roofing.
- When fasteners of a different metal are used to secure the substrate, roofing should be separated from the substrate; isolators generally used include roofing felts and bituminous coatings.

PREFORMED LAPPED METAL PANELS

■ **Roofing panels provide substrate and roofing in a single application.**
- Noninsulated, or single thickness, panels can be used where thermal resistance is not required (e.g., unheated buildings, or when used as a shelter only).
- Insulated panels may be
  - site assembled - sandwich panels insulated with glass fiber, with or without an interior liner; or
  - factory assembled - formed of sheet metal faces with a core of foamed or dense glass fiber insulation.
- For additional information on panel types and coatings, refer to CURTAIN WALLS.

■ **Lapped panels are made of:**
- **Galvanized steel** - factory coated, and roll-formed in various shapes.
- **Aluminum** - either in mill finish, or same coatings as galvanized steel. Color anodized may be available on special orders.
- **Cement bound mineral fiber** - without applied finish, corrugated only.
- **Plastic** - in various integral colors, with optional glass fiber reinforcing and/or protective weathering resistant film; corrugated, and in shapes similar to, or matching, those used for roll-forming metal.

■ Single thickness metal panels are often used for deep fascia and mansard roofing.
- Fascias and mansards are generally available as complete systems with trim pieces, gravel stops, soffit panels, framing systems, etc.
- Mansard panels and framing may also be adapted as roof top equipment screens.
- Materials and finishes are generally the same as for roofing panels.

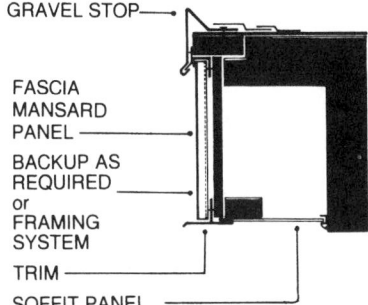

GRAVEL STOP

FASCIA
MANSARD
PANEL

BACKUP AS
REQUIRED
or
FRAMING
SYSTEM

TRIM

SOFFIT PANEL

■ General considerations during selection should include:
- To minimize the possibility of water penetration, roofing panels should have:
  - **standing seam side joints,** elevated sufficiently above the plane of the roofing.
  - **concealed fasteners** to eliminate the need of puncturing the panel. Concealed fasteners should also allow for thermal movement in the panels.
  - **minimal end joints** - placing panels in single length or in maximum length available to eliminate or minimize joints between panels perpendicular to pitch of roof.
  - **minimal flashing** because of the difficulty in providing a durable watertight seal.
  - **prefabricated curbs** to fit various configurations of ribbed roofing panels are available.
- **Fastening details should provide for thermal movement in panels and minimize the effect of through fasteners acting as thermal bridges.**

■ **Steep pitched roofs** in cold climates may accumulate ice within the ribs, which may slide off the open sides of the roof when temperatures rise, endangering passersby. Provision for their safety should be made.

B3.1 ROOFING

FLASHING, TRIM, ACCESSORIES: introduction

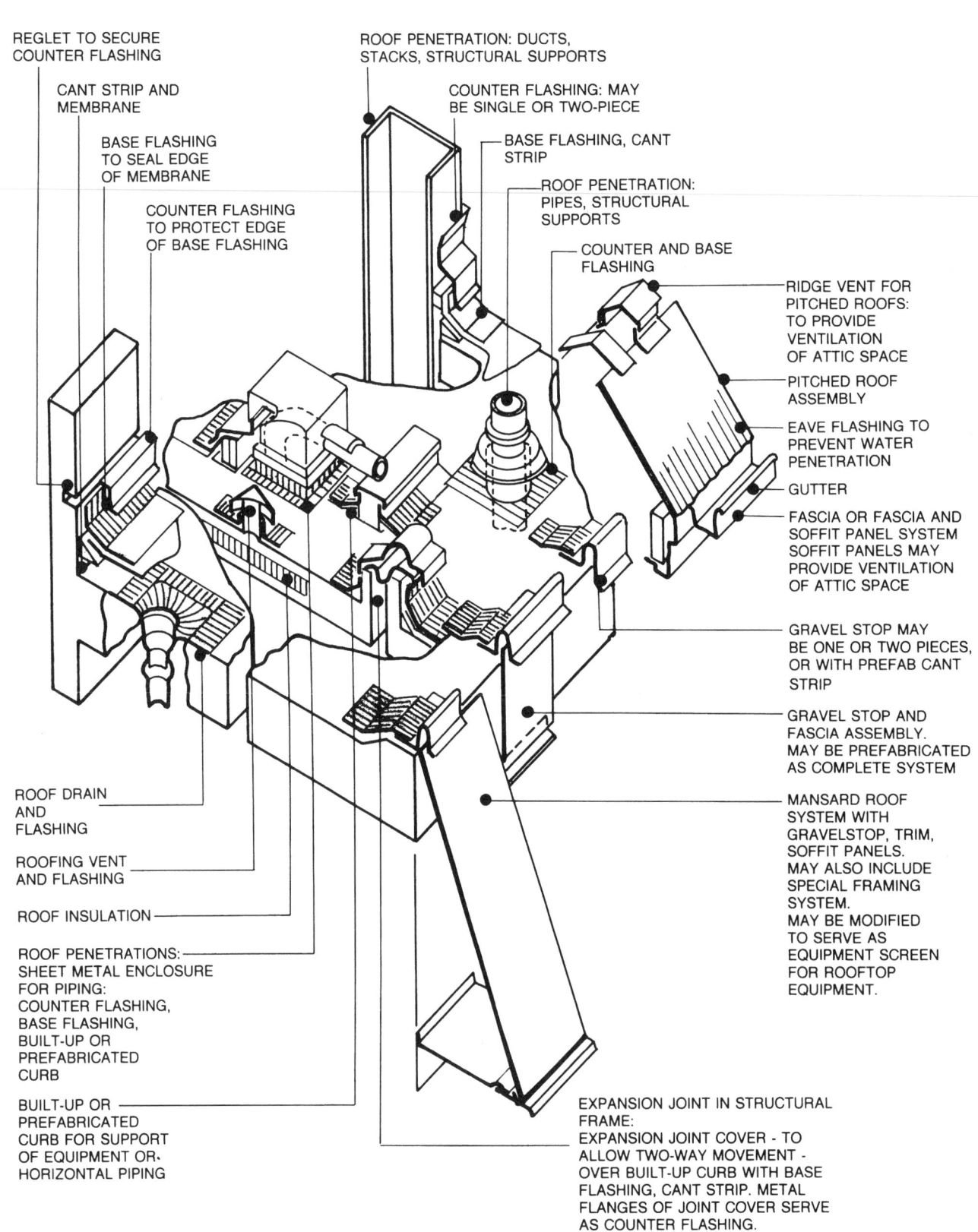

REGLET TO SECURE
COUNTER FLASHING

CANT STRIP AND
MEMBRANE

BASE FLASHING
TO SEAL EDGE
OF MEMBRANE

COUNTER FLASHING
TO PROTECT EDGE
OF BASE FLASHING

ROOF PENETRATION: DUCTS,
STACKS, STRUCTURAL SUPPORTS

COUNTER FLASHING: MAY
BE SINGLE OR TWO-PIECE

BASE FLASHING, CANT
STRIP

ROOF PENETRATION:
PIPES, STRUCTURAL
SUPPORTS

COUNTER AND BASE
FLASHING

RIDGE VENT FOR
PITCHED ROOFS:
TO PROVIDE
VENTILATION
OF ATTIC SPACE

PITCHED ROOF
ASSEMBLY

EAVE FLASHING TO
PREVENT WATER
PENETRATION

GUTTER

FASCIA OR FASCIA AND
SOFFIT PANEL SYSTEM
SOFFIT PANELS MAY
PROVIDE VENTILATION
OF ATTIC SPACE

GRAVEL STOP MAY
BE ONE OR TWO PIECES,
OR WITH PREFAB CANT
STRIP

GRAVEL STOP AND
FASCIA ASSEMBLY.
MAY BE PREFABRICATED
AS COMPLETE SYSTEM

MANSARD ROOF
SYSTEM WITH
GRAVELSTOP, TRIM,
SOFFIT PANELS.
MAY ALSO INCLUDE
SPECIAL FRAMING
SYSTEM.
MAY BE MODIFIED
TO SERVE AS
EQUIPMENT SCREEN
FOR ROOFTOP
EQUIPMENT.

ROOF DRAIN
AND
FLASHING

ROOFING VENT
AND FLASHING

ROOF INSULATION

ROOF PENETRATIONS:
SHEET METAL ENCLOSURE
FOR PIPING:
COUNTER FLASHING,
BASE FLASHING,
BUILT-UP OR
PREFABRICATED
CURB

BUILT-UP OR
PREFABRICATED
CURB FOR SUPPORT
OF EQUIPMENT OR·
HORIZONTAL PIPING

EXPANSION JOINT IN STRUCTURAL
FRAME:
EXPANSION JOINT COVER - TO
ALLOW TWO-WAY MOVEMENT -
OVER BUILT-UP CURB WITH BASE
FLASHING, CANT STRIP. METAL
FLANGES OF JOINT COVER SERVE
AS COUNTER FLASHING.

B3.1 ROOFING

TRIM, ACCESSORIES: types, uses

DESCRIPTION

GRAVEL STOP, FASCIA

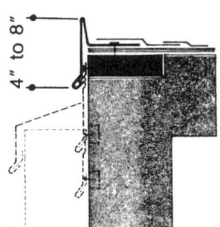

4" to 8"

MATERIALS: Galvanized steel, generally coated; aluminum, brake formed or extruded, either mill finished, coated, or anodized; stainless steel; copper; zinc alloy; terne.

INSTALLATION: Horizontal leg nailed to blocking or substrate; light gauge metal requires closely spaced nailing, heavier gauge secured at splices and at midpoint of individual sections. Length of sections generally eight to ten feet. Thermal movement accommodated by lapping ends of sections or leaving gap at splice plates. Cleats for light gauge over four inches deep generally required to prevent wind uplift.

REMARKS:
- Prefabricated scuppers and fascia sumps available.
- Systems with integrated cant strips, generally of galvanized steel, are available.

FASCIA SYSTEM

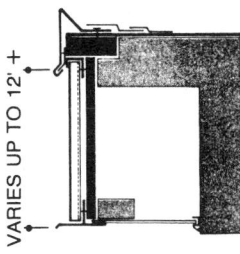

VARIES UP TO 12' +

MATERIALS:
- Prefabricated panels: galvanized steel, roll formed and coated with siliconized polymers, plastisols, or in porcelain enamel, aluminum, roll formed, generally coated with siliconized polymers or plastisols, may also be clear or color anodized, but availability may be limited; proprietary zinc alloys available. Prefabricated panels: with standing or batten seams and flat or formed pans, or roll formed in various configurations with lapped or interlocking edges. Concealed fasteners commonly used, but may be installed with exposed fasteners.
- Site fabricated panels generally of terne coated steel, terne coated stainless steel, copper. Joints used: flat lock, standing and batten seams. Secured to solid substrate/backup with concealed cleats.
- Gravel stops, trim pieces generally of the same material as panels.

MANSARD SYSTEM

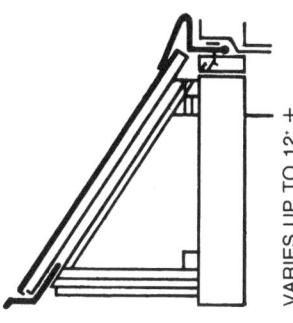

VARIES UP TO 12' +

INSTALLATION:
- Prefabricated panels generally secured to open framing; framing members may be supplied as part of the fascia/mansard system.
- Site fabricated panels require framing with nailable sheathing. Roofing felt or rosin coated paper underlayment may be required to isolate metal from sheathing.

REMARKS
- Mansard systems are available with integrated soffit panels.
- Mansard systems including framing available for use as screens for rooftop equipment.

VENTS

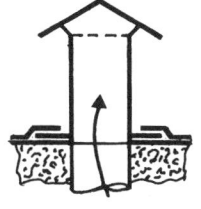

ROOFING VENT　　　　　ROOF VENT

ROOFING VENT:
- Used to allow moisture trapped within the membrane assembly to be vented to the outside.
- Sheet metal, such as galvanized steel, aluminum, stainless steel vents generally open at all times and will allow moisture to be pulled into the roof membrane as well when differences in water vapor pressure develop.
- Plastic vents available which open under outward water vapor pressure, and close to prevent backflow into membrane assembly.
- Roofing vents only effective when water vapor can migrate horizontally through membrane assembly.

ROOF VENT:
- Used to allow heated air collecting at underside of roof assembly to escape.
- Various types available.

B3.1 ROOFING

FLASHING: types, uses

DESCRIPTION

COUNTER FLASHING

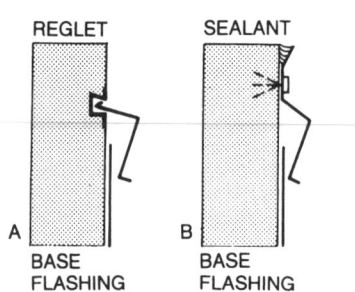

REGLET   SEALANT

A   B
BASE   BASE
FLASHING   FLASHING

MATERIALS:
- Type A = Snap-in counter flashing: aluminum, stainless steel, copper, lead coated copper. Galvanized steel not commonly used due to possibility of corrosion; available prefinished. Proprietary alloys may also be used. Aluminum type should be isolated from dissimilar materials.
- Type B = Surface mounted, either one or two-piece. Commonly available in same basic materials as type A

REMARKS:
- For information on reglets refer to BEARING WALLS.
- Type B assembly requires field installed sealant to prevent water from penetrating between wall and reglet/counter flashing.
- Counter flashing may also be built into a masonry wall, but such practice is not recommended since counter flashing would have to be bent up for installation of base flashing and thus exposed to possible damage.

BASE FLASHING, FLEXIBLE-ATTACHED

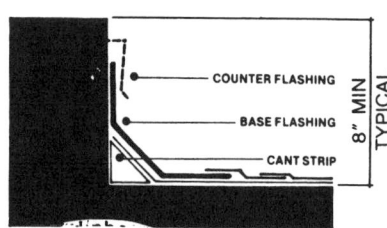

COUNTER FLASHING
BASE FLASHING
CANT STRIP
8" MIN TYPICAL

MATERIALS:
- For built-up membranes: inorganic or organic saturated and coated felts, impregnated glass fiber felts, cotton or glass fiber fabrics embedded in roofing cement in field, elastomerics such as polyvinyl chloride.
- For elastomeric and modified bitumen membranes: most are self-flashing, or use uncured polychloroprene; or chlorosulfonated polyethylene.

REMARKS:
- Bitumens for securing felted flashing to vertical surfaces must have high softening point to prevent slippage; nailing in addition to bonding is recommended.
- Some elastomers and modified bitumen may be installed with bituminous base adhesives; most will require special adhesives, generally available from manufacturer of flashing materials.

BASE FLASHING, METAL

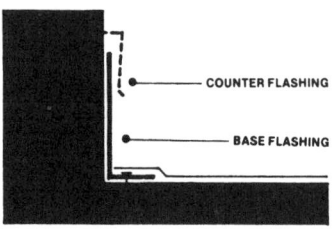

COUNTER FLASHING
BASE FLASHING

MATERIALS AND REMARKS
- Aluminum, copper, stainless steel, galvanized steel. The use of metal base flashing in conjunction with built-up roof membranes is not recommended: since the membrane has to be solidly bonded to the metal, thermal expansion/contraction between the two materials will be differential and may split the built-up membrane.
- Elastomeric coated metal base flashings are used with elastomeric single-ply membranes. Coating is generally of the same material as the membrane, and flashing is supplied by manufacturer of membrane material.
- Aluminum, copper, stainless steel, galvanized steel are used with overlapping units-shingles, shakes, tile-and lapped metal panels.

BASE FLASHING, FLEXIBLE-ISOLATED

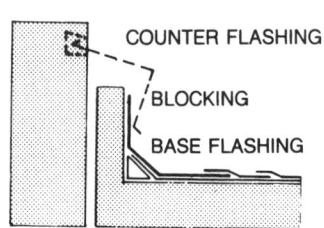

COUNTER FLASHING
BLOCKING
BASE FLASHING

MATERIALS:
- Same materials as listed under BASE FLASHING, FLEXIBLE-ATTACHED above.

REMARKS:
- Base flashings, when attached to both a vertical element such as a wall assembly abutting a roof assembly and to a roofing membrane solidly bonded to such assembly, may be subject to differential movement between the two unless the roof assembly is supported by and connected to the vertical element. Even then the exposure to solar radiation and/or the coefficients of expansion of the two may be sufficiently different for differential movement to occur.
- Base flashings for membranes over structural frame which is independent of vertical elements it abuts should be completely separated from such vertical elements.

B3.1 ROOFING

DESCRIPTION

PIPES, STRUCTURAL SUPPORTS

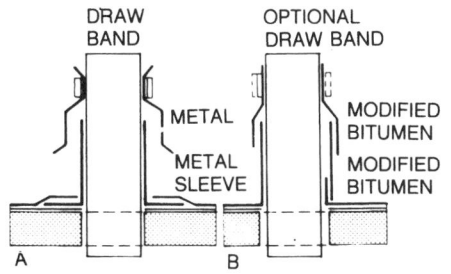

MATERIALS AND REMARKS:

- Type A: galvanized metal, aluminum, copper, stainless steel. Galvanized metal subject to corrosion; aluminum must be isolated from contact with dissimilar metals/materials. Sleeve may be of the same material or of lead. When base and/or cap flashing has to be installed in two or more sections, materials which can be field joined by soldering or welding should be selected. Wood blocking to secure base flashing may be required.
- Type B: modified bitumen. Generally heated with a torch to soften and allow it to bond to the penetration and to the membrane. Penetration must be firmly anchored to the structural frame or substrate to prevent movement in them.

EQUIPMENT SUPPORTS, CURBS

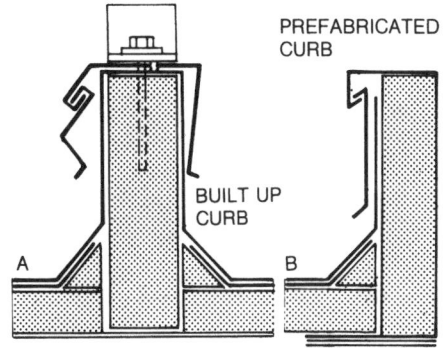

MATERIALS:

- Type A: built-up curb or equipment support generally of treated lumber. Base and counter flashing generally of the same materials as used at perimeter of membrane.
- Type B: prefabricated curbs or equipment supports generally of galvanized sheet metal covered wood frame, with or without insulation.

REMARKS:

- Curbs and equipment supports are connected to either decking or framing of the structural frame to transfer all loads to such frame; secondary framing may be required.
- Additional flashing is recommended for prefabricated curbs with counter flashing flanges to protect the base flashing.

CHANGE OF SLOPE

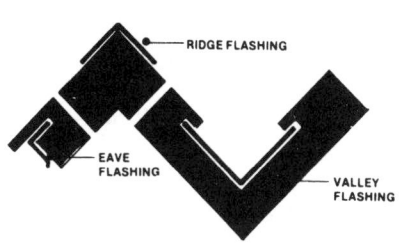

MATERIALS:

- For built-up membranes: same as base flashing.
- For elastomeric and modified bitumen single-ply membranes: membrane material or same as base flashing.
- For shingles, shakes, tile: glass fiber, or organic felts; galvanized steel, copper, stainless steel; aluminum when coated to protect it from galvanic action; elastomeric, such as polyvinyl chloride.
- For lapped metal panels: same material as panels.

REMARKS:

- Consideration must be given to differential movement between flashing and roofing; metal flashing may have to be installed using cleats to allow for thermal expansion/contraction.

EXPANSION JOINT COVER

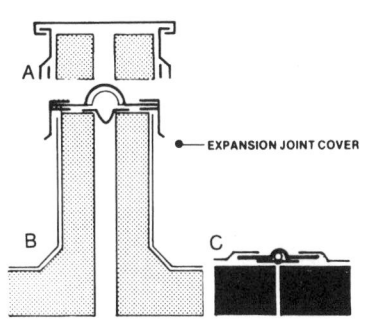

MATERIALS:

- Type A: aluminum, stainless steel, copper. Aluminum generally extruded, stainless steel and copper may be brake formed.
- Type B: elastomeric, generally with metal flanges.
- Type C: low profile joint cover, generally elastomeric.

REMARKS:

- Type A when modified, and Type B may be used at intersections of roof assemblies with vertical elements when expansion/contraction may occur at such locations.
- Type C will allow limited movement to take place; may be used when control joints in membrane between building expansion joints are required, or to repair cracks in membrane and substrate, but Type C joint covers are not recommended, due to possibility of water leakage through them.

B3.1 ROOFING

FLASHING, TRIM, ACCESSORIES:

In selecting and detailing flashing:

■ **Consideration should be given to all exposed flashing:**

- material appearance and weathering.
- layout of seams.
- joints.
- connections to roofing, wall facings, roof accessories - curbs, piping, etc.

■ **Allowance should be made for differential movement between roofing and other parts of the assembly:**

- base flashing should not be anchored to parapet walls unless the parapet wall and the substrate are continuously connected and cannot expand/contract independently.
- counterflashing should not be connected to base flashing unless the possibility of relative movement between them can be positively prevented.

■ **Flashing in-service life expectancy should at least match that of roofing selected:**

- flashings usually have to perform under more severe conditions than roofing.
- repair/replacement of flashings is costly; selecting inferior materials to realize initial savings is generally false economy.

■ **Installation and performance requirements are:**

- flexibility: to conform closely to surfaces it is attached to.
- elasticity, or tensile strength, to accommodate thermal and load induced movements.
- compatibility with roofing and adjoining materials.

■ **Flashings should be located to:**

- direct flow of water away from them; if such conditions cannot be avoided, crickets or saddles should be considered to divert water flowing towards the flashing.
- keep top of flashing above the highest level of water buildup on the roofing.
- recommended minimum for all types of flashing is eight inches.

In evaluating materials for flashings, consideration should be given to:

■ **Felted flashings:**

- are generally more durable than

fabric flashings; but
- are less puncture resistant.
- when installed with hot-mopped asphalt they require asphalt of a high softening point to avoid the possibility of sagging or slippage; this may present an inconvenience when bitumen with a low softening point is used for the membrane.
- should never be bent through a right angle as cracking at low temperatures may result.

■ **Fabric flashings, though generally less durable, are:**

- more puncture resistant.
- more easily molded to surfaces to be flashed.
- bituminous coating for fabric flashing may need periodic maintenance; if coating becomes too heavy, alligatoring may take place.

■ **Felted and fabric flashings should not be bonded to metals used in flashing systems:** differential movement between metal and fabrics or felts may cause tears or splits in fabric to develop.

■ **Elastomerics,** when compatible with bituminous materials, may not be suitable for bonding with bituminous coatings; even if they can be so bonded, special adhesives may have to be used at joints to connect individual sections of elastomerics.

■ **Metal and felted flashings** are used at roof drains; where there is need for stability and where the flashing must be well secured to the frame of the drain; lead or copper are generally used.

- in metal flashings, allowance must be made at joints to allow for thermal expansion/contraction, especially when exposed.
- all penetrations through roofing including small diameter piping should preferably be protected with base flashing and counterflashing, pitch pockets require periodic maintenance to perform satisfactorily and are not recommended.

■ **Modified bitumen** is heated using a torch to soften it so that it can be molded to conform to the shape of substrate and properly bond to such substrate.

WOOD NAILERS: provide protection for the edges of roof insulation and also a surface to which flashings can be anchored.

- nailers should be treated for decay resistance; oil borne preservatives used in some decay resistant treatments may act as solvents on roofing materials and cause bitumen drippage.
- when wood thus treated is used, it should be isolated from bituminous materials by rosin-sized paper or similar barriers.

GRAVEL STOPS

Gravel stops must perform several functions:

■ **Protect the edge** of the membrane.

■ **Serve as exposed trim,** or the top section of a fascia.

■ **Resist wind uplift** and be stiff enough to resist deformation or oil-canning between joints.

■ **Attachment of gravel stop to roofing** must be detailed to fulfill the following requirements:

- must be sufficiently frequent and secure to restrain and limit buckling of gravel stop due to its expansion and contraction which differs from that of the membrane and the substrate.
- must at the same time allow the vertical face of the gravel stop to move freely in order to prevent oil-canning and/or buckling.
- must prevent bitumen flowing out of the membrane, running down behind the gravel stop, and eventually staining the exterior wall.

GUTTERS AND ROOF DRAINS

Gutters collect water running off a pitched roof at eaves and control its flow to the ground in order to:

- prevent erosion of soil where the water would otherwise hit.
- protect people and objects in the vicinity of the building.

■ **Gutters may be:**
- suspended outside of fascia boards; or
- built-in into the roof assembly.

■ **Suspended gutters** are generally prefabricated of: galvanized steel, aluminum, polyvinyl chloride, copper.

■ **Other materials** used for the fabrication of gutters for specific applications are: zinc, terneplate, stainless steel, monel:

selection checklist

- wood gutters are infrequently used and no longer readily available.

In selecting materials for prefabricated gutters, consideration should be given to:

■ **Resistance to weathering:** is protection and/or frequent maintenance required; for example, galvanized steel may require painting or coating within a relatively short in-service period.

■ **Ease of installation:** for example, aluminum cannot be soldered in field; expansion joints need special detailing.

■ **Compatibility of materials:**

- all materials, such as fasteners, supports, brackets, leaders should be of the same or compatible metal to prevent galvanic action.
- metals should also be isolated from incompatible adjacent surfaces.

Consideration during selection should also include:

■ **Shape of the gutter:**

- should allow for expansion of any water that may freeze in it.
- vertical faces should not be parallel to one another or the gutter should be of a rounded cross section.
- the outer face should be at a lower level than the inner to prevent water spilling out or running down the fascia or wall.

■ **Size of gutter** to ensure rapid runoff; the determining factors are:

- area of roof being drained.
- size and spacing of leaders.
- intensity and duration of heaviest rainfall recorded.

■ **Flashing for built-in gutters** and connections to leaders allow access for future maintenance/replacement:

- only materials highly resistant to weathering should be used.
- installation must allow differential movement in gutter, roofing, and substrate to take place.

■ **Whether use of gutters in areas experiencing severe freezing and thawing is advisable:**

- ice may form within and on gutters in sufficient quantities to pull them down.
- ice dams may form at gutters causing water to leak through the roofing to the interior.

■ **Possibility of water running off** gutters on to adjacent surfaces and staining them:

- copper will stain adjacent surfaces if runoff is not prevented.
- galvanized steel and terneplate, when the protective coating is damaged or worn-off, will also stain adjacent surfaces.

■ **Pitching gutters** to leaders to prevent water standing in low spots.

Roof Drains

Roof drains, installed into roofing and connected to interior leaders, drain flat roofs. Roof drains are available in metal and plastic.

Roof drains should be:

■ **Flashed in properly into roofing:**
- lead or copper flashing should be used.
- drains with integral flashing collars are available.
- use of plastic drains should consider the higher coefficient of expansion of plastics and possible effects of it on adjacent roofing membrane.

■ **Located at low points of the roof to prevent water ponding on the roofing:**

- drains located near columns in a level roof may turn out to be at high points of the roof since there will be no deflection in framing at such locations.
- drains at midspan of framing for a level roof may also be at high points if framing members have been fabricated with an excessive camber.
- creep in reinforced concrete framing members and decks may, over a period of time, change the runoff pattern of an initially level roof.
- drains should be sized and spaced to ensure rapid runoff of water; smaller, closer spaced drains are preferable to larger but fewer drains.

Other considerations during selection of drains might be:

■ **Drains may be also selected for delayed drainage** when capacity of storm sewers is insufficient to take full flow. Care must then be taken to prevent excessive accumulation of water on the roof.

■ **Roofs with parapets** should, in addition to drains, have overflow scuppers in the parapet walls, this in order to prevent water accumulating on the roof to excessive depth, should roof drains be clogged or frozen.

■ **Determining factors in the sizing of roof drains are:**

- area of roof to be drained.
- intensity and duration of heaviest rainfall recorded.

■ **Connection to leaders** may have to be flexible to allow for movement in drains due to live-load-induced deflections of the roof assembly.

■ **It is recommended that positive drainage to drains be provided;** water ponding on roofing may damage or destroy the roofing membrane.

Roof Accessories

■ **Manufactured curbs** for use at larger penetrations through roofs and as supports for equipment.

■ **Roofing vents used:**

- occasionally with new roofing membranes to allow moisture to escape.
- on old roofs in the attempt to dry them out.

■ **Solar powered roof pumps** to drain water in low spots on roof.

GUTTERS AND DOWNSPOUTS

ROOF DRAINAGE OPTIONS

Direct Drainage. In many locales, it is common to allow water to drain directly off the edges of steep roofs. Except in very dry areas, direct drainage should be used in conjunction with wide overhangs and, where there is a basement, a drip bed at the ground and subsurface drainage. Direct drainage is especially common is regions subject to severe cold weather, where ice and snow may remain on roofs for long periods and where gutters and downspouts don't function in winter. Water diverters may be used to avoid discharge of water over doorways and other pedestrian walkways.

Gutters and downspouts. The most common type of roof drainage for steep roofs is a combination of gutters and downspouts (the words "leader" and "downspout" both refer to pipes conducting water down from a roof). The gutters should be designed to accept all normal roof runoff, and the downspouts should be able to discharge all water, which flows to them quickly.

Diverters. Water diverters are sometimes used in place of gutters. Instead of being mounted under the eaves, water diverters are mounted above the roof plane. While diverters do not extend to the edge of the eaves, and thus allow some dripping, they are prone to clogging.

Waterspouts. Waterspouts or gargoyles may lead the water out from the wall and allow it to drip into a pool or drip bed below.

Safety precautions. All roof drain systems in cold regions may form icicles, except internal leaders in heated buildings. Therefore one should not locate sidewalks under the eaves subject to ice formation. Alternatively, a good quality electric snow melting system in the gutters and downspouts will, when it functions, avoid icicles.

MATERIAL USED FOR GUTTERS AND DOWNSPOUTS

■ **Copper and lead-coated copper** have a long history of successful use on roofs. They are easy to form and solder. They are highly ductile, and so they tend to yield harmlessly when stressed. In those parts of the country subject to acid deposition (acid rain residue and directly-deposited acid aerosol), dew and mist may dissolve acid deposits on roofs and wash the concentrated acid onto sheet metal. See CORROSION OF METALS.

SOME PROTECTIVE METHODS INCLUDE:

• **Detail the roof** so that no water flows from other, nonreactive materials onto copper and lead-coated copper. A roof made entirely from copper is one way to accomplish this.

• **Use thicker copper,** such as 20-ounce, and plan for replacement within 15-20 years.

• **Use sheet metal not subject to acid corrosion,** such as aluminum, stainless steel, and Terne-coated stainless steel for gutters and downspouts.

• **Install sacrificial zinc anodes** at the eaves, where the acid runs into the gutters.

■ **Aluminum is vulnerable to attack by hydroxyl ions,** but concentrated hydroxyl ions are rare on roofs. Aluminum is not vulnerable to attack from the concentrations of acid, which occur on roofs. Although very thin aluminum is used on residences, the SMACNA Manual recommends the proper minimum gauge for architectural use. Aluminum is available with highly durable finishes, including fluorocarbon polymer coatings.

■ **Galvanized steel, aluminized steel, and aluminum-zinc alloy coated steel may not be satisfactory for gutters,** downspouts, valleys, and other roofing sheet metal exposed to concentrated runoff in those parts of the country with frequent rainfall and acid deposition. The protective coating may be eroded rapidly. They have been used successfully in other parts of the country, however, and if properly maintained and periodically painted, they give good service. They have the advantages of economy, strength, and low coefficient of thermal expansion. Durable finishes are available.

■ **Stainless steel and Terne®** (Terne-coated stainless steel) are highly durable metals. Terne-coated stainless steel is glossy upon first exposure, but turns matte gray with exposure.

Stainless steel is subject to stress hardening, and thus it is harder to work than copper, and it may rip rather than yield when exposed to great stress.

■ **Zinc and zinc alloys** have many of the same characteristics as copper. They are durable and easy to form and solder. Monel® is a very durable metal occasionally used for gutters and downspouts.

■ **Wood.** Gutters may be made from decay-resistant woods, and in selected uses have given decades of satisfactory use. Wooden or other gutters, lined with single-ply roof membrane material and installed to be easily replaced, give durable and beneficial service.

TYPES OF GUTTERS AND WATER DIVERTERS

Built-in gutters have been common, especially on monumental buildings with steep roofs. Built-in gutters allow little room for error; such gutters must be fabricated and installed following the recommendations of the SMACNA Architectural Sheet Metal Manual.

Metal built-in gutters. The simplest type of gutter, and generally the least troublesome, is a sheet metal hanging gutter installed under the eaves. These gutters may be half-round, rectangular, or other shapes.

• Metal built-in gutters, being stiff, require provisions for thermal expansion and shrinkage. Expansion joints should be designed between the drains, and the seams should be very firmly fastened, as with rivets as well as solder.

ALTERNATIVE DETAILS

• Many of the problems of **sheet metal built-in gutters** can be avoided by using flexible materials. These materials don't require expansion joints, and they are not subject to corrosion. They are relatively easy to repair and replace.

• **Wide, shallow gutters** do not usually become clogged with leaves and other debris; the wind helps clear them. Most gutters, however, do become clogged, and cleaning of the gutters may be expensive and difficult.

• One way to avoid the need for frequent cleaning is to **mount a screen** over the gutter. A simple screen installation.

B3.1 ROOFING

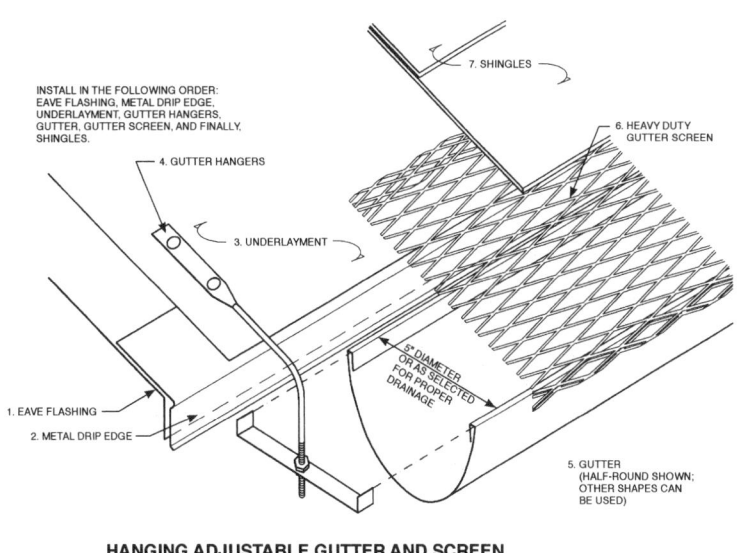

INSTALL IN THE FOLLOWING ORDER:
EAVE FLASHING, METAL DRIP EDGE,
UNDERLAYMENT, GUTTER HANGERS,
GUTTER, GUTTER SCREEN, AND FINALLY,
SHINGLES.

7. SHINGLES

6. HEAVY DUTY
GUTTER SCREEN

4. GUTTER HANGERS

3. UNDERLAYMENT

1. EAVE FLASHING

2. METAL DRIP EDGE

6" DIAMETER
OR AS SELECTED
FOR PROPER
DRAINAGE

5. GUTTER
(HALF-ROUND SHOWN;
OTHER SHAPES CAN
BE USED)

HANGING ADJUSTABLE GUTTER AND SCREEN
SCALE: 6" = 1'-0"   DWC: DETAILS\GUTTER

Fig. 3. Hanging adjustable gutter and screen (SMACNA 1993)

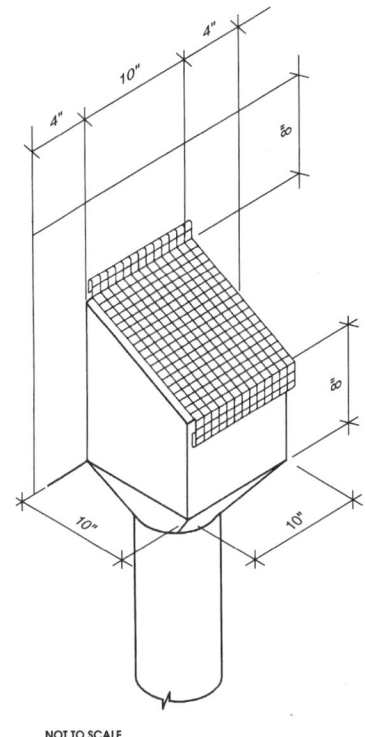

NOT TO SCALE

Fig. 4. Isometric of simple basic leader head with screen

- **Adjustable gutter hangers** allow adjustment after installation, so that the gutters will drain thoroughly.

- Gutters can make ice dams more troublesome. Eave flashing should extend behind hung gutters and, for built-in gutters, under the gutters.

Scuppers. Another type of drainage is a scupper, leader head or "conductor head," and downspout. To avoid stopping up of the gutters and upper downspouts, design an air gap between the scupper and the top of the leader head.

Water diverter. In place of gutters, water diverters may be used. They are less likely than gutters to become clogged with leaves, and they are easier to keep clear.

DOWNSPOUTS

Electrical resistance heaters in the downspouts, as well as the gutters, will melt ice and help clear the downspouts. Such equipment uses energy, and it is expended only when required by freezing conditions.

For cold climates, open front downspouts are recommended by SMACNA. Open front downspouts will develop icicles in cold weather. The icicles

will, however, melt faster than the ice inside closed downspouts.

Downspouts should not discharge directly on the ground near foundations of buildings with basements or crawl spaces. Proper methods of discharge include:

- **Boots** connected to subsurface drain lines. Do not discharge rainwater into perforated footing drains; the two should be separate. Drain boots serving open-front leaders should have sloped screens above the boots, to prevent leaves from clogging the underground drain lines.

- **Drained drip beds,** trenches, surface gutters, swales, French drains, and other surface and shallow drains.

Downspout straps and their fasteners should be heavy duty and resistant to corrosion; in cold climates they may have to support a column of ice.

REFERENCES: Donald Baerman *"Gutters and Downspouts"* (Reference 1).

SMACNA. 1993. Architectural Sheet Metal Manual - Fifth Edition. Sheet Metal and Air Conditioning Contractors' Association, 4201 Lafayette Center Drive, Chantilly, VA 20151.

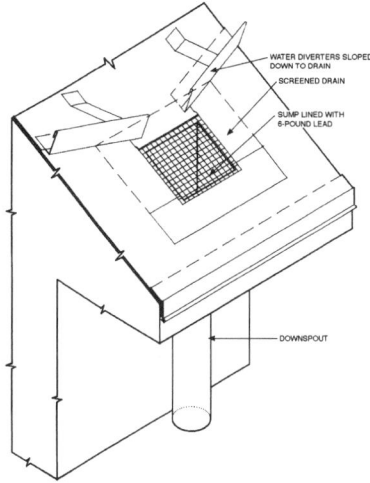

WATER DIVERTERS SLOPED
DOWN TO DRAIN

SCREENED DRAIN

SUMP LINED WITH
8-POUND LEAD

DOWNSPOUT

ISOMETRIC DETAIL OF DRAIN @ SLOPED EAVES
NOT TO SCALE

Fig. 8. Water diverter and drain.

B3.1 ROOFING

REFERENCES: Roofing

STANDARDS

ASTM Specifications for:

C222 • Asbestos-Cement Roofing Shingles.

D224 • Smooth-Surfaced Asphalt Roll Roofing (Organic Felt).

D225 • Asphalt Shingles Surfaced With Mineral Granules.

D226 • Asphalt-Saturated Organic Felt Used In Roofing and Waterproofing.

D227 • Coal-Tar Saturated Organic Felt Used In Roofing and Waterproofing.

D249 • Asphalt Roll Roofing (Organic Felt) Surfaced With Mineral Granules.

D250 • Asphalt-Saturated Asbestos Felt Used In Waterproofing.

D312 • Asphalt Used In Roofing.

D371 • Asphalt Roll Roofing (Organic Felt) Surfaced With Mineral Granules, Wide-Selvage.

D450 • Coal-Tar Pitch Used In Roofing, Dampproofing and Waterproofing.

D1227 • Emulsified Asphalt Used As a Protective Coating for Built-up Roofing.

D1863 • Mineral Aggregate Used On Built-up Roofs.

D2178 • Asphalt Glass (Felt) Mat Used In Roofing and Waterproofing.

D2626 • Asphalt-Saturated and Coated Organic Felt Base Sheet Used In Roofing.

D2822 • Asphalt Roof Cement.

D2823 • Asphalt Roof Coatings.

D2824 • Aluminum-Pigmented Asphalt Roof Coatings.

D3018 • Class A Asphalt Shingles Surfaced With Mineral Granules.

D3468 • Liquid-Applied Neoprene and Chlorosulfonated Polyethylene Used In Roofing and Waterproofing.

D3672 • Venting Asphalt-Saturated and Coated Inorganic Felt Base Sheet Used in Roofing.

D3747 • Emulsified Asphalt Adhesive for Adhering Roof Insulation.

D3909 • Asphalt Roll Roofing (Glass Mat) Surfaced With Mineral Granules.

D4434 • Poly (Vinyl/Chloride) Sheet Roofing.

D4479 • Asphalt Roof Coatings— Asbestos Free.

D4637 • Vulcanized Rubber Sheet Used in Single-Ply Roof Membrane.

ASTM Tests/Methods for:

D451 • Sieve Analysis of Granular Mineral Surfacing For Asphalt Roofing Products.

D1187 • Asphalt-Base Emulsions for Use as Protective Coatings for Metal.

E108 • Method of Fire Tests of Roof Coverings.

SELECTED REFERENCES

• An investigation into the causes of BUR Failures.
Donald F. Brotherson Research Report 61-2.
University of Illinois Small Homes Council Building Research Council.

APA — **American Plywood Association**

• APA Design/Construction Guide: Residential & Commercial

• APA Design/Construction Guide: Non-Residential Roof Systems

• Warehouse Roof Cost Estimating Sheet, and Illustrative Example

NRCC — **National Research Council of Canada**

CBD-24 • Built-Up Roofing
28 • Wind on Buildings
37 • Snow Loads on Roofs
49 • New Roofing Systems
65 • Mineral Aggregate Roof Surfacing
67 • Fundamentals of Roof Design
68 • Wind Pressures and Suctions on Roofs
69 • Flashings for Membrane Roofing
70 • Thermal Considerations in Roof Design
73 • Moisture Considerations in Roof Design
74 • Properties of Bituminous Membrane
75 • Roof Terraces
89 • Ice on Roofs
112 • Designing Wood Roofs to Prevent Decay
151 • Drainage from Roofs
176 • Venting Flat Roofs
179 • Inspection & Maintenance of Flat Roofs
181 • Shrinkage of Bituminous Roofing Membranes
193 • Estimating Snow Loads on Roofs
202 • Joints in Conventional Bituminous Roofing Systems
211 • Bituminous Roofing Membranes-Practical Considerations.

FM — **Factory Mutual**

Bulletin
1-7 • Wind Forces on Buildings
1-28 • Insulated Steel Decks
1-29 • Loose Laid Single Ply Systems
1-47 • Roof Coverings
1-48 • Repair Procedures for Built-Up Roof Coverings over Steel Decks
1-49 • Perimeter Flashing
1-50 • Zip Rib Roofing and Siding
1-52 • Field Uplift Tests
1-54 • Roof Collapse

4470 • Class I roof covers.

NBS — **National Bureau of Standards**

• Effects of Thermal Shrinkage on Built-Up Roofing

NRCA — **National Roofing Contractors**

• The NRCA Roofing & Waterproofing Manual
• Roofing Materials Reference & Guide, Vol. 3, 1983.
• The Built Up Roof
• Roof Design, Application and Maintenance M. C.

B3.1 ROOFING

Baker, distributed by
NRCA
* Roofing Spec Magazine

NTIS **National Technical
Information Service**
* Mechanical Performance
of Built-Up Roofing Mem-
brane
* Fire Endurance Test for a
Roof/Ceiling Construction
of Paper Honeycomb and
Gypsum.

RCSHSB **Red Cedar Shingle &
Handsplit Shake Bureau**
* Red Cedar Shingles and
Shakes Test Methods
for Fire Resistance of
Roof Covering Materials.
* Bulletin of Research for
Determining Wind-Uplift
Resistance of Roof
Assemblies.

UL **Underwriters Laboratories**
* Test Methods for Fire
Resistance of Roof
Covering Materials.
* Bulletin of Research for
Determining Wind-Uplift
Resistance of Roof
Assemblies.

B • BUILDING SHELL

B3.1 ROOFING

B • BUILDING SHELL

B3 • ROOFING ASSEMBLIES

B3. 2 SKYLIGHTS

SKYLIGHTS, SCUTTLES, VENTS: introduction

DAYLIGHTING, VENTILATION

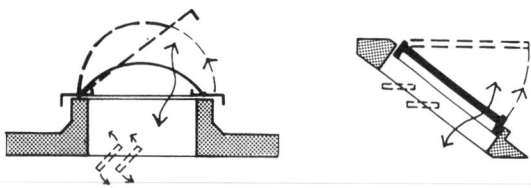

■ **Skylights provide the simplest means of introducing daylight** into a space under a flat or pitched roof assembly. Daylight entering through a skylight may be controlled by:

* selecting translucent glazing.
* addition of fixed or movable control devices.

■ **Skylights may also be hinged or pivoted to allow for air movement through them:**

* hinged types are often referred to as roof windows.

ACCESS TO ROOF

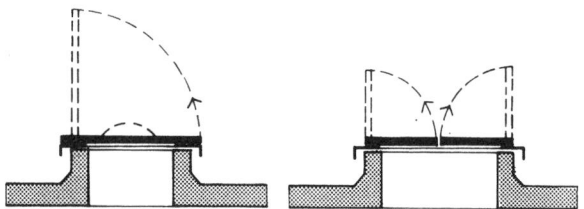

■ **Scuttles** which are also often referred to as roof hatches provide access to roof for:

* maintenance personnel only, using ladders, ship ladders, or stairs.
* emergency escape in the event of fire thus functioning as a required exit from the building.
* moving large equipment in or out of the building.

■ **May also be used as a scuttle and skylight combined when glazed.**

FIRE/EXPLOSION VENTING

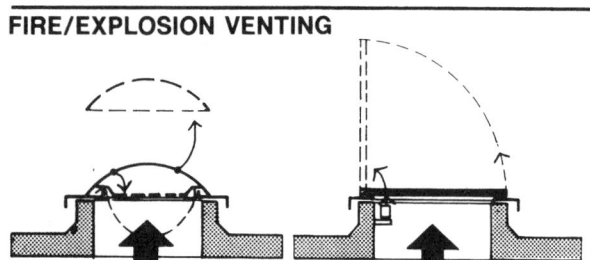

■ **Fire vents** also often referred to as smoke and fire, or smoke and heat vents function to reduce interior heat buildup during a fire by:
* opening automatically when interior temperatures rise above a predetermined level.

■ **Explosion relief vents** open automatically when interior pressure rises above a predetermined value.

■ Fire vents and explosion relief vents may also function as skylights when glazed.

SAFETY

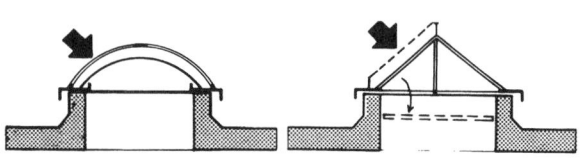

■ **Skylights and glazed fire vents** may be exposed to impact of:
* maintenance personnel accidentally falling on them.
* windblown or intentionally thrown objects crashing into them.

■ **Protection may be required to:** prevent persons from falling through the skylight, shield the glazing from falling objects, shield occupants below from falling broken glazing.

EXTERNAL FACTORS

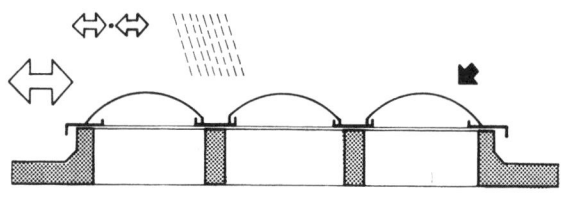

■ **Large skylights** and assemblies of skylights will be exposed to the same external factors as the roof assembly and may grow large enough to function as the total roof assembly.
* skylights will then have to resist wind pressures both postive and negative acting on the framing and glazing; rain water penetration; live loads of snow, maintenance personnel; dynamic loads of impact; provide means for draining rainwater, melting snow; safely resist thermal expansion/contraction in the framing and glazing.

■ **All skylights,** whether small units or large assemblies have to: control heat flow through them, and prevent or safely drain condensation of water vapor which may occur on them under severe temperature and/or humidity conditions.

■ **Framing and glazing** may also have to resist exposure to airborne pollutants without appreciable detrimental effect on their intended functions.

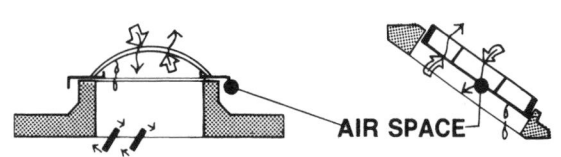

AIR SPACE

B3.2 SKYLIGHTS

SKYLIGHTS

Construction/Installation

Skylights are available as preassembled units shipped to the site ready to be installed, or as assemblies of units, or framed assemblies of stock components, prefabricated off site and then site assembled.

Scuttles and fire vents are only available as preassembled units, shipped to the site ready to be installed. Non-stock sizes may be available for scuttles on special order; when UL or FM listed fire vents are required, choice of sizes generally limited to stock sizes:

• UL or FM listings are based on tests of units of specific sizes and construction and approvals involve lengthy and costly procedures.

SKYLIGHT UNITS are available as:

■ **Self-flashing, without curb:**

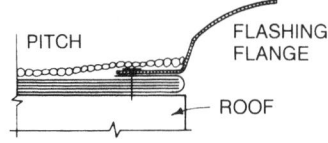

• for installation directly into roofing.
• with flashing flange integral with glazing or with added flat flashing flange.
• generally used on pitched roofs only when without a curb since flashing to prevent water penetration is difficult if not impossible to achieve on flat roofs.

■ **Self-flashing with curb.**

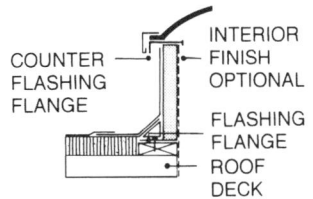

• shipped to site preassembled with a prefabricated curb, curbs generally four to twelve inches high.
• curbs commonly insulated; with flashing and counterflashing flanges.
• when counterflashing flanges are short, additional counterflashing may be required to prevent water penetration.
• where deck is field cut for skylight, trim pieces may be required to finish the exposed edges of decking.

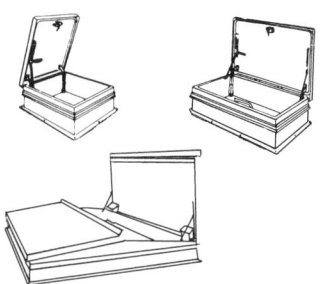

SCUTTLES are commonly available with integral curbs including flashing and counterflashing flanges, without or with insulation:
• cover generally solid, but may incorporate glazing.
• usually spring assisted opening; may also be motor operated, especially for large units.

FIRE VENTS are commonly available with integral curbs provide with flashing and counterflashing flanges. Two basic types are:
■ **Melt-down:** plastic glazing which softens and drops out of the frame when exposed to high temperature; a bar to prevent the plastic from dropping to the floor is generally incorporated into the unit:
• unit has to be replaced once exposed to fire.

■ **Automatic opening:** solid or glazed cover with springs held by fusible link, which melts when temperature rises and releases the springs for automatic opening of the cover.
• may be reused after a fire, if not damaged, by replacing fusible link.

EXPLOSION RELIEF VENTS are similar to fire vents in construction:

■ **plastic glazed units deform under rise in pressure** and are released from frame:
• may be replaced in frame if not damaged.

■ **solid or glazed cover units are also rise-in-pressure activated** releasing springs which automatically open the cover:
• may be reused if not damaged.

Fire venting and explosion relief functions may be combined in a single unit when both are required.

All skylight, scuttle, and vent units must be securely attached to the roof assembly:
• structural or miscellaneous steel frames may be required at openings in deck.

• provisions for attaching a light gauge metal flashing flange of the unit to roof substrate/decking may be required, such as wood blocking.

FRAMED ASSEMBLIES

■ **All framed skylights are custom designed** by manufacturers to meet necessary roof and/or wind loads.

■ **Framed skylights beyond a given angle** must be designed to resist wind loads similar to a curtain wall.

■ **Framed skylights require** somewhat greater mullion widths when glazed with acrylics, due to the expansion and contraction characteristics of plastics that must be taken up at the glazing connection.

■ **Mullion spacing** for framed skylights is generally limited to standard glass widths when thus glazed:
• wire glass is 60 inches.
• laminated glass is 48 inches.
• tempered glass is 72 inches.

■ **Dimensional limitations** on skylight assembly will further be imposed by:
• requirements for adequate drainage of rain/storm water from roof.
• additional requirements for condensate gutters in the body of the skylight assembly as well as around its perimeter.

■ **Mounted on built-up curb:**
• with frame and counterflashing for mounting on built-up curb.
• recommended minimum height for curb is 8 inches above roofing.
• prefabricated curbs for use in lieu of site built curbs are available without or with insulation.

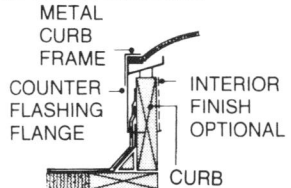

■ **Double or triple glazing** is available to minimize heat flow through the skylight:
• frames may have thermal breaks incorporated into them.

■ **Frames or screens to protect the glazing from impact,** or forced entry may be incorporated.

B3.2 SKYLIGHTS

SKYLIGHTS: units, assemblies

UNITS   **DESCRIPTION**

DOME: CIRCULAR

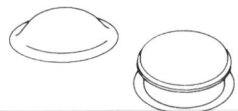

size, range	• 2'-6" to 7'-7" diameter with rise of dome from 10 to 24 inches.
glazing	• generally acrylic, single or double.
remarks	• self-flashing, with or without integral curb.
	• curbs available insulated

DOME: SQUARE, RECTANGULAR

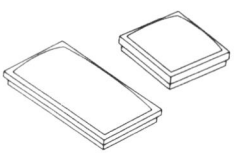

size, range	• square from 2'-0" to 10'-0", rectangular from 2'-0" to 5'-0" wide by 4'-0" to 8'-0" long, rise of dome from 5 to 18 inches.
glazing	• acrylic, polycarbonate, glass fiber reinforced polyester.
	• single, double, or triple.
remarks	• self-flashing, with or without curb.
	• self-flashing flanged, 14" x 89",available for pitched shingled roof.

DOME: PYRAMID, POLYGON

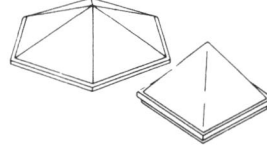

size, range	• 2'-6" to 8'-0" with rise of dome from 8 to 22 inches.
glazing	• acrylic, polycarbonate; single or double.

FLAT PANEL

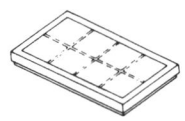 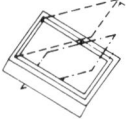

size, range	• square 1'-2" to 4'-0"; rectangular 1'-4" to 4'-0" wide, to 4'-8" long.
glazing	• insulating glass, laminated glass.
remarks	• operable sash either hinged or pivoted.
	• generally for use in pitched roofs.
	• sliding type available.

VAULT

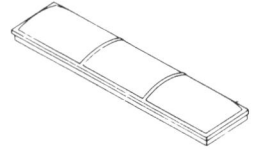

size, range	• 2'-0" to 6'-0" wide by any length; rise from 6 to 14 inches.
glazing	• acrylic, polycarbonate; single or double.
remarks	• may be installed in series; structural supports have to be then provided.

RIDGE

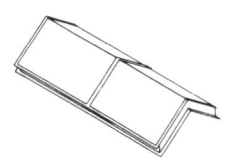

size, range	• variable pitch; generally up to 4'-0" along single slope.
glazing	• acrylic, single or double; polycarbonate, insulating or laminated glass may be available.
remarks	• may be arranged in series, structural supports have to be then provided.

B3.2 SKYLIGHTS

FRAMED ASSEMBLY	DESCRIPTION

MULTIPLE GRID

forms
- flat panels or formed units of various shapes.

spans
- custom sizes from 3'-0" x 3'-0" to 10' square factory assembled.
- grid more economical with larger units.
- uninterrupted grids of 60' x 90' not unusual.
- long-span grid networks require a perimeter gutter system to control larger water sheds.

glazing
- all types of glazing for flat panels.
- cold-formed or thermoformed plastic for curved shapes.

DOME

forms
- can be fabricated to most radii, provided rise to span ratio is minimum 22%.
- number of curved framing sections can vary within same loading conditions.
- sill and horizontals usually straight and framed into curved rafters.

spans
- 50° diameters usual; 140' available spans to 200' exist, both in static and roll-away structures.
- used with space frames.

glazing
- only plastic glazing materials used.
- cold-formed or thermoformed depending on strength requirements.

POLYGON

forms
- a variety obtainable with varying number of facets obtained by joining straight framing sections.

spans
- 45' across maximum used. Consult manufacturer before designing custom polygons.

glazing
- all types applicable.

PYRAMID

forms
- three and four sided.
- standard slopes up to 45°, maintain minimum rise of 15°.

spans
- common sizes: 20' to 30'.
- standard one piece thermoformed units up to 10' x 10' available.

framing
- standard aluminum framing available.
- all types of custom framing.

glazing
- all types applicable.

SINGLE or DOUBLE PITCH

forms
spans
- lean-to enclosure most commonly used.
- lean-to most economical for up to 15'; 20' common.
- slopes vary from 10° to 60°.

framing
- tubular or I-beam construction available.

glazing
- all types applicable; end wall glazing available.

BARREL VAULT

forms
- rises range from 10% to 50%, 22% most efficient.
- economical for skylight or covered walkway applications.

spans
framing
- up to 60 feet; span determines number of horizontal purlins required.
- rectangular base simplifies roofing, flashing and ceiling work.
- endwall panels available.

glazing
- cold formed or thermoformed plastic.

MULTIPLE RIDGE

forms
- can be used as single or multiple units, hipped gable ends also available.
- provide design flexibility and blend with aesthetics of typical roof construction.

spans
framing
glazing
- 15'-0" to 25'-0" most economical, up to 40'-0" used.
- with multiple units use structural gutter network.
- all types available.

Time-Saver Standards for Building Materials and Systems

B • BUILDING SHELL

B3.2 SKYLIGHTS

SKYLIGHT UNITS: forms, sizes, options

LEGEND

- denotes common availability in stock: custom fabrication or special model availability not included.

Form	SELF-FLASHING, no curb	SELF-FLASHING, integral curb	CURB MOUNTED, built-up curb	SINGLE GLAZING	DOUBLE GLAZING	TRIPLE GLAZING	CURB, no insulation	CURB, insulated	CURB, ventilation	HINGED FOR VENTILATION	PIVOTED FOR VENTILATION	SLIDING FOR VENTILATION	TEMPERED	LAMINATED/WIRED	INSULATING	GLASS BLOCK
			INSTALLATION		CONSTRUCTION, OPTIONS									GLAZING GLASS		
(flat panel)	•	•					•	•	•	•			•	•	•	•
(ridge)	•	•					•	•	•	•			•	•	•	
(long flat)	•			•	•							•				
(curb mounted)	•	•		•	•	•	•	•	•		•	•				
(dome)	•	•	•	•	•		•	•								
(pyramid)	•	•	•	•	•		•	•			•					

ACRYLIC	POLYCARBONATE	PVC	GRP [1]	PUSH BAR	CRANK	ELECTRIC MOTOR	SPRINGS WITH FUSIBLE LINK	PROTECTIVE FRAME, exterior	FIRE BRAND SCREEN	INTERIOR SCREEN	INSECT SCREEN	LIGHT CONTROL LOUVERS	MELT OUT FIRE VENT	EXPLOSION RELIEF VENTS	REMARKS
		●								●	●				• available hinged or pivoted for installation in pitched roofs. • type is not recommended for flat roofs.
●		●													• ridged unit may have framing member at peak or glazing may form ridge (plastic only). • muntins may be framing members or be over interior or exterior surface.
●											●	●			• end panels available flat or formed to produce dome-like appearance. • removable insect screens for sliding units only.
●	●	●	●	●	●	●	●				●	●	●	●	• operating units available also in double panel arrangement in both access/venting and fire/smoke vent applications. • ventilator units with powered fans available as well as regular gravity ventilating louvers. • built-in venetian blinds available.
●												●	●	●	• interior ceiling panels on smaller units often flat.
●	●	●										●	●	●	• rectangular units available in multi-unit assemblies.

PLASTIC — **OPERATION OPTIONS** — **OPTIONS** — **REMARKS**

[1] **Glass fiber reinforced polyester:**
• sandwich panels consisting of face panels laminated to aluminum frame are available as flat glazing.

B3.2 SKYLIGHTS

B • BUILDING SHELL

SKYLIGHTS, SCUTTLES, VENTS:

SKYLIGHTS

Daylight Admission and Control
In determining the desired form and size of the skylight unit/assembly, consideration should be given to:

■ **Environmental conditions,** including:

- orientation and the resulting solar penetration angles, winter and summer, in the given geographic location may be a critical determinant.
- prevailing winds direction and force.
- precipitation quantity and patterns.
- topography and landscaping.

■ **Views into and out of the building through clear skylights:**

- overhanging trees.
- adjacent buildings.
- nearby streetlights.
- other parts of same building.
- into building from adjacent higher areas.

■ **The more a formed plastic dome is raised,** the greater its ability to refract light of the low early morning and late afternoon sun, which:

- maximizes the use of natural light; but
- increases the solar heat gain.

GLAZING

■ **The following glazing materials are listed** in an approximate descending **order of cost:**

- formed acrylic with mar resistant finish.
- formed acrylic.
- polycarbonates.
- flat acrylic.
- laminated glass.
- tempered glass.
- clear polished wire glass.
- textured, obscure wire glass.

■ **Tinted acrylics should be limited to ¼ inch** thickness for economy. A combination fiberglass sheet and aluminum frame system which has high insulating and excellent light diffusion may be an economical alternative.

■ **Proper glazing methods have an important influence:**
- exposed gasketing is subject to

material breakdown due to ultraviolet rays of sun.
- small valleys created at bottom of sloped glazing and horizontal glazing cap will hold water.

■ **Note that:**

- domed acrylic glazing is almost self-cleaning as the sloped shapes facilitate rain washing the surface.
- normal skylight glazing is not designed to support persons. Special thicknesses and provisions are necessary if maintenance personnel is to walk on glazed surfaces.

■ **Excessive expansion and contraction of acrylic glazing** may cause "rolling" of the sealant between metal framing.

■ **The minimum rise to span on curved structures of framed skylights** for vaulted and dome shapes is 22 percent, where maximum economy is achieved.

Safety and security measures may include:

■ **Plain glass skylights** should be protected by a screen to protect occupants below in the event of breakage. In lieu of screening, codes often permit:

- laminated glass.
- tempered glass.
- wire glass.
- acrylic.
- polycarbonates.
- fiberglass.

based upon their resistance to impact and breaking characteristics.

■ **Resistance to forced entry characteristics** of a framed skylight should include:
- possibility of disassembly of framing.
- ease of removing snap-on cover.
- melting point of glazing: acrylics can be easily burned through with a torch.
- presence of metal security screens.

CONDENSATION

■ **Double glazing and thermal break framing** will prevent condensation to some extent.

■ Usually a separation is made where the glazing member is bolted into the framing member by use of a nonheat conductive material.

■ **Insulated, precured assemblies** prevent condensation buildup and energy losses.

■ **Thermalized design** will help in preventing excessive condensation buildup on the frame of domed skylight units, minimizing corrosion, staining and general maintenance.

■ **Incorporation of a contiuous condensation gutter** to collect and store moisture till it evaporates will generally minimize maintenance.

Structural Consideration

■ **Skylight units as well as scuttles, vents, should be adequately designed to resist forces of winds at roof level:** positive for steep sloping surfaces; negative for flat or low pitch.

■ **Required resistance to live loads, static and dynamic generally has to equal that of roof.**

SCUTTLES, VENTS

Selection considerations may include:

■ **If scuttles are to serve as required means of egress, consult building codes for:**

- number, size and location required.
- type of access permitted: such as ship's ladder, stair.
- force required to open unit allowable.

■ **Operable roof scuttles and skylights** must open automatically if used for smoke/fire venting. They must be able to open with a maximum snow load on them.

In evaluating quality of scuttle unit, the following may be considered:

■ **How often the unit will be used;** this will help determine:

- type of access: ladder, ship stair, regular stair.
- size of opening.
- type of operation: manual, powered, force required to open unit.
- durability of operating mechanism components, such as:
  compression spring operators
  shock absorbers
  spring latches
  hold open devices
  weatherstripping

selection checklist

■ **Fire-resistive features required may include:**

• a "label" requirement, having passed fire resistance or operating tests performed by an acceptable laboratory, such as Underwriters Laboratories or Factory Mutual fire underwriters approval.

■ **Safety features:**

• telescoping cylinder cover on the compression spring to prevent injuries from pinching or catching clothes in the spring coils.
• counterbalanced by spring operators to automatically lock the cover in an open position, preventing it from slamming shut on a person.

■ **Possibility of utilizing glazed covers for daylighting.**

■ **Quality of finishes;** such as:

• paint bond galvanized with a prime paint.
• mill finish aluminum.

■ **Ease of installation and weather-tightness:**

• whether units are preassembled with their own curbs and nailing flanges.
• whether well insulated and gasketed to prevent excessive heat loss and gain.

FIRE AND SMOKE VENTING

More loss of life results from people overcome by smoke and toxic fumes than from actual fire. Fire and smoke vents open automatically when temperatures rise in case of fire to:

■ **Permit the smoke to escape.**

■ **Lower temperatures at floor level.**

Automatic venting of fire and smoke through the use of fire vents is also a great aid in preventing loss of property and damage in single-story buildings used for manufacturing, warehousing and retailing.

Automatic opening of fire vents is accomplished either through a fusible link releasing springs held in compression which then open the cover of the vent, or through the melting of a plastic cover, which then drops out of the frame onto a special bar incorporated into the vent assembly.

■ **When fire strikes the vent or vents closest to the source of the**

fire, they open. This promptly:

• removes smoke from the building.
• confines the fire.
• reduces the heat to below the autogenous ignition temperature of the superheated gases.
• fire fighters are thus able to identify the source of the fire by observing the vent or vents which are open and to attack it from above and below.

The heat from the fire is also prevented from spreading out under the roof, causing sprinkler heads to release in areas some distance from the fire and resulting in excessive water damage.

A sufficient number of vents must be distributed over the entire roof area to assure reasonably early venting of a fire regardless of its location.

■ **The size and spacing of the vents** must be determined for each building, depending upon:

• size of the building.
• its particular use or combination of uses.
• the degree of hazard involved.

■ **Smoke venting is based upon movement of a specific number of cubic feet of air per minute through the fire vents.** Check codes for necessary capacities and compare to vent, ventilator and/or operable skylight units available.

• scuttles are available which may also serve as fire vents.

■ **Typical vent area requirements are:**

• 0.67 percent of roof area for low heat release occupancies.
• 1 percent of roof area for moderate heat release occupancies.
• 2 percent of roof area for high heat release occupancies.

■ **Roof vents are often required over:**

• stairs.
• elevator hoistways.
• high hazard occupancies to offer explosion relief.
• areas behind proscenium in theaters.

■ **In determining the number of vents** to be used to satisfy total required venting area, recognize that venting can be better accomplished by several small units than by a few larger ones. (NFPA #204) The size of the vent required is based upon its opened area, just about equal to its frame size.

The use of smoke curtains is essential to proper venting and containment of a fire in large open areas.

• curtains are usually made of sheet metal but may be constructed of many other noncombustible materials, such as cement bound mineral fiber, metal lath and plaster, or gypsum board.

■ **Consider also the spacing of vents in relation to:**

• interior spaces and their uses.
• proximity to exits.
• use to also provide daylighting.

WITHOUT VENTING

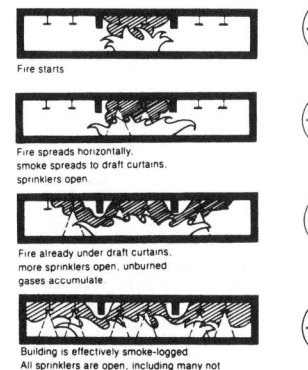

Fire starts

Fire spreads horizontally,
smoke spreads to draft curtains,
sprinklers open.

Fire already under draft curtains,
more sprinklers open, unburned
gases accumulate.

Building is effectively smoke-logged.
All sprinklers are open, including many not
over flames. Water pressure is reduced over flames.
Heat is beginning to melt steel structure.

Firemen arrive, break windows.
Sudden inrush of air causes "back draft"
explosion of unburned gases.

WITH VENTING

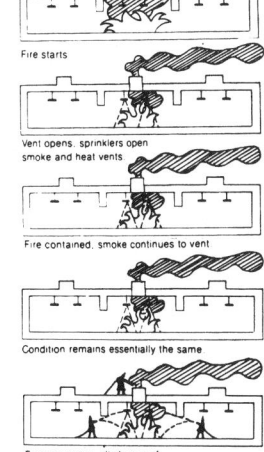

Fire starts

Vent opens, sprinklers open,
smoke and heat vents.

Fire contained, smoke continues to vent.

Condition remains essentially the same.

Firemen arrive, climb on roof,
point charge line down open
vent on fires. Others enter through
door, fight blaze from floor.

B3.2 SKYLIGHTS

SKYLIGHTS: energy notes

Skylights can be used to provide natural illumination to an enclosed space, reducing electrical lighting demand. In some cases, they can be used to admit direct solar radiation to enhance space heating. And, if openable skylights are used, natural ventilation due to convective airflow can help reduce cooling loads.

Predicting energy savings accurately requires consideration of available light, heating energy demand and light design level, and is very complex. However, for an approximate idea of the energy savings possible through the use of skylights consult the maps and charts below:

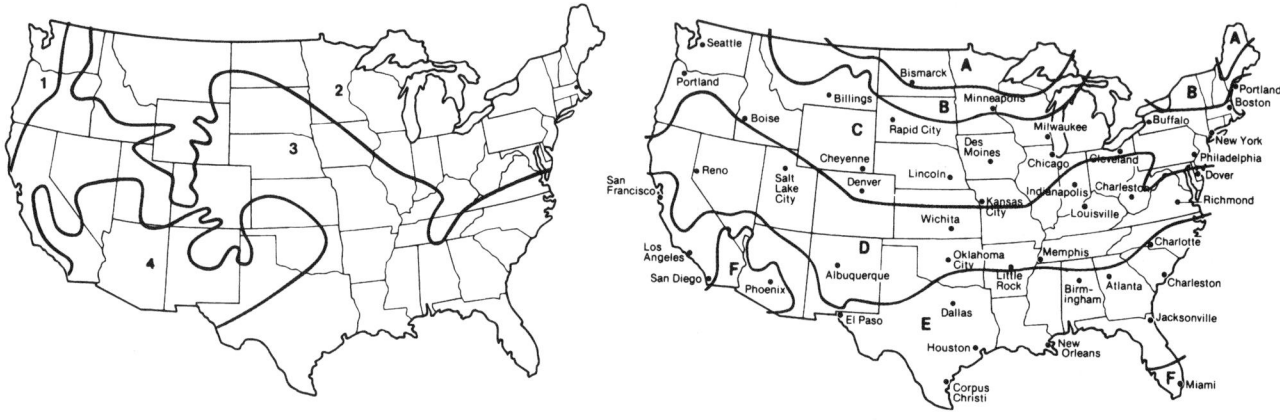

AVAILABLE LIGHT ZONES   **ENERGY ZONES**

Locate the site of the building on both the Available Light and the Energy Zone maps. The chart will indicate the percent of roof area that should be Skylights for optimum energy efficiency and the BTU's to be saved annually per sq. ft. of clear or low density white Skylights. In each case the second figure assumes air conditioning in the building, and the first assumes none.

POTENTIAL ENERGY SAVINGS WITH USE OF SKYLIGHTS.

Light Zone	Lighting Design Level	Required % Roof Area	BTU Savings in Thousands					
			Energy Zone A	Energy Zone B	Energy Zone C	Energy Zone D	Energy Zone E	Energy Zone F
1	30 f.c.	3.3	—	—	65- 70	—	—	—
	60 f.c.	5.2	—	—	102- 92	—	—	—
	120 f.c.	13.3	—	—	88- 77	—	—	—
2	30 f.c.	2.8	0	26-19	86- 73	82- 80	—	—
	60 f.c.	4.3	30-30	60-60	116-114	118-115	—	—
	120 f.c.	10.8	12-10	40-35	100- 92	83- 80	—	—
3	30 f.c.	1.8	—	—	85- 75	94- 85	175-190	300-250
	60 f.c.	3.2	—	—	135-135	135-135	225-225	225-175
	120 f.c.	6.9	—	—	120-120	120-120	192-185	320-295
4	30 f.c.	1.5	—	—	—	100- 87	195-210	315-268
	60 f.c.	2.8	—	—	—	130-140	230-225	200-160
	120 f.c.	4.0	—	—	—	140-150	235-240	290-285

B3.2 SKYLIGHTS

C • INTERIORS

C1. ❶ INTERIOR PARTITION SYSTEMS

INTERIOR WALLS/PARTITIONS: introduction

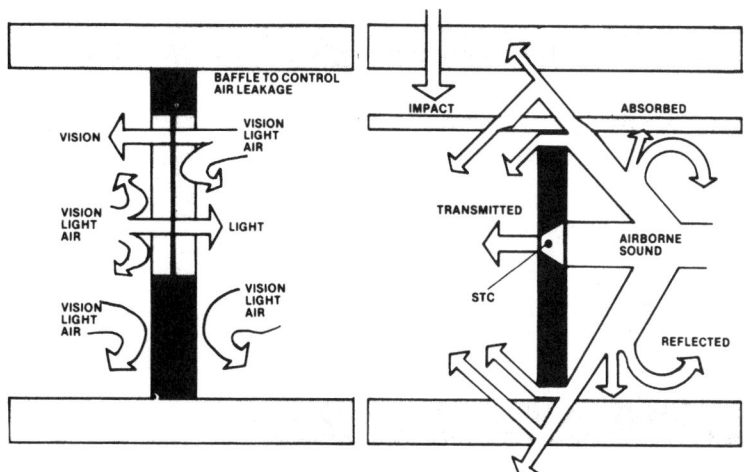

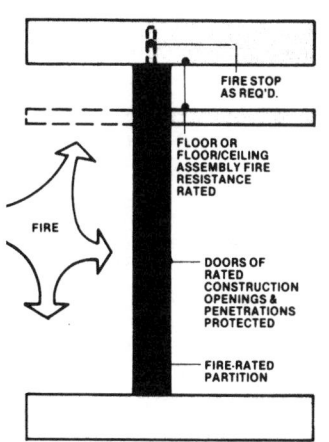

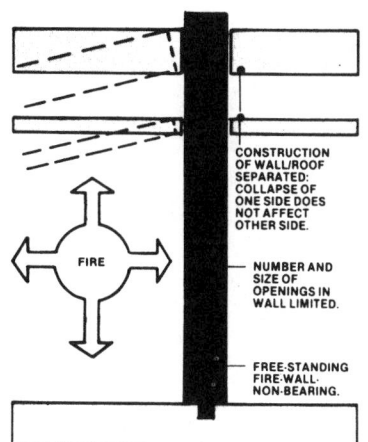

Interior walls/partitions, the means of vertical division within an enclosure, may function as:
- physical or visual barriers; and/or
- selective filters to interior environmental factors.

When serving as barriers, they may:
- control movement through and within an enclosed space.
- control passage through them from one space to another.
- separate spaces with different environments.
- isolate hazardous activities or environments within a space from adjacent ones.

When serving as selective filters, they may:
- control the flow of heat by reducing the rate of its transfer through them.
- allow or block the transmission of light.
- allow or prevent visual contact between enclosed spaces.
- control or reduce sound transmission through them.

Interior walls/partitions may serve as fire divisions to compartmentalize an enclosure to contain and/or slow down the spread of fire:
- walls and/or partitions which require a fire resistance rating must be constructed of the same materials and in the same manner as a specific tested assembly providing the necessary resistance to exposure to fire.
- they must be integrated with horizontal divisions of the enclosure to completely contain the fire within a given space.
- all openings in a fire rated wall or partition must be protected with fire-resistant rated assemblies, such as fire shutters, or doors.
- duct penetrations through a fire-resistant rated partition must be protected with fire dampers; pipe penetrations must be sealed.

Interior walls may also be used to support superimposed gravity loads of the structural frame, acting there as bearing walls; and/or to brace the structural frame against lateral displacement due to wind or seismic forces, functioning then as shear walls.

When interior walls/partitions are nonbearing, considerations of stability may include:
- ■ **Normal pressure** at 5 lbs./sq. ft. and impact assumed to act on either side, but not simultaneously on both.
- ■ **High pressure** may occur in high speed elevator shafts, pressurized rooms, e.g., clean rooms or from controlled explosions during testing.
- ■ **Deflection limits** will be governed by properties of components:
  - may range from 1/120 of span for metal or plywood to 1/360 of span for plaster.
- ■ **Cornice-high and rail-high partitions** should be braced to ceiling by posts or by cross-walls to provide lateral stability:
  - unsupported length, rather than unsupported height will then govern.
  - when no posts or cross-walls are provided, partition should be designed as a cantilever and be firmly anchored to floor.

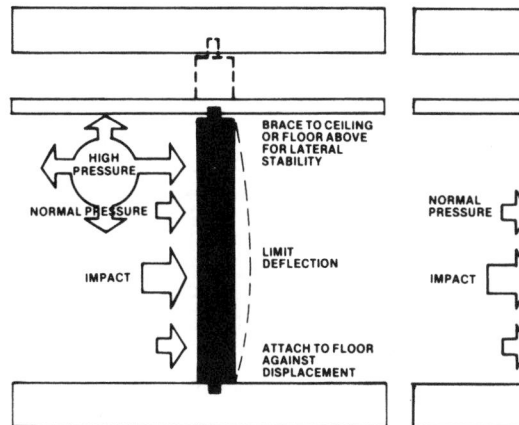

C1.1 INTERIOR PARTITION SYSTEMS

PARTITIONS: introduction

Preliminary selection of the type(s) of partitions generally includes the following:

■ **Performance requirements:**
• vision, sound, passage and fire resistance are the major variables;
• stability and control of movement are general requirements for any type of partition.

① **Comparitive Cost:**
• basis of 100 represents 2" x 4" wood studs with ½" gypsum wallboard both sides;
• no special provisions to control sound transmission.

• no fire resistance rating.
• **Cost of finishes** is not included and an average relative cost for each type is shown.

TYPES	PERFORMANCE CHARACTERISTICS	REL. COST/S.F. ①
■ Portable ■ Relocatable-Rail High ■ Fixed-Rail High: • framed • stacked	■ Vision; limited control ■ Passage: no control or limited control	600 150 180 180
■ Relocatable-Cornice High ■ Fixed-Cornice High: • framed • stacked	■ Vision: partial control ■ Sound: limited control ■ Passage: full control	250 200 200
■ Operable: • folding panels • accordion fold • coiling	■ Vision: full control ■ Sound: partial to good control ■ Passage: limited control ■ Space Division: flexible control	1300 500 1150
■ Fixed-Framed: • wire mesh	■ Vision: very limited control ■ Sound: no control ■ Passage: full control	400
■ Fixed-Framed: • glazed ■ Relocatable: • glazed-ceiling high	■ Sound: moderate to good control ■ Passage: full control	300 380
■ Fixed: • framed • stacked • solid laminated ■ Relocatable • ceiling high	■ Vision: full control ■ Sound: moderate to good control ■ Passage: full control	150 180 150 320
■ Fixed: • monolithic concrete • stacked-reinforced ■ Fire Resistant Only: • framed • solid laminated	■ Vision, Sound, Passage Control ■ Special conditions: • fire resistance • high intensity sound • high pressures • high security • radiation	300 250 220 180

C1.1 INTERIOR PARTITION SYSTEMS

PARTITIONS DEFINED

Partitions are vertical non-bearing components which, acting together with floors and ceilings, **define and organize interior space.**

■ Partitions may **divide space to various degrees,** ranging from:
• merely providing visual organization of open space with a common environment
• to creating contained units, each with its own controlled environment.

■ **Partitions may be:**
• fixed in place,
• demountable/relocatable
• operable, portable;
• may be finished in a variety of applied or attached materials.

■ Partitions **are by definition non-bearing,** but may be required to support loads attached to them in addition to their own weight, such as paintings, shelves, cabinets, equipment.

■ **Structural strength required** may range:
• from none for a suspended fabric to provide visual privacy in an open space
• to heavily reinforced concrete able to resist forced entry.

■ **Fire resistance** also varies:
• from a combustible all wood partition
• to a non-bearing fire wall, constructed to remain intact when exposed to fire for four or more hours.

■ **Sound transmission** characteristics vary:
• from none for an open mesh partition
• to high resistance for elaborate double wall construction.

PARTITION TYPES AVAILABLE

Partitions range widely in:

■ **Degree of separation** they offer:
• transparent, translucent, opaque
• rail-height, cornice-height, floor-to-ceiling, full height
• open or closed at floor level.

■ **Construction** and method of support may be:
• solid, hollow-core
• preassembled, or site-built

• free-standing, floor supported only, or braced by other partitions
• floor-supported and braced by overhead floor/ceiling assembly
• wall or ceiling hung.

Partitions may be grouped into four major types based on method of installation and/or use:

Portable partitions are prefabricated, prefinished panels which can be easily moved without the need for special equipment and include:

■ **Self-supporting** screen panels, generally up to 7 feet in height:
• individual panels with stabilizer bases
• braced interconnecting assemblies
• with spring loaded supporting posts between floor and ceiling.

■ **Full height** floor to ceiling panels supported on the floor and braced by the ceiling:
• as individual panels,
• butted for continuous runs

■ **In portable partitions:**
• attachment to floor is not required;
• fastening to ceiling is not required, or of limited extent.

Movable partitions are prefabricated, pre-finished assemblies forming a complete, continuous closure from floor to ceiling and may be:

■ **Panels butted or hinged** together, suspended from an overhead track, with or without floor guides

■ **Supported on a floor track** and guided at top.

■ **Fabric faced** folding metal frames suspended from an overhead track without floor guides.

■ **Manually** or power operated.

Relocatable or demountable partitions can be repeatedly disassembled without destroying the component parts, and re-assembled in a different location and may be:

■ **Completely prefabricated** standard size panels with interlocking joints,

supported on a continuous floor track;
• generally installed over flooring; floor track must then be secured to floor deck through the flooring.
• ceiling track generally fastened to suspension grid of acoustic tile ceiling; ceiling suspension grid may be specially shaped to receive ceiling track with minimum of fasteners

■ **Prefinished facing panels,** individually secured to a fixed framed core in a way to allow for removal of the panels at any time.

Fixed partitions are site constructed of various standard components, and generally cannot be relocated without destroying most, or all of the component parts:

■ **Framed** partitions consist of a framed core faced both sides with prefabricated panels, either prefinished or unfinished; or with various site applied facings.

■ **Stacked** partitions consist of various preformed units (generally masonry) which either combine core and facing or form a core for site-applied facings.

■ **Laminated** partitions are comprised of preformed panels for both core and facings, e.g., solid gypsumboard partitions.

■ **Monolithic** partitions are entirely site-constructed.

■ **All types** are secured or bonded to the floor and, when full height, also braced by ceilings or floor/roof assemblies.

DOORS are essentially an operable section of the partition they are part of, and are subject to the same performance requirements as the partition.

C1.1 INTERIOR PARTITION SYSTEMS

PARTITIONS: types, uses

● denotes common usage
○ denotes possible usage

TYPE		MAXIMUM UNBRACED HEIGHT range in feet	MAXIMUM UNBRACED LENGTH, range in feet	USE/CONTROL [1] SEPARATION				ISOLATION					
				VISION ONLY	SOUND ONLY	VISION AND SOUND	PASSAGE	FIRE	HEAT FLOW	HIGH INTENSITY SOUND	HIGH PRESSURE	RADIATION	FORCED ENTRY
FRAMED		16 to 18	[2] 10* to 12	●	●	●	●	●	●	●		○	
SOLID-LAMINATED		8 to 9	[2] 8* to 10	●		●	●	●					
RELOCATABLE		8 to 12	[2] 8* to 15	●	●	●	○	○					
OPERABLE PORTABLE		no limit for top-hung	no limit / 6 to 8	●		●	● [3]	○					
STACKED		12 to 16 for 4" thick	12 to 16 for 4" thick			●	●	●		○	○	○	●
MONOLITHIC		20 to 40	[4] varies			●	●	●		●	●	●	●

SOUND TRANSMISSION CLASS. range	FIRE RESISTANCE RATING highest in hours	COMPATIBLE SUB-FLOOR:			ANCHORAGE					REMARKS
		CONCRETE	WOOD/PLYWOOD	PARTICLE BOARD	NONE REQUIRED	WEDGED BETWEEN FLOOR AND CEILING	ATTACHED TO FLOOR	SUSPENDED FROM FRAME	BRACED BY CEILING	
30 to 59	2½	●	●	●			●		●	• glazed partitions will allow full vision, but will restrict sound transmission. Laminated glass is available for improved control of sound transmission. • lead lined gypsum wallboard available to isolate low level radiation; or core may be faced with sheet lead. • high degree of resistance to heat flow may be achieved by providing thermal insulation.
36 to 40	2	●	○	○			●		●	• layers of laminated gypsum wallboard. • wiring cannot be concealed; if concealed wiring is required, gypsum wallboard core may be replaced by laminated gypsum wallboard studs cut from full sheets which allow vertical runs of wiring. • used to enclose shafts or chases, or as fire-rated, minimum-thickness partitions.
30 to 50	2	●	○	○			●		●	• used when flexibility in changing layout with minimum of disruption is required. • generally installed over finished flooring. • wiring is concealed and is generally accessible for future changes. • glazed partitions available to allow full controlled vision and restricted sound transmission. • toilet partitions are included under this type.
.23 to 54 / none	2	●	●	○	●	●	●			• operating or movable partitions and portable partitions provide greatest flexibility in space subdivision. • portable partitions are prefabricated panels of all heights without any attachment to the floor, or wedged between floor and ceiling (by spring-loaded posts, continuous top caps, or screw operated continuous top caps) 3 portable partitions provide no control of passage; sound may be controlled by facing the panels with sound absorbing materials.
30 to 45	4	●					●		●	• concrete units generally used where high resistance to impact and good resistance to fire are required. • gypsum units generally used where good resistance to impact and high resistance to fire are required. • glass units used where control of vision, light and sound are required. • may be furred with lead lined gypsum wallboard for protection from low level radiation.
33 to 57	4	●					●		●	• monolithic concrete generally used only when high degree of resistance to fire and/or impact, high intensity sound, high pressures, or radiation are required. • solid gypsum plaster used when resistance to fire and/or moderate impact with minimum thickness is required. • solid Portland cement plaster used when resistance to impact and wet conditions is important. 4 depends on reinforcement

REMARKS

1 **Use/control information applies to full height or ceiling-high partitions only.** Rail-high and cornice-high partitions are available in all types except operable, but will only provide limited control of vision and passage.

Headings "Separation" and "Isolation" are arbitrary and indicate degree of control required:

• **Separation:** where no special performance requirements exist, e.g. resistance to fire, high intensity sound, pressures over 5 lbs./sq. ft.

• **Isolation:** conditions where special performance requirements must be met.

2 when top not braced.

C1.1 INTERIOR PARTITION SYSTEMS

PARTITIONS: types, materials

- denotes common usage
○ denotes possible or infrequent usage

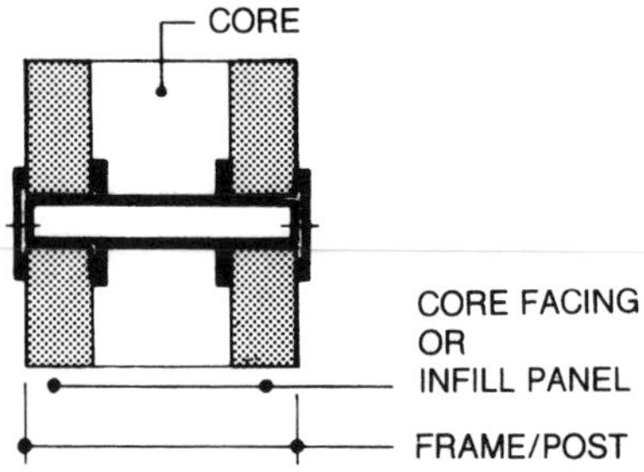

CORE

CORE FACING OR INFILL PANEL
FRAME/POST

TYPE	THICKNESS, range in inches	WEIGHT, range, in lbs./s.f.	SOLID MONO-LITHIC		PREFAB BOARDS				PREFAB UNITS			
			CONCRETE	GYPSUM & LATH	GYPSUM	PARTICLE	FIBER	RIGID INSULATION	CONCRETE	GYPSUM	CLAY	GLASS
FRAMED	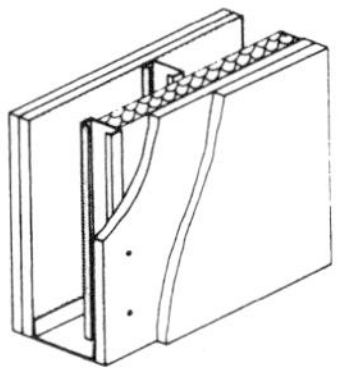 3¼ to 10¾	5 to 20										
SOLID-LAMINATED	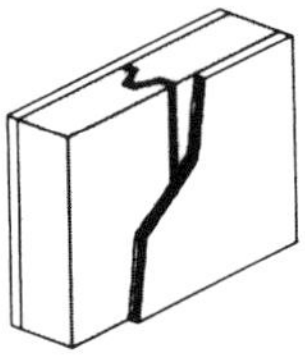 2 to 2¼	8 to 9			●							
RELOCATABLE	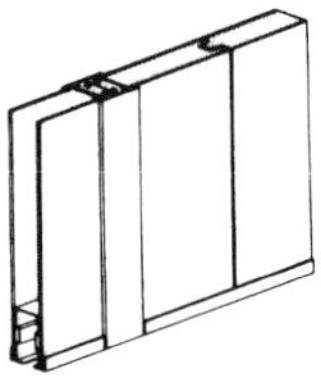 1¾ to 3¾	3 to 8				●	●	●				
OPERABLE PORTABLE	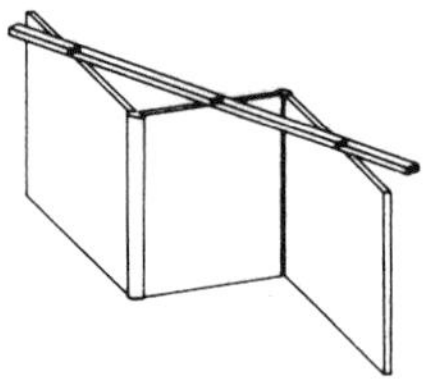 1½ to 3¾	3 to 10				●	●	○				
STACKED	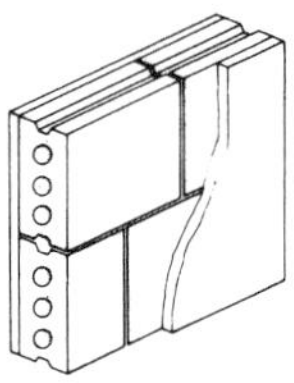 2½ to 7	20 to 45							●	●	●	●
MONOLITHIC	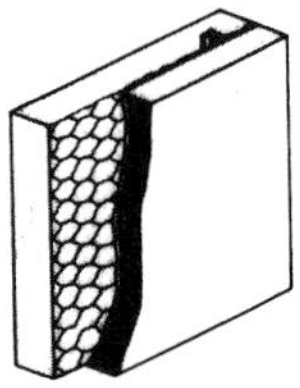 2 to 7	10 to 95	●	●								

CORE FACINGS / INFILL PANELS IN FRAMES/POSTS

	STUDS		PANELS, SLATS			CORE FACINGS									METAL		FRAMES FOR INFILL PANELS		INFILL PANELS IN FRAMES/POSTS			
	WOOD	METAL	METAL, FABRIC FACED	SLATS, WOOD, HINGED	METAL, INTERLOCKING	INTEGRAL with CORE	PLASTER FINISH COAT	GYPSUM LATH and PLASTER	METAL LATH and PLASTER	GYPSUM WALLBOARD	PREFINISHED HARDBOARD	VENEER PLYWOOD	QUARRY/TERRAZZO TILE	CERAMIC/GLAZED TILE	ON BACKING BOARD	LAMINATED to CORE	METAL	WOOD	GLASS/PLASTIC	PREFINISHED METAL	PREFINISHED PARTICLEBOARD	WIRE MESH
1	●	●						●	●	●	●	●					○	○	○	○	○	○
2								○		●												
3		●								●	○	●			●	●	●	○	●	●	●	●
4			●	●	●					○	●	●			●	●						
5						●	●	●	●	○	○	○	●	●								
6						●	●	○	○	○	○	○	●	●								

C1.1 INTERIOR PARTITION SYSTEMS

FIXED PARTITIONS: materials, limitations

● denotes common usage
○ denotes possible or
   infrequent usage

TYPE		MATERIALS	THICKNESS, range in inches	WEIGHT, range in lbs./s.f.	MAXIMUM UNBRACED HEIGHT at 5 psf pressure and 1/240 deflection	STC, sound transmission class, range	FIRE RESISTANCE RATING max. available in hours [1]	SEPARATION [2]	ISOLATION [3]
FRAMED									
	A	³/₈″ to ⁵/₈″ Gypsum Wallboard	3¼″ to 4⁷/₈″	5 to 7	9 to 14	30 to 42	1	●	
	B	2 x 4 Wood Studs or 2½″ to 3⁵/₈″ Metal Studs @ 16″ or 24″ o.c.							
	C	Rockwool or Glass Fiber Batt Insulation							
	A	¹/₁₆″ to ½″ Gypsum Plaster on Gypsum Lath or Gypsum Base	3³/₈″ to 5½″	5 to 20	9 to 16	40 to 45	1	●	
	B	2 x 4 Wood studs or 2½″ to 3⁵/₈″ Metal Studs @ 16″ or 24″ o.c.							
	C	½″ to 1″ Gypsum Plaster on Metal Lath				40 to 54	2		
	A	³/₈″ to ⁵/₈″ Gypsum Wallboard Face	4⁷/₈″ to 6½″	7 to 12	11 to 14	38 to 54	2	●	○
	B	Wood or Metal Studs at 16″ or 24″ o.c.							
	C	¼″ to ⁵/₈″ Backing Gypsum Wallboard – One or Both Faces							
	A	Gypsum Wallboard One or Two Layers	5¼″ to 5¾″	6 to 12	6 to 9	50 to 54	2	●	○
	B	2 x 4 Wood Studs at 16″ or 24″ o.c.							
	C	Resilient Furring Channel, generally 24″ o.c.							

FINISH OPTIONS

COATING, FIELD APPLIED	CHEMICAL RESISTANT	WALLCOVERING, FIELD APPLIED	FACTORY APPLIED	PANELING, PREFINISHED	CERAMIC/GLAZED FACING TILE	MARBLE/TERRAZZO UNITS	COMPARATIVE COST, range per s.f. for complete assembly
●	○	●	●	●	●	○	100 to 120
●	○	●		○	●	○	160 to 210
●	○	●	●	●	○	○	130 to 200
●	○	●	○	●	○	○	140 to 160

4 REMARKS

1 **For specific tested assemblies only**

2 **Separation:** For visual and/or speech privacy or security without special performance requirements.

3 **Isolation:** Special performance requirements in addition to separation.

4 **Basis:** ½″ Gypsum board both sides on 2 x 4 wood studs at 16″ o.c. = 100.

■ Chemical resistant coatings may require primers or surface preparation.

■ Moisture resistant gypsum wallboard should be used in all wet or damp locations. Portland cement and fine aggregate core boards available for use in wet or damp locations.

Row 100 to 120 remarks

- for fire resistance rating, gypsum wallboard should be ½″ or ⅝″ Type X.
- adding insulation increases STC by four points.
- insulation may be required in fire resistance rated assemblies.
- metal studs are pre-punched to allow wiring to be run horizontally.
- when electrical panels have to be recessed, deeper studs may be required.
- single layer of ⅜ inch thick gypsum wallboard is not recommended; if used, studs should be spaced not more than 16″ on center.
- gypsum wallboard with factory applied vinyl wallcovering is available.

Row 160 to 210 remarks

- insulation may be added between studs to improve sound resistance rating.
- trussed metal studs available. Maximum unbraced height with 3¼″ trussed studs: 20 feet.
- Portland cement plaster on metal lath available for wet locations.
- gypsum plaster and gypsum veneer plaster may be finished in several textured patterns.

Row 130 to 200 remarks

- generally used where higher fire resistance rating or improved resistance to sound transmission required.
- double layer of ½″ or ⅝″ Type X gypsum wallboard for fire rated assemblies.
- backing board ¼″ thick, also known as sound deadening, generally provided to improve resistance to sound transmission. Face layers then generally ½″ or ⅝″ thick. Half inch thick fiber sound deadening boards are available.
- insulation when added between studs will increase STC by three to four points.

Row 140 to 160 remarks

- used when high resistance to sound transmission is required.
- when fire resistance rating is also required, Type X gypsum wallboard should be used.
- insulation when added between studs will increase STC by four to six points.
- resilient channels generally used with wood framing only.

C1.1 INTERIOR PARTITION SYSTEMS

FIXED PARTITIONS: materials, limitations

● denotes common usage
○ denotes possible or
　infrequent usage

PERFORMANCE RANGE

TYPE		MATERIALS	THICKNESS, range in inches	WEIGHT, range in lbs./s.f.	MAXIMUM UNBRACED HEIGHT at 5 psf pressure and 1/240 deflection	STC, sound transmission class, range	FIRE RESISTANCE RATING max. available in hours [1]	SEPARATION [2]	ISOLATION [3]
FRAMED									
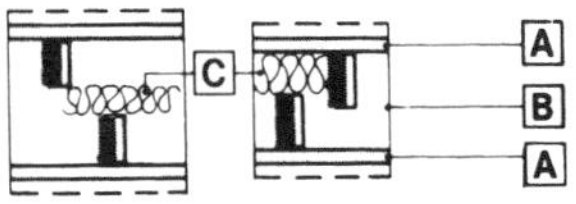	A	Gypsum wallboard One Or Two Layers, or Gypsum Plaster on Lath	6⅞″ to 8⅛″	7 to 13	12 to 16	43 to 59	2	●	●
	B	Staggered Studs 2 x 4 @ 16″ o.c. with 2 x 4 or 2 x 3 @ 16″ o.c. 1″ to 1½ ″ Apart.							
	C	Optional Insulation							
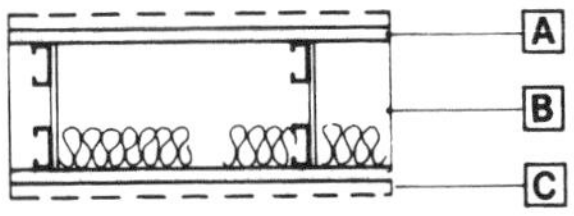	A	Gypsum Wallboard or Gypsum Plaster on Lath	6¾″ to 10¾″	6 to 10	10 to 16	47 to 52	2	●	○
	B	Spaced Metal Studs For Mechanical Chases							
	C	Optional Additional Gypsum Wallboard For Double Layer Construction							
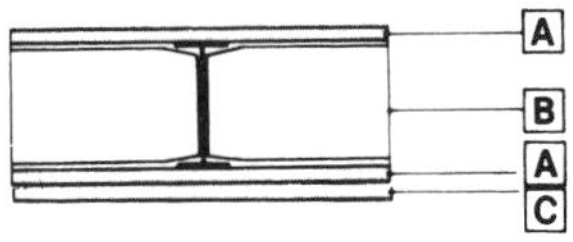	A	2-½ ″ Type X Gypsum Wallboards or 1″ Backing Board	3¼ ″ to 3½ ″	8 to 9	10 to 15	45 to 49	2	●	
	B	2″ Special Metal Studs @ 24″ o.c.							
	C	2 Layers ⅝″ Type X Gypsum Wallboard							
SOLID-LAMINATED									
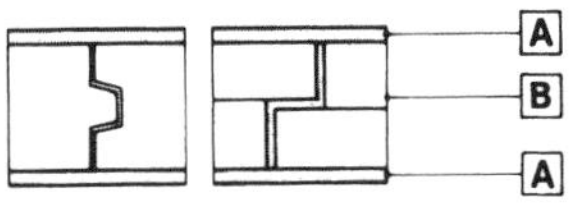	A	⅜″ to ⅝″ Gypsum Wallboard	2″ to 2¼ ″	8 to 9	9 to 11	36 to 40	2	●	
	B	1″ Gypsum Backing/Core Board / Laminated 2-½ ″ Gypsum Board Shiplap Joints							

(cont.)

FINISH OPTIONS

REMARKS

[1] **For specific tested assemblies only**

[2] **Separation:** For visual and/or speech privacy or security without special performance requirements.

[3] **Isolation:** Special performance requirements in addition to separation.

[4] **Basis:** ½″ Gypsumboard both sides on 2 x 4 wood studs at 16″ o.c. = 100.

■ Chemical resistant coatings may require primers or surface preparation.

■ Moisture resistant gypsum wallboard should be used in all wet or damp locations. Portland cement and fine aggregate core boards available for use in wet or damp locations.

COATING, FIELD APPLIED	CHEMICAL RESISTANT	WALLCOVERING, FIELD APPLIED	FACTORY APPLIED	PANELING, PREFINISHED	CERAMIC/GLAZED FACING TILE	MARBLE/TERRAZZO UNITS	COMPARATIVE COST, range per s.f. for complete assembly [4]	REMARKS
●	○	●	○	●	○	○	140 to 240	• used when very high resistance to sound transmission is required. • double studs used for party walls, with double layers of ½″ Type X wallboard on outside faces, and additional ½″ gypsum wallboard attached to the inside faces of studs. Two hour fire resistance rated. • in double layer construction, sound deadening boards, either ¼″ gypsum backing boards, or fiber sound deadening boards are used to improve resistance to sound transmission. • insulation may be added to further improve sound transmission resistance.
●	○	●	○	●	●		280 to 360	• used for mechanical/electrical chases. • metal studs generally 1⅝″ deep with gypsum board or metal channel spacers. • double layer construction for higher fire resistance rating. Type X gypsum wallboard should be used. • optional insulation between studs to improve resistance to sound transmission. • generally, metal studs are used only.
●	○	●	○	○	●	○	160 to 180	• used as shaft walls with one side finished only. • various fire resistance rated assemblies are available rated for resistance to fire on either one, or both sides, of wall. • different configurations of metal studs are available. Consult manufacturers' literature.
●	○	●	○	○	●	○	145 to 155	• used for chase walls, or when minimum thickness fire rated partition is required. • backing/core board held in place by floor and ceiling channels. • also used as double wall, with two separated cores, faced one side, when chase for mechanical/electrical services required. • gypsum wallboard spacers laminated to both face layers, instead of backing/core boards, are also used to provide space for concealed wiring.

C1.1 INTERIOR PARTITION SYSTEMS

FIXED PARTITIONS: materials, limitations

● denotes common usage
○ denotes possible or
   infrequent usage

PERFORMANCE RANGE

TYPE		MATERIALS	THICKNESS, range in inches	WEIGHT, range in lbs./s.f.	MAXIMUM UNBRACED HEIGHT at 5 psf pressure and 1/240 deflection	STC, sound transmission class, range	FIRE RESISTANCE RATING max. available in hours	[1] SEPARATION	[2] ISOLATION
STACKED									
	A	Optional: Plaster Gypsum Wallboard Paneling	3⅝″ to 5⅝″	20 to 45	9 to 12	30 to 45	2	●	○
	B	Concrete Units – Non Bearing Horizontal Joint Reinforcing Optional							
	C	Optional: Glazed Face							
	A	⅛″ to ½″ Gypsum Plaster							
	B	Hollow Clay Tile	4½″ to 7″	25 to 35	9 to 12	40 to 45	4	○	●
		Gypsum Units	2½″ to 5″	10 to 25					
MONOLITHIC									
	A	Optional: Cement or Gypsum Plaster	wash to ½″						
	B	Normal Weight Concrete	4″ to 7½″	50 to 95	20 to 40	50 to 57	4		●
		Lightweight Concrete	4″ to 6½″	40 to 65					
	A	Optional Gypsum Finishing Coat over Cement Plaster							
	B	Gypsum or Portland Cement Plaster on Metal Lath	2″ to 2¾″	10 to 12	12 to 18	33 to 39	2½	●	
		Gypsum or Portland Cement Plaster on Metal Lath & Channels							

C1.1 INTERIOR PARTITION SYSTEMS

C • INTERIORS

(cont.)

FINISH OPTIONS

REMARKS

[1] **For specific tested assemblies only.**

[2] **Separation:** For visual and/or speech privacy or security without special performance requirements.

[3] **Isolation:** Special performance requirements in addition to separation

[4] **Basis:** ½" Gypsumboard both sides on 2 x 4 wood studs at 16" o.c. = 100.

■ Chemical resistant coatings may require primers or surface preparation.

COATING, FIELD APPLIED	, CHEMICAL RESISTANT	WALLCOVERING, FIELD APPLIED	, FACTORY APPLIED	PANELING, PREFINISHED	CERAMIC/GLAZED FACING TILE	MARBLE/TERRAZZO UNITS	COMPARATIVE COST, range per s.f. for complete assembly [4]	REMARKS
●	●	○		●	●	●	117 to 258	• used where exposed to frequent impact, accidental or malicious. • available in special sound absorbing units as well as glazed both faces; for additional information refer to Stone / Masonry. • tooled mortar joints when left exposed; flush joints where faced. • may be finished with gypsum plaster, Portland cement plaster, or gypsum wallboard laminated or furred-out; or furred-out prefinished paneling. • gypsum wallboard will require furring when surface is uneven; prefinished paneling should be furred out. • concealed wiring difficult to install.
●	○	●	○	●	●		174 to 184	• more resistant to impact than gypsum wallboard partitions. • no longer in common usage; generally used when high fire resistance rating with minimum thickness required. • plaster may be applied directly over units when thin coat is used; heavier plaster coats may require gypsum lath, gypsum base, or self-furring metal lath. • for additional information refer to Stone / Masonry.
●	○	●		○	○	●		
●	●	○		○	●	●	220 to 330 / 230 to 360	• used where high impact resistance, resistance to high pressures, or high fire resistance rating is required; will also provide high resistance to sound transmission. • may be poured in place, or site precast and tilted-up into place; generally reinforced to resist shrinkage stresses, and stresses due to erection when precast. • unbraced length given is approximate only and will depend on reinforcing provided. • wallcovering requires a levelling coat of plaster; gypsum wallboard and paneling are generally installed over furring. • for additional information refer to Cement / Concrete.
●	○	●		○	●	○	130 to 200	• used where a fire resistant, moderately accidental impact resistant thin partition is required. • when used for shaft walls, one side only is finished. • wiring may be run concealed, but requires conduit and special boxes. • when Portland cement plaster is used it may be finished with Keene's cement • paneling may be laminated to smooth finished faces, especially when light-weight aggregate is used to provide nail holding strength.

C1.1 INTERIOR PARTITION SYSTEMS

DRYWALL & PLASTERING: materials

GYPSUM WALLBOARD

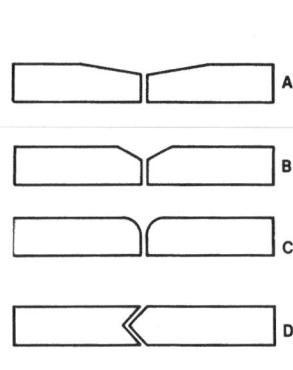

MATERIALS:
■ Incombustible core, essentially gypsum without added fibers, or with no more than 15 percent by weight of fibers; surfaced both sides with paper continuously bonded to the core.

TYPES:
■ Regular; for general use; ASTM C36 – Standard Specification for Gypsum Wallboard; also available aluminum foil backed.
■ Type X, fire-retardant; ASTM C36; used when fire resistance rating is required.
■ Water-Resistant, of water resistant core and water resistant paper; ASTM C630; used in wet locations, such as bathrooms.
• also available water-resistant as well as fire-resistant, per ASTM C36.
■ Backing Board; used as backing, base, or core material in multilayer construction; ASTM C442.
• also available aluminum foil backed, as well as fire-retardant Type X.

• backing board ¼ inch thick generally used as sound deadening layer in multi-layer partitions; or as re-surfacing board over existing plaster or drywall.

THICKNESS
• Regular: ¼ " ³⁄₈ " ½ " ⁵⁄₈ "
• Type X: ½ " ⁵⁄₈ "
• Water-Resistant: ½ " ⁵⁄₈ "
• Backing Board: ¼ " ³⁄₈ " ½ " ⁵⁄₈ " ¾ " 1"

SIZE
• Width: 2 and 4 feet.
• Length: 7 through 14 feet.
• Not all types available within above range.

EDGE DESIGN
• A = Tapered: for face layers.
• B = Square edge; for backing/core boards.
• C = Rounded or beveled edge: for vinyl fabric faced boards; or when joints between boards are not filled.
• D = Tongue and Groove: for 1" thick backing boards used as core boards.

JOINT FINISHING TREATMENT

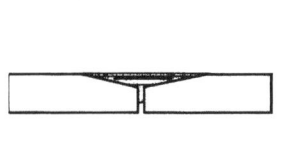

MATERIALS:
■ Joint compound consists of adhesive, with or without fillers, to cover joints in gypsum wallboard.
■ Joint tape is a strip of reinforcing material, for embedding into joint compound to reinforce the compound where it bridges the joint.
■ Finishing or topping compound is an adhesive, with or without fillers, applied over the tape and joint compound for a smooth finish.
■ Joint-Finishing compound; used for both applications.

SPECIFICATION
■ ASTM C475 – Standard Specification for Joint Treatment Materials for Gypsum Wallboard Construction.

TYPES
■ Drying compounds; with water-soluble organic adhesives or synthetic resins.
• Strength is gained through drying; which process is accompanied by shrinkage due to loss of water.
■ Setting compounds; which gain strength through chemical reaction.
• Compounds generally available either for on-site mixing or premixed.
• Pre-mixed compounds available in hand trowelling and machine consistency.
■ Tape is generally available of plain or perforated paper; with or without glass fiber mesh reinforcing; and as glass fiber open mesh.

ADHESIVES FOR GYPSUM WALLBOARD

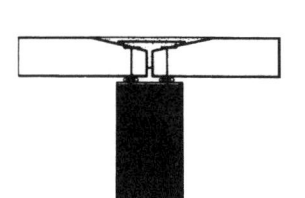

MATERIALS
■ Stud Adhesive; organic adhesive of caulking consistency. ASTM C557 – Standard Specification for Adhesives for Fastening Gypsum Wallboard to Wood Framing.
■ Dry Powder Laminating Adhesive: generally the same as Joint Compounds of the setting type.
■ Contact and Modified Contact Adhesives:
• Contact adhesive is fast drying; while
• Modified types provide up to half hour placement times.

USES
■ Stud adhesive is used to attach single layer wallboard to wood or steel studs.
• Mechanical fasteners, either nails or screws, are still required, but the number necessary can be decreased.

■ Dry Powder Laminating Adhesive is used to laminate gypsum wallboards to each other, and to masonry or concrete surfaces.
• Permanent mechanical fasteners are required at the perimeter of the boards.
■ Contact Adhesives are used to laminate gypsum wallboards to each other, or to metal studs.
• Mechanical fasteners are required at least at the perimeter of the boards.
• Adhesive will not bridge irregularities between surfaces, thus complete bonding may not be achieved.
■ Modified Contact Adhesive is used to attach gypsum wallboard to other gypsum boards, solid walls, and rigid insulation.
• Mechanical fasteners are required at least at the perimeter of the boards.
• Adhesive will bridge small irregularities between surfaces.

GYPSUM LATH

TYPES
- Gypsum Lath: incombustible core, essentially of gypsum, surfaced with paper.
  - Available as plain, perforated, and insulating. Insulating type has aluminum foil backing.
  - Also available as fire retardant – Type X.
  - Thickness: $\frac{3}{8}$" and $\frac{1}{2}$". Size 16"x48" and 24"x96".
  - Should conform to ASTM C37 – Standard Specification For Gypsum Lath.
- Gypsum Base: incombustible core, essentially of gypsum, surfaced with paper and aluminum foil backed.
  - Available fire retardant – Type X.
  - Thickness: $\frac{3}{8}$", $\frac{1}{2}$", and $\frac{5}{8}$". Size: 48" wide by 8 to 14 feet long.

USE
- Fiberboard are also used as board lath.
- Gypsum Lath used to receive base and finishing coats of plaster, about $\frac{1}{2}$ inch or more in thickness.
  - Perforated type provides for mechanical bonding of plaster; generally used in fire rated assemblies because of increased strength.
- Gypsum Base used to receive thin coat of plaster, $\frac{1}{16}$" to $\frac{3}{32}$" in thickness.
  - Less expensive than gypsum lath and plaster.
  - Joints should be reinforced with paper or glass fiber mesh tape.
  - Should conform to ASTM C588 — Standard Specification for Gypsum Base for Veneer Plaster.

METAL LATH

MATERIAL
- Expanded or punched steel: either painted with rust inhibiting coating or galvanized.
  - Weight: 2.5 to 4.0 lbs. per square yard.
- Wire lath either woven or welded: generally galvanized.
  - Weight: 1.7 to 2.5 lbs. per square yard.
  - Available with plain or waterproofed paper backing.

TYPES
- Expanded metal lath available in:
  - Diamond mesh: flat, easily bent or formed to curved surfaces. Also available in self-furring type for plastering over existing surfaces. Also available with waterproof paper or plastic backing for machine applied plaster.
  - Flat-rib lath: stiffer than diamond mesh, thus allowing wider spacing of supports. Furring over flat solid surfaces required.
  - Rib-lath: with stiffening ribs generally $\frac{1}{8}$ to $\frac{3}{8}$ inch deep. Unsuitable for contour plastering. Self-furring.
  - Self-furring lath: generally used for exterior stucco.

PLASTER

GYPSUM

MATERIALS
- Calcinated gypsum, known as plaster of Paris when 75 percent of water is removed during manufacture; and Keene's Cement, when all the water is removed.

TYPES
- Base-Coat Plaster is the portion of plaster finish which is applied to a lath base and in turn supports the Finish Coat. Base Coat Plasters, all made of plaster of Paris, are available as:
  - Gypsum-Cement Plaster, also known as hardwall or neat gypsum plaster. It comes in powder form and requires the addition of aggregate and water.
  - Sanded Gypsum Plaster requires the addition of water only. Used when good quality sand is not readily available. Equal in strength to Gypsum-Cement Plaster.
  - Gypsum Wood-Fiber Plaster contains wood fibers to give the mixture more bulk. Used where higher fire rating is required, or when good sand is not available. More expensive than sanded type, but about three times stronger in tension and compression.
  - Bond Plaster is for use over rough concrete surfaces.
  - All should conform to ASTM C28 – Specifications for Gypsum Plasters.

AGGREGATE
- Materials which are added to the base-coat mixture to minimize shrinkage, as well as to serve as fillers to add bulk, are:
  - Sand: relatively inexpensive, readily available, but heavy, at about 100 lbs. per cubic foot.
  - Lightweight aggregates: such as vermiculite, and perlite, at about 8 to 12 lbs. per cubic foot. Both will improve fire resistance rating and resistance to sound transmission.
  - All should conform to ASTM C35 – Standard Specification for Inorganic Aggregates for Use in Gypsum Plaster.
- Finish-Coat Plasters:
  - Gaging Plaster, available in quick-setting and slow-setting mixtures. Commonly used type for general plastering work. ASTM C28.
  - Molding Plaster, similar to Gaging Plaster, but finer-ground for use in ornamental work.
  - Veneer Plaster for use over Gypsum Base ASTM C587.
  - Keene's Cement; used when hard, moisture resistant surfaces are required. Is not water resistant and should not be used around bath tubs or in showers. ASTM C61.
  - Acoustic Plaster, specially formulated to provide improved sound absorption.

PORTLAND CEMENT

MATERIAL
- Mixture of Portland Cement, lime putty, and sand aggregate. For additional information refer to Cement / Concrete.

USE
- When a hard water resistant and fire retarding interior finish is required.

- May be used as scratch coat for a finishing coat of Keene's Cement.
- Should be reinforced with metal lath to control and minimize shrinkage cracking even if used over solid surfaces.

C1.1 INTERIOR PARTITION SYSTEMS

RELOCATABLE PARTITIONS: properties, materials

● denotes common usage
○ denotes possible usage

C1.1 INTERIOR PARTITION SYSTEMS

TYPE		MAXIMUM HEIGHT, range in feet	MAXIMUM LENGTH BETWEEN LATERAL SUPPORTS, range in feet	STANDARD WIDTH OF INDIVIDUAL PANELS, range in inches	THICKNESS OF PARTITION, range in inches	STC, sound transmission class	FIRE RESISTANCE RATING maximum in hours	PRIMARY SUPPORT						BRACING		
								FLOOR/CEILING TRACK	FLOOR/CEILING TRACK and POSTS/STUDS	FLOOR TRACK/TOP CAP and POSTS/STUDS	POSTS ONLY	WALL HUNG	CEILING HUNG	POSTS/STUDS TO CEILING	TOP CAP/TRIM	CROSS WALLS
CEILING HIGH																
	NONSEQUENTIAL	8	12	30	1¾	30		●	●					●		○
		to	to	to	to	to	2									
	SEQUENTIAL	12	16	60	3¾	50		●	●					●		○
CORNICE HIGH																
	REGULAR	7	10	24	1						●		●	○	●	●
		to	to	to	to											
	TOILET	10	16	60	3¾							●	●			
RAIL HIGH																
	REGULAR	3	10	24	1½						●				●	●
		to	to	to	to											
	SCREEN	5	16	60	2¼						●				●	●

MATERIALS

FACING [1] **CORE**
LAMINATED TO CORE

FINISH OPTIONS

OVER STEEL SHEETS | OVER RIGID BOARDS [2]

REMARKS

[1] **Facing** generally prefinished.

[2] **Comparative Cost:** Basis − 2 x 4 studs at 16″ on centers with ½″ gypsum wallboard facing both sides, taped and spackled = 100.

GYPSUMBOARD	HARDBOARD	SHEET STEEL	SHEET STEEL WITH BACKING	GYPSUMBOARD ATTACHED TO CORE	HONEYCOMB	RIGID INSULATION	PARTICLEBOARD	FIBERBOARD	METAL STIFFENERS	GYPSUMBOARD RIBS	BAKED-ON ENAMEL	PORCELAIN ENAMEL	VINYL COATED FABRIC	VINYL COATED FABRIC	BURLAP	CARPET	PLASTIC LAMINATE	COMPARATIVE COST, range per s.f. for complete assembly
●				●					●	●				●				120
																		to
●	●	●		●	●	●		●			●	●	●	●	●	●	●	520
●	●			●	●	●								●	●	●	●	170
																		to
	●				●		●				●						●	320
●	●			●	●	●								●	●	●	●	200
																		to
●	●	●		●	●	●					●	●	●	●	●	●	●	300

REMARKS (120 to 520):
- chalkboards and tackboards available as facings.
- stainless steel facings available.
- honeycomb available fire retardant treated.
- steel sheets generally 20 gauge; 24 gauge when laminated to core.
- prefabricated panels generally interlocking and installed in sequence; individual panels cannot be removed.
- individual panels, generally gypsum wallboard hung on a framed core, can be removed out of sequence.
- wiring for power and communication equipment generally run horizontally in base, and vertically through posts or within panel for field assembled types; base covers generally removable.

REMARKS (170 to 320):
- top cap used to brace partition; may also be used to run wiring.
- may be glazed on top, either part way, or to ceiling.
- posts may extend to ceiling, or to floor construction above, for bracing of long runs without cross-partitions.
- provisions for wiring are similar to those for ceiling-high partitions.
- toilet partitions generally available as floor-braced or ceiling-hung; marble slabs or brick masonry dividers are sometimes used, which are then supported on the floor.
- ceiling hung toilet partitions allow for easier floor maintenance.
- finishes for toilet partitions should be selected also for resistance to acid, scratches, and various types of markers.

REMARKS (200 to 300):
- glazing may be installed on top of partition; either cantilevered from top cap, or braced by extending posts.
- partitions may be closed or open at floor.
- generally braced by cross partitions; on long runs, anchorage to floor required to resist lateral forces.
- portable/relocatable screens generally used in office landscaping; may be hinged together for lateral stability.
- single screens provided with stabilizing bases or spring loaded posts.
- prefabricated cubicles available.

C1.1 INTERIOR PARTITION SYSTEMS

OPERABLE/PORTABLE PARTITIONS:

● denotes common usage
○ denotes limited or possible usage
NA not applicable

C1.1 INTERIOR PARTITION SYSTEMS

	SIZE					OPER-ATION		INSTALLATION SUP-PORT		LAYOUT					CORE			
	THICKNESS range in inches	MAXIMUM HEIGHT range in feet	MAXIMUM LENGTHS width of opening in feet	STC, sound transmission class, range	FIRE RESISTANCE RATING maximum in hours	MANUAL	POWERED	CEILING HUNG	FLOOR SUPPORTED	STRAIGHT RUN ONLY	CURVED	STACKING, SIDE	, REMOTE	, CENTER	HONEYCOMB	SOLID	METAL FRAME	WOOD/METAL STRIPS
OPERABLE																		
SINGLE TROLLEY SINGLE PANEL			no limit			●		●	●				●	●	●	●		
HINGED PAIR					1½	●		●	●	●	●		●				●	●
MULTI-PANEL HINGED	1¾ to 3¾	10 to* 36	100	23 to 54		●	●	●		●	●	●	●				●	●
DOUBLE TROLLEY SINGLE PANEL					2	●	●			●	●	●					●	●
ACCORDION FOLDING	varies	10 to 20	no limit	35 to 50	NA	●	●	●	●	●	●						●	●
SIDE COILING		10 to 14	100	NA	NA	●	●	●	●	●	●							●
PORTABLE																		
CEILING HIGH	1½ to 2	4 to 14	6 for panel	23 to 48	NA			●	●	●						●		●

properties, materials

Legend: • = applicable ○ = optional/limited

FACING		FINISH					GUIDED/SECURED				SEALS						COMPARATIVE COST, range per s.f. for completed assembly [1]	REMARKS
HARDBOARD	SHEET STEEL	VINYL COATED FABRIC	BURLAP	CARPET	WOOD VENEER	PLASTIC LAMINATE	FLOOR GUIDES	CEILING GUIDES	COMPRESSION SYSTEM	TOP/BOTTOM TRACK	FLOOR, FIXED	AUTOMATIC	TOP	JAMB, FIXED	SPRING LOADED	BETWEEN PANELS		
•	•	•	•	•	•	•	•	•			•	•	•	•	•	•	900 to 1400	• single trolley single panels can be used as continuous closure, can be spaced out, or rotated to allow passage at any point. • single trolley hinged pair can be folded to provide openings between panels for passage. • all single panels may be suspended from omni-directional track and do not require curved track at intersections. • pass-doors may be provided in most types. • chalkboards and tackboards may be substituted for other types of facings. [2] fire resistance rated partitions are of limited availability; consult manufacturer's literature. • not all available with sound transmission loss rating; consult manufacturer's literature. • pockets to receive stacked panels may be used. * larger sizes may be available.
•	•	•	•	•		•					•	•	•	•	•	•		
•	•	•	•	•		•					•	•	•	•	•	•		
•	•	•	•	•		•	•	•			•	•	•	•	•	•		
		•			•						•		•	•	○		450 to 550	• small sizes available to fit standard door openings for closets; similar in construction to operable walls, but not generally available rated for sound transmission loss. • pockets to receive door in folded position may be used.
					•	○					•		•	•	○		750 to 1550	• steel coiling type similar to steel roll-up doors. • pocket to receive partition when retracted generally required; or may be furnished as part of assembly.
•		•	•	•		•			•	•	○		○			•	250 to 800	• may be internally wired. • less-than-ceiling high types available; refer to evaluation chart for Relocatable Partitions. • ceiling high types generally secured by wedging between floor and ceiling. • also available with floor and ceiling tracks, which are secured in place, and cannot be removed without damage to flooring or ceiling construction.

[1] Comparative Cost: Basis = 2 x 4 studs at 16″ o.c. with ½″ gypsum wallboard facing both sides, taped and spackled = 100.

PARTITIONS: selection checklist

PERFORMANCE REQUIREMENTS

Stability is the one basic requirement common to partitions of all types and heights: they must resist overturning under:

■ **Normal design air pressure:** generally five pounds per square foot;

■ **Suspended loads,** such as shelves, cabinets, or equipment.

■ **Concentrated horizontal** forces, such as accidental impact.

■ Some must also resist **high or extreme air pressure,** such as may be generated in:
• high speed elevator shafts
• pressurized testing rooms, through controlled explosion.

In selecting an assembly to meet performance requirements, consideration must also include performance characteristics of **the floor the partition is supported by and of the ceiling/floor/roof assembly above,** to which the partition is connected. Some of the possibilities include:

■ **Even when sound transmission** is effectively controlled by the partition itself:
• sound may outflank the partition through the ceiling/floor/roof construction
• through closely located open windows serving two adjacent spaces.

■ **Air leakage and heat flow** may occur:
• around light fixtures recessed in the ceiling
• between acoustic tiles and their suspension system
• under and/or over the partition if not completely sealed
• at pipe or duct penetrations through the partition, or the floor/roof construction.

■ **Fire resistance rating** of a space is determined by the entire enclosing construction.

FIXED PARTITIONS

Fixed partitions, whether rail, cornice, ceiling, or full-height **offer the widest choice in materials and types of assembly.**

Framed Partitions

Framed Partitions, of wood or metal studs, are the **lowest in cost** of general purpose partitions and **can be easily modified** to meet specific performance requirements;

■ **Sound transmission resistance** may be increased by:
• providing additional layers of gypsum wallboard, generally of different thicknesses, to change frequencies of vibration
• separating studs to isolate one face of the partition from the other
• installing insulating blankets between studs to absorb sound
• resilient attachment to studs of one face layer
• combination of some or all of the above

■ **Fire resistance** may be provided and/or increased by rated materials, such as:
• Type X gypsum wallboard
• specially formulated plaster

■ **Moisture resistance** may be provided by:
• water/moisture-resistant gypsum wallboard
• Keene's cement plaster in occasionally moist locations
• Portland cement plaster in wet locations.

■ **Heat flow may be resisted by:**
• thermal insulation between studs
• split studs with the cavity filled with batt, rigid, or foamed in place thermal insulation
• layers of metal foil separated by air spaces

■ **Impact resistance** may be increased by:
• hard plaster finish, such as Keene's cement or Portland cement
• applied facings over gypsum wallboard, such as plywood or metal plates
• plaster is much easier to repair than gypsum wallboard if accidentally damaged.

■ **Sound absorption** may be provided by:
• acoustic plasters or prefabricated acoustic blocks which, however, are not resistant to impact and should be used only in protected locations
• applied sound absorbing finishes, such as felt wallcoverings, or specially woven carpeting.

Finish options for framed partitions include:

■ **Opaque coatings;** flat, semi-gloss, or gloss:
• flat coatings tend to hide surface unevenness and imperfections,
• are more porous and will pick-up and hold dirt more readily than semi-gloss or gloss coatings.

■ **Chemical resisting coatings:**
• may be used over gypsum wallboard, but:
• gypsum wallboard is too easily damaged to be used in commercial or industrial areas where such coatings may be required
• for more information see COATINGS

■ **Wallcoverings** made of vinyl fabric or paper may be installed over gypsum wallboard or over smooth finished gypsum plaster.

■ **Gypsum wallboard** that is factory-finished with vinyl wall fabric is available:
• installation is more costly than for field applied wallcovering
• skilled workmen are required.

■ **Plaster may be:**
• finished smooth
• in several textured finishes, such as sand, brush, travertine, brick, rough stone or marble imitation
• some of the latter finishes are no longer in wide usage and skilled workmen may not be readily available.

■ **Wood paneling,** prefinished or site-finished after installation, can be applied directly to studs; such use is not recommended:
• thin paneling tends to act as a sounding board without a gypsum wallboard back-up
• may also be easily damaged
• paneling is generally not used with plaster back-up because of the increased cost.

Stacked Partitions

Stacked Partitions, of concrete, burned clay, or gypsum units are generally **used when in-service conditions require:**
• a more than average resistance to impact, either accidental or malicious
• fire resistance rating.

When selecting concrete units, consideration should be given to:

■ **Sound transmission** resistance will depend upon the density of the units selected:

- porous block will absorb sound, but will not significantly resist its transmission.
- resistance of porous block to sound transmission will be improved by a heavy coating, plaster finish, or a dense applied finish.

■ **Fire resistance** may be improved by:
- lightweight aggregate used in manufacturing the units
- plaster finish on one or both faces.

Finish options for stacked partitions include:

■ **Opaque coatings:**
- for appearance and easier maintenance
- to protect the units from chemical attack.

■ **Ground exposed faces** for appearance and easier maintenance without the need for applied coatings.

■ **Glazed exposed faces** where resistance to disinfectants and cleaning agents, fatty substances, or process chemicals is required.

■ **Integrally glazed** blocks are generally less expensive than blocks with added applied glazed ceramic tile, but are harder to repair if damaged.

Solid & Monolithic Partitions

Solid and Monolithic Partitions of gypsum board, plaster, and concrete are generally **used to separate or isolate when:**

■ **Wiring** is not required to be concealed within the thickness of the partition

■ **Minimum partition thickness** and relatively high fire resistance rating are required.

■ **Concrete partitions,** in addition, will provide better resistance than other types to:
- impact
- high pressures
- wetting and drying
- forced entry

RELOCATABLE PARTITIONS

Relocatable, or demountable, partitions are either fully or partially prefabricated assemblies **capable of being completely dismounted and re-erected with all or most parts reused:**

■ Partitions are **generally installed:**
- after flooring and ceilings are in place
- **are attached to the floor through the flooring and to the finished ceiling.**

Based on method of erection, demountable partitions may be classified as:

■ **Sequential,** where completely prefabricated panels cannot be individually removed, since they are:
- either interlocking,
- or fit into special posts.

■ **Non-sequential,** where prefinished facing panels are
- secured to a framed core
- provisions are made to allow either each, or some, to be removed from either side of the wall without affecting adjacent facing panels.

■ All relocatable partitions provide **space for electrical/communications wiring;** most completely prefabricated types have removable baseboards to allow for such wiring to be quickly and easily changed.

■ Use of relocatable partitions should be carefully evaluated to determine whether the **flexibility they provide justifies the additional cost.**

PORTABLE PARTITIONS

Portable partitions are completely prefabricated units, either free standing screens, or ceiling high panels which may be butted edge to edge to form a complete partition:

■ **Portable screens may be:**
- floor braced;
- secured in place with spring-loaded posts; or by
- screw actuated top and/or bottom cap.

■ Portable partitions **may provide electrical power outlets,** wired internally to an exterior plug, which is then used to connect the panel to the building power supply.

OPERABLE PARTITIONS

Operable partitions are essentially **portable partitions designed to be movable along a fixed track.**

■ **They range:**
- from small prefabricated units similar to a multi-fold door, and used as such
- to entire walls which open to join several spaces into one.

Operable partitions may be

classified as follows:

■ **Single trolley:** each panel is supported at a single point in its center; three sub-types are seen:
- single panel: each individual panel is supported
- double panel: panels are hinged in pairs and each pair is supported at a single point
- multi-panel: all panels are hinged to each other and the entire series is supported at a single point for each panel.

■ **Double trolley:** each panel is supported at two points.

■ In considering **single-trolley vs double-trolley** partition design, note:
- all single-trolley partition panels can be rotated singly, in pairs or as a bundle, depending on the type
- this allows partial opening and thus flexibility of passage, vision and air control where desired
- double-trolley partitions generally offer:
    greater stability
    ease of operation, but
    panels cannot be rotated.

In selecting operable partitions note:

■ **Sound transmission** resistance requirements should be carefully evaluated since higher STC ratings will also significantly increase cost.

■ **Seals provided** to minimize sound transmission must be carefully detailed and installed to minimize interference with the operation of the partition.

■ **Fire resistance rated** partitions are available in specific tested assemblies only.

■ All operable partitions are only available as **completely prefabricated assemblies:**
- generally installed by the manufacturer or his representative only.
- construction is standard with each manufacturer
- finish options are generally available.

C1.1 INTERIOR PARTITION SYSTEMS

CRACKING IN PARTITIONS

Partitions may crack in any type of structure **when installed tight** to the structural frame:
- The frame may restrain movement in the partition due to changes in moisture and/or temperature,
- movement in the frames, whether caused by changes in moisture content, thermal expansion/contraction, wind loads, excessive deflection, seismic loads, or foundation settlement may impose loads upon the partition in excess of its capacity to resist them.

One particular problem is partition cracking in reinforced concrete structures. Concrete framed structures are subject not only to all the forces acting on other types of assemblies, **but also to creep,** or plastic flow in concrete. Creep is the long term deformation under a sustained load, such as the dead load of the structure itself.

Instantaneous deflection in the frame will take place right after the removal of the shoring due to the dead load of the frame.

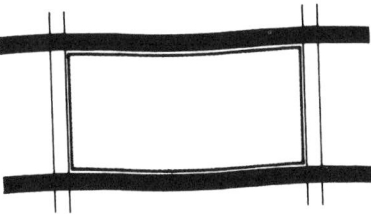

A partition is built tight to the horizontal structural elements shortly after the shoring has been removed, and **before any significant amount of creep-related deflection has taken place.**

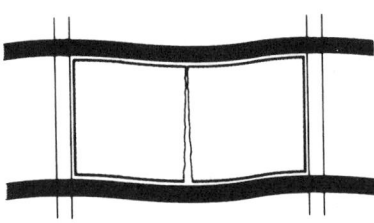

Frame continues to deform, imposing continuously increasing bending stresses in the partition which may ultimately crack it. Since the rate of creep decreases with time, cracks due to creep-related deflection may generally be successfully repaired. It should also be noted that **creep will also result in the shortening of columns,** especially of those carrying heavy, sustained loads.

C1.1 INTERIOR PARTITION SYSTEMS

C • INTERIORS

REFERENCES: Interior Partition Systems

STANDARDS

ASTM Specifications for:

C842	• Application of Interior Gypsum Plaster
C843	• Application of Gypsum Veneer Plaster
C847	• Metal Lath
C932	• Surface-Applied Bonding Agents for Exterior Plastering

ASTM Tests/Methods for:

E336	• Airborne Sound Insulation in Buildings, Test for Measurement of
E795	• Mounting Test Specimens During Sound Absorption Test

ASTM Definitions of Terms Relating to:

C11	• Ceiling and Walls
C634	• Environmental Acoustics

ASTM Classifications for:

E413	• Sound Transmission Classification

ANSI Specifications:

ANSI/AHA

A135.5-1982	• Prefinished Hardboard Paneling
A208.1-1979	• Particleboard, Mat-Formed Wood

SELECTED REFERENCES

• Paper Honeycomb Cores for Structural Building Panels: Effect of Resins, Adhesives, Fungicide and Weight of Paper on Strength and Resistance to Decay, Forest Products Laboratory, Madison.

• The Performance of a Curtain Wall Panel Faced with Ceramic Tile, test performed by Department of Engineering Research, Penn. State University.

• A Reflective Method for Testing Flatness and Thermal Buckling of Metal Panels (Study No. 4), John H. Callender, School of Architecture, Princeton University, N.J.

• Sound Insulation of Wall & Floor Construction, U. S. Department of Commerce, Building Materials and Structures Report 144, Washington.

APA American Plywood Association

F405	• APA Product Guide: Performance-Rated Panels
F800	• APA Product Guide: Panel Care & Installation

BIA Brick Institute

5A	• Sound Insulation - Clay Masonry Walls

GA Gypsum Association

600-81	• Fire Resistance Design Manual Tenth Edition
201-80	• Using Gypsum Board for Walls and Ceilings
216-82	• Recommended Specifications for the Application and Finishing of Gypsum Board
252-82	• Fire Resistant Gypsum Sheathing
218-82	• Recommended Specifications for the Application of 5/16" MH Gypsum Board to Wood Framing
504-74	• Gypsum in the Age of Man
650-80	• Gypsum Association Specifications - Covering Existing Interior Walls and Ceilings with Gypsum Products

ML/SF A Metal Lath/Steel Framing Association

No. 101	• Types of Metal Lath and Their Uses.
No. 151	• Sound Insulating Partitions and Floors.

NPA National Particleboard Association

2	• Control of Warping of Wood, Panels Faced with Plastic Laminates
4	• Particleboard Shelving Design
13	• V-Grooving With Particleboard

NRCC National Research Council of Canada

CBD 10	• Noise Transmission in Buildings.
51	• Sound Insulation in Office Buildings.
92	• Room Acoustics - Design for Listening.
139	• Acoustical Design of Open-Planned Offices.
164	• Acoustical Effects of Screen in Landscaped Offices.
186	• Office Partitions: Acoustical Requirements for Design and Construction.

OCFC Owens-Corning Fiberglass Corp.

• AIA File No. 39-3 1964 Solutions to Noise Control Problems in the Construction of Houses, Apartments, Motels and Hotels.

TCA Tile Council of America

137.1	• Recommended Standard Specifications for Ceramic Tile.

USDA U. S. Department of Agriculture

	• Basic Properties of Three Medium-Density Hardboards.
FDL 225	• Properties of Particleboards of Various Humidity Conditions.

C1.1 INTERIOR PARTITION SYSTEMS

C2. 1 STAIRCASES

STAIRCASE DESIGN

STAIRS AND STAIRWELLS
PRELIMINARY CONSIDERATIONS
Preliminary selection of stair configuration depends on:

- floor space available,
- floor-to-floor height, pedestrian movement volumes and patterns,
- groups of people who will use the stair.

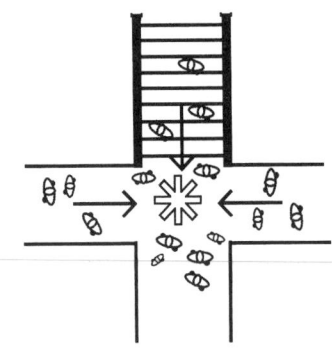

Fig. 1. Avoid conflicting pedestrian movement directions at the top or bottom of the stair

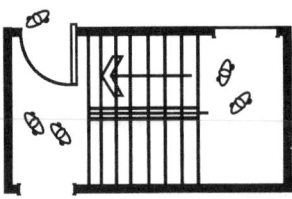

Fig. 3. Avoid entering or exit hazards

■ **Stair user group.** Potential users may be able-bodied adults, specific handicapped groups, children, the elderly, or all of these groups. The potential use of the stair may be as a means of access and egress; in a public or monumental location, with a large volume of pedestrians; or in a private location. There are special design considerations for some groups:

- **Disabled:** Provide an alternative access way so those who cannot use stairs are not denied entry.

- **Elderly and disabled:** For the elderly and some handicapped people, designs that depend only on stair access and egress should be avoided.

Fig. 2. Avoid direction, view, and illumination changes

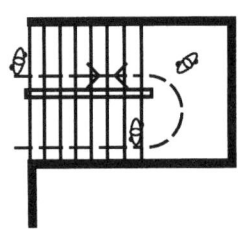

Fig. 4. Avoid keep-right conflicts

- **People carrying food, drinks, and supplies:** Avoid placing stairs where people will routinely carry food and drink that may spill on the treads. Avoid stair designs where workers must carry bulky or heavy objects, or anything else that may limit their view of the treads or affect their balance.

- **Children:** Stairs for use by children needs an additional, lower handrail. For control of toddlers, the stair may be closed by a gate.

■ **Factors that influence stair location**

- Where large crowds of people will use the stairs (theaters, stadia, fire egress for large buildings), the stair should be located so that it does not cause a hazardous bottleneck by making a sudden change of direction, such as a dog-leg stair. The stair should be located to encourage continuous direct flow.

- Avoid pedestrian movement

directional conflicts at the top and bottom of the stair, such as a stair leading directly to and across a passage with heavy traffic.

- Avoid direction, view, and illumination changes, such as a stair landing where sunlight may blind users or a fascinating view that may distract attention.

- Avoid entry- or exitway hazards, such as a door opening directly onto a landing.

- Avoid a configuration that violates the "keep-right" principle (the convention in the United States.) Helical flights that ascend by spiraling up to the right, and dog-leg and other layouts that ascend to the right, enable those ascending to keep to the right with little effort. This reduces the likelihood of conflicts with those descending the same stair.

- Avoid fire escape stairs that

continue past the ground-floor egress point down into a basement. These may mislead people during an emergency, drawing them down to a dead end.

■ **Stair types and performance**

- **Direction and flow:** The direction of travel that people will take to and from the top and bottom of the stair will be affected by the stair layout, and vice versa.

- **Stair shape, area, and performance:** The amount of floor space a staircase occupies is related to its shape. Some layouts use less space than others, and some are more effective than others for moving stretchers, furniture, crowds of people, and so on.

STAIR DIMENSIONS

Several codes and standards may govern stair design in buildings. Refer to the applicable building code, fire code, handicapped code, occupational safety and health regulations.

C2.1 STAIRCASES

The recommendations here are based on research findings, not code requirements.

■ **Width**. Stair width should be a function of comfort, capacity, and reach (handrail availability). The minimum design width of stairs for single-file use should be 29 inches (74 cm) for a public place. For comfort, about 38 inches (97 cm) is necessary. Side-by-side walking clearance dictates a width between walls of at least 56 inches (1.42 m),

■ **Riser-tread dimensions**. The **riser height** is the vertical height of a single step. The going is the horizontal distance, in plan, from the nosing edge of a step to the nosing edge of the next adjoining step. The **tread depth** is the horizontal distance, in plan, of a tread. Some codes require a nosing overhang if the treads are small. Dimensions of adjoining risers and treads must be constructed to be constant and regular.

■ **Landings.** Landings of stairs that have no change of direction are called **intermediate landings.** A change in direction of 90 degrees results in a **quarter landing** (wide L) and a change in direction of 180 degrees produces a **half-landing** (narrow U).

• Landings on long flights of stairs are necessary to provide a resting place for stair users, to form turning zones, and, in the event of a fall, to break the force of the fall. The minimum width of the landing must be equal to that of the widest flight of steps that reaches it.

■ **Headroom.** Headroom is measured vertically from the front edge of the nosing of a step to the finished ceiling.

■ **Length of Run.** A stair flight should always consist of three or more risers. The flight should not have too many steps without a landing. In many codes the maximum number of risers permitted in a flight is eighteen.

STAIR CONFIGURATION AND LAYOUT

Layout techniques vary in implementation with the configuration chosen, but in all cases, an acceptable and consistent riser and tread relationship should be established in order to provide a stair requiring minimum energy expenditure as well as maximum potential for safety. Typical configurations are:

• straight flight stair
• flight with winders and splayed steps
• spiral stair
• helical stair with an open well
• ramp is any part of a constructed pedestrian circulation way with a slope greater than 5 percent. Acceptable gradients for ramps depend on the length of ramp to be used and the location of landings

STAIR DETAILS
■ Step Shape

• **Nosings:** Abrupt nosing overhangs and any overhang that is greater than 3/4 inch (1.9 cm) should not be constructed. Where a nosing is needed, provide backward-sloping nosing overhangs (less than 3/4 inch) where this is possible.

• **Wash:** A wash is needed to throw water off external steps. It should not exceed about a 1:60 slope.

• **Dimensional regularity:** Risers, treads, and nosing projections must be constructed with a high degree of dimensional consistency, and most building codes insist on this.

■ MATERIALS

The appropriate choice of material for stair treads is influenced by several factors: structural considerations, the type and volume of traffic, appearance, resistance to wear and sometimes chemicals and the climate, ease of maintenance and cleaning, cost, slip resistance, and how comfortable it is to walk on. Avoid soft woods and stone that may erode easily.

• **Floor patterns:** Avoid the use of

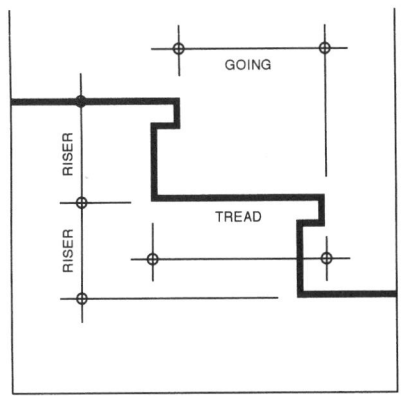

Fig. 5. Riser, tread, and going

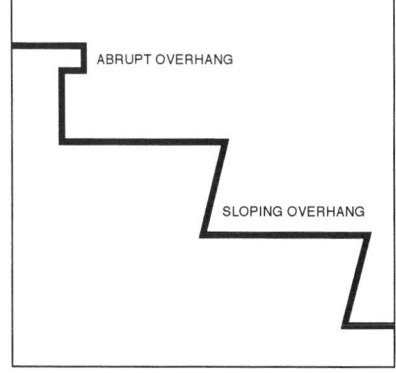

Fig. 6. Nosing overhangs

textures or patterns or nosing strips that make it difficult to discern the edge of the nosings. Use colors to emphasize the edge.

• **Carpet fixing:** Avoid step and, particularly, nosing designs that will prevent carpeting from being fixed firmly.

• **Slip resistance:** Coefficients of Friction (COF) greater than 0.3 may be adequate to prevent slips on level, dry, step surfaces in internal stairs at normal rates of climb, but coefficients of 0.5 are considered preferable and safer. The most critical area for slips is at the nosing, but it is better to ensure that the whole tread has an adequate COF rather than to add abrasive nosing strips that may be visually confusing and may cause trips.

• **Tread surface:** The surface of stair treads should be smooth, free from projecting joints stable under the loads with no tendency to shift underfoot,

C2.1 STAIRCASES

STAIRCASE DESIGN (cont.)

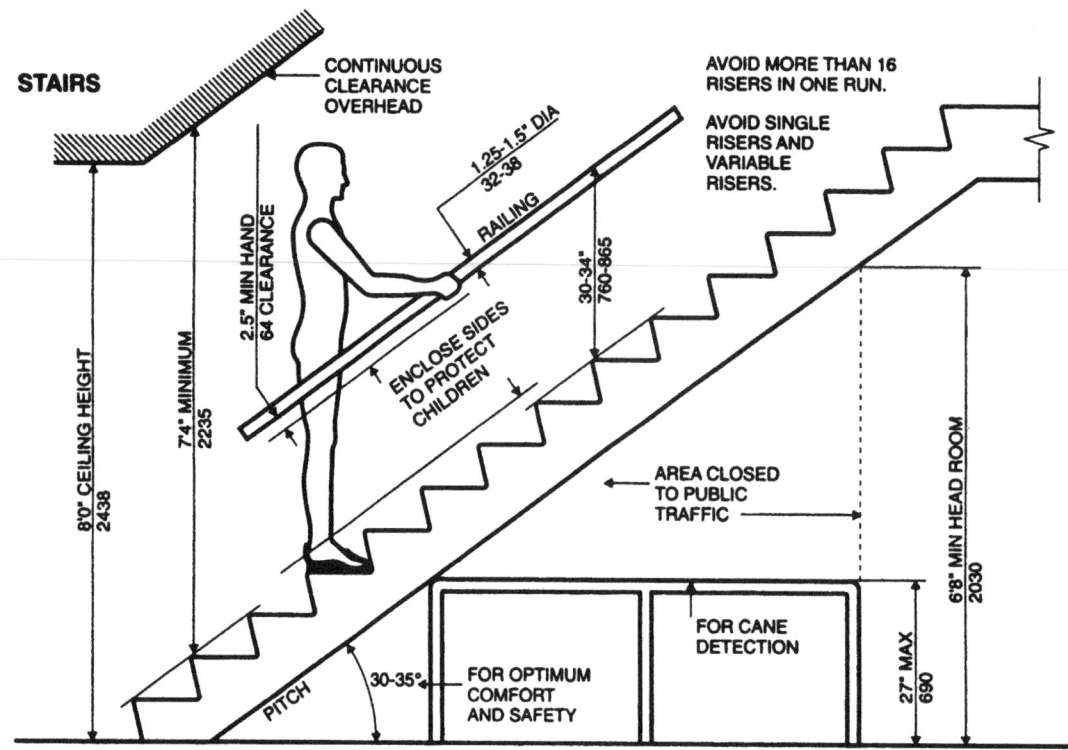

Fig. 7 Stairway dimensioning. Reproduced by permission. Henry Dreyfuss Associates. 1993. *The Measure of Man and Woman.* New York: Whitney Library of Design.

and with nothing to catch the shoe.

•   **Injurious materials:** The stair as well as the handrails and balustrading should be free from projecting elements, sharp edges and corners, and any rough surfaces, bars, rods, and other elements that have a small section. Instead, smooth, flat, impact-attenuating surfaces and gentle curves should be used.

■ HANDRAILS

•   **Location:** A stair that is 35 inches (88.9 cm) wide between the walls and with a rail on one side has the maximum feasible width if the rail must fall within the reach of adult users. A 47-inch (1.19 cm) wide stair with handrails on both sides is the maximum width for both rails to be always available to adult users. Current codes permit a single rail for stairs up to 44 inches (.12 m) and two rails for stairs up to 88 inches (2.24 m).

•   **Height:** From the forward edge of the nosing to the top of the handrail should

be 36 to 40 inches (0.91—1.02 m), but codes usually require 30 to 34 inches (76.2—86.4 cm). For children, an intermediate rail that is 21.8 to 28.7 inches (55.4—72.9 cm) should be provided.

•   **Extent:** Handrails should extend horizontally a minimum of 12 inches (30.5 cm) beyond the top of the stair and beyond the bottom riser for a distance equal to the tread width and then continue horizontally for 12 inches (30.5 cm). The handrails should not project into walkways; the ends should return to the floor or adjoining walls. Handrails should continue along at least one side of a landing.

•   **Size and shape:** A circular handrail 1-1/2 inch (3.8 cm) diameter is most effective for gripping.

•   **Spacing distance from walls:** The clearance between a handrail and an adjoining wall, assuming a circular handrail with 1-1/2 inch (3.8 cm) diameter, should not be less than 3.65

inches (9.3 cm).

•   **Materials:** Handrails should not be too slippery or too rough. Handrail materials should not conduct or retail heat to the degree that they become untouchable. Handrails must not be permitted to deteriorate so their surfaces become splintered or pitted with rust. Finally, the color of handrails should be carefully chosen so they are always highly visible. For optimal stair safety, consider that rails that are lightly padded like auto steering wheels can provide the correct range of friction characteristics, may enable better grip forces to occur, and are less likely to cause trauma if they are hit in a fall.

■ GUARDRAILS

•   **Height:** Guardrails should not be less than 42 inches (1.07 m) high unless the width is greater than 6 inches (15.2 cm). In that case, the minimum height of the guardrail should not be less than 48 inches (1.22 m) minus B, where B is the minimum

STAIRCASE DESIGN: safety checklist

width of the top surface of the guardrail. The rail height should never be less than 30 inches (76.2 cm). The design should discourage people from climbing onto a rail from which they may overbalance and fall.

• **Structural loading:** Most codes require the system to be able to withstand a test load without much deflection. A typical test requires the system to withstand a 200-pound (90.7 kg) vertical load at the midspan of the rail, with a deflection that does not exceed the length of the rail divided by 96.

• **Baluster spacing:** As a check of spacing, it should not be possible to pass a sphere of 3 1/2-inch (8.9 cm) diameter through the balustrading. This recommended dimension may be countervened by local code requirements.

STAIRWAY LIGHTING

Proper illumination of a stairway is essential for both comfort and safety. Adequate lighting must be provided under both electric lighting and daylighting conditions. The following guidelines address accident prevention and safety recommendations to reduce and eliminate accidents.

• **Adequate illumination,** either natural or electric, must be provided. The IES recommendation of 20 foot-candles (215 lux) for most applications seems to be much more realistic for stair safety than those minimum levels permitted by most building codes. A minimum of 8 foot-candles (86 lux) may be adequate.

• **Reasonably constant illumination** should be evenly spread over the whole stair.

• **Window and electric light sources** should not be placed where the stair user must include them and the steps in the same direct field of vision, and shadows on the steps should be avoided. In cases of electric lighting, attention must be given to avoid shadows that cast confusing foot and floor patterns. In cases of daylighting, attention must be given to solar glare and reflections that may occur

only at specific times of the year, falling on parts of the stair that make it difficult to see and navigate.

• **Permanent supplementary electric lighting** illumination should be provided if the stair is located where there is any risk that someone might stumble into it unexpectedly.

• **Switches that control the stair lights** should be placed sufficiently far from the stair so there is no risk of a person's falling while reaching for the switch. Three-way switches should be used at the top and bottom of the stairs.

SAFETY CHECKLIST TO REDUCE AND PREVENT ACCIDENTS

People have incidents on stairs when certain conditions are present. There are credible explanations why these conditions are dangerous. Defects can be reduced and eliminated by appropriate design, responsible construction, and vigilant maintenance. The following commentary provides an overview of conditions to be avoided. Design recommendations are highlighted as bulleted items.

Inappropriateness of stairs: Because stairs are inherently more dangerous than level walkways or ramps, there are some circumstances in which they should not be used, no matter how well designed and constructed.

■ When feasible, stairs should not be the sole means of access or egress, particularly for elderly and physically limited people and in places where alcohol is served.

■ If changes of level are inevitable, alternative means of access and egress, such as ramps and elevators, should be provided.

Inappropriate connector: Stairs are sometimes located where they will be significantly more dangerous.

■ If people traveling between two adjacent areas are likely to be carrying bulky or heavy articles, the areas should not be linked by means of steps. The change of level should be avoided, or the areas linked in some other way.

■ If there is likelihood that potentially slippery materials such as food or

drinks may be accidentally spilled on the stairs, the stairs should be isolated from the source and traffic pattern of handling the lubricant.

Hidden flights and dangerous locations: Steps are sometimes located where they may not be noticed. People may fall down stairs because they are unaware of the stair.

■ For the safety of the severely visually impaired, stairs should be located out of direct pedestrian ways so that pedestrians must make an intentional (if minor) detour to use them.

■ Doors by a landing should not open directly onto a flight of stairs.

■ Steps and stairs should be located where they may be easily seen before they are encountered. This may mean that night lights and similar permanent illumination must be provided, or it may mean that in daylight the steps must be made to contrast visually with their surroundings.

■ Single steps should not be built except at thresholds, curbs, and other places where they are customarily found. Even in some of these places, if there is any doubt whether they will be noticed, the best solution is usually to eliminate them.

■ Where it is considered likely that some severely visually impaired people or small children may fall into the stair, the stair may have to be closed off by means of a self-closing gate or some equally effective device.

More and less dangerous stair layouts: Many building codes and conventional wisdom suggest that helical stairs, long straight flights, and stairs with winders are dangerous and should be avoided or prohibited for many uses. Despite some attention paid to the hypotheses in the most significant stair studies, these conclusions are still not entirely established or substantiated.

■ Long straight flight stairs without landings are usually prohibited by building codes and should not be built.

C2.1 STAIRCASES

STAIRCASE DESIGN: safety checklist (cont.)

■ Dogleg and similar layouts with short flights are safer than straight flight stairs.

■ Dogleg, helical, and complex layouts should ascend in a clockwise direction to avoid the generation of conflicting streams of traffic.

■ For stadia, auditoriums, and other locations with substantial traffic volumes, stair layouts with abrupt turns should not be constructed to avoid the risk of people being crushed in the event of panic,

■ The stair must be large enough for the anticipated volume of pedestrians.

Too long or too short flights: Most construction codes impose limits to the number of steps permitted in one flight. Before concluding that long flights are safer, one must examine where most accidents occur. Templer (1992) cites research reports where one-third of stair incidents occurred on either the first or last step; an additional 25 percent occurred on the second or next-to-last step; and another 12 percent on the third or third-from-the-last step. That is, 70 percent of accidents occurred on the top three or bottom three steps.

■ Special attention must be paid to the design and construction of the top three and bottom three steps in any flight.

The issue of (dangerous) views from stairs: some research: If views from stairwells distract us from attending to the stairs, a misstep becomes more likely. The problem may be caused not simply as a function of views per se but by whether they distract our attention to the complete exclusion of alternative interesting scenes.

■ As a generality, anything that induces stair users to focus attention on the stair rather than the surroundings increases safety. Where this is not possible, the views should be wide open at all points on the stair, not suddenly revealed at certain places and certainly not only at the top three or bottom three steps by the landings where most accidents occur.

Where is the tread edge? Precise information is needed about the exact location of the front edge of the tread of each step. In ascent, finding the nosing edge is not usually difficult because it is close to the eyes and because both risers and treads are visible. In descent, the tread edge may be inadvertently camouflaged, even for those with perfect vision, as a result of the step design,

■ Avoid the use of flooring and carpet patterns and abrasive strips that may be visually confused with the edge of the nosing.

■ Increase the visual contrast between adjoining treads. This may be achieved by a modest change of tone.

Stairs in poor lighting conditions: Our ability to walk safely on stairs is contingent upon sufficient illumination to permit us to see the stair clearly.

■ adequate illumination, either natural or electric, must be provided. The Illuminating Engineers Society (IES) recommendation of 20 foot-candles (215 lux) for most applications seems to be much more realistic standard for stair safety than those permitted by many building codes.

■ The illumination should be reasonably constant over the whole stair. Shadows on the steps should be avoided, which in effect are equivalent to absence of illumination.

■ Window and electric light sources should not be placed where the stair user must include them, and the steps, in the same direct field of vision.

■ In terms of reflectance, the IES recommends 21 to 31 percent for floors.

■ If the stair is located where there is any risk that someone might stumble into it unexpectedly, then permanent supplementary electric lighting should be provided.

■ The switches that control the stair lights should be placed sufficiently far

from the stair so that there is no risk of a person's falling while reaching for the switch. Similarly, three-way switches should be conveniently placed at the top and bottom landings of the stairs.

■ Stairways placed near windows and doors may experience blinding sun rays or glare for selective periods of the year and thus be poorly illuminated for certain periods due to absence of light and sun control.

Risers and treads: The dimensions of risers and treads control our gait, our agility and comfort, and the probability of accidents.

■ The upper limit for riser height, the lower limit for tread depth, and the best combination of dimension for risers and treads for the most usual circumstances in buildings have been established (see accompanying Table), providing quite explicit design criteria and norms.

Projecting nosings: A nosing projection that is no more than about 11/16 inch (1.75 cm) adds a modicum of safety compared to a flight with no nosings or a flight with larger nosings. Where the nosing overhang is formed by sloping the riser back from the nosing edge, this seems to cause no difficulties, and these are generally permitted. Fixing carpeting around steps that have abrupt nosing overhangs is difficult to install and are easily loosened. Carpets are often left to bulge out around the projection without adequate fixing. The looseness of the carpet and the lack of definition of the edge of the nosing then become a new hazard.

■ Abrupt nosing overhangs, and any overhand greater than 11/16 inch (1.75 cm), should not be constructed.

■ Nosings of 11/16 inch (1.75 cm) or less seem to make steps safer.

■ Backward-sloping nosing overhangs should be used rather than abrupt nosings.

Tread surfaces and materials: The appropriate choice of a material for

C2.1 STAIRCASES

stair treads will be influenced by structural considerations, the type and volume of traffic, appearance, resistance to wear, and, sometimes, chemicals and the climate, ease of maintenance and cleaning, cost, slip resistance, and walking comfort. No materials seem to be significantly safer or less safe than any others if they are properly maintained.

■ The surface of stair treads and nosings should be smooth and even. It should be free from projecting joints or nosing strips and stable under the loads, with no tendency to shift underfoot.

■ The surface should have unacceptable coefficient of friction. (Recommended values for slip-resistance coefficient are presented in the prior section.)

The wash: The wash is the slope of the tread from riser to nosing edge. A wash is provided to throw water off the stair or to make the risers less high and therefore easier to climb.

■ The slope of the wash should not exceed about 1:60.

Handrails, guardrails, and balustrades: Handrails, guardrails, and balustrades are each needed for quite different purposes. The purpose of guardrails and balustrades is to protect people from falling over the edge of a platform, landing, balcony, stair, and so on. Handrails serve to prevent a loss of balance, to help one regain balance, to help pull oneself up a stair, and for directional guidance and stability. See discussion above for recommended dimensions.

The way the stair is built: Some of the greatest accident reductions can be realized simply by insisting that stairs be constructed according to the original design and that a reasonable level of precision be present in the finished product.

■ Risers, treads, and nosing projections must be constructed with a high degree of dimensional consistency (most building codes wisely insist on this). The differences in dimension

between the largest and smallest in a flight and between those in adjoining steps should not exceed 3/16 inch (0.48 cm). This is by no means too exacting a standard for contemporary building practices.

The way the stair is maintained: Levels of deterioration and damage that might be acceptable for level walkway surfaces cannot be tolerated on stairs; the risk of a serious injury is too great. The greatest danger comes from any structural weakness that may cause any part of the stair to break during use. Nearly as bad is the condition where the treads on the surface material are chipped, torn, splintered, loose, or excessively worn. Of as much concern are handrails and balusters that are broken, missing, or loose. Ease of stair maintenance can be helped by design that allows easy cleaning and inspection.

■ Stairs should never be used as a location for storing objects.

■ Stairs must be kept clean and free from precipitation, dust, dirt, and anything else that might act to trigger slips and trips.

■ The surface finish of treads (and carpets especially) must not be allowed to deteriorate noticeably.

■ Handrails and balustrades must be kept in good repair, firmly fixed, and structurally sound.

■ Electric illumination sources must be kept in good operating condition.

Reducing injuries from stair accidents: intelligent design, construction, maintenance, and use can reduce stair accidents, but this is not enough. Toward this end, the idea of the soft stair is proposed. Using experience with safer automobile design and by a careful choice of materials and details, stairs can be designed to obviate certain types of injuries resulting from falls, and to greatly reduce the severity of others.

■ Most injuries from stair falls are caused by collision with the steps and

landings, so the greatest rewards in injury reduction will be derived from softening these—rounding the nosings and reducing the hardness of the surfaces. These measures must be undertaken with due discretion. The rounding of the nosing must not substantially reduce the size of the tread, and the walking surfaces must not be made so soft as to interfere with a normal gait.

■ In addition to the steps and landings, impact energy-attenuating materials should be applied to all the surfaces that a victim may fall against. Walls at the side of the stair or at the end of the landing should be padded, as should any balustrading. Even the handrails should be treated like the steering wheel of many modern cars with a firm but compressible material that provides a good grip but can still reduce the impact of a blow.

■ Most cuts will come from bumping up against edges and corners such as those sometimes found on nosings or aluminum nosing inserts. Balustrades and handrails frequently are made from rectangular or square bars or wood sections, and stringers are formed from timber or plates. Frequently, these sections are left with sharp-edged corners.

REFERENCES:

John Templer *"Stair Design Checklist"* in Reference 1.

Templer, John. 1992. *The Staircase: History and Theories and Studies of Hazards, Falls, and Safer Design.* Cambridge, MA: MIT Press.

C2.1 STAIRCASES

C3. 1 CEILINGS

CEILINGS: introduction

SOUND

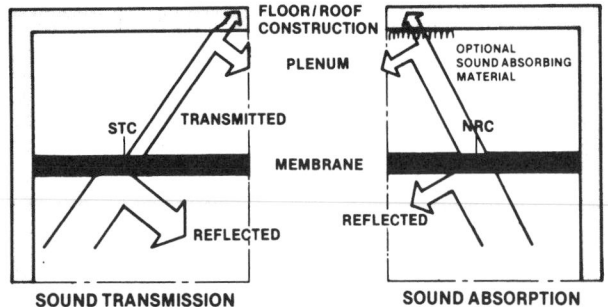

■ The control of sound may consist of:
• minimizing the transmission of sound generated within the space to and through the plenum to adjacent spaces;
• reducing the level of sound within the space in which it is generated;
• both; in this case two different materials may be required, since sound absorbing materials will not significantly resist sound transmission.

STC: Sound transmission glass, or the resistance to air borne sound transmission.

NRC: Noise reduction coefficient, or the effectiveness of material to absorb sound.

FIRE

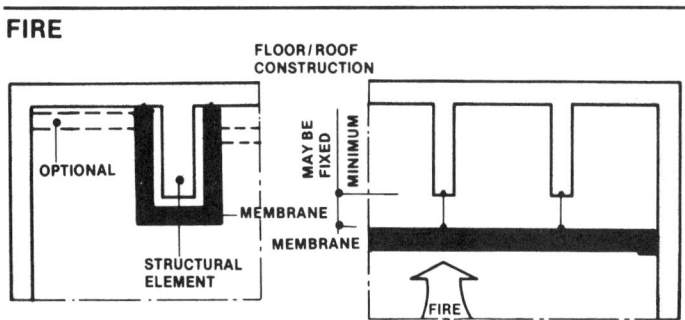

■ Ceiling membranes may:
• form a **continuous plane** suspended at a given distance below the structural framing to protect the framing and decking from exposure to effects of fire.
• **be partial,** installed to provide protection from fire only to isolated non-fire resistant structural elements of the floor/roof construction (when the construction spanning between such elements is fire resistance rated).
• fire resistance ratings are established for floor/roof assemblies, not for ceilings alone

HEAT FLOW

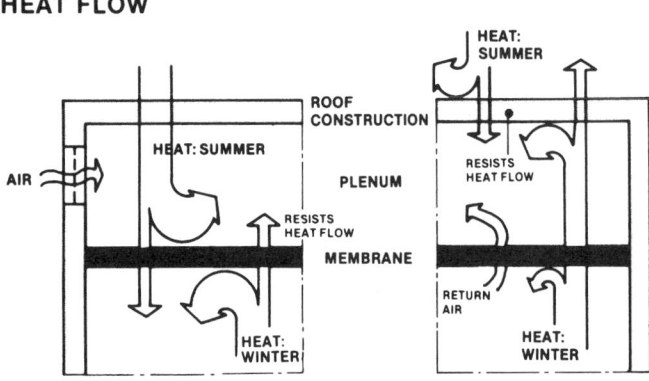

■ The flow of heat will be resisted to varying degrees by all except the open membranes;
• by the component materials alone
• by the components with thermal insulation added when high resistance to flow of heat is important.

■ Resistance to flow of heat is generally a factor in roof/ceiling assemblies only.

■ When the plenum space is ventilated and contains no services which might be affected by high or low temperatures, heat loss or gain should be controlled at the ceiling membrane level.

■ When the plenum space is used to return conditioned air from the enclosed space, or to supply conditioned air through a pressurized plenum, heat gain or loss should be controlled at the roof level.

AIR/WATER VAPOR LEAKAGE

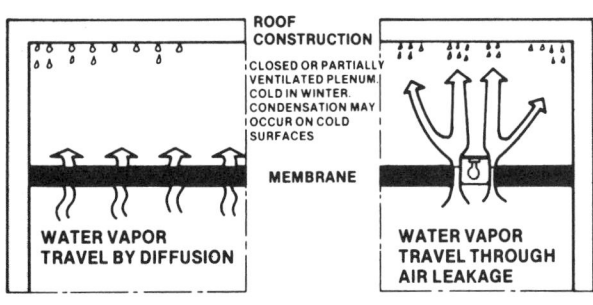

■ Water vapor will migrate by diffusion through any permeable material when differentials in pressure exist.

■ This condition is most severe:
• in cold and moderate climates
• in cold weather
• in hot climates a reverse flow may occur: water vapor may migrate from hot humid exterior into a cooled interior space.

■ Water vapor will also penetrate the membrane, carried by air leaking through poorly sealed penetrations (e.g., recessed lights) and may then condense on cold surfaces within the plenum space.

PRELIMINARY CONSIDERATIONS

A ceiling assembly generally consists of three main elements:

■ **Underside** of floor/roof construction;

■ **Plenum,** or the space immediately below the floor/roof construction; and

■ **Ceiling membrane** which may be either attached to the floor/roof construction or suspended from it, separating the plenum space, if any, from the enclosed space/room below.

■ **The type of ceiling assembly** is then determined based on whether:
• only one element is included, as is the case when the underside of construction is also the ceiling;
• a separate membrane with/without a plenum space is part of the assembly.

■ **Basic considerations** in the selection of a ceiling assembly include:
• type of the assembly itself;
• type and extent of mechanical/electrical services to be incorporated;
• type of the ceiling membrane to meet in-service performance requirements.

Types of Ceiling Assemblies

Ceiling assemblies may be classified as:

■ **No membrane/plenum:** when the exposed underside of the floor/roof construction is the only element of the assembly.

■ **Attached membrane:**
• with no plenum space when membrane is directly attached to a flat underside of floor/roof construction;
• with a limited plenum space when membrane is directly attached to framed construction.

■ **Suspended membrane:** with a plenum space of varying extent, generally based on requirements imposed by mechanical/electrical systems.

Types of Membrane

Selection of type of membrane is important in two assembly types:

the attached and the suspended and the choice may be equally applicable to both:

■ **Uninterrupted membranes** are defined as not having any or only a few small openings for mechanical/electrical services.

■ **Penetrated membranes** are those into which a limited amount of mechanical/electrical services is incorporated at random locations.

■ **Integrated membranes** are generally suspended, laid out in a grid pattern with a varying number of service outlets located within the grid.

Performance Requirements

The ceiling membrane is subject to a number of external factors which may affect its performance and the effects of which the membrane may have to resist; these include:
• **Pressure differentials**
• **Sound penetration**
• **Heat flow**
• **Air/water vapor penetration**
• **Chemical and biological attack**
• **Deflection**
• **Expansion/contraction**
• **Suspended loads**
• **Fire**

■ **All of the above factors** should be considered in the initial selection of the type of assembly and of the type of membrane, when applicable.

Attached membranes, especially of the uninterrupted and penetrated types, may not perform well if:

■ **Large deflections** are anticipated in the supporting construction to which they are directly connected, as excessive stresses may develop in the membranes.

■ Large differential **expansion/contraction** may occur between the supporting framing and the membrane.

■ **Structure borne sound** may be transmitted from one space to another through the membrane.

Suspended membranes, especially of the integrated type, may be subject to:

■ **Pressure differentials** which may lift or rattle the lay-in panels.

■ **Lateral loads** imposed upon the suspension system when partitions

are to be braced by it may exceed the strength of the suspension system to resist them.

■ **Heat flow** into a cold plenum space may also present a problem as air leaking into it will carry off heat. **Installing thermal insulation** over the membrane to minimize the flow of heat may help this problem, but also create others:
• condensation may occur within such insulation
• if recessed lighting fixtures are also covered by the insulation, heat build-up may result which will shorten the service life of the fixtures and/or lamps.

■ **Sound transmission** from one space to adjacent spaces may not be effectively controlled when soft, sound-absorbent lay-in panels are used in a suspended grid:
• sound will readily penetrate such materials
• sound may also leak through where panels are not in full contact with the suspension system.

Transmitted sound, whether air or structure borne is controlled by:

■ Sound transmission **resistant components** which enclose the receiving space: walls, partitions, doors, floor/ceiling assemblies.

■ **Sound absorbing materials** within the receiving space to control reflections of transmitted sound after it reaches the space.

■ Providing sound absorbing **flooring/furnishings** such as carpeting, drapes, upholstered furniture.

■ **Fire resistance information** has been included where relevant, but the ceiling assembly can only be evaluated as an integral part of a larger floor or roof/ceiling assembly.

C3.1 CEILINGS

CEILINGS: assembly types

A ceiling is a surface of whatever shape (level, curved, or molded) **which serves to define the upper boundary of an enclosed space.** The ceiling generally consists of:

- **underside** of floor/roof construction.
- ceiling space or **plenum** which may or may not contain mechanical/electrical service lines;
- **ceiling membrane**

Type, size, location and extent of mechanical/electrical service lines incorporated into the structure will influence the type as well as the location of the ceiling membrane:
- **[P]** denotes plumbing, HVAC, and fire protection piping runs and feeds
- **[D]** denotes HVAC ductwork runs and outlets
- **[E]** denotes electrical/communication wiring and feeds
- **[L]** denotes lighting fixtures

NO MEMBRANE/PLENUM

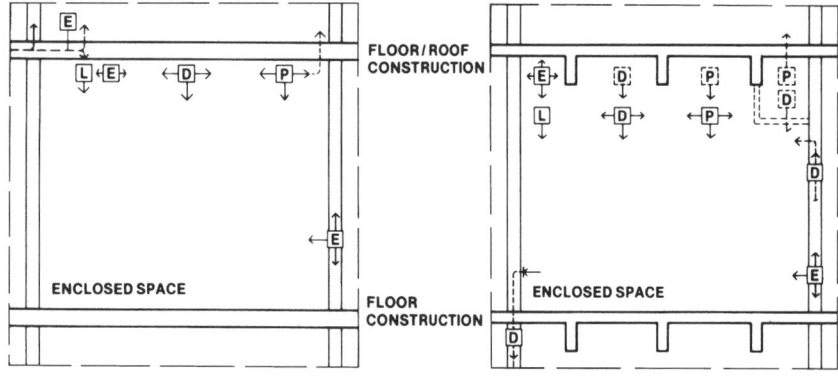

- ■ **The underside of floor/roof construction is the ceiling:**
  - exposed surfaces of floor/roof construction may be coated to improve appearance and/or minimize dust retention.
  - sound absorbing components may be incorporated into the roof construction; such as acoustical metal decking.

- ■ **Mechanical/electrical service lines are generally exposed;** except that electrical/communication wiring may be cast into concrete slabs or run within cellular metal floor decks.

- ■ **Plumbing and HVAC service lines** may be run between and sometimes through the floor/roof framing, but size, extent, and location are generally limited.

ATTACHED MEMBRANE

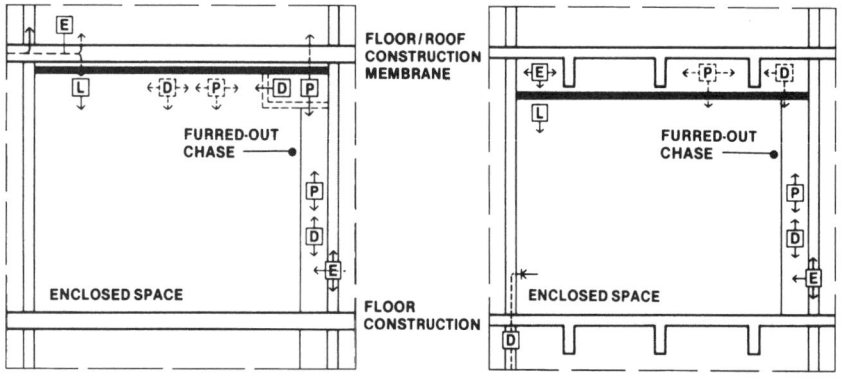

- ■ **Membrane is attached directly to, or furred down from** the underside of floor/roof construction:
  - may or may not include plenum space; when provided plenum is generally of limited extent.
  - may be of sound absorbing materials, or may have sound absorbing materials added to it: such as acoustic tile cemented to gypsum board membrane.

- ■ **Mechanical/electrical service lines may be:**
  - exposed
  - run partially or completely within the plenum space
  - grouped and concealed by furred-down-partial membrane
  - size, location, and extent of service lines within the plenum space are generally limited.

- ■ **May be a required component** of a fire resistance rated floor or roof/ceiling assembly.

SUSPENDED MEMBRANE

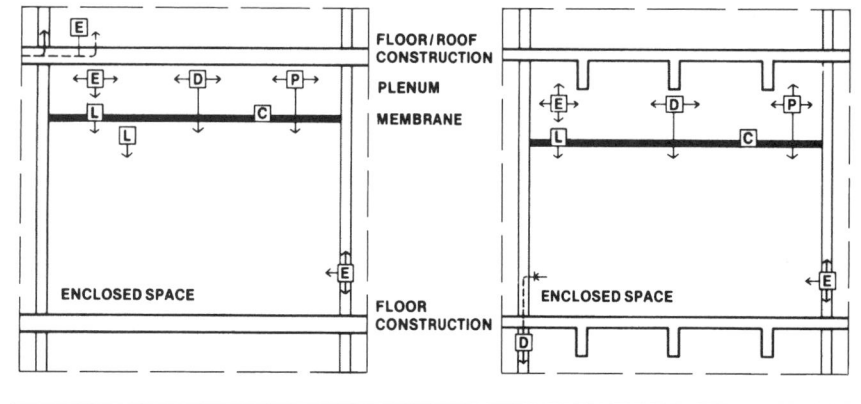

- ■ **Membrane is suspended from floor/roof construction,** generally at a distance sufficient to allow mechanical/electrical service lines to be run clear of the framing:
  - may be of sound absorbing materials, or may have such materials added to it.
  - may also be constructed in a variety of shapes: such as curved, undulating, of varying depth, coffered.

- ■ **May integrate mechanical/electrical service outlets** into the plane of the membrane either at random or as modular units within a grid.
  - may itself become source of heating, cooling, or illumination.

- ■ May serve as a required component of a fire resistance rated floor or roof/ceiling assembly.

CEILINGS: membrane types

The fabric of the ceiling membrane may be:

- a single component, selected for visual considerations and/or to perform a specific function, such as protecting the floor/roof assembly from the effects of fire;
- a number of components assembled into a membrane which in addition to visual considerations provides any/all: illumination, heating, cooling, control of sound, control of conditioned air movement, fire protection, access to service line controls.
- when selected as a component of a fire resistance rated floor or roof/ceiling

assembly, construction must conform to a tested and approved assembly.

The components and/or functions generally incorporated into a ceiling membrane are:

[L] **Light fixtures:** recessed or semi-recessed units of various shapes

[A$_S$] **Air supply:** round, rectangular, or linear air diffusers; slotted or perforated units, such as ventilating acoustic tile, or metal pan tile; housing of lighting fixtures adapted to this function.

[A$_R$] **Air return:** grilles, of metal or plastic; perforated metal pan tile; or housing of lighting fixtures.

[F] **Fire protection:** automatic sprinklers: CO_2 or Halon fire supression systems.

[S] **Sound control:** sound absorbing materials, such as acoustic plaster, acoustic tile.

[C] **Communications:** such as public address systems or masking sound systems.

The membrane may also support or incorporate:

[L] **Light fixtures:** suspended or surface mounted.

[F] **Fire protection:** surface mounted fire/smoke detection devices.

[A] **Access openings:** such as access tile, or access doors, which may be fire resistance rated for rated assemblies.

UNINTERRUPTED

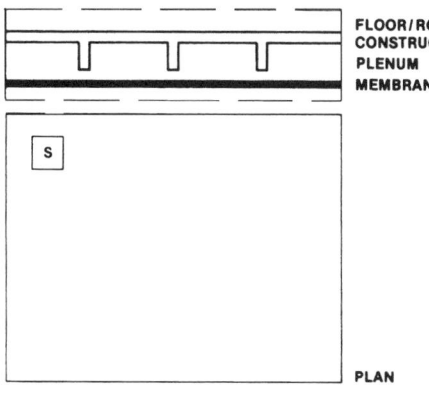

FLOOR/ROOF CONSTRUCTION
PLENUM
MEMBRANE

PLAN

■ **Membrane has no penetrations or a very limited number** of small penetrations such as:
- for sprinklers or electrical junction boxes for surface mounted light fixtures
- fire detection devices.

■ **Selected based on:**
- visual and/or economic considerations only;
- as a source of illumination, as in luminous ceiling systems;
- as a source of heat, as in radiant ceiling systems;
- when fire resistance rating for floor or roof/ceiling assembly is required.

■ **Membrane may be:**
- a monolithic hard and smooth surface, non-sound-absorbing and resistant to sound transmission;
- of sound absorbing material with poor resistance to sound transmission;
- a dense, sound transmission resistant surface with sound absorbing materials added to it.

PENETRATED

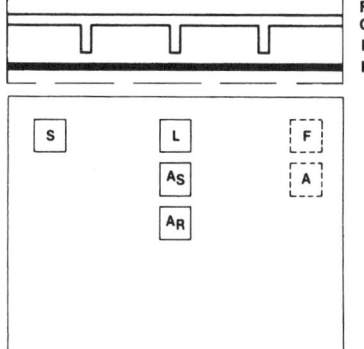

FLOOR/ROOF CONSTRUCTION
PLENUM
MEMBRANE

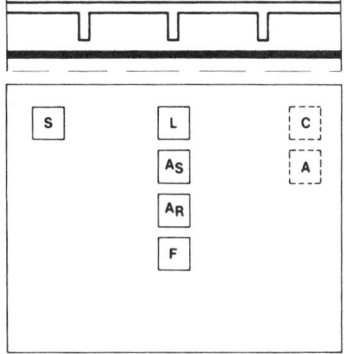

PLAN

■ **Membrane has a varying number of isolated penetrations** for mechanical/electrical systems, when such systems are not extensive and/or flexibility in locating them at any point within the membrane is a requirement.

■ **When fire resistance rating** for floor or roof/ceiling assembly is required, the number and size of penetrations may be limited, and such openings may have to be protected to maintain the integrity of the membrane; diffusers and return air grilles may require fire dampers; recessed light fixtures may have to be boxed-in with rated materials.

■ **Membrane may be dense and resistant to sound transmission:** penetration will decrease such resistance unless specifically protected.

INTEGRATED

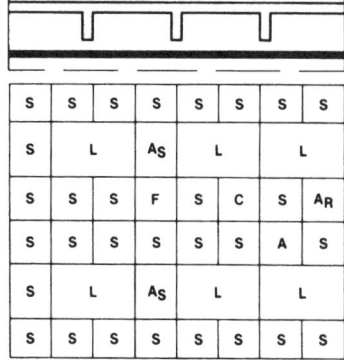

FLOOR/ROOF CONSTRUCTION
PLENUM
MEMBRANE

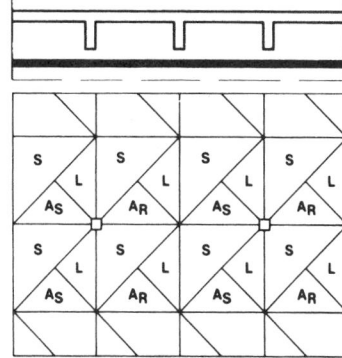

PLAN

■ **Membrane incorporates: light sources, air supply and return, materials to control sound, communication systems, and fire protection at:**
- specific fixed locations within a grid;
- throughout the entire membrane as interchangeable modules: flexibly connected to mechanical/electrical service lines to allow for future relocation.

■ **The components of the membrane** may serve dual purposes: such as light fixtures which also provide means to control air supply/return; linear diffusers which also serve as the supporting grid for other components.

■ **Membrane may also function** as a component of a fire resistance rated floor or roof/ceiling assembly.

C3.1 CEILINGS

CEILINGS: types, properties

● denotes common usage
○ denotes possible usage
P poor
F fair
G good

[1] **Basis:** 5/8 inch mineral fiber tile in exposed grid suspension system = 100

ASSEMBLY	COMPARATIVE COST RANGE [1]	WEIGHT OF ASSEMBLY, lbs./s.f.	TYPE			RESISTANCE TO: CONDITIONS WITHIN ENCLOSED SPACE						
			UNINTERRUPTED	PENETRATED	INTEGRATED	SOUND ABSORPTION	SOUND TRANSMISSION	AIR/WATER VAPOR PENETRATION	LATERAL LOADS	PRESSURE/UPLIFT	IMPACT	HIGH HUMIDITY
SUSPENDED MEMBRANE												
ACOUSTIC TILE, EXPOSED GRID	100 to 200	2 to 3	●	●	●	G	F to P	P	F to P	P to G	P to F	P to F
ACOUSTIC TILE, CONCEALED GRID	400 to 500		●	●	○	G	F to P	F to G	F	G	P	P to F
METAL PAN, CONCEALED GRID	600 to 800	.50 to 1.5	●	●	●	F	P	P	F	G	G	G
METAL SHAPES, CONCEALED GRID	1000 to 1400	1 to 2	●	●	●	F	P	P	P	G	G	G
PLASTIC PANELS, EXPOSED GRID	500 to 1000	.5 to 1	●	●	●	F	G	P	P	P to F	F	G
MODULES, INTEGRATED GRID	600 to 1000	3 to 5			●	G	G	F to P	F to P	G	P	P
PLASTER	300 to 500	4 to 6	●	●		P	G	G	G	G	P	P
GYPSUM BOARD	250 to 300	3 to 4	●	●		P	G	G	G	G	P	F to P
ATTACHED MEMBRANE												
GYPSUM BOARD	200	3 to 4	●	●		P	G	G	G	G	P	P
PLASTER	300 to 400	4 to 6	●	●		P	G	G	G	G	P	F to P
ACOUSTIC TILE, CONCEALED GRID	400 to 500	2 to 3	●	●		G	G	F to P	F to P	P	P	P to F
NO MEMBRANE												
APPLIED COATING	50 to 100		●	●		P	G				P	F to P
PLASTER FINISHING COAT	150 to 200	1 to 1.5	●	●		P	G				P	F to P
SPRAYED-ON	100 to 150	.20 to .30	●	●		G	G				P	P
SPECIAL TYPES												
BAFFLES, SUSPENDED	varies	varies				G	P					F to P
CYLINDERS, SUSPENDED						G	P					F

C3.1 CEILINGS

STEAM	CHEMICAL FUMES	CONDITIONS IN SUPPORTING CONSTRUCTION:		PREFINISHED			FINISHED IN PLACE		MECHANICAL/ELECTRICAL COMPONENTS AIR SUPPLY THROUGH:						AIR RETURN THROUGH:		
		DEFLECTION	EXPANSION/CONTRACTION	SMOOTH/PERFORATED	PATTERNED/PERFORATED	TEXTURED/PERFORATED	TEXTURED DECORATIVE COATING	PAINT	VENTILATING/SLOTTED TILE	DIFFUSED, NON-MODULAR	MODULAR	LINEAR/MAIN RUNNERS	LIGHTING FIXTURES	AIR/LIGHT/SOUND MODULES	GRILLES/TILE	LIGHT FIXTURES	AIR/LIGHT/SOUND MODULES
P to F	P to F	G	G	●	●	●				●	●		●		●	●	
P	P	G	G	●	●	●			●	●	●		●				
G	G	G	G	●					●	●	●		●		●	○	
G	G	G	G	●		●			●	●	●		●				
G	F	G	G	●	●				●								
P	P	G	G		●	●			●			●		●	●		●
P to F	P	F to P	F to P				●	●	●			●	●				
P to F	P	F to P	F to P				●	●	●			●	●				
P	P	P	P				●	●					●				
F	F	P	P				●	●			○	○	●				
P	P	F	F	○	●				●				●				
F	P	P	P				●	●									
F	P	P	P				●	●									
P	P	P	P				●						●				
P to G	P to G				●	●											
P	F				●	●											

C3.1 CEILINGS

SUSPENDED CEILINGS: exposed grid

ASSEMBLY	REMARKS

EXPOSED GRID: FLAT UNITS

SQUARE EDGE

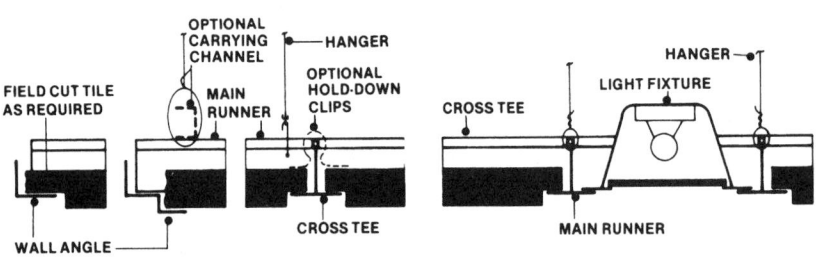

- lay-in panels should be secured in place by clips when assembly requires a fire resistance rating; also against uplift due to pressure differential.
- fixtures generally have to be boxed-in for fire resistance rating; fire dampers must be provided at all openings, such as diffusers.
- hangers secured to framing members, structural deck, or to secondary framing system.

RECESSED EDGE

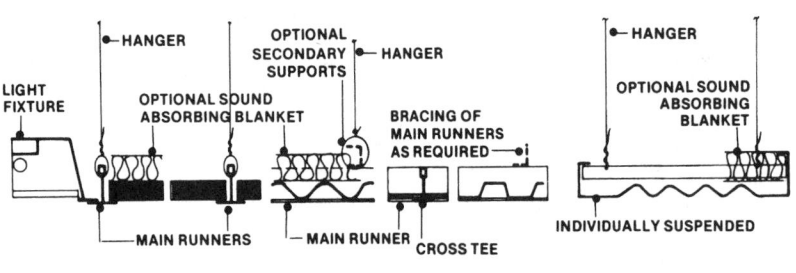

- clearance required for all lay-in panels for tilting them into place.
- suspension system used is the same as for square edge tile, but tile only available in 2x2 foot size.
- may be used in fire resistance-rated floor or roof/ceiling assemblies; clips to secure tiles in place and opening protection generally required.

EXPOSED GRID: SHAPED UNITS

INLAY PANELS, CORRUGATED, RIBBED

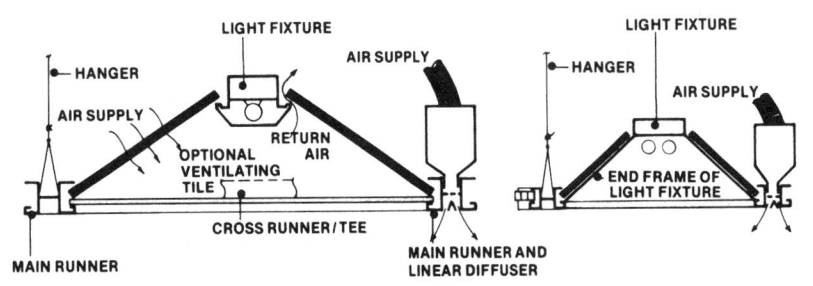

- metal panels generally perforated, with sound absorbing blankets.
- plastic panels generally solid; used in luminous ceiling installations.
- corrugated/ribbed metal or plastic panels generally used with main runners only.
- flat plastic panels generally either 2x4 or 2x2 feet in size, used with main runners and cross tees.

PRE-ASSEMBLED MODULES

- flat pre-assembled modules are also available.
- when pressurized plenum and ventilating tile are used, air return must be ducted through plenum.
- with ventilating plenum, dirt streaking may result unless the membrane is made completely air tight.
- may be used in fire resistance-rated floor or roof/ceiling assemblies.

SUSPENDED CEILINGS: concealed grid

ASSEMBLY	REMARKS

CONCEALED GRID: SHAPED UNITS

METAL PAN TILE

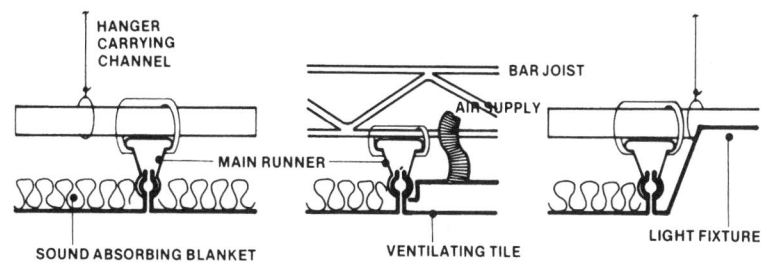

- tile may be repeatedly repainted without loss in sound absorbing characteristics.
- heating/cooling piping may be incorporated into the system.
- combination lighting/infra-red heating fixtures may be integrated into membrane.
- secondary suspension system generally required.
- tile may be used for supply/return air.

LINEAR PANELS

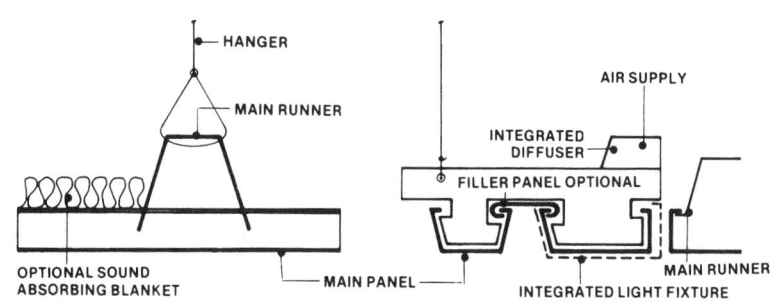

- formed prefinished metal panels in long lengths.
- air supply/return and lighting fixtures may be integrated into the system.
- may be used outdoors in protected locations, such as large soffits, canopies.
- some assemblies may be used as required components in fire resistance rated floor or roof/ceiling assemblies.
- membranes may be curved perpendicular to direction of panels.

BAFFLES

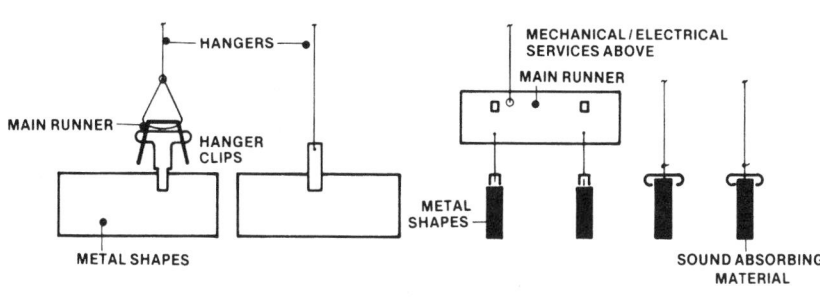

- baffles available in shaped metal, with or without sound absorbent material cores, or in faced sound absorbent material.
- various arrangements available, such as linear, radial, hexagonal.
- used to: provide additional sound absorption in selected locations; for visual interest, or to conceal mechanical/electrical services.

CONCEALED GRID: FLAT UNITS

KERFED EDGE

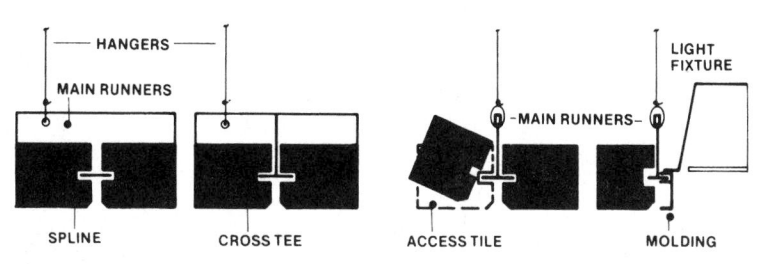

- tile, generally 12 x 12 inches in size with kerfed edges secured in place by main runners in one direction, and cross tees or splines in the other.
- secondary supports, such as carrying channels may be used to reduce spacing of hangers to framing system.
- may be used as component in fire resistance rated floor or roof/ceiling assemblies.
- special panels available to provide access to plenum.

C3.1 CEILINGS

SUSPENDED CEILINGS: exposed, concealed grid

● denotes common usage
○ denotes possible usage

ASSEMBLY	SPACING OF MAIN RUNNERS, range in feet	SPACING OF HANGERS, range in feet	TYPE UNINTERRUPTED	PENETRATED	INTEGRATED	FRAMED FLAT PLATE	STEEL BAR JOISTS	CONCRETE	STEEL BEAM	TRUSSES	TILE/PANELS MINERAL FIBER SMOOTH FACE	FISSURED	TEXTURED	PATTERNED	FACED	METAL PANEL TILE, PERFORATED	SOLID FACE
FLAT PANELS, EXPOSED GRID																	
SQUARE EDGE	3 to 5	3 to 6	●	●	●	●	●	●	●	●	●	●	●	●	●	●	
RECESSED EDGE	2	4 to 6	●	●	○	●	●	●	●	●	●	●	●	●	●		
SHAPED PANELS, EXPOSED GRID																	
CORRUGATED RIBBED	6 to 8	3 to 5	●	●		●	●	●	●	●							
MODULES	5 to 7	2 to 3			●	●	●	●	●	●							
SHAPED PANELS, CONCEALED GRID																	
METAL PAN TILE/PANELS	2 to 5	5 to 7	●	●	●	●	●	●	●	●						●	●
LINEAR PANELS	3 to 5	3 to 5	●		●	●	●	●	●	●							●
BAFFLES	5 to 7	2 to 6					●	●	●	●	●			●			●
FLAT PANELS, CONCEALED GRID																	
KERFED EDGE	2 to 5	2 to 8	●	●		●	●	●	●	●	●	●	●	●	●		

PANEL, PERFORATED	CORRUGATED, PERFORATED	RIBBED, SMOOTH, PERFORATED	GLASS FIBER, FACED	WOOD FIBER, CEMENT BINDER	FLAT, SMOOTH, TEXTURED (PLASTIC)	CORRUGATED/RIBBED (PLASTIC)	SOUND ABSORBING BLANKET	CARRYING CHANNELS	MAIN RUNNERS	CROSS TEES	SPLINES	WALL CHANNELS	END CLOSURES	HANGERS, WIRE/STRAP	CLIPS	DIFFUSERS	RETURN AIR GRILLES	LIGHTING FIXTURES	FIRE PROTECTION	ACCESS TO UTILITIES	MAIN RUNNERS (integrated)	CEILING TILES/PANELS	LIGHTING FIXTURES (integrated)	AIR RETURN AND LIGHTING	INFRARED HEAT AND LIGHTING	INTEGRAL HEATING/COOLING	LIGHT/SOUND MODULE	LIGHT/AIR/SOUND MODULE
			●	●	●		○	○	●	●		●		●		●	●	●	●	○	○		○	○				
				●	○		○	○	●	●		●		●	●	●	●	●	●	●		○	○	○				
	●	●				●	●	○	●			●		●		●		●	●	○	○							
									●	●		●		●						○		●	●	●			●	●
●							●	○	●			●		●	●	○	●	●	●			●	○		●	○		
●							○	○	●			●	●			●	●	●	●					○	○			
●								○	●					●	●													
							○	○	●	●	●	●	●	●	●	●	●	●	●	●		●	●	●				

C3.1 CEILINGS

SOUND ABSORBING COMPONENTS

● denotes common usage
○ denotes possible usage

PROPERTIES

COMPONENT	THICKNESS, range in inches	WEIGHT, range pounds per s.f.	UNIT SIZE, range in feet	RATED FOR FIRE RESISTANT ASSEMBLIES	NRC, range noise reduction coefficient	STC, range for uninterrupted ceiling	FLAME SPREAD rating
TILE/PANELS							
CELLULOSE	½ to 1	1 to 2	1 x 1 to 2 x 4		.50 to .80		26 to 75
WOOD FIBER AND CEMENT BINDER	1 to 3	1 to 4	2 x 2 to 4 x 4		.70 to .85		
MINERAL FIBER, FISSURED/PERFORATED	½ to ⅝				.50 to .60	40 to 44	
CERAMIC BONDED	⅝ only	1 to 2	1 x 1 to 2 x 4	●	.55 to .65	35 to 40	25
FACED	⅝ to ⅞					40 to 44	
GLASS FIBER, FACED	⅝ to 3		2 x 2 to 4 x 10		.85 to .95		
UNITS							
BAFFLES	1		no limit		.75 to .80		25
BLOCKS	1 to 2						
BLANKETS							
SOUND ABSORBING, GLASS FIBER	1 to 3	.05 to .35	no limit		.75 to .90		25
MINERAL WOOL	1 to 1½				.65 to .70		
APPLIED							
SPRAYED-ON-FIBERS	½ to 3			●	.65 to .85		20
ACOUSTIC PLASTER	½ to 1			●	.50 to .60		

C3.1 CEILINGS

| RESISTANCE TO | | | | [1] INSULATING VALUE thermal | SURFACE FINISH | | | | | | OPTIONAL METHOD OF ATTACHMENT | | | | | REMARKS |
IMPACT	HIGH HUMIDITY	STEAM	CHEMICAL/ACID FUMES		PAINT	GLASS FIBER FABRIC	ACRYLIC FABRIC	ACRYLIC FILM	VINYL COATING	POLYESTER FILM	EXPOSED GRID	CONCEALED GRID	FURRING CHANNELS/CLIPS	ADHESIVE	EDGE STAPLING	[1] AVERAGE C factor (at 75°F)
					●								●	●	●	• do not cement tiles to uninsulated roof-deck, where temperature differentials are likely to cause condensation or are exposed to extreme heat.
					●						●		●			• wood fiber tiles should not be used where contact with water or condensation is anticipated.
				.65	●			●			●	●	●	●	●	• mineral fiber tiles with a scrub-resistant finish are available. • special impact tiles with clips that absorb the impact are available. • when using ceramic tiles in extremely high humidity conditions, use of aluminum grid is recommended. • metal faced tiles are available.
	●	●	●	.75	●						●		●			• titles should not be directly bonded to uninsulated roof-deck, where temperature differentials are likely to cause condensation or where tiles are exposed to extreme heat.
○				.60			●	●	●	●	●	●		●		
						●			●		●			○		
●	○	○	○	.22		●							●			• used in industrial plants where the ceiling surfaces must be kept free for access to ducts, pipes, etc. • metal baffles are resistant to humidity and steam. • for effective sound absorption, use one baffle per 8 to 12 square feet based on ceiling heights of 9 to 12 feet.
					●								●	●		
													●			• up to 6″ thick available. • NRC rating depends upon the thickness and density of blanket.
													●			
																• some sprayed-on products are relatively soft and may be easily damaged. • acoustical plaster is used where curved ceilings and niches makes use of tiles impractical. • may be sprayed or hand troweled in 2 coats.

C3.1 CEILINGS

MEMBRANE CEILINGS: gypsum board, plaster

ASSEMBLY	REMARKS

GYPSUM BOARD, ATTACHED

DIRECT TO FRAMING	• secured directly to framing members or to solid furring. • most widely used in residential and light commercial construction. • two layers may be required for an improved fire resistance rating or for better resistance to sound transmission. • directly affected by deflection and/or expansion/contraction in supporting framing.

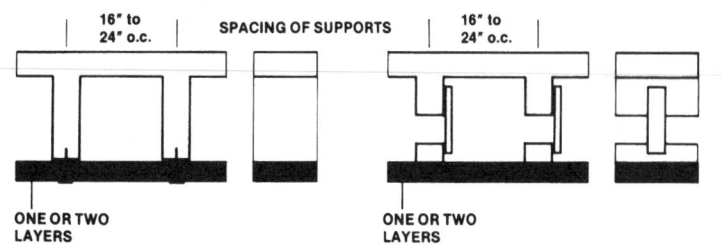

16" to 24" o.c.   SPACING OF SUPPORTS   16" to 24" o.c.

ONE OR TWO LAYERS   ONE OR TWO LAYERS

FURRED-DOWN	• hat-shaped or resilient channels may be used. • furring will minimize effects of deflection and expansion/contraction in framing upon membrane. • resilient channels also used to improve resistance to sound transmission. • furring will also minimize effects of streaking due to temperature differential which may occur with direct attachment.

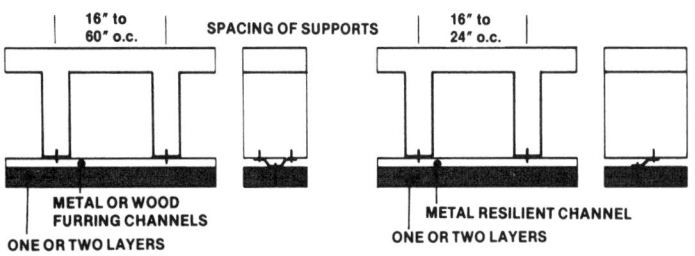

16" to 60" o.c.   SPACING OF SUPPORTS   16" to 24" o.c.

METAL OR WOOD FURRING CHANNELS   METAL RESILIENT CHANNEL
ONE OR TWO LAYERS   ONE OR TWO LAYERS

GYPSUM BOARD, SUSPENDED

PRIMARY SUPPORTS ONLY	• when framing is spaced more than 24 inches on centers, or when a plenum space for mechanical/electrical service lines is required, a suspension/support system consisting of wood or metal sections or special nailing channels is generally provided. • prefabricated metal suspension systems are available.

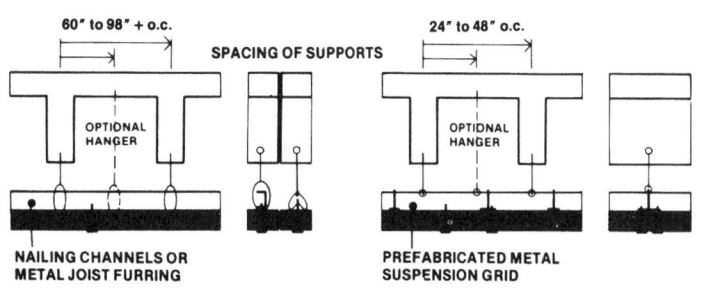

60" to 98" + o.c.   SPACING OF SUPPORTS   24" to 48" o.c.

OPTIONAL HANGER   OPTIONAL HANGER

NAILING CHANNELS OR METAL JOIST FURRING   PREFABRICATED METAL SUSPENSION GRID

PRIMARY AND SECONDARY SUPPORTS	• primary suspension system may also include a secondary system of furring channels used to align the primary system and/or to provide resilient mounting of the membrane. • it is a high cost assembly and not widely used. • resilient furring channels generally used with wood framing.

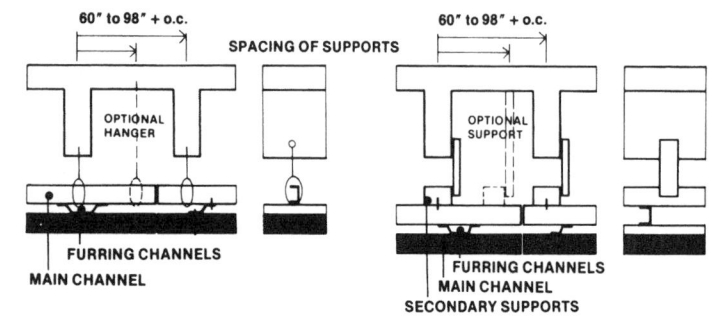

60" to 98" + o.c.   SPACING OF SUPPORTS   60" to 98" + o.c.

OPTIONAL HANGER   OPTIONAL SUPPORT

FURRING CHANNELS   FURRING CHANNELS
MAIN CHANNEL   MAIN CHANNEL
                        SECONDARY SUPPORTS

C3.1 CEILINGS

ASSEMBLY	REMARKS
PLASTER, ATTACHED	

DIRECT TO FRAMING

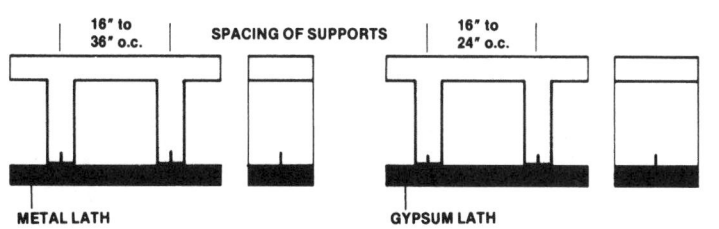

16" to 36" o.c. — SPACING OF SUPPORTS — 16" to 24" o.c.

METAL LATH   GYPSUM LATH

- metal or gypsum lath secured directly to framing.
- membrane will be directly affected by deflection and/or expansion/contraction in supporting framing.
- metal lath may be backed for machine application of plaster.
- fire resistance ratings for different assemblies have been established.

FURRED-DOWN

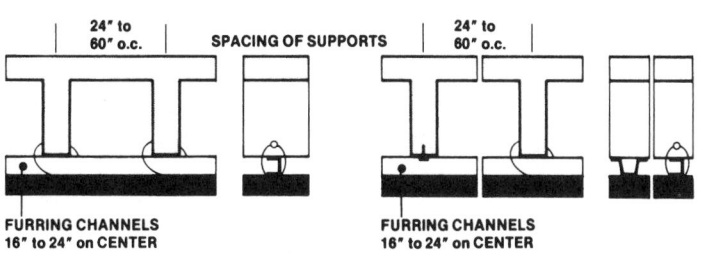

24" to 60" o.c. — SPACING OF SUPPORTS — 24" to 60" o.c.

FURRING CHANNELS 16" to 24" on CENTER   FURRING CHANNELS 16" to 24" on CENTER

- furring channels secured to framing, lath supported by furring.
- furring will minimize effects of deflection and expansion/contraction upon membrane.
- large areas of membrane should have expansion joints and should not be restrained at the perimeter.
- corners of openings in gypsum lath membranes should have metal lath reinforcing.

PLASTER, SUSPENDED	

PRIMARY SUPPORTS ONLY

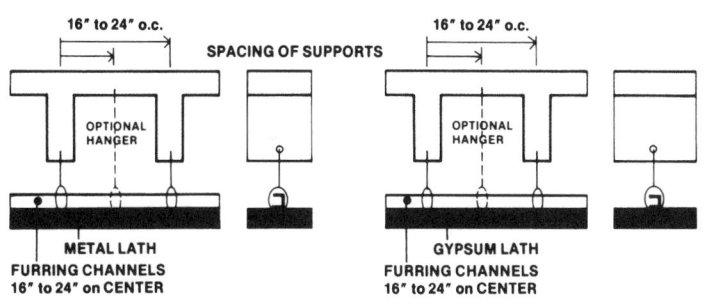

16" to 24" o.c. — SPACING OF SUPPORTS — 16" to 24" o.c.

OPTIONAL HANGER   OPTIONAL HANGER

METAL LATH FURRING CHANNELS 16" to 24" on CENTER   GYPSUM LATH FURRING CHANNELS 16" to 24" on CENTER

- suspended membrane with furring channels only is similar to furred membrane except that furring channels are suspended from, rather than directly attached to, framing members.
- suspension of membrane may be a requirement in some fire resistance rated floor or roof/ceiling assemblies.
- spacing of hangers is quite close and limits the size and/or extent of mechanical/electrical service lines in plenum space.

PRIMARY AND SECONDARY SUPPORTS

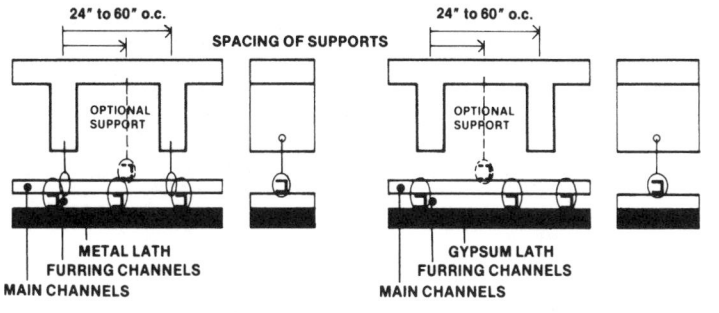

24" to 60" o.c. — SPACING OF SUPPORTS — 24" to 60" o.c.

OPTIONAL SUPPORT   OPTIONAL SUPPORT

METAL LATH FURRING CHANNELS MAIN CHANNELS   GYPSUM LATH FURRING CHANNELS MAIN CHANNELS

- when spacing of framing is wide and/or the number of hangers must be reduced, a primary support system consisting of main carrying channels may be used; the furring channels are then a secondary system, secured to such primary supports.
- for wide hanger spacing, metal joists instead of carrying channels may be used.

C3.1 CEILINGS

MEMBRANE CEILNGS: gypsum board, plaster

● denotes common usage
○ denotes possible usage

	TYPE			PLATE		FRAMED JOISTS				BEAM				TRUSSES [1]		PRINCIPAL COMPONENTS			
	UNINTERRUPTED	PENETRATED	INTEGRATED	FLAT SLAB/PLATE	WAFFLE SLAB	WOOD	METAL	STEEL BAR	CONCRETE	WOOD	ROLLED STEEL	PRE-CAST CONCRETE	TWO-WAY SLABS	SHORT SPAN, WOOD	LONG SPAN	GYPSUM BOARD	FURRING CHANNELS	RESILIENT FURRING CHANNELS	NAILING CHANNELS/T CHANNELS
GYPSUM BOARD, ATTACHED																			
DIRECT TO FRAMING	●	○				●	●							●		●			
FURRED DOWN	●	○				●	●	○	○					●		●	●	●	
GYSPUM BOARD, SUSPENDED																			
PRIMARY SUSPENSION ONLY	●	●		○	○	○	○	●	●	●	○	●	●	○	○	●	●	●	●
PRIM. AND SECONDARY SUSPENSION	●	●						○	○	●	●	●	●	●		●	●	●	○
PLASTER, ATTACHED																			
DIRECT TO FRAMING	●	●				●	●	●	●					●					
FURRED DOWN	●	●				●	●	●	●			○		●					
PLASTER, SUSPENDED																			
PRIMARY SUSPENSION ONLY	●	●		○	○	●	●	●	●	○	○			●					
PRIM. AND SECONDARY SUSPENSION	●	●		○	○	○	○	●	●	●	●	●	●	●	●			●	

C3.1 CEILINGS

(cont.)

MAIN CHANNELS	METAL LATH	GYPSUM LATH	WIRE HANGERS	STRAP HANGERS	PLASTER FINISHING COAT				MECHANICAL/ELECTRICAL		LIGHT FIXTURES				REMARKS
					GYPSUM PLASTER	ACOUSTIC PLASTER	KEENE'S CEMENT	PORTLAND CEMENT	AIR SUPPLY DIFFUSERS	AIR RETURN GRILLES	RECESSED	SURFACE MOUNTED	FIRE PROTECTION SYSTEM	ACCESS DOORS	
									○	○	○	●	○	○	• attached to wood with nails to metal joists with self-tapping screws. • screws may be used with wood framing also, and are recommended to avoid problems of nail-popping. • furring channels are nailed into wood, screw attached to metal joists and wired or clipped to steel bar joists. • resilient channels generally used with wood framing only, to minimize sound transmission. • ³/₈ inch gypsum board not recommended for joists spaced 16 inches on center: ½ inch gypsum board should be used.
									○	○	○	●	○	○	
●									●	●	●	○	●	●	• hangers secured to framing or deck by nails into wood; wired or clipped to steel joists; with self-tapping screws or clips to metal joists; to inserts cast into cement, or to power driven fasteners shot-in during installation of membrane; with clips or power driven fasteners to steel beams. • main channels are generally steel joists, 2 to 6 inches in depth. They may be supported from hangers, or furred down with short pieces of steel joists when a stiffer membrane, or supported load carrying capacity is required. • membrane may be suspended to allow mechanical/electrical service lines to pass between it and the framing or as a requirement for a floor or roof/ceiling fire resistance rated assembly.
●									●	●	●	○	●	●	
●	●	●			●	●	●	●	○	○	○	●	○	○	• metal or gypsum lath is attached to wood by nailing or screws; with self-tapping screws to metal joists. • furring channels are wired to wood and metal joists; and wire attached to inserts in concrete. Power driven fasteners to secure hangers may also be used. • spacing of furring channels generally 12 to 24″, except that ¾″ rib metal lath may be used on 36″ spacing.
●	●	●	●		●	●	●	●	●	●	○	●	○	○	
●	●	●	●	●	●	●	●	●	●	●	○	●	○	○	• screw-in metal joists or cold-rolled steel channels may be used to support gypsum lath. When metal joists are used, spacing between supports for the joists may be increased within the safe carrying capacity of furring channels is not exceeded. • providing hangers between main framing members of floor/roof construction should always be considered to minimize span, thus reducing the required size of main carrying channels. • main carrying channels may be suspended by hangers or may be furred-down using wood plank or metal joists.
●	●	●	●	●	●	●	●	●	●	●	○	●	○	○	

REMARKS

☐ **Long span trusses, rigid frames and arches may require secondary framing systems to support ceiling membranes.**

• **Vinyl fabric faced gypsum board may be used as panels where there is a possibility of exposure to damage and moisture within enclosed space.**

C3.1 CEILINGS

C • INTERIORS

GYPSUM BOARD & PLASTER: materials

GYPSUM BOARD

TYPES:

■ **Regular; for general use: ASTM C36** – Standard Specification for Gypsum board.

■ **Type X; fire retardant: ASTM C36** used when fire-resistance rating is required.

■ **Water-Resistant;** of water resistant core and water resistant paper: **ASTM C630;** used in wet locations, such as bathrooms:

- also available water-resistant as well as fire-resistant: per **ASTM C36.**
- for Joint Finishing Treatment and Adhesives, refer to PARTITIONS.

GYPSUM LATH/BASE

TYPES:

■ **Plain, perforated, and insulating:** insulation has aluminum foil backing.

- perforated type provides for mechanical bonding of plaster; generally used in fire rated assemblies because of increased strength.
- thickness: $3/8''$ and $1/2''$. Size 16"x 48" and 24" x 96".

■ **Base: incombustible core,** essentially of gypsum, surfaced with paper and aluminum foil backed.

- available fire retardant – Type X.
- for more information refer to PARTITIONS.

MATAL LATH

TYPE:

■ **Wire lath:** woven or welded; generally galvanized.

- weight: 1.7 to 2.5 lbs. per square yard.
- available with plain or water-proofed paper backing.

■ **Expanded metal lath** available in:

- diamond mesh: flat, easily bent or formed to curved surfaces.
- flat-rib lath: stiffer than diamond mesh, thus allowing wider spacing of supports; furring over flat solid surfaces required.
- rib-lath: with stiffening ribs generally $1/8$ to $3/8$ inch deep; unsuitable for contour plastering, self-furring.
- self-furring lath: generally for exterior stucco.
- rigid PVC plastic lath accessories available.

PLASTER

TYPE:

■ **Finish-Coat Plasters:**

- gaging plaster, available in quick setting and slow-setting mixtures; commonly used type for general plastering work: **ASTM C28.**
- molding plaster, similar to gaging plaster, but finer-ground for use in ornamental work.
- veneer plaster for use over gypsum base: **ASTM C587.**
- Keene's cement: used when hard, moisture resistant surfaces are required; it is not water resistant and should not be used around bath tubs or in showers: **ASTM C61.**
- acoustic plaster, specially formulated to provide improved sound absorption;
- for base-coat plaster refer to PARTITIONS

C3.1 CEILINGS

METAL & PLASTIC: materials

METAL

TYPES:

■ **Tiles** are available in:
- 24 gauge prefinished, galvanized steel.
- pre-finished .032 aluminum, perforated or smooth.
- standard sizes: 12″ x 12″, 12″ x 24″ and 12″ x 48″.
- 1 to 2 inch fiberglass sound absorbing lay-in pads may be used to increase noise reduction coefficient.
- lay-in type, suitable for an exposed grid system available.
- snap-in type also available.

■ **Panels** are available: smooth, perforated or corrugated, fabricated from;
- .025″ or .032″ prefinished aluminum.
- 24 to 28 gauge pre-finished galvanized steel.
- standard lengths: 4′, 5′, 6′, 8′ or 10′.
- widths: 23 to 55 inches.
- optional sound absorbing 1 to 3 inch fiberglass blankets used over panels.

- used with exposed grid suspension system.

■ **Shaped linear panels** are available in:
- pre-finished .020″ to .025″ aluminum.
- pre-finished 24 gauge steel.
- face width: 1½ to 7½ inches.
- length, running with the panels is generally 3 to 20 feet.
- lighting and air distribution utilizes the regressed slots between the ceiling panels.
- used with acoustic pads, recessed filler strip and with flush filler strip.
- panel carriers are suspended by wire supports and panels lock or snap on to carriers.
- linear facade components also provide exterior and interior open and closed facades.
- optional sound absorbing 1 to 3 inch fiberglass blankets may be used over panels.

PLASTIC

TYPES:

■ **Acrylic PVC or polystyrene ceiling lay-in panels** available in:
- flat, smooth and textured sheets
- also interlocking open grilles, hollow cubes and cylinders.
- variety of colors available.

- panel sizes: 12″ x 24″ to 48″ x 48″.
- length: 2 to 8 feet.
- used with concealed grid suspension systems.
- transparent polystyrene tends to yellow and weaken on exposure to ultra-violet light.

BAFFLES

TYPES:

■ **Sound absorbing baffles** available in:
- perforated .022″ pre-finished aluminum,
- 1 inch sound absorbing glass fiber core wrapped in vinyl fabric;
- 2½ inch mineral fiber, vinyl faced blocks.

- sizes: 24″ x 10″ to 48″ x 10″ also available 15″ x 15″ blocks.
- lengths: 2 to 10 feet.
- method of attachment: can be fastened mechanically, adhesively or by concealed suspension.

CEILINGS: selection checklist

DEFLECTION

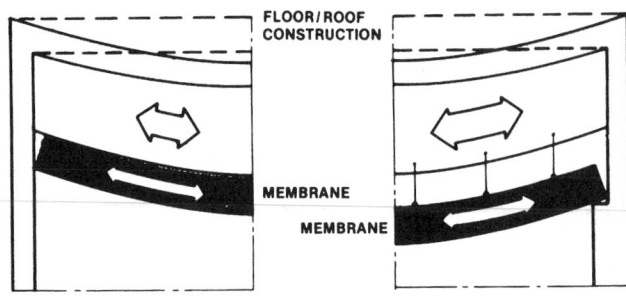

■ **Deflection in supporting floor/roof construction** will also affect the ceiling membrane, especially when the membrane is directly attached to it and is of inelastic material.

■ Deflection in the supporting construction **should be limited** so that the corresponding deflection in the ceiling material will not **exceed allowable bending stresses** for that material; e.g., the recommended deflection for a supporting member carrying an attached plaster ceiling membrane is 1/360 of the span.

■ **When deflection cannot be kept within safe limits,** consider resilient attachment of the membrane, or its suspension.

EXPANSION/CONTRACTION

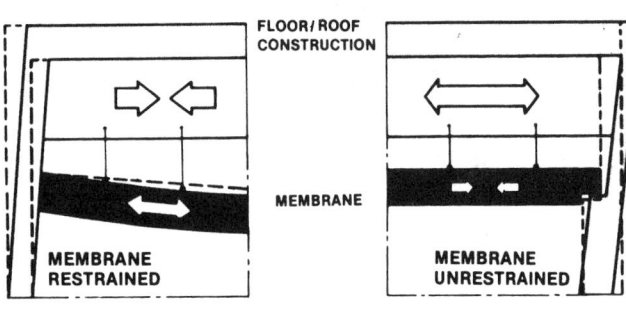

■ **Differential expansion/contraction** may occur because of differences in temperature and/or in coefficients of expansion of two materials when joined; e.g., between a roof construction exposed to solar radiation and a ceiling membrane in a cooled interior.

■ **Membranes may be:**
• restrained from free movement by adjoining vertical elements such as walls or partitions,
• separated or flexibly connected to such elements.
• differential expansion/contraction may become critical for restrained membranes, while the effects on unrestrained membranes will generally be minimal.

PRESSURE DIFFERENTIALS

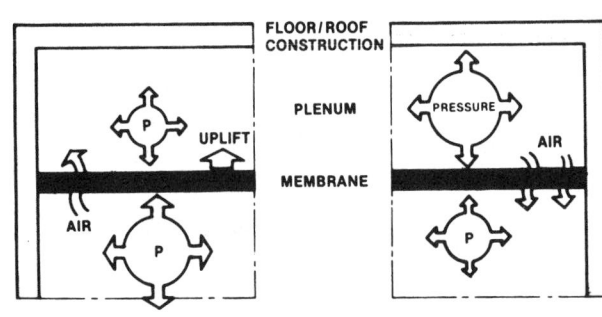

■ **Pressurized plenums,** if not completely sealed, may force air and dust through cracks in the membrane into the enclosed space.

■ **When pressure differentials such as slight negative pressure in enclosed space** exist, opening a door to the outside may result in rattling or momentary lifting of a suspended membrane.

LATERAL/SUSPENDED LOADS

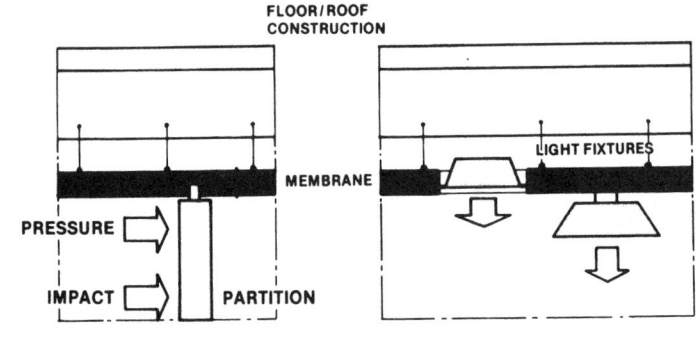

■ **Lateral loads may be imposed** upon ceiling membranes when long runs of partitions are connected to the membrane to provide them with lateral stability.

■ **The capacity of the membrane** to resist lateral loads should be investigated: some may not have the capability and the partition then should be braced to the floor/roof construction above.

■ **Vertical loads** will generally consist of light fixtures and HVAC components, e.g., air diffusers which are incorporated into, or suspended from the membrane.

C3.1 CEILINGS

In selecting a ceiling assembly, consider not only the performance requirements of the ceiling membrane itself, but also the limitations imposed by other components of the structure or by installation methods.

SUSPENDED MEMBRANES

Suspended membranes are **least affected by deflection and expansion/contraction** in the supporting construction:

■ **Suspended grid membranes** do not normally require provisions for controlling expansion in large areas: their relatively small component parts are able to accommodate expansion without much effect upon the entire assembly.

■ **Expansion in large areas of monolithic membrane** such as plaster, or in large panel membrane such as gypsum board, will be cumulative; it should be provided for by:
• allowing free movement along the perimeter
• expansion joints incorporated into the membrane itself.

In evaluating specific types within the suspended membrane group consider the following:

Acoustic Tile

The known advantages and general use of acoustic tile suggest **careful consideration of some of its limitations:**

■ **Acoustic Tile is:**
• difficult to clean
• cannot be repeatedly repainted without loss in sound absorbing properties
• where it is subject to soiling, dust accumulation or air borne moisture, the use of plastic film-faced tile is recommended, at added cost.

■ Acoustic tile is **not generally resistant to impact:** special types offering improved resistance to impact and displacement should be selected for school corridors, indoor play areas.

■ It is **not resistant to high humidity or steam,** such as found in indoor pools, commercial kitchens, or industrial spaces; special types are available for use under high humidity conditions, including suspension systems of non-corroding metal.

■ Installation of a **single suspended membrane,** especially of the exposed grid type, over several ceiling-high partitions must be carefully detailed to achieve proper closure between top of partition and membrane after the membrane is installed.

■ **Exposed grid** may restrict choice in locating lighting fixtures and/or diffusers in small spaces.

■ **Attaching wall angles/channels** directly to a framed, gypsum wallboard faced partition can seldom be successfully done: few partitions are straight enough for an angle/channel to be in continuous contact with them. The above applies to any wall angle/channel, irrespective of the type of suspension system used.

■ **Changes or repairs** to concealed grid types are difficult to make, and generally will show after completion.

In selecting metal pan tile, metal and plastic panels, note:

■ These are **generally used, when** requirements are:
• resistance to impact
• easy maintenance
• ready accessibility to utilities in plenum space

■ May be used in **protected outdoor locations.**

■ **Use of plastic panels** in certain occupancies may be limited or restricted: local building codes should be checked.

In evaluating plaster and gypsum board note:

■ **Openings in plaster** generally require plaster frames or screeds which add to the cost of the membrane.

■ **Stresses may build up** at larger openings, especially when closely spaced and cracking of the membrane may result.

■ **Plaster has:**
• poor resistance to high humidity conditions unless completely sealed
• resistance to steam and chemical fumes depends largely on the protective coating used.

■ **Sound absorption** of plaster and gypsum board membranes is poor as is that of all hard surfaces: up to 95 percent of the sound waves striking them may be reflected.

■ **Use of an acoustic plaster** finishing coat improves sound absorption, but acoustic plaster may be easily damaged and is difficult to repair successfully.

■ **Plaster on metal lath membranes** can be installed in curved or molded surfaces, but installation requires skilled labor and is high in cost.

ATTACHED MEMBRANES

Plaster and gypsum board membranes, especially when secured directly to framing, **are affected by deflections and by expansion/contraction** in the supporting construction.

■ **Attaching gypsum board** membrane directly may create the problem of nail popping unless framing members are properly aligned.

■ **Resistance to structure borne sound** of plaster and gypsum board membranes is poor, unless resiliently mounted.

■ **Attached acoustic tile** membranes are not significantly affected by normal deflections in the supporting framing.

Use of acoustic tile cemented directly to the underside of a concrete slab is generally not recommended.

■ **Moisture** may collect and condense behind the tile, causing the tile to buckle.

■ When tile is **secured to furring** channels attached to the underside of a concrete slab, provide for air to circulate between furring strips.

C3.1 CEILINGS

CEILINGS: selection checklist (cont.)

NO MEMBRANE

When the appearance of the ceiling is not an important consideration, or when the underside of construction presents a finished surface, e.g., in reinforced concrete flat plate construction, a ceiling membrane may not be provided.

Among the possible finishes then are:

■ Applied coatings and plaster finishing coat:
- will be directly affected by deflection whether under live load, dead load induced creep, or drying shrinkage.
- sprayed-on acoustical treatments will also be affected, but will not show cracks as readily as smooth coatings or plaster; some types do have a tendency to flake-off and fall.
- impact noise reduction is generally poor.

SPECIAL TYPES OF CEILING MEMBRANE

Unit acoustical absorbers - such as baffles, or cylinders - are generally suspended from the supporting construction in order to:

■ **Improve acoustical** conditions in a space.

■ **Visually isolate** mechanical/electrical services above them.

■ **Serve both** purposes, as might be the case when baffles of sound-absorbent materials are suspended over the entire ceiling area.

■ **The sound absorption** coefficients of unit sound absorbers, because of sound diffraction effects, may be greater than 1.0.

Ceilings in Open-Plan office space are another special problem:

■ **Open-plan office** space is one where individuals or groups of individuals are not separated by ceiling-high partitions.

■ Free-standing screens may be used to provide limited **visual privacy.**

■ **Speech privacy** is achieved through:
- sound absorbing materials incorporated into the ceiling membrane
- sound absorbing facings over screens and perimeter walls
- carpeting is often used as flooring to reduce impact noise of walking and scraping:

■ **Panels for the ceiling** membrane in open-plan offices should be highly absorbent in order to reduce sound reflection at a frequency range of 400 to 2000 Hz at a 45° to 60° angle of incident.

■ **Flat-lens recessed lighting fixtures,** when integrated into the membrane, tend to reflect sound from one work position to another, even when a screen is placed between them; consider the use of:
- coffered pre-assembled modules
- fixtures with special sound diffusing lenses

■ Electronically generated **masking sound system** is generally required.

■ **When isolated fully enclosed spaces,** - such as conference rooms-, are included in an open-plan office space:
- evaluate the sound attenuating properties of the membrane
- or carry the enclosing partitions to the underside of overhead construction to serve as sound baffles.

Electric Radiant Heating membranes are also in relatively frequent use. These may be:

■ **Packaged pre-assembled** systems utilizing gypsum board as the base material are available.

■ Can be **fabricated on-site,** employing a double layer of gypsum board with the heating cables embedded between them:
- installation requires special care to avoid damage to the heating cables
- face layer of gypsum board can be then finished in the usual manner.

INSULATION OVER THE CEILING MEMBRANE

Insulation may be installed directly on the suspension system or over acoustical ceiling panel to:
- improve **thermal resistance** of the assembly
- improve **sound absorbing** characteristics

■ **When insulation is added** to a ceiling assembly:
- it may interfere with accessibility to the plenum;
- adequate ventilation of the plenum should be provided to prevent high humidity conditions which may cause condensation on cold surfaces, e.g., improperly insulated ducts or cold water piping during hot weather; structural framing or other cold surfaces in winter.

SUSPENSION SYSTEMS

For information on **types of suspension systems available,** see charts. In the selection of these systems, note:

■ **Membranes are suspended** from the supporting construction by the use of:
- wire hangers, either 9 or 8 gauge for plaster, and generally 12 gauge for acoustical tile membranes.
- wire steel rods, generally 3/16 or ¼ inch in diameter
- wire steel straps, generally 1 inch by, ⅛ or 3/16 inch.

■ **Hangers should not** be connected to, or be in contact with piping and ductwork; this to prevent any vibration in such service lines from being transmitted to the membrane.

■ **When minimizing impact** noise transmission is important, resilient mounting devices should be introduced between the hanger and the suspension system of the membrane.

■ **Suspending ceiling membranes** from tabs in metal decks may present a problem under conditions where condensation collects on tabs causing corrosion and loss of their carrying capacity.

C3.1 CEILINGS

FIRE RESISTANCE RATING OF FLOOR/CEILINGS

Ceiling assemblies are non-structural and require a floor/roof construction for support; **thus the ceiling membrane is but a component of a floor/ceiling or roof/ceiling assembly.**

■ **Fire resistance tests** are conducted on full scale floor/ceiling or roof/ceiling assemblies, and fire resistance ratings are based on results of such tests.

■ In all instances **the assemblies tested are constructed of specific materials** in a specified manner, and the rating given is for that particular assembly.

■ **Changing components or methods** of installation of either the ceiling membrane or of the floor or roof construction may result in the entire assembly being no longer acceptable as a fire resistance rated assembly.

■ **Fire resistance rating** may also vary for an assembly depending on whether the elements were tested in restrained or unrestrained condition:
• restrained condition exists when expansion in a load carrying element is resisted at supports by forces external to such element.
• unrestrained condition exists when a load carrying element is free to expand at its supports.

■ **Openings for electrical outlets** and diffusers in ceiling membranes will also have an effect upon the fire resistance rating of a floor/ceiling assembly:
most Building Codes will permit unprotected openings of up to 100 square inches per 100 square feet of ceiling area, and will require protection for openings in excess of above.

■ Details of construction and sketches of a few fire resistance rated floor/ceiling assemblies:

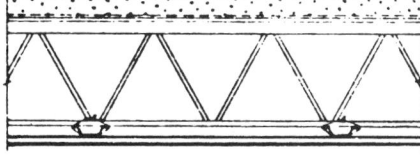

■ **Gypsum Wallboard, Steel Joists, Concrete Slab;** ½" type X gypsum wallboard or veneer base, applied across rigid furring channels 24" o.c. with 1" type S drywall screws 12" o.c.; at end joints 8" o.c. to additional pieces of drywall furring channels. Furring channel wire-tied 24" o.c. to open web steel joists which support ³⁄₈" rib metal lath or ⁹⁄₁₆" deep 28 gage corrugated steel and 2" concrete slab measured from top of flute. (Passed 90 minute fire test restrained and unrestrained.)

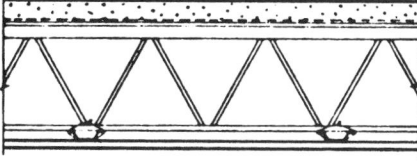

■ **Steel Joists, Concrete Slab, Gypsum Wallboard.** ⁵⁄₈" proprietary type X gypsum wallboard or veneer base applied perpendicular to rigid furring channels with 1" type S drywall screws 12" o.c. Rigid furring channels 24" o.c. attached with 18 gage wire ties to open web steel joists 24" o.c. and supporting ³⁄₈" rib metal lath and 2" concrete slab. Double channel at wallboard end joints. (Two hour restrained and unrestrained.)

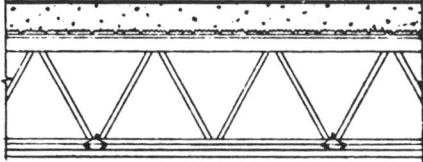

■ **Steel Joists, Concrete Slab, Metal Lath, Gypsum Wallboard.** ⁵⁄₈" proprietary typeX gypsum wallboard or veneer base screw attached with 1" type S drywall screws 12" o.c. at right angles to furring channels 24" o.c. (double channels at end joints). Furring channel wire tied to open web steel joist 24" o.c. supporting 3" concrete slab over ³⁄₈" rib metal lath. ⁵⁄₈" x 2¾" type X gypsum wallboard strips over butt joints. (Three hour restrained and unrestrained.)

C3.1 CEILINGS

REFERENCES: Ceilings

STANDARDS

ASTM Specifications for:

C28 • Gypsum Plasters
C36 • Gypsum Wallboard
C61 • Gypsum Keene's Cement.
C557 • Adhesives for Fastening Gypsum Wallboard to Wood Framing.
C587 • Gypsum Veneer Plaster.
C630 • Water-Resistant Gypsum Backing Board
C635 • Metal Suspension Systems for Acoustical Tile and Lay-In Panel Ceilings.
D1779 • Adhesive for Acoustical Materials

ASTM Tests/Methods for:

C367 • Strength Properties of Prefabricated Architectural Acoustical Tile or Lay-In Ceiling Panels.
C522 • Airflow Resistance of Acoustical Materials
C523 • Light Reflectance of Acoustical Materials by the Integrating Sphere Reflectometer
E492 • Laboratory Measurement of Impact Sound Transmission Through Floor-Ceiling Assemblies Using the Tapping Machine.
E1111 • Measuring the Interzone Attenuation of Ceiling Systems.

ASTM Recommended Practice for:

C636 • Installation of Metal Ceiling Suspension Systems for Acoustical Tile and Lay-In Panel.

Federal Specifications:

SS-S-118A(3) • Sound Controlling Blocks and Board

SELECTED REFERENCES

APA **American Plywood Association**
• APA Design/Construction: Noise-Rated Systems

CISCA **Ceilings and Interior Systems Contractors Association**
• Acoustical Ceilings Use and Practice

GA **Gypsum Association**
201-80 • Using Gypsum Board for Walls and Ceilings
216-82 • Recommended Specifications for the Application and Finishing of Gypsum Board
650-80 • Covering Existing Interior Walls and Ceilings with Gypsum Products
• Manual of Gypsum Veneer Plaster

GDCI **Gypsum Drywall Contractors International**
• Manual of Gypsum Drywall Construction

ML/SFA **Metal Lath/Steel Framing Association**
No. 101 • Types of Metal Lath and Their Uses

NBS **National Bureau of Standards**
• Technical Report on Building Materials, No. 44
• Building Materials Section, No. 17
• Building Materials & Structures Report, No. 92

NLA **National Lime Association**
• Specifications for Lime and its Uses in Plastering, Stucco, Unit Masonry and Concrete.

NRCC **National Research Council of Canada**
CBD 92 • Room Acoustics-Design for Listening
139 • Acoustical Design of Open-Planned Offices

C3.1 CEILINGS

C3. 2 FLOORING

FLOORING: introduction

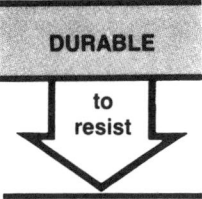

- abrasion
- indentation
- compression
- impact
- dust/dirt

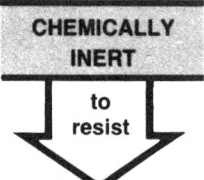

- cleaning compounds
- disinfectants
- solvents
- lubricants
- fatty substances

- fatigue of walking standing running
- impact noise of traffic

- non-slippery
- non-tripping
- non-flammable also
- conductive
- non-sparking

C3.2 FLOORING

FLOORING TYPES

Flooring today offers a considerably wider choise in materials, properties, and appearance to meet changing performance criteria.

■ **Stone** is available in a variety of colors, patterns and surface finishes. It is still the best choice in areas of very heavy traffic.

■ **Cementitious types of flooring:**
- Monolithic concete with integral finish may constitute flooring and substrate in one unit; least expensive but limited in appearance and performance characteristics.
- Concrete toppings over concrete substrate are used for industrial floors when primary importance is resistance to abrasion/wear.
- Terrazzo toppings of Portland cement and selected aggregate may be used when appearance and durability under heavy traffic are important.
- Terrazzo is also available as precast tile and precast stair treads.
- Magnesite oxychloride is similar to concrete topping, or terrazzo, but not equally wear-resistant.

■ **Burned clay** is available either unglazed or glazed, in various sizes, shapes and colors, and as bricks, pavers, ceramic and quarry tile.
- It offers generally good resistance to grease, strong cleaning solutions, abrasion/wear.
- Also available with non-slip, alkali-stain resistant, frost proof, and conductive properties.

■ **Resilient flooring** is available in sheet and tile form, of various compositions.
- Most types are resistant to moderately heavy traffic; some are also grease resistant.
- It is resistant to moisture, but may fail if the bond between it and the substrate is weakened or may be destroyed by moisture seeping in through joints between the units.

■ **Wood flooring** is produced in strips, thin blocks and planks from various hardwoods and softwoods.
- Blocks of solid end-grain provide resiliency, while wood flooring with common grain pattern provides better resistance to indentation. Acrylic impregnation improves the physical properties of wood, providing superior wear resistance.
- Natural hardwood flooring is more durable than softwood; both will

require protective surfacing and periodic maintenance.

■ **Seamless flooring** is available in formulations of a resinous matrix, fillers and/or decorative additions, which after installation and curing form a seamless finished surface.
- It has high resistance to chemicals.
- Seamless flooring is generally good for concentrated traffic areas.

■ **Carpeting** is available in a variety of yarn materials, construction types, colors, patterns and textures.
- The construction of the carpeting such as pile height, pitch and yarn count determines its resistance to wear. Carpeting reduces impact noise and provides underfoot comfort.
- It resists repetitive movement over it of large rubber wheels (such as casters on chairs), but will be quickly and severely damaged by small steel wheels.

■ **Metal plate** is available with either smooth or raised pattern; the latter to minimize the possibility of slipping.
- It is generally used over a concrete floor slab in traffic aisles only.

■ **Metal grating,** also available with concrete fill, is generally used over steel framing such that the grating spans between secondary framing members.
- Concrete fill will wear out before the steel bars of the grating, and periodic resurfacing in heavy traffic areas may be necessary.

A special type of flooring is the raised, or access floor, consisting of a system of posts supporting rigid panels surfaced with either resilient flooring or carpeting.

■ **Access floors** are used to provide space for, and easy access to, all wiring serving electronic equipment supported by the floor.
- They are also used as supply plenum for conditioned air to cool the equipment and/or to supplement the main air conditioning system of the space containing electronic equipment.

Preliminary selection of the type of flooring will not be confined to considerations of in-service conditions and appearance only; substrate condition, availability of materials, cost, and purely personal preferences will also be involved.

- Listing of flooring types is intended as a general guide only, to indicate which types of those commonly available might be used.

- Break-down by appearance is arbitrary: any type listed in either column can be used in the given in-service conditions.

SUITABLE FLOORING WHEN APPEARANCE IS AN IMPORTANT CONSIDERATION.		IN-SERVICE CONDITIONS.		SUITABLE FLOORING WHEN APPEARANCE IS A SECONDARY CONSIDERATION.
		1		
• resilient • wood strip/plank/block • carpeting • burned clay • seamless • terrazzo • stone		• light traffic • dry conditions • occasional use of mild cleaning solutions		• monolithic concrete with or without surface hardener • magnesite oxychloride
		2		
• burned clay units • seamless • terrazzo • stone		• light traffic • frequently wet conditions • frequent use of mild cleaning solutions		• monolithic concrete • concrete topping both with non-slip finish
		3		
• seamless • burned clay units		• light to moderate traffic • wet conditions • frequent use of strong cleaning solutions • chemically active substances spilled		• acid resistant protective coatings or • seamless
		4		
• some of the resilient • some of the carpeting • solid end-grain block • appearance graded • acrylic impregnated wood • seamless • terrazzo • stone		• heavy traffic • dry conditions • occasional use of mild cleaning solutions		• concrete topping • wood: solid end-grain block utility grade
		5		
• terrazzo • stone • burned clay units • seamless • acrylic impregated wood		• heavy traffic • occasionally wet • tracked-In moisture and dirt • occasional use of mild cleaning solutions		• concrete topping • seamless
		6		
• terrazzo • some of seamless • stone • burned clay units		• heavy traffic • frequently wet • freezing temperatures		• concrete topping
		7		
• brick paving • granite paving		• heavy rubber wheeled traffic • light to moderate steel wheeled traffic • dry to moderately wet conditions		• wood: solid end-grain blocks—utility grade • some of seamless
		8		
		• heavy steel wheeled traffic • dry to wet conditions		• metal: steel floor plate concrete filled steel grating

C3.2 FLOORING

C • INTERIORS

FLOORING: types, uses

● denotes common usage or suitability

○ denotes possible or limited usage or suitability

[1]   **[2]**   **USE-IN-SERVICE CONDITIONS:**

TYPE	COMPARATIVE COST, average range, base = 100 [1]	THICKNESS range in inches	WEIGHT range in lbs/sq. foot	COMFORT RATING [2]	DRY	OCCASIONALLY WET	FREQUENTLY WET	FOOT: LIGHT	FOOT: HEAVY, UNIFORM	FOOT: HEAVY, LOCALIZED	WHEELED: RUBBER or LIGHT STEEL	WHEELED: HEAVY STEEL
STONE												
MARBLE	900 / 2000	7/8 / 1¼	12 / 15		●	●	●	○	●	●	○	
SLATE	1000 / 1600	½ / 1¼	10 / 20	poor	●	●	●	○	●	●	○	
GRANITE	1600 / 3000	7/8 / 2½	10 / 40		●	●	●	○	●	●	○	
CEMEMTITIOUS												
CONCRETE, MONOLITHIC	200 / 250	4 / 6+	50 / 75+		●	●	●	●	●	○	○	
CONCRETE TOPPING	120 / 200	¾ / 1½	10 / 20	poor	●	●	●	●	●	●	●	○
TERRAZZO	400 / 500	¼ / 2½	10 / 40		●	●	●	●	●	●	○	
MAGNESITE OXYCHLORIDE	300 / 340	3/8 / 1	4 / 5	poor to fair	●	●		●	●	●		○
BURNED CLAY												
FLOOR BRICK	400 / 900	1 / 1½	20 / 40		●	●	●	●	●	●	●	
CERAMIC TILE	200 / 400	¼ / ½	4 to 6	poor	●	●	●	●	●	●	○	
QUARRY TILE	400 / 550	½ / 1½			●	●	●	●	●	●	○	
RESILIENT												
SHEET, LINOLEUM	100 / 130	1/16 / 1/8	1 to 2	good	●	●		●				
SHEET, RUBBER	150 / 160	3/32 / 3/16			●	●		●	●		●	
SHEET, VINYL	130 / 140	5/64 / 7/64		v. good	●	●		●	●		●	
TILE, RUBBER	140 / 150	5/64 / 1/8			●	●		●	●		●	
TILE, VINYL	150 / 200	5/64 / 3/32			●	●		●	●		●	
TILE, VINYL-COMPOSITION	100 / 120	5/64 / 1/8		good	●	●		●	●		○	
TILE, CORK	130 / 150	3/16 / 5/16		v. good	●			●				
TILE, ASPHALT	60 / 75	1/8 / 3/16		fair	●	●		●	●		●	○

C3.2 FLOORING

C • INTERIORS

REMARKS

[1] Base for comparative cost: vinyl-composition tile ⅛ inch thick.

[2] Comfort Rating = a measure of elasticity to absorb impact of walking and stress of prolonged standing; ranging from poor for hard surface, with no elasticity, to very good for a high degree of elasticity.

IMPACT/INDENTATION	TRACKED-IN MOISTURE/DIRT	MILD CLEANING SOLUTIONS	STRONG CLEANING SOLUTIONS	STEAM CLEANING	CHEMICALS/LUBRICANTS	ANIMAL FATS	EXPOSED TO WEATHER	PROTECTED, BELOW GRADE	PROTECTED, ON GRADE	PROTECTED, ABOVE GRADE	CONDUCTIVE	SLIP RESISTANT	WOOD/PLYWOOD/PARTICLEBOARD	CONCRETE SLAB/FILL
●	●	●		●			●	●	●	●		○	○	●
●	●	●		●			●	●	●	●		○	○	●
●	●	●		●			●	●	●	●		○	○	●
●	●	●	○	●			●	●	●	●	●			●
●	●	●	○	●			●	●	●	●	○	●		●
●	●	●	○	●			●	●	●	●	●	●	●	●
●	●	●	○	●	○			●	●	●	●	●	●	●
●	●	●	●	●	●	●	●	●	●	●		○	●	●
●	●	●	●	●	●	●	●	●	●	●		●	●	●
●	●	●	●	●	●	●	●	●	●	●		●	●	●
○	●	●							●	●		●	●	●
●	●	●							●	●		●	●	●
●	●	●	●						●	●		●	●	●
●	●	●							●	●		●	●	●
●	●	●	●						●	●		●	●	●
○	○	●	○		●	●	○		●	●		●	●	●
○	○	●							●	●		●	●	●
○	○	●				○		●	●	●		●	●	●

Remarks by group:

Marble group:
- four grades of marble available; generally only Group A suitable for exterior use.
- protective surfacing optional for marble.
- sand blasting is an optional method of cleaning.
- life expectancy of 100 plus years; but some types may wear out in relatively short time in areas subject to abrasion of concentrated high volume foot traffic.

Terrazzo/concrete/magnesite group:
- pre-cast terrazzo available. Comparative cost for terrazzo tile: 660 to 760.
- concrete will resist hard rubber wheeled traffic, and moderate steel wheeled traffic.
- concrete will resist alkalies, but must be protected from other chemicals.
- magnesite should not come in contact with metal. Use of underlayment recommended. Spark-proof.

Ceramic tile group:
- ceramic tiles are able to withstand freezing temperatures.
- may break under heavy impact loads.
- sizes and thicknesses are representative only. Other sizes, thicknesses, and various shapes are available. Consult manufacturers literature.

Resilient flooring group:
- sheets with few joints not recommended for use below grade as moisture may become trapped.
- follow manufacturer's recommendations if considering use of particle board as a substrate.
- cost listed is for vinyl composition tile ⅛" thick.
- life expectancy of 10 to 15 years.
- periodic maintenance requires the use of protective compounds to fill surface irregularities caused by abrasion and to resist further abuse.
- spillage of household grease, oil, mild solvents, and alkalis which are immediately wiped up will not affect any resilient flooring, however, if left, may damage it.

C3.2 FLOORING

C • INTERIORS

FLOORING: types, uses (cont.)

● denotes common usage or suitability
○ denotes possible or limited usage or suitability

TYPE	COMPARATIVE COST, average range, Base = 100 [1]	THICKNESS range in inches	WEIGHT range in pounds per sq. foot	COMFORT RATING [2]	DRY	OCCASIONALLY WET	FREQUENTLY WET	FOOT: LIGHT	FOOT: HEAVY, UNIFORM	FOOT: HEAVY, LOCALIZED	WHEELED: RUBBER OR LIGHT STEEL	WHEELED: HEAVY STEEL
WOOD												
HARDWOOD, STRIP/PLANK	300 500	25/32	1 to 3	fair to good	●			●	●		○	
HARDWOOD, THIN BLOCKS	300 400	3/8 to 3/4			●			●	●			
ACRYLIC IMPREGNATED: heat catalyzed	300 400	5/16			●	●		●	●	○	○	
ACRYLIC IMPREGNATED: radiation polymerized	300 400	5/16			●	●		●	●	●	●	
SOLID END-GRAIN BLOCKS	200 500	2½ to 4	10 13		●	●	●	●	●	●	●	●
SEAMLESS												
THERMOSETTING, EPOXY	300 500	1/16 to 1/8	3 to 7	fair	●	●	●	●	●	●	●	●
THERMOSETTING POLYCHLOROPRENE (NEOPRENE)	400 600	3/8 to 1/2			●	●	●	●	●	●	●	
THERMOSETTING, POLYESTER	300 1000	1/8 to 1/4			●	●		●	●	●	●	
THERMOPLASTIC, ACRYLIC	200 300	1/8 to 1/4			●	●		●	●	●	●	
THERMOPLASTIC, MASTIC	300 350	1/8 to 1½			●	●	●	●	●	○	○	
CARPETING												
ACRYLIC	150 200		½ to 1	v. good	●	○		●	●			
NYLON	100 150				●			●	●			
OLEFIN/POLYPROPYLENE	200 300	*		good	●	●	●	●	●		○	
POLYESTER	200 250			v. good	●			●	●			
WOOL	300 500				●			●	●			
METAL												
GRATING	700 2000	¾ to 2½	2 20	poor	●	●	●	●	●	●	●	●
FLOOR PLATE	200 1900	1/8 to 1	5 40		●	●	●	●	●	●	●	●
SYSTEMS												
ACCESS/RAISED	500 800		7 11	good to v. good	●			●			●	

C3.2 FLOORING

REMARKS

[1] Basis for comparative cost: **vinyl-composition** inch thick.

[2] Comfort Rating = a measure of elasticity to absorb impact of walking and stress of prolonged standing; ranging from poor for hard surface with no elasticity to very good for a high degree of elasticity.

Legend: ● = applicable / recommended ○ = limited / optional

SUBJECT TO:							LOCATION				OPTIONS		SUB-STRATE	
IMPACT/INDENTATION	TRACKED-IN MOISTURE/DIRT	MILD CLEANING SOLUTIONS	STRONG CLEANING SOLUTIONS	STEAM CLEANING	CHEMICALS/GREASE	ANIMAL FATS	EXPOSED TO WEATHER	PROTECTED, BELOW GRADE	PROTECTED, ON GRADE	PROTECTED, ABOVE GRADE	CONDUCTIVE	SLIP RESISTANT	WOOD/PLYWOOD/PARTICLEBOARD	CONCRETE SLAB/FILL
○	○								●	●		●	●	●
○	○								●	●		●	●	●
●	●	●						●	●	●		●	●	●
●	●	●						●	●	●		●	●	●
●	●				○	○			●	●		●	●	●
●	●	●	●	●	●	●	●	●	●	●	●	●	●	●
●	●	●	●	●	●	●	●	●	●	●	●	●	●	●
●	●	●	●	●	●	●	●		●	●	●	●	●	●
●	●	●	●	●	●	●	●		●	●	●	●	●	●
○	●	○		○	○	○	○	●	●	●	●	○	○	●
●	●	●							●	●	○	●	●	●
●	●	●							●	●	○	●	●	●
●	●	●	●	●	○	○	●	●	●	●	●	●	●	●
●	●	●							●	●	○	●	●	●
●	●	●							●	●	○	●	●	●
●	●						●	●	●		●	●		●
●	●	●					●	●	●		●	○		●
●	●									●	●	●		●

REMARKS (by group)

Group 1:
- strip/plank over concrete requires sleepers/nailers.
- blocks over wood: Nail to substrate or use mastic. Blocks over concrete: Use mastic.
- on grade requires vapor retarder under concrete slab and between slab and flooring.
- may last 50 plus years, but will require periodic re-surfacing and may need replacement in concentrated traffic areas.

Group 2:
- mastic can be protected with an epoxy coating.
- life expectancy varies from 10 to 20 years; generally good in concentrated traffic areas.
- for additional information refer to COATINGS.
- substrate; concrete, steel and wood; only epoxy and neoprene recommended on metal.

Group 3:
- life expectancy in commercial/institutional installations generally 5 to 10 years except in concentrated foot or wheeled traffic areas.
- carpet tiles can be moved and or replaced individually when worn.
- * For thickness see Evaluation Chart "Carpeting: types, construction".

Group 4:
- grating shop primed or galvanized steel. Also available in aluminum, plastic and stainless steel • weight range given is for aluminum and steel. • concrete-filled steel grating is available. • floor-plate is available either smooth or with a raised pattern to minimize the possibility of slipping • raised, or access floor - a system of posts supporting rigid panels surfaced with either resilient flooring or carpeting to provide access to wiring serving electronic equipment and/or as plenum for conditioned air.

C3.2 FLOORING

FLOORING: cementitious, burned clay

● denotes usage or availability
○ denotes optional usage or limited availability
■ denotes good resistance
□ denotes fair resistance
NA denotes information not available

PROPERTIES:

FLOORING TYPE	UNIT SIZE: length × width, typical in inches	THICKNESS: range in inches	GRADES AVAILABLE	[1] COMPRESSIVE STRENGTH pounds per sq. inch	[2] SHEARING STRENGTH, allowable pounds per sq. inch	[3] RUPTURE STRENGTH pounds per sq. inch	COEFFICIENT OF EXPANSION per 100°F rise	[4] THERMAL CONDUCTIVITY BTU/hr/sq. ft./°F./in.
STONE								
MARBLE	8 × 16* to 12 × 12	⅞* 1	A; also B, C, D.	7500 min	150	1,000 min	.009 to .012	
SLATE, BLUESTONE	6 × 6* to 24 × 24	½* 1¼	no grades	10,000 to 15,000	NA	9,000 and 7,200	.0036 to .0069	13 to 14
GRANITE	4 × 4* to 36 × 36	⅞* to 3	fine medium coarse	19,000 min	200	1,500 min	.0063 to .009	
CEMENTITIOUS								
CONCRETE, MONOLITHIC	not applicable	4 to 6		3,750 min	112 min			
CONCRETE, TOPPING		¾ to 2½	various mixes	4,500 to 8,000	130 to 240	650*	.007	11 to 16
TERRAZZO, MONOLITHIC		¼ 2½	various types					
TERRAZZO, TILE	8 × 8 to 16 × 16	3/16 ¼						
MAGNESITE OXYCHLORIDE	not applicable	⅜ 1	various fillers	5,000 7,000				
BURNED CLAY								
FLOOR BRICKS	8½ × 4 × 2½*	1* 1½	SW WW					
CERAMIC TILE, PORCELAIN	6 × 6*	*	standard seconds					
CERAMIC TILE, CLAY		¼ to ½	select standard commercial	3,000 to 6,000	2,300	1,600 to 2,000	.0003	5 to 9
CERAMIC TILE, MOSAIC	1 × 1* to 2 × 2		select commercial					
QUARRY TILE	6 × 6*	½* 1½	select standard seconds					
PAVERS, PORCELAIN	4 × 4* to 6 × 6	3/16* to ⅝	standard seconds					
PAVERS, CLAY			standard seconds					

C3.2 FLOORING

WEAR/ABRASION	IMPACT LOADS	INDENTATION	ALKALIES	CHEMICALS [5]	FATTY SUBSTANCES	STAINING/DIRT [6]	CHOICE OF COLOR	CHOICE OF SURFACE TEXTURE	DIRECT BONDING	MORTAR BED	ADHESIVE	SLIP PLANE RECOMMENDED	PROTECTIVE SURFACING REQUIRED	PERIODIC RESURFACING REQUIRED
■	■	■					●	●		●		○	○	
□		■	□	□			●	●		●		○	○	
■	■	■					●	●		●		○		
□	■	■	■										○	○
■	■	■	■							●				
■	■	■	□			□	●	●	●	●		●	●	○
■	■	■	□			□	●	●		●	●	○	●	○
□	■	■	□	□		□	●	○	●			○	●	○
■	■	■	■	■	■	■	●	○		●				○
■	□	■	■	□	■	■	●	●			●	●	○	
■	□	■	■	□	■	■	●				●	●	○	
■	□	■	■	□	■	■	●	●			●	●	○	
■	□	■	■	■	■	■	●	●			●	●	○	
□	□	■	■	□	□	■	●	●			●	●	○	
■	□	■	■	■	■	■				●	●	●	○	

REMARKS

[1] Compressive strength is a measure of resistance to a crushing force.

[2] Shearing strength is a measure of resistance to breaking under a load when partially supported.

[3] Rupture strength is a measure of resistance to breaking under an impact load when fully supported; or under tensile stress. For slate: high value across grain; low value along grain.

[4] Thermal conductivity will affect apparent warmth to touch.

[5] Consult manufacturer for resistance to specific chemicals.

[6] Staining by commonly used substances; soft drinks, coffee, alcoholic beverages.

- five classes of marble: calcite, dolomite, onyx, serpentine, travertine. Grades A, B, C, and D.
- slippery when wet.
- resistance to skidding depends on finish.
- resists freezing temperatures.
- resistance to staining/dirt improved by surfacing, but surfacing has to be renewed.
- * sizes and thickness approximate only; generally available in various sizes and thicknesses to meet design requirements.

- terrazzo: Use felt over wood subfloor to avoid cracks, Abrasive granules can be added to improve slip resistance. Conductive type available.
- resists staining when protective surfaced.
- skid resistant finishes available.
- for resistance to alkalies, special type cement should be selected. Refer to CEMENT/CONCRETE.
- concrete has limited resistance to lubricants.
- magnesite is no longer in wide usage.
- * Tensile strength of specially designed mixes.

- on wood subfloors use 15 lb. saturated felt.
- conductive tiles must be set in special cement.
- pavers: Available either glazed or unglazed.
- standard bricks may also be used for flooring.
- for resistance to specific chemicals consult manufacturers.
- For resistance to chemicals, lubricants and staining, special grouts required.
- protective surfacing for paving brick & pavement optional.
- in areas of concentrated foot traffic, an abrasive grain fired into the glaze is recommended.
- pregrouted ceramic tile sheets available.

- * sizes and thicknesses are representative only. Other sizes, thicknesses, and various shapes are available. Consult manufacturers literature.

C3.2 FLOORING

STONE, CEMENTITIOUS, BURNED CLAY:

STONE

Stone is **the most durable** of all materials used for flooring; it is generally suitable for indoor and outdoor locations where resistance to heavy foot traffic and appearance are important:

■ **Marble** offers the widest choice in color and markings, and is available in four grades:
- Group A includes marble with uniform and favorable working qualities; it is the only type recommended for outdoor use.
- Group B has less favorable working qualities, with faults which may require mending;
- Group C with uncertain working qualities and geological flaws requiring repair and/or reinforcing;
- Group D with larger proportion of natural faults than C; often with striking variations in color and markings.

■ Marble is **not resistant** to acid substances and alkaline salts (such as sodium carbonate and trisodium phosphate) found in some cleaning compounds, or to industrial fumes.

■ Marble flooring is **generally finished** by:
- honing, which produces a smooth, dull gloss surface;
- sand-rubbing, which produces a flat non-reflective surface; or
- grit finishing, which is an intermediate finish between honing and sand-rubbing.
- polishing is not recommended for flooring, as it results in a slippery surface.
- finish for marble used in exposed locations, or where frequently wet, should be rough enough to prevent slipping.

■ **Slate** is available in several colors, and is either:
- split to provide a natural cleft surface, or
- sawed, planed, sand-rubbed, or honed.
- honed finish is not recommended for floors as it results in a slippery surface.

■ Slate **is quite brittle,** which may limit its use in some non-residential applications where it might be subject to heavy impact loads.

■ Slate **wears very well** under heavy foot traffic, but it has a low abrasion resistance to scratches by metals, cinders, or similar hard objects.

■ **Granite** is generally more resistant to abrasion/wear than marble or slate and

may be used everywhere, including on stair treads subject to very heavy traffic. Finishes are similar to those used for marble and slate.

■ **Flagstone** is the generic name for any hard stone which splits easily into relatively thin slabs and is used as flooring or paving. It generally includes:
- slate, many varieties of sandstone, quartzite, and limestone.
- it is usually laid on a sand bed, or a setting bed over concrete substrate, in random rectangular, coursed, or irregular patterns.

■ **Stone flooring** is used mostly:
- in enclosed public areas of institutional and commercial buildings;
- occasionally in entrance halls of residential buildings;
- outdoor terraces;
- walkways, stair treads and risers.
- unless specially finished, it may be too slippery when wet for use on inclined surfaces such as ramps.

■ **Stairs and doorways concentrate** traffic and channel it into a rigid pattern, greatly increasing abrasion/wear. When durable and difficult to replace materials such as stone are selected, special consideration should be given to such areas to avoid premature failure of flooring.

In setting stone, note:

■ **Substrate** for stone must be dimensionally stable; deflection in floor decking and framing members should not be excessive to prevent cracks developing in stone flooring due to movements in substrate.

■ Stone, especially when thin slabs are used should be **fully bedded** in mortar to minimize the possibility of bending stresses developing and thus cracking the stone.

■ Synthetic rubber **surface sealers** are available, but are not generally required. They will protect the stone from soiling and wear, but have to be periodically renewed.

■ **For additional information** on stone, refer to STONE/MASONRY.

CEMENTITIOUS FLOORING

Cementitious flooring **consists of a mixture** of aggregate and cementitious binder. The type of aggregate selected will largely determine the characteristics of the flooring: it may be selected for durability only, or for both durability and appearance.

Types available include:
■ **Monolithic concrete** slabs of selected hard, wear-resistant aggregate, hard sand and cement, placed on grade or suspended:
- top surface of the slab is steel-troweled to serve also as the wearing surface, or surfacing.
- properties of the wearing surface are largely dependent on the mix of concrete used, its curing and finishing.
- the surface of smooth finished slabs may be hardened by working selected aggregate into the top layer of the freshly poured concrete, or by commercially available hardeners; coloring pigments may also be added and various decorative surface finishes are available.
- for additional information refer to CEMENT/CONCRETE.

■ **Toppings** are thin concrete slabs poured over and bonded to concrete base slabs; selected aggregate, special finishing and curing methods are used to produce a hard, dense, wear-resistant surface.

■ **Terrazzo** is a mixture of at least 70 percent of marble and/or granite chips, and 30 percent either white, gray, or pigmented Portland cement.

Both monolithic concrete slabs and concrete toppings are generally used in industrial and storage buildings, when a durable, inexpensive flooring is required.

■ **Toppings,** even though more expensive, will generally provide better service than monolithic, hardened surface slabs; a more expensive select aggregate can be used throughout without unduly increasing the cost of the mix; placement and finishing are easier and can be better controlled.

■ **Portland cement concrete** resists:
- neutral salts; most carbonates and nitrates;
- some chlorides, fluorides, silicates, petroleum products, coal tar distilates;
- but not fatty oils and other acidic products.

■ **For additional information** refer to CEMENT/CONCRETE.

Terrazzo is generally used in similar locations to stone flooring: when durability under heavy traffic and appearance are important.

■ Terrazzo may be **installed as:**
- a topping poured over and bonded to a concrete base slab;
- poured over a wood-sub-floor; or

C3.2 FLOORING

selection checklist

- an independent slab over a sand bed and a concrete sub-floor.
- it may also be placed monolithically with the base slab, but control during placement is more difficult.

■ When used as **bonded topping** or placed monolithically, terrazzo is subject to stresses developing in the base slab: shrinkage cracking in the base slab may also crack the terrazzo.

■ The recommended **method of installation,** even though more costly, is to provide a concrete sub-floor and a sand bed.

■ When **installed over a wood sub-floor,** it should be isolated to permit differential movement.

■ **Dividing strips,** set in various patterns, are used to provide control joints in terrazzo to:
- permit changes in color;
- to serve as grounds for establishing the level of the finished surface.

■ **Finishing** of terrazzo is by:
- grinding after curing,
- then sealing the surface primarily to provide protection for the cement matrix.
- abrasive aggregate may be added to the surface to provide resistance to slipping in wet locations.

■ **Conductive terrazzo,** which employs acetylene carbon black in the matrix and in the underbed for electrical conductivity, is available for such locations as hospital operating rooms.

■ **Thin-set terrazzo** is a topping, approximately ½ inch in thickness:
- poured over a base slab first coated with an organic adhesive, such as polysulfide, neoprene, vinyl, or epoxy.
- it is installed and finished similarly to conventionally placed terrazzo.
- the advantages are reduced weight and reduced curing time;
- the disadvantages are the same as for conventionally placed bonded topping.

■ **Admixtures** for thin-set terrazzo, such as acrylic or vinyl emulsions are available:
- to increase allowable tensile and flexural stresses;
- to permit reduction of the thickness of the topping to as little as ¼ inch.
- plastic matrix instead of Portland cement is also available, and is here included under Seamless flooring.

Precast terrazzo is available in the form of tiles, stair treads, risers, bases, and shower receptors; they are generally available in the same types as cast-in-place terrazzo, including abrasive

aggregate finish when required.

■ **Magnesium oxychloride** is a durable, versatile flooring of moderate cost, but is no longer in wide use; it is made by:
- adding magnesium chloride to powdered magnesium oxide and mixing it with fillers and mineral pigments.
- marble chips may be added and the surface ground to simulate terrazzo.
- abrasive granules may be used to form a slip-resistant surface; conductive and nonsparking compositions are available for use in hospital operating rooms and munition plants.

BURNED CLAY FLOORING

Burned clay flooring is **made of a variety of clays,** in diverse shapes, patterns, and colors, which may be integral with the clay body, or applied as glaze.

■ **It is suitable** for indoor and outdoor locations where resistance to wear and moisture, appearance, and low maintenance costs are important.

■ Generally **used in areas where:**
- chemicals, grease and water may be frequently spilled,
- where sanitary conditions must be maintained, such as in food processing plants, commercial kitchens, laboratories, swimming pools, toilets;
- may also be used in areas subject to freezing temperatures.

Types of burned clay flooring include brick, pavers, and ceramic tile:

■ **Bricks:**
- offer a choice of color and texture;
- may be laid in different patterns;
- have good resistance to wear, moisture, heat, and compression/ indentation.
- may be ground for a smooth surface and may be sealed and polished for easier maintenance.
- for additional information refer to STONE/MASONRY.

Ceramic Tile is available in two classes: glazed and unglazed.
- the unglazed tile only is generally used for flooring to avoid the possibility of slipping on a smooth glazed surface;
- crystalline glazed tile, because of its uneven surface, may be used for light duty floors.

■ **Ceramic tile is made** of clay or a mixture of clay and other organic materials that are fired in kilns at temperatures above red heat:
- porcelain: dense, fine-grained, and smooth, colors are usually clear and luminous.
- natural clay tile: dense, with a slightly

textured appearance.
- ceramic mosaics: glazed or unglazed.

■ **Quarry tile** is made from natural clays or shales, unglazed.

■ **Pavers:** formed of either porcelain or natural clay; unglazed.

■ **Special types** are also available such as non-slip; frostproof; acid, alkali, and stain-resistant, and conductive.

Grouts are used to fill the open joints between burned clay units and are generally the weakest part of the flooring. Various formulations are available in addition to conventional cement grout:

■ **Dry-set grout** is a mixture of Portland cement and additives; it is used for grouting floors subject to ordinary use.

■ **Latex-Portland cement** grout is a mixture of either Portland cement, sand-Portland cement, or dry-set grout with special latex additive; it is:
- more dense than regular cement grout and
- suitable for all installations subject to ordinary use and for most commercial installations.

■ **Mastic grout** is a one-part grouting composition:
- mastic hardens by coalescense and does not require damp curing as do Portland cement based grouts;
- more flexible and stain resistant than regular cement grout.

■ **Furan resin** grout system consists of furan resin and hardener portions:
- used in industrial areas requiring chemical resistance.
- use of this grout involves extra costs and special installation skills.
- primarily used for quarry tile and pavers.

■ **Epoxy grout** system employs epoxy resin and hardeners:
- especially formulated for industrial and commercial installations where chemical resistance is of primary importance.
- also provides high bond strength and impact resistance.

■ **Silicone Rubber** grout system: single component nonslumping silicone rubber;
- resistant to staining, moisture, mildew, cracking, crazing and shrinking.
- adheres well to ceramic tile, cures rapidly,
- withstands exposure to hot cooking oils, steam, oxygen as well as prolonged exposure to sub-freezing temperatures and hot humid conditions.
- for indoor use only.

C3.2 FLOORING

MORTARS, GROUTS: materials, uses, properties

● denotes property use or requirement

○ denotes optional property or use

na - information not available

MATERIAL	USED WITH: STONE/TERRAZO TILE	BURNED CLAY, BRICKS/ PAVERS	BURNED CLAY, CERAMIC/ QUARRY	RESISTANCE TO: MOISTURE	IMPACT	BACTERIAL GROWTH	NON-OXIDIZING ACIDS	ALKALIES	STAINS	SUBSTRATE: ABOVE GRADE, WOOD	ABOVE GRADE, CONCRETE	BELOW GRADE, CONCRETE
MORTAR												
PORTLAND CEMENT	●	●	●	●	●			●		●	●	●
PORTLAND CEMENT, DRY-SET	●	●	●	●	●			●		●	●	●
PORTLAND CEMENT, LATEX	●	●	●	●	●		○	●	●	●	●	●
EPOXY, TWO PART	○	●	●	●	●	●	●	●	●	●	●	●
SILICATE, TWO PART	○	●	●	●	●		●			●	●	●
FURAN, TWO PART	○	●	●	●	●		○	●	●	●	●	●
SULFUR CEMENT	○	●		●	●		●			●	●	●
PHENOLIC	○	●	●	●	●			●	○		●	●
POLYESTER	○	●	●	●	●		●				●	●
GROUT												
PORTLAND CEMENT	○	○	●	●	●			●		●	●	●
PORTLAND CEMENT, DRY-SET	○	○	●	●	●			●		●	●	●
PORTLAND CEMENT, LATEX	○		●	●	●		○	●	●	●	●	●
MASTIC	●		●	●		●		○	●	●	●	●
FURAN, ONE PART	○	●	●	●		○	○	●	●	●	●	●
EPOXY, ONE PART	●	●	●	●	●	●	●	●	●	●	●	●
SILICONE RUBBER	○		●	●		●	●	●	●		●	●

REMARKS:

- epoxy avialable in three parts, but rarely used.
- silicate is very porous, therefore needs heavy membrane vapor retarder for protection.
- ready-mix silicate available.
- some furan formulations available which will not stain adjacent materials.
- Portland cement mortar may be used to resist neutral salts, most carbonates, nitrates, weak alkaline solutions, petroleum, and coal tar distillates when contaminants are free of fatty oils and other acidic materials.
- furan and phenolic will usually not cure at below 50°F.
- Sulfur mortars may be used to resist most oils and petroleum derivatives.

- silicone is not recommended for use in kitchens.
- phenolic must be handled with care, can be toxic to sensitive skin.
- Portland cement grout requires damp curing.
- Portland cement regular and dry-set grout available formulated to resist fungus growth.
- some grout may stain or discolor ceramic tile, use of grout whose color is similar to that of tile is recommended.

C3.2 FLOORING

ADHESIVES: materials, uses, properties

- ● denotes use or property
- ○ denotes optional use, requirement or availability

[1] Most synthetic rubbers are produced by emulsion polymerization and are available in latex form, such as: styrene-butadiene, acrylonitrile-butadiene, and polychloroprene types.

Material	STONE/TERRAZO TILE	BURNED CLAY	RESILIENT, ASPHALT	RESILIENT, VINYL	CARPETING	WOOD FLOORING	MOISTURE	BACTERIAL GROWTH	STAINING	CERAMIC TILE, gloss removed	ABOVE GRADE, WOOD	ON GRADE, CONCRETE	BELOW GRADE, CONCRETE	REMARKS
	USED FOR:						RESISTANCE TO:			SUBSTRATE:				REMARKS
ADHESIVE														
NEOPRENE			●				●	○	●		●	●	○	(A) • can be used where moisture is anticipated.
RUBBER, LATEX				●	●	●	●		●	●	●	●		(B) • use where water may splash onto flooring.
RUBBER/ASPHALTIC DISPERSION			●				●				●	●	●	(C) • has tendency to stain. • rubber cement available specially formulated for clay/concrete tile.
ASPHALT, CUT-BACK, EMULSION			●	●		●	●	○			●	●	●	(D) • has tendency to stain and ooze, but is very inexpensive. • on wood substrate, felt underlayment is recommended.
SULFITE LIQUOR		●	●						●		●	○		(E) • is brittle, but inexpensive • sulfite reacts with water and alkali; not recommended where there is possibility of moisture in concrete.
SOLVENT BASED VINYL				○			●		●		●	●		
EPOXY, TWO PART	●	●		●	●	●	●	●	●	●	●	●	●	(F) • special formulations available for resilient; clay tile flooring. Used in damp, wet areas. • may be used with conductive flooring.
UNDERLAYMENT														
POLYVINYL ACETATE			●	●	●						●	●	●	(G) • properties similar to mastic. • trowel applied.
ASPHALT SATURATED FELT		●	●	●								●		(H) • secured with asphalt emulsion.
FELT AND FOAM SHEET		●	●	○								●		(I) • secured with adhesive.

RESILIENT FLOORING: types, properties

● denotes use; availability;
　or requirement.

○ denotes optional use;
　or requirement, or availability

■ denotes good resistance.

□ denotes fair resistance

na - information not available

FLOORING TYPE	SIZE: width × length in feet for sheet in inches for tile	OVERALL GAUGE, total thickness	WEARLAYER GAUGE, thickness of wearing surface	TENSILE STRENGTH psi [1]	COEFFICIENT OF ELONGATION, percent of unit length	INDENTATION allowable range pounds per sq. inch [2]	THERMAL CONDUCTIVITY BTU/hr./sq. ft./°F./in.	FLAME SPREAD RATING [3]	SMOKE DENSITY [4]
SHEET									
LINOLEUM	Width 6, 12, 15	.065 to .125	.022 to .085	na	na	75	1.5	na	na
RUBBER	Width 3 to 6	.080 to .188	.094 to .188	580 to 667	270 to 300	200	5.3	75	98
VINYL, COMPOSITION	Width 6, 12	.070 to .090	.012 to .125	993	134	75	1.4	41	215
VINYL, HOMOGENEOUS	Width 6, 9, 12	.071 to .100	.020 to .188	1,900	300	200	1.1	45 to 75	425
TILE									
LINOLEUM	9 × 9	.065 to .125	.022 to .085	na	na	75	1.5	na	na
RUBBER	9 × 9	.080 to .125	.188 to .250	580 to 667	270 to 300	200	5.3	75	98
VINYL, BACKED	9 × 9 to 18 × 18	.090 to .160	.012 to .035	993	134	75	1.4	15 to 75	215
VINYL, HOMOGENEOUS	9 × 9 12 × 12	.070 to .090	.025 to .125	1,900	300	200	1.1	45 to 75	425
VINYL-COMPOSITION	9 × 9 12 × 12	.080 to .125	.094 to .125	na	na	25 to 75	3.1	35	425
VINYL FACED CORK	9 × 9 12 × 12	.125 to .126	.02	na	na	75 to 150	.62	na	450
CORK	6 × 6 to 26 × 36	.188 to .313	na	na	na	75	.53	na	na
ASPHALT	9 × 9 18 × 18	.125 to .187	na	na	na	25	3.1	75	na

PROPERTIES:

C3.2 FLOORING

C • INTERIORS

REMARKS

[1] Tensile strength is a measure of breaking strength.

[2] Indentation is the maximum allowable load before permanent deformation takes place.

[3] Class A—Up to 25
Class B—Up to 50
Class C—Over 75

[4] Generally, allowable less than 450, when tested in accordance with ASTM E 84.

[5] Consult manufacturer for resistance to specific chemicals.

[6] Staining by commonly used substances: soft drinks, coffee, alcoholic beverages.

WEAR/ABRASION	IMPACT LOADS	ALKALIES	GREASE/OIL	CHEMICALS [5]	STAINING/DIRT [6]	CHOICE OF COLOR	CHOICE OF SURFACE TEXTURE	EMBOSSED PATTERN	CONDUCTIVE TYPE	DAMP AREAS	BELOW/ON GRADE	ABOVE GRADE ONLY	PROTECTIVE SURFACING REQUIRED	PERIODIC RESURFACING REQUIRED
□	■	□	■	□	■	●			○			●	●	●
■	■	□	□	□	■	●	●	○		●	●		●	●
■	■	■	■	□	■	●	●	●	●			●	●	●
■	■	■	■	□	■	●	●	●	●		●			○

- linoleum: Below-grade: use vapor retarder.
- rubber: Below-grade: use vapor retarder. It is a specialty floor covering for selective use.
- heat fusion of seams available in some types.

WEAR/ABRASION	IMPACT LOADS	ALKALIES	GREASE/OIL	CHEMICALS [5]	STAINING/DIRT [6]	CHOICE OF COLOR	CHOICE OF SURFACE TEXTURE	EMBOSSED PATTERN	CONDUCTIVE TYPE	DAMP AREAS	BELOW/ON GRADE	ABOVE GRADE ONLY	PROTECTIVE SURFACING REQUIRED	PERIODIC RESURFACING REQUIRED
□	■	□	■	□	■	●			○			●	●	●
■	■	□	□	□	■	●	●	○		●	●		●	●
■	■	■	■	□	■	●	●	●	●			●	●	●
■	■	■	■	□	■	●	●	●	●		●			○
■	■	■	■	■	■	●	●	●			●		○	○
■	□	■	■	□	■	●						●		○
□	□	□	□	□	■	●						●	●	●
□	□	□	■	□	■	●				●	●		●	●

- linoleum: Below-grade: use vapor retarder under concrete substrate.
- rubber: Below-grade: use vapor retarder under concrete substrate.
- generally use of 15 lb. saturated felt underlayment recommended, when there is a possibility of moisture penetrating into substrate.
- vinyl backed: Peel and stick tiles available.
- vinyl: Conductive and other sizes available.
- cork: Sheets manufactured on special orders.
- asphalt: Grease-resistant tiles available.
- leather tiles are manufactured; however, cost is very high and they are not readily available.
- periodic resurfacing depends on the amount of traffic.
- loose laid interlocking rubber tiles available.

C3.2 FLOORING

WOOD FLOORING: types, properties

- ● denotes common use or requirement
- ○ denotes optional use or requirement.
- ■ denotes good resistance.
- □ denotes poor resistance.

TYPE/SPECIES	THICKNESS range in inches	RESISTANCE TO: [1]					SUBSTRATE:				SECURED BY:	
		INDENTATION/IMPACT	MOISTURE, OCCASIONAL	MOISTURE, FREQUENT	HEAVY FOOT TRAFFIC	HEAVY INDUSTRIAL TRAFFIC	WOOD/PLYWOOD/PARTICLEBOARD	CONCRETE, MASTIC BED	CONCRETE, WOOD NAILERS	CONCRETE, METAL NAILERS	BLIND NAILING	MASTIC/ADHESIVE
STRIP/PLANK												
HARDWOOD, OAK												
HARDWOOD, MAPLE	25/32	□	□		■		●	●	●	●	●	○
HARDWOOD, PECAN												
HARDWOOD, ACRYLIC IMPREGNATED	5/16	■	■		■		●	●			●	●
THIN BLOCKS, PARQUET												
HARDWOOD, OAK												
HARDWOOD, MAPLE												
HARDWOOD, BIRCH												
HARDWOOD, PECAN	3/8 to 3/4	□	□		■		●	●			●	●
HARDWOOD, MAHOGANY												
HARDWOOD, TEAK												
HARDWOOD, WALNUT												
HARDWOOD, ACRYLIC IMPREGNATED	5/16	■	■		■		●	●	●	●	●	●
SOLID END-GRAIN BLOCKS, IMPREGNATED												
HARDWOOD, OAK												
HARDWOOD, MAPLE	2½ to 4	■	■	■	■	■		●				●
SOFTWOOD, YELLOW PINE												

FINISHING: IN FIELD　／　SURFACING OPTIONS:　／　REMARKS

SANDING	STAIN	FILLER/SEALER	PROTECTIVE SURFACING	FACTORY PREFINISHED	PENETRATING SEALER	SURFACING, SHELLAC	SURFACING, VARNISH	SURFACING, URETHANE	SURFACING, PITCH	WAX OVER SURFACING	PERIODIC RESURFACING REQUIRED	PERIODIC REFINISHING REQUIRED
●	○	○	●	●	●	●	●	●	●	●	●	●
●	○	○	●	●	●	●	●	●	●	●	●	●
●	○	○	●	●	●	●	●	●	●	●	●	●
●	○	●	●	●						●	○	
●	○		●	●				●	●		●	●

REMARKS

1. Resistance to wear, abrasion, burns, grease, chemicals is not indicated, as it will largely depend on the properties of surfacing used.

■ Flame spread average is 75 to 100 for all species.

■ Smoke density average is 250 for all species.

- coefficient of expansion for all species: average in longitudinal direction = 1.7×10^{-6} to 2.5×10^{-6} per one °F.
- of all wood flooring, acrylc impregnated has the best resistance to indentation; maple has the best resistance of natural hardwoods.
- vapor retarders should be use under concrete slabs-on-grade and between top of slab and flooring to prevent water vaper migration.
- nailers/sleepers may be installed over resilient pads to improve resiliency when flooring is used as playing surface.

- average volumetric change for wood products containing between 30 percent and 0 percent moisture: 0.4 to 0.5 percent for each one percent change in moisture content: acrylic impregnated products exhibit less variation.
- vapor retarders should be used under concrete slabs-on-grade to prevent water vaper migration.
- except for acrylic impregnated products, factory prefinished flooring is more difficult to install since it must be levelled during installation; acrylic impregnated flooring can be sanded after installation to eliminate any unevenness caused by subfloor irregularities.
- laminated blocks have highest dimensional stability within group.
- wood veneers with laminated vinyl facing and composition backing are available.
- on thin blocks as well as on strip/plank, moisture should not be allowed to remain for a long period of time.
- indentation resistance of impregnated wood is substantially better than surfaced-only wood.

- epoxy surfacings also used.

C3.2 FLOORING

RESILIENT AND WOOD FLOORING:

RESILIENT FLOORING

Resilient flooring is **produced in sheet or tile form,** of various compositions of resins, plasticizers, fibers, pigments, and fillers and formed under heat and pressure.

■ It is suitable for **indoor use** only, in dry locations subject to moderate wear of foot and rubber wheeled traffic, such as hospitals, offices, stores, residential buildings.

■ **Underfoot comfort** is greater than with stone, cementitious, and burned clay flooring; foam-backed types providing high degree of resilience are also available.

Types of resilient flooring include:

■ **Linoleum** consists of a wearing layer of oxidized linseed oil or other oleo-resinous binders, pigments, and mineral fillers over a backing of burlap or asphalt saturated felt:
• linoleum has very good resistance to grease and abrasion, but poor resistance to alkalies.
• two gauges are generally available: light-gauge for residential flooring, heavy-gauge for commercial use; conductive type is also available in heavy-gauge only.

■ **Rubber offers:**
• excellent resilience and resistance to indentation;
• good resistance to grease, alkali, and abrasion.
• it will withstand wear of spiked shoe traffic considerably longer than any other resilient flooring material; special interlocking loose laid tile is available for such areas as golf clubs, locker rooms, skating rinks.
• it is also generally used for floor runners and as stair tread covering.

■ **Homogeneous vinyls** have excellent resilience and resistance to indentation, abrasion, grease and alkalies.

■ As the **proportion of fillers to vinyl** increases, desirable properties decrease.
• backed vinyl is second to homogeneous in properties.
• vinyl composition tile is third.

■ **Slip resistant,** heat-resistant and conductive types are available in vinyl composition tile; conductive homogeneous vinyl is also available.

■ **Cork tile** is composed of compressed granulated cork bonded with a heat processed resinous binder:
• has excellent resilience, but
• poor wear resistance and should be used in light foot traffic areas only.
• protective surfacing should be used to protect the tile and periodic maintenance and resurfacing will be required.
• surface bonded vinyl facing for improved wear resistance is available.

Adhesives are of primary importance in the satisfactory performance of resilient flooring:

■ **Pastes:** water soluble, all purpose adhesive for:
• above-grade installation of felt-lined, backed sheet materials such as linoleum, cork and rubber tiles.
• not used for asphalt and vinyl composition tile.

■ **Synthetic rubber** latex-base: a water-proof-latex type of cement, used:
• over concrete surfaces or on below grade,
• for securing backed sheet materials,
• for all tiles which may be installed in these locations.
• not used for asphalt and vinyl composition tile.

■ **Epoxy:** a chemical-set, waterproof, two component adhesive with outstanding bonding characteristics, used for, on/or below-grade concrete or wood substrate.

■ Asphalt and rubber-base **brushing cement:** for all types of substrates:
• its use eliminates the need for felt lining.
• used for vinyl composition tile and asphalt tile.
• it may also be used over latex and asphalt-type levelling coats applied to uneven substrates.

■ **Cut-back asphalt:** an inexpensive, moisture resistant adhesive; it is used:
• over concrete surfaces, felt lining, and wood surfaces for asphalt and vinyl composition tile.
• should not be used over latex or asphalt-type levelling coats.

■ **Clay and asphalt-base emulsion:** a moisture-proof all-purpose adhesive for asphalt and vinyl composition tile for all locations; may be used over latex and asphalt-type levelling coats.

■ **Resin-base cement:** usually cumar or oleoresinous adhesive, used:
• on or above grade for cork or vinyl-cork;
• above grade for linoleum, vinyl sheet or tile when surface moisture is anticipated.

■ Substrate to receive the adhesive must be:
• dry, clean, free of oil or grease;
• compatible with the adhesive used,
• must allow for full bonding of the resilient flooring.

■ **The curing compound** or release agent used over concrete substrates should consist of materials that will not interfere with the installation and performance of resilient flooring; it should either be a harmless film on the concrete or be easily and economically removed prior to installation of the floor covering.

■ Among the **factors known to cause difficulties** with installation of resilient flooring are:
• solvents of low volatility that tend to remain within the concrete over long periods and can damage the tile or the adhesive after installation;
• soft, greasy, or oily compounds that interfere with the adhesive bond to the concrete;
• materials of a saponifiable nature that may soften and interfere with the bond in the presence of alkaline moisture normally present in concrete floors on- and below-grade.

WOOD FLOORING

Wood flooring is **made from various species of hardwoods,** and from certain species of softwoods.

■ It is **suitable for use** in dry indoor locations under light to moderate foot traffic conditions; special types are used for particular conditions such as gymnasium flooring.

Types and Applications

Strip/plank, generally made of hardwood, is available square edge, tongued, grooved, and end-matched:
• strips with interlocking steel splines across the ends are also available.

■ Strip/plank flooring **is secured to nailable** substrate by blind nailing; planking, because of its width, may also require face fasteners, which are then countersunk and plugged.

■ Strips are **secured to non-nailable** substrate by:
• sleepers/nailers either of wood, or metal; or
• asphalt base mastic bed; or
• rubber base adhesive.

■ **Sleepers/nailers** are secured to

C3.2 FLOORING

selection checklist

non-nailable substrate by:
- power actuated fasteners,
- bedding in mastic, or other type of adhesive,
- setting into concrete slab, or concrete fill. The last method is not recommended as sleepers cannot be shimmed after setting.

■ Sleepers/nailers **should be installed** to allow air to circulate under the finished flooring, unless the spaces between them are solidly filled with mastic; for large floor areas special provisions may be required to provide adequate ventilation.

Thin blocks generally made of hardwood, available in the following forms:
- **unit:** made in one solid piece, or short pieces of standard strip flooring, edge joined;
- **laminated:** of three to five plies of wood veneer;
- **parquet:** made of several small strips in a variety of patterns.

■ **Secured to nailable** substrate by either nailing, or by adhesive;

■ **Secured to non-nailable** substrate by fill bedding of mastic or rubber base adhesive.

Solid end-grain blocks, of impregnated hardwoods or softwoods are available square edge or grooved and generally used for heavy-duty commercial and industrial floors:

■ Also available **preassembled** in strip or parquet patterns where appearance is an important consideration.

■ **Installed** over concrete and subfloor in full bed of mastic, with joints sealed.

■ **Surface may be sanded** after installation, and should be filled and sealed, using penetrating oils, pitch base sealers, urethane, or other formulations compatible with materials used to impregnate the blocks.

■ Solid end-grain block **flooring offers:**
- durability
- improved resiliency over concrete flooring
- a non-dusting surface
- ease in maintenance and in refinishing.

In all types of wood flooring, **moisture,** whether as water spilled over the surface or as water vapor, **must be prevented from penetrating** into the wood

to avoid excessive swelling from permanently deforming the flooring:

■ **Vapor retarders** should be used with concrete slabs which are to receive directly applied wood flooring, whether installed on or below grade.

■ Vapor retarders are **also recommended** below concrete slabs and between the slab and wood flooring when installed using sleepers/nailers; this in addition to ventilation.

■ Exposed **surfaces should be sealed** to protect the wood from abrasion, moisture and stains; such protective surfacing should be maintained and renewed periodically.

Materials

Species of wood most commonly used for flooring include:

■ **Oak,** both red and white:
- is used principally as strip flooring, but also as plank, parquet, and block flooring; it is the most widely used of all species, especially in residential construction.
- available in two grades of quarter-sawed and five grades of plain-sawed strip, either square-edged or tongue-and-grooved and end-matched.
- both red and white oak are light in color; white oak having a brownish tinge and red oak a pinkish cast.

■ **Maple** is made from sugar maple; it is known in different localities by various names, including sugar tree and rock maple:
- maple's exceptional hardness, strength, resistance to abrasion and ability to take an excellent finish make it particularly desirable as flooring for areas subjected to extremely heavy wear.
- the heartwood of maple is light reddish brown; the sapwood is white with a slight brownish tinge.
- in the standard grading rules for maple flooring, the varying natural color is allowed, although a special grade provides for selected light stock.

■ **Beech and birch** are used infrequently as flooring; only two of the 15 to 20 species of birch native to the U.S. are manufactured into flooring: yellow birch and sweet birch.
- the heartwood of these woods is reddish brown, with a slight variation of color for each species

- special grades are made from all red-faced stock.

■ **Pecan** belongs to the hickory family, the hardest of all hardwoods; its heartwood is darker than that of most woods ordinarily used for flooring.

■ **Other species** of wood used for flooring include walnut, cherry, ash, hickory and East Indian teak, with all used in special custom applications.

■ **Acrylic impregnated** hardwoods are available for improved resistance to wear; they may be factory prefinished or field finished. There are two methods of impregnating natural wood for flooring.
- **Radiation polymerization** replaces the air contained in the wood cells with a liquid monomer (dyed or undyed) using vacuum and pressurization processes. The monomer is then polymerized into a hard tough acrylic plastic using irradiation (gamma rays).
- **Heat/catalyst polymerization** also uses the vacuum and pressurization processes to replace the air in the wood with liquid monomer. However, the polymerization of the monomer is effected by adding a chemical catalyst to the monomer which, upon the application of heat, breaks down into free radicals. These initiate and propogate polymerization, and the catalyst is consumed in the reaction.
- Both methods produce impregnation throughout the entire thickness of the flooring, creating continuity of color, hardness, abrasion and indentation resistance, etc.
- Radiation polymerization results in a harder, tougher, and more durable polymer than the heat/catalyst method.
- A variety of shades and finishes are available from natural through moderately colored tones all the way to black.

■ **Vinyl faced wood veneer** of various species, laminated over a composition backing, is also available.

■ **Softwoods,** such as yellow pine, are generally not sufficiently resistant to wear and impact/indentation to be used in flooring with a standard protective surfacing only; softwoods are available impregnated for use as solid end-grain block flooring.

C3.2 FLOORING

CARPETING: use, properties

● denotes usage

na - information not available

TYPE FIBER	OUTDOOR/INDOOR	INDOOR ONLY	RESILIENCE	ABRASION, range	CRUSHING, range	COMPRESSION, range	MATTING, range	HEAVY WEAR, range
	USE			**RESISTANCE TO: AS DETERMINED BY CONSTRUCTION**				
ACRYLIC/MODACRYLIC		●	fair	fair to good	good to very good	good to very good	fair to good	fair to good
NYLON		●	good	v. good to excellent	good to very good	good to very good	good to very good	fair to good
OLEFIN POLYPROPYLENE	●		poor	good to very good	fair to poor	fair to poor	good to very good	good to very good
POLYESTER		●	fair	good to very good	fair to good	fair to good	good to very good	fair to good
WOOL		●	excellent	v. good to excellent	good to very good	good to very good	good to very good	good to very good

C3.2 FLOORING

C • INTERIORS

AS DETERMINED BY FIBER MATERIAL						RATING ASTM	PER E84	REMARKS
MOISTURE	CHEMICALS/ALKALIES	MILDEW/MOTH	STAINING	SOIL/DIRT	STATIC BUILD-UP	FLAME SPREAD	SMOKE DENSITY	• smoke density shown is maximum value. • in yarn construction, weight of pile, pile height, pitch and yarn count are the critical factors in determining the service life or wear resistance of carpet.
good	fair	v. good	good	good	good	25	160	• comparable to wool as acrylics feel more like wool than any other fiber. • usually made in level-loop construction. • for good performance, the ounce weight should be the same as for wool.
fair	fair	v. good	v. good	v. good	poor	35	190	• strongest of all man-made fibers. • easy to maintain. • anti-static filaments must be added for resistance to static electricity build-up.
excellent	v. good	v. good	v. good	v. good	v. good	80	na	• most commonly used as man-made indoor/outdoor grass carpet. • available in colors, stripes and tweeds.
good	v. good	v. good	good	v. good	good	20	413	• cleans well. • tendency to fade if exposed to sun.
fair	good	fair	fair	v. good	fair	55	160	• cleans easily; crevices in carpet do not hold dirt. • treatment for moth repellancy and to prevent static build-up available.

C3.2 FLOORING

CARPETING: types, construction

● denotes usage

na - information not available

CONSTRUCTION

TYPE	PITCH	MACHINE GAUGE, range	STITCHES PER INCH, range	YARN WEIGHT, range ounces per sq. yard	TOTAL WEIGHT, range ounces per sq. yard	PILE HEIGHT, range in inches
CARPET						
ACRYLIC/MODACRYLIC	216	$5/64$ to $5/32$	8.75 to 11	25 to 42	66 to 80	$7/32$ to $7/16$
NYLON	270	$5/64$ to $3/16$	7.5 to 11.25	20 to 40	67 to 85	$1/8$ to $5/8$
OLEFIN POLYPROPYLENE	216	$5/32$	12	26	na	$5/16$
POLYESTER	na	$5/32$ to $3/16$	7 to 9	32 to 55	70 to 104	$9/16$ to $13/16$
WOOL 100 PERCENT	189	$5/32$	8 to 9	30 to 55	62 to 100	$5/16$ to $9/16$
BLEND: WOOL 80% PLUS NYLON 20%	189	$5/32$	8 to 15	30 to 37	66 to 79	$19/64$ to $11/32$
CARPET TILE						
BLEND	na	$1/10$	8	34 to 40	103 to 106	$5/32$
NYLON	216	$1/8$ to $5/32$	8.25 to 10.67	25 to 40	106 to 125	$3/32$ to $5/32$
NYLON BLEND	na	$1/8$	8.0 to 10.25	18	118	$5/16$

TUFTED	WOVEN	NEEDLEPUNCHED	FUSION BONDED	LEVEL LOOP	LEVEL LOOP-RANDOM SHEAR	CUT PILE	CUT AND LOOP	POLYESTER	JUTE	POLYPROPYLENE	COTTON/JUTE	JUTE	POLYPROPYLENE	COTTON	FOAM	UNITARY	REMARKS
		PROCESS:			SURFACE:			BACKING: PRIMARY				SECONDARY					
•	•		•	•	•	•	•			•		•					• Anti-static fibers must be added for resistance to build-up of static electricity. • most wool carpets contain added anti-static fibers and are treated for moth repellency. • olefin used as man-made grass carpets.
•	•	•	•	•	•	•	•	•	•			•	•			•	
•						•		•				•	•			•	
•				•					•	•		•	•				
•	•		•	•		•			•	•	•	•					
•	•		•			•			•	•	•	•					
	•					•		•				•					• tiles available in 19 x 19 and 24 x 24 sizes.
•		•				•		•				•					
•		•		•				•				•					

CARPETING: selection checklist

CARPETING

Considerations in the selection of carpeting will include, in addition to properties important for the satisfactory performance of any flooring, such as durability, comfort, resistance to in-service conditions, also the following:

■ **Sound absorption** qualities:
• to reduce noise levels within the space,
• to minimize impact noise transmission to spaces below.

■ **Thermal insulation** between the space and space below it.

Carpet Construction

Density: the amount of pile packed into a given volume of carpet, measured in ounces of pile yarn per unit volume. Good density pile is required for best appearance and wear and depends on the following:

■ **Pitch:** number of warp lines of yarn in a 27-inch width; the higher the pitch, the denser the carpet.

■ **Stitch:** the number of lengthwise yarn tufts in one inch of tufted carpet. The more stitches per inch, the denser the face of the carpet.

■ **Pile height:** the height of pile measured from the surface of the back to the top of the pile; generally medium high pile gives better service than very high or low.

Wearability: characterized by tight gauge, low pile and high stitch rate. A dense low pile of a given weight will often give better service than a high pile and low density of the same weight. Characteristics which affect wearability:

■ **Total weight:** weight, in ounces, of carpet pile, yarn, primary and secondary backings and coatings in a given unit of space.

■ **Yarn or face weight:** weight, in ounces, of pile yarn in a given unit of space, indicates fineness or coarseness of finished yarn.

■ **Pile:** the upright ends of yarn, whether cut or looped, that form the wearing surface of carpet; the lesser the height, the better is wearability.

■ **Gauge:** the distance between two needle points, expressed in fraction of inch. Applies to both knitted and tufted carpets: the tighter the gauge, the denser the carpet.

Resiliency: ability of a carpet yarn to spring back after being crushed or walked upon; depends also on the characteristics discussed above and on the method of carpet manufacture and its texture.

Methods of carpet manufacture include:

■ **Tufted:** a high-speed method of fabricating carpet, by which the pile yarns are inserted through a pre-woven backing fabric, leaving the stitches long enough to be cut off or left as loops.

■ **Woven:** the in-and-out interlacing of surface and backing yarns in one operation.

■ **Needlepunch:** produces carpets of limited variations in texture; used mostly when resistance to heavy traffic and abrasion required.

■ **Knitted:** a method of fabricating a carpet in one operation, as in weaving; surface and backing yarns are looped together with a stitching yarn on machines with three sets of needles.

■ **Fusion Bonded:** uses two backing fabrics that run parallel with a space between the two; the backing has an adhesive on the facing sides. By implanting a multi-fold fiber web between the backings, a sandwich is formed. A blade slices through the middle of the sandwich, producing two identical carpets.

Textures produced in carpeting include:

■ **Level Loop pile:** loops are of the same height, made by tufting, weaving, or knitting:
• variations include high and low loops.
• suitable for heavy use areas, but more difficult to clean.

■ **Cut Pile:** made from unset yarns for an even, velvety texture; or from firm-ended yarns, which look more luxurious, but also show foot steps or flaws more easily.

■ **Cut and Loop:** by combining differing loop heights, a variety of textures can be obtained.

■ **Static resistance** can be achieved by the use of special fibers, metallic wires, or chemicals to dissipate the static electricity generated by movement on carpet.

Carpet Materials

Major properties of the most commonly used fibers:

■ **Acrylics:**
• light; wool-like appearance;
• fair durability;
• fair stain and abrasion resistance;
• good crush resistance;
• higher in cost than nylon.

■ **Nylon:**
• strongest of all man-made fibers;
• economical;
• excellent abrasion resistance;
• good crush and stain resistance;
• easy to clean.

■ **Olefin** or Polypropylene: used as indoor/outdoor carpeting; "Artificial Turf" is also made from this fiber.
• resistant to fading and staining
• good abrasion resistance
• poor resiliency.

■ **Polyester:**
• feels like wool and has good abrasion resistance
• fair crush resistance
• less static-prone than other man-made fibers;
• more susceptible to oil-based stains and soil.

■ **Wool:** soft, bulky yarn
• good serviceability and resiliency
• wool is the highest priced carpet, but generally not as durable as most synthetics.

Backing is the material that forms the back of the carpet, regardless of the type of construction:

■ **Primary back:** in a tufted carpet, the material to which surface yarns are attached; may be made of jute, cotton, woven or non-woven synthetics.

■ **Secondary back:** also called "double backing". Any material: jute, woven or non-woven synthetics, foam or cushion, laminated to the primary back.
• improves resiliency
• adds dimensional stability to carpet.

■ **Separate padding,** or cushioning material:
• may be used under carpeting
• made of felted cattle hair, jute, foam rubber or plastic foam
• usually glued down over concrete floors; stapled over wood
• used to extend service life and to add resiliency to carpeting.

■ **Carpeting choices** are available that are produced with post-consumer waste and/or environmentally responsible recycling practices.

C3.2 FLOORING

SEAMLESS FLOORING: types

● denotes availability use or requirement

○ denotes optional use or requirement

NOTE: For additional information refer to COATINGS, and ROOFING.

REMARKS:
- one-component materials are inferior to two-components.
- all are directly bonded to substrate. Below/on grade protect from moisture.
- flame spread rating: 25
- Smoke developed: 32 to 40
- information on properties for thermoplastic resins not available—limited availability of material.

	FILM THICKNESS, in inches RECOMMENDED	CHOICE OF COLORS	LOCATION: DAMP AREAS	LOCATION: BELOW/ON GRADE	LOCATION: ABOVE GRADE ONLY	PROTECTIVE SURFACING REQUIRED	PERIODIC RE-SURFACING REQUIRED
THERMOSETTING RESINS							
ONE PART, EPOXY	.062 to .250	●	●	●			●
ONE PART, POLYURETHANE	.062 to .125	●			●		●
ONE PART, POLYCHLOROPRENE (NEOPRENE)	.375 to .500	●	●	●			●
ONE PART, POLYSULFIDE	.020 to .025				●		
ONE PART; POLYESTER	.125 to .250	●			●		●
TWO PART, EPOXY	.062 to .250	●	●	●			●
TWO PART, POLYURETHANE	.062 to no limit	●			●		●
TWO PART, FURAN	na	●		●			●
TWO PART, POLYCHLOROPRENE (NEOPRENE)	.062 to .375	●	●	●			●
TWO PART, POLYSULFIDE	.020 to .025				●		
TWO PART, POLYESTER	.125 to .2500	●			●		●
THERMOPLASTIC RESINS							
POLYETHYLENE	not available						
POLYSTYRENE	not available						●
POLYVINYL ACETATE	not available	●					●
POLYVINYL CHLORIDE—PVC	not available	●					●
ACRYLIC	.125 to .500	●	●				●
SILICONE	not available			●		●	
MASTIC	.125 to 1.500		●	●		●	●

THERMOSETTING RESINS remarks:
- epoxy resins will not set below 50°F.
- epoxy, polyester and polyurethane do not hide cracks.
- epoxy tiles are available.
- one-part polyester is non-industrial, decorative material.
- two-part epoxy is used for industrial flooring and for heavy traffic. Conductive material available.
- polyurethane is a more functional material rather than aesthetic. For indoor use only.
- furan: on concrete use with primer only.
- two-part polyester is for industrial use; care must be used when mixing two-part material.

THERMOPLASTIC RESINS remarks:
- polyethylene is used on metal, not recommended on concrete. Not widely used today.
- polyvinyl acetate is used as a decorative finish.
- PVC is used as top coat in multiple coats for polyvinyl acetate.
- silicone used as surface treatment, rather than coating.
- mastic is used as surfacing finish, needs primer Tends to soften over 125°F. Used as functional rather than decorative finish.

SEAMLESS FLOORING: selection checklist

SEAMLESS FLOORING

Seamless flooring—(also known as toppings, monolithic surfaces, trowelled composition floors)—consists essentially of a resinous matrix, fillers and/or decorative additions, which after installation and curing form a seamless finished surface.

Matrices, the binding materials for fillers, are the principal component of seamless flooring and largely determine its properties. They fall into two general categories: **thermosetting, and thermoplastic.**

Thermosetting Matrices:

Thermosetting matrices include materials which **cannot be melted down and reheated** after initial curing without destroying their properties:

■ **Epoxy resins:** are available in various formulations as:
• one part, which cures by solvent evaporation;
• two part, which cures by chemical reaction between the two parts once mixed just prior to installation.

■ **Epoxies:**
• are easily maintained
• have good resistance to abrasion/wear, moisture, impact/indentation and chemicals;
• the two-part formulation is superior to one-part in durability and resistance to chemicals.
• pigments may be introduced during formulation and choice of colors is available.

■ Epoxy **may be installed over** properly prepared substrate of concrete, terrazzo, marble, slate, metal, or wood; it should not be used over ceramic tile, asphalt tile, and linoleum.

■ Epoxy **should not be installed** when air temperature is below 50°F:
• generally not suitable for use in hot conditions; limited to temperatures below 140°F.
• the preparation of the substrate is critical for good performance;
• solvents given off with one-part formulation are toxic; good ventilation during installation is required.
• should not be subjected to always-wet condition.

Polyester has similar properties to epoxy; and is available in one-part and two-part formulations; the one-part is generally used for decorative, non-

industrial applications.

■ **Polyesters,** either one-part or two-part are:
• non-slippery, non-porous, non-absorbent, greaseproof;
• should not be subjected to always-wet condition;
• inert and not affected by most commonly used chemicals such as hydraulic oils, ethyl alcohol, engine oil, kerosene, lactic oil, lubricants;
• will generally withstand heavy wear without cracking, pitting, or scratching.
• will resist permanent staining, such as from tobacco, urine, coffee, carbonated or alcoholic beverages.

■ **Glass fiber rovings** may be added during spray application of polyester to improve resistance to structural cracking in the substrate.

■ Polyester **may be installed** over any clean, dry, unglazed substrate;

■ **Moisture within** the substrate, or penetrating the substrate after installation, may adversely affect or destroy the bond between the substrate and the flooring.

Polyurethane is formulated as:

■ **One-part,** which cures by oxidation or atmospheric moisture reaction; or by solvent evaporation.

■ **Two-part,** which cures by chemical reaction between the two parts when mixed.

■ Polyurethane **can be formulated** to yield a variety of finished products: from flexible foams for insulation, to clear films and rigid solids with varying properties for flooring;

■ Polyurethanes **are generally:**
• more resilient than epoxies and polyesters,
• may be expected to perform better in bridging and/or resisting cracks.
• non-slippery, grease and stain resistant,
• unaffected by most commonly used chemicals,
• has excellent resistance to scuffing, scratching, and impact/indentation.
• should not be subjected to always-wet condition.

■ One-component formulations can be applied **to a thickness** of up to 1/8 inch; thickness of two-component formulations is generally not limited and it will cure in a relatively short time.

■ **Substrate conditions** should be similar to those required for the installation of polyester or epoxy.

POLYCHLOROPRENE (neoprene) is also formulated in one- or two-component systems.

■ Both formulations **provide a high degree** of resistance to water, chemicals and oils.

■ **Two-part** formulation is used when a heavy film is required.

■ **Primers** are recommended, particularly if application is over metal.

Other thermosetting flooring materials are:

■ **Polysulfide,** formulated in one- or two-component systems:
• both formulations have outstanding resistance to oils, solvents, moisture and weathering, but
• are poor in resisting abrasion.

■ **Furan resins have:**
• excellent resistance to most chemicals and are widely employed in laboratories and chemical plants.
• they are inherently brittle, which limits the choice of substrate suitable to receive them.

Thermoplastic Matrices

Thermoplastic matrices include materials which **remain plastic after installation and curing;** they will soften when heated, and will again harden when cooled. They include:

Silicones are synthetic chemicals, produced by combining inorganic materials which contain silicone and oxygen with specially treated aggregate. The compound provides a surface texture similar to concrete.

■ **The primary use** of silicone is as a protective coating over concrete floors where water, acid or movement have caused damage.

■ **Silicone displays:**
• excellent flexibility
• good resistance to wear, abrasion and moisture
• only fair resistance to chemicals and impact.
• adhesion is only fair, unless a primer is used.

Mastics consist of asphalt emulsion, Portland cement, and stone filler, ranging from sand to coarse aggregates.

■ **Mastics are low in cost** and have:

C3.2 FLOORING

- inherent flexibility
- self-repairing capabilities
- will dampen impact noise
- are badly affected by organic solvents and mineral acids
- tend to soften under heat; therefore, should not be exposed to temperatures over 125°F.
- heavy loads should not be allowed to stand on mastic, as they will permanently indent the mastic.

■ Mastics **are commonly used** as a resurfacing material over concrete, wood, and metal surfaces:
- can be made more durable by a surface coating of epoxy
- also can be used to cover areas continuously exposed to water, such as entrance passages and loading docks.

Acrylics include a broad family of polymers based on acrylic and methacrylic acids.

■ Acrylics **are resistant to** heat, sunlight, oils, acids and alkalies and moisture, but not when continuously in contact with them.

■ **Two coats** are normally required for most applications.

■ **Recoatability** and color retention are good, and so is adhesion to concrete.

Other types of thermoplastic flooring include:

■ **Polyvinyl Acetates** are thermoplastic resins of vinyl acetate, used alone or with other compounds:
- they are resistant to grease, oils and moisture, but continuous exposure should be avoided
- are used as decorative coatings on concrete surfaces either indoors or outdoors
- weathered surfaces are readily recoated.

■ **Polyvinyl chloride** (PVC) is a water resistant coating; used as top coating for plasticized vinyl chloride acetate, to increase the chemical and abrasion resistance of the system.

■ **Polystyrene film:**
- has very good resistance to water, chemicals, oil, and grease.
- may be used outdoors, but is subject to erosion and embrittlement over a period of time, and will require periodic protective resurfacing.

■ **Polyethylene:**
- has outstanding flexibility
- high resistance to most commonly used chemicals

- generally heat fused to metal surfaces to provide a protective coating/surfacing.

Installation

Installation processes used for seamless flooring materials depend primarily on the formulation and/or viscosity of the matrix material used; and in some instances on the filler materials as well:

■ **Self-leveling,** where the matrix hardens to a smooth surface:
- requires little or no surface work;
- is generally applied in a thin film, without fillers, or with a fine particle or flake fillers only.
- will not hide cracks in substrate, and any cracks developing after installation may split the flooring.

■ **Trowel-finished,** where the matrix is quite viscous, and requires troweling to smooth out and densify the surface;
- applied in thicknesses from ⅛ to 2 inches and over with various types of fillers.
- will hide cracks in the substrate, and may resist movement in cracks; or resist cracks developing after installation, depending on thickness and type of matrix used.

■ **Surface-ground,** where the matrix is quite viscous and the filler consists of or contains marble or granite chips.
- installation and finishing are essentially the same as for conventional terrazzo, and the same limitations apply as far as substrate conditions are concerned.

■ **Substrate to receive seamless** flooring must allow for full bonding and must be:
- dry and free of dust
- glazed surfaces should be roughened by sandblasting or acid etching.
- surface oil and grease should be removed completely
- curing compounds used on concrete should not interfere with bonding of the flooring; for additional information refer to substrate requirements for resilient flooring.
- shrinkage cracks in concrete substrate generally should be patched; joints in panelized substrate should be filled in prior to installing flooring.

FLOORING ACCESSORIES

Accessories required for, or optional to the installation of flooring, include:

■ **Bases,** made of the following materials:

- rubber and vinyl: for resilient flooring and general use with other flooring materials
- terrazzo, precast; or cast in place
- burned clay: for ceramic tile, generally glazed; and for quarry tile
- wood

■ **Floor mats:**
- mats for long term service are made of rubber or vinyl strips, separated to allow dirt to collect between them.
- frames and pans are available for recessing into floors.
- frames are made of non-corroding or protected metal.

■ **Dividing strips** for cast-in-place terrazzo: made of zinc, brass, galvanized steel and plastic.

■ **Trim pieces** and edgings for carpets, generally of vinyl.

■ **Stair nosings,** treads and risers: ceramic nosings for ceramic tile faced stairs
- rubber and vinyl nosings, tread and riser coverings for cement filled steel stairs.
- precast terrazzo stair treads, with optional non-slip inserts, are available.

C3.2 FLOORING

EXPANSION JOINT COVERS: description chart

ELASTOMERIC

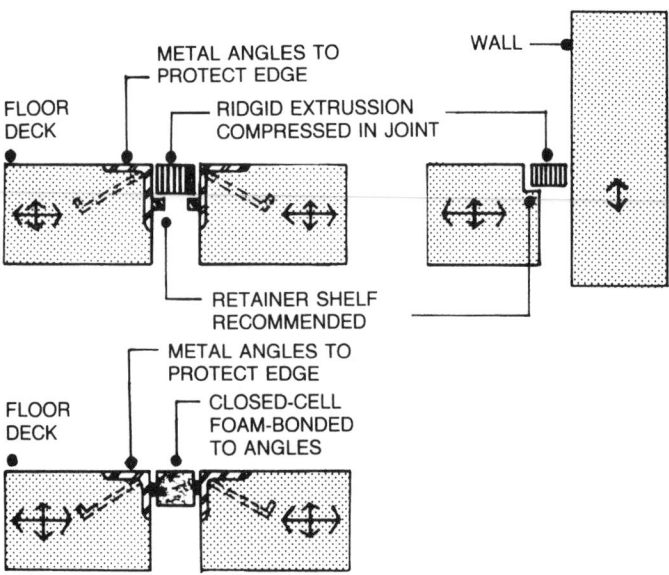

Floor and ceiling expansion joints isolate sections of enclosures to limit cumulative thermal expansion/contraction. Covers for expansion joints must accommodate such movements, which may be perpendicular, parallel or transverse to the joint due to: differences in temperature, different coefficients of expansion of materials, unequal settlement, creep and/or drying shrinkage, lateral wind or seismic forces. Expansion joint covers in floors are used to:

■ **Seal the joint** to eliminate tripping hazard and to prevent the intrusion of foreign matter, which might lock the joint and impair its proper functioning, while also:

- minimizing or preventing moisture penetration.

- minimizing air leakage.

- preventing fire penetration, when floor assembly requires a fire resistance rating.

Considerations during selection of floor expansion joint covers may include:

ELASTOMERIC

■ Rigid extrusions function by exerting pressure against the sides of the joint at all times:

- if the joint opens more than the intended design value, the extrusion may no longer be in compression and will allow moisture and air to penetrate, and may also sag out of the joint if means to retain it are not provided.

- retainer shelves will also aid in maintaining uniform depth of insertion.

- when subject to wheeled traffic, means to protect the edges of the floor deck at the joint are recommended; edge protection will also help in maintaining constant width of joint during construction.

- rigid extrusions may not function properly in preventing air and moisture penetration when joint width is not constant, especially when the changes in width are abrupt.

■ **Closed cell foams commonly function by exerting pressure,** but not as great as extrusions, against the sides of the joint:

- they may also be bonded to the sides to prevent loss of effectiveness if joint opens beyond design values, and may also resist tension in such instances.

- foams are relatively soft and when they are to be exposed to pedestrian traffic, their resistance to penetration by spiked

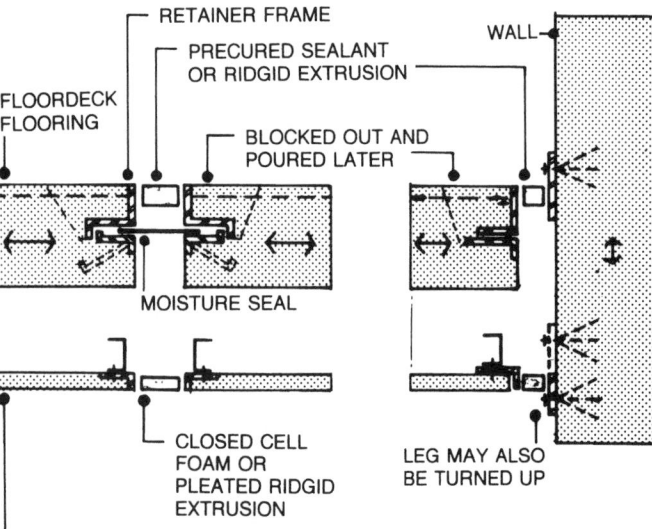

C3.2 FLOORING

METAL

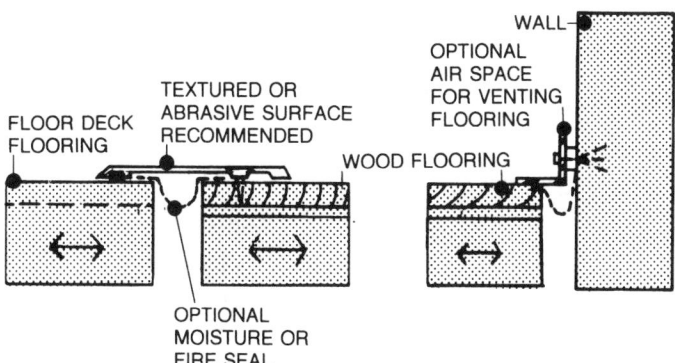

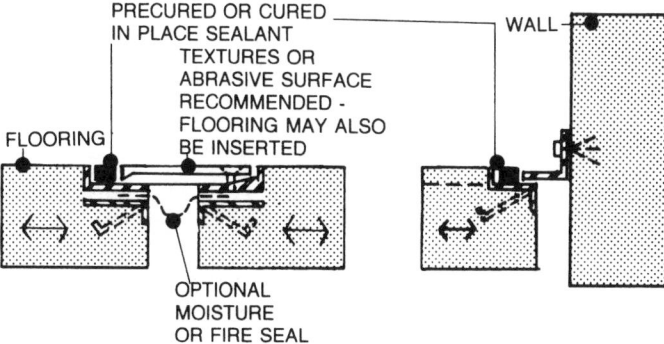

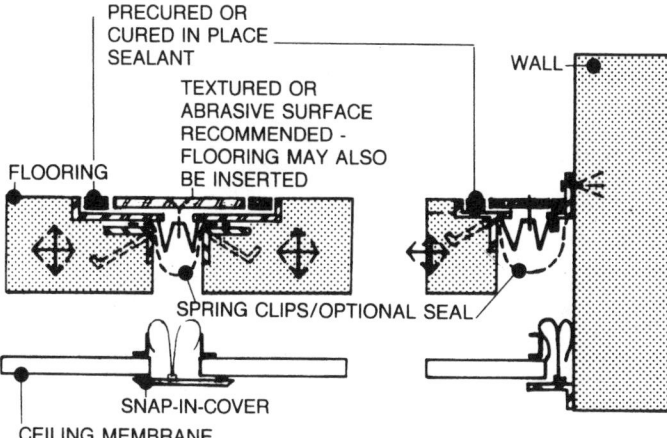

heels should be investigated to avoid creating a tripping hazard.

■ **Precured sealants,** or cured in place sealants, function by adhering to the sides only of the retaining frame:

• width of joint may be limited unless reinforcing plate across joint is provided to hold sealant in place during maximum extension.

• resistance to penetration by spiked heels should be investigated.

■ **Materials for retaining frames:** extruded aluminum, extruded bronze, stainless steel, steel:

• steel should be of corrosion resistant alloy; runoff during weathering may stain adjacent surfaces.

For additional information on materials refer to SEALANTS.

METAL

■ **Commonly available in diverse configurations for various widths of joints.**

■ **Materials used include:** extruded aluminum, extruded bronze, stainless steel:

• aluminum should be isolated from dissimilar materials.

• choice of stainless steel configurations may be limited.

■ **Sealants used may be:** cured in place, precured, rigid extrusions:

• resistance of sealants to foot traffic, especially to puncture by spiked heels, should be investigated.

■ **Seals to prevent moisture, air, and/or fire penetration** generally available.

■ **Juncture of floor and wall expansion joint covers** should be investigated for proper fit and appearance:

• for information on wall expansion joint covers refer to WALL ASSEMBLIES and COMPONENTS.

Ceiling membranes should be discontinuous through expansion joints. Expansion joints in ceilings may be incorporated into the design of the ceiling itself or an expansion joint cover may be added:

■ **Various types** such as elastomeric or metal are available:

• pressure-type extrusions should not be used to avoid damage to ceiling membrane.

C3.2 FLOORING

REFERENCES: Flooring

STANDARDS

ASTM Specifications for:

C33	• Concrete Aggregate.
C287	• Chemical Resistant Sulfur Mortar.
C395	• Chemical-Resistant Resin Mortars.
C410	• Industrial Floor Brick.
C658	• Chemical-Resistant Resin Grouts.
C902	• Pedestrian and Light Traffic Paving Brick.
D52	• Wood Paving Blocks for Exposed Platforms, Pavements, Driveways, and Interior Floors Exposed to Wet and Dry Conditions.
D1031	• Creosoted End-Grain Wood Block Flooring for Interior Use.
D3218	• Polyolefin Monofilaments.

ASTM Tests/Methods for:

C241	• Abrasion Resistance of Stone Subjected to Foot Traffic.
C323	• Chemical Analysis of Ceramic Whiteware Clays.
C482	• Bond Strength of Ceramic Tile to Portland Cement.
C483	• Electrical Resistance of Conductive Ceramic Tile.
C484	• Thermal Shock Resistance of Glazed Ceramic Tile.
C485	• Measuring Warpage of Ceramic Tile.
C499	• Facial Dimensions and Thickness of Flat, Rectangular Ceramic Wall and Floor Tile.
C501	• Relative Resistance to Wear of Unglazed Ceramic Tile by the Taber Abraser.
C502	• Wedging of Flat, Rectangular Ceramic Wall and Floor Tile.
C609	• Measurement of Small Color Differences Between Ceramic Wall or Floor Tile.

C627	• Evaluating Ceramic Floor Tile Installation Systems.
C648	• Breaking Strength of Ceramic Tile.
C650	• Resistance of Ceramic Tile to Chemical Substances.
D412	• Rubber Properties in Tension.
D418	• Woven and Tufted Pile Floor Covering.
D1335	• Tuft Bind of Pile Floor Coverings.
D1777	• Measuring Thickness of Textile Materials.
D1894	• Static and Kinetic Coefficients of Friction of Plastic Film and Sheeting.
D2394	• Simulated Service Testing of Wood & Wood-Base Finish Flooring.
D2261	• Tearing Strength of Woven Fabrics by the Tongue (Single-rip) Method.
D2646	• Backing Fabrics.
D2401	• Service Change of Appearance of Pile Floor Coverings.
D2859	• Flammability of Finished Textile Floor Covering Materials.
E84	• Surface Burning Characteristics of Building Materials.
F137	• Flexibility of Resilient Flooring Materials with Cylindrical Mandrel Apparatus.
F142	• Indentation of Resilient Floor Coverings.
F373	• Embossed Depth of Resilient Floor Coverings.
F386	• Thickness of Resilient Flooring Materials Having Flat Surfaces.
F410	• Wear Layer Thickness of Resilient Floor Covering by Optical Measurement.

ANSI Specifications:

ANSI/UL779

	• Electrical Conductive Floorings, Safety Standard for
ANSI 010.2	• Laminated Hardwood Block Flooring.
ANSI/AATCC 123	• Carpet Soiling - Accelerated Soiling Method.
122	• Carpet Soiling - Service Soiling Method.
ANSI/TCA	• Tile Council of America Recommended Standard Specifications for Ceramic Tile.
A108.1-1976	• A Collection - Including A108.1, A108.4, A108.5, A108.6, A108.7, A118.1, A118.2, A118.3, A118.4, A136.1.
A137.1-80	• Recommended Standard Practice for: Ceramic Tile.
ACI	**American Concrete Institute** • Manual of Concrete Practice Part I: Materials and Properties of Concrete Construction Practices and Inspection Pavements and Slabs.
BSI	**Building Stone Institute** • Stone Information Manual
NRCC	**National Research Council**
CBD-22	• Concrete Floor Finishes
206	• Quarry Tile and Other Floor Tiles
225	• Evaluation and Repair of Deteriorated Garage Floors
CRI	**Carpet & Rug Institute** • Carpet Specifiers Handbook, Third Edition
104	• Standard for Installation of Textile Floor Covering Materials.
	• Floor Covering Weekly Annual Handbook of Contract Floor Covering, Bart Publications
	• Commercial Carpet Maintenance Manual

C3.2 FLOORING

MFMA   **Maple Flooring Association of America**

- Spec-Data presentation of hard-maple, beech, and birch flooring.
- Recommendations for finishing and maintaining maple, beech, and birch flooring.
- Guidelines and Recommendations for the proper installation of MFMA hard maple flooring.
- Heavy Duty Finishes for maple, beech and birch floors.
- Controlling Expansion of Hard Maple Flooring.
- Specification Manual - Standard specifications and official grading rules for beech, birch and hard maple flooring.

NOFMA   **National Oak Flooring Manufacturers' Association**

- Oak Flooring Guide.
- How to Keep Oak Floors Beautiful.
- How to Install Hardwood Strip Floors.
- Architect's Specifications Manual.
- Official Flooring Grading Rules for Oak, Beech, Birch, Pecan and Hard Maple Flooring.

NTIS   **National Technical Information Service**

P380-22987

- The Effects of Some Cleaning Materials on PVC Flooring.

NTMA   **National Terrazzo & Mosaic Association**

- Terrazzo Design Data, Division 9.
- Renovation Design Ideas.

PCA   **Portland Cement Institute**

- Concrete Floors on Ground.
- Surface Treatments for Concrete Floors.
- Resurfacing Concrete Floors.

RFCI   **Resilient Floor Covering Institute**

- Recommended Installation Specification for Vinyl Asbestos and Asphalt Tile Flooring.

TCA   **Tile Council of America**

- Handbook of Ceramic Tile Installation.

USG   **Urethane Safety Group**
U-105
- Evaluation of the Fire Performance of Carpet Underlayment.

WSFI   **Wood and Synthetic Flooring Institute**

- The Care and Preservation of Your Wood Floors.
- Recommendations for the Care and Maintenance of Synthetic Sports Surfaces.
- Recommendations for Concrete Base for Synthetic Playing Surfaces.
- Recommendations for the Correct Preparation, Finishing and Testing of Concrete Subfloor Surfaces to Receive Wood Finishing.
- Specifications for Stage Floor Systems.
- Specifications for Cushioned Subfloor Construction with Laminated Subflooring and Finish Flooring.
- Specifications for Herringbone Flooring Set in Mastic.
- How to Control Cupping of Wood Floors.
- Specifications for Mastic Cushioned Construction with Nailers, Subflooring and Finish Flooring.
- Specifications for Mastic-Nailed Construction with Mastic-Set Subflooring and Nailed Finish Flooring.
- Specifications for Mosaic Wood Parquet Flooring Set in Adhesive.
- Specifications for Rubber Cushion-Sleeper

Construction with Nailers and Finish Flooring.
- Specifications for Steel Splined Continuous Strip Mastic-Set Maple Flooring.

C3.2 FLOORING

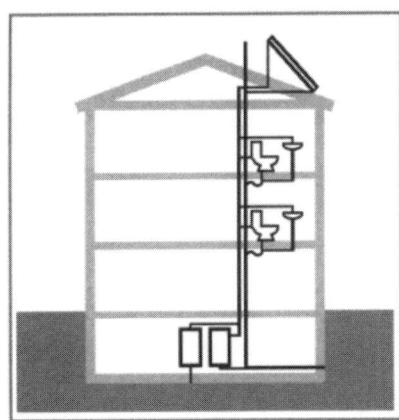

D • SERVICES

D1. ❶ CONVEYING SYSTEMS

ESCALATORS, MOVING WALKWAYS

ESCALATORS & MOVING WALKWAYS

An escalator and a moving walk are two forms of a mechanical convey-or-belt for people. If the rate of arrival is not excessive, each passenger can step on immediately without waiting, be transported to the other end, and step off immediately. This advantage of instant service is achieved at a considerable cost in other capabilities. In the usual types of escalators and walkways, these limitations are as follows:

■ Since the equipment does not stop, the passenger must accelerate to full speed in stepping on, that is, in the length of one step. This effectively limits the speed of the machine to walking pace.

■ Movement is linear, with passengers exiting in the order they entered. This means that each machine can only operate between two fixed points, unlike an elevator, which can have many intermediate stops.

■ Each machine operates in only one direction at a time, so that two-way traffic requires two separate escalators or walkways. However, in peak times one can be reversed, so that both operate in the major direction of travel.

■ The angle of incline is limited. Therefore any significant vertical rise is accompanied by a far greater horizontal movement, whether desired or not. A moving walk (that is, with a surface that does not form steps during its travel) can be built at any angle from horizontal to 15°. For escalators (which do form steps) the angle is normally 30°. Deviations from this are rare.

ESCALATORS

• Escalators are widely used to transport large numbers of people over a relatively short vertical distance.

• A single escalator has the same capacity as a large bank of elevators.

• There is little need for direction signs, since it is obvious where the escalator is going, and passengers have the benefit of an uninterrupted view during their travel.

• Escalators are also useful in a building with a large pedestrian traffic

flow, in directing that flow in the desired direction.

MOVING WALKWAYS

■ Horizontal moving walks are less common, mainly because people are more willing to walk horizontally than they are to walk up or down stairs.

• If the speed of the moving walk is the same as walking pace, there is no great time advantage in using it.

■ The use of moving walks in buildings is limited to very large horizontal buildings, such as transportation terminals, where the advantages help improve traffic flow when there are:

• Long travel distances, so that walking is tiring for many people.

• Many passengers are carrying baggage and may be fatigued from a long journey.

• There are definite entry and exit points (such as terminal lounges at an airport), and little need to stop between them.

• Passengers in a hurry can walk on the moving walkway, thereby doubling their speed.

The treads of a moving walkway travel as an endless belt, returning to the original point by a path immediately under the walkway. They can be installed in sidewalks or other public places with relatively little disruption to other traffic.

MECHANICAL DETAILS

A moving walkway requires:

• a firm, flat surface to stand on,

• close mechanical tolerances at edges and at the exit point to prevent injury,

• moving handrails to provide a means of maintaining balance.

It is acceptable to stand on a flat surface that slopes up or down as much as 15° from the horizontal (1 in 3.7), since it is not necessary to walk, and there is a handrail to hold. By way of comparison, ramps within buildings are limited by building codes to a much flatter slope, usually 1 in 10 or 1 in 12.

Escalators can be built steeper than moving walks because they form themselves into discrete horizontal steps. The industry standard is 30° from the horizontal (1 in 1.7), although in Europe some escalators are built at 35° inclination. The steps are about 14 in. (340 mm) deep (front to back) and 8 in. (200 mm) high. This is convenient for standing on, but much larger than a normal stairway. When an escalator is stopped, it forms a rather inconvenient stairway.

An escalator has the same requirements as a moving walk. In addition, the risers between the treads must be able to appear and disappear during the travel without trapping clothing or feet. The key to safety at the risers and the exits is the comb system. Both treads and risers have a grooved surface.

Modern treads are made of cast aluminum, with small and accurate grooves 4 in. (6 mm) wide and minimum clearances.

ELEVATORS

The elevator is the "batch process" of transporting people in buildings. In this respect its traffic pattern has much in common with a bus, which has a fixed route but only stops when there are passengers to get on or off.

One advantage that a modern elevator installation has over most other transport systems is that a central computer, which can assess the demands and dispatch the most appropriate car to answer each call, processes all the landing and car calls. This results in a few calls not being answered in turn, with priority being given to handling the bulk of the traffic more expeditiously.

SELECTION CRITERIA FOR ELEVATOR INSTALLATION

Elevators are called on to serve different functions according to the size and nature of the building and their location in it.

An important requirement in providing an elevator service for a building is the

& ELEVATORS

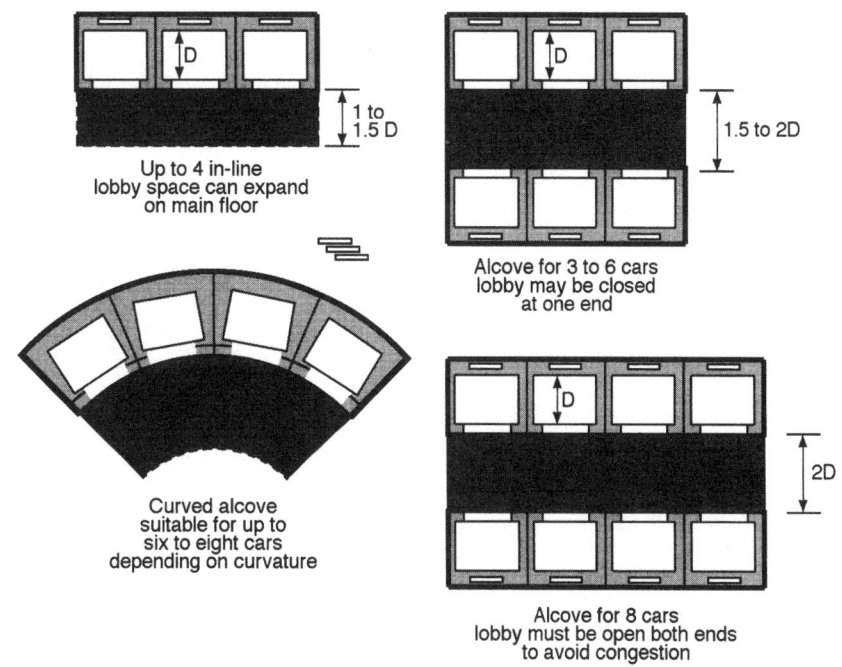

Fig. 1. Recommended lobby dimensions for various layouts of elevator groups. (Smith 1993)

Up to 4 in-line
lobby space can expand
on main floor

1 to
1.5 D

Alcove for 3 to 6 cars
lobby may be closed
at one end

1.5 to 2D

Curved alcove
suitable for up to
six to eight cars
depending on curvature

Alcove for 8 cars
lobby must be open both ends
to avoid congestion

2D

location of the elevators (in one or more groups, depending on the size of the building) in relation to access from the entrances and also in relation to the layout of the upper floors. Since elevator shafts are normally vertical, their location on plan imposes a major constraint on every floor of the building. If the building is not of uniform height, at least one elevator group must be located in the tallest portion.

The principal criteria for the selection of an elevator group can be summarized as follows:

■ Capacity to handle the passengers as they arrive, with minimum queuing, expressed as a percentage of the total building population that can be handled in the peak 5-minute period. (See accompanying TABLE for design parameters).

■ Frequency to provide an available car for arriving passengers without excessive waiting, expressed as the average time interval between cars.

■ Car size to handle the largest items required to be carried, for example, an occasional item of furniture, or a hospital bed with attendant, or a group of hotel guests with their baggage.

■ Speed or time of total trip, so that passengers do not perceive the total service as excessively slow

Fig. 2 Hydraulic elevators. (Smith 1993)

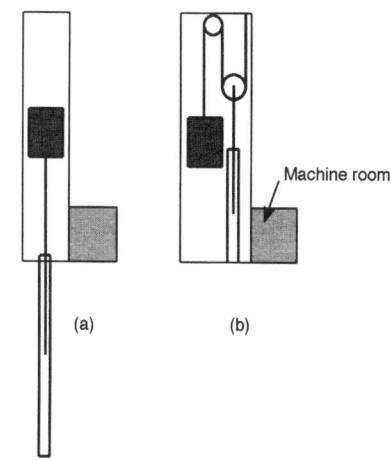

Machine room

(a)   (b)

Table 3. Design parameters for elevator selection.

$$P = F - F\left(1 - \frac{1}{F}\right)^{C}$$

where P = probable number of stops
F = number of floors served above ground
C = capacity of car on average trip (taken as no more than 80 percent of the maximum number of passengers)

D1.1 CONVEYING SYSTEMS

ELEVATORS (cont.)

In addition there are many details that should be considered for the comfort and convenience of the users:

• A means of finding the elevator lobby that serves the user's destination floor.

• Call buttons that are easy to find, and unambiguous for "up" and "down" calls. The location of the call buttons should encourage passengers to stand in a favorable position to watch all the cars in the group, and to move quickly to whichever comes first.

• Lights or indicators, and an audible indication as well, to indicate which car is arriving and in which direction it will travel.

• Enough first floor lobby space for the crowding that is expected at the up peak. (The accompanying shows minimum recommendations.)

• One set of buttons on the landing, and floor buttons in the car, should be at a level that can be reached by a person in a wheelchair, or a small child.

REFERENCES:

Peter R. Smith, Ph.D., FRAIA
"Conveying Systems" in Reference 1.

Smith, Peter R. *"The Movement of People and Goods."* in Henry J. Cowan, editor. 1991. Handbook of Architectural Technology. New York: Van Nostrand Reinhold.

D1.1 CONVEYING SYSTEMS

D2. 1 PLUMBING

PLUMBING SYSTEMS

PLUMBING SYSTEMS

Water distribution and plumbing systems are sized according to codes and occupancy, related to fixture unit count or water system demand.

System design includes:
- hot and cold water lines,
- domestic hot water systems,
- pressure and flow, and tank capacities.

Fixtures choices include a variety of dimensional, accessibility, finishes, and energy (water) conservation options.

WATER DISTRIBUTION SYSTEMS
Water Distribution Layouts

For design purposes, an architect should understand the principles of water distribution and piping system layout, in order to:
- provide necessary clearances and accessible spaces for horizontal and vertical runs of piping,
- provide for insulation to prevent condensation (and resulting dripping and staining),
- to design proper hanging and attachment, and
- provide for plumbing servicing and maintenance.

Like any system design, the best approach is to provide chases and runs for adequate water distribution, that is direct and free of unnecessary turns that may reduce flow and create locations for blockage.

The diagrammatic layouts in the accompanying figure represent typical good practice in the design of hot and cold water distribution systems for small buildings, most typical of residences. Related vents required from each fixture are not shown.

COLD WATER LINES

■ **Pipes and pipe sizes** vary depending on the fixture unit count or demand, usually represented by Gallons Per Minute (GPM) or litres/second (l/s).

■ **Codes**
- Most regional plumbing codes in the U. S. require pressure reducing valves where street main pressure exceeds 80 psi (550 kPa).
- An approved pressure-reducing valve should be installed in the water service pipe near the entrance of the water service pipe into the structure, except where the water feeds a water pressure booster.
- The pressure at any fixture normally should not exceed 80 psi (550 kPa) under no-flow conditions.

■ **Drain valves** should be installed at each of the lowest branches so that the entire system can be completely drained.

■ **Main shut-off valve** should be provided in each water supply pipe at each of the following locations:
- On the street side near the curb.
- At the entrance into the structure.
- On the discharge side of the water meter.
- On the base of every riser.
- At the supply of any equipment.
- At the connection to any fixture.

■ **Water hammer arresters or shock absorbers** will prevent water hammering throughout the piping system. They are typically located after the first fixture of a group of fixtures.

■ **Water softeners** are available in various sizes and types, all of which require a salt tank for regeneration.

■ **Pipe insulation:** Insulation of water lines is necessary to prevent undesirable condensation from cold water lines.

HOT WATER LINES

A simple form of circulation can be accomplished by connecting the hot water pipe supply after the last fixture connected to a recirculating line as a simple loop below the floor of the highest level.

Circulation serves to prevent waste of water and provides the added benefit of instantaneous hot water availability. The type of circulating hot water distribution shown is adaptable to large residences.

Hot water lines should also be insulated to avoid waste of energy due the heat loss, and for this purpose typically 1/2in. to 1in. (1.25 cm to 2.5 cm) fiber glass insulation is used. **Shut-off valves** are also required in the same fashion as indicated for cold water lines.

The variety of uses for domestic hot water and lifestyles determine system requirements and sizing. System design data are listed in the references. Special consideration should be given to facilities that have high hot-water demand, such as motels, hospitals, nursing homes, , and food service establishments.

Besides the sizing of the system, it is important to select the most efficient and economical energy source. Fuel choice will depending on local energy costs and availability. The most common sources include:
- gas fired.
- oil fired.
- electric.
- steam generated.
- solar water heaters.

Other factors to be considered at the beginning of the design include:
- storage tank requirements.
- space availability.
- peak instantaneous demand.
- water temperature requirements.
- water treatment.

The probable maximum demand is the result of the hourly hot water demand times the demand factor. The storage tank capacity is the result of probable maximum demand times the storage capacity factor.

An efficiency factor is an important

D2.1 PLUMBING

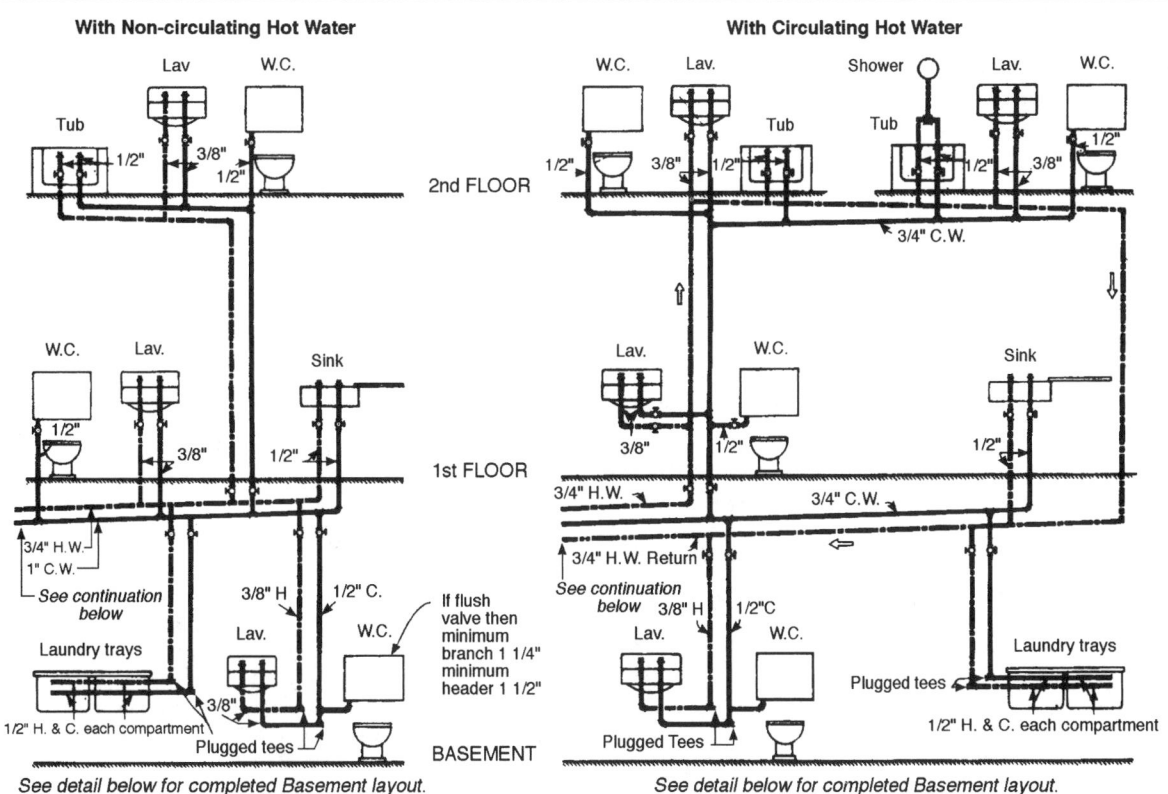

With Non-circulating Hot Water

With Circulating Hot Water

See detail below for completed Basement layout.

See detail below for completed Basement layout.

NOTE¹ *These sections are diagrammatic and drawn to no scale.*

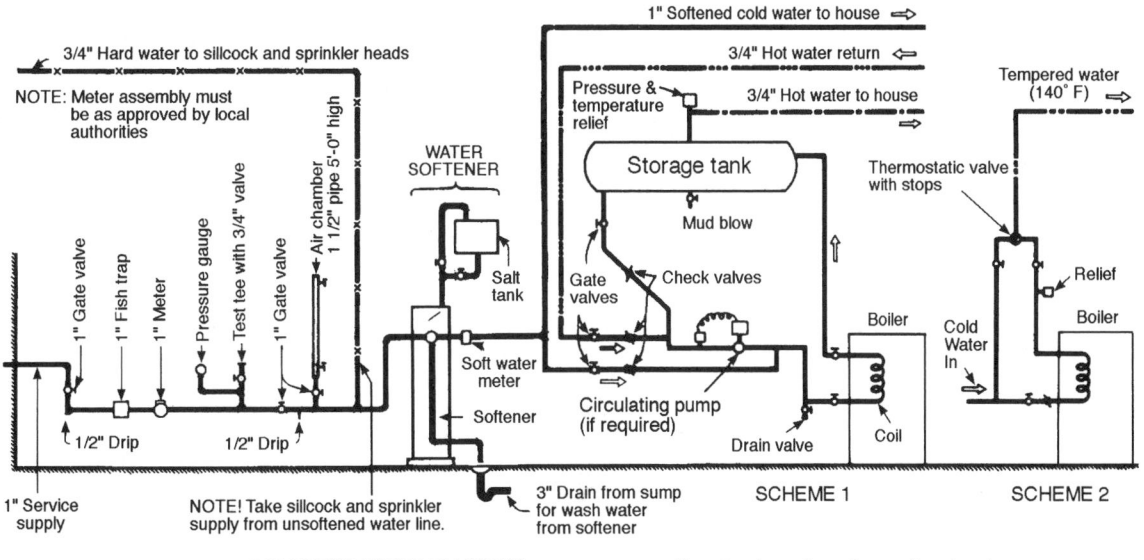

BASEMENT PIPING DIAGRAM   *(See Sections above for continuations)*

Fig. 1. Residential piping system diagrams.

D2.1 PLUMBING

PLUMBING SYSTEMS (cont.)

consideration. For example, gas water heaters can be expected to have an efficiency factor of 75% to 90%. Verification from manufacturers data and equipment performance results reported in current literature is necessary.

PRESSURE AND FLOW

■ Plumbing fixtures require certain pressure and flow to function properly. Street pressure is normally enough to satisfy the requirements of residences and two- to five-story buildings.

■ An engineer should verify the existing pressure and flow characteristics from city authorities. For structures in excess of five stories, the engineer should select any of the following means to provide adequate pressure and flow to the plumbing fixtures within the structure:

• **Booster pumps** can be a duplex system or several duplex systems supplying different zones. Booster systems can be complimented with storage tanks at the highest level of each zone.

• An alternative for pressuring the system is to provide a pneumatic booster system with the following omponents: booster pump, compressor, pneumatic tank, valves, and controllers.

• Tanks used in booster systems have also been used for the dual purpose of water supply and as a reserve for fire protection.

PIPING MATERIALS

Available water service piping materials include:

• Acrylonitrile butadiene (ABS plastic pipe).

• Brass pipe.

• Copper or Copper-alloy pipe.

• Copper or Copper-alloy tubing (Type K, WK, L, WL, M, or WM).

• Chlorinated polyvinyl Chloride (CPVC plastic pipe).

• Ductile iron water pipe.

• Galvanized steel pipe.

• Polybutylene (PE plastic pipe and tubing).

• Polyethylene (PE plastic pipe or tubing).

• Polyvinyl chloride (PVC plastic pipe).

Available water distribution piping materials include the following:

• Brass pipe

• Chlorinated polyvinyl chloride (CPVC plastic pipe and tubing)

• Copper or copper alloy pipe

• Copper or copper alloy tubing (Type K, L, or M)

• Galvanized steel pipe

• Polybutylene (PB plastic pipe and tubing)

PLUMBING FIXTURES

Selection Criteria

Before specifying plumbing fixtures, the designer should become familiar with plumbing fixture manufacturer's catalog information and associated components as fitting, faucets, toilet seats, flash valves, and supports. Each plumbing fixture must include a receptor drain or strainer, trap, cold water supply or hot and cold water supply, trim, accessories, appliances, appurtenances, equipment, and supports.

Selection criteria for plumbing fixtures include:

• **Fitting body types:** Cast brass, brass, or copper underbody with chrome-plated escutcheon, plastic underbody with chrome-plated escutcheon, plastic.

• **Finishes:** Polished chrome plated, polished brass, polished gold plated, and colored plastic finishes.

• **Handle types:** Dual handle, ornamental metal, and porcelain lever, dual three- and four-arm, dual metal or crystal knob, single lever, dual lever 4 in. (10 cm) and 6 in. (15 cm) wrist blade, push button, self closing.

• **Fixture clearances:** Proper clearances between fixtures, and between fixtures and walls must be used in the layout of washrooms and bathrooms to ensure ease of installation, ease of use, and maintenance.

• **Universal design and accessibility:** The layout of plumbing fixtures should be guided by considerations of universal design and accessibility, which is governed by the Americans with Disabilities Act (ADA). Accessible plumbing fixtures must be provided in public buildings and where required by code and authorities having jurisdiction

providing handicapped accessibility to lavatories, sinks, water closets, and water coolers.

■ **Water closets:** Two types of water closets are commonly used in new construction:

• **Wall-hanging type**, with a flush valve, are the preference in public buildings to allow toe space for wheelchair foot rest, to facilitate approach to the seat, and to allow access for cleaning below the fixture.

• **Tank type,** typically used in residential buildings. In either case, the bowl should be elongated to provide handicapped accessibility.

■ **Lavatories:** The most common types of lavatories used in new construction include wall hung, counter top, and pedestal types. Lavatories come in a variety of shapes and styles. Some common styles include back splashes, self rim, one-piece bowls, and undercounter mounted. Lavatories for handicapped access must be provided with a clear dimension of 30 in. (75 cm) under each lavatory and have insulated supply and drain lines to prevent injury and burns.

■ **Showers:** Prefabricated showers and shower compartments are manufactured in several configurations. A shower compartment should have a minimum of 900 sq. in. (5,800 sq. cm) of interior cross sectional area and have not less than 3 in. (75 cm) minimum dimension measured from its finished interior dimension, exclusive of fixture valves, shower heads, soap dishes, and safety grab bars. The minimum required area and dimension is measured from the finished interior dimension at a height equal to the top of the threshold and at a point tangent to its center line.

■ **Urinals:** Urinals are typically selected for use in mens toilet rooms intended for use by the public. They should have a visible water trap seal without strainer to permit maintenance. For the purposes of water conservation, urinals should be selected that incorporate a maximum of 1.5 gallons (5.7 l) of water per flushing cycle.

■ **Sinks:** Sinks are typically selected

D2.1 PLUMBING

for use in kitchens and laboratories. They can be provided with many different types of faucets and finishes. The most common sink finish is stainless steel because of its durability and resistance to foreign materials.

■ **Water coolers and drinking fountains:** Because of the different applications to a variety of spaces and locations, the choice of style is broad. The designer has the freedom to select shapes and colors to improve the aesthetic appearance of the units, as long as handicapped accessibility clearances are provided.

■ **Service sinks:** These fixtures are required in janitor closets or maintenance areas. The most common styles are: floor mounted, wall hung, and mop sinks. Another variety of this fixture is the laundry sink that may be used with support legs. Service sinks must be provided with hose connection and vacuum breakers.

WATER CONSERVATION

The U. S. Energy Policy Act of 1992 requires that plumbing fixtures manufactured for use in the U. S. after January 1, 1994 have the following maximum flow rates and consumption in gallons per minute (gpm) and litre/sec (l/s), gallons (gal.) and litres (l):

■ **Lavatory and sink faucet:** 2.5 gpm (0.16 l/s).

■ **Shower head:** 2.5 gpm (0.16 l/s)

■ **Water closet:** Types as follows:

• **Flushometer valve:** 1.6 gal. (6 l) per flushing cycle

• **Flushometer tank type:** 1.6 gal. (6 l) per flushing cycle

• **Commercial, tank with flush valve:** 3.5 gal. (13.2 l) per flushing cycle

• **Residential, tank with flush valve:** 1.6 gal. (6 l) per flushing cycle

■ **Urinals:** 1 gal. (3.8 l) per flushing cycle.

REFERENCES:

Arturo De La Vega *"Plumbing Systems"* in Reference 1.

Building Officials and Code Administration. BOCA National Plumbing Code. Country Club Hills, IL: BOCA. 1990.

D2.1 PLUMBING

SANITARY WASTE SYSTEMS

SANITARY WASTE SYSTEMS

WASTE SYSTEM SIZING

■ **Sanitary drainage systems** are designed to carry wastes from plumbing fixtures and floor drains to public sanitary sewers or septic tanks.

■ To apply proper sizing and slope to drainage and vent pipes, the engineer should use applicable model building codes and regulations, and consult the authority having jurisdiction.

■ System design data are available in the references, derived from the BOCA National Plumbing Code. Caution is advised that the tabulated values do not always agree with all current model, and local, building codes. Where differences exist, local requirements govern. However, the data can be used to establish limiting requirements for drainage system design.

PRELIMINARY SYSTEM DESIGN

■ To facilitate the understanding of the floor plans, and to facilitate adequate space allowance for the piping system, the designer should develop diagrams that sequentially show the system and coordinate with other engineering disciplines such as civil, structural, mechanical, electrical, communication, and fire protection.

■ The designer should also consider other sources of continuous or semi-continuous flow into the drainage system, such as from pumps, sump ejectors, and air conditioning equipment. These loads are commonly computed as one fixture unit equaling 7.5 gallons per minute (28.35 l/m).

STACK CAPACITIES

• **Waste stacks** are the vertical pipes collecting all the horizontal drainage branches from each floor.

• The collection of **pipe branches** in each level is commonly known as an "interval." A stack can take two branches from the same level with a 45-degree Y fitting or sanitary T, and that also will be an interval.

VENT REQUIREMENTS

• The size and length of **vent pipes** are directly dependent upon the volume of discharge for which the soil and waste pipe are designed.

• Unless adequate venting is provided, the flow of fixture discharges through soil or waste stacks can produce pressure variations in branches that may damage the seals of fixture traps by blowing them from positive or back pressure in lower parts of the system, or syphoning them because of negative pressure in upper sections.

SANITARY BUILDING DRAINS

• **Building drains** are typically placed at the lowest piping invert elevation in a drainage system. Building drains receive wastes from soil, waste, and other drainage pipes inside a structure and convey them into the building sewer

Critical limitations and allowances:

• No branch or fixture should be connected within 10 pipe diameters downstream from the base of the soil or waste stack.

• **Indirect waste** should not discharge through an air gap into a trapped fixture.

• **Combination waste** and vent should only be used for floor drains, standpipes, sinks and lavatories, which should discharge to a vented drainage pipe.

• **Chemical waste** should be completely separated from the sanitary drainage system.

• The minimum size for underground drainpipe should be 2 in. (5 cm).

• The size of the drainpipe should not be reduced in size in the direction of the flow.

• Drainage for future fixtures should be terminated with an approved cap or plug.

• **Dead ends** are prohibited in the installation or removal of any part of a drainage system. The application of code requirements should be accurate in this and other related matters.

• A fixture should not be connected to horizontal piping from the base of the stack within 40 pipe diameters from the stack to prevent backup of fixtures on the lower floor caused by a hydraulic jump at the base of the stack.

• **Suds pressure zones** should be considered to exist in sanitary drainage and vent systems when the piping serves fixtures on two or more floors that receive waste containing bubble bath or sudsy detergents.

PIPING MATERIALS

Available above ground drainage and vent piping materials include the following:

• Acrylonitrile butadiene styrene (ABS plastic pipe)

• Brass pipe

• Cast iron pipe

• Copper or copper alloy pipe

• Copper or copper alloy tubing (Type K, L, M, or DWV)

• Galvanized steel pipe

• Polyvinyl chloride (PVC plastic pipe type DWV)

Available underground building drainage and vent piping materials include the following:

• Acrylonitrile butadiene styrene (ABS plastic pipe)

• Cast iron pipe

• Concrete pipe

• Copper or copper alloy tubing (Type K or L)

• Polyvinyl chloride (PVC plastic pipe type DWV)

SIZING AND RATINGS

The accompanying figure illustrates types of plumbing layouts applicable to commercial buildings. Because of the wide variance in plumbing regulations, some of these diagrammatic details may be prohibited in certain localities; other details may indicate methods far in excess of mandatory requirements in other localities. All, however, reflect solutions to typical drainage problems by methods that generally constitute good plumbing practice.

RESIDENTIAL DRAINAGE SYSTEMS

• A 3 in. (7.5 cm) main soil stack is adequate for residential use in the opinion of many authorities; a 4 in. (10 cm) main soil stack is mandatory in some localities. Main house drains should never be less than 4 in. (10 cm).

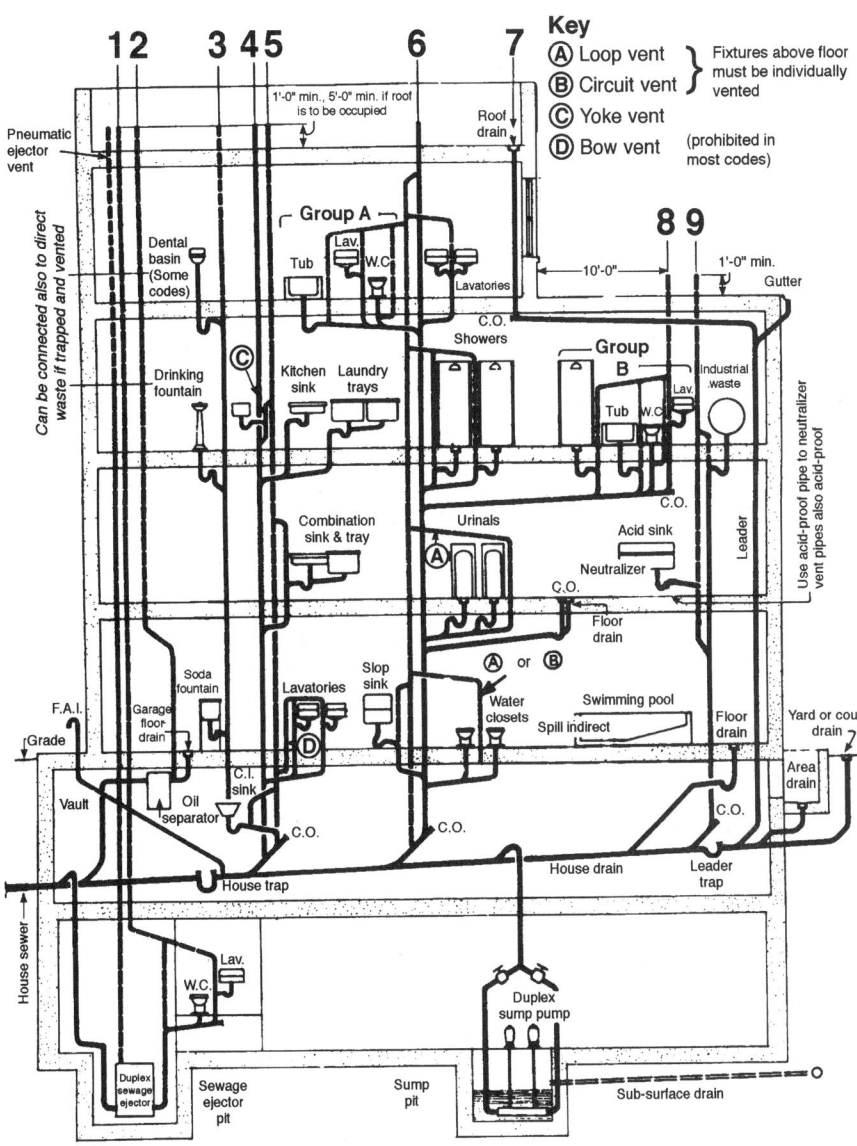

Key
- Ⓐ Loop vent ⎫ Fixtures above floor
- Ⓑ Circuit vent ⎬ must be individually
- Ⓒ Yoke vent ⎭ vented
- Ⓓ Bow vent (prohibited in most codes)

Fig. 3. Drainage system for commercial buildings

prohibited by some codes. It is generally not a desirable method of venting. If used, circuit or loop vents should not be connected to a group of more than eight fixtures in series. In a loop vent, a continuation of the branch runs up and over the fixtures to connect to the vent stack adjacent to the main soil. In a circuit vent, the connection is to a main vent stack opposite the main soil stack.

■ **Yoke Vent C:** This connects the main soil and waste stacks, with the soil at the lower end of the yoke. The connection of the fixtures as at C adds greatly to the safe capacity of soil stacks. This type of connection can be made to bathroom units in residential as well as commercial buildings.

■ **Bow Vent D:** This can be used for light discharge loads to avoid installation of an additional vent stack.

■ **Stacks 1 and 2:** These indicate the need for separate venting of the sewage ejector and the oil separator from garage drains. The vent from a pneumatic sewage ejector should not be joined to any other pipe; sewage pumps do not require any special considerations.

■ **Stack 3:** This is the vent from an indirect waste line discharging into a cast iron sink. Its fixtures must be trapped. If an indirect waste line is over 100' (30m) in developed length, it should be extended through the roof.

■ **Stacks 4, 5, and 6:** The bents should be connected into these stacks at their lower ends, so that discharge will scour the connection and thus prevent fouling. Such a connection is specifically required for cast iron because of scaling.

■ **Stack 9:** This applies to a special purpose type of installation. Corrosive wastes require acid proof pipe for waste, soil, and vent lines, for fittings, and for the house drain up to the base fitting of the next main soil stack.

• **If house sewers are connected to the septic tank** of a private sewerage disposal system, no house trap or fresh air inlet is necessary. In many communities individual venting may be eliminated and a system of wet venting or combined waste and vent system can be utilized.

COMMERCIAL DRAINAGE SYSTEMS

The figure shows a composite of drainage problems encountered in a wide range of commercial and industrial work; the installations are not typical for any specific kind of building. Soil and waste lines are shown solid; vent lines are broken.

■ **Group A:** Bathroom unit is rated for six fixture units, individually vented and connected by preferred methods to main soil and vent stacks.

■ **Group B:** Bathroom unit is rated at seven fixture units, individually vented and connected by the preferred method to a horizontal soil branch.

■ **Loop Vent A and Circuit Vent B:** Both are types of venting in which the branch drain is a "double duty" pipe carrying both air and discharge. The use of this pipe constitutes "wet venting,"

REFERENCES:

Arturo De La Vega *"Plumbing Systems"* in Reference 1.

Building Officials and Code Administration. BOCA National Plumbing Code. Country Club Hills, IL: BOCA. 1990.

D2.1 PLUMBING

SOLAR DOMESTIC WATER HEATING

SOLAR DOMESTIC WATER HEATING

Solar energy may be used to heat water for many applications, including:

- domestic water for residential uses,
- water for swimming pools and spas;
- service water for commercial use,
- industrial process water,
- water for agricultural uses.

A typical solar water heating system will provide about 75% of the annual hot water requirement of a family of four.

SOLAR COLLECTORS

Two fundamentally different types of solar collectors are in widespread use today:

- **solar electric** (often called photovoltaic) collectors convert the sun's energy directly to electricity and is ideal for producing electricity.
- **solar heat** (often called "solar thermal") collectors. convert the sun's energy directly to heat, and is thus ideal for water heating in the low temperature (below boiling) temperature range.

The solar heat collector most commonly used for domestic water heating is a flat plate collector. It consists of a glazed, metal enclosure. The enclosure frame is usually aluminum. The glazing is most often a tempered, low-iron content glass. Inside the enclosure is a metal absorber with integral flow passages, usually made of sheet metal. On the side of the absorber facing the glass is a black coating. The coating is either a black paint or a selective surface. The selective surface offers better heat retention than the black paint. The sun's heat is removed from a solar heat collector by a fluid flowing through the absorber. The fluid can be either a gas, such as air or a liquid, such as water. A layer of thermal insulation is used between the rear of the absorber and the rear of the enclosure. The collector is usually located outdoors, most often on a roof. A typical solar collector glazed with a single layer of glass weighs between 3 and 4 pounds per square foot. A well-made solar collector should last 50 years.

Flat-plate collectors may be either liquid-cooled or air-cooled. Air cooled solar thermal collectors which deliver heated air in the range of 80 to 100F are applicable to space heating systems which can reliable utilize low-temperature heated air. Liquid-cooled solar thermal collectors, utilizing water or water with an anti-freeze additive, and capable of delivering heated liquid in the range of 100 to 130F or higher are used for domestic water heating.

There are other types of solar thermal collectors, with mirrors or heliostats that track the sun's movement. These are used for high temperature (above boiling) applications. As long as energy is needed in the form of low temperature heat <160F (71°C), then a solar heat collection system is superior to a solar electric system. If the need for energy is in the form of electricity, then a solar electric collector is preferred.

DESIGN AND SELECTION FACTORS

A number of factors have a bearing on the type of solar water heating system for any given application. Some of the more significant of these factors are:

■ **Regularity of hot water demand.** When there is a daily demand for hot water that corresponds to the installed heat collection area, there is usually little chance that a solar water heating system will overheat. However, during the normal use of a dwelling, the hot water demand is rarely constant over the life of the dwelling. Ideally the solar system should be capable of enduring periods of reduced or no hot water demand. The system chosen for a given installation should be capable of operating with a minimum of owner intervention over a range of hot water demands

■ **Freeze protection:** If freezing weather occurs, even for a few weeks per year, some means is necessary of protecting the collectors from damage caused by freezing. Ideally the means of freeze protection should be automatic since an early frost may surprise a system owner. If water freezes in the absorber of most collectors, the absorber will be damaged by the

expansion of the freezing water. In regions where freezing weather is of little or no concern, a simple solar system, usually one of the open systems, is preferred.

■ **Quality of water to be heated:** Water quality may dictate the type of solar system for a given application. If the water to be heated is "hard," that is, it contains a higher than normal percentage of dissolved carbonates; or if the water is acidic (pH<7.0), then it is very desirable to isolate the solar heat collection loop components from the water to be heated. For these applications some type of closed solar heat collection system is preferred because the closed system isolates the solar collection loop from the water to be heated.

■ **Appearance of the system:** The style or form of the building on which the system is to be installed may dictate the type of solar water heating system to be used. Most passive solar water heating systems, which depend upon circulation due to a density difference between the collectors and storage, require that the collectors be located below the storage tank. On flat roofed buildings, it is relatively easy to place the collectors on the roof with the storage tank above them. Building aesthetics or local zoning codes may preclude the visibility of this type of system. A ground mount may be a possible alternative in this case if the collectors will not be shaded. Alternatively, an active system may be preferred since its use permits the storage tank to be placed below the collectors.

■ **Building structure:** The weight of the tank, or collectors plus tank, must be considered in choosing a type of system and in deciding where to place the components. The weight of the collectors alone is generally not a deciding factor in where they can be placed since the collectors seldom weigh more than 4 pounds per square foot and that is well below the design snow load in most regions. The weight of the tank is a different matter. If the tank is to be placed on a wood frame floor, or hung from roof rafters, the structure beneath it must be reinforced.

D2.1 PLUMBING

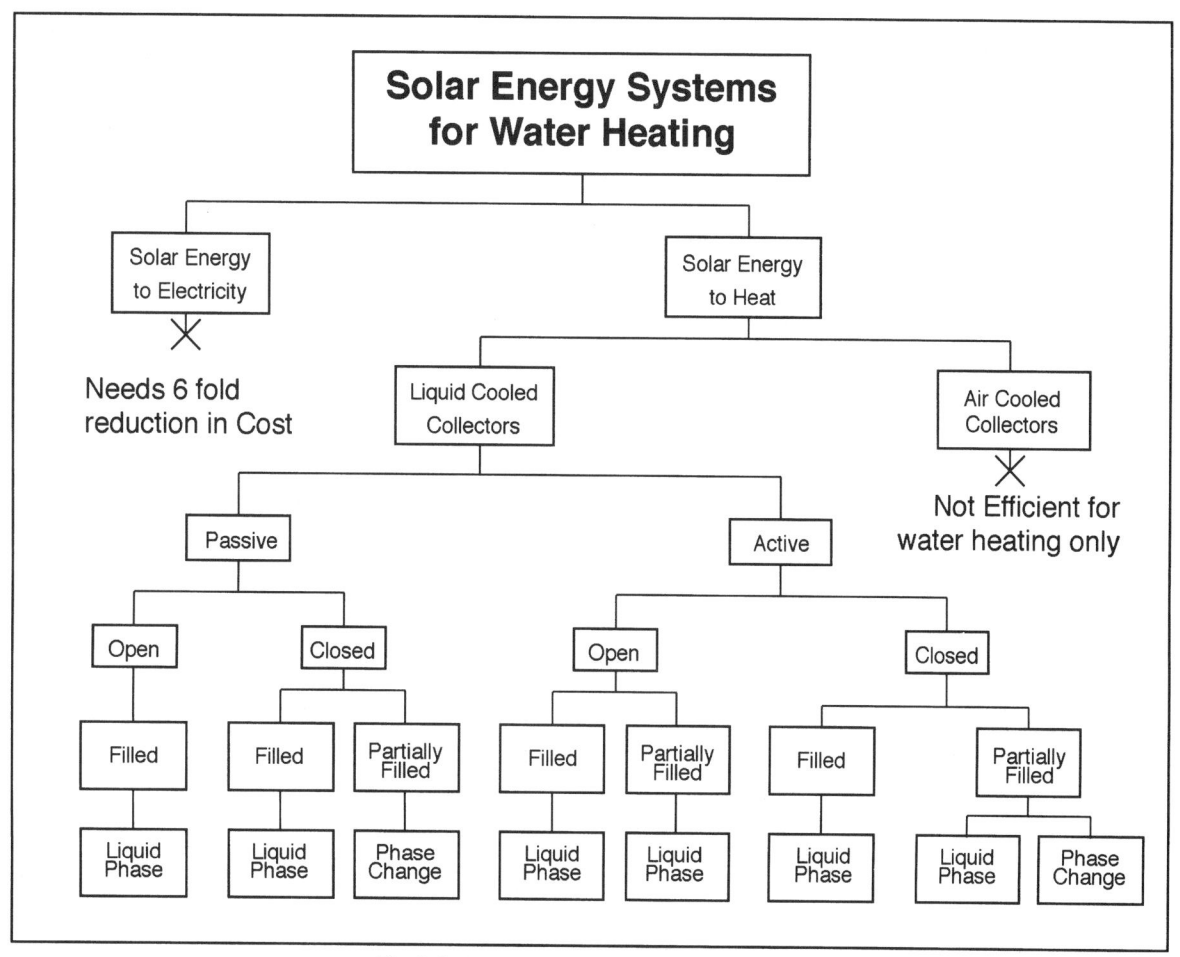

Fig. 5. Solar energy systems for water heating

From structural and maintenance standpoints, the preferred location for the storage tank is on a ground level or basement floor slab where the structure is much more likely to be able to carry the load without reinforcement.

■ **Water damage potential:** Virtually all tanks leak at some point in the life cycle. Some tanks used for residential water heating (regardless of fuel type) last no more than 7 to 12 years on average; others last 20 to 30 years. When the tanks leak they seem to have a perverse way of leaking when no one is around to discover and shut off the leak. It is not the 80 or 120 gallons of water in the tank that causes so much damage, but if continuously running, water from the well or city main will obvious cause the greatest damage.

HOT WATER STORAGE TANK

The type of solar water heating system selected can determine the location of the hot water storage tank in the building. The collectors of most solar systems are placed on the roof of the building served to raise them above surrounding shade. In general, the passive systems require that the tank be located above the solar collectors. The active systems permit the tank to be located anywhere above or below the collectors.

If a passive system is used, the tank must be placed above the collectors for the system to function properly. If this system is to be installed in a warm climate, the tank can be placed outside without risk of freeze damage to the tank or connecting piping. In that case when the tank leaks, the water will usually run off the roof with little harm done. If the passive system is to be

installed in a cold climate, then the tank should be located in a heated space. The heated space is preferred to reduce the heat loss from the tank and to minimize the risk of freezing the water lines serving the tank. The most convenient heated space is inside the dwelling. The placement of a tank high inside a building carries a significant risk of water leakage damage. The typical water tank overflow trays that are available for placement under hot water tanks have provided little protection from damage caused by ruptured tanks high in the dwelling. The tank leaks into the tray and the water runs off through the drain line unnoticed until the leak becomes a rupture and water spews out of the tank.

If a tank for a passive system must be placed inside the dwelling, in a heated space, that space must be easily accessible for inspection and tank replacement, and it should contains a

D2.1 PLUMBING

SOLAR DOMESTIC WATER HEATING (cont.)

Table 1. Which solar system to use "Daily hot water demand" implies normal occupancy and a maximum period of two weeks in which there may be no hot water demand. "Irregular hot water demand" implies intermittent occupancy with periods on no hot water demand for weeks or months at a time. Closed systems should also be used where water quality is "hard" or low pH.

		Where Storage may be above Collectors		Where storage must be Below Collectors	
		Daily hw Demand	Irregular hw Demand	Daily hw Demand	Irregular hw Demand
Non-Freezing Climate	<50% Solar	Batch	Drain Down	Basic Open Loop	Drain-Down
	>50% Solar	Thermosyphon			
Freezing Infrequent	<50% Solar	Batch	Drain Down	Closed Loop Type 1 or 2	Drain Back
	>50% Solar	Thermosyphon			
Freezing Occurs	>50% Solar	Phase Change #1	NONE	Closed Loop Type 1 or 2	Drain Back

"Daily Hot Water Demand" implies normal occupancy, and no hot water demand for a maximum of several weeks per year.

"Irregular Hot Water Demand" implies no hot water demand for weeks or months at a time.

Closed systems should also be used where water quality is 'hard' or low pH.

water runoff barrier that channels leaking water outside. If the tank cannot be located in such a manner, then an active system used.

SOLAR THERMAL SYSTEM TYPES

There are a variety of solar heat collection and storage systems in use. Solar collectors were developed commercially over one hundred years ago in the United States and were common in warm climates in the twentieth century, such as in California and Florida. Some systems heat the domestic water directly in the collector. Others isolate the collector from the domestic water. Some require no electric power to operate, while others require electric power. Some are completely filled and others only partially filled. Some are better suited for freezing climates than others. The accompanying figure provides a summary of different system types.

■ **Liquid-cooled systems:** For water

heating, the liquid-cooled heat collection systems are preferred over air-cooled systems. A liquid-cooled solar heat collection system uses a liquid such as water to transport heat from the collectors through piping to the heat storage tank. The liquid-cooled collectors are more efficient than the air-cooled collectors for this application. If an active system is used, less energy is required to collect and transport heat from a liquid-cooled collector than from an air-cooled collector. The pipes of the liquid-cooled domestic water heating systems take up far less space than do the ducts of air-cooled systems. If an open system is used, the sun's heat is more efficiently transferred to the domestic water circulating through the collector than it is in an air-cooled system where the heat must first pass through an air/liquid heat exchanger to heat the domestic water.

■ **Passive and active systems:** In addition to air-cooled and liquid-cooled systems, solar systems are further

divided into passive and active systems. There are a variety of subtypes of these, the more significant are described below.

■ **A passive solar water heating system** is one that is capable of collecting and storing the sun's heat without the use of a motor driven fan or pump. Heat transport from collector to storage in these types of systems relies either on circulation due to density difference, or on circulation due to boiling and condensation.

■ **An active solar water heating system** uses an external source of energy to power a motor driven fan or pump to force a fluid through the collectors, thereby removing the sun's heat and transporting that heat to storage. The fan or pump motor is usually turned on and off by a differential temperature thermostat. The power requirement of the pump motor is typically less than 3% of the total energy collected by the system. A solar electric collector may also power the pump motor.

D2.1 PLUMBING

■ **Open-loop and closed-loop systems:** Passive and active systems are further divided into open loop and closed loop systems. The term 'loop' refers to the flow path that permits a fluid to flow from heat storage, through the collector supply pipe, through the collectors, then back through the collector return pipe, to the heat storage tank. This circulation occurs when the collectors are warmer than the storage tank. The two systems each have merit for certain applications.

SELECTION CHECKLIST

Table 1 presents a summary of solar water heating system types appropriate for various applications. Where there is no expectation of freezing weather, then one of the open loop systems should be used because they are more efficient than closed loop systems.

■ Where freezing weather is expected, then one of the closed loop systems will offer the greatest protection from freeze damage.

■ Where there is no freezing weather and the architectural design permits, a passive unitary system such as a batch heater, or a natural circulation system is recommended. They are simple, reliable and efficient. However their use dictates that the storage be above the collectors which in many instances may not be acceptable.

■ Where the storage must be located below the collectors then some type of active system is required.

■ Where demand for hot water is expected to be intermittent, and the dwelling may be unoccupied for long periods then an active open loop drain down system or an active closed loop drain back system are preferred.

SUPPLEMENTAL HEATING

A solar water heating system is almost always used with some type of supplemental heating equipment. The supplemental heater may be located inside the solar heat storage tank, such as an electric element in the upper part of the tank; or it may be a separate appliance such as a conventional electric, oil or gas fired water heater. In many localities, with a solar system properly sized for the load, the auxiliary heater can be shut off during the summer months.

REFERENCES:

Everett M. Barber, Jr. *"Solar Domestic Water Heating"* in Reference 1.

FSEC. 1982 (Revised 1992). Solar Water and Pool Heating, Design and Installation Manual. FSEC-IN-21-82. Cocoa, FL: Florida Solar Energy Center.

D2.1 PLUMBING

DOMESTIC WATER HEATING SYSTEMS: typical

| TYPE | DESCRIPTION |

DOMESTIC WATER HEATING, liquid system

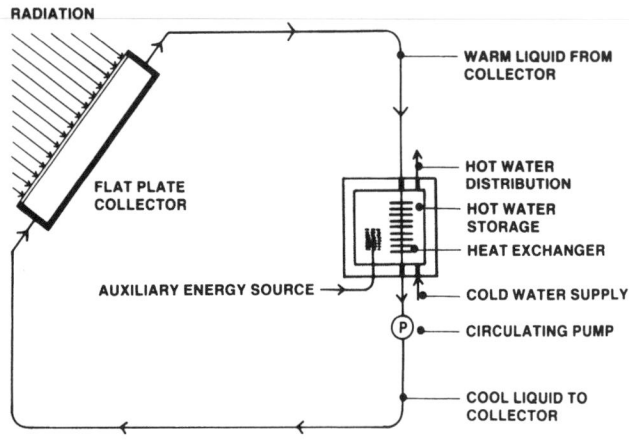

Liquid heat transport medium transfers heat from collector to domestic water storage. Typical systems include:

■ **Open flow collector** with pumped supply and gravity return of water to storage.
• collector generally with cover plate to minimize evaporative heat loss.
• flow of water over surface will erode coating, and loose particles may be picked up by the water and deposited in the hot water storage.

■ **Contained flow collector** with thermosiphoning hot water storage:
• circulating pump is not required, but storage should be located higher than collector.
• reverse thermosiphoning may be controlled by a check valve in the cool water piping.
• piping and collector must be protected from freeze-up during cloudy days and at night during winter.

■ **Contained flow collector with forced circulation:**
• liquid transport medium resistant to freezing temperatures and contained in a closed loop; as illustrated.

■ **Auxiliary energy source,** such as a heating coil, generally required for all the above systems.

DOMESTIC WATER HEATING, warm air system

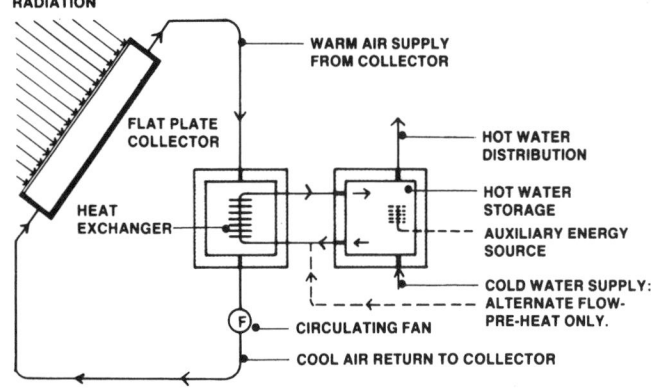

Air is used as heat transport medium to transfer heat from collector through a heat exchanger to hot water storage. Typical systems include:

■ **Forced air flow through collector picks up heat by convection and transfers it through ducts to heat exchanger. Water from storage is circulated through heat exchanger either by thermosiphoning or by a pump.**
• heat capacity of air is lower than that of water, and a large collector, in addition to ductwork, is required.
• problems of freeze-ups and corrosion of absorber plate are eliminated, but capacity of hot water storage may have to be increased over that required for liquid-type systems.

■ **Auxiliary energy source,** such as a heating coil, generally required:
• a heat exchanger may be used to preheat water and reduce energy consumption by auxiliary energy source.

SOLAR COLLECTORS

DESCRIPTION

FLAT PLATE TYPE, assembly

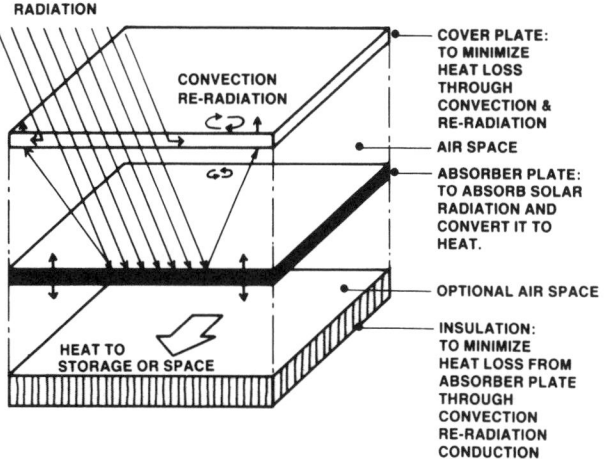

A flat-plate collector is essentially a flat surface arranged to intercept solar radiation and to allow the absorbed heat to be conveyed to a point of use remote from the absorber.

■ **The heat collecting surface may be:**
* **an array of tubing** through which a liquid heat transfer medium such as water is circulated,
* **thin metallic plates** with air as the transfer medium flowing over or between them.
* in its simplest form, a flat-plate collector may consist of the heat collecting surface only.
* generally, the heat collecting surface is enclosed to minimize heat losses to surrounding air through convection and re-radiation.

■ **Collectors should be installed at a specific angle** for maximum efficiency.

■ **Collectors may also be motorized** to follow the daily path of the sun to improve their efficiency.

FLAT PLATE TYPE, absorber plate

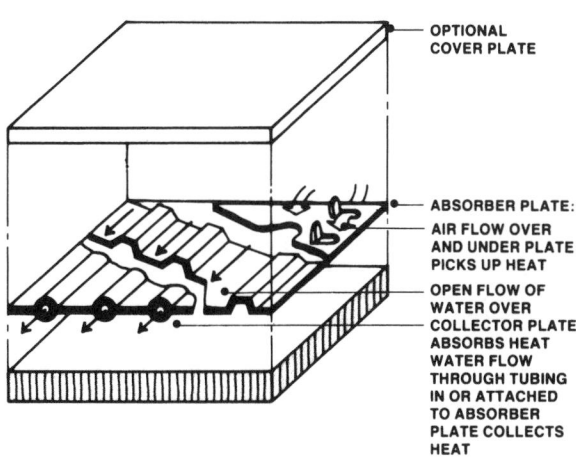

Absorber plates: the heat collecting surfaces may be classified based on the heat transfer medium used as:

■ **Liquid-type, with the liquid** such as water:
* flowing **over the surface** of the absorber plate,
* circulating **through tubing** incorporated into or attached to the plate.

■ **Air-type, with air as the heat transfer medium** flowing over the top, or top and bottom surfaces of the absorber plate.
* the **surface** of the absorber plate should be irregular to induce turbulent air flow for more efficient heat transfer.
* **for additional** information on convective heat transfer, refer to INSULATION.

■ **The degree to which radiation is absorbed** at the surface determines the efficiency of the absorber plate: the surface should have high absorptance and low emissivity.
* **related information** may also be found in INSULATION.

D2.1 PLUMBING

SOLAR COLLECTORS (cont.)

DESCRIPTION

COVER PLATE

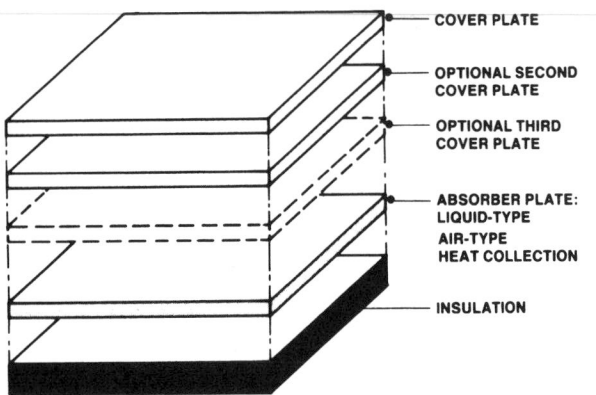

COVER PLATE

OPTIONAL SECOND
COVER PLATE

OPTIONAL THIRD
COVER PLATE

ABSORBER PLATE:
LIQUID-TYPE
AIR-TYPE
HEAT COLLECTION

INSULATION

Cover plates are transparent sheets, which allow solar radiation to penetrate to the absorber plate while minimizing heat loss from it.

■ **They may be:**
• rigid, e.g. made of glass
• flexible, e.g. made of thin plastic films.

■ **Use of several cover plates** will minimize outward flow of heat and increase collector operating temperature, but more solar radiation will be reflected, thus reducing the percentage reaching the absorber plate:

■ **At zero degree angle of incidence:**
• a **single** cover plate of clear glass may transmit up to 90 percent of solar radiation;
• **two plates** will reduce such percentage to 85;
• for **three plates** it may be about 78 percent.

■ **The reduction in transmittance may be offset** by higher collector and storage temperatures thereby increasing the effective capacity of the heat storage.

■ **The use of two cover plates** in space heating applications is generally considered the most economical choice.

EVACUATED TUBE, liquid and air types

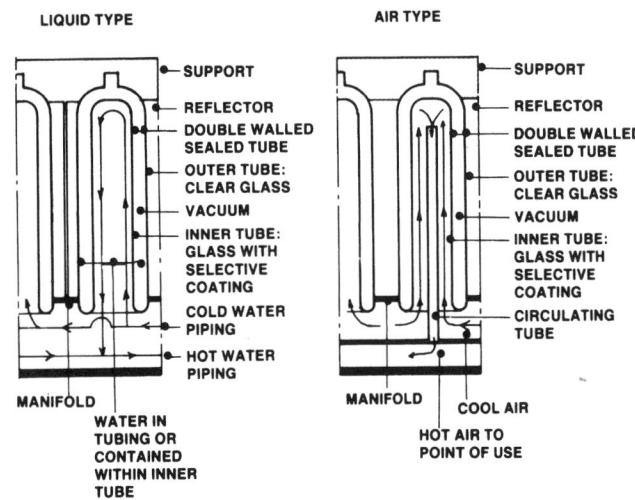

LIQUID TYPE

SUPPORT
REFLECTOR
DOUBLE WALLED
SEALED TUBE
OUTER TUBE:
CLEAR GLASS
VACUUM
INNER TUBE:
GLASS WITH
SELECTIVE
COATING
COLD WATER
PIPING
HOT WATER
PIPING

MANIFOLD
WATER IN
TUBING OR
CONTAINED
WITHIN INNER
TUBE

AIR TYPE

SUPPORT
REFLECTOR
DOUBLE WALLED
SEALED TUBE
OUTER TUBE:
CLEAR GLASS
VACUUM
INNER TUBE:
GLASS WITH
SELECTIVE
COATING
CIRCULATING
TUBE

MANIFOLD   COOL AIR
HOT AIR TO
POINT OF USE

Evacuated tube collectors are proprietary systems consisting of an array of double walled vacuum tubes connected to a manifold and generally backed with either a flat or a concentrating reflector:

■ **The outer tube is** of transparent glass and acts as a cover plate for the selectively coated inner tube which constitutes the absorber. The space between the outer and inner tubes is evacuated and sealed to eliminate convective heat loss from the absorber tube.

■ **Liquid-type collectors** have a transport medium, generally water, circulating inside the absorber tube.

■ **Air-type collectors** circulate air inside the absorber tube. Supply/return is through a double duct system.

■ **Reflectors are used** to increase heat gain of the absorber by reflecting direct radiation; the shape of the absorber contributes to efficient utilization of diffuse radiation.

■ **Evacuated tube collectors are used** generally when high operating temperatures are necessary, e.g. for cooling absorption units.

D2.1 PLUMBING

D • SERVICES

DESCRIPTION

CONCENTRATING, linear

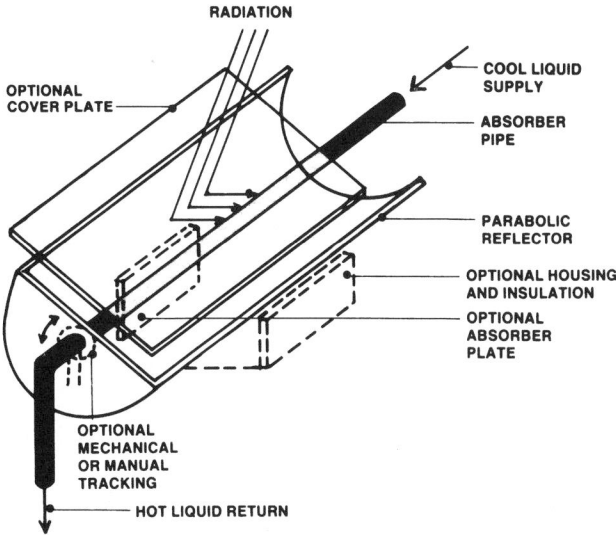

Linear concentrating collectors focus direct solar radiation on the absorber, through which the liquid transport medium is circulated:

■ **Absorber** is generally a black surfaced pipe

■ Liquid **transport medium** generally used is water or anti-freeze solution

■ Used when **high operating** temperatures are required.

■ **Linear collectors may be installed:**
• **horizontally,** with the long axis in the east-west direction;
• they may be **tilted,** with the long axis then in the north-south direction.
• to be effective, concentrating collectors **should pivot** to keep the absorber in focus:

■ **Horizontally installed** collectors have to be adjusted for changes in the altitude of the sun:
• because the change in altitude is gradual from season to season, adjustments in small installations may be made manually every few days;
• or adjustment may be automated.

■ **Tilted collectors** have to track the sun on a daily basis and automatic focusing is required.

CONCENTRATING, circular

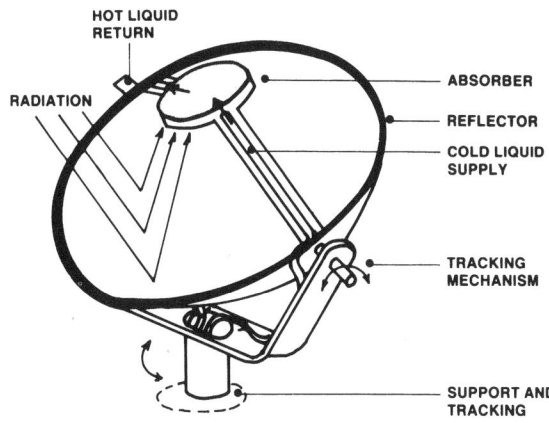

Circular concentrating collectors focus direct solar radiation on a small circular absorber through which the transport medium is circulated.

■ **Used when** very high operating temperatures are necessary.

■ Circular collectors have to be **automatically focused** to follow seasonal changes in the altitude, as well as daily changes in the compass direction of the sun.

■ **Piping containing the liquid** transport medium must accommodate two way movement of the collector.

■ **Mechanized concentrating collectors are:**
• considerably more costly than fixed flat plate collectors,
• more failure prone,
• the higher temperatures and efficiencies achieved through their use may offset their high cost.

■ All concentrating collectors with parabolic reflectors are **only efficient under direct radiation conditions;** an insignificant portion of diffuse radiation only will be caught and reflected.

D2.1 PLUMBING

D3. ❶ HEATING, VENTILATING & AIR CONDITIONING

INDOOR ENVIRONMENTS

DEFINING THERMAL COMFORT

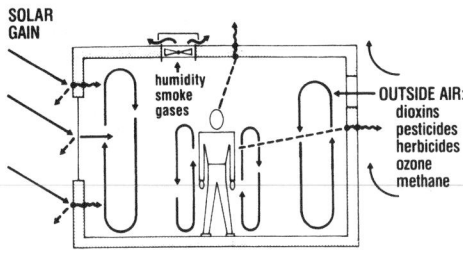

Thermal comfort is the condition of thermal neutrality when a person is not in want of warmer or cooler surroundings. The condition is considered optimum when the majority of the occupants in the space are comfortable.

■   **The success of thermal comfort design is based on defining the optimum indoor air temperature based on the comfort variables as perceived by the individual:**
- physical proximity and posture of occupant(s) within the environment;
- level of activity - sedentary, medium and high;
- amount of clothing - nude, light, medium and heavy;
- air velocity;
- relative humidity.

TEMPERATURE

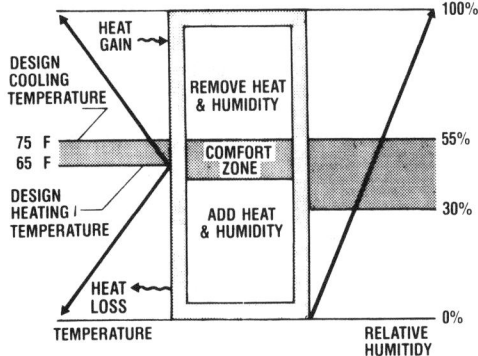

■   Three measures of temperature are used to determine thermal comfort.
- **Dry-bulb temperature:** the measure of sensible heat in ambient air, corrected to eliminate radiation. **Sensible heat** is the aspect of heat which is quantified by instruments in terms of temperature and touch.
- **Wet-bulb temperature:** the measure of air temperature with moisture; used to measure humidity at a given dry-bulb temperature.
- **Mean radiant temperature (MRT):** the weighted average of the floor, wall and ceiling temperatures used to relate to bodily comfort.

$$\text{MRT of an enclosure (°F)} = \frac{\text{sum of the individual surfaces at their temperatures surrounding the occupant (ft}^2)(°F)}{\text{sum of areas (ft}^2)}$$

■   **An ideal design condition would be to set the optimal comfort temperature equal to the MRT. Yet there are several aspects of the MRT which preclude it from use as more than a rough estimate.**
- Theoretically, if all surfaces defining the enclosure emit a uniform temperature of X°, then the MRT will equal X°. Yet, most often the surfaces will vary considerably across their surface, often creating zones of hot and cold.
- Likewise, if all surfaces reflect heat completely and the surface temperature is uniformly X°, radiated body heat of X + 1° will be reflected by the surfaces. This will cause the room to feel warmer although the MRT is the same.
- The calculation of the MRT does not consider body posture and location within the environment. Thus, the thermal comfort of a person located near a warm or heated surface may be greater than one located some distance from that or any other heated surface. Also, the thermal comfort of a reclined or seated person will be different than that of a person who is standing. For example, if a space has an eight-foot ceiling height, a radiant floor and an optimal comfort temperature of 80°F, the MRT will be reduced by increasing the ceiling height (and thus the wall height) to ten feet, but the comfort level may remain the same.
- When the MRT in the space is greater than surface body temperature, the space radiates heat to the individual and the environment may be considered too hot.
- **Heat balance of room surfaces** is attained when room surfaces are each in radiant exchange with all other surfaces, allowing convective heat exchange with the ambient air.

HUMIDITY

■   **Relative humidity is the amount of water vapor present in the ambient air as compared to the maximum amount possible at that ambient temperature.** The effects of condensation and evaporation due to humidity level will affect thermal comfort:
- Low levels of relative humidity are advantageous for thermal comfort and energy efficiency but may cause discomfort due to increased static electricity.
- For sedentary activity in low velocity air, lower relative humidity and medium clothing results in a lower optimum design temperature than light clothing and higher relative humidity.
- When relative humidity is high, a rapid change of temperature (entering or leaving a warm room when outside temperatures are cold) does not feel as extreme due to the effects of condensation or evaporation.
- **50 percent relative humidity is typically recommended for comfort and safety, as well as for preservation of artifacts (museums).**

INDOOR AIR QUALITY

■   **One of the most critical issues surrounding the conditioning of indoor environments is that of air quality.** The quality of indoor air is being implicated as a cause of persistent and drastic health problems.

● The issue of air quality is compounded by energy conservation strategies which minimize air infiltration, reduce outside air ventilation, and recirculate indoor air.

● Indoor air contaminants include formaldehyde, asbestos fibers, products of combustion, radon gas, microorganisms, and volatile organic compounds.

● Acceptable indoor air quality is defined by ASHRAE (Standard 62-1981, "Ventilation for Acceptable Air Quality") as **"air in which there are no known contaminants at harmful concentrations and with which a substantial majority of the people exposed do not express dissatisfaction."**

■   **Three basic responses to indoor air contaminants are:**

● contain or eliminate the source;

● filter out the contaminant from the air supply;

● dilute the contaminant to non-objectional levels or remove the contaminant altogether using appropriate ventilation or local exhaust.

■   **Ventilation requirements** can be determined on the basis of activity in the space or the occupant load. The basic recommended ventilation rate for acceptable indoor air quality is 15 cfm per person or as defined by local code.

● Where smoking is permitted, fresh or recycled air should be supplied at rates between 15 and 35 cubic feet per minute, and no less than 20 cfm per person.

● Offices, corridors, enclosed truck docks and textile mills generally require complete air changes every 5 to 15 minutes. Where occupants are resting or only moderately active (offices, lounges, etc.), ventilation can be as low as 5 cfm per person.

● Warehouse requirements for air changes vary greatly (every 2 minutes to every 30 minutes), depending on their contents.

MAINTAINING COMFORT

■   **The climate is defined in terms of annual average measurements:**

● Annual heating degree-days (below 65°F) define the number of °F variation between indoor design temperature (usually 65°F) and the average outdoor air temperature for one day; expressed on an annual basis. This is used for determining the heating load.

● Annual mean daily solar radiation sets the solar contribution.

● Annual wet-bulb degree-hours (below 54°F WB and 68°F DB) is used to measure relative humidity.

● Annual dry-bulb degree-hours (greater than 78°F) and annual wet-bulb degree-hours (greater than 66°F) are primary measures for cooling. The difference between outdoor and indoor wet-bulb temperatures will determine cooling and dehumidifying loads.

■   **The climate will determine the heating/cooling requirements.**

● In cold climates where the average daily temperature for the year is 50°F or less (5500 or more heating degree days), heating and ventilation are the top energy consumers, using more than half of the total energy load for the building.

● This also tends to hold true for buildings in mild climates.

● Cooling demand consumes 1/4 to 1/3 of the energy used in buildings which are located in warm climates. When the yearly average daily temperature is about 60°F or more (less than 2500 hdd), the top energy consumer is likely to be either cooling or lighting.

● Due to **solar gain** from glazed areas of southern, eastern and/or western exposures, automatic or manual controls are necessary to keep the space from overheating. Effective integration of such controls will substitute solar gain for energy consumption by heating systems.

■   **Internal factors which contribute heat must be considered.**

● **Lighting levels** play a large role in contributing to internal heat gain. With all else equal, the reduction of lighting levels may have a nearly proportional effect on cooling demand.

● Heat gain from **people and machinery** must be factored into design calculations. Many business machines contribute significant heat during the course of a day - each computer terminal may contribute as much heat as its operator.

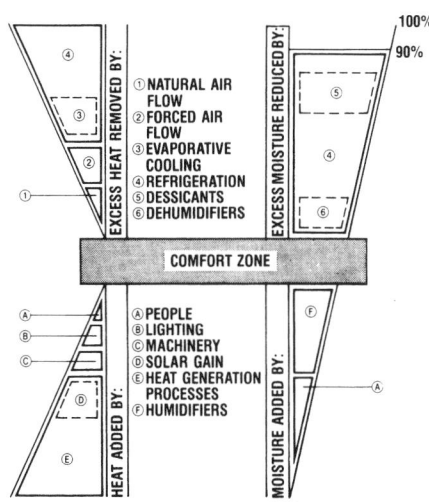

EXCESS HEAT REMOVED BY:
① NATURAL AIR FLOW
② FORCED AIR FLOW
③ EVAPORATIVE COOLING
④ REFRIGERATION
⑤ DESSICANTS
⑥ DEHUMIDIFIERS

EXCESS MOISTURE REDUCED BY:

COMFORT ZONE

HEAT ADDED BY:
Ⓐ PEOPLE
Ⓑ LIGHTING
Ⓒ MACHINERY
Ⓓ SOLAR GAIN
Ⓔ HEAT GENERATION PROCESSES
Ⓕ HUMIDIFIERS

MOISTURE ADDED BY:

100%
90%
0%

D3.1 HVAC

HEATING/COOLING: systems

HEATING

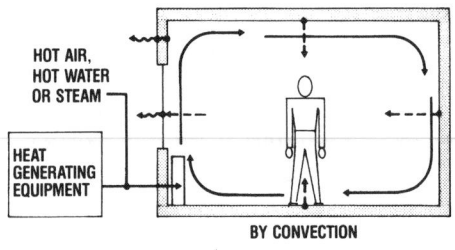

HOT AIR,
HOT WATER
OR STEAM

HEAT
GENERATING
EQUIPMENT

BY CONVECTION

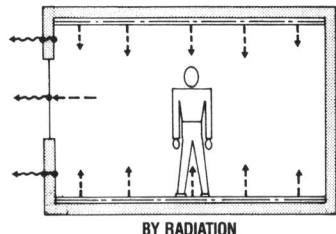

BY RADIATION

■   **Heat** is the energy transferred from one mass to another to produce an equilibrium in temperature between the two masses. The nature of heat is to pass from the warmer to the cooler mass at a rate which is directly proportional to the difference in original temperatures of those masses (the Second Law of Thermodynamics).

● Heat transfer is expressed in British thermal units (Btu), kilowatt hours (kWh), or watt hours (Wh). Heat flow is the rate of energy transfer per unit time, and it is expressed in Jules per second (J/s), watts (W), or Btus per hour. Heat flux is heat flow through a given area, expressed as J/s per m², watts/ft² or Btu/h per ft².

■   Space heating is accomplished by a combination of heat transfer mechanisms of conduction, convection and radiation, raising the temperature of the ambient air and the surfaces defining the space. The common components to any heating system are:

● **the entrance of the energy source** - fossil fuels, biofuels, solar, electric;
● **the conversion of the energy to heat** - combustion, absorption of radiation, electric resistance;
● **the transfer of heat to a thermal medium** - solid, liquid, gas; usually air, water or steam;
● **the distribution of the medium** - network of duct, pipe, or direct exposure (including electrical wire);
● **the transfer of heat into the space** - radiation, conduction, convection.

■   **Radiation is the emission of energy in wavelengths in the spectrum of infrared light or the release of stored heat.** Heat intensity on the receiving surface is inversely proportional to distance from the radiating mass. The heat may be:

● generated within the mass (e.g.; electrically);
● the result of a phase change (gas to liquid, liquid to solid); or
● the release of previously stored heat.

■   **Convective heating is the upward, cyclical/gross movement of a gas or liquid as it is heated due to contact with a surface which is at a higher temperature; may be mechanically forced or natural.**

● Natural convection is the result of the circulation movement developed when density varies due to temperature.

COOLING

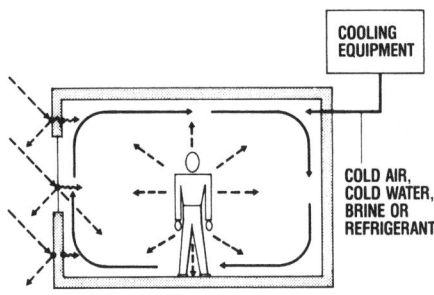

COOLING
EQUIPMENT

COLD AIR,
COLD WATER,
BRINE OR
REFRIGERANT

Cooling of a space may be accomplished by promoting natural air movement, by evaporating water into the supply air stream, by refrigeration, or by circulating cold water.

■   Thermal comfort can be attained by airflow within the environment even while temperature and relative humidity remain constant; evaporation of skin moisture cools the skin surface.

● Temperatures up to 90°F may be comfortable at 50 percent relative humidity or less; at 85 percent relative humidity, up to 85°F.
● Body movements during levels of high-activity or air movement from ventilation give a sense of cooling, reducing the optimum design temperature.

■   Evaporating water into an airstream will reduce the dry-bulb temperature but increase the relative humidity.

● Evaporative cooling is effective when dry-bulb temperatures range from 80° to 105°F, and when wet-bulb temperatures are at about 70°F or less.

■   Refrigeration comfort cooling systems operate in cycles which may include components common to heating systems in order to extract heat from the air supply to the space.

● Energy is introduced to the cycle to pressurize the refrigerant, changing it from a low-pressure vapor to a high-pressure vapor and raising its temperature.
● The refrigerant condenses from a vapor to a liquid as it passes through a stream of cooler outside/exhaust air.
● The refrigerant is then passed through a warmer thermal medium, absorbing heat as it evaporates from a liquid to a vapor again until equilibrium is achieved.
● The cooled thermal medium is then distributed to the space.
● Refrigeration systems may be self-contained and located in-space or incorporated into central distribution networks.

D3.1 HVAC

AIR HANDLING: systems

Air handling systems are of the recirculating or once-through type are used to move, remove, purify, humidify or dehumidify. The architecture of the space may be configured to encourage gross air movement or mechanical equipment may be used.

NATURAL VENTILATION

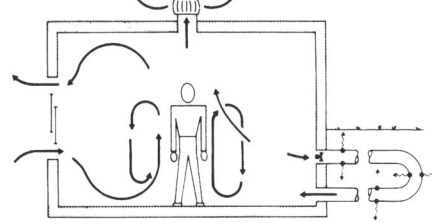

■   **Natural ventilation techniques optimize a chimney effect to move or remove indoor air.**
● Design/construction techniques may include thermal stacks/chimneys, double- or venting-skin outer walls, etc.
● Airflow can be initiated by opening a lower window (for entry of cool air) and a high window (exhausting hot air). Even opening one window 1 1/2 inches at top and bottom can satisfy the needs a small room.
● The use of wind turbines on rooftops can draw hot air from attic spaces or may be ducted directly to the living space when dampers are used.
● Solar chimneys and earth cool tubes (buried tubes used to bring cooler outside air to lower levels/basements) enhance the vertical airflow.

FORCED VENTILATION

■   Ventilation techniques, using pressurized air flow, can supply controllable amounts of fresh air and promote cooling comfort.
● Variable-speed or inlet vane controls are recommended to regulate static pressure.
● **Economizer cooling** can supply cooler outdoor air through distributing ductwork to remove hot, stale air from interior spaces without operating cooling equipment. Enthalpy controls can be integrated to improve humidity levels.
● **Low-speed, ceiling-mounted paddle fans** will provide an effective 10°F difference between air temperature and perceived comfort temperature at low humidity levels by moving air at a rate of 200 cfm. While faster air flow will increase the cooling effect, operation is not as efficient (increased power and declining benefit margin). Four-paddle units are generally quieter and more efficient.
● Rather than pushing air down, **whole area (house) fans** draw a high volume of air from open windows and doors through the space and exhaust it through an exterior wall or into the attic. This technique offers zonal cooling as windows are closed in unoccupied rooms and opened only 3 to 6 inches in occupied rooms to provide maximum air velocity. Fans should be sized to change the indoor air 30 times per hour. When exhausted into attics, roof vents must be provided to prevent recycling.

MECHANICAL EXHAUST

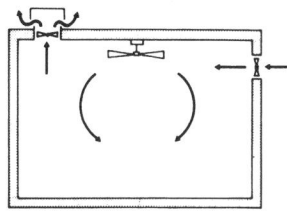

■   Exhausting indoor air creates negative pressures within the thermal envelope which need to be balanced by fresh air supply.
● **The rule-of-thumb sets exhaust air flow at about 90 percent of the supply rate so as to maintain a positive building pressure.**

■   Provide local exhaust for areas of high emissions of contaminants (e.g.; smoking areas, parking areas or spray-paint booths) or of excessive moisture and humidity.
● Air-tight construction often warrants installation of exhaust equipment to remove humidity such as moisture production in bathrooms, kitchens and laundry areas.
● **Where the enclosed space contains a homogenous ambient air quality,** the equipment selection should be on the basis of several small ventilators as opposed to using one or two large units. This promotes the uniform draw of air from the area, minimizing both potential dead spots and drafts; air from areas of heavier contamination should not be drawn across neighboring workplace.
● **Where the enclosed space contains a variety of activities and emissions of contaminants into the air,** a mixture of large and small units should be used. The size of the unit is in proportion to the activity in its zone of influence.

FILTRATION

■   Air filtration can be accomplished by forcing air through filters, using equipment such as heat exchangers and dilution control packages, or by installing electronic air cleaners.
● Chemical and mechanical filters should be employed to remove contaminants from outside air as well as recirculated indoor air.
● Air washers can be used for humidification as well as filtration.
● Pharmaceutical, electronic assembly, and other "clean room" applications require very high degrees of filtration, capturing micron size particles from the air system.

D3.1 HVAC

D • SERVICES

HEATING/COOLING: distribution systems

TYPES, USES

■ Heating/cooling systems are described by the distribution of the thermal medium (air, water or steam) from the point of energy conversion to the conditioned space:
• **In-space (direct exposure)** radiant or convective systems for single zones.
• **Centrally located** energy-conversion equipment and network distribution systems for large, single or multiple zones in a single enclosure.
• **Remotely located** conversion equipment and network distribution systems for multiple buildings, or campus environments.

■ Distribution systems connect the central plant with terminal equipment and vary according to the heat conveying medium used. Each mode contributes its idiosyncrasies to building integration issues, has particular heat loss/gain problems, and requires specific pressures to satisfy movement throughout the network. The use of appropriate terminal equipment will facilitate the conditioning of the space by adding or removing heat energy and moisture.

AIR SYSTEMS

■ Air systems are designed for low, medium or high pressure operation.
• Low pressure is used for single zone, multizone and reheat/recool systems.
• Medium and high pressure are used for induction systems, single duct constant volume systems and dual duct constant volume systems with mixing boxes (single or double motor).
• Air is heated and cooled at a fairly rapid rate, making it appropriate for applications where the space is only intermittently occupied and room temperature needs to be quickly raised.

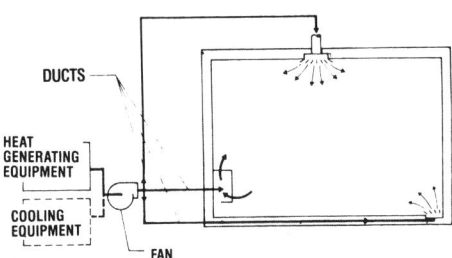

DUCTS

HEAT GENERATING EQUIPMENT

COOLING EQUIPMENT

FAN

■ **Single-duct, single-zone systems** supply air at a one set temperature (eg: 110°F winter; 60° summer) to the system, and may cover an entire building or one portion thereof. They often include air-intake filters, supply fans, heating and cooling coils and can integrate with other features such as return-air/spill-air systems, humidifiers, evaporative coolers, and various control techniques.
• Used in about 90 percent of U.S. one and two family dwellings.
• **Perimeter loop systems** are most appropriate for slabs-on-grade.
• **Radial duct systems** often use efficient, round ducts.
• **Extended plenum systems** rely on a large rectangular trunk to supply air to round or rectangular branch lines; easily adapted to frame construction.

■ **Variable-air-volume (VAV) systems** use a single duct system to supply heated or cooled air at a constant temperature to VAV boxes.
• Thermostat-controlled VAV boxes regulate the quantity of air entering each zone.
• Fan air volumes must be adjusted to compensate for changing system air quantity demands which cause pressure differentials; fans run only at full volume when all boxes are open.
• Operates at lower airflow and lower power consumption than comparable constant volume, variable temperature systems.

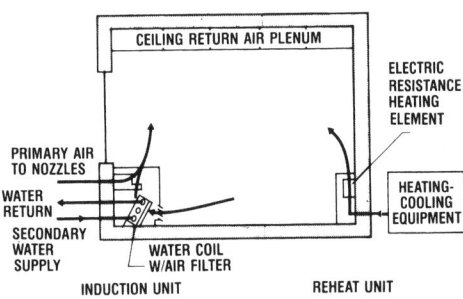

CEILING RETURN AIR PLENUM

ELECTRIC RESISTANCE HEATING ELEMENT

PRIMARY AIR TO NOZZLES

WATER RETURN

SECONDARY WATER SUPPLY

WATER COIL W/AIR FILTER

HEATING-COOLING EQUIPMENT

INDUCTION UNIT   REHEAT UNIT

■ **Terminal reheat systems** supply conditioned fresh/recirculated air through a single duct system; each terminal contains its own heating coil to heat air to the desired temperature. Central systems must cool all air to temperature and humidity requirements of the zone at its greatest load.

■ **Dual-duct systems** have two ducts supply air; one heated supply and one cooled.
• The ducts are joined at mixing boxes, then continue to the zone through one duct.

■ **Multizone systems** adjust temperature by mixing the supply of air when two supply ducts are used (one with a heating coil and one with a cooling coil) using thermostat-controlled dampers. When one supply duct is used, the duct is split just before reaching the terminal; heating and cooling coils are set into respective branches. Dampers then control incoming air temperature.

■ **Induction systems** supply conditioned air under high pressure through a single duct to the induction terminal, which is typically located on outside walls.
• Discharges conditioned air through nozzles in the induction terminal such that room air is induced back into the unit at a rate approximately four times greater than that of the primary air. The induction unit includes a water coil for conditioning of the induced room air.
• Able to handle great flux in heating/cooling loads, typical of perimeter zones.

D3.1 HVAC

D • SERVICES

WATER SYSTEMS

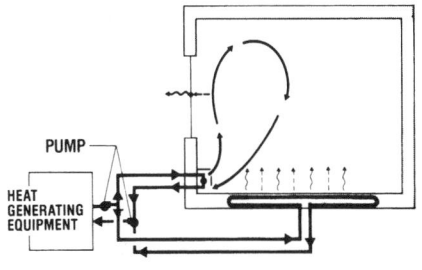

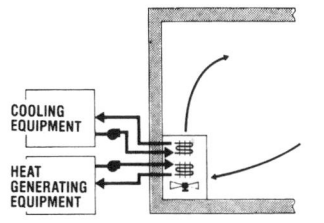

STEAM SYSTEMS

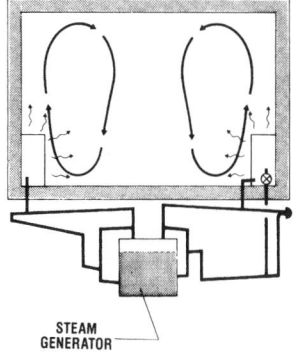

■ Water is an excellent heat transfer medium which provides the opportunity for superb temperature control. Hydronic heating systems rely on fuel combustion or electric equipment to heat water which is pumped through piping to terminal devices.
● **Hot water systems are very successful where the space is continuously occupied over several hours.** Advantages include compactness, efficiency and quietness.
● Water may be used to convey cooling only where temperatures are above its freezing point (32°F). However, the freezing point can be lowered by adding a solution of salt or antifreeze. Water-based solutions and other aqueous solutions of inorganic salts are referred to as brines. Commonly used brines are calcium chloride, ethylene glycol, and chlorinated or fluorinated hydrocarbons.

■ **Hydronic systems may be single zone, operating on a single series loop or several loops tied to supply and return main lines.**
● **One-pipe, single-zone systems:** Terminals have supply and return lines teed off the main with venturi fittings, or all terminals can be connected in series.
● **Two-pipe, single-zone systems:** Water of a constant temperature is brought to each terminal through the supply pipe. Circulated water exits the terminal, returning to the boiler through the return pipe.
● **Three-pipe, single-zone systems:** Two pipes provide separate, continuous supplies of heated and chilled water. The supplies are mixed before entering the terminal and return through a common return pipe.
● **Four-pipe, single-zone systems:** Two pipes are used to supply heated and chilled water. One supply is allowed to flow through the terminal while the other is diverted. Two pipes are provided for return.

■ **Multiple zones** are an adaptation to single-zone systems to allow control to a specific area within the system. These smaller areas, usually defined by occupancy or exposure, may have thermostat control to a zone valve. Larger areas may have thermostat control to a separate circulating pump.

■ Steam is an expansive gas produced by the vaporization of water when heated under pressure. It is not to be confused with visible condensation clouds which occur when steam is released to air of lower temperature. Steam is classified as saturated (invisible vapor at an increased temperature due to its pressure), wet (vapor at the saturation temperature and pressure containing minute water particles), dry (saturated vapor containing no moisture), superheated (at a temperature higher than at saturation) and highly superheated. They are distinguished from hot water systems by considerably higher temperatures and pressures as well as the need for specific handling of hot condensate. Steam systems may be used for power or heat and are described according to circulation pressures.
● **Low-pressure** systems supply steam at less than 15 psig; maintained by **vacuum** systems (using vacuum pumps) or **vapor** systems (rely on the pressure differential created during condensation in the heat exchanger to move the condensate).
● **High-pressure** systems operate above 15 psig.

■ Condensation occurs when steam is cooled. Steam traps are used to separate the steam from the condensate in the piping system. These thermo-mechanical devices will keep energy-rich steam in the system until it has released its latent heat of vaporization, condensing to a liquid to be returned to the boiler. **The best efficiencies are gained by returning the condensate to the boiler at the same temperature as when it condensed.**
● **Gravity return** systems rely on the weight of cooler condensate to drop while lighter steam rises. Air vents are required and must be maintained for safe operation. Terminal units must be located above the boiler water line.
● **Mechanical return** systems use pumps to dispose of, or return the condensate.
● Simple disposal of condensate should be avoided since this increases the amount of make-up water required, increasing boiler load, water treatment, etc.

■ **Piping arrangement and circuit are used to classify steam heating systems.**
● Arrangements include **one-pipe** systems (steam and condensate run in the same line) and **two-pipe** systems (provides a supply and return line).
● **Return lines can be wet** (condensate return is located below the boiler water level) **or dry** (located above water level).
● **The circuit may be divided** (multiple supply risers), **one-pipe** (single riser), **or loop** (allows placement of terminals below boiler water level).

D3.1 HVAC

HEATING/COOLING: distribution systems (cont.)

Steam Systems, continued

■ **Pressurized return systems introduce steam behind the condensate to force it back to the boiler, thus returning condensate at or near its original temperature.**

- These systems substantially reduce make-up water requirements and will reduce boiler load (Rule of Thumb: For every 11°F increase in condensate temperature, a 1 percent savings in fuel is realized).
- The components of a pressurized system include a receiver (tank for collection of condensate from the pipe runs), a three-way steam valve, a pumping chamber, and controls. Maintenance is minimal.
- Condensate is allowed to flow (due to gravity) from the receiver to the pumping chamber. When the chamber (a second collection tank) is filled, the controls signal the valves to isolate the chamber from the condensate line and open the steam valve. The condensate is forcibly discharged to the boiler. Up to 75 percent of the condensate can be recovered.
- Capacity ranges from 2,500 to 50,000 pounds of condensate per hour.
- A steam pressure of at least 15 psig is required.

DISTRICT HEATING

■ Named for the extent of distribution, district heating systems are used to supply heat to a group of buildings or campus.

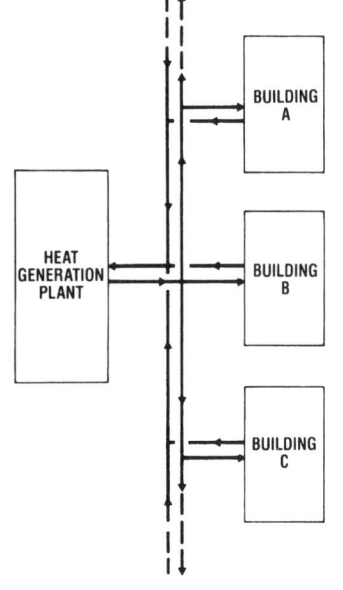

- Used in place of individual boilers, district steam allows the use of condensate return for other needs - heat make-up air, domestic hot water, space heating.
- While steam is often used, hot water is becoming a popular medium for new installations. Hot water travels farther and at lower temperatures than steam, allowing the use of smaller pipes of less expensive material.

■ **Twin pipelines, buried at a depth of 2 1/2 feet or more, provide for supply and return of steam or hot water.** Piping can range from 2 to 28 inches inside diameter.

- Foamed plastic insulation may be used for thermal protection while a high density plastic shell is used for structural purposes. The insulation layer can contain a series of wires which, by reduced wire resistance, can indicate water leakage.

TERMINAL DEVICES

Air Systems

■ In designing air distribution systems, the basic considerations for thermal comfort are air velocity, air flow and perception of draft (greater than 50 feet per minute). Therefore, the terminal device in air systems is a critical element of the overall design.

■ The basic terminal device for air systems is the diffuser. It is a register that has a damper integrated with it for the purposes of delivering balanced air flow.

- Diffusers can be linear, circular or rectangular.
- They can be integrated into ceilings, walls and floors.
- For open areas, four-way diffusers are commonly used to blow air in all directions.
- Three-way diffusers are commonly used against walls.
- Two-way diffusers are good for corner installations.

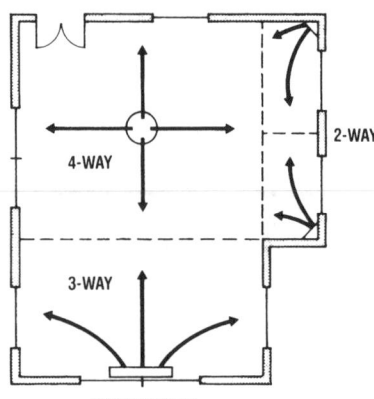

AIR DIFFUSERS

■ The diffuser is always selected so that the throw of air from the diffuser covers 75 percent of its affected zone.

- The zone may be a small, confined space or it could be only a segment of a large open area. In the latter case, the zone will be defined by lines drawn half-way between the diffuser and neighboring diffusers.

Water and Steam Systems

■ **Thermal transfer is accomplished using radiant-effect terminals, or radiators.** Types and styles vary from the old stand-up radiators to the very common fin-tube perimeter baseboard convectors to increasingly popular Eurostyle radiators.

- **Baseboard radiators** are typically convective units which operate best when they are left exposed to the room. Yet, because they often take up so many linear feet of wall space, the radiator is usually hidden behind furniture thus substantially reducing its efficiency.

■ Piping is usually copper or steel. Plastic piping can be advantageous in snow melting and radiant floor systems.

- Piping of 1 inch diameter or less is common for small scale application.
- Often, commercial buildings will require 1 1/2 or 2 inch main lines which feed smaller branch circuits. Very large buildings may have 3 inch to 12 inch diameter mains.
- It will usually be advantageous to use a single water piping for both heating and cooling.

D • SERVICES

D3.1 HVAC

HEATING/COOLING SYSTEMS: selection checklist

THERMAL COMFORT

■ **The thermoregulatory system of the human body maintains constant internal heat production in response to a given environment and thermal changes that take place within it.**

- Biological heat gain is created by metabolic rate and the level/efficiency of action.
- Cooling of the body occurs by evaporation of secreted sweat, by convective and conductive heat exchange from skin through clothing to the ambient air, and by respiration.
- For the purposes of determining the optimum design temperature, it should be noted that the temperature of the human body, measured at the surface of lightweight clothing for a person at rest, is usually about 80°F.
- Thermal comfort design is **not** significantly influenced by such factors as age, sex, body build, ethnicity, geography within climate zones, food intake, or biological cycles/rhythms.

■ **The optimum design temperature is also a function of the amount of clothing the occupant(s) can be expected to wear.**

- Conductive thermal resistance of clothing is based on the thickness of the textile layer as opposed to porosity, fiber type, etc. This involves the convection and radiation processes which occur between the skin and the outer surface of the clothing.
- Convective resistance of clothing is based on ratios of the body surface which is clothed to that which is not clothed relative to the difference between the temperature of the clothing and the ambient air.

■ **Atmosphere is usually not considered a significant factor in determining the optimum design temperature.**

- Atmospheric pressure is 29.921 inches or 760 millimeters of mercury (Hg) at sea level, which is the equivalent of 14.7 pounds per square inch.
- At sea level, the daily meteorological pressure changes only marginally affect the total air pressure in the environment.
- As elevation increases, atmospheric pressure decreases at a rate of 0.5 psi per 1000 feet of altitude above sea level; convective heat loss is also reduced.

- At altitudes significantly below sea level (or equivalent pressures), pressure is higher than normal; convective heat loss is increased; greater levels of humidity can be expected.
- Both extremes in altitude have the same effect of slightly raising optimal comfort temperatures.

■ **Thermal conditioning systems will be required to control relative humidity to ensure environmental comfort, health and safety.**

- Nearly 1000 Btus of heat are needed to vaporize one pound of water, so mechanical humidification may be expensive. Consider occupancy load and the use of waste steam condensate for increasing humidity.
- 0 to 20 percent - not recommended for human comfort, but may be required for certain scientific/manufacturing environments. At this level, energy conservation and sound absorption by air is at their greatest, but side-effects like static electricity and shock are increased.
- 20 to 30 percent - can be effective for energy conservation but may be a cause of static electricity buildup and drying of hair, skin and mucous membranes.
- 30 to 45 percent - marginal static electricity; suitable for allergy treatment. While raising relative humidity to 30 percent will reduce undesirable side-effects, it will also double energy consumption.
- 50 percent - suitable for infirmaries and other spaces where the spread of germs and organisms need to be controlled.
- 60 percent and greater - not recommended except for particular specifications.

■ High humidity is not only uncomfortable, but can affect building materials:

- Growth of mildew and mold on surfaces;
- Weakening of plaster and drywall;
- Cracking and bubbling of paint on absorptive materials;
- Condensation on structural members resulting in corrosion or rot.

HEATING/COOLING LOADS

■ **The heating/cooling load is a function of the building design, occupancy and construction relative to site and climate.**

- Basic **heat loss calculations** identify

the impact of construction materials and their details on thermal conditioning.

- **Window construction, percentage of glazing and solar orientation** are all critical to defining heat loss and gain.
- **Wind** will add structural loading as well as influence air infiltration and exfiltration. Building size and configuration can be integrated with the site to break the effect of the wind.

■ **The design heating load** is typically the basis for sizing heating systems. It is intended to define the Btus of heat needed 97.5 percent of the time during the three months of winter.

- Oversizing is often used to improve recovery time following a period of time when room temperatures are allowed to drop, such as night setback.
- A 40 percent oversizing to cover 10°F setback is recommended by some industry associations.
- Multiple equipment installations allow staging of equipment operation to match incremental imposed heating loads. The total capacity of the equipment sufficient to handle peak design loads, but that level of capacity is infrequently required. As conditions call for additional heating, additional equipment is operated at its optimum efficiency.

■ **The design cooling load is based on the standard design temperature for indoor air during the cooling season (75°F).** Equipment is typically sized to satisfy the indoor design temperature for 97.5 percent of the cooling season.

- Based on historical climate data, the 2.5 percent outdoor design temperature defines the dry-bulb temperature which includes 97.5 percent of the temperatures at which cooling is required.
- Residential applications can be estimated by rules-of-thumb based on levels of insulation and air infiltration. For tight buildings, use one ton of cooling per 1,000 square feet; one ton per 500 square feet for loose buildings.
- **As with heating equipment, oversizing should be minimized.** Oversizing uses more energy at lower efficiencies, while requiring higher first costs.

D3.1 HVAC

D • SERVICES

HEATING/COOLING SYSTEMS:

■ **Back-up systems** are employed to ensure that the design conditions are satisfied. They may take many forms:
● Standard heating equipment, such as furnaces, may be used to back-up other systems which are more economical to operate, but less reliable (e.g., heat pumps are less efficient at very low temperatures; solid fuel furnaces require more frequent attention).
● Duplicate equipment, often operated in series, can serve to satisfy peak loads or provide primary service when routine maintenance disables another system.

SYSTEM EFFICIENCY

■ **The production of heat from fuel-fired combustion equipment as rated against energy consumption is referred to as the Annual Fuel Utilization Efficiency, or AFUE.**
● **Annual Fuel Utilization Efficiency (AFUE):** the ratio of heat output to total energy consumption, including fuel conversion, electricity for motors, and thermal losses due to operational cycles and distribution.
● AFUE ratings are based on test results. Tests cover steady-state efficiency (heat production relative to fuel consumption at maximum operation), heat-up and cool-down losses occurring with each operational cycle, standby losses between cycles, jacket losses and electrical consumption for motors, etc.

■ Efficiencies of operation are also affected by factors outside of the particular piece of equipment.
● **Thermal distribution losses** must be considered.
● **Traditional system sizing** by matching peak loads can result in lower efficiency. It may be possible to regain seasonal efficiencies by sizing equipment to partial load (50 to 60 percent of full, peak load) conditions and providing back-up equipment for infrequent maximum design load conditions. The partial load should be slightly higher than a weighted average heating load.
● **Preheating combustion air** can increase efficiency at a rate of about 2 percent per 100°F rise.
● **Preheating of fuels** is necessary for certain combustion systems (such as atomization of certain oil grades).

■ Three ratios are used to gauge the efficiency of cooling equipment.
● **Energy Efficiency Ratio (EER):** the ratio of cooling capacity/output (Btu/hr) to electrical input (watts) to instantaneous cooling efficiency. Equal to the COP multiplied by 3.4.
● **Seasonal Energy Efficiency Ratio (SEER):** the average seasonal performance of cooling equipment measured by the ratio of Btu output to watts input.
● **Coefficient of Performance (COP):** the ratio of energy output to energy input to indicate operational efficiency of refrigeration systems.

$$COP = \frac{\text{heat absorbed (evaporator)}}{\text{heat rejected (condenser) -}\text{heat absorbed}}$$

$$\text{or} = \frac{\text{cooling delivered}}{\text{Btu input}}$$

● High-efficiency cooling has a COP above 2.35 (EER of 8 or better).
● COPs can be increased by increasing the evaporating temperature or reducing the condensing temperature.

Air Systems

■ **Primary equipment used to condition air should be located near the center of the distribution zone.**
● For residential and small-scale commercial applications, heat supply and return air operate best when equipment is located on lower levels.
● For larger-scale, multi-story buildings, heating and cooling functions are usually separated to take advantage of gravitational forces. Intermediate mechanical areas may be required for proper distribution.
● The disadvantages of air systems include distribution inefficiencies, noise and system-induced drafts.

■ **The use of stringent ductwork specifications can substantially increase energy efficiency.** Studies show that the actual, delivered efficiency of a heating system is often 40 to 50 percent lower than the efficiency of the heating equipment.
● The temperature of conditioned air will deteriorate over distance due to heat transfer to the duct surfaces and losses through seams.

● **Duct losses should be less than 15 percent or under 40 cfm/100 square feet of duct surface area,** allowing equipment sizing at 115 percent of the required air terminal delivery.
● Although labor and material costs will rise when construction quality standards are increased, such as caulking and taping of seams and adding insulation, duct efficiencies up to 80 percent can be economically feasible. Lesser efficiencies are tolerable for ductwork located within the conditioned space.
● The rate of flow will be affected by the size, shape and length of the duct. **To reduce resistance in the duct (static pressure), use smooth ducts of short runs and minimize dampers, elbows and fittings.** This will allow reductions in fan speed and horsepower.

■ **Insulated air ducts are likely to require a vapor barrier to keep condensation out of the insulation.**
● As a rule of thumb, when the insulation is on the inside of the duct the vapor barrier should also be on the inside, exposed to conditioned air. The duct must be sized to accomodate the insulation and maintain the proper cross-section of free air delivery.
● As a rule of thumb, when the insulation is on the outside of the duct, the vapor barrier should be on the outside of the insulation.
● Ducts built out of rigid insulation should be checked for maximum air velocities. Follow manufacturer recommendations to avoid breakoff and entrainment of insulation particles in the supply air stream.

■ Various cross-sections of ductwork are available:
● **High pressure round ductwork** is the most efficient, combining low flow resistance and typically well-sealed connections.
● **Flat oval ductwork** is not recommended for exhaust or return air systems; connections are generally difficult to seal.
● **Rectangular ductwork** is the least efficient form due to difficulties in maintaining duct shape and providing well-sealed joints.
● Thin gauge sheet metal ductwork is commonly used.

D3.1 HVAC

selection checklist (cont.)

■ **Duct losses create imbalances with air flow and pressurization.**
High, excessive duct losses cause:
● Increased fan output, results in higher static pressure which compresses ducted air, raising its temperature;
● Reduced effectiveness of ceiling air plenums, often resulting in thermal and/or quality contamination of return and supply air;
● Problems of high-pressure, hot duct air passing into lower pressure, cold air ducts where dual duct systems join for mixing.

■ Return air ductwork or plenums are critical for thermal cycling.
● Return grilles should be placed in or near the ceiling since many indoor contaminants, including water vapor, are lighter than air.
● Return grilles should be kept away from kitchen equipment or similar areas where grease, soot, etc. are likely to be drawn into the system.

AIR HANDLING

■ **Ventilation can resolve indoor air quality issues when proper designs, maintenance and operation are applied.**
● Programs that maximize the amount of outside air that is mixed with recirculated indoor air should be used in conjunction with automated dampers.
● Intake vents should be located such that air movement cannot draft contaminants from exhaust vents or other sources of pollutants into the system. Airflow around buildings must be considered to keep intakes out of areas of entrained exhaust/contaminated air.
● Ensure the proper placement and operation of supply and return air fixtures for air flow throughout the space. Be certain that the air flow is not blocked, leaving stagnate pockets of air. Avoid placement of both supply and return fixtures at floor or ceiling level as this promotes short-circuiting of air flow. The short-circuit occurs when supply air is drawn into the return system and never mixes with the air in the enclosure.

■ To ensure that outside air and recirculated indoor air are safe, the air handling systems should allow for proper filtration, adjustment, and maintenance.

● Air intakes should be designed and installed to minimize the entry of rainwater, dust, debris, and contaminants which may be discharged from nearby exhaust vents.
● Dampers should be easily inspected and adjusted.
● All equipment which handles water, creates condensation or is exposed to the weather should be installed such that all unwanted water will runoff to drains. The drains should be adequately sloped and be provided with traps and vents to eliminate backflow.
● Ducts must be designed to minimize the potential of water seepage near intakes or discharges.
● The potential for condensation within the duct must be minimized. This may be a function of over-humidification.
● Water tanks should have close-fitting covers and provision for rapid drainage and refilling.

■ Fans should be selected and installed based on the maximum allowable revolutions per minute as established by the fan manufacturer.
● Fan rpm should be sufficient to satisfy air flow needs without overloading the fan motor despite any operational conditions. For this reason, fan rpm is typically set at 50 percent of anticipated loading with clean filters in place.

MECHANICAL EXHAUST

■ **Equipment sizing** for particular applications of powered ventilation will follow the same guidelines despite the variety of equipment available.

Exhaust capacity =

$$\frac{\text{volume of enclosure to be vented}}{\text{number of air changes per hour}}$$

■ Equipment size may be limited because of **noise and vibration restrictions.**
● Larger units with slower speeds can be used to satisfy air movement where excessive noise cannot be tolerated.
● Sound attenuating and vibration isolating techniques can reduce this type of "pollution."

● Roof curbs help to weatherproof and can be used to dampen noise and vibration for roof ventilators.

■ Experiments with methods of improving air quality have resulted in new information on space planning and design.
● Providing a 4 inch clear space between the floor surface and the bottom of any continuous obstacle, such as a partition, will increase the mixing of supply air by up to 50 percent.
● Mechanical systems which provide makeup air for toilet rooms should be shut down when door louvers and undercut doors (up to 1-1/2 inch) can be employed. This causes air from outer conditioned spaces to migrate into toilet areas as makeup for the exhaust system, which should be operating at about 1 cfm per square foot of area, or as required by local code.

■ **Indoor air pressure is created by an imbalance between supply and exhaust air to a space.**
● Extreme positive air pressures can force moisture into walls and ceilings, causing condensation.
● Extreme negative air pressures increase outside air infiltration and cause energy-consuming strain on exhaust equipment.
● Preferably, air pressures should tend to the positive.

■ Interior exhaust fans can be used with ductwork, however fans must be designed for greater static pressures.
● Ceiling-mounted fans can be ducted through an attic, roof vent or soffit vent. In multi-story buildings, ductwork can be routed laterally through the ceiling structure to a wall vent.
● Fans mounted on interior walls can be ducted down the wall, then laterally through the floor structure to a wall vent.
● Fans mounted on exterior walls offer direct, ductless venting but must be carefully sealed to minimize air infiltration. Backdraft dampers can be used to reduce infiltration during periods when fans are not operated.

D3.1 HVAC

ENERGY CONSERVATION: selection checklist

RADIANT FLOOR SYSTEMS

■   Radiant floors rely on a heating element below the floor surface to provide space heating. The element may be electrical wire, but is more commonly hot water piping.
● Commonly installed in concrete slabs-on-grade, hydronic radiant floors may increase construction costs by 10 to 15 percent.
● Operating costs can run 5 to 10 percent less than other heating systems. Heat sources are maintained at temperatures of 85° to 130°F.
● Underfloor, radiant heating is quiet, clean, self-regulating, and is concealed, eliminating the space intrusion of terminals. This type of heating also minimizes drafts.
● Piping should be continuous throughout the slab, with joints and connections occurring only outside the slab perimeter.
● Plastic (polybutylene or polyethylene) pipe is frequently used, but copper, steel and neoprene piping can be used effectively.
● Radiant floors are excellent applications when the occupancy activity is at floor level, such as nursery or child-play areas.
● These have the longest recovery rate of all heating systems, but the floor mass contributes to the heat flywheel effect by providing comfort long after heat input has ceased.

ENERGY CONSERVATION STRATEGIES

■   Energy management strategies must be considered during the design stages for proper determination of thermal conditioning.
● By lowering the temperature of the space during non-use hours, the heating load can be substantially reduced. **Setbacks of 5 to 20° can be recommended, but in most cases a 10° setback is optimal.**
● Although "non-use" temperature setbacks are effective, they are further enhanced by **lowered temperatures during occupied hours** (65 to 68°F, dry-bulb).
● Many building areas are unoccupied, infrequently occupied or occupied only for short duration. These areas include corridors, stairways, lobbies, storage rooms/closets, equipment/

mechanical rooms and warehouses. The design temperature should be about 55°F during cold seasons and about 80°F during warm seasons.
● The effectiveness of these strategies will vary with the length of time and number of degrees of temperature setback.

■   **Cooling loads can also be reduced by using energy conserving strategies:**
● Like night setback for heating, cooling loads can be reduced by allowing the temperature of the unoccupied space to rise to 75° to 82°F as opposed to design temperatures around 72°F.
● Considerable energy savings can also be achieved by allowing relative humidity to climb to 55 percent.

■   **Remote thermal storage** is employed to even out the energy demand on heating and cooling systems.
● Solar heat, vented heat, recovered waste-heat and/or mechanical conditioning can be stored in a thermally isolated location, provided the efficiency of the transfer medium and the ability of the storage facility is sufficient to maintain the conditioned temperature of the medium.
● The most common technique is water storage in large capacity tanks, occasionally placed underground. The water is reintroduced to the conditioning system to supplement peak system performance in order to meet peak demand.
● Thermal storage systems can allow the calculated peak demand of the heating/cooling system to be reduced as much as 40 percent. Since the cost of energy is at its highest during the peak usage hours of the day, systems which use off-peak energy can become economically advantageous. During moderate seasons, it may be possible to heat/cool the entire building using only the thermal storage that was conditioned during off-peak hours.

■   **Solar research has supported the known effects of thermal mass exposed to ambient air.**
● Thermal mass may be inorganic matter such as rocks, concrete or brick; it may be water; it may even be a chemical phase change material.
● The benefit of thermal mass exposed directly to the ambient air is that the temperature swings between hot and cold are much more gradual, maintaining the optimum design

temperature for a longer period of time without mechanical backup.
● Thermal mass is very effective for cooling (up to 100°F) when relative humidity is at or below 50 percent. Just as in the heating mode, thermal mass will absorb heat during the course of a day and release it during the night. Night ventilation can be used to remove the stored heat, or cooling equipment may be operated during off-peak hours to minimize cooling energy costs. However, for cooling, ventilation is a better choice when greater relative humidities are encountered.

PASSIVE COOLING

■   **Roof spray systems** rely on surface evaporation to keep roof temperatures down, minimizing the effect of roof solar gain on cooling load.
● These systems are appropriate in hot climates for buildings with flat roofs of large area and low insulation levels.
● Water distribution is accomplished by a pipe grid which supply from 0.01 to 0.04 gallons per square foot per hour. The grid is elevated above the roof surface.
● Plastic or copper piping can be perforated or installed with sprinklers. Valves are incorporated to establish spray zones.
● Timers and sensors are used to control essential operating factors of temperature, time and duration.
● Clean waste water is the optimum source for these systems.

■   **Roof ponds** can be used to absorb and store daytime indoor heat which is later radiated to cooler night air.
● Roof ponds require heavy construction, flat roofs and movable insulation (for daytime protection from solar gain).
● Drainage, waterproofing, flashing and protection against freezing are common problem areas.

■   **Earth tubes** may be used for cooling by acting as a geothermal heat exchanger.
● Appropriate for short cooling seasons, buried air ducts rely on cooler ground temperatures to moderate outside air.
● These systems tend to have high first costs.
● Mold may be a concern in humid conditions and climates.

D3.1 HVAC

INTEGRATED DOMESTIC HOT WATER

■ The use of higher efficiency heat exchangers in condensing and pulse combustion burners has opened the way for hot water heating.
● Liquid-cooled furnace heat exchangers divert the coolant to heat external water tanks.
● Tankless water heaters combined with hot water storage tanks save the potential flue losses of fuel-fired water heaters.
● Systems can be equipped to produce hot water at a rate of 126 gallons per hour, over twice that of standard water heaters.

■ The combination of systems affords three primary benefits:
● Common parts, such as burners, heat exchangers and vents, are consolidated.
● Waste heat from one system is typically recovered by the other.
● First costs for both equipment and installation labor are reduced.

■ **Water quality** becomes an issue in integrated systems.
● All piping within the system will need to be rated for potable water.
● Pumps will also require brass or stainless-steel fittings and pipe.
● Requirements for part-year heating may reduce annual operating efficiency.
● High mineral content may affect heat transfer surfaces.

WASTE-HEAT RECOVERY

Energy recovery is vital to the economy of thermal conditioning. Heat energy lost from various types of energy conversion equipment can range from 100° to 2000°F. It can be considered to be a valuable fuel source, reducing system fossil fuel or electrical requirements 5 percent or more. Applications will only be useful when the amount of lost energy substantially exceeds the consumption of energy by the recovery equipment.

■ Waste-heat recovery equipment is selected on the basis of three criteria:
● thermodynamic characteristics;
● potential applications for recovered heat;
● capital payback.

■ **Waste-heat is recovered by heat exchange equipment which is incorporated into the exhaust or flue system so that escaping heat can be transferred to a separate stream of cooler supply air or to a water supply.** Equipment performance can be affected by:
● corrosive gases and liquids;
● temperature variation;
● condensation;
● flow rate.

■ Air-handling systems can be used effectively for recovery of energy from conditioned air which is being exhausted:
● When removal exceeds 4000 cfm and heating degree days are over 2500;
● When removal exceeds 4000 cfm and cooling degree hours above 78°F dry-bulb are over 8000.

■ Hot water systems provide good opportunities for waste-heat recovery. Like standard boilers, **waste heat boilers** are available in firetube and watertube versions. **These boilers produce steam or hot water from flue gases which reach temperatures between 500° and 2500°F.**
● Recovery is often from turbine/engine exhaust or cooling systems.
● Applications for recovered energy include supply air tempering, space heating, and preheating of combustion air.
● Firetube units allow the hot gases to pass rapidly through small tubing which is contained within an outer shell containing water. Capacities of up to 60,000 pounds per hour of steam are practical, though large units are quite heavy.
● Watertube units require more space as capacity increases to about 200,000 pounds per hour of steam. These units use baffles within a shell to slow the movement of hot gases over tubing containing water. This process results in higher shell temperatures, often requiring considerable insulation.

■ **Primarily used for preheating boiler water, economizers can improve system efficiency by 5 to 10 percent.** Economizers transfer about 25 to 40 percent of the heat from flue gases to boiler feedwater, raising water temperatures as much as 70° to 100°F.
● Integrated into the boiler exhaust, hot flue gases pass over finned water tubes in which feedwater (212°F or greater) from a deaerator travels to the boiler.

COGENERATION

The primary purpose of a cogeneration system is to recover the heat, steam or other by-products of the generation of electricity.

■ **Cogeneration is a viable option when:**
● Substantial electrical loads are required over long periods and recovered heat can be used.
● Displaced heating and cooling equipment costs may be taken as a credit.

■ **Fuel cell energy systems are available which are 72 to 80 percent efficient at converting fuel to electricity and heat.**
● An electrochemical process involves three primary steps:
  – Fuel processing - steam produced in step two is used to convert the fuel to a hydrogen-rich one by a catalytic steam-reforming process.
  – Power generation - the processed fuel is passed over an anode to release electrons which create a direct current electrical output; hydrogen ions move through an electrolyte solution to a cathode where the returning DC current and a supply of outside air are introduced, producing steam or hot water.
  – Power conditioning - the DC current is converted by a solid-state inverter to alternating current at specified frequency and voltage levels.
● Electrical generation is 36 to 40 percent efficient.
● Usable heat recovered from the power generation stage is transferred to the fuel processing stage and to space heating systems.
● The advantages of these systems are their relatively small size, low maintenance, high reliability, quietness, and extremely low pollution contribution.

D3.1 HVAC

FUEL CHOICES: selection checklist

■ **The selection of an energy source will be weighted to consider the economics of fuel/energy supply, convenience and operational impact:**
- Supply factors are availability, dependability, convenience of use and storage, price and cleanliness.
- Combustion/equipment factors are related to operating requirements, cost, service requirements, and ease of control.

GAS

■ Natural gas is a fossil fuel containing 80 to 95 percent methane. It is distinguished from liquefied petroleum (LP) gas by the gasoline content per 1000 cubic feet (0.1 gal. = dry; 0.1 gal. = wet).
- Gas is supplied by pipe from a central utility or is stored in 5 to 1000 gallon steel storage tanks.
- Liquid gas is highly volatile, evaporating to vapor state at ordinary temperatures.
- Gas is the cleanliest burning of the fossil fuels and requires the least complicated equipment for burning.

OIL

■ Fuel oil grades range from #1 to #6, indicating the gravity, viscosity, distillation, flash point and the percentage of ash, carbon residue, water and sediment.
- #2 has the highest ash content.
- Fuel oil is safely stored in steel tanks maintained at temperatures below flash point (110°F). It is transported within the heating system in steel or iron piping.
- Low viscosity oils, such as #1 and #2, will flow easily at ambient temperatures and will be more readily atomized for efficient combustion. Higher-viscosity oils, such as #4 and #6, will require some preheating to achieve proper atomization.
- **Viscosity is a primary element of burner design and will determine what steps will be required to prepare the oil for combustion.**
- Incomplete burning of fuel oil produces soot which is deposited at heavier rates during cooler flue temperatures (such as occurs during start-up). Buildup should be removed periodically to maintain best efficiency rating.

KEROSENE

■ The burning quality is relative to the percentage of paraffins as opposed to olefins and aromatics, all of which make up kerosene.
- A 1-K+ rating for kerosene indicates that it contains at least 78 percent paraffins, which burn cleanly.
- Kerosene is typically piped to the appliance from a steel storage tank.
- It will readily evaporate at room temperature, thus supporting combustion at any surface where kerosene is exposed to air.

SOLID FUELS

■ Although wood has been a source of heat energy since fire was discovered, precise wood-heat technology and standards are not available.
- The heat output from wood will vary by wood species, moisture content, growth region, and other factors.
- Green wood should be avoided as a fuel source, largely due to moisture content which leads to creosote deposits and poor combustion efficiency.

FUEL TYPES

Types:	Conversion: Btus/unit	Typ. Conversion Efficiency
Fuel Oil		
#1: stove, kerosene	132,000 - 137,000/gal	65 - 80%
#2: residential	137,000 - 141,000/gal	
#3:	n/a	
#4: lt. industrial	143,000 - 148,000/gal	
#5: *	n/a	
#6: hy. industrial*	151,000 - 156,000/gal	
Gas		
natural	100,000/therm, or 950 - 1200/cu.ft.	75 - 85%
LP: propane	91,500 - 93,000/gal., or 2516 - 2550/cu.ft.	
butane	approx. 101,300/gal., or 3200 - 3280/cu.ft.	
Coal	11,000 - 14,400/lb.	50 - 70%
Wood (dry)	$18(10^6)$ - $36(10^6)$/cord, or approx. 8,500/lb.	10 - 50%
Electricity	3413/kW	100%
Steam	980.3/lb. (at atmospheric pressure)	

*Must be preheated to achieve complete atomization: No. 5 = 185°F; No. 6 = 210°F.

D3.1 HVAC

AIR CONTAMINATION: selection checklist

SOURCE: OUTDOOR AIR

■ **Fresh air intakes often draw outside contaminants into buildings, where they are often recycled.** These contaminants are a complex mixture of particles and gases.

● The atmosphere is currently polluted with over 75 different contaminants, ranging from base elements (arsenic, mercury, nickel, lead, etc.) to hydrogen, nitrogen and oxide compounds.

● Some are direct emissions from sources of combustion, manufacturing or other identifiable source. Other contaminants are formed through photochemical reactions caused by airborne chemical compounds in sunlight.

● The most common are carbon dioxide, man-made ozone (O_3), methane and chloroflourocarbons. Dioxins, pesticides and herbicides are also recorded to have been drawn into buildings.

SOURCE: BUILDING/CONSUMER PRODUCTS

■ **Formaldehyde gas is a potentially harmful contaminant which is brought into buildings in many forms.**

● Formaldehyde is found in a great variety of products since its qualities include low cost, good bonding characteristics, fungicidal and preservative properties.

● **The major source of formaldehyde off-gasing is urea formaldehyde (UF) adhesive used in many pressed wood products for bonding veneers, particles and fibers.** However, other sources include carpet pads and backings, ceiling tiles, cosmetics, drapery and upholstery fabrics, paper products, permanent press fabrics, and wall coverings. Tobacco products, automobiles and fuel-burning stoves emit formaldehyde as a product of combustion.

● Formaldehyde offends most occupants when it reaches mild limits (1.5 parts per million of air), potentially causing eye and respiratory irritation, headaches, dizziness, and general malaise. **The maximum acceptable formaldehyde level is considered to be 0.1 ppm,** however even this level can cause discomfort.

■ **Asbestos is a noncombustible, nonconductive mineral which is easily pulled apart into fibers.** Its use in building construction in past years is widely documented as a health hazard.

● Abatement is generally called for when fiber density is over 10 fibers (over five micrometers in length) per cubic meter.

● **Some synthetic and manmade fibers have also been shown to be potential health hazards when inhaled.** These include long, very narrow fiberglass, ceramic and mineral fibers that can potentially penetrate deeply into lung tissue.

■ **Products of combustion, including tobacco smoke, are known to contain certain harmful gases.**

● Unvented fuel-burning appliances spew out several types of contaminants into the air (including hydrogen cyanide, carbon monoxide, and nitrogen dioxide).

● Even normal human respiration results in emission of carbon dioxide.

● **Carbon dioxide** levels in buildings have been found to be as high (0.1 percent of the air) as ten times the levels outside.

● **Carbon monoxide and nitrogen oxides are the most common emission. Nitrogen dioxide,** which has been associated with respiratory problems, can increase by up to 50 percent in enclosures where products of combustion are commonly emitted.

● Tobacco smoke is known to contain several hundred gaseous compounds, as well as other solid and liquid particles.

■ **Other air contaminants known as volatile organic chemicals (VOCs) are also present in indoor air.** Formaldehyde is the most common, but VOCs include:

● styrene (insulation);
● benzene (cleaning agent);
● methylene chloride (aerosol spray paints);
● xylene isomers (paints);
● naphthalene and paradichlorobenzene (moth balls).
● Various copying equipment is known to emit methanol, ozone and trinitrofluorenone (TNF).
● Other VOCs are contributed by cleaning products and activities, home workshop and hobby products.

SOURCE: BUILDING SITE

■ **The concern over radon gas (a naturally-occuring, colorless, odorless gas which is a product of radium decay) is centered around the carcinogenic effects of radioactivity.**

● Radon is emitted by various types of rocks and percolates through the earth to the atmosphere. The four types of rock which are considered radon-bearing are uranium, granite, phosphate/phosphorous and shale.

● **Entry into an enclosure is typically around the foundation, wherever cracks/joints occur or porous materials are used.** Some stone products used in the construction of the building may emit radon.

● Generally, an acceptable exposure is to a concentration of less than about 4 pCi/L (picoCuries per liter - a unit of radiation intensity) or 0.02 WL (working level).

● The atmosphere averages a radon level of about 0.1 pCi/L.

■ **Microorganisms, as well as pollens and other particles (aeroallergens) which are carried through air supply systems by attaching to dust and water vapor are considered to be air contaminants.**

● Areas where untreated water collects (cooling towers, drain pans, evaporative condenser units, water sumps, etc.) tend to nurture bacteria, viruses, and mold spores.

● Regular cleaning of equipment surfaces, water treatments or continuous water flow will retard bacteria and algae growth.

■ **Fresh air intaks must be located away from common sources of polluted air, such as:**

● loading docks or vehicular zones where trucks and auto emissions are present.
● Exhaust hoods from the building itself.

D3.1 HVAC

BOILERS

COMPONENTS

LOW MASS BOILERS

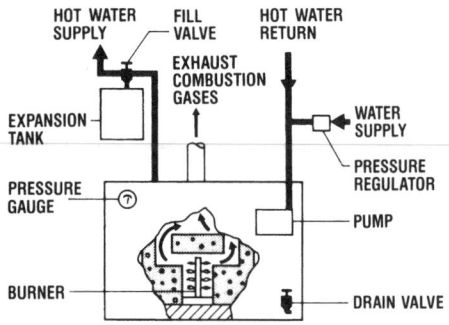

HOT WATER SUPPLY
FILL VALVE
HOT WATER RETURN
EXHAUST COMBUSTION GASES
EXPANSION TANK
WATER SUPPLY
PRESSURE REGULATOR
PRESSURE GAUGE
PUMP
BURNER
DRAIN VALVE

FIRETUBE BOILERS

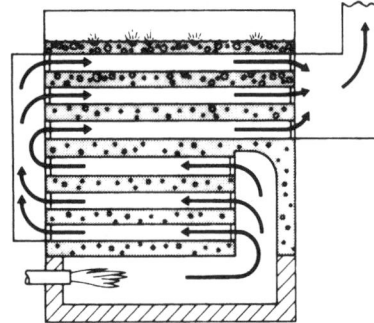

WATERTUBE BOILERS

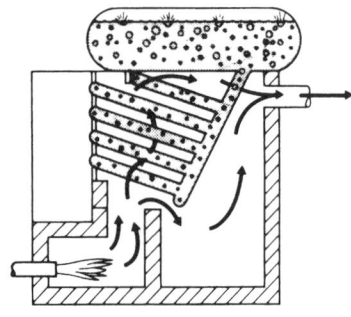

SCOTCH MARINE BOILERS

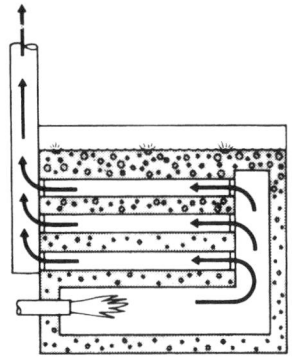

■ **Boilers are typically comprised of:**
- cast-iron or steel pressure vessel heat exchanger;
- burner or other combustion device or energy transfer means;
- fill valve; and
- controls.
- Circulating pumps and expansion tanks are used specifically for hydronic (hot water) systems.

■ **Expansion tanks are critical to any hot water system.**
- These tanks allow the expansion and contraction of water, venting the increased water volume created during the heating cycle. They also are effective in maintaining constant water pressure in the system.
- By elevating the expansion tank, the pressure in the water line is increased and the boiling point of the water rises.
- **Open** expansion tanks, used on low-pressure systems, are vented to the atmosphere.
- **Closed** expansion tanks, used on low- and high-pressure systems, are not vented.

■ **Boilers used for space heating generally fall in the classification of low pressure boilers.**
- These are constructed for maximum working pressures of 103 kPa (15 psi) steam and up to 1103 kPa (160 psi) water.
- A 30 psi operating pressure is usually specified for hot water units to compensate for water supply pressure and building height (0.434 pounds of pressure per foot above the boiler). The maximum building/system height for a 30 psi boiler is about 45 to 50 feet.
- Hot water boilers are not to exceed an operating temperature of 121°C (250°F).

■ **Low mass boilers** often come completely assembled - all piping and wiring in place - and are frequently of smaller size than other boiler types.
- These units have quick response time since the water tank may contain as little as one quart of water.
- The small water tank also allows the burner to be placed in the middle of the water storage, rather than at the bottom as in most large capacity tank systems.

■ **Firetube boilers:**
- The burner is located beneath the water storage tank, and hot combustion gases are piped through a tank of water.
- Single pass units exhaust the combustion gases after flowing through the water tank once. Two pass units force the gases through the tank twice. Three pass units have the combustion chamber located such that its heat is also transferred to the water storage.

■ **Watertube boilers:**
- Hot combustion gases flow through a chamber which contains water-filled pipes, acting as an air to water heat exchanger.
- The water storage is either remote from or above the burner.

■ **Scotch Marine boilers:**
- The combustion chamber is located within the water storage so that heat transfer takes place on all sides of the chamber.
- Combustion gases are also piped through the tank in a manner similar to fire tube boilers.

D3.1 HVAC

FURNACES

COMPONENTS

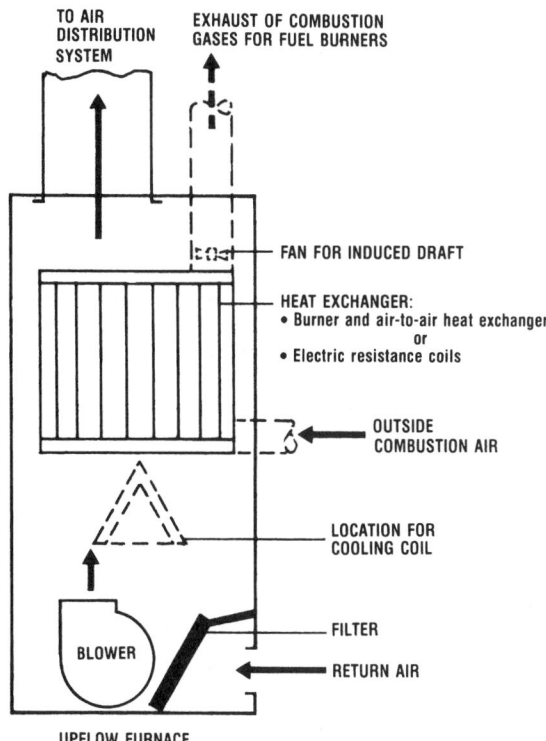

UPFLOW FURNACE

- TO AIR DISTRIBUTION SYSTEM
- EXHAUST OF COMBUSTION GASES FOR FUEL BURNERS
- FAN FOR INDUCED DRAFT
- HEAT EXCHANGER:
  - Burner and air-to-air heat exchanger or
  - Electric resistance coils
- OUTSIDE COMBUSTION AIR
- LOCATION FOR COOLING COIL
- FILTER
- RETURN AIR
- BLOWER

INSTALLATION/AIRFLOW

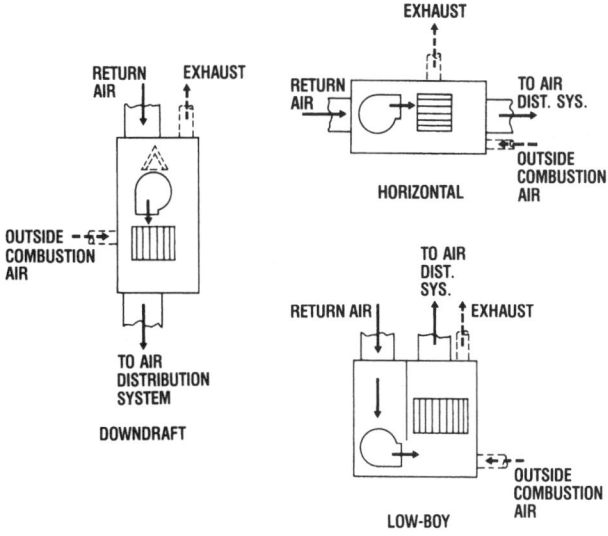

DOWNDRAFT

HORIZONTAL

LOW-BOY

■ Furnaces are appliances used to heat air. They are often classified by the method of moving air through the burner; natural (atmospheric) or powered (induced, forced or balanced). Fuel-burning furnaces are typically comprised of:

- **Burner** - standing pilot, intermittent pilot, electric/spark ignition, condensing or pulse combustion; monoport or multiport; steel, cast-iron or other corrosion-resistant materials which are compatible with the heat exchanger.
- **Heat exchanger** - steel, stainless steel, ceramic or glass coated steel; designs range from tube to baffle systems, all of which separate combustion gases from indoor air.
- **Fan/blower assembly** - usually centrifugal blowers with forward-curved blades; sized to accommodate cooling systems.
- **Air filter** - installed ahead of the heat exchanger; electric air cleaners can also be incorporated.
- **Housing** - steel, insulated or reflective-coated.
- **Controls** - ignition device, fuel valve, fan switch, limit switch.
- **Draft hood or draft diverter** - essential for natural draft systems; replaced by controls in direct vent and some forced draft systems.
- **Supply and return ductwork connectors.**
- **Flue system.**
- Many units require a sealed, outside combustion air supply.

■ Electric furnaces are similar to fuel-burning furnaces with the following exceptions:

- Modular heating elements (5 kW per module) replace the burner and are combined to determine heating capacity.
- When cooling coils are installed, they are located ahead of the blower. **The appliance is referred to as a fan-coil unit when the cooling coil is installed by the manufacturer.**
- **When the appliance is designed for outdoor installation, it is referred to as a packaged heat pump or packaged air conditioner.**

■ Equipment selection is often on the basis of installation.
- **Upflow (high-boy) -** circulating air enters below the burner/blower, moves vertically through the unit and exits the top; for basement or first floor applications.
- **Downflow (counterflow) -** circulating air enters from the top, moves vertically through the unit and exits the bottom; for second floor or perimeter duct applications.
- **Horizontal -** circulating air enters from one end, moves horizontally through the unit and exits the other end; either left- or right-hand applications in crawlspaces, attics or other areas with limited headroom; may be weatherized for outdoor installations.
- **Low-boy -** circulating air enters the top of the blower compartment, moves horizontally through the combustion chamber and exits vertically; for applications where low ceiling height restricts the use of standard upflow furnaces.
- **Gravity -** air is circulated vertically by convection across the heat exchanger.
- **Packaged units -** a horizontal furnace (gas) and an air conditioner (electric) contained in a single, weatherized housing.

D3.1 HVAC

EVAPORATIVE COOLING

COMPONENTS

■   Evaporative coolers exchange sensible heat for latent heat by evaporating water in an airstream. Components include:
● **Blower units -** generally centrifugal fans with filters.
● **Evaporative pads -** fibrous fill of resilient materials (latex-coated, glass fiber, ferrous metal, wood fiber, etc.) occasionally treated with chemicals to increase hygroscopic qualities. Pads are designed to retain sufficient moisture for evaporation.
● **Water circulation systems -** a pump supplies a measured amount of water through tubing to the moisture pads. Water overflow pans, float valves, bleedoffs and drains are included in most sump units. Water distribution is typically designed to produce a washing action to the pads. For each 10°F drop in dry-bulb temperature, estimate water usage at approximately 1.3 gallons per hour per 1000 cubic feet per minute of air.
● **Louvered housings -** permit filtering and integration with ductwork. Outside louvers should be screened.

DIRECT EVAPORATIVE COOLERS

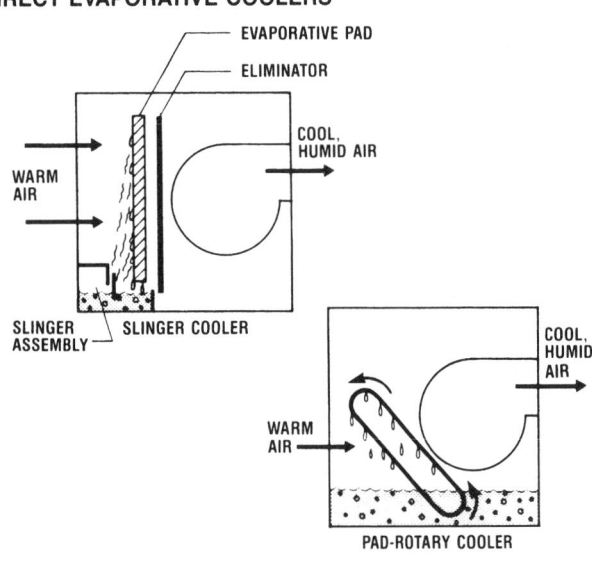

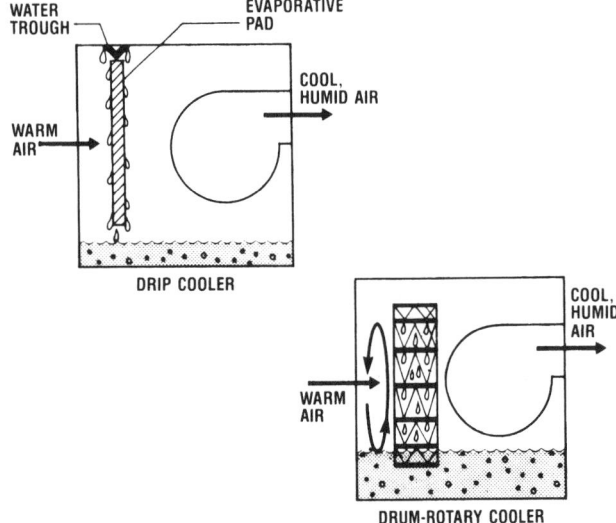

■   **Rotary coolers** are available as drum or pad types, with capacities usually between 2,000 and 12,000 cfm.
● Drum type units use axial rotors made of corrosion-resistant screening (often 2-inch thick bronze wool) are submerged up to 1/3 the fan diameter in a water tank. Air is forced through the rotor at about 300 fpm, evaporating the moisture that collected on the screen.
● Pad type units use a looped belt-type pad to move through the water storage and collect water for evaporation in the airflow section. Air flow is typically 100 to 600 fpm at 0.5 inches of water. Evaporation from the pads reduces dry-bulb temperature but raises relative humidity. This type of cooler is most appropriate for dryer climates.

■   **Spray, or slinger coolers** use a vertical disc that is partially submerged in water and spins at high speed to produce a water spray. This type of equipment is effective for cooling because it maintains relative humidity.
● A forward-curved, centrifugal fan is used to draw outdoor air through the water spray and into an evaporative filter pad followed by an entrained-moisture eliminator pad.
● The evaporative pads are usually 3/4 to 2 inches thick fiber panels, allowing air velocities of about 300 fpm.
● Cooling efficiencies of about 80 percent are available at air capacities of approximately 30,000 cfm.

■   **Drip coolers** pump water from a sump tank to the top of the unit where it is allowed to drip back down through the pads.
● Air is drawn through wet pads of aspen wood fibers or other fibrous fill. The pads are usually mounted in removable frames of corrosion-resistant material.
● Cooling efficiencies of about 80 percent are enhanced by the effective removal of airborne particles which are 10 microns and larger.
● Capacities of 2,000 to 20,000 cfm are available with pad velocities in the range of 100 to 300 cfm at a pressure drop of 0.1 inches of water at 25 Pa.

■   **Rigid-media coolers** draw air through housing louvers which have immediately adjacent water pads.
● The pads are usually 12 inches thick, comprised of rigid, fluted or corrugated, corrosion-resistant materials.
● Water is pumped from a sump at the bottom of the unit to a slotted trough system which is located over the pads. The water flows through the pads to the sump.

D • SERVICES

D3.1 HVAC

INDIRECT EVAPORATIVE COOLERS

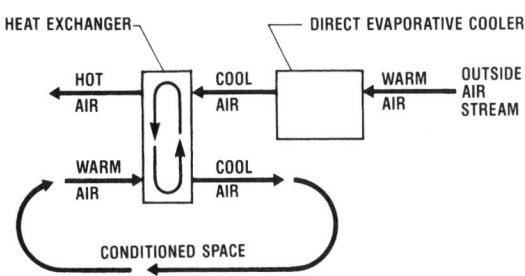

■ **Indirect evaporative coolers use heat exchangers to sensibly cool supply air as a result of direct evaporative cooling of outside air.**
● **Two airstreams are used:** Supply air, often recycled indoor air, is cooled as it passes through a heat exchanger. Outside air or exhaust air is cooled by single-stage evaporative coolers before it is passed through the heat exchanger.
● Desiccant and heat pipe technologies are used to increase cooler efficiencies.

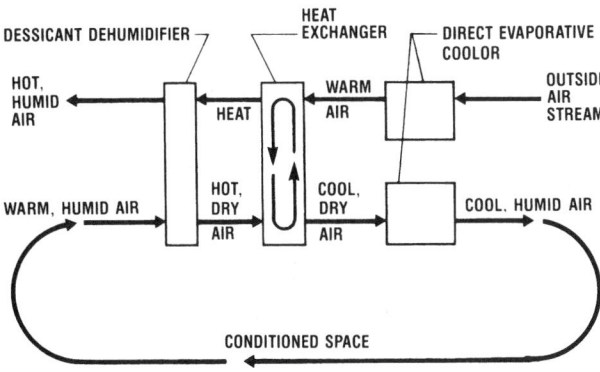

■ Desiccant dehumidifiers are being incorporated to lower the relative humidity while reclaimed heat from a heat exchanger is being used to recharge the desiccant.
● Warm, humid supply air becomes hot, dry air when passed through the dehumidifier. The hot air is cooled to warm levels by a heat exchanger, then cooled and humidified by an evaporative cooler.
● A parallel, but opposite stream of outside air is sensibly cooled by an evaporative cooler prior to receiving the heat of the indoor air via the heat exchanger. A heating element (one or any combination of steam, solar, electric, cogenerative) is set in series so as to provide hot, dry air to the dehumidifier. Thus, the dehumidifier is recharged, exhausting hot, humid air.

■ Heat pipe heat exchangers are being used in place of an evaporative cooler in order to sensibly cool outside air.
● After the heat pipe, air is routed through cooling coils to remove latent heat; moisture condenses on the coils as the air is overcooled. The cold air is passed across the other end of the heat pipe to raise its dry-bulb temperature.

■ **Packaged units contain all components in one housing.**
● They are often used as precoolers to the main cooling system. This reduces dry-bulb temperatures which improves energy efficiency and allows selection of refrigeration equipment with smaller capacities.

■ **Combination indirect/direct evaporative coolers** may be used to evaporatively cool both air streams prior to entering the heat exchanger.

COOLING TOWERS

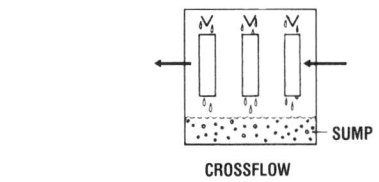

CROSSFLOW

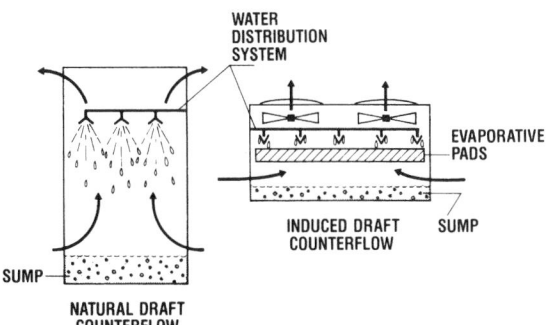

■ Cooling towers use evaporative principles to reject heat to outside air. They are classified by the direction of airflow.
● Counterflow: Air moves upward through falling water spray.
● Crossflow: Air moves horizontally through falling water spray; units tend to be smaller.
● A typically high maintenance factor is due to water handling and treatment.

■ **Cooling towers are used to lower condenser water temperature by means of direct evaporation.**
● The water is sprayed over a series of baffles and drains into a sump. The water evaporates into outside air which is drawn through the tower baffles and exhausted outside.
● The difference between the condenser water temperature and the wet-bulb temperature of the outside air is referred to as the approach temperature. Cooling towers are usually sized for a 10°F approach temperature. Efficiencies of operation rely on full flow of water and air through the tower; any reduction in flow increases the approach temperature.

D3.1 HVAC

REFRIGERATIVE COOLING

COMPONENTS

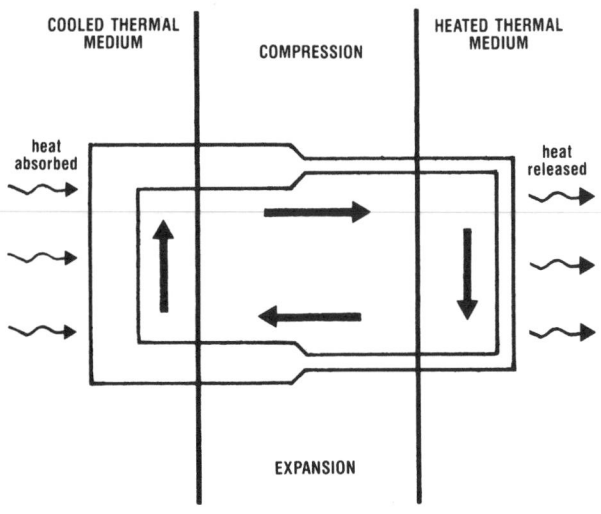

DIRECT EXPANSION SYSTEMS

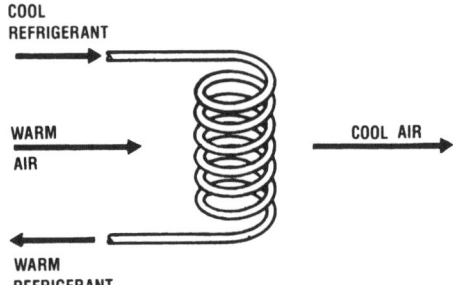

ABSORPTION SYSTEMS

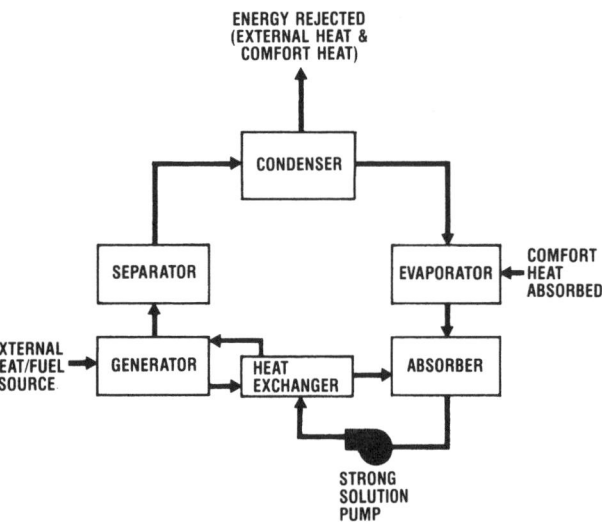

■ **Refrigerants** are liquid compounds which are sequentially evaporated, condensed or absorbed, then compressed or concentrated in solution to produce a continuous cooling effect. Each compound is designated by a number for easy reference. They are chosen for economy, applicability and according to chemical, physical and thermodynamic properties. Many types of refrigerants are available:
● **Inorganic compounds** - air, water, ammonia, carbon dioxide and sulfur dioxide.
● **Halocarbon compounds** - chlorine, fluorine and/or bromine compounds (i.e., freon).
● **Hydrocarbons** - methane, ethane, propane.
● **Azeotropes** - mixtures of two substances such that new, unique sets of properties are created and the substances cannot be separated by distillation.

■ Mechanical refrigeration system components include:
● **Compressors** - raise refrigerant pressure and, therefore, temperature (warm vapor becomes hot gas);
● **Condensers** - transfer refrigerant heat to another thermal medium to foster phase change (gas to liquid);
● **Metering devices** - restrict flow of refrigerant to evaporator coil;
● **Evaporators** - exchange the heat of air or water to the refrigerant at low circulating pressure and temperature to vaporize the refrigerant and facilitate its return to the suction side of the compressor.

■ **Also referred to as gas compression refrigeration, DX systems are mechanical refrigeration systems which include an expansion device.**
● The expansion device is located between the condenser and the evaporator. It allows the liquid refrigerant which is under high pressure to expand into a larger volume. This reduces the pressure and the boiling temperature of the refrigerant.
● Air is cooled directly by passing it through an evaporator, producing cooling and dehumidification.
● Ammonia was the refrigerant used for larger scale industrial systems, while freon is now generally used for small and large scale comfort cooling systems.

■ Absorption systems use a refrigeration generator in place of a compressor to operate a complex heating and cooling cycle:
● Water with lithium bromide (absorbent) is used as the refrigerant. The absorbent allows the water to boil at a lower temperature.
● **Generators** use steam or directly applied burners (gas or oil fired) to heat and evaporate the refrigerant, which carries the absorbent with it.
● A **separating chamber** is used to condense the absorbent into liquid form while the water remains in a vapor state.
● Water vapor is cooled to a liquid in the **condenser** and drains into a **cooling coil** that is, in effect, an evaporator. Due to its high vacuum (low absolute pressure), the coil lowers the boiling temperature of the water refrigerant, cooling and dehumidifying air that passes over it.
● The water vapor from the evaporator is piped to an **absorber**, where the absorbent is again combined with the water. Since the affinity of the absorbent to water is stronger at lower temperatures, the absorber typically employs a method of lowering the water temperature. Cooling coils are effective for this.
● The refrigerant then returns to repeat the cycle.

CHILLERS

Liquid chilling systems use the basic components of any refrigeration system to produce cold water.

■ Cooling can be delivered via cold water when the indoor air is passed over evaporative coils and blown back out into the space.
● Water temperature is between 35° and 55°F on delivery and 50° to 65°F after absorbing heat from the indoor air.

■ Chillers, like boilers, can be piped in series or parallel circuits to add cooling and dehumidification to water or steam heating systems for year-round conditioning. These units use refrigeration to cool comfort control air directly (direct expansion) or indirectly using circulated chilled water.
● Electric chiller units are commonly used in the commercial cooling market with COPs in the range of 5.4 to 6.4.
● Shell-and-tube water chillers contain the low temperature refrigerant within the insulated shell and chilled water is pumped through the tubes within that shell.

■ **Reciprocating chillers:**
● A good choice for systems with capacities up to 200 tons (700 kW).
● These units are identified by the use of reciprocating compressors.
● COPs are usually around 4.5.

■ **Centrifugal chiller packages:**
● Available with capacities ranging from 80 to 2400 tons (280 to 8400 kW).
● These units are identified by the use of centrifugal compressors.

■ **Screw chillers:**
● Use screw-type compressors to achieve capacities from 120 to 800 tons (420 to 2800 kW).

■ **Absorptive chillers:**
● Generally use a gas-fired generator/absorber, shell-and-tube condensers and one or more evaporators.
● COPs are usually 0.60 to 1.0, but will improve with lower condensing temperatures.
● Low-pressure steam systems can operate off of normally idle heating boilers. Maximum pressures are around 12 psig, but common operation is about 5 psig.

■ Double bundle condensers on chiller packages permit the use of rejected heat energy to be reclaimed.

D3.1 HVAC

HEAT PUMPS

■ **Heat pumps use vapor compression (refrigeration) to transfer heat energy from low temperature sources (air or water) to higher temperature air (or water) streams.** Air-to-air heat pumps are a good selection for moderate climates but are not appropriate in very cold climates, as COP's drop as outside air temperature drops.
● Since the components of the heat pump are essentially duplicated for each air/water stream, a thermostat-controlled reversing valve is all that is needed to change from a heating mode to a cooling mode.
● Life-cycle costing often justifies the use of heat pumps for waste-heat recovery where large quantities of waste-heat are available.

UNITARY SYSTEMS

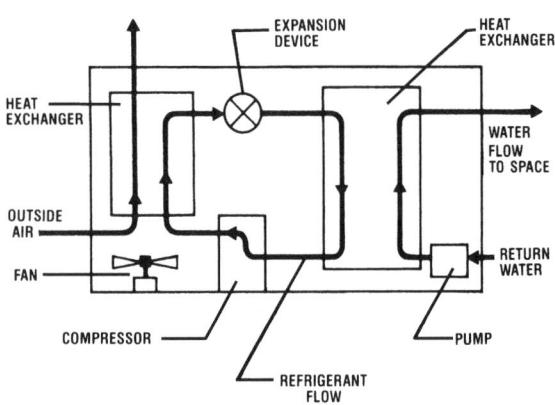

■ There are five primary components of the refrigerant cycle in the heat pump mode:
● It vaporizes as it absorbs heat at low pressure from the source when piped through a **heat exchanger or evaporator**.
● Its temperature is further increased as pressure is applied to the vapor by a **compressor**.
● Heat is transferred to another thermal medium as the refrigerant is condensed to a liquid when piped through another heat exchanger or **condenser**.
● The liquid refrigerant is reduced to its original temperature and low pressure as it passes through the **expansion valve**.

■ Often advantageous for large-scale projects (100,000 square feet and up), unitary systems can be used to transfer heat from warmer interior spaces to cooler exteriors/perimeters.
● Generally, unitary systems offer lower first costs (cost of equipment and the cost to install it).
● Equipment can be mounted in ceiling plenums (horizontally), in closets (vertically) or directly in the space along perimeter walls.
● Convenience, reliability and ease of maintenance fill out the list of advantages.

■ **Unitary heat pumps are comprised of two exchanger compartments in one housing.**
● The heat source may be outside air, recycled interior-zone air, circulating core water system, etc.
● Smaller through-wall units do not incorporate ductwork for distribution, but simply discharge the air into a single zone.

SPLIT SYSTEMS

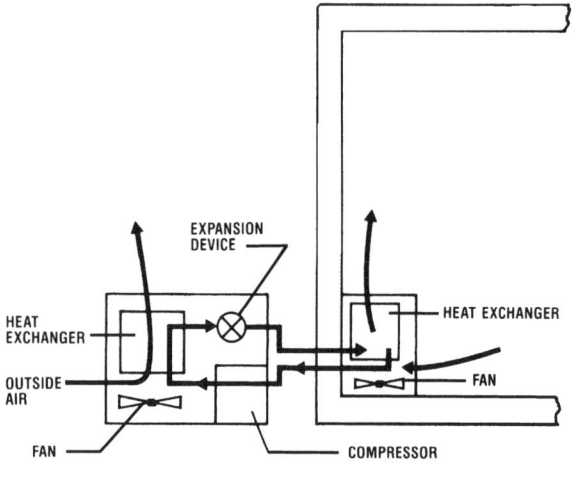

■ **Split systems operate on the same principles as unitary systems,** but the two heat exchange coils are separated.
● The compressor and a heat exchange coil are located outside the thermal envelope, while a second coil and the blower are located inside.
● The inside and outside units are connected by two insulated lines of piping which carry refrigerant back and forth for heating and cooling.

D3.1 HVAC

GEOTHERMAL SYSTEMS

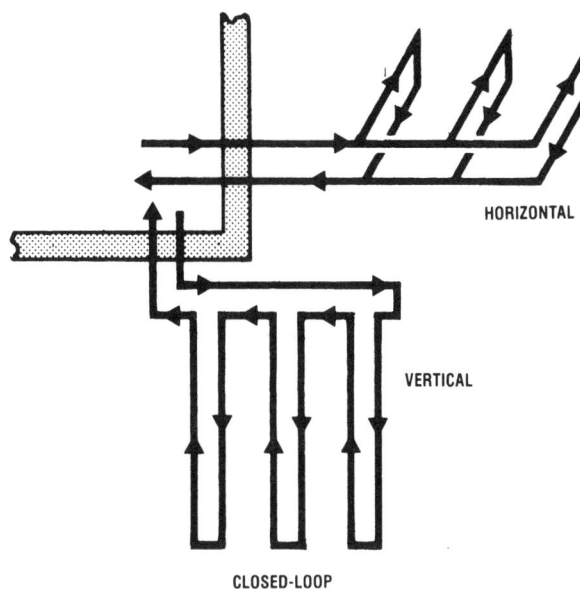

HORIZONTAL

VERTICAL

CLOSED-LOOP

■ **Small scale heat pump applications can often take advantage of geothermal sources rather than the outside air.**
• They are often referred to as water-source heat pumps.
• Any consistent source of air or water can be used effectively if held at moderate temperatures.
• Ground/well water is kept at relatively mild and constant temperatures (37° to 77°F, ranging North to South in the continental U.S.) by earth mass, and therefore is not subject to outside air temperature swings.
• Systems require water temperatures between 45° and 95°F to be effective even when air temperatures are below freezing.
• Source heat is typically delivered at lower temperatures and at a higher rate, requiring larger air-handling equipment.
• COPs range from 2.75 to 4.0 for these water source units, which rarely require defrosting.

■ **Open loop, or ground-water systems:**
• Water is drawn from well, through the system just once, and disposed into some form of water detention setup.
• Surface wells, lakes, streams or a deep recharge well can be used for disposal.
• These systems are not recommended where they can deplete or contaminate the aquifer.

■ **Closed loop, or ground-source systems:**
• A sealed, self-contained pipe loop is buried in the ground, circulating water and a corrosion inhibitor (propylene glycol or calcium chloride) to act as one of the heat exchangers.
• The circulating pump is smaller than that used in open loop systems.
• **Vertical loop systems** are suitable for heavily populated areas.
  —Sealed-casing loops, in U-shaped patterns, require 100 to 250 feet of plastic piping per ton of cooling capacity.
  —More than one run can be used in series in order to gain needed footage, but runs should be spaced at least 10 feet apart.
• **Horizontal loop systems** demand more land and are often not appropriate for densely populated areas, hilly sites, or soils which are difficult to excavate.
  —Piping is typically set four to six feet deep, with the return line placed below the line from the heat pump.
  —350 to 600 feet of piping is required per ton of cooling capacity.

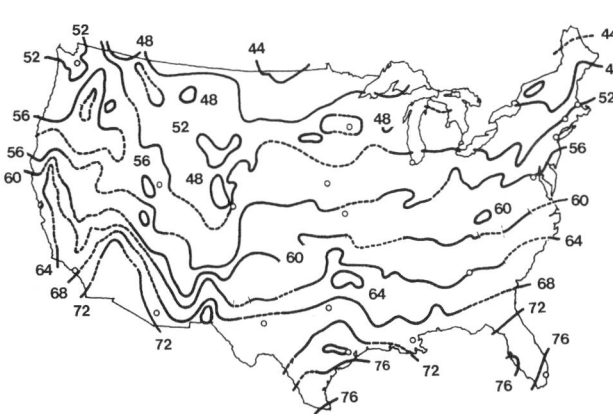

Approximate steady state of earth temperatures can be estimated from this map of shallow non-thermal well water temperature distribution (redrawn from a map prepared by the National Water Well Association).

D • SERVICES

D3.1 HVAC

IN-SPACE EQUIPMENT

DIRECT-VENTING HEATERS

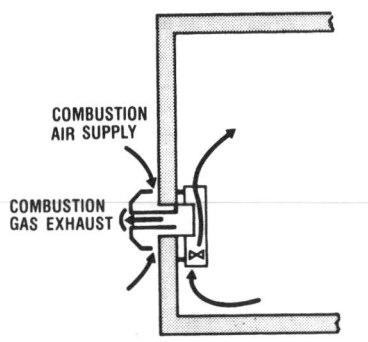

COMBUSTION
AIR SUPPLY

COMBUSTION
GAS EXHAUST

INFRARED HEATERS

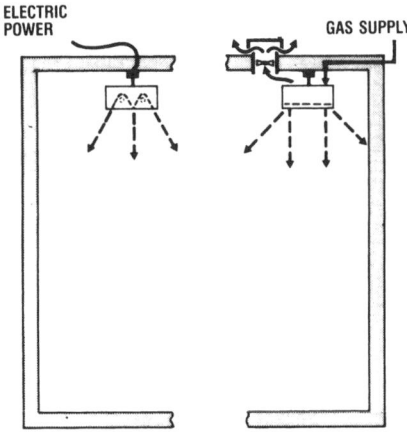

ELECTRIC
POWER   GAS SUPPLY

ELECTRIC HEATERS

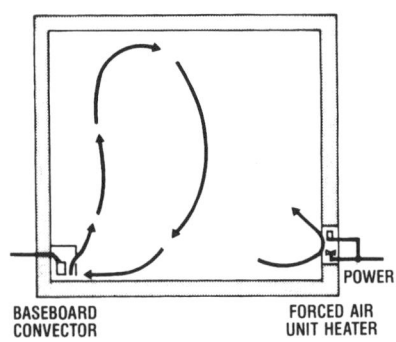

POWER

BASEBOARD   FORCED AIR
CONVECTOR   UNIT HEATER

■ **Direct-venting appliances use a double sleeve tube/flue which extends through the exterior wall to draw combustion air into the unit (outer sleeve) while exhausting gases through an inner tube.**
● Burners are sealed in the combustion chamber so that the only air flow comes from the outside.
● Some units operate with spark ignition. Others use continuously burning pilot lights.
● The heat exchanger consists of a pipe network through which the flue gases flow before being exhausted to the outside.
● A fan forces cooler room air through the heat exchanger and out into the space. Natural convection is employed in some units.
● The units are often controlled by mechanical or electronic thermostats that can be programmed for various heating schedules. They may be built-in or remote.
● Efficiencies range from 70 to 92 percent (AFUE). Output ranges from 20,000 to 50,000 Btu/hour.
● Safety features include automatic shutoff, heat sensors, indoor air quality meters, etc.

■ **Infrared heaters are available as electrically powered appliances or fuel-fired appliances.** Both are used to emit infrared radiation when convective heating is difficult to apply effectively.
● Only a small area is occupied by the unit, which may be ceiling mounted, wall mounted or portable.
● The use of reflector units allows radiation to be directed accurately to one, defined spot with little loss to adjoining spaces.
● Applications are typically in areas of large volume or high ceilings, in industrial buildings over machines or work stations, or in somewhat sheltered outdoor situations.

■ **Electric high-intensity infrared heaters** are excellent choices for beam or spot heating. Units are direct-wired or portable and are available in metal sheath, fused quartz tubing or evacuated quartz lamp types.

■ **Gas-fired infrared heaters** may be vented or unvented, and they are typically mounted near the ceiling to radiate downward.
● **Surface combustion** units burn an air-gas mixture on a metallic screen, heating it to a surface temperature of nearly 1600°F. These units are often not vented, therefore room ventilation is warranted.
● **Internally fired** units are vented and produce radiation from heat exchanger surfaces which rise in temperature to about 180°F.

■ **Electric resistance heaters generate heat from the passage of electric current through conductors.** The conductors are usually metal wires or coils installed within a system resulting in radiant heat. Resistance heaters can be effective where heating loads are small and electrical rates are low.
● The advantage of electric heaters is zone control. They are especially effective for supplemental heating or smaller, separately zoned spaces.
● Wires can be incorporated into materials so as to provide radiant surfaces.

■ **Baseboard units are considered as radiant convectors because they are designed to draw air at low velocities through the unit and into the space.**
● One or more metal element rods are integrated into a heat exchanger.
● The heat exchanger design is most often linear, but can be coiled or bent to a particular pattern.
● Efficiency is dependent on heat exchange surface; available with cast aluminum fins, concentric molded fins, baffles or vertical cells (open at top and bottom).

■ **Forced air unit heaters provide supplementary space heating from surface mounted (floor, wall, ceiling), recessed or portable units.**
● Steel housings typically enclose a heating element/exchange unit, a blower unit with filter and integrated thermostat control.
● Air supply to the blower is typically from the front or bottom, except for inverted units (top). The direction of the discharge air will be opposite or perpendicular to supply air.

ELECTRIC RADIANT HEATERS

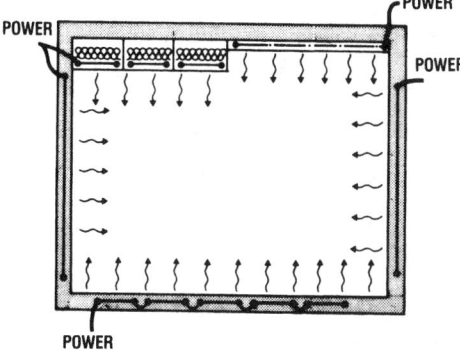

POWER

POWER

POWER

POWER

POWER

Radiant heating systems are available in a variety of types:

■ **Gypsum boards embedded with wires -** typically installed above ceiling gypsum board and between framing members.
● Applicable in or under concrete slabs; commonly finished with floor tile.
● Response time is slow, since the ceiling/floor mass must be heated before the radiant effect is realized. Therefore, these installations are best where a constant temperature needs to be maintained over a reasonable length of time.
● Surface temperatures are in the 110° to 120°F range for ceilings and walls, but only about 85°F for floor applications.
● Gypsum panels are often rated at about 25 watts per square foot and weigh 2 to 3 pounds per square foot.
● No alteration of the radiant surface by the occupant can be allowed since electrical problems may occur when holes are cut, fixtures are installed, or due to other puncturing or nailing.
● The pattern of embedded wires may become visible (mirroring) over years of use due to the reaction of gypsum to localized heat.

■ **Flexible, plastic-coated rolls/sheets** have conductors consisting of carbon graphite or metallic foil.
● Typically spread over about half of the ceiling or floor area; cable length and spacing is determined by heating requirements.
● Warm-up response time can take up to 20 minutes to reach operating temperatures of 90° to 110°F.
● Lengthy, continuous operation may cause excessive heat buildup.
● Sheets typically weigh about 2 ounces per square foot and operate at about 25 watts per square foot.

■ **Rigid panels -** heating elements backed by noncombustible, dense or rigid insulation.
● Temperatures range from 150° to 200°F with effective response to controls in about 4 minutes.
● Panel weights can vary from about one to four pounds per square foot and are rated for up to 100 watts per square foot.

■ The biggest disadvantage with radiant heating systems is the difficulty of providing even distribution. A disadvantage lies with the potential for air to become stagnate, but this can be resolved with adequate ventilation.
● Watt-density and heating output of the panels must be scaled against proximity to other surfaces to minimize "hot flashes" near radiant elements in an otherwise cool space.

COOLING APPLIANCES

■ **Self-contained unitary cooling equipment provide all of the elements of refrigeration within one, single assembly and are sized by nominal tonnage (12,000 BTU/h per ton).**
● Compressors are electric. Larger heat pumps and self-contained packaged air conditioners often use multiple compressors for capacity control.
● Window air conditioners, through-the-wall units and smaller heat pumps use reciprocating or hermetically sealed compressors.
● Sizes generally range from 2.5 to 40 tons.

■ **Self-contained direct expansion units** are made up of the minimum elements required to produce cooling. Chilled water, brine or refrigerant may be used to circulate from a remote condensing unit (condenser and compressor) to a fan-coil terminal unit. The fan-coil unit moves outdoor air and return air across the evaporator coils, supplying cooled air to the space.

■ In-space cooling units should always be mounted such that the air flow from the unit originates near the top of the space.
● Convective cooling units can use the same principles of air flow as heating units; the difference is primarily that the airflow is downward at the surface of the heat exchanger (convector) and warmer air will rise further into the space. Convective heat transfer is dependent upon the temperature difference between the convector and room air; convective cooling requires greater convector area to compensate for smaller differences in temperature.

D3.1 HVAC

BURNERS

GAS BURNERS

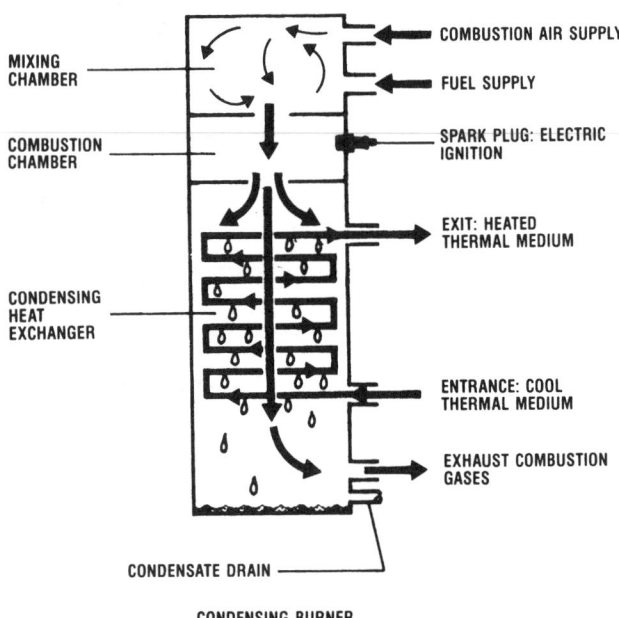

MIXING CHAMBER

COMBUSTION CHAMBER

CONDENSING HEAT EXCHANGER

COMBUSTION AIR SUPPLY

FUEL SUPPLY

SPARK PLUG: ELECTRIC IGNITION

EXIT: HEATED THERMAL MEDIUM

ENTRANCE: COOL THERMAL MEDIUM

EXHAUST COMBUSTION GASES

CONDENSATE DRAIN

CONDENSING BURNER

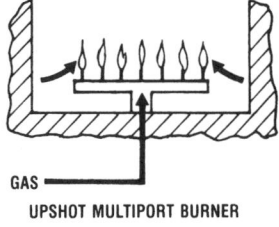

GAS

UPSHOT MULTIPORT BURNER

■ **Burners are the primary elements of all devices which convert fuel to heat energy through the process of combustion.** Complete burner and control units typically consist of fuel supply, air supply and ignition systems.

● **Air supply may be used to classify burners as atmospheric (natural draft) or power (forced/induced).**
● Power burners usually use ducted, outside air for combustion; control is achieved through fan speed and dampering.
● **Ignition systems** include standing pilot light, intermittent ignition and electric (direct spark) ignition. These become a part of more complicated burners, such as condensing and pulse combustion units.

■ **Atmospheric burners introduce primary combustion air at the throat of the gas supply to the combustion chamber.**
● The air/fuel mix is released for ignition in the secondary air of the chamber through one or more burner heads.

■ **Power burners include ring, premix, gun, condensing and pulse combustion types:**
● **Ring** - gas is introduced into a ring which allows it to enter the combustion air stream just ahead of the combustion zone.
● **Premix** - introduces air to the gas supply prior to the combustion zone.
● **Gun** - uses concentric pipes to introduce combustion air and gas simultaneously.

■ **Condensing burners gain very high operating efficiencies by recovering sensible heat from combustion exhaust as well as latent heat from vaporization.**
● Air and gas are supplied to and ignited in a sealed combustion chamber.
● The resultant hot gases are immediately exposed to a very efficient heat exchanger which is sealed within the unit.
● As heat is transferred, condensation forms on the heat exchanger and combustion gases are cooled, then exhausted.
● The condensate drops to the drain pan at the bottom of the unit and is removed.

■ **Pulse combustion burners** premix fuel and air before employing a special system of intermittent combustion to achieve greater efficiencies.
● A small blower forces air into the combustion chamber to mix with a small amount of injected gas.
● The spark plug ignites the first cycle, pressurizing combustion gases and forcing the gases through the heat exchanger.
● The positive pressure of combustion closes supply valves and opens exhaust valves.
● The release of combustion gases creates a negative pressure which reverses the valves and allows another cycle to begin.
● Retained heat provides ignition for the next cycle; cycles are perpetuated at a rate of about 60 cycles per second.

D3.1 HVAC

OIL BURNERS

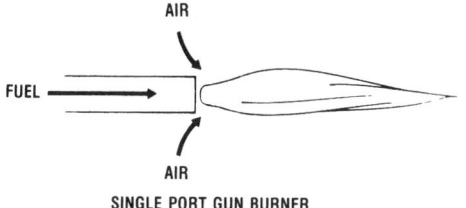

SINGLE PORT GUN BURNER

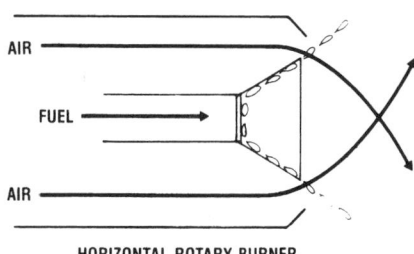

HORIZONTAL ROTARY BURNER

SOLID FUEL BURNERS

■   Oil is drawn from a storage tank through piping to a filter and pump. **It is prepared for combustion by atomization (spraying) or vaporization (heating).**

■   **Gun-type burners are used for atomization and rely on concentric tubes to deliver oil and air.** Burner designs vary by the process of mixing the oil with the air.
● **High-pressure** burners mix a forced air supply with the oil in the combustion chamber, just beyond the nozzle.
● **Low-pressure** burners mix primary air with oil at the nozzle, and secondary air from a blower is introduced just beyond the nozzle in the combustion chamber. The primary air is supplied to the nozzle in a separate channel or may be introduced in the oil supply line.

■   **Rotary type burners are used for atomization and rely on high-speed motors to spin the oil from a cup, spraying it across the combustion chamber.**
● **Horizontal** burners feed the oil directly to a relatively deep and narrow cup, which is mounted on a horizontal axis. Forced air, supplied alongside the cup, is spun in the opposite direction to the oil flow and mixes with the fuel in the chamber.
● **Vertical** burners feed the oil into a shallow cup or between rotating discs to spray the oil upward and outward across the combustion bowl (circular hearth). Air is introduced prior to and is also forced alongside the cup. The spray tends to be lateral and flat to the bowl; upon ignition, fire fills the complete bowl.
● **Vertical wall-flame** burners spin at about 1750 rpm, moving the oil through small tubes which project from the cup to an outer ring. At first, the oil is ignited along the ring by electric spark. After a few cycles, the ring temperature is high enough to vaporize the oil as it strikes the surface. Eventually, the entire wall of the burner is supporting a continuous flame.

■   **Pot-type burners heat oil to the point of vaporization.**
Oil is supplied at a low flow rate to a metal vessel which is warm enough to vaporize the oil, maintaining a pilot flame (low-fire) while the burner is not in use. On demand for heat, oil flow is increased allowing fire intensity to rise. On high, the flame commonly burns above the pot. Pot-type burners may use either forced-draft or natural draft.

■   Solid fuel, in the form of wood or coal, is converted to heat energy in a firebox.
● Coal may be pulverized, mixed with an air stream and introduced into the combustion chamber where it is ignited.
● Chain grate stokers carry a continuous supply of coal into and through the combustion chamber at a controlled rate such that it exits as ash.
● Push-feed stokers use piston mechanism to feed a supply of coal into combustion chamber. Ash falls down below the combustion bed grate and is removed.

D3.1 HVAC

HEAT EXCHANGERS

■ Heat exchangers are used transfer heat from one thermal medium/source to another with a minimal amount of contamination between them.
● They are a primary component of waste-heat recovery systems and refrigeration systems.
● Ventilation systems often use heat exchangers to remove heat from exhaust air in order to preheat supply air (recirculated or outside).
● Heat exchangers can operate using any two thermal mediums; air-to-air, air-to-liquid, liquid-to-air, liquid-to-liquid.

■ **Condensers are essentially heat exchangers which are used in refrigeration systems to cool the refrigerant to a liquid state after it has left the compressor.**
● In split systems, they are always installed in the condensing unit along with the compressors and controls. They may be cooled by air or water.

■ **Evaporators are a type of heat exchanger used in refrigeration systems to vaporize the refrigerant.**
● These heat exchanger units are essentially the same as condensers with the exception that refrigerant is heated.

■ **Air-to-air heat exchangers are used primarily for preheating fresh supply air drawn from the outside.**
● They are effective equipment selections where ventilation is important but cannot be expected to replace the specific need for localized exhaust (such as bathrooms, labs, etc.).
● **These units often consist of two isolated chambers (one for indoor airstream and one for outdoor) with blowers that are balanced to operate at the same speed.**
● **Configurations** may include sequential transfer to a slowly moving core so as to provide greater heat transfer surface area for a given air volume.
● **Core configurations** include counterflow, cross-counterflow, single-crossflow, double-crossflow, rotary and heat pipe.

SHELL AND TUBE

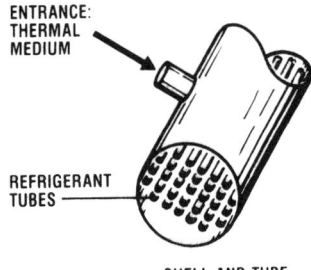

ENTRANCE: THERMAL MEDIUM

REFRIGERANT TUBES

SHELL AND TUBE

■ **Liquid-to-air/gas heat exchangers:**
● **Shell and tube** types can be horizontal or vertical.
  —A cold liquid is pumped through thin, copper tubing (often with integrated fins) which is enclosed in an airtight steel shell.
  —A warmer air/gas medium enters the shell, condensing on the tubing and dropping to a receiver.
● **Fin-tube** types use alternating positioned rows of thin wall tubes with fins on the exterior (air side).
  —Liquid or gas on the inside absorbs or rejects heat to the air flow which is usually forced through the finned tube array by a fan or blower.

■ **Flat plate heat exchangers** utilize multiple thin wall plates to separate liquid flows affording large heat transfer areas, high efficiencies, and small floor area requirements.

D3.1 HVAC

HEAT PIPES

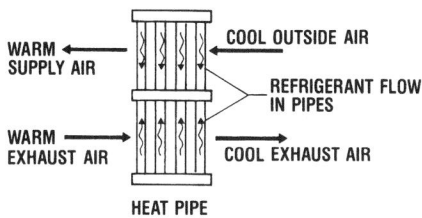

WARM
SUPPLY AIR ← / COOL OUTSIDE AIR →

REFRIGERANT FLOW
IN PIPES

WARM
EXHAUST AIR → / COOL EXHAUST AIR →

HEAT PIPE

FIXED PLATE: CROSSFLOW

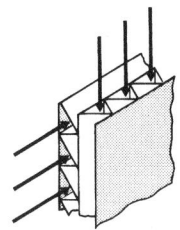

FIXED PLATE: CROSSFLOW

FIXED PLATE: COUNTERFLOW

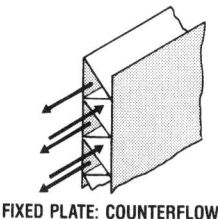

FIXED PLATE: COUNTERFLOW

ROTARY HEAT WHEELS

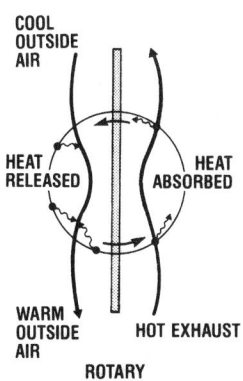

COOL
OUTSIDE
AIR

HEAT
RELEASED / HEAT
ABSORBED

WARM
OUTSIDE
AIR / HOT EXHAUST

ROTARY

■ **Liquid-to-liquid heat exchangers:**
● **Shell and coil** types circulate water both through a sealed shell and through coils set in horizontal and vertical directions in that shell.
● **Double pipe** types are assemblies comprised of a smaller inner pipe which is set into a larger outer pipe. The warmer liquid flows through the inner pipe, and the colder liquid flows through the outer pipe, or vice versa.
● **Atmospheric** types provide a washing effect as a liquid (usually water) is released over a vertical column of horizontal pipes which contain refrigerant.

■ **Heat pipes are applicable to systems with flue temperatures of 100° to 850°F.** Heat pipes are finned tubes which rely on heat energy transfer from rapidly flowing hot air across one end to vaporize refrigerant contained within the tubes.
● Vapor pressure forces the refrigerant gas to the other end of the tube which is in the midst of a separate, cooler supply air.
● Cooling the vapor causes it to condense back into a liquid which then flows toward the hot end.
● Heat can be recovered from exhaust air, flue gas, hot condensate, hot condenser water, or engine exhaust and cooling systems using heat pipes.
● Heat pipes can be used in terminal reheat systems as well as other air preheating/tempering systems.

■ **Rotary heat wheels are used to draw heat from clean/ filtered air of up to 1600°F.**
● The wheels are fluted and contain corrugated passages in the core to accept and retain hot air. They may be plastic, teflon-coated paper or honeycomb.
● The wheel, which only rotates at about one to three revolutions per minute, can be up to 50 feet in diameter and three feet thick. Smaller wheels may revolve at a faster rate (30 rpm).
● Commercial units are available with capacities of 3,000 to 60,000 cfm.

■ **Fixed plate exchangers are about 70 percent efficient operating in temperatures up to 1600°F.**
● Heat from exhaust air is conducted through the plate to a surface which is exposed to cooler air supply.
● Units may be mounted on a wall or ceiling, or placed in a basement, crawlspace, utility room or attic.

■ **Tubular (shell-and-tube) exchangers separate flue gases from supply air by setting tubes within a tube, promoting heat transfer by conduction.** One air supply flows through the small tubes, while the other air supply flows around those tubes.
● Efficiencies are typically below 50 percent and operation is usually in flue gas temperatures between 500° and 2,800°F.
● Like heat pumps, tubular exchangers can be useful for liquid to liquid, steam to liquid, and gas to liquid applications.

D3.1 HVAC

CIRCULATING EQUIPMENT

COMPRESSORS/PUMPS

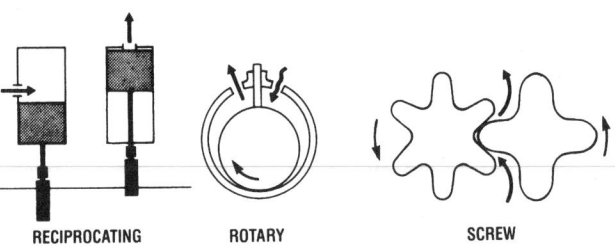

RECIPROCATING   ROTARY   SCREW

■ **Compressors may reduce the volume of the refrigerant to increase its pressure (positive displacement) or may create suction to increase static pressure (nonpositive displacement, or centrifugal compressors).** They are rated by capacity (in watts or Btu/hr) and performance factor.

● Positive displacement compressors may be reciprocating, rotary or helical rotary (screw) types. The compressor action is produced by mechanical force, usually an electric motor. Gas or diesel engines can also be used to power these units, as can steam turbines.

● Centrifugal pumps are usually used to circulate the heating/cooling medium from the central plant equipment to the terminal units.

● **Compressors may be open (driven by a separate, remote motor), hermetic (completely sealed by welding) or** semi-hermetic (sealed by bolting). Hermetic types usually incorporate an electric motor.

FANS/BLOWERS

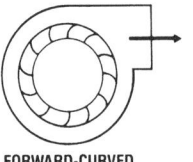 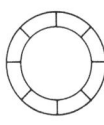 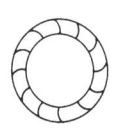 

FORWARD-CURVED   RADIAL   BACKWARD-CURVED   AIRFOIL

Fans are typically of either centrifugal or axial type and may be powered using direct-drive or belt-drive. They are usually quietest when operated at lower speeds and air volumes.

■ **Centrifugal fans are available in four types; all typically use scroll-type housings and large blades on a small hub.**
● **Airfoils** are used for general ventilation in clean air.
● **Backward-inclined/curved fans** are used for general ventilation and are appropriate in corrosive/erosion environments.
● **Radial fans** are used mainly for high-pressure industrial applications such as material handling.
● **Forward-curved fans** are used primarily as the blower in heating and cooling equipment.

■ **Axial fans are of three types and use a variety of housing types.** These fans typically have small blades on a large hub; they are placed within cylindrical mountings/housings with tight tolerances through which air is drawn.
● **Tubeaxial fans** are used for exhaust and ventilation where ductwork is designed for low and medium pressure.
● **Vaneaxial fans** incorporate guide vanes in a cylindrical housing to correct the rotary motion of the airflow to improve pressure and efficiency. These fans are used for nearly all ventilation and exhaust applications.
● **Propeller fans** are used at or near free delivery for low-pressure, high-volume ventilation. Common types of propeller fans are ceiling/paddle fans and portable fans, including window and floor units. Propeller fans usually have large blades in comparison to other axial types.

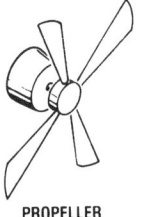  

PROPELLER   TUBEAXIAL   VANEAXIAL

VENTILATORS

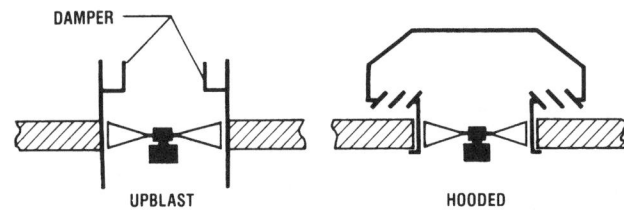

DAMPER

UPBLAST   HOODED

■ Roof ventilators may be powered or wind-driven, vane-type. They may be used for exhaust of hot or contaminated indoor air.
● They can also be used for supply, recirculation or mixture through the use of reversible fan motors and integration with certain duct and vent configurations.

■ **Powered roof ventilators are defined by the fan within them as either axial or centrifugal.**
● Axial fans are most often propeller types used to achieve high volume flow rates at low pressures.
● Centrifugal fans are airfoil or backward-inclined/curved types used without ductwork. These units typically have low first costs and low operating costs.
● Powered ventilators may be upblast (open-top cylindrical housing) or hooded.

AIR-HANDLERS

FILTERS & CLEANERS

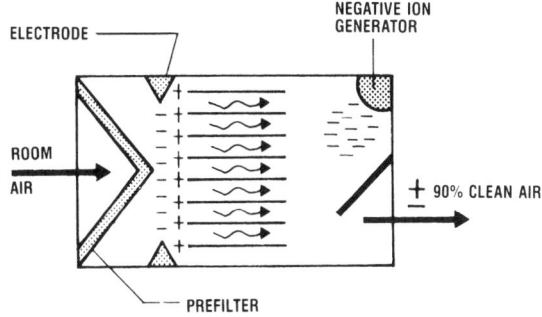

ELECTROSTATIC PRECIPITATOR

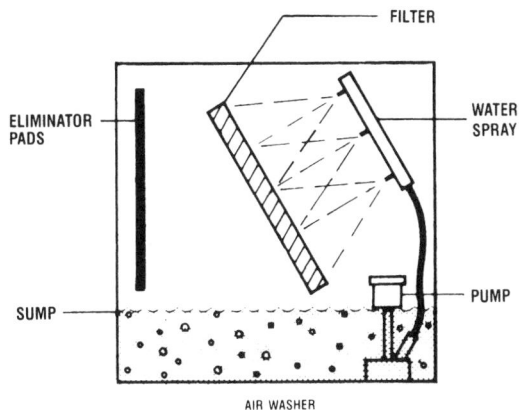

AIR WASHER

HUMIDIFIERS

DEHUMIDIFIERS

■ **Air filters are available in many forms:**
- Absorptive filters remove gases by chemical reaction.
- Activated carbon filters also remove gases and fumes.
- Foam and fiber filters remove large particles, such as lint, hair, dust and pollen.
- High-efficiency particulate air filters (HEPA) consist of a dense fiberglass web and remove smaller particles such as dust, pollen, spores, bacteria and smoke.

■ **Air cleaners** may be inexpensive fan/filter units (electric fan and filter mounted in a plastic housing) or electric units.
- **Electric air cleaners** commonly incorporate charcoal or fiber filters as well as a prefilter, an electrostatic precipitator, and a negative ion emitter.
- **Electrostatic precipitators** filter particles from indoor air by using electrodes and electrostatic cells to create an electro-magnetic bond. The electrodes charge the particles while the cells place an opposite charge on plates within the cleaner. The plates can be easily taken out for periodic cleaning.
- **Negative ion generators** emit negatively charged ions into the ambient air for the purpose of adhering to airborne particles. The particles will then stick to nearby surfaces. The generators can be effective but they typically require that surfaces be cleaned and maintained more frequently. They have been shown to be effective at removing smoke from ambient air.

■ **Air washers** use a low-velocity air flow through water spray or cell-type washer to remove dust and dirt particles from the air. These units, not effective for soot and smoke removal, have similar components and operation as evaporative coolers.

■ Humidifiers are available for integration in ductwork or as self-contained portable and room appliances. Basic types are:
- **Pan:** Heating coils (electric, low-pressure steam or hot water) maintain water temperature at 200°F or above. Vaporized water is blown out of the unit by a fan which is controlled by a humidistat.
- **Wetted element:** Many configurations of a fan and an evaporator pad are used to increase relative humidity. Units may be either bypass or duct mounted types.
- **Spray:** Spinning disc or cone units with optional filter/ eliminator can be used for humidifying or dehumidifying. Atomizing humidifiers use compressed air to produce a water mist.
- **Sonic:** High frequency sound is used to break up water into very fine particles which quickly evaporate in the airflow.

■ Dehumidifiers reduce relative humidity for purposes of thermal comfort, condensation control, or dehydration (for preservation of materials; control in manufacturing processes). They may be installed in air supply systems, may be added to heating/cooling systems to enhance performance, or may be used alone (portable units).
- Regenerative types use desiccant materials to remove moisture from the air, raising the dry-bulb temperature. Application of heat recharges the desiccant, releasing the moisture into an exhaust air stream.
- Absorption types use sorbent materials to remove moisture from the air by chemical reaction.
- Adsorption types use solid sorbent materials to hold water vapor which comes in contact with its surface.
- Refrigeration types use the principles of refrigeration, condensing humidity on cooling coils.

D3.1 HVAC

HEATING/COOLING CONTROLS

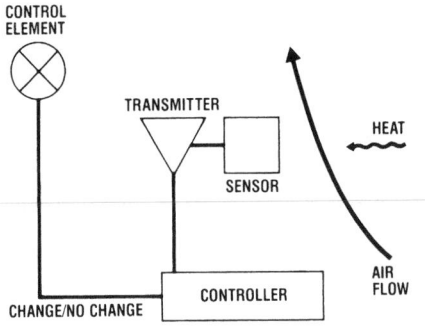

■ All control systems measure the variable being controlled using a sensor and transmitter. A controller makes a decision as to whether the measurement is at the desired value (set point) or if change is required. The decision is transmitted to the final control element (valve, regulator, actuator) which implements the action.

OPERATING CONTROLS

■ Control response systems operate the final control elements.
● **Mechanical systems** - motor driven or manually operated shafts and linkages.
● **Pneumatic systems** - transmit signals via changes in air pressure (typically 3 to 15 psig).
● **Electronic systems** - electric signals of 4 to 20 mA.
● **Digital (or discrete) systems** - use electrical binary (zeros and ones) signals.

■ The four primary measures of space conditioning can be sensed by instruments placed in the space. For general environmental information, one sensor station is sufficient, but where comfort is an issue several stations may be required.
● Temperature can be measured using a mercury thermometer, a thermocouple or an electric sensor. Electric sensors are extremely sensitive to radiant heat and require adaptation to the environment in the

form of a heat reflective shield, reduced sensor size or installation amidst increased air velocity to reduce possible measurement error.
● Air velocity is typically measured using a thermal anemometer.
● Relative humidity is measured by using a psychrometer, dew-point apparatus, hair hygrometer or electric hygrometer.
● The surface temperatures of walls, floors and ceilings for the calculation of mean radiant temperature is measured by using thermocouples or thermoradiometers.
● A globe thermometer may also be used to obtain the MRT. These thermometers contain sensors within a black spherical shell.

■ Basic thermal comfort systems operate on thermostatic control. Thermostats control heating and cooling by sensing that the ambient temperature is too hot or too cold. Thermostats can be reset manually (if automatic setback control has not been installed) or by adjusting controls to suggested temperatures (if clock, day-night, or other automatic reset controls are available).

■ **Controls for hydronic systems** use centrally-located thermostats, individual manual radiator controls, and other more advanced hydronic system controls.
● **Progressive anticipatory regulation (PAR) responds to outdoor weather conditions by regulating the temperature of the water being distributed.** Also referred to as variable ratio reset control, water is mixed in a four-way valve with an interior damper. Rather than change water temperature via boiler operation, it has been found that mixing radiator return water with hot water leaving the boiler is most efficient. Heated water is cooled to the desired level while the cool returning water is heated, thus reducing the amount of energy required by the boiler to reheat it. The valve can be operated manually, but operates best from a computer-controlled motor. The computer adjusts the valve to accommodate programmed system responses and

building envelope considerations such as insulation, air infiltration and solar contribution.
● Temperature and pressure relief valves are critical to safety and efficiency. Valve operation can be controlled so that hot water is allowed to continue to cycle through the system although the boiler is no longer burning. This reduces jacket and distribution losses.
● Automatic or manual feed valves work with expansion tanks to control system pressure.
● **Zonal controls** are often desired for hydronic systems. This is usually accomplished by installing electrically controlled zone valves at the boiler output. Conventional thermostats control the local zone valve while two-stage thermostats can control zone valves and the burner. When heat is needed, the valve is opened to admit the water already in the system. If this water does not satisfy the Btu requirement, the second stage of the thermostat will operate the burner until the water temperature is appropriate.
● **Newer zonal controls have thermostats incorporated with radiator valves.** These valves (TRVs - thermostatic radiator valves) should be used with two-pipe or single-pipe venturi systems so that distribution can continue throughout the building when the valve is closed.

■ **Electric heating controls** may be linear, input, line-voltage thermostats, low-voltage thermostats, demand/peak load controllers.
● Linear controls: A spring and diaphragm system used to control heat buildup along baseboard heaters.
● Input controls: Clock mechanisms which control line voltage to infrared units.
● **Line-voltage thermostats:** Single and double pole units that act like a switch. Also available with modulating control for two-stage heaters. Sensing elements are either enclosed within a grilled housing or are exposed to the open space. The exposed units will sense radiated heat, whereas enclosed will not.

D3.1 HVAC

- **Low-voltage thermostats:** Dual-voltage thermostats are two stage devices used to provide either full or half voltage. Thermostat-relay combinations act as a pilot device and operate a relay in the heater circuit. These are more sensitive to temperature change, but are also more expensive and require additional labor.
- **Demand or peak load controllers:** Used for sustained peak loads which are the basis for utility pricing. These devices drop out part of the heating load when a given peak is reached, or they convert the operation to half voltage.

■ **Controls for air-handling systems** guarantee fresh air, by relying on humidistats, variable speed controls and on/off switches to regulate air change rate and relative humidity. Air quality monitors are also used to measure the electrical resistance across a semiconductor to determine the level of contaminants in the air.

- Variable speed motors/drives, inlet vane controls and adjustable pitch blades can contribute to effective control of air flow.
- Where humidity is high, dehumidistat controls can be wired in series with fan on/off switches.
- Enthalpy controllers can be used to alter damper settings in ductwork. These can allow all air-handling systems to be capable of delivering 100 percent of fresh air. The rate of outdoor air versus recirculated indoor air can be controlled according to outdoor temperature with a minimum outside air setting as determined by code or good practice.
- Monitors react to carbon monoxide, carbon dioxide, formaldehyde, ethyl alcohol, cigarette smoke, and natural gas.
- Where tobacco smoke periodically fills the space, a time-of-day clock can be wired into a fan circuit. It can then be programmed for operation during occupancy by smokers; available in solid-state electronic or mechanical devices.

■ Return-air and outdoor-air dampers may provide a means of relieving air pressure.

- Reduced air flow can increase static pressure which increases energy consumption.
- Dampers are excellent sources of flow control for low first cost.

IGNITION & FLAME CONTROLS

■ **Burner controls** have three primary objectives:
- **Rate of fire:** Provide a balance between heat supply and demand.
- **Combustion efficiency:** Provide the minimum amount of fuel and the optimum air supply.
- **Supply.**

■ Flame size in larger burners is often controlled automatically.
- Using heating load as the determinant, flame size can be cycled between high and low settings respective to load.
- Modular increments can be used to set flame size to correspond to heating load.

■ Oil burners use automatic viscosity controls to mix and measure fuel oil as it is fed to the burner. Some oil burners incorporate photoelectric eye which shuts off oil flow during burner failure.

■ Furnaces and boilers prevent overheating by using high-limit switches in bonnets. These cut off air flow/combustion if temperatures exceed 250°F. Circulators will continue to operate at water temperatures over 200°F.

D3.1 HVAC

EQUIPMENT: selection checklist

BOILERS

■ **Boilers are rated by testing on the basis of output.**
● The **Net Rating** in Btus indicates the Btu per hour capacity. Net ratings should be matched to, or slightly exceed, the calculated heat loss from the building plus the estimated system heat losses.
● **Gross Output, or Heating Capacity,** determines the total Btu output produced under normal conditions for boilers with 300,000 Btu/hour input or more.

■ **Boilers may be constructed of a variety of materials responding to different needs.**
● Cast iron, stainless steel, steel and copper are used in coil construction. Larger scaled units are of cast iron or steel.
● Plate-steel and copper-tube boilers heat up more quickly than cast-iron, but may not be as durable.
● Copper and stainless steel are best for corrosion resistance. Water quality and PH should be monitored and controlled.
● The average boiler will last between 25 and 35 years, but cast-iron and steel units are known to last over 75 years.

■ Boilers are usually gas or oil-fired and must have a minimum steady state efficiency of 75 percent.
● **Gas-fired** units with atmospheric gas burners range from 77 to 85 percent efficiency. **Condensing boilers** attain AFUEs up to 95 percent because they recover an appreciable amount of heat from flue gases. **Pulse combustion steam boilers** (500,000 Btu/hr or more) may have efficiencies in the area of 85 percent.
● **Oil-fired** units have AFUEs from about 80 to 87 percent.
● Electric units may be used for terminal reheat or as localized hot water sources.

■ **Modular boilers** can be used as an energy-efficient alternative to a single, large boiler in a central system. Smaller boilers are piped in parallel or series circuits.
● Parallel systems are controlled to operate only the unit(s) necessary to respond to the heating load. Each boiler is set to fire at full rating.

—Piping should be designed to allow isolation of burners so that when one or more of the boilers are not needed, the on-line boilers maintain the temperatures in the idle boilers.

FURNACES

■ The efficiency of the furnace is set by the performance of the burner system in combination with the heat exchanger. **Many design and construction techniques air incorporated in gas furnaces to improve AFUE ratings.**
● Pilot lights are being replaced with electronic ignition systems.
● Natural draft combustion air is giving way to precise, controlled fan-forced combustion air supply. Direct vent systems can improve AFUE's by 6 percent, but the use of power burners will increase AFUE's further.
● Atmospheric burner tubes are improving flame control.
● Flue/chimney dampers are being incorporated into stacks to close them off when the equipment is not in use, thereby saving flue losses.
● Recuperative heat exchangers are being designed to condense the moisture in flue gases, thus recovering waste heat.

■ **Condensing furnaces** pass air or liquid from the combustion heat exchanger through a second heat exchanger/coil in the air flow.
● Some units incorporate a fresh water wash to flush out the bottom of heat exchangers at the end of each heating cycle. This reduces potential damage due to buildup of acidic, corrosive liquids.

■ The most common fuel burned in furnaces is natural gas. **Gas-fueled furnaces** primarily burn natural gas, but some units are also equipped to handle LPG.
● Conversion kits, generally provided by the manufacturer, allow the change from natural gas to LPG and vice versa.
● AFUE ratings of about 90 percent are available with condensing furnaces, but may be improved to 92 to 97 percent when pulse combustion is incorporated.

● Furnaces with atmospheric combustion and standing pilots generally have AFUEs in the range of 50 to 65. Intermittent ignition, direct venting, automatic vent dampers and improved heat transfer have increased efficiencies to above 80 percent.

■ Clearances are required between bonnets and combustible materials for oil and solid fuel-burning furnaces (local codes should be checked):
● Oil furnaces: Top of furnace to combustible 2 inches.
● Wood furnaces: Top of furnace to combustible 18 inches. It can be reduced to 1 inch if 28 ga. galvanized steel or aluminum heat shield is spaced away from the combustible by porcelain insulators.

■ **Oil burning furnaces** typically use pressure-atomizing, flame-retention burners and electric ignition.
● AFUE ratings range from 81 to 88 percent.

■ **Solid fuel furnaces** differ slightly in that a firebox is the burner and additional attention must be given to combustion exhaust systems.
● Chimneys and stacks require frequent maintenance to minimize the buildup of creosote which, left unattended, can result in an extremely hot chimney fire.
● The heat exchanger should be located in the bonnet and separated from the cold air return.

■ Wood furnaces are not typically rated according to accepted standards.
● Heat output is usually estimated by manufacturers.
● Efficiency ratings, via testing, have been found to fall in the range of 55 to 65 percent.
● Lower efficiencies are inherent due to the need for high flue gas temperatures which will minimize creosote buildup.

■ **Solid fuel furnaces** are integrated with indoor air systems and flue exhaust in the same way as other furnace types. In fact, many applications use these furnaces as primary heating while a standard furnace acts as backup.
● Furnaces may be set up in parallel or series arrangements, but there will be a need for two thermostats (set higher for wood furnace) and two distribution blowers.
● Humidifiers may be installed in the hot air supply duct.

- The greatest concern of combined systems is over the depositing of creosote in the exhaust stack.
- Water heating coils can be incorporated.
- Catalytic converters can be installed to reduce creosote deposits in flues.

■ **Back-up systems** are often employed to ensure that the design conditions are satisfied. They may take many forms:
- Standard heating equipment, such as furnaces, may be used to back-up other systems which are more economical to operate, but may be less reliable (e.g.: heat pumps are less efficient at very low temperatures; solid fuel furnaces require more frequent attention).
- Duplicate equipment, often operated in series, can serve to satisfy peak loads, but will largely provide back-up when maintenance routine disables one system.

EVAPORATIVE COOLERS

■ Evaporative coolers are effective options for cooling when dry-bulb temperatures range from 80° to 105°F, and when wet-bulb temperatures are at about 70°F or less.
- They are often referred to as swamp coolers because they raise relative humidity as they cool the air.
- They can be installed on roof tops or concrete pads, given consideration for unit weight and water handling.
- They may be operated individually (single-stage), in series (two-stage means two single-stage units), or in combination with other equipment (indirect).

HEAT PUMPS

■ Heat pumps are available in many sizes:
- up to 25 tons of heat transfer;
- handle temperatures up to 300°F;
- deliver heat at relatively low temperatures of 90° to 100°F;
- capacity to increase temperatures up to three times the low-heat-source temperature;
- air-to-air units improve efficiency by increasing the coil area - up to a 360° wrap around the air chamber;
- two-speed units offer greater efficiencies and control humidity better.

■ **Electricity is consumed to power the compressor and fan.**
- **Seasonal energy efficiency ratios (SEER)** rate the performance of the heat pump relative to electrical consumption; higher ratings equal better performance. SEERs of 10.5 or better are often recommended.
- Heat pumps are often economical choices for new construction when electricity costs are equivalent to fossil fuel costs or when fossil fuels are unavailable.
- Air-to-air units are the most common and inexpensive for residential and other small-scale applications.

■ **Typical performance levels:**
- SEER 8 to 14 for air units, 9 to 11.4 for water.
- COP 2.2 to 3.35 for air units, 2.9 to 4.1 for water (rated at 47°F outside temperature).
- HSPF 6.0 to 9.5 for air units.
- Balance point 20° to 30°F; it is the temperature where the heating load is equal to the output of the heat pump.

■ **The development of frost and ice in the heat exchanger during periods of low temperatures (less than 15°F) is a major hinderance to performance.**
- Electric resistance heaters are typically used to supply heat when the outside temperature falls below the balance point.
- Defrost cycles (switching to a cooling mode temporarily to heat the outdoor coil) are often employed.

D3.1 HVAC

EQUIPMENT: selection checklist (cont.)

DIRECT VENT APPLIANCES

■ Gas direct-vent units use electric or hydraulic thermostat controls.
- Burner capacities range from 12,000 to 22,000 Btu.
- Fans typically move 60 to 75 cfm.
- Sizes come in widths of 14 to 60 inches, depths of 5 to 10 inches, and heights from 13 to 31 inches.

■ Kerosene direct-vent units are available which operate on about four gallons of fuel per day. Electrical support is often required for fan and thermostat operation.
- Burner capacities range from 19,000 to 32,600 Btu.
- Fans typically move 159 to 318 cfm.
- Sizes come in widths of 30 to 32 inches, depths of 16 to 19 inches, and heights of about 25 inches.

UNVENTED APPLIANCES

Unvented, fuel-burning appliances are potential sources of considerable indoor air pollution and should be carefully researched prior to selection. All such appliances should be operated in rooms with operable windows or local supply and exhaust.

■ **Portable space heaters** usually burn kerosene to gain output of 1,000 to 15,000 Btu/hour. They have double combustion chambers and are nearly 100 percent efficient, but tests still show that carbon monoxide, carbon dioxide, nitrogen dioxide and sulfur dioxide are produced. Safety considerations incorporated into these units include:
- birdcage grilles designed to prevent contact burns;
- wick stops;
- tip-over/automatic shutoffs.

ELECTRIC HEATING

Electric space heating is considered to be 100 percent efficient because the entire energy input is converted to heat radiation. In the broader view, this is not true as generating and transmission (line) losses can consume up to two-thirds of the energy before it even reaches the electric heater. The result is often a costly alternative if not properly designed and controlled.

Radiant Heaters

■ **Radiant heaters emit energy in the far infrared range of the electromagnetic spectrum. The "invisible" energy provides lower temperature heat from larger surface radiation areas.** When the system is properly applied, thermostats can be set as much as 6° to 8°F lower to maintain equivalent thermal comfort. Other advantages include:
- no maintenance required;
- heating elements are neatly concealed;
- no restrictions are set on the placement of furniture in the space;
- quick response;
- useful for spot-heating.
- the entire ceiling or floor becomes the radiating surface, rather than a localized terminal.

■ **Radiant heating panels operate best in well-insulated enclosures with very low rates of air infiltration.** They are 20 to 50 percent less expensive to operate than convective units, adding the benefit of quiet, draft-free operation.
- Typically, the exterior side of the heating panel needs to be insulated for efficient operation.
- Higher wattage systems require less coverage of the surface area. 25-watt heaters often cover 40 to 60 percent of the area, while 100 watt heaters may be less than 20 percent.
- Their performance can be enhanced by using radiant barriers (metallic films) on envelope surfaces which are opposite the radiant surface.
- When the opposite surface is cold (noninsulated), the radiant heat is absorbed and the ambient temperature is minimally affected. In this condition, the radiant surface becomes a hot spot, causing uneven and uncomfortable sensations.
- The best control of a radiant system is provided by a thermostat which responds to both the mean radiant temperature and air temperature.
- Some panels can be surface mounted for retrofit situations.

Infrared Heaters

■ Intermediate infrared light rays provide the best source of radiation:
- These rays are generally absorbed by all colors and surfaces, with little convective loss to the ambient air.
- The effect of the radiation is felt within minutes.

■ The effectiveness depends on:
- the capacity of the equipment,
- mounting height,
- temperature of the ambient air,
- configuration of the unit,
- configuration and nature of the space/ environment.

■ Electric infrared heat appliances include:
- **Metal sheath** - emits near infrared radiation and is about 60 percent efficient.
  - Offers good resistance to impact, vibration and water.
  - Uses about 47 watts per inch of 3/8 inch diameter sheath to operate at temperatures of 1200° to 1500°F.
  - Response time: 2 to 5 minutes.
  - Reduced voltage operation is not recommended; percentage input timer controls can be effective.
- **Fused quartz tubing** - emits intermediate infrared radiation at about 80 percent efficiency.
  - Offers good resistance to water, but not to vibration and impact.
  - Uses about 34 watts per inch of 3/8 inch diameter tube to operate at temperatures of 1200° to 1800°F.
  - Response time: under 3 minutes.
  - Requires horizontal mounting.
  - Percentage input timer control can be used.
- **Evacuated quartz lamp** - emits near infrared radiation and about 8 lumens per watt of visible light.
  - Uses about 100 watts per inch of 3/8 inch diameter tubing to operate at a temperature of 4000°F and about an 80 percent efficiency.
  - Response time: less than 2 minutes.

Forced Air Unit Heaters

■ Electric heating appliances are also available in forced air units.
- Typical performance characteristics:
  - Voltage: 120 to 480.
  - Wattage: 1125 to 50000.
  - Amps: 2.7 to 139.
  - Btu/hr: 3840 to 170500.
- Up to six elements can be used, with one to five fans for distribution.
- Duct collars and manually-controlled dampers are often offered for additional control.

BURNERS

■ **Optimal designs provide complete burning and maximum efficiency by controlling the supply of oxygen.**
● **Natural draft:** Room or recirculated air is drawn into the combustion chamber by chimney-effect, maintaining air pressure lower than atmospheric. Dampers and small fans may be used for control.
● **Forced-draft:** Fans force fuel and air into the combustion chamber and maintain air pressures above atmospheric.
● **Induced-draft:** Air is drawn through the burner at pressures lower than atmospheric by fans located at the exhaust outlet.
● **Balanced-draft:** Two fans are set for equal air flow with slightly negative pressurization. One fan controls air supply, and the other regulates pressure.
● Deviation from the optimum supply of oxygen results in unburned fuel (too little) or lost heat through the chimney (too much), since air flow through the heat exchanger is too fast.
● Maximum combustion efficiency can be measured from the percentage of carbon dioxide and oxygen in the exhaust and from the temperature of exiting gases. A noticeable temperature increase is a common sign of inefficiency calling for adjustment or maintenance.

■ **Sealed combustion burner units** have been developed to improve burner operation, but have introduced other improvements as well:
● Eliminates the potential of backdrafting due to indoor/outdoor pressure differentials.
● Reduces the energy losses associated with open chimneys/stacks and make-up air.
● Reduces the air infiltration caused during operation, since room air is not cycled through the burner and out the flue.
● Often referred to as vented, these burners draw outside air in through the outer sleeve of concentric ducts while flue gases are exhausted through the center sleeve.

■ **Pilot light ignition systems, often used in atmospheric burners, burn continuously during system operation.**
● Standing pilot lights are fed by a tube which branches off the main line just after the safety control valve.
● The pilot controls the safety control valve: Located directly above the pilot is a thermocouple, which is a heat-sensitive switch. A warm thermocouple keeps the safety control valve open; a cool thermocouple closes the valve.

■ **Intermittent ignition systems burn long enough to establish the main flame.**
● These systems can improve the efficiency of an atmospheric burner with standing pilot by about 6 percent.

■ **The use of electric ignition has improved burner reliability and efficiency.** These are used in most power burners.
● Ceramic igniters are electrically heated up to 2500°F.
● Electric ignition may be provided by a high voltage current (upwards of 10,000 volts) arcing through the fuel/air mix and between two electrodes.
● Other systems use electric spark (spark plugs) for ignition.

Gas Burners

■ **Gas supply** is piped through several controls before mixing with the air.
● The supply is controlled primarily at the main shutoff valve.
● A safety control valve acts as an automatic shutoff as problems occur.
● A pressure regulator controls the supply pressure for proper mixing.
● A main gas valve responds to operating controls, such as a thermostat.

■ A blue-green inner flame denotes proper air-fuel mixing in gas burners.
● Too little air is evident by yellow tipped flames.
● Too much air shows up as a bright blue inner flame.

■ **Atmospheric burner configurations include:**
● single flame;
● long, thin ribbon flames;
● series of small flame jets;

● horizontal (inshot) - often used for Scotch-type boilers;
● vertical (upshot) - often used for firebox-type boilers.

■ Condensing burners require attention with issues of low-temperature flue gases and removal of condensation.
● **Corrosion-resistant materials** are used in construction; such as stainless steel heat exchangers, plastic flue pipes, etc.
● **Low-temperature flue gases** (less than 150°F as opposed to the usual 500°F +) allow the use of plastic pipe for venting purposes rather than chimneys or double-sleeve flues. However, the low temperature may also impact flue drafting and steps must be taken to be sure that the draft is sufficient.
● Drains made of non-corrosive materials are required to collect and remove condensate. Some units also use pumps.
● **The heat exchanger** may be an air-to-air or air-to-liquid unit. The liquid is usually a mixture of water and antifreeze.

■ Pulse combustion units may be noisier than other types of equipment, and may incorporate sound deadening systems.
● Combustion chambers are made of cast-bronze or stainless steel.
● Like condensing burners, low-temperature flue gases (110° to 150°F as opposed to initial combustion temperatures of over 1200°F) are emitted.
● The valve and spark-plug ignition system requires less fuel, thus achieving AFUEs of up to 97 percent.

■ Burner heads are available in many types including:
● drilled-port;
● slotted-port; and
● single-port.

Oil Burners

■ Atomization may be accomplished in one of three ways:
● **High-pressure atomization:** the delivery of oil under 100 psi pressure to a nozzle which sprays the oil in a fine mist.
● **Low-pressure atomization:** the delivery of oil and air under 1 to 15 psi pressures such that the air is used to produce a spray.

D3.1 HVAC

EQUIPMENT: selection checklist (cont.)

- **Centrifugal:** the delivery of oil to a cup which is rotating at speeds of 3,450 to 12,000 rpm in order to establish a thin oil film resulting in throwing off a fine spray from the edge of the cup.

■ Gun-type burners typically have good efficiencies of operation and provide a clean fuel combustion.
- This type of burner is less sensitive to viscosity, and adapts well to any grade of fuel oil.
- Retention head burners introduce air through a very small aperture to the fuel line forward of the nozzle; efficiencies range from 83 to 86 percent.
- Shell-head adapters on these burners can increase combustion efficiency.

■ The nozzle of gun-type burners is important for performing three functions:
- It produces a given spray density based on the oil pressure;
- It provides a specific quantity of oil to the combustion chamber;
- It sets the pattern (uniformity and angle) necessary for the particular burner.
- Nozzles are usually brass or steel housings with stainless or hardened steel orifices.

Solid Fuel Burners

■ Fireboxes are usually made from steel, and in large boilers may be partially enclosed with water walls. Cast iron boxes are available, but the large castings are expensive and may crack.
- Unlike the necessity of grates in coal fireboxes, wood-burning units may not include grates since wood fires are cooler and more diffuse.
- Fire temperature will determine the type of firebox linings. Coal units typically include brick linings to delay burnout.
- Unit sizes vary considerably, and the door size may be important. Wood furnaces should be sized to allow burning of a 2 foot log or larger.

■ Firebox operation is either at full blast or just smoldering.
- A thermostat-controlled damper or a small blower regulate the flow of combustion air to the firebox.

- Fireboxes should be sized so that higher fire temperatures are maintained over the majority of the heating season. This minimizes creosote generation which is greatest when the fire is smoldering. Creosote buildup is deposited on all surfaces which contact combustion gases when temperatures are around 200° to 300°F.

HEAT EXCHANGERS

■ **Air-to-air heat reclaiming exchangers** are only appropriate for recovery of heat from exhaust air and flue gas. Many units recover only sensible heat, but "enthalpy" exchangers are available for recovery of both sensible and latent heat. For minor levels of contamination, they can serve to purify the indoor air.
- **The heat exchanger core** is usually made of metal (aluminum, stainless steel, titanium), plastic or treated-paper cells. Metal heat exchangers are usually more efficient.
- Efficiency of heat transfer is often below 50 percent.
- Residential-scale units are typically designed for central air removal or forced air interface. When radon is at moderate levels or lower and the number of air changes per hour is also low, these units are the most economical solution to radon removal. They are not effective for smoke removal; air filters and cleaners can remove 50 percent of the smoke in 26 percent of the time.
- Units can be mounted in closets or basements (to the ceiling), with indoor intake through door louvers/grilles or return air ductwork. Air is exhausted through roof or wall caps. Outdoor intake (roof or wall caps) should be located 6 feet or more from exhaust vents.
- **Variable-speed fans** operating from a humidistat often produce more efficient operation. Constant volume fans are harder to balance.
- **Most units will contain some device which is designed to remedy frost/ice problems.** Common where outside temperatures fall below freezing, this is necessary at 5°F or less to ensure proper heat transfer.
- Operations can be reversed in response to heating and cooling needs due to seasonal variation.

- **Desiccants** should be used in hot, humid climates to absorb moisture from the air. For dry climates, hygroscopic materials may be used to recover water vapor. These materials can be applied in the core or on fixed sheet exchangers.
- Incidentals which improve unit design include humidistat control, backdraft dampers, override timers, condensation drains, insulation, and rodent screens at outside penetrations.

■ **Typical performance characteristics:**
- Larger heat-transfer areas tend to have greater heat recovery efficiency. The percent of sensible heat recovered varies:
  —crossflow = 48 to 71 percent;
  —counterflow = 40 to 76 percent;
  —tubular = less than 50 percent;
  —fixed plate = 22 to 89 percent;
  —rotary = 60 to 90 percent;
  —heat pipe = 47 to 64 percent.
- Only rotary types remove sensible and latent heat; at a rate of 20 to 82 percent.
- Sizes range from small window units to very large commercial units.
- Electricity is used to power blower units (wattage in the range of 20 to 500) and preheat units on some crossflow and counterflow types (1000 to 1500 watts).
- Blowers are often rated from 20 to 230 cfm. Due to static pressure in the distribution system, the actual amount of air moved through the heat exchanger may be 10 to 25 percent less than the rated fan capacity.
- Slower blower flow rates tend to increase exchange efficiency.

■ Crossflow leakage may be an important consideration in selecting heat exchangers.
- Heat pipes and tubular units have little to no leakage.
- Rotary units may allow 1 to 4 percent leakage; crossflow units may allow from less than 3 up to 12 percent; counterflow units may allow 16 to 42 percent; and fixed plate may allow about 27 percent leakage.

D3.1 HVAC

Condensers

■ **Air-cooled units** generally have lower COPs than water-cooled types.
● Metal (copper, aluminum, steel) finned coils containing hot refrigerant are arranged to promote the flow of air through them.
● Common operation is at differences of 20°F between ambient air and refrigerant temperatures.

■ **Evaporative condensers** can be designed for lower condensing temperatures than other types of condensers.
● Refrigerant is passed through a coil which is cooled by a water spray (coming from above) and a supply of cool air (from below). The air is either forced from a blower below or drawn through from a fan above.
● The evaporation of water on the coil cools the tubing and the refrigerant within.
● They are somewhat more efficient than the water- and air-cooled types because pumping and fan requirements are substantially less.
● Mineral (scale) buildup and corrosion on the coils must be controlled.

■ **Misters can be attached to cooling equipment for the purpose of cooling the outdoor condensing coils.**
● A fine mist of water is sprayed at a rate of about 0.5 gallons per hour across the coils, cooling them as the water evaporates.
● Thermostats on the coils operate the misters in temperature ranges from 95° to over 105°F.
● Mineral/scale buildup and corrosion on the coils must also be controlled here.

Evaporators

Evaporator capacity must be matched to condenser capacity. Evaporators are common to all refrigeration systems, but are available in specific types, as follows.

■ Direct-expansion (DX) evaporators are used to boil and superheat the refrigerant in the evaporator tubes for air cooling or liquid chilling.

● DX coils in HVAC duct systems or air-handling units can realize as much as 1 1/2 percent increase in efficiency for each degree rise in evaporating temperature.
● DX systems are used for small to medium sized applications.

■ Absorption type evaporators are usually generators. Systems include:
● **Single effect:** When a single generator replaces the compressor.
● **Double effect:** When one generator transfers the heat from the vaporized refrigerant so that it can be used as the lower pressure heat source for a second generator.
● **Dual cycle:** When heat from a high temperature absorption system is used by a second absorption system, while both absorption systems are housed in the same unit.

COMPRESSORS/PUMPS

■ Compressor efficiencies can be improved by reducing the resistance to refrigerant flow offered by long piping runs and other equipment in the system.
● The closer the systems components are located, the lower the compressor pressure.

■ **Reciprocating compressors** operate one or more pistons from a crank shaft. They can be either open or hermetic types.
● Compressors may be sized as small (under 5 hp; 3.7 kW), medium (5 to 25 hp; 3.7 to 18.7 kW) or large (over 25 hp; 18.7 kW).
● Open type units may have up to 16 cylinders, providing power of 0.167 hp (0.12 kW) and greater.
● Semi-hermetic units may have up to 12 cylinders, providing power of 0.5 to 150 hp (0.37 to 112 kW).
● Welded hermetic units may have up to 16 cylinders, providing power of 0.167 to 25 hp (0.12 to 18.7 kW).

■ **Rotary compressors (also known as vane compressors)** use a roller on a shaft to compress the gas within a hermetic cylinder. Two types, with very similar characteristics, are available; rolling piston and rotating vane.
● The rolling piston type has the roller mounted on an eccentric shaft so that the roller provides the pressure as it moves around the stationary cylinder. A single vane is set into the side of the cylinder and moves in and out with the action of the roller.

● The rotating vane type has two vanes inset into the roller, which is mounted concentrically on the shaft. The roller/shaft assembly is mounted eccentric to the cylinder.

■ **Screw-type compressors** use the motion of a helical-shaped screw to draw air in, compress and release it. They may be single or twin screw types.
● Single screw units have one screw-shaft with gate rotors mounted on shafts perpendicular to the screw. The gate rotors rotate in opposite directions to match the rotation of the screw.
● Twin screw units have two screw-shafts which are designed to mesh as they rotate into one another. The screws are mated by asymmetric lobes (male) and gullies (female) to produce both an axial and radial movement of gas.

FANS/BLOWERS

■ **One of the basic laws of fan operation is that slower speeds of rotation save energy.** Other laws include:
● Volume varies directly with the speed.
● Pressure varies directly with the square of the speed or volume.
● Power input varies directly with speed cubed.

Centrifugal Fans

■ Typical characteristics:
● Best for low-static-pressures (1/2 to 3/4 in. w.g.) and minimal noise production.
● Principles of operation require that the drive motor be mounted outside of the airstream.

■ Airfoils:
● Optimum operation (peak power, pressure and efficiency) is at 50 to 65 percent of maximum volume.
● Airfoils operate at the highest speeds and efficiencies of all centrifugal fans.
● Usually, 10 to 16 airfoil blades are set to curve away from direction of rotation.
● Housings are designed with close clearances for best efficiencies.

D3.1 HVAC

EQUIPMENT: selection checklist (cont.)

■ **Backward-inclined/curved:**
● Performance characteristics are nearly the same as airfoils, but not quite as efficient; also has non-overloading power characteristic.
● The 10 to 16 fan blades are sloped backward relative to the direction of rotation.

■ **Radial:**
● Efficiency is sacrificed for high mechanical strength and simplicity. Greatest efficiency is at about 30 to 40 percent of volume. Fan speeds are moderate.
● The 6 to 10 blades may be radial to the hub or sloped forward relative to direction of rotation.
● The housing is the narrowest of centrifugal fans.

■ **Forward-curved:**
● Greatest efficiency is at 40 to 50 percent of full volume. Pressure peaks before maximum efficiency is reached.
● Motor selection must consider that power input increases as free delivery is approached.
● 24 to 64 small blades are used. The blades are curved forward at both ends.

Axial Fans

■ Typical characteristics:
● Pressure - up to 1 in. w.g.
● Noise - greater than equivalent capacity centrifugal unit.
● Applications - best for static pressures under 1/2 in. w.g. where high air volume is desired.
● A better selection where variable flow and capacity to power ratios are important.

■ **Propeller:**
● Efficiency increases with power.
● Installation is via a circular ring, orifice plate or venturi and typically is not attached to ductwork.

■ **Tubeaxial:**
● Like propeller fans (but better), efficiency is greatest at free delivery, with a higher volume to pressure capability.

● Housing is typically a cylindrical tube with tight clearances between blade tips and tube.
● The hub is usually less than 50 percent of the diameter (tip to tip).
● Fan blades may number from 4 to 8.

■ **Vaneaxial:**
● Characterized by improved pressure, efficiency and noise characteristics.
● The hub is usually greater than 50 percent of the fan diameter.
● The 12 airfoil blades are either fixed or adjustable-pitch type.

Fan Drive

■ Direct drive fans are usually run by single-speed, 1750 rpm motors. Air volume can be varied by using adjustable pitch blades.
● Advantages:
—operate best at high speeds and in ambient air with low concentrations of particles;
—easy maintenance and longer operating life due to few moving parts;
—offers least obstruction of the airstream.
● Disadvantages:
—typically has only one, two or three motor speeds.

■ Belt drive motors are installed in or outside of the airstream depending on the qualities of the ambient air.
● **Motors are kept out of the airstream when contaminant concentrations are high, when the contaminants are corrosive, or when the temperature of the air is above 104°F.**
● Advantages of belt driven fans:
—most economical and quietest fan configuration at low speeds and in clean air;
—accommodates changes in air volume and speed.
● Disadvantages:
—maintenance is necessary to monitor belt life, slippage and other sources of drive loss;
—belt and bearing housings interfere with airflow.

VENTILATORS

■ **Upblast roof ventilators use dampers, a rain gutter and a wind band to weatherproof their otherwise open top.**
● Operation requires an exit velocity of 1500 fpm in order to open the dampers. Under normal conditions, this air velocity will also keep precipitation from entering the ventilator.
● **Advantages** include lower material cost, greater operating efficiency and good expulsion of unwanted air away from the building.
● **Disadvantages** include high noise factor and the potential for operational disruption by strong winds.

■ **Hooded roof ventilators eliminate most weather concerns.**
● Operation is best with low air velocity and volume.
● Disadvantages may include concerns with the flow of exhausted air that is directed back over the roof; a possible result is roof damage from or recycling of contaminants.

FILTRATION

■ **Increasing concerns with indoor air quality have resulted in the development of absorptive filters which use chemical reaction to remove contaminants.**
● Potassium permanganate impregnated alumina pellets can remove formaldehyde.
● Hot diesel exhaust gases and emissions from coal-fired boilers can have nitrogen oxides removed by passing them over cyanuric acid, a nontoxic chemical that breaks the oxides down into simple nitrogen and water.
● These filters require regeneration or replacement when noticeable increases in undesirable gases are measured.

■ Removal of formaldehyde and nitrogen dioxide gases can also be accomplished using **air cleaners with activated carbon filters.**
● Formaldehyde removal rates can range from 5 to almost 30 percent.
● Nitrogen dioxide removal rates can range from about 1 percent up to nearly 45 percent.

D3.1 HVAC

AIR FILTERS: types

Filter Type	Average Efficiency (%)	Life (hrs)	Average Pressure Drop (WG)
Cleanable 2 in. *high velocity*	8-10	600	0.24
Disposable 2 in.	10-15	480	0.20
Pleated 2 in.	36	2000	0.37
Automatic Roll *w/Bag Cart.*	20-25 50-85	3750 3750-10000	0.40 0.975-1.135
Bag Cartridge *w/prefilter*	38-97 50-97	2500-6600 3500-9000	0.525-0.74 0.55-0.74
Rigid Cartridge *w/prefilter*	55-90 55-90	2500-3500 3500-4500	0.60-0.65 0.60-0.65
Electrostatic Auto. Roll *Bag*	90 93-97	2500-4000 12000	0.45 0.775
95% rated *w/prefilter* *w/85% prefilter*	99 99	6000 24000	0.75 0.75
HEPA *w/o prefilter* *w/85% prefilter*	100 100	6000 24000	1.50 1.50

- Activated charcoal filters have a life of about 8760 hours. Periodic replacement or regeneration of filters will minimize off-gasing of previously absorbed gases.
- Average pressure drop (WG) is 0.35.

■ **Foam and fiber filters** are commonly used to catch larger particles in the indoor air, but are not generally effective for gaseous contaminants.
- These particulate filters soil rapidly. They are effective at concentrations of about 4 grams per 1000 cubic feet of air.
- Though inexpensive, fiberglass filters cannot be reused while more expensive aluminum and plastic filters can be reused after washing.

■ Filter configuration is usually best when in the shape of a "V" as opposed to perpendicular to the air flow.
- This arrangement provides more surface area and less resistance.

AIR WASHERS

■ Air washers use a water spray or cell-type washer to remove dust and dirt particles from the air. These units operate like and have similar components as evaporative coolers. They are not effective for soot and smoke removal.
- The advantage of air washers over other types of air cleaners or filters is that the filters are essentially self-cleaning.
- Capacity ranges from 2000 cfm and higher.
- Low-velocity units will provide an airflow between 300 and 600 fpm, but some units will need up to 1500 fpm. Air pressure should be limited to a drop of 0.25 to 1 inch of water.
- High-velocity units will provide an airflow between 1200 and 1600 fpm optimally, but some units operate at up to 2400 fpm. Air pressure should be limited to a drop of 0.5 to 1.5 inches of water.
- Water-spray density should be between 1 and 5 gallons per minute per square foot of eliminator area.
- Nozzle capacity (1 to 3.75 gpm), spacing (.75 to 2.5 nozzles per sq. ft. of eliminator), and pressures are critical design elements. Nozzle pressures of 20 to 40 psig provide a fine spray which is appropriate for

humidification, while dehumidification is accomplished at about 25 psig.
- Efficiencies are realized at smaller eliminator sizes and higher spray densities.

■ Cellular air washers pass water over fiber-filled screens which make up tiered cells.
- Capacities of up to 210,000 cfm are available.
- Air velocity and cell size will cause efficiency to vary from 70 to 97 percent.
- Water is usually supplied at 0.75 to 1.5 gallons per minute per 1000 cfm of airflow. Air pressure drop can range from 0.15 to 0.65 inches of water depending on water quantity and air velocity.

HUMIDIFIERS

■ **Units used in central air systems are dependent on airflow for evaporation and distribution.**
- Evaporation/output rates are published by manufacturers and are generally based on 24 hours of operation.

DEHUMIDIFIERS

■ **Regenerative/desiccant dehumidifiers:**
- Desiccants are drying agents which remove moisture from air supply.
- Desiccants may be hygroscopic salts such as calcium chloride, urea, lithium chloride or sodium chloride. Other forms of desiccants include silica or alumina gel, activated alumina, molecular sieves, or glycol solution.

■ **Absorption dehumidifiers:**
- Use liquid or solid sorbent materials.
- Rotating wheel units are generally used for commercial and industrial applications.
- Advantageous for producing extremely low moisture content of air at a given dewpoint.

■ Air washers can be used for dehumidification, using a water spray and/or evaporator pad.
- Water temperature must be below the dewpoint of the entering air.

D3.1 HVAC

SYSTEM MAINTENANCE

Air Handling Equipment

■ Regular cleaning of air handling equipment will have a substantial effect on their performance.
- Efficiencies can be maintained and friction and resistance from filters, coils, blower fan wheels and fan scroll blades can be reduced.
- Cleaning can be automatically signalled by installing pressure drop gauges which measure the resistance to air flow across filters and coils.

■ **Balancing of air flow** through ductwork requires measurement and adjustment of static pressure at various points in the system.
- Measure static pressure before and after filters, heating/cooling coils, blowers/fans, the terminal which is furthest from the source (or otherwise the most difficult terminal to supply) and at induction units.

■ **Air-to-air heat exchangers** require inspection to maintain balanced airflow.
- Varying pressures in ductwork and changing weather conditions can throw off settings established at installation.
- Problems with defrost mechanisms are known to be more frequent in colder climates resulting in ice or condensate buildup. Flow rate reductions in these units caused by frost and ice in the core and shutdown of fans during defrost cycles can be as much as 50 percent.

■ **Fan motors:**
- Check bearings for wear and resistance to movement.
- Adjust belt tension to ensure that belts are neither slipping due to weakness nor adding to motor loads due to tension. Belt tension should allow about a 3/4 inch give. A belt needs replacement when fatigue is evident -- cracking, stretching, becoming stringy, etc.

Hydronic Equipment

■ **Clean or replace strainers and filters on a regular basis to reduce the potential of chemical or mineral buildup.**
- Dirt, scale and debris must be removed from all elements which contain, distribute or filter water.

■ Water flow is key to good system maintenance.
- Motors, bearings and fan drives must be lubricated.

● Filter systems must be used to protect nozzles from plugging.
● Bleedoff techniques can be used to minimize scale accumulation in heavily mineralized water. A 10 percent bleedoff should be used for pump and spray-nozzle systems.
● Boiler blowdown can be manual or automatic. The purpose is to remove solids in the water and sludge in the boiler. Priming and carryover can be avoided by blowdown.

■ Water treatments should be used.
● Quaternary salts or other chemicals hold dissolved materials in suspension.
● Inhibitors should be used to minimize corrosion of steel.
● The pH of the water should be frequently checked and maintained above 7.0.
● Biocides can be added for control of algae, slime and bacterial growth.
● Reducing calcium hardness will often be desirable.

■ **Boiler operation** can be quickly checked:
● For hot water systems, a working circulating pump is most essential.
● For steam systems, check the water level in the sight glass. If it is too low, open the supply valve until water level is restored to the halfway mark.

■ Hydronic systems must be monitored and adjusted for proper, balanced operation.
● Fluid flow quantities, temperatures and pressures are measured to ensure balance throughout the system.
● Fluid flow can be measured by meters which are installed in the line, by calculating pressure differentials across the system components (coils, valves, etc.) or by differential between rated water temperature and measured water temperature.

■ **Cooling towers** rely on clean air and water supplies for top efficiency.
● Proper water treatment and blowdown rates will control scaling and dissolved solids.
● Cleaning with chlorine and algicide treatments need to be periodically used to control algae growth on baffles and around spray nozzles.

Fuel Combustion Equipment

■ Annual maintenance is essential for efficient **burner operation.**
● Clean and correct dirty oil nozzles and fouled gas parts.

● Check for defective gaskets, cracked brickwork, broken casings and other signs that additional air may be entering the combustion chamber.

■ Oil burners require routine maintenance (every 4 to 6 months).
● Clean nozzles for air and fuel delivery. Replace high pressure oil nozzles yearly.
● Replace filters between storage tank and pump.
● Use proven additives which will eliminate water in storage systems. Additives can also be used to reduce flashpoint temperatures and for soot control.

D3.1 HVAC

HVAC SYSTEMS: selection checklist

TABLE A. SELF-CONTAINED AIR CONDITIONERS AND AIR-SOURCE HEAT PUMPS
(window units, thru-wall units, and residential split-system units).

NOTE: This system <u>usually</u> used for perimeter space conditioning only, with separate all-air system serving large interior zones.

General building applications:	Small Residential/Light Commercial Rooms@ Exterior Wall, max. 6 stories	
System classification:	Self-Contained, All-Electric Equipment (Non-Ducted)	
System type:	**Self-Contained Air-Conditioners & Air Source Heat Pumps(ASHP)**	
Competing systems:	Modular WSHP's and GSHP's,	
	2-Pipe and 4-Pipe Fan Coil Units	

Common equipment configurations:	**Window Unit** **Through-Wall Unit**	**Split System Unit**
General System Characteristics Basic operating principle:	Constant Air Volume/Variable Air Temp.	Constant Air Volume/Variable Air Temp.
Equipment quality:	Residential "appliance" unless A.R.I. rated as PTAC (commercial)	Residential "appliance" unless A.R.I. rated as PTAC(commercial)
Cooling coil type: Heating coil type:	DX refrigerant (R-22) Electric	DX refrigerant (R-22) Electric
Typical capacity limitations:	150 - 450 cfm, 1/2 to 2 tons cooling	300 to 2000 CFM, 1/2 to 5 tons cooling [Larger units avail. in comm. units]
Number of zones/unit:	One	One to two
Comfort Considerations Degree of temperature control: RH control: Air circulation: Noise Levels:	Fair Fair Fair Poor (compressor & fan in occupied room)	Fair Fair Fair Fair (fan in occupied room)
Indoor Air Quality Considerations Continuous fresh air supply for recommended vent. rate: Air filtration capabilities: Room pressurization capabilities: Humidification capabilities: Dehumidification capabilities: IAQ maintenance issues: IAQ maintenance issues:	Good Fair Poor None Fair Many filters to replace, cond. pans to clean	None Fair Poor None Fair Many filters to replace, cond. pans to clean
Aesthetic Considerations In-room equipment: Exterior wall penetrations: Rooftop and/or On-grade equipment: Miscellaneous:	Supply fan and compressor Air-cooled condenser and Fresh air louvers None Coil condensate drains to exterior can stain wall surfaces	Supply fan Refrigerant tubing penetrations Air-cooled condenser
First Cost Considerations System first cost: Indoor equipment room space reqd.: Interior pipe/duct shaft requirements:	Low Minimum None	Low Minimum None
Operating & Maintenance Cost Considerations Maintenance staff skill level required: Maintenance hours/year required: Expected service life: Airside economizer cooling option availability: Energy efficiency of air-conditioning mode: Energy efficiency of occupied period heating mode: Energy efficiency during unoccupied period heating mode: Optional energy-efficient equipment available: Flexibility re: changes to space layouts: Adaptability to changes in space function:	Low Low, but units easily damaged 10 years (15 yrs. if PTAC) No Good Poor Poor ASHP heating Poor Poor	Low Low 15 years No Good Poor Poor ASHP heating Poor Poor

Source: "HVAC Systems for Commercial Buildings" (Ref.1)

D3.1 HVAC

TABLE B. SELF-CONTAINED SINGLE-ZONE (DUCTED) AIR CONDITIONERS
(rooftop single zone units and large split-system units).

General building applications:	Light Commercial Building
System classification:	Unitary Ducted Single Zone Air-Conditioners & ASHP's
System type:	**Single Zone Self-Contained (Ducted) Air-Conditioners**
Competing systems:	Modular WSHP's and GSHP's, Central All-Air Systems
	2-Pipe and 4-Pipe Fan Coil Units

Common equipment configurations:	Rooftop Single Zone Units	Split System Unit
General System Characteristics		
Basic operating principle:	Constant Air Volume/Variable Air Temp.	Constant Air Volume/Variable Air Temp.
Equipment quality:	Light commercial grade	Light commercial grade
Cooling coil type:	DX refrigerant (R-22)	DX refrigerant (R-22)
Heating coil type:	Electric (or optional direct gas-fired heating)	Electric(or direct gas-fired heating)
Typical capacity range:	2,000 - 10,000 CFM, 5 to 25 tons cooling	2,000 - 10,000 CFM, 5 to 25 tons cooling
Standard Number of zones/unit:	One	One
Comfort Considerations		
Degree of temperature control:	Fair	Fair
RH control:	Fair	Fair
Air circulation:	Good	Good
Noise Levels:	Fair (good vibration control reqd.)	Fair (good vibration control reqd.)
Indoor Air Quality Considerations		
Ability to provide recommended fresh air rates on continuous basis:	Good	None
Air filtration capabilities:	Fair	Fair
Room pressurization capabilities:	Poor (air flow not continuous)	Poor (air flow not continuous)
Humidification capabilities:	None	None
Dehumidification capabilities:	Fair	Fair
IAQ maintenance issues:	Filter replacement, cond. pan cleaning	Filter replacement, cond. pan cleaning
Aesthetic Considerations		
In-room equipment:	None	None(fan above clg. or in adj. space)
Exterior wall penetrations:	None	None
	None	None
Rooftop and/or On-grade equipment:	Packaged fan/air-cooled condenser/compressor	Air-cooled condenser
Miscellaneous:		
First Cost Considerations		
System first cost:	Low	Low
Indoor equipment room space reqd.:	None	Foor-Mtd. or above-ceiling fan section
Outdoor equipment space reqd.:	Roof only	Roof or on-grade
Interior pipe/duct shaft requirements:	Min. (vert. duct shaft in multi-story buildings)	None(indoor unit typ. serves 1 floor)
Operating & Maintenance Cost Considerations		
Maintenance staff skill level required:	Low	Low
Maintenance hours/year required:	Low	Low
Expected service life:	15 years	15 years
Airside economizer cooling option availability:	Yes(optional, in units > 5 ton capacity only)	No(unless field-fabricated)
Energy efficiency of air-conditioning mode:	Good	Good
Energy efficiency of occupied period heating mode:	Poor(electric heat) to Good(direct gas-fired)	Poor(gas-fired option n/a)
Energy efficiency during unoccupied period heating mode:	Poor(electric heat) to Good(direct gas-fired)	Poor
Optional energy-efficient equipment available:	Air Source Heat Pump heating	Air Source Heat Pump heating
Flexibility re: changes to space layouts:	Good	Good
Adaptability to changes in space function:	Poor	Poor

Source: "HVAC Systems for Commercial Buildings" (Ref.1)

D3.1 HVAC

TABLE C. SELF-CONTAINED DUCTED MULTI-ZONE AIR CONDITIONERS (rooftop units).

General building applications:	**Light Commercial Buildings**
System classification:	**Self-Contained All-Electric Equipment (Non-Ducted)**
System type:	**Self-Contained Ducted Multizone Air-Conditioners**
Competing systems:	Central WSHP's, Central All-Air VAV Systems,
	Multiple Self-Contained Unitary Rooftop Systems

Common equipment configurations:	Rooftop Unit

General System Characteristics	
Basic operating principle:	Constant Air Volume/Variable Air Temp.
Equipment quality:	Light Commercial
Cooling coil type:	DX refrigerant (R-22)
Heating coil type:	Electric, Direct Gas-Fired Option
Typical capacity limitations:	6,000 - 12,000 CFM; 15 to 37 tons cooling
Number of zones/unit:	8 to 12

Comfort Considerations	
Degree of temperature control:	Fair
RH control:	Poor
Air circulation:	Good
Noise Levels:	Fair(good vibration isolation reqd.; unit close to occupied space)

Indoor Air Quality Considerations	
Ability to provide recommended fresh air rates on continuous basis:	Good
Air filtration capabilities:	Good
Room pressurization capabilities:	Fair
Humidification capabilities:	None
Dehumidification capabilities:	Poor
IAQ maintenance issues:	Filter replacement

Aesthetic Considerations	
In-room equipment:	None
Exterior wall penetrations:	None
Rooftop and/or On-grade equipment:	Air-cooled condenser, fan

First Cost Considerations	
System first cost:	Low
Indoor equipment room space reqd.:	None
Interior pipe/duct shaft requirements:	Minimum(vert. duct shafts, multi-story bldg.)

Operating & Maintenance Cost Considerations	
Maintenance staff skill level required:	Moderate
Maintenance hours/year required:	Low
Expected service life:	15 years
Airside economizer cooling option availability:	Yes (optional)
Energy efficiency of air-conditioning mode:	Fair
Energy efficiency of occupied period heating mode:	Poor(elect.) to Good(gas option)
Energy efficiency during unoccupied period heating mode:	Poor(elect.) to Good(gas option)
Optional energy-efficient equipment available:	Economizer cooling
Flexibility re: changes to space layouts:	Fair
Adaptability to changes in space function:	Poor

Source: "HVAC Systems for Commercial Buildings" (Ref.1)

D3.1 HVAC

TABLE D. MODULAR WATER SOURCE HEAT PUMPS
(closed loop heat pump systems with/without central ventilation air)

General building applications:	Commercial/Institutional/Industrial Buildings	
System classification:	Central System, All-Water & Air-Water	
System type:	**Modular Water Source Heat Pumps(WSHP's)**	
Competing systems:	2-Pipe Change-Over FCU's, 4-Pipe FCU's,	
	Window/Through-Wall Air-Conditioners & ASHP's.	
	Central All-Air Systems	
Common equipment configurations:	**Wall Console Units/Unit Ventilators & Vertically-Stacked Closet Units**	**Wall Consoles, Vert. Closet Units, or Above-Ceiling Concealed Units**
Ventilation air options:	**Local Fresh Air Inlets**	**Central Ducted Ventilation System**
General System Characteristics		
Basic operating principle:	Constant Air Volume/Variable Air Temp.	Constant Air Volume/Variable Air Temp.
Equipment quality:	Commercial grade	Commercial grade
Cooling coil type:	Water-Cooled DX (R-22)	Water-Cooled DX (R-22)
Heating coil type:	Water-Cooled DX (R-22)	Water-Cooled DX (R-22)
Typical capacity range, Console & Vert. Closet Units:	200 - 800 CFM, 1/2 to 2 tons cooling	200 - 800 CFM, 1/2 to 2 tons cooling
Horiz. Cocealed Units:	n/a	200 - 1600 CFM, 1/2 to 5 tons
Unit Ventilators:	750 - 1500 CFM, 2 to 4 tons cooling	n/a
Number of zones/unit:	One	One
Comfort Considerations		
Degree of temperature control:	Good	Good
RH control:	Fair	Fair
Air circulation:	Fair	Fair
Noise Levels:	Fair (console units) to Good (closet units)	Good (unit separated from room by ceiling or closet construction)
Indoor Air Quality Considerations		
Ability to provide recommended fresh air rates on continuous basis:	Fair	Very Good
Air filtration capabilities:	Fair	Fair
Room pressurization capabilities:	Poor	Poor
Humidification capabilities:	None	None
Dehumidification capabilities:	Fair	Fair
IAQ maintenance issues:	Many filters to replace, condensate pans to clean	Many filters to replace, condensate pans to clean
Aesthetic Considerations		
In-room equipment:	Console Unit(or in-closet)	Console Unit(or concealed/ in-closet)
Exterior wall penetrations:	Fresh air louvers	None
Rooftop and/or On-grade equipment:	Cooling tower	Cooling tower
First Cost Considerations		
System first cost:	Low	Moderate
Indoor equipment room space reqd.:	Moderate	Moderate
Outdoor equipment space reqd.:	Cooling Tower	Cooling Tower
Interior pipe/duct shaft requirements:	Minimum (S/R water pipes & condensate waste lines)	Low(S/R water pipes & condensate waste lines, vert. vent. ducts)
Operating & Maintenance Cost Considerations		
Maintenance staff skill level required:	Low	Low
Maintenance hours/year required:	Moderate	Moderate; ease of access to above-ceiling units very important.
Expected service life:	19 years	19 years
Airside economizer cooling option availability:	No	No
Energy efficiency of air-conditioning mode:	Good	Good
Energy efficiency of occupied period heating mode:	Good (heat reclaim from interior zones)	Good (heat reclaim from interior zones)
Energy efficiency during unoccupied period heating mode:	Good(improved w/ thermal storage)	Good(improved w/ thermal storage)
Optional energy-efficient equipment available:	Thermal water storage for recovered heat; GSHP loop or many boiler equipment selections available; variable speed pumping.	Thermal storage for recovered heat; GSHP loop or many boiler equipment selections available; variable speed pumping, ventilation heat recovery/dessicant cooling options.
Flexibility re: changes to space layouts:	Poor	Poor
Adaptability to changes in space function:	Poor	Poor

Source: "HVAC Systems for Commercial Buildings" (Ref.1)

D3.1 HVAC

TABLE E. CENTRAL DUCTED WATER SOURCE HEAT PUMPS
(rooftop units, floor-mounted and horizontal indoor units)

General building applications:	Light Commercial Buildings
System classification:	Central Air-Water System
System type:	**Central Ducted Water Source Heat Pumps**
Competing systems:	Central WSHP's, Central All-Air VAV Systems, Self-Contained Unitary Single Zone Systems

Common equipment configurations:	Rooftop Unit	Floor-Mounted Indoor Unit & Horiz. Indoor Units
General System Characteristics		
Basic operating principle:	Constant Air Volume/Variable Air Temp.	Constant Air Volume/Variable Air Temp.
Equipment quality:	Commercial	Commercial
Cooling coil type:	Water-cooled DX	Water-cooled DX
Heating coil type:	Water-source DX	Water-source DX
Typical capacity limitations:	1,000 - 11,000 CFM; 3 to 25 tons cooling	300 - 4,000 CFM; 1 to 10 tons(Horiz. Unit) 340 - 9,000 CFM; 1 to 25 tons(Flr.-Mtd.)
Number of zones/unit:	One	One
Comfort Considerations		
Degree of temperature control:	Good	Good
RH control:	Good	Good
Air circulation:	Good	Good
Noise Levels:	Fair(good vibration isolation reqd.; unit close to occupied space)	Fair (vibration isolation reqd., init close to occupied space)
Indoor Air Quality Considerations		
Ability to provide recommended fresh air rates on continuous basis:	Good	Good
Air filtration capabilities:	Good	Good
Room pressurization capabilities:	Fair	Fair
Humidification capabilities:	Optional	Optional
Dehumidification capabilities:	Good	Good
IAQ maintenance issues:	Filter replacement	Filter Replacement, cond. pan cleaning
Aesthetic Considerations		
In-room equipment:	None	Fan, compressor
Exterior wall penetrations:	None	None
Rooftop and/or On-grade equipment:	Evap. cooler, air handler	Evap. cooler
First Cost Considerations		
System first cost:	Moderate	Moderate
Indoor equipment room space reqd.:	Small boiler, pump room	Small boiler, pump room
Outdoor equipment space reqd.:	None (roof only)	None (roof only)
Interior pipe/duct shaft requirements:	Minimum(vert. pipe shafts, multi-story bldg.)	Minimum(vert. pipe shafts, multi-story bldg.
Operating & Maintenance Cost Considerations		
Maintenance staff skill level required:	Low	Low
Maintenance hours/year required:	Low	Low
Expected service life:	20 years	20 years
Airside economizer cooling option availability:	Yes	Yes
Energy efficiency of air-conditioning mode:	Good	Good
Energy efficiency of occupied period heating mode:	Good (heat reclaim from interior spaces)	Good (heat reclaim from interior spaces)
Energy efficiency during unoccupied period heating mode:	Good (improved w/ thermal storage)	Good(improved w/ thermal storage)
Optional energy-efficient equipment available:	Thermal water storage	Thermal storage
Flexibility re: changes to space layouts:	Good	Good
Adaptability to changes in space function:	Poor	Poor

Source: "HVAC Systems for Commercial Buildings" (Ref.1)

D3.1 HVAC

TABLE F. TWO-PIPE CHANGE-OVER SYSTEMS
(fan coil units and unit ventilators, with/without central ventilation air)
NOTE: This system is sometimes used for perimeter space conditioning only, with separate all-system serving large interior zones.

General building applications:	Commercial/Institutional/Industrial Buildings	
System classification:	Central System, All-Water & Air-Water	
System type:	2-Pipe Change-Over Systems (Fan Coil Units & Unit Ventilators)	
Competing systems:	Modular Piped WSHP's/GSHP's, 4-Pipe FCU's	
	Window/Through-Wall Air-Conditioners & ASHP's	
Common equipment configurations:	**Wall Console Units & Vertically-Stacked Closet Units(FCU only)**	**Wall Consoles; Vert. Closet Units & Above-Ceiling Concealed Units(FCU)**
Ventilation air options:	**Local Fresh Air Inlets**	**Central Ducted Ventilation System**
General System Characteristics		
Basic operating principle:	Constant Air Volume	Constant Air Volume
Equipment quality:	Commercial grade	Commercial grade
Cooling coil type:	Chilled water	Chilled water
Heating coil type:	Shares ch. wa. coil; supplemental elect. heating optional	Share ch. wa. coil; supplem. elect. heating optional
Typical capacity range, Console & Vert. Closet Units:	200 - 1200 CFM, 1/2 to 4 tons cooling	200 - 1200 CFM, 1/2 to 4 tons cooling
Horiz. Cocealed Units:	n/a	200 - 1600 CFM, 1/2 to 5 tons
Unit Ventilators:	750 - 1500 CFM, 2 to 4 tons cooling	n/a
Number of zones/unit:	One	One
Comfort Considerations		
Degree of temperature control:	Poor	Poor
RH control:	Fair	Fair
Air circulation:	Fair	Fair
Noise Levels:	Fair(w/ fan in room) to Good(unit fan separated from room by ceiling construction)	Good (fan separated from room by room by ceiling construction)
Indoor Air Quality Considerations		
Ability to provide recommended fresh air rates on continuous basis:	Fair	Very Good
Air filtration capabilities:	Fair	Fair
Room pressurization capabilities:	Poor	Poor
Humidification capabilities:	None	None
Dehumidification capabilities:	Fair	Fair
IAQ maintenance issues:	Many filters to replace,	Many filters to replace,
IAQ maintenance issues:	condensate pans to clean	condensate pans to clean
Aesthetic Considerations		
In-room equipment:	Wall Console Unit (or above-ceiling unit)	None exposed
Exterior wall penetrations:	Fresh air louvers	None
Rooftop and/or On-grade equipment:	Cooling tower	Cooling tower
First Cost Considerations		
System first cost:	Low	Moderate
Indoor equipment room space reqd.:	Moderate	Moderate
Interior pipe/duct shaft requirements:	Minimum (S/R water pipes & condensate waste lines)	Low(S/R water pipes & condensate waste lines, vert. vent. ducts)
Operating & Maintenance Cost Considerations		
Maintenance staff skill level required:	Low	Low
Maintenance hours/year required:	Moderate	Moderate; ease of service access to above-ceiling units very important.
Expected service life:	20 years	20 years
Airside economizer cooling option availability:	No	No
Energy efficiency of air-conditioning mode:	Good	Good
Energy efficiency of occupied period heating mode:	Good	Good
Energy efficiency during unoccupied period heating mode:	Good	Good
Optional energy-efficient equipment available:	Wide range of chiller, heat rejection, ice storage, boiler equipment selections available; variable speed pumping possible;	Wide range of chiller, ice storage, boiler equipment selections available; variable speed pumping possible; central ventilation heat recovery & dessicant cooling options.
Flexibility re: changes to space layouts:	Poor	Poor
Adaptability to changes in space function:	Poor	Poor

Source: "HVAC Systems for Commercial Buildings" (Ref.1)

TABLE G. FOUR-PIPE FAN COIL UNITS AND UNIT VENTILATORS
(with/without central ventilation air)

NOTE: This system is sometimes used for perimeter space conditioning only, with separate all-air system serving large interior zones.

General building applications:	Commercial/Institutional Buildings	
System classification:	Central System, All-Water & Air-Water	
System type:	**4-Pipe Fan Coil Units & Unit Ventilators**	
Competing systems:	2-Pipe Change-Over FCU's, Modular Water Source Heat Pumps,	
	Window/Through-Wall Air-Conditioners & ASHP's, and Central All-Air Systems	

Common equipment configurations:	**Wall Console Units/Unit Ventilators & Vertically-Stacked Closet Units**	**Wall Consoles, Vert. Closet Units, or Above-Ceiling Concealed Units**
Ventilation air options:	**Local Fresh Air Inlets**	**Central Ducted Ventilation System**

General System Characteristics		
Basic operating principle:	Constant Air Volume/Variable Air Temp.	Constant Air Volume/Variable Air Temp.
Equipment quality:	Commercial grade	Commercial grade
Cooling coil type:	Chilled Water	Chilled Water
Heating coil type:	Hot Water or Steam	Hot Water or Steam
Typical capacity range, Console & Vert. Closet Units:	200 - 1200 CFM, 1/2 to 4 tons cooling	200 - 1200 CFM, 1/2 to 4 tons cooling
Horiz. Cocealed Units:	n/a	200 - 1600 CFM, 1/2 to 5 tons
Unit Ventilators:	750 - 1500 CFM, 2 to 4 tons cooling	n/a
Number of zones/unit:	One	One

Comfort Considerations		
Degree of temperature control:	Good	Good
RH control:	Fair	Fair
Air circulation:	Fair	Fair
Noise Levels:	Fair (console units) to Good (units in closet)	Good (unit separated from room by ceiling or closet construction)

Indoor Air Quality Considerations		
Ability to provide recommended fresh air rates on continuous basis:	Good	Very Good
Air filtration capabilities:	Fair	Fair
Room pressurization capabilities:	Poor	Poor
Humidification capabilities:	None	None
Dehumidification capabilities:	Fair	Fair
IAQ maintenance issues:	Many local filters to replace, condensate pans to clean	Many local filters to replace, condensate pans to clean

Aesthetic Considerations		
In-room equipment:	Console Unit(or in-closet)	Console Unit(or concealed/in-closet)
Exterior wall penetrations:	Fresh air louvers	None
Rooftop and/or On-grade equipment:	Cooling tower	Cooling tower

First Cost Considerations		
System first cost:	High	High
Indoor equipment room space reqd.:	Moderate	Moderate
Interior pipe/duct shaft requirements:	Minimum (S/R water pipes & condensate waste lines)	Low(S/R water pipes & condensate waste lines, vert. vent. ducts)

Operating & Maintenance Cost Considerations		
Maintenance staff skill level required:	Low	Low
Maintenance hours/year required:	Moderate	Moderate
Expected service life:	20 years	20 years
Airside economizer cooling option availability:	Unit Ventilator model only	No
Energy efficiency of air-conditioning mode:	Good	Good
Energy efficiency of occupied period heating mode:	Good	Good (heat reclaim from interior zones)
Energy efficiency during unoccupied period heating mode:	Good	Good(improved w/ thermal storage)
Optional energy-efficient equipment available:	Many chiller, heat rejection, ice storage options available; many boiler/heat source options available; var. speed pumping.	Many chiller, heat rejection, ice storage options available; many boiler/heat source equipment selections available; variable speed pumping, ventilation heat recovery/dessicant cooling options.
Flexibility re: changes to space layouts:	Poor	Poor
Adaptability to changes in space function:	Fair	Fair

Source: "HVAC Systems for Commercial Buildings" (Ref.1)

D3.1 HVAC

D • SERVICES

TABLE H. CONSTANT VOLUME/REHEAT, CENTRAL STATION AIR HANDLING SYSTEMS
(with electric, hot water or steam reheat)

NOTE: These systems limited to hospital operating rooms, laboratories, museums requiring close temp. /RH control unless reheat source is renewable/ recovered heat.

General building applications:	Commercial/Institutional Buildings
System classification:	Central System, All Air
System type:	Constant Volume Reheat, Central Station Air Handling Units
Competing systems:	Central WSHP's, Central All-Air VAV Systems, Multiple Self-Contained Unitary Rooftop Systems

Common equipment configurations:	Electric Reheat	H.W. or Steam Reheat
General System Characteristics		
Basic operating principle:	Constant Air Volume/Variable Air Temp.	Constant Air Volume/Variable Air Temp.
Equipment quality:	Commercial to Institutional	Commercial to Institutional
Expected service life:	25 years	25 years
Cooling coil type:	Chilled Water	Chilled Water
Heating coil type:	Electric or Central Hot Water/Steam	Central Hot Water/Steam
Typical capacity limitations:	1,500 - 40,000 CFM	1,500 - 40,000 CFM
Number of zones/unit:	Unlimited	Unlimited
Comfort Considerations		
Degree of temperature control:	Very Good	Best
RH control:	Very Good	Best
Air circulation:	Very Good	Very Good
Noise Levels:	Very Good	Very Good
Indoor Air Quality Considerations		
Ability to provide recommended fresh air rates on continuous basis:	Best	Best
Air filtration capabilities:	Best	Best
Room pressurization capabilities:	Best	Best
Humidification capabilities:	Yes	Yes
Dehumidification capabilities:	Very Good	Very Good
IAQ maintenance issues:	Central M.E.R. maintenance	Central M.E.R. maintenance
Aesthetic Considerations		
In-room equipment:	None	None
Exterior wall penetrations:	Central M.E.R. only	Central M.E.R. only
Rooftop and/or On-grade equipment:	Central Cooling Tower or Air-Cooled Chiller	Central Cooling Tower or Air-Cooled Chiller
First Cost Considerations		
System first cost:	High	High
Indoor equipment room space reqd.:	High	High
Interior pipe/duct shaft requirements:	High	High
Operating & Maintenance Cost Considerations		
Maintenance staff skill level required:	High	High
Maintenance hours/year required:	Low	Low
Expected service life:	25 years+(10 years for elect. reheat coils)	25 years+
Airside economizer cooling option availability:	Yes	Yes
Energy efficiency of air-conditioning mode:	Poor(Reheat reqd.)	Poor(Reheat reqd.)
Energy efficiency of occupied period heating mode:	Poor(all-elect.)	Good(depending on central heating source)
Energy efficiency during unoccupied period heating mode:	Poor(all-elect.) to Good(central HW at AHU)	Good(depending on central heating source)
Optional energy-efficient equipment available:	Central plant equipment options	Central plant equipment options
Flexibility re: changes to space layouts:	Good	Good
Adaptability to changes in space function:	Good	Good

Source: "HVAC Systems for Commercial Buildings" (Ref.1)

TABLE I.   VAV CENTRAL STATION AIR HANDLING SYSTEMS
(with electric and hot water reheat VAV boxes)

NOTE: Energy efficiency of central VAV systems strongly dependent on type of VAV terminal device used.

General building applications:	Commercial/Institutional Buildings
System classification:	Central System, All Air
System type:	VAV Central Station Air Handling Units
Competing systems:	Central WSHP's, Central Constant Volume Reheat Systems, Multiple Self-Contained Unitary Rooftop Systems

Common equipment configurations:	Electric Reheat V.A.V Boxes	H.W. Reheat V.A.V. Boxes
General System Characteristics		
Basic operating principle:	Variable Air Volume/Constant Air Temp.	Variable Air Volume/Constant Air Temp.
Equipment quality:	Commercial to Institutional	Commercial to Institutional
Cooling coil type:	Chilled Water	Chilled Water
Central AHU Heating coil type:	Electric or Central Hot Water/Steam	Central Hot Water/Steam
Typical capacity limitations:	1,500 - 40,000 CFM	1,500 - 40,000 CFM
Number of zones/unit:	Unlimited	Unlimited
Comfort Considerations		
Degree of temperature control:	Very Good	Best
RH control:	Very Good	Best
Air circulation:	Very Good	Very Good
Noise Levels:	Very Good	Very Good
Indoor Air Quality Considerations		
Ability to provide recommended fresh air rates on continuous basis:	Good	Good
Air filtration capabilities:	Best	Best
Room pressurization capabilities:	Very Good	Very Good
Humidification capabilities:	Yes	Yes
Dehumidification capabilities:	Yes	Yes
IAQ maintenance issues:	Central M.E.R. maintenance	Central M.E.R. maintenance
Aesthetic Considerations		
In-room equipment:	None	None
Exterior wall penetrations:	Central M.E.R. only	Central M.E.R. only
Rooftop and/or On-grade equipment:	Central Cooling Tower or Air-Cooled Chiller	Central Cooling Tower or Air-Cooled Chiller
First Cost Considerations		
System first cost:	High	High
Indoor equipment room space reqd.:	High	High
Interior pipe/duct shaft requirements:	High	High
Operating & Maintenance Cost Considerations		
Maintenance staff skill level required:	High	High
Maintenance hours/year required:	Low	Low
Expected service life:	25 years+(10 years for elect. reheat coils)	25 years+
Airside economizer cooling option availability:	Yes	Yes
Energy efficiency of air-conditioning mode:	Good	Very good
Energy efficiency of occupied period heating mode:	Good(all-elect.)	Good(depending on central heating source)
Energy efficiency during unoccupied period heating mode:	Poor(all-elect.) to Good(central HW at AHU)	Good(depending on central heating source)
Optional energy-efficient equipment available:	Central plant equipment options	Central plant equipment options
Flexibility re: changes to space layouts:	Good	Good
Adaptability to changes in space function:	Good	Good

Source: "HVAC Systems for Commercial Buildings" (Ref.1)

D3.1 HVAC

D4. ① FIRE PROTECTION

FIRE SAFETY DESIGN

FIRE SAFETY DESIGN

Effective fire safety in buildings requires a systemic approach to design for fire prevention, protection from fire, and fire control.

The **"Fire Triangle"** depicts fire behavior in terms of three essentials; if one is absent or removed, combustion will not occur:

■ **Oxygen** must be at least 16% of the atmosphere to sustain combustion.

■ **Fuel** (solid, liquid, or gaseous) must be present in sufficient concentration to form a combustible mixture with oxygen.

■ **Heat** must be sufficient to produce and ignite combustible gases; solid and liquid fuels must be pre-heated to distill these gases before they will ignite.

PRINCIPLES OF FIRE SAFETY

■ To understand the requirements of fire-safe buildings, consider:

• **Fire safety planning** should be based on knowledge of fire behavior in buildings and the building's behavior under fire conditions.

• **Structural fire resistance** is a prerequisite of occupant fire safety.

• **Building contents,** selection and organization must be recognized as a key influence on fire safety.

• **Integrated fire protection** should be a design consideration.

• **Redundant fire protection** (i.e., back-up provisions for key systems) should be part of an effective design.

• **Worst case "possibilities"** should parallel "probabilities" as planning tools in developing fire-safe designs.

■ **Code compliance.** Building and life safety codes define the architect's responsibilities for protecting building users from fire. Code provisions indicate:

• **Occupancy classification:** The unique fire safety requirements of buildings, based on the types of activity they accommodate.

• **Configuration:** The fire safety implications of buildings in context, including setback requirements and site access.

• **Egress and evacuation:** Requirements for exiting occupants from all spaces and moving them through the building to safety.

• **Exterior protection** from exterior-source fires, such as from adjacent buildings and vegetated/forested areas.

• **Interior protection:** Required to maintain structural stability.

• **Internal separation:** For various occupancies and functions.

• **Mechanical requirements:** To support fire prevention and control and to protect occupants and property.

The fire safety design decision tree defines design

CONCEPT OF THE "FIRE TRIANGLE"

objectives to help address three major fire safety goals. Building design factors are:

1 • PREVENT IGNITION OF BUILDING MATERIALS AND CONTENTS

• Control ignition sources
• Control fuel character
• Control fuel/heat interaction by maintaining adequate separation

2 • CONTROL FIRE DEVELOPMENT

■ **Detect fire.** Automatic fire detection include:
• Heat detectors
• Smoke detectors
• Supervised extinguishing systems.
• Flame detectors

■ **Control combustion.** Room geometry influences the movement of fire through space, a pattern called "mushrooming." Heat rises until in encounters a horizontal barrier, then moves horizontally until the fire front is able to move upward along a new vertical path or strikes a vertical barrier and is forced downward by rising fire gases behind it.

■ **Means to suppress/extinguish fire include:**
• Manual occupant suppression.
• Automatic sprinkler suppression systems.

■ **Limit spread.** Compartmentalization is the best means of fire control by dividing the building into separate fire areas by fire resistive "barriers." In order to be effective, a floor, wall, or ceiling assembly fire barrier must be:
• Physically complete
• Thermally resistant
• Structurally stable under fire conditions.

D4.1 FIRE PROTECTION

FIRE SAFETY OBJECTIVES

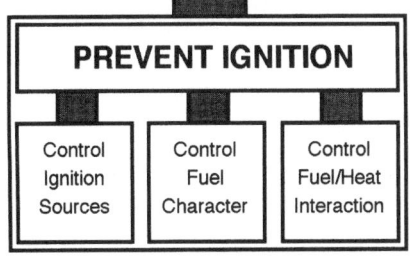

PREVENT IGNITION

| Control Ignition Sources | Control Fuel Character | Control Fuel/Heat Interaction |

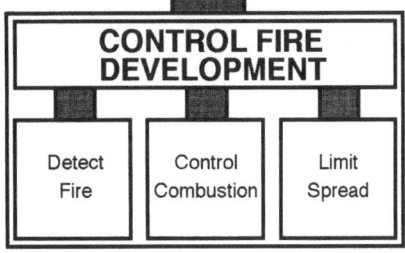

CONTROL FIRE DEVELOPMENT

| Detect Fire | Control Combustion | Limit Spread |

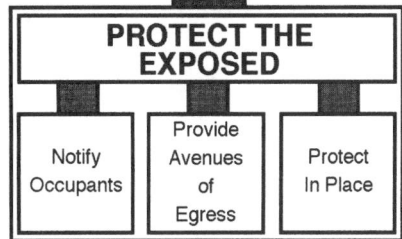

PROTECT THE EXPOSED

| Notify Occupants | Provide Avenues of Egress | Protect In Place |

Source: "Fire Safety Design" (Ref.1)

■ **Physical completeness** is generally ensured through the use of certified barrier (wall, floor, and ceiling). Attention should be given to specification of:

- walls.
- ceilings and floors.
- separation of occupied and concealed spaces.
- vertical barriers to spread between floors.
- doors.
- HVAC equipment, such as ducts, transfer grilles, etc.
- service openings, such as hose cabinets, access panels, and hatches.
- lighting equipment.
- poke-throughs for electrical, plumbing, and other utilities.

■ Thermal resistance is important in barriers to ensure they do not spread heat to adjacent areas by conduction.

3 • PROTECT THE EXPOSED BUILDING OCCUPANTS

■ **Notify occupants.** See FIRE ALARMS in this section for fire notification systems.

■ **Planning for egress** should include an effective system of building evacuation system and should provide:

- Adequate number and arrangement of exits, sufficient in capacity for the number of occupants.
- Structural integrity to ensure safety during a fire while the occupants are exiting or in an area of refuge.
- Exit continuity to ensure that exits remain clearly visible, unobstructed, and unlocked.
- Adequate marking so that exits and routes of escape are clearly marked and unambiguous under smoke conditions.

■ **Defend in place.** Refuge areas serve as an important alternative to egress systems and are required to provide safety for individuals with limited mobility or inability to quickly exit a building and for exigencies in large or tall buildings where exit passages can be blocked.

REFERENCE: Fred Malven, Ph.D. *"Fire Safety Design"* (Reference 1).

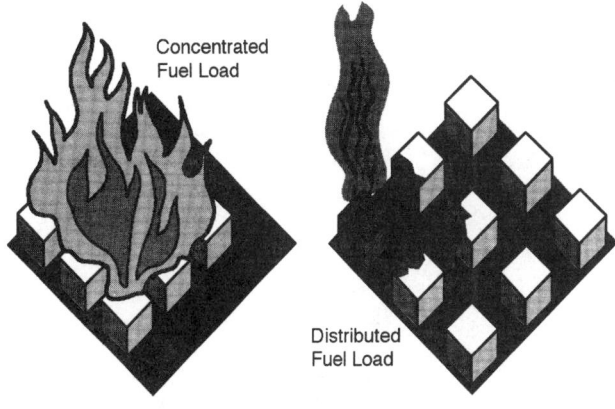

Concentrated Fuel Load

Distributed Fuel Load

TYPES OF FUEL LOADS

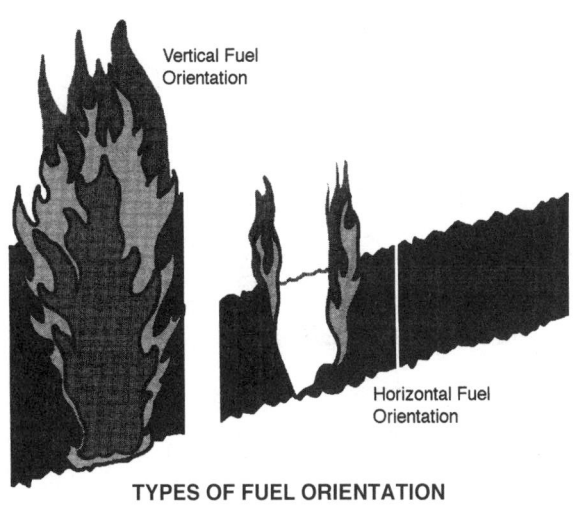

Vertical Fuel Orientation

Horizontal Fuel Orientation

TYPES OF FUEL ORIENTATION

D4.1 FIRE PROTECTION

FIRE PROTECTION: introduction

EXTERNAL FIRE CONTROL

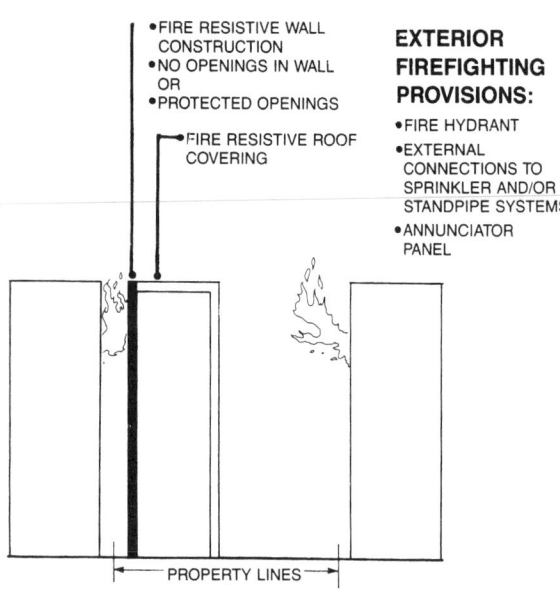

- FIRE RESISTIVE WALL CONSTRUCTION
- NO OPENINGS IN WALL OR
- PROTECTED OPENINGS
- FIRE RESISTIVE ROOF COVERING

EXTERIOR FIREFIGHTING PROVISIONS:
- FIRE HYDRANT
- EXTERNAL CONNECTIONS TO SPRINKLER AND/OR STANDPIPE SYSTEMS
- ANNUNCIATOR PANEL

PROPERTY LINES

INTERNAL FIRE CONTROL

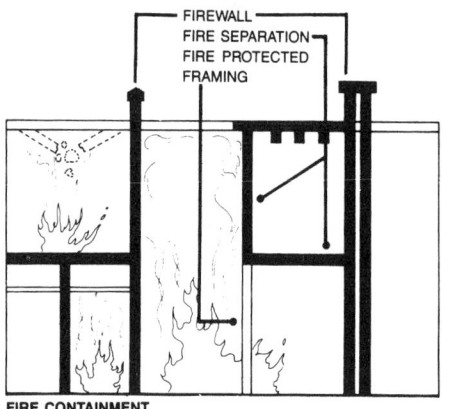

- FIREWALL
- FIRE SEPARATION
- FIRE PROTECTED FRAMING

- FIREWALLS TO REMAIN INTACT AFTER COLLAPSE OF FRAMING ON EITHER SIDE.
- FIRE SEPARATION TO BE CONTINUOUS BOTH HORIZONTALLY AND VERTICALLY WITH CONSTRUCTION OF UNIFORM FIRE RESISTANCE RATING.
- PENETRATIONS THROUGH AND GAPS BETWEEN FIRE RESISTIVE ASSEMBLIES MUST BE SEALED.

FIRE CONTAINMENT

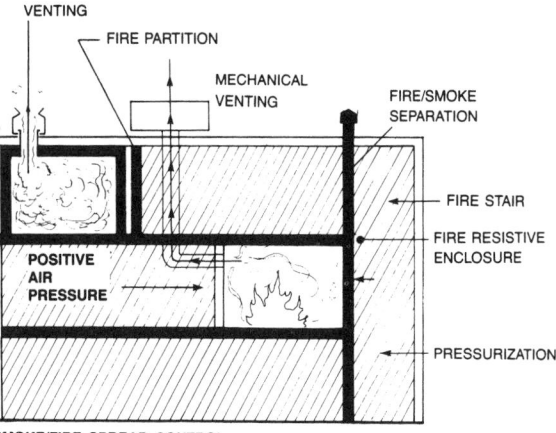

VENTING
FIRE PARTITION
MECHANICAL VENTING
FIRE/SMOKE SEPARATION
FIRE STAIR
FIRE RESISTIVE ENCLOSURE
POSITIVE AIR PRESSURE
PRESSURIZATION

SMOKE/FIRE SPREAD CONTROL

Fire-safe design is based on the assessment of:

IGNITION POTENTIAL
- fuel load.
- spread of ignition.
- heat output, etc.

SMOKE POTENTIAL
- rate of flow and production.
- quantity and content of products of combustion.

DAMAGE POTENTIAL
- horizontal and/or vertical travel of fire.
- risk to neighboring structures, etc.

■ The primary design responses to the assessed hazard are:
- **Fire-resistive construction** - building systems in which structural members are protected from the effects of fire.
- **Compartmentation** - confinement of a fire to the area of origin by using construction assemblies.
- **Life safety systems** - Fire detection and alarm systems used to identify the presence of a fire in its early stages. Another component of life safety is evacuation or **egress**, the provision for a safe means of escape within five minutes of an alarm.
- **Smoke control systems** - smoke containment and/or elimination.
- **Automatic suppression systems** - the attempt to extinguish the fire or to bring it under control until firefighters can respond.
- **Provision for firefighting** - external and internal access, water supply, hose stations, specific elevator controls, and equipment for local use.

■ **Fire control depends on how fast the fire spreads beyond its ignition point.** The selection of materials must consider the critical factors of flammability and fuel contribution.
- Ignition temperatures are attained by convective action of hot gases from a fire, by radiant heat from the fire plume, by radiant heat from other surfaces of the enclosure, and/or by the direct contact of the fire plume.
- Materials that are "noncombustible" by nature are often finished with coatings, facings, fabrics, etc. which are combustible; these finishes are a source of fire spread.
- **Noncombustible materials must be capable of withstanding continuous exposure to temperatures up to 1200°F without igniting and burning. No material with significant organic content will pass this requirement.**
- ASTM E136 is used to define material noncombustibility.

■ Smoke and fire control also relies on the building itself:
- The number, size and placement of windows and doors will affect the growth and movement of fire.
- Physical boundaries such as ceiling height, open floor area and the existence of vertical shafts are keys to the movement and density of smoke. Boundaries are required to stop fire/smoke on both horizontal and vertical paths.

■ **Construction Types have been developed based on their structural integrity during fire: The structure must remain intact while limiting the spread of ignition and flame into adjacent spaces over a prescribed period of time.**
- Construction type is selected in response to the anticipated fire hazard. It defines the required protection of exterior walls, interior walls, columns, beams/girders/trusses/arches, floor and roof construction in terms of the number of hours of fire endurance as documented by testing.

D4.1 FIRE PROTECTION

FIRE ALARM

SMOKE DETECTOR AND AUDIBLE ALARM TO WARN OCCUPANTS IN A ROOM OR ADJOINING ROOMS.

SELF-CONTAINED DETECTION AND ALARM DEVICE.

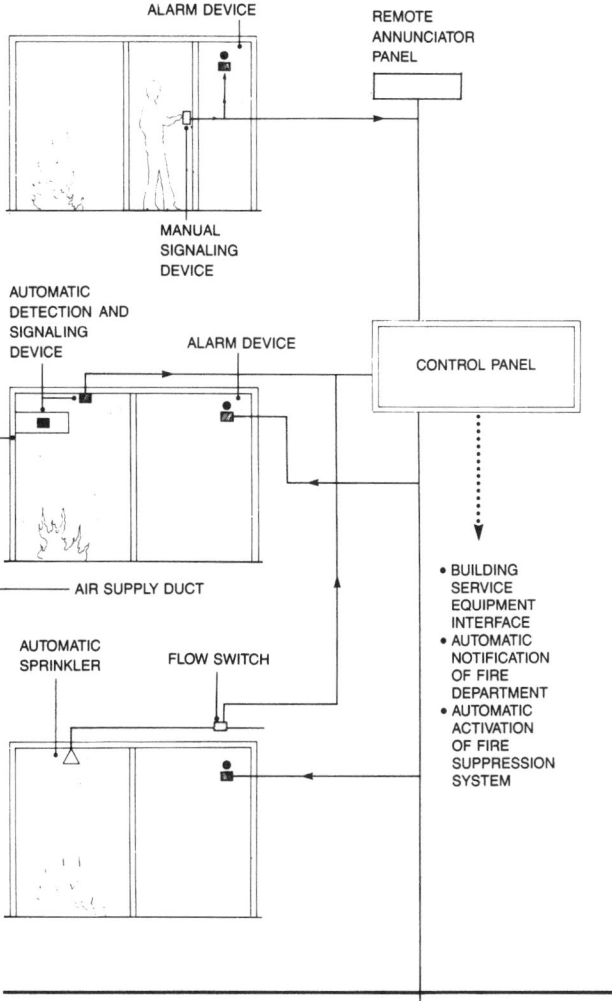

ALARM DEVICE

REMOTE ANNUNCIATOR PANEL

MANUAL SIGNALING DEVICE

AUTOMATIC DETECTION AND SIGNALING DEVICE

ALARM DEVICE

CONTROL PANEL

AIR SUPPLY DUCT

AUTOMATIC SPRINKLER

FLOW SWITCH

- BUILDING SERVICE EQUIPMENT INTERFACE
- AUTOMATIC NOTIFICATION OF FIRE DEPARTMENT
- AUTOMATIC ACTIVATION OF FIRE SUPPRESSION SYSTEM

FIRE SUPPRESSION

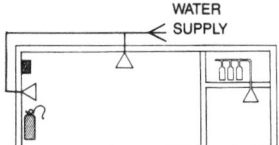

WATER SUPPLY

- FIRE EXTINGUISHER
- STANDPIPE SYSTEM
- AUTOMATIC SPRINKLER SYSTEM
- GAS SYSTEM
- FOAM SYSTEM

■ Fire detection and alarm systems function to alert the occupants of a building to the presence of a fire. A system may consist of:
- Manual signalling devices - often referred to as pull-boxes or fire stations.
- Automatic detection and signalling devices - smoke, heat or flame detectors or flow switches incorporated into automatic sprinkler systems.
- Alarm devices - bells, horns, loudspeakers, and flashing lights activated by the signalling devices.
- Annunciator panels - either centrally located within the building or at points of access to building to indicate location of fire.
- Control panel - to monitor automatic detection devices and to activate alarm devices.

■ Control panel may be designed to control the operation of the building service equipment in event of fire:
- Release hold-open devices on fire doors.
- Recall elevators automatically to ground floor.
- Pressurize fire stairs.
- Shut down recirculating air system.
- Activate smoke exhaust system.

■ Control panel may also:
- Interface with public address system.
- Alert building staff prior to sounding general alarm.
- Monitor the status of the detection and alarm devices.
- Monitor water pressure in automatic sprinkler supply or water level in water towers or tanks.

■ **Fire suppression/extinguishing relies on water or chemical agents to perform one of four basic techniques:**
- **Cooling** - the lowering of fuel surface temperatures;
  —water absorbs heat as droplets are transformed to steam; rate of heat transfer depends on immediate ambient humidity level, water temperature, etc.; typically not effective on flammable liquids and gases; optimum droplet size = 0.01 to 0.04 inches.
- **Smothering** - the displacement or replacement of available oxygen or isolation of the fuel surface from ambient air;
  —displacement of air by steam is effective for suppression of flame but not extinguishing; water with a foaming agent may be used to smother a flammable liquids fire.
- **Emulsification** - the action of creating turbulence in the fuel (usually liquids) in order to disrupt combustion and inhibit release of combustible vapors;
- **Dilution** - the addition of noncombustible matter in order to exceed or fall below the respective oxygen to fuel ratio.

IMPORTANT NOTE:
Relative to building code requirements, the topics presented herein are generalizations used for example purposes only. **Refer to sections of applicable fire/building codes for specific requirements.**

D4.1 FIRE PROTECTION

FIRE PROTECTION/SIGNALING DEVICES

D4.1 FIRE PROTECTION

■ **Detection systems are the key to life safety.**
Placement and selection are affected by;
● construction and occupancy type;
● geometry and thermal aspects of the space, including air movement patterns;
● assessed fuel load; kind, amount and location;
● flame spread characteristics;
● detector speed and reliability.

Smoke Detectors

■ **Ionization type smoke detectors respond to smoke particles that interrupt the flow of ionized air molecules moving between two electrodes within the detector.** They are:
● Most effective where fast burning materials are present or a fast response is important.
● Sensitive to very small concentrations of airborne particles.
● Reliant on the decay of radioactive matter to maintain ionization, but the amount released is roughly equal to that of a watch.
● About 10 seconds faster in flaming, hot, fast growing fires, responding in 50 seconds or less.

■ **Photoelectric type smoke detectors use a light-emitting diode to project a light beam which is reflected by smoke as it enters the detector.** The smoke is detected when light sensors receive the reflected light (scatter type) or when they sense a reduced transmission of the light beam (obscuration type). They are:
● Most effective where slow burning, smoldering fires are present.
● About 5 to 10 minutes faster in cool, smoldering fires, with response time being from 5 to 45 minutes from ignition.

■ **Smoke detectors may be powered by 9 volt battery or can be wired into the electrical circuitry.**
● Often, wired detectors are used with battery-type units as backup. The battery types are typically used in residential construction because emergency power supplies are not used.
● Most battery-powered units will "chirp" when the battery grows weak.
● Direct wired smoke detectors can easily be hooked into a central alarm system. Direct wiring may be required where commercial power is available; only overcurrent protection devices may be

installed to wiring circuits.
● **The biggest problem with smoke detectors is that they wear out over time and are generally not replaced. Testing and maintenance is critical but seldom enacted, and failure of smoke detectors is expected to become a problem.**

Heat Detectors

■ **Heat detectors are thermal sensors used to detect the heat of stage III or IV fires. They are the oldest, least accident prone type of detector. They may respond to a specific predetermined air temperature or to a specific rate of temperature increase.**
● Fixed temperature (FT) units use bimetallic contacts or fusible links to create a response.
  —Bimetallic types rely on the specific thermal expansion of two metal contacts to close an electrical circuit.
  —Fusible types use a metal with a low yield point, usually a low temperature solder, to separate two metal contacts.
● Rate-of-rise (ROR) units are designed to operate when the temperature increases 15° to 20°F in a minute's time. They may be bimetallic, thermoelectric or pneumatic tube types.
● The detectors may be spot (measuring temperature at a specific point in the space) or line type (thermally sensitive over a continuous length of wire or tubing).

Fire/Flame Detectors

■ **Flame detectors offer the fastest response by detecting the radiant energy of a fire, but they have a high potential for false alarms due to other energy radiation in the space (e.g.: a cigarette or high-intensity light).** They are most often used in high hazard locations or where the fastest response possible is needed.
● **Infrared detectors** use a light filter and lens over a photocell. They must be shielded from the sun as the solar radiation will set them off.
● **Ultraviolet detectors** use solid state devices and are not commonly activated by sunlight and artificial light.

Manual Detection

■ **Manual stations are basic to most alarm systems, allowing an occupant of a space to set off an alarm rather than wait for automatic systems to react.** Manual stations are available in several forms:
● Break-glass - a plate or rod is broken to set off the alarm.
● Prewound - the signal is sounded when the prewound mechanism is released.
● Pull lever wound - the signal is sounded when the lever is pulled, winding the mechanism.

Water-flow Detectors

■ **Water-flow switches become detection/alarm devices when installed in the supply piping of automatic sprinkler systems.**
● Vane- or paddle-type mechanisms are set in wet-pipe systems or in the wet side (before the automatic valve) of a dry system, activating an alarm via the movement of water.
● Pressure switches, set at fixed air pressure settings, are used in dry pipe systems to note pressure loss which indicates water flow.
● Another type allows water to fill a small tube, thus completing an electrical circuit.
● When located on every floor, these devices actually pinpoint the source of fire better since the sprinkler is not activated by smoke infiltrating from lower floors.

Heat-sensitive, Emissive Products

■ **New wallcovering products are being developed from special materials and treatments which, when heated, emit colorless, odorless, and harmless gases that are sensed by ionization type smoke detectors.**
● Materials are designed to reveal Stage I fires (about 300°F), before smoke/flames are visible in a room.
● Some fabric materials are designed to discolor or foam/char (intumescent) in the area being heated.
● These products are excellent for fires concealed in walls or ceiling assemblies; also good for fires ignited near walls.
● Excellent for detecting problems due to frayed, shorted, overloaded and other types of faulty wiring.

DETECTION DEVICES

1-5: RANGE OF PROPERTIES FROM BEST TO LEAST DESIRABLE

TYPE	OPERATION	COMPARATIVE PROPERTIES:				RECOMMENDED USES/PLACEMENT:
		FIRE STAGE	COST	RELIA- BILITY	RESPONSE TIME	
HEAT DETECTORS **fixed-temp.(FT)** Type I, point	bimetallic; fusible link nonrestorable	IV	1	1	5	good for small, confined spaces containing fast-burning material or where reliability is a concern; pneumatic types may be concealed from view; requires transfer of heat from air to device, thus taking longer to respond; operating temperature usually 25°F above ambient; install in grid pattern with spacing based on desired response time.
Type II, point	restorable					
Type III, line	insulated, 2-conductor cables arranged to short-circuit					
rate-of-rise (ROR)	bimetallic, pneumatic or thermoelectric	IV	1-2	1	4	
Type IV, line	insulated/nominalist tubing terminating in an orifice-vented chamber with alarm initiating contact					
Type V, point	restorable; orifice-vented chamber and alarm initiating contact					
Type VI, point	restorable					
combination		IV	2-3	1	4	
Type VII, line	restorable ROR w/integral nonrestorable FT detector(s)					
Type VIII, point	restorable ROR w/integral restorable FT detector(s)					
SMOKE DETECTORS **ionization**	interrupted ion flow between electrodes	I	3-4	2	2	on walls, 4 to 12 inches from the ceiling or corner; on ceiling, min. 4 inches from wall; locate within convective air pattern but keep away from excessive air movement, humidity or cold; maintenance is required.
photoelectric	reduced intensity of light beam or reflected light	II	4-5	2	2	
FLAME DETECTORS **ultraviolet**	filters all but UV light in modulation common to flames	III	5	3	1	shield from flickering lamps, reflections, etc.; for spaces with high ceilings, high air flow or instantaneous ignition of contents of space; maintain clear area in front of detector.
infrared	filters all but IR light in modulation common to flames	III	5	3	1	
MANUAL STATIONS **pull lever**	lever is pulled to initiate alarm	II-III	1	1	1	locate within 150-200 feet of any point on the floor; place along exit/evacuation route.
break-glass	glass is broken to initiate alarm	II-III	1	1	1	
WATER-FLOW DETECTORS **vane-, paddle-type**	passage of water in piping moves paddle/vane	I-IV	1-3	1	2	switch should be installed in wet-pipe sprinkler systems in an accessible location adjacent to piping tees to each floor.
pressure switch	senses air pressure changes in dry piping systems	I-IV	1-3	2	2	
GAS SENSING DETECTORS **semiconductor**	conductivity changes in response to oxidizing or reducing gases	II-III	2-4	2	3-4	keep away from aerosol sprays or other oxidizable vapors; locate near/on ceilings in convective air flow.
catalytic element	heat created by oxidation of combustible gases.	II-III	2-4	2	3-4	

Time-Saver Standards for Building Materials and Systems

D · SERVICES

D4.1 FIRE PROTECTION

FIRE ALARM SYSTEMS

■ **When a fire is detected, an alert is communicated to occupants:**
Alarm systems are designed to facilitate the most immediate response by integrating the following components:
● a control unit with one or more initiating device circuits; multi-circuit units use annunciators to indicate location of activated detectors.
● primary (main) and secondary (standby) power supplies.
● heat, smoke and/or flame detectors tied to the control unit.
● signal or voice/intercom alarm devices to alert occupants.
● communications link with local firefighters.

■ **The type of alarm system is defined by the extent of communications.**
● **Local systems** alert only the occupants, who must contact the fire department.
● **Proprietary systems** alert a central control panel, initiating action by building safety personnel.
● **Central station systems** alert the building occupants and a central control panel. Personnel at the control panel often supervise several buildings, notifying the fire department when alarms are activated.
● **Auxiliary systems** alert the fire department by means of direct line communications and simultaneously alert the local system. This type is frequently required in hotels, high-rise buildings and buildings in high density areas; it is required where invalid/ incapacitated occupants are regularly present.
● **Remote station systems** have a direct communications link with the fire department and a central control panel.

■ **Signal alarms use only a continuous or coded tone to direct the occupants in a space in the case of an emergency.** They are located in hallways or workspaces such that they can be readily heard by everyone on each floor.
● **Noncoded** signals are locked into continuous tone operation until manually reset.
● **Positive non-interfering** signals produce a minimum of four rounds of coded signals without interference from any other similar station on the circuit.
● **Presignal** alarms send a coded or noncoded signal to other automated

systems or the control monitor simultaneously as to the local general alarm.
● **Shunt non-interfering coded** signals produce a minimum of four rounds of coded signals without interference from any station on the circuit but further away from the control panel.
● The signal may alert the entire building (**master code system**); it may identify the location where the fire was detected (**selective code system**); or it may identify the floor or area where the fire was detected (**zone code system**).

■ **To be properly heard, signal alarms must exceed the ambient noise level (by about 15 decibels) so that all occupants may hear the alarm.** The alarm should be set 5 decibels higher than the peak noise in the space.
● Ambient noise levels can range from 30 decibels (at a whisper) to 130 decibels (the threshold of pain).
● **Buzzers and chimes** are generally effective in areas of light noise, such as offices and storage areas.
● **Bells** can be effective in areas of greater activity, such as shipping rooms or open office plans. Vibrating bells are usually louder than single stroke bells; single stoke bells are generally used when coded ringing patterns denote specific messages/ alarms/locations.
● Light machinery areas, larger assembly rooms, and operations with openings to outside often require the louder alarm of a **convention horn.**
● As mechanical activity increases, noise level increases and louder alarms are required. **Sirens, heavy duty horns, howlers and air trumpets** are generally used for such areas.

■ **Intercom, or voice alarm systems combine a coded signal with voice messages.**
● Intercom alarms are tied into the central communications system of the building.
● Often used in hotels and offices, the alarm/intercom speakers are located in each room so that occupants can be notified as to further instructions.

■ **When sound is not an effective or appropriate medium for alerting occupants or safety personnel, other systems are available.**
● At central control panels, **annunciators** provide a visual identification of alarm location.

● In occupancy groups where panic may be dangerous or where the hearing-impaired may be present, **fire lights** are most effective. Fire lights are typically high intensity strobe lights.

■ Alarm/detection systems should incorporate alarm discriminators.
● Discriminators are used to minimize false alarms due to extraneous disturbances.
● They can be adjusted to provide alarm discrimination under any job conditions.
● They may consist of a special circuit incorporated in a detector, or a specific device added to a system.

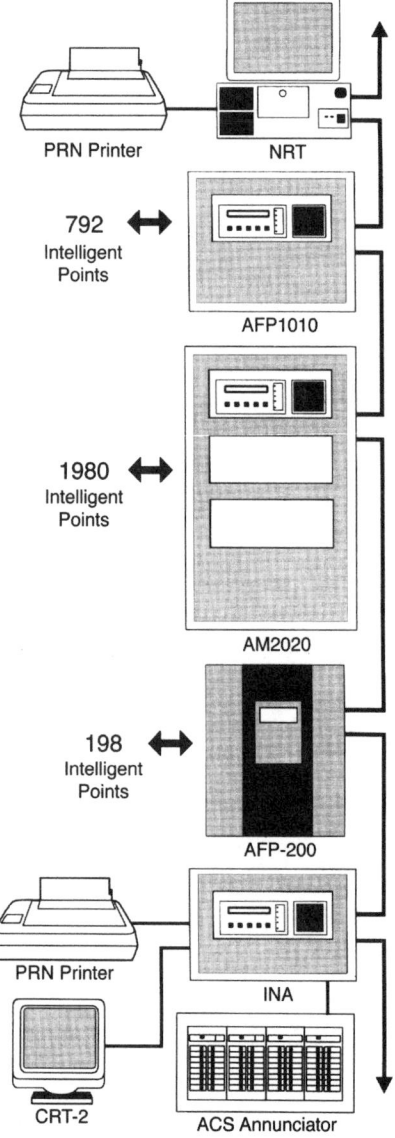

INTEGRATED NETWORK FIRE ALARM SYSTEM
Source: "Fire Alarm Systems" (Ref.1).

FIRE SUPPRESSION

■ The reliability and speed of response are critical decision points for sprinkler system design. There are several basic system alternatives:

● **Wet Pipe Systems -**
 —rapid response, 98 percent reliable;
 —relatively simple design;
 —pipes and water mains contain water under pressure at all times;
 —often chosen in new construction.
 —water damage from loosened pipe joints or damage to sprinkler heads is possible;
 —not recommended where piping passes through a space that have ambient temperatures at or below freezing (32°F) unless antifreeze is added to the water supply.

● **Dry Pipe Systems -**
 —delayed response due to the time needed to release the air in the line before the water can reach the sprinkler head;
 —dry valves allow pressurized air in the water lines to hold water back until a sprinkler head opens;
 —especially appropriate for use in unheated, low hazard occupancies; not recommended in high hazard occupancies.
 —the response delay requires that a greater quantity of water be delivered to balance fire growth; this will increase piping size and possibly the number of sprinkler heads over that for wet pipe systems.

● **Pre-action Systems -**
 —relies on a two-stage operation - notification of the hazard via the alarm and execution via sprinkler response;
 —have the advantages of the dry pipe system, but add a time delay to the fire alarm system to fill the lines with water prior to the sprinkler head opening;
 —a pre-action valve fills the pipe with water as the sprinkler heads respond to the fire.

● **Fire Cycle Systems -**
 —designed to minimize water damage while controlling the fire;
 —operate as pre-action systems, but use sensing devices that cycle sprinkler heads on and off;
 —most systems will shut off automatically when the fire is out.

● **Deluge Systems -**
 —similar to pre-action systems, except that all sprinkler heads are kept open;
 —any fire alarm will open the deluge valve, supplying water to all sprinklers on the piping system. Thus, the entire protected area is deluged with water.
 —used for high-hazard, often unheated, spaces or at multiple levels for atrium style lobbies.

SPRINKLERS

■ **Automatic sprinklers operate at certain temperatures due to the response of integrated heat sensors;** sprinklers for light and ordinary hazards are rated from 135° to 170°F (100°F ceiling temperatures) and high hazard sprinklers range from 175°F to over 650°F (150° to 625°F ceiling temperatures).

● **Fusible:** Soldered metal links keep the sprinkler head closed until the temperature reaches the melting point of the solder. When the solder yields, the connection between the links is broken and the sprinkler head opens. Fusible metal alloy pellets may also be used as the heat sensor, allowing a spring action to open the head when the pellets melt.

● **Frangible:** A breakable, transparent glass bulb contains a colored liquid which expands to a pressure which breaks the glass, opening the head. The liquid is colored to indicate temperature rating (orange/red for 135° to 170°F).

■ **Sprinkler heads are available in many types/forms which respond to the type of construction, occupancy or suppression system.**

● **Large-drop heads** discharge both large (2 mm) and small (1 mm) droplets of water simultaneously.

● **Early suppression fast response (ESFR)** heads have highly sensitive thermal sensors and a very large orifice and deflector to produce larger drops.

● **On-off** features in sprinkler heads provide sensors which allow the head to shut off when the fire is suppressed and to turn on again in case of a flare-up.

● **Residential sprinkler systems** have been developed and approved for home safety. Low temperature sprinkler heads are used, and plastic piping is generally used for water supply.

STANDPIPES & HOSE STATIONS

■ **Standpipe systems are used mostly in high-rise, institutional, assembly and some high-hazard buildings for fire suppression at close range.**

● Buildings over 75 feet in height require some sort of internal water supply for hose hookup because firefighters cannot effectively reach above that height from the street.

● Standpipes should be located in or near stair enclosures for protection of firefighters preparing to enter the floor that is burning.

■ **Standpipe systems are usually a wet pipe design, operating from a separate water reserve and/or upfeed pumping.**

● Some systems incorporate an automatic and/or a semiautomatic starting pump which responds to a fall in water pressure. Pressures should be 20 psi minimum, with 40 to 50 psi preferred.

● Standpipe systems commonly use 4 to 6 inch piping and valve networks to deliver water to the hose station.

● Dry-pipe systems can be used in unheated occupancies that are *not* considered high-hazard. Water may be supplied by fire department pumpers connected to open lines or by valves and pumps connected to central alarm systems and municipal water supply.

● Wet-pipe systems may have a constant supply of water under pressure to all hose stations or dry branch pipes with automatic devices which supply water when a hose valve is opened.

● When automatic sprinkler systems are connected to standpipe systems, required water supply for the standpipe system must be in addition to that of the sprinkler system.

D4.1 FIRE PROTECTION

FIRE SUPPRESSION (cont.)

■ **Hose stations contain fire hose and hose storage facilities, hose valves and possibly one or more accessories.** They are required to be located on each floor of a nonsprinklered building in which a standpipe system is used.
- Hose station housings/cabinets may be surface-mounted, recessed, semi-recessed or trimless. Door panels and trim styles vary.
- Hose may be rolled onto a reel which swings out for storage and ease of operation or may be folded into a rack which allows an easy pull to release hose.
- Attached hoses are 1 1/2 inches in diameter and 50 to 150 feet long (100 foot maximum usually recommended). They are made of synthetic plastic fibers or cotton with rubber lining.
- 2 1/2 inch hose valves can be integrated for hookup of fire department hoses.
- Accessories in the station may include a fire extinguisher, a fire blanket and/or a fire axe. Fire blankets are usually flameproofed wool or aluminized fabric; they are often used in laboratories for chemical fires.
- Controls include a shutoff valve and water flow device. Pressure reducing valves are often required when fire pumps are used.

■ **Hose nozzles are available in adjustable fog; spray type (30, 60 or 90 degree) and straight-stream, smoothbore type; or a combination solid-stream and spray.**
- The largest nozzle which can be safely hand-held is considered to be 1 1/8 inches, allowing a water flow over 200 gpm at 40 or 50 pounds of pressure.
- Solid-stream nozzles can deliver 120-560 gpm of water from 3/4-2 inch nozzle tips (at 40-100 pounds of pressure).
- 1 1/2 inch hoses for occupant use generally use the combination nozzle type.

Siamese Connections

■ **External hookups are essential to provide an adequate water pressure and volume.**
- Specifications for siamese connections must match municipal fire department standards for type and number per inch of connection thread, male or female, and type and size of operating nuts.

- These connections should be within 50 feet of a fire hydrant.
- The connections are usually of brass, galvanized pipe or similar materials appropriate to the use.
- Siamese connections can be mounted on the exterior of the building or may be located along a sidewalk (remote).
- A remote connection tied directly into the underground main which services the sprinkler/standpipe system should only be used when the building does not have a fire pump.

■ Siamese connections for fire department hookup to street-level hydrants should distinguish between water supply for standpipe and sprinkler systems by label or color code.
- When fire pumps are used in a fire suppression system, the fire department connection must be connected on the discharge side of the pump. This prevents pumping by the fire department through the building's fire pumps.
- Supply from hydrants will be up to 1000 gpm.

Fire Pumps

■ **Automatic and/or manually controlled fire pumps may be installed in the basement to strengthen the water supply.** Factors to be considered include:
- pressure fluctuations due to demand;
- resistance within the piping due to friction and gravity (pumping height);
- starting surges from pumps;
- pump pressure-flow characteristics;
- motor control, maintenance and back-up.

■ Self-priming centrifugal type fire pumps are most common. They may be either horizontal or vertical shaft turbine versions.
- Air-release valves are required to ensure automatic operation of split-case horizontal fire pumps.
- Pumps are powered by electric motors (on emergency power circuitry), diesel engine or steam turbine (least common).
- Pump performance varies:
  - flow = 200 to 5000 gpm.
  - pressure = 40 to 200 psi (horizontal types); 75 to 280 psi (vertical types); or 40 to 100 psi (make-up or jockey pumps).
  - rated capacity = 130 to 150%.

STANDPIPE SYSTEMS

CLASS	SYSTEM TYPE	OUTLET SIZE	INTENDED USER	STANDPIPE LOCATION	BUILDING REQUIREMENTS
I	dry	2 1/2"	fire department & trained personnel	each floor above grade; within stairwell or protected enclosure; above roof line (3 way outlet) wherever roof slope is less than 4 in 12	over 4 stories; where pipes may be subject to freezing
II	wet	1 1/2"	bldg. occupants	in corridors near stairwells such that a 100 ft. hose can reach within 30 ft. of any part of the bldg.	multi-storied occupancies
III	combo	1 1/2" & 2 1/2"	bldg. occupants; fire personnel	each floor above grade; such that a 100 ft. hose can reach within 30 ft. of any part of the building; above roof line (3 way outlet) wherever roof slope is less than 4 in 12	nonsprinklered; less than 4 stories; occupancy load of over 1000; over 20,000 square feet per floor

Fire Tanks

■ **Water tanks may be located on mechanical floors or on the roof.**
- Typically, a fire tank occupies a large space and adds considerable dead load to the structure.
- A fire reserve of 5000 gallons or more is generally recommended. Reserve water tanks rely on gravity or compressed air to provide pressure.
- **Gravity tanks** can be used when the tank can be located at least 35 feet above the top level of sprinklers. Elevated storage tanks, once common, have declined in use due to esthetics and problems with freezing. Steel gravity tanks range in size from 5,000 to 500,000 gallons, while wood tanks range from 5,000 to 100,000 gallons. The tank may be mounted on top of a building or on a pedestal/tower of steel or reinforced concrete.
- **Pneumatic tanks** force air into the top of the tank to produce the desired water pressure.
- **Ground-level suction tanks** are combined with fire pumps and range in capacity from 50,000 to 1 million gallons.

■ Combinations of pumps and tanks can be used to ensure an adequate supply of water.
- Water tanks in the basement can be fed directly from the water main. Fire pumps can then be used to move the water upward through the standpipe from the suction tanks.
- The fire pumps will also be capable of filling a gravity tank on upper levels, thus providing an emergency back-up.

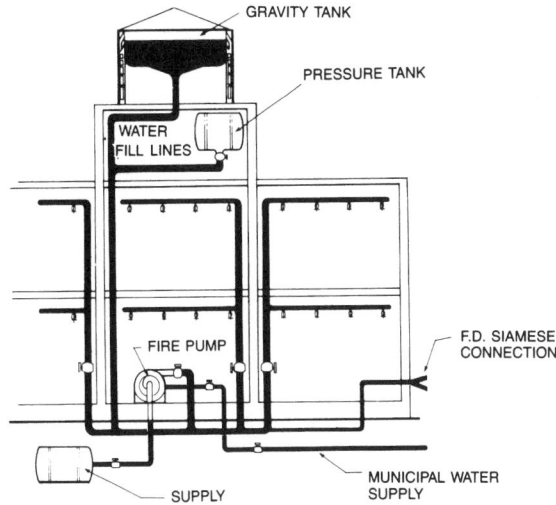

CHEMICAL SYSTEMS

■ **Chemical fire suppression systems typically use formulations specifically designed for given situations.**
- **Carbon dioxide** may be used as an extinguishing agent where water damage is to be restricted. The gas should not be used where contact may be made with oxygen-rich materials or reactive metals such as sodium or potassium. Occupants must be evacuated when the gas is deployed.
- **Dry chemical** systems are often used for electrical and flammable liquid fires. Discharge is generally forced under air pressure. While these systems are good for burning grease in kitchens, a messy residue is left that must be cleaned up. They are not appropriate for electronic equipment/equipment rooms.

- **Wet chemical** extinguishing systems, also referred to as foaming or liquid agent extinguishing systems, are often employed to control flammable liquids. Systems may be designed to inundate a surface or to fill cavities. Foaming methods include air-foaming, chemical-foaming and high density chemical foaming.
- **Dry powders** have been developed to combat burning of very specific combustible metals and chemicals.
- **Halogenated agents**, or halon, was developed due to the need to minimize water damage from sprinkler systems. Computer rooms, electronics/control rooms, or rooms which contain delicate materials (papers, fabrics, relics, etc.) are appropriate locations for halon systems.

■ **Halon suppression systems use a halogenated gas, usually halon 1301 (bromotrifluoromethane) or 1211 (bromochlorodifluoromethane), under pressure from nitrogen or halon 1301 (known as dual halon).**
- Nontoxic in its stable form, it breaks down into several potentially toxic compounds when sprayed on fire.
- The room must be contained; automatic closers are recommended for all doors; proper ventilation is required before the space can be reoccupied.
- Separate tanks of gas (100 to 500 pound capacity) are generally used for each of the protected rooms.
- Halon may be stored as a liquid at room temperature, vaporizing at higher fire temperatures; or may be stored under pressure, vaporizing on release.
- The first cost and subsequent maintenance (refilling the tanks after use) costs are generally high.

FIRE EXTINGUISHERS

■ Fire extinguishers are required to be located on each floor or in large spaces. They should be located near an exit and away from obvious fire hazards which may restrict access to the extinguisher. They should not weigh more than 50 lbs.
- Extinguishers should be located in or near any kitchen area.
- Where the fire is producing considerable smoke or is growing quickly, the first course of action is to leave the scene, not fight the fire with an extinguisher.
- Smoldering fires, such as in furniture, should never be considered as extinguished until adequate time has passed without evidence of continued burning.

■ **UL labels on the fire extinguisher indicate the relative firefighting capacity of the unit, with the number preceding the Class denoting multiples of fire size.**
- A Class 1A extinguisher is capable of putting out a blazing stack of wood (50 pieces of 20 inch long 2x2).
- A Class 1B extinguisher is capable of putting out blazing pool of flammable liquid (3 1/4 gallons contained in a 2 1/2 square foot pan).
- A 2A or 2B rating means the extinguisher can put out a fire twice that size.

■ Portable fire extinguishers are available in a variety of sizes, operations and mountings, based on extinguishing agent. Mountings include:
- wall brackets;
- wheeled carriages;
- cabinets; and
- hose stations.

D4.1 FIRE PROTECTION

EXTINGUISHING SYSTEMS

EXTINGUISHING AGENTS	EXTINGUISHING ACTION					FIRE CLASS				RESIDUAL EFFECTS						FORM: STORED/RELEASED						DELIVERY METHOD					
	COOLS	SMOTHERS	DILUTES	EMULSIFIES	REACTS CHEMICALLY	SOLIDS	LIQUIDS	ELECTRICAL	COMBUSTIBLE METALS	REIGNITION POSSIBLE	NONCONDUCTIVE	CAN BE TOXIC	PENETRATING	CORROSIVE	MESSY	POWDER/POWDER	LIQUID/LIQUID	LIQUID/FOAM	LIQUID/GAS	GAS/GAS	SALTS/FOAM	TOTAL FLOODING	LOCALIZED SPRAY	HOSE AND STANDPIPE	FIRE EXTINGUISHER	CONTAINED SUPPLY	OTHER
water: spray	●					●									●		●					●	●	●	●		
steam		●				●																					●
stream			●			●									●		●					●		●			
wet water	●					●							●		●		●					●		●			
thickened	●	●				●	●								●		●					●		●			
emulsified	●			●		●								●	●		●					●		●			
carbon dioxide	●	●					●	●		●	●	●								●		●	●		●	●	
dry chemical																											
sodium bicarbonate		●			●		●	●		●	●			●	●	●							●		●		
ammonium phosphate		●			●	●	●	●		●	●			●	●	●							●		●		
potassium bicarbonate		●			●		●	●		●	●			●	●	●							●		●		
foaming agents																											
film-forming	●	●				●	●							●	●			●					●	●			
med./high expansion	●	●				●	●					●			●			●				●					
surfactant	●	●	●			●	●								●			●				●					
halogenated agents																											
halon 1211					●	●	●	●		●	●	●	●						●						●	●	●
halon 1301					●	●	●	●		●	●	●	●							●		●	●			●	
halon 2402					●	●	●	●		●	●	●		●					●			●	●			●	
dry powder																											
graphitized coke	●	●							●							●											●
NaCl w/additives		●							●							●											●
flux: molten/crust		●							●																●		●
trimethoxyboroxine		●							●												●				●	●	

FUEL TYPES/FIRE CLASSES

COMBUSTIBLE SOLIDS:
FIRE CLASS A - pyrolytic decomposition of artificial fibers (acetate, acrylic, nylon, polyester, etc.), natural fibers (cellulosic/plant or protein/animal), plastics, wood, rubber, etc.

FLAMMABLE LIQUIDS:
FIRE CLASS B - vaporization of combustible (polishes, kerosene, heavy oil products, grease, some paints), flammable (automotive fluids, paint thinners, some paints) and extremely flammable (gasoline, acetone solvents, isopropyl alcohol, contact adhesives, some stains) liquids. Flammable liquids are always labelled.

GASES - existing in flammable (ethylene, hydrogen, methane, propane) or reactive (fluorine, chlorine) phase state. Extinguish by removing heat and/or fuel.

ELECTRICAL:
FIRE CLASS C - propagation of heat in combustible substances due to electrical current.

COMBUSTIBLE METALS:
FIRE CLASS D - metals and alloys (aluminum, lithium, magnesium, potassium, sodium, titanium, uranium, zinc, zirconium) which react severely when in contact with water.

D4.1 FIRE PROTECTION

FIRE EXTINGUISHERS

AGENT	CLASS	CAPACITY	EFFECTIVE RANGE (FT)	PRESSURE SOURCE	LIMITS ON OPERATION	DISCHARGE TIME	USES
WATER	2A	1.5 gal.	20	stored pressure	°F > +32	30 sec.	
	2A	2.5 gal.	30-40	pump			
	2A	2.5 gal.	30-40	chemical expellant		60 sec.	
	2A	2.5 gal.	30-40	stored pressure		60 sec.	
	2A	2.5 gal.	45-55	gas cartridge		30 sec.	
	3A	4 gal.	30-40	pump		120 sec.	
	4A	5 gal.	30-40	pump		2-3 min.	
	10A	25 gal.	35	CO_2 cylinder		1.5 min.	
	10A	17 gal.	50	chemical expellant		3 min.	
	20A	33 gal.	50	chemical expellant		3 min.	
	30A	45 gal.	35	CO_2 cylinder		2 min.	
	40A	60 gal.	35	CO_2 cylinder		2.5 min.	
	2A, .5-1B	2.5 gal.	30-40	stored pressure	°F > -40	60 sec.	
CARBON DIOXIDE	1-5B:C	2.5-5 lb.	3-8	self-expelling	°F > -40	8-30 sec.	
	2-10B:C	10-15 lb.	3-8			8-30 sec.	
	10B:C	20 lb.	3-8			10-30 sec.	
	10-20B:C	50-100 lb.	3-10			10-30 sec.	
HALON 1301	2B:C	2.5 lb.	4-6	stored pressure		8-10 sec.	valuables storage areas, computer rooms, etc.
1211	1-2B:C	1-2 lb.	6-10	stored pressure	°F > -40	8-10 sec.	
	5B:C	2.5 lb.	6-10			8-12 sec.	
	1A:10B:C	5.5-9 lb.	9-15			8-15 sec.	
AMMONIUM PHOSPHATE	1A:10B:C	2.25-2.5 lb.	5-12	stored pressure		8-10 sec.	garages, utility areas
	2A:10B:C	4-5 lb.	5-12	compressed gas	°F > -40	10-15.5 sec.	
	2A:20B:C	9-17 lb.	5-20	stor. pres./cart.		10-25 sec.	
	2A:80B:C	17-30 lb.	5-20	stor. pres./cart.		10-35 sec.	
SODIUM BICARBONATE	2-10B:C	1-2.5 lb.	5-8	stored pressure	°F > -40	8-12 sec.	kitchens
	5-20B:C	2.75-5 lb.	5-20	cartridge/st. pres.		8-20 sec.	
	10-160B:C	6-30 lb.	5-20	cart./stored pres.		10-25 sec.	
POTASSIUM BICARBONATE	5-20B:C	2-5 lb.	5-12	cart./stored pres.	°F > -40	8-10 sec.	
	10-80B:C	5.5-10 lb.	5-20	cart./stored pres.		8-20 sec.	
	40-120B:C	16-30 lb.	10-20	cart./stored pres.		8-25 sec.	

D4.1 FIRE PROTECTION

FIRE ALARM & SUPPRESSION:

■ **Suppression systems operate on the basis of the following:**

● **Total area flooding** - Response to fire from fixed-in-place apparatus is intended to encompass the complete space; usually a smaller, confined area. Extremely rapid detection of fire is often required.

—Occupants are often required to leave the immediate enclosure to eliminate any chance of toxic reaction when chemical agents are used. The chemical, usually in gas form, is uniformly discharged throughout the area. Chemicals are generally gas and are piped from storage tanks to nozzles.

● **Localized suppression** - Response to fire from fixed-in-place apparatus is intended to be in the immediate area of combustion.

—Method of discharge (through nozzles, tubes, troughs, chutes, etc.) depends on the phase of extinguishing agent (solid, liquid or gas).

—Chemical systems rely on the placement of concealed storage tanks above areas of specific hazard. The tanks are more easily installed and concealed than sprinkler systems, however they supply a limited amount of suppression potential.

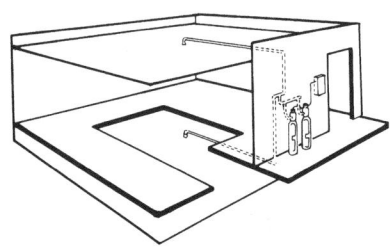

● **Standpipe systems** - Response to fire from hand-held hose lines is intended to provide direct application of extinguishing agent to the fire.

—Water systems rely on piped water supply.

—Chemical systems are not permanently connected to one supply, but are supplied by fixed or mobile containers/tanks.

● **Portable fire extinguishers** - Response with hand-held containers is intended for fires that are small enough to approach within 10 or 12

feet. At this range, application is safe and the fire has not become too severe.

■ **Two basic concepts are used to design automatic fixed-in-place applications: Fire-control and fire-suppression.**

● **Fire-control** is based on the premise that the system will contain the fire long enough for firefighters to arrive to extinguish it.

—Assumes that the operation of a large number of discharge devices (sprinklers/nozzles) at a specified minimum discharge density will minimize the spread of fire and keep ceiling temperatures low.

—Most effective for storage of low-hazard materials in low (10 foot) stacks.

—Least effective for high-hazard combustibles, stored in stacks of large quantity. High-intensity, rapid growth fires can overwhelm sprinklers before the fire can be controlled.

—Requires the availability of local firefighters to extinguish the fire.

—For water systems, a large supply of water is required to be available for up to 60 minutes.

● **Fire-suppression** is designed to extinguish the fire in its early stages.

—Assumes that the amount of water or chemical discharged onto the fire is greater than the amount required to extinguish the fire.

—Response time is critical and requires that designs reflect the assessed fire hazard.

—Response is designed to limit fire growth from reaching an intensity that will make it harder to extinguish.

AUTOMATIC SPRINKLER SYSTEMS

■ Building codes may allow trade-offs in fire prevention requirements when complete fire suppression systems are used. Allowances may include:

● Increased number of stories or floor area;

● Reduced requirements for fire resistive construction;

● Fewer firewalls and wider firewall openings;

● Reduced exit capacities, longer travel distances (150 to 200 feet), longer dead end corridors;

● Wider spacings of fire hydrants and access roads.

● In cases where sprinklers are required, they will not satisfy other requirements of construction/design.

■ **Many code issues may impact the design and installation of automatic sprinkler systems.**

● Some areas of the building may be concealed, yet contain combustible materials which are exposed within the concealed space. When such a building is sprinklered, it is often required that similar protection be provided within that concealed space.

● Although areas of refuge may be omitted in some cases because the building is sprinklered, new code requirements for the handicapped require reconsideration. A protected area of refuge may be the only choice for a wheelchair-bound occupant.

■ **Evaluation of fire suppression systems in new or existing buildings needs to consider the following:**

● What medium can be used to extinguish the fire — water or chemical?

● If water is used, what is the source and is it automatic or dependent on manual operations?

● Is the sprinkler piping to be exposed or concealed?

—In existing buildings, pipes can be hidden in soffits on either side of corridor so that branching lines can go to spaces on either side.

● What is the relation of the ceiling height to the effective detection/response height of the sprinkler head?

● What is the extent of combustibility of finishes and furnishings within the space?

● Are there any obstructions near the sprinkler that will block the sprinkler discharge from reaching the fire?

● Are special designs required because the building is located in a seismic or flood zone?

● Is the structure designed/capable of handling the additional weight of piping filled with water or chemical storage tanks?

● What conflicts exist in routing piping through the structure (i.e.: ducts, cables, other piping)?

■ **Automatic sprinkler systems are based on horizontal water distribution at ceiling level covering an area of potential fire hazard.** An important issue of sprinkler efficiency is the penetration of the water to the fire:

● Since the fire produces an upward thrust of rising gases, the downward velocity of the water droplets must be greater than that upward thrust. This is accomplished via gravity in smaller

selection checklist

fires, but requires the momentum of higher water pressure for larger, more intense fires.

- Small drops of water (1 mm diameter or smaller) used to provide a larger water surface area will absorb considerable heat, reducing the temperature of gases near the ceiling.
- Very small drops of water, more like a fine spray, are vulnerable to strong gas and air currents, potentially carrying the water away from the area of the fire. Misters, used in cooking/grease fire vents, exemplify this.
- Large drops of water can be used to penetrate high-velocity gas plumes.

■ **Spacing of sprinkler heads generally follows a layout which provides uniform distances between sprinkler and/or branch lines up to a limit of 15 feet.** Layout must ensure that no element of the building will block the discharged water spray.

- Discharge patterns vary with assessed fire hazard; coverage can vary from 100 to 200 square feet per sprinkler head in light hazard occupancies; about 90 square feet per head is used for high hazard conditions.
- The distance of a sprinkler head to a wall or structural column is usually one half of the uniform spacing. Uniform spacing of sprinkler heads varies from about 8 to 15 feet apart along the supply pipe; the pipes are usually about 10 to 14 feet apart and run at right angles to exposed beams or other construction barriers.

TYPICAL SPRINKLER LAYOUT

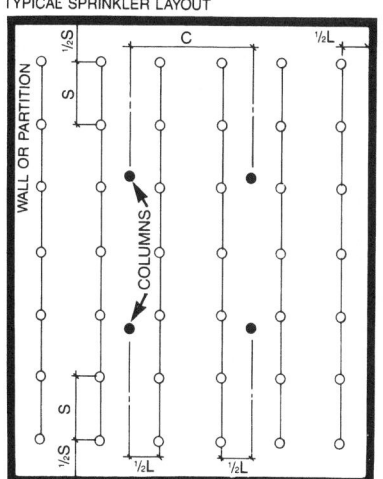

C = COLUMN SPACING
L = DISTANCE BETWEEN BRANCH LINES, LIMIT 15 FEET
S = DISTANCE BETWEEN SPRINKLERS OR BRANCH LIMIT

- Spacing varies with the fire rating of the construction, the occupancy/hazard, the type and configuration of the ceiling, the spacing of joists, and the size of the undivided area of application and the response of the water supply system. Less hazardous locations may allow low temperature heads to be spaced at greater distances or in independent zones. More hazardous locations (e.g.: warehouses containing plastics) may require high temperature heads at overlapping zones of coverage.
- Even water distribution is achieved by overlapping downward, arcing coverages, such that two sprinklers are covering any one area.

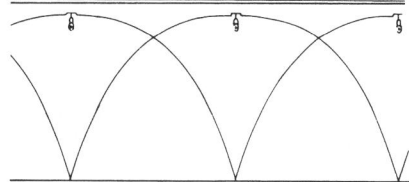

■ **Sprinkler installation will vary according to extent of coverages and exposures.**

- **Pendent -** Water is directed downward through the deflector. Often used below a finished ceiling which conceals the water piping. Heads can be recessed in the ceiling, concealed by a cover plate or mounted flush. They can be plated, polished, or colored — finishes should be factory-applied only. The umbrella-shaped pattern of the discharge helps to control ceiling temperatures.

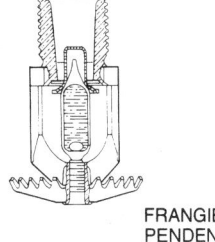

FRANGIBLE PENDENT

- **Upright -** Water is directed upward through the deflector. They are less apt to be damaged by cleaning and

maintenance personnel and are installed to operate at the highest location where the hottest gases accumulate.

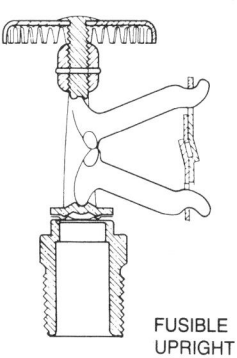

FUSIBLE UPRIGHT

- **Sidewall -** Water is distributed in a half-circle from the wall. They are located within 4 to 14 inches of the ceiling in soffits or along beams and joists, projecting about 15 feet into the protected space. Extended coverage types direct water horizontally through the deflector to produce an arc of up to 20 feet.

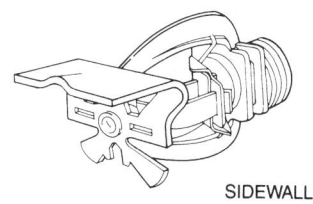

SIDEWALL

- Older sprinkler head design and installation used upward discharge (striking the ceiling, collecting and falling as large drops) as well as downward. The downward, arcing discharge enlarged the area of coverage, but spacing of the heads created small overlaps. The net result was an uneven distribution of water.

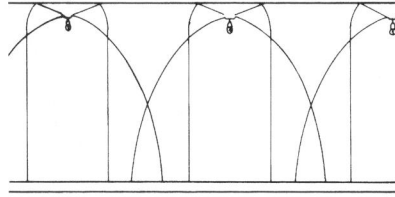

FIRE ALARM & SUPPRESSION:

■ Sprinkler operation is evaluated based on the following:

● **Response Time Index (RTI) -** an indication of sensitivity; the speed at which a temperature is measured.

● **Required Delivery Density (RDD) -** a measure (in gallons of water per minute per square foot) of the minimum quantity of water that must be delivered onto a specific burning commodity to extinguish the fire.

● **Actual Delivered Density (ADD) -** a measure of the rate that water is actually discharged from operating sprinklers and delivered onto the burning surfaces.

● Older, conventional sprinklers contain sensors that are of sufficient mass that by the time the sensor reaches 165°F, the surrounding gases can be as high as 1000°F. The Standard Time/Temperature Curve shows that this could be almost 5 minutes into the fire before the sprinkler operates; an RTI of 225 to 700. This time frame and RTI are reduced by using newer, lower-mass sensors (RTI = 48 to 50), which activate the sprinkler at considerably lower ceiling temperatures.

Sprinkler System Balance Sheet:

Attributes

● reduces flame spread
● reduces heat intensity; both fire and smoke temperature
● reduced temperatures allow faster access to fire source by firefighters

Problems

● does not effect smoke hazard
● adds to construction costs
● requires monitoring to guarantee effective operation

SPRINKLER PIPING

■ **The design of water piping for fire suppression systems is typically approached on the basis of pipe schedule rules or hydraulic calculation.**

● Pipe schedules specify the maximum number of sprinkler heads which can be installed on a pipe line of a given size. The schedule approach is more flexible to occupancy changes in the protected space.

● Hydraulic analysis sizes pipes to distribute a specified uniform water density, flow and pressure over a specific area. This design approach optimizes the available water supply and boasts a lower initial cost.

● The piping must be sized to deliver 13 to 18 gallons per minute (gpm) per sprinkler head with more than one sprinkler head open.
— If the water pressure to the building is not sufficient, a fire pump will be required for sprinkler system.

● The piping design must include the facilities to drain all water to specific discharge points.

■ Piping products used for fire sprinklers are available in steel, copper or plastic.

● The flexibility of plastic piping have improved retrofit possibilities.

● Steel is more commonly found in commercial applications.

● Common components of the distribution system include valves, flow indicators, pressure gauges and control switches.

■ **Polybutylene (PB) and chlorinated polyvinyl chloride (CPVC)** pipe and fittings have been approved for fire sprinkler piping.

● Its use is limited to indoor wet pipe systems for residential and light hazard occupancies (as defined by NFPA 13 and 13D) in locations where ambient temperatures will not exceed 120°F.

● Approved PB pipe is available in straight lengths up to 20 feet with outside dimensions matching steel pipe (IPS) or copper tubing (CTS).

● Standard fittings are approved, including adapters and fittings with threaded brass inserts for direct connection to a sprinkler head.

● Joining is accomplished by socket fusion. The pipe and fitting are joined by heat; a thin layer of PB is melted on the outside of the pipe and inside the fittings. The joint is formed when

the pipe is pushed into the fitting, and the melted polybutylene flows together, cools and resolidifies. PB pipe cannot be joined with solvent cement or glues.

● Fire protection of PB piping must be provided by one layer of 1/2 inch gypsum board, fire rated suspended ceiling assemblies or even 1/2 inch plywood soffiting.

■ **The supply of water to sprinkler heads may be provided by a riser which enters the protected space via a fire-resistive plumbing chase or may be connected to a standpipe system.** Piping designs include:

● **Tree:** Branch lines are run parallel to each other and perpendicular to the main line. The main line is teed from the riser, beginning at a large diameter (4 inch) and declining (to 2 inch) as the number of sprinkler heads supplied decreases.

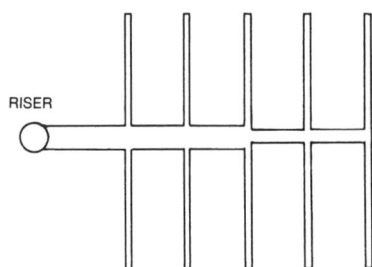

● **Loop:** The riser feeds a main loop with parallel branch lines connected perpendicular to two sides of the loop. Loop pipe diameter is smaller (2 inch) and consist throughout.

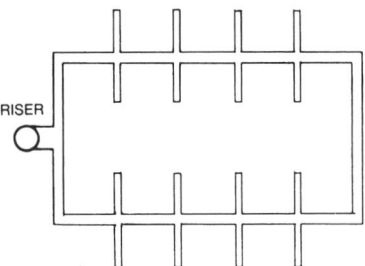

selection checklist (cont.)

• **Grid:** Allows the smallest diameter pipe (often 1 1/2 inch) to be used throughout the system since water can come from many directions to reach a sprinkler head. The supply riser may be located in the center of the loop or on one end.

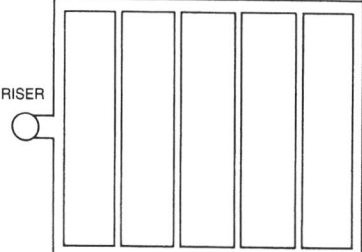

RISER

WATER SUPPLY

■ **Water supply is critical to the effectiveness of automatic sprinkler and standpipe systems.**
• Ideal building location will be sited near a water source—hydrants, public mains, rivers or reservoirs/ponds.
• Without adequate supply on site, the building must stored in tanks within the building and imported via water tank trucks.
• Public water supply is often not pressurized enough for buildings of 3 to 5 stories. Domestic water mains may also be too small to carry sufficient water volume.
• Both pumps and water tanks can be incorporated into the domestic water supply.

■ Water supply needs vary with the fire hazard:
• Light hazards = 15 psi, 500 to 750 gallons/minute for 30 to 60 minutes.
• Ordinary hazards = 15 psi or higher, 700 to 1500 gallons/minute for 60 to 120 minutes.
• High hazards, warehouses and high-rises = pressure and flow determined by specific case and authority with jurisdiction.
• Fire-alarm operated pumps in the basement and/or a reserve water tank

on a top/upper floor should be able to provide these needs until fire department equipment arrives.

■ **To improve the firefighting capacity of water, additives can be entrained into the water supply:**
• **Antifreeze -** based on solution (percentage of water by volume), it effectively lowers the freezing point of water. Types include pure glycerine (-15° to -40°F in 50 to 30 percent solution), pure propylene glycol (+9° to -60°F; 70 to 40 percent), diethylene glycol (-13° to -42°F; 50 to 40 percent; poisonous), ethylene glycol (-10° to -40°F; 61 to 47 percent; poisonous), and calcium chloride (should not be used where public water is involved).
• **Corrosion inhibitor -** sodium dichromate.
• **Wetting agents -** available in many forms, wetting agents lower the surface tension of water. They are used to improve water penetration of material surfaces, stacks or bales; not used on flammable liquids or electrical equipment.
• **Water thickeners -** available in many forms, thickeners reduce water runoff. Thickened water tends to cling to surfaces, producing a continuous coating several times thicker than water. Thickener water does not penetrate surfaces and droplet size is increased.
• **Friction reduction agents -** Polyethylene oxide is often used to reduce friction loss through piping and hoses. Significant increases in flow rate are possible and must be considered when such agents are used.

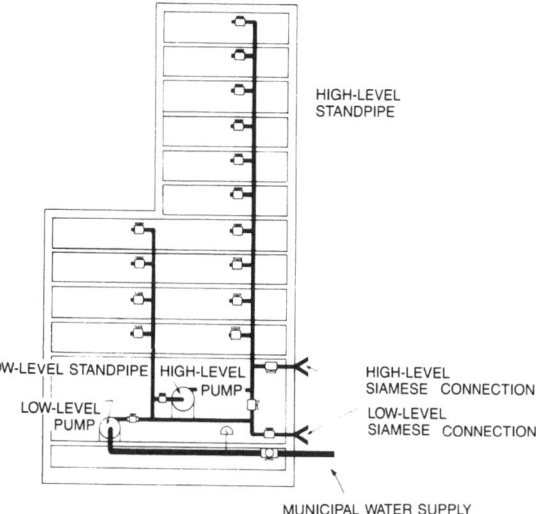

HIGH-LEVEL STANDPIPE

LOW-LEVEL STANDPIPE HIGH-LEVEL PUMP

LOW-LEVEL PUMP

HIGH-LEVEL SIAMESE CONNECTION

LOW-LEVEL SIAMESE CONNECTION

MUNICIPAL WATER SUPPLY

D4.1 FIRE PROTECTION

SMOKE CONTROL

COMPARTMENTATION

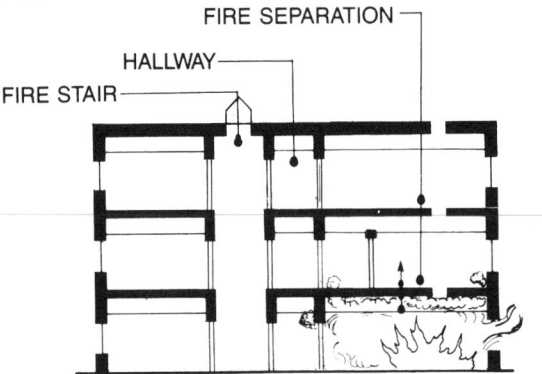

EARLY STAGE: FIRE CONTAINED IN SPACE.
         SOME SMOKE LEAKAGE INTO ADJACENT
         SPACES.

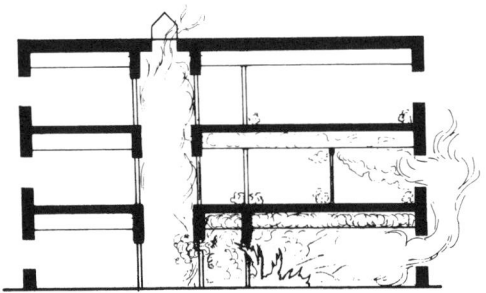

FIRE BREAKS CONTAINMENT AND SPREADS ALONG
HALLWAY AND OUT THROUGH WINDOWS.

SMOKE VENTING

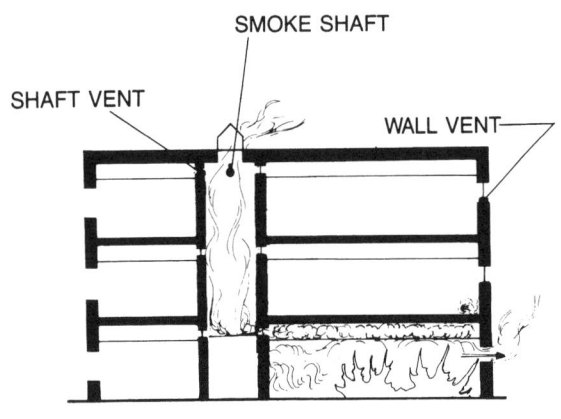

■ **Smoke control systems are used to limit the spread of smoke generated by a fire, isolating selected areas otherwise unaffected by the fire. These systems require rapid response and sealing of even the most minute openings in containment systems and assemblies.**
● The occupants of a building cannot always be expected to exit the building within a reasonable amount of time due to building height, distance to exits, obstacles in the path of egress and poor detection/alarm systems. Smoke control systems are required to perform the primary task of ensuring a smoke-free, safe environment for evacuation.

■ **The total smoke control system should be capable of both containing the smoke within the fire zone and controlling the ventilation of smoke and hot gases of combustion out of the zone. Smoke control systems generally respond to a fire by employing both passive and active techniques.**
● Passive techniques rely on the selection of material or the design of a space to limit smoke migration without mechanical assistance. Examples include compartmentation, natural ventilation and smokestopping.
● Active techniques rely on the use of air handling systems to limit smoke migration and/or create a desired pattern of movement.

PASSIVE SMOKE CONTROL

■ **Compartmentation can be used to protect occupants from smoke as well as fire.**
● Assemblies with 1-hour fire-ratings may be considered as adequate smoke barriers.
● Automatic fire doors, which are capable of containing the smoke to one side of the door only, can be used. The doors use automatic closers and are generally held open with electromagnetic locks that release as soon as a fire is detected.
● Smoke/fire seals around doors can be an important barrier to the spread of hot smoke and fumes through elevator shafts and fire escape stairwells. For example, a 1/4 inch gap around a double door can be equivalent to a 1 foot diameter hole.
● Fire dampers and doors are effective smoke barriers when equipped with devices which operate where smoke is detected. Detector units can be integrated with thermal links to respond to either the heat or smoke of a fire.

■ **Passive venting of smoke is accomplished by providing openings in the roof or upper walls. Pressure buildup is prevented by allowing expanding, combustible gases and hot, rising smoke to escape.**
● **Exterior wall vents are generally windows or wall panels that operate automatically in response to the building's alarm/detection systems.** They are useful where open floor plans allow smoke from interior spaces to move toward exterior walls. Vents need to be distributed evenly along the exterior walls to minimize the adverse effects of wind.
● **Smoke shafts, vertical shaft extending the full height of a building, are effective for smoke removal in both passive and active systems.** The shaft is open at the top to the outside and has openings (protected with automatic dampers) to each floor of the building.
● Wall and roof fire vents may open automatically by fusible link when excessive heat is detected; may have an integrated smoke or flame sensor; or may have glazing panels which melt-out due to heat.
● Venting is highly successful when combined with pressurization.

SMOKE AND FIRE VENTING

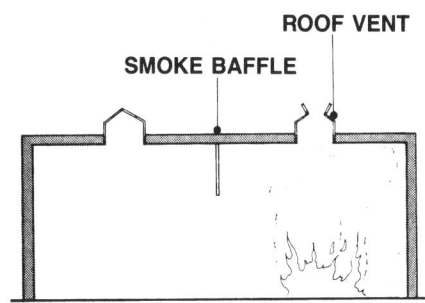

ROOF VENT

SMOKE BAFFLE

PRESSURIZATION AND VENTING

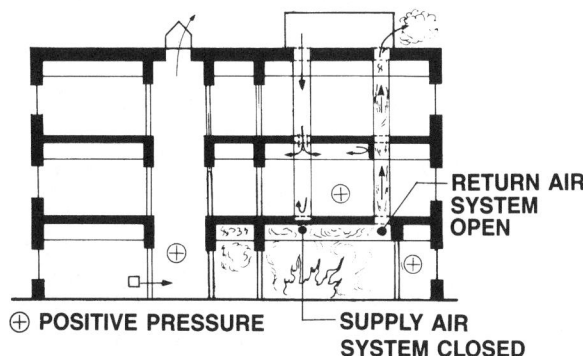

RETURN AIR SYSTEM OPEN

⊕ POSITIVE PRESSURE

SUPPLY AIR SYSTEM CLOSED

PRESSURE DIFFERENTIALS

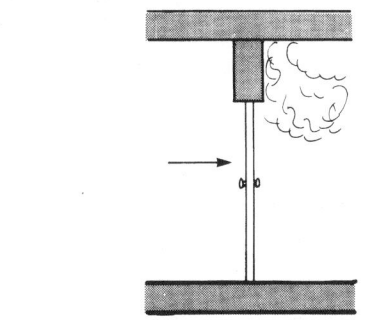

HIGH AIR PRESSURE   LOWER AIR PRESSURE

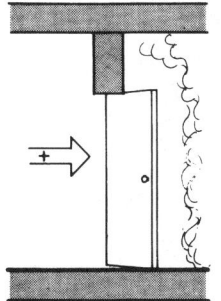

PRESSURIZED AIR THROUGH OPENING AT HIGH FLOW RATE

PRESSURIZED AIR THROUGH OPENING AT LOW FLOW RATE

■ **Fire vents are used for openings in large, flat or slightly sloped roofs.** Large, open single-story spaces and those on the top floor of a building can benefit from the use of automatic roof vents.

● Properly designed and installed, roof vents effectively improve visibility and reduce floor level temperatures to such a degree that firefighters can safely enter the space. This is due to the "chimney" effect of venting over a fire, drawing the smoke and gases straight up and out of the space.

● Roof vent performance is improved by using larger vent ratios (e.g.: from 1/140 to 1/20) and can be achieved using small vents at relatively close spacing.

Effect of Roof Venting on Sprinklers

■ Research indicates that properly installed roof vents will perform well when used in conjunction with sprinkler systems, providing an excellent backup for unexpected sprinkler failures.

● Testing has shown that perimeter venting (via doors and eaveline windows) may be detrimental to sprinkler operation. This is due to increased availability of oxygen in the fire zone and movement of hot gases across the ceiling, thus causing sprinkler heads remote from the fire zone to operate. The activation of sprinkler heads not in the immediate area of the fire will reduce the effectiveness of those above the fire.

● Sprinkler effectiveness can be improved with roof vents by increasing discharge rates (e.g.: from 15 to 25 gpm/head). This is needed due to the increased velocity of rising, escaping gases.

■ For more information on fire and smoke venting, see SKYLIGHTS, SCUTTLES, VENTS.

ACTIVE SMOKE CONTROL

■ **Active smoke control systems are intended to create an imbalance in air pressures in order to allow air leakage into the fire zone for the purposes of cooling, diluting smoke density, minimizing smoke migration, and replacement of air as smoke and hot gases are vented out to a predetermined exhaust point.** They use mechanical air handling equipment/devices to produce pressurization, exhaust venting and dilution.

● **Pressurization is the use of differential air pressures or flows to keep smoke from penetrating openings in the containment system.** Pressurization may be limited to particular areas such as stairways and elevator shafts, or may encompass entire floors or even entire buildings.

● **Exhaust systems mechanically vent smoke and expanding, hot gases to reduce smoke concentration and relieve the chance of flashover.** They rely on exhaust fans, shafts/ductwork and dampers to provide optimum exhaust from the fire floor. These systems respond to the detection of fire by shutting down normal HVAC operation, closing dampers, and operating fan units in specific locations.

● Integrated smoke venting allows only the return air fan to continue operation when smoke/fire is detected.

D4.1 FIRE PROTECTION

SMOKE CONTROL (cont.)

Exhaust fans can be used for mechanical smoke venting; supply air can be used to pressurize zones of the building; and supply air ducts can be sealed off from the fire zone so as to control the intensity of the fire through suffocation.

- Air handling plenums and equipment have traditionally been equipped with heat and/or smoke detectors to minimize the chance of circulating smoke. Recent evidence indicates that the response time for heat detectors may not be effective enough.
- Air handling/distribution systems with a capacity greater than 2,000 cfm should be equipped with duct smoke detectors. The smoke detector should be located in the return air duct before the filtering equipment and before any outside air intakes to maximize the effect of the detector.

■ **The arrangement and operation of HVAC systems requires careful planning and coordination when integrating mechanical exhaust of smoke.**

- When the HVAC system is used to exhaust the fire floor, its capacity to do so will be reduced, possibly to an ineffective level, if the system is designed to exhaust all floors simultaneously.
- The design and specification of return air shafts and plenums must consider the possibility that hot gases and fire can be drawn through with the smoke. Heat-rated fans are often required since exhaust fans may fail to operate if temperatures are too high. Noncombustible ductwork/plenum materials are recommended to reduce the spread of fire hazard via return air systems.

■ **Isolation smoke control systems** are recommended by some fire experts and codes as an alternative to fully automatic systems in high-rise buildings.

- When a fire or its smoke is detected, the HVAC system is shut down, with supply and return dampers closed to contain smoke and fire. The operation of the HVAC system is then controlled at the building's centralized security control panel.
- Controls, operated manually at the panel, include those for all supply, return and exhaust fans; all supply, return, exhaust and mixing dampers. The panel should indicate the position of all dampers and the status of fan operation.
- **Purging or dilution systems remove smoke by exhaust coupled with a supply of fresh air to replace that which is being removed.** This concept of providing additional air changes is used to reduce smoke density. This system is not often recommended because a much greater quantity of air is required over that of a pressurized system.

■ New products and systems are constantly evolving to help "buy" time until firefighters can reach trapped occupants of a building. Their rescue is often a function of available air.

- Critical to hotels, new methods of providing fresh air to trapped occupants are being tested. One such method is to use grey-water drain lines by adopting valves, hoses and masks to vent piping.

PRESSURIZATION

■ **Pressurization systems are automatically activated by the building's alarm/detection system.**

- There are several strategies which may be used to pressurize around fire zones:

—Pressurize each floor and vent the one or more floors of the smoke zone.
—Pressurize only the floors/unaffected areas adjacent to the smoke zone, vent the smoke zone, and operate the remaining floors as normal.

- Relatively low air velocity can be used to keep smoke from totally filling a passageway, but the smoke will be able to travel across the ceiling, with backflow moving into adjacent spaces through openings in the wall. Smoke traps are recommended at such openings to minimize detrimental effects.
- Relatively high air velocity can be used to eliminate backflow issues, as the smoke will be contained by the strong air flow.
- Barriers, such as fire rated doors, can be used to contain smoke when the air pressure on the unaffected side of the barrier is greater than that on the fire side. Air flow leakage through cracks in the construction prevents smoke from moving into the high pressure side.
  —Pressurization level must consider air leakage through holes in the pressurized envelope such as joints, cracks, conduit and pipe penetrations as well as open doors.

■ **Pressurized stairways provide smoke-free paths for evacuation and entry of firefighters.**

- Pressurization is provided at intervals (every 3 or 4 floors) via an air duct within the protected stairwell, rather than only at the bottom of the stairwell. This eliminates the potential for incredible pressure requirements and imbalances that can be created when the air supply is only at the bottom of the stair shaft. The imbalances are caused by air turbulence as it winds through the stairwell and by exfiltration through the stair enclosure. Pressure calculations should also include the movement of relief air at the top.
- In cold climates, the automatic pressurization of the stair tower can be tied into an emergency heating system so that subfreezing conditions of the fresh, outside are not delivered onto people entering the stair.
- Emergency generators and power supplies to air handlers in the stair tower should be protected from fire potential.
- The air supply inlet to the stair tower system must be located in an area where fresh, uncontaminated air will always be available. A smoke detector should be located at the fan to shut down the system if the air source becomes contaminated.
- A manual, master switch should also be part of the system, allowing firefighters to override automatic controls.

■ The design of pressurized stairways must consider door-opening forces:

- A force of at least 3 pounds is normally needed to close and latch a fire door.
- Most door-closing devices require more force in the opening direction than in the closing. This force often doubles the amount of force required to open the door.
- Handicap access requires that door operating force be a maximum of 5 pounds.
- The pressures cannot be so great that doors cannot be opened into the stairwell.

FIRESTOPPING

■ **The fire containment can breakdown as a result of failure to stop the spread of fire and smoke.** Firestopping techniques and materials are designed to eliminate the passage of fire/smoke through or around noncombustible or fire rated assemblies. The goals of any firestopping/safing design are:
- Seal any and all openings and voids in, around or between fire-resistive assemblies.
- Fill any void created by the disintegration of a combustible material penetrating a fire-resistive assembly.
- Remain in place for a duration equivalent to the rating of the fire-resistive assembly.

■ **Firestopping is required wherever:**
- Penetrations occur through fire resistance rated assemblies by building service lines such as piping, ductwork, conduit, etc.
- Joints occur in or voids/gaps occur around fire-rated assemblies.
- Air passages are concealed or pass vertically through floors; such as shafts, ceiling plenums and mechanical, electrical, power and communications chases.
- Concealed spaces, including plenums with very low air flow rates, with draftstopping every 3,000 square feet unless the building is sprinklered.
- Plenums concealed within a floor assembly must be firestopped over/under permanent walls and otherwise to limit clear-open spaces to 100 square feet.

FIRESTOPPING/SAFING MATERIALS

Bulk Materials

■ **Sealants/Foams:**
- One-part elastomers can be used to seal joints or gaps around ordinary metallic plumbing, cable or conduit penetrations in fire-rated wall and floor assemblies.
- Two-part elastomers are often used where more complex penetrations are involved; mixtures/multiples of metallic cables, pipes, etc.
- Intumescent materials should be used wherever assemblies are penetrated by combustible materials (plastic pipe, etc.); they can expand 5 to 10 times their original volume and are effective against fire, fumes, smoke and water.
- Mortar and grout products can be used to provide a solid, nonshrinking

seal for metallic cable and pipe penetrations.
- Other products include mastics, magnesium oxychloride coatings, silicone and polyurethane sealants and foams; silicone foam may be undesirable because of high smoke development ratings.
- In floor/roof/ceiling assemblies, openings generally require that a backing or damming material be installed before the sealant/foam is applied.

■ **Putty/putty tape:** Putty can be hand packed, trowelled, caulked or mechanically pumped.
- In floor assemblies, damming materials or backer rods are required for large openings.
- Putty tapes are wrapped around elements which penetrate a fire-rated assembly so that the tape fits inside the void or covers the void from the exposed side. The tape may have an aluminum foil backing to restrain movement in fire conditions.

■ **Fibrous:** Mineral, ceramic and cementitious-blend fiber insulations can be friction fit in voids to provide a fire barrier.
- For effective smokestopping, a flameproof sealant is applied (top side for floors, both sides for walls); fibrous insulations are porous and have been shown to allow combustible gases and smoke to penetrate when under conditions of positive pressure.
- **Rigid/semirigid sheets:** Available with galvanized steel or aluminum foil facings and filled with intumescent materials, rigid sheets are effective one-piece fire stops for large penetrations. Semirigid sheets provide more flexibility for installation and may expand and contract with cable and piping movement. Semirigid mineral fiber blankets, with or without aluminum foil facing, are effective for preventing fire and sound from penetrating or flanking the floor slab.

■ **Loose packaged:** Bulk mineral/ceramic fiber can be used for penetration seals or as a noncombustible packing material for sealants. Some products may package the fiber and other additives in fiberglass bags.

■ **Ceramic fiber cloth:** Woven ceramic fibers are available plain, silicone treated, or with an aluminized finish. They may be used as exterior wraps for

additional protection of an assembly.

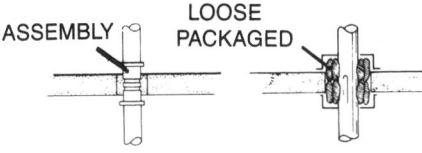

PIPE/CONDUIT/DUCT PENETRATIONS

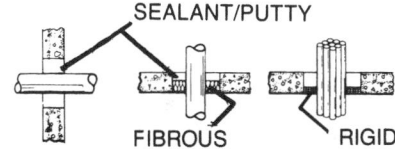

EXTERIOR WALL CONNECTIONS

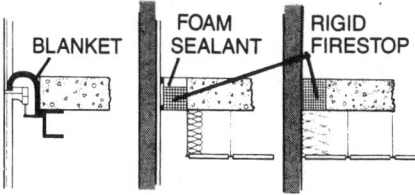

INTERIOR WALL BARRIERS

Assemblies

■ **Expansion joint covers:** Fire-rated joint covers are used to firestop voids created by expansion or isolation joints. Applications include floor to floor, floor to wall, wall to ceiling, etc.

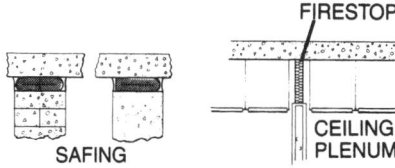

EXPANSION JOINTS

■ **Pipe penetrations:** Certain pipe sleeves are designed to be installed in poured-in-place concrete, cored holes in concrete, or gypsum walls such that piping and conduit are then connected to the sleeve. The sleeve systems can have ratings of up to 4 hours.

D4.1 FIRE PROTECTION

FIREPROOFING

■ Sprayed-on applications of mineral fiber or cementitious materials:

- Mineral fiber provides excellent protection where the surface is concealed, but requires care during and after installation. Since it is easily brushed off, damage during construction, renovation and occupant maintenance must be carefully monitored and repaired.
- Cementitious coatings are more durable to impact, but adhesion and spalling in fires have been a problem.
- Adhesion is very critical and requires careful preparation of surfaces before application. All surfaces should be free of any substance which may impair adhesion. Some sprayed-on fireproofing will require that a compatible paint/primer is used to coat the steel prior to application.
- All mechanical fasteners and hangers should be in place before spraying. Ductwork and other assemblies which can block the spraying process should be installed after the fireproofing has cured.
- Several performance standards, including impact, abrasion, bond strength, deflection and air erosion, should be known before selection.

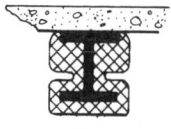

BEAMS   COLUMNS

■ Rigid board materials used to protect medium to large structural steel members are specified by board thickness and number of layers or by the required fire endurance rating.

- Assembly specifications must match applicable tested designs (per ASTM E119), following attachment of the board, corner construction, overlaps and butt joints exactly.
- Rigid board materials include fire-rated gypsum board or veneer base, molded calcium silicate boards, and mineral fiber.
- One method of attachment is to impale the board on straight or spring wire studs which are welded or clipped onto the steel member; the studs should extend through the board and be secured at its surface.

- Another method is to attach lightgage steel angles or studs to the larger steel member and use self-tapping screws to hold the board in place.
- A less desirable option is to use a laminating compound. This method is often used for applying face layers over base layers of gypsum board. It tends to be messy, requires drying time and poses potential for corrosion of metal substrates.

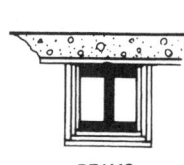 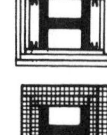

BEAMS   COLUMNS

■ Semi-rigid boards or flexible blankets are available with mineral or ceramic fiber cores.

- Semi-rigid panels can be mechanically fastened to steel beams and columns in single or double layers to provide sufficient protection.
- Flexible blankets, faced or unfaced, are available for the protection of cable trays, air and grease ducts, trash and linen chutes. They can also be compression fit to fill expansion joints. Blankets can be draped over/ around fire doors, appliances, dampers, etc.
- Facings include aluminum and stainless steel foils, plain and perforated fiberglass cloths, reinforced mylar, and foil scrim kraft paper.

■ Encapsulation of structural steel members with concrete:

- Concrete thickness must be sufficient to keep the temperature of the steel within appropriate temperature ranges and must be within heat transfer limits as defined by ASTM E119.
- Reinforced concrete slabs often incorporated the forming of concrete around the supporting steel beams.
- The cost of added formwork, concrete and dead load (resulting in larger supporting members) must be considered in the selection process.

BEAMS   COLUMNS

■ Intumescent paints can provide effective protection for short periods of time.

- Applied directly onto primed steel members, a base coat of the fireproofing paint will expand to form a layer of char (3/4 to 1 inch thick) during fire temperatures over about 200°F.
- A finish coat may be applied to provide a decorative, moisture-resistant finish.

■ A common approach to fireproofing steel joists and beams is to suspend a fire-rated ceiling from them.

- Plaster ceilings, when exposed to fire, will dry out and disintegrate. In many older buildings, this leaves the wood lath exposed, adding fuel to the fire. Metal or gypsum lath is usually specified in new construction.
- Gypsum or acoustical tiles are often set into fire-rated grid systems. The tiles may be mechanically fastened in place or set onto flanges in order resist the upward air pressure of a fire.
- Construction quality and maintenance are key elements in continued fire resistant integrity of suspended ceilings.

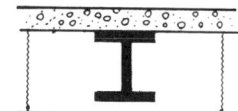

■ Water-filled assemblies around steel beams/columns have been used to transfer heat convectively away from the point of fire.

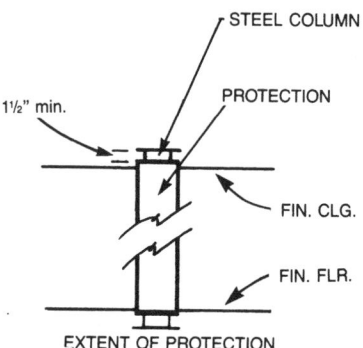

FIRE-RETARDANT TREATMENTS

■ **Fire-retardant treatments are applied to combustible materials for the purpose of retarding flame propagation and spread.** The basic methods used to treat combustibles are:
● Adding chemicals to plastics/ synthetic fibers during the manufacturing process to alter burning characteristics.
● Immersion of absorbent materials (fabrics, paper) in chemical solutions impregnates their fibers.
● Wood and similar less absorbent materials are treated by pressure impregnation.
● Fire-retardant coatings are available for many types of materials.

Fire-retardant Coatings

■ **Fire-retardant paints can be used to reduce flame spread, fuel contribution and smoke development.**
● **Nonintumescent** fire-retardant coatings may be noncombustible, insulative, or may be formulated to release water or carbon dioxide vapor to smother ignition.
● **Intumescent** coatings are formulated to expand when heated to certain temperatures. Gas is released by the chemicals in the coating, but remains contained within it so as to produce a cellular, insulative layer.

■ **Sprayed or trowelled mastics may be used to form thicker protective coatings.**
● Economy and ease of application are offset by concerns for quality control and durability.

■ **Building materials which are already assembled in place can be coated to improve fire-retardant characteristics.**
● They generally require reapplication over time when exposed to the weather. Ultraviolet radiation, wind, rain, etc., will combine to degrade the coating layers.
● Brushed-on coatings are generally suitable for interior applications.

Fire-Retardant Treated Wood

■ The behavior of untreated wood during a fire can be summarized as follows:
● Wood can ignite when exposed to 300°F for a considerable length of time. Other ignition/exposure time relationships are 600°F/5 minutes,

500°F/10 minutes, 425°F/20 minutes and 375°F/30 minutes.
● Heat release from wood averages about 8500 Btus/hr.
● Burning produces char at a rate of 1.5 inches per hour. The char jackets the wood, insulating the remaining cross-section from rapid combustion. The inner wood maintains its strength relative to the reduced cross-sectional area.
● Cell decomposition from rising temperatures produces gases which ignite when hot enough, causing flashover.

■ **Wood products can be treated or coated to improve behavior in fire.** Treatments are classified as exterior or interior; interior being of Type A or B.
● Type B chemical solutions for interior applications may leach out of the wood when relative humidity is 75 percent or greater or when the wood is in contact with water.
● Type A products have increased leach-resistance to 95 percent humidity.
● Wood treated for interior use should not be exposed directly to exterior conditions.
● Treated to be used in structural applications must have a flame-spread of 25 when burned for a 30 minute duration. The duration is reduced to 10 minutes for interior finish usages.
● Treated wood products must be labelled and/or certified as such, with special labelling for exterior use.

■ **Fire-retardant treated wood products have chemicals forced into the wood cells such that they replace moisture which was present prior to treatment.** The chemicals and their functions vary:
● Some chemicals act on the tar content of burning wood to convert it to carbon char or to lower the amount of tar produced during combustion.
● Other chemicals insulate inner wood by increasing the production of charcoal.
● Dilution of combustible gases produced by burning wood with carbon dioxide and water vapor.
● **Intumescent chemicals** will bubble out of the wood cell structure as the ambient temperatures increase, forming a thick char that isolates the wood from further damage by

sealing off the supply of oxygen. —Polyol is one type.
● **Boron** based chemicals form a glassy film on the surface of the wood when exposed to high temperatures.
● **Phosphate salts** are used but the treatments can corrode metal and they eventually wash out.
● **Aluminum trihydrate** absorbs some of the heat of combustion and lowers the temperature of the wood near the flame. It is the only chemical that can be used in the production of fiberboard.
● **Amino-resins** are leach-resistant treatments of soluble salts mixed with metal salts and are used in production of particle board products.
● Other chemicals used to treat wood include dicyandiamide, urea, melamine, and phosphoric acid. Formaldehyde makes the wood more waterproof but could be a source of unreacted formaldehyde (free formaldehyde) indoors.

■ **Although treated wood impedes the spread of flames, it is still combustible.** Other considerations of treated wood include:
● **The use of fire-retardant treated wood in a combustible assembly will not improve the fire rating of that assembly. However, treated wood can be used in certain noncombustible assemblies.**
● Many of the chemicals are toxic; handle with care and do not burn scraps.
● Impregnation of chemicals will effectively reduce the strength and fastening values of the wood by about 10 to 16 percent.
● Carbide-tipped saws are recommended for cutting.
● Galvanized or corrosion-resistant fasteners should be specified for wood treated with sulfates, ammonium phosphate, chlorides or halogens.
● Treated lumber can cost up to twice as much as untreated lumber.

■ **More information on fire-retardant treated wood can be found in** WOOD/WOOD PRODUCTS.

D4.1 FIRE PROTECTION

D • SERVICES

ELEMENTS OF FIRE CONTAINMENT

CONTAINMENT

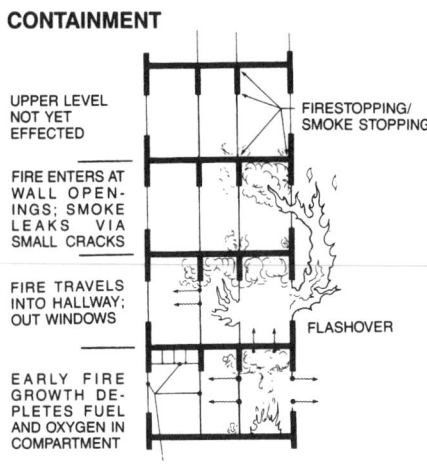

UPPER LEVEL NOT YET EFFECTED

FIRESTOPPING/ SMOKE STOPPING

FIRE ENTERS AT WALL OPEN-INGS; SMOKE LEAKS VIA SMALL CRACKS

FIRE TRAVELS INTO HALLWAY; OUT WINDOWS

FLASHOVER

EARLY FIRE GROWTH DE-PLETES FUEL AND OXYGEN IN COMPARTMENT

FIRE SEPARATION

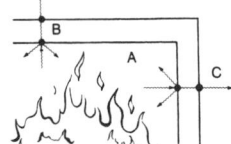

A: SURFACE DOES NOT SUPPORT COMBUSTION; REJECTS SOME HEAT RADIATION AND ABSORBS SOME.
B: THERMAL CONDUCTION VARIES WITH COM-POSITION OF ASSEMBLY; RATINGS REQUIRE STRUCTURAL INTEGRITY.
C: SURFACE RADIATES HEAT BUT DOES NOT SUP-PORT COMBUSTION.

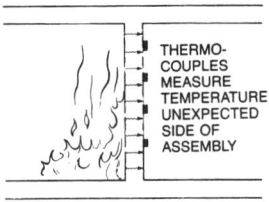

THERMO-COUPLES MEASURE TEMPERATURE UNEXPECTED SIDE OF ASSEMBLY

FIRE RESISTANCE RATINGS =
FIRE VS. TEMPERATURE RISE

THERMAL BARRIER PROTECTION

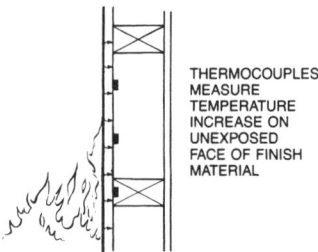

THERMOCOUPLES MEASURE TEMPERATURE INCREASE ON UNEXPOSED FACE OF FINISH MATERIAL

■ **Fire containment, or compartmentation, has a broad definition which describes the use of fire-resistive construction to limit the spread of fire and smoke through a building.** The purpose of compartmentation is to provide adequate time for orderly evacuation of occupants and to protect life and property from fire by providing adequate barriers. The elements of containment include:

- **Vertical fire separation** - fire-resistive construction (ceiling assemblies) used to provide adequate structural capacity while acting as an appropriate acoustical and fire barrier. Typical assemblies include noncombustible roof or floor/ceiling constructions (concrete slab) or combustible assemblies protected from below by layers of noncombustible materials.
- **Horizontal fire separation** - fire-resistive construction, including fire walls, used to separate and protect neighboring occupied spaces on the same floor or to minimize the spread of fire between neighboring buildings. Typical assemblies are of noncombustible materials or combustible frame with noncombustible facing materials fastened to either side.
- **Firestopping/smokestopping** - materials and methods used for protection of penetrations through fire resistive assemblies by mechanical, electrical and plumbing distribution systems.
- **Protection of openings in fire-resistive assemblies** - materials and methods used to minimize the movement of smoke and fire through doors and windows.

■ **Fire separation is accomplished by using assemblies which will have sufficient structural stability under anticipated fire conditions.** The use of these assemblies is referred to as **fire-resistive construction** and is often dictated by construction type. Such assemblies are classified as noncombustible or combustible and by their endurance, or "resistance" rating.

- **Fire resistance ratings are established from the result of tests (ASTM E119 for walls, floors and roofs; E152 for doors; E163 for windows) performed on building construction assemblies.** If the assembly consists of a homogeneous material, then it is a rating of that material.
- The rating is based on the passage of heat entirely through an assembly, measured by the rise in surface temperature on the unexposed side over time. The passage of heat and/or flame shall not be sufficient to ignite cotton waste attached on the non-fire side. The heat is measured in one direction only (fire exposure on one side of the assembly); if the assembly is symmetrical in construction, the rating is effective from either side.
- The structural integrity of the assembly or installed material must remain intact through the duration of the burn and subsequent impact of water from a fire hose.
- **The rating indicates the performance of a specific assembly with specific materials and methods of attachment. Any deviation from that specification constitutes a new assembly which must be tested to determine its rating.**
- Any alteration to an existing fire-rated assembly (by adding unrated doors, punching conduit through, etc.) voids the rating of that assembly.
- Typically, fire resistance ratings are shown as 4-hour (wall openings may not be permitted), 3-hour (wall openings are equally protected; limited in size and area), 2-hour (wall openings must be 1 1/2-hour rated), 1-hour (wall openings must be equally protected).
- When combustible materials or those with low yield points are contained in the assembly, noncombustible protective facings (thermal barriers) are required.

■ **Thermal barrier - a building assembly surface which will protect concealed spaces containing combustible materials (ie: foam plastic insulation, electrical nonmetallic tubing) for a minimum of 15 minutes.**

- The barrier must be stable, continuous and of sufficient construction to remain in place for the required time.
- The thermal barrier finish rating indicates the time it takes the side of the facing material opposite of the fire exposure to rise an average of 250°F with a temperature rise at any one point not exceeding 325°F.
- Applications of gypsum board or plaster often satisfy these requirements.

■ **The difference between the thermal barrier and the fire resistance rating is based on where temperature rise is measured — the unexposed side of the exposed material for thermal barriers; the unexposed side of the complete assembly.**

- The importance here is that it is possible for an assembly to be rated for 1-hour resistance, yet have a thermal barrier of less than 15 minutes.

FLOOR OR ROOF AND CEILING ASSEMBLIES

■ **The floor/ceiling assembly is designed for a fire-rated ceiling and a low flame spread floor. Each rated assembly is specified by material/component type, size/thickness and spacing as well as by fastening schedule; any deviation from given specifications will void the rating of the assembly.**
● Floor/ceiling assemblies are required to provide adequate structural capacity while acting as an appropriate fire and/or acoustical barrier.
● Floor/ceiling assemblies may be of combustible materials in some occupancy types and noncombustible in others while ranging from a 3/4 to 3 hour fire rating.

■ **Roof assemblies are vulnerable to both internal and external fires, but the fire rating is generally applied to the ceiling side.** Roof covering materials are rated relative to their resistance to external fire exposure.
● Roof coverings are tested according to ASTM E108 (flame spread, fire resistance, fire protection of deck, flying brand development). The roofing must not slip from position and cannot present a flying brand hazard during fire exposure. Flying brands are any piece of the roof covering which becomes separated and is blown some distance away while maintaining flaming or glowing.
● Class A roof assemblies are achieved with noncombustible decks, shingles and tiles, certain hot-applied built-up roofs and with Class B wood shake shingles over 1/2-inch Type X gypsum board over 3/4-inch plywood sheathing.
● Class B assemblies may be of metal roofing, fiber cement shingles or specific built-up roofing. Class C assemblies are typically of asphalt composition shingles, treated wood shingles or cold-applied built-up roofing.

■ **Ceiling construction may be the same for both roof and floor assemblies. The assembly may be entirely of noncombustible material or may provide noncombustible material as a layer of protection for combustible material.**
● Gypsum products are most commonly used as protective ceiling materials. They include wallboard products, panels/tiles, and plaster mixtures.
● Plaster mixtures which include insulative materials (perlite, vermiculite) usually have better fire-ratings at lower contributions to dead load.

■ **The plenum is the free air space between the ceiling and deck membranes. Considerable damage is possible from the spread of fire in this concealed space.**
● Firestopping via vertical barriers is commonly required. It must be at prescribed intervals, unpenetrated and continuous between membranes for the length of the floor span.
● Penetrations for lighting, heating, ventilating and air conditioning are the biggest cause for passage of smoke, heat, fire and hot gases into the protected plenum.
● The addition of ceilings, suspended under an otherwise fire-rated assembly are common oversights for fire protection; the new ceiling must be fire-rated as a thermal barrier (min. 15 min.) independent of the existing floor/ceiling assembly, proving that it is capable of staying in place for that period of time.
  —When supported by an exposed grid, the temperature of the fire after 15 minutes may be enough to distort or buckle supports, allowing ceiling panels to fall out or open up to expose the concealed space.
  —The number and type of supports for the ceiling assembly,

light fixtures and HVAC duct and duct openings is critical to the rating of the assembly.

■ For more information on ceiling assemblies, refer to CEILINGS.

WALL ASSEMBLIES

■ **The required wall construction and rating will depend on its function in the building.**
● **Exterior walls** should successfully perform several fire safety functions:
  —Provide a fire barrier to exposure at both inside and outside surfaces to minimize the spread of fire to or from bordering properties.
  —Provide protection to prevent the spread of fire originating from a lower floor.
  —Minimize the spread of fire between the exterior wall and a rated floor assembly.
  —Remain in place for the required duration of fire.
  —Protect openings: Any opening in the exterior wall, including doors and windows, lessens the wall's ability to resist entry or exit of fire.
● **Interior walls** are used to subdivide a floor. The relative probability of fire spread through and fire damage in a building is inversely correlated to the fire resistance of walls which define the interior spaces.
  —Interior fire (separation) walls are loadbearing and extend from top of floor slab to bottom of the next floor slab. Noncombustible requirements can be satisfied using steel studs with appropriate layers of gypsum board properly fastened.
  —Fire partitions are non-load-bearing and generally use gypsum board facing over standard framing or other support.
● **Fire separation, or party walls** are used to separate adjacent occupancies of same or different degree of hazard. They are often located over the lot line in "attached" building constructions and are usually of noncombustible construction. They must:
  —be noncombustible and remain in place despite the collapse of construction on either side.
  —extend past the roof surface (24 to 48 inches) to form a parapet unless the roof/roof deck is noncombustible.
  —extend beyond the exterior wall the same distance as any horizontal projection (overhangs, balconies, etc.) or the same as the parapet.
  —satisfy sound transmission requirements.
  —For Type IV and V construction, separation may also be consist of two individual fire rated wall assemblies. **1-hour fire walls** may consist of two assemblies, each rated at 1 hour with no openings and separated by an air space which is actually over the lot line.
● **Shaft walls** are used to enclose a vertical opening that connects two or more floors. Resistance ratings may be different for interior exposure than for exterior exposure; check local codes.

■ For further information on wall assemblies, sound and fire ratings, etc., refer to BEARING WALLS, and PARTITIONS.

D4.1 FIRE PROTECTION

PROTECTION OF OPENINGS

Fire Barriers

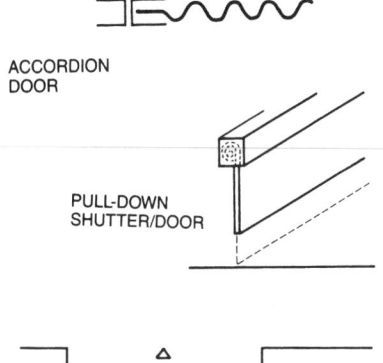

ACCORDION
DOOR

PULL-DOWN
SHUTTER/DOOR

SWINGING

■ **Openings in fire resistive assemblies are required to be protected from fire exposure for a duration equal to the that for the assembly. Protected openings, or fire barrier assemblies, consist of all hardware, fastenings, frames and operating units required for proper operation.** Fire barriers include fire rated doors (swinging, pull-down, accordion, sliding), shutters, windows and dampers. They all operate on a similar basis:

● **Automatic-closing** - fire assemblies that remain open until an increase in temperature (to 165°F) or the presence of smoke is detected. 3-hour rated fire assemblies require automatic-closing operation.

● **Self-closing** - spring actuated fire assemblies that remain closed except for normal operation; closing devices must provide sufficient force to latch and maintain an appropriate seal.

■ **Glazing is limited or restricted in protected openings because neither glass or plastic glazing will remain in place for the required duration of the fire rated assembly in which it is placed.** Allowable glazing area is limited by the hour rating of walls (window walls), doors and windows. For example, windows with 3/4-hour ratings may not exceed 12 feet in any dimension, nor 84 square feet in area. In order to accommodate such restrictions, various products have been developed:

● Standard glazing for fire rated vision panels is a minimum of 1/4 inch thick, is wire-reinforced glass, and is anchored with steel glazing angles or wire clips. Wired glass is acceptable for fire doors, exterior windows subject to moderate fire exposure, enclosures for open stairwells and for one-hour corridor walls. It distributes heat, thereby lowering thermal stresses in the glass and increasing the assembly's strength. A drawback is that the wires produce sufficient radiant heat to impair the use of the opening for egress.

● Certain glass block products can be used in various applications.

● Gel-filled double-pane glazing has been developed to resist flame, smoke and heat passage for up to 1 1/2 hours without compromising vision quality until exposed to fire (exposed gel produces an opaque protective coating).

● Multiple glazing layers, sandwiching plastic glazing materials within glass are available with ratings up to 45 minutes.

● Fire shutters can also be used, closing over openings when alarm systems operate.

Fire Dampers

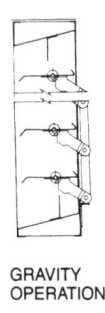

GRAVITY
OPERATION

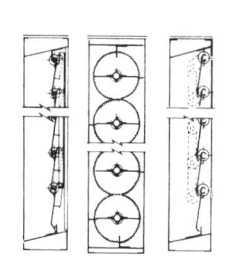

AUTOMATIC OPERATION

■ **Fire dampers, usually made of galvanized steel, are required as follows:**

● Wherever a duct passes through a two hour or greater fire resistive assembly (shaft, corridor, or exterior walls). When the opening through a protected shaft wall, one damper is sufficient. When the opening is through a fire wall, a fire door or a damper will be required on both sides of the wall.

● Wherever a duct penetrates fire resistive ceiling assembly that forms part of a total fire rated floor/ceiling or roof/ceiling assembly.

● At the fresh-air intake for the air conditioning system when that intake is located in an area assessed to be potentially hazardous.

■ **Fire dampers are integrated into any system which moves air in order to automatically seal off the system and restrict the passage of heat.**

● Fire/heat from combustion can be drawn through plenums and air conditioning duct systems by return air systems or can spread due to the presence of combustible material in the concealed space. In both cases and other similar cases, the fire is enabled to spread horizontally without being detected.

● Automatic-closing fire dampers use a fusible link to close the damper.

● Self-closing fire dampers close (by gravity) with the direction of air flow.

● Dampers may be single blade or connected multiple blades, as in a louvered damper.

FIRE DOORS

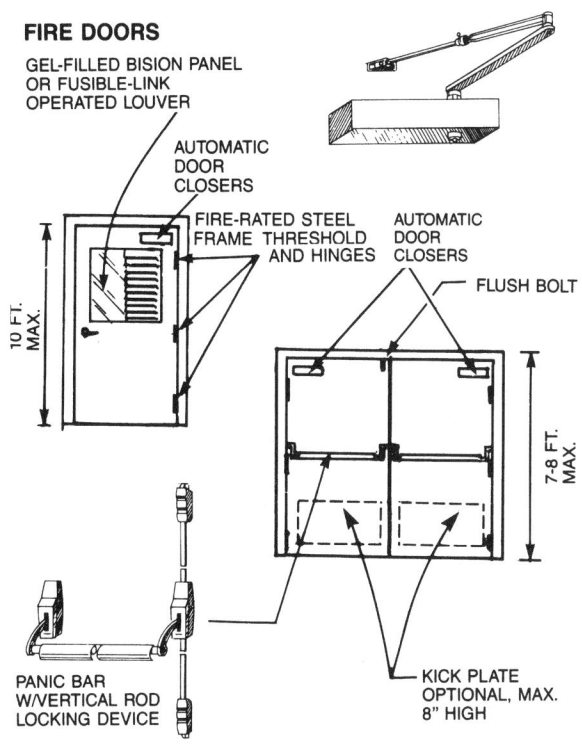

GEL-FILLED BISION PANEL
OR FUSIBLE-LINK
OPERATED LOUVER

AUTOMATIC
DOOR
CLOSERS

FIRE-RATED STEEL
FRAME THRESHOLD
AND HINGES

AUTOMATIC
DOOR
CLOSERS

FLUSH BOLT

10 FT.
MAX.

7-8 FT.
MAX.

PANIC BAR
W/VERTICAL ROD
LOCKING DEVICE

KICK PLATE
OPTIONAL, MAX.
8" HIGH

FIRE SHUTTERS

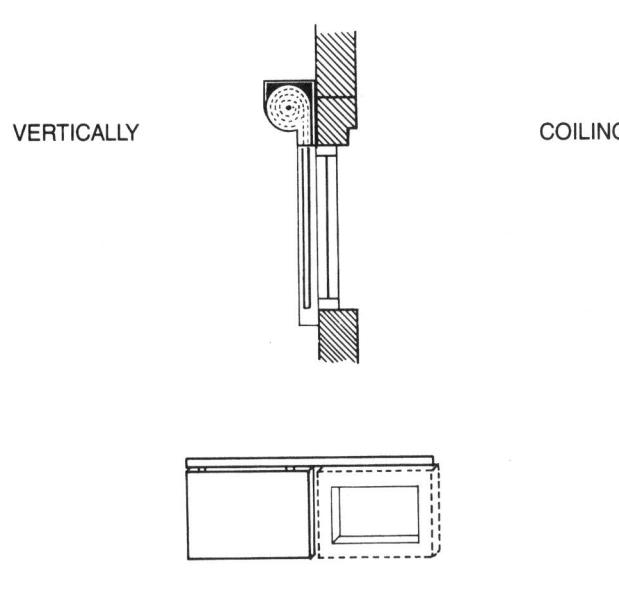

VERTICALLY   COILING

HORIZONTALLY   SLIDING

■ **Fire doors are critical elements of compartmentation, especially for corridors and vestibules to stairways and areas of refuge.** All fire doors, frames and hardware must be labelled to indicate door and frame fire rating/class.

- Once machined and labelled, no facings (other than small signs and kick plates) may be laminated onto the door without voiding the label.
- Single-swing doors are available in heights up to 10 feet; 7 or 8 feet for pairs.
- Proper operation mandates that the doors not be blocked, propped, stopped or in any other way have their normal operation hampered. The door must close to provide an airtight seal and positive latch.
- Standard equipment includes:
  —fire-rated steel hinges (standard or spring);
  —flush bolts with dust-proof strike plates;
  —automatic closing devices (when standard hinges are used), spring hinges, counterweights, etc.;
  —locksets (lever handles for handicapped);
  —panic hardware for exit doors;
  —coordinators when an astragals are used.
- Optional equipment includes:
  —silencers;
  —electronic hold-open/closer devices;
  —smoke seals;
  —door stops, acoustical gaskets and limited sizes and materials for kick/protection plates.
- For more information on door hardware, see **SD.3/En entrances/doors.**

■ **Protection of openings and doors is classified by NFₗPA according to the hourly fire protection rating of the assembly.** This classification may be combined with the former letter designations (Classes A to E) found in Appendix F of NFₗPA 80A which describe openings according to the character and location of the wall.

- **3 hour fire doors (Class A):** Rated for 3 hours; no vision panels; suitable for use in fire separation walls. They may be assembled with two doors, one on either side of the protected opening when the wall is four hour rated.
- **1 1/2 hour fire doors and shutters (Class D):** In exterior walls, severe fire exposure from outside is resisted by the door and shutters; vision panels are limited to a maximum of 100 square inches.
- **1 or 1 1/2 hour fire doors (Class B):** Suitable for 1 or 2 hour rated walls or for openings in the enclosure of stairwells and other vertical openings/shafts.
- **3/4 hour fire doors (Class C):** Generally designed for interior walls, partitions and corridor walls. Corridor walls generally require self-closing or smoke detector actuated automatic closers.
- **3/4 hour fire doors and shutters:** Rated doors and shutters are used in exterior walls with moderate or light exposures. The maximum vision panel size is a 9 square foot area and a 54 inch dimension.
- **3/4 hour fire windows:** Windows, generally with steel frame and wire glass, used to protect openings in corridor walls, partitions and exterior walls subject to moderate or light external fire exposure. For higher ratings, shutters are used.
- **1/2 hour (Class E):** Used for smoke control or protection of openings in partitions and exterior walls which have 1-hour ratings or less. 1 3/4 inch, solid core, bonded wood doors are typically 1/2 hour rated.
- **20 minute:** used for smoke control in 1-hour rated walls.

D4.1 FIRE PROTECTION

PROTECTION OF OPENINGS (cont.)

WINDOW PROTECTION

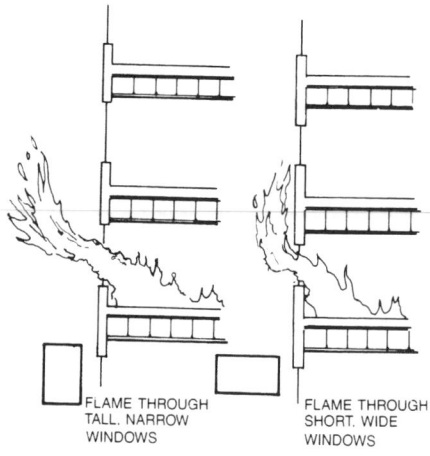

FLAME THROUGH
TALL, NARROW
WINDOWS

FLAME THROUGH
SHORT, WIDE
WINDOWS

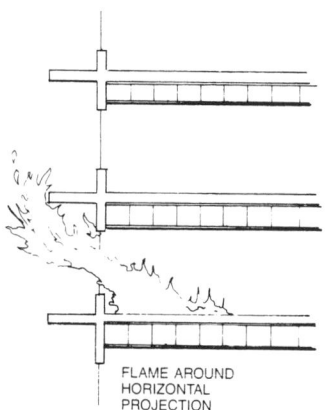

FLAME AROUND
HORIZONTAL
PROJECTION

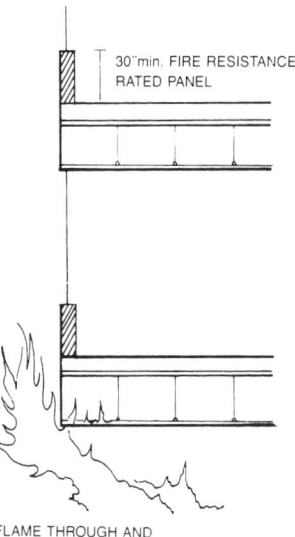

30"min. FIRE RESISTANCE
RATED PANEL

FLAME THROUGH AND
AROUND CURTAINWALLS

■   **The fire protection of windows in exterior walls is important because of the potential for fire penetration, both inward and outward.**
- The fire zone, just prior to flashover, is very hot, but does not have sufficient oxygen to burn the combustible gases released in the fire.
- Ordinary glazing (plain glass) will crack when exposed to hot gases (about 750°F) and is likely to break and fall out when temperatures reach about 930°F. Therefore, a window which breaks under thermal shock or explosion at flashover allows hot, combustible gases to flow out the upper portion of the window opening.
- The escaping, unburned gases entrain air as they exit the window, igniting into a plume of flame on the exterior face of the building. The fire plume is proportional to the fuel load in the burning space.
- Temperatures of the wall surface rise from convective heat of escaping gases and from the radiant heat of the fire plume itself.
- The shape of the flame plume is affected by window geometry. A tall, narrow window tends to eject the plume away from the building face. A short, wide window tends to allow the flame to climb the face of the structure. Smaller windows tend to produce taller plumes due to ventilation-limited burning within the building.
- The spread of fire through windows to neighboring buildings will depend on distance separation, fire-resistive rating of the exterior walls, and the respective number, size and location of windows on the facing walls. The size and number of windows will determine the amount of heat that can be radiated to the neighboring structure.

■   **One of the primary fire issues of multiple-storied buildings is the impact of windows which are stacked, or aligned on consecutive stories.**
- The fire plume can project out windows and extend 10 feet or more above window openings on the fire floor. When windows are vertically aligned in the exterior wall above the fire floor, heat transfer is likely to be sufficient to ignite furniture, draperies and other combustible items near the window above.
- This type of fire spread is accentuated when upper story windows are open or when they are broken by thermal shock. Windows in the story above the fire floor may crack within 2.5 minutes of a fire, breaking and falling out within 4 minutes, exposing combustibles above.

■   **Design techniques which can be used to minimize vertical flame spread through exterior wall openings, especially in nonsprinklered buildings, include:**
- **Window placement:** Consideration should be given to eliminating openings in exterior walls that are within 5 feet of the property line or staggering them on consecutive floors to minimize stacking. Another response is to use protected window assemblies if allowed by authorities with jurisdiction.
- **Automatic sprinkler systems:** The use of sprinklers can reduce flame and heat near an opening. These may be used to provide a curtain of water on either side of the wall.
- **Horizontal "eyebrow" projections:** Unusual, possibly expensive to build, but proven effective, horizontal flame barriers can be used to deflect flames away from the building facade. They should extend 36 inches beyond the exterior wall in the plane of the floor and be fire rated for one hour. Projections of 4 feet which are at least 3 feet wide are more likely to eliminate vertical flame spread and/or curl-back of flames toward the building.
- **Vertical spandrel walls:** One-hour fire-resistive rated panels can be used to separate windows (vertically). A minimum of 36 inches is generally required, with 30 inches or more of the panel extending above the top of the floor deck.
  —Even with spandrels, glazing on the story above the fire floor can be expected to fail due to temperatures within 15 to 30 minutes.
- **Fire-retardant textiles:** The use of Class A flame spread materials on walls, ceilings and window treatments can inhibit flame and heat intensity toward wall openings. On upper levels, such materials may be less apt to ignite from flames or radiant heat energy coming in through open windows. See Fire Retardant Treatments.

FIRE PROTECTION: selection checklist

FLAME SPREAD CONSIDERATIONS

■ **Flame- and fire-retardant claims cannot be taken at face value. Each material, its treatment, and its association with other elements of its assembly must be considered.**
● Adhesives used in glue-laminated wood products or for attachment of tiles, fabrics and other finishes can fuel a fire even though the finish material if fire-retardant.

Interior Finishes

■ The ASTM E84 Steiner Tunnel Test is the primary method of measuring the speed of flame growth across a surface, the density of smoke developed and the fuel contributed (determined by temperature rise).
● E84 is performed by lining the 20-inch wide ceiling of the 25 foot long tunnel with the material to be tested; the finished, exposed surface faces into the tunnel.
● A pilot source is located at one end, with standard draft conditions exhausting through the tunnel.
● A problem not addressed by E84 flame spread ratings is that the fuel contribution from materials varies with temperature. Some materials do not easily ignite at low temperatures and have low flame spread ratings but do contribute to fire growth at higher temperatures.
  —Materials, such as certain plastics, have low flame spread indices; however these materials have been shown to promote fast fire growth in larger scale, higher temperature fires.
  —The E84 test has been challenged in its measure of the performance of textile wall coverings.
● Thin finish materials (<1/28 inch) may be applied to wall/ceiling surfaces if their installed flame spread is not greater than that of paper of equal thickness cemented to a noncombustible backing.
● Scaled down versions, especially the Forest Products Laboratory 8 foot Tunnel Test (ASTM E286) have been proven to have very high correlation of results with the E84 tests.

■ **Another test used to measure flammability of a material finish is the radiant panel test (ASTM E162).**
● E162 is used to measure flame spread and heat release.
● A 6 by 18 inch material specimen is inclined at a 30° angle to a 12 by 18 inch radiant panel.
● Flame movement down the surface of the specimen is measured over a period of 15 minutes or until the specimen is completely destroyed. A flame spread index similar to that of the E84 test is calculated.

■ **There are many natural and synthetic textiles used as wall coverings, each with differing flame spread ratings.**
● Thin, lightweight, loose-weave fabrics ignite more easily than heavier, tightly-woven fabrics. The moisture content of a textile substantially affects its fire performance.
● The selection of adhesive is critical to the flame spread/contribution of the wall covering. Since the fire performance of adhesives vary with thickness of application, variance from the approved and specified application may result in unsatisfactory performance. High temperature mortar cement may be required.
● Flame-resistant fabrics may be cleaned without affecting fire performance by washing in phosphate-based detergents, heavy duty liquid detergents and other manufacturer specified cleaning solutions; bleach is not recommended.

■ **A flame spread rating of 25 or less is required for the use of any carpet or carpet-like material on walls and ceilings.**
● These materials may be prohibited from exit areas.
● The Class A rating limits potentially hazardous use of non-rated floor carpet as decorative wall treatment.

Flooring

■ **The flooring radiant panel test is the primary method for assessing flame spread of flooring materials in exits and corridors.**
● ASTM E648/NFPA 253 is similar to ASTM E162 in most aspects, but the sample is larger.

■ **When carpet or other textile floor covering is used, the pill test is most appropriate for fire assessment.**
● ASTM D2859 measures the ease of ignition and surface flammability of 9 inch square specimens. The test requires that 8 specimens be tested.
● The specimen is ignited by lighting a methenamine pill which is placed on each sample at the center of the frame opening.
● Results are graphed on the basis of critical radiant flux to distance (cm) with ratings ranging from 0.35 to 0.4 watts per square centimeter for red oak, 0.4 to 0.75 for wool and in excess of 1.1 w/cm^2 for vinyl. The higher the index rating, the less the hazard.

■ **The pill test (DOC FF-1-70/ carpets, DOC FF-2-70/rugs) is used to assess the surface flammability of finished textile floor covering materials.** It is specifically designed to assess horizontal material applications without concern for a specific deck or underlay.
● The charred portion of carpet/textile is not to extend within 1 inch of the holding frame edge.
● The result is rated pass or fail, and 7 out of 8 samples must pass.

Furniture

■ Upholstery fabric may be classified according to test results (Upholstered Furniture Action Council Fabric Classification Test Method) as follows:
● **Class A:** Materials that result in less than 38 mm of char when tested on glass fiberboard or cotton batting; such as wool, heavy thermoplastics, vinyl or polyurethane coated fabrics.
● **Class B:** Materials that result in 38 mm or more of char when tested on cotton batting, but less than 38 mm of char when tested on glass fiberboard; such as light weight cellulosics and medium weight thermoplastics.
● **Class C:** Materials that result in 38 to 75 mm of char when tested on glass fiberboard; such as medium weight cellulosics and light weight thermoplastics.
● **Class D:** Materials that result in more than 75 mm of char when tested on glass fiberboard; such as heavy weight cellulosics.
● Cellulosics include fabrics of natural plant fiber and man-made fibers from natural materials.

FIRE PROTECTION: selection checklist (cont.)

FLAME-SPREAD CLASSES/RATINGS

Class	Allowable Flame Spread; Smoke Developed	Required Applications By Occupancy/Hazard*	Commonly Used Building Materials: Rated By Class	Notes
A	0 to 25; 450	vertical exits = all hazards shaft walls = all hazards corridors = moderate-high	asbestos cement board (0) brick, concrete, concrete block (0) aluminum, baked enamel finish (5-10) gypsum board (10-25) mineral fiber acoustical panels (10-25) treated plywood paneling (10-25)	Special ratings may be given to materials with smoke development of 25 or less.
B	25 to 75; 450	corridors = low-moderate rooms>1500 sf = all hazards rooms<1500 sf = moderate-high	vinyl composition tile (10-50) treated red oak (35-50)	
C	76 to 200; 450	rooms<1500 sf = low-moderate	untreated red oak (100) untreated plywood paneling (75-275) untreated southern pine (130-190) cork (175)	untreated red oak is the performance standard as set by ASTM E84.
D	201 to 500; 450		linoleum (190-300)	Classes D and E are not recognized in NFPA 101.
E	over 500; 450			

* The requirements shown are for unsprinklered spaces, which are more stringent.
Some codes use Roman numerals to denote class rather than letters (A = I, B = II, C = III).

NONCOMBUSTIBLE MATERIALS USED FOR FIRE PROTECTION

TYPES / USES	protection of structural steel	thermal barrier protection	vertical fire separation	horizontal fire separation	vandal resistance	exterior/wet area	interior only
SPRAYED-ON COATINGS							
cementitious:							
calcium silicate	•	•					•
gypsum/vermiculite	•	•					•
Portland cement	•	•					
fibrous: glass	•						•
mineral	•						•
paint: intumescent	•	•					•
FACINGS							
gypsum board:							
regular	•		•	•	•	•	
Type X	•		•	•	•	•	
lath & plaster	•		•	•	•		
wire mesh & plaster	•	•	•	•	•		
mineral fiber	•			•		•	
STRUCTURAL ELEMENTS							
brick	•	•	•				
concrete	•	•	•	•		•	
concrete masonry	•	•	•				
solid gypsum/plaster	•		•		•		
steel, unprotected	•		•				

■ **Combustible construction is defined by assemblies and materials which are fuel contributing and structurally unstable under fire load conditions.**
● This definition also includes construction which is does not contribute to the fuel load but is unstable, and construction that is fuel contributing but is partially stable to the degree of fire resistance.

■ **When noncombustible construction is specified, there are limitations on the amount of combustible material that can be added to the construction assembly.**
● Combustible material, especially foam plastics, in the form of trim, ornamentation, accessories, etc., is generally limited in terms of overall percentage of coverage, flame-spread rating and thickness of application.
● Fire-retardant treated wood is allowed in interior non-load-bearing walls.
● In most cases, where a noncombustible plenum (similar to air-ceiling plenums) is specified, combustible materials are not permitted in the plenum. Exceptions to this include:
  —Combustible pipe or duct insulation, acoustical material, ceiling tile, and insulation.
  —Class 0 or 1 duct, flexible ducts, flexible connections and low-voltage wiring with insulative coatings.

■ **Common noncombustible building materials may require special consideration during selection:**
● The melting point of steel and metals may be more critical than their combustibility.
● Plaster and gypsum products are frequently used for fire protection. The high water content, nearly 21 percent by weight, is chemically released when exposed to extreme heat forming small droplets which quickly vaporize. This "weeping" is capable of absorbing heat and suppressing surface flames until the entire thickness of the material is dried, or calcinated.
● Plaster is typically formed by mixing gypsum plaster with aggregates such as sand, perlite or vermiculite. Generally, higher ratios of perlite/vermiculite to plaster improve fire resistance properties. Extreme weathering, inadequate ventilation, thermal shock and improper loading are capable of damaging plaster surfaces.
● Gypsum board is available in two formulations; regular and type X. Type X board is has better fire resistant properties because its core is reinforced with fibrous materials so as to remain in place after calcination.

D4.1 FIRE PROTECTION

FIRE PROTECTION: key issues

CONSIDERATIONS AFFECTING FIRE PROTECTION

Key Issues of Fire Protection	Behavioral Impacts	Occupancy Impacts	Construction Impacts
Fire Hazard Assessment	Arson; vandalism; pranks; improper use of electrical appliances; overloading electrical circuitry.	Storage and handling of hazardous materials; furnishings; quantity of fuel-contributing materials and proximity to other combustibles.	Noncombustible vs. combustible; extent and type of surface finish; ceiling height and floor area.
Smoke Detection	The smell of smoke may cause panic; vandalism may be a source of detector failure.	Require regular testing and cleaning of detectors; locate in/near sleeping areas, air handling systems.	Installation of detectors in convective air flow pattern; use AC and DC power unless emergency power is supplied with AC direct-wired units; ventilation via roof vents, hvac systems.
Alarm Systems	Local signals may be ignored; direct notification of fire department reduces response time; vandalized units may be inoperative.	Coded signals alert certain occupancies; intercom systems instruct via communication.	Placement and selection based on occupancy; test to assure operation.
Sprinkler Systems	Vandalism and carelessness are responsible for sprinkler non-response; continually check/monitor operation.	Extinguishing agents selected according to Fire Class and hazard; protection of electrical equipment. In special uses (museums, archives) the value of the material objects might be damaged or lost by inadvertant sprinkler operation.	Fire pumps are necessary when water supply/pressure is insufficient to suppress fires on upper floors.
Evacuation	Educate through practice drills; minimize false alarms; do not block fire doors with temporary storage, etc.; use appropriate egress hardware so building security is not compromised.	Number, size and arrangement of obvious, obstruction-free exits within safe travel distance; minimize "dead end" corridors; swing egress doors in direction of exit.	Fire doors fitted with single-action panic hardware with an automatic alarm and unlock allows egress in an emergency but prevents unauthorized personnel from using the exit. Keys and/or breakaway locks should be readily accessible. Use self-closing doors.
Fire Fighting Problems	Provide clear descriptions of spaces and occupancies to foster fast, accurate decisions during emergencies.	Fire departments must be knowledgeable of the contents of the building.	Obstructions by porches, overhangs, low electrical power lines impede firefighter access.
Structural Problems	Ensure integrity of fire barriers during design, construction, occupancy and renovation; maintain freely operating fire doors, etc.	Provide smoke and fire barriers wherever possible; provide protection for exit access and areas of refuge.	Seal all penetrations through fire rated assemblies; follow specified assemblies to the letter.

D4.1 FIRE PROTECTION

IGNITION & FIRE DEVELOPMENT

■ **Four conditions must coexist for fire to ignite and burn:**
- **Fuel -** combustible matter which will ignite when heated to a specific temperature; fire intensity depends on the nature, density, size and configuration of the fuel.
- **Oxygen -** present in air or released by chemical reaction;
- **Oxygen to fuel ratio -** the mixture of available oxygen and required to support combustion; the ratio can be diluted and ignition suppressed if air quantity/movement is too great;
- **An ignition source -** a heat source capable of raising the temperature of the fuel to its ignition point (e.g.: hot embers, sparks, open flames).
- **The aim of fire protection systems is to isolate or remove one of these four elements.**

AN EXAMPLE OF PYROLYSIS

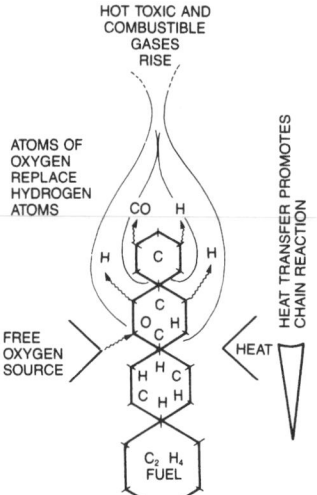

■ **Pyrolysis is the chemical decomposition of matter through the action of heat.** It progresses through several stages as the temperature of the fire increases. This progression is typically modeled by the **Standard Time-Temperature Curve** which measures temperature rise over time. The curve is used as a model to determine the endurance of building materials and assemblies in fires.
What impact does the Standard Time-Temperature Curve have on fire safety?
- Evacuation from the burning building must be completed within 5 minutes.
- Fire-resistive construction must be capable of withstanding temperatures of over 1200°F for several hours.
- Temperatures must be reduced before firefighters can get close enough to effectively try to extinguish intense fires.

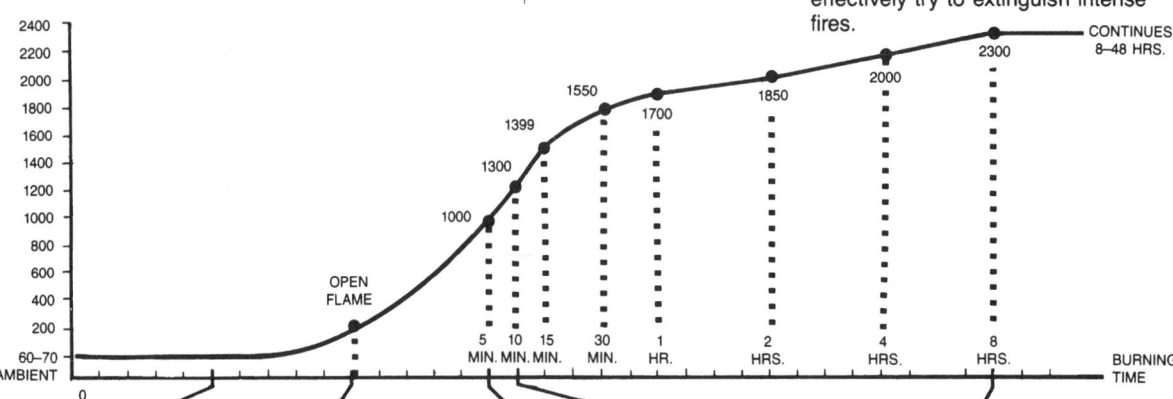

Stage I, or the Incipient Stage:

- Ignition induces heat gain, with local temperatures reaching up to 1000°F.
- This stage may progress over a period of minutes, hours or days.
- Combustion gases are released but neither visible evidence nor significant ambient temperature rise is present.

Stage II, or the Smoldering Stage:

- Smoke develops from combustion at low levels of heat. A flame may appear and increase in size with the heat of combustion; smoke production increases as well.
- In this stage, ambient room temperatures can rise to over 200°F. Consider that water boils at 212°F, but may scald skin at 115°F. Computer data is lost at 150°-200°F.

Stage III, or the Initial Developing (Flame) Stage:

- Within 5 minutes of flame development, a rapidly growing, flaming fire will raise ambient temperatures as high as 1000°F (possibly reaching flashover).
- At these room temperatures, wood chars (300°F), steel loses half its strength (1000°F) and plastic polymers self-ignite (650°F).

Flashover:

- An explosion caused by the pressurization of hot, combustible gases expanding within a confined space.
- The entrance of oxygen into the space and the hot gases into outside areas allows fire to spread/grow.
- The explosion may occur when radiant flux at floor level reaches 20kW/m² or when ambient upper air temperatures reach 1100°F.

Stage IV, or the Fully Developing (Heat) Stage:

- The space is totally engulfed in fire, burning at a steady rate.
- In 10 minutes, fire temperatures range from 1100°F to 1300°F. Steel will yield to structural loads at 1100°F.
- Rising to 1399°F within 15 minutes, the rate of temperature increase finally begins to slow (1550°F at 30 min.; 1700°F at one hour).

Cool-down Stage:

- The return to ambient temperature is the longest fire stage - a matter of hours or days.

D4.1 FIRE PROTECTION

SMOKE PRODUCTION & TOXICITY

■ **There are five major static test methods and three major dynamic test methods being used to determine the toxicity of smoke.**
- They use rats or mice as the test sample and measure from one to nine responses.
- The tests measure density of smoke, rate of off-gasing, amount and accumulation of smoke, and the contents of smoke.
- These toxicity tests produce significant differences in their findings, thus there has not been any agreement on a measurement method for a standard. Results have produced differences in rankings or ratings of similar materials.
- The fact that fire/toxic hazard is not measured creates another problem.

■ The result of burning various synthetic or natural materials is not significantly different in terms of the gases produced. Test results in this area are not comprehensive, but more than 400 gases can be attributed to combustion. The gases must be significantly toxic in significant concentrations to be considered deadly.
- Many codes use wood or paper as a standard for allowable products of combustion; no material shall be more toxic in point of concentration.
- Carbon monoxide is recognized as the primary killer; 1.3 percent in lungs can cause death.

■ **The "N-gas model" was developed by the National Bureau of Standards to measure smoke toxicity.** Its premise is that a small number of gases present in smoke are responsible for fire fatality.
- On the basis of chemical analysis of smoke from a material, an "LC_{50}" value is assigned to indicate time of exposure verses the smoke dosage.
- It represents the lethal content concentration of toxic gases at which 50 percent of test animals exposed to smoke for a 30 minute duration succumb.

■ **Smoke from a fire spreads throughout a building by entering through broken windows, HVAC distribution networks, plumbing chases, stairways, elevator shafts, seismic joints, etc. Any airflow through baseboards, electrical boxes, etc., can also allow smoke to spread. The dynamics of smoke movement rely on the heat component of its associated air and gases.**
- In Stage I and II of a fire, hot smoke (at temperatures as low as 100°F) is rising nearly vertically through cool, ambient air. This movement is conducive to being directed out of a space and away from the fire.
- When the smoke layer is over 3 feet deep and hot enough to set off a sprinkler head, it is buoyant enough to resist the downward pull of water from those heads.
- At flashover, a fire will produce 100,000 cfm of smoke or more at very high temperatures (up to 1500°F).
- The broad variance in smoke temperatures at different stages of fire requires that smoke control be effective in response to the entire range of temperatures.

■ **Building codes have adopted regulations to control smoke by defining the smoke developed rating of interior finishes.** The ratings are based on the measure of optical density.
- Properties such as moisture content, thermal conductivity, density and size will be determinants as to smoke development.

- The potential rate of production and concentration of smoke, expressed as parts per million or weight per unit volume, resulting from the burning of a material is referred to as the smoke load.

■ **Smoke movement is a function of the volume and geometry of the space, the heat of the fire, air leakage characteristics and pressurization.**
- Smoke moves rapidly, rising from the fire, collecting and spreading across the ceiling. The rate at which the smoke fills a space is controlled by the ceiling height, the ceiling area, leakage from the space and smoke contributed by the combustion of the fuel.
- The temperature of the fire causes upward movement due to the expansion of air molecules, reducing the air density but its increasing volume. If an opening in the containment system exists, the hot gases are forced out, carrying smoke fumes and combustible molecules with them. When a vertical shaft condition exists, the heat of the fire will carry the smoke up the shaft and into any space which is open to that shaft (stack effect).
- The temperature difference between inside and outside can also cause a thermal stack effect of smoke movement. The greater the difference in indoor and outdoor temperatures, the greater the rate of air flow.
- The effect of wind/wind pressure on the building can cause smoke to move horizontally. Pressure differences caused by wind can move smoke vertically movement in shafts or through the opening in any barrier where a significant pressure difference exists.
- Air conditioning distribution networks may be a path for smoke movement, even when fans are shut down. Adequate smokestopping is required within ductwork to minimize such migration.

SMOKE DEVELOPMENT RATINGS FOR INTERIOR FINISHES

SPACE/OCCUPANCY	ALLOWABLE RATING
• exits; corridors	25
• where evacuation of all occupants is a problem	50
• where occupant density is 10 sf/person or less	100
• where combustible finishes cover more than 20% of the exposed surface area	450

TOXIC GASES

Gas Present	Concentration		Effect of Inhalation
carbon monoxide	>1000	ppm	D<60 min.
	>4000	ppm	F<60 min.
	>10000	ppm	F< 1 min.
	>60000	ppm*	F = 1 breath
hydrogen cyanide	>350	ppm	F<10 min.
nitrogen dioxide	>200	ppm	F<10 min.
ammonia	>1000	ppm	F<10 min.
hydrogen chloride	>500	ppm	F<10 min.
sulfur dioxide	>500	ppm	F<10 min.

D = discomfort; F = fatal.
* At flashover, carbon monoxide can reach 60-80,000 ppm.

D4.1 FIRE PROTECTION

FUEL PROPERTIES & FUEL LOAD

■ **Fuel may be simply defined as any combustible substance. The substance may be in the form of a solid, liquid or gas.** These substances are reduced by decomposition and vaporization (by rising temperatures) to release flammable vapors and gases which diffuse in oxygen. Four chemical categories of fuel exist:
- oxidizable nonmetals (arsenic, carbon, phosphorus, and sulfur);
- hydrocarbons - composition based on carbon and hydrogen molecules;
- organic compounds - composition based on carbon, hydrogen and oxygen (alcohol, cellulose, lignin/protein - wood, vegetation, etc.);
- combustible metals/alloys (sodium, potassium, magnesium, aluminum, zinc, titanium, zirconium, uranium).

■ **The most common forms of combustible materials are organic compounds and hydrocarbons.**
- Organic compounds include oxygen in their make-up and release it during burning as well as combustible hydrogen gas. This oxygen source can perpetuate the burning of the material and may be a cause of smoldering.
- The basic products of combustion for an organic material are carbon dioxide and water. Of the basic elements in an organic compound, only hydrogen is combustible. When heat is applied to the compound, the hydrogen (as well as nitrogen and oxygen) molecules are released (known as free radicals), leaving carbon as a residue.
- Hydrocarbons also include fossil fuel products.
- Solid, combustible materials are measured by rate of ignition, flame spread and heat release through testing. Test standards are listed in the REFERENCES. Fire is controlled by using noncombustible or fire resistive materials and coatings, and by using suppression systems.

■ **Areas which contain/store flammable liquids must be assessed for the hazard of extremely dangerous fumes produced when flammable liquids evaporate.**
- These liquids are typically hydrocarbon compounds (made of carbon and hydrogen) requiring oxygen to maintain combustion.
- Flammable liquids are measured in terms of flashpoint and volatility. Where flammable liquids are stored or used, fire control is provided by adequate ventilation. This reduces hazardous concentrations of vapor.
- The vapors can be readily ignited from a spark, small flame (pilot light), a glowing electric heating element or electric motors, flashing back to the liquid.
- Some vapors, such as gasoline, can be heavier than air. They will drift at floor level and down stairs/openings, igniting on low sources. Others may be lighter, moving along ceilings and down wall surfaces.
- **The flammable/explosive limits of vapors are expressed as the percent of vapor in air at atmospheric conditions and define the minimum and maximum vapor concentrations at which flame can be propagated.** A mixture below the lower flammable limit is considered "too lean" and will not burn or explode; "too rich" when above the upper limit. The limits vary with ambient temperatures, pressures and relative humidities.

■ **Hazardous gases** are measured by limits of ignitability/flammability, density and diffusion coefficient. Safe storage and handling is paramount, requiring storage in atmospheres of inert gases ventilation controlled areas.

■ **The extent or duration of a fire can be measured in terms of the fuel load. It is the quantity and contribution (measured in units of heat produced) of fuel contained in a given area.** The fire load equals the sum of all contents in a particular area (in square feet) with equivalent weights set to fuel having a heat value of 8000 Btu/lb.
- A fire can be estimated to consume fuel at a rate of 10 psf/hour up to 30 psf (40 psf = 4.5 hrs; 50 psf = 6 hrs; 60 psf = 7.5 hrs).

■ **The fuel load is not only a function of fuel quantity, but also a function of the fuel array.**
- The size of the individual particle (sawdust vs. wood chips vs. lumber) and its storage arrangement (horizontal, vertical, cribbed, piled) define the fuel array.
- Many ordinary materials become dangerous/explosive fuels because of the way they are handled and stored.
  - Spontaneous combustion can occur in improperly ventilated stored materials which contain or are saturated or treated with certain oils (linseed, tung, turpentine).
  - Dust can act as an explosive or result in a flash fire when allowed to reach certain pressures and concentrations. Ease of ignition increases as particle size gets smaller.

■ **The contents of a building, therefore, are considered in terms of all combustible surfaces and their arrangement in the fire zone.** Flashover is reached sooner when highly flammable finishes are present.
- Interior finishes are those surfaces exposed to the interior space and potential fires. Walls and ceilings are the primary concern, but surfaces of some fixed-in-place furniture may also be considered as interior finishes.
  - Room corner tests are used to model interior finishes more accurately.
- **Furnishings often account for a high percentage of the fuel content of a room.** Upholstered furniture produces considerable smoke and toxic fumes when burning or smoldering; cigarettes can smolder on upholstered furniture for nearly 30 minutes and account for about 80 percent of such fires. Upholstered furniture should be labeled to indicate resistance to ignition by cigarettes.

■ Combustible materials vary in response to fire due to density, moisture content, etc. **It is possible, however, to impede the normal combustion process by treating the material with fire-retardant chemicals.** Fire-retardants are odorless and colorless salts which typically inhibit combustion by:
- decomposing endothermically under high temperatures;
- emission of a blanket of inert/nonflammable gas over the material surface;
- charring or producing an inert, glazed coating over the material surface;
- interfering with the chain reaction of fire by releasing free radicals;
- breaking down to small particles which change combustion properties.

MEANS OF EGRESS

SIZING EXITS

1. Determine Occupant Design Loads (X); floor area divided by net floor area per occupant allowed by applicable codes per occupancy type. For stairs, add the occupant load from upper floors; for exits at grade be sure to cover the total occupant load of the building.
2. Determine Exit Unit Capacity (C); refer to chart below.
3. Determine Number of Exit Units; X/C.
4. Determine Exit Size; divide X/C by the number of exits designed into the space and match unit quantities to sizes per the following:

Exit Width:

Units	1	1.5	2	2.5	3	3.5	4	4.5
Size (in.)	22	34	44	56	66	78	88	100

EXIT AND ACCESS REQUIREMENTS

	High Hazard	Commercial	Residential*
Max. Travel Distances (ft)			
Unsprinklered	NA	100 - 200	150
Sprinklered	150	150 - 300	200
Dead-end corridors	NA	20 - 50	20 - 40
Min. Corridor Width (in)	36	36, 44, 66 or 96	36
Capacity^c (C)			
Interior exit doors	50	30 - 80	40
Exterior exit doors	40	30 - 100	50
Vertical exits	30	15 - 60	30
Horizontal exits	50	30 - 100	50
Ramps (>1:10 slope)	37	22 - 75	37
Max. Occupant Load:			
Spaces with 1 door	10	15 - 75	20

* Does not include one- and two-family dwellings.
^{c} Number of persons per unit of width.
NA = not allowed.

■ **Egress is the means to escape from a burning building and is facilitated by the proper access to exits, adequate exits per occupant load, and proper discharge from the exits.** Exit components include doors (with panic hardware), stairs, ramps, passageways, escalators/moving walks and fire escapes.

● Exit access is any space, passage and/or corridor between the occupied space and the exit itself. It should be free from any obstruction, should not require passage through another occupied space (i.e.: office, storeroom, etc.) and should provide protection from fire.

● Exits should be arranged within the space in sufficient number and width so as to minimize gathering of escaping occupants. They should be obvious, within short travel distance and provide protection from smoke and fire. **A critical goal of egress design is to minimize occupant panic.**

—Revolving doors, elevators, and stairs/escalators which are open to an atrium are not acceptable exits.
—Windows are acceptable for egress in certain cases.
—Fire escapes may not be appropriate due to the threat of vertical flame spread and smoke through windows.
—Exit ladders may be used to serve occupant loads of 10 or less, single dwellings or guest rooms in buildings of three stories or less. They must be accessible from an egress window or a balcony.

● Exit discharge is the space outside of the exit which allows occupants to get away from a building. Discharge areas must be free from obstruction and of sufficient width.

● Existing/historic buildings may not meet current egress requirements: Restoration to original condition of spaces, finishes and egress capacities is likely to place a restriction on the maximum allowable occupant load.

■ **The occupant load may be critical to life safety planning.** It is based on the net floor area (sf) per occupant as established by codes/standards to ensure adequate space to perform the particular function in that occupancy type.

● The occupant load is additive for successive floors or adjoining spaces; as the number of occupants with access to an exit increases, so will the number of units of width.

■ **Exit signs and emergency lighting are required in all but residential occupancies and are to be located such that the travel path to the exit is visibly and clearly marked.** Directional signs may be necessary.

● Exit signs may be internally or externally lighted, daylighted in buildings not provided with electrical lighting, or may be electroluminescent (will glow without an electrical power source). **Select signs for legibility under smoke conditions.**

■ Exit and access requirements vary with occupancy type, building design and building regulations, therefore requiring designers to carefully review the applicable code.

● Egress from one- and two-family detached dwellings is provided by a minimum of two exterior doors and one window in each bedroom which is of sufficient openable area to allow passage in a relatively quick manner.

● Egress from high-rise buildings generally involves the exit from the fire floor through protected corridors and stairways. Floors above grade are usually required to have at least two separate exits; more are required with higher occupancy loads (3 over 500; 4 over 1000).

—Loop corridor arrangements offer two directions to reach an exit and eliminate dead end corridors.

D4.1 FIRE PROTECTION

ROLE OF TESTING

- ■ Fire statistics document that:
- ● **With decreases in the fire resistive characteristics of construction, loss of life and property increases as does the percentage of out-of-control fires.**
- ● Fire safety, therefore, relies on the use of proven materials and assemblies. For most purposes, full scale and reduced scale models are tested.

- ■ For many years, fire tests/standards have been developed for the purpose of modeling fire in various constructions. The many test methods are listed in REFERENCES.
- ● Developed by many research organizations, most fire test methods have become established as consensus industry standards through ASTM or NF₁PA.
- ● **Most existing fire tests are designed to produce models of performance in fire exposures.** New fire research and test standards are stressing full scale mock-ups of hazardous conditions for more comprehensive understanding of interactive forces of fuel contribution, spacial aspects and air movement.
- ● **These standards and test methods are typically adopted by building codes to regulate materials and assemblies used in buildings.**

- ■ When a particular material and/or assembly is tested in accordance with prescribed standards by recognized testing agencies, the results are referred to as accepted fire performance ratings.
- ● Each rated assembly is identified by a reference number which is assigned by the independent testing agency. Trade associations, such as the Gypsum Association or the defunct American Insurance Association, also have published tested and rated assemblies. Some product manufacturers also publish rated assembly information.
- ● Any deviation in composition, application and/or installation of the assembly/material from the test specifications will void the fire rating.

Terminology

- ■ Several terms are commonly used in the construction industry to describe the nature of materials when exposed to fire but are not technically accurate.

- ● **Fireproof:** Nothing is truly fireproof because the heat of fire, given a high enough temperature over a sufficient period of time, is capable of altering the structural properties of most materials. Those that are not altered transfer sufficient heat to promote ignition on the nonfire side. Preferred terms include fire resistant and noncombustible.
- ● **Flame retardant:** This term is used to describe materials which will not contribute to the growth of the fire. However, many flame retardant materials are only effective in low temperature fires; when at higher temperatures in fire stages beyond the incipient stage, these materials are likely to contribute fuel to the fire. These materials should be described in terms of their flame-spread ratings only.

Other Considerations

- ■ **The use of particular materials and assemblies in buildings can indicate potential fire hazard.**
- ● In wood frame construction, lightweight metal connecting plates are commonly used for trusses, both for floors and roofs. Within 10 or 15 minutes of a fire, these plates can be heated to the point where they char the wood and fall out, causing roofs and floors to fail.
- ● Concealed materials and assemblies should be investigated. Remodeling frequently covers over materials that can burn, even though the facing assembly is fire-resistive. Examples include:
  - —suspended ceilings that meet code requirements but are hung below a combustible finish or assembly;
  - —grid ceilings which meet code but conceal a fire may be destroyed and fall when opened to expose a fire -- flashover may occur when the burst of air is allowed to reach the fire;
  - —low ceiling areas may have one or more ceilings above them, each creating a void where fires can start without being detected;
  - —untreated fiberboard, used in furniture, cabinetry and as sheathing, burns tenaciously when ignited;
  - —insulated ceilings with steel joists and/or a tin ceiling finish tend to retain heat and conduct it to other parts of the assembly.

- ■ **Essential to fire-resistive construction is the protection of structural members from the effects of heat and fire.**
- ● Reinforced concrete is noncombustible and is not likely to experience structural fatigue due to excessive heat. Exposed reinforcing steel loses 85 percent of its yield strength at 1200°F and 95 percent at 1800°F.
- ● Concrete exhibits a loss of about 1/3 its strength when exposed to temperatures above 1200°F.
- ● While steel and iron are noncombustible, their strength is greatly affected by heat. Steel actually gets stronger at 500°F, it softens and loses about 1/2 its strength at 1000 to 1400°F. Protection from heat is required.
- ● Large wood members are used for beams, girders and interior columns in Heavy Timber construction. It is combustible, charring when exposed to fire/heat. Although surface wood is burned away, the charring effect allows the remaining wood to maintain its structural integrity.

- ■ **Protection of structural steel is designed to insulate the steel member from reaching its yield point temperature (1100°F) over a duration of time given by design/code parameters.** This protection is referred to as "fireproofing."
- ● Heavy gauge steel beams are fireproofed on the three exposed sides while members used in columns must be protected on four sides (such that the steel does not reach 1022°F - point of diminished strength).
- ● Lightgage steel framing requires protection from a thermal barrier in order to obtain fire-resistive ratings.
- ● Reinforcing steel loses half its strength at 1000°F while prestressing steel does so at 850°F; both may be protected by applying sufficient thickness of concrete.
- ● **When there is insufficient fuel to produce high fire temperatures, bare steel may not be structurally damaged.**

REFERENCES: Fire protection

STANDARDS

ASTM Test Methods/Methods for:

D1230 • Flammability of Clothing Textiles.

D1360 • Fire Retardancy of Paints (Cabinet Method).

D1929 • Ignition Properties of Plastics.

D2633 • Thermoplastic Insulations and Jackets for Wire and Cable.

D2843 • Density of Smoke from the Burning or Decomposition of Plastics.

D2859 • Flammability of Finished Textile Floor Covering Materials.

D2863 • Measuring the Minimum Oxygen Concentration to Support Candle-Like Combustion of Plastics (Oxygen Index).

D3675 • Surface Flammability of Flexible Cellular Materials Using a Radiant Heat Energy Source.

D3713 • Measuring Response of Solid Plastics to Ignition by a Small Flame.

D3806 • Small-scale Evaluation of Fire-Retardant Paints (2-ft. Tunnel Method).

D3894 • Evaluation of Fire Response of Rigid Cellular Plastics Using a Small Corner Configuration.

D4108 • Thermal Protective Performance of Materials for Clothing by Open-Flame Method.

D4151 • Flammability of Blankets.

E84 • Surface Burning Characteristics of Building Materials.

E108 • Fire Tests of Roof Coverings.

E119 • Fire Tests of Building Construction and Materials.

E136 • Behavior of Materials in a Vertical Tube Furnace at 750, C.

E152 • Fire Tests of Door Assemblies.

E160 • Combustible Properties of Treated Wood by the Crib Test.

E162 • Surface Flammability of Materials Using a Radiant Heat Energy Source.

E163 • Fire Tests of Window Assemblies.

E286 • Surface Flammability of Building Materials Using an 8-Ft. (2.22-M) Tunnel Furnace.

E605 • Thickness and Density of Sprayed Fire-Resistive Material Applied to Structural Members.

E648 • Critical Radiant Flux of Floor Covering Systems Using a Radiant Heat Energy Source.

E662 • Specific Optical Density of Smoke Generated by Solid Materials.

E736 • Cohesion/Adhesion of Sprayed Fire-resistive Materials Applied to Structural Members.

E814 • Fire Tests of Through-Penetration Fire Stops.

E859 • Air-Erosion of Sprayed Fire-Resistive Materials Applied to Structural Members.

E906 • Heat and Visible Smoke Release Rates for Materials and Products.

E970 • Critical Radiant Flux of Exposed Attic Floor Insulation Using A Radiant Heat Energy Source.

P190 • Proposed Test Method for Heat and Visible Smoke Release Rates for Materials and Products Using an Oxygen Consumption Calorimeter.

ASTM Terminology Relating to:

D4391 • Burning Behavior of Textiles.

E176 • Fire Standards.

ASTM Guides/Practices for:

D4205 • Flammability and Combustion Testing of Rubber.

E800 • Measurement of Gases Present or Generated During Fires.

E931 • Assessment of Fire Risk by Occupancy Classification (Commentary).

NFPA Standards/Standard Methods for:

13 • Installation of Sprinkler Systems.

13D • Installation of Sprinkler Systems in One- and Two-Family Dwellings and Mobile Homes.

14 • Installation of Standpipe and Hose Systems.

17A • Liquid Agent Extinguishing Systems.

261 • Tests and Classification System for Cigarette Ignition Resistance of Components of Upholstered Furniture.

UL Standards for:

8 • Foam Fire Extinguishers.

19 • Safety for Woven-Jacketed Rubber Lined Fire Hose.

38 • Manually Actuated Signaling Boxes for Use with Fire Protection Signaling Systems.

154 • Carbon Dioxide Fire Extinguishers.

193 • Alarm Valves for Fire-Protection Services.

199 • Automatic Sprinklers for Fire Protection Service.

203 • Pipe Hanger Equipment for Fire-Protection Service.

217 • Single and Multiple Station Smoke Detectors.

219 • Safety for Lined Fire Hose for Interior Standpipes.

228 • Door Closers-Holders, and Integral Smoke Detectors.

260 • Dry Pipe, and Deluge Valves for Fire-Protection Service.

262 • Gate Valves for Fire-Protection Service.

268 • Smoke Detectors for Fire Protective Signaling Systems.

268A • Smoke Detectors for Duct Application.

299 • Dry Chemical Fire Extinguishers.

312 • Check Valves for Fire-Protection Service.

346 • Waterflow Indicators for Fire Protective Signaling Systems.

405 • Fire Department Connections.

448 • Pumps for Fire-Protection Service.

521 • Heat Detectors for Fire Protective Signaling Systems.

D4.1 FIRE PROTECTION

REFERENCES: Fire protection (cont.)

539 • Single and Multiple Station Heat Detectors.

626 • 2¹/₂ Gallon Stored Pressure Water Type Fire Extinguishers.

668 • Hose Valves for Fire-Protection Service.

711 • Rating and Fire Testing of Fire Extinguishers.

753 • Alarm Accessories for Automatic Water-Supply Control Valves for Fire-Protection Service.

864 • Control Units for Fire-Protective Signaling Systems.

985 • Household Fire Warning System Units.

1058 • Halogenated Agent Fire Extinguishing System Units.

1093 • Halogenated Agent Fire Extinguishers.

1468 • Direct-Acting Pressure-Reducing and Pressure-Control Valves for Fire-Protection Service.

1474 • Adjustable Drop Nipples for Sprinkler Systems.

1478 • Fire Pump Relief Valves.

1480 • Speakers and Amplifiers for Fire Protective Signaling Systems.

1486 • Quick Opening Devices for Dry Pipe Valves for Fire-Protection Service.

1626 • Safety for Residential Sprinklers for Fire Protection Service.

1767 • Early Suppression Fast Response Type Sprinklers.

SELECTED REFERENCES

AISI **American Iron and Steel Institute**
• Designing Fire Protection for Steel Columns, 1975.
• Designing Fire Protection for Steel Trusses, 1976.
• Fire-Resistance Ratings of Load-Bearing Steel Stud Walls with Gypsum Wall-Board Protection with or without Cavity Insulation, 1981.
• Fire-Safe Structural Steel - A Design Guide, 1980.

ASHRAE **American Society of Heating, Refrigerating and Air Conditioning Engineers**
• Design of Smoke Control Systems for Buildings.

ASTM **American Society for Testing and Materials**

STP 614 • Fire Standards and Safety; A. Robertson, ed., 1977.

STP 685 • Design of Buildings for Fire Safety; Harmathy, T.Z. and Smith, E.E., ed., 1979.

STP 762 • Fire Risk Assessment; Castino/Harmathy, ed., 1982.

STP 816 • Behavior of Polymeric Materials in Fire; E.L. Schaffer, ed., 1983.

STP 826 • Fire Resistive Coatings; Lieff/Stumpf, ed., 1984.

STP 882 • Fire Safety: Science and Engineering; Harmathy, T.Z., ed., 1985.

CFR **Center for Fire Research, National Bureau of Standards**

NBSIR 81-2427 • Calculation of the Heat Release Rate by Oxygen Consumption for Various Applications.

NBSIR 85-3267 • A Summary of the NBS Literature Reviews on the Chemical Nature and Toxicity of the Pyrolysis and Combustion Products from Seven Plastics: Acrylonitrile-Butadiene-Styrenes (ABS), Nylon, Polyesters, Polyethylenes, Polystyrenes, Poly(Vinyl Chlorides) and Rigid Polyurethane Foams.

NBSIR 86-3372 • Fire Research Publications, 1985.

NBSIR 87-3535 • Ceiling Jet-Driven Wall Flows in Compartment Fires.

NBSIR 87-3550 • Mixing in Variable Density, Isothermal Turbulent Flow and Implications for Chemically Reacting Turbulent Flows.

NBSIR 87-3555 • Fire Research Publications, 1986.

NBSIR 87-3560 • An Engineering Analysis of the Early Stages of Fire Development - The Fire at the Dupont Plaza Hotel and Casino - December 31, 1986.

NBSIR 87-3562 • FIREDOC Users Manual

NBSIR 87-3567 • Comparisons of NBS/Harvard VI Simulations and Full-Scale, Multi-Room Fire Test Data.

NBSIR 87-3572 • Diffusion-Controlled Reaction in a Vortex Field.

NBSIR 87-3591 • EXITT - A Simulation Model of Occupant Decisions and Actions in Residential Fires: Users Guide and Program Description.

NBSIR 87-3626 • An Overview of Smoke Control Technology.

NBSIR 87-3633 • Test Results and Predictions for the Response of Near-Ceiling Sprinkler Links in a Full-Scale Compartment Fire.

GA **Gypsum Association**
• Fire Resistance Design Manual.

NFPA **National Fire Protection Association**

92A • Manual for Recommended Practice for Smoke Control.

101 • Life Safety Code.

UL **Underwriters' Laboratories**
• Fire Resistance Directory.
• Fire Protection Equipment List.

OTHER SOURCES:

• Room Fire Experiments of Textile Wall Coverings; Fisher, Fred L., MacCracken, Bill and Williamson, Robert Brady; Service to Industry Report No. 86-2; March, 1986; Fire Research Laboratory, University of California, Berkeley, California.
• Colburn, Robert E., **Fire Protection & Suppression,** McGraw-Hill, NY 1975

PERIODICALS

• Factory Mutual Record.
• Fire Journal, NFPA.
Fire Technology.
• U.S. Fire Sprinkler Reporter.

D4.1 FIRE PROTECTION

D5. 1 ELECTRICAL

ELECTRICAL SYSTEMS

ELECTRICAL SYSTEMS

There are three major components of a building's electrical power system:

• wiring, including conductors and raceways of all types.

• power-handling equipment, including transformers, switchboards, panelboards, large switches, and circuit breakers.

• control and utilization equipment, such as lighting, motors, controls, and wiring devices and receptacles.

The National Electric Code (NEC) of the National Fire Protection Association (NFPA) defines the fundamental safety measures that must be followed in selection, construction, and installation of all electrical equipment.

ELECTRICAL EQUIPMENT RATINGS.

The ratings that describe electrical performance are t voltage and current.

■ **Voltage.** The voltage rating (V) of electrical equipment is the maximum voltage that can be applied safely to the unit continuously. Frequently, it corresponds to the voltage applied in normal use.

■ **Current.** The maximum operating temperature at which electrical components can operate properly and continuously determines the current rating. That rating in turn depends on the type of insulation used.

The function of any **interior wiring system** is to conduct electricity from one point to another and may be either:

• **Electric power system** to distribute electrical energy.

• **Electric signal or communications system** to transmit information.

AN ELECTRICAL DISTRIBUTION SYSTEM

is designed to safely provide the energy required at the locations required. The principal types of interior wiring systems are:

■ exposed insulated cables

■ insulated cables in open raceways

■ insulated conductors in closed raceways

■ combined conductor and enclosure: This category describes all types of fac-

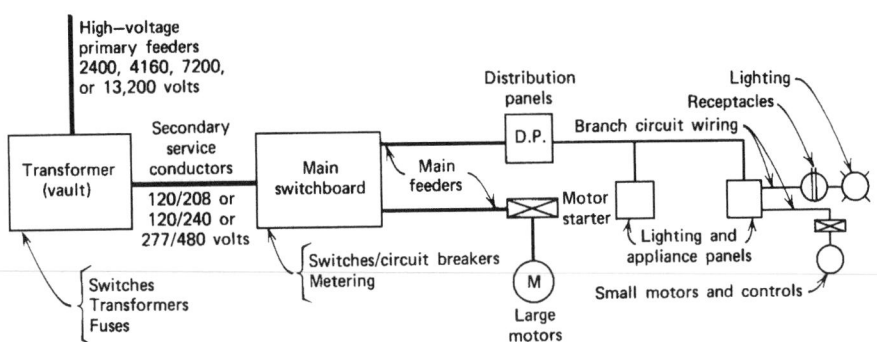

BLOCK DIAGRAM OF AN ELECTRICAL SYSTEM

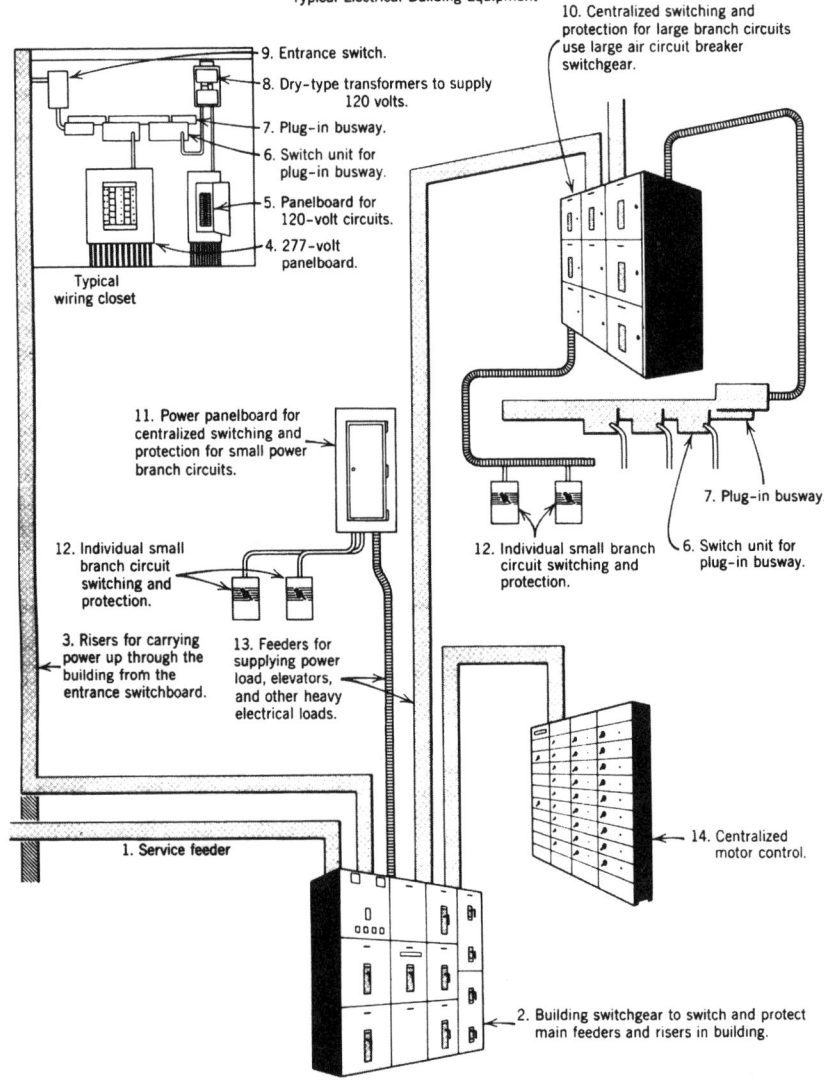

DEPICTION OF A BASIC ELECTRICAL SYSTEM (courtesy of General Electric).

Source: Electrical Wiring Systems (Ref. 1)

D5.1 ELECTRICAL

tory-prepared and factory-constructed integral assemblies of conductor and enclosure, including:

• flat cable intended for under-carpet installation

• flat cable assemblies

• lighting track,

• manufactured wiring systems

• all types of busway, busduct and cable bus

The components of these types of electrical systems and their application to electric wiring layout include:

Busway / busduct. A busway (bus-duct) ils an assembly of copper or aluminum bars, in a rigid metallic housing. It is normally the preferred economical choice when:

• it is necessary to carry large amounts of current (power) and,

• it is necessary to tap onto an electric power conductor at frequent intervals along its length.

Cablebus is similar to ventilated busduct, except that it uses insulated cables instead of busbars. These cables are rigidly mounted in an open space-frame.

Light-duty busway, flat cable assemblies and lighting track. Special construction assemblies that act as light-duty (branch circuit) plug-in electrical feeders are widely used because of simplicity of installation and their flexibility due to its plug-in mode of connection.

■ **Light-duty plug-in busway** may be used either for feeder or branch circuit wiring application, with restrictions when applied as branch circuit wiring.

■ **Lighting track** is a factory-assembled channel with conductors for one to four circuits permanently installed in the track.

Cable tray is a continuous open support for approved cables. When used as a general wiring system the cables must be self-protected. The advantages of this system are free-air rated cables, easy installation and maintenance, and relatively low cost. The disadvantages are bulkiness and the required accessibility.

Raceways. Types and characteristics of closed wiring raceways include:

■ **Steel conduit.** The purpose of conduit is to:

• protect the enclosed wiring from mechanical injury and corrosion.

• provide a grounded metal enclosure for the wiring in order to avoid shock hazard.

• provide a system ground path.

• protect surroundings against fire hazard as a result of overheating or arcing of the enclosed conductors.

• support the conductors.

The NEC generally requires that all wiring be enclosed in a metallic raceway. Metal electrical conduits and associated fittings must be corrosion resistant.

■ **Steel conduit is available as:**

• heavy-wall steel conduit, also referred to as "rigid steel conduit."

• intermediate metal conduit, referred to as "IMC."

• electric metallic tubing, known as "EMT" or "thin-wall conduit."

■ **Aluminum conduit.** The use of aluminum conduit has increased due to its light weight and resulting labor cost savings.

■ **Flexible metal conduit** is used principally for motor connections or other locations where vibration is present, where movement is encountered, or where physical obstructions make its use necessary.

■ **Nonmetallic conduit.** For use above ground, this conduit must be flame retardant, tough, and resistant to heat distortion, sunlight, and low-temperature effects.

■ **Surface metal raceways** are normally installed in exposed conditions and in places not subject to physical injury.

■ **Floor raceways.** The NEC recognizes three types of floor raceways:

• underfloor raceways

• cellular metal floor raceways

• cellular concrete floor raceways

All three types are applicable to all types of structures. None may be used in corrosive or hazardous areas. The fundamental difference between them is that underfloor raceways are added on to the structure, whereas cellular floor raceways are part of the structure itself—and therefore have a direct effect on the building design.

■ **Underfloor raceways.** Underfloor ducts may be cast into the structural slab in lieu of being in fill or topping, but the slab must be designed to

accommodate them. The use of a fill or topping on the structural slab for underfloor duct has these advantages:

• Ducts can be run in any direction, without conflict to structural elements.

• Formwork and construction sequence are simplified.

• Finishing is simplified.

To justify the relatively higher cost of underfloor raceways, the building should meet these conditions:

• Open floor areas, with a requirement for outlets at locations removed from walls.

• Outlets from ceiling systems is unacceptable.

• Frequent rearrangement of furniture and other items requiring electrical and signal service.

■ **Cellular metal floor raceway** may be provided by a cellular (metal) floor that is an integrated structural/electrical system, fully or partially electrified.

■ **Precast cellular concrete floor raceways.** This structural concrete system is similar to a cellular metal floor.

■ **Full access floor construction** is applicable to spaces with very heavy cabling requirements, particularly if frequent recabling and reconnection is required. It provides for instant and complete access to an underfloor plenum.

■ **Under-carpet wiring system**

originally developed as an inexpensive alternative to an underfloor or cellular floor system and as a means for providing a flexible a flexible floor-level branch circuit wiring system.

■ **Ceiling raceway systems.** The need for electrical flexibility in facilities with limited budgets has encouraged the use of equivalent over-the-ceiling systems.

■ **Ceiling raceways and modular wiring systems** are more flexible than underfloor counterparts, because they energize lighting as well as provide power and telephone facilities.

■ **Boxes and cabinets** including pull boxes, splice boxes, and outlet boxes. All boxes must be equipped with tightly fitting, removable covers.

REFERENCE: Stein, Benjamin and John S. Reynolds. 1992. *Mechanical and Electrical Equipment for Buildings.* Eighth Edition. John Wiley & Sons.

D5.1 ELECTRICAL

ILLUMINATION: introduction

THE NATURE OF LIGHT

As vision is the primary sensory tool of the human species, light is the most important form of energy we encounter. Light provides the stimulus which allows recognition, organization and evaluation of visual information perceived within an environment. Light defines form, color; creates mood, atmosphere; emphasizes direction and movement. **Therefore, the use of light should be considered to be an art as well as a science.**

■ **The energy of light is a small band of wavelengths in the electromagnetic spectrum which ranges from radio frequencies to cosmic rays.**

■ **Infrared (IR) radiation is non-visible light characterized by its relation to heat:**
- it can dehydrate many dyes and fibers;
- it is generally used for drying and heat therapy;
- it can be the source of skin or eye tissue burns and has been related to cataracts of the eye.

■ **Visible light is recognized by the human eye as white light but is actually the combination of the colors of the visible spectrum; red, orange, yellow, green, blue and violet.**

■ **Ultraviolet (UV) radiation is non-visible light that can have both harmful and beneficial effects:**
- it can burn or embrittle paper and textiles;
- it can fade or discolor dyes and oils;
- it is generally associated with the aging processes of skin and eyes;
- it benefits the generation of Vitamin D in the body;
- it can enhance certain types of illumination such as black light effects.

There are three types of visible light sources:

■ **Natural phenomena** - radiation of the complete spectrum of light from the sun or stars; the reflection of sunlight off the moon, sky, clouds, earth and bodies of water; lightning; Aurora Borealis, Aurora Australis; and bioluminence (light produced from the oxidation of chemical compounds by plants and animals).

■ **Incandescence or temperature radiation** - as an element is heated, its molecules become increasingly active.
- Until the temperature of the element reaches about 500°C, only heat

radiation is released. From 500° to 600 °C, infrared light is also produced. At 600°C, visible light of long, red wavelengths is radiated. As the temperature rises, the wavelengths of light become shorter, progressing through the spectrum of visible light.
- For example, as electricity is introduced to a steel bar, the current of energy gradually heats the bar. As the energy current is increased, the bar turns from red to yellow and eventually to white. If energy is applied only to one end of the bar, it will be possible to see this spectrum by looking from the cooler end toward the energy source.
- Typically, incandescent sources emit infrared light along with visible light and heat.

■ **Luminescence** - when an electric current is introduced through a gas or solid which is composed of single valence atoms, it causes the valence electrons to temporarily rise to a higher level of energy. When the electron resumes its natural state, it radiates energy in the form of light.

- **Photoluminescence or gaseous discharge** is the principle which describes the response of single valence electrons when their atoms are in a gaseous state. The collision of electrons induced by the introduction of electrical current (free electrons) within a confined gas generates an arc discharge, ionizing the vapor molecules. The wavelengths of the emitted light will vary with the type of gas vapor.
- **Fluorescence** is a means of producing light from gaseous discharge. In this case, the confined gas is one which produces a relatively high percentage of ultraviolet light. The confining glass tube is coated with phosphor crystals which absorb and convert the UV radiation into longer wavelengths in the visible spectrum.
- **Phosphorescence** describes the action of organic fluorescent materials (rare earth phosphors) which retain the radiant energy for short periods of time.
- **Electroluminescence** is the conversion of electricity to light using

WAVELENGTHS

10^5 meters	RADIO FREQUENCIES
	AM Radio
	Short Wave Radio
	FM Radio
1 meter	Television
	Radar
	Microwave
10^{-3} m (1 millimeter)	LIGHT FREQUENCIES
1400 millimicrons	Far Infrared
	Near Infrared
10^{-6} m (1 micron)	
770 millimicrons	Visible Light
400	
380 millimicrons	Ultraviolet
30 millimicrons	
10^{-9} m (1 millimicron)	
10^{-10} m (1 angstrom unit)	X-RAY FREQUENCIES
10^{-11} m	
10^{-12} m	
10^{-13} m	GAMMA RAY FREQUENCIES
10^{-14} m	COSMIC RAY FREQUENCIES
10^{-16} m	

770 millimicrons
Red — 620
Orange — 590
Yellow — 560
Green — 490
Blue — 430
Violet — 380 millimicrons

D5.1 ELECTRICAL

the passage of alternating or direct current through special conductors of solid matter producing visible light; includes AC capacitive lamps and light emitting diodes (direct current).
- Other luminescent phenomena are created chemically, during crystallization or oxidation, from radio-activity, etc.

■ **The distribution of light in a given environment will often influence the activity within it. There are three basic forms of distribution which are used to control the effect of illumination:**
- **direct lighting** - provided by the introduction of the light source immediate to the environment, creating illumination, definite shadows, contrast, and depth of field; sunlight on a clear day.
- **indirect lighting** - when the light source is located in but shielded from the environment such that all light is allowed only to radiate away from the environment, illumination is available only from the surfaces which define the environment; produces a general, even illumination nearly void of shadow and contrast; moonlight on a clear night.
- **diffuse lighting** - provided by the introduction of an element between the light source and the environment such that light radiation is allowed to pass through in limited or controlled amounts, resulting in scattered light and soft shadows; sunlight on an overcast day.

Basic Lighting Physics

The control of light distribution is established using the principles which define the effect of light on objects and surfaces: Absorption, reflection, refraction and transmission.

■ **When light strikes an object, it is either absorbed or reflected.** The amount of light absorbed will depend on the surface color, contour and composition. The absorption of all light will result in darkness, as the absence of light is black. **The absorption process converts the radiant energy of light into thermal energy and releases it as heat.**

■ **The light which is not absorbed is reflected in a pattern which is dependent upon surface characteristics:**
- **specular** or regular reflection - when the reflected light rays are bounced

off the surface at the same angle at which they approached and struck the surface (the angle of incidence); characteristic of a mirror;

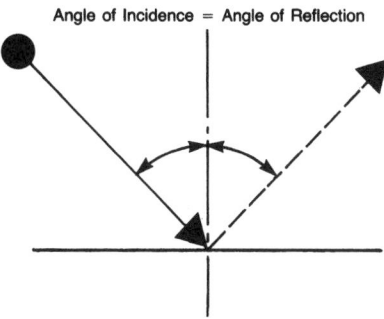

Angle of Incidence = Angle of Reflection

- **spread** reflection - when the reflected light is distributed in all directions, but allows the greatest intensity of light to be sent off at, or near the angle of incidence; for surface highlight;

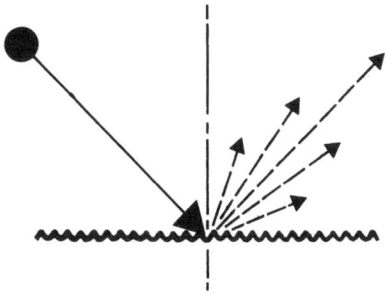

- **diffuse** or scattered reflection - when the reflected light rays are distributed in all directions with maximum intensity normal to the surface; characteristic of a matte finished surface; promotes uniform color and luminance.

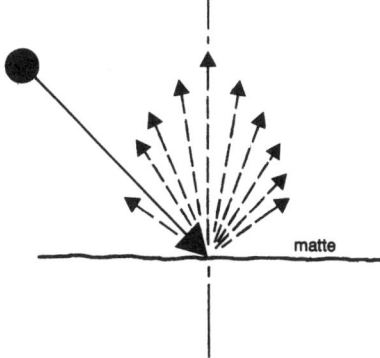

matte

■ **Refraction is when the light beam undergoes a change or shift in**

direction as it passes through a medium, but is otherwise unaffected.

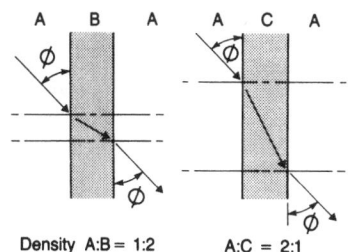

Density A:B = 1:2   A:C = 2:1

■ **The transmission of light is the allowable passage of light beams through a medium (solid, liquid or gas).**
- **Direct** transmission describes the response of a transparent medium which allows the light beam to pass through and emerge unaffected by the transmission.

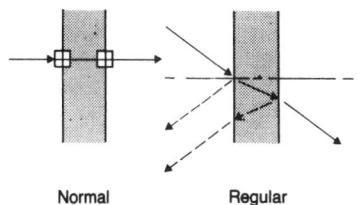

Normal   Regular

- **Spread** transmission is when the medium refracts only a portion of the light beam; often used to conceal a light source while still providing surface highlights (sparkle).

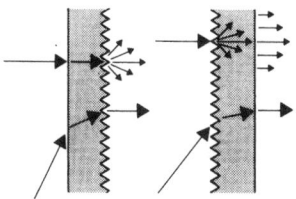

- **Diffuse** transmission describes the passage of light through a translucent medium which disperses the light beam within itself such that the images behind the medium cannot be distinguished.

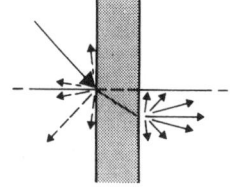

D5.1 ELECTRICAL

ELEMENTS OF LIGHTING DESIGN

THE MEASURE OF LIGHT

■   There are many terms associated with the measurement of light. For specific definition, more detail or instructions for use in design calculations, reference should be made to complete engineering documents such as the current editions of the IES Lighting Handbook Reference and Application Volumes.

Terminology

- **Amperes:** The unit of measure of electrical current.
- **Bulb:** The glass enclosure of a lamp designed to contain inert gases, protect inner elements of the lamp and occasionally to determine distribution (diffuse or reflective); also used as the lay term for 'lamp'.
- **Candela:** The metric unit of measure of light intensity.
- **Candlepower:** The intensity of light from a source in a certain direction and measured in candelas.
- **Coefficient of Utilization (CU):** The ratio of illuminance to the lumens radiated for the light source.
- **Efficacy:** the ratio of the approximate initial lumens produced by a light source divided by the necessary power to produce them, expressed as percentage (lumens/watt; formerly 'efficiency').
- **Footcandle (fc):** A unit of measurement in the English system used to gauge the amount of light falling on a surface; the number of lumens that are incident on each square foot of work surface. 1 fc = 10.76 lx.
- **Footlambert (fl):** A unit of measurement in the English system used to gauge the amount of light leaving a surface; footcandles times reflectance.
- **Illuminance:** The light falling on a surface; measured in footcandles or lux.
- **Lamp:** The mechanism which converts electricity into light by means of incandescent filament or gaseous discharge; also used as the common term for portable luminaires.
- **Lens:** An element of a luminaire which is used to alter or redirect light distribution using diffusion, refraction or filtration.
- **Light Loss Factor (LLF):** The design factor that accounts for atmosphere dirt depreciation, normal degrading of lamp lumens over the life of the lamp and other factors that add to the fact that less light is available over time.

- **Lumen:** A measure of total light-producing output of a source; the quantity of visible light emitted.
- **Luminaire:** An assembly used to house one or more light sources (lamps), connect light and power sources and distribute light (also referred to as 'fixture').
- **Luminance:** The emitted or reflected light from a surface measured in footlamberts (commonly 'brightness').
- **Lux (lx):** A unit of measurement in the metric system used to gauge the amount of light falling on a surface; the number of lumens incident on each square meter of work surface. 1 lx = 0.0929 fc.
- **Ohms:** The unit of measure of resistance.
- **Visual acuity:** a measure of the ability to distinguish fine details.
- **Volts:** The unit of measure of electrical force;
- **Watts:** The unit of measure of electric power; the power required to keep a current of one ampere flowing under the pressure of one volt.

Calculating Illumination

Basic techniques available for illumination design are briefly summarized below merely for the purpose of demonstrating the important elements of illumination design. These simple calculations are general design tools and guidelines; within a given space the incidence of the calculated footcandle level is not likely to occur.

■   **The Lumen Method** (also referred to as the Zonal Cavity Method) is useful for general lighting design when the average illuminance on a horizontal work surface is to be estimated.

- For average maintained footcandle values, use the following equation:

$$fc = \frac{(\# \text{ of luminaires}) \times (\# \text{ lamps per luminaire}) \times (\text{initial lumens per lamp}) \times CU \times LLF}{\text{area (in square feet)}}$$

- The CU for a particular luminaire can be selected from tables in the manufacturer's catalog following the calculation of the room cavity ratio (RCR) and the determination of wall and ceiling surface reflectance. Floor reflectance is typically given at 20 percent. The formula for calculating the RCR is:

$$RCR = \frac{5 \times (\text{mounting height of luminaire above task}) \times 1/2 \ (\text{perimeter of room})}{\text{area}}$$

- The LLF can be calculated, but for general purposes, the following values can be used: .60 to .65= poor, .70 to .75 = average, .80 to .85 = good.

■   **The Point Method** is used for street lighting, spotlighting and floodlighting.

- For the footcandle value, at a point straight down, use the following equation:

$$fc = \frac{\text{candlepower}}{(\text{mounting height})^2}$$

where the mounting height is the distance from the work surface to the luminaire.

- For footcandle values at a point away from nadir (straight down):

$$fc = \frac{\text{candlepower} \times \text{cosine(beam angle/2)}}{(\text{mounting height})^2}$$

- For the area of coverage of the projected light pattern from a given source with a specific beam pattern use trigonometric functions:

Beam = tan(beam angle/2)
Dia.   x mounting height x 2

Note: 'tan'= a tangent of an angle of a right triangle = the length of the opposite side divided by the length of the adjacent side. The mounting height is taken at a right angle to the work surface and bisects the beam angle.

- Candlepower and beam pattern can be found in manufacturer catalogs.

Illuminance

■   **Various measurements of footcandles/lux are used as illumination guidelines for different tasks and applications.** A detailed listing of these guidelines can be found in the **IES Lighting Handbook Application Volume**; a summary is as follows:

- **Four categories cover general lighting throughout the environment:**

— 2, 3 or 5 fc for open areas where orientation is the predominant visual need or where subdued activities are enjoyed;

— 5, 7.5 or 10 fc for circulation (corridors, stairwells) or for simple visual task of short duration;

applicable to lobbies, reception rooms, dining rooms, etc;

— 10, 15 or 20 fc for those areas where a broad range of visual tasks are performed, but where no one task is carried out for a substantial length of time; applicable to stores, offices, etc;

— 20, 30 or 50 fc for an area where a broad range of visual tasks are carried out continuously or where other close visual tasks are performed on a background of high reflectance.

- **Five categories describe the specific illuminance on or for the visual task:**

— 50, 75 or 100 fc may be appropriate for work with considerable detail and moderate contrast; should be used with general lighting of 20 to 50 fc;

— 100, 150 or 200 fc should be used for visual tasks of extreme detail and poor contrast; should be used with general lighting of 30 to 50 fc, with maximum contrast of 5:1, task to general lighting.

— 200, 300 or 500 fc levels are used where the performance of the task or the inspection of the product is very difficult due to low contrast and small size;

— 500, 750 or 1000 fc levels are typically used for very fine tasks of exacting visual acuity;

— 1000, 1500 or 2000 fc levels are necessary for inspection of cloth and clothing products.

- **Visual tasks which are of prolonged duration, fine detail, or close work on tasks of low reflectance require high levels of general illumination with supplemental local illumination.**
- The lowest two categories (1 to 10fc) meet most residential needs, and attractive general lighting is accommodated by levels of 3 to 15fc.

■ **The effect of the lighting system will be highly responsive to the incidence of light.**
- The placement of the light and distribution from it will produce certain surface reflection.
- Illuminance on the horizontal plane (work or circulation) promotes awareness of people, movement and surroundings.
- Luminance on the vertical plane highlights the environment through contrast and silhouette, setting a more introspective, intimate, passive mood.

PLANNING VARIABLES

The quantity and quality of illumination are generally the paramount factors in lighting design. Yet, these factors must be tempered by other planning variables which will play their part in the design process.

Intended Use

■ **The most basic input to the design process is the identification of the intended use of the environment.**
- **The type of activity:** From restful to highly demanding visual tasks.
- **Quality or accuracy:** Product quality can be enhanced through the use of appropriate illumination for the visual task.
- **The motivation of the occupant:** Some activities require high energy levels (quickness, alertness, awareness) associated with greater levels of illumination. Others can afford subdued lighting associated with relaxing, intimate or passive activities.

Occupancy

■ **Psychological elements play a large role in lighting an environment** since the visual sense introduces the images that formulate knowledge. This knowledge, along with experience, emotion and expectation, forms the **perceptions** of what is seen - the order, pattern figure, background, color, size, shape and other visual sensitivities.

■ **Biological needs: About 75% of human activity depends upon the eyes for guidance.** Therefore, eyesight impairment and various other disorders which may exist must be considered as well as potential eye strain and fatigue.
- **Age** affects the ability to see because the eye gradually requires greater stimulation to respond.
- **Orientation:** Motor skills rely on references to horizon, depth of field, obstructions, contours, etc., for proper response. Time orientation is critical to lighting as the biological clock is relative to the solar clock.
- **Physical security:** Illuminating or defining the environment reduces the unknown.
- **Relaxation:** Induced by using lighting techniques to produce focal interest at the workplace while providing a calm, interesting environment for leisure.

Environmental Definition

■ **The limits of space are defined by the existence and nature of reflective surfaces.**
- Surfaces of light color value are highly reflective and will produce a brighter

environment with more general diffusion of light; less contrast and shadow.
- Finishes of dark value have low reflectance and will absorb great quantities of light, resulting in the impression of a dark space. This is the same effect as if no walls existed.
- Rough surfaces and matte coatings effectively break down the directional qualities of light, and may be more difficult to maintain due to dust collection; typical of flat paint, plaster, etc.

■ **Buildings provide specific environments where the geometry and color of the room/space will affect the amount of light necessary to illuminate an area or task.** Given similar reflective characteristics of surfaces, the losses in lighting efficiency due to room proportions are given by the ratio of floor area to ceiling height.
- In a large, low room, a substantial portion of the light reaches the work plane without reflectance.
- In a high, narrow room, a higher proportion of light strikes the walls and other surfaces, resulting in reduced lighting efficiency.
- The geometry of the space can be factored into the design of the lighting system by using the **room cavity ratio (RCR)**, which is a component of the coefficient of utilization. The RCR relates the various dimensions of a room in order to establish a weighted number on a scale of 1 to 10. It assumes that the work surface is typically 30 inches above the floor and measures from that point to the luminaire plane (the bottom of the luminaire).

■ **Buildings must be analyzed for structure and space.**
- The location and size of columns, bay size, wall height, etc., must be defined.
- The ceiling must be classified for height, location and symmetry of roof trusses/girders. This is relevant to ease of wiring, control and luminaire placement.

■ **Air quality, particularly dust, dirt, smoke/soot and moisture will produce a filtering effect between the light source and the visual task surface.**
- Often, greater levels of direct lighting are required for proper illumination of dusty, smoky or foggy environs.
- The efficient operation of the light source may be affected as a coating of airborne particles collects on the lighting elements.

D5.1 ELECTRICAL

ELEMENTS OF LIGHTING DESIGN (cont.)

■ **Provision for the thermal contribution by the activity and the lighting must be integrated into the environmental design.**
- Ambient temperatures can effectively reduce the efficiency of the light source if temperatures rise too high.
- The operation of light sources will contribute to the heat radiation in the environment, adding to or subtracting from the efficiencies of HVAC systems.

■ **All surfaces, especially ceilings, that have integrated lighting systems must recognize the need for sound control techniques.** The relationship of acoustics and lighting should be examined for its influence on the selection and placement of materials, surfaces and contours.

■ **The proper design of any space, indoor or outdoor, coordinates the transition from daylight to electrical light (and vice versa).** It should be as gradual as that of day to night, maintaining the lighting design parameters. It is important to minimize the adjustment by the user's eye to the gradual change in the angle, direction and color of the light as the source changes.
- Daylighting design must consider the entrance and distribution of light as well as the potential for space heating. Like overheating, overlighting is an undesirable design trait which can be avoided by employing principles of reflection and diffusion.
- Normal outdoor illumination will measure over 10,000 fc in sunlight on a clear day; 1000 fc in the shade or on an overcast day; about 0.025 to 0.00.01 fc in moonlight.
- Window arrangements can permit high levels of natural light near the window, with only about 10% of that illumination reaching further into the space. Designs which consider this impact can reduce the need for perimeter lighting by incorporating electric light controls. Shading controls at the windows can reduce contrast.

Economics

■ **The economics of lighting extend beyond the initial (first) costs of materials and installation to the operational considerations of maintenance and energy costs which continue throughout the life of the lighting system.** Planning must account for these costs as well as the potential for future alterations and the flexibility of the lighting system to meet the needs of the evolving environment.

■ **First costs:** The purchase and installation of the lighting components.
- Material and labor costs are optimized when the lighting equipment is properly coordinated with the available power supply and the other elements of the environment (power, electronics, communications, air handling, sprinkler systems, etc.).
- Installation costs will respond to architectural form and structure.

■ The cost to add or relocate sources of illumination will define the **system flexibility.** Ceiling lighting has less flexibility for physical modifications, while furniture lighting is easily moved or replaced but is limited to locations near a power source. However, ceiling systems promote flexibility in use and activity, while furniture lighting is more specific in area. Fixed power sources such as conduit and wire may often be inappropriate; flat cable and surface floor outlets may be expensive.

■ **Maintenance** costs will vary according to three primary functions; the frequency of the need to replace the expended light source, the repair or replacement of auxiliary parts, and the cleaning schedule.

■ **Energy** costs reflect the consumption of wattage over time. The design and specification of lighting systems considers the amount of illumination provided and the expenditure of energy necessary to accomplish the design.
- Lighting controls contribute to controlling energy costs.
- Energy concerns are being addressed by building code officials and state regulatory agencies as they come to realize that illumination consumes up to 25% of the power used in a typical office building. Some codes are limiting power for lighting to 1.5 watts per square foot.
- Energy costs, like first costs, are reduced when the selection of lighting equipment is matched to the available power supply. For further details and requirements for circuitry, wire sizing and so on, refer to the National Electric Code.

VISUAL COMFORT

The visual comfort of an environment and the efficiencies of illumination are the paramount concerns of lighting. Electrical lighting design can be expanded beyond simple concepts of general lighting and quantitative response to function, task, focus, orientation and guidance. Design dynamics respect the relationship of forms, spaces, color, contrasts and textures as well as typical concerns of intensity, distribution and glare.

Contrast

■ **Contrast is the perceived and/or measured variation of luminances; a difference of shade, texture or other effect from reflection.**

■ **The application of contrast in luminance value is paramount in lighting design.** Proper contrasts can lower levels of illumination (thus saving energy), can reduce distraction from the visual task and can alleviate sources of fatigue. The closest ratio of these measures which can be discerned by the human eye is 2:1.
- Standard guidelines describe maximum ratios so that the ill-effects of excessive contrast may be avoided:
  — 3:1 between the task and immediate adjacent areas (working plane or visual surround);
  — 5:1 between the task and general background;
  — 10:1 between the task and remote areas;
  — 20:1 between the light source and its background;
  — 40:1 between any two luminous intensities anywhere within the normal field of vision should not be exceeded.
  — 50:1 will highlight objects to the exclusion of everything else.

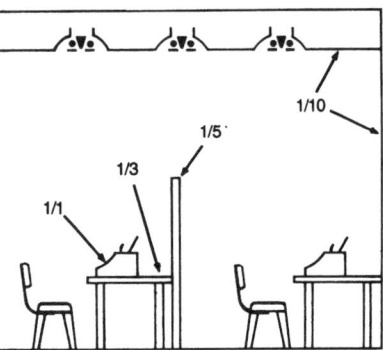

D5.1 ELECTRICAL

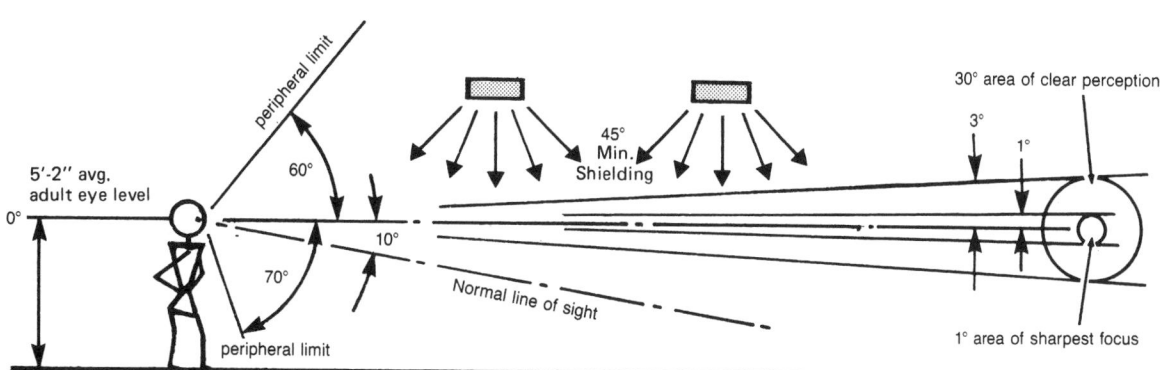

DEFINING THE VISUAL FIELD

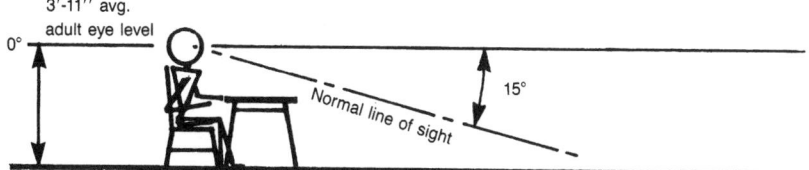

- **Visual acuity is influenced by contrast.** When the surround has a ratio in the range of 1:1 to 1:10 with the background, acuity will be optimized.
  — With a background of low luminosity, the object of the visual task must be larger in order to be visible, but as luminosity increases, objects of smaller size can be discernible. This is referred to as the minimum perceptible contrast.
- **Diffuse lighting techniques will provide low contrast environments** where difficult and sustained visual tasks are free of luminous distraction.
- **Direct lighting promotes high contrasts** due to the creation of shadow and highlight. Bold contrasts using highlighting are appropriate for directing attention while lesser highlight contrasts provide orientation.

■ **The higher the contrast of image or task to background, the easier it is to perceive the image and the lower the footcandle requirement.**

- Excessive contrast (black, sharp shadows due to overlighting) will produce eye strain and fatigue.
- Vision becomes increasingly accurate as the luminosity approaches an upper limit of about 1000 fc.
- Other factors affected by increased levels of illumination are the sensitivity

to contrast, visual acuity, and the amount of time required to perform a task (less time needed for study, perception and effort).
- Illuminance requirements on a light (good reflectance) task will be approximately 10% of those for lighting a dark task.

Glare

■ **The intrusion of glare is probably the most potent cause of visual disability and discomfort.**

■ **Direct glare is the result of excessive luminance of a light source. It can be controlled or reduced using one or more techniques.**
- **Increase the area of luminance:** Small, concentrated light sources produce a higher average luminance per surface area. The use of a globe or lens to broaden that area will reduce the average luminance, thus reducing glare.
- **Moderate the contrast of the source to its background.**
- **Position the source for proper distribution relative to the line of sight:** Substantial quantities of luminance can be distributed at a right angle (90°) to the line of sight because the structure of the forehead and brow shade the eyes. Extreme glare exists where the light source, unshielded, lies on the visual horizon.
- **Shielding is effective for limiting**

the amount of light emitted in the direction of the eye: Shielding devices are used to conceal the light source from direct view. Some options will cover the source completely; others are designed to cut off visibility of the source at 45° from horizontal.

■ **Veiling reflections (or reflected glare) are excessive, uncontrolled luminances reflected from objects or surfaces in the field of view.** When on the periphery, the glare may be referred to as indirect. Veiling reflections are also specular reflections which are superimposed upon the diffuse reflections of an object such that the details of the image are obscured.
- **Ideally, the light source would be positioned to either side or to the rear of the observer.** This, however, causes shadows which may be detrimental.
- **Work surfaces (both task and surround) should be finished to reduce glossiness.** In some cases, such as with computer CRT screens, it is not possible to avoid glossy surfaces to eliminate reflections from overhead lighting. In this case, a lower level of illumination (7.5 fc) is recommended to reduce disability.
- **Overlighting may be described as a glare effect.** The excessive illumination and resultant reflection over a large area produces the same ill-effects.

D5.1 ELECTRICAL

ELEMENTS OF LIGHTING DESIGN (cont.)

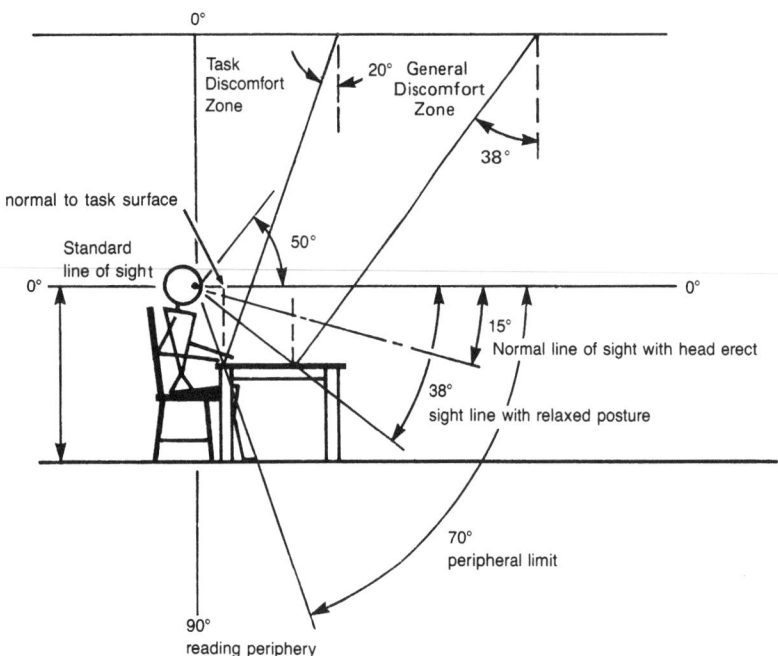

VEILING REFLECTION

COLOR

■ **The human eye is sensitive to the full spectrum of visible light.**
- When all of the wavelengths of that spectrum are present in equal proportion, the sensation is of white light.
- When variation exists in the presence and/or energy level (luminance) of the wavelengths, the sensation is of a color of light referred to as **hue**. The extreme in variation is when only one wavelength exists, producing the purest sensation of the hue (monochrome).
- The strength or intensity of the hue is relative to its purity, or the **saturation** of light radiation within a particular band of wavelengths.
- Color value graduates from pure color (high value) to white (producing tints of lower value) or to black (producing shades of lower value) or to gray (producing tones of lower value). Gray is itself a tone which varies between pure white and pure black.
- **Color contrast** might be considered to be a comparison of saturations and hues since saturated blues, greens and reds will stand out in white light while yellow must be considerably brighter (higher in luminance) to be equally contrasting.

■ **The integration of lighting and environmental surfaces requires an elemental understanding of the relationship of primary colors.**
- The primary colors are any set of three unique colors such that no combination of two colors will produce a match with the third.
- **The primary colors of light are red, green and blue**; equal proportions of the three produce white light.
- **The secondary colors of light are magenta, cyan and yellow**; adding equal proportions of the three produce black, or the subtraction of all color. Magenta is the equal combination of blue and red, cyan is the combination of blue and green, while yellow is produced from red and green.
- **The primary colors of dye and pigment** are those which absorb one of the primary colors of light and emit (reflect or transmit) the other two; thus they are the same as the secondary colors of light.
- **The perceived color of an object** is the color of the transmitted or reflected light.
- **Complimentary colors** are any two colors which will produce white when mixed in proper proportions (green

Texture

■ **Texture in lighting design refers to patterns of shadow and highlight.** Texture is important in the creation of visual interest; it is used to direct attention, affect mood or atmosphere, respond to visual expectations.

- **Shade and shadow provide depth and form in an environment.** It is most effective when the lines of shadow are softened.

- **Shadow is not desirable where specific tasks are being performed or for general lighting.**

- **Light pattern depends on the direction of light as well as its luminance and contrast.** Natural or conventional patterns are formed by concentrating light from the front (straight on, above) or rear, creating expected impacts. Perceptions can be upset by using silhouette (diffuse light from the rear) or by reversing natural patterns (concentrated light from below).

- The use of **grazing light** is effective for creating surface texture and sculptural relief. Grazing light accentuates surface irregularities when the light source is located to distribute light across the surface, creating sharp highlight and shadow. Where

grazing is not desirable, provide illumination from opposing directions to minimize shadows. Light distributed straight onto the surface will create flat images of truer color.

■ **One form of lighting texture which is usually not desirable, but often difficult to eliminate is scallop shadows.** These uneven patterns of shadow and light are due to improper placement of light sources in relation to surfaces and each other. If the overlapping of distribution patterns does not eliminate the scallop effect, it can be minimized by reducing the contrast of light and shadow through the use of a darker, low reflectance surface in the area of pattern.

■ **Sparkle is a technique which employs contrasts to create visual stimulation.**
- Direct sparkle is produced from the exposure of small, low-intensity lights, or pinhole exposures through a device used to shield a larger light source.
- Reflected sparkle is created through specular reflection from irregular, patterned or other textures of surfaces.
- Transmitted sparkle is the result of refraction through smaller prisms.

and magenta, red and cyan, blue and yellow).

Color Rendition

■   **The conditioned response of the occupant of an environment is to perceive illumination in comparison to natural light, but few artificial light sources produce the full spectrum of light that the sun does. Therefore the color, as well as the intensity, of light radiation will vary the perception of environment. This concept is referred to as color rendering.**
• Color rendering is also affected by the 'overlapping' of color due to the reflections of two or more adjacent colored surfaces when viewed simultaneously.

■   **When radiated light is composed of a narrow band of wavelengths, the color of the exposed surface may fall into one of three renderings:**
• If the surface is white or another neutral color, the emitted light will be mostly reflected (80% and greater for white, 60-70% for off-whites, 20% for medium gray), giving the perception that the surface is actually that color;
• If the surface color is not contained in the radiated light, the light will be mostly absorbed, dulling or graying the surface colors (good for identifying slight variations in surface color);
• If the surface color is contained in the radiated light, the light will be almost entirely reflected, accentuating the surface color. (Light sources which produce ultraviolet light as well as visible light may create a fluorescent effect on the reflecting surface.)

Color Reference Standards for Light Sources

■   **Color temperature refers to the appearance of emitted light as compared to white light.** Measured in Kelvins (K), color temperature relates to the physics of light (Planck's radiation law) rather than heat radiation.
• Low color temperatures are correlated to yellowish light, giving the sense of warmth. The lowest color temperature of a light source would be about 1900K for the rich yellow glow of candlelight.
• High color temperatures are correlated to bluish light, giving the sense of coolness. Higher color temperatures of light sources are over 9000K, such as the reflected sunlight of a clear blue sky.
• The range of color temperatures can be said to start with reddish light,

gradually changing to orange, yellow, white, and blue as more blue light is added. Thus, when the radiated light is in balance, such as direct sunlight, the light is at its whitest.

■   **The color rendering index (CRI) is an international standard that rates the color rendering properties of light sources.**
• The index is based on a scale of 0 to 100, with 100 representing the standard for comparison.
• The standard for comparison may be any light source, but is typically representative of similar color temperature or chromaticity.
• Lower CRIs indicate predominate light color (approaching monochromaticity) and grayed or dulled appearance of rendered surfaces which do not reflect that color.
• High CRIs indicate a fuller spectrum of light color approaching the white light of daylight.

■   **Spectral energy distribution (SED) refers to the measure of the relative energy radiated across the spectrum of light wavelengths for a specific light source.** The measure of this distribution is taken by a device which is called a spectroradiometer.
• The information obtained is generally plotted on a graph which has wavelengths graduated on the x-axis and some common reference (output per lumen, watts per 10 nanometers, etc.) on the y-axis.
• Incandescent light sources typically plot as a smooth curve which climbs from initial radiance in UV wavelengths and peaks in near-infrared, declining gradually as wavelength increases.
• Gaseous discharge light sources typically emit dominant wavelengths which are plotted as bars, and the radiance in other wavelengths appears under an erratic curve of lower value.

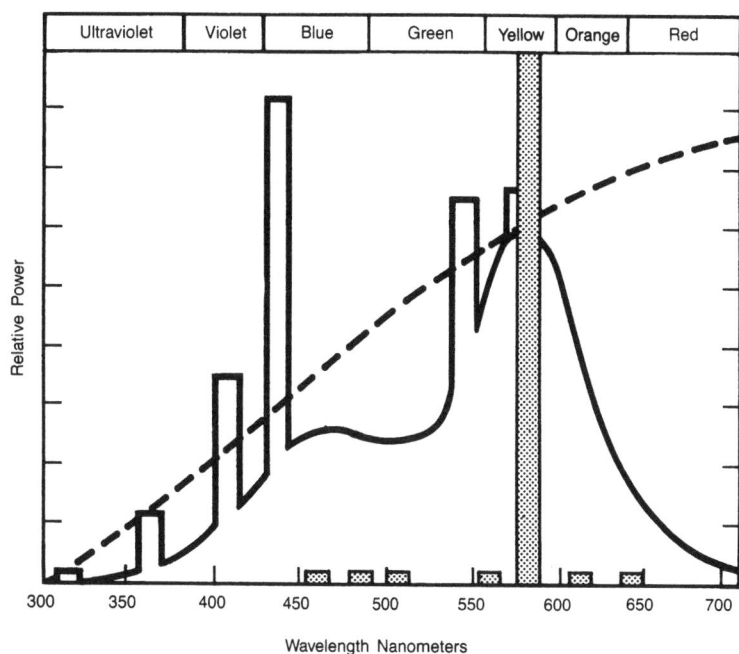

— — — Typical Incandescent Lamp

———— Cool White Florescent Lamp

▨▨▨ Low Pressure Sodium Lamp

D5.1 ELECTRICAL

LIGHTING APPLICATIONS

The essence of lighting design is to provide a lighting system which will establish focus and ambience while addressing the functional and economic needs of the occupants.

GENERAL LIGHTING

■ **Ambient, or area lighting responds to the general illumination requirements of the total space.**

■ **Uniform distribution of indoor area lighting is most valuable for flexibility of use of the illuminated space.**
- Appropriate for common areas, open office plans, other large open areas, and for traffic circulation through the space.
- Supplementary illumination may be desirable at the perimeter of the area.
- **The illumination needs of interior area lighting vary.** Private offices, conversation areas and entrance lobbies require lower levels of illuminance and are often accompanied by periodic use of complimentary lighting (focal or accent). Often, an environment will require more than one type of illumination to suit the various functions performed in the space. For example, in conference rooms where luminance will vary according to task, dimmers or separately switched systems will allow high task luminance or low-level orientation to compliment focused activities (presentation, displays).
- **Interior gardens** can be maintained under artificial light as well as natural light, given the proper distribution of radiance in the spectrum of wavelengths and adequate illuminance levels. Photosynthesis responds well to blue light while chlorophyll synthesis responds to red light. Illumination ranges in which common plant groups will flourish include low light plants (50 to 75 fc), medium light plants (75 to 200 fc) and high light plants (above 200 fc). Most colored plants or blooming flowers need up to 1000 fc.

■ **In order to provide adequate levels of illumination without creating glare, general area lighting often employs diffuse or indirect distribution.** This type of design provides little contrast or shadow and therefore does not enhance depth perception and scale.

- The primary light source is the generation of light radiation.
- The secondary light source is the surface which is reflecting the primary light.
- Indirect lighting systems rely on highly reflective surfaces for their effectiveness.
- Ceiling mounted diffuse lighting systems will provide good distribution regardless of surface reflectance, but low reflectance finishes will create the perception of lower levels of light.
- The shadowless atmosphere of indirect lighting can be made more interesting with the effective use of accent lighting, perimeter lighting or decorative lighting that provides some 'sparkle'.

■ **Where the primary light source is to illuminate the working plane (visual task), direct light distribution may be appropriate for general lighting.**
- Ceiling mounted light sources with direct distribution must be properly spaced (according to beam spread and distance above the floor or work surface) to provide even distribution.
- With high reflectance finishes, the walls and/or work surface will reflect light to contribute to the general illuminance.
- When low reflectance finishes are prevalent, concentrated beams will highlight a particular area or object, but contribute very little to general illumination.
- The spacing of light sources should provide a slight overlap of distribution on the horizontal plane to minimize uneven illumination. Such distribution may cause a scallop effect on vertical surfaces if the source is spaced too closely to those surfaces.

■ **Outdoor area lighting** also regards the average maintained level of footcandles to be important over the entire area, but uniformity variances are accepted especially in roadway lighting.
- Most roadways require a 3:1 ratio of average minimum uniformity and a 12:1 maximum to minimum uniformity ratio.
- Residential and local roadways allow a 6:1 ratio of average minimum uniformity and a 15:1 maximum to minimum uniformity ratio.
- Assymetrical light distribution is often used for roadway, security and building lighting.
- Maximum to minimum uniformity in building lighting should be 10:1.

- **Exterior area lighting** typically has values in the area of 5 fc, but will increase to 10 fc where mechanical conveyance or other equipment is in use. Levels of 20 fc and 30 fc are necessary for more dangerous or visually specific tasks, and levels up to 50 fc are used in sports lighting (tennis courts and stadiums). Footpaths are adequately illuminated at a level of 0.5 fc.
- **In parking areas** lighting plays a large role of providing comfort and convenience to the public, providing for safe vehicular activity and a sense of pedestrian security. Three levels of illumination are generally recommended which are based on traffic flow: 2 fc for heavy traffic (major community complexes), 1 fc for moderate traffic (commuter lots, hospitals, fast food franchises), or 0.5 fc for light traffic (neighborhood activities, schools).

TASK LIGHTING

■ **Illumination designed to relate to a specific task, location and orientation is referred to as task lighting. It is intended to respond to the specific needs of visual acuity.**
- Applications should avoid the glare potential of localized high luminance.
- Glare and shadow are undesirable effects of task lighting which is not properly aimed. Task lighting is best placed in front and to one side of the work surface with the angle of incidence away from the viewer's line of sight. Then, area lighting should be used to fill in the shadows and to achieve proper contrast ratios.

■ **Task lighting distribution is typically direct.**
- Linear sources or diffused point sources are often used for task lighting so that less distinct shadows and minimal glare are produced. Projecting light with point sources should be limited to high ceiling areas to allow wider spread distribution.
- Light from multiple locations or from a diffuse source is best for working surfaces because irritating (or hazardous) shadows are reduced. o Diffuse surface reflection is also preferred. Optimum surface reflectances in work areas under task lighting are: 70% to 90% for ceilings, 45% to 70% for walls, 10% to 40% for floors, and 25% to 50% for furniture, machines, desks, benches and so on.

PERIMETER LIGHTING

■ **Perimeter lighting may be both utilitarian and aesthetic:**
- It provides a continuity of intensity to the boundaries of area lighting applications.
- At night, it replaces the entrance of daylight at window walls.
- It provides ambient light to passageways while guiding the traveller through.

■ **Perimeter lighting involves both point and line sources of light.**
- Point sources are effective for wall-washing or for accent lighting where non-uniformity is desired. For uniformity, wall-washing luminaires with point sources should be spaced the same distance apart as the distance from the wall.
- When point sources are used for indirect light from ceiling reflection, a technique called cross-lighting should be used. Cross-lighting is when the light beam is directed at a 15° to 20° angle in order to cover about 2/3 of the ceiling span. Vertical surfaces may also be cross-lit.
- Line sources are appropriate for soft distribution but cannot be projected as far as point sources. Light level fall-off is generally more prominent from the top of the wall to the bottom. Reflectors are not as effective with linear sources.

ACCENT & DECORATIVE LIGHTING

■ **Lighting used primarily to draw attention to particular points of interest, special structural features or to add highlights to the lighting composition is known as accent, decorative or architectural lighting.**
- It is primarily an aesthetic approach, but must respond to all design planning variables.
- Lighting equipment must be either unobtrusive, nearly camouflaged or represent a distinct decorative element.
- Accent lighting can be the same as perimeter or display lighting when used to create a high contrast between focal point and background.
- It is also used to define contour, line and composition of architecture. Indirect lighting on high ceilings, an illuminated handrail, stairway lights, and more free-form light rays are types of architectural lighting.
- Decorative lighting typically describes chandeliers, wall sconces or tiny sparkle lights, used as aesthetic accents or as decorative objects.

■ **Landscape lighting techniques** are used to imitate the effects of natural light sources, both sunlight and moonlight.
- A key to landscape lighting is to produce the lighting effect without exposing the light source.
- Several lighting techniques are effective means of producing dramatic images. Downward lighting resembles moonlight when mounted high within a tree. Upward lighting is often used to define the tree canopy. Another technique is to illuminate a vertical surface such that nearby trees and shrubs are silhouetted.
- Spread lighting techniques provide illumination over larger ground areas and are effective for paths, decks and gardens.
- Color rendition and reflection from plants, walkways and other surfaces are major design factors for the selection and placement of light sources.
- Flower gardens are not often included in landscape lighting because a high luminance is required in order to properly display the colors.

■ **Facade lighting often uses a grazing illumination to accentuate the irregular or patterned surfaces of a building.** If the surface is smooth and uniform, the spacing and distribution of luminaires may be used to create texture through light patterns.
- Coordinate building lighting with wall surfaces:
  — Window walls are often best illuminated from indoor perimeter lighting systems with no additional indoor or outdoor lighting (better for energy conservation);
  — Opaque walls and exterior detail are perceptible only in silhouette.
- Wall washing produces a low-key impression or a focused, specific accent (entrances).
- Energy codes often impact building illumination, limiting it to a maximum of 2% of the total interior lighting load.

■ **Underwater lighting** can be effective, indoor or out, for highlight in a moderate to low level of illumination.
- Geysers of aerated water can be used as a transmitting medium, becoming radiant columns when the light sources are concentrated, direct beams from below.
- Light sources may be mounted below the water level, dispersing their light horizontally. For uniform illumination, the light source should be recessed into the wall of the pool.

DISPLAY LIGHTING

■ **Display lighting is typically high luminance focused on a single object or group of objects.** In some cases, such as store show windows, the illumination is indeed high (100 to 500 fc), but many effective displays may be illuminated with lower level, high contrast, direct concentrated beams.
- Fixed focal centers of direct lighting, high contrast and subdued background are used to dominate the environment, thereby directing and holding the viewer's attention and interest.

■ **Showroom and merchandise displays** often create a sense of drama.
- The light source is typically kept from view, using shielded or recessed equipment.
- The illumination on the display must be greater than any surrounding luminance. When glossy surfaces are present in the display, care must be taken to avoid reflected glare.

■ **Museum lighting** should combine natural and artificial light to provide ambient lighting and illumination of exhibited objects. Yet, light must be controlled since it is one of many environmental factors which can degrade materials.
- The balance between illuminating the exhibit for the viewer's appreciation and minimizing light to protect the exhibit is delicate.
- Both UV and IR light will degrade materials. Fluorescent sources can be used when a UV filter is applied.
- Sunlight can be used for illumination when the light is repeatedly reflected from white surfaces.
- Limiting the duration of exposure to light is an alternative technique. Controls may be employed to switch lighting for certain viewing times, for viewing of only certain items, for viewing only on demand or upon the presence of a viewer near the piece.
- As certain materials can degrade when exposed to light, the maximum year-round average illumination may be calculated as a preventative guideline for an exhibition. This means that, for example, a maximum illumination of 30 fc in the summer months must be offset by winter illumination below 10 fc and the remaining months below 15 fc in order to average 15 fc or less for the year.

D5.1 ELECTRICAL

LIGHTING APPLICATIONS (cont.)

ILLUMINANCE CATEGORIES FOR SEVERAL TYPES OF VISUAL TASKS

Type of Activity	Illuminance Category	Ranges of Illuminances Lux	Footcandles
Public spaces with dark surroundings	A	20 - 30 - 50	2 - 3 - 5
Simple orientation for short Temporary visits	B	50 - 75 - 100	5 - 7.5 - 10
Working spaces where visual tasks are only occasionally performed	C	100 - 150 - 200	10 - 15 - 20
Performance of visual tasks of high contrast or large size	D	200 - 300 - 500	20 - 30 - 40
Performance of visual tasks of low medium contrast or small size	E	500 - 750 - 1000	50 - 75 - 100
Performance of visual tasks of low contrast or very small size	F	1000 - 1500 - 2000	100 - 150 - 200
Performance of visual tasks of low contrast and very small size over a prolonged period	G	2000 - 3000 - 5000	200 - 300 - 500
Performance of very prolonged and exacting	H	5000 - 7500 - 10000	500 - 750 - 1000
Performance of very special visual tasks of extremely low contrast and small size	I	10000 - 15000 - 20000	1000 - 1500 - 2000

Adapted from **IESNA LIGHTING HANDBOOK** (1993)
SOURCE: "LIGHTING" (REF. 1)

SIGNS & LIGHTING GRAPHICS

■ **Sign illumination levels must be sufficient for the viewer to be able to read or interpret the sign or vertical surface at an adequate distance in an appropriate amount of time.**
- Consider uniformity, contrast and reflectance of the sign and its background.
- Surface to surrounding luminance ratio = 2:1.

■ **Exit signs** are very important elements of emergency egress systems. They should be evenly illuminated and are often required (by code) to be illuminated to 5 fc on the face of the sign.
- Lower luminance can be used in lower general illumination (theatres, etc.). Higher luminance is necessary in areas of difficult visual tasks.

- Exit signs should be evaluated for legibility under smoke conditions.

■ Backlighting through translucent materials or glow lights are often used for signs, commercial messages, decoration or as an art medium.
- Be careful that the shape of the background does not communicate a message which is other than the intended/printed one.
- Backlit or internally lit signage provides excellent visibility for nighttime driving conditions (way-finding.)

D5.1 ELECTRICAL

LAMPS: types, properties

INCANDESCENT LAMPS

■ **Incandescent filament lamps operate on the principle of temperature radiation. Incandescent glow increases relative to wattage of input.** The filament produces a very small, luminous point source of radiation.

• Incandescent lamps are a source of IR light. They are a less efficient method for converting power to visible light: About 72% of energy is radiated as infrared heat, up to 18% is converted to heat which is distributed by conduction or convection, leaving only 6% to 12% of energy converted to light.

• Immediate full output response to power and unimpeded operation irrespective of ambient temperatures are common traits.

• The lamp temperature at peak light radiation will vary from 100° to over 500°F and is proportional to wattage input.

• These lamps generally produce a 'warm' light (2700-3200K) which is often preferred in environments of low luminance or where illumination is below 20 fc. The yellowish white light enhances red, oranges and yellows, while dulling blues.

• Increased voltage will produce slightly greater lumen production, but lamp life will be substantially reduced (a lamp rated at 120 volts will produce 130 percent greater output if operated at 130 volts, but its life is reduced to 40 percent). Reduced voltage will significantly prolong life while lumen production is reduced and the color of the light will be yellower (a lamp rated at 130 volts will last three times its rated life if operated at 120 volts, but its output is reduced to 78 percent).

• Extended service lamps offer filament temperatures, resulting in an average life up to 2500 hours, sacrificing 15% of light output and lamp efficacy.

■ Lamps are labeled to denote shape and size. The many styles of lamp are identified by the letter prefix. The numbers identify the size.

• B, C and F lamps are commonly used for decorative lighting. Type G lamps are often left exposed, but are used for general, low level illumination as are Type A and T. Type S lamps are often used for appliances and special applications. Type PS are used for higher levels of general illumination.

• Tubular lamps are used to produce a linear source of light as opposed to the point source of other incandescents. They are best for showcase/display lighting.

• **Bulb diameter** is given in terms of 1/8-inch increments of diameter (i.e.: a 2 3/8' diameter bulb = 19/8; A-19). The light center length is from tip of base to centerline of filament.

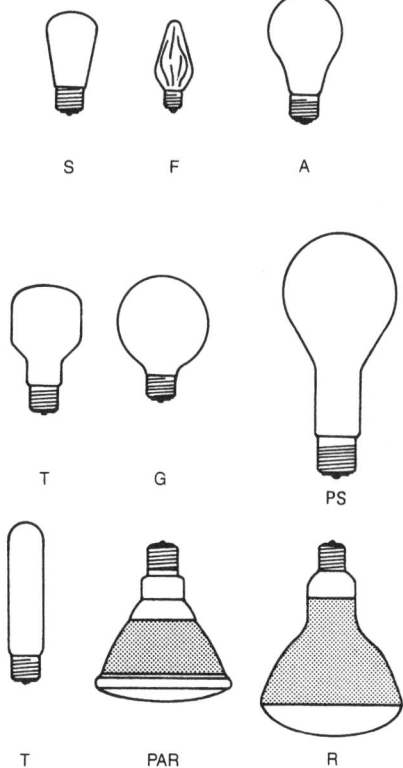

■ Reflector lamps (R, ER, PAR) are classified as narrow (0-15°), medium (15-30°) or wide (over 30°) beam spreads and are used for spotlighting or floodlighting.

• The beam spread is defined by a line which marks 10% of maximum lamp candlepower and indicates the width of the projected light pattern.

• At 90° to the illuminated surface (0° aiming angle), the area of coverage will be in a circle or ellipse. The spotlight will generally produce a 60% tighter beam with higher candlepower than that of a floodlight with comparable wattage.

■ **Tungsten-halogen lamps** (often called quartz lamps because of their glass envelope) use the introduction of an electronegative 'halogen' gas (bromine, chlorine, fluorine, iodine, etc.) to regenerate the filament. This extends lamp life to about twice that of standard incandescent lamps, as the gas reacts with the evaporated tungsten particles and redeposits them on the filament. This gas also allows higher filament temperatures; thus a 'whiter' light (3190K) which includes more UV radiation.

• The sizes and shapes of halogen lamps vary, and low-voltage miniaturized lamps (requiring a transformer) are gaining acceptance. They can be single or double ended.

• Handling of halogen lamps is critical as fingermarks and scratches may promote failure. Other precautions are necessary to screen the lamp in the case of bulb failure, which could cause an explosion of the glass.

• Halogen lamps produce an extremely intense and small point source. Direct viewing should be avoided. Heat radiation is also quite high, requiring proper ventilation and possibly heat resistant fittings. Lamps with dichroic coatings and glass reflectors will also emit IR radiation toward the lamp base.

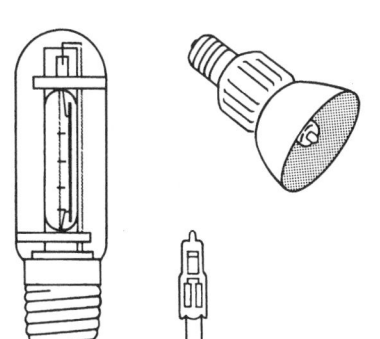

■ The incandescent lamp consists of few parts, thus requiring little additional maintenance other than lamp cleaning and replacement.

• The source of light is **the filament**. Usually a tungsten wire which is straight, coiled or a coiled coil, the filament is held in place by support wires, current and lead-in wires which connect it to the base. Lamp burnout is often caused by a break in the filament due to evaporation.

continued on next page

D5.1 ELECTRICAL

LAMPS: types, properties (cont.)

INCANDESCENT

Category	Type		BEAM TYPE/SPREAD	PROJECTION RANGE: S = short = 0'-8' M = 8'-12' L = long = over 12'	EFFECT ON SURFACE PERPENDICULAR TO BEAM: C = circular E = eliptical	LUMENS (measured within primary distribution)	APPROX: INITIAL EFFICACY (lumens/watt)	LUMEN DEPRECIATION (%)	WATTAGE RANGE	BALLAST WATTAGE
General Service	Type S		Broad	S	C,E	44-80	8-30	80-90	6,10	N/A
General Service	Type A		Broad	S-M	C,E	125-4000	8.3-20	79-93	15-200	N/A
General Service	Type G		Broad	S	C	125-35500	8-22	80-85	25-1000	N/A
General Service	Type PS		Broad	S-L	C,E	2660-34000	13.9-22.6	78-90	150-1500	N/A
General Service	Tubular		Linear	S-M	E	244-800	8.5-27	80-85	25-75	N/A
General Service	Decorative		Point	S	C,E		6-10	80-85	6-60	N/A
Reflector	Type R-20	Flood	85° / 90°	S	C	150 / 430	4-8	80-85	30 / 50	N/A
Reflector	Type R-30	Spot / Flood	30° / 130°	S / S-M	C	400 / 610	5-9	80-85	75 / 175	N/A
Reflector	Type R-40	Spot	37° / 35° / 60°	L / L / M-L	C	835 / 1800 / 3300	5-11	80-85	150 / 300 / 500	N/A
Reflector	Type R-40	Flood	110° / 115° / 120°	M / M / M	C	1550 / 3000 / 5700	5-11	80-85	150 / 300 / 500	N/A
Ellipsoidal Reflector	Type R-52	Flood	90° / 110	M-L	E	7750, 13000 / 16300	15 / 17	80-85	500,756 / 1000	N/A
Ellipsoidal Reflector	Type RB-52	Flood	130°	M	E	18900	18	80-85	1000	N/A
Ellipsoidal Reflector	Type ER-30	Flood	Medium	M	C,E	420-950	1.5-4.0	80-85	250	N/A
Ellipsoidal Reflector	Type ER-40	Flood	Medium	M	C,E	1475	6	80-85	250	N/A
Parabolic Aluminized Reflector	PAR-36	Spot / Flood	Narrow / Wide	S / L	C,E			80-85	25,50	N/A
Parabolic Aluminized Reflector	PAR-38	Spot / Flood	30° × 30° / 60° × 60°	M-L / S-M	C	465-110 / 570-1350	6-9	80-85	75,100 / 150	N/A
Parabolic Aluminized Reflector	PAR 46	Spot / Flood	17° × 23° / 20° × 40°	S-L	E	1200 / 1300	6-8	80-85	200	N/A
Parabolic Aluminized Reflector	PAR-56	Narrow Spot / Medium Flood / Wide Flood	15° × 20° / 20° × 35° / 30° × 60°	S-L	E	1800 / 2000 / 2100	6-7	80-85	300, 500	N/A
Parabolic Aluminized Reflector	PAR-64	Narrow Flood / Medium Flood / Wide Flood	13° × 20° / 20° × 35° / 35° × 65°	M-L	E	3000 / 3400 / 3500	6-7	80-85	500, 1000	N/A
Tungsten-Halogen		Double Ended / Single Ended	Linear	M	E	3460-35800 / 1400-22400	17.3-24.5 / 17.5-25	93-96	200-1500 / -1000	N/A
Tungsten-Halogen	PAR-38	Spot / Medium Flood	60° × 60° / 26° × 26°	L / M	C	1600 / 2400	6-10	93-96	250	N/A
Tungsten-Halogen	PAR-56	Narrow Spot / Medium Flood / Wide Flood	15° × 32° / 20° × 42° / 34° × 66°	L	E	4900 / 5700 / 5725	9-12	93-96	500	N/A
Tungsten-Halogen	PAR-64	Narrow Flood / Medium Flood / Wide Flood	14° × 31° / 22° × 45° / 45° × 72°	L	E	8500 / 10000 / 13500	8.5-13.5	93-96	1000	N/A

D5.1 ELECTRICAL

		BULB DESCRIPTION		COATINGS									BASE DESCRIPTION														
VOLTAGE	AVERAGE LIFE (HRS.)	DIAMETERS (INCHES)	LENGTHS (INCHES)	CLEAR	FROSTED	SILICA	WHITE/WHITE BOWL	COLORED/ENAMELED	SILVER BOWL/SEMI-SILVERED/REFLECTORIZED	PHOSPHOR: STANDARD	PHOSPHOR: COLOR IMPROVED	PHOSPHOR: DAYLIGHT	JACKETED/OUTDOOR/LOW-TEMP.	HEAT RESISTANT	PRISMATIC	MINIATURE/BAYONET/CANDELABRA	SCREW: MEDIUM/MED. SKIRTED	SCREW: MOGUL	END/SIDE PRONG	RECESSED: SINGLE CONTACT	RECESSED: DOUBLE CONTACT	BIPIN: MIN	BIPIN: MEDIUM	BIPIN: MOGUL	FOUR PIN/RECESSED BIPIN	SINGLE PIN	CAP: FERULE, SLIMLINE, CLOVERLEAF
115-125	1500	1 3/4	3 1/2	●	●			●									●										
115-277	750-2500	1 7/8, 2 3/8, 2 5/8, 2 7/8	3 1/2-6 5/16	●	●	●	●	●	●								●										
115-277	200-800	3 9/16-8	3 9/16-8	●	●			●									●										
115-277	750-2500	3 1/8, 3 3/4, 4 3/8, 5, 6 1/2	6 15/16 - 13 1/16	●	●	●	●	●	●								●	●									
115-277	1000,1500	1 3/16, 1, 1 1/4	3 1/8 - 17 3/4	●	●		●	●	●								●										
115-125	750,1000,1500		1-4 7/16	●				●								●											
120	2000	2 1/2	3 15/16		●			●	●								●										
					●			●	●								●										
120	2000	3 3/4	5 3/8	●				●	●								●										
					●			●	●								●										
120	2000	5	6 1/2	●				●	●								●										
			6 7/8, 6 1/2, 7 1/4	●	●			●	●								●	●									
			7 1/4	●				●	●									●									
120	2000	5	6 1/2		●			●	●								●										
			6 7/8, 6 1/2, 7 1/4		●			●	●								●	●									
			7 1/4		●			●	●									●									
120	2000	6 1/2	11 3/4		●				●									●									
					●				●									●									
120	2000	6 1/2	12 3/4		●				●									●									
120	2000	6 3/8	6 3/8		●				●																		
120	2000	7 3/8	7 3/8		●				●																		
120	2000			●					●								●										
				●					●								●										
120	2000	4 3/4	4 5/16, 5 5/16	●					●								●		●								
					●				●						●		●		●								
120	2000	5.6	4	●					●																		
									●																		
120	2000	7	5	●					●																		
									●						●												
									●						●												
120	2000	8	6						●						●												
									●						●												
									●						●												
120-277	500-4000	3/8, 1/2,	3 1/8 - 10 1/16						●											●							
12-130	1000-3000	3/4	2 1/4 - 9 1/2						●							●											
120-277	4000	5.6	5 5/16	●	●				●								●										
				●	●				●								●										
120-277	4000	7	5	●					●										●								
									●						●				●								
									●						●				●								
120-277	4000	8	6	●					●										●								
									●						●				●								
									●						●				●								

D5.1 ELECTRICAL

LAMPS: types, properties (cont.)

- The glass casing which encloses the filament is called **the bulb**. Silverized coatings reflect light in a certain direction - coating the bulb over the filament will send the light backwards for indirect distribution, while coatings from the neck of the bulb and outward tend to concentrate the light into a conical beam. Coatings may reduce lamp efficiency by the following percentages:
  - clear & silica coated glass-l00% inside frosted - 99% blue/daylight glass - 65% white outside enamel - 85% ivory outside enamel - 73% flametint outside paint - 58% tinted ceramic enamels - 40-85%
- The bulb is often sealed tight to protect a vacuum-effect in low-wattage lamps. Gases are sealed inside higher-wattage lamps (a combination of argon and nitrogen, krypton or halogen) in order to impede filament evaporation.
- Fuses are often used in high-wattage lamps to protect against filament arcs. Mica or ceramic discs may be incorporated to keep heated gases away from the base of the lamp where excessive heat can degrade bulb seals.
- The connection of lamp to the electrical system depends on **the lamp base**. Miniature bayonet and miniature screw bases are used for low wattage and series burning lamps. Recessed contact (single and double), medium screw (lower wattages) and mogul screw (300 watt and greater) bases are more common. Side prong, end prong and screw terminal are common to PAR lamps and bipost bases are used in many high wattage lamps.

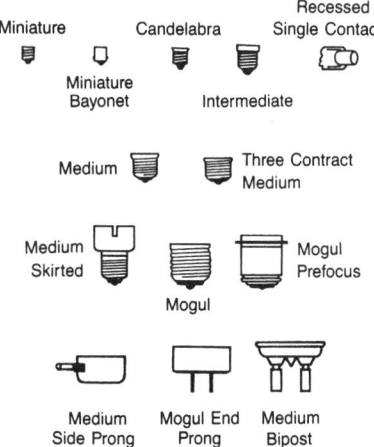

Miniature | Candelabra | Recessed Single Contact

Miniature Bayonet | Intermediate

Medium | Three Contract Medium

Medium Skirted | Mogul | Mogul Prefocus

Medium Side Prong | Mogul End Prong | Medium Bipost

FLUORESCENT LAMPS

■ **Fluorescent lamps operate on the principle of gaseous discharge. They rely on the production of ultraviolet light from the ionization of mercury to charge fluorescent materials which absorb the UV rays and transform them into longer wavelengths in the range of visible light.** The color of the light and the efficacy of the lamp are affected by the selection of the phosphor compound.

- Due to the inherently diffuse nature of the lamp, fluorescent lighting has very poor beam control characteristics. An effective line of light is produced however.
- These lamps compare favorably to the performance of incandescent sources: 3 to 5 times greater lumens per watt, from 7 to 20 times greater lamp life; considerably less waste heat produced.
- The operation of fluorescent lamps can be affected by the temperature of surrounding air. With special ballasts, the lamps will operate down to -20°F. The lamp temperature at peak light radiation is 104°F.
- Efficacy is affected by the selection of phosphor coatings, lamp length (the longer the lamp, the longer the arc, the greater the lumens per watt) and lamp diameter (the optimum efficacy is at a 1″ or 1 1/2″ diameter, declining to the least efficacies at 1/2 inch and 2 1/8 inches). Operating voltage rises as lamp length increases.
- More than half of the input energy used by fluorescent lamps is used to produce UV light. Yet, due to the fluorescence of the phosphor compounds, the distribution of energy output is modified; less than 20 percent UV, up to 37 percent IR, about 40 percent heat radiation and less than 5 percent visible light.
- Lamp designations often denote wattage (**F60**=60 watts), shape (F60**T**=tubular) and bulb diameter (F60T**I7**=I7/8″=2 I/8″).

■ As a source of UV radiation, fluorescent lamps use coatings (and occasionally filters) to improve visible light production.

- The coatings are identified by lamp names which are linked to the light radiation in Kelvin temperature.
- **Daylight** lamps (6250K) produce a bluish white light, enhancing greens and blues and dulling reds and oranges. These are often used for color matching due to the predominance of blue light.

- **Cool-white** lamps (4250K) produce a white light that enhances oranges, yellows and blues, but that grays and dulls reds.
- **Deluxe cool-white** lamps (4050K) are used where color rendition is important. They produce a white light, enhancing all colors and having very little dulling effect on any color.
- **White** lamps (3450K) produce a pale yellowish white light that enhances oranges and yellows, but dulls reds, greens and blues.
- **Warm-white** lamps (3020K) are efficient lamps which produce a yellowish white light. Their light enhances oranges and yellows while dulling reds, greens and blues.
- **Deluxe warm-white** lamps (2940K) are the least efficient of fluorescent lamps producing a pinkish white light which enhances reds and oranges; greens and blues are grayed.
- Several other types of fluorescent lamps are also produced. Many types have been developed for specific uses (plant growth, germicidal, etc.) and others are intended to duplicate daylight.
- Dual-coat phosphor technology has improved energy efficiency by adding a **triphosphor** coating to the standard phosphor. With excellent color rendering, the triphosphor lamp uses lower wattage (e.g.: 34 vs. 40) and nearly the same light output. Triphosphor lamps are available in 3000, 3500 and 4100K.

The method of generating the arc in fluorescent lamps distinguishes the three primary types.

■ **Preheat lamps require starters and ballasts which control electrical current in order to sufficiently raise the temperature of the filament ('hot') cathode to produce electron emission.**

- The starter operates manually (push-button) or automatically, permitting momentary heating of cathodes before starting. Automatic starters may damage other luminaire components when they continue to function but the lamp has failed. Starter types include thermal switch, glow-switch and cut-out starters. A delay of 2 to 4 seconds from initial power to full light output is common.
- Typical usage of preheat lamps is limited to applications where lamp length is three-foot (30 watts nominal) or less.

- **Cut-out starters** are glow-switch starters that automatically (or manually) reset, disconnecting the circuit when the lamp fails to start after several tries.
- **Trigger start ballasts** can be used with I4, I5 and 20 watt pre-heat lamps to eliminate the need for starters. This, in effect, converts the pre-heat unit into a rapid-start unit.

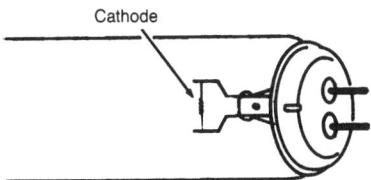
Cathode

■ **Instant-start lamps are started by an initial surge of 400 to I000 volts to the electrode to strike the arc and produce full light output immediately.**
- Because the electrode is not heated for starting purposes, hot cathode instant-start lamps only require a single pin base.
- Cold cathode instant-starting lamps have considerably longer life, but have a lower efficacy and lower output per foot of length than hot cathode. They have ferrule, slimline or cloverleaf bases.

■ **Rapid-start lamps use ballasts that operate without starters and that allow dimming and flashing.**
- These lamps require grounding strips within close proximity to and the full length of the lamp.
- Their operation is based on the use of hot cathodes, which require double contact or bipin bases.
- The ballasts heat the cathodes throughout the operation of the lamp, using additional wattage over and above the lamp requirements. These lamps require 1 to 2 seconds to reach full light output.
- **Preheat rapid-start lamps** are available which are interchangeable with either preheat or rapid start systems. They can be adapted to instant-start, but will have very short life. Starting characteristics are consistent with the existing systems.

■ **All fluorescent lamps typically include a small amount of inert gas contained in the bulb to aid in starting the arc.** Argon is the most common, since low levels of electricity will cause the argon to glow, initiating excitement of the low pressure mercury

vapor. Other inert gases include a mixture of argon and krypton, argon and neon, or argon plus neon plus xenon.

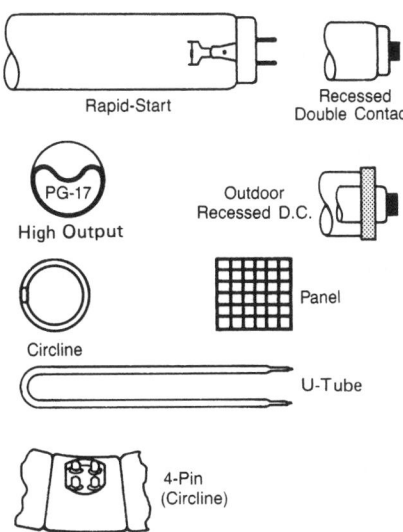
Rapid-Start

Recessed Double Contact

PG-17

High Output

Outdoor Recessed D.C.

Circline

Panel

U-Tube

4-Pin (Circline)

- **The starting method and the number of starts will be the paramount factor in determining lamp life.** Except for cold cathode lamps, lamp life is measured on the basis of 3 hours of operation per start. If the lamp is started once and left on continuously, the life of the lamp could be doubled.

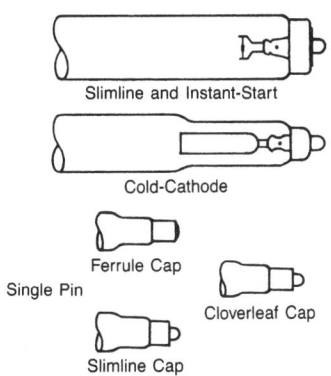
Slimline and Instant-Start

Cold-Cathode

Ferrule Cap

Single Pin

Cloverleaf Cap

Slimline Cap

■ Fluorescent lighting can also employ other techniques to save energy:
- **Energy saving retrofit lamps** use krypton gas instead of argon to reduce lamp wattage (e.g.: 34 vs. 40), but lose 12 percent in light output. They will operate on standard ballasts and certain energy saving ballasts.
- **Reactive impedance retrofit lamps** are designed to replace one of the two lamps in a two-lamp series rapid-

start circuit. Wattage is reduced 33% or 50%, as is output.
- In a series circuit, one of the two lamps can be eliminated by using **lamp impedance** devices. Wattage will be reduced about 65% and output reduction is nearly 70%.

■ **Many types of compact, or miniature fluorescent lamps** are available, offering all of the benefits of modern fluorescent lighting in addition to smaller size, lower cost and easier installation.
- Some lamps are designed for fluorescent lampholders and auxiliary ballasts, while others are available with adapters or fittings that allow installation in ordinary incandescent sockets. The adapters contain the appropriate ballasts for lamp operation.
- Other lamps are available with all components contained within a single assembly so that operation is as easy as with incandescent lamps, but service life is nearly ten times longer.
- Compact fluorescents are intended to be long life, efficient sources of general light. They are not effective substitutes for incandescent lamps where point sources are specified.
- Lamp wattages are typically quite low (5, 7, 9, 13, or 18 watts) with lumen output of 250 to 1250.
- Estimated life ranges from 7,500 to 10,000 hours.
- Operation is independent of burning position, however use is effectively limited to indoor applications because of the need for above-freezing ambient temperatures.

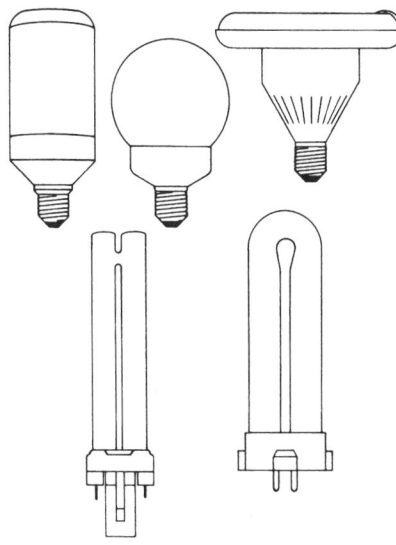

D5.1 ELECTRICAL

LAMPS: types, properties (cont.)

Fluorescent Lamp Ballasts

■ **All fluorescent lamps require ballasts. They are typically encased in metal, are a component of the luminaire and, like starters, can burn out. Ballasts regulate the amount of current passing through a lamp and the voltage required to start and operate the lamp.**

- The **power factor** refers to the differential in phase relation between voltage and current. High power factors indicate a reduction in the amount of current used by a particular unit, reducing operating and installation costs. Most common ballasts operate on an 0.90 power factor, but some are as low as 0.35.
- The **sound rating** refers to the production of a low humming sound due to current passing through the ballast. Sound ratings are derived from test results and are not always available. Typically, preheat system ballasts tend to have the best ratings, followed by low-wattage rapid-start, instant-start, and high-wattage rapid start systems. Relative ratings are A (most quiet) to F (loudest). F-rated ballasts may require isolated location to reduce distraction.
- **Ballast selection must be made on the basis of lamp, power and ballast specifications.** Certified Ballast Manufacturers (CBM) labels indicate that certain basic manufacturing and safety standards have been met. Underwriters Laboratories (UL) also approve and label ballasts as to thermal protection. UL's Class-P rating indicates that the ballast/lamp will turn off automatically if the ballast overheats due to misapplication, failure or other cause.
- Typically, **standard ballasts last about 12 years** under medium use.
- **Energy saving ballasts** can reduce input of wattage for systems using standard lamps, but generally should not be used with energy saving lamps.
- **Solid state ballasts** use high frequency output to increase lumen production in fluorescent lamps without changing input power. They use 75% less electricity than standard ballasts and are adjustable.
- Other ballast types include common dimming ballasts for low dimming ratios (up to 59:I) and peaking dimming ballasts for high dimming ratios.

■ **The operation of the fluorescent lamp produces a cyclic variation at twice the frequency of the electrical** current input (common 60 Hz alternating current results in 1220 Hz output variation). **This variation is referred to as flicker and produces a stroboscopic effect.** Flicker has been reduced to an unnoticeable level through phosphorescence.

HIGH INTENSITY DISCHARGE LAMPS

■ **There are four types of high intensity discharge (HID) lamps which differ primarily due to the respective gases pressurized within them. Common to HID lamps are sealed inner arc tubes which contain the gases and electrodes; the outer bulb, usually glass, which may be clear or coated to control wavelength and intensity of light; and a single base with a medium or mogul screw for ease of installation. Many characteristics of HID lamps are common to all.**

- The short arc length, pressurized gases, and an initial surge of electricity produce an extremely intense and small point of light. Unlike fluorescent lamps which only rely on the arc to produce the ionization of mercury vapor, the arc in HID lamps continues throughout the lamp operation and is the actual source of light radiation. Like incandescent sources, the small original light source lends itself to good control of distribution.
- Typical of gaseous discharge light production, HID lamps have inherently long life. Lamp failure is generally due to deterioration of electrodes caused by starting.
- Ballasts are required for starting and operation, using high input wattage but operating at 10% to 20% of lamp wattage.
- The outer-glass jacket, or bulb, protects the inner arc tube so that high operating temperatures can be maintained. Lamp temperature radiation is typically 300° to 400°F at peak output.
- Bulb types include standard (BT), elliptical (E), tubular (T), globe (G) and reflector (R).

■ **The four types of HID lamps are mercury, metal halid, high pressure sodium and low pressure sodium.**

■ **Mercury HID lamps** have a third electrode in the quartz arc tube which is used to ionize argon gas. The activity of the argon raises the temperature and pressure within the tube, vaporizing the mercury for arc generation.

- Arc tubes may also contain neon and krypton.
- An inert fill gas (nitrogen) is sealed between the arc tube and the outer bulb to prevent oxidation of internal parts. The outer bulb may be coated on the inside with various phosphors to improve color rendering.
- Mercury lamps convert about 124% of input energy into visible light radiation. They are the least efficient (lumens/watt) of the HID sources typically.
- Indoor uses are usually commercial or industrial, but outdoor applications are more varied (building illumination, landscape lighting, floodlighting, etc.). Applications are limited by ambient temperature as special ballasts are required for operation down to -20°F. Operation is also affected at 105°F or above.
- Color rendition varies between clear, high-efficiency phosphor coatings and color-improved phosphor coatings. With clear lamps (5710K), yellows, blues and greens are enhanced by the greenish-bluish white light, while reds and oranges are grayed. The color-improved phosphor produces a yellower light which dulls only blues (4430K).
- The time to reach full light output will range from 3 to 7 minutes.

■ **Metal halide HID lamps** are smaller than comparable wattage mercury lamps but are otherwise quite similar in construction and operation. They are distinguished by the addition of metal

D5.1 ELECTRICAL

halides (sodium, thallium and indium iodides; sodium and scandium iodides; or dysprosium and thallium iodides) which exhibit up to twice the lumens per watt of mercury lamps and match the improved color rendering of the phosphor-coated mercury lamps. Metal halide lamps are also available with phosphor coatings.

- Nearly 21% of metal halide electrical consumption is converted into visible light.
- Color rendition can be altered by varying the content of sodium (orange), thallium (green), indium (blue), lead (UV), etc. These lamps (3720 to 4200K) typically emit a greenish white light that enhances yellows, blues and greens but dulls reds. Lower color temperatures approaching incandescent can be produced.
- Common outdoor uses include sports lighting, floodlighting, parking area and sign illumination.
- Most indoor applications where high levels of illumination are required can be satisfied with metal halide lamps.
- Compared to mercury HID sources, metal halides have greater efficacy and color, but shorter life.
- Full light output is attained within 5 minutes of starting (restart time of 15 minutes), but special ballasts will be required to start at low temperatures (-20°F).
- Double ended lamps that appear similar to tungsten-halogen lamps are available as low as 70 watts with high color rendering (CRI = upper 80's) and low color temperature (3000K).

■ **High pressure sodium (HPS) HID lamps** produce light by passing the arc through a combination of sodium, mercury and xenon (for starting).
- The arc tube is made of a sodium resistant ceramic (polycrystalline alumina). It is contained in a vacuum which is protected by the outer bulb.
- As sodium pressure is increased, color rendition is improved but life and efficacy are reduced.
- HPS is more efficient than mercury or metal halide, producing levels of visible light up to 26% of input wattage.
- HPS lamps can operate at temperatures down to -30°F with appropriate ballasts.
- The yellowish or golden-white light of the HPS lamp (2100K) enhances yellows, oranges and greens, while blues and reds appear dulled or grayed.

- HPS lamps require 4 to 8 minutes to warm up and may take up to 15 minutes to restart after warm.
- Bulb temperatures can reach 400°C, with base temperature of 210°C.
- The HPS lamp is the most popular choice for roadway lighting. When color rendition is not critical, they are good for sports lighting and most other outdoor applications.

■ **Low pressure sodium lamps (LPS)** produce light by passing an electric charge through sodium vapor. A neon starting gas is ionized, producing a red glow and warming the sodium. When vaporized, the sodium emits its characteristic yellow color.
- The light of the LPS lamp is the most efficient and the most monochromatic of all light sources. The yellow-orange light (1740K) renders all surfaces yellow, orange or some shade of brown.
- Nearly 35% of the input wattage is converted to visible radiation.
- LPS lamps typically require 7 to 15 minutes to reach full light output levels.
- Ambient temperatures have little effect on lamp production. The lamp itself will be about 500°F at peak light radiation.

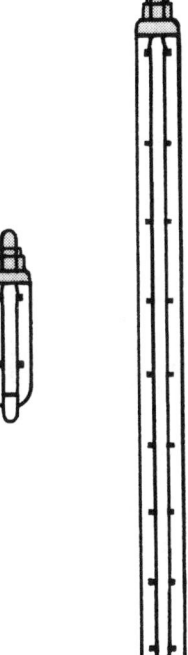

■ **Ballasts for HID lamps provide the correct starting current and voltage in order to initiate and maintain the arc.** They also regulate the flow of current and compensate for low power factor.
- The reactor ballast is the smallest, most economical and most efficient. It is used only when the line voltage is greater than the minimum required lamp starting voltage. It is not recommended where line voltage fluctuations exceed 5%.
- The high reactance ballast, or autotransformer, is often used in conjunction with the reactor ballast to provide proper starting voltage when the line voltage is either above or below the required lamp starting voltage.
- The constant wattage autotransformer ballast (CWA) is used mainly with mercury and metal halide lamps to reduce voltage fluctuations, therefore reducing wattage.
- HPS lamps use three types: A constant wattage autotransformer ballast is used to provide wattage regulation for changes both in line voltage and lamp wattage; the reactor ballast and reactance ballast are used for high power factor lamp types and have a starting device that is required to start the lamp.
- HID lamps must be operated by ballasts specific to their lamp type and wattage. The ballasts cannot be interchanged among varying lamps.

■ Some mercury lamps are self-ballasted through internal circuitry and may be used to replace incandescent lamps where sockets, wiring and power supply are appropriate.

■ Transformers and reactor ballasts or high power factor autotransformer ballasts are required for current protection for LPS lamps.

continued on page D5.1-24

LAMPS: types, properties (cont.)

FLUORESCENT, HIGH INTENSITY DISCHARGE

Types			DISTRIBUTION: BEAM TYPE/SPREAD	DISTRIBUTION: PROJECTION RANGE: S = short = 0′-8′ M = 8′-12′ L = long = over12′	DISTRIBUTION: EFFECT ON SURFACE PERPENDICULAR TO BEAM: C = circular E = eliptical	OPERATION: LUMENS (measured within primary distribution)	OPERATION: APPROX: INITIAL EFFICACY (lumens/watt)	OPERATION: LUMEN DEPRECIATION (%)	OPERATION: WATTAGE RANGE	OPERATION: BALLAST WATTAGE
Preheat		Type T5	Linear	S	E	85-880	22-63	67-75	4-13	2,5
Preheat		Type T8	Linear	S	E	610-2255	22-63	79	15,30	4.5, 10.5
Preheat		Type T12	Linear	S	E	476-3185	22-63	79-85	14-40	4.5-12
Preheat		Type T17	Linear	S	E	5525-6400	22-63	82	90-100	20
Preheat	Energy-saving	Type T12	Linear	S	E	2800-2900	82-85		34	
Preheat	Energy-saving	Type T17	Linear	S	E	5400-6400	64-78		82-84	
Rapid Start	Circline	Type T9 Type T10	Diffuse	S	C	630-2650	60-71	72-82	20,22.5 33,41.5	
Rapid Start	U-shaped		Linear	S	E	1480-2965	60-71	84	40.5-41	13
Rapid Start	Type T10	low medium high	Linear	S	E	2270-3252 4690-14000	60-71	66	41 105-205	12-13
Rapid Start	Type T12	low medium high	Linear	S	E	1505-3250 1400-9200 4900-15250	60-71	81-84 77-82 69-72	32.4,41 37-113 116-215	10.5-13
Rapid Start	Type PG17		Linear	S	E	5200-16000	60-71	69	116-215	
Rapid Start	Panel		Diffuse	S	C	3200-4800	60-71		55-96	
Rapid Start	Energy-saving	Type T12	Linear	S	E	1940-9100	46-78		25-195	
Rapid Start	Energy-saving	Type PG17	Linear	S	E	6550-14900	69-81		95-185	
Instant Start		Type T12	Linear	S	E	2560-3150	52-85	83	40.5	23.5-24
Instant Start		Type T17	Linear	S	E	1990-2940	52-63	89	42	24
Instant Start	Energy-saving	Type T8	Linear	S	E	3725-3795	93		40-41	
Instant Start	Energy-saving	Type T12	Linear	S	E	2550-6000	85-95		30-60	
Instant Start	Slimline	Type T6	Linear	S	E	1265-3050	52-63	76-77	25.5, 38.5	13,16
Instant Start	Slimline	Type T8	Linear	S	E	2100-4265	52-63	83-89	38-51	16
Instant Start	Slimline	Type T12	Linear	S	E	1010-6365	52-63	78-91	21.5-75	16
Instant Start		Cold-Cathode	Linear	S	E	950-3400	38-52		26-65	9-19
HIGH INTENSITY DISCHARGE										
	Mercury	Type H	Broad	S-L	C,E	580-63,000	30-65	52-89	40-1075	7-570
	Metal Halide	Type M	Broad	S-L	C,E	8300-210,000	69-125	66-75	100-3500	460-1080
	High-Pressure Sodium	Type S	Broad	S-L	C,E	2150-140,000	60-140	83-88	35-1000	85-1100
	Low-Pressure Sodium		Broad	S	E	1800-33,000	up to 183	91-95	18-180	
	Self-Ballasted Mercury		Broad	S-L	C,E	1700-10,880	20-25		160-250	

		BULB DESCRIPTION		COATINGS									BASE DESCRIPTION														
VOLTAGE	AVERAGE LIFE (HRS.)	DIAMETERS (INCHES)	LENGTHS (INCHES)	CLEAR	FROSTED	SILICA	WHITE/WHITE BOWL	COLORED/ENAMELED	SILVER BOWL/SEMI-SILVERED/REFLECTORIZED	PHOSPHOR STANDARD	PHOSPHOR COLOR IMPROVED	PHOSPHOR DAYLIGHT	JACKETED/OUTDOOR/LOW-TEMP.	HEAT RESISTANT	PRISMATIC	MINIATURE/BAYONET/CANDELABRA	SCREW MEDIUM/MED. SKIRTED	SCREW MOGUL	END/SIDE PRONG	RECESSED SINGLE CONTACT	RECESSED DOUBLE CONTACT	BIPIN MIN	BIPIN MEDIUM	BIPIN MOGUL	FOUR PIN/RECESSED BIPIN	SINGLE PIN	CAP: FERULE, SLIMLINE, CLOVERLEAF
29-95	6000-7500	5/8	6,9,12,21							●	●	●										●					
55-99	7500	1	18,36							●	●	●											●				
40-101	7500, 9000, 20000	1 1/2	15,18,24, 28,33,48							●	●	●											●				
65	9000	2 1/8	60							●	●	●												●			
84	15000	1 1/2	48							●	●												●				
	9000	2 1/8	60							●	●	●												●			
48,61	12000	1 1/8	6 1/2,8 1/4 ∅							●	●	●													●		
81,108		1 1/4	12,16 ∅							●	●	●													●		
100-103	12000	1 1/2	24							●	●												●				
104	24000		48							●	●												●				
		1 1/4								●											●						
80-160	9000		48,72,96							●			●								●						
81-101	15,000-20,000		36,48							●	●												●				
41-153	9000-12000	1 1/2	24-96							●		●									●						
84-163	9000		48-96							●			●								●						
84-163	9000	2 1/8	48,72,96							●											●						
	7500	1 1/2	11 5/8 × 11 5/8							●											●				●		
64-137	10,000-20,000	1 1/2	36,48,96							●	●	●									●		●				
64-144	12000	2 1/8	48,96							●											●						
104	7500-12000	1 1/2	48							●	●	●											●				
107	7500-9000	2 1/8	60							●	●	●												●			
	7500	1	96							●	●															●	
153	9000-12000	1 1/2	48,96							●	●	●														●	
150,233	7500	3/4	42,64							●	●	●														●	
220-295	7500	1	72,96							●		●														●	
53-197	7500-12000	1 1/2	24-96							●		●														●	
240-450	12500-30000	1	45-93								●	●															●
90-265	12000-24000 +	3 1/8 - 7	4 11/32 - 17 7/8	●					●	●	●			●			●	●									
130-270	2000- 20000	3 1/2, 4 5/8, 7	6 1/2- 19 5/16						●	●	●			●				●		●							
162-706	16000 - 24000	2 1/4, 3 1/8, 3 1/2, 4 5/8	4 5/16- 15 1/16						●	●	●	●					●	●		●							
57-240	10000- 20000	2 1/8 , 2 5/8	9 3/8,20, 36	●																						●	
120-277	8000- 20000		6 1/2- 15 3/8	●					●	●	●			●			●	●									

D5.1 ELECTRICAL

LAMPS: types, properties (cont.)

OTHER LAMPS FOR ILLUMINATION

■ **Short-arc lamps** are gaseous discharge type of an extremely concentrated point source used for high luminance, long projection range and/or specific beam control. These lamps operate at very high internal pressures, requiring protective enclosure for safety purposes. Discharge gases may be mercury and argon, mercury and xenon or xenon. The xenon vapor produces a white light (5920K) which enhances reds and dulls blues, while the mercury vapors produce bluish white light which brightens blues, greens and yellows but dulls reds.

■ **Glow lamps** are used for art, signage and other graphic communication. Glow lamps are inexpensive and easy to manufacture, using available and stable gases such as neon and argon.

■ **Zirconium concentrated arc lamps** use direct current arcs to create very small concentrated point sources of light of very high luminance. At peak light radiation, these lamps can range from 140° to 520°F; used primarily for photochemical applications and UV production.

■ **Pulsed xenon arc lamps** provide immediate start, daylight color rendition, and high levels of light output with little flicker; used for graphic arts.

■ **Custom-fabricated lamps** are often produced to form continuous lines of light (coves, arches, stairs and walks) and for lighting graphics (light sculpture, signs, etc.).
- The lamps are of cold cathode type, with tube diameter ranging from 1/2 inch to 1 inch. The tube may be straight, curved or bent. Up to 120 feet of lamp tubing can be connected to only two leads.
- Remote transformers provide the high voltage requirements of the lamp and dimming can be accomplished without special dimming ballasts (use low power factor transformers with SCR dimmers).
- The lamp color is based on the gas contained in the lamp (neon = orange-red, argon-mercury = blue, helium = pinkish white) and the tube coloring or coating (phosphors provide bluish white light with mercury vapor). Altogether, over 25 colors are common.

- The lamps typically produce a low, no-glare luminance as light discharge is spread over the length of the tube. A 1-inch diameter tube typically produces about 300 lumens per foot. Lamp life is up to 25,000 hours.
- Installation and replacement is easy, but loose connections can cause flickering.

D5.1 ELECTRICAL

LUMINAIRES

Luminaire or Fixture?

The lighting assembly which is designed for the purpose of housing one or more lamps, connecting them to the power source and distributing their light is properly termed a luminaire (but in common usage is a fixture). The two terms are synonymous on most levels, since 'fixture' connotes annexation or integration of the assembly to the building - legally, chattel that becomes a fixed part of the property. Further complications come from the common (and acceptable) use of the term 'lamp' in exchange for portable luminaire.

LUMINAIRE DESIGN

The most important element of luminaire design is luminance. But that is followed closely by safety, appearance and durability. Each component in the assembly of a luminaire affects these factors. They are the key to good luminaire design - from the lamp and pull-chain closet light to the most complex integrated ceiling system.

Light Distribution

■ **Luminaire design and selection is based on six types of light distribution.** Each type of distribution may be used solely or in conjunction with other types to form a lighting system which is appropriate for a given environment. Each type is differentiated by the amount of light sent upwards or downwards, and the nature of the light beam.

● Each luminaire is measured for candlepower distribution from the source in vertical and horizontal directions. The measurement is referred to as photometric data.

● Since distribution is axial about the centerline of the lamp and/or luminaire (through the light center) the measure of candlepower distribution can be plotted on a graph to indicate half, or 180°, of the distributed light.

● The intensity distribution curve is a critical element of luminaire selection and is essential for calculating the illumination within the environment. However, beware that photometric data can vary as much as 15% due to variance in light distribution and output from one luminaire to the next.

Variation between supply-line voltage and theoretical design voltage will also be a factor.

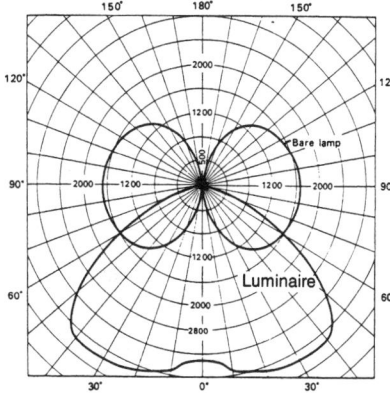

■ **Direct lighting systems distribute 90% or more of their luminance downward, producing sharp shadows.**
● Distribution is from 0° to nearly 90° typically (spread).
● When distribution is limited from 0° to 45° or less, it is referred to as concentrated..

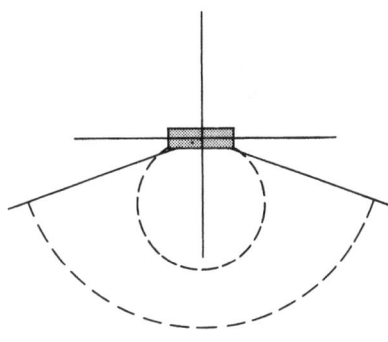

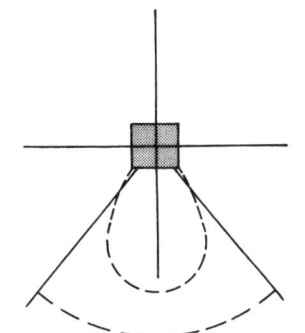

■ **Semi-direct lighting systems distribute 60% to 90% of the luminance downward.**
● The purpose of this luminaire type is to soften the contrasts and shadows created by the direct distribution. 10% to 40% of the luminance is distributed upward, typically through a diffusing element.

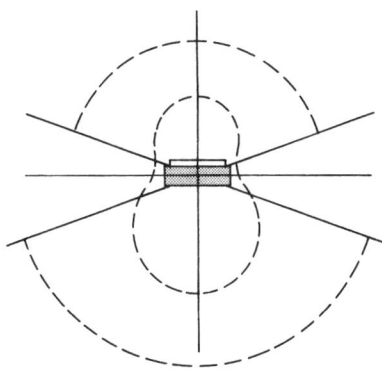

■ **General diffusing lighting systems distribute light in nearly equal amounts in all directions.**
● Shadow and contrast are minimized.

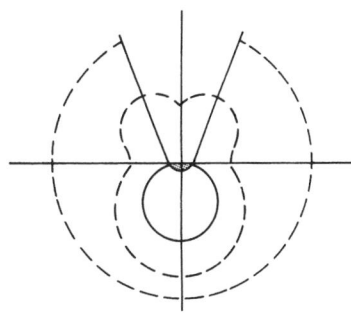

■ **Direct/indirect systems typically distribute half of their luminance downward and half upward.**
● The two-directional quality of the light reduces contrast and provides localized general illumination.

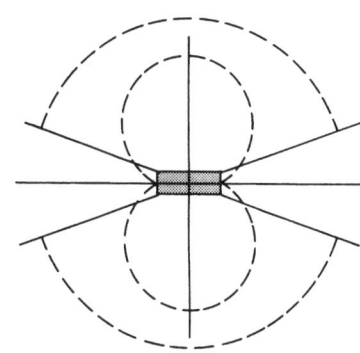

D5.1 ELECTRICAL

LUMINAIRES (cont.)

■ **Semi-indirect lighting systems distribute 10% to 40% of the luminance downward and 60% to 90% upward.**
- Downward distribution is typically diffuse in order to soften contrasts and shadows created by the upward element of light.

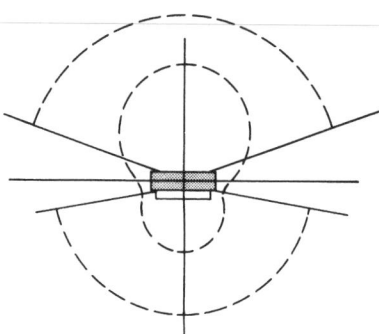

■ **Indirect lighting systems distribute 90% or more of the luminance upward.**
- Distribution is typically from 90° to 180° (spread), producing low-contrast general illumination and emphasizing structural form.
- The angle of the distribution can be narrowed to create a concentrated beam of light when highlight, texture and contrast are desirable.
- Indirect lighting systems rely on the nature and proximity of reflecting surfaces to determine the level of general illumination.

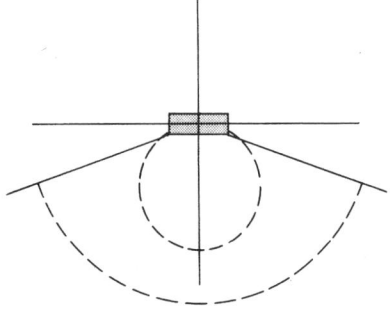

■ **Two factors are used to indicate the potential of luminaire performance.**
- The luminaire efficiency is the ratio of lumens emitted by the luminaire to lumens produced by the light source. It is dependent on materials used, type of lamp, internal finishes and contours, beam type and distribution.
- The coefficient of utilization (CU) is the ratio of lumens received on the intended surface to lumens emitted by the luminaire. The CU varies with the shape of the room/space, the reflectance of surfaces (typically walls and ceilings) and the luminaire efficiency.

Lamping Considerations

■ Lampholders and receptacles are designed for the many respective lamp base types, providing a variety of shapes, sizes, materials, controls and options for connecting the light source to the power source.

■ **The type(s) of lamp to be used in a particular luminaire will dictate certain aspects of its design - size, shape, material type - because of the dimensions, light distribution and heat generation of the lamp.**
- Fluorescent lamps may require grounding strips within close proximity to the lamp and along its entire length.
- Most filament, HID and short arc lamps produce a **point source** of light which allows the luminaire design to define the light distribution.
- Some filament and HID lamps (types R, ER and PAR) are internally reflectorized to concentrate and direct the beam.
- Fluorescent lamps and tubular lamps of other types produce a **linear source** of light which typically allows limited beam control.
- The heat generated by the lamp must be vented away from the lamp and luminaire. Materials used in luminaire design must be selected on the basis of expansion, melting point and proximity to the lamp as well as properties of reflection, transmission and weight.

■ The relative temperatures of ambient air and lamp bulb may have a significant impact on lamp performance.
- Incandescent lamps produce considerable internal convection as the filament is heated to as high as 5000°F or more to produce light. This convection protects the filament from the effects of ambient air temperature, thereby allowing consistent lumen production in any temperature.
- Both mercury vapor and phosphor coatings are sensitive to ambient air temperature. Deviation from optimum bulb temperatures of 100° to 120°F will result in distorted processes of producing ultraviolet light and converting it to visible light. Thus, fluorescent lamps lose lumen production (reduced efficacy) and vary in color rendition.

- Lamps ranked by bulb temperature during operation (coolest to hottest): Fluorescent, HID, and incandescent.
- Temperatures also affect ballast operation; when the metal ballast casing reaches about 194°F, the capacitor inside the ballast is likely to break down, allowing full flow to the transformer. When the transformer overheats, the ballast will cease operation.

Luminaire Housings and Reflectors

The luminaire enclosure may be used for protection from the environment, for shielding or diffusing/enlarging the light source or for directing and controlling the light beam.

■ **The enclosure, or housing, can be of three basic types: opaque, transmitting or reflecting.**

■ Opaque housings are primarily used to conceal the bare lamp and lampholder. Depending on the nature of the lamp's beam direction, the housing may also be used to define luminous distribution.
- Materials typically include various metals, wood, porcelain and even stone/masonry.
- Opaque housings which conceal the light source have high reflectance inner surfaces to maximize lumen output, minimize absorption and therefore reduce contained heat.
- Transmitting housings and shades are made of glass, plastic, fabric or other such material. They are used to protect the light source, diffuse the light radiation and to filter the light to produce certain colors. Shades are often more elaborately decorated than opaque housings. High reflectance inner surfaces are beneficial.

■ **Reflectors are a type of opaque housing which are designed for specific types of beam control.** Smaller reflectors are often used within reflector housings in order to further direct light.
- **Reflectors use geometric contours and principles of reflection to control light distribution.** Reflectors are most accurate with the smallest point source of light; bulb coatings will enlarge the source and reduce beam control. Greater candlepower is produced in the beam when the reflector extends beyond the source, shielding it from view, but is achieved at the expense of efficiency. Several types of reflectors are used:

D5.1 ELECTRICAL

- **Parabolic section** - used to produce a concentrated beam by setting the focus of all reflected light ahead of the point source of light (such as floodlights), at or behind the point source of light (such as spotlights).

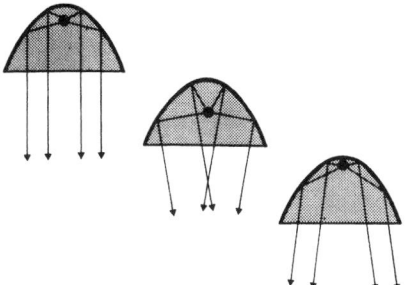

- **Elliptical section** - used to produce a divergent beam where the emitted light is concentrated on a second point or focus. Elliptical reflectors are appropriated for projection beams.

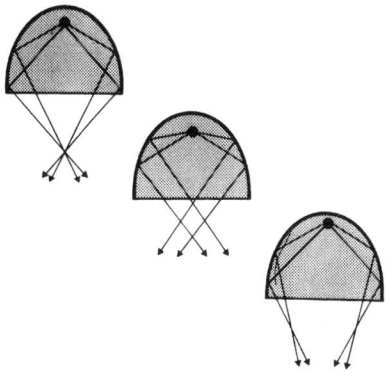

- **Circular section** - used to optimize lumen output by returning a ray of light which originated at the focal point back through that same point; reflected light can cause excessive temperatures on lamp parts.

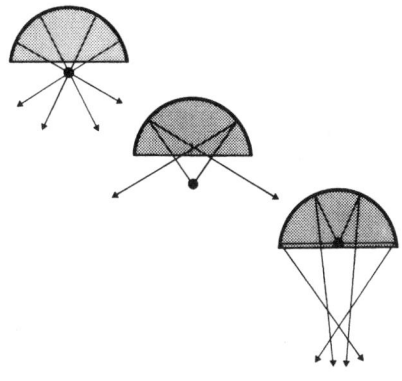

- **Hyperbolic section** - used to produce a divergent beam similar to an elliptical section; more difficult to manufacture.

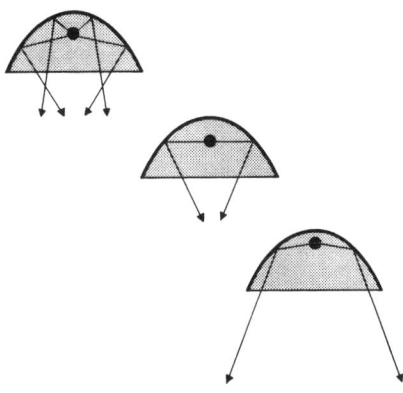

- **Compound sections** - used to improve beam control and distribution; often combining circular and parabolic curves (paracyl) to shield the light source from view while focusing the beam.
- **Angular sections** - used for more economical solutions where beam control does not need to be as precise.

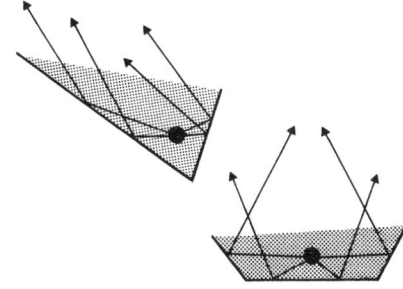

■ **Supplemental reflectors are used to control glare and to improve indirect distribution of light.**
- The reflector intercepts stray light and redirects it toward the primary reflector.
- Silver-bowl lamps, socket attachments and other similar forms of reflectors are available.

■ **Reflector materials may be specular or diffuse, but should always be resistant to heat, scratching and marring.**
- Specular materials include silver, silver deposited on glass (mirror glass), electroplated metals, film coating, aluminum and totally-reflecting prismatic glassware.

- Diffuse materials include plastics, porcelain enamel and coated/painted materials.
- High reflectance coatings, paints and electroplating are applied to base materials which are chosen for their strength, weight, cost, durability, ease of manufacture and other characteristics.

■ **Although opaque and reflector housings are very common, portable and decorative luminaires often employ light transmitting shades or diffusers.**
- Shades and diffusers can be made of fabric, spun glass, plastic, perforated metal, paper, parchment.
- Primary concerns of shade selection are the spread/distribution of light, depth (to conceal socket and bulb), and density of material.
- Care should be taken that the diffusers conceal the primary source of light and hardware from normal viewing angles.

■ **The thermal impact of lighting is a concern in luminaire design and specification.** Three types of thermal gain must be considered: Convective heat, conductive heat and radiant heat transfer.
- **Convective heat** raises ambient temperature as air moves from warmer to cooler surfaces in and about the operating luminaire. The effect of convection will be controlled by the size and placement of opening in the housing. Convection may also occur within the lamp, distributing hot vapors to the uppermost part of the lamp. Lamp performance is often defined by the position of the lamp within the installed luminaire because this convection may cause glass blisters or base seal failure.
- The interception and absorption of light by luminaire components produces **conductive heat**. Conductive heat radiates into the space to raise ambient temperature, but can also be a cause of improper luminaire operation as thermally sensitive materials may be adversely affected. Excessive conductive heat can lead to reduced lamp efficacy and eventual failure.
- The production of infrared light and the absorption of light by reflective surfaces produce another type of lighting heat - **radiant transfer**.

D5.1 ELECTRICAL

LUMINAIRES (cont.)

■ **The housing design is critical to the control of thermal gain from lighting.** Housings absorb and often trap heat radiating from the lamp, becoming a heat source themselves.
- Heat buildup from high wattage and/or poor ventilation can be a potential fire hazard.
- Housings should be provided with vents or slots such that a minimum quantity of air flows at low velocity across the lamp surface. A chimney-effect should be avoided so that the base seals of the lamp are not exposed to heated air. Baffles within the housing can provide adequate protection.
- Heat control can be accommodated by integrating the luminaire with other systems; air conditioning or water systems can be used to cool housings through ventilation or conduction.

Luminaire Lenses and Filters

■ **Lenses are elements with light transmission qualities which are used to modify the light emitted from the luminaire, to enclose the luminaire or to protect the lamp or modify the light radiation.** The selection of the lens material based on transmission properties will affect luminaire heat generation from absorbed and reflected light.
- Gasketed lenses are occasionally incorporated in luminaires to provide waterproof or dust/dirt-proof properties. Luminaire designs which allow proper maintenance without disturbing the lens seal are more desirable since the quality of the seal is secure.

■ Lens material may vary in density, color, shape and surface contour.
- Density and color are used to control the percentage of light transmitted. The density of the material relates to the transparency or translucence of the lens.
- The lens shape can be curved or flat, with consistent thickness and parallel surfaces or formed with opposing surfaces. Parallel surfaces provide regular transmission while opposing surfaces create positive (concentrating) or negative (spreading) light distribution.
- The surface may be smooth, prismatic, fluted, beveled or otherwise contoured to create the desired distribution of light. These contour variations use reflection and refraction

to direct or spread the light beam. The epitome of surface contour variation is the Fresnel lens which uses beveled and curved surfaces on one side with a flat surface on the other.
- The most common lens material is acrylic, because of its durability, light weight and ability to retain its original color. Configured panels for spread transmission are usually made of acrylics formed into diamonds, pyramids, ribs or ripples. Prismatic lenses are available. The disadvantage of acrylic lenses is their expense.
- Polystyrene is also used as a lens material, but it discolors (yellows) over time and is restricted from use by some building codes. Prismatic lenses can also be made with polystyrene.
- Like all plastic lenses, acrylics and polystyrenes conduct static electricity, therefore attracting dust.
- Optical glass comes in various formulations (soda-lime, lead-alkali, brosilicate, aluminosilcate, 96% silica or vitreous silica), can be produced several ways (blown, pressed, rolled, drawn, cast or sagged) and may have various surface characteristics (polished, sandblasted, chemically-etched, stained or coated). Although glass breaks and is relatively heavy, it will not change color, and it is inert. Spread transmission can be promoted by using configured glass panels which are pebbled, beaded, hammered, ribbed, stippled, fluted, prismatic or square patterned.
- Prismatic diffusers can reduce beam spread, directing more light downward. These lenses are 70% to 80% efficient, creating surface highlights (sparkle). Overlighting causes glare above about ll50 lumens/square foot.

■ **Light beams can be modified using filters.** Filters are primarily used to allow only selected wavelengths of visible light to radiate from the luminaire. Selected colors of light are absorbed by the filter and converted to heat (absorbing filter) or are reflected back into the luminaire (interference filter).
- Absorbing lenses tend to maintain higher surface temperatures than clear glass. The amount and type of light transmitted depends upon the hue (color saturation) of the filter; the greater the saturation, the more selective the absorption. Materials of absorbing filters include:
  — Gelatines - these rapidly fade, and will shrink from heat;

  — Colored glass - smooth, flat, stippled, molded or prismatic;
  — Split-glass - separate strips of glass set in a metal rim or frame;
  — Liquid applied - spray painted, enameled and inside coated bulbs;
  — Film coatings.
- Interference filters are generally used to reflect UV and IR radiation. The most common is the dichroic (two-color) filter which transmits one color and reflects the complimentary color.
- The result of a filter will depend on the color and intensity of the light source. Spectral distribution will affect luminaire efficiency; filters over white light sources are most efficient, followed by a match of the filter hue to the strongest wavelengths of light radiated, and the lowest efficiency is when the filter hue is the compliment of the color omitted by the light source. Saturated blues, reds and greens will always reduce transmission drastically, but increased levels of source luminance will change the transmitted color. This is not true with yellow filters.

■ **A distinct set of lens-type materials have been developed to diffuse light: Diffusers and diffuser panels.**
- For good illumination, a diffuser panel should allow 40% to 60% transmittance while minimizing absorption and maximizing reflectance. Reflectance from the diffuser surface will be reflected again from the inner surfaces of the luminaire. When these surfaces are of high reflectance, overall transmission can be increased to 70% to 80%, increasing luminaire efficiency.
- Sizes and shapes of diffuser panels vary considerably. Illumination characteristics are a function of pattern and surface treatment. Flat sheets are available with polished or matt surfaces in natural, colorless, white and a wide range of transparent and translucent colors. Pebbled, hammered and other configured lenses are also used for diffusion. Domed panels are typically translucent white and can be installed either dome up or down.
- Luminance uniformity from the diffuser is best when the lamp is spaced no more than l l/2 times the height measured from light center to panel.
- Decorative plastic sheets are available with smooth, polished, mat and rough surface textures.

Luminaire/Lamp Shielding

■ **Shielding or baffling is a means of control of direct glare. It is properly applied when the shielding angle (the maximum angle that the eye can be raised above the horizontal without viewing the light source) is between 45 and 60.** Shielding devices include baffles, louvers, grilles, screens and supplemental reflectors.

■ **Baffles are opaque or translucent materials installed parallel to one another and located over the light source to provide shielding in one direction. Perimeter baffles may be used on small luminaires to limit beam spread.**

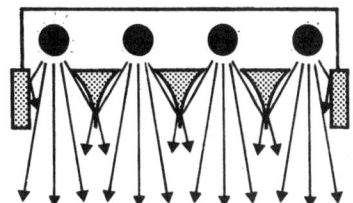

 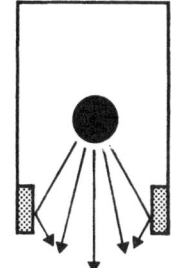

- Appropriate for low ceilings, baffles separate rows of lights so that the viewer sees the luminance of the vertical baffle surface rather than the light source.
- Baffles are usually made of metal or plastic, and occasionally of wood.
- Because of their application, baffles provide little surface area where dust can build up; lumen production is not affected by dust collection on baffles (as opposed to diffusers).
- The direct distribution of light permitted by baffles may be the source of veiling reflections or direct

glare. If glossy surfaces are necessary beneath the luminaire, diffusers should be considered.

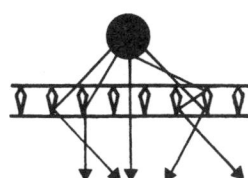

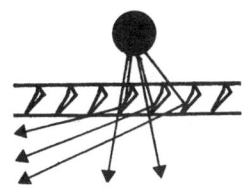

- Luminaire designs can combine baffles and diffusers to control glare. The diffuser panel may rest on the baffles, or a one-piece panel can be created.

■ **Louvers are a series of baffles arranged in a geometric pattern to provide shielding from many directions.**

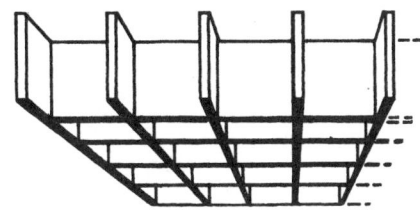

- Louvers will reduce luminaire efficiency in proportion to the percentage of open area. Large cell parabolic louvers are more efficient than lenses or diffusers.
- Louvers are generally an integral part of the luminaire. When they are part of a ceiling system with lighting above, they are supported on four sides, using t-bar or hanger clip supports. Some panels are designed to overlap, eliminating the need for end support between panels.
- Cell patterns include parabolic, square, grid, eggcrate, hexagonal and circular. Cell sides can be flat or curved

(wedge shaped or parabolic). Concentric patterns can be nested.

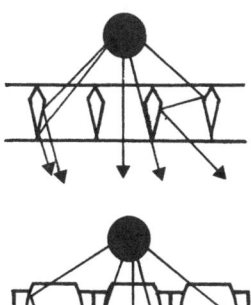

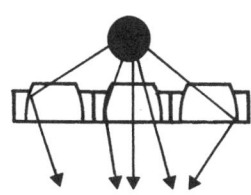

- Smaller cell size promotes diffuse luminance qualities.
- Materials used include aluminum, steel, plastic and wood.
- The control of light by louvers is similar to that of baffles. Matte black surfaces provide the best cutoff shield, but specular parabolic surfaces add directional control to good shielding qualities.

■ Control of transmission and reflection are essential for adequate shielding.
- Slight spread transmission can be used to create sparkle for decoration, but transmission should be diffuse. Likewise, reflective qualities should be carefully chosen.
- High reflectance finishes improve luminaire efficiency but should be of spread or diffuse nature. Specular finishes are the most efficient (best at low brightness) and can be contoured to direct light for certain applications. However, they must be maintained carefully and as often as necessary to remain specular.
- Low reflectance baffles are good for cut-off and dramatic effect. Colored finishes will reflect colored light causing some color overlap, but general illumination will not be affected.

■ Grilles and screens are used most often as decorative space dividers (movable partitions, etc.), however they are also effective shielding devices. These assemblies are appropriate for luminance reduction at oblique angles.
- Metal grilles include expanded metal mesh, stamped metal, perforated metal, and various densities of common screening. Material selection are steel, aluminum, brass, copper, bronze, tin and gold.

D5.1 ELECTRICAL

LUMINAIRES (cont.)

- Plastics are used to create fabrics and grilles of varying density, pattern, thickness and cost.
- Wood products are used to create roll screens, slat shades, fiber mesh, and screen framing.

Support and Installation

■ Logical and efficient luminaire design facilitates proper installation, easier maintenance, and therefore better illumination.

- Luminaires should have hinged access on parts which do not need removal except for vigorous cleaning.
- Parts which are commonly replaced should be detachable without removing the whole unit.
- Lowering equipment may be helpful on luminaires used in very high ceiling applications.

■ Incandescent lamping usually allows various types of mounting and support conditions due to the point source lighting qualities.

- Mountings can range from outlet boxes (supported by structural elements) to self-supporting or non-supporting wireways to pipe clamps to a simple cord and plug.

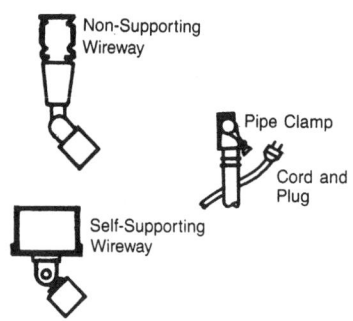

Non-Supporting Wireway

Pipe Clamp

Cord and Plug

Self-Supporting Wireway

- Recessed luminaires (cans) can be supported by the ceiling structure or can be suspended before the ceiling is installed. The luminaire and associated hardware is commonly limited by building codes to a weight of 6 pounds when supported by the ceiling.
- Portable luminaires typically rely on cord and plug mountings, supporting the lamp, housing, reflector and lampholder on a stand or base.
- Outlet boxes usually support ceiling mounted luminaires. When the luminaire is suspended, a canopy is used to finish the connection of luminaire and box, covering the ceiling around the box opening. The

lampholder, lamp, reflector and housing (globe, basin, bowl) are supported by hanger rod, chain, or fitter. Otherwise, the housing attaches directly to the box and ceiling.

- Track mounting systems are typical of self-supporting wireways. Track systems can be surface, recessed or pendant mounted.

■ Fluorescent lamps are typically set into lampholders which are connected to a metal channel. The channel houses the starter, ballast, filters and other controlling equipment. It also serves as the grounding strip which is required for many lamps.

- The channel can be mounted directly to the ceiling, or can be suspended below it using threaded support rods, suspension stirrups or other methods.
- The channel may also connect to trunking which is supported by other means. Trunking allows the lighting system to be independent of ceiling structure or height.

■ Louvers and diffusers are typically supported by the luminaire housing or are suspended below the lamp system.

■ Roadway, parking and other outdoor area lighting systems require appropriate elevation of the luminaire to meet allowable spacing and illumination needs. Mounting on structures is occasionally adequate, but most systems rely on posts (shorter lengths; up to 12') and poles for height.

- The shape, profile and sometimes the length of the post/pole is a function of the style chosen and the material used. Wood poles will have a round cross-section and a taper from butt to tip of 1/16 to 1/8 inches per foot of length. Shorter wood posts can be shaped to form multiple profiles for decorative purposes. Metal posts/poles can be formed to various cross-sections and the shape can be consistent or modified over its length. Multiple sections of metal poles can be used, whereas wood and concrete cannot be practically joined. Concrete poles also offer a variety of shapes and profiles.
- Metal poles offer an opportunity to conceal all wiring, ballasts and other equipment within them. Concrete poles can be formed with a conduit in them for internal wiring, but wood poles involve external assemblies.
- Posts/poles may be held in place by embedding in the ground or concrete, or by mechanical fastener or welding

to an anchoring base, pedestal or foundation.

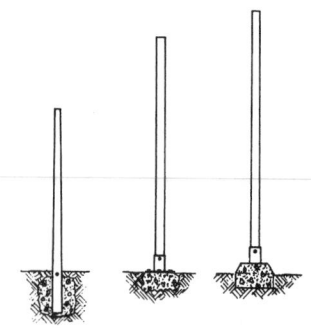

- Resistance to the outdoor environment includes consideration of rust, atmospheric corrosion, moisture, insect attack (wood), freeze-thaw cycles, delamination, vandalism, cracking, chipping, impact/denting, wind and gust factors, etc.
- Posts/poles should be selected based on ease of installation, of maintenance (access, relamping, cleaning) and mounting of luminaire.
- Luminaires may be mounted on brackets, struts, or post tops. Bracket and strut designs will be based on maximum supported load for a given span/extension.
- Finishes of posts/poles should be weather-resistant, of diffuse reflection and non-toxic.
- High mast lighting is typically an arrangement of high intensity luminaires mounted 80 to 100 feet above grade.

Luminaire Construction

■ Construction of the luminaire must be of sufficient quality to meet the intended use and life of the lighting system. All aspects of the system rely

D5.1 ELECTRICAL

on serviceability, electrical safety, operating temperatures, dimensional consistency, ease of assembly and installation, etc.

■ One measure of construction quality is indicated by labelling. The Underwriters Laboratories (UL) labels luminaires to indicate their compliance with appropriate electrical safety standards.
- Refer to appropriate electrical codes to be certain of wiring specifications such as total current, voltage, insulation ground, application and exposure.
- Check sockets for suitability with standard lamp bases.
- Check clearance for standard lamp sizes.
- Be sure that power circuits, lamps, ballasts, and other components all match in wattage and voltage.

■ Mechanical, or structural considerations include matters of strength, durability and heat resistance of materials and connections.
- Tolerances between luminaire components must consider lamp sizes, thermally induced expansion rates of different materials - both for luminaire components and adjacent surroundings. Industry and ANSI standards address these issues.
- The assembly should be essentially complete and ready for installation, with appropriate instructions enclosed. Materials should be strong enough to endure shipping and installation handling, not only the elements to be met in the environment (weight, impact, moisture).
- In many cases, special housings, enclosures or supports are required. Examples include submerged luminaires, enclosures around thermally-sensitive short-arc lamps, tamperproof requirements, and protection against accidental lamp bulb fracture.
- Connections within and supporting the luminaire must be sufficient to resist vibration.
- Fastening technique (welding, riveting, screwing) should respond to permanent, temporary, accessible and operable connections.
- All materials should be protected from corrosion.
- Details at joints and connections should be designed to eliminate light leakage or lines of shadow.

Efficiency

■ Power is measured in dollars per kilowatt hour. Thus the dollars spent on power will be minimized by reducing the number of hours that the system operates and by optimizing the use of the input wattage through lighting system efficiencies.

■ Assuming that appropriate controls are installed to monitor the length of operating time, that the proper level of illumination (in footcandles) has been determined and that the environment to be illuminated has been defined (area in square feet), only efficiency factors can reduce the number of watts used. The efficiency factors are:
- Lamp efficacy;
- Light loss factor;
- Luminaire efficiency;
- Coefficient of utilization.

■ Luminaire efficiency is the ratio of the lumens emitted by the luminaire to the lumens emitted by the lamp(s) contained within the luminaire.

■ The **coefficient of utilization** (CU) depends on the type of luminaire, the spatial geometry of the environment, and the reflectance of adjacent surfaces. It is a rating of the fraction of useful illumination received on the visual task. The CU responds to:
- The **shape and size of the environment**: The larger and more open the environment, the higher the CU. Building enclosures with large floor areas and moderate ceiling heights allow less light to be absorbed by wall surfaces. Smaller floor areas and/or higher ceilings will allow greater quantities of light to strike wall surfaces (wall surface area increases as ceiling height increases or floor area decreases).
- The **reflection factor of ceilings**: percentage of light reflected.
- The **reflection factor of walls**: percentage of light reflected.
- The **distribution of light from a luminaire**.

■ For photometric data (the amount and distribution of emitted light) and coefficient of utilization, refer to manufacturer catalogs for specific luminaires.

D5.1 ELECTRICAL

LUMINAIRES: types, uses

Architectural/Built-in

■ General Notes:
- Best uniformity with linear source or large number of small point sources;
- Lenses can be shaped to imitate the structural form or create new forms.

■ Light Shelf Type:
- Use structural elements to support luminaire and shield light source;
- Similar to cove lighting, but lenses are used as the finished ceiling as opposed to building elements.

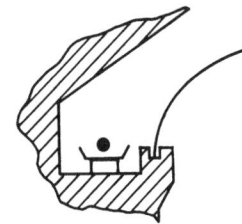

■ Flush Panel Type:
- Diffuse lenses;
- Single or double rows of lamps;
- Cylindrical or rectangular cavities.

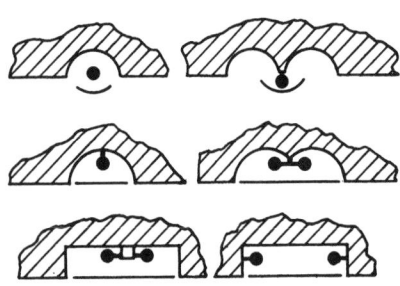

■ Corner Panel Type:
- Typically used vertically;
- Flat or quarter cylinder diffuse lenses.

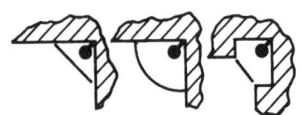

■ Projecting Type:
- Usually used for wall texture;
- Single or double rows of lamps;
- Diffuse or spread lenses.

■ Coffer Type:
- Recessed cavity that is open to the space with deeply concealed lamps producing indirect light;

- Common shapes include circular, rectangular, irregular/free form;
- Usually set in ceiling, but some walls may also have coffers for prominence;
- Lamps may be arranged in a grid or in parallel rows;
- Concealment of lamps is provided by structural component or by housings (metal, plastic, wood);
- The more concentrated the light source, the greater the distance from reflective surface to lamp center (distance = 1/6 lamp spacing for diffuse; = 1/3 lamp spacing for concentrated).

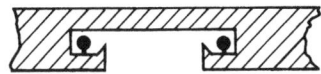

■ Cove Lighting Type:
- Good for high ceiling room; set high on wall;
- Keep light source 10 to 15 inches from reflecting surfaces to maximize uniform distribution;
- Cove (trough) conceals lamp from direct view and is made of wood, plastic, metal or plaster;
- External shields (attached to wall and set over lamp) can be used to keep light off the wall;
- Recessing the luminaire provides an integral shield which limits distribution;
- Usually the luminaire consists of a fluorescent lamp with a reflector or custom formed cold cathode lamping;
- Effective for cross-lighting;
- Using a glass or plastic bottom would allow downward light for accent.

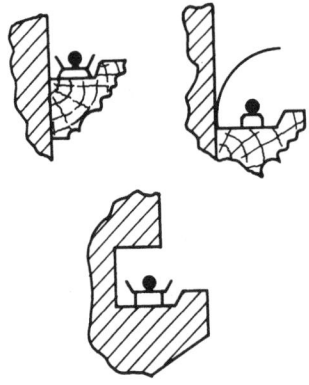

■ Cornice Lighting Type:
- Indirect lighting distributed downward across the wall surface;
- Located at ceiling line;

- Wood or plastic baffle used for lengthwise shielding.

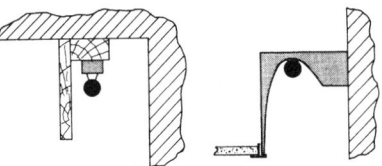

■ Valance Lighting Type:
- Used over windows, draperies, etc.;
- Should not be less than 6 inches deep nor 6 inches from the wall or window nor 10 inches from the ceiling;
- Lamps are typically linear;
- Provides indirect lighting both upward and downward.

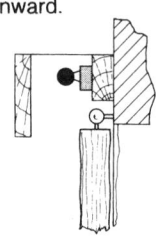

Recessed

■ General Notes:
- Interior reflectance is critical to luminaire efficiency.

■ Downlight Type:
- Baffles are used to control luminance;
- Venting is necessary with Type R, ER, and PAR lamps;
- Reflector may be parabolic or elliptical with general service lamps and no lens.

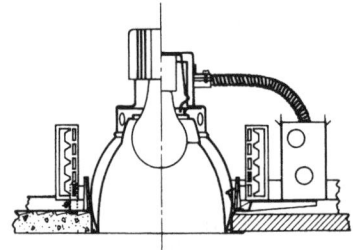

■ Totally Enclosed Type:
- Housing is completely enclosed and often has an additional reflector surface to define distribution;
- Lens may be transparent, translucent (diffusing) or refracting.

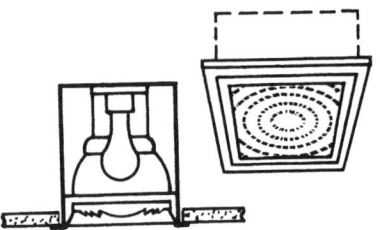

Wall Washer Type:
- Reflectors can be parabolic or elliptical;
- Spread lenses may be used for beam control;
- Reflective louvers may be used to control luminance;
- Recessed housings with specular reflectors and general service lamps;
- Recessed housings with reflector lamps and specular reflectors;
- Ceiling mounted PAR lamps with a maximum spacing of 0.8 times the distance from luminaire to wall;
- Ceiling mounted reflector lamps with a maximum spacing of 1.0 times the distance from luminaire to wall.

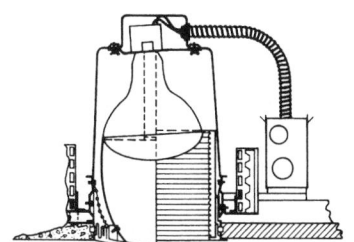

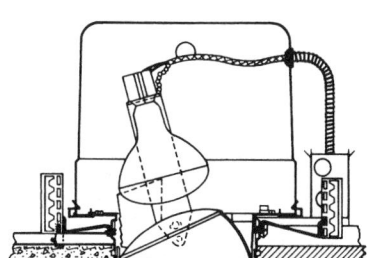

Luminous Ceilings

General Notes:

- Characterized by a network of louvers and/or transmitting lenses suspended below a lamping system to provide uniform, diffuse illumination;
- Lamps should be below and at right angles to ducts, sprinkler pipes and all other obstructions to avoid shadows;
- Ceilings should be washable with 80% to 85% reflectance (reflectors or painted) to increase efficiency;
- Lamp spacing should not exceed the mounting height (light center to louver); lamp spacing to wall should be 1/2 of standard spacing;
- Glare is reduced by using small cell opaque louvers, diffusing or prismatic lens panels; opaque sides;

- Use broad beam lamps of the same lumen output;
- Luminaire design and application requires attention to acoustical control.

Suspended Type:
- Luminaire and diffuser attached to channel framework which is suspended from ceiling structure;
- Side panels are used to minimize shadows on wall.

Ceiling Mounted Type:
- Luminaire mounted on surface, diffuser gridwork suspended below;
- Translucent side panels.

Baffled Type:
- Baffles provide lamp shielding and beam cutoff as well as support for diffusers;
- Luminaire is surface mounted with baffles set between.

Modular Types:
- Pre-assembled and installed separately or in groups;
- Either suspended or flush-mount installation;
- Common module sizes of 2'x4', 2'x6', 2'x8', 4'x4', 4'x6', 4'x8', 6'x6', 6'x8'.

Plenum Ceilings

General Notes:
- Air handling, lighting and often fire suppression systems are incorporated to make one assembly.

Air Return Type:
- Open louvers and vented housings allow low velocity air to move from the illuminated space to an exhaust duct, removing lighting heat;
- Allows cooling load to be unaffected by lamp heat;
- Exhaust air may be raised 10 to 25°F as it passes through.
- Fan driven exhaust ducts aid the convection process;

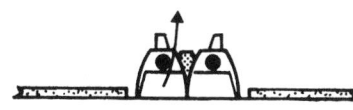

Air Supply Type:
- Conditioned air is forced over the housing and out through a slot in the ceiling;
- The duct is created by the lamp housing and the upper dust cover;
- Heat from luminaire will affect the temperature of the air being distributed over the housing;

- Luminaires tend to be expensive, but the installation and the appearance of the ceiling system is simplified.

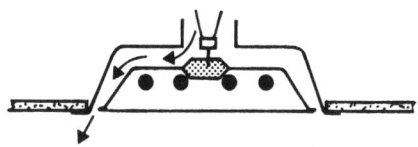

Ceiling Mounted

Surface/Flush Type:
- Distribution is direct or general diffuse;
- Available for exterior use.

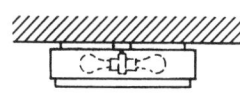

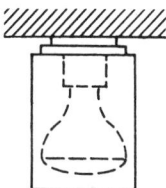

Pendant Type:
- Typically of general diffusing type.

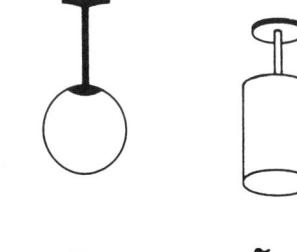

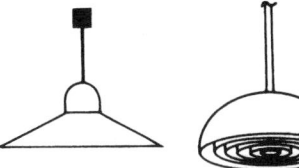

LUMINAIRES: types, uses (cont.)

■ Suspended Type:
- Luminaire is supported by rods, chains or adjustable chords;
- All distribution types are available;
- See also Luminous Ceilings.

■ Track Type:
- Direct concentrating illumination.

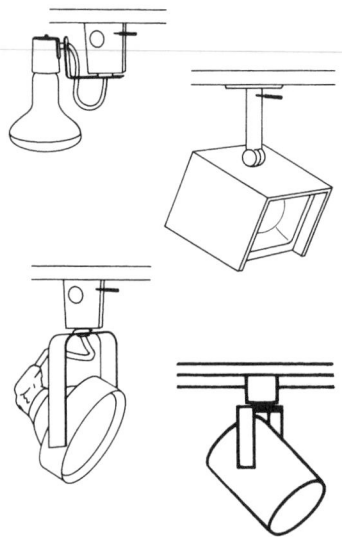

■ Chandelier Type:
- A decorative luminaire which can be used for low-level general illumination.

Wall Mounted

■ General Notes:
- Typically shielded by shades, diffusers or baffles to reduce glare and avoid excessive luminance on adjacent wall surface;
- Usually for local illumination or highlight;
- Often available as a portable luminaire.

■ High Bracket Type:
- Same as valance but on interior or windowless wall;

- Shielding can be a fixed faceboard on brackets or supported on ends; can also be on adjustable brackets to change shielding and distribution angles.

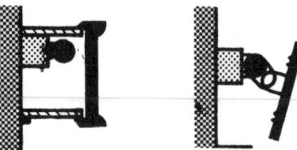

■ Low Bracket Type:
- For lighting specific task areas;
- Height varies by use, whether seated or standing positions;
- Glass or plastic shelving necessary if upward lighting is permitted.

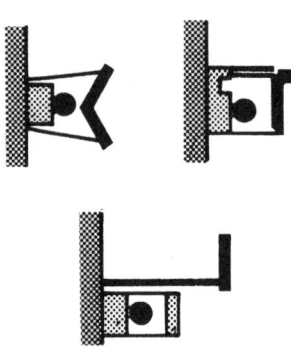

■ Indirect Reflector (Wall Urn) Type:
- Indirect distribution.
- May have decorative or utilitarian housing.

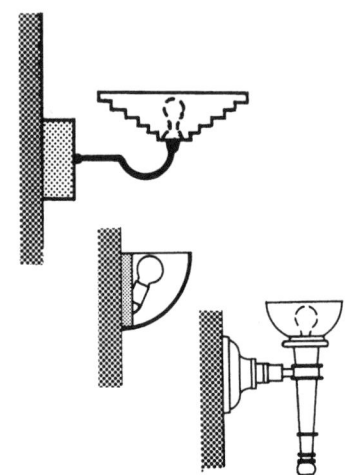

■ Colonial Type: chimney-style, candle-type.
- Strictly decorative;

- Sconce types have reflectors or chimneys to control the light from the unshielded lamp.

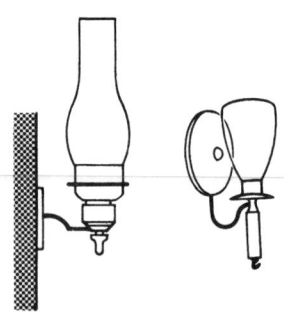

■ Pendent Type:
- For utilitarian direct task lighting or decoration.

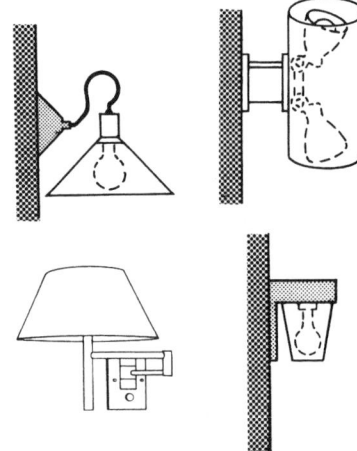

■ Lantern Type:
- Mainly for exterior use.

Portable

■ General Notes:
- Easily relocated and readily connected or disconnected at the nearest outlet;
- Flexibility of location of controls;
- Shades, reflectors or both used;
- Common base materials include pottery, metal, wood, leather, glass and plastic.

■ **Decorative Type:**
- Small luminaire with low wattage lamps;
- Strictly aesthetic; night light; background.

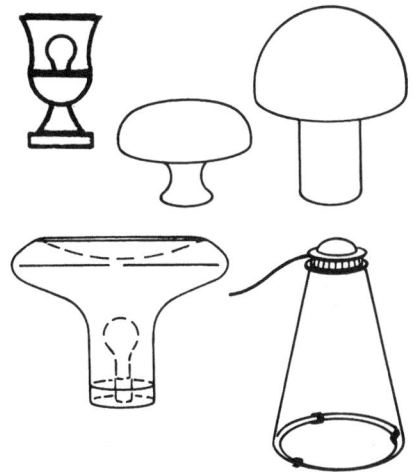

■ **Table Type:**
- Single, two or three socket with standard or three-way lamps;
- General lighting provided by open shades, top and bottom, with no reflector;
- Diffuse downward lighting, produced using reflectors, is recommended for prolonged difficult visual tasks;
- End table, dresser and vanity applications use 18- to 26-inch tall units; not appropriate for reading;
- Desk, reading and study units typically 18 to 30'' tall.

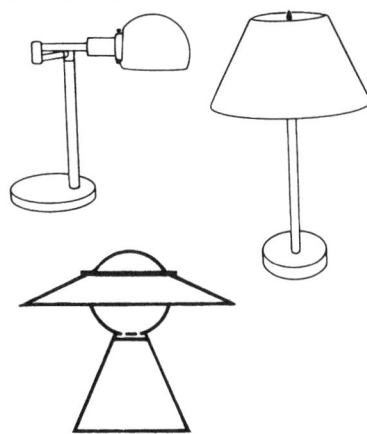

■ **Floor Type:**
- Free-standing, adjustable, direct, general diffuse or indirect (torchere);
- Single, two or more sockets; standard or three-way lamps;
- Semi-indirect lighting provided by reflectors;

- Multiple switching allows major wattage lamps for task lighting or low wattage lamps for decorative general lighting;
- Torchere provides indirect lighting.

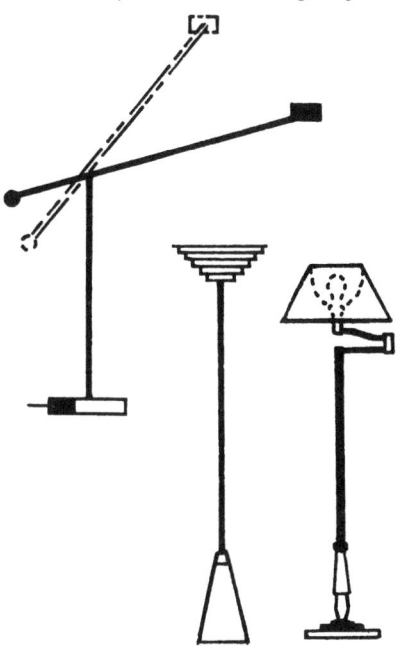

Floodlights/Spotlights

■ **General Notes:**
- Louvers can be used to control beam spread, but efficiency is reduced about 10%;
- Flexible mounting specifications (height and spacing);
- Type of non-cutoff luminaire; for indoor and outdoor applications.

■ **Round Spun Type:**
- Low cost;
- Single piece spun aluminum reflector, modular ballast assembly, distinct support assembly, hinged glass lenses;
- Available with weather-protective aluminum covers.

■ **Cast Heavy-Duty Type:**
- More expensive than round spun;
- Lower operating and maintenance costs; durable;
- Permits all classifications of beam spreads.

■ **Cutoff Site Type:**
- Can produce precise beam control; little waste light or spillover.

■ **Projector Type:**
- Usually weatherproof; lamp, reflector and cover glass with trunnion mounting;
- General service or PAR lamps.

■ **Recessed Adjustable Type:**
- Recessed housings with specular reflectors, general service lamp and an asymmetric lens with offset adjustment;
- Recessed housings with PAR lamps, spread diffuser, vertical adjustment and 360° rotation.

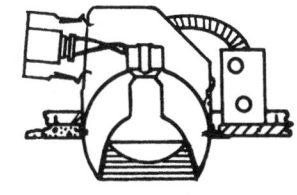

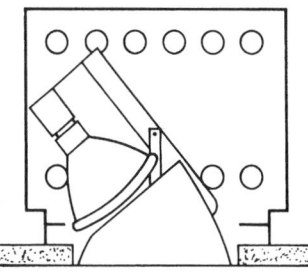

Pole Mounted

■ **General Notes:**
- Opaque cutoff housing, characterized by restriction of light radiation to 90° or less from vertical.
- Pleasing aesthetics combined with excellent beam control, little glare, little spillover.

■ **Cutoff Type:**
- Lessens glare with a 70° cutoff; sharp cutoff is down to 36°;
- Extremely distinct and predictable light patterns;
- May require greater mounting height or closer pole spacing to eliminate dark spots;
- Tend to maintain symmetrical distribution but asymmetric is available;
- Adjustable types are appropriate for irregularly shaped areas; good precision, lower height and wider spacing of asymmetric illumination.

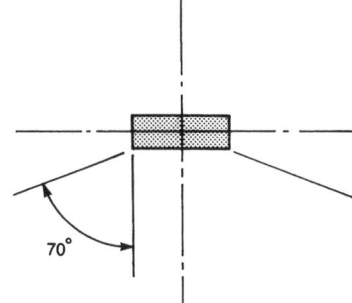

70°

D5.1 ELECTRICAL

LUMINAIRES: types, uses (cont.)

■ Shoebox Cutoff Type:
- Lens is set parallel to the surface being illuminated;
- Housing is totally enclosed and extended beyond lens to prevent direct glare;
- Good for roadway lighting.

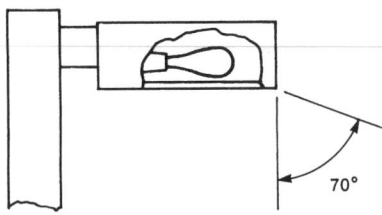

■ Lantern Type:
- Old style lantern coach lights used for greater aesthetic qualities;
- Mounts up to 15 ft. above the ground;
- Typically with non-cutoff housings;
- Higher luminances controlled with prismatic refractors;
- Available with high wattage lamp, clear lenses and sharp cutoff housing.

■ High Mast Type:
- Applications: Large parking areas, roadways;
- For mounting heights of 80 to 150 feet;
- For use with HID lamps;
- Distribution varies between Types I through Type V.

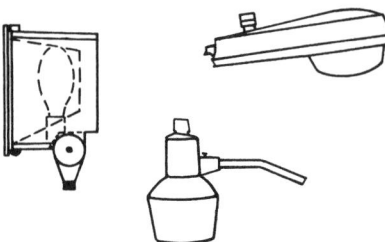

Landscape

■ Spread Type:
- Unobtrusive designs which typically provide circular patterns of light from lower than 4 feet above grade;
- Direct light is shielded above horizontal;
- Reflective louvers can provide indirect light above horizontal;
- Allows lower wattage lamps, lower mounting and/or wider spacing.

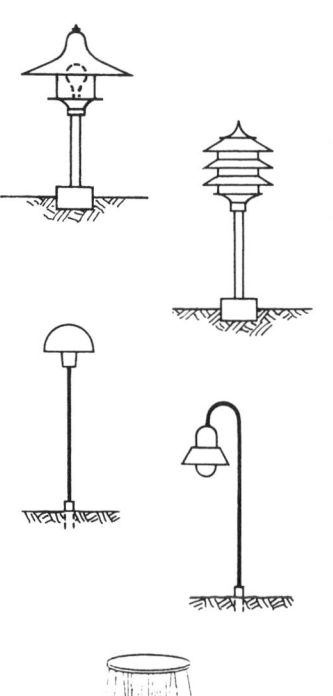

■ Accent Type:
- Typically surface mounted at grade, in trees or on facades;
- Effective for uplighting;
- Distribution is directed away from observers;
- Commonly with adjustable base for angle correction.

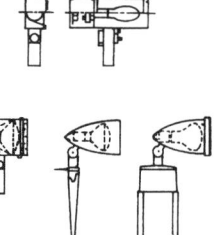

■ Well Type:
- Effective for uplighting with PAR or reflector lamp;
- In-ground applications with tilt adjustment inside housing;
- Louver grates protect lamp and control glare;
- Luminaires not designed for submersion should have all wiring and fittings well sealed and should have drainage provided to prevent flooding.

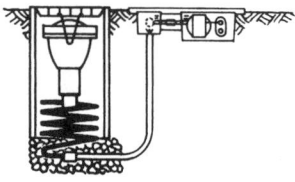

■ Step Light Type:
- Indirect lighting concealed within stair sidewalls, under stair nosings, under handrails, etc.

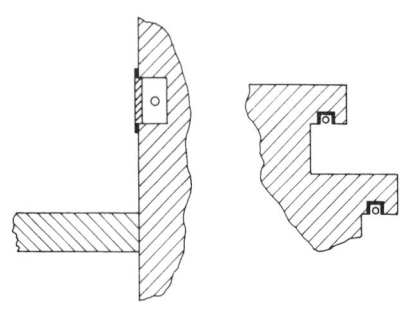

■ Bollards:
- Low, closely spaced for borders, pedestrian control and/or path lighting;
- Often designed to be vandal-resistant.

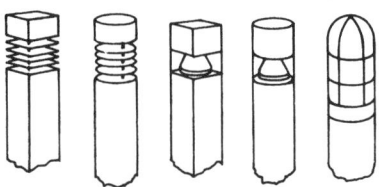

D5.1 ELECTRICAL

LIGHTING: controls & maintenance

LIGHTING CONTROLS

■ Lighting controls are an essential part of lighting design since they allow the flexibility to satisfy the varied requirements of use, function, time of day and other client needs. Several control techniques are also available to optimize lighting energy consumption. Two basic concepts are to limit the rate of work (wattage) and to limit the duration of work (time). Although the efficiency of luminaire components can be optimized (lamps, ballasts, luminaires), power savings are keyed more to the length of time of operation of lighting systems.
• Use controls to turn off lights in areas which are not in use;
• Use daylighting;
• The intensity of general illumination can be reduced when area function changes or when additional task lighting is employed.

■ The lighting control techniques that have evolved in response to task, security, safety and energy needs can save 30% to 50% of the energy otherwise expended on lighting by providing illumination only where it is needed and turning off other luminaires. Since many codes and standards limit the number of watts used for lighting per square foot of floor area, the reduction or elimination of electrical lighting where it is not required can be balanced by increasing the average wattage per square foot where it is needed.

• **Manual switching:** Two levels of switches allow an occupant to reduce light output by 50% by switching alternative rows of lamps in luminaires to reduce illumination. Luminaires which are close to windows should be on a separate circuit from the rest of the system to allow the luminaires to be switched on or off respective to time of day.

• **Daylight maintenance** provides a constant level of illumination by using a photoelectric cell connected to a dimming system, allowing output to be automatically changed gradually or in steps. 20% to 30% savings can be taken by using daylight sensors on circuits or individual luminaires located near windows.

• **Lumen maintenance** reduces light level initially to designed maintained level. For example, if a system is designed to be 60 fc maintained,

initially it will be 75 fc. This can be accomplished by using devices which measure the light output. When lumen output is reduced by age, dust, etc., the device will increase the power levels to compensate.

• **Time of day scheduling** can be provided by manual time-switch or automated/microprocessor programmable lighting timers to shut off lights when the space is not in use (i.e.: after working hours).

• **Load shedding** saves energy by turning off non-essential lamps during peak energy demand hours.

• **Occupancy sensors** are appropriate for more localized and less predictable lighting needs. Infrared/heat and sound/microwave systems can be used to save about a third of potential energy use.

■ **Dimmers provide variable control of lighting.** There are several types available, but electronic technology has rapidly outdated all but the solid-state. Solid-state dimmers generally use about 2% of the energy supplied during lamp operation, but will increase lamp life and decrease lamp wattage proportional to the percentage of dimming.
• Solid-state dimmers commonly use a small, light weight thyristor as the solid state device that 'chops' the wave, thus dimming the lamp. It requires a 'firing circuit' and an output filter to eliminate radio frequency interference (RFI) noise and to reduce lamp filament vibration (buzzing). These devices are provided integral to the dimmer.
• Higher wattage systems often use a triac or silicon controlled rectifier (SCR). These were originally common to television and theatre lighting. Instant response and quiet operation are typical, but control capability and system quality vary tremendously.

■ Power control provided by solid-state dimmers allows a 0% to 100% output range for standard, low-voltage and quartz incandescent lamps; 0% to 100% output range for fluorescent; down to 2% for mercury lamps, and down to 40% for other HID lamps.
• Dimming of incandescent lamps will cause a dramatic color shift, becoming more yellow.
• Quartz/tungsten halogen lamps should not be dimmed below 60% for

extended time periods or the lamp will blacken and burn out sooner than rated life. The lamp can be restored by bringing it to full voltage for a period of time.
• Low voltage incandescent lamps (eg: MR 16, PAR 36, tubelights, striplights) can be dimmed if the dimmer interacts with a transformer which will reduce the voltage from 120 volts to the voltage required by the lamp (6v, 12v, 24v). Newer solid state transformers may cause interference if used with electronic dimmers.
• Dimming ballasts are required with rapid-start fluorescent lamps on fluorescent dimmers.
• Some high frequency electronic ballasts are also capable of dimming fluorescent lamps.
• Some standard-ballast fluorescent dimmers can provide energy savings while providing dimming capabilities as low as 30%.
• Use of dimmers with gaseous discharge lamps can allow the arc to be maintained at a low level of light output, eliminating the need for restarting and lengthening the life of such lamps.
• Cold-cathode lamps require a low power factor transformer between the dimmer and the lamps.
• Metal halide lamps are not often dimmed because of expense, lack of infinite range and color shift with phosphor-coated lamps when voltage is decreased.
• Technology has been developed to dim high pressure sodium lamps, but it is not yet readily available.

■ **Variable auto-transformers** are still available but are usually specified only in areas where RFI may be a problem such as recording studios or theatres with remote microphone systems.
• They are up to 95% efficient, low heating, insensitive to load and of moderate cost, but they do not provide energy savings.
• They can be motorized for remote control. Response time can vary from a few seconds to nearly one minute.

■ **Automatic light detectors** are often used to operate electrical lighting to coordinate with daylighting or to be operated by the presence of a person or object. Operation based on movement or pressure is often used for security lighting and for energy-conserving lighting design. There are two basic types of light sensors (photodetectors):

D5.1 ELECTRICAL

LIGHTING: controls & maintenance (cont.)

- **Thermal detectors** absorb light radiation (all or only selected wavelengths), raising the temperature of the detector with output from the detector being proportionate to temperature gain over time.

- **Photon detectors** absorb light radiation, converting the absorbed light according to principles of quantum physics.

■ **Automatic light detectors should be selected by their affect on the lighting circuit.** Evaluation should be based on:
- Responsivity - the ratio of output to input;
- Spectral response - amps per lumen of visible light;
- Response time - exponential rise or fall relative to input;
- Detector area.

■ Automatic occupancy sensors can be thermally sensitive, air pressure sensitive or vibration sensitive. These occupancy sensors are applicable to energy management plans for small office areas, security lighting or in random use areas such as museum displays.
- Thermally sensitive sensors can be used in small areas or immediate proximity to measure temperature rise due to body heat.
- Pressure sensitive sensors rely on the force placed on them, such as the weight of an object or person. These sensors may also be used to measure displacement.
- Vibration sensitive sensors are often tied into systems where sound is generated by another source. The source can be a foreign element entering the environment, microwave transmission or another source of sound waves which is beyond the human threshold of hearing.

■ Programmable systems are available which coordinate sensor response, microprocessing and light output control. Several types are available:
- Building-wide management systems that control other building systems as well (i.e.: security, communications);
- Memory dimmer controls (hardware logic and programmable software systems);
- Automatic, programmable timers.

MAINTENANCE CHECKLIST

Light Loss Factors

The maintenance of any lighting system is critical to the satisfaction of the lighting design.

■ Luminance is a function of lamp performance and luminaire design. However, illumination is estimated by considering several environmental and maintenance factors which effectively reduce the incidence of light. Some factors which affect or can be affected by an appropriate maintenance program are listed below.
- Lamp Lumen Depreciation (LLD), measured at 70% of rated life (see LAMPS (properties for average lamp life):
  — incandescent = 0.78 to 0.90;
  — tungsten halogen = 0.93 to 0.96;
  — fluorescent (3 hours per start average) = 0.66 to 0.9l;
  — mercury HID (5 hours per start average) = 0.70 to 0.75;
  — high pressure sodium (5 hours per start average) = 0.83 to 0.88.
- Voltage drop ratings indicate the effect on incandescent lamps when socket volts are lower than lamp voltage rating:
  — 2.5% lower voltage = 0.94;
  — 5.0% lower voltage = 0.87;
  — 7.5% lower voltage = 0.76;
  — l0.0% lower voltage = 0.69 (0.95 and greater for ballasted lamps).
- Dirt and dust depreciation (considering that luminaires are cleaned at each lamp replacement) is rated for fairly clean or fairly dirty locations respectively.
  — incandescent = 0.8, 0.7;
  — fluorescent = 0.7, 0.6;
  — HID = 0.7, 0.6.
The depreciation of light due to dust and dirt includes reduced lumen production from lamps, reduced luminaire efficiency, reduced room surface reflections and the potential for filtering by airborne particles.
- Luminaire ambient temperature may affect output:
  — incandescent = l.0;
  — fluorescent = 1.0 at l00°F, 0.87 at 80°F, 0.84 at l20°F, 0.70 at l40°F, 0.l8 at 40°F.
  — HID = l.0.
- Ballast factors:
  — low power factor = 0.40;
  — high power factor = 0.90;
  — energy saving ballast factor = 0.99.

■ Luminaire components will also affect luminance. The properties of these components may impact the maintenance program.
- Plastic components conduct static electricity which attracts dust;
- Aluminum components are not conductive;
- Open cell parabolic louvers allow air circulation around lamps, have less surface (vs. diffusers, for example) for collection of dirt and therefore have a higher maintenance factor;
- Luminaires can be gasketed to minimize dust collection on the inside of lenses and on reflector surfaces.

■ Maintenance should be scheduled so that strict adherence will result in proper illumination of the environment at all times. Schedules should address:
- Relamping;
- Luminaire cleaning; and
- Inspection of auxiliary units.

■ Relamping may be scheduled on a regular basis depending on lamp wattage, rated life and lumen depreciation or it can occur only when the lamp burns out. Responsibilities include replacement of all burned out, broken or missing lamps.
- Group relamping produces more efficient use of labor time and better illumination. Scheduling strategies should determine the frequency of relamping, the percentage of the lamps to be replaced and the pattern within the system which constitutes that percentage.
- Spot burnout replacement often creates a period of time when lighting levels are deficient. Also, the blinking effect of a burned out fluorescent lamp can eventually cause damage or failure of starters and ballasts.
- Lamp schedules are an important control to assure that each fixture is relamped with the proper type and wattage of lamp. Fluorescent lamp color should be consistent without affecting the light output. Generally, the visual effect of multiple lamp colors is undesirable.

■ Luminaire cleaning frequency may vary due to soiling conditions. The percentage of light output will decline over time due to:
- deterioration of luminaire surfaces;
- dirt on room surfaces; and
- dirt on luminaires.

■ Cleaning of room surfaces is considered a lighting maintenance issue because of the contribution of reflected light to the environment.

■ Cleaning of the luminaire:
• Luminaires with appropriate ventilation will often reduce dust and dirt accumulation, allowing less frequent cleaning.
• Sealed housings require cleaning in a clean environment to eliminate the chance of trapping dust within the luminaire or potential degradation of the seal.
• Inherently cleaner operation is promoted by using direct downward lighting, vented housings with diffuse reflectance and no diffuser, louver or baffle.
• Great care must be taken in selecting cleaning materials. Most polished surfaces are easily scratched, many surfaces are not resistant to abrasive cleansers, and chemical compounds can degrade the composition of many plastics.
• Soiling conditions must be evaluated on the basis of particles per million (ppm) as very clean, clean, medium, dirty or very dirty.
• Soiling conditions will vary due to the quantity and frequency of dust/ dirt generation within immediate and/or surrounding environments, the typical quality of the ambient air, the removal of particles by air handling systems or other ventilation, and the nature of the particles relative to the ambient air (ranging from no adhesive collection to heavy adhesion due to static, humidity or chemical composition).
• Replace broken or damaged refractors as warranted.

■ Inspection of auxiliary units, such as ballasts, starters and programmable controllers, should be performed periodically to ensure proper performance. Wiring and circuitry should be checked when systems are used outdoors or underwater. Poles, standards and other structural components must be inspected for proper support and durability.

D5.1 ELECTRICAL

LIGHTING: selection checklist

ILLUMINATION

General Guidelines

■ The design of lighting systems involves several universal considerations for selection of equipment. Together they form the general guidelines for effective illumination.

■ **Illumination of the visual task is of premier importance.**
- Provide adequate illumination at the plane of work, on the object;
- Provide proper lighting for passages, corridors, staircases and other means of circulation;
- Ensure proper contrast of surround and background, not to exceed basic illumination ratios;
- Selection of lamp should be based on lamp efficacy, lumen production, lumen maintenance and wattage;
- Provide coordination with natural lighting arrangements;
- Provide auxiliary lighting as needed for safety, security, emergency, etc.

■ **Distribution of light within the environment imparts the quality of illumination.**
- The suitability of distribution relies on the position of the lighting equipment within the environment;
- The location of the light source should be chosen for surface brightness uniformity, for shadow and for lighting texture;
- Provide good room appearance with ambient lighting by eliminating spotty or excessively uneven distribution;
- Choose the light source for appropriate function - for precise beam concentration, for large area diffuse lighting, for wall wash, for general horizontal illumination;
- Provide combinations of complimentary lighting for spatial distribution, continuity and interest;
- Minimize direct glare;
- Maximize diffuse-type high reflectance surfaces and diffuse transmission materials for general lighting;
- Choose surface finishes, type of distribution and angle of incidence to minimize veiling reflection.

■ **Visual interest is a function of illumination quality.**
- Variable distribution can be used to provide highlight;
- The distribution and incidence of light on a surface can be used to create certain perceptions or reveal qualities;
- Use specular and spread reflection/transmission to reduce contrast.

■ **Correct color of light may be essential.**
- Select the light source with regard to requirements of work or process, psychological satisfaction, and combination with natural light;
- Be aware of color rendition for impact on environmental surfaces - color in decorations, on machines and equipment, emergency and safety color identifications.

■ **Operation, control and maintenance are a factor of illumination quality.**
- Energy conservation is relative to lamp wattage and coverage as indicated by the following comparison: Given a particular height and distribution, a luminaire with a 46-watt mercury lamp might illuminate 1000 sq. ft. of area to 1 fc. Energy savings could be achieved by switching to a 28-watt metal halide, 20-watt HPS or 17-watt LPS without loss in illumination.

APPEARANCE

■ Several guiding questions can be used to define the aesthetics of the luminaire:
- How can the architecture of the environment be described? Style? Finishes? Purpose?
- What styles and finishes will be compatible with the architecture?
- Is the luminaire to be exposed or concealed?
- Is the luminaire to be subordinate or dominant in the environment?
- Recessed, shielded, on the periphery? Or suspended, portable, within the visual plane?
- Will the luminaire be independent of or integrated into other electrical, mechanical or architectural systems?
- Are the decorative qualities of the luminaire appropriate when lighted? When unlighted?
- Does the luminaire enhance, maintain or subdue sound transfer?
- Are protective finishes necessary? Vandal resistant? Corrosion resistant? Impact resistant?

■ Luminaire selection is greatly affected by component materials and their finishes. The decision to use a certain luminaire should be based on the following criteria:
- Illumination quality and efficiency - luminance, distribution, color;
- Illuminating properties - transmission, reflection, absorption, light stability;
- Maintenance properties - cleaning, handling, access, durability;
- Safety - labelling, fire resistance, thermal properties, power loading;
- Cost - first, operating.

■ The integration of lighting equipment with the environment involves many considerations:
- **Architecturally exposed** equipment should be selected relative to prominent pattern, color, scale, form and so on; it should be coordinated with the major design element. Selection should evaluate appearance during operation on and off.
- **Architecturally concealed** equipment should be chosen on the basis of the emphasis of major architectural surfaces and the prominence of the equipment as a design element.
- Architectural forms can be defined using luminaire components; for example, bent or curved lenses can be incorporated with recessed luminaires and a high reflectance cavity to create an illuminated arch.
- **Mechanical integration** of lighting systems with air handling, acoustical control and fire suppression systems is available for large, open floor areas.
- Baffles, louvers, diffusers and other components of luminaires can be produced to offer creative opportunities of architectural and illuminating natures.

■ Luminaire selection should be weighted heavily as to the nature of its distribution.
- Is the light distribution to be subordinate to the environment; accent or perimeter lighting, indirect, diffusing, shielded?
- Is the light distribution to be prominent in the environment; direct concentrating, direct task lighting?

MAINTENANCE

■ Incandescent lamps, due to their conductive heating characteristics, require certain considerations.
- Lamps should be properly matched to voltage and wattage specifications as both over-voltage operation and use of lamp wattage exceeding recommended levels can result in damage to the bulb seal, socket and wiring.
- Higher wattage lamps, compact lamps and reflectors may require certain types of glass or may limit bulb position in operation; based on the maximum safe operating temperature of the bulb glass.

- Gas-filled lamps must be protected from rain, snow, contact with cooler metallic pieces or other sources of great temperature variation in small areas; the resultant thermal shock is a cause of stress within the glass bulb resulting in breakage.
- Lamp base deterioration due to heat is reduced within the lamp by using mica or ceramic discs in the neck of the lamp to deflect convective radiation and shield the base when white and silver-bowl coatings are present; deterioration is reduced outside the lamp by providing adequate ventilation to the lamp, yet minimizing the flow of cool air immediate to the neck and base of the lamp.
- Non-metallic housings are susceptible to charring and/or combustion when they are too close to the lamp, poorly ventilated or components in a high-wattage luminaire.
- Lampholder materials should be appropriate for the intended use. Refer to UL ratings and manufacturer specifications. Examples include the limitations on porcelain sockets for spotlights (use only in recessed or totally enclosed luminaires); nickel-plated brass or other high-heat tolerant materials (little expansion) for tungsten-halogen and other high-wattage lamps.

■ Fluorescent lamp output is adversely affected by high and low ambient temperatures.
- Low temperature operation may require glass jackets which retain lamp heat and special ballasts for lower temperature starting (below 50°F).
- High temperature operation (above 80° F) often requires ventilation around the lamp.
- Ventilation should be limited to low velocities in order to optimize lamp output. High velocities typically reduce lamp efficacy and shift the optimum ambient air temperature to about 90° to 100°F.

■ HID lamps rely on relatively precise arcs, which operate at high temperatures. Therefore, precautions are taken to insure proper operation.
- Over-voltage operation is detrimental to the arc tube; voltage surges can also raise the arc temperature beyond the melting point of the tube.
- Selection of HID lamps should be made with respect to burning position. HID lamps are designated for base up, base down, horizontal or universal burning.

■ Ballasts should be mounted such that the internal vibration of the ballast is minimized.
- This vibration causes a humming noise which may be intensified when the ballast is in contact with luminaire components or structural material that can act as a resonator. Components which are not securely fastened in place will tend to produce additional noise.
- Consult manufacturer specifications for sound ratings.

■ Systems with dimmers require certain considerations.
- All such circuits should be grounded.
- Check lamp requirements for special ballast specifications.
- Cold-cathode fluorescent lamps should be operated at 20% of lamp load for dimming to 10% of light output.
- Constant wattage transformers cannot be used in dimmer circuits.
- Ensure that all dimming, ballast and lamp equipment is suitably matched.

■ Materials used for luminaire construction or placed adjacent to the luminaire should be resistant to light and heat degradation.
- **Deformation** due to thermal variation can cause seal and/or connection failure as well as create unsightly appearances.
- Some materials and finishes are susceptible to **discoloration** from ultraviolet, infrared and thermal radiation.
- **Aging** of materials can be accelerated in hot, dry environments; some materials may become increasingly brittle or may experience breakdown in composition.

■ **Luminaire design should facilitate easy maintenance** (relamping and cleaning).
- Assembly should present a logical progression of parts, with connections and fastenings accessible for successive parts.
- Diffusers, louvers and baffles should be supported for easy removal without the use of fixed/mechanical connection.

■ Although the visual components of a luminaire should be suited to the appearance of the overall environment, **it is very important to use basic components in luminaire assemblies**.
- The use of basic, or standard components creates flexibility of parts usage and relieves inventories in the maintenance department.

- Standards include wattage, lamp type, lampholder/base type, etc.

■ **A major consideration for the selection of luminaires is its durability within the environment given a specific application.**
- Vandal-resistant housing and lens designs and materials are recommended for any application which places the luminaire accessible to the public. Ground fault circuitry is required for damp locations in areas where the public comes in contact. Protection for restroom luminaires and ground-mounted floodlights are typical examples.
- Corrosion and chemical degradation must be evaluated.
- Applications, such as industrial lighting, where movement, vibration or impact occur should be fitted with luminaires which are designed to accommodate it.

INDOOR APPLICATIONS

■ Flexibility of intended use of the space is the biggest benefit of area lighting, and it relies on uniform illumination on the horizontal plane. Uniform distribution can be accomplished by using certain arrangements of like-luminaire systems or by combining different types of systems.
- Systems which use the same luminaires throughout are typically ceiling mounted. Examples include linear arrangements of fluorescent luminaires, modular pattern of fluorescent luminaires, symmetric/regular pattern (rows and columns) of incandescent/HID luminaires, or patterns of incandescent/HID luminaires
- Combination systems will use direct, diffuse and/or indirect lighting to provide general illumination for the entire area and task lighting in defined areas.

■ **A concern of area lighting systems is acoustical control.** Although sound waves are of a different physical nature, they respond to surfaces in a similar manner, being absorbed or reflected. The amount of absorption or reflection depends on the contour of the surface, its density and hardness.
- Airborne transmission is promoted by large open areas with hard, dense surfaces.
- Porous materials are effective for sound absorption; typical of acoustical tiles, baffled and louvered systems.

D5.1 ELECTRICAL

LIGHTING: selection checklist (cont.)

■ Office lighting can be satisfied with recessed, surface-mounted, pendant-mounted or portable luminaires. It is often most effective when combined with task lighting.
• Conference rooms generally require high luminance to horizontal surfaces for certain activities and low luminance at the perimeter or a wall wash for others. Display lighting is often necessary at the front of the room. Dimmer controls are usually desirable for all three systems.

■ Typical area lighting seldom provides sufficient illuminance to support growth of plants in interior landscaping designs. Ground covers and flowering plants typically require substantial light.
• Proper illumination for growth can be provided by using extremes of footcandle production during non-business hours or when the space is not otherwise occupied. Timing controls on precise, concentrated luminaires can be used. HID lamps can produce the high luminance (about 600 fc) at ground level.

■ Indoor industrial tasks include a variety of visual tasks often in one large, open space. Building definition can vary greatly and movement of equipment and material is constant. Two basic applications are typical:
• High bay lighting is defined by applications where the ceiling line (defined often by the lower chord of roof structure/trusses) is 25 feet or more above the ground/floor. This is necessary for material handling and manufacture of heavy/large assemblies. Direct downward concentrating ceiling luminaires with point light sources (usually HID) are most effective.
• Low bay lighting includes all manufacturing/processing areas having a ceiling height of less than 25 feet (average of 12 feet). High coefficients of utilization and little glare can be gained by using 90% direct concentrating/10% indirect ceiling luminaires with high wattage lamps at close spacing over areas of difficult visual task, allowing spillover light to illuminate travel lanes.

■ Sports lighting typically requires illumination beyond the limits of the area of play. Illumination levels should remain constant for a certain distance, possibly into the stands, in order to relieve the potential of high contrast which may distract the athletes and to provide adequate illumination for TV cameras.

• Color rendition is very important and may limit the illuminance to avoid overlighting (washing out colors).

■ When direct and/or indirect luminaires are installed on vertical surfaces or near corners their distribution may create scallops of shadow.
• Scallops and other undesirable patterns of darkness can be eliminated by careful selection and installation of line sources of light. Fluorescent lighting is frequently used because of energy and maintenance factors. Specify wiring channels that permit back-to-back lampholders or tandem channels that accommodate 2 or more lamps. Multiple rows of lamps with staggered lampholders can also be effective. A single row of lamps set on a slight diagonal to overlap sockets will also minimize shadows.
• With point sources of light, the distribution, beam spread, distance and angle of incidence will determine coverage. Wall wash incandescent luminaires are often used to reduce shadow patterns.

■ Wall lighting from ceiling mounted sources can be guided by standard placement ratios.
• The distance from the luminaire to the wall will determine the effect of the distribution.
• When using broad beam sources, a distance from wall to luminaire equal to 1/4 of the mounting height will provide uniform illumination, with luminance at the bottom being 1/10 of that at the top.
• Moving the luminaire closer to the wall will increase the grazing effect, emphasizing surface texture and sculptural relief.
• If luminaires are located at both floor and ceiling levels, the placement ratio can be 1/6 of the ceiling height. The ratio can be even less when more concentrated, precise beams are used.

■ Two aspects of wall lighting are important in museum exhibit lighting; the light on the wall on which the art is hung/set, and the light on the art itself.
• Wall-washer luminaires can be surface mounted (ceiling) incandescent units with reflectors and opaque housings or recessed fluorescent lamps with low brightness louvers.
• Illumination of the art achieved with a surface mounted spotlight (adjustable or fixed). Wall-mounted picture lights (tubular lamps in a curved reflector)

are not recommended as they tend to produce uneven distribution across the surface of the piece; cause of 'hot spots' and glare.

OUTDOOR APPLICATIONS

■ The basic design and application of outdoor luminaires follow the same guidelines as for indoor luminaires. However outdoor environments, exterior architecture and aesthetics are considerably different. The primary design elements remain:
• All underground, submerged or embedded (in concrete) wiring should be run in conduit and must be grounded.
• The light source should be properly shielded to prevent glare.
• Continuous density and color characteristics should be maintained using only one lamp type in a given application.
• Control and direction of light at the source is emphasized to minimize glare and spill light to areas outside of the intended area of coverage. Distraction to passers-by may be hazardous, while neighbors may be annoyed by the excessive luminance.
• The control of outdoor lighting differs from indoor systems in that a central control station or light actuated photocell control may be preferable as is a system of separate lighting circuits. This allows for control of the basic lighting functions (security, building illumination, parking, landscape, decorative) as needed.
• Luminaire selection should consider lit and not lit aesthetics, shape, configuration, height, style, finish, pole spacing, poles, standards and mountings. Luminaires should be gasketed, sealed and fitted with an air-filter breather and have safety glass lenses for protection.

■ **Security lighting is employed to monitor entrance/exit from an enclosed area, and discourage thievery and vandalism.**
• Illumination of fences should:
— enable the security guard stationed within the fence to see a person outside the fence and to apprehend that person if he/she penetrates the fence;
— light the fence inside and out;
— be greater in the area around the entrance gates.
• Distribution should illuminate or silhouette intruders without producing glare in the direction of the guard;

COMMON LIGHT SOURCE CHARACTERISTICS. (Compiled by R. M. Harrold)

Incandescent including properties/ specifications	tungsten halogen	Fluorescent	Mercury - vapor (self - ballasted)	Metal halide	High pressure sodium (improved color)	Low pressure sodium
			High - intensity discharge			
Wattages (lamp only)	15 - 1,500	15 - 219	40 - 1,000	175 - 1,000	70 - 1,000	60 - 180
Life (hours)	750 - 12,000	7,500 - 24,000	16,000 - 24,000	1500 - 15,000	24,000 (7500)	16,000
Efficacy, lumens per watt (lamp only)	15-25	55-100	50-60 (20 - 25)	80-100	75-140 (67 - 112)	Up to 180
Lumen maintenance	Fair to excellent	Fair to excellent	Very good (good)	Good	Excellent	Excellent
Color rendition	Excellent	Good to excellent	Poor to excellent	Very good	Fair (very good)	Poor
Light direction control	Very good to excellent	Fair	Very good	Very good	Very good	Fair
Source size	Compact	Extended	Compact	Compact	Compact	Extended
Relight time	Immediate	Immediate	3 - 10 minutes	10 - 20 minutes	Less than 1 minute	Immediate
Comparative fixture cost	Low—simple fixtures	Moderate	Higher than incandescent and fluorescent	Generally higher than mercury	High	High
Comparative operating cost	High—short life and low efficacy	Lower than incandescent	Lower than incandescent	Lower than mercury	Lowest of HID types	Low
Auxiliary equipment needed	Not needed	Needed— medium cost	Needed— high cost	Needed— high cost	Needed— high cost	Needed— high cost

D5.1 ELECTRICAL

LIGHTING: selection checklist (cont.)

asymmetric patterns accentuate shadows while maximizing coverage.
- Color rendition is not a factor.
- Passers-by and neighbors should not be affected by glare or spillover light.
- Distribution and illumination may need to be within the sensitivity range of camera monitoring systems.
- Building mounted luminaires are effective when located on facades and parapets.
- Ground installed luminaires will set enlarged, dark silhouettes against a well-lit building facade; can be broken due to accessibility.
- Illuminate all pathways and the areas adjacent to them. Keep the trees and shrubs trimmed and illuminated to minimize hiding places.
- Minimize dark areas for security.

■ **Exterior illumination of buildings is often accomplished using floodlighting techniques**. The building's surface and ornamentation (detailing) are the keys to effective applications as well as level of illumination and angle of incidence.
- Illumination levels vary with surface reflectance and background luminance; high reflectance surfaces in a dark setting require low illuminance of about 5 fc. The Illuminating Engineering Society recommends 50 fc when a low reflectance surface is illuminated amidst other exterior lighting.
- Floodlighting is not generally recommended for specular reflectance materials (aluminum or glass curtain walls) since they may reflect an image of the source.
- The incident light will create shadow and reflection. Therefore, the angle of incidence should be consistent in order to keep the shadows in the same pattern. A 30° angle of incidence is the minimum for effecting good texture. Reflection should compliment the viewing angle; the straight-on distribution which is effective for passing motorists will be uninteresting to the pedestrian as opposed to a grazing effect.

■ **Parking areas** have six factors which affect illumination, with most being dependent on the height of light source.
- Luminaires set below eye level do not illuminate faces, a basic security need. A height of 15 feet is about the minimum effective height.
- The higher the luminaire, the greater the spacing, the lower the initial cost of installation, the higher the lamp wattage. However, lower mounting

heights are often more sympathetic to the environment.
- The other factors are pole spacing, the nature of the light source, the distribution from the luminaire, the reflectance of the paving material, and the desired level of illumination.

■ **Garden and landscape lighting** should be dramatic.
- General lighting is not effective except over large lawn or deck areas.
- Maintain low levels of illumination; overlighting tolerance is low.
- Shield all light sources from normal night viewing.
- Camouflage all luminaires by painting, recessing or concealing in plant growth unless the luminaires are intended to be decorative.
- Individual luminaires present more interest through highlight and shadow by illuminating only certain spots (a path, a tree, etc.).
- Effective luminaire types include pole-mounted, tree-mounted, building mounted, pathway luminaires and bollards.

■ **Underwater lighting** may be used for outdoor or indoor applications.
- The luminaire should be adjustable by means of flexible conduit.
- Dimmers should be used.
- Two basic types of luminaires are used for underwater lighting. One is permanently set in the wall below the water line and requires water circulation for cooling. The water must be drained to a level below the luminaire in order to replace a burned-out lamp. The other luminaire is a sealed unit which can be lifted to dry deck areas for relamping.
- The lens must be submerged a minimum of 2 in. to prevent overheating and eventual breakage.
- All luminaires, especially exposed, should have tamperproof, vandal-resistant fittings or metal grille enclosures are recommended.
- The luminaire must be labelled and should be grounded in accordance with the National Electrical Code or prevailing local code.

■ **Flagpole lighting** is accomplished with independently mounted spotlights in the ground, on the flag pole or on the building.
- Upward spotlighting mounted on flagpoles is often housed in round or square aluminum canisters and wired internally through the pole shaft.

Index

INDEX

INDEX

INDEX

INDEX

Conversion Tables

Inches/Feet to Millimetres

inches	milli-metres	inches	milli-metres	inches	milli-metres	inches	milli-metres	ft.	in.	milli-metres
1/64	0.3969	1 27/32	46.8313	4 21/32	118.269	8 15/16	227.012	3	7	1092.20
1/32	0.7938	1 7/8	47.6250	4 11/16	119.062	9	228.600	3	8	1117.60
3/64	1.1906	1 29/32	48.4188	4 23/32	119.856	9 1/16	230.188	3	9	1143.00
1/16	1.5875	1 15/16	49.2125	4 3/4	120.650	9 1/8	231.775	3	10	1168.40
5/64	1.9844	1 31/32	50.0063	4 25/32	121.444	9 3/16	233.362	3	11	1193.80
3/32	2.3813	2	50.8000	4 13/16	122.238	9 1/4	234.950	4	0	1219.20
7/64	2.7781	2 1/32	51.5938	4 27/32	123.031	9 5/16	236.538	4	1	1244.60
1/8	3.1750	2 1/16	52.3875	4 7/8	123.825	9 3/8	238.125	4	2	1270.00
9/64	3.5719	2 3/32	53.1813	4 29/32	124.619	9 7/16	239.712	4	3	1295.40
5/32	3.9688	2 1/8	53.9750	4 15/16	125.412	9 1/2	241.300	4	4	1320.80
11/64	4.3656	2 5/32	54.7688	4 31/32	126.206	9 9/16	242.888	4	5	1346.20
3/16	4.7625	2 3/16	55.5625	5	127.000	9 5/8	244.475	4	6	1371.60
13/64	5.1594	2 7/32	56.3583	5 1/32	127.794	9 11/16	246.062	4	7	1397.00
7/32	5.5563	2 1/4	57.1500	5 1/16	128.588	9 3/4	247.650	4	8	1422.40
15/64	5.9531	2 9/32	57.9438	5 3/32	129.381	9 13/16	249.238	4	9	1447.80
1/4	6.3500	2 5/16	58.7375	5 1/8	130.175	9 7/8	250.825	4	10	1473.20
17/64	6.7469	2 11/32	59.5313	5 5/32	130.969	9 15/16	252.412	4	11	1498.60
9/32	7.1438	2 3/8	60.3250	5 3/16	131.762	10	254.000	5	0	1524.00
19/64	7.5406	2 13/32	61.1188	5 7/32	132.556	10 1/16	255.588	5	1	1549.40
5/16	7.9375	2 7/16	61.9125	5 1/4	133.350	10 1/8	257.175	5	2	1574.80
21/64	8.3344	2 15/32	62.7063	5 9/32	134.144	10 3/16	258.762	5	3	1600.20
11/32	8.7313	2 1/2	63.5000	5 5/16	134.938	10 1/4	260.350	5	4	1625.60
23/64	9.1281	2 17/32	64.2938	5 11/32	135.731	10 5/16	261.938	5	5	1651.00
3/8	9.5250	2 9/16	65.0875	5 3/8	136.525	10 3/8	263.525	5	6	1676.40
25/64	9.9219	2 19/32	65.8813	5 13/32	137.319	10 7/16	265.112	5	7	1701.80
13/32	10.3188	2 5/8	66.6750	5 7/16	138.112	10 1/2	266.700	5	8	1727.20
27/64	10.7156	2 21/32	67.4688	5 15/32	138.906	10 9/16	268.288	5	9	1752.60
7/16	11.1125	2 11/16	68.2625	5 1/2	139.700	10 5/8	269.875	5	10	1778.00
29/64	11.5094	2 23/32	69.0563	5 17/32	140.494	10 11/16	271.462	5	11	1803.40
15/32	11.9063	2 3/4	69.8500	5 9/16	141.288	10 3/4	273.050	6	0	1828.80
31/64	12.3031	2 25/32	70.6438	5 19/32	142.081	10 13/16	274.638	6	1	1854.20
1/2	12.7000	2 13/16	71.4375	5 5/8	142.875	10 7/8	276.225	6	2	1879.60
33/64	13.0969	2 27/32	72.2313	5 21/32	143.669	10 15/16	277.812	6	3	1905.00
17/32	13.4938	2 7/8	73.0250	5 11/16	144.462	11	279.400	6	4	1930.40
35/64	13.8906	2 29/32	73.8188	5 23/32	145.256	11 1/16	280.988	6	5	1955.80
9/16	14.2875	2 15/16	74.6125	5 3/4	146.050	11 1/8	282.575	6	6	1981.20
37/64	14.6844	2 31/32	75.4063	5 25/32	146.844	11 3/16	284.162	6	7	2006.60
19/32	15.0813	3	76.2000	5 13/16	147.638	11 1/4	285.750	6	8	2032.00
39/64	15.4781	3 1/32	76.9938	5 27/32	148.431	11 5/16	287.338	6	9	2057.40
5/8	15.8750	3 1/16	77.7875	5 7/8	149.225	11 3/8	288.925	6	10	2082.80
41/64	16.2719	3 3/32	78.5813	5 29/32	150.019	11 7/16	290.512	6	11	2108.20
21/32	16.6688	3 1/8	79.3750	5 15/16	150.812	11 1/2	292.100	7	0	2133.60
43/64	17.0656	3 5/32	80.1688	5 31/32	151.606	11 9/16	293.688	7	1	2159.00
11/16	17.4625	3 3/16	80.9625	6	152.400	11 5/8	295.275	7	2	2184.40
45/64	17.8594	3 7/32	81.7563	6 1/16	153.988	11 11/16	296.862	7	3	2209.80
23/32	18.2563	3 1/4	82.5500	6 1/8	155.575	11 3/4	298.450	7	4	2235.20
47/64	18.6531	3 9/32	83.3438	6 3/16	157.162	11 13/16	300.038	7	5	2260.60
3/4	19.0500	3 5/16	84.1375	6 1/4	158.750	11 7/8	301.625	7	6	2286.00
49/64	19.4469	3 11/32	84.9313	6 5/16	160.338	11 15/16	303.212	7	7	2311.40
25/32	19.8438	3 3/8	85.7250	6 3/8	161.925	12	304.800	7	8	2336.80
51/64	20.2406	3 13/32	86.5188	6 7/16	163.512	13	330.200	7	9	2362.20
13/16	20.6375	3 7/16	87.3125	6 1/2	165.100	14	355.600	7	10	2387.60
53/64	21.0344	3 15/32	88.1063	6 9/16	166.688	15	381.000	7	11	2413.00
27/32	21.4313	3 1/2	88.9000	6 5/8	168.275	16	406.400	8	0	2438.40
55/64	21.8281	3 17/32	89.6938	6 11/16	169.862	17	431.800	8	1	2463.80
7/8	22.2250	3 9/16	90.4875	6 3/4	171.450	18	457.200	8	2	2489.20
57/64	22.6219	3 19/32	91.2813	6 13/16	173.038	19	482.600	8	3	2514.60
29/32	23.0188	3 5/8	92.0750	6 7/8	174.625	20	508.000	8	4	2540.00
59/64	23.4156	3 21/32	92.8688	6 15/16	176.212	21	533.400	8	5	2565.40
15/16	23.8125	3 11/16	93.6625	7	177.800	22	558.800	8	6	2590.80
61/64	24.2094	3 23/32	94.4563	7 1/16	179.388	23	584.200	8	7	2616.20
31/32	24.6063	3 3/4	95.2500	7 1/8	180.975	24	609.600	8	8	2641.60
63/64	25.0031	3 25/32	96.0438	7 3/16	182.562	25	635.000	8	9	2667.00
1	25.4000	3 13/16	96.8375	7 1/4	184.150	26	660.400	8	10	2692.40
1 1/32	26.1938	3 27/32	97.6313	7 5/16	185.738	27	685.800	8	11	2717.80
1 1/16	26.9875	3 7/8	98.4250	7 3/8	187.325	28	711.200	9	0	2743.20
1 3/32	27.7813	3 29/32	99.2188	7 7/16	188.912	29	736.600	9	1	2768.60
1 1/8	28.5750	3 15/16	100.012	7 1/2	190.500	30	762.000	9	2	2794.00
1 5/32	29.3688	3 31/32	100.806	7 9/16	192.088	31	787.400	9	3	2819.40
1 3/16	30.1625	4	101.600	7 5/8	193.675	32	812.800	9	4	2844.80
1 7/32	30.9563	4 1/32	102.394	7 11/16	195.262	33	838.200	9	5	2870.20
1 1/4	31.7500	4 1/16	103.188	7 3/4	196.850	34	863.600	9	6	2895.60
1 9/32	32.5438	4 3/32	103.981	7 13/16	198.438	35	889.000	9	7	2921.00
1 5/16	33.3375	4 1/8	104.775	7 7/8	200.025	36	914.400	9	8	2946.40
1 11/32	34.1313	4 5/32	105.569	7 15/16	201.612	37	939.800	9	9	2971.80
1 3/8	34.9250	4 3/16	106.362	8	203.200	38	965.200	9	10	2997.20
1 13/32	35.7188	4 7/32	107.156	8 1/16	204.788	39	990.600	9	11	3022.60
1 7/16	36.5125	4 1/4	107.950	8 1/8	206.375	40	1016.00	10	0	3048.00
1 15/32	37.3063	4 9/32	108.744	8 3/16	207.962	41	1041.40	11	0	3352.80
1 1/2	38.1000	4 5/16	109.538	8 1/4	209.550	42	1066.80	12	0	3657.60
1 17/32	38.8938	4 11/32	110.331	8 5/16	211.138			13	0	3962.40
1 9/16	39.6875	4 3/8	111.125	8 3/8	212.725			14	0	4267.20
1 19/32	40.4813	4 13/32	111.919	8 7/16	214.312			15	0	4572.00
1 5/8	41.2750	4 7/16	112.712	8 1/2	215.900			16	0	4876.80
1 21/32	42.0688	4 15/32	113.506	8 9/16	217.488			17	0	5131.60
1 11/16	42.8625	4 1/2	114.300	8 5/8	219.075			18	0	5486.40
1 23/32	43.6563	4 17/32	115.094	8 11/16	220.662			19	0	5791.20
1 3/4	44.4500	4 9/16	115.888	8 3/4	222.250			20	0	6096.00
1 25/32	45.2438	4 19/32	116.681	8 13/16	223.838			21	0	6400.80
1 13/16	46.0375	4 5/8	117.475	8 7/8	225.425			22	0	6705.60

Millimetres to Inches/Feet

milli-metres	inches	milli-metres	inches	milli-metres	inches	milli-metres	inches	milli-metres	inches
1	0.0394	91	3.5827	181	7.1260	271	10.6693	361	14.2126
2	0.0787	92	3.6221	182	7.1654	272	10.7087	362	14.2520
3	0.1181	93	3.6614	183	7.2047	273	10.7480	363	14.2913
4	0.1575	94	3.7008	184	7.2441	274	10.7874	364	14.3307
5	0.1969	95	3.7402	185	7.2835	275	10.8268	365	14.3701
6	0.2362	96	3.7795	186	7.3228	276	10.8661	366	14.4094
7	0.2756	97	3.8189	187	7.3622	277	10.9055	367	14.4488
8	0.3150	98	3.8583	188	7.4016	278	10.9449	368	14.4882
9	0.3543	99	3.8976	189	7.4409	279	10.9843	369	14.5276
10	0.3937	100	3.9370	190	7.4803	280	11.0236	370	14.5669
11	0.4331	101	3.9764	191	7.5197	281	11.0630	371	14.6063
12	0.4724	102	4.0158	192	7.5591	282	11.1024	372	14.6457
13	0.5118	103	4.0551	193	7.5984	283	11.1417	373	14.6850
14	0.5512	104	4.0945	194	7.6378	284	11.1811	374	14.7244
15	0.5906	105	4.1339	195	7.6772	285	11.2205	375	14.7638
16	0.6299	106	4.1732	196	7.7165	286	11.2598	376	14.8031
17	0.6693	107	4.2126	197	7.7559	287	11.2992	377	14.8425
18	0.7087	108	4.2520	198	7.7953	288	11.3386	378	14.8819
19	0.7480	109	4.2913	199	7.8347	289	11.3780	379	14.9213
20	0.7874	110	4.3307	200	7.8740	290	11.4173	380	14.9606
21	0.8268	111	4.3701	201	7.9134	291	11.4567	381	15.0000
22	0.8661	112	4.4095	202	7.9528	292	11.4961	382	15.0394
23	0.9055	113	4.4488	203	7.9921	293	11.5354	383	15.0787
24	0.9449	114	4.4882	204	8.0315	294	11.5748	384	15.1181
25	0.9843	115	4.5276	205	8.0709	295	11.6142	385	15.1575
26	1.0236	116	4.5669	206	8.1102	296	11.6535	386	15.1969
27	1.0630	117	4.6063	207	8.1496	297	11.6929	387	15.2362
28	1.1024	118	4.6457	208	8.1890	298	11.7323	388	15.2756
29	1.1417	119	4.6850	209	8.2284	299	11.7717	389	15.3150
30	1.1811	120	4.7244	210	8.2677	300	11.8110	390	15.3543
31	1.2205	121	4.7638	211	8.3071	301	11.8504	391	15.3937
32	1.2598	122	4.8032	212	8.3465	302	11.8898	392	15.4331
33	1.2992	123	4.8425	213	8.3858	303	11.9291	393	15.4724
34	1.3386	124	4.8819	214	8.4252	304	11.9686	394	15.5118
35	1.3780	125	4.9213	215	8.4646	305	12.0079	395	15.5512
36	1.4173	126	4.9606	216	8.5039	306	12.0472	396	15.5906
37	1.4567	127	5.0000	217	8.5433	307	12.0866	397	15.6299
38	1.4961	128	5.0394	218	8.5827	308	12.1260	398	15.6693
39	1.5354	129	5.0787	219	8.6221	309	12.1654	399	15.7087
40	1.5748	130	5.1181	220	8.6614	310	12.2047	400	15.7480
41	1.6142	131	5.1575	221	8.7008	311	12.2441	401	15.7874
42	1.6535	132	5.1969	222	8.7402	312	12.2835	402	15.8268
43	1.6929	133	5.2362	223	8.7795	313	12.3228	403	15.8661
44	1.7323	134	5.2756	224	8.8189	314	12.3622	404	15.9055
45	1.7717	135	5.3150	225	8.8583	315	12.4016	405	15.9449
46	1.8110	136	5.3543	226	8.8976	316	12.4409	406	15.9832
47	1.8504	137	5.3937	227	8.9370	317	12.4803	407	16.0236
48	1.8898	138	5.4331	228	8.9764	318	12.5197	408	16.0630
49	1.9291	139	5.4724	229	9.0158	319	12.5591	409	16.1024
50	1.9685	140	5.5118	230	9.0551	320	12.5984	410	16.1417
51	2.0079	141	5.5512	231	9.0945	321	12.6378	411	16.1811
52	2.0472	142	5.5906	232	9.1339	322	12.6772	412	16.2205
53	2.0866	143	5.6299	233	9.1732	323	12.7165	413	16.2598
54	2.1260	144	5.6693	234	9.2126	324	12.7559	414	16.2992
55	2.1654	145	5.7087	235	9.2520	325	12.7953	415	16.3386
56	2.2047	146	5.7480	236	9.2913	326	12.8346	416	16.3780
57	2.2441	147	5.7874	237	9.3307	327	12.8740	417	16.4173
58	2.2835	148	5.8268	238	9.3701	328	12.9134	418	16.4567
59	2.3228	149	5.8661	239	9.4095	329	12.9528	419	16.4961
60	2.3622	150	5.9055	240	9.4488	330	12.9921	420	16.5354
61	2.4016	151	5.9449	241	9.4882	331	13.0315	421	16.5748
62	2.4409	152	5.9843	242	9.5276	332	13.0709	422	16.6142
63	2.4803	153	6.0236	243	9.5669	333	13.1102	423	16.6535
64	2.5197	154	6.0630	244	9.6063	334	13.1496	424	16.6929
65	2.5591	155	6.1024	245	9.6457	335	13.1890	425	16.7323
66	2.5984	156	6.1417	246	9.6850	336	13.2283	426	16.7716
67	2.6378	157	6.1811	247	9.7244	337	13.2677	427	16.8110
68	2.6772	158	6.2205	248	9.7638	338	13.3071	428	16.8504
69	2.7165	159	6.2599	249	9.8031	339	13.3465	429	16.8898
70	2.7559	160	6.2992	250	9.8425	340	13.3858	430	16.9291
71	2.7953	161	6.3386	251	9.8819	341	13.4252	431	16.9685
72	2.8347	162	6.3780	252	9.9213	342	13.4646	432	17.0079
73	2.8740	163	6.4173	253	9.9606	343	13.5039	433	17.0472
74	2.9134	164	6.4567	254	10.0000	344	13.5433	434	17.0866
75	2.9528	165	6.4961	255	10.0393	345	13.5827	435	17.1260
76	2.9921	166	6.5354	256	10.0787	346	13.6220	436	17.1654
77	3.0315	167	6.5748	257	10.1181	347	13.6614	437	17.2047
78	3.0709	168	6.6142	258	10.1575	348	13.7008	438	17.2441
79	3.1102	169	6.6535	259	10.1969	349	13.7402	439	17.2835
80	3.1496	170	6.6929	260	10.2362	350	13.7795	440	17.3228
81	3.1890	171	6.7323	261	10.2756	351	13.8189	441	17.3622
82	3.2284	172	6.7717	262	10.3150	352	13.8583	442	17.4016
83	3.2677	173	6.8110	263	10.3543	353	13.8976	443	17.4409
84	3.3071	174	6.8504	264	10.3937	354	13.9370	444	17.4803
85	3.3465	175	6.8898	265	10.4331	355	13.9764	445	17.5197
86	3.3858	176	6.9291	266	10.4724	356	14.0157	446	17.5591
87	3.4252	177	6.9685	267	10.5118	357	14.0551	447	17.5984
88	3.4646	178	7.0079	268	10.5512	358	14.0945	448	17.6378
89	3.5039	179	7.0472	269	10.5906	359	14.1339	449	17.6772
90	3.5433	180	7.0866	270	10.6299	360	14.1732	450	17.7165

Conversion Tables

The following tables are furnished to assist you in utilizing the information provided by building product manufacturers.

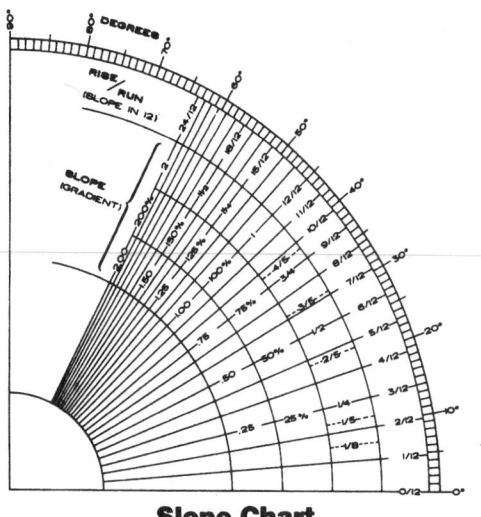

Slope Chart

Basic SI Units (Systéme International d'Unités)

physical quantity	name of unit	symbol for unit
length	metre	m
mass	kilogramme	kg
time	second	s
electric current	ampere	A
thermodynamic temperature	kelvin	K
luminous intensity	candela	cd
area	square metre	m^2
volume	cubic metre	m^3
density	kilogramme per cubic metre	kg/m^3
velocity	metre per second	m/s
angular velocity	radian per second	rad/s
acceleration	metre per second squared	m/s^2
pressure	newton per square metre	N/m^2
kinematic viscosity, diffusion coefficient	square metre per second	m^2/s
dynamic viscosity	newton second per square metre	$N s/m^2$
electric field strength	volt per metre	V/m
magnetic field strength	ampere per metre	A/m
luminance	candela per square metre	cd/m^2

Symbols for units do not take a plural form.

Distance

Imperial	Metric		Metric		Imperial
1 inch	= 2.540 centimetres		1 centimetre	=	0.3937 inch
1 foot	= 0.3048 metre		1 decimetre	=	0.3281 foot
1 yard	= 0.9144 metre		1 metre	=	3.281 feet
1 rod	= 5.029 metres			=	1.094 yard
1 mile	= 1.609 kilometres		1 decametre	=	10.94 yards
			1 kilometre	=	0.6214 mile

Weight

1 ounce (troy)	= 31.103 grams		1 gram	= 0.032 ounce (troy)
1 ounce (avoir)	= 28.350 grams		1 gram	= 0.035 ounce (avoir)
1 pound (troy)	= 373.242 grams		1 kilogram	= 2.679 pounds (troy)
1 pound (avoir)	= 453.592 grams		1 kilogram	= 2.205 pounds (avoir)
1 ton (short)	= 0.907 tonne*		1 tonne	= 1.102 ton (short)

*1 tonne = 1000 kilograms

Capacity

Imperial			U.S.		
1 pint	= 0.568 litre		1 pint (U.S.)	=	0.473 litre
1 gallon	= 4.546 litres		1 quart (U.S.)	=	0.946 litre
1 bushel	= 36.369 litres		1 gallon (U.S.)	=	3.785 litres
1 litre	= 0.880 pint		1 barrel (U.S.)	=	158.98 litres
1 litre	= 0.220 gallon				
1 hectolitre	= 2.838 bushels				

Area

1 square inch	= 6.452 square centimetres
1 square foot	= 0.093 square metre
1 square yard	= 0.836 square metre
1 acre	= 0.405 hectare*
1 square mile	= 259.0 hectares
1 square mile	= 2.590 square kilometres
1 square centimetre	= 0.155 square inch
1 square metre	= 10.76 square feet
1 square metre	= 1.196 square yard
1 hectare	= 2.471 acres
1 square kilometre	= 0.386 square mile

*1 hectare = 1 square hectometre

Volume

1 cubic inch	= 16.387	cubic centimetres
1 cubic foot	= 0.0283	cubic decimetres
1 cubic yard	= 0.765	cubic metre
1 cubic centimetre	= 0.061	cubic inch
1 cubic decimetre	= 35.314	cubic foot
1 cubic metre	= 1.308	cubic yard

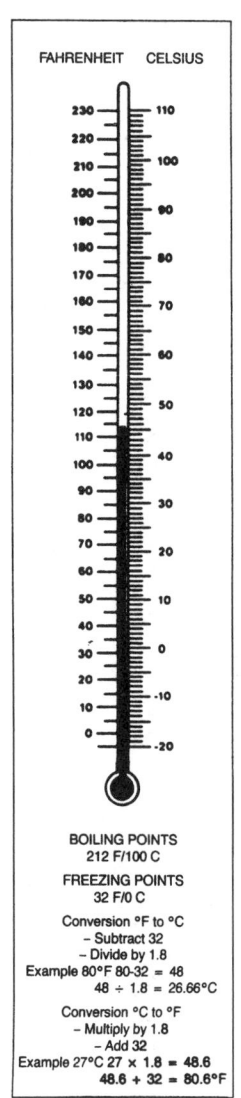

BOILING POINTS
212 F/100 C

FREEZING POINTS
32 F/0 C

Conversion °F to °C
– Subtract 32
– Divide by 1.8
Example 80°F 80-32 = 48
48 ÷ 1.8 = 26.66°C

Conversion °C to °F
– Multiply by 1.8
– Add 32
Example 27°C 27 x 1.8 = 48.6
48.6 + 32 = 80.6°F

NOTES

NOTES

NOTES